HIGHWAY ENGINEERING HANDBOOK

Other McGraw-Hill Books of Interest

HIGHWAY ENGINEERING HANDBOOK

Building and Rehabilitating the Infrastructure

Roger L. Brockenbrough, P.E.   Editor

R. L. Brockenbrough & Associates, Inc.
Pittsburgh, Pennsylvania

Kenneth J. Boedecker, Jr., P.E.   Editor

W. R. Grace & Company
Charleston, South Carolina

McGRAW-HILL

New York   San Francisco   Washington, D.C.   Auckland   Bogotá
Caracas   Lisbon   London   Madrid   Mexico City   Milan
Montreal   New Delhi   San Juan   Singapore
Sydney   Tokyo   Toronto

Library of Congress Cataloging-in-Publication Data

Highway engineering handbook : building and rehabilitating the
  infrastructure / Roger L. Brockenbrough, editor, Kenneth J.
  Boedecker, Jr., editor.
          p.   cm.
  Includes bibliographical references and index.
  ISBN 0-07-008777-6 (acid-free paper)
    1. Highway engineering—United States—Handbooks, manuals, etc.
  I. Brockenbrough, R. L.   II. Boedecker, Kenneth J.
  TE23.H484   1996
  625.7′0973—dc20   95-48283
                                                            CIP

McGraw-Hill

*A Division of The **McGraw·Hill** Companies*

2 3 4 5 6 7 8 9 0   DOC/DOC   9 0 1 0 9 8 7 6

ISBN 0-07-008777-6

*The sponsoring editor for this book was Larry S. Hager, the editing supervisor was
Stephen M. Smith, and the production supervisor was Pamela A. Pelton. It was set in
Times Roman by Donald A. Feldman of McGraw-Hill's Professional Book Group
composition unit.*

Printed and bound by R. R. Donnelley & Sons Company.

This book is printed on acid-free paper.

CONTENTS

Chapter 4. Bridge Engineering   *Walter J. Jestings, P.E.*   **4.1**

Chapter 5. Culverts, Drainage, and Bridge Replacement   **5.1**
Paul W. Cotter, P.E.

Chapter 6. Safety Systems   *Roger L. Brockenbrough, P.E.*   **6.1**

Chapter 7. Signing and Roadway Lighting

Chapter 8. Retaining Walls *A. J. Siccardi, P.E.* 8.1

Chapter 9. Noise Walls   *James J. Hill, P.E., and Roger L. Brockenbrough, P.E.*   **9.1**

Chapter 10. Value Engineering and Life Cycle Cost   *Harold G. Tufty, CVS, FSAVE*   **10.1**

VALUE ENGINEERING ON HIGHWAY PROJECTS

FUNDAMENTALS OF VALUE ENGINEERING

LIFE CYCLE COST

EXAMPLES OF SUCCESSFUL VE HIGHWAY STUDIES

Index follows Chapter 10

CONTRIBUTORS

Brian L. Bowman, P.E. *Associate Professor, Auburn University, Auburn, Alabama* (Chap. 7)

Roger L. Brockenbrough, P.E. *R. L. Brockenbrough & Associates, Inc., Pittsburgh, Pennsylvania* (Chaps. 6, 9)

Cheryl Bly Chester, P.E. *Senior Engineer, Levine-Fricke, Inc., Roseville, California* (Chap. 1)

Paul W. Cotter, P.E. *Underground Structures Engineer, California Department of Transportation, Sacramento, California* (Chap. 5)

Roger L. Green, P.E. *Pavement Research Engineer, Ohio Department of Transportation, Columbus, Ohio* (Chap. 3)

James J. Hill, P.E. *Structural Engineer, Minnesota Department of Transportation, Roseville, Minnesota* (Chap. 9)

Walter J. Jestings, P.E. *Bridge Engineer, Parsons Brinkerhoff, Quade & Douglas, Inc., Atlanta, Georgia* (Chap. 4)

Aric A. Morse, P.E. *Pavement Design Engineer, Ohio Department of Transportation, Columbus, Ohio* (Chap. 3)

Larry J. Shannon, P.E. *Standards and Geometrics Engineer, Ohio Department of Transportation, Columbus, Ohio* (Chap. 2)

A. J. Siccardi, P.E. *Formerly, Staff Bridge Engineer, Colorado Department of Transportation, Denver, Colorado* (Chap. 8)

Harold G. Tufty, CVS, FSAVE *Editor and Publisher,* Value Engineering and Management Digest, *Washington, D.C.* (Chap. 10)

C. Paul Watson, P.E. *State Electrical Engineer, Alabama Department of Transportation, Montgomery, Alabama* (Chap. 7)

PREFACE

The *Highway Engineering Handbook* has been developed by knowledgeable engineers to serve as a comprehensive reference source for those involved in highway design. This handbook is broad in scope, presenting information on topics ranging from environmental issues to value engineering, from the design of culverts, lighting, and noise walls to the design of safety systems, retaining walls, and bridges. In addition, such fundamental subjects as location and pavement design are fully discussed.

This volume should be useful to a wide range of personnel involved in highway design and construction, including consulting engineers; engineers employed by departments of transportation in federal, state, and local governments; those involved with turnpike authorities; and engineering educators. Both experienced practitioners and serious students will find the information presented here useful and easy to apply. It should enable the engineer to create a design that fulfills the requirements of the highway user: a safe, smooth, durable, aesthetically pleasing, environmentally sensitive, and economical highway system.

Contributors to this handbook are experienced highway engineers, consultants, or educators. They are leading authorities in their subject areas. The guiding principle of this book is to present practical information that has direct application to situations encountered in the field. Efforts were made to coordinate the information with that of the American Association of State Highway and Transportation Officials (AASHTO). Metric units are used where feasible to ease the transition to that system.

The material in this book follows a logical sequence. It begins with a discussion of environmental issues, a fundamental consideration in modern highway design. This is followed by a chapter on location, design, and traffic that includes extensive examples of typical standard treatments. A subject critical to building and maintaining durable systems, pavement design and rehabilitation, is then presented. Following this, aspects of bridge engineering are discussed to aid in the selection of bridge type and material for a durable design. The essentials of culvert design are then offered, as well as information on the various culvert types available. Next, a discussion of roadway safety addresses the latest options for providing for errant vehicles that leave the traveled way. A wealth of information follows on signing and lighting highways, subjects that also are closely related to highway safety. A comprehensive chapter next addresses the selection and design of retaining walls and considers both generic and proprietary systems. Walls to reduce traffic noise and screen unsightly areas are then considered. Finally, a chapter on value engineering and life cycle cost presents fundamental insights into these areas, as well as application examples, to encourage cost-effective design.

The contributors and editors are indebted to their colleagues and a variety of sources for the information presented. Credit is given in references throughout the text to the extent feasible.

The reader is cautioned that independent professional judgment must be exercised when information given in this handbook is applied. Anyone making use of this information assumes all liability arising from such use.

Roger L. Brockenbrough, P.E.
Kenneth J. Boedecker, Jr., P.E.

FACTORS FOR CONVERSION TO SI UNITS OF MEASUREMENT

Multiply	By	To find
Length:		
Inches (in)	25.400	Millimeters (mm)
Feet (ft)	0.3048	Meters (m)
Yards (yd)	0.9144	Meters (m)
Miles (mi)	1.6093	Kilometers (km)
Area:		
Square inches (in^2)	645.16	Square millimeters (mm^2)
Square feet (ft^2)	0.09290	Square meters (m^2)
Square yards (yd^2)	0.8361	Square meters (m^2)
Square miles (mi^2)	2.5900	Square kilometers (km^2)
Acres (ac)	0.4047	Hectares (ha)
Mass:		
Ounces (oz)	28.350	Grams (g)
Pounds (lb)	0.4536	Kilograms (kg)
Tons, short (T)	0.9072	Megagrams (Mg), or tonnes
Volume:		
Ounces, fluid (oz)	29.574	Milliliters (mL)
Gallons (gal)	3.7854	Liters (L)
Cubic feet (ft^3)	0.02832	Cubic meters (m^3)
Cubic yards (yd^3)	0.07646	Cubic meters (m^3)
Velocity:		
Miles per hour (mi/h)	1.6093	Kilometers per hour (km/h)
Force:		
Pound (lb)	4.4482	Newton (N)
Kip	4.4482	Kilonewton (kN)
Stress:		
Pounds per square inch (lb/in^2)	6.8948	Kilopascal (kPa)
Kips per square inch (kips/in^2)	6.8948	Megapascal (MPa)
Kips per square foot (kips/ft^2)	47.880	Kilopascal (kPa)

CHAPTER 1
ENVIRONMENTAL ISSUES

Cheryl Bly Chester, P.E.

Senior Engineer
Levine-Fricke, Inc.
Roseville, California

Environmental issues play a major role in the planning, design, construction, and rehabilitation of highways. This chapter provides an overview to lend guidance in understanding the environmental issues, and information to aid in the process of implementing the various requirements in this area. It begins with an overall review of the
numerous federal laws and regulations that must be considered. A thorough discussion
is provided of the requirements of the National Environmental Policy Act and the
preparation of environmental impact statements. Information is presented on the
important topics of storm water pollution prevention and lead-based paint removal and
containment. The chapter concludes with a discussion of resource recovery and the
use of waste material, including the recycling of hazardous wastes.

1.1 FEDERAL LAWS AND REGULATIONS

Federal environmental laws and regulations may affect planning, design, construction,
and rehabilitation of the infrastructure. The objective of environmental legislation is to
protect the health and welfare of the general public. Primary legislation has been
developed by the federal government, and the authority for implementing the laws has
been given to various federal agencies. State-level environmental legislation has, in
general, followed or expanded upon federal objectives and programs. Increasingly,
states are being given powers to implement federal programs. This has led to further
state involvement in promulgating laws, regulations, and judicial interpretations.

Federal legislation, and, therefore, requirements for compliance, have been steadily
evolving. The information in this section is intended to provide the highway engineer
and other interested parties with a basic understanding of the legal framework in
which environmental compliance requirements were developed, what they are intended to address, and how to find current information on the constantly changing federal,
state, and local agency regulations specific to a proposed project. Federal laws are
codified in the U.S. Code (USC) and are enacted by Congress. Federal regulations are
codified in the Code of Federal Regulations (CFR). Department of Transportation reg-

ulations are codified in Title 23 CFR-DOT regulations. The environmental regulations discussed in this chapter are found in Title 40 CFR-EPA regulations. Executive orders are issued by the President of the United States to mandate policy on specific issues. A few executive orders relevant to highway engineering have been included in this discussion. Executive orders designate the implementing agency, which generally responds by promulgating guidelines or issuing directives to address the stated policy. Table 1.1 indicates the environmental laws discussed in this chapter, the USC designations that apply, and which federal agency has the implementing authority.

1.1.1 National Environmental Policy Act (NEPA)

The National Environmental Policy Act (NEPA) of 1970 (with amendments) is the most important environmental legislation to be considered by highway engineers prior to planning or implementing highway projects. NEPA applies to all federal government agencies and the programs and projects that they manage, permit, or fund. NEPA mandates that the environmental implications of all federally funded programs or projects be evaluated and disclosed to the public. The purpose, foundation, and major provisions of NEPA are discussed in the following paragraphs. Implementing procedures are discussed in Art. 1.2.

The purpose of NEPA is "to declare a national policy which will encourage productive and enjoyable harmony between man and his environment; to promote efforts which will prevent or eliminate damage to the environment and biosphere and stimulate the health and welfare of man; to enrich the understanding of the ecological systems and natural resources important to the Nation; and to establish a Council on Environmental Quality."

Under Title I, Section 101, of NEPA, Congress declares that it is the "...policy of Federal Government, in cooperation with State and local governments, and other concerned public and private organizations, to use all practicable means and measures, including financial and technical assistance, in a manner calculated to foster and promote the general welfare, to create and maintain conditions under which man and nature can exist in productive harmony, and fulfill the social, economic, and other requirements of present and future generations." Section 101 also states that this policy should be "consistent with other essential considerations of national policy."

NEPA, Title I, Section 102, states that, among other things, "a systematic and interdisciplinary approach" should be utilized to ensure "the integrated use of the natural and social sciences and the environmental design arts in planning and decision making." A major provision of NEPA under Title I, Section 102(2)(C), sets forth requirements for environmental impact statements (EISs), which are required for every federal project to determine and document that the purpose of NEPA is fulfilled. NEPA, Title I, Section 102(2)(C), established the framework for the guidelines and requirements in preparing an EIS that were issued by the Council on Environmental Quality (CEQ). That section requires that the environmental impacts of the proposed action be stated. An EIS must identify adverse environmental effects that cannot be avoided, along with any irreversible and irretrievable commitments of resources that would be involved if the proposal is implemented. Alternatives to the proposal should also be identified. The EIS must evaluate and show the relationship between local short-term uses of the environment and the maintenance and enhancement of future environmental productivity.

The CEQ is established as environmental advisory body under Title II of NEPA. The CEQ's primary mandate is to oversee federal efforts in complying with NEPA.

TABLE 1.1 Summary of Environmental Laws, USC Designators that Apply, and Implementing Authority

---

National Environmental Policy Act, 1970 (42 USC §4321 et seq.)—Executive Office of the President, Council on Environmental Quality

Clean Air Act (42 USC §7401 et seq.)—U.S. Environmental Protection Agency (EPA)

Noise Control Act, amended 1978 (42 USC §§4901–4918)—U.S. EPA

Clean Water Act, 1977 (33 USC §1251 et seq.)—U.S. EPA, Army Corps of Engineers

Safe Drinking Water Act (SDWA; 42 USC §300)—U.S. EPA

Resource Conservation and Recovery Act (RCRA), 1974, amended 1984 (42 USC §6901 et seq.)—U.S. EPA

Toxic Substances Control Act (TSCA), 1976–1986 (15 USC §2601 et seq.)—U.S. EPA

Comprehensive Environmental Response, Compensation, and Liability Act (CERCLA), 1980 (42 USC §9601 et seq.)—U.S. EPA

Superfund Amendments and Reauthorization Act (SARA), 1986 (42 USC §6991 et seq.)—U.S. EPA

Intermodal Surface Transportation Efficiency Act (ISTEA), 1991 (23 USC §109, Section 1038)—U.S. Department of Transportation (DOT)

Farmland Protection Policy Act, 1981 (73 USC §4201 et seq.)—U.S. Department of Agriculture

Floodplain Management, 1977, Executive Order 11988

Federal Coastal Zone Management Act, 1984 (16 USC §§1451–1464)—U.S. Department of Commerce

Wild and Scenic Rivers Act (16 USC §§1271–1287)—U.S. Department of the Interior (DOI)

Protection of Wetlands, 1977, Executive Order 11990

Fish and Wildlife Coordination Act (16 USC §§661–666)—U.S. DOI, Fish and Wildlife Service

Federal Endangered Species Act (ESA) (16 USC §£1531–1543)—U.S. DOI, Fish and Wildlife Service

Department of Transportation Act, Section 4(f) [49 USC §1653(f)]—U.S. DOT

Rivers and Harbor Act, 1899 (33 USC §401, et seq.)

National Historic Preservation Act (16 USC §470 et seq.)—Advisory Council on Historic Preservation

Historic Sites and Buildings Act of 1935 (16 USC §£461–471)

Protection and Enhancement of the Cultural Environment, 1971, Executive Order 11593

Reservoir Salvage Act, 1960 (16 USC §469)

The Archaeological and Historical Preservation Act, 1974 (16 USC §469)

Archaeological Resources Act, 1979 (16 USC §470 et seq.)

CEQ regulations set forth procedures and considerations for preparing several environmental documents, including:

- Notice of intent (NOI)
- Environmental assessment (EA)
- Draft environmental impact statement (DEIS)
- Final environmental impact statement (FEIS)
- Finding of no significant impact (FONSI)
- Record of decision (ROD)

Each federal agency has set forth its procedures for implementing CEQ requirements and guidelines. The Federal Highway Administration (FHWA) regulations for implementing NEPA are set forth under 23 CFR 771, and the Department of Transportation's NEPA procedures are set forth in Order 5610.1C. An important aspect of NEPA is public involvement. In accordance with CEQ regulations, each agency must make diligent efforts to involve the public in preparing and implementing its NEPA procedures. Public involvement is discussed in Part 771.111 of the Federal Highway Administration NEPA regulations.

1.1.2   Clean Air Act (CAA)

In 1970, the Clean Air Act (CAA) provided the nation with a new approach to controlling air quality over the then existing Air Quality Act of 1967. The CAA established a national policy "to protect and enhance the quality of the Nation's air resources so as to promote public health and welfare and the productive capacity of its population" [42 USC, §1857(b)(1) 1970]. The CAA was amended in 1977, and was significantly amended again in 1990 with the enactment of the 1990 Clean Air Amendments (CA 90). CA 90 included provisions for stricter mobile source emissions (specifically, tailpipe emissions), as well as emissions linked to stationary sources such as hazardous or toxic pollutants.

The U.S. Environmental Protection Agency (EPA) has overall administrative authority for the implementation of CAA requirements. Every state must promulgate regulations by developing state implementation plans (SIPs). According to CAA provisions, federal facilities must comply with the SIP of the state in which they are located. CAA provides for the U.S. EPA to establish primary and secondary national ambient air quality Standards (NAAQSs). The goal of the primary standards is to protect public (human) health. Secondary standards are set with the intent of protecting the public welfare. These secondary standards consider deterioration of or harm to vegetation (including crops), visibility, aesthetics, and property. Many states' SIPs have set air quality goals that exceed federal requirements and carry their own set of penalties and fines for noncompliance.

The 1970 CAA requires review and approval of any new highway on which the anticipated average daily volume within 10 years is more than 20,000 vehicles, or the modification of a highway that will foster increased average daily traffic of more than 10,000 vehicles within 10 years. In addition, the legislation requires review and approval of any new parking facility for 1000 cars or more or modification of a parking facility to increase capacity by 500 cars or more, or any other indirect source of air pollution.

There are additional regulations promulgated under CAA that deal with air pollution created directly or indirectly by highway construction and use. Current provisions of CAA and CA 90 relevant to highway engineering are Titles I (Attainment and Maintenance of NAAQS), II (Mobile Sources), and VII (Enforcement). Title I addresses air pollution control requirements for "nonattainment areas," which are those metropolitan areas in the United States that have failed to meet NAAQSs. Because ozone is the most widespread pollutant in nonattainment areas, the requirements focus on controlling the volatile organic compounds (VOCs) and nitrogen oxides that contribute to ground-level ozone formation. Title II deals with revised tailpipe emission standards for motor vehicles, requiring automobile manufacturers to reduce carbon monoxide, hydrocarbon, and nitrogen oxide emissions. Provisions for enforcement under Title VII include fines and terms of imprisonment. Federal violations prosecuted by EPA may result in civil penalties of up to $25,000 per day and criminal enforcement if the violator fails to abate on notice [42 USC §7413(b)].

If a SIP does not demonstrate sufficient progress in achieving compliance with NAAQSs in a nonattainment area, EPA may prepare an implementation plan of its own and/or impose construction bans on stationary sources and/or withhold EPA-approved federal funds (such as transportation improvement grants) targeted for the state.

Federal Regulations under 40 CFR 50–99. CA 90 designated 189 hazardous air pollutants. Legislation prior to CA 90 provided for hazardous air pollutants to be regulated under the National Emission Standards for Hazardous Air Pollutants (NESHAP) program. NESHAP set emissions limitations for asbestos, benzene, beryllium, inorganic arsenic, mercury, radionuclides, and vinyl chloride, which will stay in effect unless specifically modified by CA 90. Each will be reviewed and revised, if appropriate, by November 15, 2000.

1.1.3 Noise Control Act (NCA)

The Noise Control Act (NCA) was enacted to control noise emissions that result from human activity. Among the objectives of NCA are two that are relevant to highway engineering: (1) developing state and local programs to control noise, and (2) controlling the sources of noise of surface transportation and construction activities. Under the 1978 amendments to NCA, greater responsibility for noise control was delegated to state and local agencies.

Under NCA, Section 7, one objective of the amendments was to develop and implement a national noise environmental assessment program, including evaluating the effectiveness of noise abatement measures, and to set ambient noise levels.

Determining the effect of an increase in noise levels due either to construction activities or to traffic is an important element of the EIS process. See DOT Order 5610.1C, Attachment 2, Section 9(c), Standards for Noise, Air, and Water Pollution.

1.1.4 Legislation to Protect Freshwater Sources

Clean Water Act (CWA). The Clean Water Act (CWA), which was enacted in 1977 as amendments to the Federal Water Pollution Control Act of 1972, constitutes the principal water pollution control program in the United States. The purpose of CWA is to "restore and maintain the chemical, physical, and biological integrity of the Nation's waters." The aspect of CWA that most influences the design, construction, or rehabilitation of highways is the mandate that the EPA (or the implementing state agency) develop a permit program "for the discharge of any pollutant, or combination of pollutants to the Nations waters." Waters of the United States have been defined in 40 CFR 122.2 as navigable waters, tributaries of navigable waters, and wetlands, including those adjacent to waters of the United States.

The requirement in CWA to establish a permitting program has led to development of the National Pollutant Discharge Elimination System (NPDES) (42 USC §§1251–1389). Two types of permitted discharges are especially important to the highway engineer: (1) discharges due to highway and related construction activities, and (2) discharges due to storm water runoff from highways and related facilities. Specific procedures to comply with NPDES permitting requirements are addressed in Art. 1.3.

Another provision of the CWA that affects highway planning decisions is the "404" permit process, applicable to the protection of national wetlands. The Army Corps of Engineers is responsible for implementing the regulations and issuing these permits. CWA water quality goals and standards may also influence highway planning, design, construction, and rehabilitation decisions.

Some regulations promulgated as a result of CWA, and the areas that they cover, are as follows:

- 40 CFR §§100–149, water programs and enforcement
- 40 CFR §122, wastewater discharges of waste to land and waters (federal permits)
- 40 CFR §§122–124, storm water runoff regulations

Safe Drinking Water Act.   The Safe Drinking Water Act was enacted at the end of 1974 to protect the nation's drinking water supply and protect public health through appropriate water treatment technologies. The act establishes maximum containment levels (MCLs), or standards for the maximum safe levels of specific constituents in potable water. Most important to highway engineers is the provision of this act that mandates protection of sources of drinking water.

1.1.5   Legislation to Control Hazardous and Nonhazardous Waste

Resource Conservation and Recovery Act (RCRA).   The Resource Conservation and Recovery Act (RCRA) was enacted in 1974, and amended in 1984, to address growing concerns related to solid waste disposal, both hazardous and nonhazardous. RCRA is the foundation of the hazardous waste management system in the United States. RCRA requires states to develop their own EPA-approved hazardous waste management plans and encourages options other than landfill disposal for final disposition of hazardous waste. A major objective of RCRA is to conserve and protect environmental resources, including the land resource that is lost to other uses when it is filled with solid waste.

RCRA established:

- A system for defining hazardous waste
- A method to determine whether hazardous waste has been generated
- Guidelines on how to store, handle, or treat hazardous waste
- Standards for proper disposal of waste
- Methods to track hazardous waste to its ultimate disposition

Resource recovery, which is an important research area mandated by RCRA, covers several materials used in highway construction, such as recycled glass, scrap tires, and recycled construction materials. Some hazardous materials can be treated and recycled for use in highway construction. RCRA also covers issues of "use constituting disposal" for projects that seek to use embankments or road subbase as disposal areas for hazardous waste, if suitability can be demonstrated. Some of the research and demonstration projects in the area of resource recovery that are applicable to highways are discussed in Art. 1.5.

Toxic Substances Control Act (TSCA). The Toxic Substances Control Act (TSCA) sets the policy for researching and testing suspected toxic substances to evaluate persistence in the environment and the effect on exposed humans (acute toxicity levels and/or carcinogenic effects). This act also regulates toxic substances not regulated by RCRA such as asbestos-containing materials (ACM) and polychlorinated biphenyls (PCBs).

Comprehensive Environmental Response, Compensation, and Liability Act (CERCLA). The Comprehensive Environmental Response, Compensation, and Liability Act (CERCLA) was passed in 1980. It established national policy and procedures for identifying and cleaning up sites that are found to be contaminated with hazardous substances, along with procedures for containing and removing releases of hazardous substances. CERCLA was amended and expanded by the Superfund Amendments and Reauthorization Act (SARA) of 1986. CERCLA established a hazard ranking system. Sites with the highest ranking have been placed on the National Priorities List (NPL) and are eligible for money from the substantial fund established for the environmental cleanup under CERCLA.

CERCLA provides for "joint and several liability," which means that any party identified as responsible for contamination of a site is considered equally responsible for cleanup costs with all other parties identified, and can be held 100 percent financially responsible in the event that other parties do not pay. Recovering costs from nonpaying parties is then the burden of the paying party and is pursued through the judicial system. Potentially responsible parties (PRPs) may be current or past owners and/or operators of a site where hazardous substances have been released, or persons who arranged for disposal or treatment of hazardous substances at the site. In addition, any person who knowingly accepted hazardous substances for transport to the site may be considered a PRP. Liability under CERCLA may also be retroactive to an era when the practices leading to the contamination were accepted industry standards. Petroleum is excluded from CERCLA unless mixed with other hazardous substances, and then the entire mixture is considered hazardous. Provisions have been established under SARA for an Underground Storage Tank Trust Fund that will address petroleum releases.

Another environmental concept mandated by CERCLA is "cradle-to-grave" responsibility for hazardous substances. Liability for a hazardous substance begins when it is accepted on the site or formulated at the site and continues even after it is disposed off-site at a legally permitted facility.

CERCLA is important to the highway planning process primarily in the acquisition of right-of-way. Accepting financial liability for contaminated property may adversely affect the economic analysis of a project and therefore its financial feasibility. In addition, if significant cleanup must take place before highway construction can begin, substantial delays to the project can be anticipated. Contaminated properties identified during route planning can be grave hindrances to project development and may be the crucial element in selecting alternative routes. Careful evaluation of the nature and extent of the contamination as well as the cleanup alternatives, costs, schedule, and ongoing liability is warranted on all sites with an identified release within the planned right-of-way purchase.

Superfund Amendments and Reauthorization Act of 1986 (SARA).   Title III of the
Superfund Amendments and Reauthorization Act of 1986 (SARA), Emergency
Planning and Community Right-to-Know (Public Law 99-499, Title III; 42 USC
§11001), established mandatory federal standards for community right-to-know pro-
grams and for reporting toxic chemical release by manufacturers under Section 313 of
USC §11001.

Intermodal Surface Transportation Efficiency Act (ISTEA).   Section 1038 of the
Intermodal Surface Transportation Efficiency Act (ISTEA) addresses the use of recycled paving materials. Section 1038(a) states, "a patented application process for recycled rubber shall be eligible for approval under the same conditions that an unpatented
process is eligible for approval." The subsection also provides that the U.S. EPA shall
evaluate the human health and environmental impacts of asphalt pavement that contains recycled rubber, determine the percentage of the pavement that can practicably
be composed of recycled rubber and the comparative performance of such pavement,
and conduct a study of the uses and performance of recycled materials in highways to
determine the environmental impacts and benefits of recycling such materials as
reclaimed asphalt and asphalt containing recycled glass and/or plastic. The study must
contain an economic cost-benefit analysis and an estimate of the environmental savings in terms of reduced air emissions, conservation of natural resources, and reduced
landfill waste.

Under Section 1038(d), ISTEA sets forth requirements to the states for use of scrap
tires in paving materials. By the year 1997, 20 percent of the highway paving material
applied in each state must be composed of recycled rubber.

1.1.6   Legislation to Govern Special Land Use

Farmland Protection Policy Act (FPPA).   The Farmland Protection Policy Act
(FPPA) of 1981 (73 USC §4201 et seq.) requires that the lead agency on a project
evaluate the effects a federal project may have on farmland before that agency can
approve any action that may result in the conversion of farmland from agricultural use
to nonagricultural use. If there are adverse effects, then alternatives to lessen them or
eliminate them must be considered in the evaluation.

Floodplain Management.   Executive Order 11988, Floodplain Management (May
24, 1977), directs all federal agencies to discourage development within floodplains
and to avoid long- and short-term modification of floodplains. Attachment 2 of DOT
Order 2610.1C, Section 11, Floodplain Management Evaluation, provides for NEPA
compliance related to floodplains and refers to DOT Order 5650.2, Floodplain
Management and Protection, for compliance criteria.

Federal Coastal Zone Management Act.   The federal Coastal Zone Management Act
of 1972 (16 USC §§1451–1464) provides for states with coastlines to develop and
implement federally approved coastal zone management programs (CZMPs). Once a
state has an approved management program, federal projects or federally permitted
development affecting the coastal zone must conform to the requirements of the state
program "to the maximum extent practicable." A determination of consistency with
the approved CZMP is required from the state before federal approval can be granted.

DOT Order 5610.1C, which sets forth the federal agency's procedures for complying with NEPA, covers coastal zones in Attachment 2 under subsection 12(d).

1.1.7 Legislation to Protect Natural Resources and Recreation Lands

Federal Wild and Scenic Rivers Act. The federal Wild and Scenic Rivers Act (16 USC §§1271–1287) provides that selected rivers that meet specified requirements and their immediate environments shall be preserved in free-flowing condition, and that they and their immediate environments shall be protected for the benefit and enjoyment of present and future generations. If a selected river is placed in the Wild and Scenic River System, no such designated river, or proposed designated river, may be degraded in its wild and scenic value by a federal project or agency. Any proposed federal construction projects on the river or in its immediate environment must be brought before Congress with an explanation of how the act will continue to protect the river despite the proposed construction activity.

Protection of Wetlands. Executive Order 11990, Protection of Wetlands (May 24, 1977), directs all federal agencies to refrain from assisting in or giving financial support to projects that encroach upon public or private wetlands unless the agency determines that there are no practicable alternatives to such construction and that the proposed action includes all practicable measures to minimize harm to wetlands that may result from such use.

As mentioned under the Clean Water Act subsection 101(4)(f), the Corps of Engineers is given permit authority to protect or receive resource compensation for wetland impacts. DOT Order 5610.1C, which sets forth the federal agency's procedures for complying with NEPA, covers wetlands in Attachment 2 under subsection 12, Considerations Relating to Wetlands or Coastal Zones.

Fish And Wildlife Coordination Act. The Fish and Wildlife Coordination Act (16 USC §§661–666) requires coordination and consultation among (1) the agency proposing the highway project, (2) the U.S. Fish and Wildlife Service of the Department of the Interior, and (3) the state agency responsible for protecting wildlife resources whenever the waters of any stream or other body of water are proposed to be impounded, diverted, or otherwise modified. Full consideration and evaluation of the cost and benefit on a resource and public welfare scale must be performed including proposed mitigation measures for potential impacts.

Federal Endangered Species Act. The Federal Endangered Species Act of 1973 (16 USC §§1531–1543) provides a means whereby the ecosystems upon which endangered species and threatened species depend may be conserved. It also provides a program for the conservation of such endangered and threatened species. Section 7 requires each federal agency, in consultation with, and with the assistance of, the Secretary of the Interior, to ensure that actions authorized, funded, or carried out by such agency do not jeopardize the continued existence of any endangered or threatened species or result in the destruction or adverse modification of habitat of such species unless such agency has been granted an exemption for such action.

For federal highway projects, a request is made to the Fish and Wildlife Service regarding whether any species listed or proposed as endangered are present in the project area. If so, a biological assessment must be completed and reviewed by the Fish and Wildlife Service. The Fish and Wildlife Service will make a determination as to the impacts on critical habitat or on the species itself and whether the impacts can be mitigated or avoided. In order to proceed with a project where impacts to endangered species have been identified, an exemption from the Endangered Species Act must be obtained.

The Department of Transportation Act, Section 4(f).   One significant environmental provision appears in the Department of Transportation Act, which established the Department of Transportation and mandated its mission. This is Section 4(f), which is a duplication of the language under Section 138 of the Federal-Aid Highway Act (23 USC §138). The provision states that "...the Secretary shall not approve any program or project which requires the use of any publicly owned land from a public park, recreation area, or wildlife and waterfowl refuge of national, State, or local significance as determined by the Federal, State, or local officials having jurisdiction thereof, or any land from an historic site...unless (1) there is no feasible and prudent alternative to the use of such land, and (2) such program includes all possible planning to minimize harm to such park, recreational area, wildlife and waterfowl refuge, or historic site resulting from such use."

According to regulations set forth in 23 CFR 771, a Section 4(f) evaluation must be prepared when a project will require the use of 4(f) land. The EIS process generally incorporates this evaluation. The final evaluation must include sufficient information to support a determination that the requirements of the act have been met. Section 4(f) is relevant only when there is actual taking or use of land from a federal park or site that is included in the National Register of Historic Places. The Advisory Council on Historic Preservation's procedures (36 CFR 800) must also be complied with when a project involves historic resources.

Rivers and Harbor Act.   The Rivers and Harbor Act (33 USC 401 et seq.) was enacted in 1899 and later amended to protect navigation and the navigable capacity of the nation's waters. The two provisions of the act most significant to highway projects proposed in or around U.S. harbors or rivers are:

- Section 9, which requires a permit for the construction of bridges or causeways across navigable waters of the United States
- Section 10, which requires a permit for various types of work performed in navigable waters including stream channelization, excavation, and filling

As stated under the Clean Water Act, the Army Corps of Engineers has permitting and enforcement jurisdiction for construction activities performed in a stream or on the shore of waters of the United States.

1.1.8   Legislation to Protect Historical and Cultural Resources

National Historic Preservation Act.   The purpose of the National Historic Preservation Act is to protect the historical and cultural foundations of the nation. The act provides for review by the Advisory Council on Historic Preservation (ACHP) of federal projects that may affect a historic site. The act mandates (in Section 106) that federal agencies take into account the effect of an undertaking on a property which is included in, or eligible for inclusion in, the National Register of Historic Places. Impacts on historic and culturally significant sites are considered in the EIS process under NEPA and implemented under DOT Order 5610.1C, Attachment 2, Section 5, entitled Properties and Sites of Historic and Cultural Significance.

The National Historic Preservation Act has itself not been a significantly controversial statute in highway projects. However, it is important to note the quasi-environmental implications: "...any Federal agency having direct or indirect jurisdiction over a proposed Federal or federally assisted undertaking...shall take into account the effect of the undertaking on any district, site, building, structure, or object that is included in the National Register."

Historic Sites and Buildings Act.   The Historic Sites and Buildings Act of 1935 (16 USC §§461–471) authorized the Historic American Buildings Survey, the Historic American Engineering Record, and the National Survey of Historic Sites. It authorized the establishment of national historic sites and the designation of national historic landmarks. It also mandated and encouraged interagency, intergovernmental, and interdisciplinary efforts for the preservation of cultural resources.

Protection and Enhancement of the Cultural Environment.   Executive Order 11593, Protection and Enhancement of the Cultural Environment (May 13, 1971), directs federal agencies to ensure the preservation of cultural resources in federal ownership. Each agency must also institute procedures to ensure that federal projects and programs contribute to the preservation and enhancement of non-federally owned sites that are of cultural significance. Each federal agency must provide for recording of National Register properties that will be unavoidably altered or destroyed as a result of federal action.

The DOT regulations, under Order 5610.1C, address EIS procedures for historic and culturally significant sites under Section 5, Properties and Sites of Historic and Cultural Significance.

Reservoir Salvage Act and the Archaeological and Historical Preservation Act.   The Reservoir Salvage Act of 1960 (16 USC §469) provided for the recovery and preservation of historical and archaeological data that might be lost or destroyed as a result of the construction of dams, reservoirs, and attendant facilities. This act was amended by the Archaeological and Historical Preservation Act of 1974 (also known as the Moss-Bennett Act) to include all proposed federal construction projects that threaten the loss or destruction of significant scientific, historic, or archaeological data. The act requires that the proposing agency notify the secretary of the interior of the threat. The federal agency may undertake the survey or recovery of data, or it may request that the secretary of the interior do so. If the agency itself undertakes the survey and recovery of data, it must provide the secretary of the interior with a report. The FHWA historic preservation procedures under the National Historic Preservation Act (Section 106) provide similar protection, and so this act is not applied if federal funding or other involvement is used in a highway project.

Archaeological Resources Act.   To protect archaeological resources on public lands, the Archaeological Resources Act requires the issuance of permits in order to excavate or remove any archaeological resources. Unauthorized activities are punishable by fine, imprisonment, or both. Rules and regulations concerning this act are printed under 43 CFR 7.

1.1.9   Additional Information Sources

In 1993 alone, more than 11,000 pages of new federal environmental legislation was promulgated. The volume has been escalating annually and can be expected to continue to do so as the environmental field matures. Even legal analysts sometimes have difficulty keeping abreast of all the changes; the highway engineer and other interested parties should defer to legal counsel for advice on the legislation applicable to a particular project and information on the implications of the latest proposed legislation, as well as current applicable state and local laws and regulations.

1.2 NATIONAL ENVIRONMENTAL POLICY ACT (NEPA)

The National Environmental Policy Act (NEPA) goals have been created to ensure that governmental actions promote the general welfare of people and fulfill the social, economic, and environmental requirements of present and future generations. Executive Order 11991 (May 24, 1977) directed the Council on Environmental Quality (CEQ) to issue regulations to federal agencies for implementing NEPA. CEQ regulations were published on November 29, 1978. In response to the CEQ regulations, the U.S. Department of Transportation (U.S. DOT) issued DOT Order 5610.1C on October 1, 1979. The DOT order supplements the CEQ regulations and establishes general procedures and requirements for the consideration of environmental impacts by agencies within DOT. Supplementary guidance and procedures for federal highway projects have been issued by the Federal Highway Administration (FHWA).

The purpose of Order 5610.1C is to establish procedures for consideration of environmental impacts in decision making on proposed U.S. DOT actions. It provides environmental impact information to the public in the form of environmental impact statements, assessments, or findings of no significant impact.

The order implements the mandate of NEPA, as defined and elaborated upon by CEQ regulations, within the programs of the U.S. DOT. The intent is to provide for a single process, set forth under this order, to meet requirements for environmental studies, consultations, and reviews in as many projects as possible.

It is U.S. DOT policy to integrate national environmental objectives. Through its missions and programs, DOT aims to:

• Restore or enhance environmental quality to the fullest extent practicable

• Preserve the natural beauty of the countryside and preserve public park and recreation lands, wildlife and waterfowl refuges, and historic sites; and to preserve, restore, and improve wetlands

• Improve the urban physical, social, and economic environment (e.g., increase access to opportunities for disadvantaged persons)

• Utilize a systematic, interdisciplinary approach in planning that may have a positive impact on the environment

FHWA regulatory procedures are contained in 23 CFR 771, and further guidance is provided in FHWA Technical Advisory T6640.8A. Title 23 CFR 771 embodies most of the requirements of other federal laws and regulations to which the lead agency on a federal action must conform before obtaining various approvals for federal highway projects. The requirements for environmental documents under NEPA also include subjects in other areas of environmental legislation and implementing regulations. Laws and regulations regarding air, noise, water, historic preservation, parklands, the coastal zone, farmlands, hazardous wastes, wildlife, and plants must be closely followed to meet the expressed public policy.

An outline of the steps in the NEPA process is presented in Table 1.2. Each step is discussed in detail in subsequent articles of this chapter. Figure 1.1 illustrates the environmental review process.

1.2.1 Proposal

Highway and related projects are usually initiated by either state or local agencies. If federal funding or approval will be required, the project is considered a federal action

TABLE 1.2 Outline of Steps in the National Environmental Policy Act (NEPA) Process

Step	Description
Proposal	The requirements of NEPA are initiated when a project or program (an "action" under NEPA) is proposed that is either conducted, sponsored, funded, assisted, or approved by a federal agency. A proposal under NEPA does not exist until a goal has been established and the proposing agency has begun taking steps toward deciding among different methods to achieve the goal.
Categorical exclusion	After the proposal is made, the proposing agency must evaluate whether the action falls under a categorical exclusion (CE). A CE is a category of actions that do not individually or cumulatively have a significant effect on the environment. Categorical exclusions are set forth in 23 CFR 771.117.
Environmental assessment	If the action is not categorically excluded, the agency initiates an environmental assessment (EA). An EA should be a concise report that provides information and analysis for determining whether an action will or will not have significant environmental effect, or whether certain environmental issues warrant further study.
Finding of no significant impact	If the EA findings indicate that no significant adverse environmental effects will result from implementation of the proposed action, then a finding of no significant impact (FONSI) is prepared and approved by FHWA. If an EA determines that environmental impacts may result from the action, the environmental impact statement (EIS) processes are initiated.
Notice of intent	The first filing in the EIS processes, published by the proposing agency in the *Federal Register,* is the notice of intent (NOI). The NOI is a public notice that an EIS will be prepared.
Scoping	After the NOI is filed, the scoping process begins. Scoping determines the scope of issues to be addressed in an EIS and identifies issues to be addressed in an EIS.
Draft EIS	Once the scope of the EIS has been established, preparation of the draft EIS (DEIS) begins.
Review	Following a period of public and agency review and comment on the DEIS, the agencies begin to prepare the final EIS (FEIS).
Record of decision	If the FEIS is approved, a record of decision (ROD) is filed and the proposing agency can move forward with the action. The ROD is a formal, written statement, required under NEPA, wherein a federal lead agency must present the basis for its decision to approve a selected project alternative, summarize mitigation measures incorporated into the project, and document any required Section 4(f) approval.

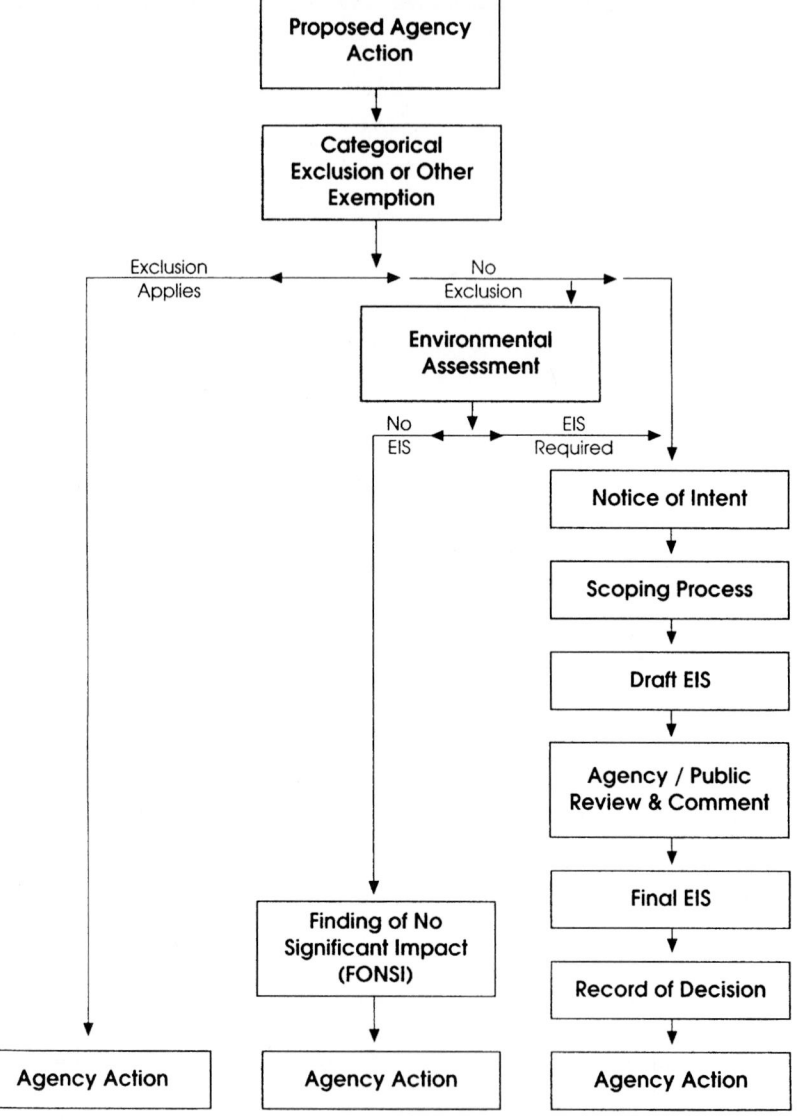

FIGURE 1.1 Overview of NEPA environmental review process. (*From R. E. Bass and A. I. Herson, Mastering NEPA: A Step-by-Step Approach, Solano Press Books, Point Arena, Calif., 1993, with permission.*)

under NEPA. For highway projects, formal approval by the FHWA is required under any of the following conditions:

- Federal funds will be used for engineering, construction, or acquisition of right-of-way.
- Revisions will be made to interstate highways. This does not include modifications of a facility or a new structure over or under an interstate highway or the construction over an interstate highway.
- Modifications will be made on noninterstate access-controlled highways so that access control will be affected or right-of-way previously financed with federal funds will be either disposed of or relinquished.

It should not be assumed, however, that lack of federal funding necessarily means no federal involvement. Certain federal approvals and permits may still be required. For example, an Army Corps of Engineers Section 404 permit is required for any dredging or fill operations in navigable waters.

If the proposal involves a federal action, a determination is also made regarding compliance with NEPA and other environmentally related federal requirements. This determination will indicate which of the related federal requirements are involved and whether the project:

- Requires preparation of an EIS (environmental impact statement, class I action)
- Is a CE (categorical exclusion, class II action)
- Requires preparation of an EA (environmental assessment, class III action)

For federal actions, the selection of the type of environmental document to prepare for a project must be made in consultation with the FHWA transportation engineer. For those projects that are not exempt (see Article 1.2.2), a preliminary environmental evaluation of the items listed in Table 1.3 should be conducted to provide the basic information needed at this stage of project development. The purpose of this initial evaluation is to identify environmental resources and issues that may be adversely affected. These issues would be evaluated in relation to the proposed project development process in order to determine:

- Whether additional alternatives should be studied to avoid or minimize significant environmental impacts
- The amount of time and resources likely to be needed to perform and document environmental studies, including time and resources for conforming to special-purpose environmentally related requirements
- Necessary mitigation measures, a range of costs, and an estimate of the time needed to negotiate with permitting agencies
- Whether the project has unusual problems or issues that are controversial, and therefore whether it may require legal review

1.2.2 NEPA Exempt Actions

NEPA provides for certain specific actions to be exempt from the environmental review process as described in the following paragraphs.

TABLE 1.3 Topics for Preliminary Environmental Evaluation

Aesthetics	Housing
Affected environment	Hydrology
Agricultural lands	Land use
Air quality	Noise
Archaeological resources	Parks
Biological environment	Population
Climate	Schools
Cultural resources	Seismicity
Dust	Social effects
Economic effects	Soils
Employment	Topography
Endangered species	Traffic
Energy	Urban development
Environmental setting	Utilities
Erosion	Vegetation
Floodplains	Water quality
Geology	Wetlands
Growth	Wildlife
Historic resources	

Categorical Exclusions. The NEPA regulations under 40 CFR 1508.4 provide for each responsible agency to identify types of federal actions under its purview that routinely do not individually or cumulatively carry significant environmental impacts. These projects, designated as categorical exclusions (CEs), are exempt from the requirements to prepare an environmental assessment or an EIS. The intent of this provision is to reduce paperwork, delays, and expense on federal actions that are relatively small and occur often enough that the environmental effects are known and predetermined as not significant.

FHWA sets forth highway-related actions that are CEs in two groups. The first group, CEs with no impacts, are found in 23 CFR 771.1179(a) and are listed in Table 1.4. The second group, CEs that generally have no significant environmental impact except under certain circumstances, are found in 23 CFR 771.1179(b). A CE determination is recommended by the proposing agency and approved by FHWA. The determination is documented by completing the CE form.

Federal actions, including CEs, must comply with a number of environmentally related federal laws. Compliance may require added studies and documentation. The CE determination for a federal highway or related action will be made by the FHWA transportation engineer. When satisfied that the project meets the exclusion criteria and that other environmentally related requirements have been met, the FHWA transportation engineer will indicate approval by signing the CE form. A copy of any documentation required to back up this determination should be sent to the FHWA transportation engineer.

In many cases, FHWA has reached agreement with a proposing agency on the treatment of very routine, repetitive projects with little or no environmental impact

TABLE 1.4 Categorical Exclusions Defined by the FHWA

1. Activities that do not involve or lead directly to construction
2. Approval of utility installations along or across a transportation facility
3. Construction of bicycle and pedestrian lanes, paths, and facilities
4. Activities included in the state's highway safety plan under 23 USC §402
5. Transfer of federal lands pursuant to 23 USC §317 when the subsequent action is not an FHWA action
6. Installation of noise barriers or alterations to existing publicly owned buildings to provide for noise reduction
7. Landscaping
8. Installation of fencing, signs, pavement markings, small passenger shelters, traffic signals, and railroad warning devices where no substantial land acquisition or traffic disruption will occur
9. Emergency repairs under 23 USC §125
10. Acquisition of scenic easements
11. Determination of payback under 23 CFR §480 for property previously acquired with federal-aid participation
12. Improvements to existing rest areas and truck weigh stations
13. Ride-sharing activities
14. Bus and railcar rehabilitation
15. Alterations to facilities or vehicles in order to make them accessible for elderly and handicapped persons
16. Program administration, technical assistance activities, and operating assistance to transit authorities to continue existing service or increase service to meet routine changes in demand.
17. Purchase of vehicles by the applicant where the use of these vehicles can be accommodated by existing facilities or by new facilities which themselves are within a categorical exclusion
18. Track and railbed maintenance and improvements when carried out within the existing right-of-way
19. Purchase and installation of operating or maintenance equipment to be located within the transit facility and with no significant impacts off the site
20. Promulgation of rules, regulations, and directives

Source: Adapted from 23 CFR 771.1179(a).

implications. Such projects may be processed on a programmatic CE basis if certain specified conditions are met. Use of this programmatic process is subject to annual review. A CE classification does not exclude a project from the requirements of federal environmentally related processes. These requirements must be met before FHWA will make the exclusion determination.

For example, CE actions where National Register–eligible properties are present or are potentially present (such as an archaeologically sensitive area) require early public involvement. Projects that have been defined as having minimal areas of potential effect (APEs) and therefore do not fall within the Section 106 definition of undertaking, and where no known historic resources are present, are excepted from this requirement. Generally, an APE project consists of the operation, repair, maintenance, or minor alteration of existing highways and streets (within already-established rights-of-way), sidewalks, gutters, bicycle and pedestrian trails, and similar facilities. Table 1.5 lists types of highway-related projects with minimal APEs. Interested parties are to be notified of the conclusions of studies. When minimal-APE projects have known historic resources, the public involvement process will begin when resources have been identified.

Express Statutory Exemptions.   Congress has the authority to override NEPA requirements. Congress may, at its discretion, exempt a specific federal project or program from NEPA through legislation.

Statutory Conflicts.   If Congress passes legislation relevant to an action that is in direct conflict with requirements in NEPA, the action could be exempt from NEPA on the basis of statutory conflicts. This may occur if legislation requires an agency to act within a time frame that would preclude the NEPA process.

Emergency Actions.   Emergency projects such as road repair following a flood are statutorily exempt under 23 CFR 771.1179. Emergency repair applies to work necessary to reopen a road or bridge as a result of a slide or slipout or other storm damage. The regulations define *emergency* as a sudden, unexpected occurrence, involving a clear and imminent danger, demanding immediate action to prevent or mitigate loss of, or damage to, life, health, property, or essential public services. Emergency includes such occurrences as fire, flood, or earthquake or other soil or geologic movements, as well as such occurrences as riot, accident, or sabotage. Depending on the nature and extent of the emergency, FHWA may modify or waive certain environmental requirements as long as the emergency response is directly related to controlling the effects of the emergency.

1.2.3 Public Involvement

Under the CEQ regulations, public involvement is an essential element of the NEPA process, and the proposing agency must make sincere efforts to encourage and provide for early and continuing public participation in the decision-making process [40 CFR 1506(a)]. Opportunities for public involvement are provided at several stages during the development of NEPA documents, including the notice of intent (NOI), public scoping meetings, public contribution and comments on EAs, public comment periods on the draft EIS, public hearings for the final EIS, and approval period for the final EIS. Opportunities for the public to review and comment on documents occur when a notice of availability is published. A notice of availability is a formal public notice under NEPA announcing the availability of a completed EA, DEIS, or FEIS. Such notice is to be published in local newspapers or other local print media, provided to the state clearinghouse, presented in special newsletters, provided to community and business associations, placed in legal postings, and presented to interested Native American tribes, if appropriate. For an EIS, publication of such notice is also required in the *Federal Register*. Notices and other public announcements regarding the project should be sent individually to anybody expressing an interest in a specific action.

Early incorporation of public input on issues dealing with social, economic, and environmental impacts helps in deciding whether to prepare an EA or an EIS, the scope of the document, and the important or controversial issues related to the project or program. When impacts involve the relocation of individuals, groups, or institutions, special notification and public participation efforts should be undertaken.

TABLE 1.5   Project Types Having Minimal Areas of Potential Effect (APEs) on Cultural Resources

1. Pavement reconstruction, resurfacing, and placement of seal coat.

2. Work on bridge structures and appurtenant facilities, such as traffic control devices or toll collection facilities.

3. Addition of lanes in the median of a divided highway.

4. Repair and maintenance of the highway and all its appurtenant facilities, including replacement of damaged or inadequate facilities.

5. Removal and/or replacement of distinctive roadway markings such as painted stripes, raised pavement markers, or thermoplastic tape.

6. Landscaping within highway right-of-way and on department-owned property.

7. Bridge maintenance painting when performed in conformance with the requirements of air pollution control and water quality control agencies having jurisdiction.

8. Abandonment, removal, reconstruction, or alteration of railroad grade crossings or grade crossing protection.

9. Additions to existing structures, provided that the addition will not result in an increase of more than 50 percent of the floor area of the structure before the addition, or 2500 ft^2, whichever is less; included herein is work on items such as office and equipment buildings, warehouses, roadside rests, weight and inspection stations, toll facilities, and state-owned rentals.

10. Addition or replacement of devices such as glare screen, median barrier, fencing, guiderail, safety barriers, energy attenuators, guideposts, markers, safety cables, ladders, lighting hoists, or signs.

11. Installation of noise barriers and alteration to existing buildings to provide for noise attenuation.

12. Projects to eliminate hazards within the operating areas or the operating right-of-way.

13. Modifying existing features such as curbs, dikes, headwalls, slopes, or ditches within the right-of-way to improve roadside safety.

14. New copy on existing on- and off-premise signs.

15. Maintenance of existing landscaping, native growth, and water supply reservoirs (excluding the use of defined economic poisons); included herein is work on such items as treatment, maintenance, and replacement of all vegetative material, native or planted, on state-owned property, including highway rights-of-way, building sites, and rental unit. Work will include such items as watering, fertilizing, weed control by hand or mechanical means, trimming and cutting by hand or mechanical means, tree trimming, and tree removal when required for traffic safety or because of disease infestation or for pest control.

16. Maintenance of fish screens, fish ladders, wildlife habitat areas, artificial wildlife waterway devices, stream flows, springs and waterholes, and stream channels (clearing of debris) to protect fish and wildlife resources.

17. Minor widening of less than a lane width and/or adding paved shoulders.

18. Minor operational improvements, such as median and side ditch paving or drainage facilities.

19. Modification of traffic control systems and devices, including addition of new elements, such as signs, signals, and controllers.

20. Installation, removal, or modification of regulatory, warning, and informational signs.

21. Minor widening of less than half width and/or adding paved shoulders.

22. Addition of auxiliary lanes when required for traffic separation, such as truck lanes, or for lane changing between adjacent interchanges.

TABLE 1.5   Project Types Having Minimal Areas of Potential Effect (APEs) on Cultural Resources

---

23. Modification, upgrading, alteration, or relocation of railroad grade crossing protection and the construction of bus and truck stop lanes at railroad grade crossings.

24. Alteration or widening of existing grade separation structure where the primary function and utility remains unaltered.

25. Restoration or rehabilitation of deteriorated or damaged structure, facilities, or mechanical equipment to meet current standards of public health and safety, unless it is determined that the damage was substantial and resulted from an environmental hazard such as earthquake, landslide, or flood and there is an option to relocate.

26. Installation of new traffic control systems, such as signs and/or signals, channelization of intersections, or pavement striping.

27. Addition of nonmotorized trails to separate such use from motorized traffic.

Source: Technical Advisory T6640.8A, FHWA.

The proposing agency must provide for one or more public hearings or the opportunity for hearings to be held at a convenient time and place for federal actions that require significant amounts of right-of-way acquisition, substantially change the layout or function of connecting roadways or of the facility being improved, have substantial adverse impact on abutting properties, or otherwise have a significant social, economic, or environmental effect [23 CFR 771.111(2)(iii)].

During public hearings, the public should receive information on the project's purpose and reason and be told how it is integrated with local planning goals. The public should be provided with information on the major design features of the project, potential impacts, and available alternatives under consideration. Processes of special interest to the public, such as relocation procedures and right-of-way acquisition, should be carefully explained, as should the agency's procedures and timing for receiving oral and written public comments [23 CFR 771.111(2)(v)]. The public comment period for a draft EIS is at least 45 days, except in rare circumstances determined by the Environmental Protection Agency (EPA). Public hearings must be documented, including providing FHWA with a copy of the transcript from the hearing.

Under the Freedom of Information Act (5 USC §552), an agency must make documentation, including interagency comments, available to the public at no cost.

1.2.4   Environmental Assessment

An environmental assessment (EA) is conducted for a project that is not under a CE and does not require an EIS. The purpose of the EA is to identify the environmental impacts of the project and to provide the necessary information of significant impacts. The EA process includes studies and procedures required by environment-related laws and regulations. An environmental assessment may also be conducted for projects requiring an EIS. This should be completed in order to coordinate, at the earliest appropriate time, with local, regional, state, and federal agencies having jurisdictional authority regarding the project. It would also serve to focus, at an early stage, on the significant environmental issues deserving study, evaluate their potential impact, begin to identify alternatives and measures that might mitigate adverse environmental impacts, and eliminate study and discussion of nonpertinent issues. At any point during the EA process, FHWA can determine that the preparation of an EIS is required. However, if the study indicates that there will be no significant impact, a finding of no significant impact (FONSI) will be prepared by FHWA.

Significance under NEPA. The NEPA regulations define the context of the project and the intensity of its impacts. The context is the setting (physical, social, political, and geographic) within which the proposed project should be considered. Under the regulations, several potential contexts must be considered for the assessment of resources. These include society as a whole, an affected region, affected interests, or a locality.

Significance also varies with the setting of the proposed action. For instance, noise increases in an undeveloped area may be evaluated to be of less significance than in an urban area because there are fewer affected receptors. In addition, both short- and long-term effects are relevant in determining context.

Intensity refers to the severity of an impact. Measures of intensity vary according to resource type. In socioeconomic issues, intensity would vary with the number of displaced homes or businesses. Measures may diminish the intensity of project effects, and in many situations, intensity will be reduced to the point where the impacts will not be significant.

Determining the significance of an impact can be done in four systematic steps:

- Eliminating obviously insignificant effects
- Performing studies to obtain more information if the impacts are not clear
- Applying the context and intensity criteria to evaluate significance
- If an EIS or a FONSI is prepared, holding a public review for the determination of significance

Using a systematic approach and including public involvement will result in fewer court challenges, gain a greater number of cooperating agencies, build consensus earlier, screen out inappropriate alternatives, and determine whether a FONSI or an EIS needs to be prepared.

Upon completion of an EA, formal FHWA approval of the study is required before it can be made available to the public. The determination of environmental effect is not made by FHWA until after the public availability period.

Format of an Environmental Assessment. The contents of an EA are determined by the number of environmental resources identified in the preliminary environmental evaluation: the effect on each needs to be evaluated. A preliminary review will identify potentially affected resources and what level of analysis may be necessary. It will also identify those resources that are clearly without the potential for significant adverse effect. This will allow for effective and efficient use of staff and resources. Early coordination with other involved agencies and the public can also assist in determining which assessments are necessary.

The amount of detail in the EA is to be the absolute minimum, conveying key issues and the significant potential adverse impacts. The discussion must be focused on substantiating whether the proposed project will have a significant effect on the environment. It is not necessary to discuss those impacts where the effect is not significant. The following elements should be included in an EA document:

- Cover sheet
- Table of contents
- Statement of project purpose
- Description of the project
- Description of the affected environment

- Environmental evaluation
- Impacts and proposed mitigation measures
- Consultation and coordination
- List of preparers

Purpose of an Environmental Assessment. The purpose and need for the project should be succinctly explained at the beginning of the EA. The need for the project is based on an objective and analytical evaluation of current information and future anticipated conditions. Statements regarding the need for the project should be factual, and, to the extent feasible, quantifiable. Table 1.6 shows various types of information related to need for highway projects.

Description of the Project. The project description should be written in clear, non-technical language. A glossary or footnotes should be used to define or explain technical terms. Exhibits are also essential for a clear understanding of the project features.

The description of the project should include the scope of the project; location and limits; major design features; typical sections (where appropriate); a location map (district, regional, county, or city map showing state highways, major roads, and well-known features to orient the reader to the project location); a vicinity map (detailed map showing project limits and adjacent facilities); current status of the project including integration into regional transportation plans, regional transportation improvement programs, congestion management plans, and the state transportation improvement program; proposed construction date; funding source; and the status of other projects or proposals in the area. For projects involving more than one type of improvement, the major design features of each type should be included. For instance, a freeway widening may also involve a median barrier, ramp metering, sound barriers, and park-and-ride lots. The project will also include necessary borrow or disposal sites, as well as detours and contractor equipment yards.

TABLE 1.6 Basic Information Requirements to Establish Need for Highway Projects

Problem or deficiency	Information requirements
Capacity problems	(1) Hours of congestion being experienced and reasons for congestion; (2) present and projected average daily traffic (ADT), peak-hour volumes (PHV), and level of service; (3) basis for projected ADT, e.g., development in areas served by facility, traffic diversion, or increasing recreational traffic
Safety problems	(1) Personal injury and fatality rates vs. statewide rates; (2) reasons why accidents are happening (e.g., poor geometrics, turning movements, lack of merging or weaving distance, curves, lack of shoulders); (3) safety benefits of the project
Operational deficiencies (ramp metering and auxiliary lane projects)	Explicit description of the problem(s), such as excessive demand or steep grades
Structural deficiencies	(1) Results of field survey or report on condition of pavement, bridges, drainage capacity, etc.; (2) maintenance cost vs. statewide rate for the same type of facility

Alternatives to the project, including the no-project option and nonhighway alternatives, are also to be discussed in this section. Project alternatives can be classified into two types: viable, and those studied but no longer under consideration. Viable alternatives are to be described in detail to compare their effectiveness against the proposal in meeting the project purpose and need, potential impacts, and cost. Alternatives no longer under consideration should be explained briefly.

Description of the Affected Environment.   The setting should be written in a clear manner. Visual displays can be used to eliminate verbiage and clearly describe the setting. Most of the setting will be written after the impacts have been identified to focus on the appropriate details for impact evaluation. Beyond the general contextual background, discussion should include only that portion of the setting that is affected. For example, if there are no effects on riparian habitat, then the setting would not include a description of riparian habitat beyond that needed for background information.

Environmental Evaluation.   The environmental evaluation section should list technical studies or backup reports used in making the environmental evaluation. It should indicate where reports are available. Initial or preliminary environmental evaluations should form the basis for deciding what information and subjects need to be obtained and discussed in the EA. It is not intended that each item be supported by detailed studies or discussion, but only that there be enough information in the EA to allow a reasonable determination of effect and significance. The evaluation should be completed to assess the effects of the proposal and of viable alternatives to the proposed project as described. If more than one alternative is involved, the evaluation discussion must identify the alternative(s) associated with each specific impact being discussed. The following subject areas should each be addressed and either discussed or set aside as not being a significant issue.

- *Topography.*   Briefly describe the terrain in the project area.
- *Geology.*   Note any geological features that would affect or be affected by the project. If the project involves known unstable slopes or a structure, then the characteristics of any faults in the area should be described. This will include the location, name, and maximum credible earthquake associated with each fault.
- *Hazardous wastes.*   Describe the results of the initial site assessment.
- *Soils.*   Identify the existence of any prime or unique farmland, including any agriculture preserve (under the Williamson Act), and any slide- or erosion-prone soils.
- *Hydrology.*   Describe existing drainage patterns, floodplains, and any downstream or adjacent bodies of water.
- *Wetlands.*   Describe the limits and significance of any area that meets the Corps of Engineers' criteria of wetlands.
- *Vegetation.*   Describe existing trees, brush, and landscaping. Identify the presence of any rare or endangered species and significant habitat characteristics.
- *Fish and wildlife.*   Describe species important to the area and any endangered or threatened species present.
- *Land use.*   Generally describe the existing land use in the area and any signs of pressure for growth, decline, or substantial change in character.
- *Planning.*   Briefly describe any aspects of community, local district, local government, and regional plans that are relevant to the effects that the project may cause. The plans of state or federal agencies may also be relevant. The most important

information comes from goals and policies, specific proposed actions, and progress to date in plan implementation. When the project gives access to undeveloped land or adds capacity, discuss the attitude of agencies toward growth and development adjacent to the proposed project.

- *Natural resources.* Discuss any changes in the use of natural resources as a result of the project.
- *Aesthetics.* Scenic resources should also be considered, together with the effect on planned local or regional aesthetic goals.
- *Social factors.* Briefly describe relevant ethnic and income data for the census tracts affected, the character of adjacent communities, and the value and availability of housing if any would have to be taken.
- *Economic factors.* Briefly describe the job situation, local businesses, and any other aspects of the local economy that might be affected significantly.
- *Public services and facilities.* Describe parks, schools, or similar facilities and police, health, or other public services. Comment whether they might be adversely affected by right-of-way requirements, noise, construction activities, traffic diversion, changes in use, or changes in taxes and revenue base.
- *Historic and cultural resources.* If the project has been categorized as having a minimal area of potential effect and historic studies are not needed, include a statement that the project has a minimal area of potential effect and no historic properties are involved. If eligible properties are found, describe their location and significance in terms of National Register criteria. Please note that the exact locations or trinomial designations of archaeological sites are not to be described or displayed on exhibits. Discussions of history or ethnography of a region are not necessary. This section should document the public involvement process that occurred as part of the identification and evaluation of resources. If consultation with the state historic preservation officer has taken place, it should be documented here.

Impacts and Proposed Mitigation Measures of the Environmental Assessment. Mitigation measures proposed are to be clearly identified as mitigation measures. Proposed measures are to be presented as commitments and not just actions that could or should be done. As part of the proposal, they should be included in the project description. Though the following is not a complete list, many highway-related projects will have impacts in the following areas:

- *Erosion and siltation.* Many projects have cut or fill slopes that require mitigation of potential erosion. Mitigation measures should be spelled out. An example would be seeding and mulching, placing of straw, or other deterrents to slope erosion. Siltation of streams is also a concern. Use of special dams or catch basins to prevent silt from getting into streams should be mentioned.
- *Hazardous waste.* If any initial studies or preliminary environmental evaluations identify known or potential hazardous waste sources, alternatives to avoid the site must be explored. If the site cannot be avoided, an assessment including sampling and possibly a characterization of the problem should be conducted. Findings and recommendations are to be discussed. If a hazardous waste site is identified (through information referenced to the address of the site or the county assessor's number), specify the type of regulatory actions it is subject to and any environmental databases or lists that it appears on along with regulatory identification number.
- *Floodplains.* If the project involves several encroachments, the description can include a discussion of the potential flood risks and floodplain impacts with a sum-

mary for each encroachment placed in an appendix. Floodplain impacts include the support of probable incompatible floodplain development and impacts on natural and beneficial floodplain values. Discussion should include measures to minimize floodplain impacts and restore and preserve floodplain values impacted by the action.

- *Air quality.* A carbon monoxide air quality study should be performed on highway projects. This may vary from an extensive study with a formal report to a brief investigation with an informal memo to the file. The results of an air study may be presented here in their entirety, if one page, or summarized if lengthy. Quantified microscale results are frequently required.

- *Noise.* A noise study is needed for almost every highway project. The results should be summarized. Both existing and projected exterior noise levels at sensitive receptors should be tabulated for each alternative. If there are no sensitive receptors in the area of impact (e.g., undeveloped or rural areas), it may be adequate to state this fact without preparing a formal noise study and report. A noise study is not required for projects unrelated to traffic noise such as lighting, signing, landscaping, or safety. Mitigation should be proposed when the predicted traffic noise levels approach or exceed the FHWA noise abatement criteria, or when predicted levels substantially exceed existing levels (23 CFR 772). Where mitigation is not proposed, the reasons must be given. Acceptable reasons include frequent access openings that negate attenuation, cost out of proportion to the benefits, aesthetics, and local opposition.

- *Consistency with community policies and goals.* The extent to which project effects are consistent with community goals should be noted. Identify the RTP and FTIP in which the project is included. A determination must be made as to the consistency of project effects with any adopted local or regional conservation or preservation plans.

- *Historic sites.* The effects, as described in 36 CFR 800, of the various alternatives on historic properties need to be explicitly stated. If there are no effects, include a statement that FHWA determined that the project will have no effect on historic properties. If there are effects but they are not adverse, any specific data leading to that finding should be stated. Documentation of public involvement efforts must be included along with consultation with the Advisory Council on Historic Preservation (ACHP) if it has occurred. If the effect is adverse, describe surveys undertaken, the number and type of historic properties to be affected, and the proposed treatment of these properties. A Section 4(f) evaluation is to be included as a separate section in the EA if there will be involvement with a Section 4(f)-protected resource.

1.2.5 Finding of No Significant Impact

A finding of no significant impact (FONSI) is a document issued by a federal agency briefly presenting the reasons why a federal action, not otherwise excluded, will not have a significant effect on the human environment and therefore does not require the preparation of an environmental impact statement (EIS). If the conclusion of an environmental assessment is that a proposed project will not result in any significant impacts, the EA will be sent to FHWA with a recommendation for approval and the preparation of a FONSI. Agencies that decide not to prepare an EIS after completion of an EA must obtain a FONSI from the FHWA.

If the analysis clearly indicates that the proposed action will not have a significant adverse impact, the EA will be attached to a FONSI to support the determination of no

TABLE 1.7 Evaluation of Mitigated FONSIs

Advantages	Criticisms
Encourage environmentally acceptable projects	Lack of uniform standards for public notice and review
Save time and cost	Lack of public involvement in decision-making process Absence of means to enforce mitigation Excessive judicial scrutiny No evaluation of alternatives

significant effect. If, during the course of the analysis, acceptable mitigation is identified and incorporated into the project that clearly reduces any potential impact to below a level of significance, the EA will be attached to a FONSI to support the determination of no significant effect with appropriate mitigation.

If there is to be a FONSI, the EA must show that each potential environmental issue was adequately reviewed and each will have no effect, no significant effect, or a potentially significant effect that will be mitigated. A mitigated FONSI is one in which significant environmental effects have been determined but can and will be mitigated to a level of no significant impact. For a mitigated FONSI to be effective, the evaluation of the issue and potential mitigation measures must be thorough, mitigation measures must be very specific and provide a convincing solution to the environmental concern, and the proposing agency must be committed to the solution. Table 1.7 outlines some of the criticisms and advantages of the mitigated FONSI.

Federal procedures do not allow FHWA to make a determination of no significant impact until after a public availability period. There is no record of decision (ROD) for FONSIs, and there is no formal distribution of the FONSI. Normally, copies are sent to individuals, groups, and agencies who have made substantive comments for the EA or who have expressed concern.

1.2.6 Notice of Intent (NOI)

If it is determined that a NEPA environmental impact statement is required, the proposing agency submits a notice of intent (NOI) to FHWA for publication in the *Federal Register*. The format and content of the notice of intent are shown in Appendix A-2 of FHWA Technical Advisory T6640.8A. In addition to those instructions, it is recommended that a brief statement be added, under Supplementary Information, soliciting the views of agencies that may have knowledge about historic resources potentially affected by the proposal or may be interested in the effects of the project on historic properties.

1.2.7 Scoping

Scoping is NEPA terminology for the process of developing the scope of issues to be addressed in an EIS and for identifying significant issues to warrant in-depth study. The process of scoping involves mailing a notice of preparation to initiate early consultation and coordination with cooperating agencies (federal agencies that have jurisdiction over the project). Other federal agencies with relevant expertise may also be asked to serve as cooperating agencies. This latter, nonmandatory request may be similar to the notice of preparation, except that the statements regarding permit or approval requirements do not apply. In addition to cooperating agencies, any other affected agency or party should be invited to participate in the scoping process.

There should be an early and open scoping process involving the public and state and local governments. Preparation of an EIS requires consideration of an array of environmental issues in order to comply with NEPA requirements. The intention of the scoping process is to emphasize important or controversial concerns early in the process: real problems are identified early and receive primary focus, and time and resources are not expended on insignificant issues.

1.2.8 Preparation of the Draft Environmental Impact Statement (DEIS)

Preparation of the draft EIS should begin at the earliest practical time. A key element is the early exploration of alternatives and their associated environmental effects. This will assist in early identification of avoidance alternatives and allow early coordination efforts with cooperating and responsible agencies, aimed at identifying and reaching agreement on appropriate mitigation measures. The draft EIS is prepared in consultation with FHWA. The purpose of the EIS is to provide information on the environmental implications of an action or a program. Responsible decisions can be then made regarding possible alternatives or mitigation measures intended to decrease the adverse environmental consequences. In the spirit of NEPA's purpose, the EIS is to be used as an important and necessary tool before decision making as opposed to a backfilling justification for decisions already made.

Concise and succinct statements, evaluations, and conclusions are important in preparing the EIS. Lengthy, encyclopedic discussions of the subject matter diffuse the focus of the document from its analytical purpose. The document should be written to emphasize the significant environmental impacts. Discussions of less significant impacts should be brief, but sufficient to demonstrate that due consideration was given and more detailed study is not warranted. By law and necessity, the document will be prepared using an interdisciplinary approach that integrates the use of natural sciences, social sciences, and the design arts.

Format of the Draft EIS. The objective in establishing a format is to present information in a consistent manner and sequence so that public issues and environmental impacts can be tracked throughout the document. This consistency must be reflected not only within each section but among the various sections where it is necessary to revisit some topics. An example of this concept would be the discussion of project alternatives that are treated in various sections of an environmental document. The desired product is a concise, easily understandable document that can fully inform decision makers and the public about the environmental effects of a proposal and its alternatives. In accordance with 40 CFR 1502.10, the following outline is used unless compelling reasons to do otherwise are given by the proposing agency:

- Cover sheet
- Summary
- Table of contents
- Purpose of and need for action
- Alternatives
- Affected environment

- Environmental consequences
- Mitigation measures
- List of preparers
- List of who received copies
- Appendixes
- Index

An environmental impact statement is an analysis of environmental aspects of a proposed project that are anticipated to have a significant effect on the human environment. EISs follow the instructions and content as described in FHWA Technical Advisory T6640.8A, Guidance Material for the Preparation of Environmental Documents. The technical advisory provides guidance on the format and content of EISs needed to meet the requirements of FHWA regulations for implementing NEPA (23 CFR 771). The regulations and technical advisory are in Appendix A-2 of that document. The Council on Environmental Quality (CEQ) regulations stress the importance of focusing on significant issues and avoiding detailed discussion of less important matters in an EIS. CEQ regulations emphasize brevity. Normally, EISs should be less than 150 pages, or less than 300 pages if unusual in scope and complexity.

The preparation of a DEIS calls for the exercise of judgment in dealing with questions of selection, relevance, emphasis, and organization. Exhibits (charts, tables, maps, and other graphics) are useful in reducing the amount of narrative required. They should be technically accurate and of high quality. Brevity in the text, consistent with the extent of the impact and the scale of the project, should be the objective. It is emphasized that the elements listed in the following paragraphs require discussion only if they are pertinent or are not adequately covered in another section. The adequacy of a DEIS is measured by its functional usefulness in decision making, not by its size or amount of detail.

Cover Sheet.   The cover sheet must clearly indicate that the document is a draft EIS and include the information indicated in Table 1.8.

TABLE 1.8   Information That Should Be Included in a Draft Environmental Impact Statement (DEIS)

Item	Information
Title	A title, location of the proposed action, and a report number are required. The FHWA division office will add this number to the EIS cover sheet when the document is approved for circulation.
Agencies	The proposing (or lead) agency must be clearly identified, including the name, address, and telephone number of a primary contact person. In addition, list all agencies that were requested to serve as a cooperating agency. These will include agencies that have special expertise with respect to any environmental issue addressed in the statement.
Abstract	The abstract is a one-paragraph description of the proposed project, alternatives, significant impacts, and major mitigation measures.
Comments due date	The ending date that comments on the draft EIS must be received should be clearly visible and easily identifiable.

Summary. The summary should be an overview of the entire statement and should run about 10 to 15 pages. The summary must include the following information:

- Briefly describe the proposed project, including the route, termini, type of facility, number of lanes, length, county, city, and state, along with significant appurtenances, as appropriate.

- List other federal actions required because of this project, such as permit approvals. Also describe any significant actions proposed by other governmental agencies in the same geographic area as the proposed project.

- Summarize the major alternatives considered.

- Summarize the significant environmental impacts of each alternative, both beneficial and adverse.

- Identify proposed mitigation measures.

- Briefly describe any areas of concern (including issues raised by agencies and the public) and any important unresolved issues.

As with most environmental documents, replacing discussion with matrices, charts, maps, tables, or other graphics is preferred in the interest of brevity. This is especially applicable in the summary, where items relating to alternatives and their impacts can often be presented in a matrix format, with impacts quantified where possible, thus keeping the text discussion to a minimum.

Table of Contents. A table of contents must be prepared for the document. The table of contents should closely follow the document outline that is established through the scoping process. It should include a lists of tables and figures and a list of appendixes.

Purpose of and Need for Action. A statement of purpose and need is an important section of an EIS. Unless the need for a facility is clearly and convincingly stated, the remainder of the EIS will fail in its purpose. The discussion of the purpose and need of a project for the EA presented in Art. 1.2.4 is applicable to the draft EIS. It is important to be specific and to present quantifiable information about the current and projected circumstances that are driving the decision to go forward with the project. This would include such things as accident data, existing and projected traffic data, and the ability of existing transportation facilities to provide for future transportation needs.

Alternatives. The purpose of the alternatives section is to describe reasonable alternatives and identify them as either the no-project alternative, an alternative under consideration, or an alternative that has been withdrawn from consideration. The DEIS is to state that alternatives currently under consideration are being given equal treatment at this stage of the EIS process and that a preferred alternative will be identified following the review and public hearing process.

It is in the best public interest to make the presentation of project alternatives as objective and neutral as possible. The purpose is to present the decision maker with the most relevant, comprehensive, and factual database possible that fairly covers project alternatives and their associated beneficial and adverse impacts. It is largely from this database that the decision maker will develop the rationale for making the selected alternative decision. Implementation of this concept must be recognizable in the draft environmental document and must be carried through and retained in the final environmental document.

No-project alternatives. The no-project alternative is always a viable alternative and is to be discussed and evaluated in this section with equal consideration to the other potential alternatives. The discussion is to identify the transportation, economic, and social consequences of not building a project including the tradeoffs in terms of benefits. This will inform the decision maker and the public of the choices available. A helpful reference for developing this discussion of the proposed project is *Impact Assessment Guidelines: The Role of the No-Build Alternative in the Evaluation of Transportation Projects,* National Cooperative Highway Research Program Report No. 8-11, prepared by David A. Crane and Partners et al.

Alternatives under consideration.   This section is to describe the physical features of the alternative proposals under consideration that provide feasible solutions to the transportation problem. Sufficient description of the scope of each proposal should be briefly presented without lengthy detail beyond that needed for evaluation and review of the anticipated environmental impacts. The discussion should be augmented with matrices, maps, and other graphics to help clarify each proposal. The description of each proposal is to contain data pertinent to the following:

- *Location.*   Boundaries (reference to map); type of facility (freeway, expressway, conventional, conversion to freeway, etc.); number of lanes; termini; length; and whether the alignment is new or existing.
- *Major design features and possible variations for each proposal.*   Predominant right-of-way width and access control; general horizontal and vertical alignment; interchanges, separation structures, at-grade intersections, river crossings, etc.; and design options that will not alter the environmental impacts.
- *Current status.*   Brief summary of any technical, social, and economic studies made, route adoptions, agreements, conferences, meetings, and notifications to local governing bodies; estimated cost of each proposal; construction period; and discussion of phases, extensions, and incremental stages of improvement where applicable for each proposal.

Alternatives withdrawn from consideration.   These are alternatives that were investigated during early stages of project development but were considered infeasible or otherwise inappropriate and were therefore dropped after due consideration. These alternatives should be briefly described, and a brief, credible discussion of the reasons for their elimination should then be given.

Affected Environment.   The next section describes in general terms the environmental setting for the entire area of impact of the alternative proposals. The setting limits should embrace areas that may be influenced by the proposed project. Refer to page 17 in FHWA Technical Advisory T6640.8A for a discussion of the content of this section. As is mentioned in Art. 1.2.4, the discussion of the setting should be brief with graphics used in lieu of verbiage. Only relevant aspects of the setting should be discussed in detail, with other descriptions limited to that necessary to provide context.

Environmental Consequences.   It is assumed that benefits will accrue during the life of a project. At the same time there will usually be costs, side effects, and loss of natural resources that have long-term productive value. The significant and potentially significant impacts of the project are to be discussed in detail in the environmental consequences section and will form the scientific and analytical foundation for evaluation of the merits of each alternative. The treatment of the various alternatives and their impacts in this section will probably be the best indicator of the objectivity and neutrality of the document. The text should only attempt to present, and quantify where possible, the various impacts of the individual alternatives and commit to mitigation as appropriate. The following discussion supplements the guidance contained in FHWA Technical Advisory T6640.8A.

The environmental consequences section should be a summary of positive and unavoidable adverse environmental affects, any tradeoffs caused by the proposed project that would lead to short-term (economic) gains at the expense of long-term (natural) productivity, and any irretrievable or irreversible commitments of resources. Short-term gains include such benefits as improved transportation, better safety, lowered energy use, better public services, more efficient economic activities, and improved development potential. Short-term costs include construction materials consumed, disrupted community or economic activities, and existing homes or businesses removed. Long-term productivity refers to valuable uses for existing environment (e.g., wetlands, open space, recreation areas, floodplains, wildlife habitat, groundwater recharge, areas that support rare species, or existing urban living and working places) and renewable resources (e.g., agriculture, timber, fisheries, ranching, or water supply). Long-term productivity also refers to environmental quality, such as low noise levels, clean air, pure water, and low levels of other kinds of pollutants. The section should discuss the extent that each alternative precludes future options for long-term productivity (such as loss of land suitable to be farmed) and any long-term risks of damage to productivity (such as making floods, landslides, or polluted groundwater more likely). This section must also discuss cumulative impacts on long-term productive values from other projects, if applicable. If the losses to long-term productivity would be significant, the reasons why the project should go forward now, rather than be left as an option for future consideration, must be explained.

The environmental consequences should also include a discussion of significant resources that would be used or removed by the project, including the following:

- The materials, labor, and energy needed to build the project or consumed in maintenance and operation of the project
- Land, and present uses of that land, directly taken to make way for the project (e.g., agricultural land, housing, wildlife habitat)
- Environmental conditions degraded or destroyed by the project (e.g., polluted waters, reduced wildlife populations, noisier communities)
- Properties indirectly used by the project (e.g., fill disposal sites, borrow sites, sediment basins)
- Public service capacities used up by the project (e.g., available water supply, storm sewer capacity, or police patrol time committed)

For historic and cultural resources, if the project is of a type that has been categorized as having a minimal area of potential effect and one for which historic studies are not needed, include a statement that the project has a minimal area of potential effect and no historic properties are involved. A negative historic property survey report (HPSR) must be prepared; include a statement that historic studies were performed by qualified persons and no historic resources were identified within the area of potential effects. If eligible properties were found, describe their locations and significance in terms of National Register criteria. Please note that the exact location or trinomial designations of archaeological sites are not to be described or displayed on exhibits. Also, lengthy discussions of history or ethnography of a region are not necessary. This section should document the public involvement process that occurred as part of the identification and evaluation of resources.

Mitigation Measures. Mitigation measures are the steps that can be taken to lessen the adverse environmental impacts of an action. In order for an impact to be successfully mitigated, the mitigation measure must be economically and practically viable and the agency has to be capable of and committed to implementing it. Under CEQ regulations, mitigation can be achieved by avoiding the adverse impact, minimizing the adverse effect by reducing the scope of the project, rectifying (repairing) the damage caused by the action, implementing a program to reduce the impact over time, or compensating for the impact by replacing or providing substitute resources.

List of Preparers. List the names and appropriate qualifications (professional license, academic background, certification, professional working experience, and special expertise) of the persons who were principally responsible for preparing the draft EIS or significant background papers. These latter will include but not be limited to those primarily responsible for the air quality and noise studies, biological studies, historic and cultural resource studies, and social and economic studies. The list should include any FHWA personnel who had primary responsibility for preparation or review of the draft EIS. If a consulting firm was used for any part of the document, then the consultant's primary author and project manager should be listed by name. Casual contacts for expert opinion made during the preparation of these reports should not be listed. An optional item that may be included in the document is a separate list giving recognition to people and organizations outside the lead agency and FHWA that assisted in its preparation.

List of Who Received Copies. The draft EIS must list the name and address of each agency and organization that was sent a copy of the document for review purposes.

Appendixes. The appendixes contain the reports and documents that support the findings of the draft EIS. Detailed technical discussions and analyses that substantiate the concise statements within the body of the draft EIS are most appropriately placed in the appendixes. Therefore, the appendixes will include such details of each study as the technical approach and scope of the study, protocols and procedures, specific modeling or testing parameters, data generated, interpretation of the data, and findings and limitations of the study. Appendixes must either be circulated with the draft EIS or be readily available for public review.

Index. Each draft EIS must contain an index that is sufficient to assist the reader in locating topics of interest and importance with ease. The list in Table 1.9 is not intended to be comprehensive, but to present an example of topics that might be found in a draft EIS index.

1.2.9 Supporting Documents

Federal actions must comply with a number of environmentally related federal laws and regulations described in Art. 1.1. This requirement applies to categorical exclusions as well as to projects requiring an environmental document. This article presents a summary of the documentation and processing requirements for some of those laws and regulations that are frequently applicable to highway projects. These include Section 4(f), Section 106, and wetlands, floodplains, endangered species, hazardous waste, and farmland protection considerations.

Section 4(f). Section 4(f) clearance is required whenever a federal action involves the use of a publicly owned park or recreation area, a wildlife or waterfowl refuge, or

TABLE 1.9 Example of Topics in a Draft EIS Index

Accidents	Hydrology
Adverse impacts	Land use
Aesthetics	List of preparers
Affected environment	Location map
Agricultural lands	Long-term impacts
Air quality	Mitigation measures
Alternatives	Need for project
Appendixes	No-build alternative
Archaeological resources	Noise
Biological environment	Noise attenuation
Circulation list	Parks
Climate	Permits
Comments	Population
Cooperating agencies	Preferred alternative
Coordination	Public facilities
Costs	Relocation
Cover sheet	Responsible agencies
Cultural resources	Right-of-way
Description of project	Schools abstract
Detours	Seismicity
Dust	Short-term impacts
Economic effects	Social effects
Employment	Soils
Endangered species	Summary
Energy	Table of contents
Environmental impacts	Topography
Environmental setting	Traffic
Erosion	Urban development
Floodplains	Utilities
Geology	Vegetation
Glossary	Visual assessment
Growth	Water quality
Historic resources	Wetlands
Housing	Wildlife

land from a historic site. Section 4(f) clearance requires that the following additional steps be included in the environmental process:

- Include in the early coordination and consultation process agencies that have jurisdiction over the 4(f) property.
- Take possible planning actions to avoid the use of 4(f) property or to minimize harm to any 4(f) property affected by the project. Each project proposal must include a 4(f) avoidance alternative.
- Prepare a draft Section 4(f) evaluation as a section in the draft environmental document, whether it is an EA, an EIS, or any similar document.
- Circulate the draft environmental document to agencies involved in the 4(f) property, and request that the agency review and comment on the 4(f) evaluation. If there is no draft environmental document, the draft 4(f) evaluation will be circulated separately. The comment period must be a minimum of 45 days.
- Evaluate the 4(f) comments and prepare a final 4(f) evaluation as a separate document or as a separate section in the final environmental document.
- The regional FHWA administrator will indicate Section 4(f) approval either in approval of the final environmental document or in the record of decision for other actions processed with EISs.

Section 4(f) evaluation documents the considerations, consultations, and alternatives. Alternatives may be considered not feasible or prudent for any of the following reasons:

- Not meeting the project purpose and need
- Excessive cost of construction
- Severe operational or safety problems
- Unacceptable adverse social, economic, or environmental impacts
- Serious community disruption
- A combined effect of several of the foregoing types of factors though each is present in lesser degree

If a proposed alternative involves more than one 4(f) resource, a separate 4(f) evaluation must be prepared for each such resource involved. Where a project proposal results in only minor impact upon a 4(f) resource and certain specified conditions can be met, a *programmatic* section 4(f) evaluation process is available. For routine involvements, a 4(f) evaluation will be a separate section of the environmental document. Including 4(f) evaluations within environmental documents serves two key purposes: (1) they are subject to essentially the same public review process as an EIS, and (2) Section 4(f) approval will be automatic with approval of the environmental document. In the case of an EIS, the 4(f) approval is summarized in the record of decision (ROD). This practice is an excellent example of the benefit to be gained by implementing environmental processes concurrently rather than consecutively.

Additional instructions are contained in FHWA Technical Advisory T6640.8A.

Section 106 (Historic Resources).   The Advisory Council on Historic Preservation (ACHP) has established implementing regulations for the protection of historic properties (36 CFR 800). These procedures must be followed for federal undertakings. An *undertaking* is defined as any project, activity, or program that can result in changes in the character or use of historic properties, if any such historic properties are located in the area of potential effects (APE). Following is a summary of the major steps in the project development process when historic properties (whether eligible or potentially eligible for the National Register of Historic Places) may be present in the APE.

An opportunity for early public involvement must be provided for federal actions during the identification phase of the project development process. For CE projects when National Register–eligible properties are present or are potentially present (such as in an archaeologically sensitive area), there must be early public involvement. Projects are excepted from this requirement if (1) they have been defined as having a minimal APE and therefore do not fall within the Section 106 definition of *undertakings* and (2) no known historic resources are present (Table 1.5). For projects that are CEs and are not listed in Table 1.5, the appropriate stages during the planning process to involve the public will have to be established by the FHWA transportation engineer. These are generally at the identification, evaluation, and consultation stages. Interested parties are to be notified of the conclusions of the studies. When projects listed in Table 1.5 have known historic resources, the public involvement process would begin when resources have been identified.

For those actions requiring either a FONSI or an EIS, any notices (notice of initiation of studies, notice of preparation, notice of intent, public hearing, opportunity for a hearing, and notice of availability) must state whether any alternatives could potentially involve historic properties, if the answer is known. If this potential is not known, then the notices must request the names of those persons who may have information relating to historic properties that may be affected or who may be interested in the effects of the undertaking on historic properties. At any hearing, the presence of any historic properties and the effects of any alternatives on those properties must be identified. For categorically excluded projects, a public mailing inviting written comments must be made. This notice need not be made for those projects that have been defined as having minimal APEs unless resources have been identified during project studies or the area is known to be sensitive for resources.

For projects where an EIS or a FONSI has been prepared, documentation of completion of the Section 106 process for projects with resources present is in the form of the signature of the transportation engineer on the historic property survey report (HPSR) form. For categorically excluded projects, Section 106 documentation is completed separately when resources have been identified. A positive HPSR is prepared which would conclude that the resources involved were found not to be eligible. Where no resources have been identified, FHWA concurrence is evidenced by the transportation engineer's signature on the negative HPSR, or by the transportation engineer's signature on the CE form for projects with minimal APEs.

Wetlands Involvement (Executive Order 11990).   The following procedures must be followed for any federal action that involves wetlands.

- An opportunity for early public involvement must be provided for actions involving wetlands. For those actions requiring either a FONSI or an EIS, any notices for a public hearing, or an opportunity for a hearing, must indicate if any alternatives are located in wetlands. At any hearing, the location of wetlands must be identified. For CEs, a newspaper notice inviting written comments must be published.

- Alternatives that would avoid wetlands must be considered, and if avoidance is not possible, measures to minimize harm to wetlands must be included in the action. Documentation of the above must be included in the environmental assessment or

EIS. Documentation for CEs should be reviewed and placed in the project files with
a copy to the FHWA area engineer.

- A wetlands-only-practicable-alternative finding must be prepared for actions
requiring a FONSI or an EIS (see FHWA Technical Advisory T6640.8A for content). Findings are not required for CEs.

Floodplain Involvement (Executive Order 11988).   The following additional steps
in the environmental process are required for actions that encroach on base floodplains:

- As with wetlands and historic properties, the public must be given the opportunity
for early review and comment, and notices must reference potential encroachments
on the base floodplain. Also, all public hearings must provide an opportunity to
address any floodplain encroachments.

- A floodplain-only-practicable-alternative finding must be prepared for actions
involving a significant encroachment (see FHWA Technical Advisory T6640.8A).
The finding is included in the final environmental document.

- A technical report entitled Floodplain Evaluation is to be prepared and summarized
in the environmental document in accordance with federal regulation 23 CFR 650,
Subpart A. If floodplain encroachments are significant, then alternatives should be
developed jointly among an interdisciplinary team that includes design, hydraulics,
and environmental analysis professionals. The detail in the report is to be kept to a
minimum commensurate with the significance of potential flooding, environmental
impact, and risk of economic loss, private or public. The report should contain a
project description, including a map of the project showing the base floodplain and
all project encroachments, as well as alternatives to encroachment. Include a discussion of the practicality of alternatives that would avoid longitudinal or significant encroachments, or the support of incompatible floodplain development and
risk assessment, which are part of the location hydraulic and/or design studies
called for in 23 CFR 650, Subpart A; impacts of the project, including the support
of probable incompatible floodplain development and impacts on natural and beneficial floodplain values; and mitigation measures to minimize floodplain impacts
including measures to restore and preserve floodplain values impacted by the
action. A matrix should be used to summarize the risk assessment and impacts.

The floodplain evaluation report should be summarized in the environmental document. The portions of the evaluation pertaining to fish and wildlife, vegetation, wetlands, growth inducement, etc., are to be included in the respective sections of the
document. Summaries involving floodplains in general, as well as hydraulics and risk,
are to be included in a section entitled Floodplains.

Section 7 (Endangered Species).   All projects must be evaluated to determine
whether any endangered or threatened species may be affected. If it is determined that
a species could be expected or is known to occur in the project area, further study and
evaluation will be required. This additional effort may include formal consultation
with the U.S. Fish and Wildlife Service and preparation of a biological assessment.
The level of involvement with the endangered species process can vary widely from
project to project and will generally involve the following steps:

- Establish an area of potential environmental impact (APEI) and potential for conflict with endangered species.

- Once preliminary alternatives are selected, determine whether a request for a species list from the U.S. Fish and Wildlife Service (FWS) is required and then request a list, through FHWA, if required.
- Perform a biological assessment and write up results as appropriate.
- If there are no species present or there will be no effect, obtain FWS concurrence through FHWA before circulating the draft environmental document.
- If the preferred alternative affects species, request conference or consultation with the U.S. Fish and Wildlife Service through FHWA, which must be completed before the final environmental document can be approved.

Hazardous Waste.   As discussed earlier, it is prudent to avoid acquisition of right-of-way that has been contaminated with hazardous waste. Generally, every project that includes purchase of new right-of-way, excavation, and/or structure demolition or modification should be evaluated to determine whether there is any known or potential hazardous waste within the proposed project limits. The discovery of hazardous waste after a project has gone to construction will usually result in a long and costly project delay while required cleanups are being completed. Consequently, early detection, evaluation, and remediation of hazardous waste is essential. Where hazardous substances are involved, adequate protection must be ensured for employees, workers, and the community prior to, during, and after construction. Typical materials that may constitute hazardous waste include pesticides, organic compounds, heavy metals, industrial waste, or other compounds injurious to human health and the environment. If site evaluation does not identify a known or potential hazardous waste site within the project area, a statement to that effect should be made in the environmental document.

The field of preliminary site assessment for hazardous substances has been steadily becoming more uniform, with several organizations establishing procedures and protocols for these studies. Generally, the studies are conducted in two stages referred to as either phases I and II or levels I and II. Phase I (level 1) is a research and visual observation stage to identify concerns and evaluate the likelihood that hazardous substances have affected the property. The purpose of phase II (level II) is to collect additional information (preferably quantifiable data through sampling and laboratory analysis) to confirm that contamination is present or to set any concerns aside. Phase I generally consists of historical research to evaluate current and past land uses and operations with a focus on what hazardous substances may have been introduced into the soil or water (including groundwater at the site); a search of regulatory records to evaluate whether the site or adjacent properties are listed in files as having violations, recorded hazardous substances releases or incidents, or a history of storing, handling, using, transporting, or disposing of hazardous substances; physical description of the soil geology and of surface water and groundwater, in order to evaluate the potential for migration of contaminants from the source to another property; and a site walk to observe the site conditions and operations as well as those of the neighboring properties. Phase II is a specifically designed sampling and laboratory analysis program that effectively addresses the concerns raised in the phase I study, if any. Phase II should be designed to collect sufficient information to establish that a valid concern exists and to indicate what level of effort may be required to address the concern.

The American Society for Testing and Materials (ASTM) has developed standards for phase I and phase II reports that effectively cover how to conduct these studies and prepare appropriate reports. These reports should be presented along with the draft environmental document.

Farmland Protection Policy Act, Amended 1984.   The Farmland Protection Policy Act (FPPA) and the National Environmental Policy Act (NEPA) require that before any federal action that would result in conversion of farmland is taken or approved, the Department of Agriculture must examine the effects of the action using the criteria set forth in the acts. If it is determined that there are adverse effects, alternatives to lessen them must be considered. This process requires an inventory, description, and classification of affected farmlands as well as early consultation with the U.S. Soil Conservation Service (SCS) within the U.S. Department of Agriculture. Processing of Form AD 1006 (Farmland Conversion Impact Rating) is also necessary. The FPPA must be part of the NEPA process in accordance with 7 CFR 658.4(e).

Where the right-of-way required for a transportation project is clearly not farmland (i.e., where it is rocky terrain, sand dunes, etc.) and the project would not indirectly convert farmland, the FPPA does not apply. The evaluation of land for agricultural use includes productivity, proximity to other land uses, urban and rural, impacts on remaining farmland after the conversion, and indirect or secondary effects of the project on agricultural and other local factors. Reports and findings about agricultural land are to be included with the draft environmental document.

1.2.10   Review

FHWA Review.   Timely federal approvals are an outgrowth of the ongoing participation of the FHWA division transportation engineer in preparation of the document. As a result of document preparation, the transportation engineer should have a good working knowledge of the project. The proposing agency should also have received FHWA comments on individual sections of the document during their preparation. This will minimize the rounds of comments and responses to comments made between FHWA and the district after formal submittal to FHWA. Sixteen copies of the draft document and two copies of each technical report should be sent to FHWA for review and comment. The FHWA division and region term this a *predraft review.* It is a full interdisciplinary and legal review by FHWA division, region, and headquarters (Washington, D.C.). The FHWA division will ensure that these reviews are concurrent, although consolidation and reconciliation of FHWA comments will always precede their return to the district. Copies of backup reports must be available for FHWA prior to the start of its review.

FHWA has adopted specific review periods for drafts. These periods should be viewed as planning tools only; specific review periods will always depend upon FHWA's priorities. The specified periods of time can be shortened considerably if the transportation engineer has been involved throughout document preparation. Once comments have been addressed to the satisfaction of FHWA, the original signed cover sheet, two copies of the DEIS, and two copies of any revised backup reports are to be sent to FHWA for approval to circulate. However, as a result of the "predraft" review, FHWA may require additional copies.

FHWA will assign an EIS number to the document and enter this number on the cover sheet. When comments from FHWA have been resolved and the FHWA division administrator is in agreement with the scope and content of the DEIS, the original cover sheet will be signed and returned along with authorization to circulate. FHWA approval may be contingent on some revisions to the DEIS prior to circulation or revisions to be made later in a FEIS. Before distribution, the district will add to the cover sheet the date by which public comments must be received (approximately 60 to 70 days from the date copies are sent to the FHWA during the circulation period). This will allow for mail time, as well as the required 45 days for comment after publication in the *Federal Register.* Any comments received from FHWA are to be considered in-house review comments. The commenting letter is not to be placed in the comments-received section of the EIS.

Public Review and Comment. If a public hearing is to be held on the project for project approval purposes, the draft environmental document (DED) must be circulated for comment at least 15 days prior to the hearing. Environmental issues raised, alternatives proposed, and substantive comments of an environmental nature made at the public hearing are to be included in the comments and reply section of the DED along with responses. The comments do not have to be presented verbatim but can be paraphrased and grouped together for brevity.

All letters received on substantive environmental issues related to the project as a result of the public hearing process are to be included in the comment section of the final environmental document (FED) along with letters received during circulation of the DED. There is no requirement to respond to letters received on nonsubstantive issues. Receipt of comments may be acknowledged. This acknowledgment can be by letter or postcard. DOT comments, however, are to be included. Responses to comments will be made in the final environmental document. Comments received on the DED must be available to the public for review in accordance with public information procedures.

The lead agency is to make copies of the draft environmental document available for public review in such places as public libraries and city halls. At the same time the document is circulated for comment, a notice announcing its availability should be published in a newspaper having general circulation in the vicinity of the proposed project and in any newspapers having a substantial circulation in the area concerned, such as foreign-language and community newspapers. If there will be impacts to wetlands, historic sites, or floodplains, the public notice must identify this fact and give the approximate acreage involved. For CE projects, a separate public notice must be published. The locations of public copies are to be noted in the newspaper notices, as well. Send a copy of the notice, along with a listing of the other publications, by name of newspaper and the date on which the notice appeared, to FHWA immediately after publication.

Sufficient copies should be printed to be available to fill requests. Normally, charges will not be made for individual copies, or for those distributed to persons and organizations that may be affected by the project. Charges are appropriate, however, for requests for multiple copies of reports and for requests from persons outside the project area. FHWA has specified that the fee may not be more than the actual printing cost. Technical study reports (usually called *backup reports*) such as noise, air, water, cultural, and biological study reports, are to be summarized in the draft environmental document. Copies need not be included in the DED, but should be referenced and made available upon request. Charges for the actual printing cost may also be made for these reports.

1.2.11   Final EIS

A final EIS, or FEIS, will be prepared in basically the same format and have the same supporting information required for a DEIS. In fact, if few substantive comments are received on the draft EIS, only minor changes to it will be necessary in order to produce a final EIS suitable for the decision maker. The major difference between the draft EIS and the final EIS is that a separate section is added to the alternatives section that identifies a preferred alternative and provides the supporting rationale for its selection. In general, the need, affected environment, and environmental consequences sections should not change. However, other changes that might cause revisions in these sections are choice of an alternative that is a composite of several alternatives; additional information that is pertinent; results of new studies; and corrections and updating of the text as a result of comments received from circulation of the draft EIS, the public hearing process, or changed circumstances due to the passage of time. Table 1.10 outlines the content of the final EIS.

As an aid to the reviewer of a final EIS, a vertical line should be made in the outside margins of the text showing added and revised portions from the draft EIS. Any unresolved environmental issues should be described along with a description of the efforts to resolve them through further consultation or otherwise. For instance, when an agency comments that a draft EIS has inadequate analysis or that the agency has reservations concerning the impacts, or believes that the impacts are too adverse for the project to proceed, the final EIS should reflect efforts to resolve the issue and set forth any action that will result. If the preferred alternative is in conflict with regional or local plans, convincing arguments must be presented on why the action should proceed and how the conflict will be resolved.

If the preferred alternative will involve the use of a 4(f) resource, a final Section 4(f) evaluation must be prepared and included as a separate section in the FEIS or as a separate document. If the proposed project is a federal action and is located in wetlands and/or encroaches on a floodplain, a wetlands-only-practicable-alternative finding and/or a floodplain-only-practicable-alternative finding must be added as a section.

Record of Decision. For federal actions documented with an EIS, FHWA must prepare and sign a record of decision (ROD) as part of the final project approval. At the time the FEIS is prepared, a draft of the ROD must also be prepared for the federal process (see Appendix A-2, FHWA Technical Advisory T6640.8A, for format and content). The final EIS is reviewed by the appropriate specialists before approval. Send eight copies of the final EIS (12 copies if prior concurrence by FHWA headquarters is needed) plus two copies of any new or revised technical reports, two copies of the draft ROD, and a keyed list of the disposition of FHWA comments on the draft EIS to FHWA for review ("prefinal" EIS).

Following the required reviews and approvals, the signed final EIS is distributed to all responsible agencies, those who made substantive comments on the draft EIS, and other appropriate parties. A notice of availability will be published in the *Federal Register* and in local newspapers at the time of distribution. CEQ regulations require that no administrative action be taken on a federal project until 30 days after publication of a notice of availability in the *Federal Register* by EPA. The filing date for final EISs received by EPA in Washington, D.C., during a given week (Monday through Friday) will be the Friday of the following week. Following this 30-day review period and the successful resolution of any objections raised, the FHWA regional administrator will approve the project by issuing the ROD.

Distribution of the Final EIS. After FHWA approval of the final EIS is received, the document is to be distributed to the public. To ensure that the statement is recognized as the formal distribution of the FHWA-adopted FEIS, specifically state in the document transmittal letter that distribution is being made on behalf of FHWA in accordance with 23 CFR 771. A dated copy of the transmittal letter must be sent to the Environmental Division. The district is to publish a notice of availability at this time in a newspaper having general circulation. Upon completion of distribution, the district will notify FHWA, which will furnish the final EIS to EPA and request publication of a notice of availability in the *Federal Register.*

TABLE 1.10 Development of Final Environmental Impact Statement (EIS) from Draft EIS

Item	Content
Cover sheet	The cover sheet must reflect that this is a *final* EIS.
Summary sheet	The summary sheet is revised to be consistent with the final EIS and to incorporate the preferred alternative identification and information. Briefly and concisely describe all mitigation measures that are part of the preferred alternative. Monitoring and enforcement measures are also listed, where applicable, for any mitigation.
Alternatives	The preferred alternative is identified and described in a separate subsection. In addition, this subsection must develop and present in detail a defensible rationale for the preferred alternative selection decision. This rationale must reflect a comparison of the strengths and weaknesses of the various alternatives considered. The discussions relating to alternatives that were not selected should generally remain unchanged.
Effects on sites of historical and cultural significance	When it has been determined that the preferred alternative will produce an effect on properties included in or eligible for inclusion in the National Register of Historic Places, this discussion must be revised as necessary to demonstrate that all requirements of 36 CFR 800 have been met. If the effects are not adverse, any specific conditions leading to that finding should be stated. Documentation of public involvement efforts and the views of the Advisory Council on Historic Preservation (ACHP) must be included. Include a statement that FHWA has made a finding of no adverse effect with the concurrence of the ACHP. Include a copy of the ACHP concurrence letter as an exhibit (do not include the FHWA concurrence letter). If the effect is adverse, describe the results of the Section 106 process, including surveys undertaken, the number and type of historic properties to be affected by the project, and the proposed treatment of these properties. A copy of the memorandum of agreement that has been signed by the consulting agencies and reflects their agreement for avoiding, mitigating, or accepting the adverse effects on the historic properties must be attached as an exhibit.
Mitigation and other commitments summary	The final EIS includes a separate added section summarizing all mitigation measures, and any other conditions, committed to in previous sections of the document. This summary listing is to include a page reference where each commitment is discussed.
Distribution list	Indicate on the list those entities commenting.
Comments and coordination	This section is to include a copy of all comments from public agencies, all substantive comments of an environmental nature (or summaries if comments are particularly voluminous) from others, a summary of substantive comments of an environmental nature received at public hearings or presentations, responses to those substantive comments needing answers and/or references to the text where the comment has been answered, and a summary or categorization of nonsubstantive comments received with a list of these commentators' names.

1.2.12   Environmental Reevaluation

A formal, written environmental reevaluation (ER) is prepared only on projects with federal involvement for which an EIS has been prepared when any of the following circumstances occur:

- An acceptable FEIS is not submitted to FHWA within three years from the date of draft EIS circulation.
- No major steps have been taken to advance a project [e.g., allocation of a substantial portion of right-of-way or construction funding, or request for FHWA Plans, Specifications and Estimates (PS&E) approval] within three years from the date of final EIS approval.
- FHWA requests an ER on EIS projects or on large and complex projects with lengthy periods of inactivity between major steps to advance the project.

An environmental reevaluation (ER) will be performed prior to the next major state or federal approval step (e.g., an action such as allocation of right-of-way or construction funding, or a request for FHWA PS&E approval) if the time between the filing of the notice of determination (or a previous ER) and the major approval is over 12 months. If circumstances warrant it, the reevaluation will be performed even if less than 12 months has elapsed. Examples of this would be a project that is located in a rapidly developing area, or in position of a new retroactive regulation that has a bearing on the project. The purpose of the reevaluation is to determine whether there has been a substantial change in the social, economic, and environmental effects of the proposed project. This could result from changes in the project itself or from changes in the circumstances under which the project is to be undertaken.

A supplemental draft and final EIS are prepared when there are changes that result in significant impacts not previously evaluated. A draft EIS or a final EIS may be supplemented or amended at any time and must be supplemented or amended when (1) changes to the proposed project would result in significant environmental impacts that were not evaluated in the EIS or (2) new information or circumstances relevant to environmental concerns and bearing on the proposed project or its impacts would either bring to light or result in significant environmental impacts not evaluated in the EIS. The supplemental statement need only address those subjects in the EIS affected by the changes or new information. For example, if changes in the proposed project significantly change the number of houses involved in right-of-way acquisition, only this issue and the associated effects need be discussed. The required sections of a supplemental EIS will follow the format of an EIS. There is no set ER format, but the ER should include the project title and description, previous environmental approvals, date of FHWA approval, filing date of the notice of determination, and project changes including changes in setting or circumstances. Describe changes and identify and assess the significance of any effects. If a supplemental EIS requires the preparation of additional studies—air quality, noise, water quality, etc.—these should be treated the same as for an EIS.

In addition, where any type of environmental document (FONSI, or even CE) has been approved for a project, some form of reevaluation is required at each major milestone, such as commencement of right-of-way acquisition or approval of PS&E. The environmental branch chief must document in the project files, to the degree necessary, that an environmental reevaluation has been made. This must be done with the signed endorsement of the FHWA transportation engineer.

For older FONSIs and EISs, backup documentation of predicted effects should be reviewed in light of new retroactive regulations (such as for historic resources) and major advancements in the state of the art regarding analytical methodologies for such elements as air, noise, water, and social effects.

The results of the environmental reevaluation, conducted after major milestones during the design phase, are documented in a memo to the project file. This normally need only indicate that there are no substantial changes to the project itself or to the project's impacts. When a formal reevaluation is required for projects with an EIS, two copies of the environmental reevaluation are submitted to FHWA. To prevent delay in project approval, the reevaluation must be submitted at least three months prior to the next approval action.

1.3 STORM WATER POLLUTION PREVENTION

1.3.1 Background

The Clean Water Act (CWA, 33 USC §251) prohibits storm waters from being used to transport or collect wastes and requires that standards for water pollution be established that do not diminish the uses of the water. EPA is given the authority to develop a framework of regulation that is fully delegated to states once the EPA has approved their regulatory program. The federal act requires states to establish a policy of nondegradation that protects and preserves water. (J. T. Dufour, *California Environmental Compliance Handbook,* California Chamber of Commerce, Sacramento, 1993, pp. 72–74.)

Amendments to the Clean Water Act in 1972 separated point sources (known industrial and wastewater treatment outfall locations) from nonpoint sources (general urban and agricultural discharges that cannot be associated with a specific location). The amendments provided for controlling the level of pollutants discharged into bodies of water by way of storm water drainage. Initial programs concentrated on point sources, since they were seen as the most immediate threat to the environment and are easily identifiable. In 1987, further amendments to CWA added Section 402(p), which established authority for regulating municipal and industrial storm water discharges. Thus the National Pollutant Discharge Elimination System (NPDES) was established.

1.3.2 Municipal NPDES Permits for State-Owned Highways

Under Section 402(p)(2)(C) and (D) of the Clean Water Act, storm water permits are necessary for discharges from a municipal separate storm sewer system serving an incorporated or unincorporated area with a population over 100,000. The EPA definition of *municipal separate storm sewer* is "a conveyance or system of conveyances (including roads with drainage systems, municipal streets, catch basins, curbs, gutters, ditches, manmade channels, or storm drains)." The definition goes on to specify that the system of conveyances may be owned by any of a number of types of municipal governing bodies and specifically includes states (elsewhere in CWA, states are not regulated as municipalities), that the conveyances must be specifically designed for the purpose of collecting or conveying storm water, and that they are not to be part of a combined sewer or part of a publicly owned treatment works (POTW).

In the preamble of the November 1990 amendments, EPA explains its decision to include state-owned highways as municipal separate storm sewers. EPA identifies discharges from state highways as a significant source of runoff and pollutants and as one of the "issues and concerns of greatest importance to the public" (*Federal Register,* Part

II, Environmental Protection Agency 40 CFR Parts 122, 123, and 124, November 16, 1990, p. 48039). One of the primary traffic-related contaminants in stormwater is lead particulate from the exhaust of vehicles. Certain forms of lead are fairly soluble. Ionic lead precipitates in soil as lead sulfate and has a low solubility, thus remaining relatively immobile (K. W. Olson and R. K. Skogerboe, "Identification of Soil Lead Compounds from Automotive Sources," *Environmental Science and Technology,* vol. 9, no. 3, March 1975, pp. 227–230). In 1978, the EPA established the National Urban Runoff Program (NURP) to evaluate average concentrations of pollutants in storm water runoff. Studies on specific catch basins evaluated for NURP confirm that lead is a primary traffic-related contaminant (*Results of the Nationwide Urban Planning Runoff Program,* U.S. Environmental Protection Agency, Water Planning Division, Washington, D.C., 1983).

Sediment and water samples (both groundwater and surface water) collected near a major highway interchange in Miami, Florida, revealed that lead concentrations in the water were very low. Sediment concentrations of lead were relatively high, and it was believed that lead had precipitated out as a result of interaction with bicarbonates. (T. R. Beaven and B. F. McPherson, "Water Quality in Borrow Ponds near a Major Dade County, Florida, Highway Interchange, October–November 1977," U.S. Geological Survey Open File Report, in Review, 1978.)

In order to avoid the problems associated with multiple permittees for systemwide discharges, the CWA regulations include a method whereby interconnected systems owned and operated by local agencies and state-owned highways in areas of medium to high population may be combined into a single permit. The regulations allow the state transportation agency to be named as a co-permittee in a systemwide permit, or to be named in a separate municipal permit. (E. M. Jennings, "Coverage of State Highways under Municipal Storm Water Permits," Memorandum, Office of Chief Counsel, California State Water Resources Control Board, Sacramento, November 2, 1992, p. 31.)

Storm water systems owned by state highway departments in low-population areas (under 100,000) are not required to be permitted. Appendixes to Part 122 indicate the incorporated areas and unincorporated counties in the United States with sufficient population to require municipal storm water permits. This information is shown in Table 1.11.

Application requirements for a group permit for discharges from large and medium municipal storm sewers fall into two parts. Also, an annual report must be submitted, as discussed later.

Part 1 of the Application. For Part 1, the application must include the following items [Environmental Protection Agency, 40 CFR 122.26(d)(1), Application Requirements for Large and Medium Separate Storm Sewer Discharges, Part 1, *Federal Register,* vol. 55, no. 222, November 16, 1990, pp. 48068–48069]:

General information, including the applicant's name and status as a state or local entity and the name, address, and telephone number of the contact person

Legal authority to control discharges to the municipal separate storm sewer system

Source identification that includes:

1. U.S. Geological Survey (USGS) 7.5-minute topographic maps that show the surrounding area extending one mile beyond the service boundaries of the municipal storm water system covered in the permit application along with the location of known municipal storm sewer outfalls

2. A description of land use activities; population growth projections; the estimated average runoff coefficient associated with the population, the location of

TABLE 1.11 Population Areas That Require Storm Water Permits

A. Incorporated places with populations of 250,000 or more	
State	Incorporated place
Alabama	Birmingham
Arizona	Phoenix, Tucson
California	Long Beach, Los Angeles, Oakland, Sacramento, San Diego, San Francisco, San Jose
Colorado	Denver
District of Columbia	
Florida	Jacksonville, Miami, Tampa
Georgia	Atlanta
Illinois	Chicago
Indiana	Indianapolis
Kansas	Wichita
Kentucky	Louisville
Louisiana	New Orleans
Maryland	Baltimore
Massachusetts	Boston
Michigan	Detroit
Minnesota	Minneapolis, St. Paul
Missouri	Kansas City, St. Louis
Nebraska	Omaha
New Jersey	Newark
New Mexico	Albuquerque
New York	Buffalo, Bronx Borough, Brooklyn Borough, Manhattan Borough, Queens Borough, Staten Island Borough
North Carolina	Charlotte
Ohio	Cincinnati, Cleveland, Columbus, Toledo
Oklahoma	Oklahoma City, Tulsa
Oregon	Portland
Pennsylvania	Philadelphia, Pittsburgh
Tennessee	Memphis, Nashville/Davidson
Texas	Austin, Dallas, El Paso, Fort Worth, Houston, San Antonio
Virginia	Norfolk, Virginia Beach
Washington	Seattle
Wisconsin	Milwaukee

B. Incorporated places with populations greater than 100,000 and less than 250,000	
State	Incorporated place
Alabama	Huntsville, Mobile, Montgomery
Alaska	Anchorage
Arizona	Mesa, Tempe
Arkansas	Little Rock
California	Anaheim, Bakersville, Berkeley, Concord, Fremont, Fresno, Fullerton, Garden Grove, Glendale, Huntington Beach, Modesto, Oxnard, Pasadena, Riverside, San Bernardino, Santa Ana, Stockton, Sunnyvale, Torrance
Colorado	Aurora, Colorado Springs, Lakewood, Pueblo
Connecticut	Bridgeport, Hartford, New Haven, Stamford, Waterbury
Florida	Fort Lauderdale, Hialeah, Hollywood, Orlando, St. Petersburg
Georgia	Columbus, Macon, Savannah

TABLE 1.11 Population Areas That Require Storm Water Permits (*Cont.*)

State	Incorporated place
Idaho	Boise
Illinois	Peoria, Rockford
Indiana	Evansville, Fort Wayne
Iowa	Cedar Rapids, Davenport, Des Moines
Kansas	Kansas City, Topeka
Kentucky	Lexington-Fayette
Louisiana	Baton Rouge, Shreveport
Massachusetts	Springfield, Worcester
Michigan	Ann Arbor, Flint, Grand Rapids, Lansing, Livonia, Sterling Heights, Warren
Mississippi	Jackson
Missouri	Independence
Nebraska	Lincoln
Nevada	Las Vegas, Reno
New Jersey	Elizabeth, Jersey City, Paterson
New York	Albany, Rochester, Syracuse, Yonkers
North Carolina	Durham, Greensboro, Raleigh, Winston-Salem
Ohio	Akron, Dayton, Youngstown
Oregon	Eugene
Pennsylvania	Allentown, Erie
Rhode Island	Providence
South Carolina	Columbia
Tennessee	Chattanooga, Knoxville
Texas	Amarillo, Arlington, Beaumont, Corpus Christi, Garland, Irving, Lubbock, Pasadena, Waco
Utah	Salt Lake City
Virginia	Alexandria, Chesapeake, Hampton, Newport News, Portsmouth, Richmond, Roanoke
Washington	Spokane, Tacoma
Wisconsin	Madison

C. Counties with unincorporated urbanized areas with population of 250,000 or more

State	County
California	Los Angeles, Sacramento, San Diego
Delaware	New Castle
Florida	Dade
Georgia	De Kalb
Hawaii	Honolulu
Maryland	Anne Arundel, Baltimore, Montgomery, Prince Georges
Texas	Harris
Utah	Salt Lake
Virginia	Fairfax
Washington	King

D. Counties with unincorporated urbanized areas with population
greater than 100,000 and less than 250,000

State	County
Alabama	Jefferson
Arizona	Pima
California	Alameda, Contra Costa, Kern, Orange, Riverside, San Bernardino

TABLE 1.11 Population Areas That Require Storm Water Permits (*Cont.*)

State	County
Florida	Broward, Escambia, Hillsborough, Orange, Palm Beach, Pinellas, Polk, Sarasota
Georgia	Clayton, Cobb, Richmond
Kentucky	Jefferson
Louisiana	Jefferson
Nevada	Clark
North Carolina	Cumberland
Oregon	Multnomah, Washington
South Carolina	Greenville, Richland
Virginia	Arlington, Chesterfield, Henrico
Washington	Pierce, Snohomish

Source: Adapted from *Federal Register,* vol. 55, no. 222, November 16, 1990, pp. 48073, 48074.

landfills, the location of industrial discharges to the system that have been issued NPDES permits, the location of major structural controls (e.g., retention basins, detention basins, and major infiltration devices); and the location of publicly owned parks, recreation areas, and open spaces

3. Existing quantitative data describing the volume and quality of discharges from the municipal storm sewer, including a description of the outfall samples, sampling procedures, and analytical methods
4. A list of water bodies that receive discharges from the municipal storm sewer, including a brief description of water quality impacts with information on whether these water bodies are: (a) assessed by the state on the basis of CWA attainment goals for fishable and swimmable waters, (b) listed as water bodies that are not expected to meet water quality standards or water quality goals, (c) listed in state nonpoint source assessments, (d) identified and classified according to eutrophic condition in state reports, (e) designated an Area of Concern of the Great Lakes, (f) designated as an estuary in the National Estuary Program, (g) highly sensitive or valuable water bodies, (h) wetlands, or (i) found to have pollutants in bottom sediments, fish tissue, or biosurvey data
5. Field screening for illicit connections and illegal dumping—along with a narrative description that includes the location of field screening points in accordance with requirements set forth in CWA Part 122.26 (d)(1)(iii)(D), Screening Methods—and findings
6. A characterization plan that includes the location and description of the selected outfalls and why they are representative, the seasons when sampling is intended, and a description of sampling equipment

Management program to control pollutants discharged from the municipal separate storm water system, including existing structural and source controls, along with their associated operation and maintenance measures; and a description of the existing program to identify illicit connections and prevent illicit discharges

Fiscal resources currently available to complete Part 2 of the permit application, including the budget for existing storm water programs, overall indebtedness, and assets and sources of funding for the storm water program

Part 2 of the Application. Part 2 of the application consists of the following items (Ibid., Part 2, pp. 48069–48071):

Demonstration of adequate legal authority to control storm sewer discharges by industrial facilities into the system; prohibit illicit discharges to the system; control spills, dumping, or disposal of materials other than storm water; require compliance; and carry out inspections.

Source identification of locations of outfalls and an inventory (organized by watershed) of facilities that may discharge to the system.

Characterization data (quantitative data) collected in accordance with 40 CFR 122.21(g)(7) and including data from representative outfalls based on information presented in Part 1. Outfalls selected for screening will be sampled during discharges from three storm events occurring at least one month apart and shall include a description of the event. Pollutant samples at a minimum shall include total suspended solids, total dissolved solids, chemical oxygen demand, biological oxygen demand, oil and grease, fecal coliform, fecal streptococcus, pH, total Kjeldahl nitrogen, nitrate plus nitrite, dissolved phosphorus, total ammonia plus organic nitrogen, and total phosphorus. Characterization shall also include estimates of the annual pollutant load of the cumulative discharges from all identified municipal outfalls and of the event mean concentrations of the seasonal discharges; a proposed schedule and estimate of each major outfall of the seasonal pollutant load; an estimate of the mean concentration of a storm event; and a monitoring plan.

Proposed management program covering the duration of the permit and including comprehensive planning with public participation and intergovernmental coordination, with the intent of reducing the discharge of pollutants using management practices, control techniques, engineering methods, and other practical methods. The program should include information on available staff, equipment, and resources. The management program should be based on:

1. A description of the structural and source control measures to reduce pollution from runoff, along with an estimate of the expected reduction of pollutant loads and a proposed schedule for implementing such controls
2. A description of the program, including a schedule to detect and remove illicit discharges and improper disposal into the storm sewer
3. A description of the program to monitor and control pollutants in storm water discharges to municipal systems from municipal landfills, hazardous waste treatment, disposal and recovery facilities, and industrial facilities
4. A description of the program to implement and maintain structural and nonstructural best management practices (BMPs) to reduce pollutants in storm water runoff from construction sites to the municipal storm water sewer system

Assessment of controls, including estimated reductions in loading of pollutants from discharges of municipal storm sewer systems expected as the result of the management program.

Fiscal analysis for each year of the permit, including analysis of the necessary capital, operation, and maintenance costs necessary for the activities of the entire program and sources of funds and legal restrictions on the use of those funds.

Clear identification of roles of co-permittees, along with the responsibilities of each and how they will be effectively coordinated.

Director exclusions, if applicable. In some instances, the director of the EPA has discretion to exclude systems from permit requirements. If director exclusions apply, they should be stated in the application.

Annual Report.   The operator must submit an annual report by each anniversary date of the permit. The report shall include:

1. The status of each component of the storm water management program, as established in the permit
2. Proposed changes to the storm water management program
3. Proposed revisions to the assessment of controls and the fiscal analysis reported in the permit application
4. A summary of data, including monitoring data, accumulated throughout the reporting year
5. Projected annual expenditures and budget for the year following each annual report
6. A summary describing the number and nature of enforcement actions, inspections, and public education programs
7. Identification of water quality improvements or degradation

1.3.3 Storm Water Permits for Construction

The U.S. EPA, as administrator of CWA, in 1990 established final regulations for controlling storm water runoff from specific categories of industries and activities (40 CFR 122, 123, and 124). Any discharger of, or person who proposes to discharge, a waste other than to a sewer system, or changes the character of a current discharge, is required to report this activity to the local enforcement agency (Dufour, op. cit., p. 75). Once reported, the agency will evaluate the discharge and may:

- Issue waste discharge requirements
- Waive discharge requirements for insignificant discharges such as well testing or construction dewatering (no waiver is permitted if the discharge is to surface waters)
- Prohibit the discharge if sufficiently protective discharge requirements cannot be met by the discharger

Permit applications are called *notices of intent* (NOIs). Waste discharge requirements are issued through an NPDES permit that specifies conditions the discharger must meet. The conditions are based on the established water quality objectives and the capacity of the existing storm water drainage system or receiving waters to assimilate the discharge. Discharge limitations are usually expressed as a combination of quantitative and procedural specifications (Ibid.). CWA provides for three types of NPDES permits: individual, group, and general. Issuance of waste discharge requirements must be noticed for public comment and approved at a hearing of the local authority (Ibid., p. 74).

Industrial facilities are covered under a general permit for discharges associated with industrial activities according to the applicable SIC (Standard Industrial Classification) codes. These general permits are not driven by discharge standards or numerical receiving water limits, but rather are requirements for the facility operator to establish a storm water pollution prevention plan (SWPPP) that the operator certifies will be effective and within the facility's financial means. (L. D. Duke, "Industrial Storm Water Runoff: Pollution Practices," *AIPE Facilities,* January–February 1995, p. 49.)

The primary industrial category relevant to building and maintaining highways is construction activities. Construction activities, in this context, include clearing, grading, and excavating that result in the disturbance of five acres or more of land that is not part of a larger (nonhighway) construction project. Construction sites were targeted because studies showed that the runoff from construction sites has high potential for serious water quality impacts. Sediment runoff from construction sites may be 10 to 20 times that from agricultural lands. Non-point-source pollutants from construction sites include sediment, metals, oil and grease, nitrates, phosphates, and pesticides.

In order to obtain an NPDES construction permit, a notice of intent (NOI) must first be filed requesting permit coverage. The NOI contains the following information:

1. Owner of the site (legal name and address) and contact person's name, title, and telephone number

2. Construction site information—whether the construction is part of a larger project, and the portion of the site that is impervious both before and after construction activities

3. Scheduled beginning and ending dates of construction

4. Identification of the receiving body of water, and storm water drainage information, including a site map

5. Type of construction activity: transportation should be indicated if the project is a roadway; utilities should be indicated for the installation of sewer, electric, and telephone systems

6. Material handling and management practices indicating the type of material to be stored and handled on-site and the management practices to be used to control storm water pollution

7. Regulatory status of the site, including approval status of the erosion or sediment control plan

8. Signature of the owner of the site certifying that the information is accurate

Most statewide NPDES permits for general construction activities require the permit holder to develop and implement a storm water pollution prevention plan (SWPPP) using either best available technology economically achievable (BAT), best conventional technology (BCT), or best management practices (BMP) to control pollutant discharge both during and after construction activities. Once prepared, the SWPPP will be maintained at the construction site by the highway department representative and made available on request by the local enforcement agency. All contractors and subcontractors working at the site are responsible for implementing the plan. The plan will generally include the following components:

1. Location, including a $\frac{1}{4}$-mi vicinity map that shows nearby surface water bodies, drainage systems, wells, general topography, and location where storm water from the construction activities will be discharged.

2. A site map that indicates the total site area and total area to be disturbed. This map should indicate the location of the control practices to be implemented, areas where wastes and soils will be stored, drainage patterns for the site both before and after construction activity, areas of soil disturbance, areas of surface water, potential soil erosion areas, existing and planned paved areas, vehicle storage areas, areas of existing vegetation, and areas of postconstruction controls.

3. A narrative description of the construction site, project, and activities. This should include a description of the fill material and native soils at the construction site and the percentage of site surface area that is impervious both before and after construction activities.

4. A narrative description of toxic material used, treated, or disposed of at the construction site.

5. Identification of potential sources of storm water pollution, and name of receiving water.

6. Proposed controls and best management practices (BMPs) during construction, including description of:
   - State and local erosion sediment control requirements
   - Source control practices intended to minimize contact between the construction equipment and materials and the storm water being discharged
   - Erosion and sediment control procedures to be implemented
   - Plan to eliminate or reduce discharge of other materials into the storm water

7. Proposed postconstruction waste management and disposal activities and planned controls, including a description of state and local erosion and sediment postclosure control requirements.

8. Estimated runoff coefficient for the site, estimated increase in impervious area following the construction, nature of fill, soil data, and quality of discharge.

9. List of the contractors and their subcontractors who will be working at the construction site.

10. Employee training.

11. Maintenance, inspection, and repair activities.

Control measures for sediment include grading restrictions, runoff diversion, straw bales and filter fabric, revegetation requirements, and retention basins. Control measures for other pollutants include roof drains, infiltration trenches, grassy swales to detain storm water to allow sediments to settle out, oil/grit separators, detention basins, and proper management practices such as the proper application of fertilizers and pesticides.

1.3.4 Comprehensive Permits for Highway Departments

Another approach to implementing the NPDES program for state highway agencies is to issue comprehensive permits for all relevant highway construction, maintenance, and operations activities in areas meeting the population requirements outlined in 40 CFR 122.26. The benefit of a comprehensive permit is the management efficiency of administering the permit from both the regulatory agencies' and the highway department's perspectives.

In California, for instance, the California Regional Water Quality Control Board (RWQCB) is responsible for issuing storm water discharge permits. The RWQCB in the San Francisco Bay area has issued a comprehensive NPDES permit for storm water discharged directly or through municipal storm drain systems to lakes, water supply reservoirs, groundwaters, the Pacific Ocean, San Francisco Bay, San Pablo Bay, Suisun Bay, the Sacramento River Delta, or tributary streams or watercourses and contiguous water bodies in the San Francisco Bay region (District 4 and portions of District 10 of the California Department of Transportation, or Caltrans).

Provisions of the permit cover maintenance operations and include requirements to submit plans for maintenance activities that affect storm water discharges and to improve practices that will result in reduction of pollutants in discharges. Road sweeping plans, storm drains, catch basins, inlet and channel maintenance, and vegetation control plans are required. Caltrans must prepare storm water pollution prevention plans for maintenance facilities that cover such activities as vehicle and equipment maintenance, cleaning, fueling practices, and storage and handling of construction materials, fertilizers, pesticides, paints, solvents, and other chemicals.

TABLE 1.12   Examples of Activities Included in Identification and Control Procedures for Storm Water Permits

Water line flushing
Landscape irrigation
Diverted stream flows
Rising groundwaters
Uncontaminated groundwater infiltration
Uncontaminated pumped groundwater
Discharges from potable water sources
Foundation drains
Air-conditioning condensate
Irrigation water
Springs
Water from crawl space pumps
Footing drains
Flows from riparian habitats and wetlands
Discharges or flows from emergency firefighting activities

Under the provisions of the permit, Caltrans must rank construction activities on the basis of their potential impacts on receiving waters from pollutants in storm water discharges. Plans must be developed for erosion control, chemical and waste management, and postconstruction permanent features. Training is a key component of these plans.

The permit also encompasses permanent control measures for the management of storm water draining from Caltrans rights-of-way in areas meeting the population criteria. Consideration must be given to high-risk areas where spills may occur and must include a plan to reduce the pollutants discharged into the system over time. This portion of the permit requires Caltrans to develop mechanisms to control illegal dumping, to respond to accidental discharges and to identify and control procedures for discharge in a category not expressly prohibited by the permit. Activities included are listed in Table 1.12.

The RWQCB included specific provisions to assist in meeting water quality goals. For example, requirements of the permit include specific measures to reduce the mass load of copper in storm water discharges.

Monitoring plans and annual reports are also required in the NPDES permit and are generally consistent with these provisions in standard construction and/or municipal storm water permits.

1.4   LEAD-BASED PAINT REMOVAL

In 1992, it was estimated that 80 percent of all state-maintained steel bridges and 35 to 40 percent of all industrial steel structures were coated with lead-based paint. Steel bridges have been coated with lead-based paint for more than 40 years. The coating systems have an expected effective life of 15 to 25 years, and those on many bridges are now deteriorating. Life extension and overall protection of the bridges from corrosion are dependent on refurbishing deteriorating coatings.

Steel bridges are protected from corrosion with long-lasting, durable coating systems. Use of lead in corrosion-resistant painting systems continued until the early 1980s, for several reasons:

1. Lead-based paint systems are efficient in preventing corrosion.
2. They do not require extensive surface preparation before painting (especially an advantage around the inaccessible angles and connections of steel structures).
3. Lead-based paints are inexpensive to formulate.

Through studies, the public has become increasingly aware that lead (and other metals contained in paint) can represent a significant human health and environmental threat. When intact and in good condition, the paint does not pose a threat. It is when paint is removed to prepare the surface for coating replacement, or as the paint deteriorates, that the release of toxic components is likely.

Many highway structures are located in urban areas where lead-based paint removal has the potential to affect adjacent commercial or residential properties and to expose their occupants to hazardous concentrations of lead. Bridges are often constructed over water bodies where lead-containing dust from removal operations can affect water quality and the aquatic environment.

1.4.1 Biohazards of Lead

A bioaccumulative substance such as lead can be stored in various organs and tissues of the body. As lead-storing tissues are consumed by larger organisms in the food chain, a cumulative effect occurs in each subsequent organism. For example, a fish in a lead-affected environment is exposed to lead not only in the water, but also in the insects that it eats, which have accumulated lead from their food source, and so on down the chain. Organisms at the top of the food chain are therefore exposed to higher concentrations of lead.

In humans, long-term exposure can result in brain and nerve disorders, anemia, blood pressure problems, reproductive problems, decreases in red blood cell formation, and slower reflexes. In high enough doses or after long-term bioaccumulation, lead exposure can cause death. The Occupational Safety and Health Administration's (OSHA's) Interim Final Rule on Lead Exposure in Construction (29 CFR 1926.62) describes long-term overexposure effects of lead.

The primary methods of exposure to toxic levels of lead are inhalation and ingestion. For example, paint removal workers inhale leaded dust or—in the absence of proper cleaning and preventative measures—ingest it after it has settled on food, cigarettes, utensils, or other items that come to their mouths.

1.4.2 Regulatory Framework

Hazardous waste is regulated under the Resource Conservation and Recovery Act (RCRA) if more than 100 kg (220 lb) is generated each month, as is the case in most bridge paint removal projects. RCRA defines the concentrations of a waste that should be considered hazardous and establishes procedures for handling and disposing of hazardous waste. Disposing of waste is the responsibility of the waste generator. The lead-based paint and blasting grit recovered in bridge paint removal projects most likely contain concentrations of lead sufficient to classify the waste as hazardous, and in all instances, the owner of the structure is considered the generator (in some states the contractor removing the paint may be considered a cogenerator). Subtitle C under RCRA is relevant to lead-removal activities. Table 1.13 provides a listing of the pertinent RCRA regulations.

Methods of testing wastes for toxicity characteristics for the purpose of determining whether the waste is hazardous are described in 40 CFR 261. Appendix II of that regulation describes the toxicity characteristic leaching procedure (TCLP, Method 1311) that must be used to analyze hazardous constituents such as lead. Leachable levels of various elements that will establish waste as hazardous are found in Table 1 of 40 CFR 261.24 and are presented here in Table 1.14. Under 40 CFR 261, those wastes with any of the listed characteristics would be considered hazardous. For example, using the toxicity characteristic leaching procedure (TCLP) testing method, if 5.0 mg/liter or more of lead can be extracted, lead paint debris is considered toxic and hazardous.

Through the Comprehensive Environmental Response, Compensation, and Liability Act (CERCLA) and the Superfund Amendments and Reauthorization Act (SARA), EPA regulates the amount of hazardous substances and waste that can be released into the environment. Thus, the owner is required to contain the lead paint removed from a structure. A response could be initiated at a paint removal project if improper containment of dust or debris resulted in a release of lead to the environment. A reportable quantity of released leaded waste is 1 lb at this time, but a proposed change in the regulations would increase that amount to 10 lb. The report must

TABLE 1.13   Pertinent Regulations of the Resource Conservation and Recovery Act (RCRA)

RCRA regulation	Description of regulation
40 CFR 260	Hazardous waste management system
40 CFR 262	Standards applicable to generators of hazardous waste
40 CFR 261	Identification and listing of hazardous waste
40 CFR 268	Land disposal restrictions (land ban)
40 CFR 263	Standards applicable to transporters of hazardous waste

TABLE 1.14   RCRA Toxicity Characteristics and Waste Limits

RCRA waste number	Characteristic	Waste limit, ppm*
D001	Ignitability	
D002	Corrosivity	pH<2; pH>12.5
D003	Reactivity	
D004	Arsenic toxicity	5.0
D005	Barium toxicity	100.0
D006	Cadmium toxicity	1.0
D007	Chromium toxicity	5.0
D008	Lead toxicity	5.0
D009	Mercury toxicity	0.2
D010	Selenium toxicity	1.0
D011	Silver toxicity	5.0

*Corrosivity is measured in pH units.
Source: Based on Table 1 of 40 CFR 261.24.

TABLE 1.15 Example Calculation of Surface Area Required to Generate a Unit Weight of Lead

---

Assumptions:

  Lead in paint = 1% (10,000 ppm)

  Dry film thickness (DFT) = 10 mil (0.010 in)

  Density of dried paint = 1.5 g/cm^3 (can range from 1.1 to 2.5)

Calculations:

  1. Calculate volume of paint in 1 ft^2 (1 ft^2 = 929 cm^2):

      Volume = 929 cm^2 (DFT × 2.54 cm/in)

                = 929 cm^2 (0.010 in × 2.54 cm/in)

                = 23.60 cm^3

  2. Calculate weight of paint in 1 ft^2:

      Paint weight = density × volume

                  = 1.5 g/cm^3 × 23.60 cm^3

                  = 35.4 g

  3. Calculate weight of lead in 1 ft^2 of paint (1 ppm = 1 μg/g):

      Lead weight = ppm lead × paint wt/ft^2

                = 10,000 μg/g × 35.4 g/ft^2

                = 354,000 μg/ft^2

  4. Calculate square feet required to generate 1 lb of lead (1 lb = 454 g × 1,000,000 μg/g = 454,000,000 μg/g):

      Area = 1 lb ÷ wt of lead/ft^2

           = 454,000,000 μg ÷ 354,000 μg/ft^2

           = 1282 ft^2

Source: Adapted from K. A. Trimbler, *Industrial Lead Paint Removal Handbook,* 2d ed., Steel Structures Painting Council/KTA-Tator, Inc., Pittsburgh, 1993.

be made to the National Response Center [(800) 424-8802] and to state and local regulatory authorities within 24 hours. The calculations presented in Table 1.15 demonstrate how to estimate what unit area of paint on a bridge surface equates to a reportable CERCLA release of lead.

CERCLA and SARA regulations are found in 40 CFR 300 through 373. Discharges into the air and water are also regulated by the Clean Air Act (CAA) and Clean Water Act (CWA), respectively. EPA has mandated enforcement and interpretations of regulations to the states, leading to nonuniformity in the procedures to be followed and the stringency of enforcement. Permits for blasting are required in some states but not others.

As mentioned previously, the generator of the waste is responsible for it onward from the time the paint is removed from the structure, and is not released from that responsibility even when the waste is disposed of in a landfill, regardless of whether it is properly permitted, manifested, tested, and buried in a fully approved and permitted facility. If the waste site operations are under cleanup orders, the generators of the waste disposed of at the facility are liable for the cleanup along with the operator.

Also, because of the joint and several liability provision of CERCLA, it is possible that any one generator (or responsible party) may be liable for the entire waste disposal site cleanup. This is true even if there is no negligence on the part of the highway agency or its contractors. Regulatory agencies do not recognize contractual obligations among responsible parties and will seek financial compensation from whoever has funds and can be connected to the contamination.

In addition to these regulations, OSHA has established several regulations applicable to worker protection during paint removal as summarized in Table 1.16.

TABLE 1.16 Regulations for Worker Protection during Paint Removal

RCRA regulation	Description
29 CFR 1926	Safety of health regulations for construction
29 CFR 1926.33	Access to employee exposure and medical records
29 CFR 1926.51	Sanitation
29 CFR 1926.59	Hazard communication
29 CFR 1926.62	Lead
29 CFR 1926.63	Cadmium
29 CFR 1926.103	Respiratory protection

1.4.3 Approaches

Work technology is advancing to develop methods that effectively contain blasting debris and to minimize the amount of waste generated (vacuum blasting and recycling of grit). These methods are discussed in Art. 1.4.5. However, because there are so many existing steel bridges with lead-based coatings and highway funding is being stretched, highway departments are faced with the fiscal problem of prioritizing structures for coating replacement as part of the bridge maintenance program. Funding is a crucial element in scheduling coating replacement. The problem is planning the lead-based paint removal program and distributing funds on a time schedule that allows for preservation of the existing infrastructure.

The problem has been approached by either deferring maintenance, overcoating the existing paint, fully removing the paint and repainting the structure, or removing and replacing steel members. The problems with each are as follows:

Deferring maintenance does not serve to protect the bridge, which represents an enormous investment by the public.

Overcoating consists of applying new layers of nonleaded paint over lead-based paint with the intent of extending the coating system for another five years or so. This method may reduce short-term costs and buy the maintenance agency more time while new innovations in lead paint removal are being developed. However, worker safety and environmental issues still remain with the structure until the lead-based paint is removed. For example, the volume of unleaded paint increases with each coat, and thus a greater quantity of lead-contaminated paint must be disposed of as hazardous waste in many cases. Additionally, performance of the overcoating products has been highly variable, depending on operator skill and experience, application conditions, existing paint that is being overcoated, and product consistency.

Removal and repainting entails using abrasive blasting or other means to completely remove the existing paint, followed by application of a high-technology coating system. This is expected to provide the most durable and effective protection. However, its cost-effectiveness has greatly diminished because the spent paint and blasting grit must be collected and disposed of as hazardous waste. Also, worker safety during removal is a significant consideration.

Removing and replacing steel members involves removing portions of the bridge during major rehabilitation efforts; removing the paint from them within an enclosed workplace such as a fabricating shop, where the members are also repainted; and then putting the members back into place. Containment of the lead paint and blasting grit is more easily achieved. This method could approach cost-effectiveness only on major rehabilitation projects.

1.4.4 Worker Protection

Workers involved in removal, containment, or other work dealing with lead paint must be protected against lead hazards. Blood poisoning has historically been a serious job hazard in the bridge painting industry and is likewise dangerous in removing lead-based paint. Lead enters the body primarily through inhalation and ingestion. If not properly protected and trained, workers can be affected by lead while they are cleaning power tools, sanding, scraping, and performing abrasive blasting. Enclosing the work area to capture the blasting grit and waste paint creates a confined area for the workers, increasing the concentrations of lead to which they are potentially exposed.

The guidance document developed by the U.S. Occupational Safety and Health Administration (OSHA) entitled *Lead in Construction* sets forth proper health and safety procedures to be observed by painting contractors. The methodology generally requires training of employees, enclosure of the work area, decontamination of workers, the use of personal protection and monitoring equipment, and decontamination of personnel and equipment when leaving the work space. The structure maintenance engineer should be aware of these procedures so that appropriate schedule adjustments and budgetary planning can be included to cover contractor health and safety compliance.

It should be recognized that unconfined removal of paint regardless of lead content presents environmental and health and safety concerns. It will result in unacceptable deposition of dust and debris in roadways, streams, and communities as well as presenting an excessive dust hazard to workers.

1.4.5 Removal Methods and Containment

Lead paint removal methods have been improving in recent years, especially with respect to reduced impacts on the environment. Each method developed has characteristics that may make it more desirable in certain applications. K. A. Trimbler described and compared 17 methods of lead paint removal. His findings are summarized in Table 1.17 and described below. (K. A. Trimbler, *Industrial Lead Paint Removal Handbook,* 2d ed., Steel Structures Painting Council/KTA-Tator, Inc., Pittsburgh, 1993.)

Open Abrasive Blast Cleaning with Expendable Abrasives.   In the first method described by Trimbler, compressed air propels blasting grit against the coated surface. The spent blasting grit is then collected for disposal. The major advantages of this method are that contractors are familiar with this long-practiced method, it is very effective in creating a superior surface preparation, it reaches areas difficult to access, and it is quick (if containment considerations are not included in the evaluation). The major disadvantage is that it creates a high level of leaded dust and large quantities of debris that typically must be disposed of as hazardous waste. The additional containment requirements, hygiene training, and personal protection equipment drive the cost of the operation up. Blasting grit additives are currently being investigated that may render the grit nonhazardous.

Open Abrasive Blast Cleaning with Recyclable Abrasives.   In a similar method, metallic abrasives are used that can be separated from the debris (paint, rust, mill scale) and reused. The volume of dust and debris is reduced as compared with open

TABLE 1.17 Comparison of Paint Removal Methods

| Method and name | Equipment investment[a] | Quality of preparation | | | | | Debris created | | | Production rate[e] |
| | | Paint removal[b] | | Rust/mill scale removal[b] | | Quality for painting[c] | Dust generation[d] | Volume of debris[d] | Containment required[d] | |
		Flat	Irregular	Flat	Irregular					
Method 1. Open abrasive blast cleaning with expendable abrasives	2–4[f]	5	5	5	5	5	1	1	1–2	5
Method 2. Open abrasive blast cleaning with recyclable abrasives	1	5	5	5	5	5	3	4	1	5
Method 3. Closed abrasive blast cleaning with vacuum	1	5	3–4	5	3–4	5	4–5	4	4	2
Method 4. Wet abrasive blast cleaning	2–3[f]	5	5	5	5	4–5	4–5	1	2–3	4
Method 5. High-pressure water jetting	2	3–4	2–3	1	1	3–5	5	2–4	2–4	3
Method 6. High-pressure water jetting with abrasive injection	2	5	4–5	5	4–5	4–5	5	2–3	2–4	3–4
Method 7. Ultrahigh-pressure water jetting	1	4–5	3–4	1	1	3–5	5	2–4	2–4	4
Method 8. Ultrahigh-pressure water jetting with abrasive injection	1	5	4–5	5	4–5	4–5	5	2–3	2–4	4
Method 9. Hand-tool cleaning	5	1–2	1–2	1	1	1–3	4–5	4	4	2
Method 10. Power-tool cleaning	4	2–3	2	1–2	1–2	1–3	3–4	4	4	2
Method 11. Power-tool cleaning with vacuum attachment	3	2–3	2	1–2	1–2	1–3	4–5	4	4–5	2
Method 12. Power-tool cleaning to bare metal	4	4–5	2–3	4–5	2–3	4–5	3	4	3–4	1–2

Method										
Method 13. Power-tool cleaning to bare metal with vacuum attachment	3	4–5	2–3	4–5	2–3	4–5	4–5	4	4–5	1–2
Method 14. Chemical stripping	3–4	3–4	3	1	2–5	5	5	2–3	3–4	1
Method 15. Sponge jetting	2–3	5	5	4–5	5	4	4	3–4	3–4	2–3
Method 16. Sodium bicarbonate blast cleaning	2–3	3	2–3	1–2	3–4	4–5	2–4	3–4	2–4	2–3
Method 17. Carbon dioxide blast cleaning	1	2–3	2–3	1	3–4	4–5	4	4	2–4	1–2
Method 18. Combinations of removal methods	Ratings dependent upon combinations of methods used.									

[a] 5, very inexpensive; 4, inexpensive; 3, moderately expensive; 2, expensive; 1, very expensive.

[b] 5, highly effective; 4, effective; 3, moderately effective; 2, poor; 1, very poor (ineffective).

[c] 5, excellent; 4, good; 3, marginal; 2, poor; 1, very poor.

[d] 5, no/none; 4, little/low; 3, moderate; 2, sizable; 1, substantial.

[e] 5, very high; 4, high; 3, moderate; 2, low; 1, very low.

[f] Most contractors already own much of this equipment. Therefore, even though the purchase price is high, little additional investment may be needed.

Source: From K. A. Trimbler, *Industrial Lead Paint Removal Handbook*, 2d ed., Steel Structures Painting Council/KTA-Tator, Inc., Pittsburgh, 1993, with permission.

abrasive blast cleaning with expendable abrasives, but the effectiveness and the ability to reach inaccessible areas are the same. Additional disadvantages are contractors' unfamiliarity with the method and the special additional care that must be taken to keep the blasting grit moisture-free to avoid rusting and clumping. Should the abrasive dust escape containment, it may cause rust spots on the surfaces where it settles. Additionally, because the grit is recycled, higher concentrations of airborne lead dust within the containment area will have to be considered for worker safety.

Closed Abrasive Blast Cleaning with Vacuum.   A third method is a compressed-air propellant method using a nozzle fitted with a localized containment assembly that employs a vacuum. The recycled metallic grit, dust, and debris are vacuumed as the surface is blasted. This method is rated as highly effective, both in surface preparation and in containment of dust and debris, but the rate of cleaning is slow. The greatest limitation of this method is that the containment mask must be held tightly to the surface of the structure, reducing the method's effectiveness on irregular and inaccessible surfaces. The containment method confines the blast spray pattern so that only small surface areas are being blasted at any one time. This, along with maintaining a tight seal, is arduous and leads to operator fatigue.

Wet Abrasive Blast Cleaning.   In the wet abrasive method, water is injected into a stream of slag abrasive propelled by compressed air. An additional step after paint removal to flush the surface is required using this method. Water, grit, and debris are collected for disposal. This method is effective both in dust control and in the quality of surface preparation; however, the amount of waste produced is substantial and difficult to clean up. Inhalation hazard is greatly reduced, but ingestion hazard still exists.

High-Pressure Water Jetting.   Highly pressurized water (20,000 lb/in²) propelled against the surface is effective without the use of grit. This method reduces dust to negligible levels; however, ingestion hazard still exists. The water is voluminous in quantity and difficult to capture in containment. The method does remove paint, but is not effective in inaccessible areas or in removing mill scale. A rust inhibitor is usually used, which may affect the applied coating.

High-Pressure Water Jetting with Abrasive Injection.   Combining the previous method with abrasive injection results in all the disadvantages of the previous methods but with the additional complication of having grit in the disposal water. It is considered highly effective in removing mill scale and paint from inaccessible areas.

Ultrahigh-Pressure Water Jetting.   Even more highly pressurized water (up to 40,000 lb/in²) can be propelled against the surface without the use of grit. This method is more efficient in removing paint than the high-pressure water jetting method; however, the main advantages and disadvantages still apply.

Ultrahigh-Pressure Water Jetting with Abrasive Injection.   The ultrahigh-pressure water jet method can also be enhanced by the addition of disposable abrasives to the jet stream. The result is rated highly effective, with advantages and disadvantages similar to those of the previously described water jetting methods.

Hand-Tool Cleaning.   Manually operated impact tools and scrapers can be used to remove paint and mill scale. This method is very inexpensive with regard to the contractor's investment, but it is rated poor in efficiency and effectiveness. Also, since only small amounts of localized dust and debris are created, workers are given a false sense of security about exposure, thus making it difficult to enforce personal protective equipment requirements.

Power-Tool Cleaning.   Power tools such as chippers, needle guns, descalers, wire brushes, sanding disks, and grinding wheels can be used to remove paint, rust, and scale from the bridge surface. This is a labor-intensive method. The quality of preparation of the surface may be inadequate, depending on the condition of the coating being removed. Airborne dust is generated, and workers must be properly protected.

Power-Tool Cleaning with Vacuum Attachment.   In another version of the previous method, a vacuum attachment is added around power tools and debris. This has the additional disadvantage that accessibility in tight areas is further reduced because of the shroud and vacuum attachment. On irregular surfaces, a seal may be difficult to maintain, and airborne leaded dust may be present. Because a seal typically minimizes dust, workers may not be aware when it has slipped and they thus require additional respiratory protection.

Power-Tool Cleaning to Bare Metal.   Power tools can also be used to clean to the bare metal. This method adds such tools as scarifiers (rotary peening tools) to the power-tool set and can achieve a generally higher level of surface preparation. More dust is created, and higher levels of worker protection and training are required. Productivity is low, and a high quality of surface preparation may not be achieved in inaccessible or heavily pitted areas.

Power-Tool Cleaning to Bare Metal with Vacuum Attachment.   A modification of the previous method contains dust and debris using a shroud and a vacuum attachment around the scarifying power tools, creating a seal with the bridge surface. This has the same disadvantages as the method of power-tool cleaning with vacuum attachment, but with additional training required on the equipment and greater cost to achieve bare-metal standards.

Chemical Stripping.   Chemical stripping agents can be applied to the surface, left in place for several hours, and then scraped off along with paint, rust, and scale. The surface must then be flushed with water and the chemical agent neutralized. The rinsates must be contained and disposed of properly. This method virtually eliminates airborne debris. Personal protective clothing should be worn during the removal process to prevent dermal contact with leaded debris. Not all chemicals are effective on all paints, and few will remove all the rust and scale.

Sponge Jetting.   In the sponge jetting method, compressed air is used to propel polyurethane particles (sponge) that may be seeded with abrasives against the bridge surface. The debris and sponges are collected and sorted. The sponges can then be reused. The quality of surface preparation is similar to that from other blast cleaning methods, but the productivity is lower. The amount of debris is significantly reduced because of the recycling of the sponges. Visible dust is reduced, although containment and personal protection gear must be maintained as in other blasting methods. Costs of the equipment and abrasives are high.

Sodium Bicarbonate Blast Cleaning.   Either jetted water or compressed air can be used to propel water-soluble sodium bicarbonate against the bridge surface. This method does not remove mill scale or rust effectively. Dust is significantly reduced when jetted with water, and so hazards of lead inhalation are reduced (not eliminated), but lead ingestion remains a hazard. Containment of the water is difficult. It may be demonstrated on a case-by-case basis that the sodium bicarbonate serves to stabilize lead in the paint so that it does not leach into the water in concentrations great enough to render the blasting water a hazardous waste. There is no grit waste. This method requires inhibitors to prevent flash rust from forming when the paint is removed. Sodium bicarbonate does neutralize chlorine contamination of the steel.

Carbon Dioxide Blast Cleaning.   Small pellets of dry ice can be propelled, using compressed air, against the bridge surface. This method does not remove mill scale or heavy rust, and production is slow. This method reduces the volume of waste to only the actual paint being removed. It also greatly reduces sparking risk, and dust is reduced. Worker exposure is reduced, though it must still be controlled. The equipment and materials for this method are quite expensive.

Combinations of Removal Methods.   Combining methods, if done effectively, will achieve a specific advantage such as reducing the volume of waste or increasing productivity or the quality of surface preparation. The objective is to select methods that are complementary. An example would be first using a chemical stripper, which yields low dust and minimizes the need for containment. The chemicals will remove the leaded paint but not the mill scale. Once the hazardous substances are removed, another method, such as wet blasting, can remove the mill scale and rust without necessitating further hazardous waste disposal.

1.4.6   Containment Considerations

Design of proper containment requires specialists in structural engineering, coatings, ventilation, and exhaust. The following aspects of containment design should be considered:

- The environmental media (air, water, soil) that are vulnerable as a result of the method being used, and the containment methods that will provide the best protection
- Durability, for the sake of worker safety and containment integrity
- Compatibility with the selected removal method, and noninterference with the productive removal of the paint, mill scale, and rust and the application of a new coat of paint
- Ease of construction and disassembly, and of moving from one area of the structure to another
- Design for the natural elements such as wind and rain loads
- Continued usability of the structure and proximity of nearby structures and people
- Cost-effectiveness
- Compliance with applicable regulations

Materials used to construct containments are either rigid panels or flexible materials such as tarpaulins and windscreens. Materials should be fire-retardant because of the sparking hazard, high dust, and high ventilation requirements. For flexible materials, other considerations would include resistance to ultraviolet radiation, solvents, abrasion, and bursting.

The checklist in Table 1.18 is a guideline prepared by Trimbler to follow in designing an appropriate containment.

Various debris-recovery assessment methods are being developed. Some, such as air monitoring and analysis of soil and water samples to evaluate whether lead content has increased as a result of paint removal activities, have been codified in regulations. A method for calculating debris recovery that has been used by the South Carolina Department of Transportation is found in the Structural Steel Painting Council (SSPC)

TABLE 1.18   Containment Design Checklist

1. Review drawings and specifications for project familiarity.

2. Investigate OSHA and EPA regulations affecting worker protection and control over emissions.

3. Determine method of surface preparation to be employed.

4. Examine the structure to be prepared:
   - Confirm that the selected method of preparation is suitable.
   - Determine if any coats of paint will be applied in containment.
   - Assess the load-bearing capacity of the structure to support containment.
   - Examine the structure for attachment points for the containment.
   - Divide large structures into logical containment units according to size and configuration. Consider the air movement requirements and the need to have a large enough area for productive surface preparation and painting.
   - Determine if a working platform should be used on elevated projects. Determine how far ground covers should extend beneath or around the removal operation.
   - When working over water, determine if a barge is going to be used for spent abrasive collection or staging, and assess the need for water booms to minimize problems due to inadvertent spills. Determine the need for U.S. Coast Guard approval and navigation restrictions.
   - Determine methods for conveying the debris for recycling or disposal.

5. Determine project-specific ventilation requirements.
   - Consult *Industrial Ventilation: A Manual of Recommended Practice* (Committee on Industrial Ventilation, American Conference of Government Industrial Hygienists, Cincinnati; 20th ed., 1988) for engineering guidance.
   - Select the air velocity (air speed) throughout the work area and exhaust volume required.
   - Determine the necessary transport velocity through the exhaust ductwork required to avoid dropout of debris.
   - Lay out the ductwork as short as practical with as few bends as possible. Do not use bends with a centerline radius less than 2 times the duct diameter. Include the use of exhaust hoods or plenums within containment.
   - Select the air-cleaning device (dust collector) on the basis of the volume of air and dust loading of the airstream (air-to-cloth ratio).
   - Select the fan that will provide an adequate volume of air, and that is able to overcome the resistance throughout the system.
   - Provide adequate makeup air (supply air), properly distributed to provide a uniform airflow. Include properly balanced forced air if required.
   - Confirm that all of the above will provide ample airflow throughout the work area. If not, consider the use of localized ventilation and exhaust.

6. Obtain and review equipment manufacturers' technical information.

7. Complete the design package. Utilize the expertise of structural and mechanical engineers, industrial hygienists, coatings specialists, and equipment specialists.

Source: From K. A. Trimbler, *Industrial Lead Paint Removal Handbook,* 2d ed., Steel Structures Painting Council/KTA-Tator, Inc., Pittsburgh, 1993, with permission..

Guide 61, *Guide for Containing Debris Generated during Paint Removal Operations* (SSPC 92-07, March 1992). The following equation is used in the SSPC guideline for calculating debris recovery:

$$\text{RE} = \frac{W_d}{W_a + W_p} \times 100$$

where RE = efficiency of recovery
W_d = dry weight of abrasive and paint debris collected
W_a = dry weight of abrasive used
W_p = calculated weight of paint to be removed

This method has limitations such as not weighing the various release media (air, soil, water), which have a great deal of bearing on the effectiveness of a containment. A 1 percent debris loss into the soil is not as significant as a 1 percent debris loss into the air. Care must be taken when using this method to measure only the abrasive and paint from the project and not to measure soil that may have come into contact with the debris.

The project designer should require an environmental monitoring plan to evaluate the effectiveness of the containment methods. Reporting and record keeping within the plan should include the following, at a minimum:

• Name and location of the site, along with a site plot plan
• Identification of the individual or company that is conducting the monitoring
• Name and qualifications of the analytical laboratory used
• Criteria and rationale for selecting monitoring and sampling sites and duration of sampling
• Descriptions of sampling and monitoring methods
• Quality assurance and quality control plans
• Examples of reporting forms
• Acceptance criteria
• Reporting procedures and corrective actions if acceptance criteria are not met

1.4.7 Community Relations

Bridges are public structures, exposed to public scrutiny. Lead poisoning, caused in large part by lead-based paint, has come to the forefront of public awareness. Any inconvenience to the public revolving around bridge maintenance activities calls attention to the structure and ongoing operations. If not handled well, lead paint removal from bridges can become a volatile community issue. The highway agency should be prepared with complete, accurate, and current documentation on the safety procedures that are implemented to protect the public health and the environment.

The agency's position is very strong if it is able to demonstrate that it has an effective technical program with stringent monitoring and inspection procedures in place for hazard containment. Soliciting regulatory agreement with the methodology and establishing a good working relationship with the regulators will also be valuable in smoothing public perception problems that may arise. Containments can be designed to minimize the appearance of risk to the public in addition to fulfilling their intended purpose. Adjusting the timing for paint removal activities to be conducted during off-peak hours also serves to diminish the attention that the operation receives.

1.4.8 Specifications Guidelines

The specification for scraping or blasting lead-based paint from structures should be written with worker safety and environmental issues in mind, so that qualified contractors who can adhere to a high level of quality and compliance are selected for the project. These specifications should:

- Describe the extent of surface preparation and the degree of containment required and let the contractor propose how to accomplish this.
- Identify key health and safety and environmental regulations to ensure that the contractor is aware of these regulations and plans compliance strategies in the bid.
- Clearly state that the paint to be removed is lead-based. The highway department should have had the paint tested prior to contract bid if there is any doubt whether the paint is lead-based. The cost differential is too great to make the assumption or let a contract with the lead concentration factor as an unknown.
- Specify how the waste is to be treated, tested, handled, and disposed of.
- Identify the worker protection standards and requirements that the contractor's health and safety plan must meet at a minimum.

1.4.9 Management of Industrial Lead Paint Removal Projects

The following steps have been developed as guidelines for managing lead paint removal projects and are set forth in Trimbler's *Industrial Lead Paint Removal Handbook:*

Initial project evaluation. In the initial project evaluation, the owner or specifier must determine whether the coating contains lead-based paint either by reviewing earlier plans and specifications for the structure or by sampling and analysis.

Prebid assessment of paint removal methods and debris generated. The owner or specifiers should estimate how much waste will be generated by methods evaluated to be appropriate to the size and circumstances of the project. Designing a testing program to evaluate the toxicity of waste generated may be appropriate for large paint removal projects.

Understanding the regulations before preparing the specifications. The regulations regarding air quality, water quality, soil cleanup, unauthorized releases, worker protection, and hazardous waste generators should be thoroughly understood. How these regulations are enforced should be discussed with both state and local officials.

Preparing the project specifications. Both painting and lead removal requirements should be addressed in the specifications. These should identify the methods for surface preparation and the coating system to be applied. The relevant regulations, the degree of containment, and the evaluation of performance criteria should all be specified.

Developing a worker protection plan. Prior to start-up, the contractor should provide a worker safety plan that addresses exposure monitoring, the compliance program, the respiratory protection program, personal protective equipment, housekeeping issues, hygiene facilities and procedures, medical surveillance, employee removal for exposure to lead, employee training, signage, record keeping, and employees' right to observe and review monitoring information.

Preparing environmental protection monitoring plans. The procedures developed to verify environmental protection should include high-volume air samplers, tests for visible air emissions (opacity), personal air quality monitors, measurement (and reporting requirements) of unauthorized releases, and pre- and postproject soil quality and water quality sampling.

Developing procedures for the control and handling of hazardous waste. Assuming that hazardous waste is to be generated, plans should be developed for identifying the waste, obtaining a hazardous waste generator identification number from the EPA, preparing for proper notification and certifications with each shipment, preparing waste manifests, packaging and labeling waste, implementing contingency plans, conducting waste treatment and analysis for on-site handling, and record keeping.

Designing a containment and ventilation plan. The contractor should develop detailed plans to select appropriate support structures and containment, address ventilation and other worker safety issues, provide emissions control, achieve water and soil protection and debris recovery, and verify the integrity of the containment structure.

Monitoring the project. The project manager should develop a plan to monitor the adequacy of all of the control measures; visually monitor the project regularly and use approved testing methods to evaluate adequacy of controls; regularly monitor the ventilation system and the integrity of the containment; regularly examine waste storage facilities, and the handling and transportation methods and procedures; and verify worker protection and hygiene procedures. OSHA standards must be observed.

Figure 1.2 illustrates a decision tree to aid in the management of lead paint removal.

1.5 RESOURCE RECOVERY AND USE OF WASTE MATERIAL

Solid waste production in the United States exceeds 4.5 billion tons annually for nonhazardous waste alone. Dividing the waste by classification, agricultural waste accounts for 46 percent of the total, mineral waste 40 percent, industrial waste less than 10 percent, and domestic waste less than 5 percent. Because of the vast amount of building materials required to construct and maintain the transportation infrastructure in the United States, our highway systems represent a tremendous potential for beneficial use of reclaimed and recycled resources.

The generation, handling, and safe disposal of solid wastes are a growing national concern that is reflected in the promulgation of increasingly comprehensive and restrictive land disposal legislation. Increased cost of complying with this legislation and other environmental laws has enhanced the appeal of recycling and resource management.

Because solid waste material is not as uniform as raw materials, the characteristics, performance, cost of preparing, and therefore appropriate uses vary with the source of the material. Results in highway applications vary considerably and depend on several parameters including climate differences, composition, material handling practices, and construction procedures. Factors to be considered when recycling a waste to a highway construction end use include the following [National Cooperative Highway

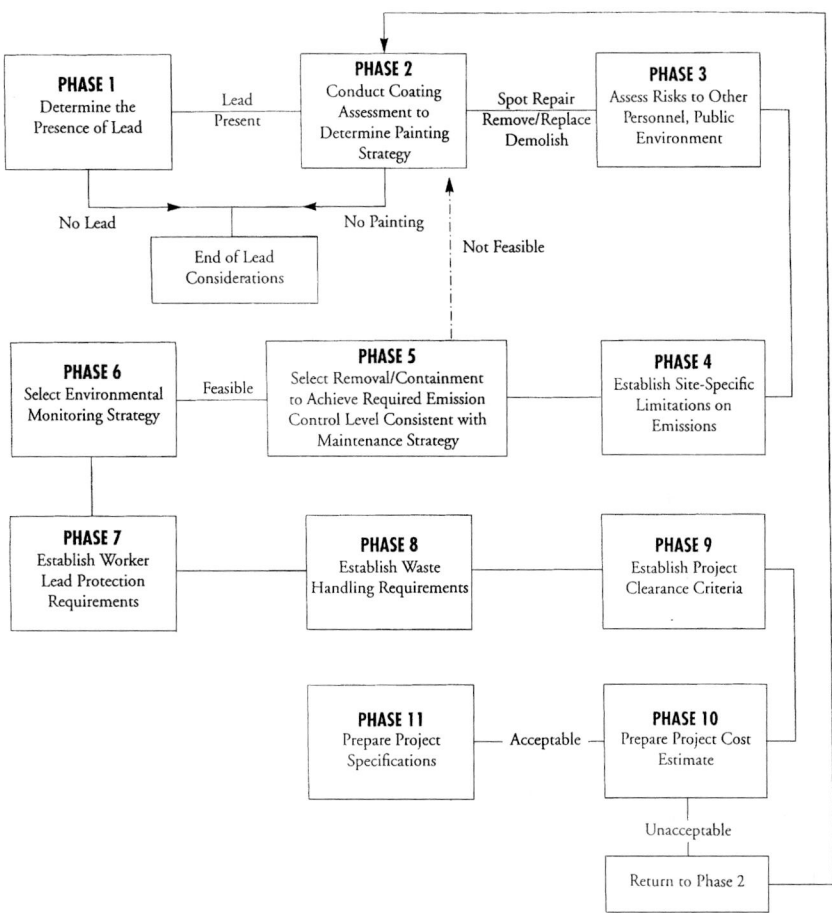

FIGURE 1.2   Decision chart for management of lead paint removal. (*From K. A. Trimbler,* Industrial Lead Paint Removal Handbook, *2d ed., Steel Structures Painting Council/KTA-Tator, Inc., Pittsburgh, 1993, with permission.*)

Research Program (NCHRP), Transportation Research Board, Synthesis 199, *Recycling and Use of Waste Materials and By-Products in Highway Construction: A Synthesis of Highway Practice,* Washington D.C., 1994]:

> *Environmental threats and benefits.*   Along with the considerable environmental benefit of reducing the landfill burden, potential threats to the environment caused by the use of recycled material must be considered and compensated, mitigated, or otherwise overcome before using recycled material is feasible.
>
> *Regulatory requirements, guidelines, and restrictions.*   The federal and state legislation regarding recycled materials reflects reduced landfill capacity in the United States and the recognition that there is a net benefit to producing resources from waste. Some states now have recycling mandates in place. Permits are often required when conducting recycling activities, and/or when effectively creating an

authorized disposal site in using certain wastes as embankment fill. If the recycling activity falls under the RCRA classification of "use constituting disposal," additional regulations apply (see the discussion on recycling hazardous waste, Art. 1.5.10).

Economic cost and benefit. Economic considerations are often the driving force behind recycling efforts on the county and city level, because of the increased landfill costs and increasingly limited capacity. Recycling for highway departments may become more attractive as budget cuts increase and the price for recycled waste materials decreases. In some cases, recycled materials extend the service life of highway components, making the life-cycle costs of using such materials attractive.

Engineering properties and technical performance. Because of inconsistencies in the constitution of some waste materials, performance results for some end products may vary significantly and careful evaluation of appropriate applications is necessary. The primary question is, Does the performance of the material compare favorably with the same material constituted from raw materials? In some instances, the use of waste material has consistently improved performance. For instance, silica fume use in portland cement results in higher compressive strength and higher resistance to corrosion of steel reinforcement due to the increased density and reduced porosity of the resulting concrete. ("Silicon," *Minerals Yearbook,* U.S. Bureau of Mines, Washington, D.C., 1989.)

Construction materials shortages and alternative resource availability. Millions of tons of aggregate are used each year in the construction of highways. Resources from existing quarry mining are being depleted and the new sources are often not accessed because of restrictive regulation and preferred uses of the land. In areas experiencing shortages, recycling construction materials, waste minerals, and other products into aggregate is more cost-effective than shipping aggregate from distant quarries. Steel is one of the most widely recycled materials used in highways. Steel reinforcement can be composed completely of recycled scrap steel, and steel girders can contain as much as 25 percent recycled scrap steel. Recycling scrap steel greatly reduces reliance on foreign sources for raw materials in the steel industry. (NCHRP Synthesis 199, p. 6.)

1.5.1 Legislation

Under RCRA, the approach was established to classify solid waste management facilities and practices, and to direct states to develop comprehensive state plans for solid waste management. (Dufour, op. cit., p. 99). The solid waste disposal emphasis of RCRA called attention to the growing landfill capacity problem and the need to develop approaches to handling wastes. In the preamble of RCRA, attention was called to the millions of tons of recoverable materials that are placed in landfills and to the fact that the recovery or conservation of many of these materials would benefit the United States by reducing projected landfill capacity requirements, retaining and expanding on our national resources, reducing the country's dependence on foreign resources, and thus also reducing the balance-of-payments deficit.

In reference to recycled materials, Section 6002 of RCRA requires that federal agencies and state or local agencies receiving funds from the federal government must procure supplies and other items composed of the highest practical percentage of recovered or recycled materials, consistent with maintaining satisfactory levels of:

- Product quality
- Technical performance

- Price competition
- Availability (appropriate quantities for the required project)

Also, under RCRA, specifications cannot be written to discriminate against materials with recycled constituents. In addition, EPA was authorized to prepare guidelines for recycling, and resource recovery guidelines addressing procurement practices and information on research findings about the uses and availability of recycled materials. Guidelines covering coal fly ash in portland cement, recycled paper, retreaded tires, building insulation, and re-refined oil have been developed. While not specifically required by EPA, the guidelines have encouraged most state highway programs to prepare specifications allowing the substitution of fly ash in concrete.

The Intermodal Surface Transportation Efficiency Act of 1991 (ISTEA) authorized DOT to coordinate with EPA and state programs in developing information on the economic savings, technical performance qualities, and environmental and public health threats and benefits of using recoverable resources in highway construction. ISTEA specifically calls out requirements for the percentage of asphalt pavement containing recycled rubber from scrap tires. In an escalating scale, this percentage is to reach 20 percent by 1997.

In addition to RCRA and ISTEA, an executive order has been signed with the intent of promoting cost-effective waste reduction and recycling, developing policy options and procurement practices for recycled materials within the federal government, and encouraging the use of such materials through procurement specifications favoring such practices. The executive order also establishes a federal recycling coordinator and recycling coordinators for every federal agency.

State legislation has been developing to promote both research into the performance and viability of recycled materials and the procurement of such materials. The results of a study conducted by the National Solid Wastes Management Association indicated that 42 states have passed laws to stimulate recycling markets by encouraging state agencies to purchase products with recycled content and 33 states had enacted other legislation related to recycling (*Recycling in the States: 1990 Review*, National Solid Wastes Management Association, Washington, D.C., 1990). In a study performed by the National Cooperative Research Program of the National Research Council Transportation Research Board, 26 of the 45 responding states surveyed indicated that state legislation had been passed requiring the state department of transportation and other state agencies to investigate waste material use. Seventeen of these have established mandatory recycling laws. Projections from 18 of the states responding indicated that they do not expect to have sufficient landfill capacity in the next 5 to 10 years. Thirty of the states responding have used wastes or waste by-products in their highway programs.

1.5.2   Waste Material Generated

Generated waste material can be categorized as either construction wastes, industrial wastes, mining or mineral wastes, agricultural wastes, or domestic wastes (of which scrap tires are a significant subset). Many advanced recycling programs have been established to make use of these wastes, such as requiring identifying codes for the base resin in plastic products to enable more refined recycling of plastics. Some of these wastes are not suitable for or do not make a significant recycling contribution to highway use. For example, only an insignificant amount of the total crop waste (estimated to be about 9 percent of all the total nonhazardous solid waste generated each year in the United States) has a beneficial highway use. These uses are as an asphalt extender or portland cement additive and are experimental. (NCHRP Synthesis 199, p. 7.) The following articles contain brief descriptions of the types of wastes generated that have at least been researched for use in highway projects (Ibid., pp. 7–25):

1.5.3 Construction Waste

Much construction demolition debris consists of wastes with little recycling value for highways, such as wood and plaster. However, demolition debris also includes concrete, glass, metal, brick, and asphalt, most of which can be reused in highways as aggregate. In order to be a viable resource and meet the standard specifications as aggregate when crushed, the construction and/or demolition rubble must be separated from the other debris and cleaned of detritus. Construction wastes generated and the associated annual tonnage produced are presented below. These tonnage estimates were collected from numerous sources and presented in NCHRP Synopsis 199.

Reclaimed Asphalt.   The asphalt pavement from the demolition of parking lots, roads, and highways can be reclaimed. Most states are making at least some use of reclaimed asphalt pavement (RAP) in highways, with recycling back into asphalt pavement the most prevalent use. Although 18 states see their stockpiles of RAP growing, there are so many uses for RAP that state transportation agencies do not consider this a problem. Estimated tonnage is 50 million tons annually. Because the use of RAP interferes with the ability to control hot-mix temperatures during formulation, asphalt mixtures can contain only between 20 and 50 percent RAP. Achieving 50 percent RAP content is practical only in a laboratory setting, where thorough blending of the RAP and new aggregate can be controlled. When plant efficiency is a concern, 50 percent RAP in hot mix is not practical. The differential between the temperature of the discharged gases and the discharged asphalt mix reaches 70°F. High exhaust gas temperatures can lead to premature corrosion of plant equipment. Thus the percentage of RAP that can be incorporated efficiently is based on the plant efficiency that can be maintained. (U.S. Army Corps of Engineers, *Hot-Mix Asphalt Paving Handbook,* AC 150/5370-14, July 1991, Appendix 1, pp. 1-21, and 2-45.)

Reclaimed Concrete Pavement (RCP).   Recycling concrete pavement began in this country more than 20 years ago, first as unbound aggregate, then in asphalt-wearing surfaces, and now as concrete aggregate. Improved methods of breaking up concrete and separating out the rebar have made the use of RCP more cost-competitive. At least 16 states now recycle concrete pavements either as new concrete or as aggregate in subbase material or base course. This does not include demolition debris of concrete structures.

Roofing Shingles.   Scrap from and leftover materials from composite shingle manufacturing operations amount to an estimated 10 million tons of waste annually. The waste includes fragments, asphalt binder, and granules. These wastes can be recycled as asphalt paving material. Shingle waste from roofing contractors and demolition operations is less viable because of possible contamination.

Sandblasting Residue.   Many uses of sandblasting grit are possible if the removed paint was not lead-based. If the paint was lead-based or contained other metals, the debris would have to be analyzed to determine if it was nonhazardous before it would be deemed suitable for recycling use.

Demolition Debris. Demolition debris is estimated to produce 25 million tons of waste annually. Much of this debris is restricted from landfills, and alternatives are being sought. In order to be viable for recycling, the debris has to be separated into homogeneous materials. Rubble material has many recycling uses in highways. Wood debris can be chipped and used for lightweight fill and mulch, but only if it is untreated. Disposing of asbestos-containing material (ACM, prevalent in buildings constructed before 1979) is difficult because chrysotile asbestos fibers are known to increase cancer risk if inhaled. If demolition of buildings with ACM from state transportation right-of-way is required in a project, it is possible in some states to arrange for on-site disposal in a state monitored landfill.

1.5.4 Industrial Waste

Approximately 150 million tons of the types of industrial waste that can be reused to some degree in highway projects is produced annually in the United States. Dredge spoils are not included in this estimate. Very little of this waste can be landfilled. Many kinds of industrial wastes are not suitable for highway use because they are hazardous or because once they are in place, leachate or breakdown products from these materials are a threat to the environment. Through treatment, some industrial wastes otherwise deemed a threat to the environment may be rendered usable. Petroleum-contaminated soils, for instance, once thermally treated, can be used as fill material and have been used in asphalt mixtures as fine aggregates. Petroleum-contaminated soils are not currently being recycled into highway projects but have been used on road and street construction at the local level. The principal recoverable wastes from industrial activities are described below.

Coal Ash By-Products. NCHRP Synthesis 199 cites an American Coal Ash Association publication, *1992 Coal Combustion By-Product Production and Consumption,* when noting that 66 million tons of coal ash are produced annually from the 420 coal-burning power plants across the country. Only six states do not have coal-burning power plants. Coal is either anthracite, bituminous, or lignite (subbituminous); the particular form has a bearing on the characteristics of the by-products.

Fly Ash. The American Society for Testing and Materials (ASTM) divides fly ash into two classes: class F, from anthracite coal; and class C, from lignite coal. Anthracite coal is mined from areas east of the Mississippi River and is more plentiful and has a higher sulfur content than lignite coal. Its fly ash reacts with calcium and water at ordinary temperatures to form a cementlike compound. Class C fly ash has a higher lime content than class F fly ash and therefore can be self-setting. To be usable as a cementitious substitute for portland cement, the fly ash must meet quality standards established by ASTM (Standard C-618). Approximately 25 percent of the fly ash produced meets this standard, yet only about half of the viable resource is being used.

Bottom Ash and Boiler Slag. Bottom ash and boiler slag are also by-products of coal burning, amounting to approximately 18 million tons of waste produced annually. These by-products are being researched for use in embankments, unbound aggregate, and asphalt paving and antiskid material.

Blast-Furnace Slag. The slag that is the by-product of producing iron in a blast furnace is nonferrous and consists of silicates and aluminosilicates of lime. Of the three types of slag produced from blast furnaces (expanded, granular, and air-cooled), about

90 percent of that recovered for use in construction is air-cooled. Air-cooled slag is porous and suitable for use as aggregate in lightweight concrete, in asphalt, in roadway bases, and in fill material. Granulated slag can be finely ground as slag cement, and expanded slag can be used as aggregate in lightweight concrete. The primary barrier in the past to using slag that accumulated is that it was not separated into homogeneous piles and it was mixed with steel slag.

Steel Slag.   Steel slag is the product of lime flux reacting with products in a steel furnace such as pig iron. Steel slag consists of calcium, iron, unslaked lime, and magnesium. It can be very expansive if not properly treated with water to "age" it. Because of its characteristics of being very hard, stable, and abrasion-resistant, it is used in paving material and snow control. It is heavier than most aggregate and has been used as fill material and as railroad ballast. However, some concern has developed recently that the leachate from these two uses clogs drains and can affect receiving waters. About 7.9 million tons of steel slag is sold in the United States annually.

Nonferrous Slag.   Slag from the smelting operations for other ores such as copper, lead, zinc, nickel, and phosphates is grouped together under a single heading for discussion only. Each must be evaluated and treated separately because of the varying properties these slags possess. Phosphate slag, copper oxide blasting slag, and zinc slag have been used as aggregate in paving mixtures. Aluminum slag was used experimentally for asphalt paving aggregate, but the material was not durable and is no longer used.

Foundry Wastes.   Foundry wastes are estimated at just under 3 million tons produced annually in the United States, including furnace dust, arc furnace dust, and sand residue. There are more than 2300 foundries operating in the United States, concentrated primarily in the Great Lakes states. Foundry dust is often disposed of as hazardous material because of its high concentration of metals. Foundry sand, however, is not generally hazardous and has been used as fill material, pipe bedding, and fine aggregate in paving mixtures. Tests must be conducted on the material prior to reuse to determine the properties of the leachate and to ensure that it is environmentally safe. Research into the use of foundry sand is being conducted by departments of transportation in five states, and limited use has begun with limited success. The permanence of foundry sand as pipe bedding in Illinois, however, was not considered acceptable.

Flue Gas Desulfurization Sludge (FGD).   Flue gas desulfurization sludge (FGD) is the product of wet scrubbing of flue gases at coal-burning plants and consists of calcium sulfate or sulfite slurry. These slurries are generally landfilled through various processes. By dewatering FGD (especially the sulfate slurries) and blending it with a reactant like portland cement, or cement fly ash, the mixture can be used as stabilized base material or as fill material. FGD has also been used as a dust control palliative, and additional uses are being investigated.

Paper Mill Wastes.   Inorganic paper mill sludge has been used occasionally for dust control on highway projects. Although some research has indicated that spent sulfide liquor from the paper milling process may have application in soil stabilization, it is believed that a higher level of use exists for the material within the paper industry. The ash residue from burning bark at paper mills, when pulverized with coal and burned, has been shown to be as effective a portland cement substitute as class F fly ash and is being considered for use in highway projects.

1.5.5 Mining Waste

Coal Refuse. Coarse coal refuse from mining operations is produced at a rate of 120 million tons per year. Coarse material is banked, while fine coal refuse is put into a silt-sized slurry mix and placed in impoundments. It is estimated that between 3 and 4 billion tons of coal mining refuse has accumulated in the United States. Concern about spontaneous combustion and leachate of the material (composed of slate and shale with sandstone and clay mixed in) has impeded in-depth studies of the use of coal waste. It is currently being evaluated for use in embankments and as subbase material, two applications that reportedly have been used in the past.

Quarry Wastes. Fairly consistent wastes are produced at any one quarry consisting of fines from stone washing, crushing, and screening and wet, silty clay from washing of sand and gravel. Most quarry waste materials are not generated as reusable material and are therefore not sized within standard specifications, but are stockpiled in ponds. Reclamation through dewatering and segregating coarse and fine materials would be necessary to use the 175 million tons of quarry waste produced each year, or any of the approximately 4 billion tons that have accumulated in the United States. The mineral properties and characteristics of the waste differ from quarry to quarry, limiting the beneficial end use, but quarry wastes have been used as fill and borrow material, flowable fill, and cement-treated subbase.

Mill Tailings. Mill tailings are the remains left after processing ore to concentrate it. Large amounts of mill tailing are generated from copper, iron, lead, zinc, and uranium ores. They have been used as fill materials, in base courses, and in asphalt mixtures for years in areas where they are abundant and conventional sources are limited. Because of the metal content in the mill tailings, the stockpiles must be carefully analyzed for leachate properties before use of the material is deemed appropriate.

Waste Rock. Surface mining operations and subsurface mining operations produce an estimated billion tons of waste rock annually in the United States. Some has been used as construction aggregate and in embankments; however, transportation costs from remote mines to construction areas often render the use of the rock economically infeasible. Where transportation is reasonable, waste rock can be used as stone fill for embankments or as riprap, or crushed for aggregate. These uses have been shown to be successful. Environmental considerations of leachate, low-level radiation, and sulfuric acid content should be investigated before use is deemed appropriate.

1.5.6 Agricultural Waste

Recycled agricultural waste has a high potential for use in applications not related to highways. Uses of agricultural wastes (with a few notable exceptions) in highways are usually in landscaping applications as an extension of their use in the agricultural industry. It is estimated that more than 2 billion tons of agricultural wastes are produced each year in the United States annually. This represents about 46 percent of the total waste produced in the United States each year.

Animal Manure. Animal manure is produced at a rate of 1.6 billion tons annually in the United States. Other than its use as fertilizer or as composting material for landscaping rights-of-way, it has little recycling value for highways.

Crop (Green) Waste.   Of the 400 million tons of crop waste produced annually from harvesting operations and grain processing, the potential to use rice husk ash to increase compressive strength in concrete is the most promising highway use. Research has also been conducted into converting cellulose waste to an oil appropriate as an asphalt extender.

Logging and Wood Waste.   It is estimated that about 70 million tons of lumber waste from logging and milling operations is produced each year. Only about one-third of the wood from logged trees is used as lumber. Much of this is used in other industry applications. Uses in highways include mulching and lightweight fill material for embankments or to repair slides. Application as lightweight fill material has been well documented and proven to be successful. Life expectancy of such embankments is estimated at 50 years.

1.5.7   Domestic Waste

It is estimated that approximately 4 pounds of domestic refuse is generated every day for every person in the United States; about 3 pounds per day goes to domestic landfills and only 11 percent is recycled. It is estimated that about 185 million tons of domestic waste is generated per year in the United States. Several of these wastes have a potential for reuse in highways.

Refuse.   Landfill refuse is not sought for its reuse potential in highway construction. There is no homogeneity among landfills, and so a great deal of expensive analysis must be performed on individual landfills to determine the potential future use. However, there have been occasions when a highway right-of-way traverses a landfill. In such cases, analysis to find appropriate on-site placement of the refuse instead of costly relocation and disposal has been found to be cost-effective. The refuse was spread in thin layers and compacted into embankment material or used for raised medians.

Paper and Paperboard.   Approximately 40 percent of the domestic waste generated is paper or cardboard. Approximately 25 percent of the wastepaper products are recycled each year and used primarily in making more paper, cardboard, and related materials. A highway use of wastepaper, particularly slick paper such as magazine paper, is in the production of mulch material.

Yard Waste and Compost.   There are approximately 1400 yard waste composting stations in the country now, and this number is growing. Yard waste is banned completely from landfills in 17 states, and this trend is also growing. Compost material must meet pathogen control, pH, metal concentration, nitrogen ratio, water-bearing capacity, maturity, particle size, and nutrient content control standards set by the EPA. Compost materials are used for mulching, soil amendment, fertilizers, and erosion control. Concerns related to leaching potential, odors, worker health and safety, long-term exposure, and public acceptance have limited use in highways to the experimental stage, except in landscape use. Studies are ongoing in several states.

Plastics.   The amount of plastic waste generated each year is growing. Recycling plastics is complicated by the fact that plastics are developed from at least six different resin bases, that must be sorted for the most effective recycling. About 30 percent of the plastics made from polyethylene terephthalate (PET), the resin base of soda bottles, is recycled. One use of PET is as a geotextile. Low-density polyethylene (LDPE) resin from film and trash bags can be recycled into pellets for use as an asphalt modifier in paving mixes. High-density polyethylene (HDPE) from milk jugs has been used in manufacturing plastic posts. Reuse of commingled plastics is more difficult but has been applied in fencing and posts. Such plastics have also been used as traffic delineators.

Glass. The amount of glass containers produced each year is declining, but about 12.5 million tons of glass is disposed of as domestic waste each year. To be reused in glass manufacturing, the glass must be sorted according to color. Uses in highways include fine aggregate in unbound base courses, as pipe bedding, as aggregate in asphalt mixes, and as glass beads in traffic paint.

Ceramics. Ceramic waste consists of factory rejects as well as discarded housewares and plumbing fixtures. Only in infrequent instances are large quantities of waste ceramics available for reuse in large projects, such as highway projects. In California, crushed porcelain has been used as an unbound base course aggregate. Crushed porcelain has been found to meet or exceed quality requirements for concrete aggregate.

Incinerator Ash. Incinerator ash results from burning municipal waste. About 26 million tons of incinerator ash is produced each year, of which 90 percent is bottom ash and the remainder is fly ash. The fly ash separately often exceeds regulatory limits for concentrations of lead and cadmium; however, fly ash is most often mixed in with the bottom ash, and this mixture generally does not contain sufficient concentrations of metals to render it hazardous. Incinerator ash has been used successfully as a partial replacement of coarse aggregate in asphalt mixtures, as roadway fill, and in base course construction when stabilized with portland cement. Concerns on the part of the EPA about the leaching of heavy metals have initiated several studies. While these studies are in progress, no field applications of incinerator ash in highways are anticipated.

Sewer Sludge Ash. The more than 15,000 municipal wastewater treatment plants in the country produce more than 8 million tons of dry solids of sewage sludge. Following dewatering, sludge cake has from 18 to 24 percent solids consisting mostly of nitrogen and phosphorus, but may be contaminated from various wastewater streams. Much of this sludge cake is incinerated, producing about 1 million tons of ash a year. Sludge ash has the potential for use as an asphalt filler and use in brick manufacturing. Studies indicate that with heat treatment, the ash can produce lightweight pellets that can increase concrete compressive strength by 15 percent when replacing aggregate. Sewage sludge ash has been used experimentally as a mineral filler in asphalt paving. Sewage sludge can be composted for agricultural uses such as soil amendments, compost, or fertilizer. Recycled municipal sewage sludge can be a health and safety concern for highway workers using it in landscaping.

Scrap Tires. In 1994, the National Cooperative Highway Research Program (NCHRP) published findings of a five-year review and synthesis of all of the states' highway practices involving the use of waste tires. This document, entitled *Uses of Recycled Rubber Tires in Highways,* is the result of a vast compilation of over 500 sources of information on the topic. The discussion in this section is a synopsis of the information provided in that document. For those interested in more detailed information on the uses of recycled tires, a copy of the document with a listing for specific references can be obtained through the Transportation Research Board of the National Research Council, 2101 Constitution Avenue, NW, Washington, D.C. 20418.

It is estimated that 2 to 3 billion waste tires have accumulated in the United States, about 70 percent of which are dumped illegally throughout the countryside or disposed of in unauthorized, uncontrolled stockpiles. Also, scrap tires account for about 2 percent of the solid waste that is disposed in regulated landfills. Each year an additional 242 million more scrap tires add to the nation's solid waste dilemma. Scrap tires are regulated under RCRA Subtitle D as a nonhazardous waste. However, if they are burned, the resulting residue, which may consist of oils, carbon black, and metal-concentrated ash, may be hazardous. In addition, leachate from tire-based products may also be a hazardous or toxic concern. Potential uses of scrap tires in highways and related facilities are numerous.

Table 1.19 details the uses of tires in transportation facilities in several states. The environmental implications of the use of scrapped tires in pavement are issues of emissions from the manufacture and placement of rubber asphalt. Leachate is also a major concern, particularly of metals (arsenic, barium, cadmium, chromium, lead, selenium, and zinc) and PAHs (polyaromatic hydrocarbons). A Minnesota study conducted in wetland areas concluded that the use of waste tires in asphalt-rubber pavements may affect groundwater quality. The study's results were comparable to two other studies with regard to metal leachates, but PAH leachate concentrations were not confirmed by the other studies. Mitigation measures suggested in the Minnesota study would be to place tire materials only in unsaturated zones of the subgrade or fill areas and not below the water table or within surface water boundaries. A Wisconsin study found [through EP TOX (extraction procedure for toxicity characteristics) analysis] that scrap, shredded, and crumbed tires were not hazardous, nor did they release significant amounts of priority pollutants. Several studies have indicated that the emissions in asphalt-rubber operations are not significantly higher than with conventional asphalt concrete. The one exception to this may be the release of methyl isobutyl ketone, which appears to be consistently slightly higher than with the conventional mixture. It is important to point out the difficulty and limitations presented by the large number of variables within the studies. First of all, the tires from which asphalt rubber is made are not of the same chemical composition, and are continuing to change. The rubber-asphalt formulation process also varies significantly, changing the emissions and leachable properties of the asphalt rubber. Comparison difficulties are compounded in that the composition and formulation processes for the conventional concrete asphalt that is being used for a standard vary tremendously also. Common and innovative uses of scrap tires are summarized in Table 1.20.

1.5.8 Research and Failed Attempts

Because highways represent such a large potential for use of reclaimed and recycled materials, the building and maintenance of highways can be integrated with national solid waste reduction goals. Our ability to recycle materials and reduce the growing amount of waste collecting in piles, landfills, and ponds across the United States will not improve without dedicated research.

All state transportation agencies have been actively involved with either research and development of innovative solid waste uses or the testing and practical application of new technologies. Research in the different states tends to vary, understandably, with the type of waste generated in highest abundance within the state. Some waste problems, such as scrap tire accumulation, are nationwide problems that have been addressed in virtually every state. Aside from the accepted technologies described above, several research projects and technology developments are currently under way and show a great deal of promise. Additionally, several end uses of recycled wastes have been shown to be unacceptable either in performance of the material or because

TABLE 1.19 Uses of Scrap Tires in Transportation Facilities

Type of use	State	Description of use	Advantages	Concerns
Erosion use	California	Shoulder reinforcement Channel slope protection	Disposal Low cost Erosion control	Visual acceptance by public Labor intensive Cost
	Louisiana	Windbreaks Slope reinforcement	Availability of tires Disposal	Pull out values
	Pennsylvania	Pending project		
	Vermont	Side slope fill	Disposal Flatten side slope	Unloading Leachate Cost
	Wisconsin	Experimental project		
Retaining wall	California	Anchored timber walls		
	North Carolina	Experimental retaining wall		
	Rhode Island	Experimental retaining wall		
Membrane	Arizona	Membrane to control expansive subgrade soils Shoulder membrane Ditch membrane	Less moisture fluctuations Seal out moisture Prevent cracking Ride quality Lower maintenance cost	
	California	Routine use		
	Oregon	Routine use		
	Washington	Routine use on bridge decks		
	Wisconsin	Experimental use		
Safety hardware	Colorado	Experimental project		Tires become projectiles
	Connecticut	Tire-sand inertial barrier	Disposal Low cost Maintenance	Debris Deceleration of vehicle

TABLE 1.19 Uses of Scrap Tires in Transportation Facilities (*Cont.*)

Type of use	State	Description of use	Advantages	Concerns
Safety hardware (*cont.*)	Oregon	Bases for tubular markers		
	Pennsylvania	Pending projects		
	Texas	Bases for vertical panel supports		
Railroad crossings	Oregon	Routine use	Ease of installation Smooth Reduced maintenance Potential reuse	
Valve box coverings	Pennsylvania	Experimental only		
	Oregon		Ease of installation Reduced maintenance Easy to adjust Durability	
Planks and posts	California Ontario	Laminated tires for planks and posts Sound barrier walls	Strength Durability Lightweight Sound loss	Burning Smoke
Drainage material	Pennsylvania	Aggregate drain rock replacement	Water-draining Stable roadway	Leachate
Culvert	Vermont	Whole tires bound together to form culvert	Cost	
Interlocking block	Minnesota	Erosion control, safety barriers, retaining walls, dikes, levees	Ease of installation Shock absorbing Resist chemical damage Durability	

Source: Adapted from *Uses of Recycled Rubber Tires in Highways,* National Cooperative Highway Research Program (NCHRP), Transportation Research Board, Washington, D.C., 1994.

TABLE 1.20 Highway Uses of Scrap Tires

Common uses	Innovative uses
Fills and embankments	Railroad grade crossing
Erosion control	Valve box
Shoulder stabilization	Drainable mat
Channel slope protection	Planks and posts
Windbreak	Culverts
Side slope fill	Interlocking blocks
Slope reinforcement	
Retaining wall	
Membranes	
Safety hardware	
Tire-sand inertial barrier	

Source: Based on National Cooperative Highway Research Program (NCHRP), Transportation Research Board, *Uses of Recycled Rubber Tires in Highways,* Washington, D.C., 1994.

of environmental or health threats that develop with use of the material. Failed attempts are also valuable. The information that is gathered can assist in developing evaluation criteria for other materials in similar uses and perhaps can redirect the research on the particular waste toward overcoming the difficulties or toward more appropriate uses. The following is a list of waste products that are currently being researched for use in highways or have been found to be unacceptable in a particular highway end use.

Spent Oil Shale.   During the oil crisis of the 1970s, several pilot programs were initiated to extract oil from shale deposits in Colorado. Though discontinued, several million tons of shale ash remains at the processing sites. Through research, it has been determined that the use of shale ash to replace 10 percent of the asphalt-concrete binder improved the stability, cohesion, stiffness, and resistance to stripping of asphalt mixtures. No use of this material has been reported in highways in the United States.

Cement and Lime Kiln Dust.   Much of the cement kiln dust produced (60 percent) is recycled at the cement batch plant. Approximately 8 million tons per year is not recycled at the plant. Successful nonhighway uses of cement kiln dust include substitution for agricultural lime and use as a stabilizer for municipal sludge. Experimental uses of cement kiln dust as embankment material, soil stabilization, stabilized base material, and mineral filler were all deemed unsuccessful. Limekiln dust can be considerably reactive because of the high lime content and has also been used as a stabilization material for sewage sludge and in agricultural uses. One reported successful highway application of limekiln dust as the reagent in stabilized base material was reported in Kentucky.

Advanced Sulfur Dioxide Control By-Products.   Wastes are developed when burning coal in advanced sulfur dioxide control processes. Examples of such coal-burning include fluidized bed combustion, spray drying, and dry furnace injection. This technology is often used in cogeneration plants.

Carpet Waste. It is estimated that 2 million tons of carpet wastes are disposed of each year. Research on the use of polypropylene fibers as reinforcement of concrete is being performed. Adding polypropylene fibers to concrete mixtures improves flexural strength and toughness but compromises the compressive strength somewhat. Carpet waste is not currently being used, but is promising.

Carbide Lime Waste. Carbide lime waste is a by-product of the acetylene manufacturing process. It comes either in powdered form, or in a slurry, depending on the processing method. Research has been conducted indicating that carbide lime as a mineral filler in asphalt paving mixtures improves viscosity and temperature susceptibility and otherwise performs satisfactorily. No field use in highway projects or current research has been reported.

Washery Rejects (Alumina Mud). Washery rejects are phosphate slimes from processing phosphatic clay and alumina mud from extracting alumina from bauxite. Because they have very low solids content, only settling out to higher concentrations after many years, and because of the difficulties in handling the slurries or slimes, no highway use has been found for them. Lack of strength and durability have rendered attempts at using alumina brown mud as base and subgrade material unsuccessful.

Phosphogypsum. Phosphogypsum is a by-product of finely ground phosphate rock. Approximately 35 million tons is produced each year. Much research and some effective uses have been carried out, mainly in Florida, Louisiana, and Texas, where total accumulations of phosphogypsum stacks are estimated at 700 million tons. In 1989, concerns about radon emanations from phosphogypsum stacks were the reason EPA imposed a ban on the use of and research on uses for phosphogypsum, pending studies on the health and environmental safety implications of the radon emissions.

1.5.9 End Uses in Highways

It is apparent that there are many uses of recycled materials in highway construction and related applications. Table 1.21 provides a summary of these uses for reference.

1.5.10 Recycling Hazardous Wastes

Under Subtitle C of RCRA, EPA has the authority to regulate recyclable hazardous waste material. If the intent is to recycle a waste, what the waste is and how it is to be recycled must be known to determine whether it is regulated under Subtitle C. The definition of solid waste under Section 261.2 identifies four types of recycling activities for which recycled wastes may be subject to Subtitle C regulation: use constituting disposal, burning waste-derived fuels for energy recovery, reclamation, and speculative reclamation.

Use Constituting Disposal. Use constituting disposal is defined as placing or applying a solid waste or a material contained in a product that was a solid waste on the land in a manner constituting disposal. In this case, land disposal regulations under RCRA Parts 264 and 265 apply. Use constituting disposal may include the following uses involved in the construction of highways or maintenance of highway landscaping: fill material, cover material, fertilizer, soil conditioner, dust suppressor, asphalt additive, and foundation material.

TABLE 1.21 Applications of Recycled Materials in Highways

Asphalt: Crop waste and other cellulose material may be reduced to an oil suitable for asphalt extender.

Asphalt paving aggregate: Incinerator ash.

Asphalt mineral filler: Sewage sludge ash, fly ash, baghouse fines, cement kiln dust, lime waste.

Asphalt-rubber binder: Scrap tires.

Asphalt stress-absorbing membranes: Scrap tires.

Asphalt rubberized crack sealant: Scrap tires.

Asphalt aggregate: Mill tailings, phosphogypsum, slag.

Asphalt fine aggregate: Glass and ceramics.

Asphalt cement modifier: Plastic waste.

Asphalt plant fuel: Used motor oil.

Asphalt paving: Bottom ash, boiler slag, blast furnace slag, steelmaking slag, nonferrous slag, reclaimed asphalt pavement, foundry sand, roofing shingle waste, petroleum-contaminated soils (after thermal treatment).

Base course: Glass and ceramic waste, construction and demolition debris, nonferrous slags, reclaimed asphalt pavement, reclaimed concrete pavement, mill tailings.

Pipe bedding: Foundry sand, glass, and ceramic waste.

Borrow material: Quarry waste, construction and demolition material.

Slope stabilization and erosion control: Sawdust and wood waste.

Mulch: Wood waste, paper waste (especially slick, magazine-type paper), compost.

Fertilizer: Animal manure and farm waste.

Embankments: Lumber and wood waste, sawdust and wood chips, recycled sanitary landfill refuse, fly ash, bottom ash, construction and demolition waste, sulfate waste, waste rock, mill tailings, coal refuse.

Cement stabilized base: Incinerator ash, fly ash, bottom ash, advanced SO_2 control by-products, cement kiln dust, reclaimed asphalt pavement, petroleum-contaminated waste (after thermal treatment), coal refuse, and rice husk ash may be used as supplementary cementing material.

Concrete: Incinerator ash from sewage sludge cake as vitrified aggregate or palletized aggregate.

Lightweight fill material: Wood waste, sawdust, chipped wood, scrap tires.

Geotextile: Plastic waste.

Sealant: Scrap tires.

Safety hardware, fencing, signposts: Plastic wastes.

Flowable fill and grout: Quarry waste, fly ash.

Soil stabilization: Fly ash, advanced SO_2 control by-product, cement kiln dust, lime waste.

Antiskid material: Bottom ash, steelmaking slag.

Blasting grit: Nonferrous slags.

Burning and Blending of Waste Fuels. Burning and blending would be the applicable method for recycling used oil for fuel in asphalt plants. Used oil is not currently considered a hazardous waste unless it has a characteristic of ignitability, corrosivity, reactivity, or extraction procedure toxicity (ICRE characteristic). If the used oil is mixed with a hazardous waste, it is regulated as a hazardous waste fuel under RCRA, Part 266, Subpart D. Specifications for nonhazardous used oil fuel are described in Table 1.22. Used oils that do not meet one or all of these specifications and are not mixed with hazardous waste may still be burned in industrial boilers, but they must have an EPA identification number for this activity and must meet a higher standard of reporting than used oil meeting the specifications. A burner of either specification or off-specification used oil fuel must notify EPA of its used-oil fuel activities and state the location and a general description of the used-oil management activities.

TABLE 1.22 Specification Levels for Used Oil Fuels

Specification	Maximum allowable level
Arsenic concentration	5 ppm
Cadmium concentration	2 ppm
Chromium concentration	10 ppm
Lead concentration	100 ppm
Flash point	1000°F
Total halogen concentration (unmixed)	4000 ppm
Total halogen concentration (mixed)	1000 ppm

Source: Adapted from Travis Wagner, *Complete Handbook of Hazardous Waste Regulation,* Perry-Wagner Publishing, Brunswick, Maine, 1988, p. 46.

Copies of invoices and waste analysis conducted on the used oil must be maintained for at least three years.

Reclamation. Reclamation is the recovery of materials with value from a waste material and involves regeneration of waste material from the reclamation activities. Recovering precious metals from a waste stream (such as silver from x-ray film) is an example of reclamation. When the lead plates from lead-acid batteries are recovered, the activity is regulated under RCRA as reclamation. Use of material as feed stocks or ingredients in the production of a new product is not considered reclamation.

Speculative Accumulation. Any hazardous secondary material is considered a solid waste if accumulated before recycling unless 75 percent of the stockpile is recycled during a calendar year.

CHAPTER 2
HIGHWAY LOCATION, DESIGN, AND TRAFFIC

Larry J. Shannon, P.E.

Standards and Geometrics Engineer
Ohio Department of Transportation
Columbus, Ohio

This chapter begins with a description of the overall transportation development process, and then presents comprehensive information on the various elements of highway location and design. Included is the determination of horizontal and vertical alignment, with attention to obtaining proper sight distance and superelevation. The design of roadway cross sections, intersections, ramps, and service roads is addressed. Traffic aspects include an introduction to intelligent vehicle highway systems and the use of high-occupancy vehicle lanes. A presentation on preparation of highway construction plans and organizing CADD drawings is also provided. A list of references, which are noted in the text, concludes the chapter. Some design issues related to roadside safety are also discussed in Chap. 6.

2.1 TRANSPORTATION DEVELOPMENT PROCESS

2.1.1 Statewide Systems Planning

The beginnings of any roadway project involving government money are found in a statewide transportation planning program. The state transportation department develops a set of goals and objectives which take into account social, economic, environmental, and developmental goals of other state, federal, and local agencies. Based on these goals and objectives, the department identifies transportation improvement needs throughout the state. The approach is from a multimodal standpoint; that is, not just highways are considered, but all forms of transportation, including public transportation, railroads, water, aviation, bikeways, and pedestrian ways (Ref. 1).

2.1.2   Transportation Programming Phase

In order to evaluate various projects from various parts of the state, information is collected consisting of the following items: transportation inventories, traffic analyses,
modal forecasts, future system requirements, levels of service, population data and
forecasts, land use inventories, public facilities plans, and basic social, economic, and
environmental data. This information comes from various sources, both public and
private, is updated on a regular basis, and is used in developing the state's transportation improvement program.

The statewide fiscal program is also considered in developing the plan.
Transportation investment, fiscal forecasts, and consideration of expenditure tradeoffs
between modes are some of the financial considerations affecting the project selection
process.

Public input is sought from regional to local levels. Local and regional planning
organizations, as well as private individuals, have a chance to express opinions and
provide input to the project selection process. Once all factors have been evaluated,
the state announces and publishes its recommended transportation improvement plan.
This usually consists of a one-year plan and a five-year plan, with remaining projects
grouped under long-range plans.

2.1.3   Project Evaluation

Once projects reach the selected lists, the next phase is project evaluation. This phase
will determine which projects can advance to detail design and which will require a
more detailed evaluation in preliminary development.

Projects that can advance directly to design phase meet the following criteria:

- No additional right-of-way (permanent or temporary) will be required to accomplish the work and there will be no adverse effect on abutting real properties.
- No major changes in the operation of access points, traffic volumes, traffic flows,
  vehicle mix, or traffic patterns.
- No involvement with a live stream or an intermittent stream having significant
  year-round pools, upstream or downstream, in the immediate vicinity.
- No involvement with a historic site.

Examples of these types of improvement are:

- Restoration and/or reconstruction of existing pavement surfaces
- Modernization of an existing facility by adding or widening shoulders
- Modernization of existing facilities by adding auxiliary lanes or pavement widening
  to accomplish a localized purpose (weaving, climbing, speed change, protected
  turn, etc.)
- Intersection improvements
- Reconstruction or rehabilitation of existing grade separation structures
- Reconstruction or rehabilitation of existing stream crossings which do not involve
  any modification of a live stream or otherwise affect the water quality
- Landscaping or rest area upgrading projects
- Lighting, signing, pavement marking, signalization, freeway surveillance and con-
  trol systems, railroad protective devices, etc.

- Minor safety-type improvements, such as guiderail replacement or installation of breakaway sign hardware
- Outdoor advertising control programs
- Bicycle or pedestrian facilities provided within existing right-of-way

All projects that do not fall into the above categories must undergo additional evaluation in a preliminary development phase.

2.1.4 Preliminary Development Phase

Two types of projects are considered here: (1) projects that involve studies outside the existing corridor or where a facility for more than one alternate mode of transportation may be involved, and (2) projects where feasible alternatives are limited to the existing corridor but did not qualify to pass directly to the design phase. The main difference between the two as far as processing is concerned is that the first group has not yet narrowed its alternatives down to feasible alternatives.

In each case, a project inventory is developed. This information includes historical sites; public recreational facilities; school, church, fire, and police districts; proposed development; land use; existing and other proposed transportation facilities; preliminary traffic assignments; and other similar social, economic, and environmental features which are pertinent to the area under study. Using this information as a guide, all preliminary alternatives are developed together with documentation of the anticipated effects on community, preliminary cost estimates, and other technical considerations. Advantages and disadvantages of each alternative are studied. Where appropriate, coordination with other modes is considered. The "no-build" alternative is also considered and provides a reference point for defining potential beneficial and adverse impacts. Public hearings are held to gain input from the local public in the affected areas. Following an evaluation of all input received, alternatives are weighed and only those considered to be feasible are forwarded to the next step. From this point on, all projects in the preliminary development phase are on the same path.

Among the environmental concerns which must be considered for each alternative are the following (see also Chap. 1):

Air quality. A study of the effect of a proposed transportation improvement on the quality of the air

Historic or prehistoric. A study of the effect of the proposed transportation improvement on historic or prehistoric objects or on lands or structures currently entered into the National Register or which may be eligible for addition to the National Register

Endangered species. A study of the effect of the proposed transportation improvement on rare or endangered plants or animals having national or state recognition

Natural areas. A study of the effect of the proposed transportation improvement on natural areas designated as having regional, state, or national significance

Parks and recreation. A study of the effect of the proposed transportation improvement on publicly owned parks, recreation areas, or wildlife and waterfowl refuge designated as having national, state, or local significance

Prime farmlands. A study of the effect of the proposed transportation improvement on farmlands with high productivity due to soil and water conditions or having other unique advantages for growing specialty crops

Scenic rivers. A study of the effect of the proposed transportation improvement on any scenic rivers of state or national significance

Streams and wetlands. A study of the effect of the proposed transportation improvement on streams and wetlands on project and abutting land areas

Water quality. A study of the effect of the proposed transportation improvement on the quality of live streams or bodies of water

The next step is a refinement of feasible alternatives. This requires additional work sufficient to prepare an environment document. This could include such items as approximate construction costs; alignment and profile studies; typical section development; preliminary designs for geometric layout, drainage, right-of-way, and utilities; location of interchanges, grade separations, and at-grade intersections; preliminary bridge designs at critical locations; channel work; air, noise, and water studies; flood hazard evaluations; and other supplemental studies and right-of-way information. Once again, input is sought from the public sector through advertisement and public hearings.

Figure 2.1 shows the corridors for the feasible alternatives for an 11-mile relocation of U.S. 30 in Ohio (Ref. 13). The map is part of a study evaluating crossroad treatment for each alternative. Figures 2.2 and 2.3 show the projected crossroad treatments for the various alternatives. The options are (1) interchange, (2) grade separation, or (3) closing roads with cul-de-sacs. Since the proposed segment will be a limited-access highway, the option of at-grade intersection was not considered. Figures 2.4 and 2.5 show current and 20-year projected traffic volumes for all roadways. These are examples of maps used in the study of feasible alternatives.

After consideration of all the input and comparing the benefits and disadvantages of each alternative, the next step is to make a selection of the recommended alternative. This selection is certified by the state's transportation director. Following approval of the environmental document, the project may proceed to the design phase.

2.1.5 Detail Design Phase

During the detail design phase, various design elements are finalized and construction plans are developed. Project development in this phase can include many intermediate reviews prior to final plan submission. These may include some or all of the following, depending on the complexity of the plan:

Traffic request/validation

Traffic signal warrant analysis

Airway-highway clearance study

Alignment, grade, and typical section review

Conceptual maintenance of traffic review

Structure type study

Retaining wall justification

Service road justification

Preliminary drainage review

Preliminary right-of-way review

Bridge type, size, and location study

Drive review

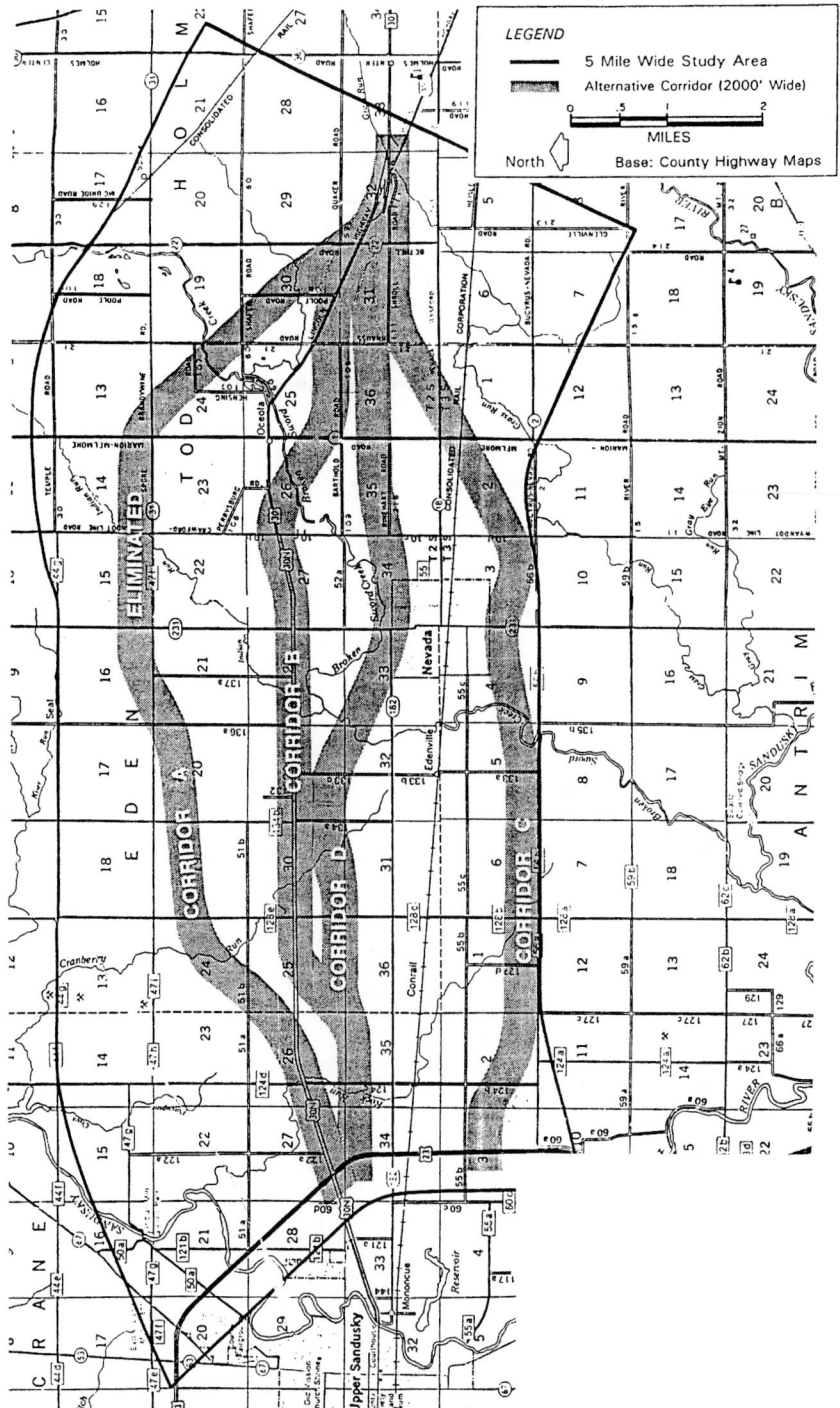

FIGURE 2.1 Example of map used in study of alternate routes showing four possible corridors. (*From* Justification Study for Crossroad Grade Separations, US 30, *by Balke Engineers for Ohio Department of Transportation, with permission*)

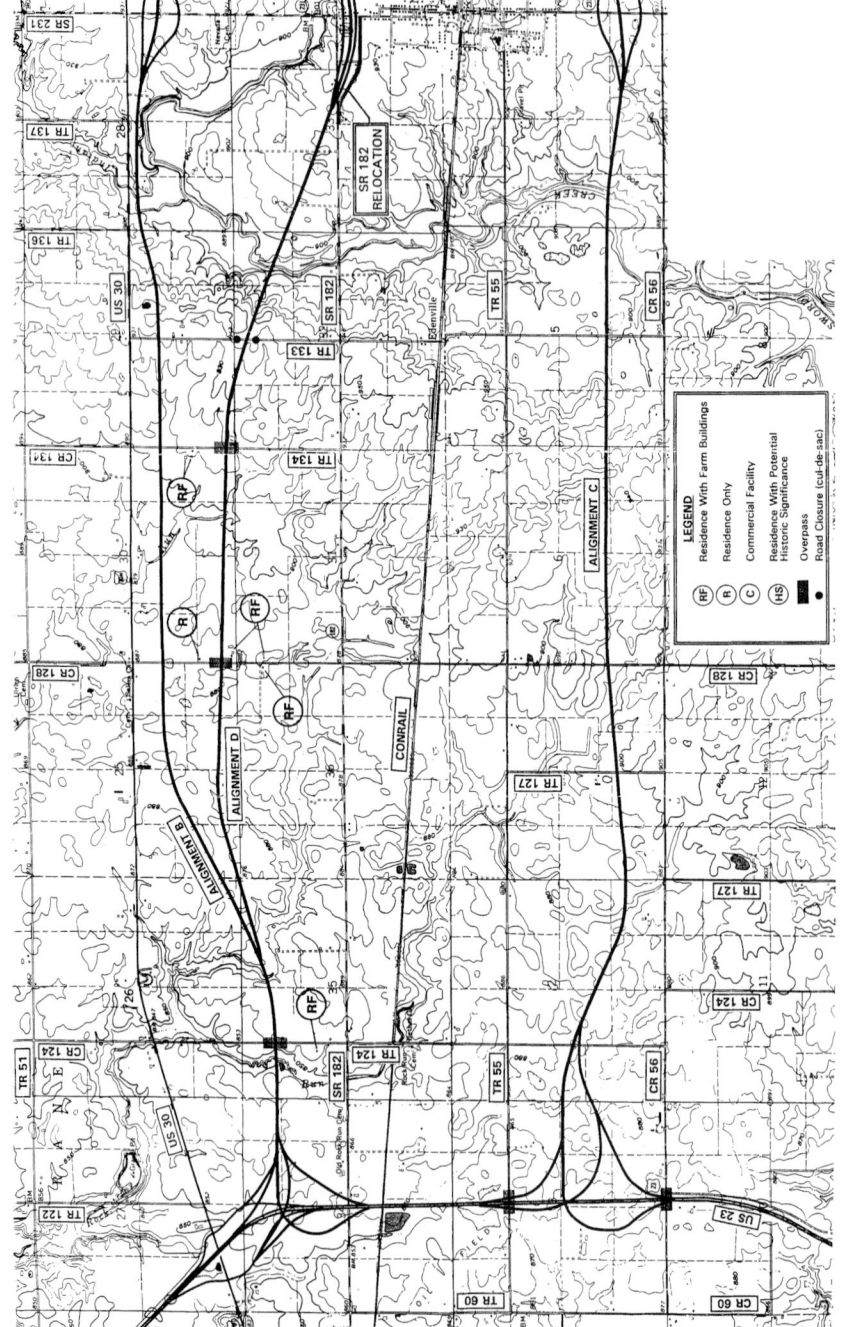

FIGURE 2.2 Example of map used in study of alternate routes showing projected crossroad treatments for western portion of corridor D. (*From Justification Study for Crossroad Grade Separations, US 30, by Balke Engineers for Ohio Department of Transportation, with permission*)

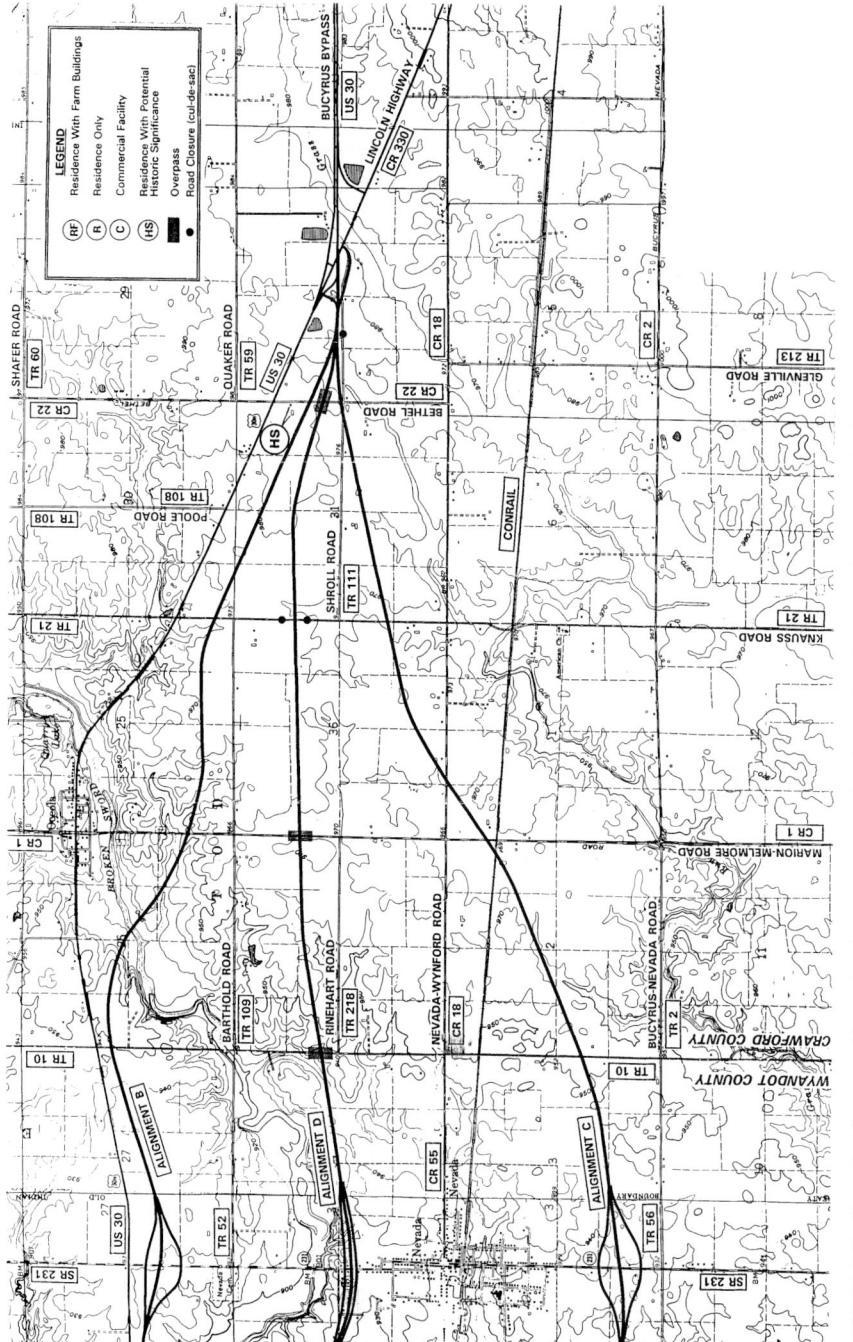

FIGURE 2.3 Example of map used in study of alternate routes showing projected crossroad treatments for eastern portion of corridor D. (*From Justification Study for Crossroad Grade Separations, US 30, by Balke Engineers for Ohio Department of Transportation, with permission*)

2.7

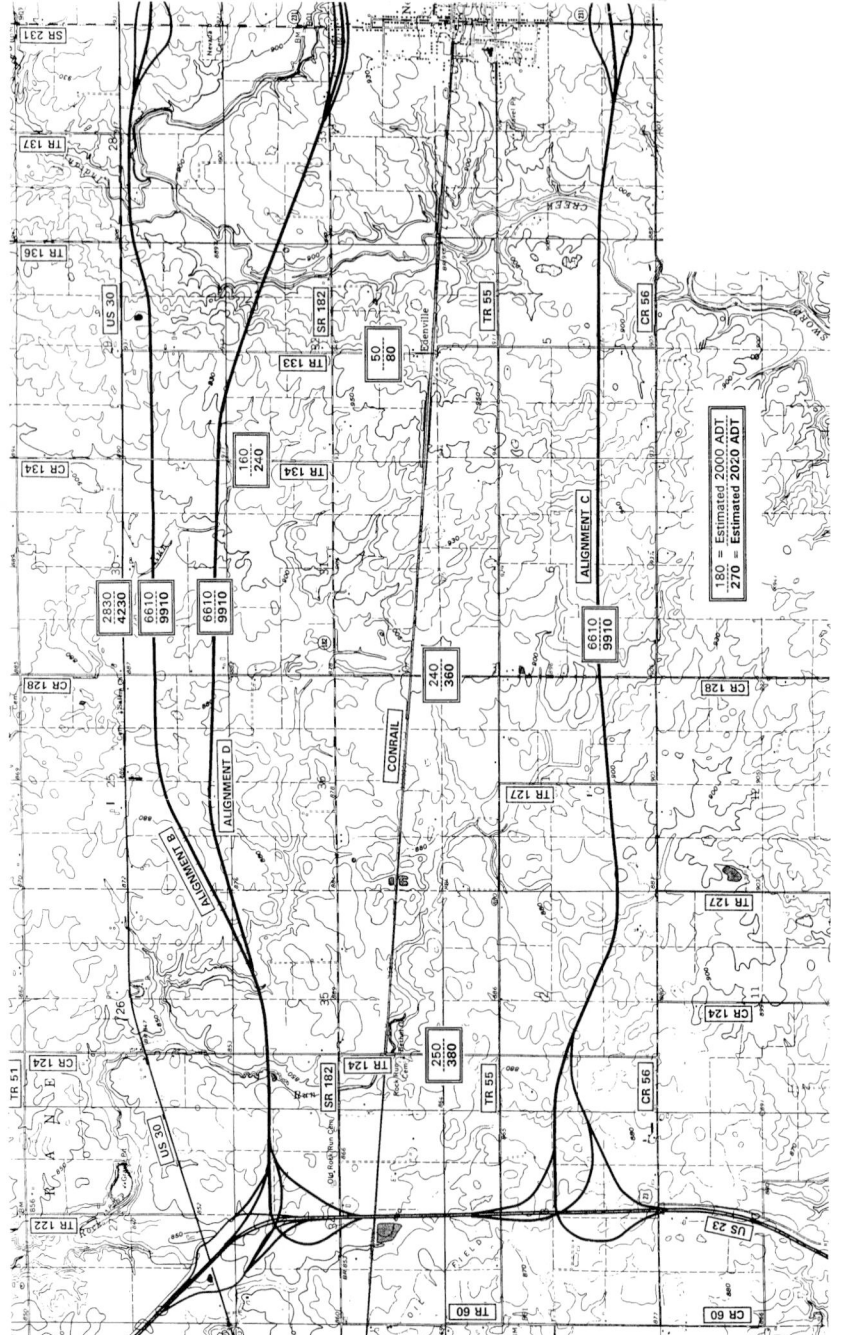

FIGURE 2.4 Example of map used in study of alternate routes showing projected traffic (ADT, average daily traffic, in years 2000 and 2020) for western portion of corridor D. (*From Justification Study for Crossroad Grade Separations, US 30, by Balke Engineers for Ohio Department of Transportation, with permission*)

2.8

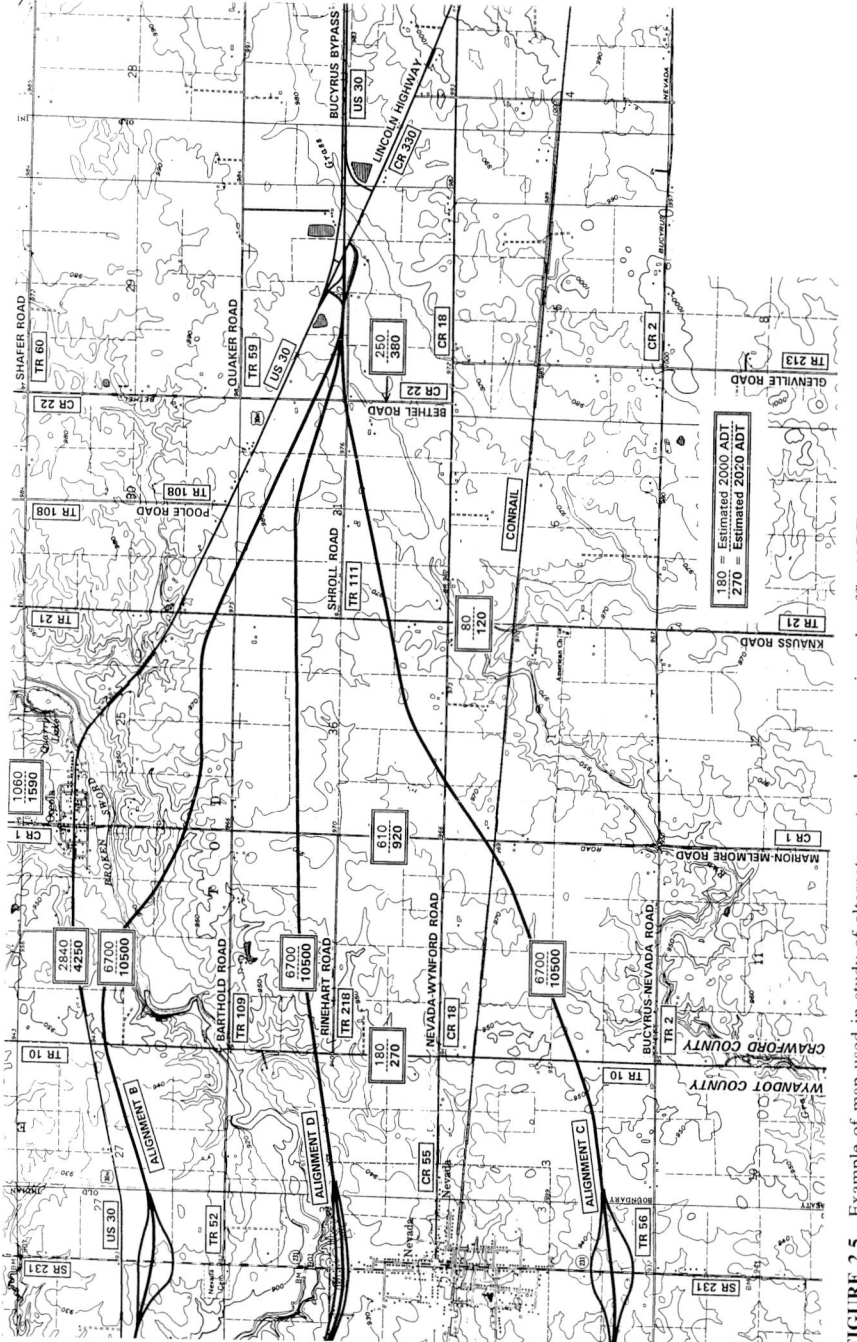

FIGURE 2.5 Example of map used in study of alternate routes showing projected traffic (ADT, average daily traffic, in years 2000 and 2020) for eastern portion of corridor D. (*From Justification Study for Crossroad Grade Separations, US 30, by Balke Engineers for Ohio Department of Transportation, with permission*)

2.9

Slope review

Traffic control

Lighting

Waterline

Sanitary sewer

Final roadway, field and office check

This is not intended to be an all-inclusive list. The designer should contact the government agency having review and final acceptance authority to see what reviews are required during this phase of plan development.

Following acceptance of the final plans, specifications, and estimates, the project is processed for letting. Any necessary consent legislation is obtained. The project is then advertised, bids are taken, and the construction contract is awarded.

2.2 GEOMETRIC DESIGN

2.2.1 Design Controls

Once a route has been selected for a new highway, or a decision has been made to perform major work on an existing facility, the next step is to establish the design controls. The various factors considered for design controls may be generally grouped into four categories: functional classification, traffic data, terrain and locale, and design speed.

Functional classification is a way of grouping roadways together by the character of service they provide. The initial division is between urban and rural roadways. The urban classification may be defined differently in various parts of the country, but one definition is incorporated areas having a population of 5000 or more (Ref. 7). Rural areas are those areas outside of urban areas.

Each of these may be further subdivided into other classifications defined as follows:

Interstate. Roadways on the federal system with the highest design speeds and the highest design standards.

Freeway. An expressway with full access control and no at-grade intersections.

Expressway. A divided arterial highway with full or partial control of access and generally having grade separations at major intersections.

Arterial. A facility primarily used for through traffic, usually on a continuous route.

Collector. An intermediate roadway system which connects arterials with the local road or street systems.

Local road or street. A road whose primary function is to provide access to residences, businesses, or other abutting properties.

Traffic data are an important foundation in highway design. The information used in design is usually a future forecast on the basis of existing traffic counts and expanded on the basis of normal expected growth in the area or enhanced by estimates of future business, commercial, or residential development. Most highway design is based on what traffic demands will be 20 years from the current year. Shorter time periods, such as 10 years, may apply to resurfacing projects or other minor repair projects. It is important that, within the same jurisdiction, traffic data be forecast using the same methods and techniques, in order to ensure similar designs for similar type roadways. This is especially true for roadways in a given state jurisdiction.

The following types of traffic numbers are used most frequently in design:

Average daily traffic (ADT). The average number of vehicles using a roadway in a 24-hour period.

Design hourly volume (DHV). The estimated number of vehicles using the roadway in the 30th highest hour of the year. This number is generally 8 to 12 percent of the ADT and is used extensively in determining lane widths and shoulder characteristics of the roadway cross section.

Directional design hourly volume (DDHV). The estimated number of vehicles traveling in one direction of a two-way roadway in the 30th highest hour of the year. This number must be at least 50 percent of the DHV and is usually in the range of 50 to 60 percent. A higher value would indicate that the roadway is a major link in the commuter network, carrying a heavy inbound load in the morning and reversing that flow in the evening.

Truck percentage (T). The portion of the ADT which consists of *B* and *C* trucks. Traffic counts are usually separated according to vehicle type:

P = passenger cars (%)

A = commercial (%), consisting of light delivery trucks, panel trucks, and pickup trucks

B = commercial (%), consisting of semitrailer and truck-trailer combinations

C = commercial (%), consisting of buses or dual-tired trucks having single or tandem rear axles

Traffic counts sometimes group the *P* and *A* vehicles together and the *B* and *C* together.

Terrain is a factor that can significantly influence design features, especially in rural areas. Various categories of terrain are level, rolling, and hilly. They are further described as follows:

Level terrain. Any combination of grades and horizontal and vertical alignment permitting heavy vehicles to maintain approximately the same speed as passenger cars. Grades are generally limited to 1 or 2 percent.

Rolling terrain. Any combination of grades and horizontal and vertical alignment causing heavy vehicles to reduce their speeds substantially below those of passenger cars, but not to operate at crawl speeds.

Hilly terrain. Any combination of grades and horizontal and vertical alignment causing heavy vehicles to operate at crawl speed.

Heavy vehicles are defined as any vehicle having a weight (pounds) to horsepower ratio of 300 or more. Crawl speed is defined as the maximum sustained speed heavy vehicles can maintain on an extended upgrade, or 35 mi/h for design purposes.

Locale affects the design of roadways in urban areas. It can be divided into commercial, industrial, and residential areas, which, for the purposes of this text, are self-explanatory.

TABLE 2.1 Relationship between Design Controls and Design Features

Design features	Design controls			
	Functional classification	Traffic data	Terrain locale	Design speed
Lane width, rural	X	X		X
Lane width, urban	X		X	
Rural shoulder width, type	X	X		
Urban shoulder width, type	X		X	
Guiderail offset	X	X		
Degree of curve				X
Grades	X		X	X
Bridge clearances (horizontal and vertical)	X	X		
Stopping sight distance				X
Passing and intersection sight distance				X
Decision sight distance				X
Superelevation				X
Widening on curves				X
Rural design speeds	X	X	X	
Urban design speeds	X		X	

Source: *Location and Design Manual,* Vol. 1, *Roadway Design,* Ohio Department of Transportation, with permission.

Design speed is defined as "the maximum safe speed that can be maintained over a specific section of highway when conditions are so favorable that the design features of the highway govern" (Ref. 1). When designing new or reconstructed roadways, the design speed should always equal or exceed the proposed legal speed of the roadway.

Table 2.1 shows the relationship of the functional classification, traffic data, terrain and locale, and design speed to the various geometric design features listed on the chart.

It should be noted that there are situations when it will not be possible or reasonable to meet the design standard for a particular feature in a given project. When this occurs, the designer must bring this to the attention of the reviewing authority for approval of what is being proposed, or suggestions on what other course of action to take. A design exception must be approved by the reviewing authority when a substandard feature is allowed to remain as part of the design. In this way, it can be documented that this was not an error or oversight on the part of the designer and that every effort has been made to provide the best design possible in the given situation.

2.2.2 Sight Distance

A primary feature in the design of any roadway is the availability of adequate sight distance for the driver to make decisions while driving. In the articles that follow, the text contains conclusions based on information contained in Ref. 1. Derivation of formulas and references to supporting research are contained in that document and will not be repeated here. The reader is encouraged to consult that document for more detailed background information. The following paragraphs discuss various sight distances and the role they play in the design of highways.

Stopping Sight Distance. Stopping sight distance is the distance ahead that a motorist should be able to see so that the vehicle can be brought safely to a stop short of an obstruction or foreign object in the road. This distance will include the driver's reaction or perception distance and the distance traveled while the brakes are being applied. The total distance traveled varies with the initial speed, the brake reaction time, and the coefficient of friction for wet pavements and average tires. Two sets of data have been derived for stopping sight distance. In one case, called "minimum," the assumption is made that under wet pavement conditions, the driver will make necessary adjustments to lower speed for conditions. In the other case, called "preferred," the assumption is made that the driver will not reduce speed for wet pavement conditions. Table 2.2 shows both sets of values for design speeds ranging from 20 to 70 mi/h. Design speed values were used to calculate the preferred values, while "average running speed" (Ref. 1) values were used to calculate the minimum values. When using stopping sight distance values in design, the designer is encouraged to use the preferred values whenever possible.

TABLE 2.2   Basic Minimum and Preferred Stopping Sight Distance for Design Speeds from 20 to 70 mi/h

Design speed, mi/h	Minimum, ft	Preferred, ft	Design speed, mi/h	Minimum, ft	Preferred, ft
20	125	125	46	340	415
21	130	130	47	355	430
22	135	135	48	370	445
23	140	140	49	385	460
24	145	145	50	400	475
25	150	150	51	410	490
26	160	160	52	420	505
27	170	170	53	430	520
28	180	180	54	440	535
29	190	190	55	450	550
30	200	200	56	465	570
31	205	210	57	480	590
32	210	220	58	495	610
33	215	230	59	510	630
34	220	240	60	525	650
35	225	250	61	530	665
36	235	265	62	535	680
37	245	280	63	540	695
38	255	295	64	545	710
39	265	310	65	550	725
40	275	325	66	565	750
41	285	340	67	580	775
42	295	355	68	595	800
43	305	370	69	610	825
44	315	385	70	625	850
45	325	400			

Minimum stopping sight distance is based upon average running speed and a driver reaction time of 2.5 s. Preferred stopping sight distance is based upon design speed and a driver reaction time of 2.5 s.

Source: *Location and Design Manual,* Vol. 1, *Roadway Design,* Ohio Department of Transportation, with permission.

When considering the effect of stopping sight distance, it is necessary to check both the horizontal and the vertical stopping sight distance. Horizontal sight distance may be restricted on the inside of horizontal curves by objects such as bridge piers, buildings, concrete barriers, guiderail, cut slopes, etc. Figure 2.6 shows a diagram describing how horizontal sight distance is checked along an extended curve. Both formulas and a nomograph are provided to enable a solution. Many times, where the curve is not long enough or there are a series of roadway horizontal curves, a plotted-out "graphic" solution will be required to determine the available horizontal sight distance.

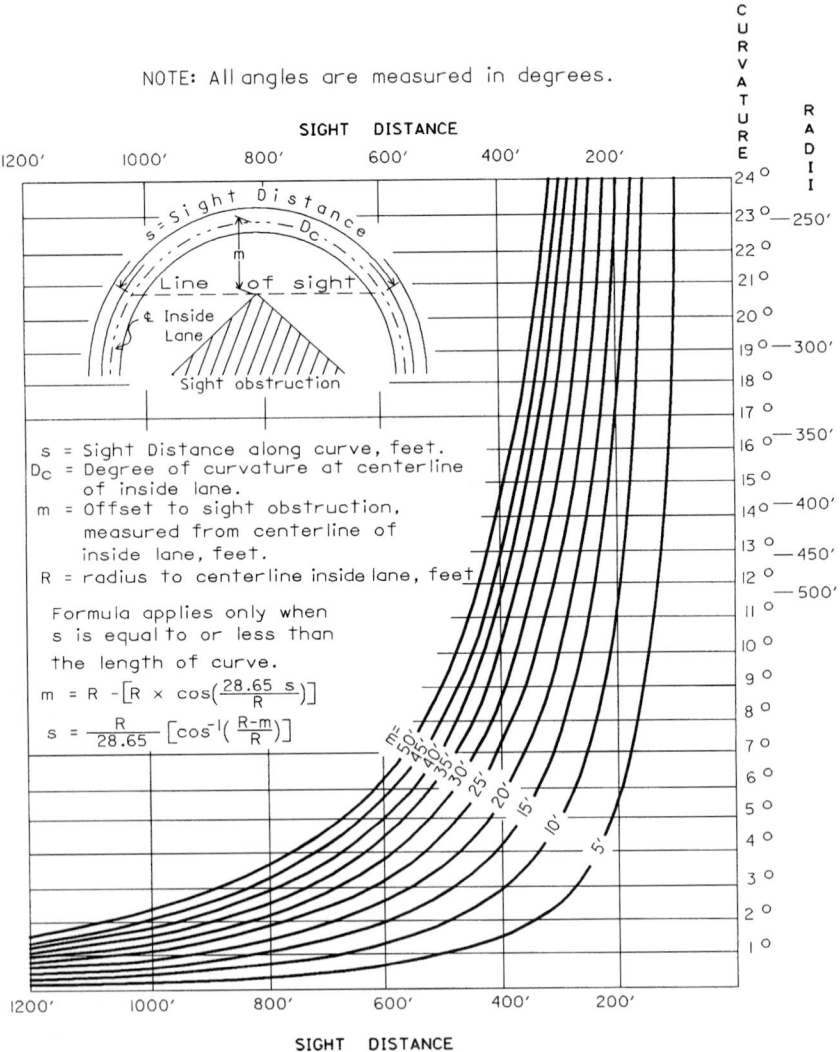

FIGURE 2.6 Horizontal sight distance along curve. (*From* Location and Design Manual, *Vol. 1, Roadway Design, Ohio Department of Transportation, with permission*)

When a cut slope is the potential restriction, the offset should be measured to a point on the backslope having the same elevation as the average of the roadway where the driver is, and the location of the lane downstream where a potential hazardous object lies. In this way, an allowance of 2 ft of vegetative growth on the backslope can be made, since the driver's eye is assumed to be 3.5 ft above the pavement and the top of a 6-in hazardous object downstream may still be seen.

Vertical sight distance may be restricted by the presence of vertical curves in the roadway profile. The sight distance on a crest vertical curve is based on a driver's ability to see a 6-in-high object in the roadway without being blocked by the pavement surface. The height of eye for the driver used in the calculations is 3.5 ft.

The sight distance on a sag vertical curve is dependent on the driver's being able to see the pavement surface as illuminated by headlights at night. The height of the head-light is assumed to be 2.0 ft, and the height of the object is 0.0 in. The upward divergence angle of the headlight beam is assumed to be 1°.

Intersection Sight Distance. A motorist attempting to enter or cross a highway should be able to observe traffic at a distance that will allow safe movement. There are three possible maneuvers for a motorist stopped at an intersection to make. The motorist can (1) cross the intersection by clearing oncoming traffic on both the left and right of the crossing vehicle, (2) turn left into the crossing roadway after first clearing traffic on the left and then making a safe entry into the traffic stream from the right, or (3) turn right into the crossing roadway by making a safe entry into the traffic stream from the left. Option 2 requires the longest distances and is the one usually chosen for design purposes. In Table 2.3, the "Intersection" column shows the distances required for this movement for the various design speeds listed. If these distances cannot be obtained, the minimum sight distance provided should not be less than the stopping sight distance for the through roadway. This would allow a driver on the through roadway adequate time to bring a vehicle to a stop if the waiting vehicle started to cross the intersection and suddenly stopped or stalled. If this distance cannot be provided, additional safety measures must be provided. These could include, but are not limited to, advance warning signs and flashers and/or reduced speed limit zones in the vicinity of the intersection.

Figure 2.7 shows a diagram in which the horizontal and vertical components of intersection sight distance are illustrated. Figure 2.7*a* shows the line of sight of the waiting driver for traffic approaching from the left and the right. Any obstacles blocking this line of sight should be removed if possible. For design purposes, the waiting vehicle is assumed to be 10 ft off the through road edge of pavement and the left edge of the through vehicle approaching from the left is assumed to be 9 ft from the pavement edge. The tabular data at the bottom of Fig. 2.7 show the effect of an intervening obstacle on the line of sight. A clearance distance *B* is given, which varies with the obstacle offset distance and the design speed of the through roadway. In a similar manner, a vehicle approaching from the right is assumed to be 2 ft off the opposite edge of pavement. An intervening obstacle at an offset of *A* feet from the near pavement edge must be at least two-thirds the *B* distance away from the waiting driver. Finally, in considering the horizontal elements of intersection sight distance, Fig. 2.7*c* illustrates the effect of guiderail on the through roadway located near the intersection. It should be flared back from the through roadway from its normal offset so that it is at least 15 ft away from the through pavement edge at the intersection radius return. The table at the bottom of Fig. 2.7 specifies a flare rate that varies with design speed to be used to transition back to the normal guiderail offset.

Figure 2.7*b* shows the vertical component of intersection sight distance. The design of a crest vertical curve is based on a height of a waiting driver's eye of 3.5 ft and a

TABLE 2.3   Minimum Sight Distance for Passing (MPSD) and at Intersections (MISD) for Design Speeds from 20 to 70 mi/h

Height of eye, 3.50 ft; height of object, 4.25 ft

	Minimum sight distance			
	Passing		Intersection	
Design speed, mi/h	MPSD, ft	K, ft/%	MISD, ft	K, ft/%
70	2500	2021	950	292
65	2300	1710	875	248
60	2100	1426	825	220
55	1950	1230	750	184
50	1800	1048	700	158
45	1650	880	625	126
40	1500	728	575	107
35	1300	547	500	81
30	1100	391	450	66
25	950	292	375	46
20	800	207	300	29

Using S = minimum passing or intersection sight distance, ft
    L = length of crest vertical curve, ft
    A = algebraic difference in grades, %
    K = rate of crest vertical curvature, ft per % of grade change
- For a given design speed and A value, the calculated length $L = KA$.
- To determine S with a given L and A, use the following:
  For $S < L$: $S = 55.61\sqrt{K}$ where $K = L/A$
  For $S > L$: $S = 1546.36/A + L/2$

   Source: *Location and Design Manual,* Vol. 1, *Roadway Design,* Ohio Department of Transportation, with permission.

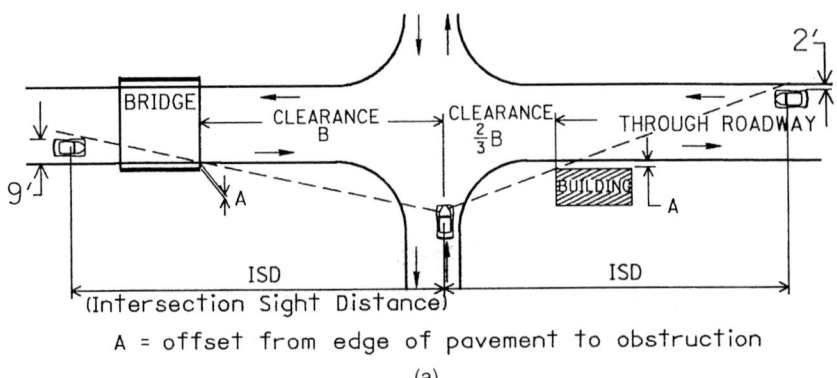

A = offset from edge of pavement to obstruction

(a)

FIGURE 2.7   Typical intersection sight distance conditions. (*a*) Horizontal components. (*b*) Vertical components. (*c*) Guardrail flare. (*d*) Dimensional data. (*From* Location and Design Manual, *Vol. 1,* Roadway Design, *Ohio Department of Transportation, with permission*)

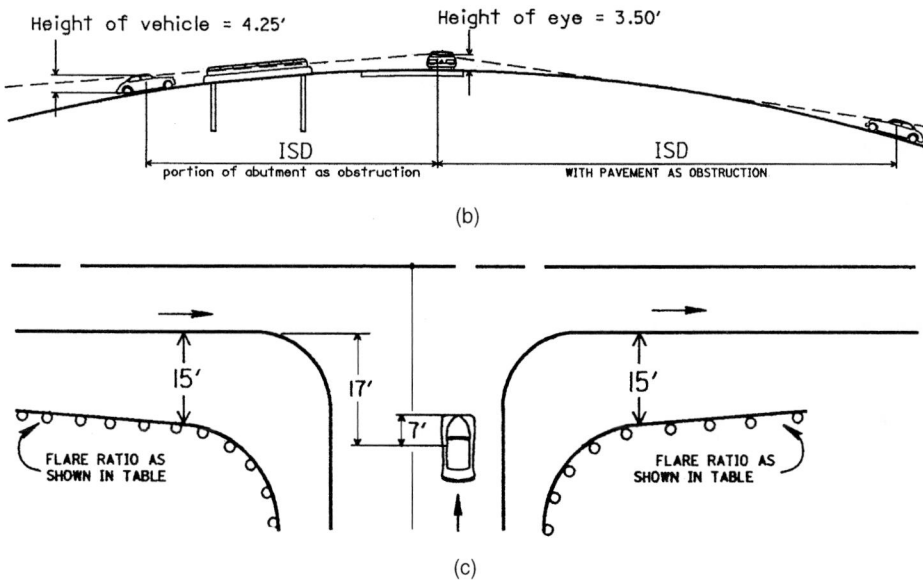

(b)

(c)

DESIGN SPEED (mi/h)	ISD (ft.)	CLEARANCE B (Obstruction to vehicle; ft) Offset A =						GUIDERAIL FLARE RATIO
		2 ft	4 ft	6 ft	8 ft	10 ft	12 ft	
70	950	550	475	400	330	255	185	45:1
65	875	510	445	375	305	240	170	45:1
60	825	475	415	350	285	225	160	40:1
55	750	440	380	320	260	205	145	40:1
50	700	405	350	295	240	190	135	35:1
40	575	330	290	245	200	155	110	30:1
30	450	260	225	190	155	120	85	25:1

(d)

FIGURE 2.7   (*Continued*)

height of an approaching vehicle of 4.25 ft. Table 2.3 shows the calculated K values for each of the design speeds listed (K being the rate of vertical curvature in feet per percent of grade change). Although the values listed assume that the line of sight is being obstructed by the pavement surface, this is not always the case. Figure 2.7*b* shows a case where the bridge parapet is providing the obstruction, in which case a graphical solution may be required to determine the available sight distance.

Passing Sight Distance.   In Table 2.3, the "Passing" column lists the distances required for passing an overtaken vehicle at various design speeds. These distances are applicable to two-lane roadways only. Among the assumptions that affect the required distance calculations are that (1) the passing vehicle averages 10 mi/h faster than the vehicle being passed, (2) the vehicle being passed travels at a constant speed and this speed is the average running speed (which is less than the design speed), and (3) the oncoming vehicle is traveling at the same speed as the passing vehicle. Table 2.3 contains K values for designing crest vertical curves to provide passing sight distance. These values assume that the height of the driver's eye is 3.5 ft and the height of the oncoming vehicle is 4.25 ft. The equations at the bottom of the table provide mathematical solutions for sight distance on the crest curves.

On two-lane roadways, it is important to provide adequate passing sight distance for as much of the project length as possible to compensate for missed opportunities due to oncoming traffic in the passing zone. On roadways where the design hourly traffic volume exceeds 400, the designer should investigate the effect of available passing sight distance on highway capacity using procedures outlined in the 1985 Transportation Research Board "Highway Capacity Manual" (Ref. 10). If the available passing sight distance restricts the capacity from meeting the design level of service requirement, then adjustments should be made to the profile to increase the distance. If the problem cannot be resolved in this manner, then consideration should be given to providing passing lane sections or constructing a multilane facility.

Decision Sight Distance.   Stopping sight distances are usually sufficient to allow reasonably competent drivers to come to a hurried stop under ordinary circumstances. However, these distances may not be sufficient for drivers when information is difficult to perceive, or when unexpected maneuvers are required. In these circumstances, the decision sight distance provides a greater length for drivers to reduce the likelihood of error in receiving information, making decisions, or controlling the vehicle.

The following are examples of locations where it is desirable to provide decision sight distance: (1) exit ramps, (2) diverging roadway terminals, (3) intersection stop bars, (4) changes in cross section, such as toll plazas and lane drops, and (5) areas of concentrated demand where there is apt to be "visual noise" (i.e., where sources of information compete, such as roadway elements, traffic, traffic control devices, and advertising signs).

Table 2.4 shows decision sight distances based on design speed. If a crest vertical curve is involved, a list of K values and related formulas is also provided. Where conditions call for the use of the decision sight distance in design but it cannot be provided, every effort should be made to provide the preferred stopping sight distance values from Fig. 2.7. Consideration should also be given to using suitable traffic control devices to provide advance warning of the unexpected conditions that may be encountered.

2.2.3   Horizontal Alignment and Superelevation

The horizontal alignment of a roadway should be designed to provide motorists with a facility for driving in a safe and comfortable manner. Adequate stopping sight distance should be furnished. Also, changes in direction should be accompanied by the use of curves and superelevation when appropriate in accordance with established guidelines. Some changes in alignment are slight and may not require curvature. Table 2.5 lists the maximum deflection angle which may be permitted without the use of a horizontal curve for each design speed shown. It is assumed that a motorist can easily negotiate the change in direction and maintain control over the vehicle without leaving the lane.

When centerline deflections exceed the values in Table 2.5, it is necessary to introduce a horizontal curve to assist the driver. Curves are usually accompanied by super-

TABLE 2.4   Minimum and Preferred Decision Sight Distance (DSD) for Design Speeds from 30 to 70 mi/h

Height of eye, 3.50 ft; height of object, 0.50 ft

Design speed, mi/h	Minimum		Preferred	
	DSD, ft	K, ft/%	DSD, ft	K, ft/%
30	450	152	625	294
40	600	271	825	512
50	750	423	1025	791
60	1000	752	1275	1223
70	1100	910	1450	1582

Using S = decision sight distance, ft
   L = length of crest vertical curve, ft
   A = algebraic difference in grades, %
   K = rate of crest vertical curvature, ft per % of grade change
• For a given design speed and A value, the calculated length $L = KA$.
• To determine S with a given L and A, use the following:
   For $S < L$: $S = 36.45\sqrt{K}$ where $K = L/A$
   For $S > L$: $S = 664.58/A + L/2$

Source: Location and Design Manual, Vol. 1, Roadway Design, Ohio Department of Transportation, with permission.

elevation, which is a banking of the roadway to help counteract the effect of centrifugal force on the vehicle as it moves through the curve. In addition to superelevation, centrifugal force is also offset by the side friction developed between the tires of the vehicle and the pavement surface. The relationship of the two factors when considering curvature for a particular design speed is expressed by the following equation:

$$e + f = \frac{V^2}{15R} \tag{2.1}$$

where e = superelevation rate, ft per ft of pavement width
   f = side friction factor
   V = design speed, mi/h
   R = radius of curve, ft

   In developing superelevation guidelines for use in designing roadways, it is necessary to establish practical limits for both superelevation and side friction factors. Several factors affect the selection of a maximum superelevation rate for a given highway. Climate must be considered. Regions subject to snow and ice should not be superelevated too sharply, because the presence of these adverse conditions causes motorists to drive slower, and side friction is greatly reduced. Consequently, vehicles tend to slide to the low side of the roadway. Terrain conditions are another factor. Flat areas tend to have relatively flat grades, and such conditions have little effect on superelevation and side friction factors. However, mountainous regions have steeper grades, which combine with superelevation rates to produce steeper cross slopes on the pavement than may be apparent to the designer. Rural and urban areas require different maximum superelevation rates, because urban areas are more frequently subjected to congestion and slower-moving traffic. Vehicles operating at significantly less

TABLE 2.5 Maximum Centerline Deflection Not Requiring a Horizontal Curve

Design speed, mi/h	Maximum deflection*
25	5°30′
30	3°45′
35	2°45′
40	2°15′
45	1°15′
50	1°15′
55	1°00′
60	1°00′
65	0°45′
70	0°45′

Based on the following formulas:
  Design speed 45 mi/h or over: $\tan \Delta = 1.0/V$
  Design speed under 45 mi/h: $\tan \Delta = 60/V^2$
    where V = design speed, mi/h
      Δ = deflection angle

Note: The recommended *minimum* distance between consecutive horizontal deflections is:

  200 ft where design speed > 40 mi/h

  100 ft where design speed ≤ 40 mi/h

  *Rounded to nearest 15 minutes.

  Source: *Location and Design Manual,* Vol. 1, *Roadway Design,* Ohio Department of Transportation, with permission.

than design speeds necessitate a flatter maximum rate. Given the above considerations, a range of maximum values has been adopted for use in design. A maximum rate of 0.12 or 0.10 may be used in flat areas not subject to ice or snow. Rural areas where these conditions exist usually have a maximum rate of 0.08. A maximum rate of 0.06 is recommended for urban high-speed roadways (50 mi/h or greater), while 0.04 is used on low-speed urban roadways and temporary roads.

Various factors affect the side friction factors used in design. Among these are pavement texture, weather conditions, and tire condition. The upper limit of the side friction factor is when the tires begin to skid. Highway curves must be designed to avoid skidding conditions with a margin of safety. Side friction factors also vary with design speed. Higher speeds tend to have lower side friction factors. The result of various studies leads to the values listed in Table 2.6, which shows the side friction factors by design speed generally used in developing superelevation tables (Ref. 1).

Taking into account the above limits on superelevation rates and side friction factors, and rewriting Eq. (2.1), it follows that for a given design speed and maximum superelevation rate, there exists a minimum radius of curvature that should be allowed for design purposes:

$$R_{min} = \frac{V^2}{15(e + f)} \qquad (2.2)$$

To allow a lesser radius for the design speed would require the superelevation rate or the friction factor to be increased beyond the recommended limit.

TABLE 2.6 Friction Factors for Design Speeds from 20 to 70 mph Used in Developing Superelevation Tables

Design speed, mi/h	Side friction factor f
20	0.17
30	0.16
40	0.15
50	0.14
55	0.13
60	0.12
65	0.11
70	0.10

Highway design using English units defines horizontal curvature in terms of *degree of curve* as well as radius. Under this definition, the degree of curve is defined as the central angle of a 100-ft arc using a fixed radius. This results in the following equation relating R (radius, ft) to D (degree of curve, degrees):

$$D = \frac{5729.6}{R} \tag{2.3}$$

Substituting in Eq. (2.2) gives the maximum degree of curvature for a given design speed and maximum superelevation rate:

$$D_{max} = \frac{85,660(e + f)}{V^2} \tag{2.4}$$

Before presenting the superelevation tables, one final consideration must be addressed. Because for any curve, superelevation and side friction combine to offset the effects of centrifugal force, the question arises how much superelevation should be provided for curves flatter than the "maximum" allowed for a given design speed. The following five methods have been used over the years (Ref. 1):

Method 1. Superelevation and side friction are directly proportional to the degree of curve.

Method 2. Side friction is used to offset centrifugal force in direct proportion to the degree of curve, for curves up to the point where f_{max} is required. For sharper curves, f_{max} remains constant and e is increased in direct proportion to the increasing degree of curvature until e_{max} is reached.

Method 3. Superelevation is used to offset centrifugal force in direct proportion to the degree of curve for curves up to the point where e_{max} is required. For sharper curves, e_{max} remains constant and f is increased in direct proportion to the increasing degree of curvature until f_{max} is reached.

Method 4. Method 4 is similar to method 3, except that it is based on average running speed instead of design speed.

Method 5. Superelevation and side friction are in a curvilinear relationship with the degree of curve, with resulting values between those of method 1 and method 3.

Figure 2.8 shows a graphic comparison of the various methods. Method 5 is most commonly used on rural and high-speed (greater than 40 mi/h) urban highways. Method 2 is used on low-speed urban streets and temporary roadways.

Recommended superelevation rates for horizontal curves in relation to other design elements are given in Tables 2.7 through 2.11, where each table represents a different maximum superelevation rate. Table 2.7 shows values for a maximum rate of 0.04; Table 2.8, for 0.06; Table 2.9, for 0.08; Table 2.10, for 0.10; and Table 2.11, for 0.12. Method 5 was used to assign the appropriate superelevation rate to each degree of curve listed. Transition runoff lengths are discussed subsequently.

The superelevation rates on low-speed urban streets are set using method 2 described above, in which side friction is used to offset the effect of centrifugal force up to the maximum friction value allowed for the design speed. Superelevation is then introduced for sharper curves. The design data in Table 2.12, based on method 2 and a maximum superelevation rate of 0.04, can be used for low-speed urban streets and temporary roads.

In attempting to apply the recommended superelevation rates for low-speed urban roadways, various factors may combine to make these rates impractical to obtain. These factors include wide pavements, adjacent development, drainage conditions, and frequent access points. In such cases, curves may be designed with reduced or no superelevation, although crown removal is the recommended minimum.

Effect of Grades on Superelevation.   On long and fairly steep grades, drivers tend to travel somewhat slower in the upgrade direction and somewhat faster in the downgrade direction than on level roadways. In the case of divided highways, where each pavement can be superelevated independently, or on one-way roadways such as ramps, this tendency should be recognized to see whether some adjustment in the superelevation rate would be desirable and/or feasible. On grades of 4 percent or greater with a length of 1000 ft or more and a superelevation rate of 0.06 or more, the designer may adjust the superelevation rate by assuming a design speed 5 mi/h less in the upgrade direction and 5 mi/h greater in the downgrade direction, provided that the assumed design speed is not less than the legal speed. On two-lane, two-way roadways and on other multilane undivided highways, such adjustments are less feasible, and should be disregarded.

Superelevation Methods.   There are three basic methods for developing superelevation on a crowned pavement leading into and coming out of a horizontal curve. Figure 2.9 shows each method. In the most commonly used method, case I, the pavement edges are revolved about the centerline. Thus, the inner edge of the pavement is depressed by half of the superelevation and the outer edge raised by the same amount. Case II shows the pavement revolved about the inner or lower edge of pavement, and case III shows the pavement revolved about the outer or higher edge of pavement. Case II can be used where off-road drainage is a problem and lowering the inner pavement edge cannot be accommodated. The superelevation on divided roadways is achieved by revolving the pavements about the median pavement edge. In this way, the outside (high side) roadway uses case II, while the inside (low side) roadway uses case III. This helps control the amount of "distortion" in grading the median area.

Superelevation Transition.   The length of highway needed to change from a normal crowned section to a fully superelevated section is referred to as the *superelevation transition*. This length is shown as X in Fig. 2.9, which also shows the various other elements described below. The superelevation transition is divided into two parts: the tangent runout, and the superelevation runoff.

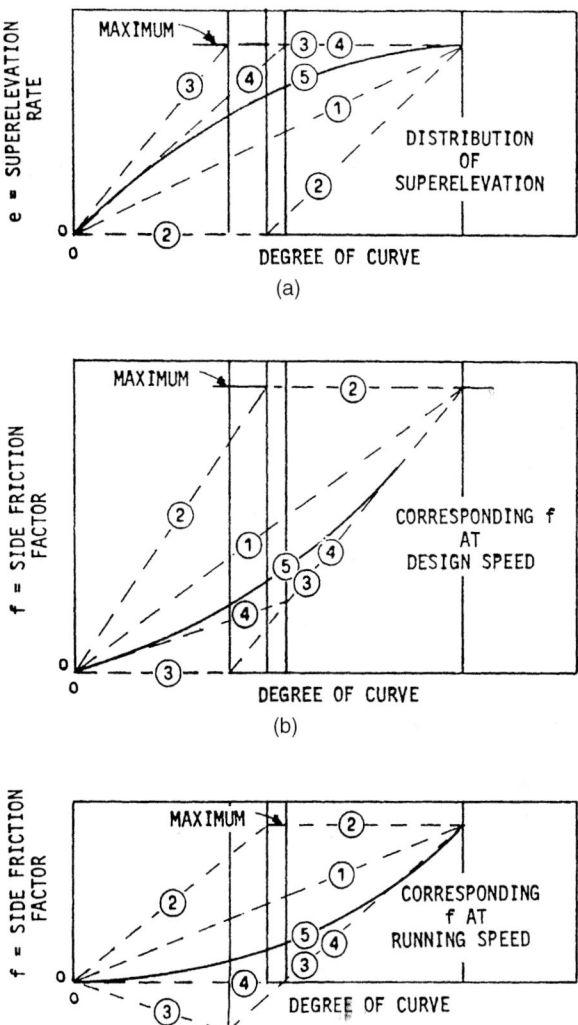

FIGURE 2.8 Methods of distributing superelevation and side friction. (*a*) Superelevation. (*b*) Corresponding friction factor at design speed. (*c*) Corresponding friction factor at running speed. (*From* A Policy on Geometric Design of Highways and Streets, *American Association of State Highway and Transportation Officials, Washington, D.C., 1990, with permission*)

TABLE 2.7 Recommended Superelevation Rates (to $e_{max} = 0.04$) and Runoff Lengths for Design Speeds from 30 to 60 mi/h and Curves up to 19°

D	R, ft	V = 30 mi/h e	L, ft Two lanes	L, ft Four lanes	V = 40 mi/h e	L, ft Two lanes	L, ft Four lanes	V = 50 mi/h e	L, ft Two lanes	L, ft Four lanes	V = 55 mi/h e	L, ft Two lanes	L, ft Four lanes	V = 60 mi/h e	L, ft Two lanes	L, ft Four lanes
0°15′	22,918	NC	0	0	NC	0	0	NC	0	0	NC	0	0	NC	0	0
0°30′	11,459	NC	0	0	NC	0	0	NC	0	0	NC	0	0	NC	0	0
0°45′	7,639	NC	0	0	NC	0	0	NC	0	0	RC	160	240	RC	175	265
1°00′	5,730	NC	0	0	NC	0	0	RC	150	225	.021	160	240	.023	175	265
1°30′	3,820	NC	0	0	RC	125	190	.024	150	225	.026	160	240	.029	175	265
2°00′	2,865	RC	100	150	.022	125	190	.027	150	225	.030	160	240	.033	175	265
2°30′	2,292	RC	100	150	.025	125	190	.030	150	225	.033	160	240	.037	175	265
3°00′	1,910	.020	100	150	.027	125	190	.033	150	225	.036	160	240	.039	175	265
3°30′	1,637	.022	100	150	.028	125	190	.035	150	225	.038	160	240	.040	175	265
4°00′	1,432	.024	100	150	.030	125	190	.037	150	225	.039	160	240			
5°00′	1,146	.026	100	150	.033	125	190	.039	150	225						
6°00′	955	.028	100	150	.035	125	190	.040	150	225						
7°00′	819	.030	100	150	.037	125	190									
8°00′	716	.031	100	150	.039	125	190									
9°00′	637	.033	100	150	.040	125	190									
10°00′	573	.034	100	150	.040	125	190									
11°00′	521	.035	100	150												
12°00′	477	.036	100	150												
13°00′	441	.037	100	150												
14°00′	409	.038	100	150												
16°00′	358	.039	100	150												
18°00′	318	.040	100	150												
19°00′	302	.040	100	150												
		$D_{max} = 19°00′$			$D_{max} = 10°00′$			$D_{max} = 6°00′$			$D_{max} = 4°45′$			$D_{max} = 3°45′$		

2.24

D = degree of curve

R = radius of curve

V = assumed design speed

e = rate of superelevation

L = minimum length of runoff (does not include tangent runout)

NC = normal crown section

RC = remove adverse crown, superelevate at normal crown slope

Note: Lengths rounded in multiples of 25 or 50 ft permit simpler calculations. In recognition of safety considerations, use of e_{max} = 0.04 should be limited to urban conditions.

Source: *A Policy on Geometric Design of Highways and Streets*, American Association of State Highway and Transportation Officials, Washington, D.C., 1990, with permission.

2.25

TABLE 2.8 Recommended Superelevation Rates (to e_{max} = 0.06) and Runoff Lengths for

		V = 30 mi/h			V = 40 mi/h			V = 50 mi/h		
			L, ft			L, ft			L, ft	
			Two	Four		Two	Four		Two	Four
D	R, ft	e	lanes	lanes	e	lanes	lanes	e	lanes	lanes
0°15′	22,918	NC	0	0	NC	0	0	NC	0	0
0°30′	11,459	NC	0	0	NC	0	0	NC	0	0
0°45′	7,639	NC	0	0	NC	0	0	RC	0	00
1°00′	5,730	NC	0	0	NC	0	0	.020	150	150
1°30′	3,820	NC	0	0	.020	125	125	.028	150	150
2°00′	2,865	RC	100	100	.025	125	125	.035	150	150
2°30′	2,292	.020	100	100	.030	125	125	.040	150	150
3°00′	1,910	.023	100	100	.034	125	125	.045	150	160
3°30′	1,637	.026	100	100	.038	125	125	.048	150	170
4°00′	1,432	.029	100	100	.041	125	130	.052	150	180
5°00′	1,146	.034	100	100	.046	125	140	.056	150	200
6°00′	955	.038	100	100	.050	125	160	.059	150	210
7°00′	819	.041	100	110	.053	125	170			
8°00′	716	.043	100	120	.056	125	180		D_{max} = 6°45′	
9°00′	637	.046	100	120	.058	125	180			
10°00′	573	.048	100	130	.059	125	190			
11°00′	521	.050	100	140	.060	130	190			
13°00′	477	.052	100	140						
13°00′	441	.054	100	140		D_{max} = 11°15′				
14°00′	409	.055	100	150						
16°00′	358	.058	100	160						
18°00′	318	.059	110	160						
20°00′	286	.060	110	160						
21°00′	273	.060	110	160						

$$D_{max} = 21°00′$$

D = degree of curve
R = radius of curve
V = assumed design speed
e = rate of superelevation
L = minimum length of runoff (does not include tangent runout)
   Source: *A Policy on Geometric Design of Highways and Streets,* American Association of State Highway and Transportation Officials, Washington, D.C., 1990, with permission.

   The tangent runout (T, in Fig. 2.9) is the length required to remove the adverse pavement cross slope. As is shown for case I of Fig. 2.9, this is the length required to raise the outside edge of pavement from a normal cross slope to a half-flat section. The superelevation runoff (L in Fig. 2.9) is the length required to raise the outside edge of pavement from a half-flat section to a fully superelevated section. The length of transition required to remove the pavement crown (R in Fig. 2.9) is generally equal to twice the T distance.
   The minimum superelevation transition length X should be equal in feet to 3 times the design speed in miles per hour. This includes the tangent runout (T) as previously described. The reason to specify this minimum is to avoid the appearance of a "kink"

Design Speeds from 30 to 60 mi/h and Curves up to 21°

	V = 55 mi/h			V = 60 mi/h			V = 65 mi/h			V = 70 mi/h	
	L, ft			L, ft			L, ft			L, ft	
e	Two lanes	Four lanes	e	Two lanes	Four lanes	e	Two lanes	Four lanes	e	Two lanes	Four lanes
NC	0	0	NC	0	0	NC	0	0	NC	0	0
NC	0	0	NC	0	0	RC	190	190	RC	200	200
RC	160	160	.021	175	175	.024	190	190	.026	200	200
.023	160	160	.027	175	175	.030	190	190	.033	200	200
.032	160	160	.037	175	175	.041	190	190	.046	200	200
.040	160	160	.045	175	180	.050	190	210	.055	200	230
.045	160	170	.051	175	200	.056	190	230	.059	200	260
.050	160	190	.055	175	220	.059	190	250			
.054	160	210	.058	175	230	.060	190	250	D_{max} = 2°45′		
.057	160	220	.060	175	240						
.060	160	230				D_{max} = 3°30′					
			D_{max} = 4°15′								
D_{max} = 5°15′											

NC = normal crown section
RC = remove adverse crown, superelevate at normal crown slope
Note: Lengths rounded in multiples of 25 or 50 ft permit simpler calculations.

in the roadway that a shorter transition would provide. The distance is approximately equal to that traveled by a vehicle in 2 s at design speed. This requirement does not apply to low-speed roadways, temporary roads, superelevation transitions near intersections, or transitions between adjacent horizontal curves (reverse or same direction) where normal transitions would overlap each other. In these cases, the minimum transition length is determined by multiplying the edge of pavement correction by the equivalent slope rate (G) shown in Table 2.13. The rate of change of superelevation should be constant throughout the transition X. Some agencies use a flatter rate of transition through the T or R sections than that recommended in Table 2.13, an acceptable but unnecessary practice.

TABLE 2.9   Recommended Superelevation Rates (to $e_{max} = 0.08$) and Runoff Lengths for

| | | V = 30 mi/h | | | V = 40 mi/h | | | V = 50 mi/h | | |
| | | | L, ft | | | L, ft | | | L, ft | |
D	R, ft	e	Two lanes	Four lanes	e	Two lanes	Four lanes	e	Two lanes	Four lanes
0°15′	22,918	NC	0	0	NC	0	0	NC	0	0
0°30′	11,459	NC	0	0	NC	0	0	NC	0	0
0°45′	7,639	NC	0	0	NC	0	0	RC	150	150
1°00′	5,730	NC	0	0	NC	0	0	.021	150	150
1°30′	3,820	NC	0	0	.021	125	125	.030	150	150
2°00′	2,865	RC	100	100	.027	125	125	.038	150	150
2°30′	2,292	.021	100	100	.033	125	125	.046	150	170
3°00′	1,910	.025	100	100	.038	125	125	.053	150	190
3°30′	1,637	.028	100	100	.043	125	140	.058	150	210
4°00′	1,432	.031	100	100	.047	125	150	.063	150	230
5°00′	1,146	.038	100	100	.055	125	170	.071	170	260′
6°00′	955	.043	100	120	.062	130	190	.077	180	280′
7°00′	819	.048	100	130	.067	140	210	.080	190	280
8°00′	716	.053	100	140	.071	150	220			
9°00′	637	.056	100	150	.075	160	240	$D_{max} = 7°30′$		
10°00′	573	.060	110	160	.078	160	240			
11°00′	521	.063	110	170	.079	170	250			
12°00′	477	.065	120	180	.080	170	250			
13°00′	441	.068	120	180						
14°00′	409	.070	130	190	$D_{max} = 12°15′$					
16°00′	358	.074	130	200						
18°00′	318	.077	140	210						
20°00′	286	.079	140	210						
22°00′	260	.080	140	230						

$$D_{max} = 22°45′$$

D = degree of curve
R = radius of curve
V = assumed design speed
e = rate of superelevation
Note: Lengths rounded in multiples of 25 or 50 ft permit simpler calculations.
Source: *A Policy on Geometric Design of Highways and Streets,* American Association of State Highway and Transportation Officials, Washington, D.C., 1990, with permission.

The values given for *L* in Tables 2.7 through 2.12 are based on two and four lanes revolved about the centerline. Table 2.13 provides a summary of ways to adjust the table values of *L* for cases involving other lane widths or where more than one lane is being revolved about the centerline or baseline. Note that when using the formula $W(S + N)G$ from Table 2.13 to calculate the transition length for revolving multiple lanes, the result is considered to be the minimum transition length for interstate or freeway-type design. It is the desirable length for all other roadways. The minimum length for these roadways is calculated similarly; however, the *G* value from the table is selected for a speed 15 mi/h less than the design speed. The 1990 AASHTO "Green Book" provides other suggestions for shortening these transitions (Ref. 1).

Design Speeds from 30 to 70 mi/h and Curves up to 22°

| | V = 55 mi/h | | | V = 60 mi/h | | | V = 65 mi/h | | | V = 70 mi/h | |
| | L, ft | | | L, ft | | | L, ft | | | L, ft | |
e	Two lanes	Four lanes	e	Two lanes	Four lanes	e	Two lanes	Four lanes	e	Two lanes	Four lanes
NC	0	0	NC	0	0	NC	0	0	NC	0	0
NC	0	0	RC	175	175	RC	190	190	RC	200	200
RC	160	160	.022	175	175	.025	190	190	.028	200	200
.025	160	160	.029	175	175	.032	190	190	.036	200	200
.035	160	160	.041	175	175	.046	190	200	.051	200	240
.045	160	170	.051	175	210	.058	190	250	.065	200	290
.053	160	200	.061	175	240	.068	190	300	.075	220	330
.060	160	230	.068	180	270	.075	210	320	.080	230	350
.067	170	260	.074	200	300	.079	220	350			
.071	180	270	.078	210	310				$D_{max} = 3°00'$		
.078	200	300		$D_{max} = 4°45'$			$D_{max} = 3°45'$				

$D_{max} = 6°00'$

L = minimum length of runoff (does not include tangent runout)
NC = normal crown section
RC = remove adverse crown, superelevate at normal crown slope

Superelevation Position.   Figure 2.9 shows the recommended positioning of the proposed superelevation transition in relationship to the horizontal curve. For those curves with spirals, the transition from adverse crown removal to full superelevation should occur within the limits of the spiral. In other words, the spiral length should equal the L value, usually rounded to the nearest 25 ft.

For simple curves without spirals, the L transition should be placed so that 50 to 70 percent of the maximum superelevation rate is outside the curve limits (point of curvature PC to point of tangency PT). It is recommended that whenever possible, two-thirds of the full superelevation rate be present at the PC and PT. See the case diagrams in Fig. 2.9 for a graphic presentation of the recommended positioning.

TABLE 2.10 Recommended Superelevation Rates (to $e_{max} = 0.10$) and Runoff Lengths for

			$V = 30$ mi/h			$V = 40$ mi/h			$V = 50$ mi/h	
			L, ft			L, ft			L, ft	
D	R, ft	e	Two lanes	Four lanes	e	Two lanes	Four lanes	e	Two lanes	Four lanes
0°15′	22,918	NC	0	0	NC	0	0	NC	0	0
0°30′	11,459	NC	0	0	NC	0	· 0	NC	0	0
0°45′	7,639	NC	0	0	NC	0	0	RC	150	150
1°00′	5,730	NC	0	0	NC	0	0	.021	150	150
1°30	3,820	NC	0	0	.021	125	125	.031	150	150
2°00′	2,865	RC	100	100	.028	125	125	.040	150	150
2°30′	2,292	.021	100	100	.034	125	125	.049	150	180
3°00′	1,910	.025	100	100	.040	125	125	.057	150	210
3°30′	1,637	.029	100	100	.046	125	140	.065	160	240
4°00′	1,432	.033	100	100	.051	125	160	.072	180	260
5°00′	1,146	.040	100	110	.061	130	190	.083	200	300′
6°00′	955	.046	100	120	.070	150	220	.092	220	330′
7°00′	819	.053	100	140	.078	160	240	.098	240	350′
8°00′	716	.058	110	160	.084	180	260	.100	240	360′
9°00′	637	.063	120	170	.089	190	280			
10°00′	573	.068	120	180	.094	200	290		$D_{max} = 8°15′$	
11°00′	521	.072	130	200	.097	200	310			
12°00′	477	.076	140	210	.099	210	310			
13°00′	441	.080	140	220	.100	210	320			
14°00′	409	.083	150	220						
16°00′	358	.089	160	240		$D_{max} = 13°15′$				
18°00′	318	.093	170	250						
20°00′	302	.097	170	260						
22°00′	260	.099	280	270						
24°00′	239	.100	180	270						

$$D_{max} = 24°45′$$

D = degree of curve
R = radius of curve
V = assumed design speed
e = rate of superelevation
Note: Lengths rounded in multiples of 25 or 50 ft permit simpler calculations.
Source: *A Policy on Geometric Design of Highways and Streets,* American Association of State Highway and Transportation Officials, Washington, D.C., 1990, with permission.

Profiles and Elevations. Breakpoints at the beginning and end of the superelevation transition should be rounded to obtain a smooth profile. One suggestion is to use a "vertical curve" on the edge of the pavement profile with a length in feet equal to the design speed in mi/h (i.e., 45 ft for 45 mi/h). The final construction plans should have the superelevation tables or pavement details showing the proposed elevations at the centerline, pavement edges, and, if applicable, lane lines or other breaks in the cross slopes. Pavement or lane widths should be included where these widths are in transition. Pavement edge profiles should be plotted to an exaggerated vertical profile within the limits of the superelevation transitions to check calculations and to determine

Design Speeds from 30 to 70 mi/h and Curves up to 24°

| | V = 55 mi/h | | | V = 60 mi/h | | | V = 65 mi/h | | | V = 70 mi/h | |
| | L, ft | | | L, ft | | | L, ft | | | L, ft | |
e	Two lanes	Four lanes	e	Two lanes	Four lanes	e	Two lanes	Four lanes	e	Two lanes	Four lanes
NC	0	0	NC	0	0	NC	0	0	NC	0	0
NC	0	0	RC	175	175	RC	190	190	RC	200	200
RC	160	160	.023	175	175	.025	190	190	.028	200	200
.025	160	160	.030	175	175	.033	190	190	.037	200	200
.037	160	160	.043	175	190	.048	190	220	.054	200	260
.048	160	180	.055	175	230	.062	190	290	.070	220	330
.056	160	210	.067	190	280	.075	220	330	.085	260	390
.067	170	250	.077	210	320	.087	250	380	.096	280	420
.075	190	290	.086	350	.095	.095	270	400	.100	300	450
.083	210	320	.093	250	380	.099	280	420			
.094	240	360	.098	270	400					$D_{max} = 3°30'$	
.099	250	380					$D_{max} = 4°15'$				
	$D_{max} = 6°30''$			$D_{max} = 5°15'$							

L = minimum length of runoff (does not include tangent runout)

NC = normal crown section

RC = remove adverse crown, superelevate at normal crown slope

the location of drainage basins. Adjustments should be made to obtain smooth profiles. Special care should be taken in determining edge elevations in a transition area when the profile grade is on a vertical curve.

Superelevation between Reverse Horizontal Curves. When two horizontal curves are in close proximity to each other, the superelevation transitions calculated independently may overlap each other. In these cases, the designer should coordinate the transitions to provide a smooth and uniform change from the full superelevation of the first curve to the full superelevation of the second curve. Figure 2.10 shows two dia-

TABLE 2.11   Recommended Superelevation Rates (to $e_{max} = 0.12$) and Runoff Lengths for

			V = 30 mi/h			V = 40 mi/h			V = 50 mi/h	
			L, ft			L, ft			L, ft	
			Two	Four		Two	Four		Two	Four
D	R, ft	e	lanes	lanes	e	lanes	lanes	e	lanes	lanes
0°15′	22,918	NC	0	0	NC	0	0	NC	0	0
0°30′	11,459	NC	0	0	NC	0	0	NC	0	0
0°45′	7,639	NC	0	0	NC	0	0	RC	150	150
1°00′	5,730	NC	0	0	NC	0	0	.022	150	150
1°30′	3,820	NC	0	0	.022	125	125	.032	150	150
2°00′	2,865	RC	100	100	.029	125	125	.042	150	150
2°30′	2,292	.022	100	100	.035	125	125	.051	150	180
3°00′	1,910	.026	100	100	.042	125	140	.060	150	220
3°30′	1,637	.030	100	100	.048	125	160	.069	170	250
4°00′	1,432	.034	100	100	.054	125	180	.077	190	280
5°00′	1,146	.041	100	110	.065	140	210	.092	220	330′
6°00′	955	.049	100	130	.075	170	250	.104	250	370
7°00′	819	.055	100	150	.085	180	280	.113	270	410′
8°00′	716	.062	110	170	.094	200	300	.119	290	430
9°00′	637	.068	120	180	.101	220	320			
10°00′	573	.074	130	200	.107	230	340	$D_{max} = 9°00′$		
11°00′	521	.079	140	210	.112	240	360			
12°00′	477	.084	150	230	.116	240	370			
13°00′	441	.089	160	240	.119	250	370			
14°00′	409	.093	170	250	.120	250	380			
16°00′	358	.101	180	270						
18°00′	318	.108	190	290	$D_{max} = 14°30′$					
20°00′	286	.113	200	310						
22°00′	260	.116	210	310						
24°00′	239	.119	210	320						
26°00′	220	.120	220	320						

$$D_{max} = 26°45′$$

D = degree of curve
R = radius of curve
V = assumed design speed
e = rate of superelevation
Note: Lengths rounded in multiples of 25 or 50 ft permit simpler calculations.
   Source: *A Policy on Geometric Design of Highways and Streets,* American Association of State Highway and Transportation Officials, Washington, D.C., 1990, with permission.

grams suggesting ways in which this may be accomplished. In both diagrams each curve has its own L value (L_1, L_2) depending on the degree of curvature, and the superelevation is revolved about the centerline.

   The top diagram involves two simple curves. In the case of new or relocated alignment, the PT of the first curve and the PC of the second curve should be separated by enough distance to allow a smooth, continuous transition between the curves at a rate not exceeding the G value for the design speed (Table 2.13). This requires that the distance be not less than 50 percent nor greater than 70 percent of $L_1 + L_2$. Two-thirds is

Design Speeds from 30 to 70 mi/h and Curves up to 26°

	V = 55 mi/h L, ft			V = 60 mi/h L, ft			V = 65 mi/h L, ft			V = 70 mi/h L, ft	
e	Two lanes	Four lanes	e	Two lanes	Four lanes	e	Two lanes	Four lanes	e	Two lanes	Four lanes
NC	0	0	NC	0	0	NC	0	0	NC	0	0
NC	0	0	RC	175	175	RC	190	190	RC	200	200
.020	160	160	.023	175	175	.026	190	190	.029	200	200
.026	160	160	.030	175	175	.034	190	190	.038	200	200
.038	160	160	.044	175	175	.050	190	220	.056	200	250
.049	160	190	.058	175	230	.065	190	290	.073	220	330
.060	160	230	.070	190	280	.080	230	350	.090	270	410
.071	180	270	.082	220	330	.093	270	410	.106	320	480
.081	210	310	.094	250	380	.106	310	470	.118	350	530
.090	230	340	.104	280	420	.116	350	510		$D_{max} = 3°45'$	
.106	270	400	.117	310	470	$D_{max} = 4°45'$					
.116	290	440	$D_{max} = 5°45'$								
.120	300	460									

$D_{max} = 7°00'$

L = minimum length of runoff (does not include tangent runout)

NC = normal crown section

RC = remove adverse crown, superelevate at normal crown slope

the recommended portion. When adapting this procedure to existing curves where no alignment revision is proposed, the transition should conform as closely as possible to the above criteria. When the available distance between the curves is less than 50 percent of $L_1 + L_2$, the transition rate may be increased and/or the superelevation rate at the PT or PC may be set to less than 50 percent of the full superelevation rate.

The lower diagram involves two spiral curves. Where spiral transitions are used, the spiral-to-tangent (ST) point of the first curve and the tangent-to-spiral (TS) transition of the second curve may be at, or nearly at, the same location, without causing

TABLE 2.12 Recommended Superelevation Rates (to $S = 0.04$) and Runoff Lengths for Design Speeds from 20 to 40 mi/h and Curves up to 72° for Urban Streets and Temporary Roads

		Design speed								
		$V =$ 20 mi/h		$V =$ 25 mi/h		$V =$ 30 mi/h		$V =$ 35 mi/h		$V =$ 40 mi/h	
D	Radius	S	L	S	L	S	L	S	L	S	L
4°00′	1432	NC	—	NC	—	NC	—	NC	—	NC	—
5°00′	1146	NC	—	NC	—	NC	—	NC	—	.017	50
6°00′	955	NC	—	NC	—	NC	—	NC	—	.020	60
7°00′	819	NC	—	NC	—	NC	—	.017	50	.024	70
8°00′	716	NC	—	NC	—	NC	—	.019	55	.027	75
9°00′	637	NC	—	NC	—	NC	—	.022	60	.031	85
10°00′	573	NC	—	NC	—	.016	45	.024	65	.034	95
11°00′	521	NC	—	NC	—	.018	50	.026	70	.038	105
11°30′	498	NC	—	NC	—	.018	50	.028	75	.039	110
12°00′	477	NC	—	NC	—	.019	55	.029	75	$D_{max} = 11°45′$	
13°00′	441	NC	—	NC	—	.021	55	.031	80		
14°00′	409	NC	—	NC	—	.022	60	.034	85		
15°00′	382	NC	—	NC	—	.024	60	.036	90		
16°00′	358	NC	—	.016	45	.026	65	.039	100		
16°30′	347	NC	—	.016	45	.027	65	.040	100		
18°00′	318	NC	—	.018	50	.029	70	$D_{max} = 16°30′$			
20°00′	286	NC	—	.020	55	.032	75				
22°00′	260	NC	—	.022	55	.035	80				
24°00′	239	NC	—	.024	60	.038	85				
24°45′	231	NC	—	.025	60	.040	90				
26°00′	220	NC	—	.026	60	$D_{max} = 24°45′$					
28°00′	205	NC	—	.028	65						
30°00′	191	.016	45	.030	65						
35°00′	164	.019	50	.035	70						
40°00′	143	.022	55	.040	85						
50°00′	116	.027	60	$D_{max} = 40°00′$							
72°45′	78	.040	70								
$D_{max} = 72°45′$											

See superelevation notes, Table 2.13.

S = superelevation rate, ft/ft

L = runoff length, ft

Source: *Location and Design Manual,* Vol. 1, *Roadway Design,* Ohio Department of Transportation, with permission.

superelevation problems. In these cases, the crown should not be reestablished as shown in Fig. 2.9, but instead, both pavement edges should be in continual transition between the curves, as shown in the lower diagram of Fig. 2.10. The total superelevation transition length is the distance between the curve-to-spiral (CS) point and spiral-to-curve (SC) point.

Spiral Transitions. When a motor vehicle enters or leaves a circular horizontal curve, it follows a transition path during which the driver makes adjustments in steering to account for the gain or loss in centrifugal force. For most curves, the average driver can negotiate this change in steering within the normal width of the travel lane.

NOTE: The diagrams below show positioning of the superelevation transition for both simple curves and spiral curves. Only one of these conditions would exist for a given transition.

LEGEND: X = Length of superelevation transition.
L = Length of superelevation runoff.
T = Tangent runout
R = Crown removal
G = Equivalent slope rate of Change of outside pavement edge compared to the control line in each case. (See Table 2.13 for values.)
N = Normal cross slope
S = Full superelevation rate

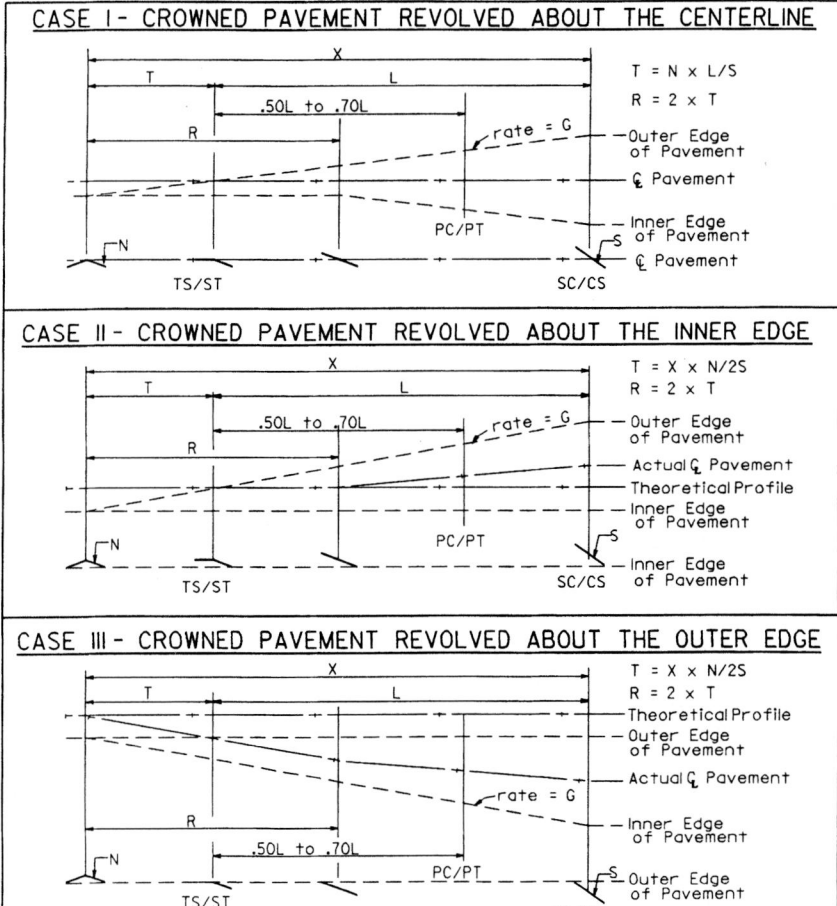

FIGURE 2.9 Superelevation transition between tangent and simple or spiral curves for three cases. (*From* Location and Design Manual, *Vol. 1,* Roadway Design, *Ohio Department of Transportation, with permission*)

TABLE 2.13 Superelevation Notes for Adjusting Runoff Lengths in Tables 2.6 to 2.12

(a) Maximum relative gradients for profiles between the edge of pavement and the centerline or reference line

Design speed, mi/h	Relative gradient	Equivalent slope rate G
20	0.75	133:1
25	0.71	141:1
30	0.67	150:1
35	0.625	160:1
40	0.58	175:1
45	0.54	185:1
50	0.50	200:1
55	0.475	210:1
60	0.45	222:1
65	0.425	235:1
70	0.40	250:1

(b)

Calculate X	Case I: $X = W(S + N)G$	Cases II, III: $X = WSG$
If $X > 3D$	$L = WSG$	$L = W(S - N/2)G$
	$T = WNG$	$T = WN/2G$
If $X < 3D$	$X = 3D$	$X = 3D$
	$L = 3D\left(\dfrac{S}{N + S}\right)$	$L = 3D\left(\dfrac{2S - N}{2S}\right)$
	$T = 3D\left(\dfrac{N}{N + S}\right)$	$T = 3D\left(\dfrac{N}{2S}\right)$

1. The L values in Tables 2.7 to 2.12 are based on a two-lane, 24-ft-wide pavement revolved about the centerline.

2. Adjustments to L for varying pavement widths can be made by direct proportion. For a 20-ft pavement revolved about the centerline, $L' = L(20/24)$.

3. As shown in Fig. 2.9, the minimum X for each case is dependent on design speed and is approximately the distance traveled in 2 s. This number has been rounded off to a figure in feet to 3 times the design speed. The numbers that appear in the L columns of Tables 2.7 to 2.12 represent those situations where the minimum length controls and L plus T would equal 3 times the design speed.

4. Determination of X, L, and T when more than one lane is revolved about the centerline (or other reference line, such as a baseline or edge of pavement) is shown in part (b). In part (b), $W =$ pavement width (ft) from point of rotation to farthest edge, and $D =$ linear-foot equivalent of design speed in mi/h (example: 55 ft for 55 mi/h). See Fig. 2.9 for other symbol definitions and case descriptions.

5. The L value is also the recommended spiral length where spirals are used.

Source: *Location and Design Manual,* Vol. 1, *Roadway Design,* Ohio Department of Transportation, with permission.

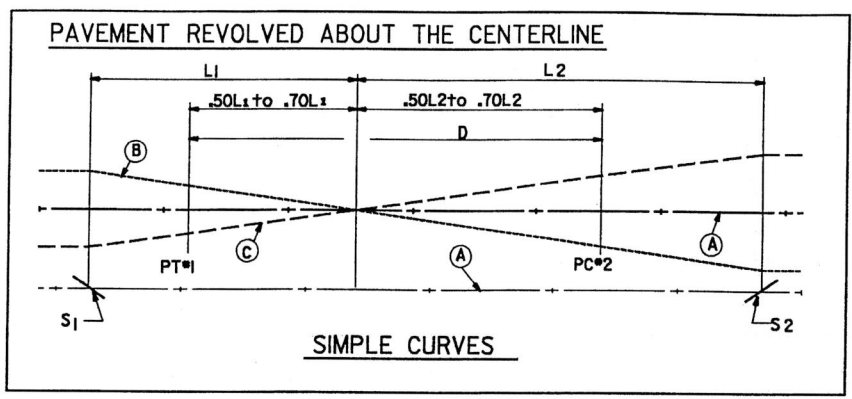

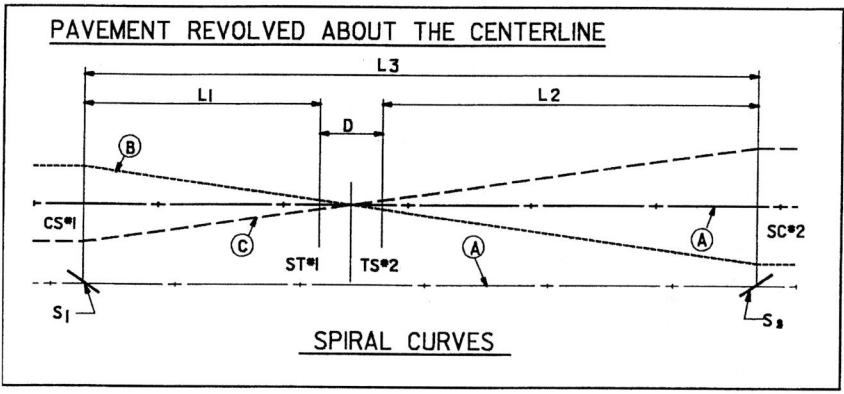

LEGEND:

Ⓐ – Centerline Pavement

Ⓑ – Outside E.P. Curve 1,
   Inside E.P. Curve 2   E.P.=Edge of Pavement

Ⓒ – Inside E.P. Curve 1, Outside E.P. Curve 2

S_1, S_2 = Superelevation Rates: Curves 1 & 2

L_1, L_2 = Superelevation Transition Lengths: Curves 1 & 2

D = Distance between Curves

L_3 = Total Superelevation Transition between Spiral Curves

FIGURE 2.10 Superelevation transition between reverse horizontal curves, simple or spiral. (*From Location and Design Manual, Vol. 1, Roadway Design, Ohio Department of Transportation, with permission*)

However, combinations of higher speeds and sharper curvature may cause the driver to move into an adjacent travel lane while accomplishing the change. To prevent this occurrence, the designer should use spirals to smooth out transitions.

There are several advantages to using spiral transitions for horizontal curves:

- They provide an easy-to-follow path for the driver to negotiate.
- They provide a convenient area in which to place the superelevation transition.
- They provide an area where the pavement width can be transitioned when required for curve widening.
- They provide a smoother appearance to the driver.

The Euler spiral is the one most commonly used in highway design. The degree of curve varies gradually from zero at the tangent end to the degree of the circular arc at the curve end. By definition, the degree of curve at any point along the spiral varies directly with the length measured along the spiral. In the case where a spiral transition connects two simple curves, the degree of curve varies directly from that of the first circular arc to that of the second circular arc. As a general guideline, spirals should be used on roadways where the design speed is greater than 40 mi/h and the degree of curvature is 1°30′ or greater.

Horizontal Alignment Considerations. The following items should be considered when establishing new horizontal alignment:

- The alignment should be as directional as possible while still consistent with topography and the preservation of developed properties and community values.
- Maximum allowable curvature should be avoided whenever possible.
- Consistent alignment should be sought.
- Curves should be long enough to avoid the appearance of a kink.
- Tangents and/or flat curves should be provided on high, long fills.
- Compound curves should be used only with caution.
- Abrupt alignment reversals should be avoided.
- Two curves in the same direction separated by a short tangent (broken-back or flat-back curves) should be avoided.

2.2.4 Vertical Alignment

The design of the vertical alignment of a roadway also has a direct effect on the safety and comfort of the driver. Steep grades can slow down large, heavy vehicles in the traffic stream in the uphill direction and can adversely affect stopping ability in the downhill direction. Grades that are flat or nearly flat over extended distances will slow down the rate at which the pavement surface drains. Vertical curves provide a smooth change between two tangent grades, but must be designed to provide adequate stopping sight distance.

Tangent Grades. The maximum percent grade for a given roadway is determined by its functional classification, surrounding terrain, and design speed. Table 2.14 shows how the maximum grade can vary under different circumstances. Note that relatively flat grade limits are recommended for higher functional class roadways and at higher

TABLE 2.14 Maximum Grades as Determined by Function, Terrain, and Speed, %

Functional classification	Terrain	Design speed					
		20 to 29	30 to 39	40 to 49	50 to 59	60 to 69	70 and over
Urban:							
Interstate,* other freeways, and expressways	Level				4	3	3
	Rolling				5	4	3†
	Hilly				6	6	5
Arterial street‡	Level		8	7	6	5	
	Rolling		9	8	7	6	
	Hilly		11	10	9	8	
Collector streets‡	Level	9	9	9	7	6	
	Rolling	12	11	10	8	7	
	Hilly	14	12	12	10	9	
Local streets‡	Level	10	9	9	8	7	
	Rolling	13	12	11	10	8	
	Hilly	15	15	14	12	10	
Rural:							
Interstate,* other freeways, and expressways	Level				4	3	3
	Rolling				5	4	3†
	Hilly				6	6	5
Arterials‡	Level				4	3	3
	Rolling				5	4	4
	Hilly				7	6	5
Collectors‡	Level	7	7	7	6	5	
	Rolling	10	9	8	7	6	
	Hilly	12	10	10	9	8	
Local roads‡	Level	8	7	7	6	5	
	Rolling	11	10	9	8	6	
	Hilly	15	14	12	10	8	

*Grades 1% steeper may be used for extreme cases where development in urban areas precludes the use of flatter grades. Grades 1% steeper may also be used for one-way down-grades except in hilly terrain.

†Grades 1% steeper may be used for 70 mph design on non-Interstate freeways and expressways in rolling terrain.

‡Grades 1% steeper may be used for short lengths (less than 500 ft.) and on one-way down-grades. For rural highways with current ADT less than 400, grades may be 2% steeper.

Source: *Location and Design Manual,* Vol. 1, *Roadway Design,* Ohio Department of Transportation, with permission.

design speeds, whereas steeper grade limits are permitted for local roads and at lower design speeds.

Concerning minimum grades, flat and level grades may be used on uncurbed roadways without objection, as long as the pavement is adequately crowned to drain the surface laterally. The preferred minimum grade for curbed pavements is 0.5 percent, but a grade of 0.3 percent may be used where there is a high-type pavement accurately crowned and supported on firm subgrade.

Critical Length of Grade.   Freedom and safety of movement on 2-lane highways are adversely affected by heavily loaded vehicles operating on upgrades of sufficient lengths to result in speeds that could impede following vehicles. The term *critical length of grade* is defined as the length of a particular upgrade which reduces the operating speed of a truck with a weight-to-horsepower ratio of 300 to 10 mi/h below the operating speed of the remaining traffic. Assuming an entering speed of 55 mi/h, the critical lengths corresponding to various upgrades are as listed in Table 2.15.

If after an investigation of the project grade line, it is found that the critical length of grade must be exceeded, an analysis of the effect of the long grades on the level of service of the roadway should be made. Where speeds resulting from trucks climbing up long grades are calculated to fall within the range of service level *D* or lower, consideration should be given to constructing added uphill lanes on critical lengths of grade. Refer to the "Highway Capacity Manual" (Ref. 10) for methodology in determining level of service. Where the length of added lanes needed to preserve the recommended level of service on sections with long grades exceeds 10 percent of the total distance between major termini, consideration should be given to the ultimate construction of a divided multilane facility.

Vertical Curves.   A vertical curve is used to provide a smooth transition between vertical tangents of different grades. It is a parabolic curve and is usually centered on the intersection point of the vertical tangents. One of the principles of parabolic curves is that the rate of change of slope is a constant throughout the curve. For a vertical curve, this rate is equal to the length of the curve divided by the algebraic difference of the grades. This value is called the *K* value and represents the distance required for the vertical tangent to change by 1 percent. The *K* value is useful in design to determine the minimum length of vertical curve necessary to provide minimum stopping sight distance given two vertical grades.

Allowable Grade Breaks.   There are situations where it is not necessary to provide a vertical curve at the intersection of two vertical grades because the difference in grades is not large enough to provide any discomfort to the driver. The difference varies with the design speed of the roadway. At 25 mi/h, a grade break of 1.85 percent without a curve may be permitted, while at 55 mi/h the allowable difference is only 0.40 percent. Table 2.16 lists the maximum grade break permitted without using a vertical curve for various design speeds. The equation used to develop the distances is indicated as well as a recommended minimum distance between consecutive grade breaks. Where consecutive grade breaks occur within 100 ft for design speeds over 40 mi/h, or within 50 ft for design speeds at 40 mi/h and under, this indicates that a vertical curve may be a better solution than not providing one.

Crest Vertical Curves.   The major design consideration for crest vertical curves is the provision of ample stopping sight distance for the design speed. Calculations of available stopping sight distance are based on the driver's eye 3.5 ft above the road-

TABLE 2.15 Critical Length of Grade Based on Entering Speed of 55 mi/h

Grade, %	Critical length, ft
2.0	2500
2.5	1800
3.0	1400
3.5	1150
4.0	1000
4.5	875
5.0	775
5.5	700
6.0	625
7.0	525

TABLE 2.16 Maximum Change in Vertical Alignment Not Requiring a Vertical Curve

Design speed, mi/h	Maximum grade change, %*
25	1.85
30	1.30
35	0.95
40	0.75
45	0.55
50	0.45
55	0.40
60	0.30
65	0.30
70	0.25

Based on the following equation:

$$A = \frac{46.5L}{V^2} = \frac{1162.5}{V^2}$$

where A = maximum grade change, %
L = length of vertical curve, ft; assume 25
V = design speed, mi/h

Note: The recommended *minimum* distance between consecutive deflections is 100 ft where design speed > 40 mi/h and 50 ft where design speed ≤ 40 mi/h.

*Rounded to nearest 0.05%.

Source: *Location and Design Manual,* Vol. 1, *Roadway Design,* Ohio Department of Transportation, with permission.

way surface with the ability to see an object 6 in high in the roadway ahead over the top of the pavement. Table 2.17 lists the minimum and preferred stopping sight distance values and the corresponding K values for design speeds from 20 to 70 mi/h in 1-mi/h increments. The values shown are based on the assumption that the curve is longer than the sight distance. In those cases where the sight distance exceeds the vertical curve length, a different equation is used to calculate the stopping sight distance provided. The equations are shown in the table.

Another consideration in designing crest vertical curves is passing sight distance, especially when dealing with two-lane roadways. This has already been discussed under "Passing Sight Distance" earlier in this chapter. Also, in addition to being designed for safe stopping sight distance, crest vertical curves should be designed for comfortable operation and a pleasing appearance whenever possible. To accomplish this, the length of a crest curve in feet should be, as a minimum, 3 times the design speed in miles per hour.

Sag Vertical Curves.   The main factor affecting the design of a sag vertical curve is headlight sight distance. When a vehicle traverses an unlighted sag vertical curve at night, the portion of highway lighted ahead is dependent on the position of the headlights and the direction of the light beam. For design purposes, the length of roadway lighted ahead is assumed to be the available stopping sight distance for the curve. In calculating the distances for a given set of grades and a length of curve, the height of the headlight is assumed to be 2 ft and the upward divergence of the light beam is considered to be 1°. Table 2.18 lists the minimum and preferred stopping sight distance values and the corresponding K values for design speeds from 20 to 70 mi/h in 1-mi/h increments. As was the case with crest curves, the values shown are based on the assumption that the curve is longer than the sight distance. In those cases where the sight distance exceeds the vertical curve length, a different equation is used to calculate the actual stopping sight distance provided as indicated in the table.

Note for sag curves, when the algebraic difference of grades is 1.75 percent or less, stopping sight distance is not restricted by the curve. In these cases, the equations in Table 2.18 will not provide meaningful answers. Minimum lengths of sag vertical curves are necessary to provide a pleasing general appearance of the highway. To accomplish this, the minimum length of a sag curve in feet should be equal to 3 times the design speed in miles per hour.

Vertical Alignment Considerations.   The following items should be considered when establishing new vertical alignment:

- The profile should be smooth with gradual changes consistent with the type of facility and the character of the surrounding terrain.
- A "roller-coaster" or "hidden dip" profile should be avoided.
- Undulating grade lines involving substantial lengths of steeper grades should be appraised for their effect on traffic operation, since they may encourage excessive truck speeds.
- Broken-back grade lines (two vertical curves—a pair of either crest curves or sag curves—separated by a short tangent grade) should generally be avoided.
- Special attention should be given to drainage on curbed roadways where vertical curves have a K value of 167 or greater, since these areas are very flat.
- It is preferable to avoid long, sustained grades by breaking them into shorter intervals with steeper grades at the bottom.

TABLE 2.17 Minimum and Preferred Stopping Sight Distance (SSD) for Crest Vertical Curves at Design Speeds from 20 to 70 mi/h

Height of eye, 3.50 ft; height of object, 0.50 ft

Design speed, mi/h	SSD, ft Min	SSD, ft Pref	Rate K, ft/% Min	Rate K, ft/% Pref	Design speed, mi/h	SSD, ft Min	SSD, ft Pref	Rate K, ft/% Min	Rate K, ft/% Pref
20	125	125	12	12	46	340	415	87	130
21	130	130	13	13	47	355	430	95	139
22	135	135	14	14	48	370	445	103	149
23	140	140	15	15	49	385	460	112	159
24	145	145	16	16	50	400	475	120	170
25	150	150	17	17	51	410	490	126	181
26	160	160	19	19	52	420	505	133	192
27	170	170	22	22	53	430	520	139	203
28	180	180	24	24	54	440	535	146	215
29	190	190	27	27	55	450	550	152	228
30	200	200	30	30	56	465	570	163	244
31	205	210	32	33	57	480	590	173	262
32	210	220	33	36	58	495	610	184	280
33	215	230	35	40	59	510	630	196	299
34	220	240	36	43	60	525	650	207	318
35	225	250	38	47	61	530	665	211	333
36	235	265	42	53	62	535	680	215	348
37	245	280	45	59	63	540	695	219	363
38	255	295	49	65	64	545	710	223	379
39	265	310	53	72	65	550	725	228	396
40	275	325	57	79	66	565	750	240	423
41	285	340	61	87	67	580	775	253	452
42	295	355	65	95	68	595	800	266	482
43	305	370	70	103	69	610	825	280	512
44	315	385	75	112	70	625	850	294	544
45	325	400	79	120					

Using S = stopping sight distance, ft
    L = length of crest vertical curve, ft
    A = algebraic difference in grades, %
    K = rate of vertical curvature, ft per % change
- For a given design speed and A value, the calculated length $L = KA$.
- To determine S with a given L and A, use the following:
  For $S < L$: $S = 36.45\sqrt{K}$ where $K = L/A$
  For $S > L$: $S = 664.58/A + L/2$

Source: *Location and Design Manual*, Vol. 1, *Roadway Design*, Ohio Department of Transportation, with permission.

TABLE 2.18 Minimum and Preferred Stopping Sight Distance for Sag Vertical Curves at Design Speeds from 20 to 70 mi/h

Height of headlight = 2.00 ft

Upward light beam divergence = 1°00′

Design speed, mi/h	SSD, ft		Rate K, ft/%		Design speed, mi/h	SSD, ft		Rate K, ft/%	
	Min	Pref	Min	Pref		Min	Pref	Min	Pref
20	125	125	19	19	46	340	415	71	93
21	130	130	20	20	47	355	430	77	97
22	135	135	21	21	48	370	445	81	101
23	140	140	22	22	49	385	460	85	105
24	145	145	23	23	50	400	475	89	109
25	150	150	24	24	51	410	490	92	114
26	160	160	27	27	52	420	505	94	118
27	170	170	29	29	53	430	520	97	122
28	180	180	31	31	54	440	535	100	126
29	190	190	34	34	55	450	550	103	130
30	200	200	36	36	56	465	570	107	136
31	205	210	38	39	57	480	590	111	141
32	210	220	39	41	58	495	610	115	147
33	215	230	40	44	59	510	630	119	152
34	220	240	41	46	60	525	650	123	158
35	225	250	43	49	61	530	665	125	162
36	235	265	45	53	62	535	680	126	166
37	245	280	48	57	63	540	695	127	171
38	255	295	50	61	64	545	710	129	175
39	265	310	53	65	65	550	725	130	179
40	275	325	56	69	66	565	750	134	186
41	285	340	58	71	67	580	775	138	193
42	295	355	61	77	68	595	800	143	200
43	305	370	63	81	69	610	825	147	207
44	315	385	66	85	70	625	850	151	214
45	325	400	69	89					

Using S = stopping sight distance, ft
    L = length of sag vertical curve, ft
    A = algebraic difference in grades, %
    K = rate of vertical curvature, ft per % change
- For a given design speed and A value, the calculated length $L = KA$
- To determine S with a given L and A, use the following:

For $S < L$: $S = \dfrac{3.5L + \sqrt{12.25L^2 + 1600AL}}{2A}$

For $S > L$: $S = (AL + 400)/(2A - 3.5)$

Source: *Location and Design Manual,* Vol. 1, *Roadway Design,* Ohio Department of Transportation, with permission.

2.2.5 Coordination of Horizontal and Vertical Alignments

When designing new roadway projects, the following items should be considered to coordinate the horizontal and vertical alignments:

- Curvature and tangent sections should be properly balanced. Normally, horizontal curves will be longer than vertical curves.
- It is generally more pleasing to the driver when vertical curvature can be superimposed on horizontal curvature. In other words, the PIs (points of intersection) of both the vertical and horizontal curves should be near the same station or location.
- Sharp horizontal curves should not be introduced at or near the top of a pronounced crest vertical curve or at or near the low point of a pronounced sag vertical curve.
- On two-lane roadways, long tangent sections (horizontal and vertical) are desirable to provide adequate passing sections.
- Horizontal and vertical curves should be as flat as possible at intersections.
- On divided highways, the use of variable median widths and separate horizontal and vertical alignments should be considered.
- In urban areas, horizontal and vertical alignments should be designed to minimize nuisance factors. These might include directional adjustment to increase buffer zones and depressed roadways to decrease noise.
- Horizontal and vertical alignments may often be adjusted to enhance views of scenic areas.

2.3 CROSS-SECTION DESIGN

This article provides information to assist the designer in determining lane widths, pavement cross slopes, shoulder widths, interchange cross-section elements, medians, curbs, pedestrian facilities, and grading and side slopes. The number of lanes for a given roadway facility is best determined using principles and procedures contained in the "Highway Capacity Manual" (Ref. 10). This manual analyzes roadways to determine an appropriate "level of service," by which a letter value (A through F) is assigned depending on the volume of traffic and other geometric features. Table 2.19 provides a design guide for level of service for various facilities by functional classification and terrain or locale. The table includes a brief description of the characteristics of each level of service.

2.3.1 Roadway Criteria

Lane Widths and Transitions. When considering the physical characteristics of cross sections, the values selected will depend on location (rural or urban), speed, traffic volumes, functional classification, and, in urban areas, the type of adjacent development. Tables 2.20, 2.21, and 2.22 provide values currently used in Ohio. Lane width is dependent on design speed, especially in rural areas. Widths may be as narrow as 9 ft for a local, low-volume road. In urban areas, lane widths can be as narrow as 10 ft, if the road is primarily a residential street. The maximum lane width is generally accepted to be 12 ft in all locales.

TABLE 2.19 Guide for Selecting Design Service Level As Determined by Function and Terrain or Locale

| | Minimum level of service for area and terrain or locale | | | |
| | Rural | | | Urban and |
Functional classification	Level	Rolling	Hilly	suburban
Interstate, other freeways, and expressways	B	B	B	C
Arterial	B	B	C	C
Collector	C	C	D	D
Local	D	D	D	D

A: Free flow, with low volumes and high speeds.

B: Stable flow, speeds beginning to be restricted by traffic conditions.

C: In stable flow zone, but most drivers are restricted in freedom to select own speed.

D: Approaching unstable flow; drivers have little freedom to maneuver.

E: Unstable flow; short stoppages may occur.

F: Forced or breakdown flow.

Source: *Location and Design Manual,* Vol. 1, *Roadway Design,* Ohio Department of Transportation, with permission.

In some cases it may be necessary to widen the pavement on sharp curves to accommodate off-tracking of larger vehicles. Table 2.23 provides a chart of recommended pavement widening based on degree of curvature and design speed. The widened portion of the pavement is normally placed on the inside of the curve. Where curves are introduced with spiral transitions, the widening occurs over the length of the spiral. On alignments without spirals, the widening is developed over the same distance that the superelevation transition occurs. The centerline pavement marking and the center joint (if applicable) should be placed equidistant from the pavement edges.

Whenever the driver's lane is being shifted—for example, when lanes are being added or eliminated—the shifting rate should be controlled using the following equations:

$$L = WS \quad \text{for design speeds over 40 mi/h} \tag{2.5}$$

$$L = W\frac{S^2}{60} \quad \text{for design speeds up to 40 mi/h} \tag{2.6}$$

where L = approach taper length, ft
W = offset width, ft
S = design speed, mi/h

Where lanes are being added but the driver is not being "forced" to follow the actual transition (such as in adding right turn lanes), the transition can occur in 50 ft on most roadways or 100 ft on freeway designs.

TABLE 2.20   Guide for Selecting Lane Width for Rural Areas

Functional classification	Current ADT	DHV	Minimum lane widths, ft,* for design speed in mi/h					
			20 to 29	30 to 39	40 to 49	50 to 59	60 to 69	70 and over
Interstate, other freeways, and expressways	All	All	—	—	—	12	12	12
Arterials	—	>200	—	—	—	12	12	12
	≥400	<201	—	—	—	12	12	—
	<400	—	—	—	—	12†	12	—
Collectors	—	>400	—	—	12	12	12	—
	—	201–400	—	—	11	12	12	—
	≥400	<201	—	10	11	11	11	—
	<400	—	10	10	10	10	11	—
Local roads	—	>400	—	12	12	12	12	—
	—	201–400	—	11	11	12	12	—
	>400	<201	—	10	11	11	11	—
	250–400	—	10	10	10	10	11	—
	<250	—	9	9	10	10	10	—

There may be locations having rural functional classifications that are urban in character. An example would be a village where adjacent development and other conditions resemble an urban area. In such cases, urban design criteria may be used.

*The number of lanes should be determined by capacity analysis.

†May be 11 ft on nonfederal projects if design year ADT includes fewer than 250 B and C truck units.

Source: *Location and Design Manual,* Vol. 1, *Roadway Design,* Ohio Department of Transportation, with permission.

Pavement Cross Slopes.   Roadways on tangent or relatively straight alignments where no superelevation is required are normally crowned (peaked) in the middle. Cross slopes are usually in the range of 0.015 to 0.020 ft/ft. Urban areas with curbed pavements are more likely to have a slope near the upper limit, while rural roadways tend to have a little flatter cross slope. The following guidelines are applicable to the location of the crown point:

- Crowns should be located at or near lane lines.
- For pavements with three or four lanes, no more than two should slope in the same direction.
- Undivided pavement sections should be crowned in the middle when the number of lanes is even, and at the edge of the center lane when the number is odd.
- Narrow raised median sections should be crowned in the middle, so that the majority of the pavement will drain to the outside.

Shoulders.   A shoulder is the area adjacent to the roadway that (1) when properly designed, can provide lateral support to the pavement, (2) is available to the motorist in emergency situations, and (3) can be used to maintain traffic during construction. *Graded shoulder width* is the width of the shoulder measured from the edge of the pavement to the intersection of the shoulder slope and the foreslope. *Treated shoulder*

TABLE 2.21 Guide for Selecting Shoulders for Rural Areas

Functional classification	Traffic		Graded width, ft			Type[a]	Rounding, ft, for design speed in mi/h[b]		Normal barrier offset, ft
	Current ADT	DHV	With barrier or foreslope steeper than 6:1	Without barrier, 6:1 or flatter foreslope	Treated width, ft		>50	<50	
Interstate, other freeways, and expressways	All	All	15 right, 9 median[c]	10 right, 4 median	10 right,[d] 4 median[e]	Paved	10	—	12 right, 6 median[f]
Arterials[g]	—	>400	14	10	10	Paved[h,i]	8	4	12
	—	201–400	12	8	8	Paved[h,i]	8	4	10
	—	100–200	10	8	6	Bituminous surface treatment[h]	8	4	8
	≥400	<100	10	8	6	Bituminous surface treatment[h]	4	4	8
	<400	—	8	8	4	Stabilized aggregate	4	4	6
Collectors[g]	—	>400	12	8	8[j]	Bituminous surface treatment[h]	8	4	10[k]
	—	201–400	10	8	4	Bituminous surface treatment[h]	8	4	8[k]
	—	100–200	8	6[l]	4	Stabilized aggregate	8	4	6[k]
	≥400	<100	6	4[m]	4	Stabilized aggregate	4	4	4
	<400	—	6	4[n]	4	Stabilized aggregate	4	4	4
Local roads	—	>400	12	8[o]	8[j]	Bituminous surface treatment[h]	8	4	10[k]
	—	201–400	10	8[o]	4	Bituminous surface treatment[h]	8	4	8[k]
	—	100–200	8	6[m]	4	Stabilized aggregate	8	4	6[k]
	≥400	<100	6	4[n]	4	Stabilized aggregate	4	4	4
	<400	—	6	4[n]	4	Stabilized aggregate	4	4	4

Treated shoulder width for both right and median sides should be 12 ft if used for maintenance of traffic lanes. When 12 ft treated width is used, the following applies:

12-ft graded shoulder width (6:1 or flatter foreslopes without barrier)

17-ft graded shoulder width (foreslopes steeper than 6:1 or with barrier)

14-ft normal barrier offset

There may be locations having rural functional classifications that are urban in character. An example would be a village where adjacent development and other conditions resemble an urban area. In such cases, urban design criteria may be used.

[a]Turf shoulders may be used on non–state-maintained roads at option of local government if current year ADT includes fewer than 250 B and C trucks. Turf shoulders are not to be used on state-maintained roads.

[b]Rounding should be 4 ft where the foreslope begins beyond the clear zone or where guiderail is installed and foreslope is steeper than 6:1. No rounding is required when foreslope is 6:1 or flatter.

[c]If six or more lanes, use 15 ft.

[d]Consider increasing treated shoulder width to 12 ft on interstate when truck traffic exceeds 250 DDHV. If 12-ft treated width is used, graded width with barrier or foreslope steeper than 6:1 and normal barrier offset would increase an additional 2 ft.

[e]If six or more lanes, use 10 ft. Consider increasing treated shoulder width to 12 ft on interstate when truck traffic exceeds 250 DDHV. If 12-ft treated width is used, graded width with barrier or foreslope steeper than 6:1 and normal barrier offset would increase an additional 2 ft.

[f]Use 12 ft if footnote e applies.

[g]The median shoulder width criteria for interstates, other freeways, and expressways shall apply to the medians of divided arterials and divided collectors.

[h]Use stabilized aggregate on state-maintained roads if design year ADT includes less than 250 B and C truck units. Paved may be used if design year ADT includes over 1000 B and C truck units.

[i]Use bituminous surface treatment if design year ADT includes between 250 and 1000 B and C truck units.

[j]Use 6 ft if design year ADT includes fewer than 501 B and C truck units. If 6-ft treated width is used, graded width may be reduced to 10 ft and normal barrier offset will be 8 ft.

[k]Whenever a design exception is approved for graded shoulder width, the guiderail offset may be reduced but shall not be less than 4 ft.

[l]Turf shoulders may use 6 ft regardless of foreslope.

[m]Turf shoulders may use 4 ft regardless of foreslope.

[n]Turf shoulders may use 2 ft regardless of foreslope.

[o]Graded shoulder is 8 ft regardless of foreslope.

Source: *Location and Design Manual*, Vol. 1, *Roadway Design*. Ohio Department of Transportation, with permission.

TABLE 2.22 Guide for Selecting Lane Width and Shoulders for Urban Areas[a]

Functional classification	Locale	Lane width, ft		Minimum curbed shoulder width,[b] ft	
		Minimum	Preferred	Without parking lane	With parking lane[c]
Interstates, other freeways, and expressways	All	12	12	10 right paved, 4 median paved[d,e]	—
Arterial streets	Free-flowing, over 40 mi/h	12	12	10 each side paved[e,f]	—
	Controlled-flow, 40 mi/h or less	12 12[g]	12	1–2 paved	10–12 paved
Collector streets	Commercial or industrial	11	12	1–2 paved	9–10 paved
	Residential	11	12	1–2 paved	7–10 paved
Local streets	Commercial or industrial	11	12	1–2 paved	9 paved
	Residential	10[h]	11	1–2 paved	7 paved

[a]Use rural criteria (Tables 2.20 and 2.21) for uncurbed shoulders. Rural functional classification should be determined after checking the urban route extension into a rural area.

[b]The median shoulder width for divided arterials should follow the median criteria for interstates, other freeways, and expressways.

[c]Use minimum lane width if, in the foreseeable future, the parking lane will be used for through traffic during peak hours or continuously.

[d]Use 10-ft median shoulder on facilities with six or more lanes.

[e]Minimum curbed shoulder width for both right and median sides should be 11 ft unless otherwise justified.

[f]May be reduced to 8 ft if DHV is less than 250.

[g]Lane width may be reduced to 11 ft where right-of-way is limited and current truck ADT is less than 250; however, on all federal aid primary projects, at least one 12-ft lane in each direction is required.

[h]Lane width may be 9 ft where right-of-way is limited and current ADT is less than 250.

Source: *Location and Design Manual,* Vol. 1, *Roadway Design,* Ohio Department of Transportation, with permission.

width is that portion of the graded shoulder that has been improved to at least stabilized aggregate or better. Figure 2.11 illustrates these definitions.

Four basic types of shoulder are used: (1) paved, (2) bituminous surface treated, (3) stabilized aggregate, and (4) turf. Paved shoulders may be rigid (concrete) or flexible (asphalt). Turf shoulders are usually used on low-volume, uncurbed, local roads. Tables 2.21 and 2.22 provide recommended shoulder widths and types based on functional classification and traffic volumes or locale.

Whenever practical, shoulders should be designed to be wide enough and strong enough to accommodate temporary traffic, especially on high-volume roadways. Figures 2.12, 2.13, and 2.14 provide information on recommended cross slopes and allowable grade breaks depending on the type of shoulder chosen.

TABLE 2.23 Recommended Pavement Widening on Horizontal Curves, ft

	Pavement width on tangent, ft											
	24					22				20			
	Design speed, mi/h					Design speed, mi/h				Design speed, mi/h			
Degree of curve	30 to 39	40 to 49	50 to 59	60 to 69	70 and over	30 to 39	40 to 49	50 to 59	60 and over	30 to 39	40 to 49	50 to 59	60 and over
1°00′	0	0	0	0	0	0.5	0.5	0.5	1.0	1.5	1.5	1.5	2.0
2°00′	0	0	0	0.5	0.5	1.0	1.0	1.0	1.5	2.0	2.0	2.0	2.5
3°00′	0	0	0.5	0.5	1.0	1.0	1.0	1.5	1.5	2.0	2.0	2.5	2.5
4°00′	0	0.5	0.5	1.0	1.0	1.0	1.5	1.5	2.0	2.0	2.5	2.5	3.0
5°00′	0.5	0.5	1.0	1.0		1.5	1.5	2.0	2.0	2.5	2.5	3.0	3.0
6°00′	0.5	1.0	1.0	1.5		1.5	2.0	2.0	2.5	2.5	3.0	3.0	3.5
7°00′	0.5	1.0	1.5			1.5	2.0	2.5		2.5	3.0	3.5	
8°00′	1.0	1.0	1.5			2.0	2.0	2.5		3.0	3.0	3.5	
9°00′	1.0	1.5	2.0			2.0	2.5	3.0		3.0	3.5	4.0	
10°00′	1.0	1.5				2.0	2.5			3.0	3.5		
11°00′	1.0	1.5				2.0	2.5			3.0	3.5		
12°00′	1.5	2.0				2.5	3.0			3.5	4.0		
13°00′	1.5	2.0				2.5	3.0			3.5	4.0		
14°00′	1.5	2.0				2.5	3.0			3.5	4.0		
14°30′	1.5	2.0				2.5	3.0			3.5	4.0		
15°00′	2.0					3.0				4.0			
18°00′	2.0					3.0				4.0			
19°00′	2.5					3.5				4.5			
21°00′	2.5					3.5				4.5			
22°00′	3.0					4.0				5.0			
25°00′	3.0					4.0				5.0			
26°00′	3.5					4.5				5.5			
26°30′	3.5					4.5				5.5			

Note: Values less than 2.0 ft may be disregarded. Multiply table values by 1.5 for three lanes and by 2.0 for four lanes.

Source: *Location and Design Manual,* Vol. 1, *Roadway Design,* Ohio Department of Transportation, with permission.

2.3.2 Grading and Side Slopes

This section is concerned with the design of the slopes, ditches, parallel channels, and interchange grading. It incorporates into the roadside design the concepts of vehicular safety developed through dynamic testing. Designers are urged to consider flat foreslopes and backslopes, wide gentle ditch sections, and elimination of barriers.

Slopes. Several combinations of slopes and ditch sections may be used in the grading of a project. Details and use of these combinations are discussed in subsequent paragraphs. In general, slopes should be made as flat as possible to minimize the necessity for barrier protection and to maximize the opportunity for a driver to recover control of a vehicle after leaving the traveled way. Regardless of the type of grading used, projects should be examined in an effort to obtain flat slopes at low costs. For

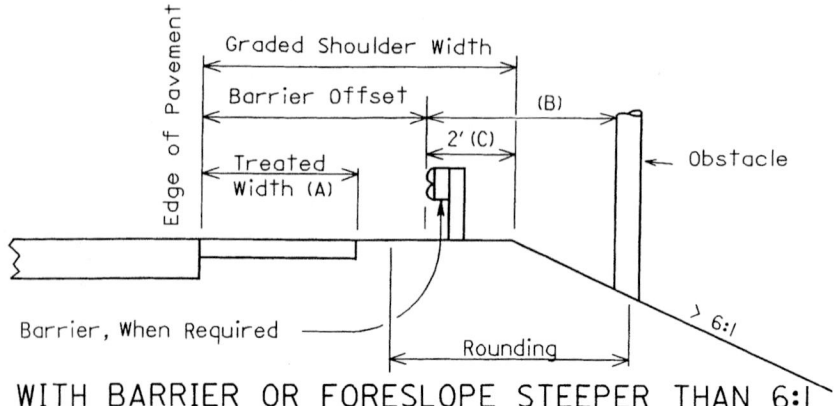

WITH BARRIER OR FORESLOPE STEEPER THAN 6:1

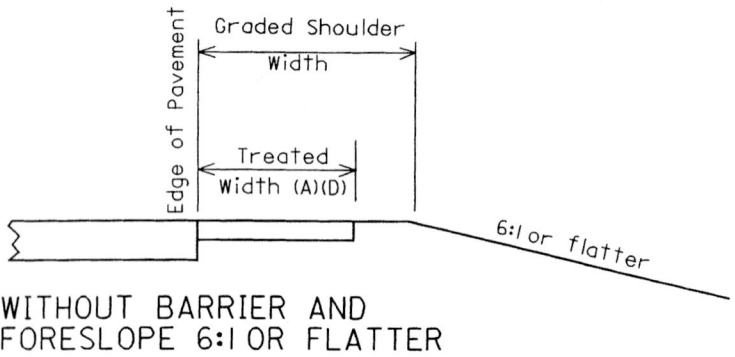

WITHOUT BARRIER AND
FORESLOPE 6:1 OR FLATTER

(A) Treated width includes that portion of the shoulder improved
    with stabilized aggregate or better.

(B) Minimum barrier clearance.

(C) 3' on interstate, other freeways and expressways.

(D) Treated shoulder width may equal graded shoulder width in some cases.

FIGURE 2.11 Cross sections of shoulders showing graded and treated shoulder widths. (*From* Location and Design Manual, *Vol. 1,* Roadway Design, *Ohio Department of Transportation, with permission*)

example, fill slopes can be flattened with material that might otherwise be wasted, and backslopes can be flattened to reduce borrow.

To better understand the various types of grading, it is necessary to become familiar with the concept of a clear zone. *Clear zone* is defined as the border area along a highway, outside the pavement edges, available for use by an errant vehicle (Ref. 2). Within this area, most motorists will be able to recover control of their vehicle in a safe manner. In the following paragraphs, four types of roadside grading are described. The designer must select the appropriate one for the roadway being designed.

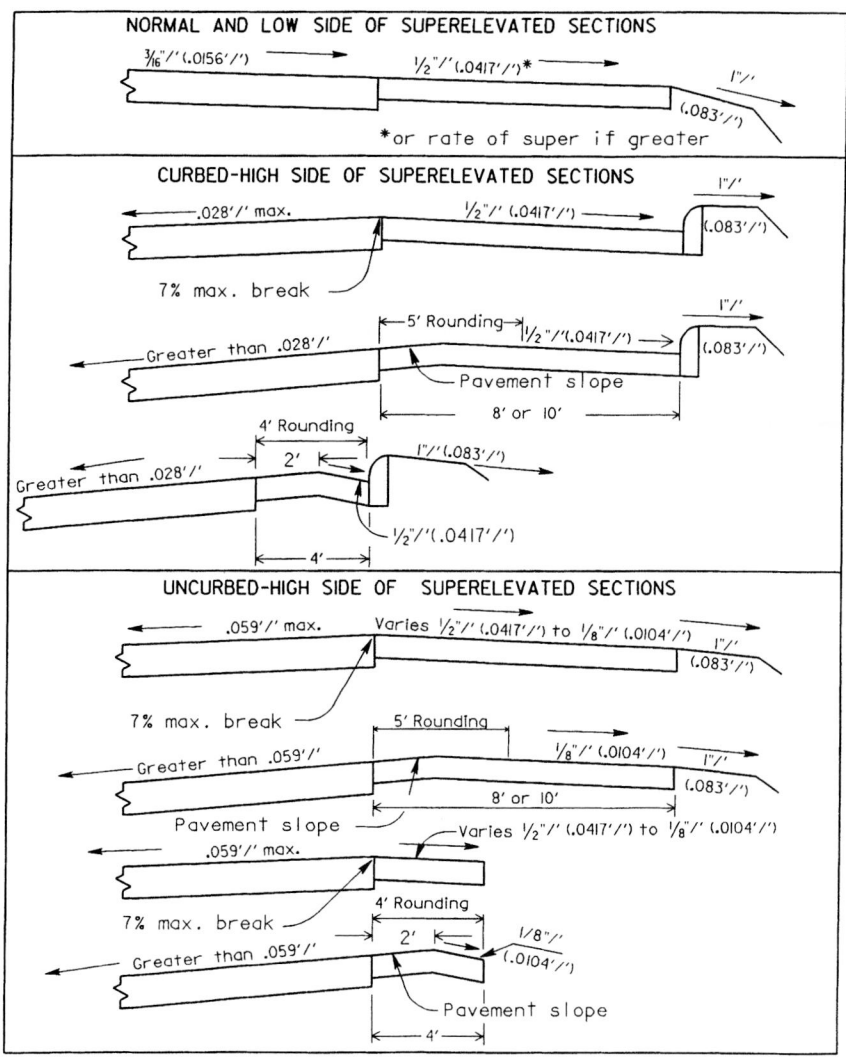

FIGURE 2.12 Recommended cross slopes and grade breaks for paved shoulders. (*From* Location and Design Manual, *Vol. 1,* Roadway Design, *Ohio Department of Transportation, with permission*)

Safety grading is the shaping of the roadside using 6:1 or flatter slopes within the clear zone area, and 3:1 or flatter foreslopes and recoverable ditches beyond the clear zone. Safety grading is used on interstate highways, other freeways, and expressways. Figures 2.15 and 2.16 show many of these details.

Clear zone grading is the shaping of the roadside using 4:1 or flatter foreslopes and traversable ditches within the clear zone area. Foreslopes of 3:1 may be used but are not measured as part of the clear zone distance. Clear zone grading is recommended for undivided rural facilities where the design speed exceeds 50 mi/h, the design hourly volume is 100 or greater, and at least one of the following conditions exists:

NORMAL AND LOW SIDE OF SUPERELEVATED SECTIONS

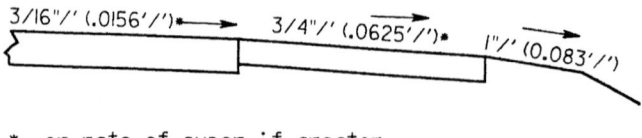

* or rate of super if greater

HIGH SIDE OF SUPERELEVATED SECTIONS

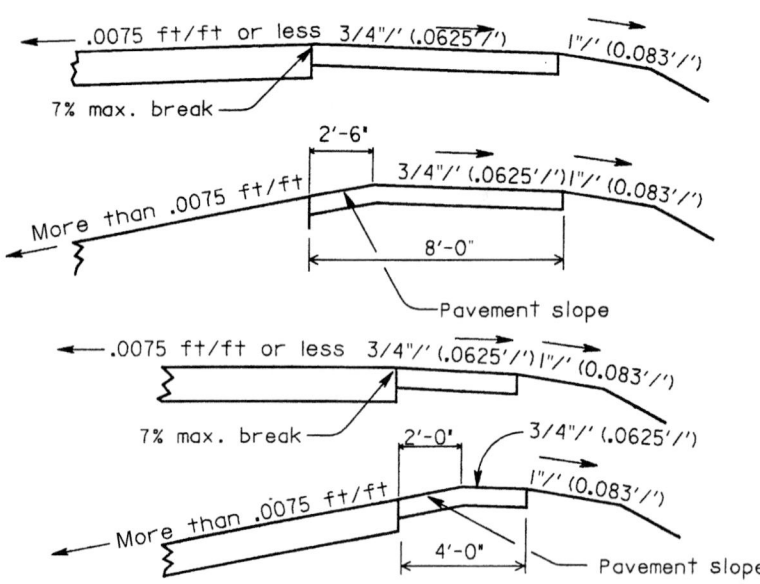

FIGURE 2.13 Recommended cross slopes and grade breaks for bituminous surface treated or stabilized aggregate shoulders. (*From* Location and Design Manual, *Vol. 1,* Roadway Design, *Ohio Department of Transportation, with permission*)

- The wider cross section is consistent with present or future planning for the facility.
- The project is new construction or major reconstruction involving significant length.
- The wider cross section can be provided at little or no additional cost.

Figure 2.17 shows examples of clear zone grading and traversable ditches.

Standard grading is the shaping of the roadside using 3:1 or flatter foreslopes and normal ditches. Standard grading is used on undivided facilities where the conditions for the use of safety grading or clear zone grading do not exist. The designer should ensure that any obstacles within the clear zone receive proper protection. Figure 2.18 shows examples of standard grading and normal ditches.

NORMAL AND LOW SIDE (INNER SIDE) SUPERELEVATED SECTIONS

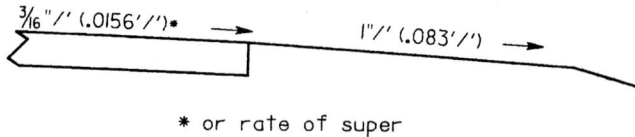

* or rate of super

RISING SIDE (OUTER SIDE) OF SUPERELEVATED SECTIONS IN TRANSITION

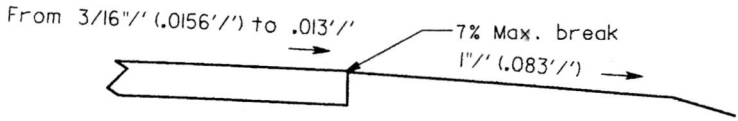

HIGH SIDE OF SUPERELEVATED SECTIONS

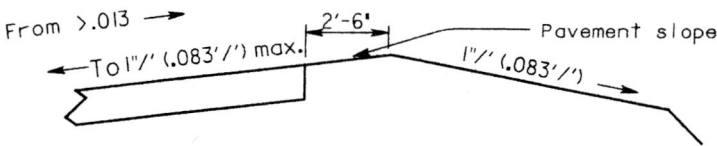

The break at the edge of the pavement shall not exceed 7%.

FIGURE 2.14 Recommended cross slopes and grade brakes for turf shoulders. (*From* Location and Design Manual, *Vol. 1,* Roadway Design, *Ohio Department of Transportation, with permission*)

Barrier grading is the shaping of the roadside when a barrier is required for slope protection. Normally, 2:1 foreslopes and normal ditch sections are used. Figure 2.18 includes an example of barrier grading.

Rounding of Slopes. Slopes should be rounded at the break points and at the intersection with the existing ground line to reduce the chance of a vehicle's becoming airborne and to harmonize with the existing topography. Rounding at various locations is illustrated in Figs. 2.15, 2.17, and 2.18.

Special Median Grading. Figure 2.19*c* shows some examples of median grading when separate roadway profiles are used.

Rock and Shale Slopes. In rock or shale cuts, the maximum rate of slope should be determined by a soils engineer. In deep rock or shale cuts where slopes are steeper than

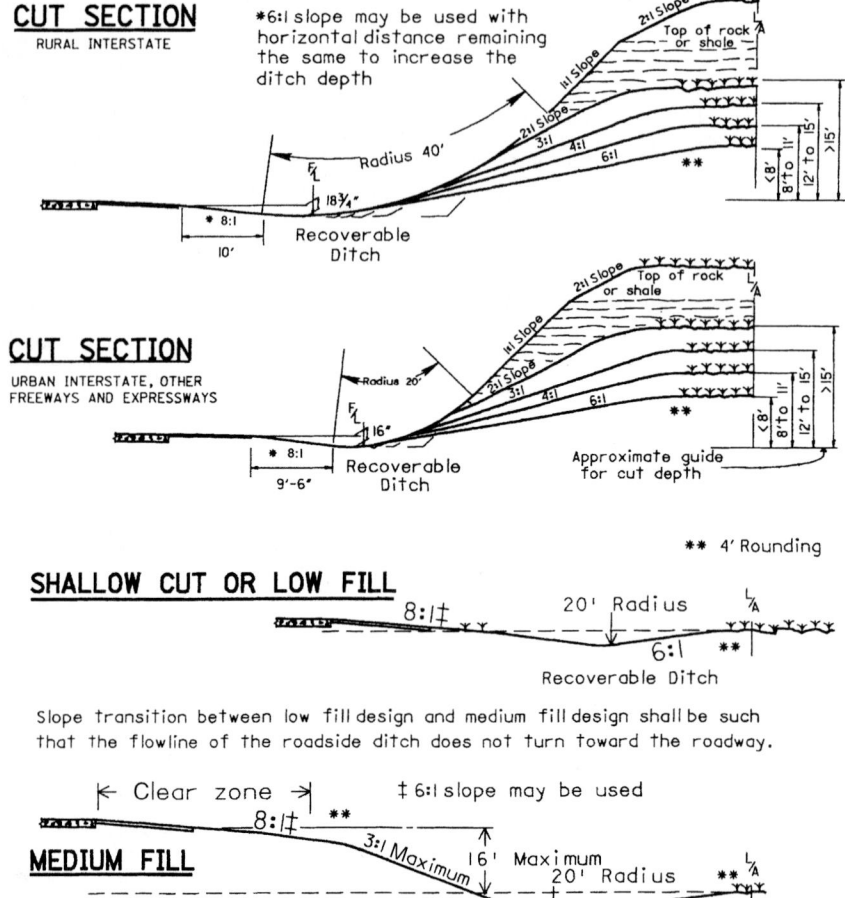

FIGURE 2.15 Cross sections showing safety grading for four different conditions. (*From* Location and Design Manual, *Vol. 1,* Roadway Design, *Ohio Department of Transportation, with permission*)

1:1, a 10-ft-wide bench should be provided between the top of the ditch backslope and the toe of the rock face as illustrated in Fig. 2.19a. In shale cuts, the designer should not use backslopes steeper than 2:1 unless excessive waste would result. In any event, 2:1 slopes should be used for all shale cut sections less than 20 ft in depth, and the bench should be omitted. In this discussion, depth of cut is measured from the top of shale or rock to the ditch flow line. Backslopes steeper than 2:1 should not be used in rock cuts until the depth exceeds 16 ft. In such cases the bench may be omitted.

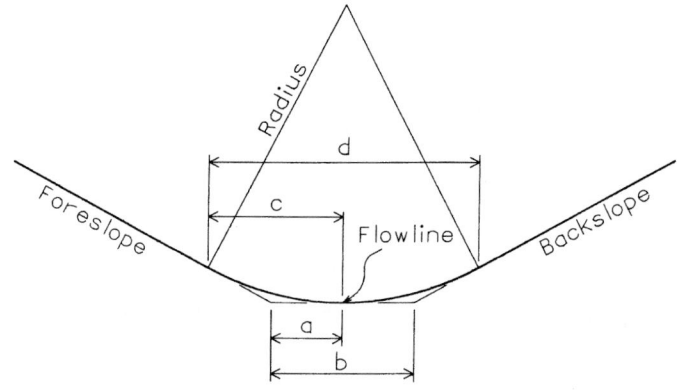

Fore-slope	40' RADIUS		Backslope									
			6:1		4:1		3:1		2:1		1:1	
	a	c	b	d	b	d	b	d	b	d	b	d
8:1	2'6"	5'0"	6'0"	11'6"	7'6"	14'8"	9'0"	17'7"	12'0"	22'10"	19'0"	33'3"
6:1	3'4"	6'7"	6'8"	13'2"	8'4"	16'3"	9'10"	19'3"	12'9"	24'6"	20'0"	34'10"

Fore-slope	20' RADIUS		Backslope									
			6:1		4:1		3:1		2:1		1:1	
	a	c	b	d	b	d	b	d	b	d	b	d
8:1	1'3"	2'6"	3'0"	5'9"	3'9"	7'4"	4'6"	8'10"	6'0"	11'5"	9'6"	16'8"
6:1	1'8"	3'3"	3'4"	6'7"	4'2"	8'2"	4'11"	9'7"	6'5"	12'3"	10'0"	17'5"
4:1	2'6"	4'10"	4'2"	8'2"	5'0"	9'8"	5'9"	11'2"	7'3"	13'9"	10'9"	19'6"
3:1	3'3"	6'4"	4'11"	9'7"	5'9"	11'2"	6'6"	12'8"	8'0"	15'4"	11'6"	20'6"

FIGURE 2.16 Details of ditch rounding for safety grading. (*From* Location and Design Manual, *Vol. 1,* Roadway Design, *Ohio Department of Transportation, with permission*)

Curbed Streets. Figure 2.20 shows typical slope treatments next to curbed streets.

Driveways and Crossroads. At driveways or crossroads, where the roadside ditch is within the clear zone distance and where clear zone grading can be obtained, the ditch and pipe should be located as shown on Fig. 2.21.

Ditches. When the depth or velocity of the design discharge accumulating in a roadside or median ditch exceeds the desirable maximum established for the various highway classifications, a storm sewer will be required to intercept the flow and carry it to a satisfactory outlet. If right-of-way and earthwork considerations are favorable, a deep, parallel side ditch (see Fig. 2.19*b*) may be more practical and should be considered instead of a storm sewer. In some cases where large areas contribute flow to a highly erodible soil cut, an intercepting ditch may be considered near the top of the cut to intercept the flow from the outside and thereby relieve the roadside ditch. Constant-depth ditches (usually 18 in deep) are desirable. Where used, the minimum pavement profile grades should be 0.24 to 0.48 percent. Where flatter pavement

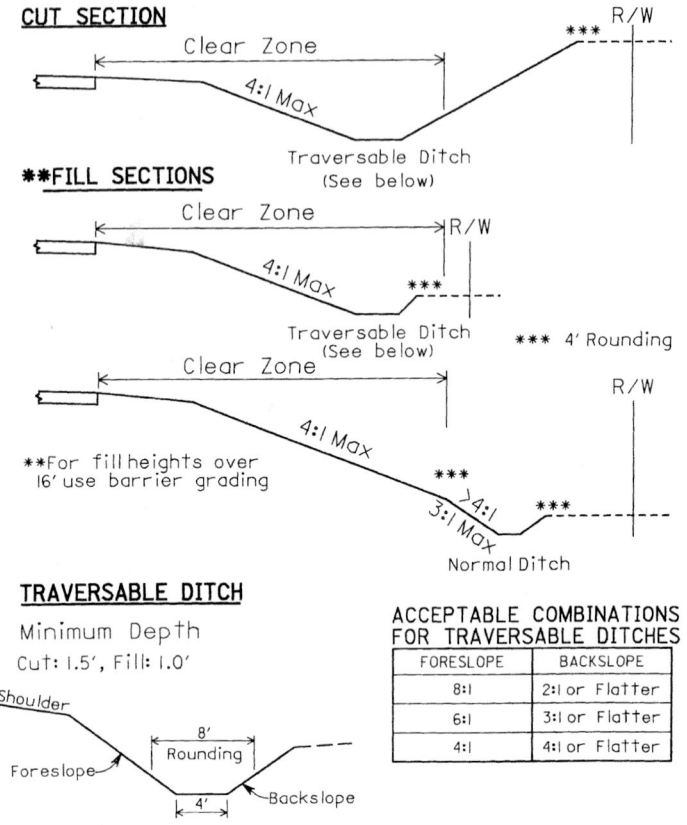

FIGURE 2.17 Examples of clear zone grading and traversable ditches. (*From Location and Design Manual, Vol. 1,* Roadway Design, *Ohio Department of Transportation, with permission*)

grades are necessary, separate ditch profiles are developed and the ditch flow line elevations shown on each cross section.

Parallel Channels.   Where it is determined that a stream intercepted by the roadway improvement is to be relocated parallel to the roadway, the channel should be located beyond the limited access line (or highway easement line) in a separate channel easement. This arrangement locates the channel beyond the right-of-way fence, if one is to be installed. Figure 2.19*b* shows a parallel channel section. This does not apply to conventional intercepting erosion control ditches located at the top of cut slopes in rolling terrain.

In areas of low fill and shallow cut, protection along a channel by a wide bench is usually provided. Fill slope should not exceed 6:1 when this design is used, and maximum height from shoulder edge to bench should generally not exceed 10 ft. If it should become necessary to use slopes steeper than 6:1, guiderail may be necessary and fill slopes as steep as 2:1 may be used. In cut sections 5 ft or more in depth, earth barrier protection can be provided. This design probably affords greater protection where very deep channels are constructed and requires less excavation. Where the sec-

STANDARD GRADING

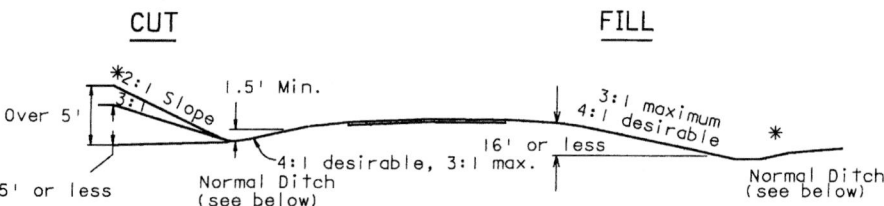

BARRIER GRADING

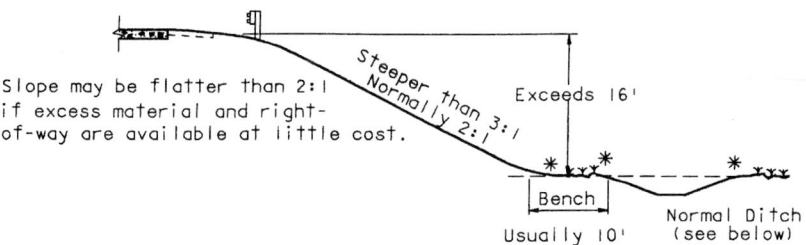

NORMAL DITCH SECTIONS

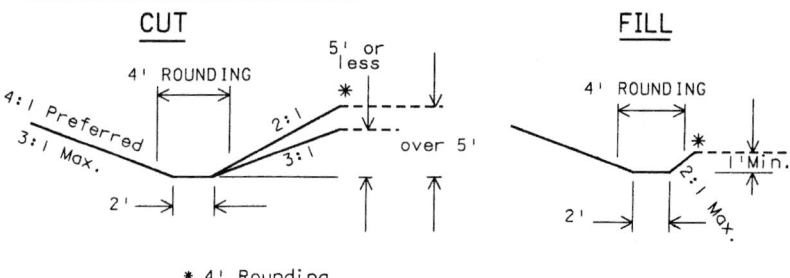

* 4' Rounding

FIGURE 2.18   Examples of standard grading and normal ditches. (*From* Location and Design Manual, *Vol. 1,* Roadway Design, *Ohio Department of Transportation, with permission*)

tions alternate between cut and fill and it is desired to use but a single design, earth barrier protection is less costly if waste excavation material is available. Likewise, bench protection is less costly if borrow is needed on the project as a whole.

Earth bench or earth barrier protection provided adjacent to parallel channels should not be breached for any reason other than to provide an opening for a natural or relocated stream that requires a drainage structure larger in rise than 42 in. Outlet pipes from median drains or side ditches should discharge directly into the parallel channel.

Channels and toe-of-slope ditches, used in connection with steep fill slopes, are both removed from the normal roadside section by benches. The designer should establish control offsets to the center of each channel or ditch at appropriate points that govern alignment so the flow will follow the best and most direct course to the outlet. Bench width should be varied as necessary.

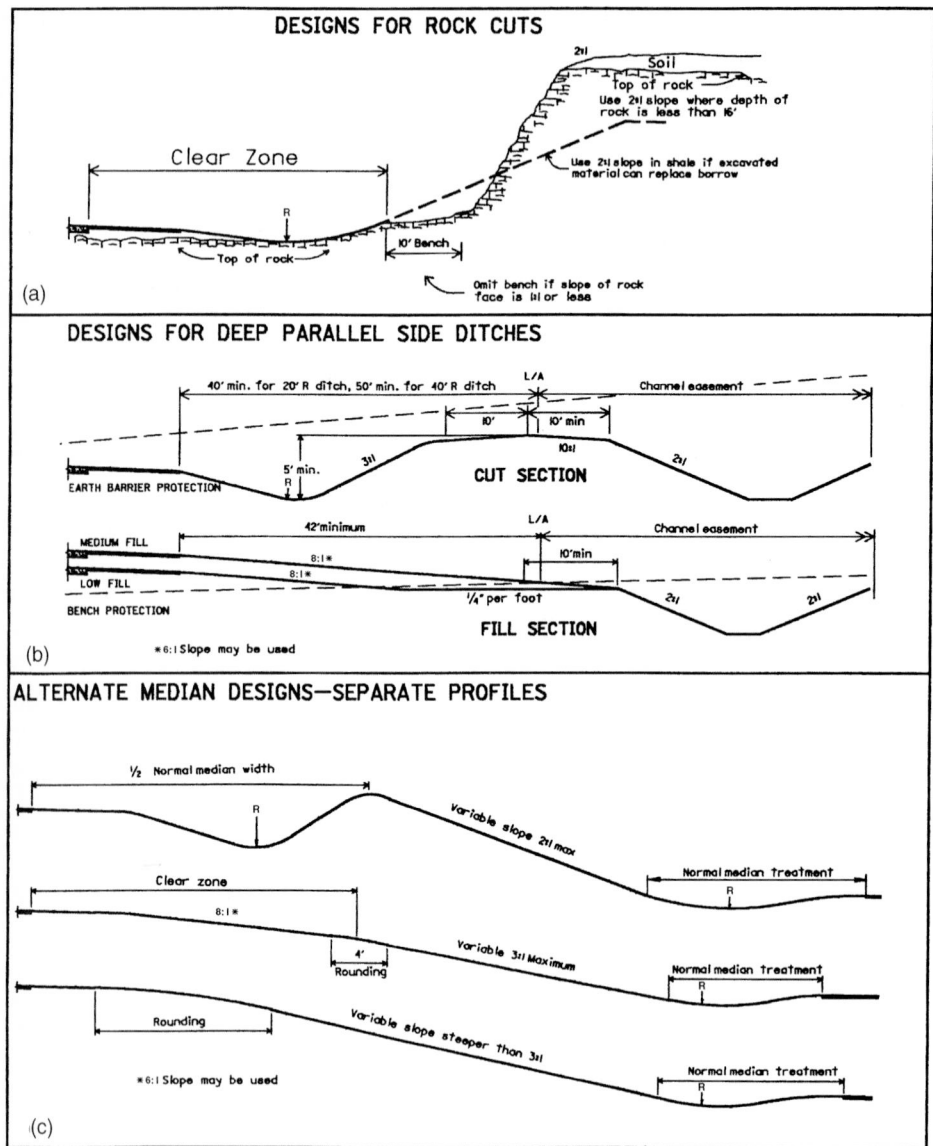

FIGURE 2.19 Examples of special designs for grading. (*a*) Designs for rock cuts. (*b*) Designs for deep, parallel side ditches. (*c*) Alternate median designs. (*From* Location and Design Manual, *Vol. 1,* Roadway Design, *Ohio Department of Transportation, with permission*)

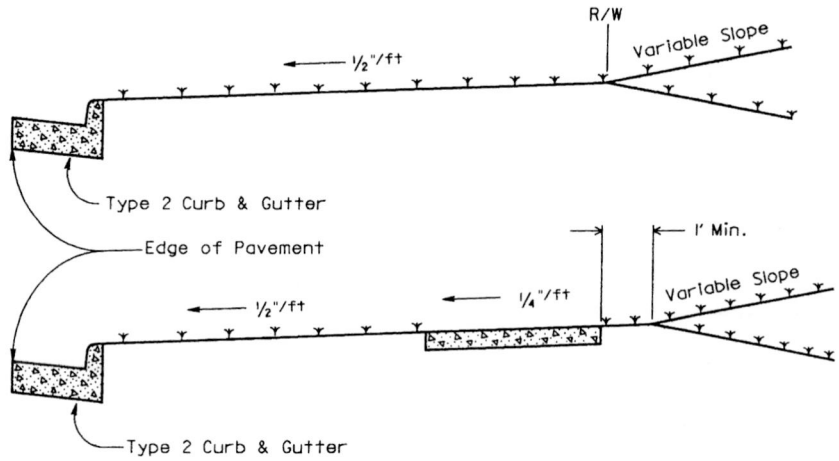

FIGURE 2.20 Examples of slope treatment adjacent to curbed streets. (*From* Location and Design Manual, *Vol. 1,* Roadway Design, *Ohio Department of Transportation, with permission*)

Design speed 50 mi/h or more

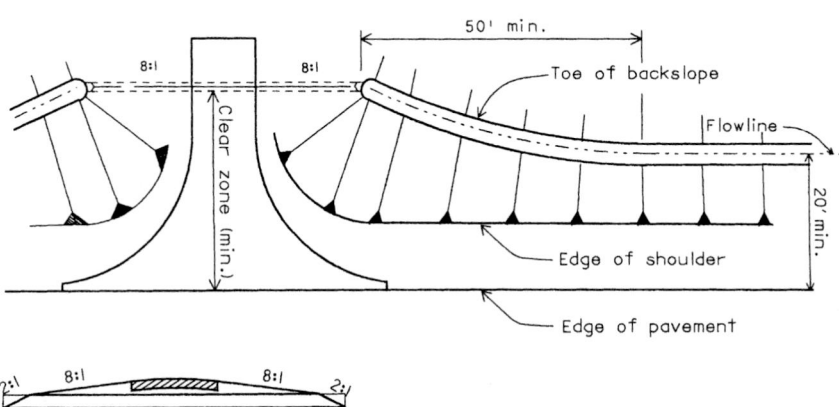

To be used on clear zone grading projects where the roadside ditch flowline is located within the clear zone distance

FIGURE 2.21 Slopes and ditches at driveway and crossroad in cut or low fill for use on clear zone grading projects where ditch is within clear zone distance. (*From* Location and Design Manual, *Vol. 1,* Roadway Design, *Ohio Department of Transportation, with permission*)

Interchange Grading.   Interchange interiors should be contour-graded so that maximum safety is provided and the least amount of guiderail is required. Figures 2.22 and 2.23 show examples. The generous use of flat slopes (6:1 or flatter) will also be easier for maintenance crews to work with. Sight distance is critical for passenger vehicles on ramps as they approach entrance or merge areas. Therefore, sight distance should be unobstructed by landscaping, earth mounds, or other barriers on the merging side of the vehicle.

Crossroads.   At a road crossing within an interchange area, bridge spill-through slopes should be 2:1, unless otherwise required by structure design. They should be flattened to 3:1 or flatter in each corner cone and maintained at 3:1 or flatter if within the interior of an interchange. Elsewhere in interchange interiors, fill slopes should not exceed 3:1.

Ramps.   Roadside design for ramps should be based on Fig. 2.15 or 2.16, depending on the mainline grading concept.

Gore Area.   Gore areas of trumpets, diamonds, and exteriors of loops adjacent to the exit point should be graded to obtain slopes of 6:1 or flatter, which will not endanger a vehicle unable to negotiate the curvature because of excessive speed.

Trumpet Interiors.   Interior areas of trumpets (Fig. 2.22) should be graded to slopes not in excess of 8:1, sloping downward from each side of the triangle to a single, rounded low point. Roadside ditches should not be used. Exteriors should be graded in accordance with mainline or ramp standards.

Loop Interiors.   In cut, the interior of a loop should be graded to form a normal ditch section adjacent to the lower part of the loop, and the backslope should be extended to intersect the opposite shoulder of the upper part of the loop. This applies unless the character and the amount of material or the adjacent earthwork balances indicate that the cost would be prohibitive. Roadside cleanup and landscaping should be provided in undisturbed areas of loop interiors. If channels are permitted to cross the loop interior, slopes should not be steeper than 4:1. Figure 2.23 shows an example.

Diamond Interiors.   If the location of the ramp intersection at the crossroad is relatively near the main facility, a continuous slope between the upper roadway shoulder and the lower roadway ditch will provide the best and most pleasing design. If the ramp intersection at the crossroad is located a considerable distance from the main facility, then both ramp and mainline roadsides should have independent designs, until the slopes merge near the gore.

   If the quadrant is entirely, or nearly so, in cut, the combination of a 3:1 backslope at the low roadway ditch and a gentle slope down from the high roadway shoulder will provide the best design in the wide portion of the quadrant. Approaching the gore, the slopes should transition to continuous 4:1 and 6:1 or flatter slopes. Quadrants located entirely in fill areas should have independently designed roadways for ramp, mainline, and crossroad. Each should be provided with normal slopes not greater than 3:1, with the otherwise ungraded areas sloped to drain without using ditches. If the quadrant is located part in cut and part in fill, the best design features a gentle fill slope at the upper roadway and a gentle backslope at the lower roadway, joined to a bench at the existing ground level that is sloped to drain. The combination of a long diamond ramp having gentle alignment with a loop ramp in the same interchange quadrant is not to be treated as a trumpet. Each ramp should be designed independently of the other in accordance with the suggested details set forth above.

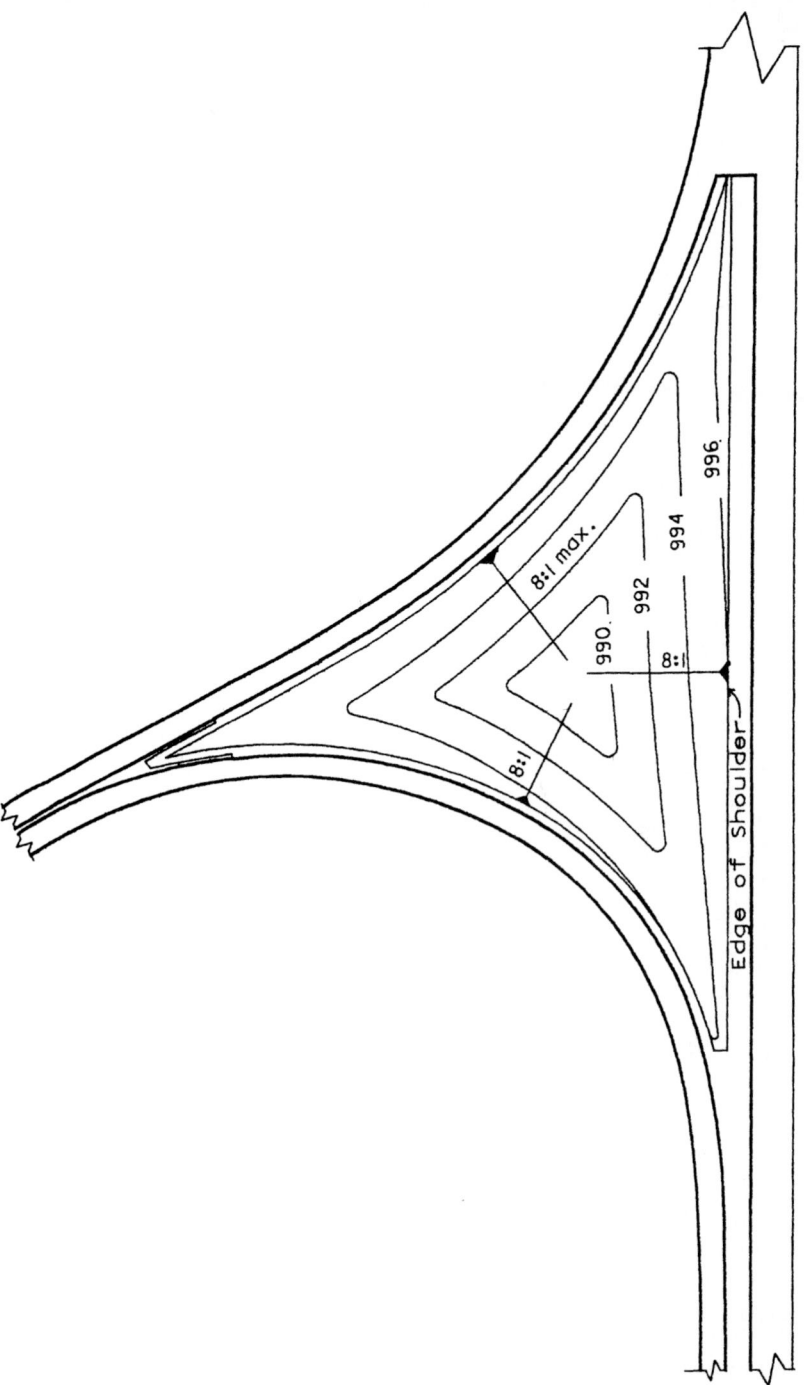

FIGURE 2.22 Contour grading of trumpet interior at interchange. (*From* Location and Design Manual, *Vol. 1,* Roadway Design, *Ohio Department of Transportation, with permission*)

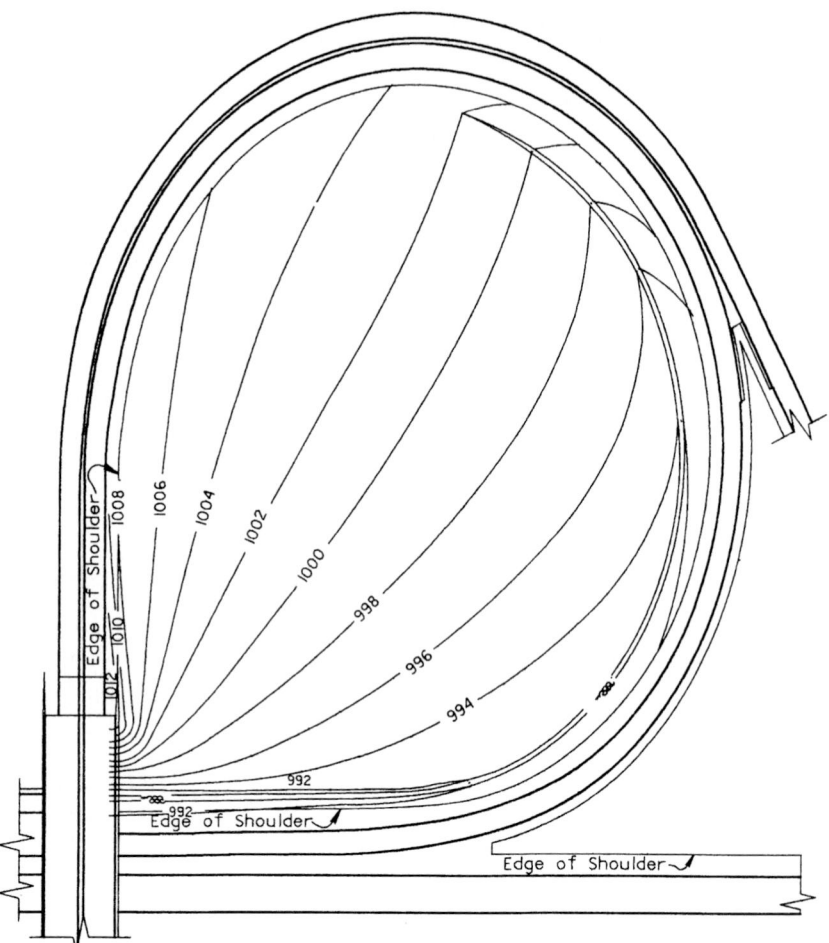

FIGURE 2.23   Contour grading of loop interior in cut section at interchange. (*From* Location and Design Manual, *Vol. 1,* Roadway Design, *Ohio Department of Transportation, with permission*)

2.3.3   Bridge Criteria

Although bridge engineering is discussed in Chap. 4, information on pertinent physical dimensions is presented here. Lateral clearance at underpasses and vertical clearance over roadways, as used in Ohio, are given in Table 2.24 for new and reconstructed bridges. The table notes provide a good insight into when variations from the standards are allowed.

2.3.4   Interchange Elements

Cross-section information pertaining to interchange elements, such as ramps and directional roadways, is given in Fig. 2.24. This information includes pavement and shoulder dimensions for acceleration-deceleration lanes, one- and two-lane directional roadways, and medians between adjacent ramps. Notice that for a single-lane ramp, the shoulder and guiderail offset distances are greater on the driver's right-hand side than on the left. This is to provide more width for drivers to pull over in emergencies and to allow people a better opportunity to go around disabled vehicles.

2.3.5 Medians

A median is a desirable element on all streets or roads with four or more lanes. The principal functions of a median are to prevent interference of opposing traffic, to provide a recovery area for out-of-control vehicles, to provide areas for emergency stopping and left turn lanes, to minimize headlight glare, and to provide width for future lanes. A median should be highly visible both day and night and in definite contrast to the roadway.

Width.   The width of a median is the distance between the inside edges of the pavement. See Fig. 2.25 for examples of various medians. The width depends upon the type of facility, topography, and available right-of-way. In rural areas with flat or rolling terrain, the desirable median width for freeways is 60 to 84 ft. Although the minimum median width is normally 40 ft, narrower medians may be used in rugged terrain. A constant-width median is not necessary, and in fact, variable-width medians and independent profiles may be used for the two roadways. Narrow medians with a barrier (barrier medians) are normally used in urban areas. Under normal design, the median width will vary depending on the width of the barrier and the shoulder width required (Table 2.22).

Types.   Medians are divided into types depending upon width and treatment of the median area and drainage arrangement. In general, raised or barrier medians are applicable to urban areas, while wide, depressed medians apply to rural areas. Figure 2.25 shows examples. Medians in rural areas are normally depressed to form a swale in the center and are constructed without curbs. The type of median used in an urban area depends on the traffic volume, speed, degree of access, and available right-of-way. On major streets with numerous business drives, a median consisting of an additional lane, striped as a continuous two-way left turn lane, is appropriate. A solid 6-in-high concrete median may be used in low-speed areas (where the design speed is 40 mi/h or less) and where an all-paved section is desired and a wider median cannot be justified. Barrier medians are normally recommended for urban facilities when the design speed is over 40 mi/h. However, care must be exercised when barrier medians are used on expressways with unsignalized at-grade intersections because of sight distance limitations and end treatments of the barrier.

U-Turn Median Openings.   U-turn median openings may be provided on expressways, freeways, or interstate highways with nonbarrier medians where space permits and there is a need. U-turns may be needed for proper operation of police and emergency vehicles, as well as for equipment engaged in physical maintenance, traffic service, and snow and ice control. U-turn crossings should not be constructed in barrier-type medians. When U-turn median openings are permitted, it is intended they be spaced as close to 3-mi intervals as possible. Crossings should be located at points approximately 1000 ft beyond the end of each interchange speed change lane.

An example of a typical U-turn median opening is shown in Fig. 2.26, which indicates geometric features applicable to crossings located in medians of widths ranging

TABLE 2.24 Recommended Lateral and Vertical Clearances for New and Reconstructed Bridges

Functional class	Traffic Current ADT	Traffic DHV	Lateral clearances, ft (in addition to approach pavement) On bridge[a] Rural Minimum	On bridge[a] Rural Preferred	On bridge[a] Urban, minimum	Under bridge[b] Minimum	Under bridge[b] Preferred	Minimum design loading	Vertical clearance over surfaced roadway, ft[c] Minimum	Preferred
Urban interstate	All	All	Right, 10[d,e] Left, 4[e,f]	Right, 12[e] Left, 6[e,g]		i		HS-20	16.5[l]	17.0
Rural interstate	All	All	Right, 19[d,e] Left, 4[e,f]	Right, 12[e] Left, 6[e,g]				HS-20	16.5[l]	17.0
Freeways and expressways	All	All	Right, 10[d,e] Left, 4[e,f]	Right, 12[e] Left, 6[e,g]		i, j		HS-20	16.5[l]	17.0
Arterials	—	>400	10[d]	12		j		HS-20	16.5[l]	17.0
	—	201–400	8[d]	10						
	≥400	≤200	6[d]	8						
	<400	—	4[d]	6						
Collectors	—	>400	8[m]	10		j		HS-20	14.5	15.0
	—	201–400	4[m]	8						
	—	100–200	4[m,n]	6						
	≥400	<100	4[m,n]	4						
	<400	—	4[m,o]	4						
Locals	—	>400	8[m]	10		j		HS-20	14.5	15.0
	—	201–400	3	8						
	—	100–200	3	6						
	>400	<100	3	4						
	≤400	—	2	4						

Under bridge notes:

Clear zone width[k]

Curbed treated shoulder width (Fig. 2.22) plus barrier clearance[h]

Urban, minimum notes:

For curbed shoulders, use shoulder width, Table 2.22.

For uncurbed shoulders, use rural criteria at left.

2.66

[a]Distance measured to face of curb or railing if no curb is provided.

[b]Distance measured from edge of traveled lane to face of walls or abutments and piers. Regardless of rural or urban condition, the absolute minimum clearance is 4 ft.

[c]To minimize structure cost, design tolerances for clearances are plus 4 in, minus 0 in. Sign supports and pedestrian structures have a 1-ft additional clearance. Clearances shown are over paved shoulder as well as pavement.

[d]If bridge is considered to be a major structure having a length of 200 ft or more, the lateral clearance may be reduced, subject to economic studies, but to no less than 3 ft (4 ft on interstates).

[e]When 12-ft treated shoulder width is provided on the approach roadway. Use 14 ft on left and/or right side as appropriate.

[f]Use 10-ft width if 10-ft paved shoulder is provided on the approach roadway.

[g]Use 12-ft width if 10-ft paved left shoulder is provided on the approach roadway.

[h]Minimum barrier clearance.

[i]In locations with restricted right-of-way, may be reduced to a clearance of 8.0 ft right side, 4.5 ft median side, plus barrier clearance, except where footnote *j* applies.

[j]May be reduced to a clearance of 2 ft plus barrier clearance on urban streets with restricted right-of-way and a design speed less than 50 mi/h.

[k]Clear zone width is defined in Art. 6.2.

[l]A 15.5-ft minimum clearance may be used in highly developed urban areas if attainment of 16.5-ft clearance would be unreasonably costly and if there is an alternate freeway route or bypass which provides a minimum 16.0-ft vertical clearance.

[m]May be 3-ft width if bridge length exceeds 100 ft.

[n]May be 3-ft width if turf shoulder is used.

[o]May be 2-ft width if turf shoulder is used.

Source: *Location and Design Manual*, Vol. 1, *Roadway Design*, Ohio Department of Transportation, with permission.

INTERCHANGE ELEMENT	TOTAL PVMT. WIDTH (ft)	GRADED SHLDR. WIDTH (ft)				PAVED SHLDR. WIDTH (ft)		NORMAL ROUNDING (ft) (D)	NORMAL BARRIER OFFSET	
		LEFT		RIGHT						
		WITH BARRIER OR FORE-SLOPE STEEPER THAN 6:1	W/O BARRIER SLOPES 6:1 OR FLATTER	WITH BARRIER OR FORE-SLOPE STEEPER THAN 6:1	W/O BARRIER SLOPES 6:1 OR FLATTER	LT	RT		LT	RT
RAMP	16 (A)	9 (B)	6	11 (B)	8	3	6	10	6	8
1-LANE DIRECTIONAL ROADWAY	16 (A)	9 (B)	6	11 (B)	8	4	6	10	6	8
2-LANE DIRECTIONAL ROADWAY	24	9 (B)	6	15 (C)	10 (C)	4	10	10	6	12 (C)
ACCEL/DECEL LANE OR COMBINED	VARIABLE	NA	NA	13 (C)	8 (C)	NA	8	10	NA	10 (C)

(A) Use 18' when inside pavement edge radius is less than 200'.

(B) May be reduced 1' if the face of the mainline barrier is 2' from the outside edge of the graded shoulder.

(C) Or match mainline dimension if lesser.

(D) Rounding is 4' when barrier is used. No rounding is required when foreslope is 6:1 or flatter.

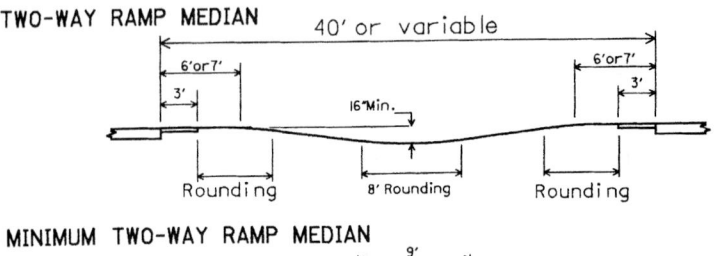

TWO-WAY RAMP MEDIAN

MINIMUM TWO-WAY RAMP MEDIAN

FIGURE 2.24 Cross-section information for interchange elements—pavement, shoulders, and medians. (*From* Location and Design Manual, *Vol. 1,* Roadway Design, *Ohio Department of Transportation, with permission*)

BARRIER MEDIAN

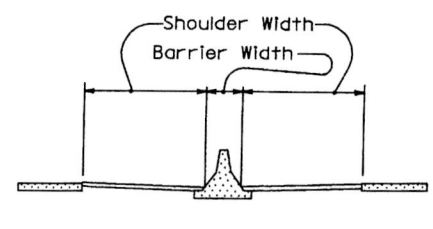

DEPRESSED MEDIANS

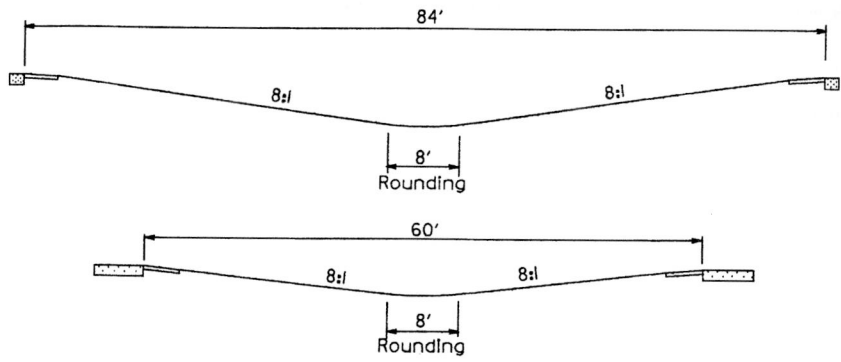

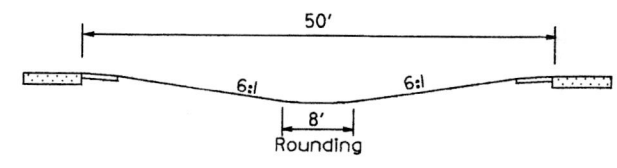

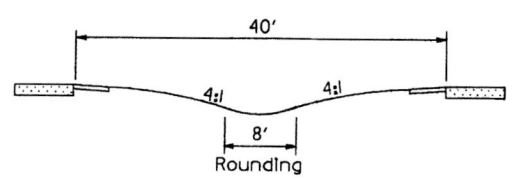

FIGURE 2.25 Typical designs for medians. (*From* Location and Design Manual, *Vol. 1,* Roadway Design, *Ohio Department of Transportation, with permission*)

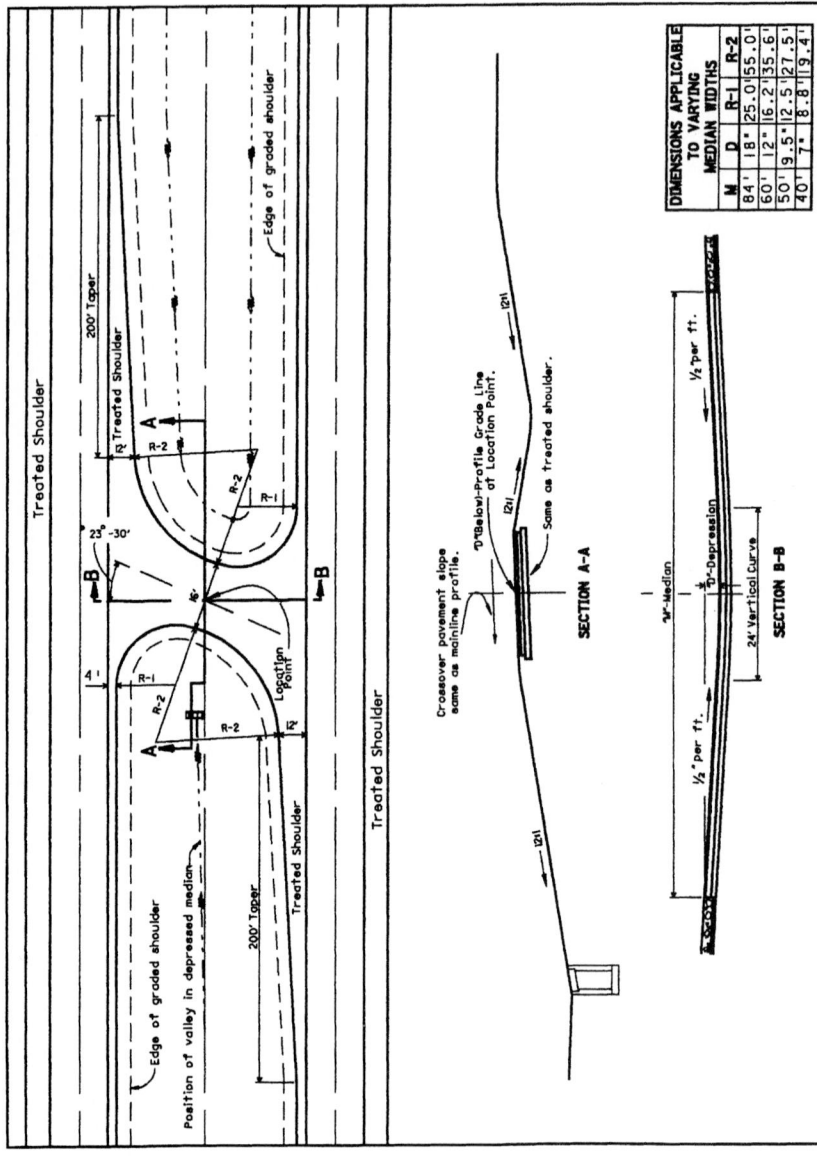

FIGURE 2.26 Design for U-turn median opening. (*From Location and Design Manual, Vol. 1, Roadway Design, Ohio Department of Transportation, with permission*)

from 40 to 84 ft. Turning radius should be modified proportionately for medians of varying widths. Tapers should be 200 ft in length for all median widths. The profile grade line should normally be an extension of the cross slope of the shoulder paving, rounded at the lowest point.

2.3.6 Curbs

The type of curb and its location affect driver behavior patterns, which, in turn, affect the safety and utility of a road or street. Curbs, or curbs and gutters, are used mainly in urban areas. They should be used with caution where design speeds exceed 40 mi/h. Following are various reasons for justifying the use of curbs, or curbs and gutters:

- Where required for drainage
- Where needed for channelization, delineation, control of access, or other means of improving traffic flow and safety
- To control parking where applicable

Types of Curb.   There are two general categories of curbs: barrier curbs and mountable curbs. Barrier curbs are relatively high (6 in or more) and steep-faced. Mountable curbs are 6 in or less in height and have flatter, sloping faces so that vehicles can cross them with varying degrees of ease. Figure 2.27 (Ref. 14) shows various curb designs that are commonly used on roadways. Types 1, 3, and 4 are examples of mountable curbs and are used for channelizing traffic, especially in islands and medians. Types 2 and 6 are barrier curbs used along pavement edges in urban areas and are designed to handle drainage more efficiently. Types 7 and 8 are tall barrier curbs (10 in and 8 in) designed to provide a more positive traffic barrier than the others. Type 7 is used as an alternate for guiderail in low-speed urban situations.

Position of Curb.   Curbs are normally used at the edge of pavement on urban streets where the design speed is 40 mi/h or less. Curbs at the edge of pavement have an effect on the lateral placement of moving vehicles. Drivers tend to shy away from them. Therefore, all curbs should be offset at least 1 ft and preferably 2 ft from the edge of the traffic lane. Where curb and gutter are used, the standard gutter width is 2 ft.

On roads where the design speed exceeds 40 mi/h, curbs should be used only in special cases. Special cases may include, but are not limited to, the use of curb to control surface drainage or to reduce right-of-way requirements in restricted areas. When it is necessary to use curbs on roads where the design speed is over 40 mi/h, they should not be closer to the traffic than 4 ft or the edge of the treated shoulder, whichever is greater.

Curb/Guiderail Relationship.   If curbs are used in conjunction with guiderail on roads having a design speed in excess of 40 mi/h, the face of curb should preferably be located either at or behind the face of guiderail. Under no conditions should the face of curb be located more than 9 in in front of the face of rail. This restriction is necessary to prevent a vehicle from "vaulting" over the rail or striking it too high to be contained. Although guiderail is not normally used on curbed roadways having design speeds of 40 mi/h or less, the same criteria used for higher-speed roadways should apply. Where this is not feasible or practical, the curb may be placed in front of the rail. Regardless of the design speed of the roadway or the placement of the curb, the face of guiderail should not be located closer than 4 ft to the roadway.

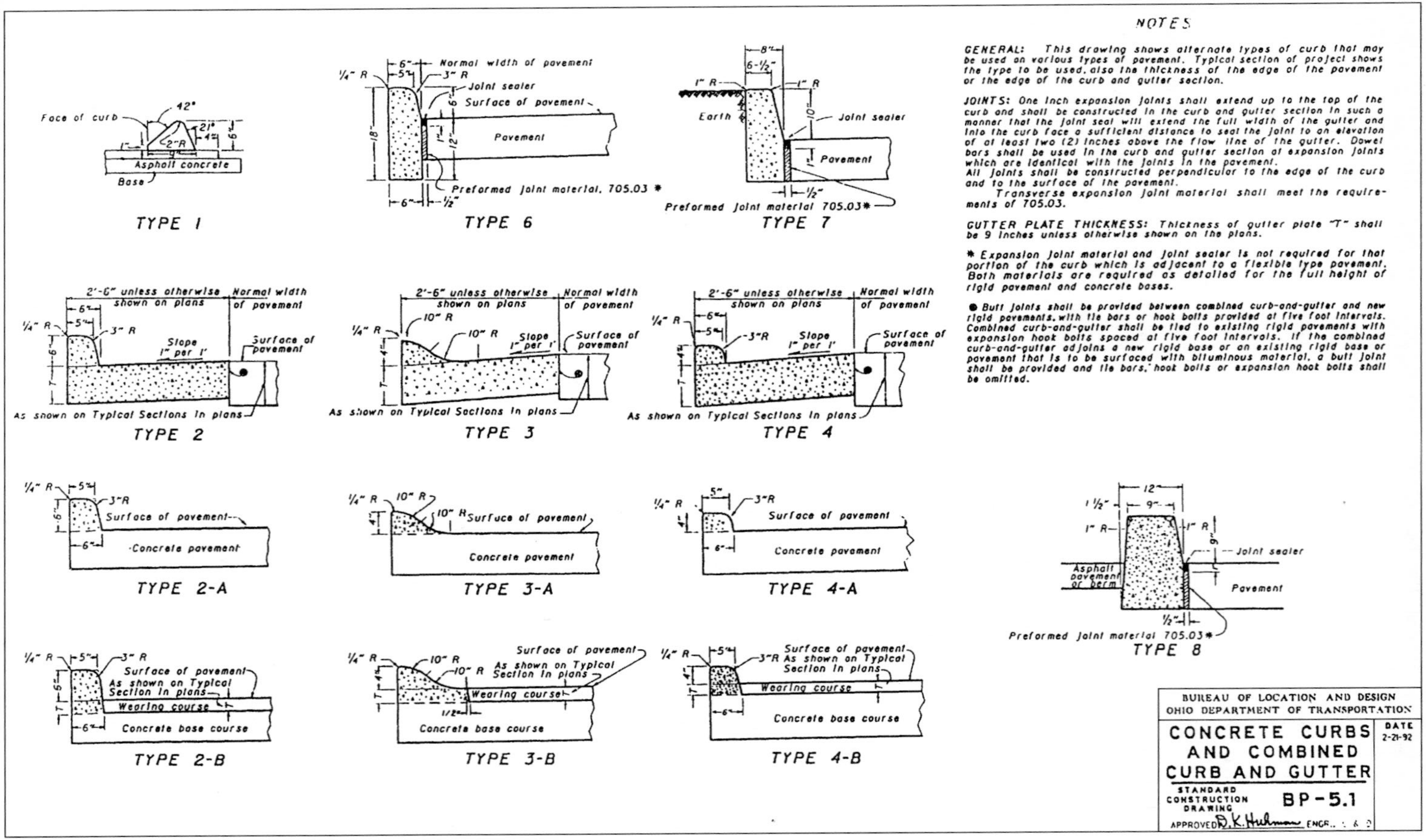

FIGURE 2.27 Examples of several designs for curbs. (*From Standard Construction Drawings,* Bureau of Location and Design, Ohio Department of Transportation, with permission)

Curb Transitions. Curb and raised median beginnings and endings should be tapered from the curb height to 0 in in 10 ft. When an urban-type section with curbs at the edge of pavement changes to a rural-type section without curbs, the curb should be transitioned laterally at a 4:1 (longitudinal:lateral) rate to the outside edge of the treated shoulder, or 3 ft, whichever is greater. When a curbed side road intersects a mainline that is not curbed, the curb should be terminated no closer to the mainline edge of pavement than 8 ft or the edge of the treated shoulder of the mainline, whichever is greater.

2.3.7 Pedestrian Facilities

When pedestrian facilities are to be constructed or reconstructed as part of project plans, the facilities should be designed to accommodate the handicapped. Guidance in design of pedestrian facilities with access for the handicapped is available (Ref. 11).

Walks. Walks should be provided in urban areas where pedestrian traffic currently exists or is planned in the future. Walks may be provided in rural areas where they will have sufficient use in relation to cost and safety. Walks are usually made of concrete, although asphalt or gravel may be used under special circumstances. Concrete walks are usually 4 in thick. At drive locations, the thickness is increased to 6 in, or the drive thickness, whichever is greater. Asphalt or gravel walks are mostly used in parks, rest areas, etc., where there is low usage. Asphalt walks consist of 2 in of asphalt and 5 in aggregate base, while gravel walks are constructed of 4 in compacted aggregate base.

Walk Design. The normal width of walks is 4 ft for residential areas and 6 ft for commercial areas or major school routes. In downtown areas, the walk width normally extends from the curb to the right-of-way or building line. Transverse slopes should be $\frac{1}{4}$ in per foot. The grade of the walk is normally parallel to the curb or pavement grade, but may be independent. The walk and the "tree lawn" (see next section) normally slope toward the pavement. Care should be taken in setting the pavement curb grade so that the sidewalk and the curb will not trap water or otherwise preclude usability of the adjoining property. The back edge of the walk should be located 2 ft inside the right-of-way line, unless grading, utilities, or other considerations require a greater dimension.

Tree Lawn. The tree lawn is defined as the area between the front of the curb and the front edge of the sidewalk. Grass is usually provided in the tree lawn, although in some urban areas the tree lawn is paved. As shown in Fig. 2.28, the desirable tree lawn width is 8 ft or more. The 8-ft width provides an area for snow storage and for traffic signs, and an adequate distance for elevation changes at drives. Tree lawn widths of less than 5 ft result in locating of signposts close to pedestrians using the walk, and steep grades on drive profiles. The minimum tree lawn width is 2 ft.

Border Area. In an urban area where a walk is not provided, the area between the face of curb and the right-of-way line is often referred to as a *border*. As indicated in Fig. 2.28*d*, the border width in residential areas should be at least 8 ft and preferably 14 ft. In commercial areas, the minimum border width is 10 ft, while a 16-ft width is preferable.

Walks on Bridges. Walks should be provided on bridges located in urban or suburban areas having curbed sections under two conditions: (1) where there are existing

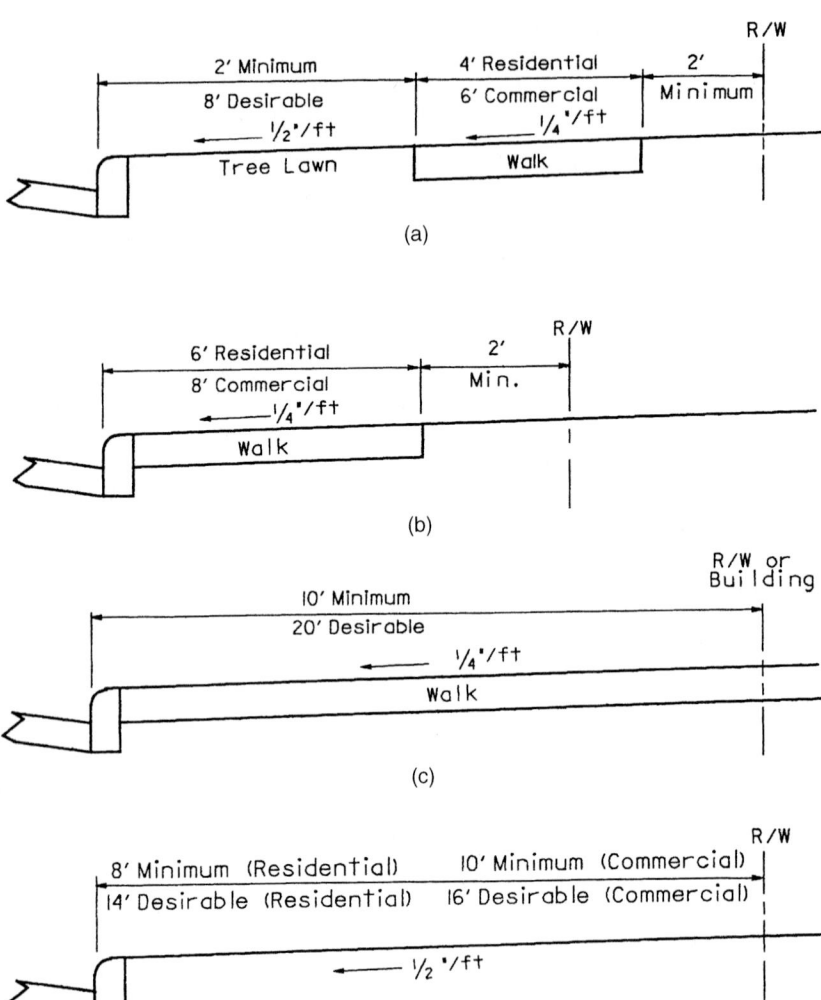

FIGURE 2.28 Examples of walk designs. (*a*) Walk with tree lawn. (*b*) Walk with no tree lawn. (*c*) Walk in downtown area. (*d*) Border area with no walk. (*From* Location and Design Manual, *Vol. 1, Roadway Design, Ohio Department of Transportation, with permission*)

walks on the bridge and/or bridge approaches, or (2) where evidence can be shown through local planning processes, or similar justification, that walks will be required in the near future (5 to 10 years). Anticipated pedestrian volumes of 50 per day justify a walk on one side, and 100 per day justify walks on both sides. Walks on bridges should preferably be 6 ft wide in residential areas and 8 ft wide in commercial areas measured from the face of curb to face of parapet. Widths, however, may be as much as 12 ft in downtown areas. The minimum bridge walk width is 5 ft.

Walks under Bridges. The criterion for providing walks at underpasses is basically the same as described above for walks on bridges. An exception is that in areas where there are no approach walks, space will be provided for future walks but walks will not be constructed with the project unless there is substantial concurrent approach walk construction. Where the approach walks at underpasses include a tree lawn, the tree lawn width may be carried through the underpass wherever space permits.

Curb Ramps. A curb ramp is a portion of the walk that is modified to provide a gradual elevation transition through the face of the adjoining curb. It is designed to provide safe and convenient curb crossings for the handicapped in wheelchairs, but it can also be used by others. Examples include wheeled vehicles maneuvered by pedestrians and bikeway traffic, when such use is permitted. Curb ramps should be provided where curb and walks are being constructed at intersections and other major points of pedestrian curb crossing such as mid-block crosswalks. When a curb ramp is built on one side of a street, a companion curb ramp is required on the opposite side of the street. The basic requirement is that a crosswalk must be accessible via curb ramps at both ends, not one end only. In most cases, curb ramps will be installed in all quadrants of an intersection. Curb ramps should be located within crosswalk markings to permit legal street crossings. The ramp location must be coordinated with drainage structures, utility poles, etc. The normal gutter profile should be continued through the ramp area, except the profile may be altered to avoid a location conflict between the ramp and a drainage structure. Drainage structures should not be located in the ramp or in front of the ramp. Catch basins should be placed upstream from the ramp.

2.4 INTERSECTION DESIGN

An intersection is defined as an area where two or more roadways join or cross. Each roadway extending from the intersection is referred to as a *leg*. The intersection of two roadways has four legs. When one roadway ends at the intersection with another roadway, a three-leg intersection, or T intersection, is formed. Some intersections have more than four legs, but this design should be avoided, since the operation of traffic movements is usually inefficient. There are three general types of intersections: (1) at-grade, where two or more roadways cross in the same vertical plane, (2) grade-separated, where one roadway is bridged over or tunneled under the other roadway but no turning movements are allowed, and (3) interchanges, a special type of grade-separated intersection where turning movements are accommodated by ramps connecting the two roadways.

2.4.1 At-Grade Intersections

At-grade intersections should be designed to promote the safe movement of traffic on all legs with a minimal amount of delay to drivers using the intersection. The amount of delay a driver experiences is the measure of effectiveness for signalized intersections as used in capacity analysis. Factors to be considered in designing an intersection are:

- Traffic volumes on all legs, including separate counts for turning vehicles
- Sight distance
- Traffic control devices
- Horizontal alignment
- Vertical alignment
- Radius returns
- Drainage design
- Islands
- Left turn lanes
- Right turn lanes
- Additional through lanes
- Recovery areas
- Pedestrians
- Bicycles
- Lighting
- Development of adjacent property

Traffic Volumes.   No intersection can be properly designed without first obtaining accurate traffic counts and reliable projections for the design year of the project. Traffic counts are best determined from actual field counts, including all turning movements, and are broken down by vehicle type. Vehicle types are divided into two groups. The first group includes passenger cars and type A commercial vehicles (pickup trucks and light delivery trucks not using dual tires). The second group includes type B commercial vehicles (tractor, semitrailer, truck-trailer combinations) and type C commercial vehicles (buses, dual-tired trucks with single or tandem rear axles). Adjustments are made to field counts to allow for day of the week, month of the year, time of day, and other site-related factors that may have a significant effect on the counts. Most urbanized areas have regional planning agencies that either provide or certify the traffic data used in intersection design.

Traffic Control.   There are four basic types of traffic control at at-grade intersections:

- Cautionary, or nonstop, control
- Stop control for minor traffic
- Four-way stop control
- Signal control

In discussing at-grade intersections, the terms *major roadway* and *minor roadway* are sometimes used to distinguish between the two roads. The major roadway usually has a higher functional classification and a greater volume of traffic.

Cautionary, or nonstop, control is used only in special circumstances, such as an at entrance terminal on a freeway. Stop control for the minor roadway is one of the most common treatments found in practice. In these cases, the traffic volumes on the minor roadway are light enough that a signal is not required. The major roadway apparently has volumes low enough to allow gaps for the minor road traffic to enter or cross the intersection. Four-way stop control is effective in situations where the roadways have nearly equal traffic volumes but not great enough volume to justify installing a signal. Finally, signal control is used for intersections where volumes are large enough to preclude using one of the other types.

Sight Distance. Adequate sight distance is an important consideration when designing an at-grade intersection. The alignment and grade on the major roadway should, as a minimum, provide stopping sight distance as given in Table 2.2. The criteria for intersection sight distance (Table 2.3) should also be met wherever possible. Figure 2.7 illustrates the lines of sight involved in intersection design.

Horizontal Alignment Considerations. It is best to avoid locating an intersection on a curve. Since this is often impossible, it is recommended that intersection sites be selected where the curve superelevation is ½ inch per foot (0.0417 ft/ft) or less. It is also recommended that intersections be located where the grade on the major roadway is 6 percent or less, with 3 percent the desirable maximum. Intersection angles of 70 to 90° are provided on new or relocated roadways. An angle of 60° may be satisfactory if right-of-way is to be purchased for a future grade separation and the smaller angle will avoid reconstruction of the intersecting road. In such cases, it may be desirable to locate the intersection so the separation structure can be constructed in the future without disrupting the intersection operation.

Relocation of the minor road is often required to meet the desired intersection location, to avoid roadway segments with undesirable vertical alignments, and to adjust intersection angles. Horizontal curves on the minor roads should be designed to meet the design speed of the road. The minor road alignment should be as straight as possible. Figure 2.29 shows the alignment for a typical rural crossroad relocation.

Vertical Alignment Considerations. On roadways with stop control at the intersection, the portion of the intersection located within 60 ft of the edge of the mainline pavement is considered to be the *intersection area.* The pavement surface within this intersection area should be visible to the driver within the limits of the minimum stopping sight distance listed in Table 2.2. By being able to see the pavement surface (height of object of zero), the driver (height of eye of 3.5 ft) can observe the radius returns and pavement markings and recognize an approaching intersection. Figure 2.30 shows acceptable practice for design of the intersection area.

Combinations of pavement cross slopes and profile grades may produce unacceptable edge of pavement profiles in the intersection area. For this reason, edge of pavement profiles should be plotted and graphically graded to provide a smooth profile. Profile grades within the intersection area for stop conditions are shown in Figs. 2.30 and 2.31. The grade outside the intersection area is controlled by the design speed of the crossroad. Normal design practices can be used outside the intersection area with the only restriction on the profile being the sight distance required as discussed above.

Grade breaks are permitted at the edge of the mainline pavement for a stop condition. If these grade breaks exceed the limits given in note 3 of Fig. 2.30, they should be treated according to note 3 of Fig. 2.31. Several examples are shown in Fig. 2.31 of the use of grade breaks or short vertical curves adjacent to the edge of through pavement.

Signalized intersections require a more sophisticated crossroad profile. Whenever possible, roadway profiles through the intersection area of a signalized intersection

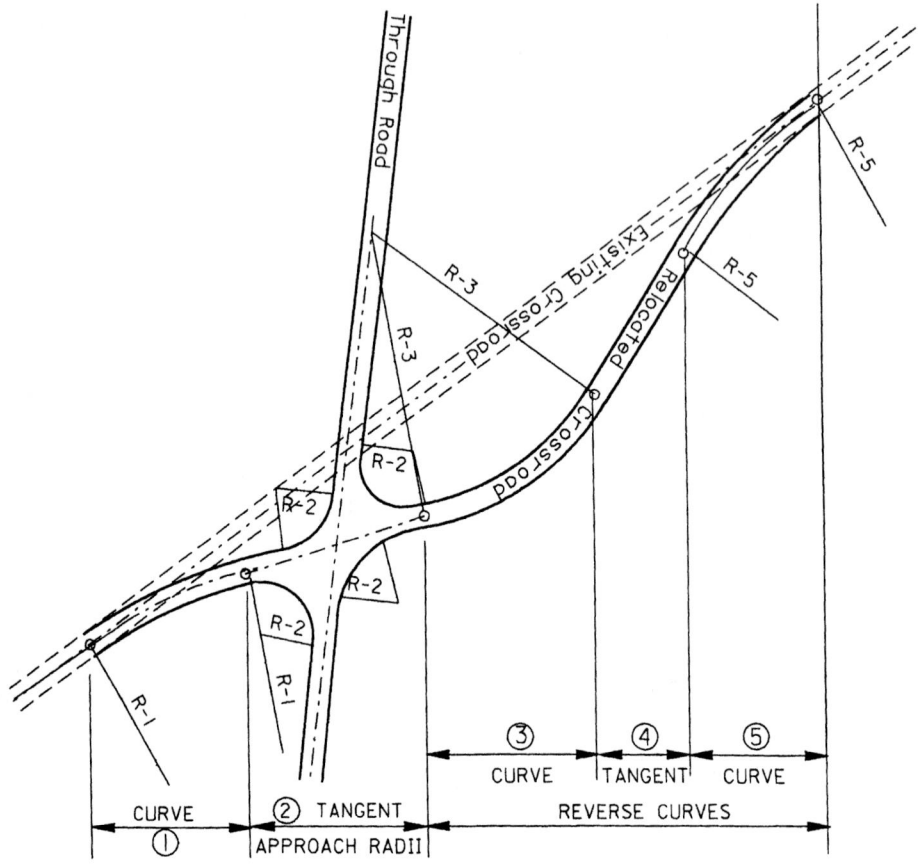

FIGURE 2.29   Typical rural crossroad relocation. (*From* Location and Design Manual, *Vol. 1,* Roadway Design, *Ohio Department of Transportation, with permission*)

should be designed to meet the design speed of the roads. Grade breaks at signalized intersections should be in accordance with Table 2.16. Since the grade break across a normal crowned pavement is usually 3.12 percent, it should be noted that the crown must be flattened. This will allow vehicles on the crossroad to pass through the intersection on a green signal safely without significantly adjusting their speed. The sight distance requirements within the intersection area that were discussed for stop-controlled roadways are also applicable for signalized intersections. Figure 2.32 shows examples of crossroad profiles through a signalized intersection.

Radius Returns at Intersections.   Intersection radii in rural areas should normally be 50 ft, except that radii less than 50 ft (minimum 35 ft) may be used at minor intersecting roads if judged appropriate for the volume and character of turning vehicles. Radii larger than 50 ft, a radius with a taper, or a three-center curve should be used at any intersection where the design must routinely accommodate semitrailer truck turning movements. Truck turning templates should be used to determine proper radii and stop

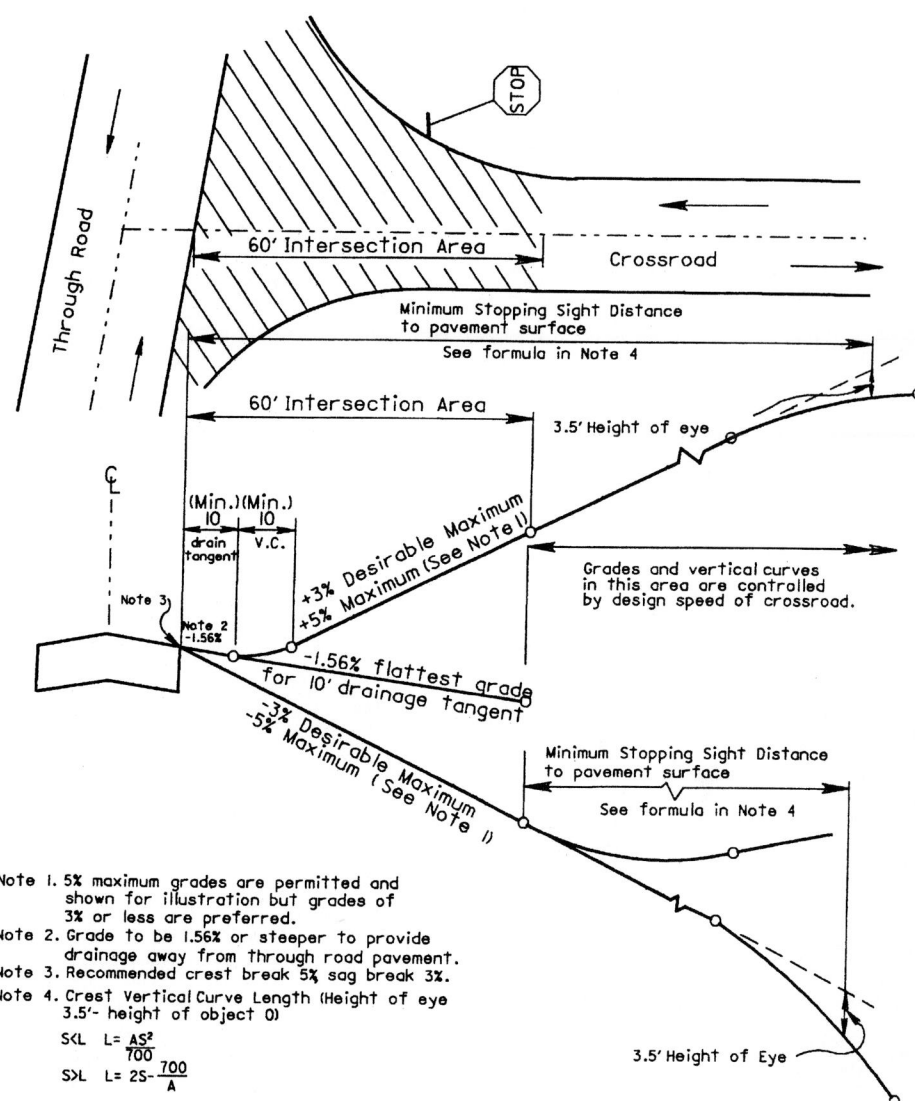

FIGURE 2.30 Crossroad profile for stop condition where through road has normal crown. (*From* Location and Design Manual, *Vol. 1,* Roadway Design, *Ohio Department of Transportation, with permission*)

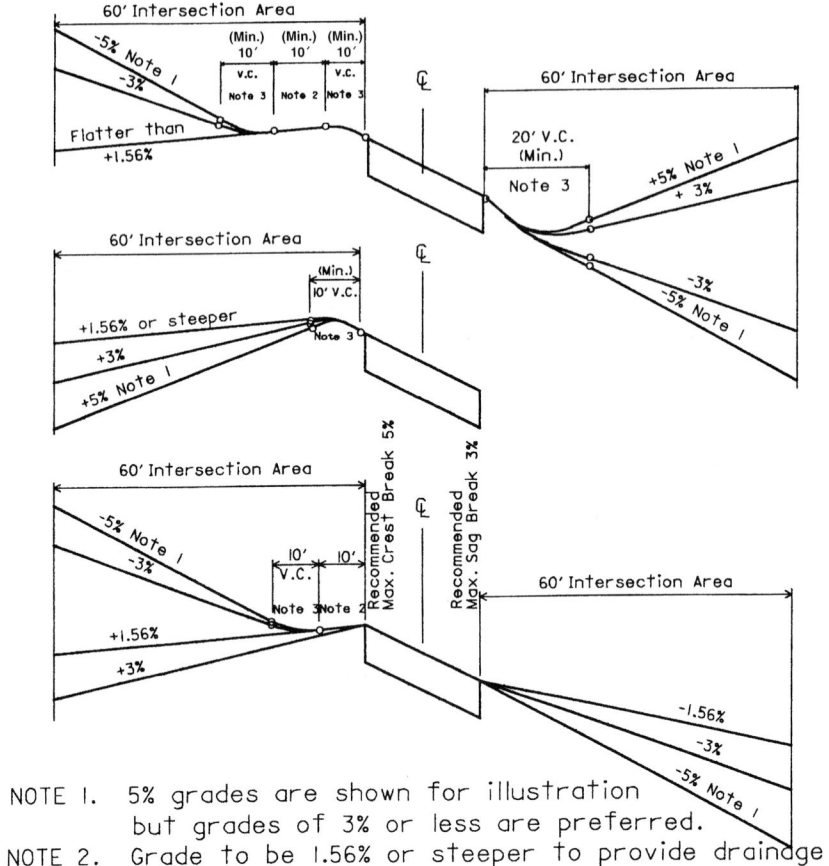

NOTE 1. 5% grades are shown for illustration but grades of 3% or less are preferred.

NOTE 2. Grade to be 1.56% or steeper to provide drainage.

NOTE 3. Crest breaks exceeding 5% should be rounded using vertical curves having a K of 1 or greater.

Sag breaks exceeding 3% should be rounded using vertical curves having a K of 1.5 or greater.

FIGURE 2.31 Crossroad profile for stop condition where through road is superelevated. (*From Location and Design Manual, Vol. 1,* Roadway Design, *Ohio Department of Transportation, with permission*)

bar location. Figure 2.33 shows an example of a turning template for a WB-50 semi-trailer truck. Complete sets of turning templates may be obtained from the Institute of Traffic Engineers (Ref. 12). When truck turning templates are used, a 2-ft clearance should be provided between the edge of pavement and the closest tire path.

Corner radii at street intersections in urban areas should consider the right-of-way available, the intersection angle, pedestrian traffic, approach width, and number of lanes. The following should be used as a guide:

- Radii of 15 to 25 ft are adequate for passenger vehicles and may be provided at minor cross streets where there are few trucks or at major intersections where there are parking lanes.

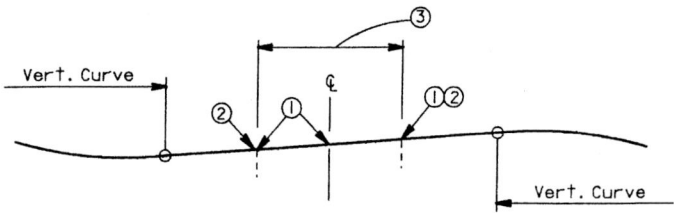

Example A – Crossroad Profile Tangent through Intersection

① Location of permissible grade break

② Edge of pavement of intersecting roadway extended through the intersection

③ Width of intersecting roadway

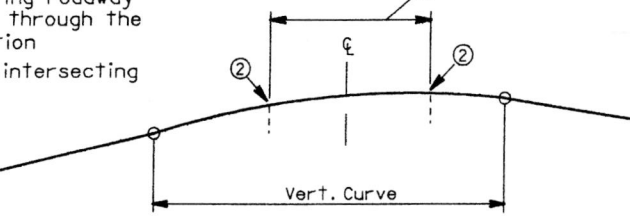

Example B – Crossroad Profile on Vertical Curve through Intersection

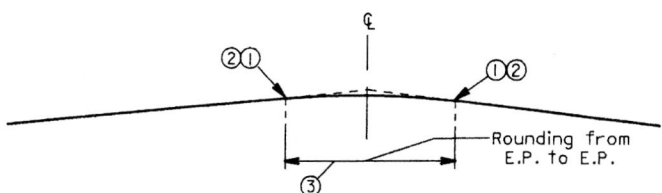

Example C – Crossroad Profile Fitted to a Normal Crown on the Mainline Road

FIGURE 2.32 Examples of crossroad profile through signalized intersection. (*From* Location and Design Manual, *Vol. 1,* Roadway Design, *Ohio Department of Transportation, with permission*)

- Radii of 25 ft or more should be provided at minor intersections on new or reconstruction projects where space permits.
- Radii of 30 ft or more should be used where feasible at major cross street intersections.
- Radii of 40 ft or more, three-centered compound curves, or simple curves with tapers to fit truck paths should be provided at intersections used frequently by buses or large trucks.

Drainage Considerations. Within the intersection area, the profile of the crossroad should be sloped wherever possible so the drainage from the crossroad will not flow

THIS TURNING TEMPLATE SHOWS THE TURNING PATHS OF THE AASHTO DESIGN
VEHICLES. THE PATHS SHOWN ARE FOR THE LEFT FRONT OVERHANG AND THE
OUTSIDE REAR WHEEL. THE LEFT FRONT WHEEL FOLLOWS THE CIRCULAR CURVE;
HOWEVER, ITS PATH IS NOT SHOWN.

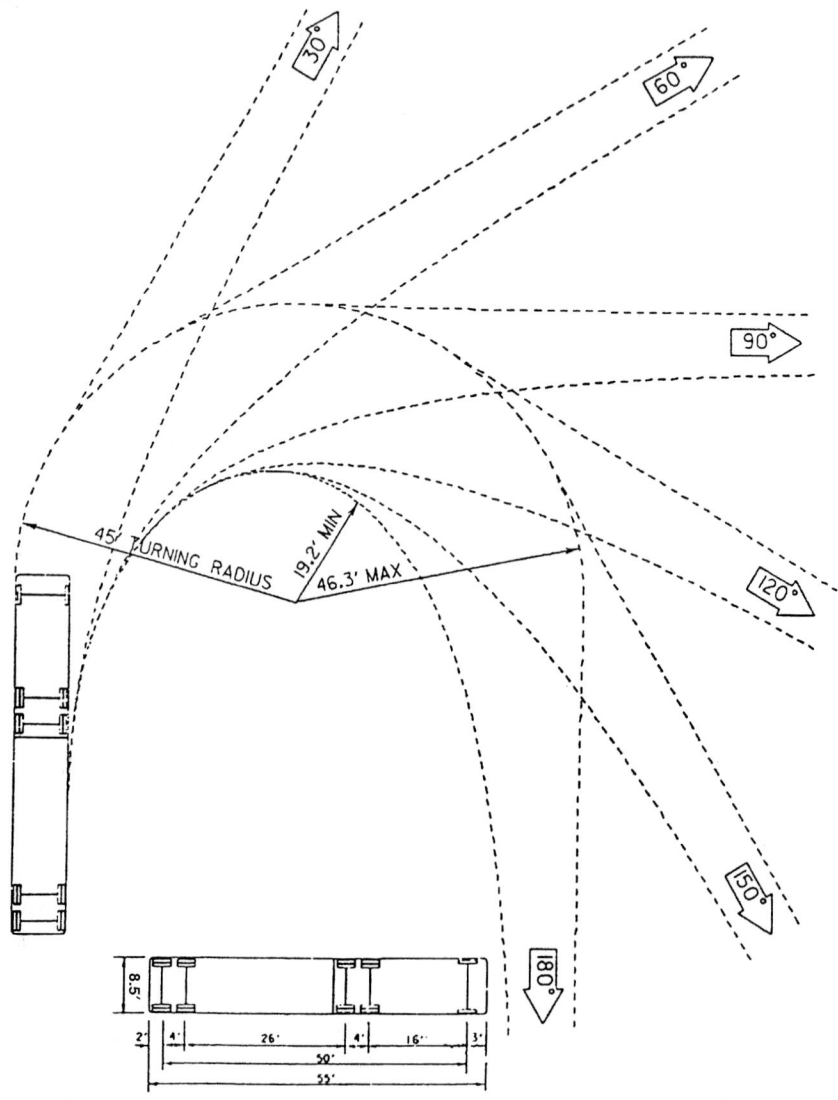

FIGURE 2.33 Template of minimum turning path for WB-50 semitrailer truck. (*From* A Policy on
Geometric Design of Highways and Streets, *American Association of State Highway and
Transportation Officials, 1990, with permission*)

across the through road pavement. For a stop condition, the 10 ft of crossroad profile adjacent to the through pavement is normally sloped away from the through pavement, using at least a 1.56 percent grade, as shown in Fig. 2.30. The profiles of curbed radius returns within the intersection may be adjusted to accommodate location of catch basins. It is recommended that exaggerated profiles be used to make adjustments. To ensure smooth transitions around the returns, plot the pavement edges for at least 25 ft going away from the returns for each leg of the intersection.

Islands at Intersections. In intersection design, an *island* is defined as an area between traffic lanes that has been delineated to control traffic movements through the intersection. An island may be curbed or uncurbed. It may be concrete, grass, or the same material as the traffic lanes. Islands may be used at intersections for the following reasons:

- Separation of conflicts
- Control of angle of conflict
- Reduction in excessive pavement areas
- Favoring a predominant movement
- Pedestrian protection
- Protection and storage of vehicles
- Location of traffic control devices

Although certain situations require the use of islands, they should be used sparingly and avoided wherever possible. Curbed islands are most often used in urban areas where traffic is moving at relatively low speeds (40 mi/h or less) and fixed-source lighting is available. Curbed islands with an area smaller than 50 ft^2 in urban locations and 75 ft^2 in rural areas should generally not be used. An area of 100 ft^2 is preferred in either case. Where pedestrian traffic will be using curbed islands, the islands must be provided with curb ramps. Islands delineated by pavement markings are often preferred in rural or lightly developed areas, when approach speeds are relatively high, where there is little pedestrian traffic, where fixed-source lighting is not provided, or where traffic control devices are not located within the island. Nonpaved islands are normally used in rural areas. They are generally turf and are depressed for drainage purposes.

Left Turn Lanes. Probably the single item having the most influence on intersection operation is the treatment of left-turning vehicles. Left turn lanes are generally desirable at most intersections. However, cost and space requirements do not permit their inclusion in all situations. Intersection capacity analysis procedures should be used to determine the number and use of all lanes. Left turn lanes are generally required under two conditions: (1) when left turn design volumes exceed 20 percent of total directional approach design volumes, and (2) when left turn design volumes exceed 100 vehicles per hour in peak periods.

Opposing left turn lanes should be aligned opposite each other because of sight distance limitations. They are developed in several ways depending on the available width between opposing through lanes. Figure 2.34*a* shows the development required when additional width must be generated. The additional width is normally accomplished by widening on both sides. However, it could be done all on one side or the other. In Fig. 2.34*b,* the median width is sufficient to permit the development of the left turn lane. Figure 2.35 shows the condition where an offset left turn lane is required to obtain adequate sight distance in wide medians.

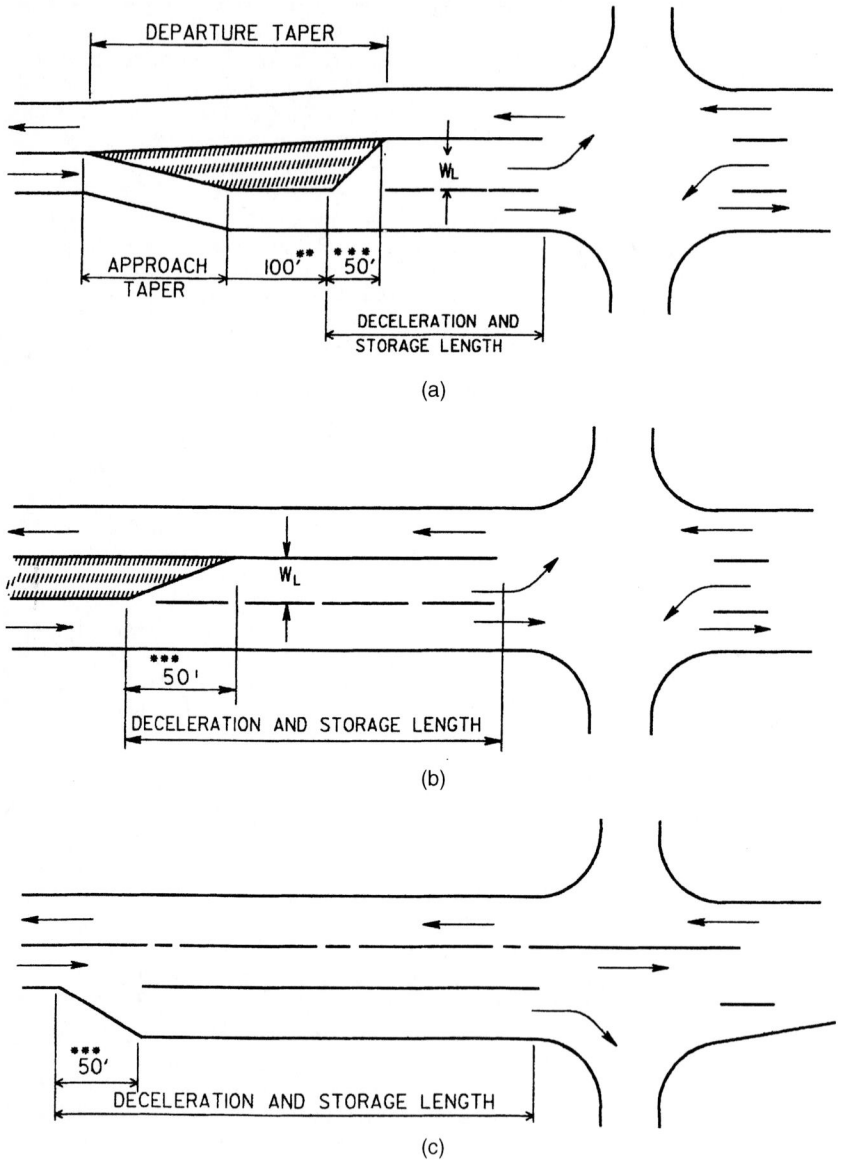

FIGURE 2.34 Turning lane designs showing roadway taper. (*a*) Left turn lane with no median or median width less than W_L. (*b*) Left turn lane with median wider than W_L. (*c*) Right turn lane. (*From Location and Design Manual, Vol. 1,* Roadway Design, *Ohio Department of Transportation, with permission*)

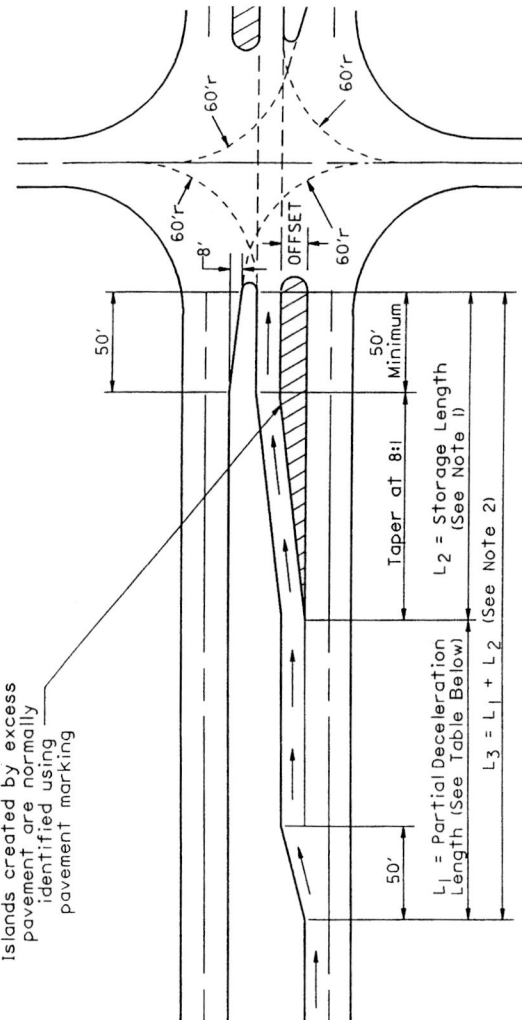

Islands created by excess pavement are normally identified using pavement marking

L_1 = Partial Deceleration Length (See Table Below)

L_2 = Storage Length (See Note 1)

$L_3 = L_1 + L_2$ (See Note 2)

Taper at 8:1

OFFSET

Design Speed (mi/h)	L_1 (ft)	L_3 (ft)
40	111	125
45	125	175
50	143	225
55	164	285
60	181	345

NOTES

1. Determine a minimum length for L_2 by multiplying 8 times the lateral offset and adding 50 ft. Using the design turning volume, obtain the required storage length from Table 2.26. Use the greater of these two numbers for L_2. If the length from Table 2.26 is the greater value, increase the tangent area at the intersection.

2. Using the value of L_2 from Note 1, add the value for L_1 from table at left to obtain L_3. If this value is less than the table value at left, increase one or both of the tangent areas of L_1 and L_2 to make up the difference.

FIGURE 2.35 Offset left turn lane for sight distance at wide median. (*From Location and Design Manual, Vol. 1, Roadway Design, Ohio Department of Transportation, with permission*)

In developing turn lanes, several types of tapers may be involved as shown in Figure 2.34:

Approach taper. An approach taper directs through traffic to the right. Approach taper lengths are calculated using Eq. (2.5) or (2.6).

Departure taper. The departure taper directs through traffic to the left. Its length should not be less than that calculated using the approach taper equations. Normally, however, the departure taper begins opposite the beginning of the full-width turn lane and continues to a point opposite the beginning of the approach taper.

Diverging taper. The diverging taper is the taper used at the beginning of the turn lane. The recommended length of a diverging taper is 50 ft.

Tables 2.25 and 2.26 have been included to aid in determining the required lengths of left turn lanes at intersections. After determining the length of a left turn lane (Table 2.25), the designer should also check the length of storage available in the adjacent through lane(s) to ensure that access to the turn lane is not blocked by a back-up in the through lane(s). To do this, Table 2.26 may be entered using the average number of through vehicles per cycle, and the required length read directly from the table. If two or more lanes are provided for the through movement, the length obtained

TABLE 2.25 Determination of Length of Left Turn Lanes

	Turn demand volume for design speed, mi/h					
Type of traffic control	30–35		40–45		50–60	
	High	Low*	High	Low*	High	Low*
Signalized	A	A	B or C†	B or C†	B or C†	B or C†
Unsignalized stopped crossroad	A	A	A	A	A	A
Unsignalized through road	A	A	C	B	B or C†	B

Condition A: storage only:

Length = 50 ft (diverging taper) + storage length

Condition B: high-speed deceleration only:

Design speed, mi/h	Length (including 50-ft diverging taper), ft
40	125
45	175
50	225
55	285
60	345

Condition C: moderate-speed deceleration and storage:

Design speed, mi/h	Length (including 50-ft diverging taper), ft
40	111 ⎫
45	125 ⎪
50	143 ⎬ + storage length
55	164 ⎪
60	181 ⎭

*Low is considered 10% or less of approach traffic volume.

†Whichever is greater.

Source: *Location and Design Manual*, Vol. 1, *Roadway Design*, Ohio Department of Transportation, with permission.

TABLE 2.26 Storage Length at Intersections

Average no. of vehicles per cycle*	Required length, ft	Average no. of vehicles per cycle*	Required length, ft
1	50	17	600
2	100	18	625
3	150	19	650
4	175	20	675
5	200	21	725
6	250	22	750
7	275	23	775
8	325	24	800
9	350	25	825
10	375	30	975
11	400	35	1125
12	450	40	1250
13	475	45	1400
14	500	50	1550
15	525	55	1700
16	550	60	1850

*Average vehicles/cycle = [DHV (turning lane)]/(cycles/hour)

If cycles/hour are unknown, assume:

Unsignalized or 2-phase—60 cycles per hour

3-phase—40 cycles per hour

4-phase—30 cycles per hour

Source: *Location and Design Manual,* Vol. 1, *Roadway Design,* Ohio Department of Transportation, with permission.

should be divided by the number of through lanes to determine the required storage length.

It is recommended that left turn lanes be at least 100 ft long, and the maximum length be no more than 600 ft. The width of a left turn lane should desirably be the same as the normal lane widths for the facility. A minimum width of 11 ft may be used in moderate- and high-speed areas, while 10 ft may be provided in low-speed areas. Additional width should be provided whenever the lane is adjacent to a curbed median as discussed previously under "Position of Curb."

Double Left Turn Lanes.   Double left turn lanes should be considered at any signalized intersection with left turn demands of 300 vehicles per hour or more. The actual need should be determined by performing a signalized intersection capacity analysis. Fully protected signal phasing is required for double left turns. When the signal phasing permits simultaneous left turns from opposing approaches, it may be necessary to laterally offset the double left turn lanes on one approach from the left turn lane(s) on the opposing approach to avoid conflicts in turning paths. Figure 2.36 provides an example. All turning paths of double left turn lanes should be checked with truck turning templates allowing 2 ft between the tire path and edge of each lane. Expanded throat widths are necessary for double left turn lanes as illustrated in Fig. 2.37.

Right Turn Lanes.   Exclusive right turn lanes are less critical in terms of safety than left turn lanes. However, right turn lanes can significantly improve the level of service of signalized intersections. They also provide a means of safe deceleration for right-

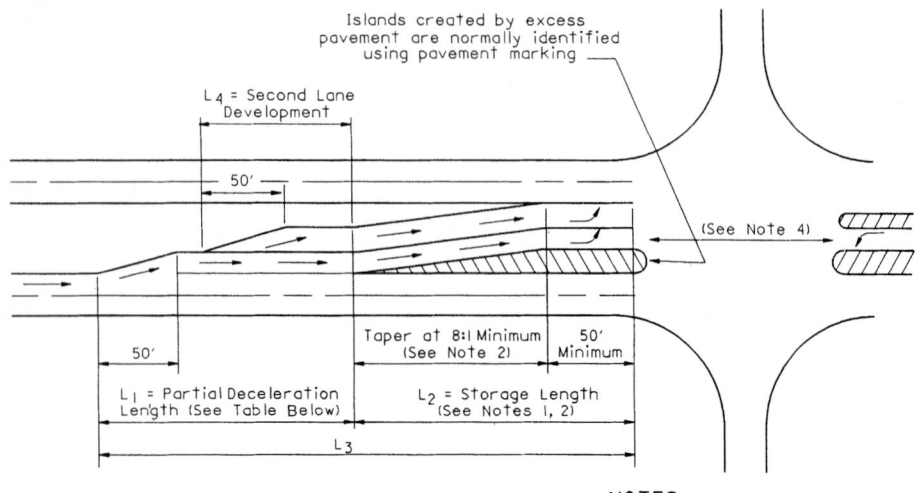

NOTES

1. Using the design turning volumes, obtain the required storage length from Table 2.26 and divide by 2. This figure becomes L_2.

2. If the lanes are being shifted laterally, the first portion of L_2 should be used to accomplish the transition. The minimum taper rate should be 8:1, although flatter rates may be used for better appearance. The last 50 ft at the intersection should not be tapered. Increase L_2 if necessary to meet above requirements.

3. Using the value of L_2 from Notes 1 and 2, add the value of L_1 from the table at left to obtain L_3. If this value is less than the table value at left, increase one or both of the tangent areas of L_1 and L_2 to make up the difference.

4. If the opposite approach has one left turn lane, these lanes should line up.

Design Speed (mi/h)	L_1 (ft)	L_3 (ft)	L_4 (ft)
40	111	125	61
45	125	175	75
50	143	225	75
55	164	285	75
60	181	345	75

FIGURE 2.36 Layout for double left turn lanes with lateral offsets. (*From* Location and Design Manual, *Vol. 1*, Roadway Design, *Ohio Department of Transportation, with permission*)

turning traffic on high-speed facilities and separate right-turning traffic from the rest of the traffic stream at stop-controlled or signalized intersections. As a general guideline, an exclusive right turn lane should be considered when the right turn volume exceeds 300 vehicles per hour per lane.

Figure 2.34*c* shows the design of right turn lanes. Table 2.26 may be used in preliminary design to estimate the storage required at signalized intersections. The recommended maximum length of right turn lanes at signalized intersections is 800 ft, with 100 ft the minimum length.

The blockage of the right turn lane by the through vehicles should also be checked using Table 2.26. With right-turn-on-red operation, it is imperative that access to the right turn lane be provided to achieve full utilization of the benefits of this type of operation.

The width of right turn lanes should desirably be equal to the normal through lane width for the facility. In low-speed areas, a minimum width of 10 ft may be provided. Additional lane width should be provided when the right turn lane is adjacent to a curb.

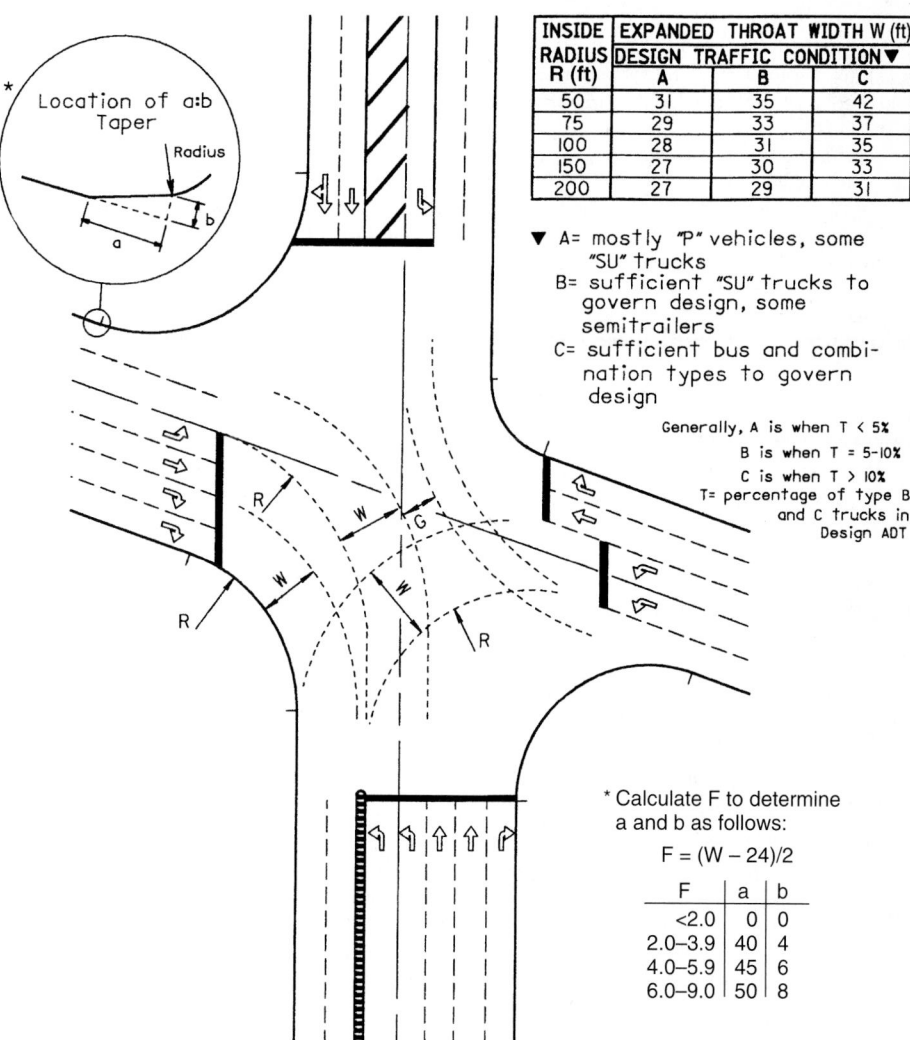

INSIDE RADIUS R (ft)	EXPANDED THROAT WIDTH W (ft) DESIGN TRAFFIC CONDITION▼		
	A	B	C
50	31	35	42
75	29	33	37
100	28	31	35
150	27	30	33
200	27	29	31

▼ A= mostly "P" vehicles, some "SU" trucks
B= sufficient "SU" trucks to govern design, some semitrailers
C= sufficient bus and combination types to govern design

Generally, A is when T < 5%
B is when T = 5-10%
C is when T > 10%
T= percentage of type B and C trucks in Design ADT

* Location of a:b Taper
Radius

* Calculate F to determine a and b as follows:

$$F = (W - 24)/2$$

F	a	b
<2.0	0	0
2.0–3.9	40	4
4.0–5.9	45	6
6.0–9.0	50	8

FIGURE 2.37 Layout for double left turn lanes showing expanded throat width required. (*From Location and Design Manual, Vol. 1, Roadway Design, Ohio Department of Transportation, with permission*)

Double Right Turn Lanes.   Double right turn lanes are rarely used. When they are justified, it is generally at an intersection involving either an off-ramp or a one-way street. Double right turn lanes require a larger intersection radius (usually 75 ft or more) and a throat width comparable to a double left turn (Fig. 2.37).

Additional Through Lanes.   Normally, the number of through lanes at an intersection is consistent with the number of lanes on the basic facility. Occasionally, through lanes are added on the approach to enhance signal design. As a general suggestion, enough main roadway lanes should be provided that the total through plus turn volume does not exceed 450 vehicles per hour per lane.

Recovery Area at Curbed Intersections.   When a through lane becomes a right-turn-only lane at a curbed intersection, an opposite-side tapered recovery area should be considered. The taper should be long enough to allow a trapped vehicle to escape, but not so long as to appear like a merging lane. Taper lengths may vary from 200 to 250 ft depending on design speed.

Pedestrians.   Whenever sidewalks approach a curbed intersection, curb ramps must be provided, lining up with the crosswalks. At signalized intersections, when pedestrians are moving concurrently with traffic on one of the phases, sufficient time must be provided on the phase to allow pedestrians to cross the intersection. This is especially significant on intersections with large radii or multiple through lanes. There may be situations where pedestrian volumes will require a separate phase of the signal to be dedicated to their passage.

Other Considerations.   On designated bikeway routes, bicycles may have their own lane approaching the intersection. This will require special handling in cases where right-turning vehicles may be crossing the path of through bicycle traffic. The designer should consult with a bicycle facility design reference before proceeding with the intersection design.

Providing fixed-source lighting at intersections, especially in urban areas, is always a safety benefit to drivers. It is particularly important for large-area intersections and channelized intersections, since turning paths may be difficult to determine at night.

The development of adjacent property can sometimes have a detrimental effect on intersection design, since driveways accessing the development may be located too close to the intersection. Whenever possible, accesses to adjacent properties should be located far enough from the intersection so as not to interfere with turn lane design.

2.4.2   Two-Way Left Turn Lanes

A two-way left turn lane may be considered a special type of "intersection" design, since its purpose is to provide a separate lane for traffic in both opposing lanes to slow down and turn out of the traffic stream in front of opposing traffic. Rather than concentrate the left turners at a single crossroad intersection, the two-way left turn lane spreads out the turning movements over a continuous stretch of roadway. Mid-block left turns are often a serious problem in urban and suburban areas. They can be a safety problem due to angle accidents with opposing traffic as well as rear-end accidents with traffic in the same direction. Mid-block left turns also restrict capacity. Two-way left turn lanes (TWLTLs) have proven to be a safe and cost-effective solution to this problem. TWLTLs should be considered whenever actual or potential mid-block conflicts occur. This is particularly true when accident data indicate a history of mid-block left turn–related accidents. Closely spaced driveways, strip commercial development, or multiple-unit residential land use along the corridor are other indicators of the possible need for a TWLTL. Some guidelines that may be used to justify the use of TWLTLs are listed below:

- 10,000 to 20,000 vehicles per day for four-lane highways
- 5000 to 12,000 vehicles per day for two-lane highways
- 70 mid-block turns per 1000 ft during peak hour
- Left turn peak hour volume 20 percent or more of total volume
- Minimum reasonable length of 1000 ft or two blocks

Widths for TWLTLs are preferably the same as through lane widths. Lane widths may be reduced to as little as 10 ft in restricted areas. Care should be taken not to make a TWLTL wider than 14 ft, since this may encourage shared side-by-side use of the lane.

2.4.3 Interchanges

An *interchange* is defined as a system of interconnecting roadways in conjunction with one or more grade separations that provides for the movement of traffic between two or more roadways or highways on different levels. Interchanges are utilized on freeways and expressways, where access control is important. They are used on other types of facilities only where crossing and turning traffic cannot be accommodated by a normal at-grade intersection.

Interchange Spacing.   Interchanges should be located close enough together to properly discharge and receive traffic from other highways or streets, and far enough apart to permit the free flow and safety of traffic on the main facility. In general, more frequent interchange spacing is permitted in urbanized areas. Minimum spacing is determined by weaving requirements, ability to sign, lengths of speed change lanes, and capacity of the main facility. Interchanges within urban areas should be spaced not closer than an average of 2 mi, in suburban sections an average of not closer than 4 mi, and in rural sections an average of not closer than 8 mi. In consideration of the varying nature of the highway, street, or road systems with which the freeway or expressway must connect, the spacings between individual adjacent interchanges may vary considerably. In urban areas, the minimum distance between adjacent interchanges should not be less than 1 mi, and in rural areas not less than 2 mi.

Interchange Type.   The most commonly used types of interchanges where two routes cross each other are the diamond, cloverleaf, and directional interchanges. When one route ends at an interchange with another route, a T or Y interchange can be used. Figures 2.38 and 2.39 show schematic examples of various types of interchanges. The diamond interchange is the most common type where a major facility intersects a minor facility. The capacity is limited by the at-grade intersections at the crossroad. Cloverleaf or partial cloverleaf designs may be used in lieu of a diamond when development or other physical conditions prohibit construction in a quadrant, or where heavy left turns are involved. A continuous-flow design is required where two major facilities intersect. In this case, a cloverleaf interchange is the minimum design that can be used. The designer should consider collector-distributor roads in conjunction with cloverleaf interchanges to minimize weaving problems. Directional inter-

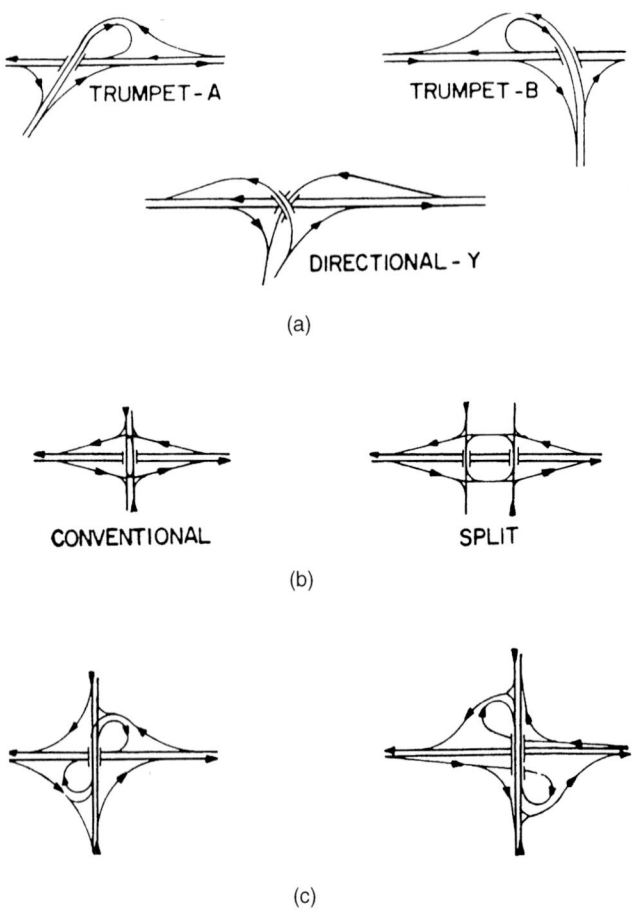

FIGURE 2.38 Common types of interchanges. (*a*) T and Y. (*b*) Diamond. (*c*) Partial cloverleaf. (*From* A Policy on Geometric Design of Highways and Streets, *American Association of State Highway and Transportation Officials, 1990, with permission*)

changes are the highest type and most expensive. They permit vehicles to move from one major highway to another major highway at relatively fast and safe speeds. The T and Y interchanges may take the form of a "trumpet" interchange, in which at least one movement has a loop ramp design. A directional Y interchange is designed to allow vehicles to maintain a relatively high speed while negotiating the ramps.

2.5 INTERCHANGE RAMP DESIGN

An interchange ramp is a roadway that connects two legs of an interchange. Ramp cross-section elements are discussed in Art. 2.3, Cross-Section Design. Elements con-

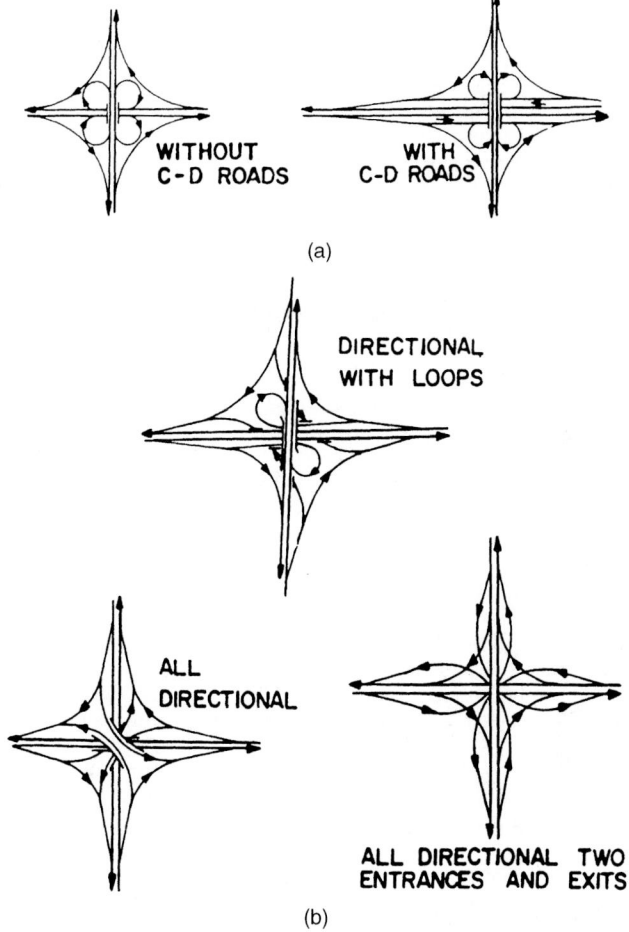

WITHOUT
C-D ROADS

WITH
C-D ROADS

(a)

DIRECTIONAL
WITH LOOPS

ALL
DIRECTIONAL

ALL DIRECTIONAL TWO
ENTRANCES AND EXITS

(b)

FIGURE 2.39 Common types of interchanges. (*a*) Full cloverleaf. (*b*)
Directional. (*From* A Policy on Geometric Design of Highways and Streets,
*American Association of State Highway and Transportation Officials, 1990,
with permission*)

tributing to horizontal and vertical alignments are designed similar to any roadway
once the ramp design speed has been determined.

2.5.1 Ramp Design Speed

To design horizontal and vertical alignment features, a design speed must be determined for each ramp. Since the driver expects a speed adjustment on a ramp, the
design speed may vary within the ramp limits. Table 2.27 includes three ranges of
ramp design speeds that vary with the design speed of the mainline roadway. The
ramp design speed range is determined by judgment based on several conditions:

TABLE 2.27 Guide for Selecting Ramp Design Speed

Mainline design speed, mi/h	30	35	40	45	50	55	60	65	70
Ramp design speed, upper range, mi/h	25	30	35	40	45	45	50	55	60
Ramp design speed, middle range, mi/h	20	25	30	30	35	35	40	45	50
Ramp design speed, lower range, mi/h	15	15	20	20	25	25	30	30	35

Source: Location and Design Manual, Vol. 1, *Roadway Design,* Ohio Department of Transportation, with permission.

- The types of roadways at each end of the ramp and their design speeds
- The length of the ramp
- The terminal conditions at each end
- The type of ramp (diamond, loop, or directional)

Diamond ramps normally have a high-speed condition at one end and an at-grade intersection with either a stop or a slow turn condition at the other. Upper- to middle-range design speeds in Table 2.27 are normal near the high-speed facility. Middle- to lower-range design speeds are usually used closer to the at-grade intersection. Loop ramps may have a high-speed condition at one end and either a slow- or a high-speed condition at the other. Loop ramps, because of their relatively short radius, usually have lower-range design speeds in the middle- and slow-speed end of the ramp, and upper- to middle-range design speeds nearer the high-speed terminal(s). Directional ramps generally have high-speed conditions at both ends. They are normally designed using an upper-range design speed, and the absolute minimum design speed should be from the middle range.

2.5.2 Single-Lane Ramp Terminals

A ramp terminal is that portion of a ramp adjacent to the through lane on the mainline. It includes both the taper and speed-change lane. Ohio uses three basic ramp terminal classifications:

Class I terminals (Fig. 2.40).   Class I terminals are intended for use on all rural interstate highways. They may also be used on other limited-access freeways or expressways having similar design standards. Design notes are provided in Table 2.28.

Class II terminals (Fig. 2.41).   Class II terminals are intended for use on urban interstate highways and all other limited-access freeways and expressways where class 1 terminals are not designated or required. Design notes are provided in Table 2.29.

Class III terminals. (Fig. 2.42).   Class III terminals are intended for use on highways that have little or no access control except through an interchange area. Many of the features of class III terminals are applicable to a terminal of one ramp with another ramp. Class III terminals are also used with collector-distributor roads. Design notes are provided in Table 2.30.

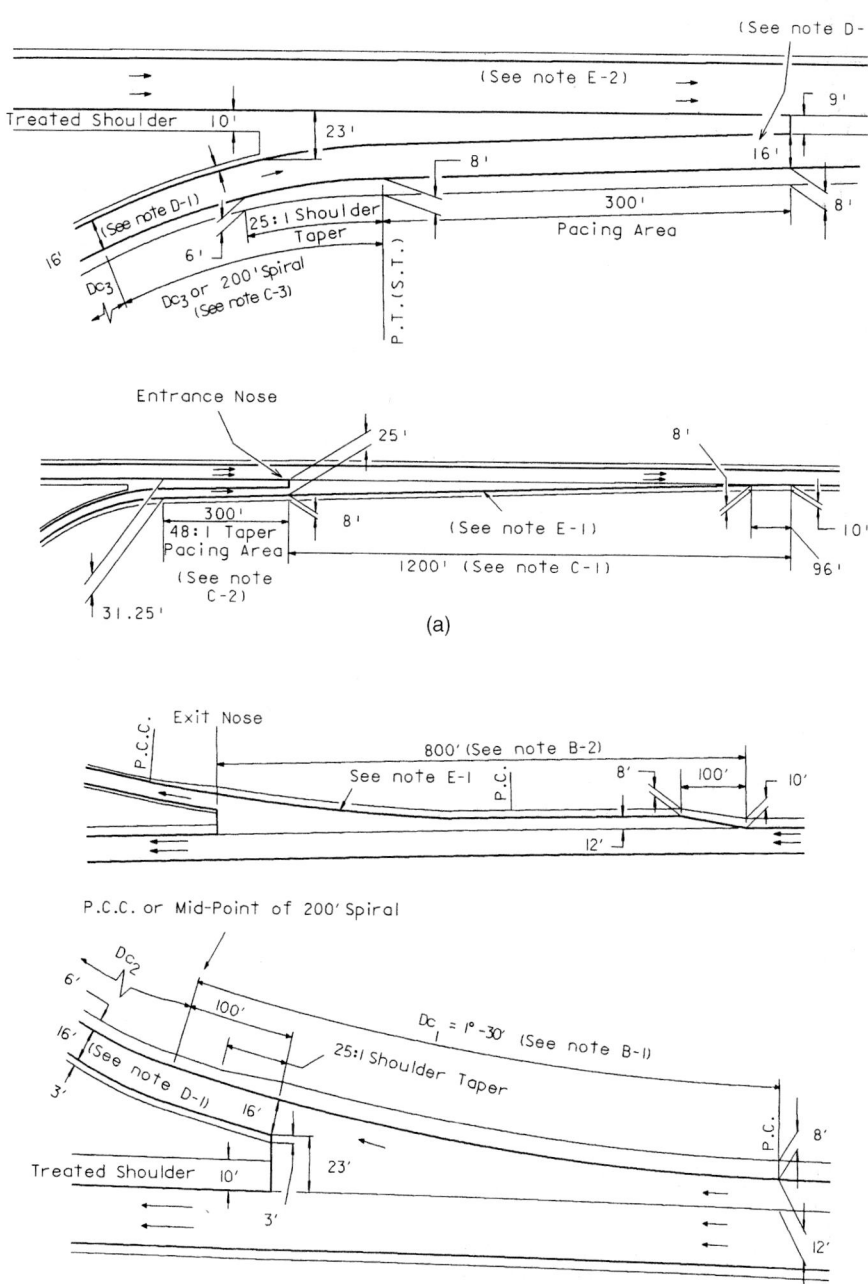

FIGURE 2.40 Class I ramp terminals. (*a*) Entrance terminals. (*b*) Exit terminals. See Table 2.28 for notes. (*From* Location and Design Manual, *Vol. 1,* Roadway Design, *Ohio Department of Transportation, with permission*)

TABLE 2.28   Design Notes for Class I Entrance and Exit Terminals

See Fig. 2.40

A. *General*
   1. Class I terminals are intended for use on all rural interstate highways. They will be used on such other limited access facilities as may be approved by the Bureau of Location and Design.

B. *Exit terminal*
   1. Exit curve Dc_1 shall be 1°30′ where the mainline is on tangent alignment. Where the mainline is on a curving alignment, the maximum differential between Dc_1 and the mainline curve shall normally be 1°30′. This differential, however, may vary by as much as one degree in order to avoid a tangent exit alignment.
   2. The 800-ft deceleration lane length may be reduced to 600 ft if such reduction would eliminate the need for bridge widening.
   3. When ramp curve Dc_2 does not exceed 8°, exit curve Dc_1 may be compounded directly with Dc_2 at a PCC 100 ft beyond the nose. When Dc_2 does exceed 8°, a spiral should be placed between Dc_1 and Dc_2 and the beginning of the spiral (CS) should be at the exit nose.

C. *Entrance terminal*
   1. The acceleration lane shall be a uniform taper (48:1) relative to the mainline pavement edge for either tangent or curving alignment.
   2. The right edge of the 300-ft pacing area shall be a rearward extension of the right edge of the acceleration lane (i.e., on a 48:1 taper relative to the mainline).
   3. When ramp curve Dc_3 does not exceed 8°, the PT shall be at the beginning of the 300-ft pacing area. When Dc_3 does exceed 8°, a 200-ft spiral shall be placed between Dc_3 and the pacing area.

D. *Ramp width*
   1. Normally single lane ramps will have a width of 16 ft. The width shall be increased to 18 ft when the ramp radius is less than 200 ft. When an 18-ft-wide ramp is used, the 39-ft exit and 25-ft entrance terminal widths shall be retained and the 23-ft and 9-ft widths reduced 2 ft.

E. *Treated shoulder*
   1. If the treated shoulder is less than 8 ft in width along the mainline, this lesser width shall be used along the speed change lanes.
   2. The 9- to 23-ft variable-width treated shoulder of the entrance terminal shall be sloped for 10 ft as required for mainline design (usually ½ in/ft), except for the last 100 to 200 ft at the 9-ft end, which is to be sloped as required for proper terminal grading.

F. *Left side terminals*
   1. A left side exit or entrance terminal shall be designed similarly to the drawing shown, but of opposite hand. Treated shoulder widths shall be in accord with notes E-1 and E-2.

Source: Location and Design Manual, Vol. 1, *Roadway Design,* Ohio Department of Transportation, with permission.

2.5.3 Superelevation at Terminals

Superelevation at ramp terminals should be developed using the following guidelines. The rate of superelevation at the entrance and exit nose should be selected on the basis of the design speed of the ramp at the nose. All transverse changes or breaks in superelevation should be made at joint lines in concrete pavement. In the case of bituminous pavement, the superelevation breaks should occur in the same locations as they would in concrete pavement. For class I and class II terminals, the transverse breaks in superelevation cross-slope should not exceed a differential of 0.032 ft/ft at the mainline pavement edge or 0.050 ft/ft at other locations. When a double break occurs on

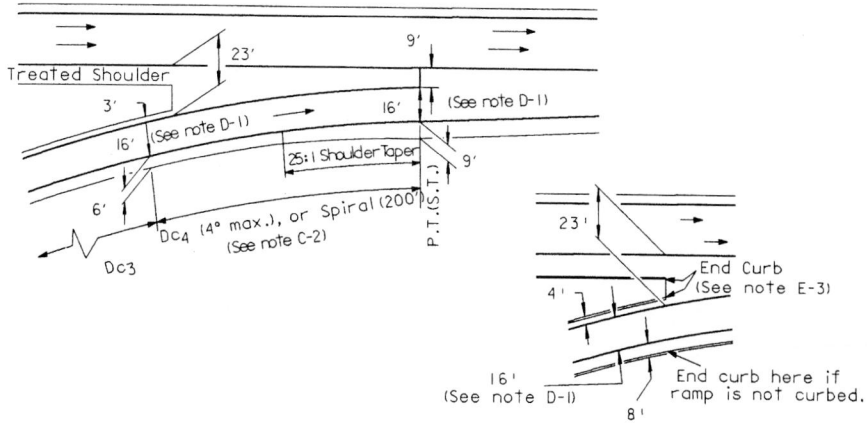

CURBED ENTRANCE TERMINAL

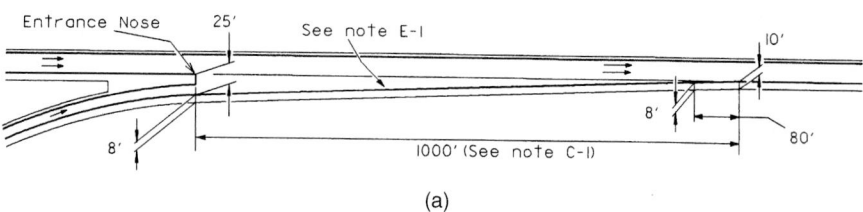

(a)

FIGURE 2.41 Class II ramp terminals. (*a*) Entrance terminals. (*b*) Exit terminals. See Table 2.29 for notes. (*From* Location and Design Manual, *Vol. 1,* Roadway Design, *Ohio Department of Transportation, with permission*)

longitudinal joints less than 6 ft apart, it should not exceed a total differential of 0.032 ft/ft, if adjacent to the mainline, or 0.050 ft/ft elsewhere. On class III terminals, the transverse breaks in superelevation cross slope should not exceed a differential of 0.05 to 0.06 ft/ft. For class I and II terminals, the rate of rotation of a superelevated ramp pavement or speed change lane pavement should be in accordance with rates from Table 2.13. Where possible, the terminal area pavement and shoulder should slope away from the mainline pavement so that a minimum amount of water drains across the mainline pavement.

2.5.4 Terminals on Crest Vertical Curves

Mainline crest vertical curves in the vicinity of ramp terminals should be designed using preferred stopping sight distances. Where a crest vertical curve occurs on an exit ramp at or near the nose, the crest vertical curve should be designed using the "upper-range" design speeds of Table 2.27.

2.5.5 Ramp At-Grade Intersections

Ramp at-grade intersections are designed using many of the same criteria as outlined in Art. 2.4.1. However, one of the basic differences is the one-way nature of ramps

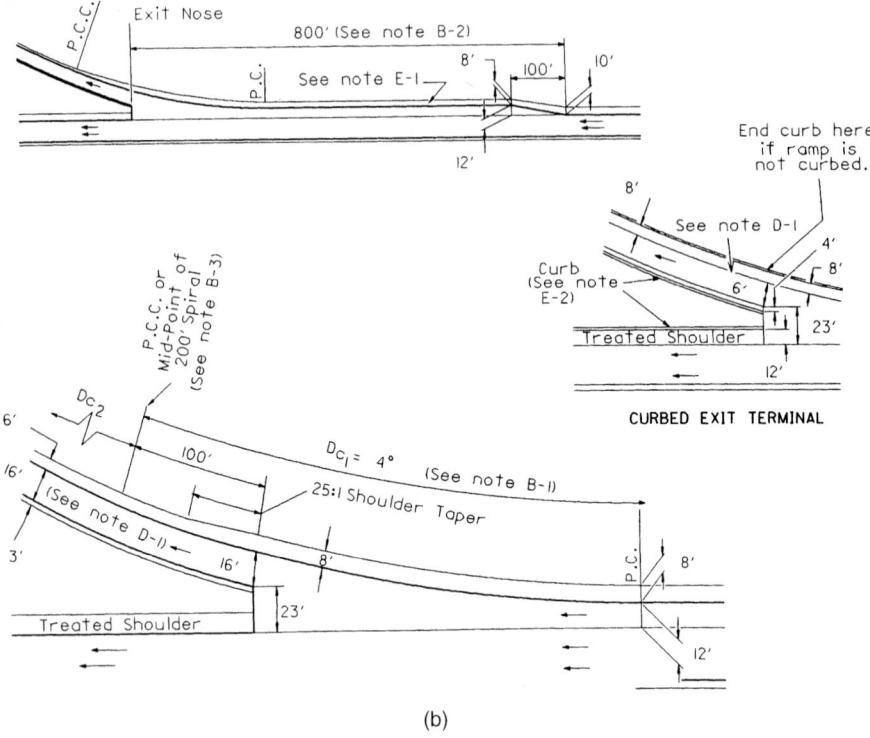

(b)

FIGURE 2.41 *(Continued)*

and the fact that most traffic at ramp intersections is turning. Figure 2.43 shows the design of a typical uncurbed ramp intersection. Curbed returns are normally used in urban areas where space is more restricted.

2.6 COLLECTOR-DISTRIBUTOR ROADS

Collector-distributor (C-D) roads are used to minimize weaving problems and reduce the number of conflict points (merging and diverging) on the mainline. C-D roads may be used within a single interchange, through two adjacent interchanges, or continuously through several interchanges.

2.6.1 Design of C-D Roads

When a C-D road is provided between interchanges, a minimum of two lanes should be used. Either one or two lanes may be used on C-D roads within a single interchange. The cross-section elements for one- and two-lane C-D roads should be in accordance with the criteria for one-lane and two-lane directional roadways provided in Fig. 2.24. The separation between the mainline and C-D road pavements should be

TABLE 2.29   Design Notes for Class II Entrance and Exit Terminals

See Fig. 2.41

A. *General*
1. Class II terminals are intended for use on all urban interstate highways and all other limited access facilities where the use of class I is not designated or required.

B. *Exit terminal*
1. Exit curve Dc_1 shall be 4°00′ maximum where the mainline is on tangent alignment. Where the mainline is on a curving alignment, the maximum differential between Dc_1 and the mainline curve shall normally not exceed 4°. This differential, however, may vary by as much as one degree in order to avoid a tangent exit alignment.
2. The 800-ft deceleration lane length may be reduced to 600 ft if such reduction would eliminate the need for bridge widening.
3. When ramp curve Dc_2 does not exceed 8°, exit curve Dc_1 may be compounded directly with Dc_2 at a PCC 100 ft beyond the nose. When Dc_2 does exceed 8°, a 200-ft spiral should be placed between Dc_1 and Dc_2 and the beginning of the spiral (CS) should be at the exit nose or on the ramp.

C. *Entrance terminal*
1. The acceleration lane shall be a uniform taper (40:1) relative to the mainline pavement edge for either tangent or curving alignment.
2. The design of the entrance terminal curvature shall be based on the following:
   (*a*) *Ramp curve* Dc_3 *of 8° or less.*   When the mainline is on a tangent or a curve to the right, Dc_4 shall be a 200-ft-long simple curve of a degree such that the differential between it and the mainline will not exceed 4°. When the mainline is on a curve to the left. A 200-ft tangent shall be substituted for Dc_4.
   (*b*) *Ramp curve* Dc_3 *greater than 8°.*   A 200-ft spiral shall be substituted for Dc_4. When the mainline is on a curve to the left, a 100-ft tangent shall be inserted between the 200-ft spiral and the entrance nose.

D. *Ramp width*
1. Normally single lane ramps will have a width of 16 ft. The width shall be increased to 18 ft when the ramp radius is less than 200 ft. When an 18-ft wide ramp is used, the 39-ft exit and 25-ft entrance terminal widths shall be retained and the 23- and 9-ft widths reduced 2 ft.

E. *Treated shoulder*
1. If the treated shoulder is less than 8 ft in width along the mainline, this lesser width shall be used along the speed change lanes.
2. When needed, taper curb to required offset along ramp or mainline pavement using a 25:1 taper.
3. At all curb ends use a 6-in to 0-in height reduction in 10 ft.

F. *Left side terminals*
1. A left side exit or entrance terminal shall be designed similarly to the drawing shown, but of opposite hand. Treated shoulder widths shall be in accord with note E-1.

Source: *Location and Design Manual,* Vol. 1, *Roadway Design,* Ohio Department of Transportation, with permission.

designed to prevent, or at least discourage, indiscriminate crossovers. As a minimum, the separation should be wide enough to provide normal shoulder widths for both the mainline and C-D road roadways plus a suitable median. Normally, a standard concrete barrier median is used, since C-D road separation often involves obstructions such as bridge parapets, piers, or overhead sign supports. There may be isolated cases where a lesser-type median may be used.

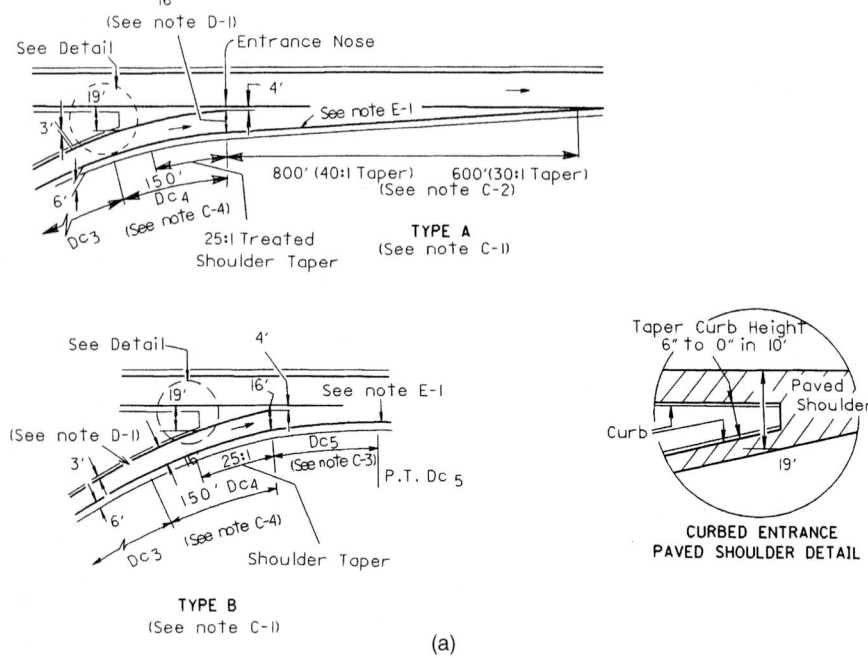

FIGURE 2.42   Class III ramp terminals. (*a*) Entrance terminals. (*b*) Exit terminals. See Table 2.30 for notes. (*From* Location and Design Manual, *Vol. 1,* Roadway Design, *Ohio Department of Transportation, with permission*)

2.6.2   C-D Road Entrance and Exit Terminals

Figure 2.44 shows both class I and class II C-D road entrance terminals. The class I collector-distributor entrance terminal is intended for use on rural interstate highways and other freeways where class I design has been designated. The class II collector-distributor entrance terminal is intended for use on all other freeways. Three exit terminal lane conditions are shown on Fig. 2.45. These terminal designs are to be applied to highways using either class I or class II terminals.

Superelevation at C-D terminals should be developed similar to that prescribed for standard ramp terminals.

2.7   MULTILANE RAMP AND ROADWAY TERMINALS AND TRANSITIONS

When two roadways converge or diverge, the less significant roadway should exit or enter on the right. Left-hand exits or entrances are contrary to driver expectancy and should be avoided wherever possible.

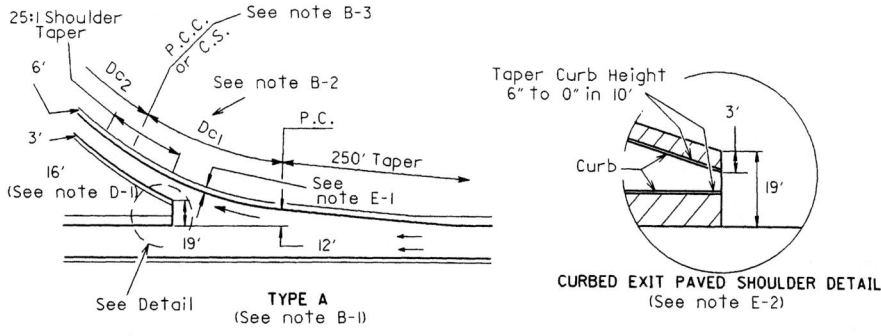

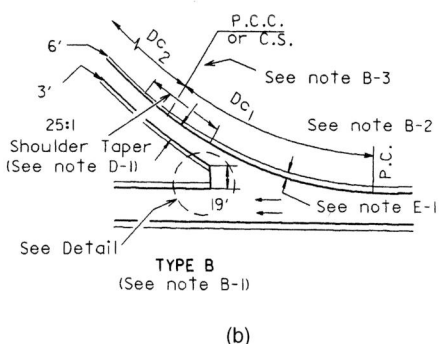

(b)

FIGURE 2.42   *(Continued)*

2.7.1   Multilane Entrance Ramps and Converging Roadways

Figure 2.46 shows recommended designs to be used for multilane entrance ramps and converging roadways. Converging roadways are defined as separate and nearly parallel roadways or ramps that combine into a single continuous roadway or ramp having a greater number of lanes beyond the nose than the number of lanes on either approach roadway. Class I, II, and III entrance terminals should be used in lieu of converging roadway drawings when applicable. Class I and II converging roadways should be used when either or both of the converging roadways are mainline roadways of an expressway or a freeway. Class III should be used at the convergence of directional ramps within an interchange or at the convergence of interchange ramps with non-limited-access roads or streets. In general, class III is applicable at all locations other than those requiring the use of class I or II.

Lane Balance and Continuity.   To avoid inside merges, the number of mainline lanes plus converging lanes approaching the nose must be equal to the resultant number of lanes leaving the nose. To make this possible, it is often necessary to carry additional mainline lanes past the nose for an adequate distance prior to tapering back to the desired number of lanes. These details are shown in Fig. 2.46.

TABLE 2.30   Design Notes for Class III Entrance and Exit Terminals

See Fig. 2.42

A. *General*
   1. Class III terminals are intended for use on highways which have little or no access control except through an interchange area. Many of the features of Class III terminals are applicable to a terminal of one ramp with another ramp in a freeway interchange.

B. *Exit terminal: type A and type B*
   1. Type A shall normally be used on highways (including off-system highways) having design speeds of 50 or 60 mi/h; however, type B may be used where substantial savings in bridge or right-of-way cost would result. Type B shall normally be used on a highways having design speeds of 40 mi/h or less.
   2. The curve differential between the through roadway and exit curve Dc_1 may vary from a minimum of 4° to the maximum allowable differential.
   3. Exit curve Dc_1 may be either compounded or spiraled into ramp curve Dc_2.

C. *Entrance terminal: type A and type B*
   1. Type A is preferred and shall normally be used; however, when a ramp enters as an added lane or as a combined acceleration-deceleration lane, type B may be used if its use would result in a substantial savings in cost (i.e., reduced bridge width).
   2. The acceleration lane of type A shall be a uniform taper relative to the through pavement edge for either tangent or curving alignment. A 40:1 taper shall be used for design speeds of 50 or 60 mi/h and a 30:1 taper shall be used for design speeds of 40 mi/h or less.
   3. The curve differential between the through roadway and entrance curve Dc_5 of type B shall be 4°.
   4. The design of the entrance terminal shall be based on the following:
      (a) *Ramp curve* Dc_3 *of 8° or less.*   When the through roadway is on a tangent or a curve to the right, Dc_4 shall be a 150-ft-long simple curve of a degree such that the differential between it and the through roadway will not exceed 4°. When the through roadway is on a curve to the left, a 150-ft tangent shall be substituted for Dc_4.
      (b) *Ramp curve* Dc_3 *greater than 8°.*   A 150-ft spiral shall be substituted for Dc_4.

D. *Ramp width*
   1. Normally single-lane ramps will have a width of 16 ft. The width shall be increased to 18 ft when the ramp radius is less than 200 ft. When an 18-ft-wide ramp is used, the 35-ft exit and 20-ft entrance terminal widths shall be retained and the 19- and 4-ft widths reduced 2 ft.
   2. When it is necessary to provide curbing on a ramp without treated shoulders, the width of pavement shall be 20 ft face-to-face of curb.

E. *Treated shoulder*
   1. The width of the treated shoulders along the speed change lane shall be as shown in Fig. 2.24.
   2. If the ramp or through roadway has a curb offset greater than 6 ft (or 3 ft) the greater width shall be used at the terminal. Retain the 19-ft width.
   3. The special detail drawings shall apply when the through roadway is curbed.

F. *Left side terminals*
   1. Left side entrance and exits shall be designed similarly to the drawing shown, but of opposite hand.

Source: Location and Design Manual, Vol. 1, Roadway Design, Ohio Department of Transportation, with permission.

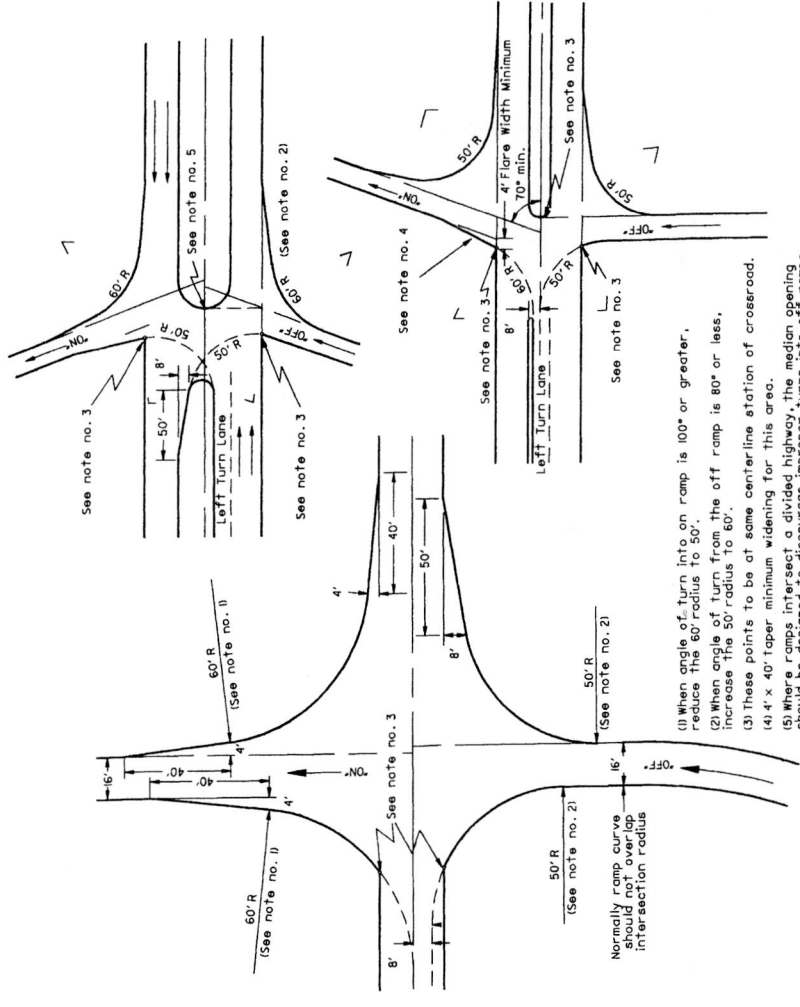

FIGURE 2.43 Design for ramp intersection without curbs. *(From Location and Design Manual, Vol. 1, Roadway Design, Ohio Department of Transportation, with permission)*

(1) When angle of turn into on ramp is 100° or greater, reduce the 60' radius to 50'.

(2) When angle of turn from the off ramp is 80° or less, increase the 50' radius to 60'.

(3) These points to be at same centerline station of crossroad.

(4) 4' × 40' taper minimum widening for this area.

(5) Where ramps intersect a divided highway, the median opening should be designed to discourage improper turns into off ramps.

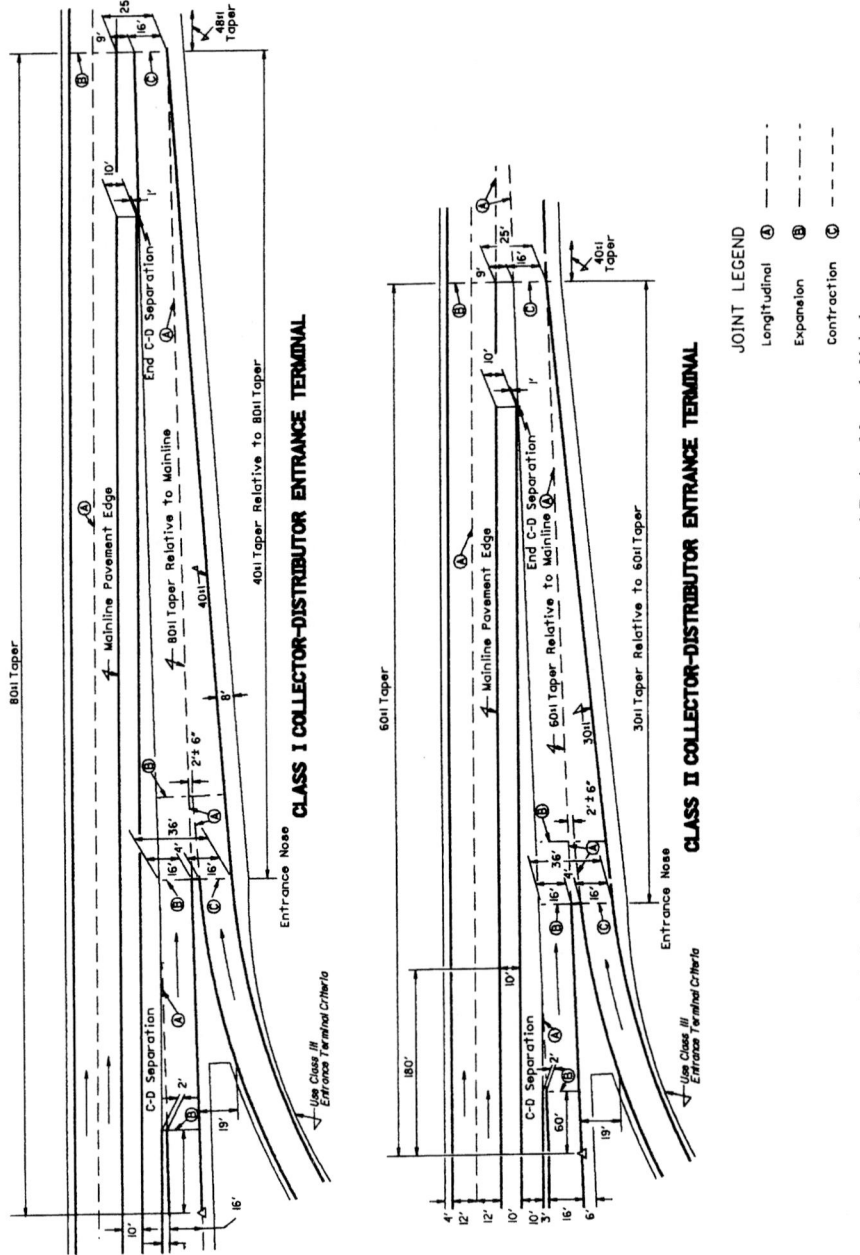

FIGURE 2.44 Entrance terminals for collector-distributor roads. (*From* Location and Design Manual, *Vol. 1,* Roadway Design, *Ohio Department of Transportation, with permission*)

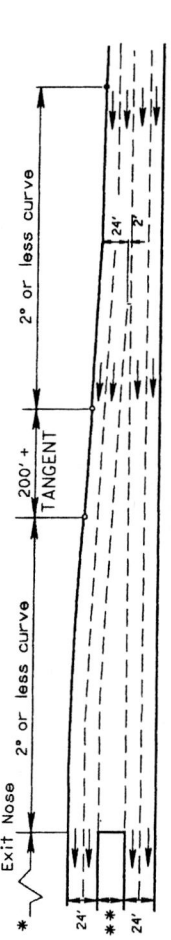

2 LANE EXIT FROM 4 LANES

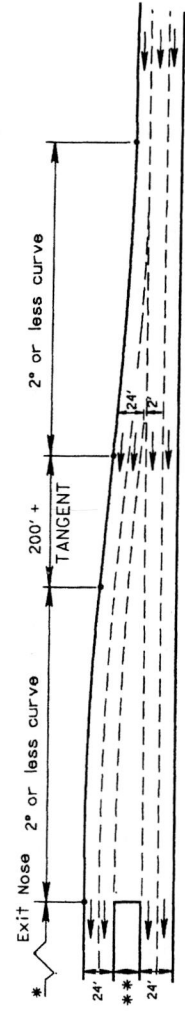

2 LANE EXIT FROM 3 LANES

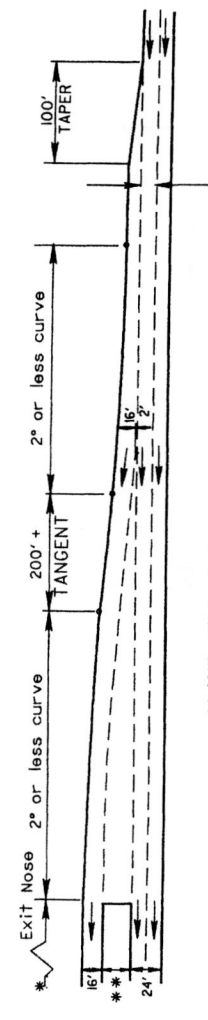

1 LANE EXIT FROM 2 LANES

* Distance to first ramp exit nose 600' minimum with up to
1000' desirable especially for 2 lane C-D roads and for Class 1 Design

** Width determined by type of C-D separation chosen

FIGURE 2.45 Exit terminals for collector-distributor roads. *(From Location and Design Manual, Vol. 1, Roadway Design, Ohio Department of Transportation, with permission)*

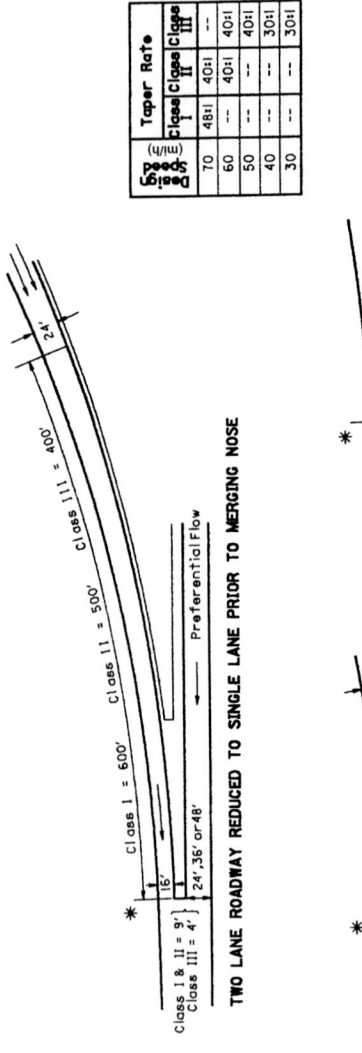
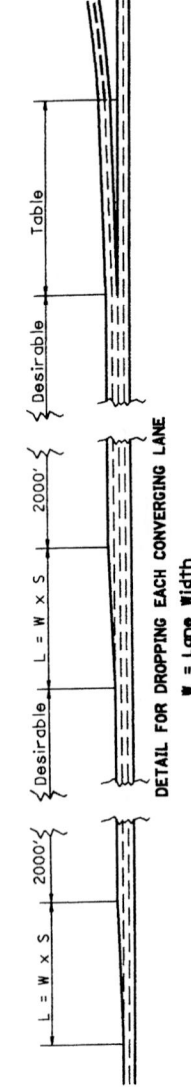

FIGURE 2.46 Designs for multilane entrance ramps and converging roadways. (*From* Location and Design Manual, *Vol. 1,* Roadway Design, *Ohio Department of Transportation, with permission*)

Preferential Flow. In Fig. 2.46, one roadway in each design is labeled "preferential flow." This indicates the more important of the two approaching traffic flows. In selecting the preferential flow, a designer must consider the effect of traffic volumes, number of lanes, the continuity and importance of signed routes, vehicle speeds, and roadway alignment. Lanes carrying the preferential flow are given the higher design treatment. When it is necessary to reduce a number of converging lanes or where an angular change in direction must occur, the design should favor the preferential flow.

Horizontal Curvature. Horizontal curves of roadways approaching the terminal nose should conform to mainline roadway criteria in the case of mainline roadways and to ramp entrance terminal criteria in the case of ramps.

Crest Vertical Curves. Crest vertical curves on constant-width roadways approaching the merging nose should be designed to provide sight distance consistent with the design speed of the roadway. Crest vertical curves from the merging nose forward to a point where pavement convergence ceases, and to the converging portion of an approaching roadway where the number of lanes is being reduced in advance of the nose, should be designed using the preferred stopping sight distance shown in Table 2.17. When design speeds differ on approaching roadways, the higher of the two design speeds should be used in designing the crest vertical curve beyond the merging nose.

Superelevation and Joint Location. Superelevation in the terminal area should be designed in accordance with the guidelines given for single-lane ramp terminals. Longitudinal joints should be located so they will coincide with and define the lane lines.

2.7.2 Multilane Exit Ramps and Diverging Roadways

Figure 2.47 shows recommended designs for multilane exit ramps and diverging roadways. A diverging roadway is defined as a single roadway that branches or forks into two separate roadways without the use of a speed change lane.

Class I and II diverging roadways should be used when either or both the diverging roadways are mainline roadways of an expressway or a freeway. Class III diverging roadways should be used at the divergence of directional ramps within an interchange or at the divergence of ramps with non-limited-access roads or streets. In general, class III is applicable at all locations other than those requiring class I or class II.

Lane Balance and Continuity. To have lane continuity, the number of mainline lanes leaving the diverging nose must be equal to the number of mainline lanes approaching the nose. The total number of lanes leaving the diverging nose (mainline lanes plus diverging lanes) must be 1 greater than the total number of lanes approaching the nose to obtain lane balance. The purpose of obtaining lane continuity and lane balance is to avoid a drop lane situation.

It may be necessary to obtain lane balance by adding additional lanes upstream from the diverging nose. The length of each additional lane should be 2500 ft and should be introduced using a 0- to 12-ft taper of 100 ft as recommended in Fig. 2.47 for the approach roadway class and design speed. There may be conditions off the mainline, such as on collector-distributor roads or within interchanges, where lane balance and continuity are less important. In such cases, the special diverging roadway design shown in Fig. 2.47b may be used.

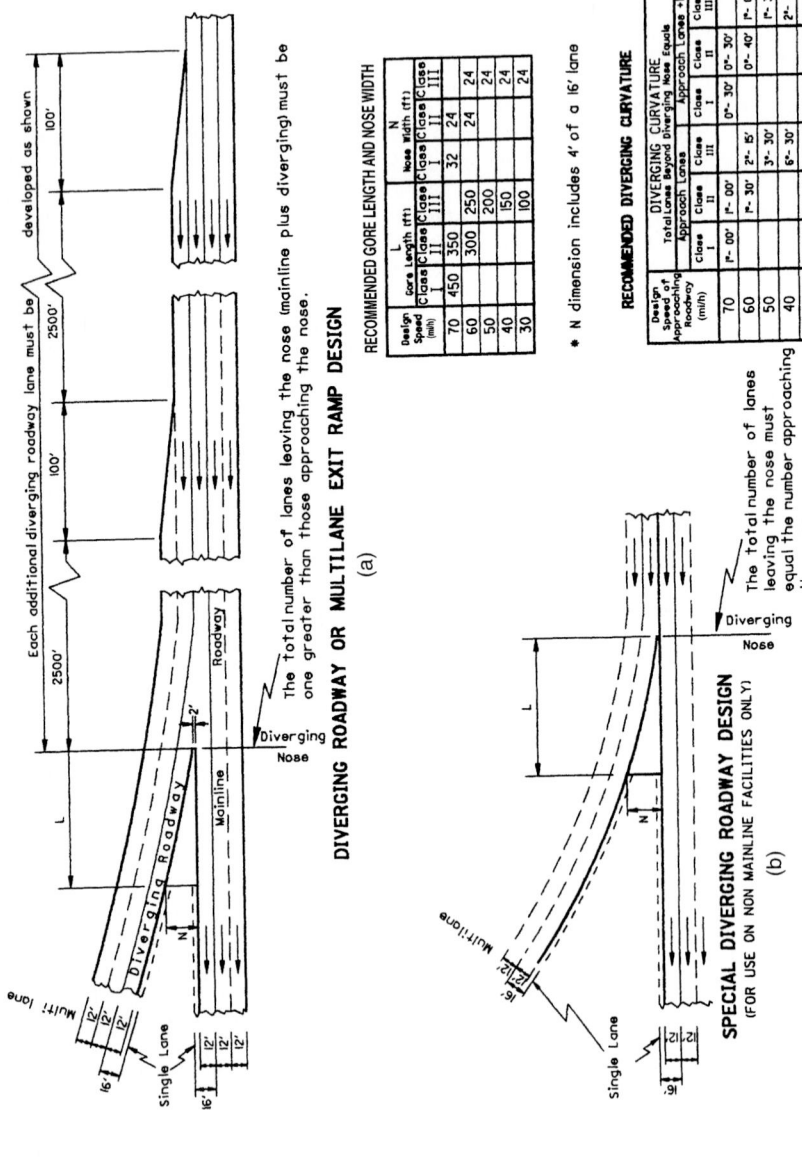

FIGURE 2.47 Designs for multilane exit ramps and converging roadways. (*From* Location and Design Manual, *Vol. 1, Roadway Design, Ohio Department of Transportation, with permission*)

Terminal Design.   The design of diverging roadway terminals is determined by the class and design speed of the approach roadway, and is based on the required neutral gore length L and the required nose width N. Figure 2.47 includes recommended length L and nose width N for various design speeds in diverging roadway classes.

Horizontal Curvature.   The inset table in Fig. 2.47 lists recommended values for the diverging curvature (curve differential) between the outer pavement edges of diverging roadways. These values apply only when the alignment between the diverging nose and the PC of the diverging curvature is on tangent or simple curvature. When compounded or spiral curvature is used in the diverging area, it will be necessary to design diverging roadway alignments individually to provide the proper. L and N for the approach roadway class and design speed.

Crest Vertical Curves.   When a diverging nose is located on a crest vertical curve, the curve should be designed using the design speed of the approach highway and preferred stopping sight distance K value from Table 2.17.

Superelevation and Joint Location.   The superelevation rate should be based on the design speed of the approach roadway. Superelevation in the terminal area should be designed in accordance with the guidelines given for single-lane ramp terminals (Art. 2.5.2). Longitudinal joints should be located so they will coincide with and define the lane lines.

2.7.3   Four-Lane Divided to Two-Lane Transition

Figure 2.48 shows a reversed curve design (types A and B), a tapered design (type C), and a design for a transition on a curve (type D) for achieving a four-lane divided to two-lane transition. The pavement transition should be located in an area where it can easily be seen. Intersections or drives should be avoided in the transition area. Vertical or horizontal curves should provide preferred stopping sight distance. Reverse curve transitions should normally be used for median widths of 20 ft or wider. Taper lengths are based on the design speed of the mainline and are calculated from Eq. (2.5).

2.8   *SERVICE ROADS*

Service roads, or *frontage* roads, as they are sometimes called, are used to enhance capacity on the mainline, control access, serve adjacent properties, or maintain traffic circulation. They permit development of adjacent properties while preserving the through character of the mainline roadway. Service roads may be either one-way or two-way, depending on where they are located and the purpose they are intended to serve.

Although the alignment and profile of the mainline may have an influence, service roads are generally designed to meet specific criteria based on functional classification (usually "local"), traffic volumes, terrain or locale, and design speed. Two features, however, are unique to service roads and are further discussed below. They are (1) the separation between the service road and mainline and (2) the design of the crossroad connection.

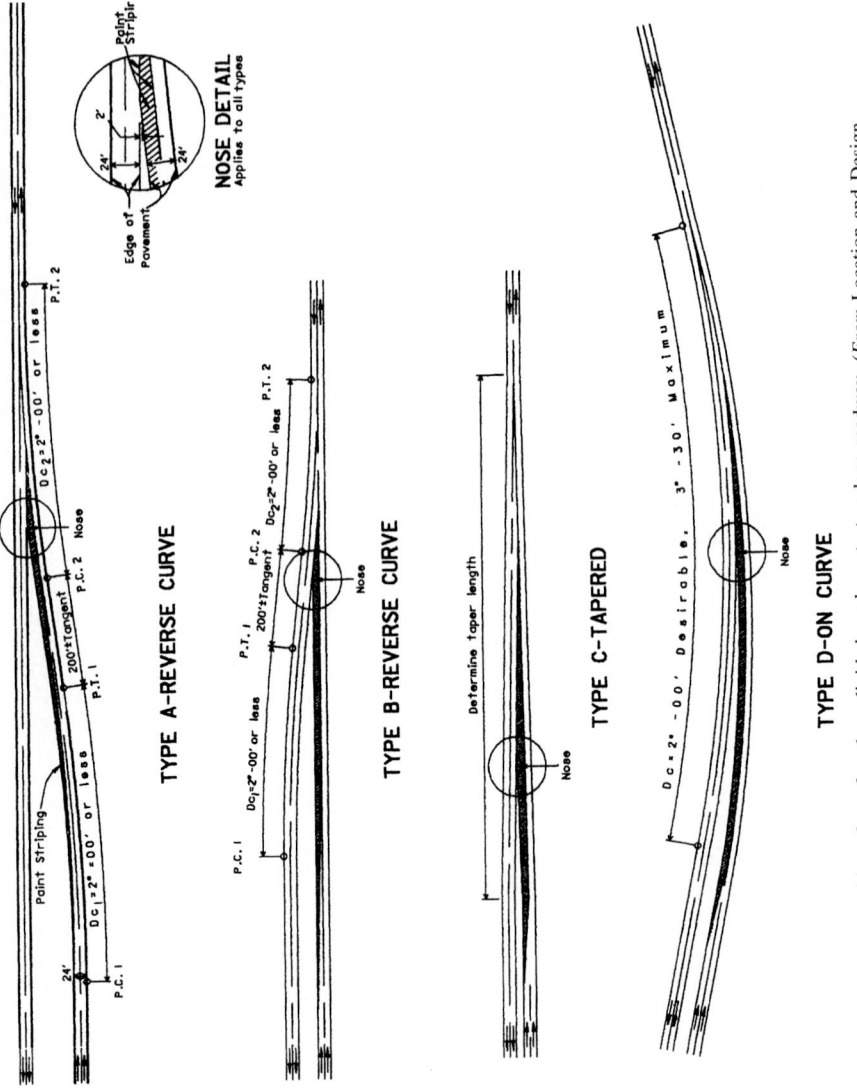

FIGURE 2.48 Transitions from four-lane divided roadway to two-lane roadway. (*From* Location and Design Manual, *Vol. 1*, Roadway Design, *Ohio Department of Transportation, with permission*)

The farther the service road is located from the mainline, the less influence the two facilities will have on each other. A separation width that exceeds the clear zone measurement for each roadway is desirable. However, the separation should be at least wide enough to provide normal shoulder widths on each facility, and also to accommodate surface drainage and a suitable physical traffic barrier. Glare screen is desirable to screen headlights when the service road is two-way.

At intersections with crossroads to the mainline, the distance between the mainline and service road becomes critical. This distance should be great enough to provide adequate storage on the crossroad approach lanes to both the mainline and service road. The recommended minimum distance between the mainline and service road pavement edges is 150 ft in urban areas and 300 ft in rural areas. In addition, the designer should check the adequacy of stopping sight distance on the crossroad as well as intersection sight distance at the service road.

2.9 ACCESS TO PUBLIC ROADS

2.9.1 Defining Access Control

Control of access is the condition where the right of owners or occupants of abutting land to access a highway is fully or partially controlled by public authority. Access control is usually defined by regulations of the authority having jurisdiction over the roadway. The purpose of establishing access control is to provide for the safe and expeditious movement of vehicles on the street or highway, while upgrading the level of service and safety to those living near and using the facility. Regulations may be categorized as full control of access, partial control of access, and driveway and approach regulations.

Full control of access is the means by which preference is given to through traffic by providing access connections only at selected public roads and by prohibiting at-grade crossings and direct private driveway connections. Partial control of access still gives preference to through traffic, but allows some at-grade crossings and some driveway connections. Driveway or approach regulations may apply where no control of access is obtained. Each abutting property is permitted access to the street or highway, but the location, number, and geometrics of the access points may be governed by the regulations.

2.9.2 Establishing Access Control

Access control may be exercised and established by statute—through zoning ordinances, driveway controls, and turning and parking regulations—and by geometric design. Control by statute is used where full access control or a high degree of access control is required. Direct driveway connections may be prohibited, and at-grade intersections may be allowed only with major crossroads. This may be employed for a major urban arterial.

Zoning can control the type of property development adjacent to the roadway, and thereby influence the amount and type of traffic generated in the area. Property uses can be limited to those that attract very few people, excluding those that would generate significant volumes of traffic during hours of peak movement. Zoning regulations can require off-street parking provisions as a condition for permit approval.

Driveway controls can be effective in preserving the functional character of the roadway. On arterials in built-up urban areas, it is important to establish minimum spacing requirements for driveways, as well as the minimum distance from a driveway to the nearest intersection.

An example of geometric design to control access is the use of a frontage road to provide indirect access of abutting properties to a major arterial. Also, the use of a raised concrete median strip in the center of the road can effectively prohibit left turns into or out of driveways.

2.10   DRIVEWAY DESIGN

2.10.1   Location

Part of the process in obtaining a driveway permit is to determine where the driveway will be located. The following guidelines may be used to establish this location.

Wherever possible, drives should be located in accordance with the intersection sight distance criteria (see Table 2.3). Special consideration should be given to the location of drive access to high-volume traffic generators such as shopping centers and industrial plants and parks, as well as other types of development having similar traffic characteristics. These should be treated as standard intersections with appropriate spacing to the nearest intersection. A driveway serving all directions of traffic should be located a minimum of 600 ft from the nearest major highway or street intersection. A new driveway should not be located where it will create an offset intersection opposite an existing street, highway, or major commercial driveway.

2.10.2   Rural Driveway Geometrics

Rural residential drives and field drives should normally conform to the type 1 design shown in Fig. 2.49 (Ref. 14). New drives should intersect the highway at an angle between 70 and 90°. In some cases, however, it may be necessary to retain existing drive angles that vary from these desirable angles.

If the project involves existing drives, the existing width is normally retained unless it is less than 12 ft. In that case, it should be widened to provide a 12-ft throat width. In the case of new drives, the width should normally be 12 ft. If the new driveway is a combined drive between two properties, the width should normally not exceed 24 ft. Also, a wider field drive may be used if it will keep a farm equipment operator from encroaching on the opposing traffic lane when entering or exiting the highway.

The radius of the type 1 driveway should normally be 25 ft. The radius may be increased on field drives if it is deemed that the larger values will improve driveway operation and reduce the hazard to motorists and farm equipment operators.

Driveways abutting uncurbed highways may be curbed. However, the curb should not extend closer to the through pavement edge than 8 ft or the treated shoulder width, whichever is greater. This is recommended to avoid curb obstruction for vehicles, snowplows, etc., using the shoulder.

2.10.3   Urban Residential Drives

Either type 1 or type 2 drives (Fig. 2.49) may be used in urban areas. If used in urban areas, the radius and flare dimensions may be reduced so that the apron does not

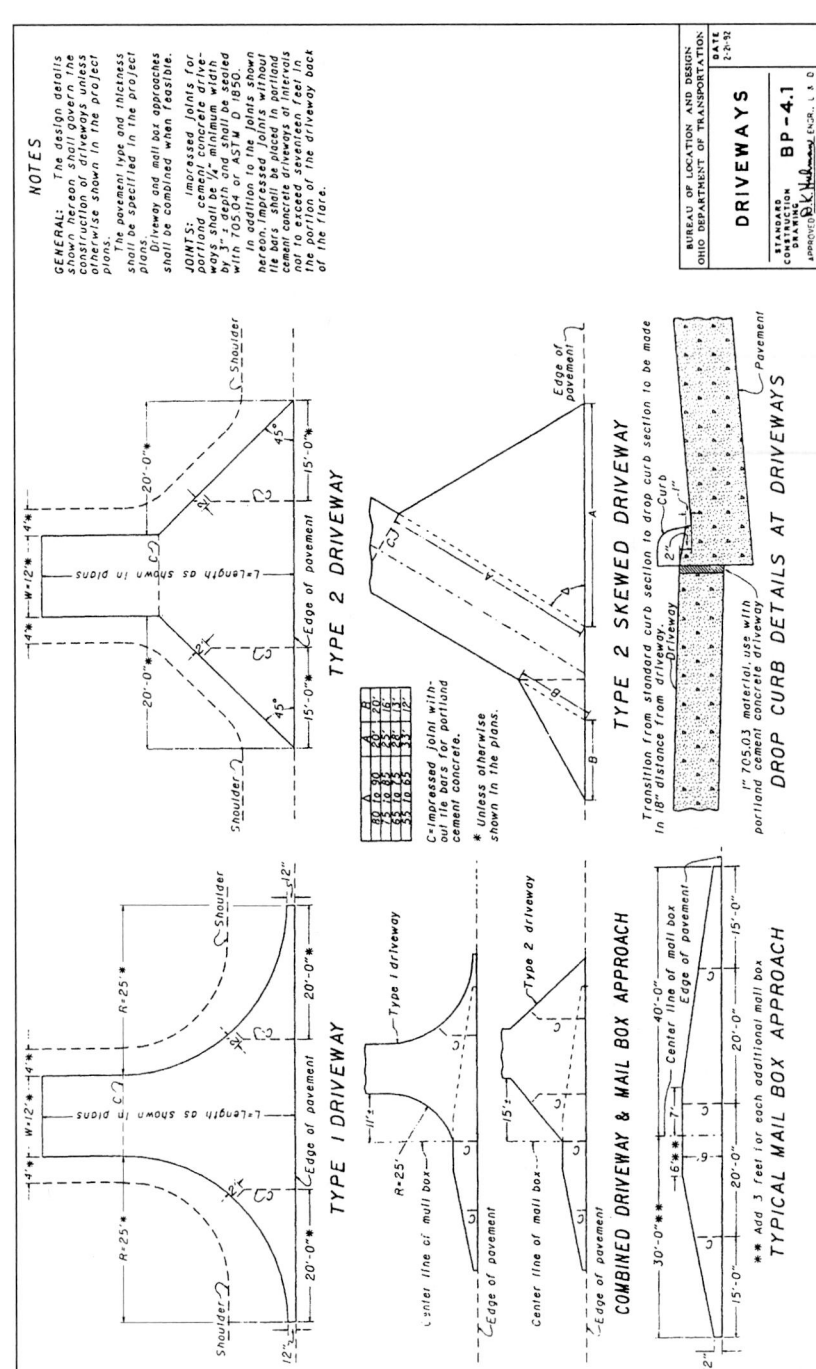

FIGURE 2.49 Driveway designs. *(From Standard Construction Drawings, Bureau of Location and Design, Ohio Department of Transportation, with permission)*

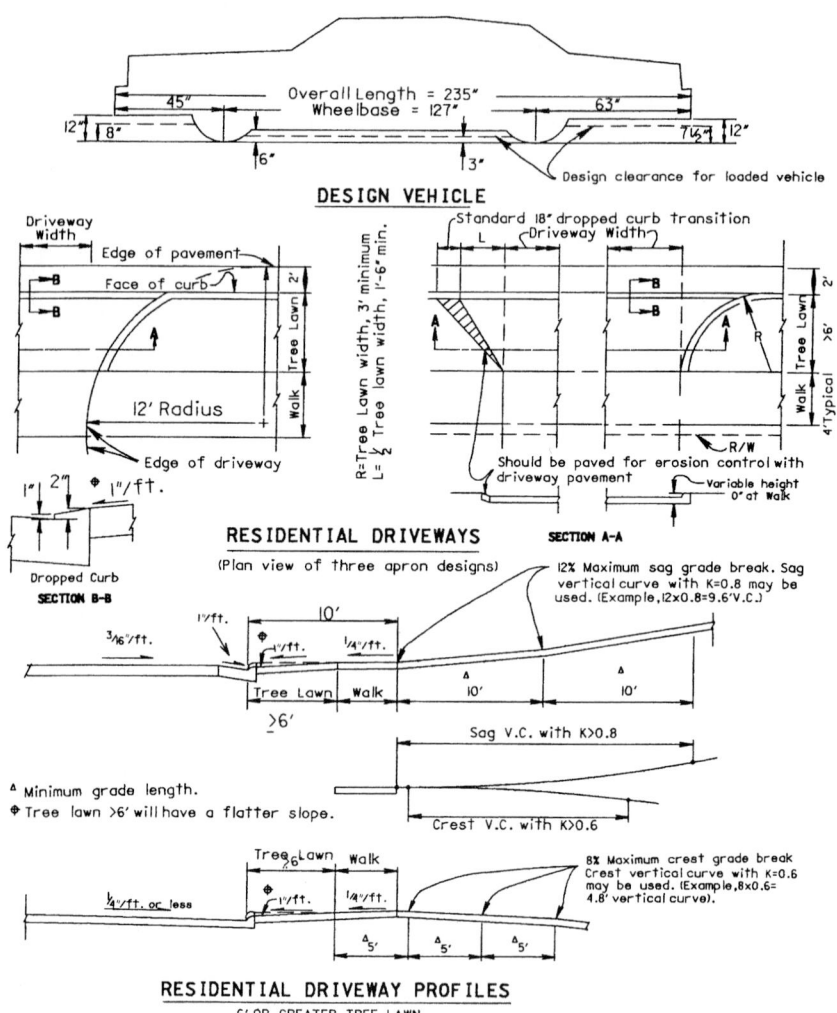

FIGURE 2.50 Details of driveway designs. (*From* Location and Design Manual, *Vol. 1,* Roadway Design, *Ohio Department of Transportation, with permission*)

extend past the back of the sidewalk, or past the right-of-way line if there are no sidewalks. The desirable minimum radius for type 1 drives, when the through highway is curbed, is 15 ft.

Three methods are shown in Fig. 2.50 for designing driveways between the curb line and sidewalk to provide for turning vehicles. Other designs may be used if they are approved for use by the local governmental agencies responsible for maintenance of the project. Additional details are shown in Fig. 2.51 when the tree lawn is less than 6 ft. Residential drives on curbed streets should use a dropped curb as shown in section B-B of Fig. 2.50.

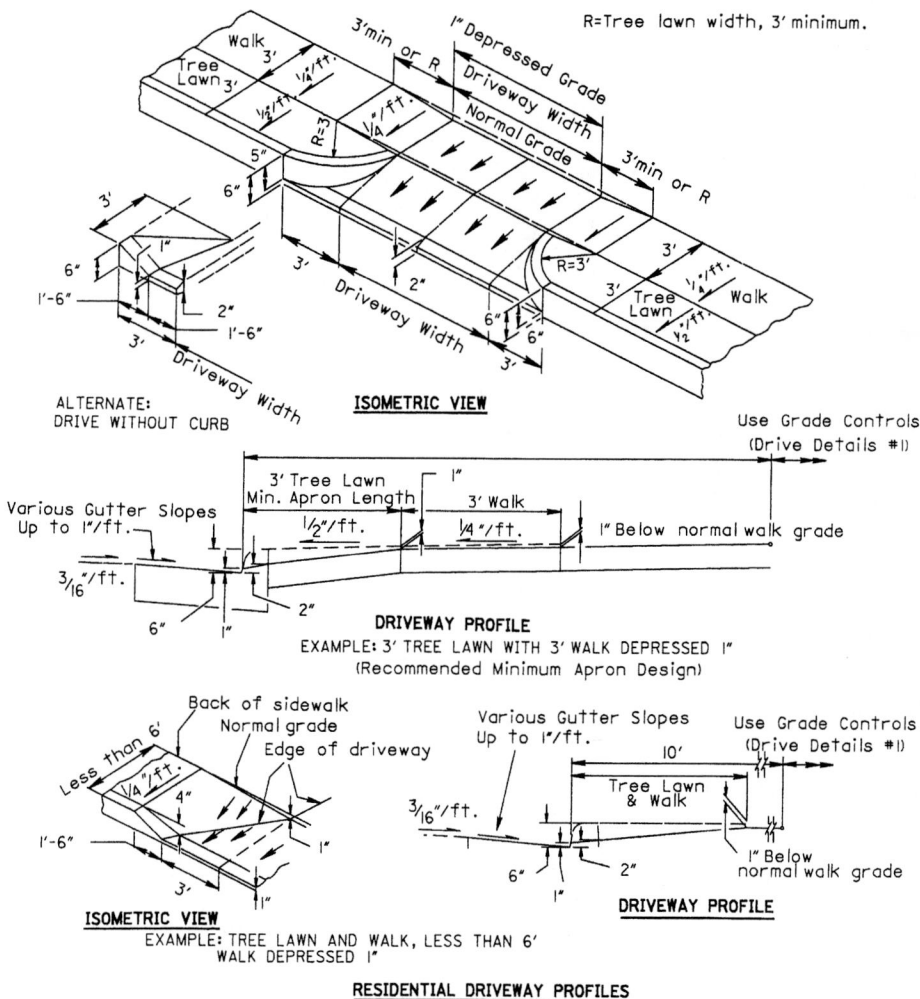

FIGURE 2.51 Additional details of driveway designs. (*From* Location and Design Manual, *Vol. 1,* Roadway Design, *Ohio Department of Transportation, with permission*)

2.10.4 Service Station Drives

Service station drive approach geometry is probably the most complex of any drive design. Many of the geometric features may be used in the design of other commercial and industrial drives. Figures 2.52 through 2.55 illustrate service station approach designs under varying conditions.

The location and angle of an approach in relation to an adjacent highway intersection should be such that a vehicle entering or leaving the site may turn out of or into the nearest lane of traffic moving in the desired direction and be channeled within this lane before entering the intersection or proceeding along the highway. The interior

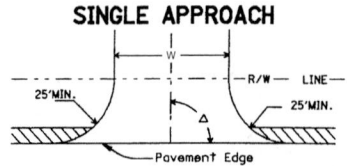

L 15' or greater

W 35' maximum

Δ 70° to 90° (for approach with two-way operation)

θ 45° to 90°

R' Nonturning Radius, 5' min. to 10' max

R'' Turning Radius, 15' min, 25' to 50' Desirable

r Permissible Rounding 15' max

▨▨ Treated Shoulder

DUAL APPROACHES & INTERMEDIATE ISLAND

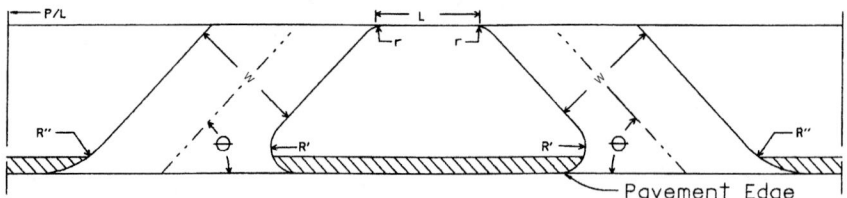

DUAL APPROACHES WITH RETURN FLOW & INTERMEDIATE ISLAND

(For use on Crossroads in the Vicinity of Interstate Routes or Freeways)

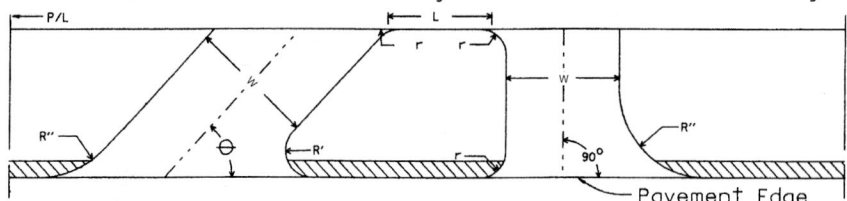

FIGURE 2.52 Design for service station drives with uncurbed roadway and uncurbed approach. (*From* Location and Design Manual, *Vol. 1,* Roadway Design, *Ohio Department of Transportation, with permission*)

angle between the axes of dual approaches and the centerline of the roadway should fall between 45 and 90°. This interior angle should fall between 70 and 90° for single approaches designed for two-way operation. The width of all approaches should not be greater than 35 ft in the throat of the approach measured at right angles to the axis of the approach. Where public alleys adjoin the service station property, approaches may begin at the far side of the alley, and if so used, the width of the alley should be included as part of the approach opening.

Approach radii on uncurbed highways should be as follows:

Turning radii: 15 ft minimum, 25 to 50 ft desirable

Nonturning radii: 5 ft minimum, 10 ft maximum

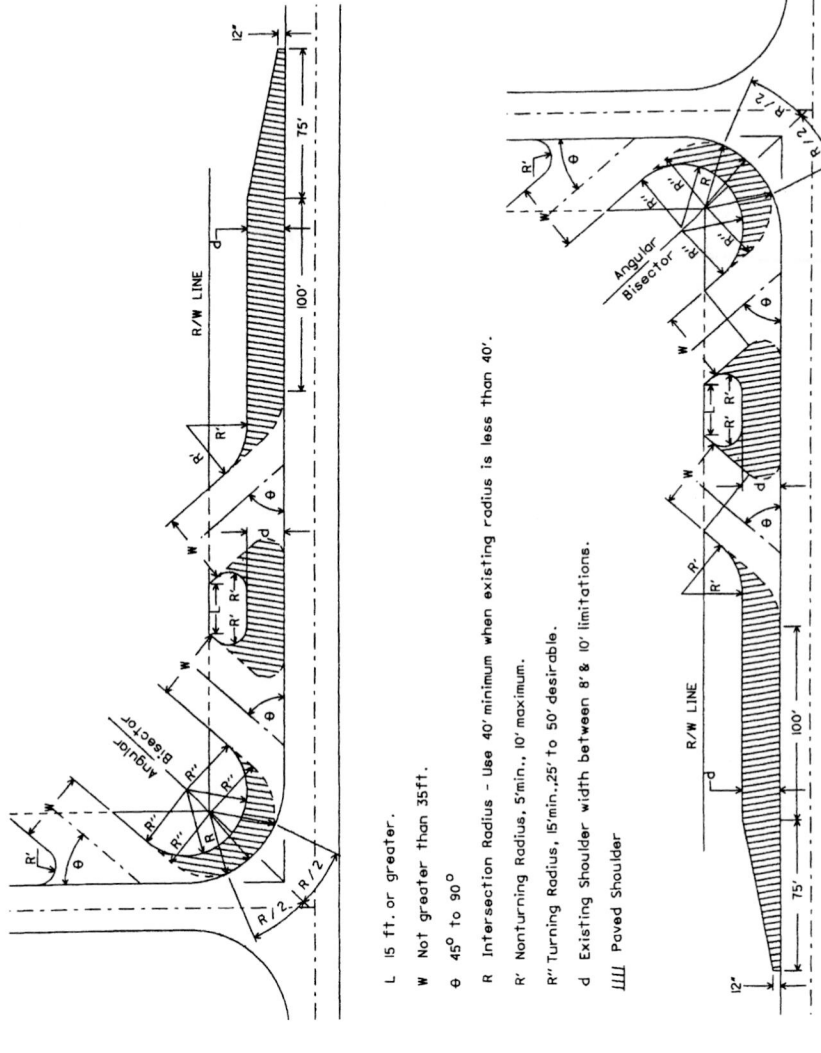

L 15 ft. or greater.

W Not greater than 35ft.

θ 45° to 90°

R Intersection Radius - Use 40' minimum when existing radius is less than 40'.

R' Nonturning Radius, 5'min., 10' maximum.

R'' Turning Radius, 15'min.,25' to 50' desirable.

d Existing Shoulder width between 8' & 10' limitations.

⁄⁄⁄⁄ Paved Shoulder

FIGURE 2.53 Design for service station drives with special paved shoulder detail. (*From Location and Design Manual, Vol. 1, Roadway Design, Ohio Department of Transportation, with permission*)

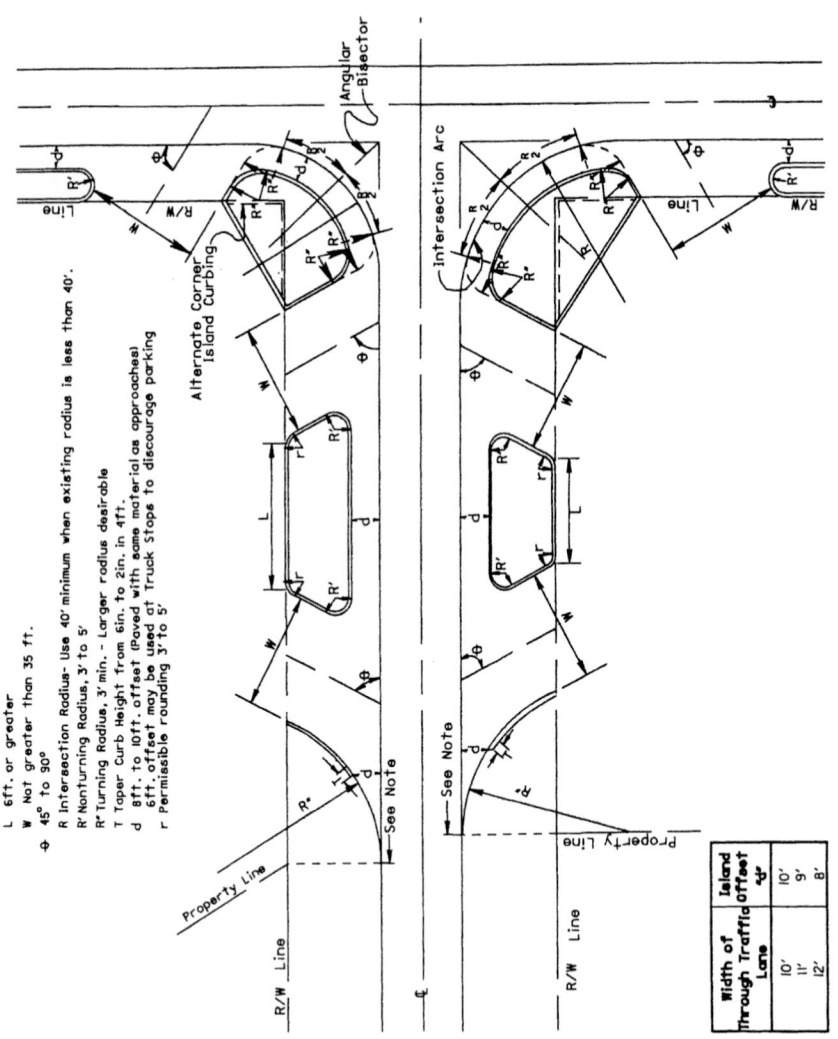

FIGURE 2.54 Design for service station drives with uncurbed roadway and curbed approach. (*From Location and Design Manual, Vol. 1, Roadway Design, Ohio Department of Transportation, with permission*)

L 6ft. or greater
W Not greater than 35 ft.
φ 45° to 90°
R Intersection Radius- Use 40' minimum when existing radius is less than 40'.
R' Nonturning Radius, 3' to 5'
R" Turning Radius, 3' min. - Larger radius desirable
T Taper Curb Height from 6in. to 2in. in 4ft.
d 8ft. to 10ft. offset (Paved with some material as approaches)
  6ft. offset may be used at Truck Stops to discourage parking
r Permissible rounding 3' to 5'

Width of Through Traffic Lane	Island offset "d"
10'	10'
11'	9'
12'	8'

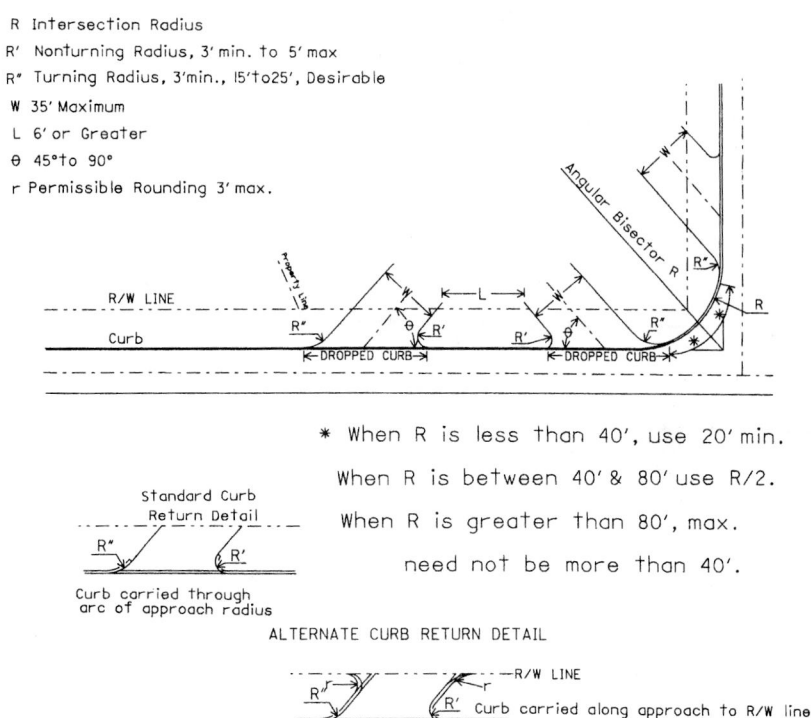

R Intersection Radius
R' Nonturning Radius, 3'min. to 5'max
R″ Turning Radius, 3'min., 15'to25', Desirable
W 35' Maximum
L 6' or Greater
θ 45°to 90°
r Permissible Rounding 3' max.

* When R is less than 40', use 20' min.

When R is between 40' & 80' use R/2.

When R is greater than 80', max.

need not be more than 40'.

Standard Curb
Return Detail

R″   R′

Curb carried through
arc of approach radius

ALTERNATE CURB RETURN DETAIL

R/W LINE
r
R′   R′ Curb carried along approach to R/W line.

FIGURE 2.55 Design for service station drives with curbed roadway and curbed approach. (*From* Location and Design Manual, *Vol. 1*, Roadway Design, *Ohio Department of Transportation, with permission*)

Approach radii on curbed highways should be as follows:

Turning radii: 3 ft minimum, 15 to 25 ft desirable

Nonturning radii: 3 ft minimum, 5 ft maximum

Where the approach radius controls the turning radius of a right-turning vehicle entering the service station from the adjacent traffic lane of the roadway, the radius of that edge should be as long as practical to provide a free and safe movement.

2.10.5 Commercial Drives

The access requirements of most commercial developments can be served by driveways having standard design characteristics. The exceptions are driveways having high traffic volumes, those being used by large vehicles, or those serving businesses that engender unique traffic patterns. For standard commercial drive designs, see Fig. 2.56. The recommended radii are (1) 15 ft minimum, when the through highway is curbed, and (2) 25 ft minimum, when the through highway is uncurbed. The maximum width is 35 ft. A dropped curb should be used on curbed streets as shown in section B-B in Fig. 2.50.

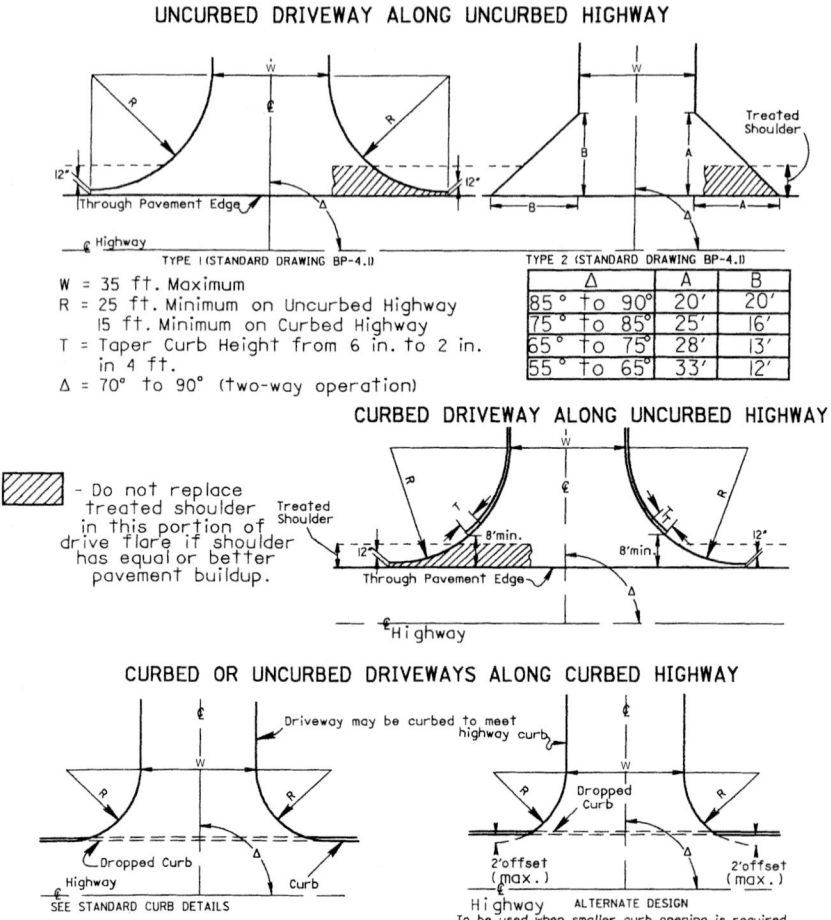

FIGURE 2.56 Designs for commercial drives. (*From* Location and Design Manual, *Vol. 1,* Roadway Design, *Ohio Department of Transportation, with permission*)

Where access requirements are such that a nonstandard driveway is necessary, the design may approximate the design of shopping center driveways, to be discussed in Art. 2.10.6, or that of a public road intersection. Specially designed radii and a width greater than 35 ft may be permitted, as necessary, to accommodate the type of vehicle using the driveway. For example, a truck stop may require two one-way driveways, or a single drive with width greater than 35 ft, and radii as great as 75 ft to facilitate turning movements.

2.10.6 Shopping Center and Industrial Drives

Figure 2.57 shows two typical driveway designs to be used as a guide for the design of driveways serving high-volume traffic generators such as shopping centers, industrial plants, industrial parks, and other types of developments having similar traffic charac-

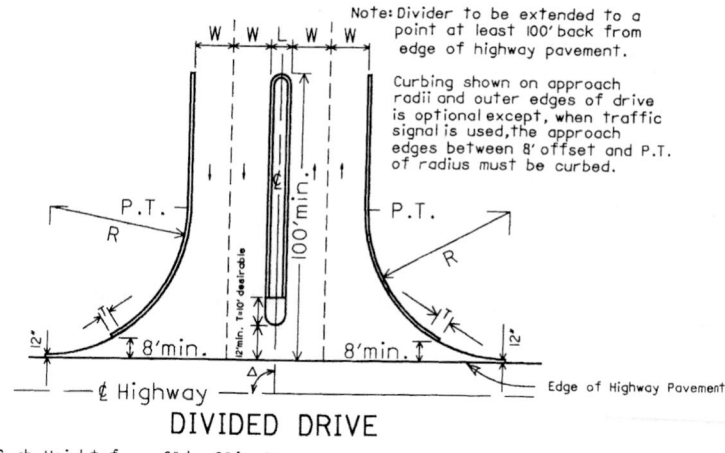

DIVIDED DRIVE

T = Taper Curb Height from 6″ to 2″ in 4′ or greater.
W = 10′ to 14′ per single traffic lane.
R = 35′ Minimum, 50′ Desirable.
Δ = 70° to 90°
L = Median Width, 6′ Minimum.
   (Median must be curbed for 6′ to 15′ widths)

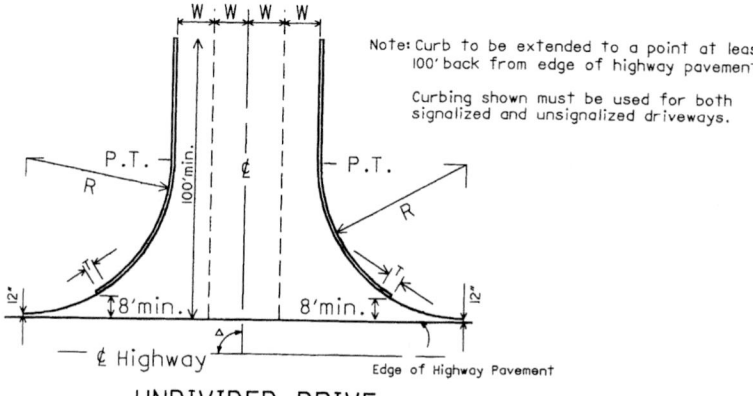

UNDIVIDED DRIVE

FIGURE 2.57 Designs for shopping center and industrial drives. (*From* Location and Design Manual, *Vol. 1,* Roadway Design, *Ohio Department of Transportation, with permission*)

teristics. Many of the design features discussed in Art. 2.41, At-Grade Intersections, are applicable here. Geometric considerations are as follows:

- Driveways should intersect the highway at an angle between 70 and 90°.
- Each driveway traffic lane should have a minimum width of 10 ft, with 12 ft preferred.
- Major driveways in shopping centers should be constructed to prevent cross movement of internal traffic within 100 ft of the entrance approach. This may be accomplished by use of a raised divider 6 in high, 6 ft wide (minimum) and 100 ft long, and/or by use of curbing, sidewalk, or other barrier along the drive edges for a length of 100 ft.

- Driveways designed for traffic signal operation should have curbed radii and should provide a minimum of two lanes for vehicles entering the highway.

2.10.7 Drive Profile Design

Drive profiles on uncurbed roadways should slope down and away from the pavement edge at the same slope as the graded shoulder. Any vertical curve should be developed outside the normal graded shoulder width. Vertical curve lengths should be 10 to 20 ft, depending on the grade differential. Under normal circumstances, rural drive grades should not exceed 10 percent, with 8 percent the preferred maximum.

The drive profiles for curbed roadways were developed using the design vehicle described in Fig. 2.50. The profile criterion shown provides clearance for this vehicle when its springs are completely compressed. If conditions of a particular driveway do not meet the cross-section criteria listed below, a template of the design vehicle can be used to design the driveway profile.

For tree lawns 6 ft or wider, the ramp grade from the gutter to the edge of the sidewalk should be 1 in/ft or less for normal cross-section design. Figure 2.50 shows this condition for the following cross-section conditions:

- Sidewalk and tree lawn slope of ¼ in/ft

and

- A 6-in curb height with pavement slope of ³⁄₁₆ or ¼ in/ft

or

- Type 2 curb and gutter with pavement slope of ³⁄₁₆ in/ft

If the cross-section design does not meet the above conditions (has sharper grade breaks), the profile should be designed using a template of the design vehicle.

For tree lawns less than 6 ft wide, Fig. 2.51 shows the profile treatment. Clearance for the design vehicle is achieved by depressing the sidewalk 1 in at the driveway. The sidewalk cross slope of ¼ in/ft is retained. The design may be used directly with curbed highways having cross-section criteria as listed above and the profile conditions of Fig. 2.50. For other cross sections, a template of the design vehicle may be used to design the profile.

Figure 2.51 shows an isometric view and profile for a driveway where only a 3-ft tree lawn is available. This design is shown not because it is desirable, but because right-of-way width and property development may require this type of design. Whenever feasible, the tree lawn should be 8 ft or wider. Where the total width of tree lawn and sidewalk is less than 6 ft, the minimum 3-ft apron designs are inappropriate and cannot be used, as they extend curb or sharp flares into the sidewalk area. For this condition, the sidewalk and curb are transitioned to meet the drive profile as shown on the lower portion of Fig. 2.51. The profile of the drive meets the 1-in depressed grade of the sidewalk, as shown in the drive profile.

The tree lawn and walk design shown in Figs. 2.50 and 2.51 will keep storm water, flowing at the curb design height or less, from flowing over the sidewalk. If it is necessary to lower the curb and sidewalk more than 1 in, the drainage condition should be checked thoroughly.

Commercial drive profiles usually use a dropped curb across the approach. However, some commercial drives serving large traffic generators may be designed as at-grade intersections, without dropped curbs, because of their high traffic volumes.

Figure 2.58 shows the recommended grade controls for commercial driveways. The grade should be as flat as possible and still meet drainage requirements. The 20-ft length between grade breaks is required by the low clearance and the long axle spacing of the commercial design vehicle shown in Fig. 2.59. Tree lawn profile design

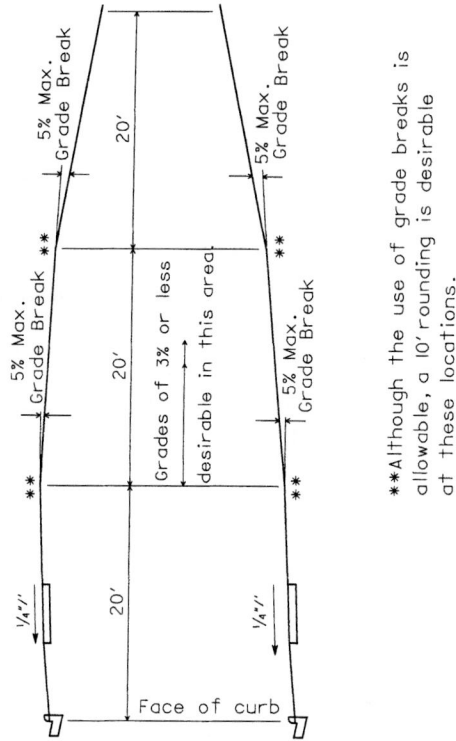

FIGURE 2.58 Profiles for commercial drives. (*From Location and Design Manual, Vol. 1,* Roadway Design, *Ohio Department of Transportation, with permission*)

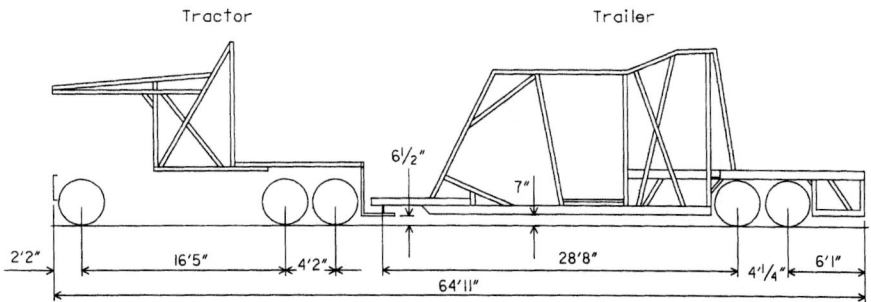

FIGURE 2.59 Commercial design vehicle showing wheel spacings and clearances. (*From* Location and Design Manual, *Vol. 1,* Roadway Design, *Ohio Department of Transportation, with permission*)

should be in accordance with Figs. 2.50 and 2.51. The grade break at the face of the curb is critical for some commercial vehicles, and the cross-section requirements for residential drives on curbed streets should be used.

2.11   THE COST OF CONGESTION

Congestion, when applied to traffic, refers to that condition which occurs when drivers experience a noticeable delay in completing a trip because of inability to maneuver through the traffic stream. This condition is characterized by slow travel speeds, increased travel times, increased accident frequencies, erratic stop-and-go driving, increased vehicle operating costs, and other undesirable circumstances leading to driver dissatisfaction (Ref. 4).

Congestion on urban freeways is of two types—recurring, and nonrecurring. Congestion that occurs regularly at particular locations during certain time periods is said to be recurring. On the other hand, congestion that occurs as a result of irregular events, such as accidents, disabled vehicles, or other similar happenings, is said to be nonrecurring. Both can cause driver dissatisfaction, but drivers usually expect the recurring congestion and make adjustments in their travel plans to accommodate it.

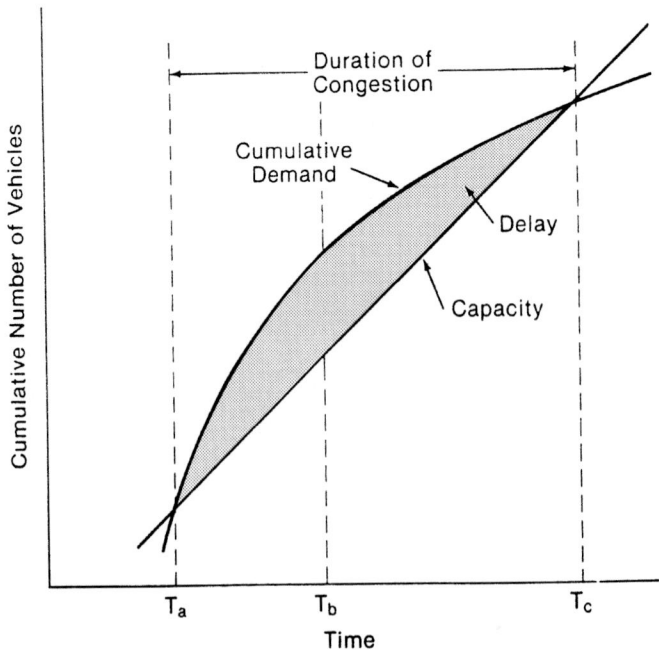

FIGURE 2.60   Illustration of relationships among demand, capacity, and congestion. (*From* Traffic Engineering Handbook, *Institute of Transportation Engineers, 1992*)

The most common example of recurring congestion is the morning and afternoon "rush hour" periods, when traffic demands can exceed the capacity of the freeway.

Figure 2.60 shows a graph illustrating what happens when demand exceeds capacity. The straight line represents the capacity of a section of freeway at a particular time (i.e., the number of vehicles getting past the point under prevailing roadway conditions). As long as traffic demand, or the number of vehicles arriving at that point (shown by the curved cumulative demand line), is less than or equal to the capacity of that section of the freeway, there is little congestion. However, once the arrival rate begins to exceed the capacity at time T_a, a bottleneck is formed and vehicles begin to accumulate upstream until time T_b, when the demand once again falls below the capacity. Congested conditions continue until time T_c, when the accumulated traffic at the bottleneck dissipates. The area between the capacity and demand curves during congested conditions is the delay resulting from the congestion (Ref. 4).

The cost of congestion is the sum of individual costs of items that represent an increase over normal operating costs directly attributable to the congestion. Among these items are fuel and oil consumption, tire wear caused by the frequency and severity of speed changes, other maintenance items affected by the speed changes, and increased idling times. There are other costs that may not be readily apparent to the individual driver but are real costs affecting the general public, such as inefficient movement of commercial vehicles and the increased level of pollutant emissions.

Where congestion on freeways is of the recurring kind, the usual solutions to try and solve the problem are geometric in nature. The most logical solution in many cases is lane addition. In urban areas, congestion frequently occurs downstream from entrance ramps when the combination of traffic entering the freeway and the traffic already present exceeds the capacity of that segment. In other cases, the existing horizontal alignment may contain one or more "sharp" curves, which result in lower capacity. Ramp designs may have a detrimental effect on freeway capacity if their merge or diverge areas are too short, or if they are too closely spaced, creating weaving problems for traffic entering and exiting the freeway traffic stream. Other problems involving physical features include unconventional interchanges, inadequate shoulders, narrow medians, poor surface quality, and poor signing.

The next article discusses a new approach to the problem that takes advantage of evolving technology: intelligent vehicle highway systems.

2.12   INTELLIGENT VEHICLE HIGHWAY SYSTEMS

Intelligent vehicle highway systems (IVHS) refers to transportation systems that involve integrated applications of advanced surveillance, communications, computer, display, and control process technologies, both in the vehicle and on the highway (Ref. 4).

In 1991, Congress passed the Intermodal Surface Transportation Efficiency Act (ISTEA), which included an authorization of $660 million to create an IVHS program for the nation. The goals for IVHS were defined as follows: to improve safety, to reduce congestion, to enhance mobility, to minimize environmental impact, to save energy, and to promote economic productivity. Research studies and demonstration projects to accomplish these goals are in progress.

The IVHS program is not limited to urban areas. It should result in benefits for both urban and rural drivers, and for both younger and older drivers. People who use public transportation will also benefit, and others will be persuaded to join them.

A planning process was undertaken following the adoption of the act, which identified 28 user services in six categories (Ref. 5):

Travel and traffic management

- Pretrip travel information
- En route driver information
- Traveler services information
- Route guidance
- Ride matching and reservation
- Incident management
- Travel demand management
- Traffic control

Public transportation management

- En route transit information
- Public transportation management
- Personalized public transit
- Public safety security

Electronic payment

- Electronic payment services

Commercial vehicle operations

- Commercial vehicle electronic clearance
- Automated roadside safety inspection
- Commercial vehicle administrative processes
- Onboard safety monitoring
- Commercial fleet management
- Hazardous material incident notification

Emergency management

- Emergency vehicle management
- Emergency notification and personal security

Advanced vehicle safety systems

- Longitudinal collision avoidance
- Lateral collision avoidance
- Intersection collision avoidance
- Vision enhancement for crash avoidance
- Safety readiness
- Precrash restraint deployment
- Automated vehicle operation

For those services listed under *travel and traffic management,* the emphasis will be upon providing real-time data to help the driver make the best decisions during a trip or even make last-minute changes in itinerary prior to departure. This category also encourages the use of high-occupancy vehicles and provides traffic control procedures and mechanisms to deal with situations as they occur.

The services under *public transportation management* will improve the efficiency, safety, and effectiveness of public transportation systems for users and providers alike. Again, the emphasis is on gathering and relaying real-time information to the users of the systems. It provides for automation of operations, planning, and management functions of public systems. It will also be able to monitor the environment in public station areas, including bus stops, and parking lots, to generate alarms when necessary and increase public safety.

Electronic payment services will promote intermodal travel by providing a common electronic payment medium for all transportation modes and functions, including tolls, transit fares, and parking. One "smart card" could be used for several different modes of transportation.

Under *commercial vehicle operations,* trucks and buses equipped with transponders could have their safety status, credentials, and weight checked at mainline speeds. Vehicles passing the check would not have to pull over into the inspection/weigh facility. Automated safety inspections would allow "real-time" access at the roadside to the performance record of carriers, vehicles, and drivers. By using sensors and diagnostic equipment, vehicle systems and even driver alertness can be checked without stopping the vehicle.

Under *emergency management,* the capabilities of fleet management, route guidance, and signal priority can be used for emergency vehicles. Police, fire, and medical units can be directed over the most expeditious route to an incident site using real-time information. Driver and personal security systems will allow the user to initiate distress signals for incidents like mechanical breakdowns. Automatic collision notification would send information regarding location, nature, and severity of the incident to emergency personnel.

Concerning those services under *advanced vehicle safety systems,* a series of collision avoidance systems would be developed. For potential longitudinal, lateral, and intersection collisions, the systems would be able to sense impending trouble, warn the driver, and temporarily control the vehicle. On attempted lane changes, the driver's blind spot would be monitored, and the vehicle prevented from making the switch if a vehicle was present. Another possibility is vision enhancement for the driver, in which the roadway and roadside are continuously scanned for potential hazards and the driver is made aware of situations when necessary. In-vehicle equipment can be used to monitor the driver's condition and issue appropriate warnings. In employing precrash restraint deployment, the velocity, mass, and direction of the vehicles and objects involved in a potential crash are identified and the number, location, and physical characteristics of occupants are determined. This information in turn is used to trigger responses, such as tightening of lap-shoulder belts, arming and deploying air bags at optimal pressure, and deploying roll bars. Another program being investigated is automated vehicle operation. This would ultimately provide an accident-free environment on the roadway. Drivers would be able to buy a vehicle already equipped to drive under these conditions, or purchase instrumentation and have it installed on an existing vehicle.

2.13   *HIGH-OCCUPANCY VEHICLE LANES*

Another method that is being increasingly used to relieve congestion on urban freeways is the establishment of *high-occupancy vehicle (HOV)* lanes. Although the first instances of use in California in the early 1970s met with much public resistance, the idea was revisited and accepted more readily during the mid-1980s and continues to grow in acceptance in highly congested urban traffic areas (Ref. 9). The concept is to provide a separate lane or lanes for high-occupancy vehicles such as buses, carpools, vanpools, and other ride-sharing modes of transportation. This, in turn, provides a positive incentive for the general public to seek out ride-sharing transportation modes, both public and private. The overall goal is to move more people in fewer vehicles.

2.13.1 Planning Considerations

The following transportation system goals can be achieved by proper development and use of HOV lanes (Ref. 3):

- To maximize the person-moving capacity of roadway facilities by providing improved level of service for high-occupancy vehicles, both public and private
- To conserve fuel and to minimize consumption of other resources needed for transportation
- To improve air quality
- To increase overall accessibility while reducing vehicular congestion

Designing and implementing HOV lanes should be limited to those cases where extreme congestion occurs on a regular basis. They should be used in conjunction with other programs that will promote the use of ride-sharing modes, such as park-and-ride lots, park-and-pool lots, and information services to facilitate bus and ride-share needs.

The following guidelines should be used to determine when an HOV lane should be implemented:

- Compatibility with other plans

    HOV lanes should be part of an overall transportation plan.

    Community support should be obtained for developing HOV lanes.

    Intense, recurring congestion should be occurring on the freeway general-purpose lanes.

    Peak-period traffic per lane should be approaching capacity (1700 to 2000 vehicles per hour).

    During peak periods, average speeds on the freeway main lanes during nonincident conditions should be less than 30 mi/h over a distance of about 5 mi or more.

    Compared with using the freeway general-purpose lanes, the HOV lanes should offer a travel time savings of at least 5 to 7 min during the peak hour.

- Coordination with travel patterns that encourage ridesharing

    Significant volume of peak-period trips (e.g., more than 6000 home-based work trips during the peak hour) on the freeway should be destined to major activity centers or employment areas in or along the freeway corridor.

    At least 65 to 75 percent of peak-period freeway trips to major activity centers should be 5 mi or more in length.

    Resulting ride-share demand should be sufficient to generate HOV volumes that are high enough to make the facility appear to be adequately utilized; volumes may vary by type and location of facility.

- A design that allows for safe, efficient, and enforceable operation

2.13.2 Operational Considerations

Management of HOV facility operations may be accomplished by a range of technological and personnel means. Minimum control may consist of passive signing and delineation. Maximum control may involve sophisticated surveillance, vehicle detection with computer integration, and dynamic, real-time signing or delineation.

A determination of the level of vehicle restriction must be made on the basis of traffic characteristics and how much the HOV lane is used. Restricting use of the HOV lane to vehicles with three or more passengers (3+) may give the appearance that the lane is underused. On the other hand, restricting the lane to vehicles with only two or more passengers (2+) sometimes results in lane utilization approaching capacity. The best rule of thumb is to start with the 2+ restriction, and go to the 3+ restriction when the level of service of the HOV lane is approaching capacity.

A proper level of enforcement is necessary to ensure that the HOV lane will operate efficiently. Absence of enforcement defeats the purpose of the lane, since single-passenger vehicles will see no need to stay out of the lane. Detection and apprehension of violators, issuance of citations to violators, and effective prosecution of violators is essential.

The hours of operation that an HOV lane is in effect are also an important consideration. Twenty-four-hour operation is preferred over peak-hour operation, simply because there is less chance for driver confusion and violations tend to be lower.

2.13.3 Freeway Design Considerations

There are three types of HOV lane patterns—separated lanes, concurrent flow lanes, and contraflow lanes. Regardless of which pattern is chosen, consideration should be given to traffic operations at interchanges and on-ramps, pedestrian access to on-line stations, the availability of parking areas at or near the stations, and the possible use of HOV lanes during freeway maintenance of traffic operations. Design speeds should generally be the same as for the mainline facility. Recommended lane and shoulder widths can be seen in the next group of referenced figures.

A separated HOV lane may be located in the median or on the outside of the general lanes, or follow an independent alignment. See Figs. 2.61, 2.62, and 2.63 for examples of cross sections. Figures 2.64 and 2.65 show two examples of how separated HOV lanes tie in with the general main lanes of travel. Figure 2.66 shows sample signing and pavement marking used in connection with HOV lanes. Note the diamond symbol that signifies an HOV lane.

Concurrent flow lanes are located adjacent to traffic lanes and are not physically separated from them. Figures 2.67 and 2.68 are examples of typical sections for concurrent HOV lanes.

Contraflow lanes provide an exclusive lane for HOVs traveling in the peak direction by removing a lane from service in the off-peak direction. These may be used in areas where traffic volumes in the off-peak directions are such that the level of service is not seriously affected. Some kind of buffer zone or device is strongly recommended for obvious safety considerations. Figure 2.69 provides examples of cross sections for contraflow HOV lanes.

2.13.4 Arterial Design Considerations

There are two general categories of HOV lanes for use on surface arterial streets: (1) those which assign exclusive use of designated lanes for HOV use and (2) those which give preferential treatment or special privileges to HOVs through traffic control measures. The first category includes concurrent and contraflow reserved lanes, reversible median or center lanes, and streets devoted to HOV use. The second includes such

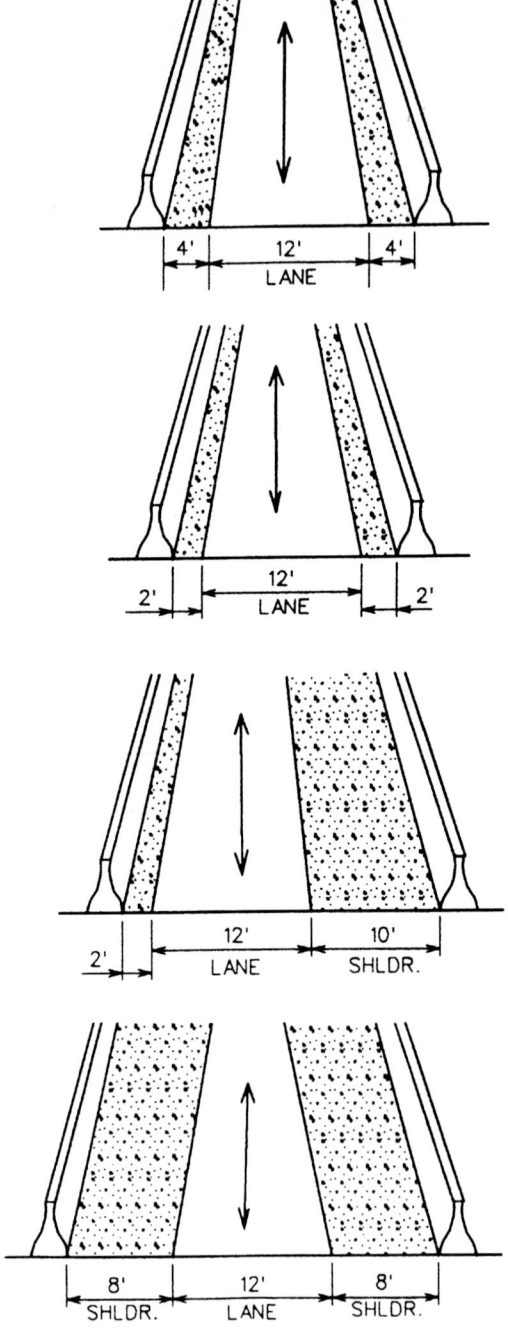

FIGURE 2.61 Cross sections for high-occupancy–vehicle single lane on separated roadway; one-way or reversible. (*From* Guide for the Design of High Occupancy Vehicle Facilities, *American Association of State Highway and Transportation Officials, Washington, D.C., 1992, with permission*)

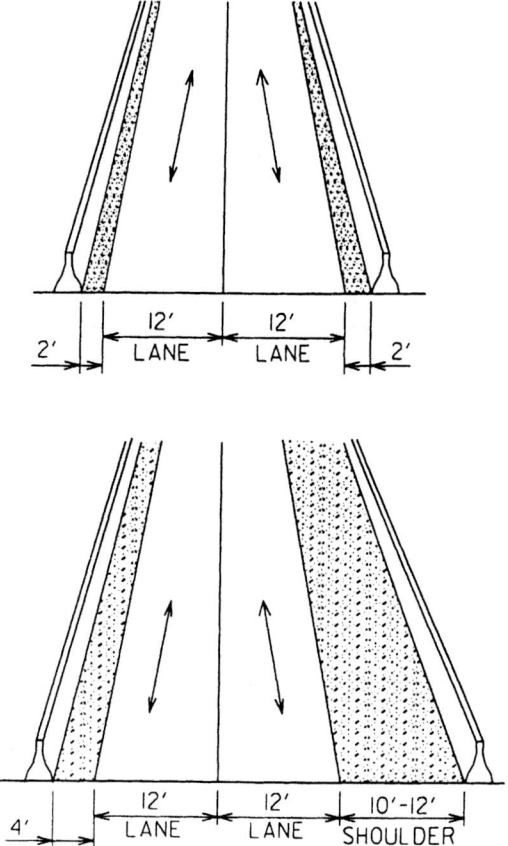

FIGURE 2.62   Cross sections for high-occupancy–vehicle double lanes on separated roadway; one-way or reversible. (*From* Guide for the Design of High Occupancy Vehicle Facilities, *American Association of State Highway and Transportation Officials, Washington, D.C., 1992, with permission*)

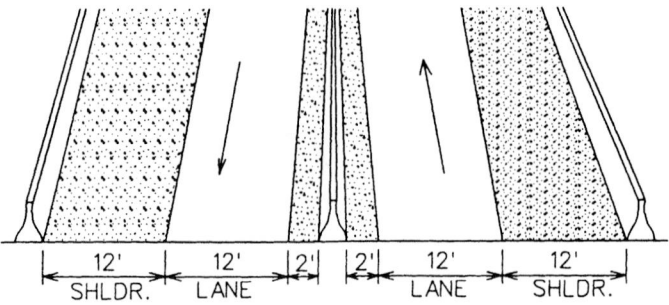

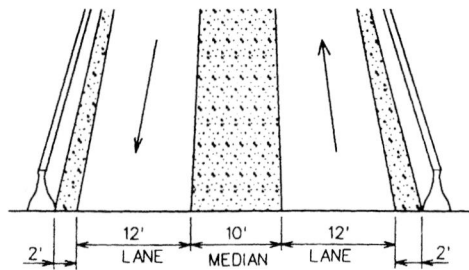

FIGURE 2.63 Cross sections for high-occupancy–vehicle two-way separated lanes on separated roadway. (*From* Guide for the Design of High Occupancy Vehicle Facilities, *American Association of State Highway and Transportation Officials, Washington, D.C., 1992, with permission*)

measures as traffic signal preemption systems for buses, and special traffic provisions that allow HOVs to make turns or other maneuvers that are prohibited for other traffic.

Regardless of the type of treatment, the geometric design and traffic control features should accommodate all vehicles that might ultimately use the HOV lane. Since the primary vehicle type using the urban HOV lanes will be buses, special consideration should be given to designing for the vehicle's dimensions and turning pattern.

Figure 2.70 shows two examples of center lane HOV use. Note the location of passenger loading areas in Fig. 2.70*b*. The advantage of a center HOV lane over other schemes is that it can be made reversible. Figure 2.71 shows various ways these HOV center lanes are developed.

Figure 2.72 shows the more commonly seen concurrent HOV lane developed in the curb lane of an urban street. The advantage of this type of HOV lane is that it is the simplest and least costly to implement. This usually involves only changing signs and pavement markings and coordinating traffic signals.

Contraflow HOV lanes may be used on one-way or two-way streets. On one-way streets, the HOV lane may be either the right or the left lane, while on two-way streets it can be either the right lane or the inside lane adjacent to the median or centerline of the arterial street. Two examples of contraflow lanes are shown in Fig. 2.73.

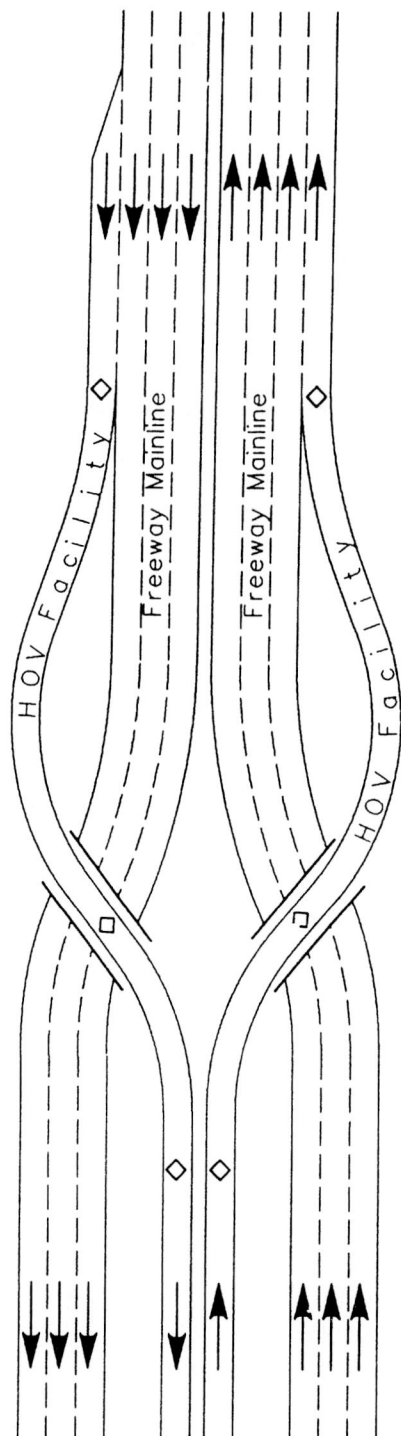

FIGURE 2.64 Connection of HOV terminal mainline lanes to freeway median with flyovers. (*From Guide for the Design of High Occupancy Vehicle Facilities, American Association of State Highway and Transportation Officials, Washington, D.C., 1992, with permission*)

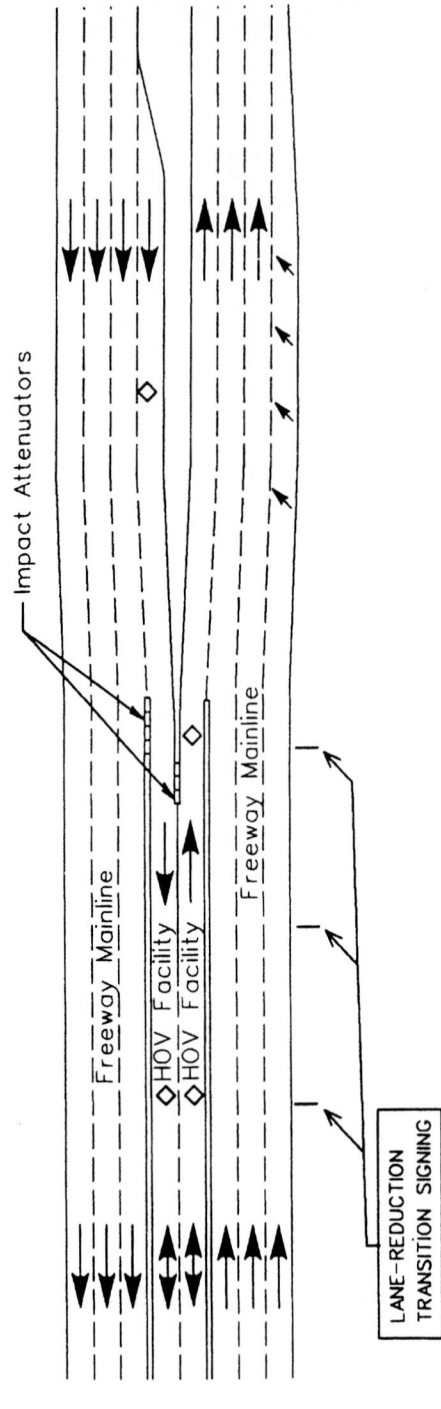

FIGURE 2.65 Connection of HOV terminal mainline lanes to freeway median with slip ramps. (*From Guide for the Design of High Occupancy Vehicle Facilities, American Association of State Highway and Transportation Officials, Washington, D.C., 1992, with permission*)

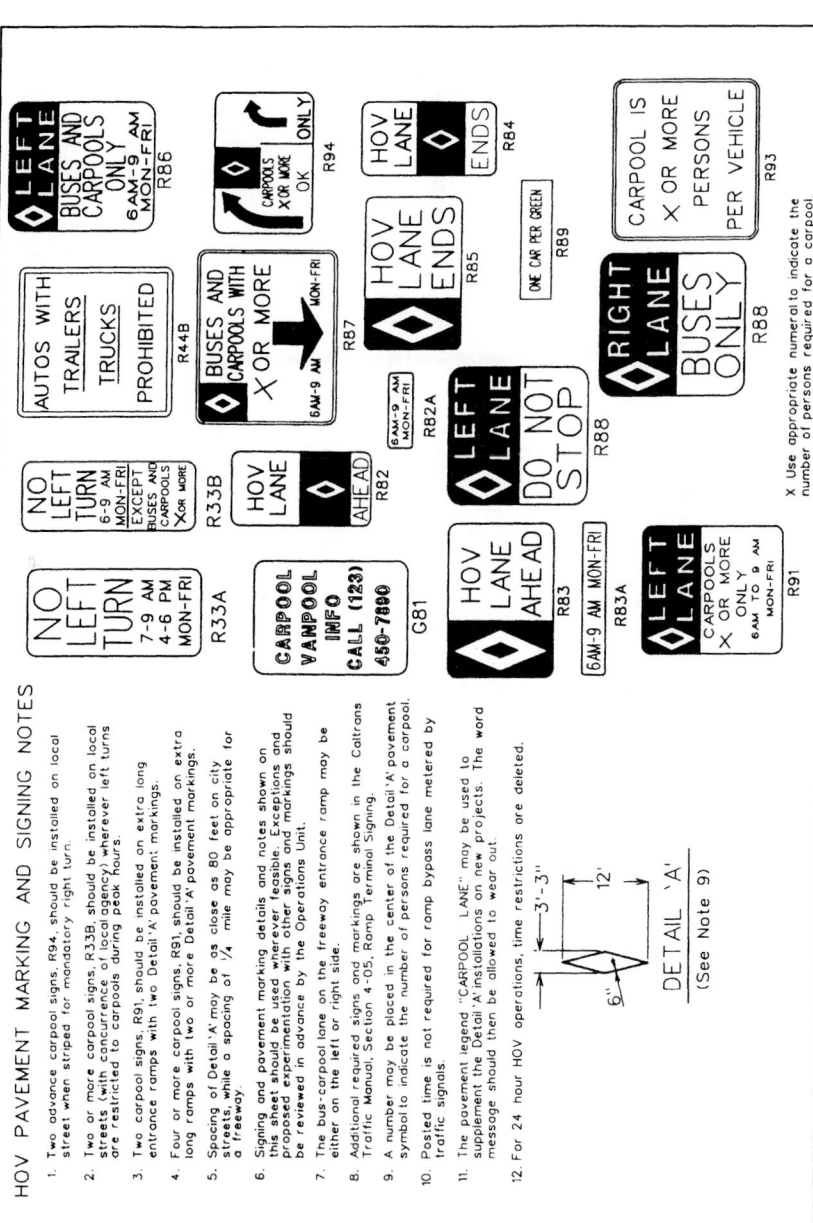

FIGURE 2.66 Sample signing and pavement markings for HOV lanes. *(From Guide for the Design of High Occupancy Vehicle Facilities, American Association of State Highway and Transportation Officials, Washington, D.C., 1992, with permission)*

2.135

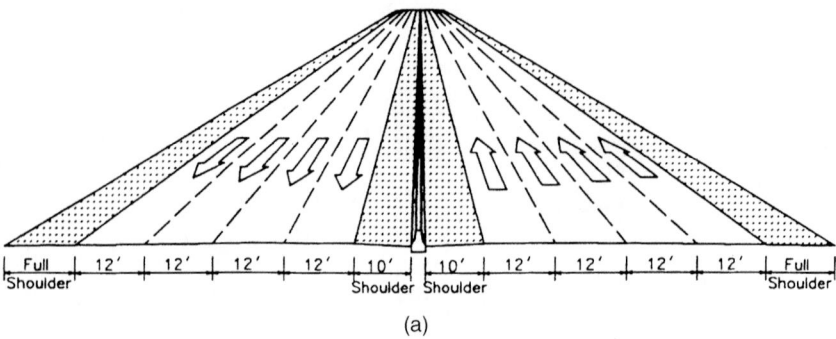

(a)

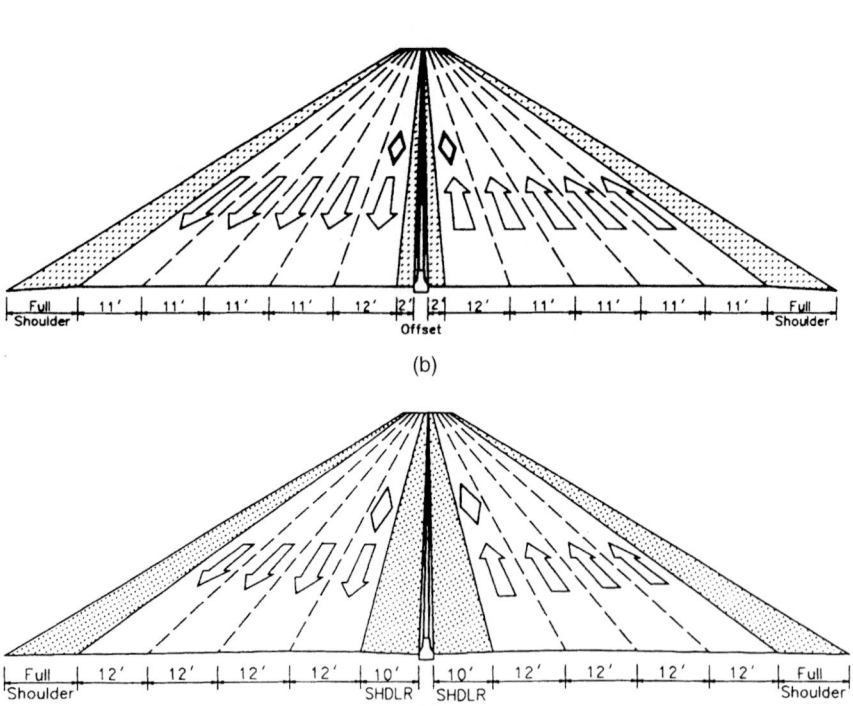

(b)

(c)

FIGURE 2.67 Contiguous concurrent HOV lanes. (*a*) Before adding HOV lanes. (*b*) After adding HOV lanes. (*c*) Alternative for HOV lanes with shoulders. (*From* Guide for the Design of High Occupancy Vehicle Facilities, *American Association of State Highway and Transportation Officials, Washington, D.C., 1992, with permission*)

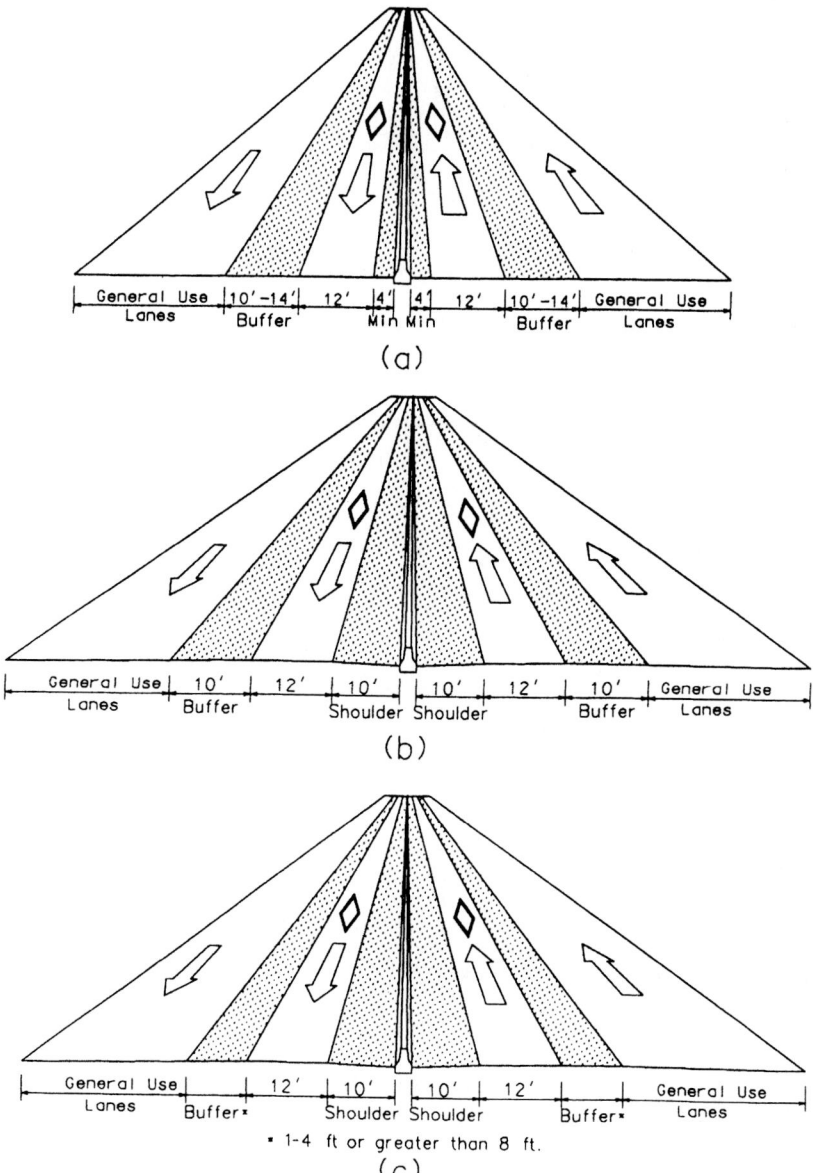

FIGURE 2.68 Examples of concurrent HOV lanes with buffer separation. (*a*) Without shoulders. (*b*) With 10-ft-wide buffers. (*c*) With other buffer widths. (*From* Guide for the Design of High Occupancy Vehicle Facilities, *American Association of State Highway and Transportation Officials, Washington, D.C., 1992, with permission*)

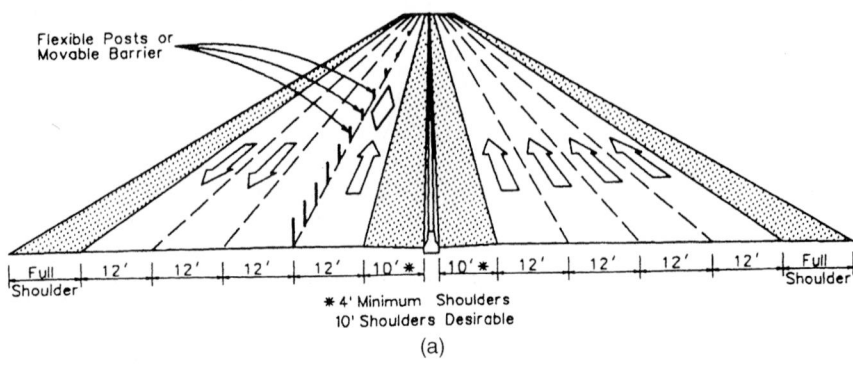

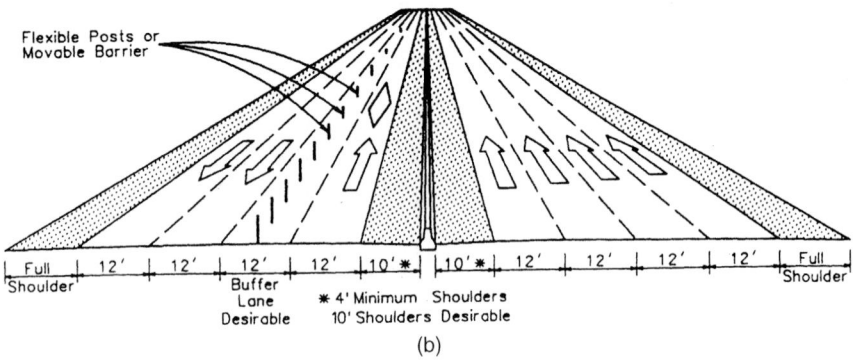

Note: Cones or flexible posts in predrilled holes may be moved toward
      the normal flow lanes increasing the contraflow lane width and
      providing an additional shoulder where a minimum inside shoulder
      exists. If cones are used they should be deployed with the traffic
      flow and removed against the traffic flow for safety. If a buffer
      lane design is not employed, the posts will be placed along the
      lane line in the gaps between the traffic stripes.

FIGURE 2.69 Examples of contraflow HOV lanes. (*a*) With posts on lane line and no buffer. (*b*) With posts in lane providing buffer. (*From* Guide for the Design of High Occupancy Vehicle Facilities, *American Association of State Highway and Transportation Officials, Washington, D.C., 1992, with permission*)

2.14 *HIGHWAY CONSTRUCTION PLANS*

2.14.1 Plan Preparation

The purpose of a set of highway construction plans is to delineate the proposed work with sufficient design details, supplemented with notes, calculations, and summary of quantities, so that it can be clearly and uniformly interpreted by engineers and contrac-

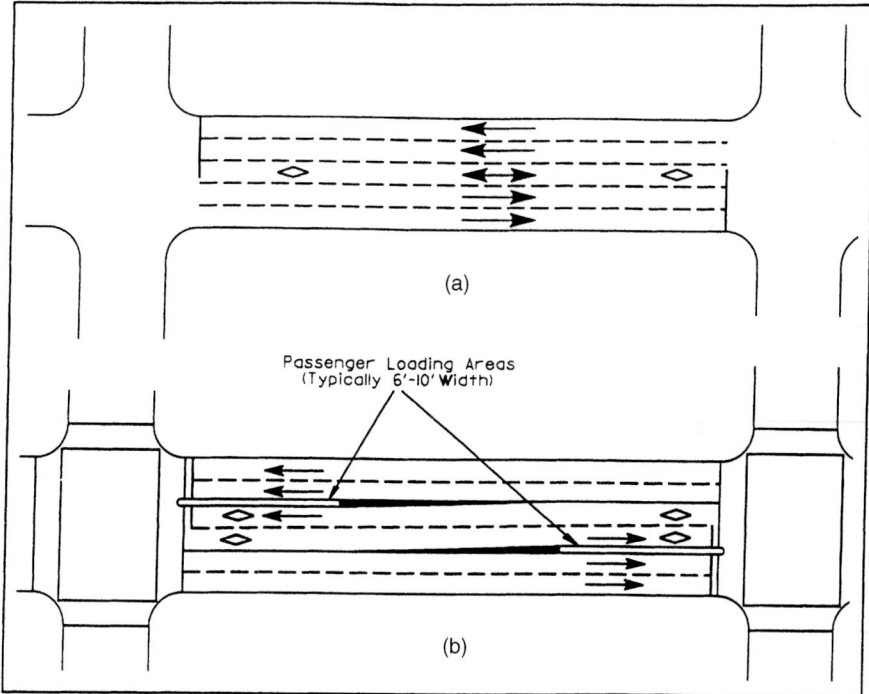

FIGURE 2.70 Examples of median HOV lanes for buses. (*a*) Reversible lanes for express buses. (*b*) Two-way lanes for local buses. (*From* Guide for the Design of High Occupancy Vehicle Facilities, *American Association of State Highway and Transportation Officials, Washington, D.C., 1992, with permission*)

tors (Ref. 8). Sufficient data must be provided to enable the contractor to make an intelligent bid and to perform the work as intended. Clarity, completeness, and conciseness are essential so as to avoid misinterpretation. Unnecessary details should be avoided.

The original tracings serve as a permanent record of the project. They must be prepared on a material acceptable to the agency responsible for maintaining the plans as a record. A currently widely recommended material is the polyester film Mylar with a thickness of 4 mil (3 mil minimum), double- or single- (top side) matted. The surface should not be highly reflective. Only black ink should be used, although grid lines may be colored. Materials that are usually not acceptable include negatives, sepias, vellum, old sheets, dark background, pencil, paste-ons, stick-ons, or bond paper. Original tracings are usually about 22 by 34 in. The designer should prepare the plans keeping in mind that the drawings will most likely be reduced to quarter size (i.e., 11 by 17 in) prior to distribution.

2.14.2 Computer-Aided Design and Drafting (CADD)

The use of computer-aided design and drafting (CADD) techniques is an increasingly acceptable and encouraged method of preparing construction plans. It is important for the designer to consult with the government agency that the plans are being prepared for, to determine the acceptable methods and procedures to be used in developing the

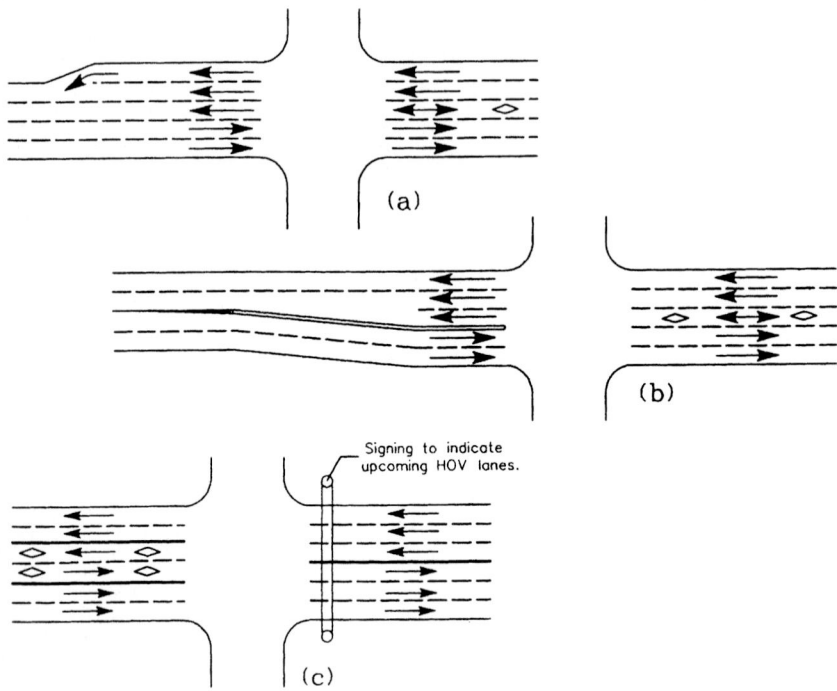

FIGURE 2.71 Typical transitions for median or center lane HOV. (*a*) With reversible HOV lane and outer transition. (*b*) With reversible HOV lane and inner transition. (*c*) With two one-way HOV lanes. (*From* Guide for the Design of High Occupancy Vehicle Facilities, *American Association of State Highway and Transportation Officials, Washington, D.C., 1992, with permission*)

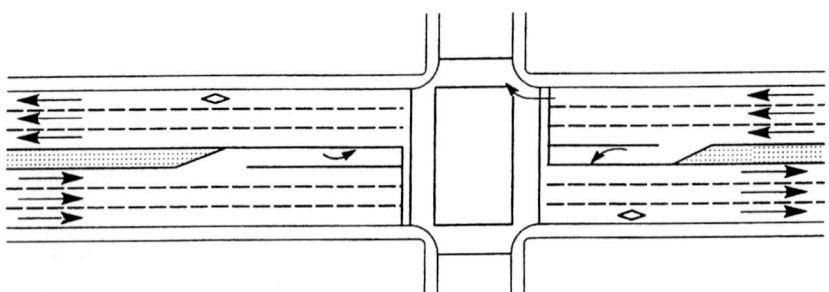

FIGURE 2.72 Common transition for concurrent HOV lanes. (*From* Guide for the Design of High Occupancy Vehicle Facilities, *American Association of State Highway and Transportation Officials, Washington, D.C., 1992, with permission*)

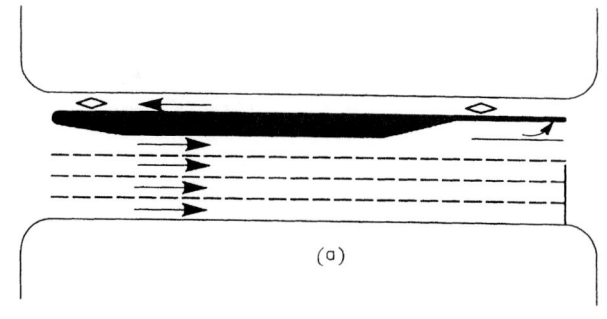

(a)

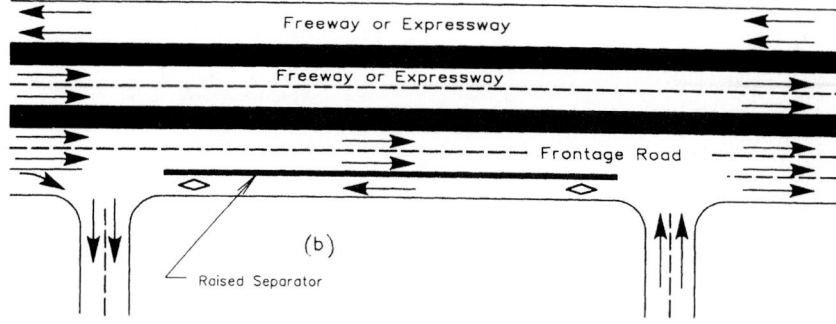

(b)

FIGURE 2.73 Typical contraflow HOV lane arrangements. (*a*) Right curb lane. (*b*) Left curb lane. (*From* Guide for the Design of High Occupancy Vehicle Facilities, *American Association of State Highway and Transportation Officials, Washington, D.C., 1992, with permission*)

plans using CADD format. The following information is presented as an example and represents the current acceptable policies in Ohio.

Any CADD information to be submitted or exchanged should be in either DWG (Autocad) or IGDS (Intergraph) format. CADD-generated plan sheets should conform to the general plan specifications of the agency. This includes sheet size and material, drafting conventions (symbols, line weights, line styles, etc.), plan format, and other applicable specifications.

Filename Conventions. The following filename conventions are used:

- Filename extension (.ext) should be as follows:

  Autocad drawing files   .DWG

  Intergraph design files   .DGN

  Cell libraries   .CEL

- Filenames should consist of eight characters using the following guidelines:

  ppppppaab.ext

where   ppppp   is a 5-digit project identification number

         aa   is a 2-letter code designating the drawing type (see Table 2.31 for codes)

         b   is a 1-letter code (A–Z) indicating the number of drawings of the same type

TABLE 2.31 Code Designations for Drawing Types for Use in Assigning CADD Filenames

Base:		**Maintenance of traffic:**	
BM	base map	MD	MOT details
		MN	MOT notes
Drainage:		MP	MOT plan
DD	drainage details	MS	MOT subsummary
DE	erosion control plan		
DF	drainage profile	**Right-of-way:**	
DN	drainage notes	RC	centerline survey plat
DP	drainage plan	RD	right-of-way details
DS	drainage subsummary	RM	property map
		RP	right-of-way schematic plan
General:		RR	railroad plat
GA	pavement area details	RS	summary of additional rights-of-way
GB	subsummary	RT	right-of-way title sheet
GC	computations		
GD	detour plan	**Structures:**	
GE	superelevation tables	SD	structure details
GG	general summary	SG	structure general summary
GM	miscellaneous details	SN	structure notes
GN	general notes	SP	structure site plan
GP	general plan and profile	SS	structure subsummary
GR	guiderail and barrier details		
GS	schematic plan	**Utilities:**	
GT	title sheet	UD	utilities details
GX	cross sections	UE	utilities elevation views
GY	typical sections	UG	utilities general summary
GZ	general plan and profile	UN	utilities notes
	(2d series)	UP	utilities plan
		US	utilities subsummary
Landscaping:			
PD	landscaping details	**Traffic control:**	
PN	landscaping notes	TD	traffic control details
PP	landscaping plan	TE	traffic control elevation views
PS	landscaping subsummary	TG	traffic control general summary
		TN	traffic control notes
Lighting:		TP	traffic control plan
LC	lighting circuit diagrams	TS	traffic control subsummary
LD	lighting details		
LG	lighting general summary	**Waterwork:**	
LP	lighting plan	WD	waterwork details
LN	lighting notes	WN	waterwork notes
LS	lighting subsummary	WP	waterwork plan
		WS	waterwork subsummary

Source: Location and Design Manual, Vol. 3, *Highway Plans,* Ohio Department of Transportation, with permission.

Working Units.   When submitting or exchanging CADD design files in the IGDS (Intergraph) format, the working units utilized in the file should be FT, 12 IN, 1000 positional units.

Levels and Layers.   It is critical that an organized leveling and layering system be utilized. Table 2.32 provides recommended leveling and layering schemes for use with both IGDS (Intergraph) and DWG (Autocad) formats. The table also addresses screen colors, element weights, and line codes (IGDS only), which, when coupled with the

(Text continues on page 2.155.)

TABLE 2.32 Information for CADD Leveling and Layering Schemes in IGDS and DWG Formats

Level no.	Layer name	Feature name	IGDS			DWG		Symbol character (font 103)	Pattern name
			Style	Color	Weight	Style	Color		
1	SHEETS	BORDER	0	4	4	Cont.	4	—	—
		TRIMLINE	3	4	0	Dashed	6	—	—
		TITLE BOX	0	4	4	Cont.	4	—	—
		PROJECT DATA	0	0	2	Cont.	2	—	—
		SHEET DATA	0	0	2	Cont.	2	—	—
2	MAJOR_GRID	MAJOR GRID LINE	0	4	1	Cont.	7	—	—
3	MINOR_GRID	MINOR GRID LINE	0	10	0	Cont.	6	—	—
4	STATIONS_20	STATION TICKS	0	4	1	Cont.	7	—	—
		STATION NUMBERS	0	4	1	Cont.	7	—	—
5	STATIONS_50	STATION TICKS	0	4	1	Cont.	7	—	—
		STATION NUMBERS	0	4	1	Cont.	7	—	—
6	STATIONS_MISC	STATION TICKS	0	4	1	Cont.	7	—	—
		STATION NUMBERS	0	4	1	Cont.	7	—	—
7	SURVEY_1	TACKED HUB	—	4	0	—	6	x	—
		P.K. NAIL FOUND	—	4	0	—	6	Z	—
		MINE/R.R. SPIKE FOUND	—	4	0	—	6	B	—
		CHISLED B/MARK	—	4	0	—	6	7	—
		BENCHMARK	—	4	0	—	6	1	—
		U.S.G.S DISK	—	4	0	—	6	Z	—
		U.S.G.S DISK SET	—	4	0	—	6	B	—
8	SURVEY_2	IRON PIN FOUND	—	0	1	—	7	w	—
		IRON PIPE FOUND	—	0	1	—	7	w	—
		STONE FOUND	—	0	1	—	6	z	—
		IRON PIPE SET	—	0	1	—	7	B	—
9	DIVISIONS	STATE LINE	6	3	3	Phantom	3	STATE	STATE
		COUNTY LINE	7	1	2	Center	2	—	—
		TOWNSHIP LINE	2	2	1	Dashed	7	—	—

TABLE 2.32 Information for CADD Leveling and Layering Schemes in IGDS and DWG Formats *(Continued)*

Level no.	Layer name	Feature name	IGDS Style	IGDS Color	IGDS Weight	DWG Style	DWG Color	Symbol character (font 103)	Pattern name
9	DIVISIONS (cont.)	SECTION LINE	2	5	0	Dashed	6	—	—
		LOT LINE	3	4	0	Dashed	6	—	—
		CORPORATION LINE	7	0	0	Center	6	8	CORP
		PROPERTY LINE	0	0	0	Cont.	6	r	—
		SECTION CORNER	—	0	0	—	6	s	—
		SECTION HALF	—	0	0	—	6	T	—
10	EX_GEOM	EX P.I.	—	4	0	—	6	Z	—
		EX P.C.	—	4	0	—	6	Z	—
		EX P.T.	—	4	0	—	6	Z	—
		EX T.S.	—	4	0	—	6	Z	—
		EX S.C.	—	4	0	—	6	Z	—
		EX C.S.	—	4	0	—	6	Z	—
		EX S.T.	—	4	0	—	6	Z	—
		EX CENTERLINE	7	4	0	Center	6	—	CL20
		EX BASELINE	7	4	0	Center	6	—	CL20
11	EX_HIGHWAY	EX EOP MAJ HIGHWAY	3	0	0	Dashed	6	—	—
		EX EOP SEC HIGHWAY	2	3	0	Dashed	6	—	—
		EX CURB	5	10	0	Hidden	6	—	—
		EX SHOULDER	5	6	0	Hidden	6	—	—
12	EX-ROW	EX RIGHT OF WAY	3	5	0	Dashed	6	—	—
		EX LIMITED ACCESS	3	13	0	Dashed	6	—	—
		EX L/A AND R/W	3	13	0	Dashed	6	—	—
		EX EASEMENT	2	5	0	Dashed	6	—	—
		EX MONUMENT BOX	—	0	0	—	6	U	—
		CONC MONUMENT FOUND	—	0	0	—	6	q	—
13	EX_STRUCTURES	EX CULVERT	1	1	0	Hidden	6	—	—
		EX HEADWALL	1	10	0	Hidden	6	—	—

		Layer				Linetype			
14	EX_ROADWAY	EX BRIDGE SUPERSTRUCTURE	5	10	0	Hidden	6		
		EX ABUTMENT	1	10	0	Hidden	6		
		EX PIER	1	10	0	Hidden	6	A	
		EX WINGWALL	5	10	0	Hidden	6		
		EX GUIDERAIL	5	9	0	Hidden	6		
		EX GUIDERAIL RIGHT	5	9	0	Hidden	6		EG20R
		EX GUIDERAIL LEFT	3	10	0	Dashed	6		EG20L
		EX CONC. BARRIER	5	9	0	Hidden	6		
		EX ANCHOR ASSEMBLY	2	8	0	Dashed	6		
		EX FENCE	0	8	0	Cont.	6		EFENCE
		EX GATE	1	10	0	Hidden	6		
		EX WALL	0	5	0	Cont.	6		
		EX RAILROAD	0	9	0	Cont.	6		RR20
		EX POST	—	9	0	—	6	Z	
		EX PILING	—	8	0	—	6	Z	
		LARGE ROCK	6	10	0	—	6	6	
15	EX_HYDRO	EX STREAM/RIVER	6	11	0	Phantom	6		
		EX POND/LAKE	6	11	0	Phantom	6		
		EX DITCH	6	11	0	Phantom	6		
		EX SWAMP AREA	3	11	0	Dashed	6		
		EX ROCK CHANNEL PROT.	—	11	0	—	6		
		EX RIPRAP	—	11	0	—	6		
16	EX_MISC	EX WALK	5	10	0	Hidden	6		
		EX STEPS	1	10	0	Hidden	6		
		EX DRIVEWAY	5	0	0	Hidden	6		
		EX FIELD DRIVE	5	2	0	Hidden	6		
		EX PARKING LOT	5	0	0	Hidden	6		
		EX CONCRETE PADS	1	10	0	Hidden	6		
		EX HOUSE	0	7	1	Cont.	7		
		EX BUILDING	0	7	1	Cont.	7		
		EX STRUCTURE MISC	0	7	1	Cont.	7		
		EX MOBILE HOME	0	7	1	Cont.	7		
		EX MAILBOX	—	9	0	—	6	d	

TABLE 2.32 Information for CADD Leveling and Layering Schemes in IGDS and DWG Formats (*Continued*)

Level no.	Layer name	Feature name	IGDS Style	IGDS Color	IGDS Weight	DWG Style	DWG Color	Symbol character (font 103)	Pattern name
17	EX_TREES	EX TREE	—	2	0	—	6	b	—
		EX SHRUB	—	2	0	—	6	b	—
		STUMP	—	2	0	—	6	o	—
		EX TREE LINE	0	2	0	Cont.	6	—	TREEE
		EX SHRUB LINE	0	2	0	Cont.	6	—	SHRUBE
18	EX_UTILITIES	UNKNOWN MANHOLE	—	0	0	—	6	A	—
		EX TELEPHONE MANHOLE	—	6	0	—	6	A	—
		EX ELECTRIC MANHOLE	—	3	0	—	6	A	—
		EX WATER MANHOLE	—	1	0	—	6	A	—
		EX STORM MANHOLE	—	5	0	—	6	A	—
		EX SANITARY MANHOLE	—	2	0	—	6	A	—
		EX P CATCH BASIN	—	5	0	—	6	e	—
		EX L CATCH BASIN	1	5	0	Hidden	6	—	—
		EX GAS LINE	5	4	0	Hidden	6	—	—
		EX TELEPHONE LINE	5	6	0	Hidden	6	—	—
		EX ELECTRIC LINE	5	3	0	Hidden	6	—	—
		EX WATER LINE	5	1	0	Hidden	6	—	—
		EX STORM LINE	5	5	0	Hidden	6	—	—
		EX SANITARY LINE	5	2	0	Hidden	6	—	—
		GUYWIRE ANCHOR	—	0	0	—	6	Z	—
		UNKNOWN POLE	—	0	0	—	6	D	—
		EX LIGHT POLE	—	3	0	—	6	E	—
		EX TELEPHONE POLE	—	6	0	—	6	F	—
		EX POWER POLE	—	3	0	—	6	G	—
		EX TELE/LIGHT POLE	—	6	0	—	6	H	—
		EX TELE/POWER POLE	—	6	0	—	6	h	—
		EX LIGHT/POWER POLE	—	3	0	—	6	I	—
		EX TELE PEDESTAL	—	6	0	—	6	z	—

		Layer							
		EX TELE BOOTH	—	6	0	—	6	z	—
		TELE LINE MARKER	—	6	0	—	6	3	—
		POWER LINE MARKER	—	3	0	—	6	3	—
		EX GAS METER	—	4	0	—	6	P	—
		EX GAS VALVE	—	4	0	—	6	S	—
		GAS LINE MARKER	—	4	0	—	6	3	—
		EX GAS LINE VENT	—	4	0	—	6	z	—
		EX GASOLINE PUMP	—	4	0	—	6	e	—
		EX OIL TANK	1	4	0	Hidden	6	—	—
		EX WATER METER	—	1	0	—	6	Q	—
		EX WATER VALVE	—	1	0	—	6	S	—
		EX WATER WELL	—	1	0	—	6	Z	—
		EX UNDERDRAIN	1	5	0	Hidden	6	—	—
		EX FIRE HYDRANT	—	1	0	—	6	R	—
		FLOOR DRAIN	—	0	0	—	6	Z	—
		TANK FILLER CAP	—	2	0	—	6	Z	—
		EX SEPTIC	—	2	0	—	6	Z	—
19	EX_CONTOURS	EX LEACH BED AREA	6	2	0	Phantom	6	—	—
		EX MINOR CONTOUR LINE	3	0	0	Dashed	6	—	—
		EX MAJOR CONTOUR LINE	3	7	1	Dashed	7	—	—
20	P_GEOM	PROP P.I.	—	3	0	—	6	T	—
		PROP P.C.	—	3	0	—	6	Z	—
		PROP P.T.	—	3	0	—	6	Z	—
		PROP T.S.	—	3	0	—	6	Z	—
		PROP S.C.	—	3	0	—	6	Z	—
		PROP C.S.	—	3	0	—	6	Z	—
		PROP S.T.	—	3	0	—	6	Z	—
		PROP CENTERLINE	7	3	0	Center	6	Z	CL20
		PROP BASELINE	7	3	0	Center	6	—	CL20
21	P_HIGHWAY	PROP EOP SEC HIGHWAY	0	6	1	Cont.	7	—	—
		PROP EOP MAJ HIGHWAY	0	0	1	Cont.	7	—	—
		PROP CURB	0	10	1	Cont.	7	—	—
		PROP SHOULDER	0	6	1	Cont.	7	—	—

TABLE 2.32 Information for CADD Leveling and Layering Schemes in IGDS and DWG Formats (*Continued*)

Level no.	Layer name	Feature name	IGDS			DWG		Symbol character (font 103)	Pattern name
			Style	Color	Weight	Style	Color		
22	P_ROW	PROP RIGHT OF WAY	0	1	2	Cont.	2	—	—
		TEMP R/W OR EASE	0	1	2	Cont.	2	—	—
		PROP LIMITED ACCESS	0	9	2	Cont.	2	—	—
		PROP L/A AND R/W	0	9	2	Cont.	2	—	—
		C/SL/SE/UTIL EASE	0	7	2	Cont.	2	—	—
		PROP MONUMENT BOX	—	0	1	—	7	a	—
		PROP CONC. MONUMENT	—	0	1	—	7	q	—
23	P_STRUCTURES	PROP CULVERT	0	1	1	Cont.	7	—	—
		PROP HEADWALL	0	10	1	Cont.	7	—	—
		PROP BRIDGE SUPERSTRUCTURE	0	10	1	Cont.	7	—	—
		PROP ABUTMENT	0	10	1	Cont.	7	—	—
		PROP PIER	0	10	1	Cont.	7	Z	—
		PROP WINGWALL	0	10	1	Cont.	7	—	—
24	P_ROADWAY	PROP GUIDERAIL	0	9	1	Cont.	7	—	—
		PROP GUIDERAIL RIGHT	0	9	1	Cont.	7	—	PG20R
		PROP GUIDERAIL LEFT	0	9	1	Cont.	7	—	PG20L
		PROP CONCRETE BARRIER	0	10	1	Cont.	7	—	—
		PROP ANCHOR ASSEMBLY	0	9	1	Cont.	7	—	—
		EX/PROP GUIDERAIL	0	9	1	Cont.	7	—	—
		EX/PROP GUIDERAIL RIGHT	0	9	1	Cont.	7	—	BG20R
		EX/PROP GUIDERAIL LEFT	0	9	1	Cont.	7	—	BG20L
		PROP FENCE	0	8	1	Cont.	7	—	PFENCE
		PROP GATE	0	8	1	Cont.	7	—	—
		PROP WALL	0	10	1	Cont.	7	—	—
		PROP RAILROAD	0	5	1	Cont.	7	—	RR20
		PROP POST	0	6	0	—	6	B	—
25	P_HYDRO	RELOC STREAM/RIVER	6	11	0	Phantom	6	—	—
		RELOC POND/LAKE	6	11	0	Phantom	6	—	—

No.	Group	Name							
26	P_MISC	PROP DITCH	6	11	1	Phantom	7	—	—
		PROP JUTE MATTING	0	0	1	Cont.	7	—	—
		PROP ROCK CHAN. PROT.	0	0	1	Cont.	7	—	—
		PROP RIPRAP	0	0	1	Cont.	7	—	—
		PROP WALK	0	10	1	Cont.	7	—	—
		PROP STEPS	0	10	1	Cont.	7	—	—
		PROP DRIVEWAY	0	0	1	Cont.	7	—	—
		PROP FIELD DRIVE	0	2	1	Cont.	7	—	—
		PROP PARKING LOT	0	0	1	Cont.	7	—	—
		PROP HOUSE	0	7	1	Cont.	7	—	—
		PROP BUILDING	0	7	1	Cont.	7	—	—
		RELOC MOBILE HOME	0	7	1	Cont.	7	—	—
27	P_TREES	PROP MAILBOX	—	1	0	—	7	k	—
		PROP TREE	—	2	0	—	6	b	—
		TREE REMOVED	—	2	0	—	6	c	—
		PROP SHRUB	—	2	0	—	6	b	—
		SHRUB REMOVED	—	6	0	—	6	c	—
		STUMP REMOVED	—	2	0	—	6	—	—
		PROP TREE LINE	0	2	0	Cont.	6	—	TREEP
		PROP SHRUB LINE	0	2	0	Cont.	6	p	SHRUBP
28	P_UTILITIES	PROP TELE MANHOLE	—	6	1	—	7	B	—
		TELE MANHOLE ADJUSTED	—	6	1	—	7	C	—
		PROP ELECTRIC MANHOLE	—	3	1	—	7	B	—
		ELEC MANHOLE ADJUSTED	—	3	1	—	7	C	—
		PROP WATER MANHOLE	—	1	1	—	7	B	—
		WATER MANHOLE ADJUSTED	—	1	1	—	7	C	—
		PROP STORM MANHOLE	—	5	1	—	7	B	—
		STORM MANHOLE ADJUSTED	—	5	1	—	7	C	—
		PROP SANITARY MANHOLE	—	2	1	—	7	B	—
		SANITARY MH ADJUSTED	—	2	1	—	7	C	—
		PROP P CATCH BASIN	—	5	1	—	7	g	—
		PROP L CATCH BASIN	0	5	1	Cont.	7	—	—
		PROP CB ADJUSTED	—	5	1	—	7	f	—

TABLE 2.32 Information for CADD Leveling and Layering Schemes in IGDS and DWG Formats (*Continued*)

Level no.	Layer name	Feature name	IGDS			DWG		Symbol character (font 103)	Pattern name
			Style	Color	Weight	Style	Color		
28	P_UTILITIES (*cont.*)	PROP GAS LINE	0	4	1	Cont.	7	—	—
		PROP TELEPHONE LINE	0	6	1	Cont.	7	—	—
		PROP ELECTRIC LINE	0	3	1	Cont.	7	—	—
		PROP WATER LINE	0	1	1	Cont.	7	—	—
		PROP STORM LINE	0	5	1	Cont.	7	—	—
		PROP SANITARY LINE	0	2	1	Cont.	7	—	—
		PROP LIGHT POLE	—	3	1	—	7	K	—
		PROP TELEPHONE POLE	—	6	1	—	7	L	—
		PROP UNDERDRAIN	0	5	1	Cont.	7	—	0
		PROP POWER POLE	—	3	1	—	7	M	—
		PROP TELE/LIGHT POLE	—	6	1	—	7	N	—
		PROP TELE/POWER POLE	—	6	1	—	7	i	—
		PROP POWER/LIGHT POLE	—	3	1	—	7	O	—
		PROP TELE PEDESTAL	—	6	1	—	7	u	—
		PROP TELE BOOTH	—	6	1	—	7	u	—
		PROP GAS METER	—	4	1	—	7	V	—
		PROP GAS VALVE	—	4	1	—	7	Y	—
		PROP GAS LINE VENT	—	4	1	—	7	B	—
		PROP GASOLINE PUMP	—	4	1	—	7	g	—
		PROP OIL TANK	0	4	1	Cont.	7	—	—
		PROP WATER METER	—	1	1	—	7	W	—
		PROP WATER VALVE	—	1	1	—	7	Y	—
		PROP WATER WELL	—	1	1	—	7	B	—
		PROP FIRE HYDRANT	—	1	1	—	7	X	—
		PROP SEPTIC	—	2	1	—	7	B	—
		PROP LEACH BED AREA	0	2	1	Cont.	7	—	—
29	P_CONTOURS	PROP MINOR CONTOUR LINE	0	4	2	Cont.	2	—	—
		PROP MAJOR CONTOUR LINE	0	6	3	Cont.	3	—	—

#	Layer	Item				Linetype		
30	LIMITS	CONSTRUCTION LIMITS	1	10	0	Hidden	6	—
		WORK AGREE LIMIT	1	1	0	Hidden	6	—
31	EX_SIGNALS	EX SIGNAL HEAD	0	4	0	—	6	—
		EX SIGNAL SUPPORT	0	0	0	Cont.	6	—
		EX SPAN WIRE	1	3	0	Cont.	6	—
		EX WIRING	1	3	0	Dashed	6	—
		EX CONTROLLER	1	3	0	Dashed	6	—
		EX PULL BOX	—	4	1	Dashed	7	—
32	PR_SIGNALS	PROP SIGNAL HEAD	0	0	1	—	7	—
		PROP SIGNAL SUPPORT	0	3	1	Cont.	7	—
		PROP SPAN WIRE	0	0	1	Cont.	7	—
		PROP WIRING	0	0	1	Cont.	7	—
		PROP CONTROLLER	0	0	1	Cont.	7	—
		PROP PULL BOX	0	0	1	Cont.	6	—
33	EX_PMARKINGS	EX LANE LINE	3	0	0	Dashed	6	—
		EX EDGE LINE	0	4	0	Cont.	6	—
		EX CENTER LINE	1	4	0	Hidden	6	—
		EX CHANNELIZING LINE	0	0	2	Cont.	7	—
		EX STOP BAR	0	0	2	Cont.	2	—
		EX SYMBOL	0	0	0	—	6	—
		EX WORD	—	0	0	—	7	—
34	PR_PMARKINGS	PROP LANE LINE	3	4	1	Dashed	7	—
		PROP EDGE LINE	0	4	1	Cont.	7	—
		PROP CENTER LINE	1	0	1	Hidden	7	—
		PROP CHANNELIZING LINE	0	0	2	Cont.	2	—
		PROP STOP BAR	0	0	3	Cont.	3	—
		PROP SYMBOL	0	0	1	—	7	—
		PROP WORD	—	0	0	—	7	—
35	SIGNING	EX FOUNDATION	1	0	0	Hidden	6	—
		EX 1-POST SIGN	—	2	0	—	6	3
		EX 2-POST SIGN	—	2	0	—	6	4
		EX 3-POST SIGN	—	2	0	—	6	5
		EX TRUSS OVERHEAD	2	1	1	Dashed	7	—

TABLE 2.32 Information for CADD Leveling and Layering Schemes in IGDS and DWG Formats (*Continued*)

Level no.	Layer name	Feature name	IGDS			DWG		Symbol character (font 103)	Pattern name
			Style	Color	Weight	Style	Color		
35	SIGNING (*cont.*)	EX CANT OVERHEAD	2	1	1	Dashed	7	—	—
		EX STRUCTURE MOUNT	2	1	1	Dashed	7	—	—
		PROP FOUNDATION	0	0	1	Cont.	7	—	—
		PROP 1-POST SIGN	—	2	1	—	7	3	—
		PROP 2-POST SIGN	—	2	1	—	7	4	—
		PROP 3-POST SIGN	—	2	1	—	7	5	—
		PROP TRUSS OVERHEAD	0	1	2	Cont.	2	—	—
		PROP CANT OVERHEAD	0	1	2	Cont.	2	—	—
		PROP STRUCTURE MOUNT	0	1	2	Cont.	2	—	—
36	EX_LIGHTING	EX SUPPORT	1	1	0	Dot	6	—	—
		EX CANT SUPPORT	2	1	0	Hidden	6	—	—
		EX FOUNDATION	1	0	0	Dot	6	—	—
		EX PULL BOX	1	4	0	Dot	6	—	—
		EX CIRCUITRY	2	3	0	Dashed	6	—	—
37	PR_LIGHTING	PROP SUPPORT	0	1	1	Cont.	7	—	—
		PROP CANT SUPPORT	0	1	1	Cont.	7	—	—
		PROP FOUNDATION	0	0	1	Cont.	7	—	—
38	LUMIN_PATTERN	INTENSITY LEVEL 1	0	1	1	Cont.	2	—	—
		INTENSITY LEVEL 2	0	1	2	Cont.	7	—	—
39	PR_CIRCUITRY	PROP PULL BOX	0	4	1	Cont.	7	—	—
		PROP CONTROL CENTER	0	3	1	Cont.	7	—	—
		PROP WIRING	0	4	1	Cont.	7	—	—
40	LIGHT_ANNOT	PROP LIGHTING ANNOT	—	1	1	—	7	—	—
41	MOT PHASE_1	DRUMS	—	6	1	—	7	z	—
		PORT. CONCRETE BARRIER	0	0	2	Cont.	2	—	—
		TEMP EDGE LINE	0	0	1	Cont.	7	—	—
		TEMP CENTER LINE	1	4	1	Hidden	7	—	—
		TEMP LANE LINE	3	0	1	Dashed	7	—	—

No.	MOT Phase	Item							
		TEMP CHANNELIZING LINE	0	0	2	Cont.	2		—
		TEMP STOP BAR	0	0	3	Cont.	3		—
		TEMP SIGNAL	0	2	1	Cont.	7		—
42	MOT PHASE_2	DRUMS	—	6	1	—	7	z	—
		PORT. CONCRETE BARRIER	0	0	2	Cont.	2		—
		TEMP EDGE LINE	0	4	1	Cont.	7		—
		TEMP CENTER LINE	1	0	1	Hidden	7		—
		TEMP LANE LINE	3	0	1	Dashed	7		—
		TEMP CHANNELIZING LINE	0	0	2	Cont.	2		—
		TEMP STOP BAR	0	0	3	Cont.	3		—
		TEMP SIGNAL	0	2	1	Cont.	7		—
43	MOT PHASE_3	DRUMS	—	6	1	—	7	z	—
		PORT. CONCRETE BARRIER	0	0	2	Cont.	2		—
		TEMP EDGE LINE	0	4	1	Cont.	7		—
		TEMP CENTER LINE	1	0	1	Hidden	7		—
		TEMP LANE LINE	3	0	1	Dashed	7		—
		TEMP CHANNELIZING LINE	0	0	2	Cont.	2		—
		TEMP STOP BAR	0	0	3	Cont.	3		—
		TEMP SIGNAL	0	2	1	Cont.	7		—
44	MOT PHASE_4	DRUMS	—	6	1	—	7	z	—
		PORT. CONCRETE BARRIER	0	0	2	Cont.	2		—
		TEMP EDGE LINE	0	4	1	Cont.	7		—
		TEMP CENTER LINE	1	0	1	Hidden	7		—
		TEMP LANE LINE	3	0	1	Dashed	7		—
		TEMP CHANNELIZING LINE	0	0	2	Cont.	2		—
		TEMP STOP BAR	0	0	3	Cont.	3		—
		TEMP SIGNAL	0	2	1	Cont.	7		—
45									
46									
47									
48									
49									
50									

TABLE 2.32 Information for CADD Leveling and Layering Schemes in IGDS and DWG Formats (*Continued*)

Level no.	Layer name	IGDS			DWG		Symbol character (font 103)	Pattern name
		Style	Color	Weight	Style	Color		
51								
52	(Blank)							
53								
54								
55								
56								
57								
58								
59								
60	SCRATCH_1							
61	SCRATCH_2							
62	SCRATCH_3							
63	SCRATCH_4							

Source: *Location and Design Manual*, Vol. 3, *Highway Plans*, Ohio Department of Transportation, with permission.

TABLE 2.33 CADD Plotting Conventions to Use with Information in Table 2.32

DWG			
Color	Hue	IGDS: weight	Full-size sheet line width, in
1	Red	1	0.014
2	Yellow	2	0.020
3	Green	3	0.024
4	Cyan	4	0.031
5	Blue	5	0.039
6	Magenta	0	0.010
7	White	1	0.014
8	Dark gray	*	*
9	Dark red	1	0.014
10	Dark brown	2	0.020
11	Dark green	3	0.024
12	Dark cyan	4	0.031
13	Dark blue	5	0.039
14	Dark magenta	0	0.010
15	Light gray	*	*

*Designer may use any pen at own discretion.

Source: *Location and Design Manual,* Vol. 3, *Highway Plans,* Ohio Department of Transportation, with permission.

appropriate plotting conventions shown in Table 2.33, will produce acceptable plan sheet plots.

Pen Tables. Table 2.33 is included to organize and standardize plotting to achieve the appropriate appearance of plan sheets. The values for colors (DWG) and weights (IGDS) from Table 2.32 are referenced to indicate pen numbers, pen sizes, and line weights that should be used.

Fonts. The following fonts are recommended for use:

- Straight font

  IGDS: 35

  DWG: romans

- Slanted font

  IGDS: 83

  DWG: romans (7° slant)

- Gothic font

  IGDS: 8

  DWG: gothic

- Filled font

  IGDS: 70

- Symbol font

   IGDS: 103

   DWG: font103

- Default font

   IGDS: 0

   DWG: STANDARD

- IGDS font library

   FONTLIBA

- Autocad font file

   gothic.shx,

   romans.shx

Feature Tables. Feature tables used with survey data collectors and IGDS coordinate geometry software implement level, color, line weight, and line code information to produce or add to base map information. The feature tables listed below, when used as such, will produce output data in conformance with the recommended data characteristics discussed under "Levels and Layers."

The feature tables are shown with the drawing scale to which they are to be designed.

Feature table name	Scale
ODOT10	10
ODOT20	20
ODOT50	50

Sheets and Files. Each sheet in the plan should be contained in a separate design or drawing file. An exception can be made for cross-section sheets. Cross-section design or drawing files may contain more than one plan sheet when utilizing the leveling and layering schemes outlined in Table 2.32.

2.14.3 Plan Components

Typical highway construction plans are made up of several individual components. The paragraphs that follow will present a brief discussion of various types of plan sheets that make up a set. Except for major projects, seldom will all of the components be required in a plan. However, when required, they are usually placed in the order discussed.

The *title sheet* is the first in the set and contains a brief description of the project and indication of its length. It displays the title of the project in large, bold letters. It lists the specifications under which the project is to be built, states whether traffic is to be maintained or detoured, gives an index of all plan sheets, lists standard construction drawings and supplemental specifications, and contains the signatures of approval by the appropriate officials. See Fig. 2.74 for an example.

The *schematic plan* shows the geometric location of proposed roadway segments in relation to existing roadway segments and other major topographic features (rivers, streams, railroads, high-voltage lines, pipelines, etc.). See Fig. 2.75 for an example.

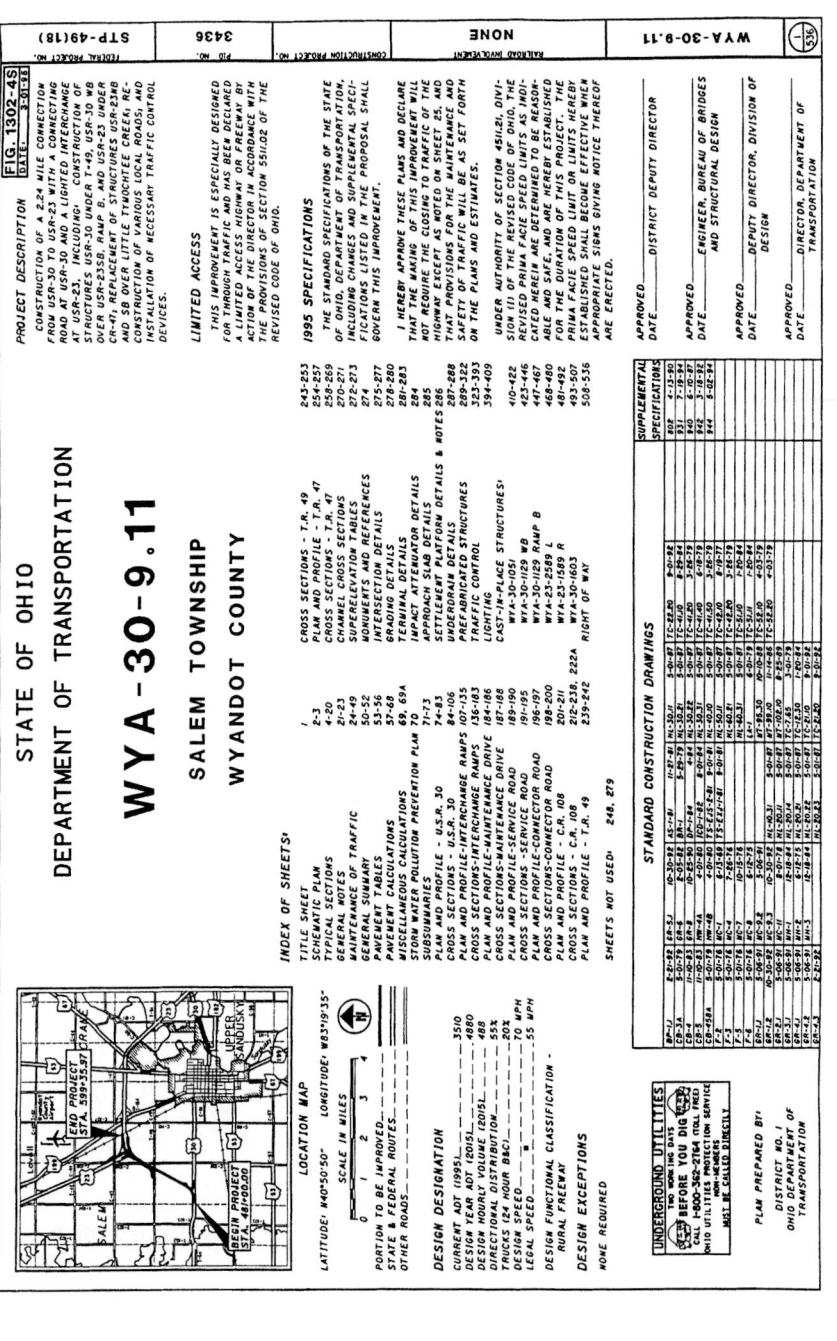

FIGURE 2.74 Typical highway construction plans: example of title sheet showing index to various types of plans. (*From* Location and Design Manual, *Vol. 3*, Highway Plans, *Ohio Department of Transportation, with permission*)

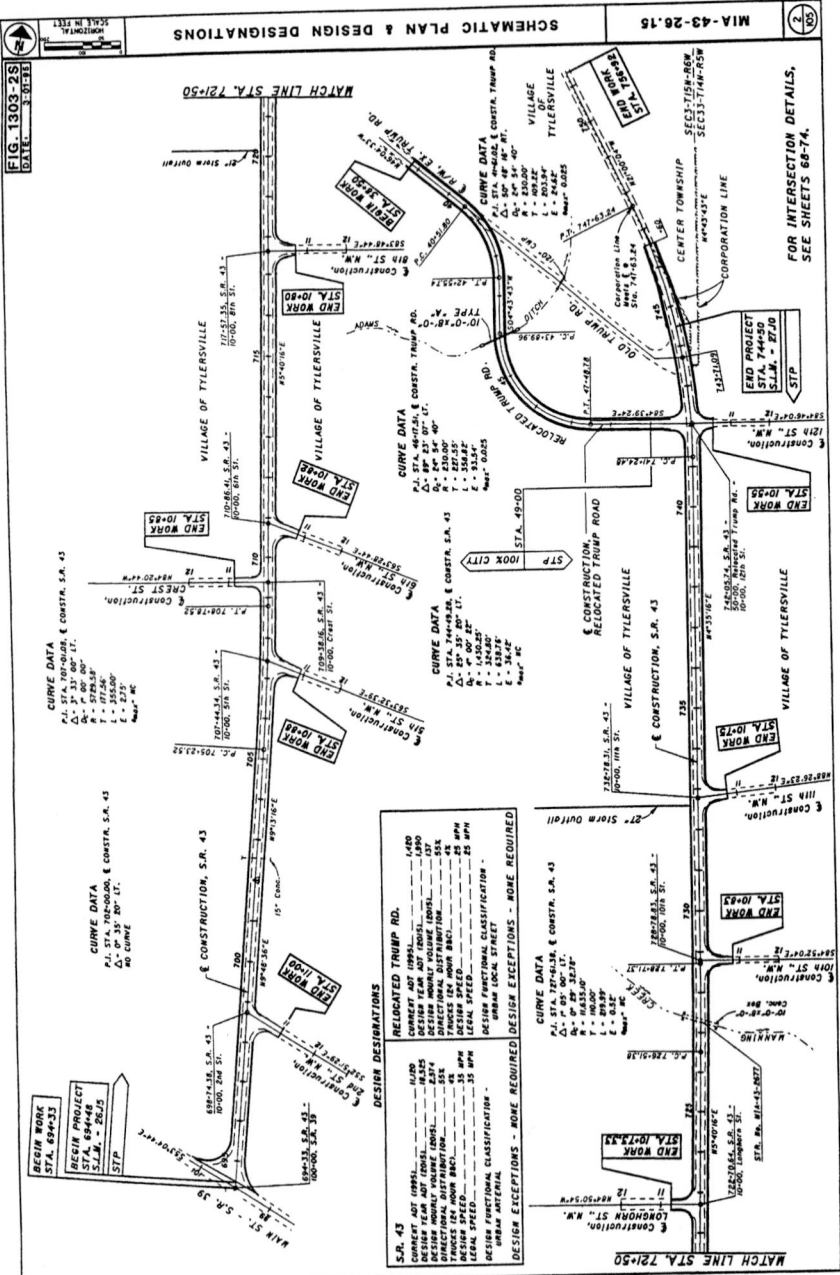

FIGURE 2.75 Typical highway construction plans: example of schematic plan showing location. (*From Location and Design Manual, Vol. 3, Highway Plans, Ohio Department of Transportation, with permission*)

The *typical sections* sheet is a dimensioned cross-sectional view of how the roadway will appear after construction is completed. These sheets generally show lane widths, shoulder widths, pavement buildup, ditch design, foreslope and backslope recommendations, and tie-ins to existing ground lines. Each section is accompanied by a set or sets of station limits identifying to which portion of the roadway it applies. See Fig. 2.76 for an example.

The *general notes* sheets contain plan notes to clarify construction items that are not satisfactorily covered by the specifications or plan details. They may be used to modify standard construction drawings.

The *maintenance of traffic* sheets may include plan view sheets showing location of temporary roads, temporary pavement widening, or detour routes, as well as sheets providing specific notes and instructions regarding sequential construction phases. Details included may be transverse sections showing relationships between the maintenance roadway and the construction area, as well as placement of channelizing devices and lateral construction limits.

The *general summary* sheets contain the itemized list of quantities on which the contractor bid and eventual payment will be based. Any items not listed on these sheets will contain a sheet number reference where they are listed elsewhere in the plans. Each item will usually have a sheet number reference indicating where the item may be found in the plan, or where a subsummary of this information may be found. See Fig. 2.77 for an example.

The *calculations* sheet provides a record of how quantity pay items were calculated. The sheet provides a way of checking these quantities and also usually indicates by station references where these items are used in the plans.

A *storm water pollution prevention plan* may be required by some agencies, depending on how much surface area is disturbed by construction. The threshold limit in Ohio is currently 5 acres. The purpose of these sheets is to provide information on how storm water runoff is to be controlled during construction. Details shown will include the location of existing streams, lakes, wetlands, springs, etc., within 250 ft of the construction area.

The *plan and profile* sheets show what an area looks like before and after construction of the project. In addition, they show quantities, dimensions, and other items required to lay out and construct the project. The sheet is normally divided into three areas—plan view, profile, and quantities. See Fig. 2.78 for an example.

The *cross sections* sheets contain a series of section "slices" of the roadway taken at regular intervals and are used primarily to determine the amount of earthwork and seeding required on the project. They may also be used to locate ditches, show proposed drainage features, design driveways, and establish limits of proposed right-of-way. See Fig. 2.79 for an example.

The *miscellaneous details* sheets are a section of the plans that serve as a "catchall" for items that do not fit under other headings. Items that may appear on these sheets include approach slab details with elevations, driveway details, grading plans at intersections or interchanges, guiderail details, impact attenuator details, intersection details with elevations, linear grading details, pavement details showing elevations, superelevation tables, and noise barriers.

The *drainage details* sheets provide details for prefabricated structures and other drainage items that cannot be adequately shown on other sheets. These sheets include culvert details—not only the structure details, but also the grading plan in the vicinity.

Other specialized sheets that may be part of the plan are as follows:

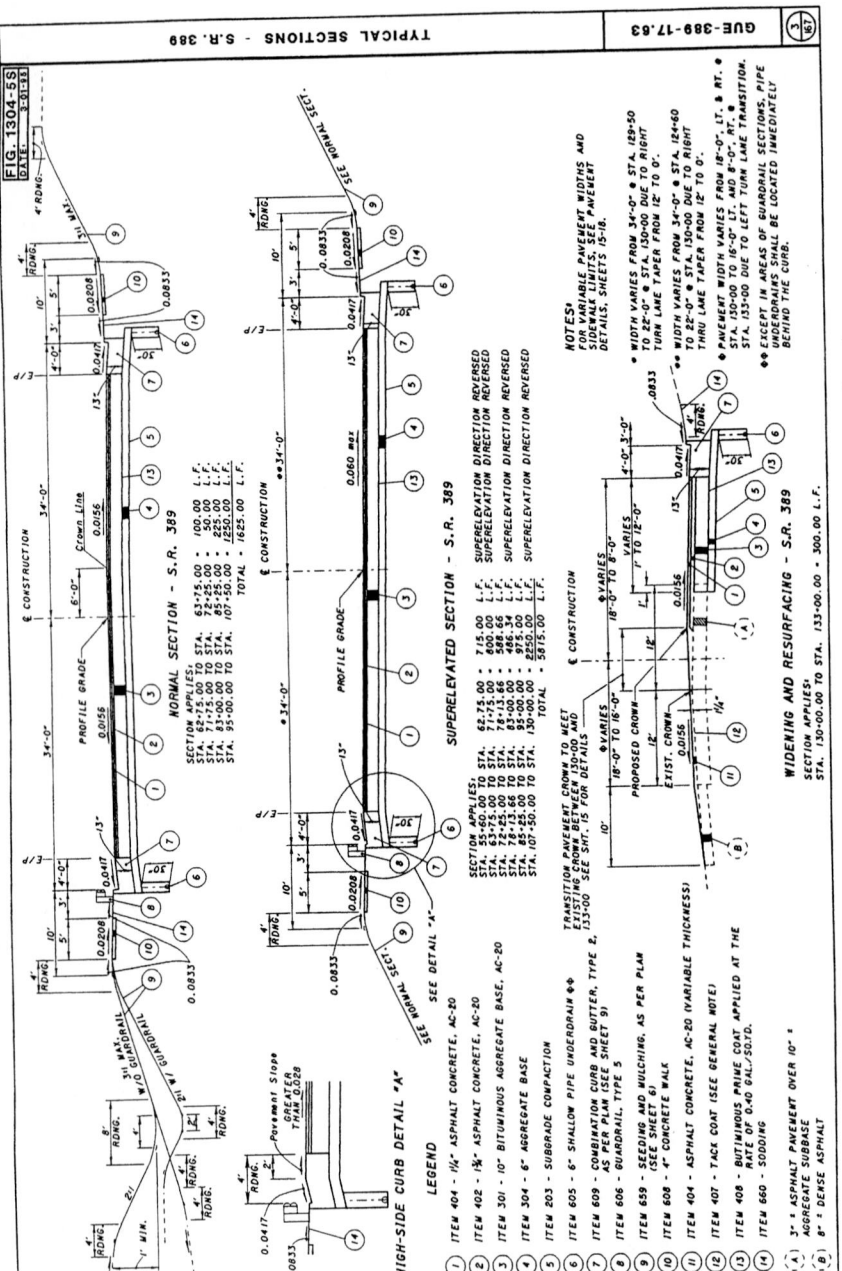

FIGURE 2.76 Typical highway construction plans: example of typical roadway cross sections. (*From Location and Design Manual, Vol. 3, Highway Plans, Ohio Department of Transportation, with permission*)

	SHEET NUMBER				FUNDING								
32	64	68	165	175	100% CITY	NH	STP	ITEM	ITEM EXT.	GRAND TOTAL	UNIT	DESCRIPTION	SEE SHEET NO.
												PAVEMENT	
1844							1844	254	01000	1844	SQ. YD.	PAVEMENT PLANING, BITUMINOUS	
1265							1265	301	00002	1265	CU. YD.	BITUMINOUS AGGREGATE BASE, AC-20	
2627						986	1641	304	20000	2627	CU. YD.	AGGREGATE BASE	
5333						5333		305	12000	5333	SQ. YD.	8" CONCRETE BASE	
1048						1048		310	12000	1048	CU. YD.	SUBBASE, TYPE 1, GRADING A	
497						268	229	402	20000	497	CU. YD.	ASPHALT CONCRETE, AC-20	
38							38	403	20000	38	CU. YD.	ASPHALT CONCRETE, AC-20	
415						226	189	404	20000	415	CU. YD.	ASPHALT CONCRETE, AC-20	
18							18	404	25000	18	CU. YD.	ASPHALT CONCRETE, AC-20 (DRIVEWAYS)	
533						533		407	10000	533	GALLON	TACK COAT	
3114						1056	2048	408	00000	3114	GALLON	BITUMINOUS PRIME COAT	
57						36	21	409	12000	57	CU. YD.	SEAL COAT COVER AGGREGATE, No. 8	
1411						802	609	409	14000	1411	GALLON	SEAL COAT BITUMINOUS MATERIAL	
4000						4000	4000	609	12000	4000	LIN. FT.	COMBINATION CURB AND GUTTER, TYPE 2	
73						73	73	611	10000	73	SQ. YD.	REINFORCED CONCRETE APPROACH SLAB (T=12")	
35							35	612	14000	35	SQ. YD.	6" CONCRETE TRAFFIC ISLAND	
												WATER WORK	
	896	896			896			SPECIAL 63860400		896	LIN. FT.	12" DUCTILE IRON WATER PIPE AND FITTINGS (COL. 801)	
	14	14			14			SPECIAL 63860900		14	EACH	6" VALVE AND APPURTENANCES (COL. 802)	
	10	10			10			SPECIAL 63863800		10	EACH	1½" WATER SERVICE TAP, COMPLETE (COL. 805)	
	8	8			8			SPECIAL 63865502		8	EACH	SERVICE BOX REMOVED AND RESET (COL. 807)	
	4	4			4			SPECIAL 63866602		4	EACH	FIRE HYDRANT, TYPE A (COL. 809)	
												SANITARY SEWER	
	12						12	202	58700	12	EACH	MANHOLE ABANDONED	
	200						200	603	00900	200	LIN. FT.	6" CONDUIT, TYPE B, 706.01 OR 706.08 WITH 706.11 OR 706.12 JOINTS	
	284						284	603	02000	284	LIN. FT.	8" CONDUIT, TYPE C, 706.08 WITH 706.12 JOINTS	
	273						273	603	04400	273	LIN. FT.	12" CONDUIT, TYPE B, 706.03 2750 D-LOAD WITH 706.11 JOINTS	
	28						28	603	05900	28	LIN. FT.	15" CONDUIT, TYPE B, 706.03 WITH 706.11 JOINTS	
	230						230	603	07400	230	LIN. FT.	18" CONDUIT, TYPE B, 706.03 3250 D-LOAD WITH 706.11 JOINTS	
	5						5	604	31500	5	EACH	MANHOLE, No. 3 WITH 706.11 JOINTS	
	8						8	604	34500	8	EACH	MANHOLE ADJUSTED TO GRADE	
	3						3	604	35500	3	EACH	MANHOLE RECONSTRUCTED TO GRADE	
												LIGHTING	
				2				202	75403	2	EACH	LIGHT POLE REMOVED FOR STORAGE, AS PER PLAN	130
				2				202	75500	2	EACH	LIGHT POLE FOUNDATION REMOVED	
				2				202	75505		EACH	LUMINAIRE REMOVED FOR STORAGE, AS PER PLAN	130
				1				625	00500	1	EACH	CONNECTOR KIT, TYPE II	
				1				625	00600	1	EACH	CONNECTOR KIT, TYPE III	
	2			2				625	14100	2	EACH	LIGHT POLE FOUNDATION, 24" X 8" DEEP	
	80			80				625	23200	80	LIN. FT.	No. 4 AWG 5000 VOLT DISTRIBUTION CABLE	
	40			40				625	29002	40	LIN. FT.	TRENCH, 24" DEEP	
	2			2				625	35001	2	EACH	REERECT EXISTING LIGHT POLE, AS PER PLAN	131
	2			2				625	35501	2	EACH	REERECT EXISTING LUMINAIRE, AS PER PLAN	131
												LANDSCAPING	
			5				5	657	10000	5	SQ. YD.	RIPRAP FOR TREE PROTECTION	
			25				25	658	10000	25	CU. YD.	TREE ROOT AERATION	
			400				400	659	10000	400	SQ. YD.	SEEDING AND MULCHING	
			6				6	662	00000	6	EACH	EVERGREEN SHRUB, 15"-18" HEIGHT, JUNIPERUS HORIZONTALIS, 10" B&B	
			15				15	666	10000	15	EACH	PRUNING EXISTING TREE, 8"-16" DIAMETER	

FIGURE 2.77 Typical highway construction plans: example of general summary sheet showing itemized quantities. *(From Location and Design Manual, Vol. 3, Highway Plans, Ohio Department of Transportation, with permission)*

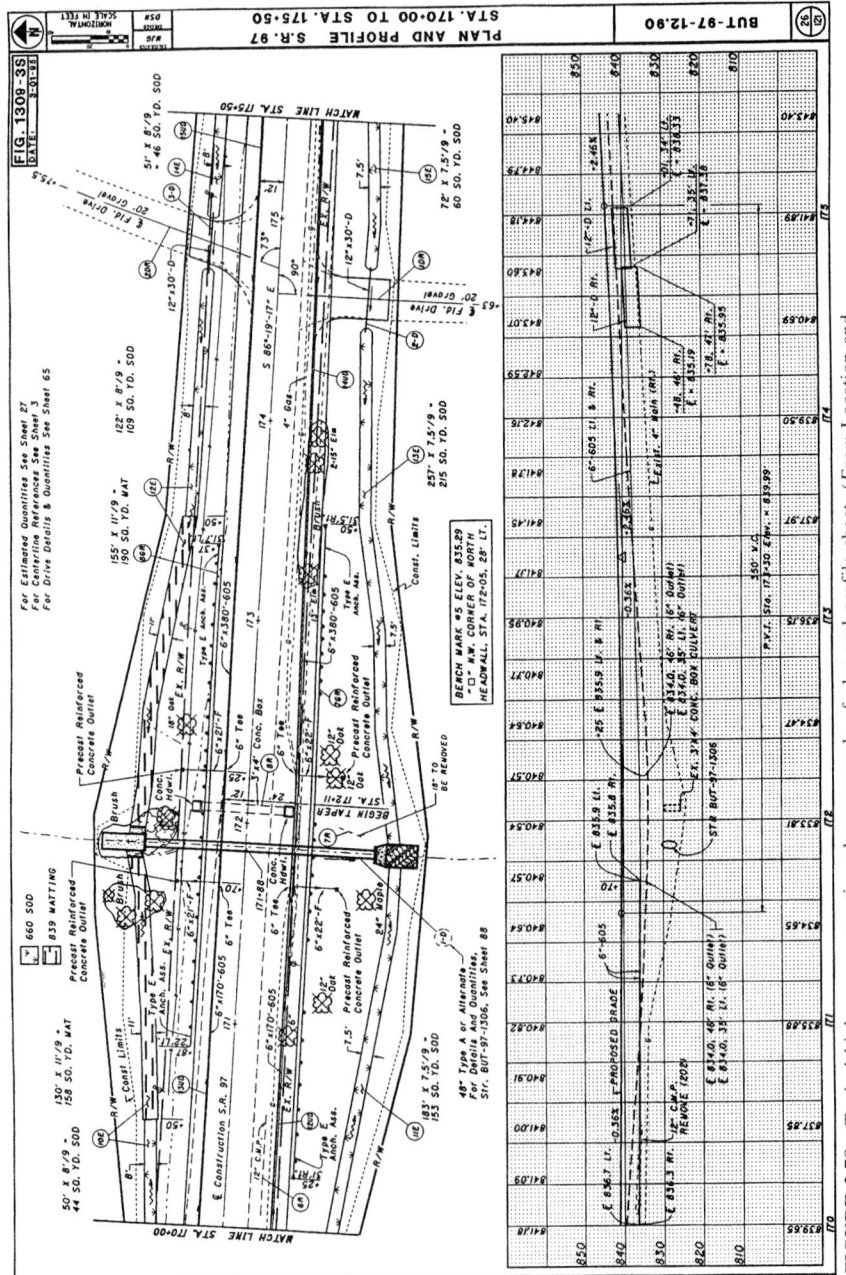

FIGURE 2.78 Typical highway construction plans: example of plan and profile sheet. (*From Location and Design Manual, Vol. 3, Highway Plans, Ohio Department of Transportation, with permission*)

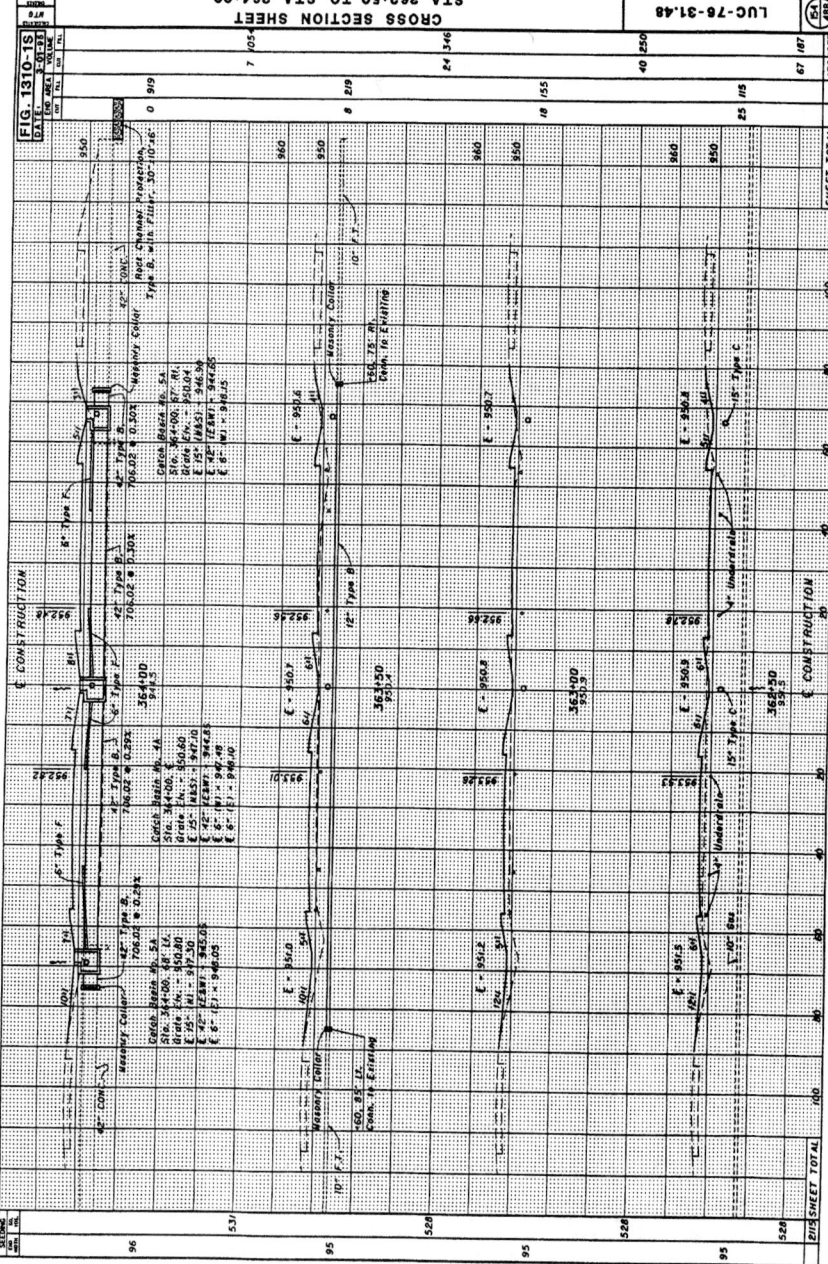

FIGURE 2.79 Typical highway construction plans: example of general cross sections at specific locations. (*From Location and Design Manual, Vol. 3, Highway Plans, Ohio Department of Transportation, with permission*)

Prefabricated structures

Sanitary sewers

Water lines

Traffic control (includes proposed signing, striping, and traffic signals)

Lighting

Landscaping

Cast-in-place structures (includes bridges, retaining walls)

Fence plan (refers to right-of-way fencing on limited-access projects)

Right-of-way (listing all affected property owners, parcel numbers, and required right-of-way to be purchased)

Soil profile and foundation investigation

2.15 REFERENCES

1. *A Policy on Geometric Design of Highways and Streets,* AASHTO, Washington, D.C., 1990.

2. *Roadside Design Guide,* AASHTO, Washington, D.C., 1989.

3. *Guide for the Design of High Occupancy Vehicle Facilities,* AASHTO, Washington, D.C., 1992.

4. *Traffic Engineering Handbook,* Institute of Transportation Engineers, Washington, D.C., 1992.

5. Intelligent Vehicle-Highway Society of America, *IVHS Architecture Development Program, Interim Status Report,* IVHS America, Washington, D.C., 1994.

6. *The Transportation Development Process,* Ohio Department of Transportation, Columbus, 1983.

7. *Location and Design Manual,* Vol. 1, *Roadway Design,* Ohio Department of Transportation, Columbus, 1992.

8. *Location and Design Manual,* Vol. 3, *Highway Plans,* Ohio Department of Transportation, Columbus, 1993.

9. State of California Department of Transportation, "High Occupancy Vehicle (HOV) Guidelines for Planning, Design, and Operations," U.S. Department of Transportation, DOT-T-91-17, Washington, D.C., 1991.

10. "Highway Capacity Manual," Special Report 209, Transportation Research Board, National Research Council, Washington, D.C., 1985.

11. Architectural and Transportation Barriers Compliance Board, "Americans with Disabilities Act (ADA) Accessibility Guidelines: State and Local Government Facilities," pub. *Federal Register,* vol. 57, no. 245, Monday, December 21, 1992; Architectural and Transportation Barriers Compliance Board, Washington, D.C., 1991, 1992.

12. Jack E. Leisch, *"Turning Vehicle Templates,"* Institute of Transportation Engineers, Washington, D.C., 1988.

13. Balke Engineers, *Justification Study for Crossroad Grade Separations, US 30 (Lincoln Highway), from Upper Sandusky to Bucyrus, Wyandot and Crawford Counties, Ohio,* Ohio Department of Transportation, District 1, Lima, 1994.

14. *Standard Construction Drawings,* Bureau of Location and Design, Ohio Department of Transportation, Columbus, 1994 (rev.).

CHAPTER 3
PAVEMENT DESIGN AND REHABILITATION

Aric A. Morse, P.E.

Pavement Design Engineer
Ohio Department of Transportation
Columbus, Ohio

Roger L. Green, P.E.

Pavement Research Engineer
Ohio Department of Transportation
Columbus, Ohio

The movement of people and goods throughout the world is primarily dependent upon a transportation network consisting of roadways. Most, if not all, business economies, personal economies, and public economies are the result of this transportation system. Considering the high initial and annual costs of roadways, and since each roadway serves many users, the only prudent owner of roadways is the public sector. Thus it is the discipline of civil engineering that manages the vast network of roadways.

The surface of these roadways, the *pavement,* must have sufficient smoothness to allow a reasonable speed of travel, as well as ensure the safety of people and cargo. Additionally, once the pavement is in service, the economies that depend upon it will be financially burdened if the pavement is taken out of service for repair or maintenance. Thus, pavements should be designed to be long lasting with few maintenance needs.

The accomplishment of a successful pavement design depends upon several variables. The practice of pavement design is based on both engineering principles and experience. Pavements were built long before computers, calculators, and even slide rules. Prior to more modern times, pavements were designed by trial-and-error and commonsense methods, rather than the more complicated methods being used currently. Even more modern methods require a certain amount of experience and common sense. The most widely used methods today are based on experiments with full-scale, in-service pavements that were built and monitored to failure. Empirical information derived from these road tests is the most common basis for current pavement design methods. More recently, with the ever expanding power of personal computers, more mathematically based pavement design methods such as finite element analysis and refined elastic layer theory have been introduced. These methods require extensive training to use and are not developed for the inexperienced.

Types of pavements can be broadly categorized as rigid, flexible, or composite. The characteristics of these types are reviewed in the following articles.

3.1   RIGID PAVEMENT

Rigid pavement can be constructed with contraction joints, expansion joints, dowelled joints, no joints, temperature steel, continuous reinforcing steel, or no steel. Most generally, the construction requirements concerning these options are carefully chosen by the owner or the public entity that will be responsible for future maintenance of the pavement. The types of joints and the amount of steel used are chosen in concert as a strategy to control cracking in the concrete pavement. Oftentimes the owner specifies the construction requirements but requires the designer to take care of other details such as intersection jointing details and the like. It is imperative that a designer understand all of these design options and the role each of these plays in concrete pavement performance.

The category of rigid pavements can be further broken down into those with joints and those without. Jointed reinforced concrete pavement (JRCP) and jointed plain concrete pavement (JPCP) are the two basic types of jointed concrete pavement. Continuously reinforced concrete pavement (CRCP) has no joints. JRCP is designed for maximum joint spacing permitting cracking between joints and requires temperature steel. JPCP is designed for no cracking between joints; thus, joint spacing is minimized and temperature steel is eliminated. Historically, many jointed pavements were constructed without dowelled joints. Past performance of undowelled jointed pavements—with the exception of warm, dry climates or low-volume roadways—has been poor. Where there are more than a few trucks per day, dowels should be considered at contraction joints. However, low-volume roadways that do not carry significant trucks, such as residential streets, may perform satisfactorily without dowelled joints.

3.1.1   Jointed Rigid Pavement

Jointed rigid pavements tend to crack at 13- to 25-ft lengths because of (1) initial shrinkage after placement as excess water evaporates, (2) temperature-induced expansion and contraction resisted by friction with the subgrade, (3) curling and warping caused by temperature and moisture differences between the top and bottom of the slab, and (4) load-induced stresses.

As slabs contract as a result of seasonal temperature changes, cracks form and widen, or formed joints widen, allowing incompressible materials into the cracks or joints. Subsequently, expansion is hindered and pressure is built up in the pavement. This pressure can result in pressure spalling or even blowups. To control this, partial depth saw cuts are made at regular intervals which induce concrete to crack at these locations. The timing and depth of these saw cuts is critical to ensure that the pavement cracks at the controlled location. Saw cuts should be made as soon as the pavement can support the weight of the saw and operator. The saw cuts should be made at a depth of one-third of the slab thickness for longitudinal joints, and one-fourth of the slab thickness for transverse joints. These saw cuts are then sealed with some type of joint sealer to prevent intrusion of incompressibles. If the saw cut interval (joint spacing) is short enough, intermediate cracks are eliminated. If longer intervals are used, intermediate cracks will form (Fig. 3.1).

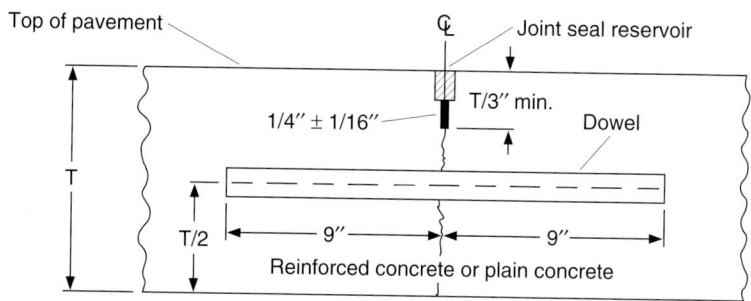

FIGURE 3.1 Typical contraction joint in rigid pavement with dowel for positive load transfer.

Load transfer is the critical element at joints and cracks. In undowelled, unreinforced pavements, any load transfer must be provided by aggregate interlock. Aggregate interlock is lost when slabs contract and the joints or cracks open up. Also, interlock is slowly destroyed by the movement of the concrete as traffic passes over. Given large temperature variations and heavy trucks, aggregate interlock is ineffectual, and faulting is the primary result.

To provide load transfer at the joints, dowels are used which allow for expansion and contraction. Figure 3.1 illustrates a typical doweled joint with saw cut and joint seal. Figure 3.2 shows a similar joint without the dowel to provide load transfer.

Where a long joint spacing is used and intermediate cracks are expected, steel reinforcement is added to hold the cracks tightly closed (JRCP). This allows the load transfer to be accomplished through aggregate interlock without the associated problems described above. Contraction joints do not provide for expansion of the pavement unless the same amount of contraction has already taken place. This contraction will initially be from shrinkage due to concrete curing. Later changes in the pavement length are due to temperature changes.

Where fixed objects such as structures are placed in the pavement, the use of an expansion joint is warranted. Expansion joints should be used sparingly. The pavement will be allowed to creep toward the expansion joint, thus opening the adjacent contraction joints. This can cause movement in the adjacent contraction joints in excess of their design capabilities and result in premature failures. Figure 3.3 shows a detail for a typical expansion joint.

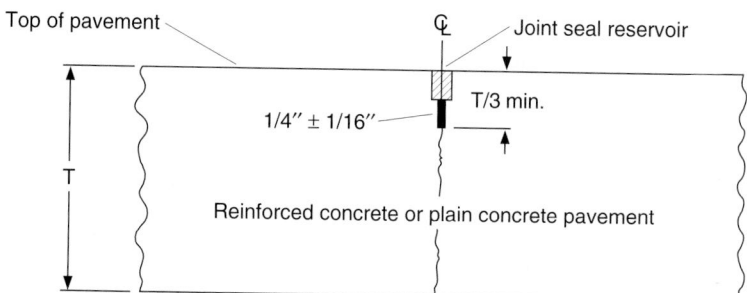

FIGURE 3.2 Typical contraction joint in rigid pavement without dowel for load transfer.

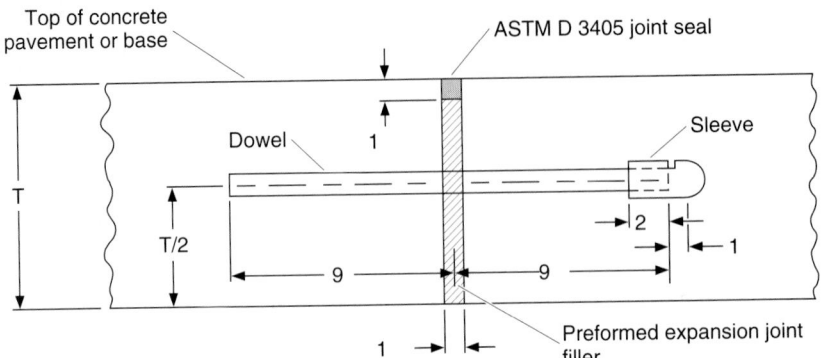

FIGURE 3.3 Typical expansion joint in rigid pavement.

The design of reinforcing steel is a function of seasonal temperature change, sub-base friction, and the weight of the slab. Inadequate reinforcing will not be able to hold the cracks together, and faulting will result. The amount of reinforcing needed to hold cracks together is traditionally calculated using a relationship based on the friction between the subgrade and the bottom of the slab. This relationship assumes that for a crack to open enough to fail the aggregate interlock, the slab will have to slide along the subbase. The current AASHTO recommendation is based on this traditional approach (*Guide for Design of Pavement Structures*, American Association of State Highway and Transportation Officials, 1993):

$$P_s = \frac{LF}{2f_s}(100) \tag{3.1}$$

where P_s = percent steel required per foot

L = intended length of slab, ft

F = frictional coefficient between subgrade and slab
f_s = allowable stress in steel, lb/in²

Another form of this relationship can be expressed as follows:

$$A_s = \frac{WFL}{2f_s}(100) \tag{3.2}$$

where A_s = area of steel required, in²/ft of width

W = weight of slab, lb/ft²

F = coefficient of resistance between slab and subgrade (1.5 unless otherwise known)

L = length of slab, ft
f_s = allowable stress in steel, lb/in²

Although these relationships are accepted by most leading authorities and are referred to in almost every reference on the subject, it is important to understand that they make many assumptions about physical quantities that are seldom consistent throughout a length of pavement. For instance, the friction factor can be affected by something as insignificant as a large footprint in the base course prior to paving. Additionally, the environment can play an important role as water and salt erode the steel, thus reducing the sectional area of the steel.

Where reinforcement is not desired, slab lengths must be chosen so that intermediate transverse cracks are eliminated. The most current theory used to determine allowable slab lengths involves a very old concept developed by Dr. H. M. Westergaard. Westergaard defined a constant called the *radius of relative stiffness* as an algorithm that relates the modulus of subgrade reaction to the flexural stiffness of the slab. The most current theory is based on research that indicates that cracking can be expected when the ratio between the slab length and the radius of relative stiffness is greater than 5. The radius can be calculated from the following equation (Federal Highway Administration Technical Advisory T 5040.30, November 30, 1990):

$$l = \left[\frac{Eh^3}{12k(1 - \nu^2)} \right]^{0.025} \tag{3.3}$$

where l = radius of relative stiffness, in

E = modulus of elasticity of concrete, lb/in^2

h = pavement thickness, in

k = modulus of subgrade reaction, (lb/in^2)/in
ν = Poisson's ratio (0.15)

3.1.2 Rigid Pavement Jointing Details for Intersections

The following basic principles must be observed in developing a correct jointing detail:

1. Never taper concrete down to less than 2 ft in width.

2. Depending upon the amount of transverse reinforcing steel, be careful of the number of lanes that are tied together. In JPCP, tying more than three 12-ft lanes together may result in uncontrolled longitudinal cracking.

3. Always design the secondary (intersecting) route as independent in movement from the primary route. Thus, as the primary route expands and contracts, no unnecessary forces will be created in the secondary route.

4. Provide for expansion wherever payment is interrupted in its longitudinal direction.

5. Terminate joints at 90° to any intersecting joints, obstructions, or edges of pavement.

6. Where possible, lay out lane widths of the same dimension. This permits the contractor to pave all the lanes without changing the paving machine setup dimensions.

7. Unless unavoidable, all joints should be in a straight line. Curved joints are difficult to saw and generally require additional forming.

8. For plain (nonreinforced) concrete pavement, keep the slab length/slab width ratio at a maximum of 2:1.

Intersection details should always be included in construction plans. A proper jointing layout ensures that cracking occurs at locations where load transfer exists (contraction joints) and away from wheel paths (longitudinal joints). The jointing detail should be a separate detail in the plan to eliminate confusion and allow field personnel to easily lay out the intersection without construction delay. Figures 3.4 and 3.5 show jointing layouts that have been used for typical intersections.

3.1.3   Rigid Pavement Joint Sealing

Joint sealing is done to prohibit the infiltration of water into the pavement base and to prevent incompressibles from lodging within the joint cavity. The advantages of keeping water out from under a pavement are documented extensively in the AASHTO *Pavement Design Guide* and in various articles in this chapter. With an unsealed joint, contraction under cooler temperatures allows joint cavities to open up and become filled with sand, stone, and other incompressible material. When warmer temperatures try to expand the length of the pavement, the joints are unable to close, compressive stresses develop, and spalling may result.

The purpose of a sealant reservoir (Figs. 3.1, 3.2, and 3.6) is to prevent water and incompressibles from entering the joint cavity. The design criteria for the sealant reservoir ensure that the sealant stays in place. The ability of the sealant to expand and contract with the movement of the joint is a function of the material properties of the sealant (defined by the manufacturer's specifications) and the expected movement of the joint. Joint movement can be calculated using the following relationship:

$$\Delta L = CL(A \ \Delta T + Z) \tag{3.4}$$

where ΔL = joint opening created by changes in temperature and loss of moisture during curing (joint movement), ft

C = constant used to adjust for friction between bottom of slab and the material that directly supports the pavement (0.65 for granular material, 0.80 for stabilized material)

L = joint spacing, ft

A = thermal coefficient of concrete, $10^{-6}/°F$. If the type of coarse aggregate is quartz, the thermal coefficient is 6.6; sandstone, 6.5; gravel, 6.0; granite, 5.3; basalt, 4.8; limestone, 3.8.

ΔT = difference in minimum temperature pavement will be subjected to and temperature at which pavement was placed

Z = drying shrinkage coefficient of the portland cement concrete (PCC) slab, in/in. If the indirect tensile strength (lb/in^2) is 300, the shrinkage coefficient is 0.0008; 400, 0.0006; 500, 0.00045; 600, 0.0003; 700, 0.0002.

(See "AASHTO Design Procedures for New Pavements," FHWA Report HI-94-023, ERES Consultants, Inc., February 1994; and FHWA Technical Advisory T 5040.30, November 30, 1990.)

There are two categories of joint sealants. The field-molded sealant and the preformed compression seal are used extensively in rigid pavements. Also, field-molded sealants are gaining acceptance and being used in flexible pavements.

For field-molded sealants, the design is very simple and is controlled by the following relationship:

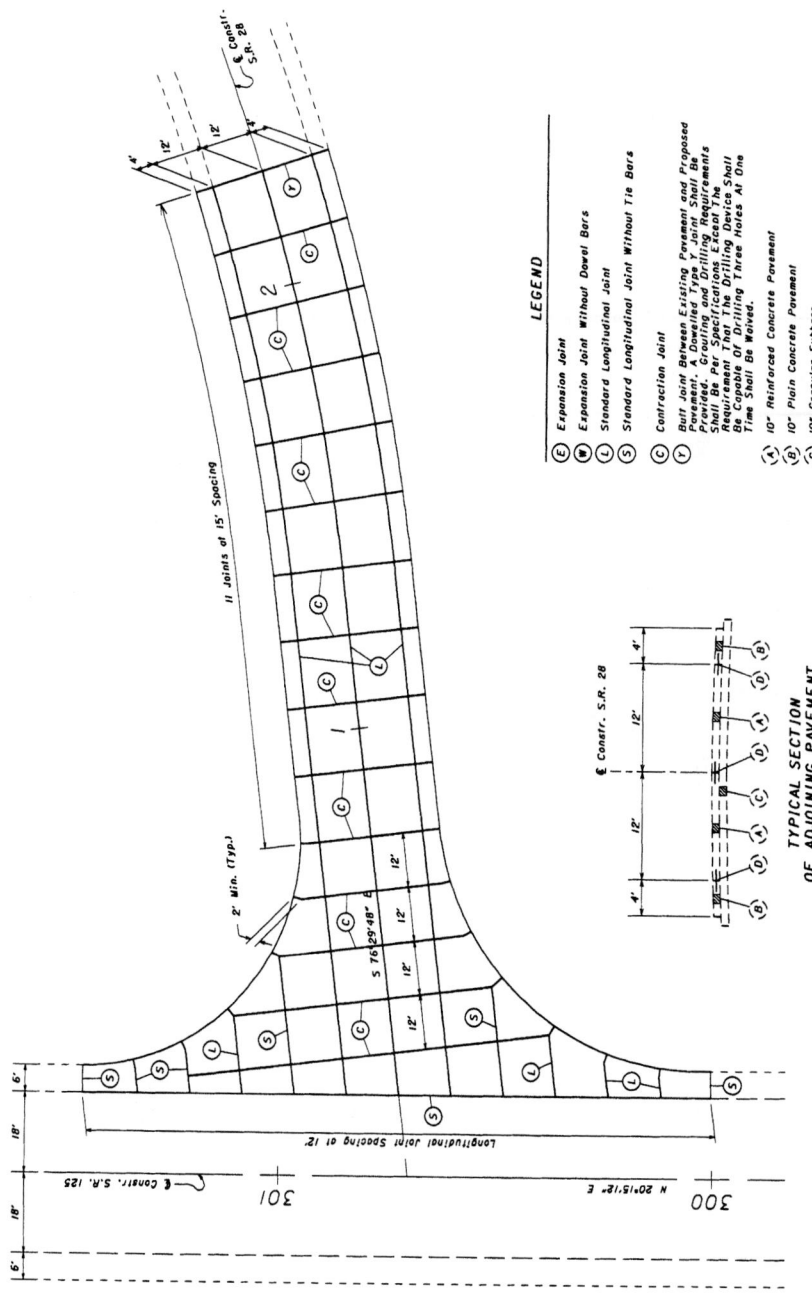

FIGURE 3.4 Layout of joints in rigid pavement at skewed intersection.

3.7

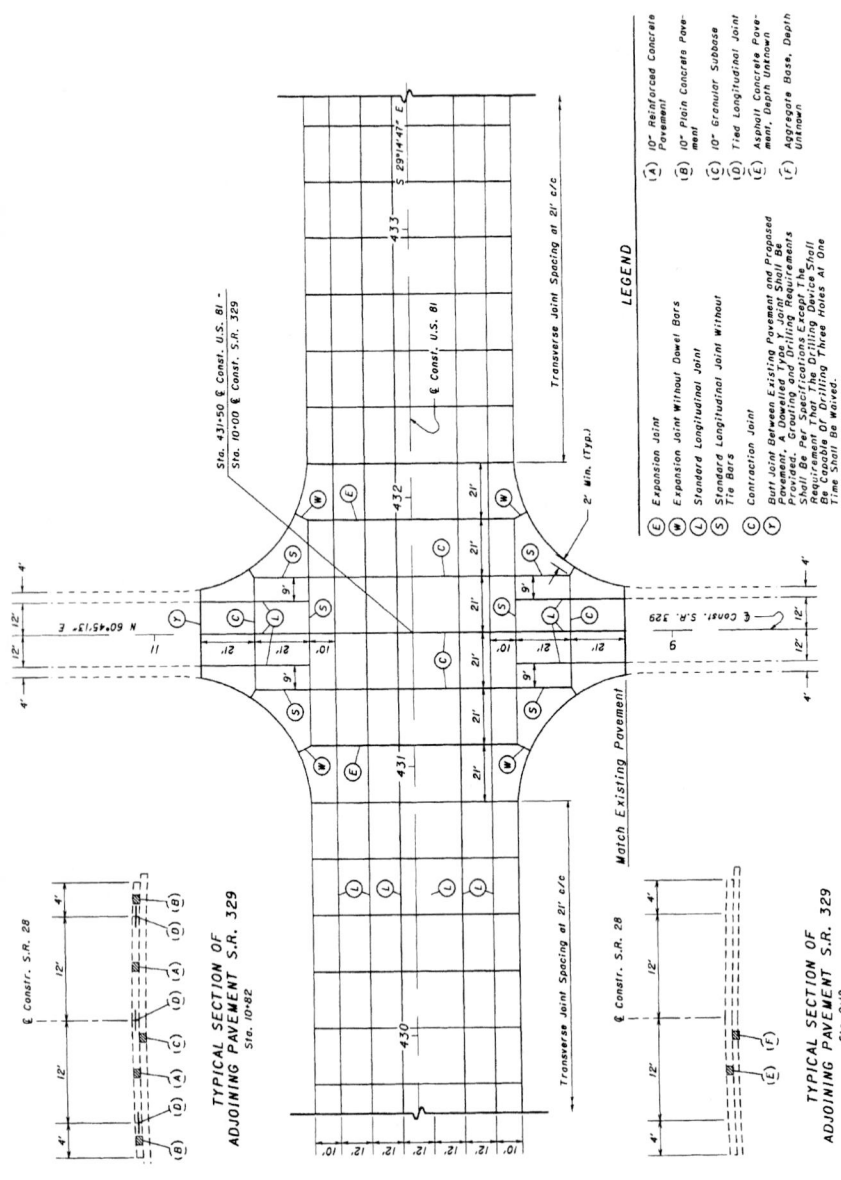

FIGURE 3.5 Layout of joints in rigid pavement at right-angle intersection.

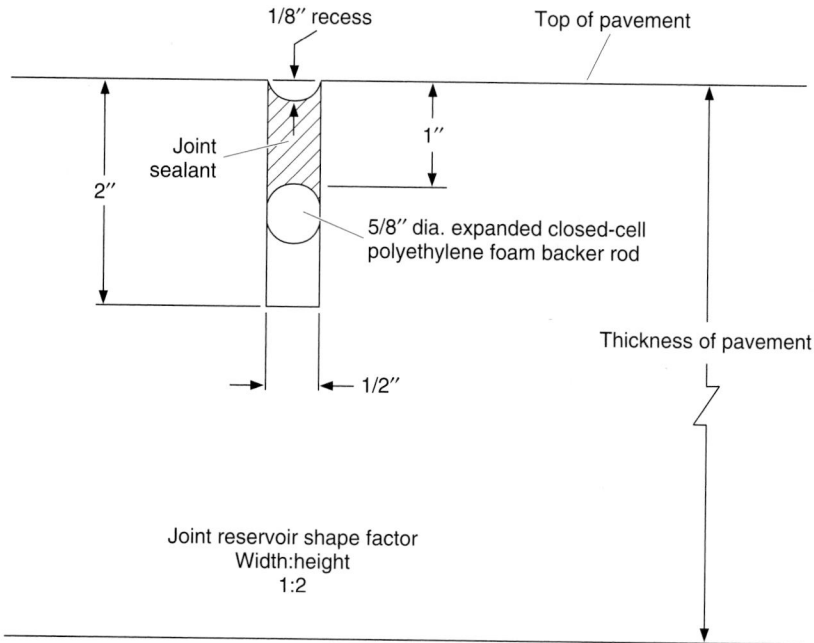

FIGURE 3.6 Typical field-formed joint seal in rigid pavement.

$$W = \frac{\Delta L}{S} \tag{3.5}$$

where W = design width of transverse contraction joint, in
S = allowable strain specified by sealant manufacturer (typically 25 to 50 percent for field-molded sealants)

To control the strain in field-molded sealants, manufacturers recommend a reservoir shape factor (width to depth), and the use of a backer rod as illustrated in Fig. 3.6. The purpose of the backer rod is to prevent bond at the bottom of the sealant reservoir where the actual crack in the pavement exists. It is at this crack that the greatest strain will occur. Typical joint sealants are either asphalt-based or silicone-based.

For preformed compression seals, the uncompressed width of the compression seal should be chosen according to manufacturer's specifications, as the material response characteristics are of primary importance. The calculated movement of the joint, normalized by the width of the uncompressed seal, should be less than or equal to the allowable movement of the compression seal, as determined by the manufacturer. Figure 3.7 shows a typical preformed seal installed in a pavement joint.

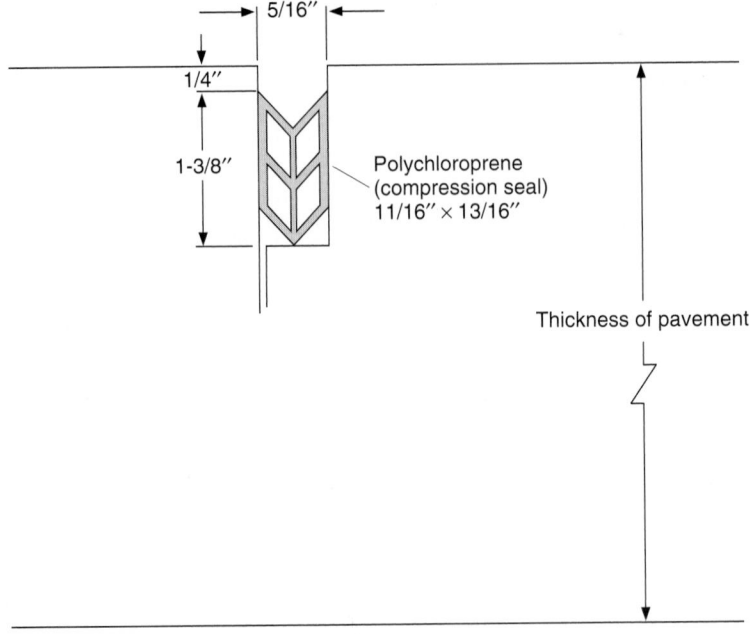

FIGURE 3.7   Typical preformed joint seal in rigid pavement.

3.1.4   Continuously Reinforced Rigid Pavement

As the name implies, continuously reinforced concrete (CRC) pavement is a rigid pavement constructed with continuous longitudinal reinforcement. No transverse joints are installed. Instead, the pavement is allowed to develop random transverse cracks, and the steel reinforcement holds the cracked sections together. The size and spacing of the cracks is influenced by the percentage of reinforcing steel used. Current practice calls for 0.5 to 0.7 percent of the slab cross-section area. The design of the reinforcement is covered in the AASHTO *Pavement Design Guide*. The thickness of the slab is determined the same way as for other concrete pavements. There is currently about 30,000 lane miles of CRC pavement in the United States, most of which has been installed since the 1960s.

3.2   FLEXIBLE PAVEMENT

Asphalt concrete pavement, also referred to as *flexible pavement,* is a mixture of sand, aggregate, a filler material, and asphalt cement combined in a controlled process, placed, and compacted. The filler material can range from quarry crushing dust and asphalt-plant baghouse fines to wood fibers (cellulose). There are many additives that can be used in asphalt concrete mixes to encourage thicker cement coatings, more elastic mixes, stiffer mixes, and less temperature-sensitive mixes. Flexible pavements can be of a type constructed on a prepared subgrade, which is called *full-depth asphalt concrete pavement (FDACP)*, or of a type built on an untreated granular base, which is not as carefully identified by the industry but is referred to herein as *deep-strength asphalt concrete pavement (DSACP)*. (See Art. 1.5.3 and 1.5.5.)

Flexible pavements are designed to bend and rebound with the subgrade. The design concept is to place sufficient layers of base and intermediate courses of pavement so as to control the deflections in the subgrade so that no permanent deflections result. Loading of an asphalt pavement requires the stiffest layers to be placed at the surface with successively weaker layers down to the subgrade. The types and thicknesses of subbase materials placed above the subgrade should be selected with consideration of the strength of the subgrade. Very weak subgrades, after compaction, can lose compaction when very stiff aggregate bases are placed above. It is often advantageous to place a granular subbase, which is much weaker than an aggregate base, above weak subgrades to ensure that compaction is sustained. Most, if not all, flexible pavement design procedures are based on a combination of elastic layer theory and experience. The elastic layer theory is used to calculate stresses in each of the layers so as to ensure that excessive deflections will not occur. The experience is related to performance parameters that predict the number of times the pavement can bend (loadings) until cracking results.

3.3 COMPOSITE PAVEMENT (OVERLAYS)

Rigid pavement constructed with an asphalt overlay is referred to as *composite pavement*. The advantage of constructing an asphalt overlay on a rigid pavement is solely in the areas of ridability and noise. Rigid pavements are considered by most to create more road noise inside a vehicle than flexible pavements. This phenomenon is largely due to the surface texture specified for rigid pavements to ensure proper skid resistance. By specifying an asphalt overlay with the rigid base, surface texture requirements can be relaxed and noise can be reduced.

There are few documented composite pavement design procedures available to determine the proper thickness ratios between the rigid base thickness and the flexible surface thickness. One way to determine an equivalent composite thickness buildup can be done using elastic layer theory. A convenient computer program called ELSYM5 (public domain) can be used for analysis of different layer combinations, provided the designer is willing to make some assumptions. By accepting the assumptions, the designer is getting results that are only approximate but relative. ELSYM5 is based on elastic layer theory, which is not entirely appropriate for rigid pavement, since rigid pavement is not continuous and elastic in all directions. The procedure involves calculating the required rigid slab thickness for the conditions present where the composite pavement will be constructed. This is done using the AASHTO *Pavement Design Guide* or another method. The second step is to analyze the required rigid slab using ELSYM5 under the conditions designed for by calculating the deflections predicted under the maximum legal loading configuration. Finally, using a trial-and-error procedure, replace up to 3 in of the rigid slab with enough thickness of asphalt to achieve the same deflections under the same loading scheme. A rule of thumb is to replace the first 1 in of slab thickness with 3 in of asphalt concrete. However, there is not a linear relationship of 1 in of PCC to 3 in asphalt concrete; for additional reductions in the rigid slab thickness, the elastic layer theory is relied upon to calculate equivalent deflections.

Because a composite pavement behaves more like a rigid pavement, special treatment is required for the transverse joints. Reflective cracking, cracks that propagate from the rigid pavement joint through the asphalt overlay, can be an intolerable dis-

FIGURE 3.8   Joint in composite pavement that has been sawed and sealed.

tress which induces a rough riding pavement. The reflective crack allows water to enter, which induces stripping in the asphalt and slow deterioration into a spalled pothole. The suggested treatment to counter the reflective cracking is to saw and seal the asphalt concrete overlay directly above the concrete joint. The joints should be sawed as soon as the asphalt overlay is placed, and the joint sealant reservoir should be constructed the same way as discussed previously. Figure 3.8 shows a joint in a composite pavement that has been properly sawed and sealed.

3.4 DEVELOPMENT OF AASHTO PAVEMENT DESIGN EQUATIONS

Perhaps the most widely used pavement design method in the United States and throughout the world is that presented in the AASHTO *Guide for Design of Pavement Structures.* A long history of pavement studies has led to the current (1993) edition.

The developments leading to the current AASHTO design procedure began with the Bates Experimental Road, which was constructed in 1922 near Springfield, Illinois. The purpose of the experimental road was to determine what factors affected pavement performance. The researchers found that pavement performance could be correlated with truck loading. No further major research was conducted over the ensuing 25 years.

The changes in truck configuration and expansion of the highway network resulting from World War II brought pavement performance to the forefront again. In 1949, the Council of State Governments held a meeting in Columbus, Ohio. At this meeting, highway officials decided there was a "need for more factual data concerning the effects of axle loads of various magnitudes on pavements." The effort to advance the science of pavement design was led by the American Association of State Highway Officials (AASHO, which later became AASHTO). The regional AASHO associations decided to construct test pavements in each region. The first of these test roads was constructed by the Southeastern AASHO states. Named Road Test One, two test loops were constructed in 1950 near La Plata, Maryland, each loop containing two 12-ft-wide pavement lanes. All sections constructed were concrete with a pavement thickness of 7 in thickening to 9 in at the edge of the pavement. Each lane of a loop carried only one loading and axle configuration.

A second regional test road was constructed by the Western Association of State Highway Organizations (WASHO). Named the WASHO Road Test, two test loops were constructed in 1952 near Malad, Idaho, each consisting of two 12-ft lanes. All pavement was comprised of asphalt concrete on a crushed aggregate base, constructed on subbases from 0 to 16 in thick. Each lane of a loop carried only one loading and axle configuration. Because of the limited number of sections constructed in Maryland and Idaho, a rational design procedure could not be developed.

In 1951, support was growing within AASHO for an expanded road test. This led to the construction of the AASHO Road Test near Ottawa, Illinois, which contained six loops with two 12-ft lanes. The AASHO Road Test contained 468 asphalt sections and 368 concrete sections. Each lane of a loop carried only one loading and axle configuration. A total of 1,114,000 load applications were applied over a two-year period.

Unpublished preliminary results from the road test were released to the states in 1961 and 1962. The AASHO *Interim Guide for Design of Pavement Structures* was published in 1972. Chapter 3 of the interim guide was revised in 1981. The first edition of the AASHTO *Guide for Design of Pavement Structures* (1986) introduced many new concepts including the reliability concept. It was published in two volumes, the first giving design procedures and the second providing documentation and explanatory information. The second edition of the Guide was published in 1993.

The rigid pavements in the AASHO Road Test were concrete slabs ranging in thickness from 2½ to 12½ in thick. The slabs were placed either on a granular subbase or directly on the subgrade. Flexible pavements at the AASHO Road Test consisted of asphalt pavements placed on a base and/or subbase. As confirmed by these tests, rigid pavements carry traffic loads through beam action whereas flexible pavements carry traffic loads by spreading the stress through the underlying layers.

3.5 PARAMETERS FOR AASHTO PAVEMENT DESIGN

The AASHTO pavement design equations have some variables that are common to both rigid and flexible pavements, including serviceability, traffic loading, reliability, overall standard deviation, and roadbed soil resilient modulus. These parameters are discussed in the following articles. Subsequently, the design procedure is presented for rigid pavements in Art. 3.6 and for flexible pavements in Art. 3.7.

3.5.1 Serviceability

The AASHTO design equations are developed around the concept of serviceability, which serves as the pavement performance parameter by which a pavement's condition is valued. *Present serviceability* is defined as the momentary ability of a pavement to serve traffic. The present serviceability rating (PSR) was developed to measure serviceability. PSR is a rating of pavement ride based on a scale of 0, for impassible, to 5, for perfect. For the development of the original AASHO equation, individuals (the raters) would ride the pavements and assign a PSR value. To avoid riding and rating every pavement to determine serviceability, a relationship is usually developed between PSR and measurable pavement attributes. The value determined by this relationship is called the *present serviceability index (PSI)*. At the AASHO Road Test, the PSI was found to be related to slope variance, cracking, and patching for concrete pavements, and related to slope variance, rutting, cracking, and patching for asphalt pavements. The relationship between pavement thickness and serviceability index is defined by the AASHTO pavement design equations.

3.5.2 Traffic Loading

Perhaps the most important step in designing a pavement is the estimation of the design traffic. Overestimation of the design traffic results in a thicker pavement than necessary with associated higher costs. Underestimation of traffic results in a thin pavement that will fail prematurely, resulting in higher maintenance and user costs. If the proposed pavement will be used to replace an existing pavement, the design traffic could be a projection of the existing traffic. If the proposed pavement is a new location, the design traffic will have to be estimated on the basis of the proposed use of the pavement. For design purposes, all traffic is equated to an equivalent 18-kip single-axle load, or ESAL. Each vehicle in the expected design traffic volume is converted to an ESAL by an equivalency factor. The equivalency factor is a function of the axle loading, pavement thickness, axle configuration, and terminal serviceability. As discussed in Art. 3.6, the terminal serviceability is an index of the serviceability of a pavement immediately before rehabilitation is needed.

The equivalency factors as given by the AASHTO *Pavement Design Guide* are presented here for flexible pavements in Tables 3.1 through 3.9, and for rigid pavements in Tables 3.10 through 3.18. For each pavement type, the tables are arranged by axle configuration and terminal serviceability p_t. Factors are included for single-axle, tandem-axle, and triple-axle configurations, and for p_t values of 2.0, 2.5, and 3.0. In the tables for flexible pavements, the pavement strength is characterized by a pavement structural number (SN), which is defined in Art. 3.7. The use of the tables is illustrated by the following example.

Consider a 30,000-lb transit bus that has a single front axle load of 10,000 lb and a tandem rear axle load of 20,000 lb. Before the ESAL can be determined, the pavement thickness or structural number must be known, as well as the terminal serviceability. In an initial design, this necessitates assumptions, and very likely an iteration after the thickness or structural number has initially been determined. In this example, the ESAL is to be determined for a rigid pavement 7 in thick and for a flexible pavement with a pavement structural number of 4. The p_t is taken as 2.5. The tables show that, for this case, the equivalency factor for rigid pavement is 0.089 for the front axle (Table 3.13) and 0.220 for the rear axle (Table 3.14). The equivalency factor for flexible pavement is 0.102 for the front axle (Table 3.4) and 0.141 for the rear axle (Table 3.5). Each bus equals $0.089 + 0.220 = 0.309$ ESAL for rigid pavement and $0.102 + 0.141 = 0.243$ ESAL for flexible pavement. A similar analysis would be completed for each vehicle type. A worksheet for making the calculations is provided in Table 3.19, and an example for using the worksheet is presented in Table 3.20.

The traffic supplied to the designer is usually the total traffic in both directions and all lanes. This traffic needs to be distributed by direction and lane to determine the required pavement thickness. The pavement is first divided by direction by multiplying by the directional factor. In most cases, this factor is equal to 0.5, assuming the

TABLE 3.1 Axle Load Equivalency Factors for Flexible Pavements, Single Axles, and p_t of 2.0

Axle load, kips	Pavement structural number (SN)					
	1	2	3	4	5	6
2	0.0002	0.0002	0.0002	0.0002	0.0002	0.0002
4	0.002	0.003	0.002	0.002	0.002	0.002
6	0.009	0.012	0.011	0.010	0.009	0.009
8	0.030	0.035	0.036	0.033	0.031	0.029
10	0.075	0.085	0.090	0.085	0.079	0.076
12	0.165	0.177	0.189	0.183	0.174	0.168
14	0.325	0.338	0.354	0.350	0.338	0.331
16	0.589	0.598	0.613	0.612	0.603	0.596
18	1.00	1.00	1.00	1.00	1.00	1.00
20	1.61	1.59	1.56	1.55	1.57	1.59
22	2.49	2.44	2.35	2.31	2.35	2.41
24	3.71	3.62	3.43	3.33	3.40	3.51
26	5.36	5.21	4.88	4.68	4.77	4.96
28	7.54	7.31	6.78	6.42	6.52	6.83
30	10.4	10.0	9.2	8.6	8.7	9.2
32	14.0	13.5	12.4	11.5	11.5	12.1
34	18.5	17.9	16.3	15.0	14.9	15.6
36	24.2	23.3	21.2	19.3	19.0	19.9
38	31.1	29.9	27.1	24.6	24.0	25.1
40	39.6	38.0	34.3	30.9	30.0	31.2
42	49.7	47.7	43.0	38.6	37.2	38.5
44	61.8	59.3	53.4	47.6	45.7	47.1
46	76.1	73.0	65.6	58.3	55.7	57.0
48	92.9	89.1	80.0	70.9	67.3	68.6
50	113.	108.	97.	86.	81.	82.

Source: *Guide for Design of Pavement Structures,* American Association of State Highway and Transportation Officials, Washington, D.C., 1993, with permission.

loads are distributed equally in both directions. In some cases, the directional factor may be greater than 0.5. An example would be an industry where material is hauled in by truck and shipped out by rail. In this case, loaded trucks would be going into the plant and empty trucks would be exiting the plant. The next factor is the lane distribution factor. As more lanes are added to a section of road, the traffic will be more distributed among these lanes. However, trucks tend to use the outermost lane, so the distribution of ESALs is not in proportion to the number of lanes added. Many of the state DOTs have developed lane distribution factors for use in pavement design. The AASHTO *Pavement Design Guide* presents a range of factors used for lane distribution as given below. It should be noted that for the same traffic, the thickness design will be greater for the pavement with the smaller number of lanes.

Number of lanes in both directions	Percent of 18-kip ESAL traffic in design lane
1	100
2	80–100
3	60–80
4 or more	50–75

TABLE 3.2 Axle Load Equivalency Factors for Flexible Pavements, Tandem Axles, and p_t of 2.0

Axle load, kips	Pavement structural number (SN)					
	1	2	3	4	5	6
2	0.0000	0.0000	0.0000	0.0000	0.0000	0.0000
4	0.0003	0.0003	0.0003	0.0002	0.0002	0.0002
6	0.001	0.001	0.001	0.001	0.001	0.001
8	0.003	0.003	0.003	0.003	0.003	0.002
10	0.007	0.008	0.008	0.007	0.006	0.006
12	0.013	0.016	0.016	0.014	0.013	0.012
14	0.024	0.029	0.029	0.026	0.024	0.023
16	0.041	0.048	0.050	0.046	0.042	0.040
18	0.066	0.077	0.081	0.075	0.069	0.066
20	0.103	0.117	0.124	0.117	0.109	0.105
22	0.156	0.171	0.183	0.174	0.164	0.158
24	0.227	0.244	0.260	0.252	0.239	0.231
26	0.322	0.340	0.360	0.353	0.338	0.329
28	0.447	0.465	0.487	0.481	0.466	0.455
30	0.607	0.623	0.646	0.643	0.627	0.617
32	0.810	0.823	0.843	0.842	0.829	0.819
34	1.06	1.07	1.08	1.08	1.08	1.07
36	1.38	1.38	1.38	1.38	1.38	1.38
38	1.76	1.75	1.73	1.72	1.73	1.74
40	2.22	2.19	2.15	2.13	2.16	2.18
42	2.77	2.73	2.64	2.62	2.66	2.70
44	3.42	3.36	3.23	3.18	3.24	3.31
46	4.20	4.11	3.92	3.83	3.91	4.02
48	5.10	4.98	4.72	4.58	4.68	4.83
50	6.15	5.99	5.64	5.44	5.56	5.77
52	7.37	7.16	6.71	6.43	6.56	6.83
54	8.77	8.51	7.93	7.55	7.69	8.03
56	10.4	10.1	9.3	8.8	9.0	9.4
58	12.2	11.8	10.9	10.3	10.4	10.9
60	14.3	13.8	12.7	11.9	12.0	12.6
62	16.6	16.0	14.7	13.7	13.8	14.5
64	19.3	18.6	17.0	15.8	15.8	16.6
66	22.2	21.4	19.6	18.0	18.0	18.9
68	25.5	24.6	22.4	20.6	20.5	21.5
70	29.2	28.1	25.6	23.4	23.2	24.3
72	33.3	32.0	29.1	26.5	26.2	27.4
74	37.8	36.4	33.0	30.0	29.4	30.8
76	42.8	41.2	37.3	33.8	33.1	34.5
78	48.4	46.5	42.0	38.0	37.0	38.6
80	54.4	52.3	47.2	42.5	41.3	43.0
82	61.1	58.7	52.9	47.6	46.0	47.8
84	68.4	65.7	59.2	53.0	51.2	53.0
86	76.3	73.3	66.0	59.0	56.8	58.6
88	85.0	81.6	73.4	65.5	62.8	64.7
90	94.4	90.6	81.5	72.6	69.4	71.3

Source: Guide for Design of Pavement Structures, American Association of State Highway and Transportation Officials, Washington, D.C., 1993, with permission.

TABLE 3.3 Axle Load Equivalency Factors for Flexible Pavements, Triple Axles, and p_t of 2.0

Axle load, kips	Pavement structural number (SN)					
	1	2	3	4	5	6
2	0.0000	0.0000	0.0000	0.0000	0.0000	0.0000
4	0.0001	0.0001	0.0001	0.0001	0.0001	0.0001
6	0.0004	0.0004	0.0003	0.0003	0.0003	0.0003
8	0.0009	0.0010	0.0009	0.0008	0.0007	0.0007
10	0.002	0.002	0.002	0.002	0.002	0.001
12	0.004	0.004	0.004	0.003	0.003	0.003
14	0.006	0.007	0.007	0.006	0.006	0.005
16	0.010	0.012	0.012	0.010	0.009	0.009
18	0.016	0.019	0.019	0.017	0.015	0.015
20	0.024	0.029	0.029	0.026	0.024	0.023
22	0.034	0.042	0.042	0.038	0.035	0.034
24	0.049	0.058	0.060	0.055	0.051	0.048
26	0.068	0.080	0.083	0.077	0.071	0.068
28	0.093	0.107	0.113	0.105	0.098	0.094
30	0.125	0.140	0.149	0.140	0.131	0.126
32	0.164	0.182	0.194	0.184	0.173	0.167
34	0.213	0.233	0.248	0.238	0.225	0.217
36	0.273	0.294	0.313	0.303	0.288	0.279
38	0.346	0.368	0.390	0.381	0.364	0.353
40	0.434	0.456	0.481	0.473	0.454	0.443
42	0.538	0.560	0.587	0.580	0.561	0.548
44	0.662	0.682	0.710	0.705	0.686	0.673
46	0.807	0.825	0.852	0.849	0.831	0.818
48	0.976	0.992	1.015	1.014	0.999	0.987
50	1.17	1.18	1.20	1.20	1.19	1.18
52	1.40	1.40	1.42	1.42	1.41	1.40
54	1.66	1.66	1.66	1.66	1.66	1.66
56	1.95	1.95	1.93	1.93	1.94	1.94
58	2.29	2.27	2.24	2.23	2.25	2.27
60	2.67	2.64	2.59	2.57	2.60	2.63
62	3.10	3.06	2.98	2.95	2.99	3.04
64	3.59	3.53	3.41	3.37	3.42	3.49
66	4.13	4.05	3.89	3.83	3.90	3.99
68	4.73	4.63	4.43	4.34	4.42	4.54
70	5.40	5.28	5.03	4.90	5.00	5.15
72	6.15	6.00	5.68	5.52	5.63	5.82
74	6.97	6.79	6.41	6.20	6.33	6.56
76	7.88	7.67	7.21	6.94	7.08	7.36
78	8.88	8.63	8.09	7.75	7.90	8.23
80	9.98	9.69	9.05	8.63	8.79	9.18
82	11.2	10.8	10.1	9.6	9.8	10.2
84	12.5	12.1	11.2	10.6	10.8	11.3
86	13.9	13.5	12.5	11.8	11.9	12.5
88	15.5	15.0	13.8	13.0	13.2	13.8
90	17.2	16.6	15.3	14.3	14.5	15.2

Source: *Guide for Design of Pavement Structures,* American Association of State Highway and Transportation Officials, Washington, D.C., 1993, with permission.

TABLE 3.4 Axle Load Equivalency Factors for Flexible Pavements, Single Axles, and p_t of 2.5

Axle load, kips	Pavement structural number (SN)					
	1	2	3	4	5	6
2	0.0004	0.0004	0.0003	0.0002	0.0002	0.0002
4	0.003	0.004	0.004	0.003	0.002	0.002
6	0.011	0.017	0.017	0.013	0.010	0.009
8	0.032	0.047	0.051	0.041	0.034	0.031
10	0.078	0.102	0.118	0.102	0.088	0.080
12	0.168	0.198	0.229	0.213	0.189	0.176
14	0.328	0.358	0.399	0.388	0.360	0.342
16	0.591	0.613	0.646	0.645	0.623	0.606
18	1.00	1.00	1.00	1.00	1.00	1.00
20	1.61	1.57	1.49	1.47	1.51	1.55
22	2.48	2.38	2.17	2.09	2.18	2.30
24	3.69	3.49	3.09	2.89	3.03	3.27
26	5.33	4.99	4.31	3.91	4.09	4.48
28	7.49	6.98	5.90	5.21	5.39	5.98
30	10.3	9.5	7.9	6.8	7.0	7.8
32	13.9	12.8	10.5	8.8	8.9	10.0
34	18.4	16.9	13.7	11.3	11.2	12.5
36	24.0	22.0	17.7	14.4	13.9	15.5
38	30.9	28.3	22.6	18.1	17.2	19.0
40	39.3	35.9	28.5	22.5	21.1	23.0
42	49.3	45.0	35.6	27.8	25.6	27.7
44	61.3	55.9	44.0	34.0	31.0	33.1
46	75.5	68.8	54.0	41.4	37.2	39.3
48	92.2	83.9	65.7	50.1	44.5	46.5
50	112.	102.	79.	60.	53.	55.

Source: *Guide for Design of Pavement Structures,* American Association of State Highway and Transportation Officials, Washington, D.C., 1993, with permission.

Abbreviated procedures for determining ESALs have been developed by several states. These procedures usually involve grouping classifications of trucks into several categories and assigning an average equivalency factor to these categories. For example, Ohio groups trucks into two categories, single, or C units; and tractor-trailer, or B combinations. The average equivalency factors used by Ohio for these two categories are shown in Table 3.21.

3.5.3 Reliability and Overall Standard Deviation

Rarely does the actual traffic loading to failure equal the predicted traffic loading; the difference is due to the deviations that exist. These deviations include (1) lack of fit of the AASHTO design equations, since these are empirical equations; (2) variations in construction, which cause variations in the equation input factors such as the strength and thickness of pavement layers; and (3) variations in the predicted traffic (see App. EE of Vol. 2 of the AASHTO *Guide for Design of Pavement Structures,* August 1986 edition). The AASHTO equations account for these variations by multiplying the predicted traffic by a safety factor. The safety factor is determined by the reliability

TABLE 3.5 Axle Load Equivalency Factors for Flexible Pavements, Tandem Axles, and p_t of 2.5

Axle load, kips	Pavement structural number (SN)					
	1	2	3	4	5	6
2	0.0001	0.0001	0.0001	0.0000	0.0000	0.0000
4	0.0005	0.0005	0.0004	0.0003	0.0003	0.0002
6	0.002	0.002	0.002	0.001	0.001	0.001
8	0.004	0.006	0.005	0.004	0.003	0.003
10	0.008	0.013	0.011	0.009	0.007	0.006
12	0.015	0.024	0.023	0.018	0.014	0.013
14	0.026	0.041	0.042	0.033	0.027	0.024
16	0.044	0.065	0.070	0.057	0.047	0.043
18	0.070	0.097	0.109	0.092	0.077	0.070
20	0.107	0.141	0.162	0.141	0.121	0.110
22	0.160	0.198	0.229	0.207	0.180	0.166
24	0.231	0.273	0.315	0.292	0.260	0.242
26	0.327	0.370	0.420	0.401	0.364	0.342
28	0.451	0.493	0.548	0.534	0.495	0.470
30	0.611	0.648	0.703	0.695	0.658	0.633
32	0.813	0.843	0.889	0.887	0.857	0.834
34	1.06	1.08	1.11	1.11	1.09	1.08
36	1.38	1.38	1.38	1.38	1.38	1.38
38	1.75	1.73	1.69	1.68	1.70	1.73
40	2.21	2.16	2.06	2.03	2.08	2.14
42	2.76	2.67	2.49	2.43	2.51	2.61
44	3.41	3.27	2.99	2.88	3.00	3.16
46	4.18	3.98	3.58	3.40	3.55	3.79
48	5.08	4.80	4.25	3.98	4.17	4.49
50	6.12	5.76	5.03	4.64	4.86	5.28
52	7.33	6.87	5.93	5.38	5.63	6.17
54	8.72	8.14	6.95	6.22	6.47	7.15
56	10.3	9.6	8.1	7.2	7.4	8.2
58	12.1	11.3	9.4	8.2	8.4	9.4
60	14.2	13.1	10.9	9.4	9.6	10.7
62	16.5	15.3	12.6	10.7	10.8	12.1
64	19.1	17.6	14.5	12.2	12.2	13.7
66	22.1	20.3	16.6	13.8	13.7	15.4
68	25.3	23.3	18.9	15.6	15.4	17.2
70	29.0	26.6	21.5	17.6	17.2	19.2
72	33.0	30.3	24.4	19.8	19.2	21.3
74	37.5	34.4	27.6	22.2	21.3	23.6
76	42.5	38.9	31.1	24.8	23.7	26.1
78	48.0	43.9	35.0	27.8	26.2	28.8
80	54.0	49.4	39.2	30.9	29.0	31.7
82	60.6	55.4	43.9	34.4	32.0	34.8
84	67.8	61.9	49.0	38.2	35.3	38.1
86	75.7	69.1	54.5	42.3	38.8	41.7
88	84.3	76.9	60.6	46.8	42.6	45.6
90	93.7	85.4	67.1	51.7	46.8	49.7

Source: Guide for Design of Pavement Structures, American Association of State Highway and Transportation Officials, Washington, D.C., 1993, with permission.

TABLE 3.6 Axle Load Equivalency Factors for Flexible Pavements, Triple Axles, and p_t of 2.5

Axle load, kips	Pavement structural number (SN)					
	1	2	3	4	5	6
2	0.0000	0.0000	0.0000	0.0000	0.0000	0.0000
4	0.0002	0.0002	0.0002	0.0001	0.0001	0.0001
6	0.0006	0.0007	0.0005	0.0004	0.0003	0.0003
8	0.001	0.002	0.001	0.001	0.001	0.001
10	0.003	0.004	0.003	0.002	0.002	0.002
12	0.005	0.007	0.006	0.004	0.003	0.003
14	0.008	0.012	0.010	0.008	0.006	0.006
16	0.012	0.019	0.018	0.013	0.011	0.010
18	0.018	0.029	0.028	0.021	0.017	0.016
20	0.027	0.042	0.042	0.032	0.027	0.024
22	0.038	0.058	0.060	0.048	0.040	0.036
24	0.053	0.078	0.084	0.068	0.057	0.051
26	0.072	0.103	0.114	0.095	0.080	0.072
28	0.098	0.133	0.151	0.128	0.109	0.099
30	0.129	0.169	0.195	0.170	0.145	0.133
32	0.169	0.213	0.247	0.220	0.191	0.175
34	0.219	0.266	0.308	0.281	0.246	0.228
36	0.279	0.329	0.379	0.352	0.313	0.292
38	0.352	0.403	0.461	0.436	0.393	0.368
40	0.439	0.491	0.554	0.533	0.487	0.459
42	0.543	0.594	0.661	0.644	0.597	0.567
44	0.666	0.714	0.781	0.769	0.723	0.692
46	0.811	0.854	0.918	0.911	0.868	0.838
48	0.979	1.015	1.072	1.069	1.033	1.005
50	1.17	1.20	1.24	1.25	1.22	1.20
52	1.40	1.41	1.44	1.44	1.43	1.41
54	1.66	1.66	1.66	1.66	1.66	1.66
56	1.95	1.93	1.90	1.90	1.91	1.93
58	2.29	2.25	2.17	2.16	2.20	2.24
60	2.67	2.60	2.48	2.44	2.51	2.58
62	3.09	3.00	2.82	2.76	2.85	2.95
64	3.57	3.44	3.19	3.10	3.22	3.36
66	4.11	3.94	3.61	3.47	3.62	3.81
68	4.71	4.49	4.06	3.88	4.05	4.30
70	5.38	5.11	4.57	4.32	4.52	4.84
72	6.12	5.79	5.13	4.80	5.03	5.41
74	6.93	6.54	5.74	5.32	5.57	6.04
76	7.84	7.37	6.41	5.88	6.15	6.71
78	8.83	8.28	7.14	6.49	6.78	7.43
80	9.92	9.28	7.95	7.15	7.45	8.21
82	11.1	10.4	8.8	7.9	8.2	9.0
84	12.4	11.6	9.8	8.6	8.9	9.9
86	13.8	12.9	10.8	9.5	9.8	10.9
88	15.4	14.3	11.9	10.4	10.6	11.9
90	17.1	15.8	13.2	11.3	11.6	12.9

Source: *Guide for Design of Pavement Structures,* American Association of State Highway and Transportation Officials, Washington, D.C., 1993, with permission.

TABLE 3.7 Axle Load Equivalency Factors for Flexible Pavements, Single Axles, and p_t of 3.0

Axle load, kips	Pavement structural number (SN)					
	1	2	3	4	5	6
2	0.0008	0.0009	0.0006	0.0003	0.0002	0.0002
4	0.004	0.008	0.006	0.004	0.002	0.002
6	0.014	0.030	0.028	0.018	0.012	0.010
8	0.035	0.070	0.080	0.055	0.040	0.034
10	0.082	0.132	0.168	0.132	0.101	0.086
12	0.173	0.231	0.296	0.260	0.212	0.187
14	0.332	0.388	0.468	0.447	0.391	0.358
16	0.594	0.633	0.695	0.693	0.651	0.622
18	1.00	1.00	1.00	1.00	1.00	1.00
20	1.60	1.53	1.41	1.38	1.44	1.51
22	2.47	2.29	1.96	1.83	1.97	2.16
24	3.67	3.33	2.69	2.39	2.60	2.96
26	5.29	4.72	3.65	3.08	3.33	3.91
28	7.43	6.56	4.88	3.93	4.17	5.00
30	10.2	8.9	6.5	5.0	5.1	6.3
32	13.8	12.0	8.4	6.2	6.3	7.7
34	18.2	15.7	10.9	7.8	7.6	9.3
36	23.8	20.4	14.0	9.7	9.1	11.0
38	30.6	26.2	17.7	11.9	11.0	13.0
40	38.8	33.2	22.2	14.6	13.1	15.3
42	48.8	41.6	27.6	17.8	15.5	17.8
44	60.6	51.6	34.0	21.6	18.4	20.6
46	74.7	63.4	41.5	26.1	21.6	23.8
48	91.2	77.3	50.3	31.3	25.4	27.4
50	110.	94.	61.	37.	30.	32.

Source: *Guide for Design of Pavement Structures,* American Association of State Highway and Transportation Officials, Washington, D.C., 1993, with permission.

desired and the amount of total variation or the overall standard deviation. AASHTO recommends the following reliability based on the functional classification of the road:

Functional classification	Recommended level of reliability	
	Urban	Rural
Interstate/freeway	85–99.9	80–99.9
Principal arterials	80–99	75–95
Collectors	80–95	75–95
Local	50–80	50–80

Overall standard deviation values recommended by AASHTO are 0.30 to 0.40 for rigid pavements and 0.40 to 0.50 for flexible pavements. The lower values are used when traffic predictions are more reliable. Values derived from the AASHTO Road Test are 0.39 for rigid pavements and 0.49 for flexible pavements.

TABLE 3.8 Axle Load Equivalency Factors for Flexible Pavements,
Tandem Axles, and p_t of 3.0

Axle load, kips	Pavement structural number (SN)					
	1	2	3	4	5	6
2	0.0002	0.0002	0.0001	0.0001	0.0000	0.0000
4	0.001	0.001	0.001	0.000	0.000	0.000
6	0.003	0.004	0.003	0.002	0.001	0.001
8	0.006	0.011	0.009	0.005	0.003	0.003
10	0.011	0.024	0.020	0.012	0.008	0.007
12	0.019	0.042	0.039	0.024	0.017	0.014
14	0.031	0.066	0.068	0.045	0.032	0.026
16	0.049	0.096	0.109	0.076	0.055	0.046
18	0.075	0.134	0.164	0.121	0.090	0.076
20	0.113	0.181	0.232	0.182	0.139	0.119
22	0.166	0.241	0.313	0.260	0.205	0.178
24	0.238	0.317	0.407	0.358	0.292	0.257
26	0.333	0.413	0.517	0.476	0.402	0.360
28	0.457	0.534	0.643	0.614	0.538	0.492
30	0.616	0.684	0.788	0.773	0.702	0.656
32	0.817	0.870	0.956	0.953	0.896	0.855
34	1.07	1.10	1.15	1.15	1.12	1.09
36	1.38	1.38	1.38	1.38	1.38	1.38
38	1.75	1.71	1.64	1.62	1.66	1.70
40	2.21	2.11	1.94	1.89	1.98	2.08
42	2.75	2.59	2.29	2.19	2.33	2.50
44	3.39	3.15	2.70	2.52	2.71	2.97
46	4.15	3.81	3.16	2.89	3.13	3.50
48	5.04	4.58	3.70	3.29	3.57	4.07
50	6.08	5.47	4.31	3.74	4.05	4.70
52	7.27	6.49	5.01	4.24	4.57	5.37
54	8.65	7.67	5.81	4.79	5.13	6.10
56	10.2	9.0	6.7	5.4	5.7	6.9
58	12.0	10.6	7.7	6.1	6.4	7.7
60	14.1	12.3	8.9	6.8	7.1	8.6
62	16.3	14.2	10.2	7.7	7.8	9.5
64	18.9	16.4	11.6	8.6	8.6	10.5
66	21.8	18.9	13.2	9.6	9.5	11.6
68	25.1	21.7	15.0	10.7	10.5	12.7
70	28.7	24.7	17.0	12.0	11.5	13.9
72	32.7	28.1	19.2	13.3	12.6	15.2
74	37.2	31.9	21.6	14.8	13.8	16.5
76	42.1	36.0	24.3	16.4	15.1	17.9
78	47.5	40.6	27.3	18.2	16.5	19.4
80	53.4	45.7	30.5	20.1	18.0	21.0
82	60.0	51.2	34.0	22.2	19.6	22.7
84	67.1	57.2	37.9	24.6	21.3	24.5
86	74.9	63.8	42.1	27.1	23.2	26.4
88	83.4	71.0	46.7	29.8	25.2	28.4
90	92.7	78.8	51.7	32.7	27.4	30.5

Source: Guide for Design of Pavement Structures, American Association of State
Highway and Transportation Officials, Washington, D.C., 1993, with permission.

TABLE 3.9 Axle Load Equivalency Factors for Flexible Pavements, Triple Axles, and p_t of 3.0

Axle load, kips	Pavement structural number (SN)					
	1	2	3	4	5	6
2	0.0001	0.0001	0.0001	0.0000	0.0000	0.0000
4	0.0005	0.0004	0.0003	0.0002	0.0001	0.0001
6	0.001	0.001	0.001	0.001	0.000	0.000
8	0.003	0.004	0.002	0.001	0.001	0.001
10	0.005	0.008	0.005	0.003	0.002	0.002
12	0.007	0.014	0.010	0.006	0.004	0.003
14	0.011	0.023	0.018	0.011	0.007	0.006
16	0.016	0.035	0.030	0.018	0.013	0.010
18	0.022	0.050	0.047	0.029	0.020	0.017
20	0.031	0.069	0.069	0.044	0.031	0.026
22	0.043	0.090	0.097	0.065	0.046	0.039
24	0.059	0.116	0.132	0.092	0.066	0.056
26	0.079	0.145	0.174	0.126	0.092	0.078
28	0.104	0.179	0.223	0.168	0.126	0.107
30	0.136	0.218	0.279	0.219	0.167	0.143
32	0.176	0.265	0.342	0.279	0.218	0.188
34	0.226	0.319	0.413	0.350	0.279	0.243
36	0.286	0.382	0.491	0.432	0.352	0.310
38	0.359	0.456	0.577	0.524	0.437	0.389
40	0.447	0.543	0.671	0.626	0.536	0.483
42	0.550	0.643	0.775	0.740	0.649	0.593
44	0.673	0.760	0.889	0.865	0.777	0.720
46	0.817	0.894	1.014	1.001	.920	.865
48	0.984	1.048	1.152	1.148	1.080	1.030
50	1.18	1.23	1.30	1.31	1.26	1.11
52	1.40	1.43	1.47	1.48	1.45	1.43
54	1.66	1.66	1.66	1.66	1.66	1.66
56	1.95	1.92	1.86	1.85	1.88	1.91
58	2.28	2.21	2.09	2.06	2.13	2.20
60	2.66	2.54	2.34	2.28	2.39	2.50
62	3.08	2.92	2.61	2.52	2.66	2.84
64	3.56	3.33	2.92	2.77	2.96	3.19
66	4.09	3.79	3.25	3.04	3.27	3.58
68	4.68	4.31	3.62	3.33	3.60	4.00
70	5.34	4.88	4.02	3.64	3.94	4.44
72	6.08	5.51	4.46	3.97	4.31	4.91
74	6.89	6.21	4.94	4.32	4.69	5.40
76	7.78	6.98	5.47	4.70	5.09	5.93
78	8.76	7.83	6.04	5.11	5.51	6.48
80	9.84	8.75	6.67	5.54	5.96	7.06
82	11.0	9.8	7.4	6.0	6.4	7.7
84	12.3	10.9	8.1	6.5	6.9	8.3
86	13.7	12.1	8.9	7.0	7.4	9.0
88	15.3	13.4	9.8	7.6	8.0	9.6
90	16.9	14.8	10.7	8.2	8.5	10.4

Source: *Guide for Design of Pavement Structures,* American Association of State Highway and Transportation Officials, Washington, D.C., 1993, with permission.

TABLE 3.10 Axle Load Equivalency Factors for Rigid Pavements, Single Axles, and p_t of 2.0

Axle load, kips	Slab thickness D, in								
	6	7	8	9	10	11	12	13	14
2	0.0002	0.0002	0.0002	0.0002	0.0002	0.0002	0.0002	0.0002	0.0002
4	0.002	0.002	0.002	0.002	0.002	0.002	0.002	0.002	0.002
6	0.011	0.010	0.010	0.010	0.010	0.010	0.010	0.010	0.010
8	0.035	0.033	0.032	0.032	0.032	0.032	0.032	0.032	0.032
10	0.087	0.084	0.082	0.081	0.080	0.080	0.080	0.080	0.080
12	0.186	0.180	0.176	0.175	0.174	0.174	0.173	0.173	0.173
14	0.353	0.346	0.341	0.338	0.337	0.336	0.336	0.336	0.336
16	0.614	0.609	0.604	0.601	0.599	0.599	0.598	0.598	0.598
18	1.00	1.00	1.00	1.00	1.00	1.00	1.00	1.00	1.00
20	1.55	1.56	1.57	1.58	1.58	1.59	1.59	1.59	1.59
22	2.32	2.32	2.35	2.38	2.40	2.41	2.41	2.41	2.42
24	3.37	3.34	3.40	3.47	3.51	3.53	3.54	3.55	3.55
26	4.76	4.69	4.77	4.88	4.97	5.02	5.04	5.06	5.06
28	6.58	6.44	6.52	6.70	6.85	6.94	7.00	7.02	7.04
30	8.92	8.68	8.74	8.98	9.23	9.39	9.48	9.54	9.56
32	11.9	11.5	11.5	11.8	12.2	12.4	12.6	12.7	12.7
34	15.5	15.0	14.9	15.3	15.8	16.2	16.4	16.6	16.7
36	20.1	19.3	19.2	19.5	20.1	20.7	21.1	21.4	21.5
38	25.6	24.5	24.3	24.6	25.4	26.1	26.7	27.1	27.4
40	32.2	30.8	30.4	30.7	31.6	32.6	33.4	34.0	34.4
42	40.1	38.4	37.7	38.0	38.9	40.1	41.3	42.1	42.7
44	49.4	47.3	46.4	46.6	47.6	49.0	50.4	51.6	52.4
46	60.4	57.7	56.6	56.7	57.7	59.3	61.1	62.6	63.7
48	73.2	69.9	68.4	68.4	69.4	71.2	73.3	75.3	76.8
50	88.0	84.1	82.2	82.0	83.0	84.9	87.4	89.8	91.7

Source: *Guide for Design of Pavement Structures,* American Association of State Highway and Transportation Officials, Washington, D.C., 1993, with permission.

3.5.4 Roadbed Soil Resilient Modulus

The *resilient modulus* is a measure of the ability of a soil or granular base to resist permanent deformation under repeated loading. Many soils are stress-dependent. As the stress level increases, these soils will behave in a nonlinear fashion. Fine-grain soils tend to be stress-softening, whereas granular soils tend to be stress-hardening. Laboratory procedures for determining resilient modulus have been published by the Strategic Highway Research Program (SHRP) as Protocol P46 and by AASHTO as test methods T292-91 and T294-92. A typical setup for the laboratory test is shown in Fig. 3.9. The stress due to the repeated load applied through the load actuator is the deviator stress and is intended to duplicate the effect of loads passing over a section of pavement. The confining stress within the chamber is intended to duplicate the confinement of the soil within the subgrade. A typical load-response curve is shown in Fig. 3.10. As shown, the resilient modulus (M_R) is the ratio of deviator stress to strain in the elastic range.

The laboratory procedures for determining resilient modulus are complex and time-consuming. Many equations have been developed relating the resilient modulus to soil properties that are more easily determined. One such property is the California Bearing Ratio (CBR). Common equations using CBR to calculate resilient modulus values include the following:

TABLE 3.11 Axle Load Equivalency Factors for Rigid Pavements, Tandem Axles, and p_t of 2.0

Axle load, kips	Slab thickness D, in								
	6	7	8	9	10	11	12	13	14
2	0.0001	0.0001	0.0001	0.0001	0.0001	0.0001	0.0001	0.0001	0.0001
4	0.0006	0.0005	0.0005	0.0005	0.0005	0.0005	0.0005	0.0005	0.0005
6	0.002	0.002	0.002	0.002	0.002	0.002	0.002	0.002	0.002
8	0.006	0.006	0.005	0.005	0.005	0.005	0.005	0.005	0.005
10	0.014	0.013	0.013	0.012	0.012	0.012	0.012	0.012	0.012
12	0.028	0.026	0.026	0.025	0.025	0.025	0.025	0.025	0.025
14	0.051	0.049	0.048	0.047	0.047	0.047	0.047	0.047	0.047
16	0.087	0.084	0.082	0.081	0.081	0.080	0.080	0.080	0.080
18	0.141	0.136	0.133	0.132	0.131	0.131	0.131	0.131	0.131
20	0.216	0.210	0.206	0.204	0.203	0.203	0.203	0.203	0.203
22	0.319	0.313	0.307	0.305	0.304	0.303	0.303	0.303	0.303
24	0.454	0.449	0.444	0.441	0.440	0.439	0.439	0.439	0.439
26	0.629	0.626	0.622	0.620	0.618	0.618	0.618	0.618	0.618
28	0.852	0.851	0.850	0.850	0.850	0.849	0.849	0.849	0.849
30	1.13	1.13	1.14	1.14	1.14	1.14	1.14	1.14	1.14
32	1.48	1.48	1.49	1.50	1.51	1.51	1.51	1.51	1.51
34	1.90	1.90	1.93	1.95	1.96	1.97	1.97	1.97	1.97
36	2.42	2.41	2.45	2.49	2.51	2.52	2.53	2.53	2.53
38	3.04	3.02	3.07	3.13	3.17	3.19	3.20	3.20	3.21
40	3.79	3.74	3.80	3.89	3.95	3.98	4.00	4.01	4.01
42	4.67	4.59	4.66	4.78	4.87	4.93	4.95	4.97	4.97
44	5.72	5.59	5.67	5.82	5.95	6.03	6.07	6.09	6.10
46	6.94	6.76	6.83	7.02	7.20	7.31	7.37	7.41	7.43
48	8.36	8.12	8.17	8.40	8.63	8.79	8.88	8.93	8.96
50	10.00	9.69	9.72	9.98	10.27	10.49	10.62	10.69	10.73
52	11.9	11.5	11.5	11.8	12.1	12.4	12.6	12.7	12.8
54	14.0	13.5	13.5	13.8	14.2	14.6	14.9	15.0	15.1
56	16.5	15.9	15.8	16.1	16.6	17.1	17.4	17.6	17.7
58	19.3	18.5	18.4	18.7	19.3	19.8	20.3	20.5	20.7
60	22.4	21.5	21.3	21.6	22.3	22.9	23.5	23.8	24.0
62	25.9	24.9	24.6	24.9	25.6	26.4	27.0	27.5	27.7
64	29.9	28.6	28.2	28.5	29.3	30.2	31.0	31.6	31.9
66	34.3	32.8	32.3	32.6	33.4	34.4	35.4	36.1	36.5
68	39.2	37.5	36.8	37.1	37.9	39.1	40.2	41.1	41.6
70	44.6	42.7	41.9	42.1	42.9	44.2	45.5	46.6	47.3
72	50.6	48.4	47.5	47.6	48.5	49.9	51.4	52.6	53.5
74	57.3	54.7	53.6	53.6	54.6	56.1	57.7	59.2	60.3
76	64.6	61.7	60.4	60.3	61.2	62.8	64.7	66.4	67.7
78	72.5	69.3	67.8	67.7	68.6	70.2	72.3	74.3	75.8
80	81.3	77.6	75.9	75.7	76.6	78.3	80.6	82.8	84.7
82	90.9	86.7	84.7	84.4	85.3	87.1	89.6	92.1	94.2
84	101.	97.	94.	94.	95.	97.	99.	102.	105.
86	113.	107.	105.	104.	105.	107.	110.	113.	116.
88	125.	119.	116.	116.	116.	118.	121.	125.	128.
90	138.	132.	129.	128.	129.	131.	134.	137.	141.

Source: *Guide for Design of Pavement Structures,* American Association of State Highway and Transportation Officials, Washington, D.C., 1993, with permission.

TABLE 3.12 Axle Load Equivalency Factors for Rigid Pavements, Triple Axles, and p_t of 2.0

Axle load, kips	Slab thickness D, in								
	6	7	8	9	10	11	12	13	14
2	0.0001	0.0001	0.0001	0.0001	0.0001	0.0001	0.0001	0.0001	0.0001
4	0.0003	0.0003	0.0003	0.0003	0.0003	0.0003	0.0003	0.0003	0.0003
6	0.0010	0.0009	0.0009	0.0009	0.0009	0.0009	0.0009	0.0009	0.0009
8	0.002	0.002	0.002	0.002	0.002	0.002	0.002	0.002	0.002
10	0.005	0.005	0.005	0.005	0.005	0.005	0.005	0.005	0.005
12	0.010	0.010	0.009	0.009	0.009	0.009	0.009	0.009	0.009
14	0.018	0.017	0.017	0.016	0.016	0.016	0.016	0.016	0.016
16	0.030	0.029	0.028	0.027	0.027	0.027	0.027	0.027	0.027
18	0.047	0.045	0.044	0.044	0.043	0.043	0.043	0.043	0.043
20	0.072	0.069	0.067	0.066	0.066	0.066	0.066	0.066	0.066
22	0.105	0.101	0.099	0.098	0.097	0.097	0.097	0.097	0.097
24	0.149	0.144	0.141	0.139	0.139	0.138	0.138	0.138	0.138
26	0.205	0.199	0.195	0.194	0.193	0.192	0.192	0.192	0.192
28	0.276	0.270	0.265	0.263	0.262	0.262	0.262	0.262	0.261
30	0.364	0.359	0.354	0.351	0.350	0.349	0.349	0.349	0.349
32	0.472	0.468	0.463	0.460	0.459	0.458	0.458	0.458	0.458
34	0.603	0.600	0.596	0.594	0.593	0.592	0.592	0.592	0.592
36	0.759	0.758	0.757	0.756	0.755	0.755	0.755	0.755	0.755
38	0.946	0.947	0.949	0.950	0.951	0.951	0.951	0.951	0.951
40	1.17	1.17	1.18	1.18	1.18	1.18	1.18	1.18	1.19
42	1.42	1.43	1.44	1.45	1.46	1.46	1.46	1.46	1.46
44	1.73	1.73	1.75	1.77	1.78	1.78	1.79	1.79	1.79
46	2.08	2.07	2.10	2.13	2.15	2.16	2.16	2.16	2.17
48	2.48	2.47	2.51	2.55	2.58	2.59	2.60	2.60	2.61
50	2.95	2.92	2.97	3.03	3.07	3.09	3.10	3.11	3.11
52	3.48	3.44	3.50	3.58	3.63	3.66	3.68	3.69	3.69
54	4.09	4.03	4.09	4.20	4.27	4.31	4.33	4.35	4.35
56	4.78	4.69	4.76	4.89	4.99	5.05	5.08	5.09	5.10
58	5.57	5.44	5.51	5.66	5.79	5.87	5.91	5.94	5.95
60	6.45	6.29	6.35	6.53	6.69	6.79	6.85	6.88	6.90
62	7.43	7.23	7.28	7.49	7.69	7.82	7.90	7.94	7.97
64	8.54	8.28	8.32	8.55	8.80	8.97	9.07	9.13	9.16
66	9.76	9.46	9.48	9.73	10.02	10.24	10.37	10.44	10.48
68	11.1	10.8	10.8	11.0	11.4	11.6	11.8	11.9	12.0
70	12.6	12.2	12.2	12.5	12.8	13.2	13.4	13.5	13.6
72	14.3	13.8	13.7	14.0	14.5	14.9	15.1	15.3	15.4
74	16.1	15.5	15.4	15.7	16.2	16.7	17.0	17.2	17.3
76	18.2	17.5	17.3	17.6	18.2	18.7	19.1	19.3	19.5
78	20.4	19.6	19.4	19.7	20.3	20.9	21.4	21.7	21.8
80	22.8	21.9	21.6	21.9	22.6	23.3	23.8	24.2	24.4
82	25.4	24.4	24.1	24.4	25.0	25.8	26.5	26.9	27.2
84	28.3	27.1	26.7	27.0	27.7	28.6	29.4	29.9	30.2
86	31.4	30.1	29.6	29.9	30.7	31.6	32.5	33.1	33.5
88	34.8	33.3	32.8	33.0	33.8	34.8	35.8	36.6	37.1
90	38.5	36.8	36.2	36.4	37.2	38.3	39.4	40.3	40.9

Source: *Guide for Design of Pavement Structures,* American Association of State Highway and Transportation Officials, Washington, D.C., 1993, with permission.

TABLE 3.13 Axle Load Equivalency Factors for Rigid Pavements, Single Axles, and p_t of 2.5

Axle load, kips	Slab thickness D, in								
	6	7	8	9	10	11	12	13	14
2	0.0002	0.0002	0.0002	0.0002	0.0002	0.0002	0.0002	0.0002	0.0002
4	0.003	0.002	0.002	0.002	0.002	0.002	0.002	0.002	0.002
6	0.012	0.011	0.010	0.010	0.010	0.010	0.010	0.010	0.010
8	0.039	0.035	0.033	0.032	0.032	0.032	0.032	0.032	0.032
10	0.097	0.089	0.084	0.082	0.081	0.080	0.080	0.080	0.080
12	0.203	0.189	0.181	0.176	0.175	0.174	0.174	0.173	0.173
14	0.376	0.360	0.347	0.341	0.338	0.337	0.336	0.336	0.336
16	0.634	0.623	0.610	0.604	0.601	0.599	0.599	0.599	0.598
18	1.00	1.00	1.00	1.00	1.00	1.00	1.00	1.00	1.00
20	1.51	1.52	1.55	1.57	1.58	1.58	1.59	1.59	1.59
22	2.21	2.20	2.28	2.34	2.38	2.40	2.41	2.41	2.41
24	3.16	3.10	3.22	3.36	3.45	3.50	3.53	3.54	3.55
26	4.41	4.26	4.42	4.67	4.85	4.95	5.01	5.04	5.05
28	6.05	5.76	5.92	6.29	6.61	6.81	6.92	6.98	7.01
30	8.16	7.67	7.79	8.28	8.79	9.14	9.35	9.46	9.52
32	10.8	10.1	10.1	10.7	11.4	12.0	12.3	12.6	12.7
34	14.1	13.0	12.9	13.6	14.6	15.4	16.0	16.4	16.5
36	18.2	16.7	16.4	17.1	18.3	19.5	20.4	21.0	21.3
38	23.1	21.1	20.6	21.3	22.7	24.3	25.6	26.4	27.0
40	29.1	26.5	25.7	26.3	27.9	29.9	31.6	32.9	33.7
42	36.2	32.9	31.7	32.2	34.0	36.3	38.7	40.4	41.6
44	44.6	40.4	38.8	39.2	41.0	43.8	46.7	49.1	50.8
46	54.5	49.3	47.1	47.3	49.2	52.3	55.9	59.0	61.4
48	66.1	59.7	56.9	56.8	58.7	62.1	66.3	70.3	73.4
50	79.4	71.7	68.2	67.8	69.6	73.3	78.1	83.0	87.1

Source: *Guide for Design of Pavement Structures,* American Association of State Highway and Transportation Officials, Washington, D.C., 1993, with permission.

E (lb/in^2) = 1500CBR   (Shell Oil Co.)

E (lb/in^2) = 5409CBR$^{0.711}$ (U.S. Army Waterway Experiment Station)

E (lb/in^2) = 2550CBR$^{0.64}$ (Transport and Road Research Laboratory, England)

(See "Pavement Deflection Analysis," FHWA Report HI-94-021, NHI, February 1994.) More detailed equations have been developed by correlating laboratory results with fundamental soil properties. R. F. Carmichael III and E. Stuart ("Predicting Resilient Modulus: A Study to Determine the Mechanical Properties of Subgrade Soils," Transportation Research Record 1043, Transportation Research Board, National Research Council, Washington, D.C., 1985) developed the following models for the U.S. Forest Service:

TABLE 3.14 Axle Load Equivalency Factors for Rigid Pavements, Tandem Axles, and p_t of 2.5

Axle load, kips	Slab thickness D, in								
	6	7	8	9	10	11	12	13	14
2	0.0001	0.0001	0.0001	0.0001	0.0001	0.0001	0.0001	0.0001	0.0001
4	0.0006	0.0006	0.0005	0.0005	0.0005	0.0005	0.0005	0.0005	0.0005
6	0.002	0.002	0.002	0.002	0.002	0.002	0.002	0.002	0.002
8	0.007	0.006	0.006	0.005	0.005	0.005	0.005	0.005	0.005
10	0.015	0.014	0.013	0.013	0.012	0.012	0.012	0.012	0.012
12	0.031	0.028	0.026	0.026	0.025	0.025	0.025	0.025	0.025
14	0.057	0.052	0.049	0.048	0.047	0.047	0.047	0.047	0.047
16	0.097	0.089	0.084	0.082	0.081	0.081	0.080	0.080	0.080
18	0.155	0.143	0.136	0.133	0.132	0.131	0.131	0.131	0.131
20	0.234	0.220	0.211	0.206	0.204	0.203	0.203	0.203	0.203
22	0.340	0.325	0.313	0.308	0.305	0.304	0.303	0.303	0.303
24	0.475	0.462	0.450	0.444	0.441	0.440	0.439	0.439	0.439
26	0.644	0.637	0.627	0.622	0.620	0.619	0.618	0.618	0.618
28	0.855	0.854	0.852	0.850	0.850	0.850	0.849	0.849	0.849
30	1.11	1.12	1.13	1.14	1.14	1.14	1.14	1.14	1.14
32	1.43	1.44	1.47	1.49	1.50	1.51	1.51	1.51	1.51
34	1.82	1.82	1.87	1.92	1.95	1.96	1.97	1.97	1.97
36	2.29	2.27	2.35	2.43	2.48	2.51	2.52	2.52	2.53
38	2.85	2.80	2.91	3.03	3.12	3.16	3.18	3.20	3.20
40	3.52	3.42	3.55	3.74	3.87	3.94	3.98	4.00	4.01
42	4.32	4.16	4.30	4.55	4.74	4.86	4.91	4.95	4.96
44	5.26	5.01	5.16	5.48	5.75	5.92	6.01	6.06	6.09
46	6.36	6.01	6.14	6.53	6.90	7.14	7.28	7.36	7.40
48	7.64	7.16	7.27	7.73	8.21	8.55	8.75	8.86	8.92
50	9.11	8.50	8.55	9.07	9.68	10.14	10.42	10.58	10.66
52	10.8	10.0	10.0	10.6	11.3	11.9	12.3	12.5	12.7
54	12.8	11.8	11.7	12.3	13.2	13.9	14.5	14.8	14.9
56	15.0	13.8	13.6	14.2	15.2	16.2	16.8	17.3	17.5
58	17.5	16.0	15.7	16.3	17.5	18.6	19.5	20.1	20.4
60	20.3	18.5	18.1	18.7	20.0	21.4	22.5	23.2	23.6
62	23.5	21.4	20.8	21.4	22.8	24.4	25.7	26.7	27.3
64	27.0	24.6	23.8	24.4	25.8	27.7	29.3	30.5	31.3
66	31.0	28.1	27.1	27.6	29.2	31.3	33.2	34.7	35.7
68	35.4	32.1	30.9	31.3	32.9	35.2	37.5	39.3	40.5
70	40.3	36.5	35.0	35.3	37.0	39.5	42.1	44.3	45.9
72	45.7	41.4	39.6	39.8	41.5	44.2	47.2	49.8	51.7
74	51.7	46.7	44.6	44.7	46.4	49.3	52.7	55.7	58.0
76	58.3	52.6	50.2	50.1	51.8	54.9	58.6	62.1	64.8
78	65.5	59.1	56.3	56.1	57.7	60.9	65.0	69.0	72.3
80	73.4	66.2	62.9	62.5	64.2	67.5	71.9	76.4	80.2
82	82.0	73.9	70.2	69.6	71.2	74.7	79.4	84.4	88.8
84	91.4	82.4	78.1	77.3	78.9	82.4	87.4	93.0	98.1
86	102.	92.	87.	86.	87.	91.	96.	102.	108.
88	113.	102.	96.	95.	96.	100.	105.	112.	119.
90	125.	112.	106.	105.	106.	110.	115.	123.	130.

Source: *Guide for Design of Pavement Structures,* American Association of State Highway and Transportation Officials, Washington, D.C., 1993, with permission.

TABLE 3.15 Axle Load Equivalency Factors for Rigid Pavements, Triple Axles, and p_t of 2.5

Axle load, kips	Slab thickness D, in								
	.6	7	8	9	10	11	12	13	14
2	0.0001	0.0001	0.0001	0.0001	0.0001	0.0001	0.0001	0.0001	0.0001
4	0.0003	0.0003	0.0003	0.0003	0.0003	0.0003	0.0003	0.0003	0.0003
6	0.001	0.001	0.001	0.001	0.001	0.001	0.001	0.001	0.001
8	0.003	0.002	0.002	0.002	0.002	0.002	0.002	0.002	0.002
10	0.006	0.005	0.005	0.005	0.005	0.005	0.005	0.005	0.005
12	0.011	0.010	0.010	0.009	0.009	0.009	0.009	0.009	0.009
14	0.020	0.018	0.017	0.017	0.016	0.016	0.016	0.016	0.016
16	0.033	0.030	0.029	0.028	0.027	0.027	0.027	0.027	0.027
18	0.053	0.048	0.045	0.044	0.044	0.043	0.043	0.043	0.043
20	0.080	0.073	0.069	0.067	0.066	0.066	0.066	0.066	0.066
22	0.116	0.107	0.101	0.099	0.098	0.097	0.097	0.097	0.097
24	0.163	0.151	0.144	0.141	0.139	0.139	0.138	0.138	0.138
26	0.222	0.209	0.200	0.195	0.194	0.193	0.192	0.192	0.192
28	0.295	0.281	0.271	0.265	0.263	0.262	0.262	0.262	0.262
30	0.384	0.371	0.359	0.354	0.351	0.350	0.349	0.349	0.349
32	0.490	0.480	0.468	0.463	0.460	0.459	0.458	0.458	0.458
34	0.616	0.609	0.601	0.596	0.594	0.593	0.592	0.592	0.592
36	0.765	0.762	0.759	0.757	0.756	0.755	0.755	0.755	0.755
38	0.939	0.941	0.946	0.948	0.950	0.951	0.951	0.951	0.951
40	1.14	1.15	1.16	1.17	1.18	1.18	1.18	1.18	1.18
42	1.38	1.38	1.41	1.44	1.45	1.46	1.46	1.46	1.46
44	1.65	1.65	1.70	1.74	1.77	1.78	1.78	1.78	1.79
46	1.97	1.96	2.03	2.09	2.13	2.15	2.16	2.16	2.16
48	2.34	2.31	2.40	2.49	2.55	2.58	2.59	2.60	2.60
50	2.76	2.71	2.81	2.94	3.02	3.07	3.09	3.10	3.11
52	3.24	3.15	3.27	3.44	3.56	3.62	3.66	3.68	3.68
54	3.79	3.66	3.79	4.00	4.16	4.26	4.30	4.33	4.34
56	4.41	4.23	4.37	4.63	4.84	4.97	5.03	5.07	5.09
58	5.12	4.87	5.00	5.32	5.59	5.79	5.85	5.90	5.93
60	5.91	5.59	5.71	6.08	6.42	6.64	6.77	6.84	6.87
62	6.80	6.39	6.50	6.91	7.33	7.62	7.79	7.88	7.93
64	7.79	7.29	7.37	7.82	8.33	8.70	8.92	9.04	9.11
66	8.90	8.28	8.33	8.83	9.42	9.88	10.17	10.33	10.42
68	10.1	9.4	9.4	9.9	10.6	11.2	11.5	11.7	11.9
70	11.5	10.6	10.6	11.1	11.9	12.6	13.0	13.3	13.5
72	13.0	12.0	11.8	12.4	13.3	14.1	14.7	15.0	15.2
74	14.6	13.5	13.2	13.8	14.8	15.8	16.5	16.9	17.1
76	16.5	15.1	14.8	15.4	16.5	17.6	18.4	18.9	19.2
78	18.5	16.9	16.5	17.1	18.2	19.5	20.5	21.1	21.5
80	20.6	18.8	18.3	18.9	20.2	21.6	22.7	23.5	24.0
82	23.0	21.0	20.3	20.9	22.2	23.8	25.2	26.1	26.7
84	25.6	23.3	22.5	23.1	24.5	26.2	27.8	28.9	29.6
86	28.4	25.8	24.9	25.4	26.9	28.8	30.5	31.9	32.8
88	31.5	28.6	27.5	27.9	29.4	31.5	33.5	35.1	36.1
90	34.8	31.5	30.3	30.7	32.2	34.4	36.7	38.5	39.8

Source: *Guide for Design of Pavement Structures,* American Association of State Highway and Transportation Officials, Washington, D.C., 1993, with permission.

TABLE 3.16   Axle Load Equivalency Factors for Rigid Pavements, Single Axles, and p_t of 3.0

Axle load, kips	Slab thickness D, in								
	6	7	8	9	10	11	12	13	14
2	0.0003	0.0002	0.0002	0.0002	0.0002	0.0002	0.0002	0.0002	0.0002
4	0.003	0.003	0.002	0.002	0.002	0.002	0.002	0.002	0.002
6	0.014	0.012	0.011	0.010	0.010	0.010	0.010	0.010	0.010
8	0.045	0.038	0.034	0.033	0.032	0.032	0.032	0.032	0.032
10	0.111	0.095	0.087	0.083	0.081	0.081	0.080	0.080	0.080
12	0.228	0.202	0.186	0.179	0.176	0.174	0.174	0.174	0.173
14	0.408	0.378	0.355	0.344	0.340	0.337	0.337	0.336	0.336
16	0.660	0.640	0.619	0.608	0.603	0.600	0.599	0.599	0.599
18	1.00	1.00	1.00	1.00	1.00	1.00	1.00	1.00	1.00
20	1.46	1.47	1.52	1.55	1.57	1.58	1.58	1.59	1.59
22	2.07	2.06	2.18	2.29	2.35	2.38	2.40	2.41	2.41
24	2.90	2.81	3.00	3.23	3.38	3.47	3.51	3.53	3.54
26	4.00	3.77	4.01	4.40	4.70	4.87	4.96	5.01	5.04
28	5.43	4.99	5.23	5.80	6.31	6.65	6.83	6.93	6.98
30	7.27	6.53	6.72	7.46	8.25	8.83	9.17	9.36	9.46
32	9.59	8.47	8.53	9.42	10.54	11.44	12.03	12.37	12.56
34	12.5	10.9	10.7	11.7	13.2	14.5	15.5	16.0	16.4
36	16.0	13.8	13.4	14.4	16.2	18.1	19.5	20.4	21.0
38	20.4	17.4	16.7	17.7	19.8	22.2	24.2	25.6	26.4
40	25.6	21.8	20.6	21.5	23.8	26.8	29.5	31.5	32.9
42	31.8	26.9	25.3	26.0	28.5	32.0	35.5	38.4	40.3
44	39.2	33.1	30.8	31.3	33.9	37.9	42.3	46.1	48.8
46	47.8	40.3	37.2	37.5	40.1	44.5	49.8	54.7	58.5
48	57.9	48.6	44.8	44.7	47.3	52.1	58.2	64.3	69.4
50	69.6	58.4	53.6	53.1	55.6	60.6	67.6	75.0	81.4

Source: *Guide for Design of Pavement Structures,* American Association of State Highway and Transportation Officials, Washington, D.C., 1993, with permission.

Cohesive soils:

$$M_R = 37.431 - 0.4566(\text{PI}) - 0.6179(\%\text{W}) - 0.1424(\text{S200}) +$$
$$0.1791(\text{CS}) - 0.3248(\text{DS}) + 36.422(\text{CH}) + 17.097(\text{MH}) \qquad (3.6)$$

where M_R = resilient modulus, kips/in^2
  PI = plasticity index
  %W = percentage water
  S200 = percentage passing the no. 200 sieve
  CS = confining stress, lb/in^2
  DS = deviator stress, lb/in^2
  $\text{CH} = \begin{cases} 1 \text{ for CH soil (Unified Soil Classification, Art. 8.3.2)} \\ 0 \text{ otherwise} \end{cases}$
  $\text{MH} = \begin{cases} 1 \text{ for MH soil (Unified Soil Classification)} \\ 0 \text{ otherwise} \end{cases}$

TABLE 3.17 Axle Load Equivalency Factors for Rigid Pavements, Tandem Axles, and p_t of 3.0

Axle load, kips	Slab thickness D, in								
	6	7	8	9	10	11	12	13	14
2	0.0001	0.0001	0.0001	0.0001	0.0001	0.0001	0.0001	0.0001	0.0001
4	0.0007	0.0006	0.0005	0.0005	0.0005	0.0005	0.0005	0.0005	0.0005
6	0.003	0.002	0.002	0.002	0.002	0.002	0.002	0.002	0.002
8	0.008	0.006	0.006	0.006	0.005	0.005	0.005	0.005	0.005
10	0.018	0.015	0.013	0.013	0.013	0.012	0.012	0.012	0.012
12	0.036	0.030	0.027	0.026	0.026	0.025	0.025	0.025	0.025
14	0.066	0.056	0.050	0.048	0.047	0.047	0.047	0.047	0.047
16	0.111	0.095	0.087	0.083	0.081	0.081	0.081	0.080	0.080
18	0.174	0.153	0.140	0.135	0.132	0.131	0.131	0.131	0.131
20	0.260	0.234	0.217	0.209	0.205	0.204	0.203	0.203	0.203
22	0.368	0.341	0.321	0.311	0.307	0.305	0.304	0.303	0.303
24	0.502	0.479	0.458	0.447	0.443	0.440	0.440	0.439	0.439
26	0.664	0.651	0.634	0.625	0.621	0.619	0.618	0.618	0.618
28	0.859	0.857	0.853	0.851	0.850	0.850	0.850	0.849	0.849
30	1.09	1.10	1.12	1.13	1.14	1.14	1.14	1.14	1.14
32	1.38	1.38	1.44	1.47	1.49	1.50	1.51	1.51	1.51
34	1.72	1.71	1.80	1.88	1.93	1.95	1.96	1.97	1.97
36	2.13	2.10	2.23	2.36	2.45	2.49	2.51	2.52	2.52
38	2.62	2.54	2.71	2.92	3.06	3.13	3.17	3.19	3.20
40	3.21	3.05	3.26	3.55	3.76	3.89	3.95	3.98	4.00
42	3.90	3.65	3.87	4.26	4.58	4.77	4.87	4.92	4.95
44	4.72	4.35	4.57	5.06	5.50	5.78	5.94	6.02	6.06
46	5.68	5.16	5.36	5.95	6.54	6.94	7.17	7.29	7.36
48	6.80	6.10	6.25	6.93	7.69	8.24	8.57	8.76	8.86
50	8.09	7.17	7.26	8.03	8.96	9.70	10.17	10.43	10.58
52	9.57	8.41	8.40	9.24	10.36	11.32	11.96	12.33	12.54
54	1.13	9.8	9.7	10.6	11.9	13.1	14.0	14.5	14.8
56	13.2	11.4	11.2	12.1	13.6	15.1	16.2	16.9	17.3
58	15.4	13.2	12.8	13.7	15.4	17.2	18.6	19.5	20.1
60	17.9	15.3	14.7	15.6	17.4	19.5	21.3	22.5	23.2
62	20.6	17.6	16.8	17.6	19.6	22.0	24.1	25.7	26.6
64	23.7	20.2	19.1	19.9	22.0	24.7	27.3	29.2	30.4
66	27.2	23.1	21.7	22.4	24.6	27.6	30.6	33.0	34.6
68	31.1	26.3	24.6	25.2	27.4	30.8	34.3	37.1	39.2
70	35.4	29.8	27.8	28.2	30.6	34.2	38.2	41.6	44.1
72	40.1	33.8	31.3	31.6	34.0	37.9	42.3	46.4	49.4
74	45.3	38.1	35.2	35.4	37.7	41.8	46.8	51.5	55.2
76	51.1	42.9	39.5	39.5	41.8	46.1	51.5	56.9	61.3
78	57.4	48.2	44.3	44.0	46.3	50.7	56.6	62.7	67.9
80	64.3	53.9	49.4	48.9	51.1	55.8	62.1	68.9	74.9
82	71.8	60.2	55.1	54.3	56.5	61.2	67.9	75.5	82.4
84	80.0	67.0	61.2	60.2	62.2	67.0	74.2	82.4	90.3
86	89.0	74.5	67.9	66.5	68.5	73.4	80.8	89.8	98.7
88	98.7	82.5	75.2	73.5	75.3	80.2	88.0	97.7	107.5
90	109.	91.	83.	81.	83.	88.	96.	106.	117

Source: *Guide for Design of Pavement Structures,* American Association of State Highway and Transportation Officials, Washington, D.C., 1993, with permission.

TABLE 3.18 Axle Load Equivalency Factors for Rigid Pavements, Triple Axles, and p_t of 3.0

Axle load, kips	Slab thickness D, in								
	6	7	8	9	10	11	12	13	14
2	0.0001	0.0001	0.0001	0.0001	0.0001	0.0001	0.0001	0.0001	0.0001
4	0.0004	0.0003	0.0003	0.0003	0.0003	0.0003	0.0003	0.0003	0.0003
6	0.001	0.001	0.001	0.001	0.001	0.001	0.001	0.001	0.001
8	0.003	0.003	0.002	0.002	0.002	0.002	0.002	0.002	0.002
10	0.007	0.006	0.005	0.005	0.005	0.005	0.005	0.005	0.005
12	0.013	0.011	0.010	0.009	0.009	0.009	0.009	0.009	0.009
14	0.023	0.020	0.018	0.017	0.017	0.016	0.016	0.016	0.016
16	0.039	0.033	0.030	0.028	0.028	0.027	0.027	0.027	0.027
18	0.061	0.052	0.047	0.045	0.044	0.044	0.043	0.043	0.043
20	0.091	0.078	0.071	0.068	0.067	0.066	0.066	0.066	0.066
22	0.132	0.114	0.104	0.100	0.098	0.097	0.097	0.097	0.097
24	0.183	0.161	0.148	0.143	0.140	0.139	0.139	0.138	0.138
26	0.246	0.221	0.205	0.198	0.195	0.193	0.193	0.192	0.192
28	0.322	0.296	0.277	0.268	0.265	0.263	0.262	0.262	0.262
30	0.411	0.387	0.367	0.357	0.353	0.351	0.350	0.349	0.349
32	0.515	0.495	0.476	0.466	0.462	0.460	0.459	0.458	0.458
34	0.634	0.622	0.607	0.599	0.595	0.594	0.593	0.592	0.592
36	0.772	0.768	0.762	0.758	0.756	0.756	0.755	0.755	0.755
38	0.930	0.934	0.942	0.947	0.949	0.950	0.951	0.951	0.951
40	1.11	1.12	1.15	1.17	1.18	1.18	1.18	1.18	1.18
42	1.32	1.33	1.38	1.42	1.44	1.45	1.46	1.46	1.46
44	1.56	1.56	1.64	1.71	1.75	1.77	1.78	1.78	1.78
46	1.84	1.83	1.94	2.04	2.10	2.14	2.15	2.16	2.16
48	2.16	2.12	2.26	2.41	2.51	2.56	2.58	2.59	2.60
50	2.53	2.45	2.61	2.82	2.96	3.03	3.07	3.09	3.10
52	2.95	2.82	3.01	3.27	3.47	3.58	3.63	3.66	3.68
54	3.43	3.23	3.43	3.77	4.03	4.18	4.27	4.31	4.33
56	3.98	3.70	3.90	4.31	4.65	4.86	4.98	5.04	5.07
58	4.59	4.22	4.42	4.90	5.34	5.62	5.78	5.86	5.90
60	5.28	4.80	4.99	5.54	6.08	6.45	6.66	6.78	6.84
62	6.06	5.45	5.61	6.23	6.89	7.36	7.64	7.80	7.88
64	6.92	6.18	6.29	6.98	7.76	8.36	8.72	8.93	9.04
66	7.89	6.98	7.05	7.78	8.70	9.44	9.91	10.18	10.33
68	8.96	7.88	7.87	8.66	9.71	10.61	11.20	11.55	11.75
70	10.2	8.9	8.8	9.6	10.8	11.9	12.6	13.1	13.3
72	11.5	10.0	9.8	10.6	12.0	13.2	14.1	14.7	15.0
74	12.9	11.2	10.9	11.7	13.2	14.7	15.8	16.5	16.9
76	14.5	12.5	12.1	12.9	14.5	16.2	17.5	18.4	18.9
78	16.2	13.9	13.4	14.2	15.9	17.8	19.4	20.5	21.1
80	18.2	15.5	14.8	15.6	17.4	19.6	21.4	22.7	23.5
82	20.2	17.2	16.4	17.2	19.1	21.4	23.5	25.1	26.1
84	22.5	19.1	18.1	18.8	20.8	23.4	25.8	27.6	28.8
86	25.0	21.2	19.9	20.6	22.6	25.5	28.2	30.4	31.8
88	27.6	23.4	21.9	22.5	24.6	27.7	30.7	33.2	35.0
90	30.5	25.8	24.1	24.6	26.8	30.0	33.4	36.3	38.3

Source: *Guide for Design of Pavement Structures,* American Association of State Highway and Transportation Officials, Washington, D.C., 1993, with permission.

TABLE 3.19 Worksheet for Calculating 18-kip Equivalent Single-Axle Load (ESAL) Applications

Analysis period = _____ years

Location _____

Assumed SN or D = _____ in

Vehicle types	Current traffic (A)	Growth factors (B)	Design traffic (C)	ESAL factor (D)	Design ESAL (E)
Passenger cars Buses					
Panel and pickup trucks Other 2-axle/4-tire trucks 2-axle/6-tire trucks 3 or more axle trucks All single-unit trucks					
3-axle tractor semitrailers 4-axle tractor semitrailers 5+ axle tractor semitrailers All tractor semitrailers					
5-axle double trailers 6+ axle double trailers All double trailer combos					
3-axle truck-trailers 4-axle truck-trailers 5+ axle truck-trailers All truck-trailer combos					
All vehicles				Design ESAL	

Source: *Guide for Design of Pavement Structures,* American Association of State Highway and Transportation Officials, Washington, D.C., 1993, with permission.

Granular soils:

$$\log M_R = 0.523 - 0.0225(\%W) + 0.544(\log T) + 0.173(SM) + 0.197(GR) \qquad (3.7)$$

where M_R = resilient modulus, kips/in^2

$\%W$ = percentage water

T = bulk stress = DS + 3CS

$SM = \begin{cases} 1 \text{ for SM soil (Unified Soil Classification, Art. 8.3.2)} \\ 0 \text{ otherwise} \end{cases}$

$GR = \begin{cases} 1 \text{ for GR soil (Unified Soil Classification)} \\ 0 \text{ otherwise} \end{cases}$

TABLE 3.20   Worksheet for Calculating 18-kip Equivalent Single-Axle Load (ESAL) Applications

Location _____ Example 1 _____

Analysis period = _____ 20 _____ years

Assumed SN or D = _____ 9 _____ in

Vehicle types	Current traffic (A)	Growth factors (B)	Design traffic (C)	ESAL factor (D)	Design ESAL (E)
		2%			
Passenger cars	5,925	24.30	52,551,787	0.0008	42,041
Buses	35	24.30	310,433	0.6806	211,280
Panel and pickup trucks	1,135	24.30	10,066,882	0.0122	122,816
Other 2-axle/4-tire trucks	3	24.30	26,609	0.0052	138
2-axle/6-tire trucks	372	24.30	3,299,454	0.1890	623,597
3 or more axle trucks	34	24.30	301,563	0.1303	39,294
All single-unit trucks					
3-axle tractor semitrailers	19	24.30	168,521	0.8646	145,703
4-axle tractor semitrailers	49	24.30	434,606	0.6560	285,101
5+ axle tractor semitrailers	1,880	24.30	16,674,660	2.3719	39,550,626
All tractor semitrailers					
5-axle double trailers	103	24.30	913,559	2.3187	2,118,268
6+ axle double trailers	0	24.30			
All double trailer combos					
3-axle truck-trailers	208	24.30	1,844,856	0.0152	28,042
4-axle truck-trailers	305	24.30	2,705,198	0.0152	41,119
5+ axle truck-trailers	125	24.30	1,108,688	0.5317	589,489
All truck-trailer combos					
All vehicles	10,193		90,406,816	Design ESAL	43,772,314

Source: *Guide for Design of Pavement Structures,* American Association of State Highway and Transportation Officials, Washington, D.C., 1993, with permission.

TABLE 3.21   Equivalency Factors for Determining ESAL

Function classification	Rigid pavement		Flexible pavement	
	B	C	B	C
Rural interstate	2.27	0.914	1.60	0.735
All other rural	2.16	1.02	1.44	0.777
Urban interstate, freeway, and expressway	2.50	1.48	1.74	1.13
All other urban	1.61	0.673	1.11	0.534

Notes: B = tractor-trucks with semitrailers and trucks with trailers. C = single-unit trucks (2 axles, 6 tires or more).

Source: Ohio Department of Transportation, *Location and Design Manual,* Vol. 1, *Roadway Design,* December 1990, revised October 1992, with permission.

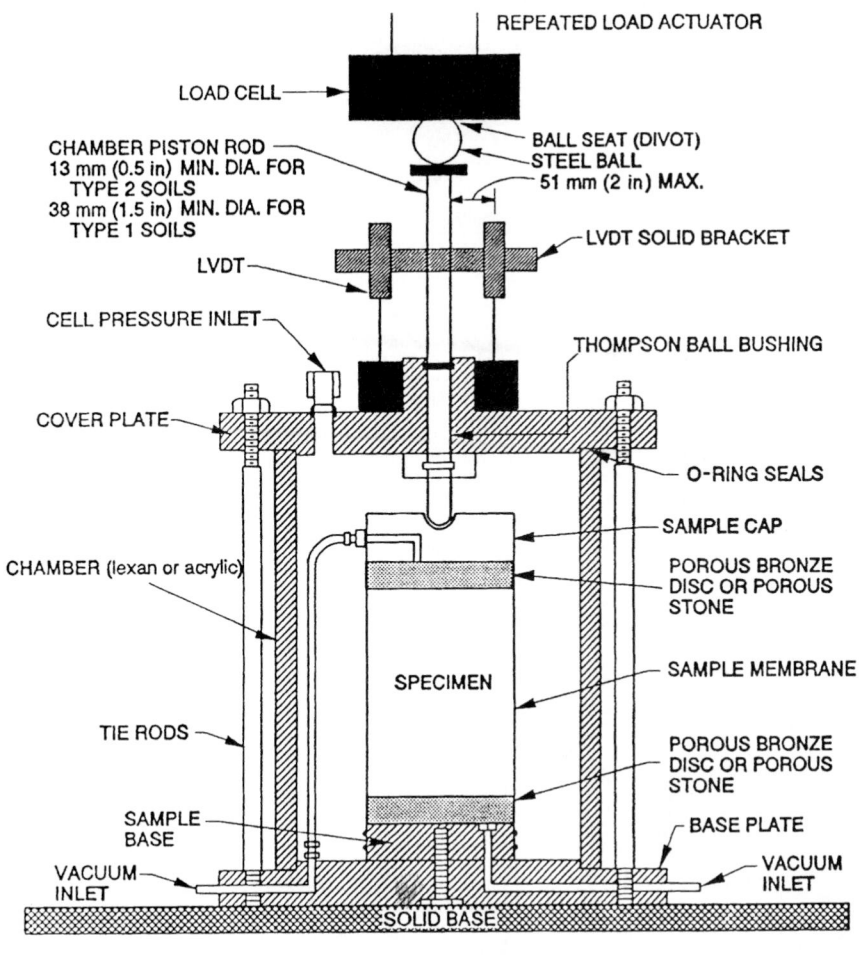

SECTION VIEW

Note: LVDT tips shall rest on the triaxial cell itself or on a plate/bracket which is rigidly attached to the triaxial cell.

NOT TO SCALE

FIGURE 3.9 Triaxial test chamber for determining resilient modulus of soil specimen. (*From NRC Operational Guide No. SHRP-LTPP-OG-004, "SHRP-LTPP Interim Guide for Laboratory Material Handling and Testing," with permission*)

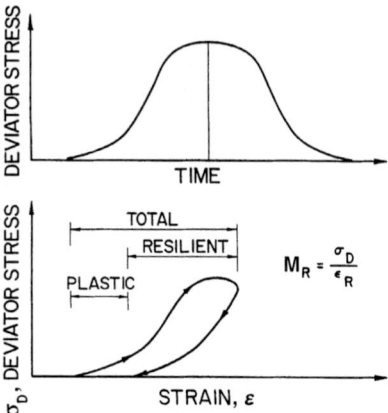

FIGURE 3.10   Load-response curve used to determine resilient modulus. (*From M. Thompson, "Factors Affecting the Resilient Modulus of Soils and Granular Materials,"* Proceedings of Workshop on Resilient Modulus Testing, *Oregon State University, Corvallis, 1989, with permission*)

3.6 RIGID PAVEMENT DESIGN PROCEDURE

The steps involved in designing a rigid pavement using the AASHTO design equations are as follows:

1. Determine the effective modulus of subgrade reaction.
2. Select the material properties for the concrete pavement.
3. Determine the drainage coefficient for the pavement.
4. Select the design serviceability loss.
5. Estimate the total number of 18-kip equivalent single-axle loads for the design period.
6. Select a level of reliability and the overall standard deviation.
7. Determine slab thickness and steel reinforcement.

Determine Effective Subgrade Modulus.   The first step in designing the thickness of a rigid pavement is the determination of the effective modulus of subgrade reaction. The effective modulus (or composite modulus) is the modulus of subgrade reaction after correction for use of subbase, seasonal variation in subgrade and subbase strength, rigid foundation within 10 ft of the surface, and loss of support. Figure 3.11 is used to determine the composite modulus of subgrade reaction when a subbase will be used under the concrete pavement. If the pavement will be placed directly on the subgrade, the AASHTO *Pavement Design Guide* recommends a composite modulus of subgrade reaction of:

Example:

D_{SB} = 6 in

E_{SB} = 20,000 lb/in²

M_R = 7,000 lb/in²

Solution: k_∞ = 400 lb/in³

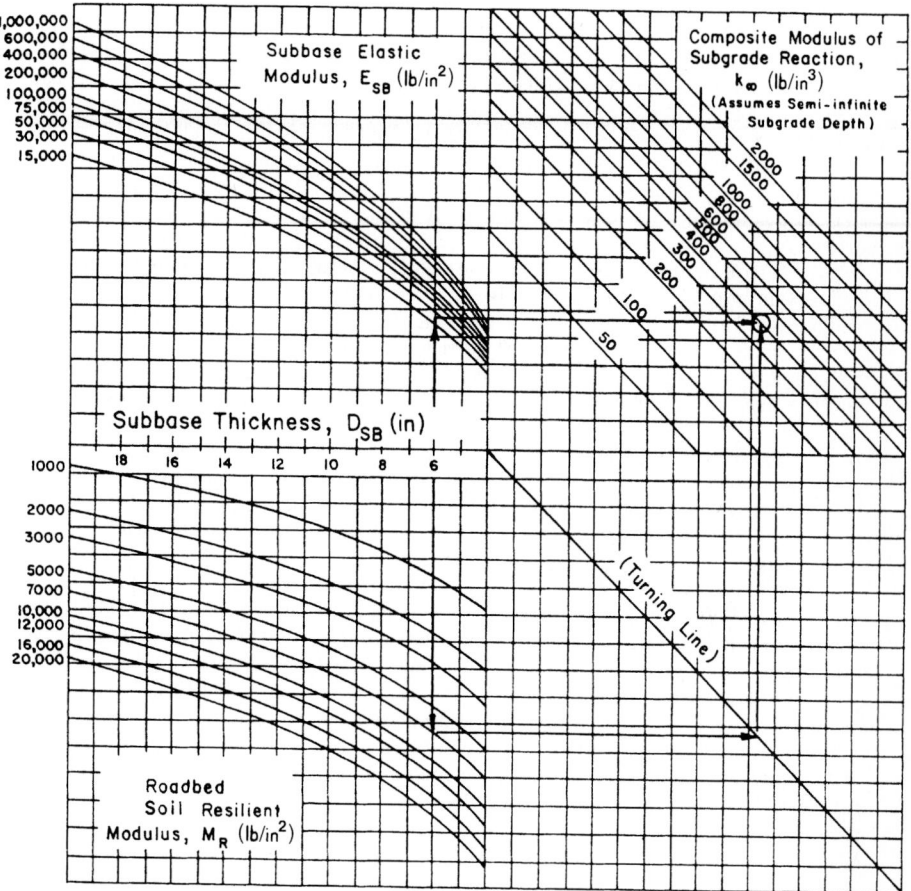

FIGURE 3.11   Chart for estimating composite modulus of subgrade reaction k_∞, assuming a semi-infinite subgrade depth; defined as over 10 ft below subgrade surface. (*From* Guide for Design of Pavement Structures, *American Association of State Highway and Transportation Officials, Washington, D.C., 1993, with permission*)

$$k = \frac{M_R}{19.4} \tag{3.8}$$

where k is in lb/in^3 (pci) and M_R is in lb/in^2.

When a stiff layer (bedrock, etc.) is located within 10 ft of the surface, the stiff layer will provide additional support for the pavement. Figure 3.12 is used to correct the composite modulus of subgrade reaction for this additional support.

In regions where large moisture variations, freeze and thaw, etc., will affect the strength of the subgrade soils and subgrade, AASHTO provides a procedure to modify the composite modulus of subgrade reaction. Table 3.22 provides a worksheet, and Table 3.23 shows an example. The seasonal variation in strength is determined using laboratory procedures or nondestructive testing (NDT). The seasonal strength of the subbase and subgrade is entered in columns 2 and 3 of Table 3.22. The composite modulus of subgrade reaction is determined using Fig. 3.11 and entered in column 4. If a rigid foundation is present within 10 ft of the surface, the k value is corrected using Fig. 3.12 and entered in column 5. The corrected k value is then used in Fig. 3.13 to determine the seasonal or relative damage factor, which is entered in column 6. The sum of relative damage is divided by the total number of periods to determine the average relative damage factor. This value is entered in Fig. 3.13 to determine the average composite modulus of subgrade reaction for the year. Many concrete pavements fail as a result of pumping or loss of support under the slab. Figure 3.14 is provided to correct the effective modulus of subgrade reaction for loss of support. This figure lowers the k so that the stress in the slab will be the same as for a slab with a void. Although the AASHTO procedure includes design for loss of support, it is recommended that a pavement base be designed to prevent or reduce loss of support, especially under pavements supporting a large number of heavy loads. The cost of providing a base to resist loss of support may be less than the cost of restoring support in the future.

Select Pavement Material Properties. The next step in the design of a rigid pavement is to select material properties. The reliability level and overall standard deviation consider the variation in material properties. Therefore, average material property values must be used in design. The concrete material values needed for design are the average concrete modulus of elasticity and the average concrete modulus of rupture. These values are not known until after construction of the pavement unless the plans for the pavement contain a performance specification. Material properties from past pavement construction may be used for design purposes provided a similar mix will be used. American Society for Testing and Materials "Test Method for Static Modulus of Elasticity and Poisson's Ratio of Concrete in Compression," ASTM C469, details the laboratory test method for determining the concrete modulus of elasticity. ASTM C78, "Test Method for Flexural Strength of Concrete (Using Simple Beam with Third-Point Loading)," details the laboratory test method for determining the concrete flexural strength (modulus of rupture).

Determine Drainage Coefficient. The drainage coefficient is used to modify the design thickness for drainage conditions. Moisture affects the pavement performance by decreasing the strength of the subgrade and subbase material and affects the warping and curling behavior of the concrete slabs. The intent of the drainage coefficient is to allow performance prediction for pavements without a proper drainage system. Increasing the pavement thickness should not be used in lieu of a properly designed

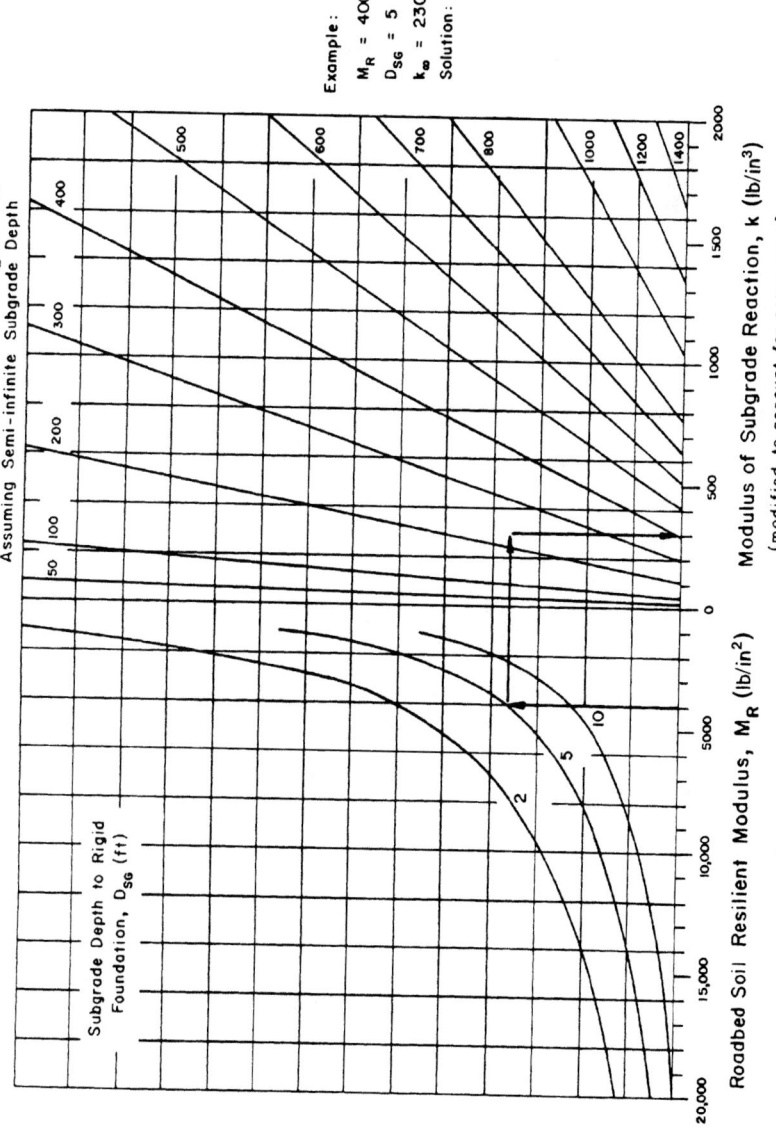

FIGURE 3.12 Chart for modifying modulus of subgrade reaction to consider effect of rigid foundation near the surface. (*From Guide for Design of Pavement Structures, American Association of State Highway and Transportation Officials, Washington, D.C., 1993, with permission*)

TABLE 3.22 Table for Estimating Effective Modulus of Subgrade Reaction

Trial subbase: Type _____   Depth to rigid foundation, ft _____

Thickness, in _____   Projected slab thickness, in _____

Loss of support, LS _____

Month (1)	Roadbed modulus M_R, lb/in² (2)	Subbase modulus, E_{SB}, lb/in² (3)	Composite k value, lb/in³ (Fig. 3.11) (4)	k value, lb/in³ on rigid foundation (Fig. 3.12) (5)	Relative damage u_r (Fig. 3.13) (6)
January					
February					
March					
April					
May					
June					
July					
August					
September					
October					
November					
December					

Summation: $\Sigma u_r =$

Average: $\bar{u}_r = \dfrac{\Sigma u_r}{n} =$ _____

Effective modulus of subgrade reaction k (lb/in³) = _____

Corrected for loss of support (Fig. 3.14): k (lb/in³) = _____

Source: Guide for Design of Pavement Structures, American Association of State Highway and Transportation Officials, Washington, D.C., 1993, with permission.

TABLE 3.23 Example Application of Method for Estimating Effective Modulus of Subgrade Reaction

Trial subbase: Type _____Granular_____   Depth to rigid foundations, ft ___5___

Thickness, in ____6____   Projected slab thickness, in ___9___

Loss of support, LS ___1.0___

Month (1)	Roadbed modulus M_R, lb/in^2 (2)	Subbase modulus, E_{SB}, lb/in^2 (3)	Composite k value, lb/in^3 (Fig. 3.11) (4)	k value, lb/in^3 on rigid foundation (Fig. 3.12) (5)	Relative damage u_r (Fig. 3.13) (6)
January	20,000	50,000	1,100	1,350	0.35
February	20,000	50,000	1,100	1,350	0.35
March	2,500	15,000	160	230	0.86
April	4,000	15,000	230	300	0.78
May	4,000	15,000	230	300	0.78
June	7,000	20,000	410	540	0.60
July	7,000	20,000	410	540	0.60
August	7,000	20,000	410	540	0.60
September	7,000	20,000	410	540	0.60
October	7,000	20,000	410	540	0.60
November	4,000	15,000	230	300	0.78
December	20,000	50,000	1,100	1,350	0.35

Summation: Σu_r = 7.25

Average: $\bar{u}_r = \dfrac{\Sigma u_r}{n}$ = ___0.60___

Effective modulus of subgrade reaction k (lb/in^3) = __540__

Corrected for loss of support (Fig. 3.14): k (lb/in^3) = __170__

Source: Guide for Design of Pavement Structures, American Association of State Highway and Transportation Officials, Washington, D.C., 1993, with permission.

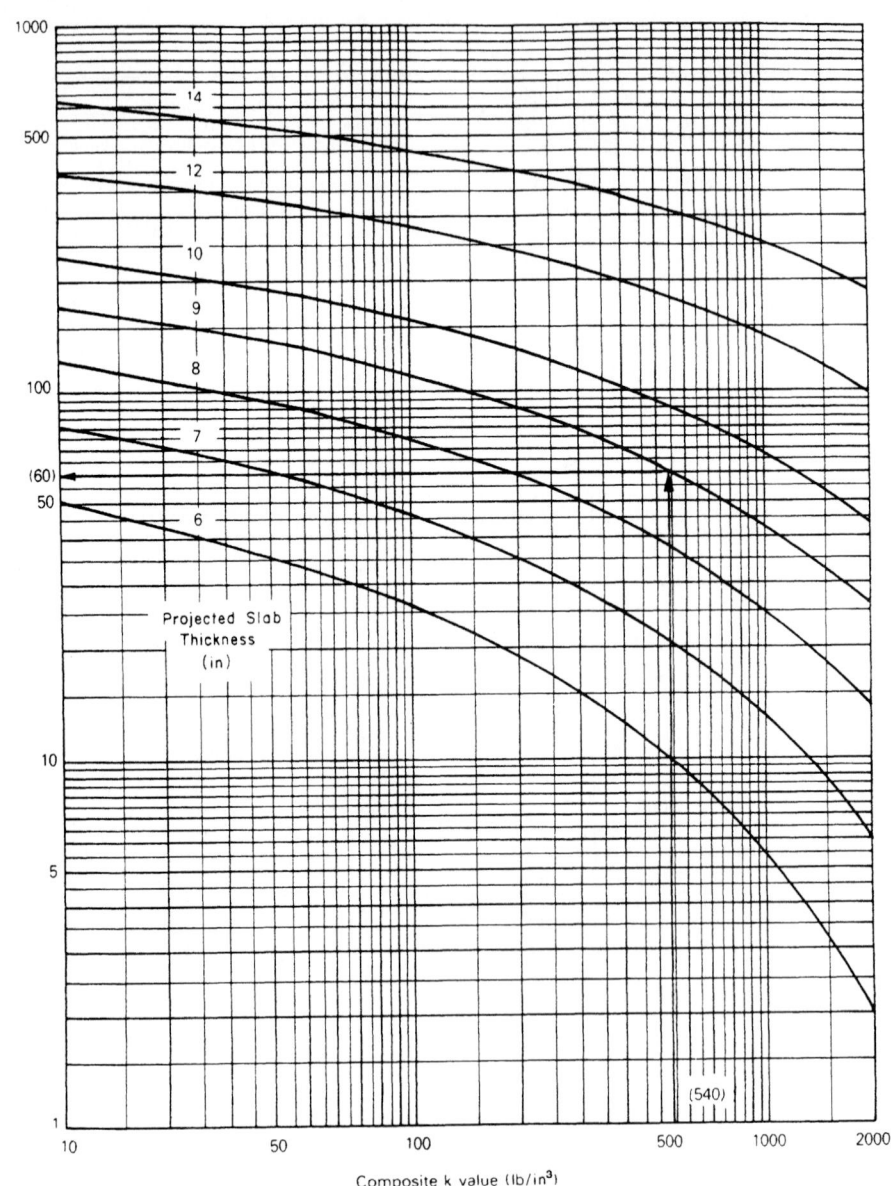

FIGURE 3.13 Chart for estimating relative damage to rigid pavements based on slab thickness and underlying support. (*From* Guide for Design of Pavement Structures, *American Association of State Highway and Transportation Officials, Washington, D.C., 1993, with permission*)

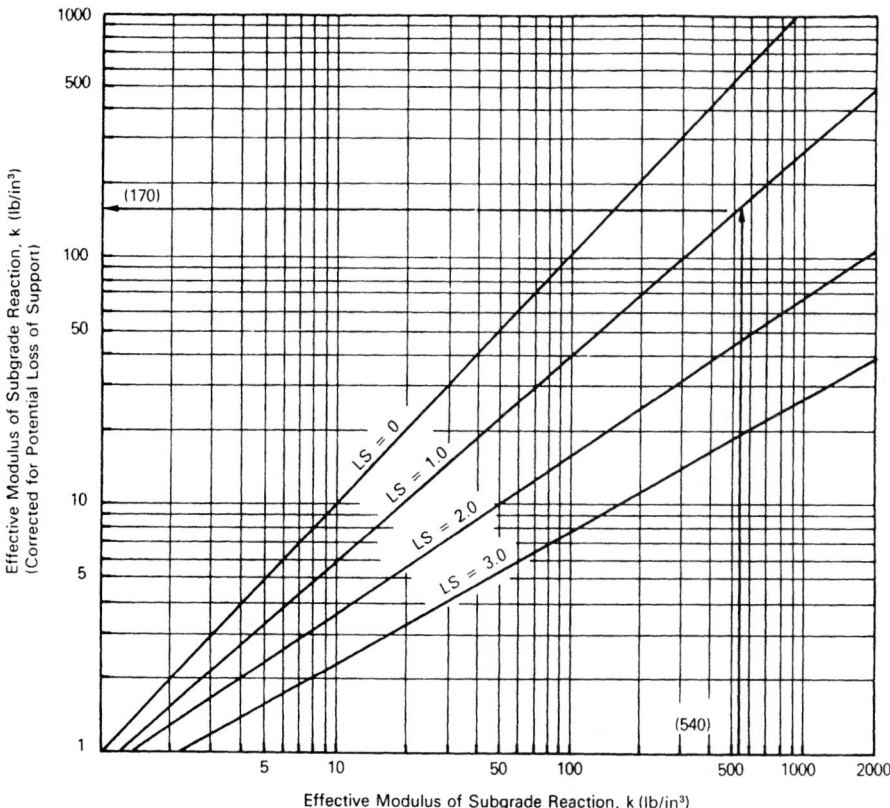

FIGURE 3.14 Chart for correction of effective modulus of subgrade reaction for potential loss of subbase support. (*From* Guide for Design of Pavement Structures, *American Association of State Highway and Transportation Officials, Washington, D.C., 1993, with permission*)

drainage system. Recommended values for the drainage coefficient are given in Table 3.24. The Federal Highway Administration's "Highway Subdrainage Design Manual," Report No. FHWA-TS-80-224, provides a procedure that may be used to determine drainage times for base material. (See Art. 5.4.5 and 5.4.6.)

Select Design Serviceability Loss. The design serviceability loss is the amount of serviceability loss the agency will tolerate before rehabilitation. To select a design serviceability loss, the designer needs to know the initial serviceability and the terminal serviceability of the pavement. The initial serviceability is the serviceability immediately after construction. Since this value is unknown at the time of construction, the designer will usually use the average initial serviceability of previously constructed pavements. The terminal serviceability is the serviceability of a pavement immediately before rehabilitation. The terminal serviceability is a function of traffic volume and speed. A low-volume road with low speeds may be allowed to deteriorate to a lower serviceability than a high-volume freeway, since the associated user costs will be lower. The terminal serviceability used by an agency is a policy decision. Common terminal serviceabilities are 2.5 for high-volume roads and 2.0 for low-volume roads.

TABLE 3.24 Recommended Values of Drainage Coefficient C_d for Rigid Pavement Design

Quality of drainage	Percent of time pavement structure is exposed to moisture levels approaching saturation			
	Less than 1%	1–5%	5–25%	Greater than 25%
Excellent	1.25–1.20	1.20–1.15	1.15–1.10	1.10
Good	1.20–1.15	1.15–1.10	1.10–1.00	1.00
Fair	1.15–1.10	1.10–1.00	1.00–0.90	0.90
Poor	1.10–1.00	1.00–0.90	0.90–0.80	0.80
Very poor	1.00–0.90	0.90–0.80	0.80–0.70	0.70

Source: Guide for Design of Pavement Structures, American Association of State Highway and Transportation Officials, Washington, D.C., 1993, with permission.

Estimate ESALs. The daily ESAL loadings are determined as outlined in Art. 3.5.2. The total number of ESAL loadings for design is the cumulative number of ESAL loadings expected over the design life of the pavement. This value can be determined by assuming a growth rate or, if the pavement is being built on an existing alignment, by extrapolating past traffic patterns.

Select Level of Reliability and Standard Deviation. The level of reliability and overall standard deviation can be selected using the guidelines discussed in Art. 3.5.3.

Determine Slab Thickness and Reinforcement. The design slab thickness is determined by using the design values as outlined above in the nomograph shown in Fig. 3.15. The design thickness is usually rounded up to the nearest $\frac{1}{2}$ or 1 in depending on the local practice for specifying slab thickness. As mentioned in Art. 3.5.2, if the design thickness varies significantly from the thickness used to determine the equivalency factors, the equivalency factors should be recalculated and the thickness design checked. Determination of steel reinforcement content, if used, is detailed in Art. 3.1.1.

3.7 FLEXIBLE PAVEMENT DESIGN PROCEDURE

The steps involved in designing a flexible pavement using the AASHTO design equations are as follows:

1. Determine the effective resilient modulus of the subgrade.
2. Select the design serviceability loss.
3. Estimate the total number of 18-kip equivalent single-axle loads for the design period.
4. Select a level of reliability and the overall standard deviation.
5. Determine the pavement structural number.
6. Select the layer material type and determine the layer thickness.

Determine Effective Resilient Modulus. The first step in designing the thickness of a flexible pavement is the determination of the effective resilient modulus of the subgrade. The resilient modulus may be determined as described in Art. 3.5.4. In regions

Nomograph solves:

$$\log_{10} \frac{W}{18} = Z_R * S_o + 7.35 * \log_{10}(D+1) - 0.06 + \frac{\log_{10}\left(\frac{\Delta \text{ (lb/in}^2)}{4.5-1.5}\right)}{1 + \frac{1.624 * 10^7}{(D+1)^{8.46}}} + (4.22 - 0.32p_t) * \log_{10}\left[\frac{S'_c * C_d\left(D^{0.75}-1.132\right)}{215.63 * J\left(D^{0.75} - \frac{18.42}{(E_C/k)^{0.25}}\right)}\right]$$

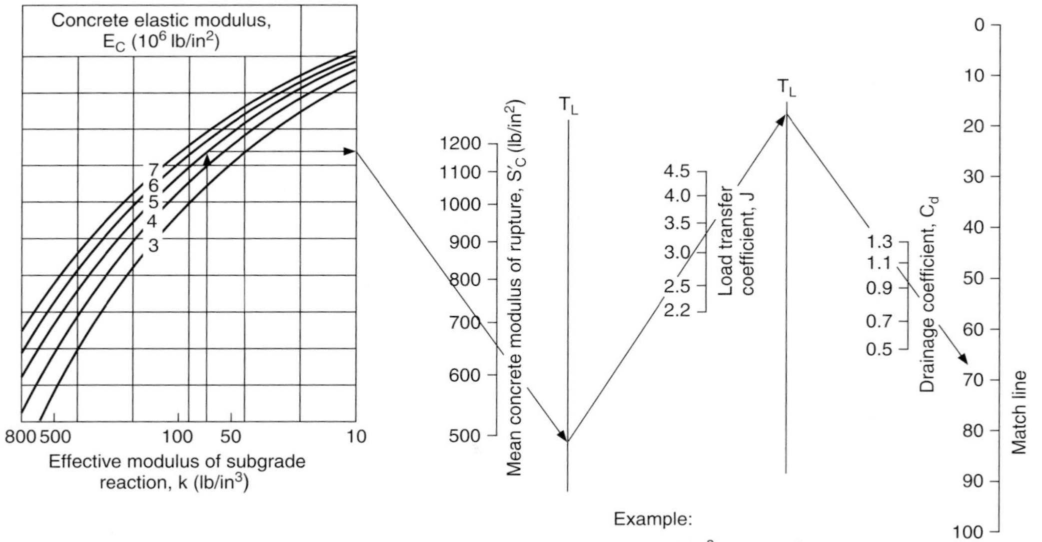

Match line

0

10

20

30

40

50

60

70

80

90

100

Drainage coefficient, C_d

1.3
1.1
0.9
0.7
0.5

T_L

Load transfer coefficient, J

4.5
4.0
3.5
3.0
2.5
2.2

T_L

Mean concrete modulus of rupture, S'_c (lb/in^2)

1200
1100
1000
900
800
700
600
500

Concrete elastic modulus, E_C (10^6 lb/in^2)

7 6 5 4 3

800 500

100

50

10

Effective modulus of subgrade reaction, k (lb/in^3)

Example:

k = 72 lb/in^3
E_C = 5 × 10^6 lb/in^2
S'_c = 650 lb/in^2
J = 3.2
C_d = 1.0

S_o = 0.29
R = 95% (Z_R = −1.645)
Δ (lb/in^2) = 4.2 − 2.5 = 1.7
W_{18} = 5.1 × 10^6 (18-kip ESAL)
Solution: D = 10.0 in.
(nearest half-inch, from part b)

FIGURE 3.15a Design chart for rigid pavements based on using mean values for each input variable. (*From Guide for Design of Pavement Structures, American Association of State Highway and Transportation Officials, Washington, D.C., 1993, with permission*)

3.45

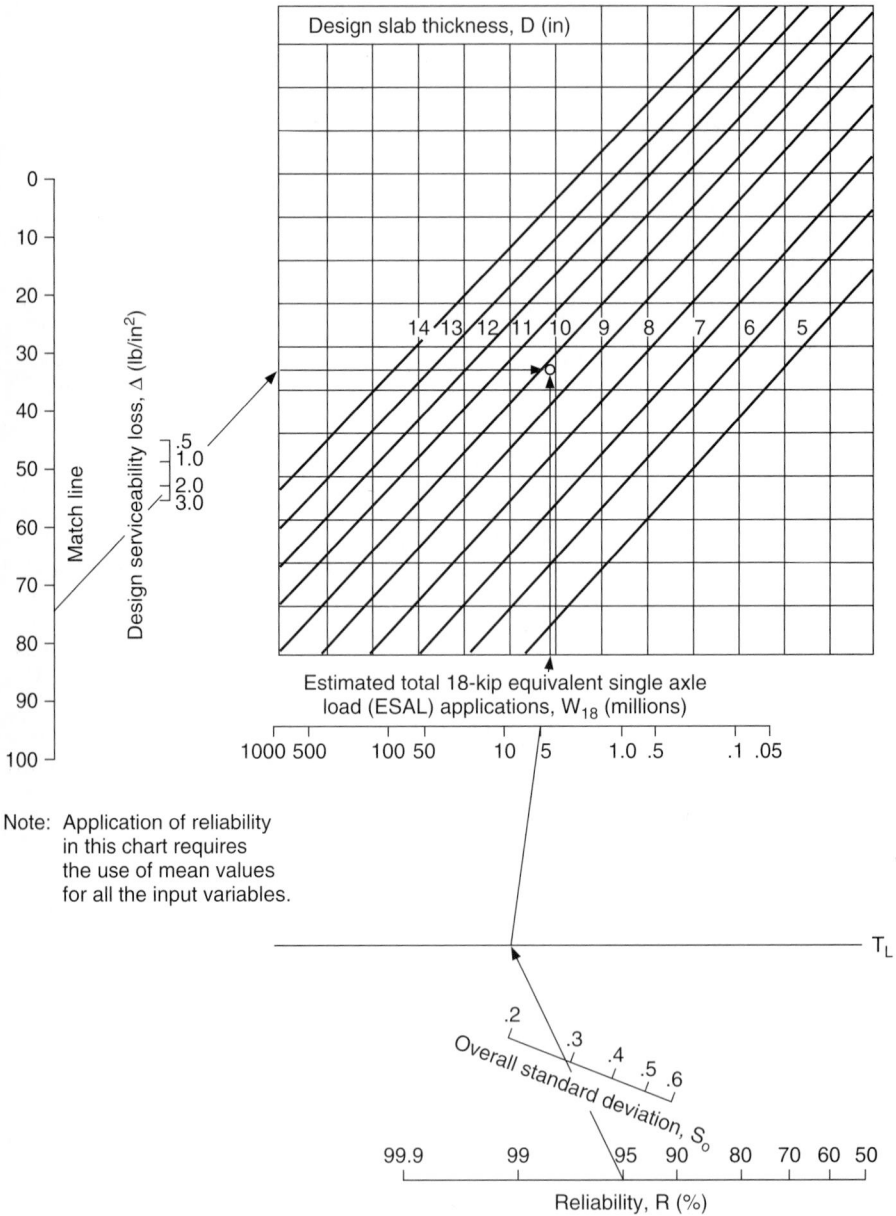

FIGURE 3.15b (*Continued*)

where large moisture variations, freeze and thaw, etc., will affect the strength of the subgrade soils and subgrade, AASHTO provides a procedure to modify the composite modulus of subgrade reaction. The seasonal variation in strength is determined using laboratory procedures or through the use of nondestructive testing (NDT). The seasonal strength of the subgrade is entered in Fig. 3.16. The relative damage is determined from the nomograph and the sum of relative damage divided by the total number of periods to determine the average relative damage factor. This factor is then entered in the nomograph to find the effective resilient modulus.

Select Design Serviceability Loss.   Considerations in the selection of the design serviceability loss are the same as discussed for rigid pavements, Art. 3.6.

Estimate ESALs.   The daily ESAL loadings are determined as outlined in Art. 3.5.2. The total number of ESAL loadings for design is the cumulative number of ESAL loadings expected over the design life of the pavement. This value can be determined by assuming a growth rate or, if the pavement is being built on an existing alignment, by extrapolating the past traffic patterns.

Select Level of Reliability and Standard Deviation.   These selections are made from the guidelines discussed in Art. 3.5.3.

Determine Pavement Structural Number.   The flexible pavement process involves the calculation of a pavement structural number. This is an abstract number reflecting the relative strength contribution of all layers in the pavement buildup. The structural number SN is calculated using the design values determined as outlined above in the nomograph shown in Fig. 3.17. The design thickness for each layer is determined by the following equation:

$$SN = a_1 D_1 + a_2 D_2 m_2 + a_3 D_3 m_3$$

where a_1, a_2, a_3 = structural coefficients of surface, base, and subbase, respectively

   D_1, D_2, D_3 = thickness of surface, base, and subbase, respectively
   m_2, m_3 = drainage coefficients for base and subbase (see Table 3.25)

The structural coefficients of the asphalt layer, granular base, and subbase can be estimated using Figs. 3.18, 3.19, and 3.20, respectively, or can be estimated from Table 3.26.

Select Layer Material and Thickness.   Once the structural coefficients are known, the thickness of the individual layers is determined by varying D_1, D_2, and D_3 until the calculated SN is equal to or greater than the required SN. Unbound bases are commonly specified to the nearest inch, and asphalt concrete is normally specified to the nearest ¼ inch. The procedure shown in Fig. 3.21 illustrates one method recommended for determining layer thickness. This procedure designs the upper layers to protect the lower layers.

3.8   PAVEMENT MANAGEMENT

Project-level pavement management is responsible for continuous evaluation of pavement present serviceability, monitoring of the pavement loading rate, determination of

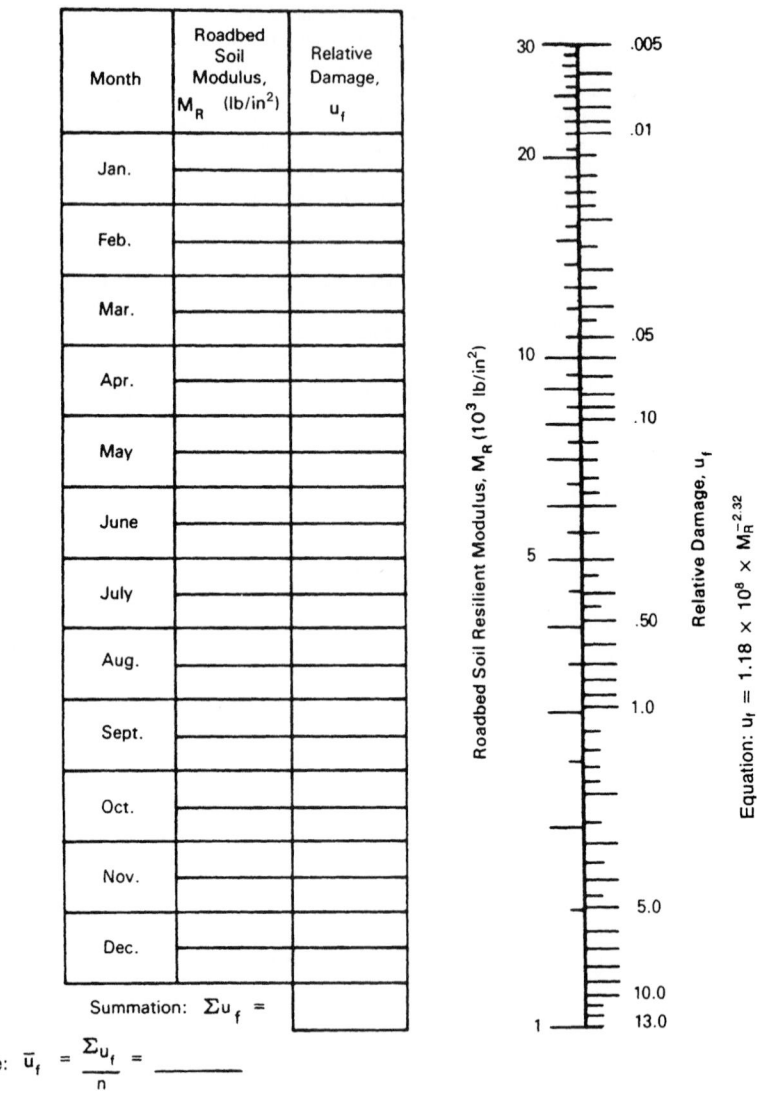

FIGURE 3.16 Chart for estimating effective roadbed soil resilient modulus for flexible pavements designed using the serviceability criteria. (*From* Guide for Design of Pavement Structures, *American Association of State Highway and Transportation Officials, Washington, D.C., 1993, with permission*)

Nomograph solves:

$$\log_{10} W_{18} = Z_R * S_o + 9.36 * \log_{10}(SN + 1) - 0.20 + \frac{\log_{10}\left(\frac{\Delta\ (\text{lb/in}^2)}{4.2 - 1.5}\right)}{0.40 + \frac{1094}{(SN + 1)^{5.19}}} + 2.32 * \log_{10} M_R - 8.07$$

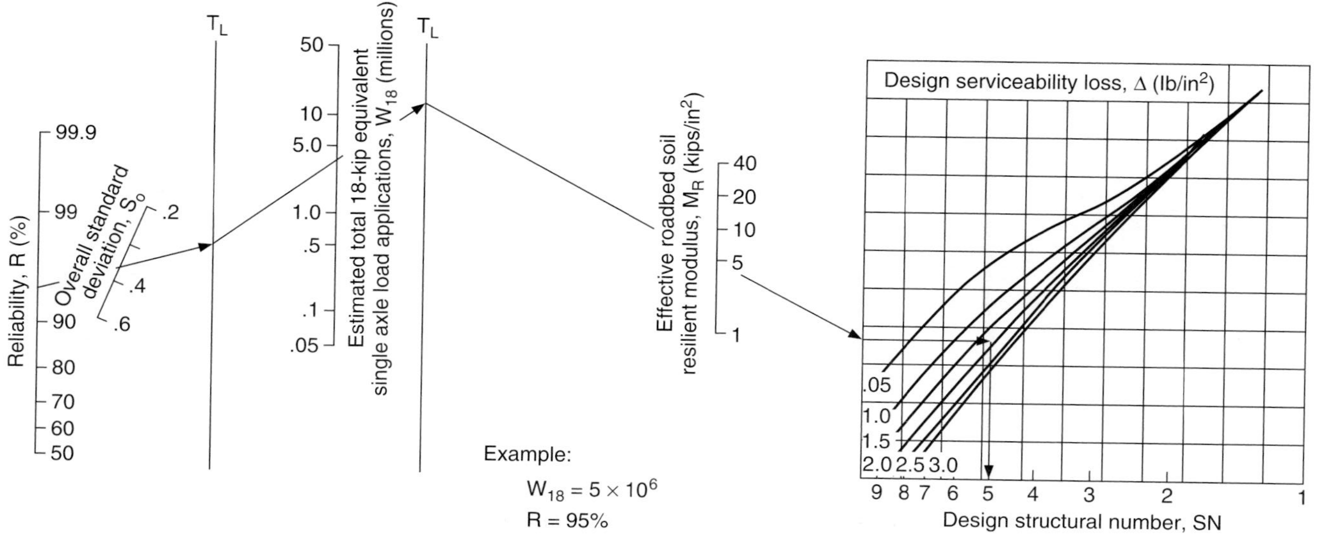

Reliability, R (%)
99.9
99
90
80
70
60
50
Overall standard deviation, S_o
.2
.4
.6
T_L
50
10
5.0
1.0
.5
.1
.05
Estimated total 18-kip equivalent single axle load applications, W_{18} (millions)
T_L
Effective roadbed soil resilient modulus, M_R (kips/in^2)
40
20
10
5
1
Design serviceability loss, Δ (lb/in^2)
.05
1.0
1.5
2.0
2.5
3.0
9 8 7 6 5 4 3 2 1
Design structural number, SN

Example:

$W_{18} = 5 \times 10^6$

$R = 95\%$

$S_o = 0.35$

$M_R = 5000$ lb/in^2

Δ (lb/in^2) 1.9

Solution: SN = 5.0

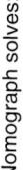

FIGURE 3.17 Design chart for flexible pavements based on using mean values for each variable. (*From* Guide for Design of Pavement Structures, *American Association of State Highway and Transportation Officials, Washington, D.C., 1993, with permission*)

TABLE 3.25 Recommended m_i Values for Modifying Structural Layer Coefficients of Untreated Base and Subbase Materials in Flexible Pavements

	Percent of time pavement structure is exposed to moisture levels approaching saturation			
Quality of drainage	Less than 1%	1–5%	5–25%	Greater than 25%
Excellent	1.40–1.35	1.35–1.30	1.30–1.20	1.20
Good	1.35–1.25	1.25–1.15	1.15–1.00	1.00
Fair	1.25–1.15	1.15–1.05	1.00–0.80	0.80
Poor	1.15–1.05	1.05–0.80	0.80–0.60	0.60
Very poor	1.05–0.95	0.95–0.75	0.75–0.40	0.40

Source: *Guide for Design of Pavement Structures,* American Association of State Highway and Transportation Officials, Washington, D.C., 1993, with permission.

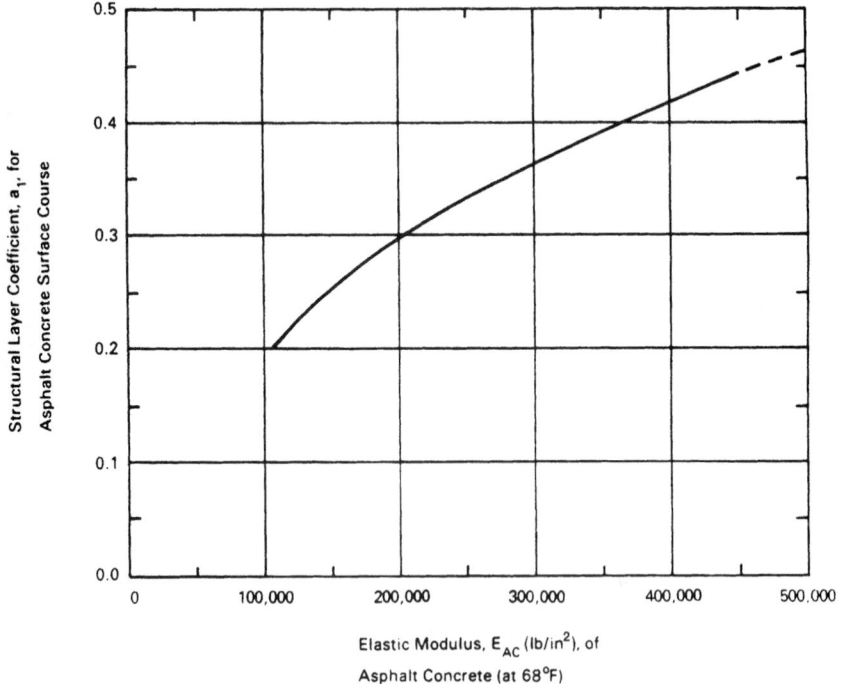

FIGURE 3.18 Chart for estimating structural layer coefficient (a_1) of dense-graded asphalt concrete based on the resilient modulus. (*From* Guide for Design of Pavement Structures, *American Association of State Highway and Transportation Officials, Washington, D.C., 1993, with permission*)

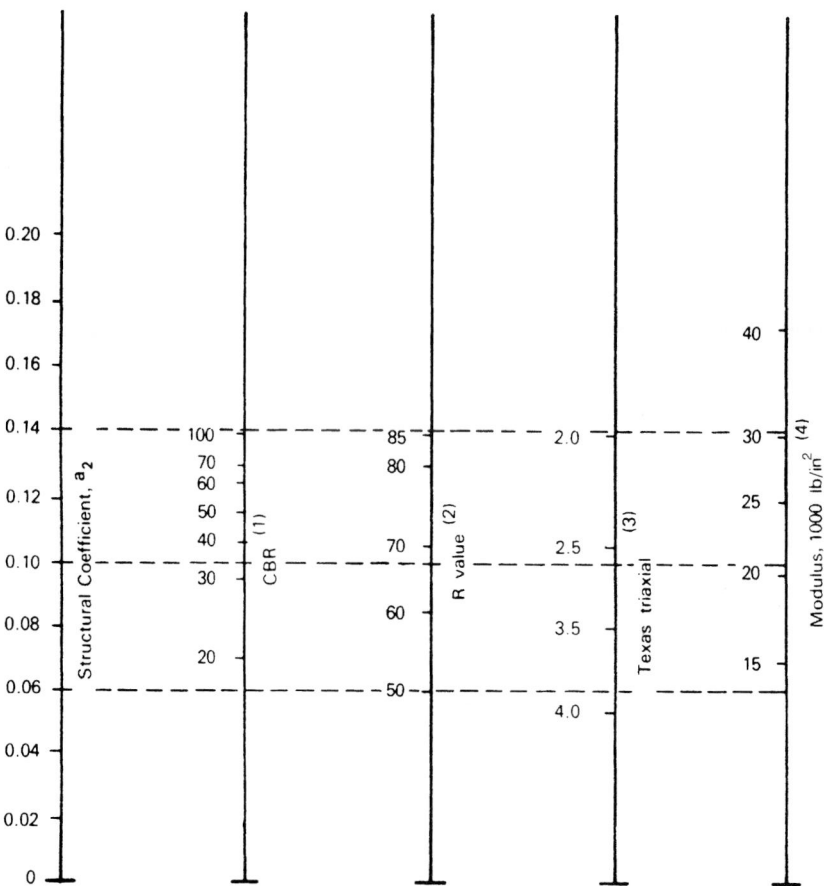

(1) Scale derived by averaging correlations obtained from Illinois.
(2) Scale derived by averaging correlations obtained from California, New Mexico, and Wyoming.
(3) Scale derived by averaging correlations obtained from Texas
(4) Scale derived on NCHRP project.

FIGURE 3.19 Variation in granular base layer coefficient (a_2) with various base strength parameters. (*From* Guide for Design of Pavement Structures, *American Association of State Highway and Transportation Officials, Washington, D.C., 1993, with permission*)

the cause and rate of pavement deterioration, prediction of optimal time for intervention, and evaluation of the most economical rehabilitation strategy.

Pavement management can be applied at the project level or at the network level. Although both levels are very dependent upon one another, they are seldom applied for the same purpose. The network level applies to the whole system in a global sense. Network refers to systemwide averages and is used for system budgeting and performance modeling. This chapter addresses only the project-level aspects. Project-level pavement management is considered to be more complicated and more important than pavement design. Pavement management is applied throughout the life of a pavement, whereas pavement design is completed and forgotten once the pavement is initially in service.

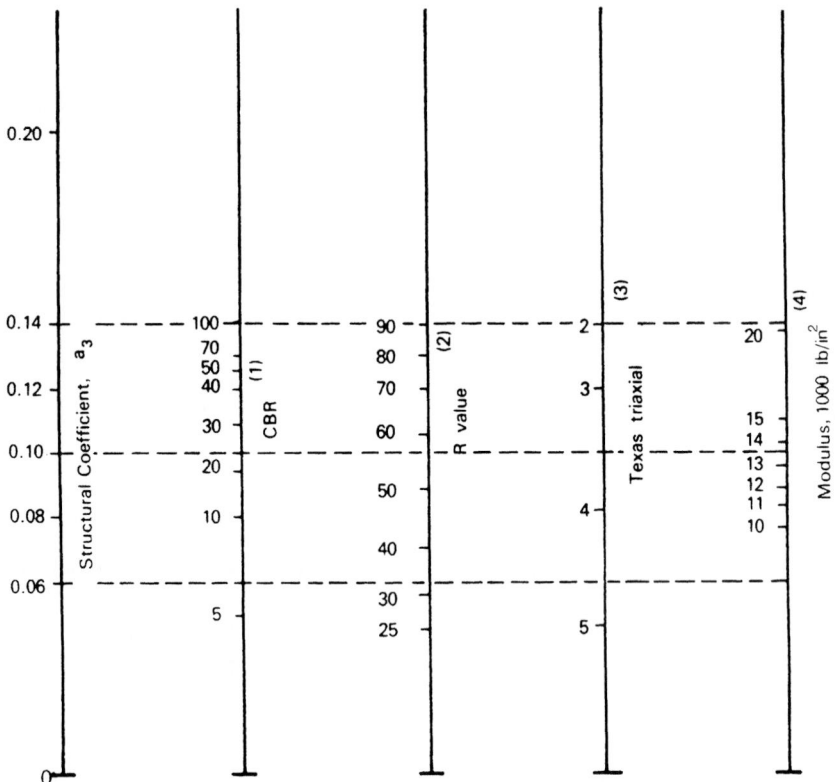

(1) Scale derived from correlations from Illinois.

(2) Scale derived from correlations obtained from the Asphalt Institute, California, New Mexico, and Wyoming.

(3) Scale derived from correlations obtained from Texas.

(4) Scale derived on NCHRP project.

FIGURE 3.20 Variation in granular base layer coefficient (a_3) with various subbase strength parameters. (*From* Guide for Design of Pavement Structures, *American Association of State Highway and Transportation Officials, Washington, D.C., 1993, with permission*)

3.8.1 Pavement Deterioration

Pavement deterioration or distress can be classified into two basic categories for all pavement types—structural and functional. The most serious category is structural. Structural deterioration results in reduced ability to carry load and a decreased pavement life. Functional deterioration can lead to and accelerate structural deterioration, but it is only related to ride quality and frictional characteristics. A third and less accepted type of pavement deterioration is environmental deterioration, which most pavement engineers lump with functional and structural deterioration. Environmental deterioration affects only the pavement materials and will generally exhibit itself as either functional or structural deterioration.

Pavement deterioration is an important measurement for a pavement engineer. To determine the remaining life of a pavement, or the amount of pavement repair required

TABLE 3.26 Pavement Coefficients for Flexible Section Design, Louisiana

	Strength*	Coefficient
I. Surface course†		
Asphaltic concrete		
Types 1, 2, and 4 BC and WC	1000+	0.40
Types 3 WC	1800+	0.44
BC	1500+	0.43
II. Base course		
Untreated‡		
Sand clay gravel—grade A	3.3−	0.08
Sand clay gravel—grade B	3.5−	0.07
Shell and sand-shell	2.2−	0.10
Cement-treated§		
Soil-cement	300+	0.15
Sand clay gravel—grade B	500+	0.18
Shell and sand-shell	500+	0.18
Shell and sand-shell	650+	0.23
Lime-treated‡		
Sand-shell	2.0−	0.12
Sand clay gravel—grade B	2.0−	0.12
Asphalt treated†		
Hot-mix base course (type 5A)	1200+	0.34
Hot-mix base course (type 5B)	800+	0.30
III. Subbase course‡		
Lime-treated sand clay gravel—grade B	2.0−	0.14
Shell and sand-shell	2.0−	0.14
Sand clay gravel—grade B	3.5−	0.11
Lime-treated soil	3.5−	0.11
Old gravel or shell roadbed (8-in thickness) (200 mm)	—	0.11
Sand (R-value)	55+	0.11
Suitable material—A−6 (PI = 15−)	—	0.04
IV. Coefficients for bituminous concrete overlay		
Base course		
Bituminous concrete pavement		
New		0.40
Old		0.24
Portland cement concrete pavement		
New		0.50
Old, fair condition		0.40
Old, failed		0.20
Old, pumping		0.10
Old, pumping (to be undersealed)		0.35

*Refer to the following footnotes for strength designations. See the AASHTO guide referenced below for further details.

†Marshall stability number.

‡Texas triaxial values.

§Compressive strength, lb/in^2

Source: *Interim Guide for Design of Pavement Structures,* American Association of State Highway and Transportation Officials, Washington, D.C., 1972 (rev. 1981), with permission.

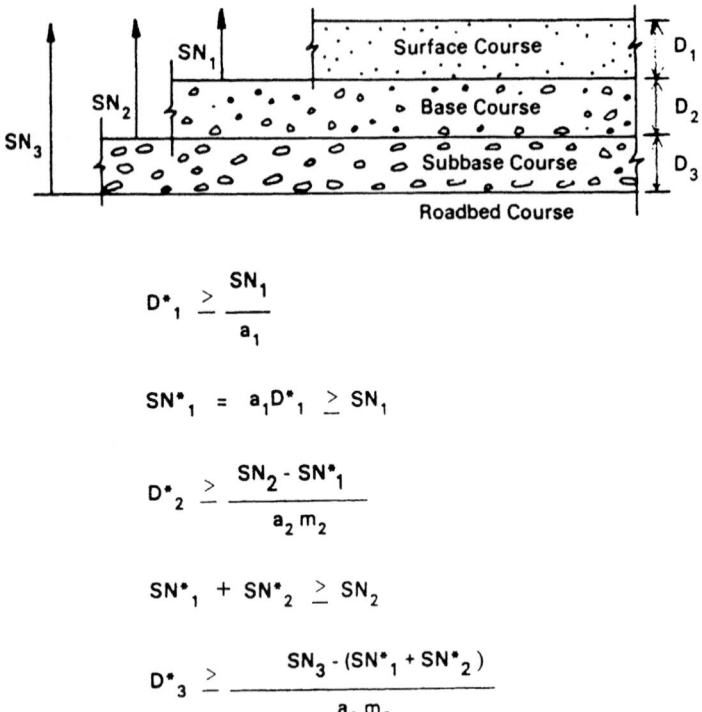

$$D^*_1 \geq \frac{SN_1}{a_1}$$

$$SN^*_1 = a_1 D^*_1 \geq SN_1$$

$$D^*_2 \geq \frac{SN_2 - SN^*_1}{a_2 m_2}$$

$$SN^*_1 + SN^*_2 \geq SN_2$$

$$D^*_3 \geq \frac{SN_3 - (SN^*_1 + SN^*_2)}{a_3 m_3}$$

FIGURE 3.21 Procedure for determining thicknesses of layers using a layered analysis approach. *a,* *D, m,* and SN are defined in the text and are minimum required values. An asterisk indicates that the value actually used is represented; this value must be equal to or greater than the required value. (*From Guide for Design of Pavement Structures, American Association of State Highway and Transportation Officials, Washington, D.C., 1993, with permission*)

to extend a pavement life for a given time period, or the most appropriate time for pavement repair, the amount and type of deterioration in a pavement must be measured. Methods of measurement of pavement deterioration vary, but most are similar in that they all require a visual inspection of the pavement and a somewhat subjective distress rating. More in-depth pavement evaluation methods include pavement coring to help the engineer visualize what is below the surface, and nondestructive testing using a Dynaflect, Falling Weight Deflectometer (FWD), or Road Rater to allow the engineer to measure a portion of the pavement stress-strain relationship. Chapter 5 of the AASHTO *Pavement Design Guide* provides several methods for evaluating the remaining life of a pavement.

3.8.2 Jointed Rigid Pavement Distress—Visual Rating

Jointed rigid pavement deterioration is exhibited in any combination of the following distresses:

Surface Deterioration. Surface deterioration (Fig. 3.22) is the result of loss of cement at the surface of the slab (scaling). It is generally caused by excessive surface water and finishing practice, or the loss of both small aggregates and cement caused

FIGURE 3.22   Example of surface deterioration in jointed rigid pavement.

by abrasion from tires. Surface deterioration affects the noise level of a pavement and cannot be repaired. Surface deterioration by itself is generally of little concern.

Popouts.   Figure 3.23 shows a typical popout. Popouts are generally due to high steel placement, but also may be the result of poor-quality aggregate, which disintegrates, causing cavities at the surface of the slab. Popouts affect the noise level of a pavement and cannot be repaired. Popouts by themselves are generally of little concern.

Pumping.   Pumping is defined as the ejection of subbase or subgrade materials from under a pavement through a joint or crack and out onto the pavement and shoulder. The loss of subbase or subgrade material causes loss of support and leads to corner breaks and faulting. The existence of pumping can be determined visually by the presence of soil stains at the joints or cracks on the adjacent shoulder.

Faulting.   Faulting is a result of the loss of load transfer across a joint or crack, which causes the slab on one side of the joint or crack to be at a lower elevation than the slab on the other side. Faulting (Fig. 3.24) is generally a result of pumping. Faulting affects the noise level and the smoothness of a pavement. It is generally considered excessive when faulting exceeds $\frac{1}{4}$ in. Faulting can be corrected by pavement grinding, joint or crack repair, or slab jacking. However, unless load transfer is established across the joint or crack and any existing voids under the joint or crack are filled, faulting can be expected to return.

Settlement.   Settlement is the result of poor construction practice. It may be due to either poor compaction over a utility, poor grade control during the final grading of the subgrade, or possible localized soil conditions that cannot resist additional over-

FIGURE 3.23 Example of popout in jointed rigid pavement.

burden or increased loading. Settlement, which is displayed by a depression in the profile of the pavement, affects smoothness. Repair methods consist of replacement to the corrected profile, or an overlay of some type. Settlements are generally of little concern unless they are numerous and severely affect the ride of the pavement.

Joint Spalling.   Figure 3.25 shows typical joint spalling, defined as deterioration of the concrete slab around transverse or longitudinal joints. The deterioration is generally only to partial depth and is visible from the surface of the slab. Joint spalling may result from poor-quality aggregates (D cracking); improperly placed dowels, tie-bars, or dowel baskets; or excessive expansion of the concrete (pressure). Repair of spalled joints can be accomplished by either partial-depth joint repairs or full-depth joint repairs.

Transverse Cracking.   A significant transverse crack is depicted in Fig. 3.26. Transverse cracking severity varies from hairline cracks to cracks sufficiently wide to completely separate the slab into two distinct pieces. Hairline cracks are expected in reinforced concrete and pose no expected problems. In plain concrete pavement, a hairline crack can be a sign of future problems. Without reinforcing mesh to hold the crack together, the long-term performance of the slab is questionable; however, as long as the crack is tightly closed (hairline), it poses no problem. Regardless of whether the pavement contains mesh, cracks that have separated by a distance greater than one-half of the largest aggregate diameter are generally considered to be failed.

Longitudinal Cracking.   Longitudinal cracking, such as shown in Fig. 3.27, may be caused by excessive lane widths, longitudinal joints that were not sawed properly, or local conditions that increase the stress level along the pavement. Longitudinal crack-

FIGURE 3.24 Example of faulting at joint in rigid pavement; pavement on right is about ½ in lower than that on left.

ing is primarily a concern when it occurs within the wheel track. Where a longitudinal crack is faulted, spalled, pumping, or working and is in the wheel path, it can become a safety hazard.

Corner Breaks. As illustrated in Fig. 3.28, corner breaks are cracks found at the corner of the slab. They usually propagate from the transverse joint to the longitudinal joint. Corner breaks are full-depth cracks and are generally the result of loss of support under the corner of the slab.

3.8.3 Continuously Reinforced Rigid Pavement Distress—Visual Rating

Continuously reinforced concrete (CRC) pavement deterioration is exhibited by the same distresses discussed for jointed concrete pavement along with the following additional considerations.

FIGURE 3.25   Example of joint spalling in rigid pavement.

Settlement.   As previously stated, settlement as displayed by a depression in the profile of the pavement affects the smoothness of a pavement. It may be the result of poor construction practice such as poor compaction over a utility, poor grade control during final grading of the subgrade, or localized soil conditions that cannot resist additional overburden or increased loading. Repair methods consist of replacement to the corrected profile or an overlay. However, settlements are more predominant in CRC pavement, because transverse cracks are inherently more numerous. With the transverse cracking at a very close spacing (5 to 8 ft), the pavement is able to bend more freely and does not bridge weak foundations as effectively.

Transverse Cracking.   Although CRC pavement is designed to have transverse cracks, the cracks should be spaced properly. Transverse cracks spaced too closely (less than 3 ft, as illustrated in Fig. 3.29) have a good chance of interconnecting, because they do not form uniformly straight and perpendicular to the centerline. Thus, as they interconnect, spalling will occur and pavement failures will result. On the other hand, transverse cracks spaced too far apart create higher stresses than the reinforcement can tolerate, and this can also result in pavement failures. Although incorrect transverse crack spacing is not a distress by itself, it must be monitored to help pavement engineers predict failures. Once failures are evident, they must be repaired by full-depth pavement removal and replacement. It is important to reestablish continuity of the reinforcement within the repair.

Punchouts.   Figure 3.30 shows a punchout in a CRC pavement. A punchout is formed by the combination of intersecting transverse and longitudinal cracks over an area of weak foundation.

FIGURE 3.26 Example of transverse crack in jointed rigid pavement.

FIGURE 3.27 Example of longitudinal crack in jointed rigid pavement.

FIGURE 3.28 Example of corner break in jointed rigid pavement.

3.8.4 Flexible Pavement Distress—Visual Rating

Flexible pavement deterioration is exhibited in any combination of the following distresses.

Raveling.   Raveling, as shown in Fig. 3.31, is the result of loss of small aggregates from the pavement surface. Raveling can be caused by oxidation of the mix, improper mix design, segregation, or lack of compaction.

Bleeding.   Bleeding is the flushing of excess asphalt cement to the surface of the pavement, as evident in Fig. 3.32. Asphalt cement concrete mixtures are more prone to bleed with hotter pavement surface temperatures. Bleeding is a result of excess asphalt cement in the mix and/or low air voids in the mix.

Potholes.   One of the most common problems is the development of a pothole (Fig. 3.33). Potholes are small, localized, but deep pavement failures characterized by a round shape. Potholes are caused by weak and wet subbase and/or subgrade. In freeze-thaw environments, potholes are generally formed during the thaw.

Rutting.   Rutting (Fig. 3.34) is the longitudinal deformation of the pavement structure within the wheel tracks. Where found only in the uppermost portions of the pavement, it is caused by poor mixture design and lack of stability. Where rutting is deep-seated and found throughout the depth of the pavement structure, it is caused by inadequate pavement structure above the founding layers or by a weak, wet subgrade.

Corrugation.   Corrugations (Fig. 3.35) are transverse waves in the pavement profile, which are found most generally at stop lights, at stop signs, or on hills. Corrugations are found in the wheel track and are the result of acceleration and deceleration of heavy trucks in a regular pattern on the roadway surface. The stability of the asphalt mix can also be a contributing factor.

FIGURE 3.29   Example of transverse cracks spaced too closely in continuously reinforced rigid pavement.

FIGURE 3.30   Example of punchout in continuously reinforced rigid pavement.

FIGURE 3.31   Example of raveling in flexible pavement.

Longitudinal Cracking.   Longitudinal cracking, such as shown in Fig. 3.36, is most often found at paving joints established during construction. The construction joint is most generally specified at lane lines. As weathering of the pavement takes place, the longitudinal joint ravels and eventually spalls. Longitudinal cracks found at locations other than paving joints are due to thermal shrinkage from seasonal temperature changes.

Transverse Cracking.   As illustrated by Fig. 3.37, transverse cracking is best described by cracks that form across the pavement perpendicular to the centerline. Transverse cracking is caused by thermal shrinkage from seasonal temperature changes and age hardening of the binder.

Block Cracking.   Block cracking is the combination of longitudinal and transverse cracking, as shown in Fig. 3.38. As the cracks worsen with time as a result of weathering, they join each other and form block cracking.

Wheel Track Cracking.   Wheel track cracking is shown in Fig. 3.39. It can be described as mostly longitudinal cracks found at the surface of the pavement within a 3-ft-wide strip considered to be the wheel track. Wheel track cracking ranges from a single longitudinal crack to a series of interconnected longitudinal cracks, also referred to as alligator cracking. Wheel track cracking is commonly considered to be the most alarming distress found in a flexible pavement. This type of cracking starts at

FIGURE 3.32   Example of bleeding in flexible pavement.

the bottom of the pavement structure and is transmitted to the surface. By the time alligator cracking can be detected by visual inspection, the pavement is generally considered to be failed.

Edge Cracking.   Edge cracking, as shown in Fig. 3.40, is a series of short longitudinal or irregular-shaped cracks at the outer 15 in of the pavement. Edge cracking is a result of lack of support outside the pavement edge.

3.8.5   Composite Pavement Distress—Visual Rating

Composite pavement deterioration is exhibited in a combination of some flexible pavement distresses and some rigid pavement distresses. The most prominent composite pavement distresses, which were defined under flexible or rigid pavement, are raveling, bleeding, rutting, corrugations, pumping, and various slab distresses.

FIGURE 3.33   Example of small pothole in flexible pavement.

FIGURE 3.34   Example of rutting in flexible pavement.

3.8.6 Pavement Coring

Without question, the simplest and most reliable method of identifying pavement deterioration is pavement coring. Pavement coring can be used to investigate many different pavement distress factors, from rigid joint deterioration to stripping in asphalt concrete pavement layers. The following are examples of pavement cores taken in various investigations.

FIGURE 3.35   Example of corrugations in flexible pavement.

FIGURE 3.36   Example of longitudinal cracking in flexible pavement.

Figure 3.41 shows a core of a composite pavement taken at a transverse joint. The core reveals a tight joint with aggregate interlock and little or no deterioration. The asphalt overlay is left intact. However, during the coring operation, the asphalt portion of the core should be inspected for delaminations between paving layers, rutting of any layers, or stripping of the asphalt from the aggregate.

Figure 3.42 shows a core hole in the pavement taken at a midpanel transverse crack. A wealth of information can be obtained by inspection of the core hole. The

FIGURE 3.37   Example of transverse cracking in flexible pavement.

FIGURE 3.38   Example of block cracking in flexible pavement.

FIGURE 3.39   Example of wheel track cracking in flexible pavement.

core hole reveals aggregate interlock to be questionable. A close inspection revealed the reinforcing mesh to be rusted and broken, not cut by the coring operation. Because the core hole indicates most of the aggregate interlock is lost, this crack can be considered a working crack and should be repaired.

Figures 3.43 and 3.44 show the remains of cores taken at transverse joints. It is obvious that these joints need a full-depth repair.

Figure 3.45 shows a core of an asphalt pavement that indicates a delamination approximately 3 in from the surface. Several cores should be taken to verify the extent of the flaw. A delamination found in an asphalt pavement such as this could result in debonding of the surface layer. If warranted, asphalt milling may be required to a depth sufficient to remove the delamination.

3.9 METHODS OF PAVEMENT REHABILITATION

Once a pavement is determined to have unacceptable smoothness or has lost its ability to properly transport goods, it is reasonable to determine the best strategy to return the

FIGURE 3.40   Example of edge cracking in flexible pavement.

pavement to its original intended function. Many of the decisions that define the point where corrective action should be taken are management decisions and can be addressed properly only in a comprehensive study of pavement management. Many considerations must be addressed before determining a list of good rehabilitation options. Leading rehabilitation techniques are reviewed in the following articles.

3.9.1   Rehabilitation of Rigid Pavement

CPR.   The most common method of restoration for jointed pavement, both reinforced and nonreinforced, is termed *concrete pavement restoration (CPR)*. CPR includes joint removal and replacement, construction of rigid shoulders (if not already present), profile grinding to reestablish smoothness, and usually the resealing of the joints. The CPR technique is only used when nondestructive testing measurements indicate that an asphalt overlay is not needed for the future design traffic. The disadvantage of this type of treatment is that, if the joints are not repaired properly, they will fail prematurely. Then, joint repair quantities are difficult to estimate, because joints continue to fail between the time the rehabilitation was designed and the time construction begins.

FIGURE 3.41 Pavement core taken at transverse joint in composite pavement.

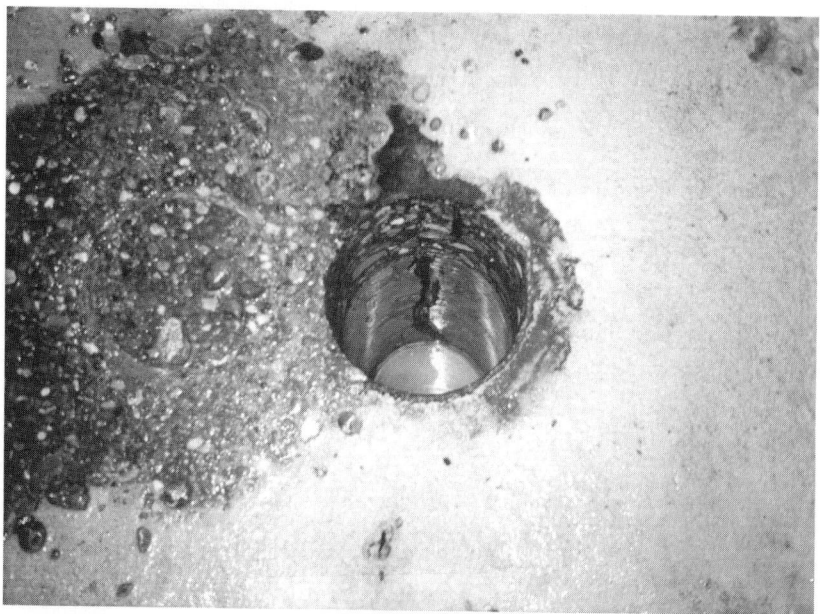

FIGURE 3.42 Hole in pavement after core was drilled at transverse crack in composite pavement.

FIGURE 3.43 Crumbled core taken from transverse joint in rigid pavement.

The advantage of this type of treatment is that it utilizes the strength of existing pavement rather than an overlay, so overhead clearance problems are postponed or eliminated.

Repair and Overlay. When nondestructive testing measurements indicate that the existing slab thickness is insufficient to carry future design traffic, a common technique is to repair failed joints or pavement and add an asphalt overlay. Generally, rigid repairs are preferred over flexible repairs. Flexible repairs in a rigid pavement do nothing to reestablish load transfer across the failed joint. Flexible repairs also have a tendency to heave because they are weak in compression and the rigid pavements need to expand during hot weather. Flexible repairs allow joints to open up beyond the design of the joint sealant, causing the joint sealant to fail. Finally, flexible repairs reduce pressure in a pavement and allow midpanel cracks to open up and lose aggregate interlock. The advantages of flexible repairs are the favorable cost and construction time. Disadvantages of rigid repairs include the construction complexity and time. It is important to realize that the biggest drawback of the repair and overlay strategy is the inability to estimate the amount of repair required at each pavement failure and, for jointed pavements, the number of joints that need repair. Designed overlays are usually thin (3 to 6 in). In cold climates, joints usually reflect through the overlay after one or two winters. Joint reflection cracking can be addressed by sawing and sealing a

FIGURE 3.44 Crumbled core taken from transverse joint in composite pavement.

FIGURE 3.45 Core from flexible pavement indicating delamination about 3 in from surface.

joint in the asphalt overlay at the exact same location as the joint in the underlying rigid pavement. Failure to align the flexible joint with the rigid joint will result in premature joint spalling of the asphalt layer.

Bonded Concrete Overlay.   Another technique to increase pavement structural capacity is to bond additional concrete to the surface of the existing concrete pavement. The required overlay thickness is determined by subtracting the effective thickness, determined by nondestructive testing of the pavement, from the thickness required for a new pavement. Cracks in the underlying pavement will reflect through the overlay. Therefore, all joints and working cracks must be established in the overlay directly over joints and cracks in the existing pavement. For CRC pavement, this is generally not a concern. The existing pavement must be cleaned to ensure a proper bond. This technique is advised only for pavements that are in sound condition with little distress. Any areas showing deterioration must be repaired prior to the overlay.

Break and Seat for JRCP.   The break and seat method for jointed reinforced concrete pavement is accomplished by breaking the long slabs into shorter slabs to distribute the expansion and contraction movement of the pavement over more cracks or joints. This reduces the strains in the asphalt overlay over the cracks or joints to the point where reflective cracking is retarded. The smaller slabs are seated in the subgrade by rolling to reduce vertical deflections. The overlay is designed as a new flexible pavement section with the broken and seated pavement as a base. The broken and seated pavement is given a structural coefficient as determined by nondestructive testing. One disadvantage of this technique is that, to fail the reinforcing steel, tremendous breaking effort is required, and this results in a weak and nonuniform base. Where the reinforcing steel is not failed, large slabs continue to behave as large slabs, causing the joints to reflect through the overlay. Additionally, breaking does not correct problems at joints. Failed joints continue to be weak points in the pavement and usually heave, creating a hump in the overlay. The advantage to this technique is that broken and seated pavements tend to require thick overlays and maintain a high level of serviceability. Additionally, reflective cracking is of low severity when compared with cracking in thin asphalt overlays.

Crack and Seat for PCC.   The crack and seat method for plain concrete pavement (nonreinforced) is accomplished by producing several transverse cracks in each slab, thus transforming the long slabs into shorter slabs to distribute the expansion and contraction movement. This reduces the strains in the asphalt overlay over the joints to the point where reflective cracking is retarded, and the smaller slabs are seated in the subgrade to reduce vertical deflections. By definition, crack and seat produces a crack visible when the pavement is wetted with water. As with break and seat, the overlay is designed as a new flexible pavement section with the cracked and seated pavement as a base, and with a structural coefficient as determined by nondestructive testing. The disadvantage of this method is that the cracking does not correct problems at joints. Joints that have failed continue to be weak points in the pavement and usually heave, creating a hump in the overlay. The advantage of this method is that cracked and seated pavements with thick overlays (7 in or more) exhibit a high level of serviceability, and reflective cracking is of low severity when compared with cracking in thin asphalt overlays.

Rubblize and Roll.   Rubblize and roll is applicable for all types of rigid pavement. This method is accomplished by breaking the existing pavement into 6-in size or less using a resonant beam breaker. The crushed concrete is compacted with a roller and used as a base for a new pavement. The overlay is designed as a new flexible pavement section with the rubblized and rolled pavement as a base. The rubblized pavement is given a structural coefficient based on nondestructive testing. One disadvantage to this technique is that rubblizing weakens the pavement and thereby increases the required overlay thickness. Areas with soft subgrade require removal of the pavement and undercutting; otherwise, the rubblization process cannot be achieved properly. The geometry of the equipment prohibits breaking near portable barriers used for traffic control. Another disadvantage of this technique is that, because the resulting overlay is thick, elevation transitions at bridges require pavement replacement. One advantage of this technique is the complete utilization of the existing pavement as a uniform base without discontinuities. For reinforced concrete pavement, the technique serves to completely debond the steel from the concrete.

Thick Asphalt Overlay with No Repairs.   A thick asphalt overlay with no repairs is a quick and inexpensive rehabilitation strategy that can be used on any rigid pavement beyond economical repair. As the overlay thickness is increased, vertical deflection is decreased as a result of the increased structure. Horizontal movements in the slab are decreased because of lower temperature variations. This decreases the strain at the interface of the overlay and pavement, which retards reflective cracking. The overlay is designed as a new flexible pavement section with the existing pavement as a base. The existing pavement is given a structural coefficient based on deflection testing. A disadvantage of this strategy is that problems at joints are not corrected. Joints that have failed continue to be weak points in the pavement. Another disadvantage is that the thick overlay necessitates pavement replacement to make elevation transitions at bridges. The advantages of this strategy are the low initial cost and ease of construction. Reflective cracking is of low severity when compared with cracking in thin asphalt overlays.

Unbonded Concrete Overlay.   The purpose of breaking the bond between the old pavement and the proposed overlay is to separate the distresses in the old pavement from the new concrete overlay. Thus, the concrete overlay can be treated as a separate pavement, and the existing distressed pavement as a uniform base. There is little benefit derived from repairing the existing pavement prior to placing the overlay, as the bondbreaker will provide uniform support and interface for the concrete overlay. The bondbreaker is placed as a thin (1- to 3-in) asphalt overlay on the existing pavement, and the concrete overlay is placed on the bondbreaker. The thickness required for the concrete overlay can be determined using the following modified version of an equation developed by the Army Corps of Engineers:

$$T = \sqrt{(RT)^2 - (ET)^2} \tag{3.9}$$

where T = the required thickness of the concrete overlay, in

RT = required thickness of new concrete pavement on the existing subgrade and for the anticipated truck loading, in; the existing subgrade strength can be determined from original construction and design records or from nondestructive testing

ET = effective thickness of existing concrete pavement as determined by nondestructive testing, in

This technique is most efficient if the entire width of the roadway is available for overlay at the same time, but this makes maintenance of traffic difficult. However, the strength of the existing pavement is utilized, and the performance can be expected to be similar to that of a new pavement.

3.9.2 Rehabilitation of Flexible Pavement

Asphalt Overlay.   Without question the most common method of rehabilitation for flexible pavement is an asphalt overlay. There are many variations of this technique ranging from pavement planing and a thick asphalt overlay to a thin skin patch placed infrequently along a pavement. The existing condition of the asphalt pavement and the results of nondestructive testing dictate the most economical strategy. The pavement can be designed as a layered system.

Whitetopping.   The construction of a concrete pavement on an existing asphalt pavement is termed *whitetopping*. An asphalt pavement provides an excellent base for a rigid pavement. The concrete pavement is designed as if it were a new pavement constructed on an asphalt base. The AASHTO design procedure can be used to design the concrete pavement, and the strength of existing pavement is utilized. A concrete overlay is an acceptable rehabilitation technique for flexible pavements beyond economical repair. However, construction is difficult unless lane lines are shifted permanently, and the thickness of the overlay makes elevation transitions at bridges difficult.

3.10   LIFE CYCLE COST ANALYSIS OF PAVEMENTS

It is seldom readily apparent which is the most economical rehabilitation method for a particular pavement. Each rehabilitation strategy has unique initial construction costs, performance expectations, and future maintenance needs. What is most economical for one pavement may not be for another. Local costs may differ from one location to another, and material performance expectations may be different from region to region. The only rational way to compare one rehabilitation strategy and another is to perform an economic analysis of the alternative strategies. The method used for such a study is the life cycle cost analysis (see Chap. 10).

It is not good practice to compare a minor pavement rehabilitation strategy and a complete pavement replacement strategy. Even when comparing a new rigid pavement and a new flexible pavement, difficult choices must be made concerning the expected performance of each pavement type. Table 3.27 shows a hypothetical example of life cycle cost analysis assuming a 35-year performance period for both alternatives with no salvage values at the end of the period. It is not the intent to show that one pavement type has an economical advantage over another, as many hypothetical assumptions were made in the example. The intent is to indicate the level of information needed to make a life cycle cost analysis, and the information an analysis presents.

Probably the most important consideration in a life cycle cost analysis is the selection of the discount rate used to evaluate the time value of money. It is sometimes defined as the difference between the market interest rate and the rate of inflation. (Article 10.17.2 provides further discussion on this subject.) Because costs are incurred at different times over the life of a pavement, the discount rate is used to convert these costs occurring at different times to equivalent costs in present dollars. In the example shown in Table 3.27, the discount rate was unrealistically assumed as

TABLE 3.27 Life Cycle Cost Analysis Comparing Rigid (ALT1) and Flexible (ALT2) Pavements

Length: 3.16 mi
Lane number: 5
Lane width: 12 ft   All sections in curb—no shoulders
Lane widths vary; average width of roadway = 60 ft

Item	Dimension, in	Unit	Quantity analysis		Price	Cost analysis	
			ALT1	ALT2		ALT1	ALT2
Main lane							
AC surface course	1.25	CY		3,862	$46.00		$178,000
AC intermed. course	1.75	CY		5,407	$44.00		$238,000
Bituminous base	7	CY		21,628	$39.00		$844,000
Aggregate base	6	CY	18,539	18,539	$18.00	$334,000	$334,000
JRCP	9	SY	111,232		$22.00	$2,447,000	
Asphalt prime coat		SY		44,493	$1.50		$67,000
				Subtotal		$2,781,000	$1,661,000
Future maintenance							
10 years							
Pavement milling	1.50	SY		111,232	$1.20		$133,000
AC surface course	1.25	CY		3,862	$46.00		$178,000
AC intermed. course	1.75	CY		5,407	$44.00		$238,000
20 years							
Pavement milling	3.00	SY		111,232	$1.75		$195,000
AC surface course	1.25	CY		3,862	$46.00		$178,000
AC intermed. course	1.75	CY		5,407	$44.00		$238,000
Bituminous base	3	CY		9,269	$39.00		$362,000
Joint repair, 3%		SY	3,337		$35.00	$117,000	
Pavement sawing		LF	1,430		$1.20	$2,000	
Diamond grinding		SY	111,232		$2.00	$222,000	
Transverse joint reseal		LF	49,101		$1.50	$74,000	
Longitudinal joint reseal		LF	66,739		$1.50	$100,000	

AC = asphalt concrete; JRCP = jointed reinforced concrete pavement; CY = cubic yards; SY = square yards; LF = linear feet.

TABLE 3.27 Life Cycle Cost Analysis Comparing Rigid (ALT1) and Flexible (ALT2) Pavements (*Continued*)

| Item | Dimension, in | Unit | Quantity analysis | | Price | Cost analysis | |
			ALT1	ALT2		ALT1	ALT2
Future maintenance (*cont.*)							
30 years							
Pavement milling	1.50	SY		111,232	$1.20		$133,000
AC surface course	1.25	CY	3,862	3,862	$46.00		$178,000
AC intermed. course	1.75	CY	5,407	5,407	$44.00		$238,000
					Subtotal	$515,000	$2,071,000
					Grand total	$3,296,000	$3,732,000

AC = asphalt concrete; JRCP = jointed reinforced concrete pavement; CY = cubic yards; SY = square yards; LF = linear feet.

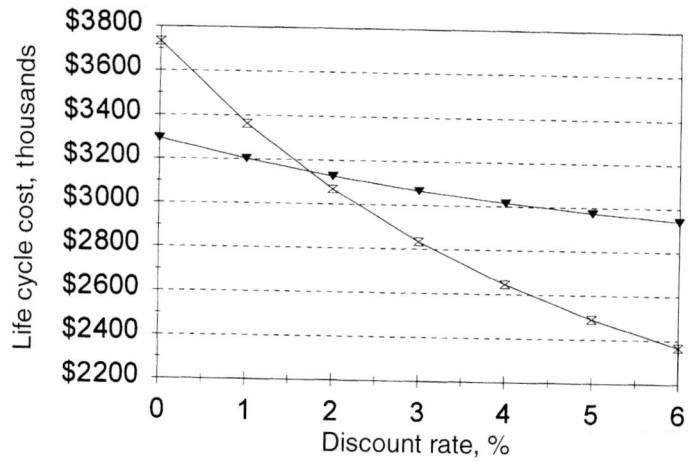

FIGURE 3.46   Sensitivity analysis showing effect of discount rate selection on life cycle cost of pavement alternatives.

zero. Figure 3.46 shows the effect of discount rates from 0 to 6 percent. As is typically the case, the analysis is very sensitive to the discount rate. In this example, the rigid pavement provides the lower life cycle cost when the discount rate is less than about 1.7 percent, and the flexible pavement when the rate is higher. It is apparent that the discount rate must be selected with great care.

REFERENCE MATERIAL

The following reference sources were helpful in developing this chapter.

AASHTO, *Guide for Design of Pavement Structures,* Vol. 2, Washington, D.C., 1986, App. EE.

AASHTO, *Guide for Design of Pavement Structures,* Washington, D.C., 1993.

"AASHTO Design Procedures for New Pavements," FHWA Report HI-94-023, National Highway Information, ERES Consultants, Inc., Washington, D.C., February 1994.

FHWA Technical Advisory T 5040.30, "Concrete Pavement Joints," November 30, 1990.

FHWA Technical Paper 89-04, "Preformed Compression Seals for PPC Pavement Joints," Washington, D.C., September 11, 1989.

FHWA Highway Subdrainage Design Manual, FHWA Report No. TS-80-224, Washington, D.C., August 1980.

Implementation and Revision of Developed Concepts for ODOT Pavement Management Program, Vol. II, *Pavement Conditions Rating Manual,* February 1987, Resource International, Inc., Engineering Consultants, Columbus, Ohio.

Ohio Department of Transportation, *Location and Design Manual,* Vol. 1, *Roadway Design,* Columbus, Ohio, December 1990, revised October 1992.

"Pavement Deflection Analysis," FHWA Report HI-94-021, National Highway Information, Washington, D.C., February 1994.

"Pavement Design for Federal, State and Local Engineers," U.S. Department of Transportation, FHWA, National Highway Information, Contract No. DOTFH 61-86-C-00038, Washington, D.C., 1986.

"Pavement Design: Principles and Practices," FHWA, National Highway Information, ERES Consultants, Inc., Washington, D.C., September 1987.

"SHRP Distress Identification Manual for the Long Term Pavement Performance Project," SHRP-P-338, Strategic Highway Research Program, Washington, D.C., May 1993.

"SHRP-LTPP Interim Guide for Laboratory Material Handling and Testing," NRC Operational Guide No. SHRP-LTPP-OG-004, Strategic Highway Research Program, Washington, D.C., July 1993.

R. F. Carmichael III and E. Stuart, "Predicting Resilient Modulus: A Study to Determine the Mechanical Properties of Subgrade Soils," Transportation Research Record 1043, Transportation Research Board, National Research Council, Washington, D.C. 1985.

CHAPTER 4
BRIDGE ENGINEERING

Walter J. Jestings, P.E.

Bridge Engineer
Parsons Brinkerhoff, Quade & Douglas, Inc.
Atlanta, Georgia

This chapter is directed at practical issues of importance in the design and rehabilitation of traditional bridge types for short and medium spans. Subjects addressed include characteristics of various bridge types, considerations in their selection, and suggestions for economical design; materials for bridges and bridge decks; bridge deck design, construction, and maintenance; deflection and expansion joints; and bridge bearings. The issues are addressed from a general viewpoint, with the emphasis on what is generally done and why. Detailed design methods are available in other publications. (See R. L. Brockenbrough and F. S. Merritt, *Structural Steel Designer's Handbook*, McGraw-Hill, and E. H. Gaylord and C. N. Gaylord, *Structural Engineering Handbook*, McGraw-Hill.)

The author has attempted to confine discussion to his own realm of experience, from his background as a former state engineer of bridges and as a practicing manager of the structural department of a transportation engineering consulting firm. It is the author's hope that the combination of these two perspectives will provide some helpful insights, to engineers employed by agencies and to consultants and those working for consultants alike.

4.1 CLIENT-CONSULTANT RELATIONSHIP

State departments of transportation, bridge and turnpike authorities, and other agencies often require the services of a consulting engineering firm. This may be because the agency chooses not to maintain an engineering staff of its own, because its workload is greater than its staff can handle, or because expertise in special kinds of bridges is needed. Consultants can fill these needs.

Where only routine types of bridges are involved and the agency has an engineering staff, the best that a consultant can be expected to do, usually, is only as good a job as the agency's engineers can do. The agency's staff may include veteran engineers who have become extremely proficient in design of routine and not so routine bridges, and who also know exactly how to prepare plans in the proper format and sheet sequence preferred by the agency, as well as how to use exact pay item descriptions and to refer to pertinent proposal notes and special provisions. A consultant, in this instance, is like a temporary employee who knows the basics but needs to be trained in local procedures.

When a consultant serves a client for many years, however, that consultant can become as proficient as the agency's staff. Long-term contracts for continuing or on-call services eliminate the need to train a new consultant. However, they can be seen as showing favoritism in an environment where other consultants expect an opportunity to compete for contracts. For this reason, and because a long-term contract may allow a consultant to become complacent, the client may limit the term of the contract and, upon expiration, issue a request for proposals (RFP) to perform the services. The original contract holder may or may not be eligible to respond to this RFP, at the agency's discretion.

Consultant's Responsibilities. To serve the client in a professional and efficient manner, the consultant should:

- Deliver the product promised in the contract scope of services
- Deliver the product on time
- Conform to accepted codes and standards
- Develop economical designs
- Use time-tested materials, avoiding purely experimental materials and systems
- Confirm in writing to the client any verbal understandings
- Keep the client informed of project status
- Avoid issues that could involve the client in litigation
- Not make statements to the public or to the media without the client's knowledge and authorization

Client's Responsibilities. Just as the consultant has a responsibility to the client, the client has a responsibility to the consultant. Responsibilities include:

- Executing a contract with the consultant that includes adequate hours to perform the work, recognizing any unique requirements, and not applying standard allowances for nonstandard work.
- Performing reviews in a timely manner.
- Performing reviews either concurrently or sequentially, but not expecting the consultant to make changes required by one branch of the client's office only to be subsequently countermanded by another branch. In other words, the client should transmit consolidated review comments.
- Not interrupting the work unless absolutely necessary.
- Paying invoices in a timely manner. Contracts with subconsultants often stipulate that the subconsultant's invoices will not be paid until payment is received from the prime consultant's client, so a delay in payment from the client results in delay of payment to the subconsultant.
- Being frank with the consultant about any dissatisfaction the client may have with the consultant's performance so that corrective action can be taken immediately.

If the client and consultant meet their respective responsibilities, the relationship will be a partnership that benefits both parties.

4.2 AESTHETIC CONSIDERATIONS

While highway bridges are utilitarian structures, they are visible to the public and therefore should be pleasing to the eye. At the outset of design, one should be conscious of the aesthetic qualities of the structure, lest one end the project saying (after Shelley), "Look on my works, ye Mighty, and despair!"

Some basic guidelines that were adopted by the Ohio Department of Transportation (DOT), Bureau of Bridges and Structural Design, and are included in its *Bridge Design Manual* illustrate the commonsense approach that can be taken to apply this consciousness:

> Aesthetics. Each structure should be evaluated for aesthetics. Normally it is not practical to provide cost premium aesthetic treatments without a specific demand, however careful attention to the details of the structure lines and forms will generally result in a pleasing structure appearance.
>
> Some basic guidelines that should be considered are as follows:
>
> *a.* Avoid mixing structural elements, for example concrete slab and steel beam superstructures or cap and column piers with wall type piers.
>
> *b.* In general, continuous superstructures shall be provided for multiple span bridges. Where intermediate joints cannot be avoided, the depth of spans adjacent to the joints preferably should be the same. Avoid the use of very slender superstructures over massive piers.
>
> *c.* Abrupt changes in beam depth should be avoided where possible. Whenever sudden changes in the depth of the beams in adjacent spans are required, care should be taken in the development of details at the pier.
>
> *d.* The lines of the structure should be simple and without excessive curves and abrupt changes.
>
> *e.* All structures should blend in with their surroundings.
>
> One of the most significant design factors contributing to the aesthetic quality of the structure is unity, consistency, or continuity. These qualities will give the structure an appearance of a design process that was carefully thought out.
>
> The aesthetics of the structure can generally be accomplished within the guidelines of design requiring only minimum special designs and minor project cost increase. As special situations arise preliminary concepts and details should be developed and coordinated with the Bureau of Bridges and Structural Design.

Some states have adopted, in principle if not in writing, a similar philosophy in regard to aesthetics of their bridges. California, for example, is known and admired for applying some degree of architectural attention to all of its bridges. Some agencies, however, seem to neglect aesthetics, particularly in regard to the very visible piers of grade-separation bridges. Here the primary objectives seem to be standardization of shape to facilitate computer design, and emphasis on straight, flat lines to obtain minimum cost of forming. These objectives are achieved at a price—ungraceful substructures not in keeping with the lines of the superstructure.

4.3 BRIDGE DESIGN SPECIFICATIONS

AASHTO Specifications.   For many years, the basic manual for design of highway bridges has been the *Standard Specifications for Highway Bridges* adopted by the American Association of State Highway and Transportation Officials (AASHTO). These specifications permit use of either allowable stress design or load factor design.

In 1994, however, AASHTO published a completely new alternative volume, *LRFD Bridge Design Specifications*. Based on the *load and resistance factor design method,* the *LRFD Specifications* represent a major step in improved bridge design and analysis methods. It is anticipated that usage of the new specifications will lead to bridges with improved serviceability, long-term maintainability, and more uniform levels of safety. The new volume resulted from a five-year research effort conducted under AASHTO's National Cooperative Highway Research Program. Independent consultants, technical representatives from various industries, AASHTO members, and other engineers participated in the effort to develop a draft document. Then the provisions were tested in trial designs at fourteen AASHTO member departments before final specifications were adopted. It is expected that the *LRFD Specifications* will become the mandatory AASHTO document in the near future. One of the most useful features included is a detailed commentary that explains the specification provisions and gives references for further study.

AASHTO specifications are developed under the direction of the AASHTO Highway Subcommittee on Bridges and Structures. This subcommittee consists of all bridge engineers of states of the United States and of Canadian provinces and officials of selected turnpike and bridge authorities. The specification development process is a deliberate one. Nevertheless, changes are made on a regular basis (some would say too frequently for the average bridge designer to stay abreast of them). Between new editions, revisions are published under the title of *Interim Specifications.* When identifying the AASHTO specifications used for design on plans, some states refer to "AASHTO *Standard Specifications for Highway Bridges,* Current Edition." A better practice is to refer to the specific edition, by number and year, along with any interims that were in effect at the time of design.

Unless there is a cogent reason for not meeting the minimum requirements of the AASHTO Specifications, engineers designing bridges where they are in effect should apply and conform to them. Any exceptions should be noted on the plans. In case of litigation, one would have to explain why these recognized standards were not met.

Other AASHTO Specifications. A recent edition of *AASHTO Publications* listed over two dozen specifications, manuals, and commentaries related to bridges and structures. A bridge designer should be aware of the availability of these publications and should use them where applicable.

In this chapter, references to the "AASHTO Specifications" or "AASHTO" will be to the AASHTO *Standard Specifications for Highway Bridges* unless otherwise noted.

Bridge Design Manuals. Many state departments of transportation publish bridge design manuals, which they develop for guidance of their own staff and consultants. States that do not have manuals often publish design memoranda. Before starting a bridge design project, a consultant should determine which of these aids are available, acquire and become familiar with them, and apply them in designing and preparing plans. Some state bridge design manuals are quite explicit, and are almost textbooks on bridge design.

4.4 BRIDGE GEOMETRICS

Bridge Width. Roadway width on bridges is the inside measurement to the bottom of the sidewalk curb or the bottom of the safety barrier. For bridges on roads where sidewalks are not provided, the bridge width is made equal to the approach roadway width including shoulders, so that the bottom of the barrier curb or the near face of the railing is aligned with the face of the barrier rail at the outside edge of the shoulder.

In the past, policy did not always permit full shoulders to be accommodated on bridges. Often the roadway was made narrower, particularly on longer bridges. This was done strictly to reduce bridge cost. From the traffic operations standpoint, however, it was an unwise practice. Disabled vehicles could not find refuge on the shoulder, and a full shoulder was not available for temporary maintenance of traffic during road rehabilitation or repaving. It is now recognized that a bridge is an integral part of a highway system when it comes to roadway width.

Bridge Horizontal Clearance. For bridges over streams, the location of substructure units, and therefore the length of spans, is controlled by hydraulic requirements and by navigation clearance requirements established by agencies such as the U.S. Coast Guard and the U.S. Army Corps of Engineers.

For crossings of highways, the bridge columns or pier walls should clear the traveled way, shoulders, ditches where required, barrier rail, and any additional width required to provide a safe clear zone from edge of pavement. A minimum clearance of 30 ft from edge of pavement is required except where this clearance is impractical, in which case the pier or wall may be placed closer to the edge of pavement, with barrier rail 2'-0" minimum from edge of shoulder, and pier or wall 2'-0" minimum from face of barrier rail. The barrier rail offset from face of pier or wall will be further controlled by the dynamic deflection of the particular system used. (See Chap. 6 for additional information.)

For crossings over railroads, the horizontal clearance requirements are usually set by the railroad company or by the state public utilities commission. In addition to clearance for safe operation of trains, including allowance for accidentally overhanging cargo, railroad companies are cognizant of the importance of trackside drainage and require that drainage ditches be accommodated where present. In addition, a maintenance roadway for off-track equipment is often required. A horizontal clearance of 25 ft from the centerline of the track is desirable and will obviate the need for pier crash walls.

If a pier adjacent to a railroad track is located closer than what is considered to be an adequate distance to prevent derailed cars from striking the pier (generally 25 ft from centerline of railroad track), the pier is required to be of heavy construction, or a substantial crash wall is required to be constructed to protect the pier and prevent catastrophic collapse of the bridge. This wall should be aligned with the pier. For additional details, refer to the American Railway Engineering Association (AREA) *Manual for Railroad Engineering.*

Bridge Vertical Clearance. Generally, a clearance of 16 ft plus an allowance for resurfacing should be provided over major state, U.S., and interstate highways, over the entire width of roadway. Over less important highways, a clearance of 14 ft should be provided. These are AASHTO requirements. Published state standards, if different from AASHTO, should be followed.

The above vertical clearances apply to vehicular bridges. Because pedestrian bridges are narrower and lighter in weight, and therefore more vulnerable to major damage or collapse in the event of collision from overheight vehicles passing under the bridge, states are beginning to require an additional clearance of 1 ft for pedestrian bridges. This additional clearance is also recommended for overhead sign structures.

Vertical clearance requirements over railroads, like horizontal clearances, are set by the railroad company or state public utilities commission. A minimum clearance of 23 ft above high rail is common for new bridges. Widened or rehabilitated bridges will generally be allowed to maintain the existing clearance, but no less.

4.5 BASIC BRIDGE MATERIALS

The basic materials most often used to construct bridges are concrete and steel. Timber is occasionally used for deck construction and sometimes for short-span bridges.

4.5.1 Concrete

High strength is desirable for bridge concrete to reduce member size and weight, but durability is equally or more important. Component materials must be compatible with each other, and the concrete must have low permeability.

A long-term destroyer of concrete from within is alkali-silica reaction. While material specifications for concrete have been developed to preclude use of cement and aggregates that will produce alkali-silica reaction, the best prevention of this problem is the use of cements and aggregates from sources that have a known history of absence of this problem.

Given that required strength can be obtained by mix design with relative ease, low permeability becomes one of the most desirable properties, because bridge concrete is reinforced or prestressed and prevention of corrosion of the embedded steel is essential for long-term durability. Reduced permeability will also reduce carbonation, alkali-silica reaction, and freeze-thaw damage. Permeability can be reduced by proper mix design, by maintenance of a low water/cement ratio during concrete placement, by use of admixtures, by compactive effort, by use of specialty concretes, or by application of concrete sealers or coatings. Often a combination of these procedures is employed.

Another desirable quality for the prevention of reinforcing steel corrosion, along with low permeability, is resistance to cracking. The cracking of bridge decks and of bridge deck overlays has been a persistent problem. The use of shrinkage-compensating concrete using type K cement has been found to be effective in reduction of cracking, and some agencies mandate its use in construction of bridge decks. One property of shrinkage-compensating concrete that is different from regular concrete is the need for adequate amounts of mix water to cause the chemical reaction necessary for the development of the expansion. The normal rules of low water/cement ratio do not apply and must not be enforced. Another difference is that bleed water cannot be expected to appear. Waiting for bleed water to appear will result in the start of concrete hardening, making finishing very difficult. Agencies that have adopted shrinkage-compensating concrete have also adopted strict specifications for the production, placement, and curing of the concrete. These specifications address the requirements peculiar to shrinkage-compensating concrete, but also include many requirements that are applicable to normal concrete as well. Perhaps the most important requirement is that prior to placement of the deck concrete, a preconstruction meeting be held, and that *all* participants in the cement manufacturing and concrete mixing, delivery, placement, finishing, curing, and inspection be *required* to take part. This meeting gives the type K cement manufacturer the opportunity to instruct the other participants in special requirements, and to correct any misconceptions that exist. Such meetings in themselves go a long way toward improving the quality of the concrete.

Admixtures.   Various admixtures are available to enhance the properties of concrete made with the basic ingredients: coarse aggregate, fine aggregate, portland cement, and water.

Air-entraining admixtures produce a distribution of bubbles that become permanent tiny voids in the concrete. This system of voids makes the concrete resistant to scaling, a surface failure that became frequent when deicing salts came into use. Air entrainment has virtually eliminated scaling. The use of air entrainment is recommended even with high-strength concrete.

Water-reducing admixtures make concrete mixtures workable at a lower water/cement ratio than is possible with use of "water of convenience" (water in excess of that required for hydration of the cement) alone. High-range water reducers provide great workability at very low water/cement ratios, and have been developed to provide reasonable control of duration of extra fluidity.

Dense concrete is a concrete developed by the Iowa Department of Transportation for overlayment of new and existing concrete bridge decks. It has a low water/cement ratio and relies on special compactive effort imparted by vibrating screeds to produce a dense concrete with reduced permeability.

Latex-modified concrete uses an admixture of latex, generally liquid styrene butadiene with a minimum solids content of 40 percent. It achieves a reduced permeability equivalent to dense concrete at a lesser thickness. This quality is important to the viability of latex-modified concrete as an alternative option to dense concrete because the cost of latex-modified concrete is higher on a volume unit measure basis.

Silica fume concrete, or microsilica concrete, incorporates extremely fine particles of microsilica. Added to concrete in powder or liquid form, it densifies the concrete, increases strength, and reduces permeability. Where silica fume concrete has been allowed as a contractor's alternative option to latex-modified or dense concrete, it has rapidly replaced these other specialty concretes.

Calcium nitrite concrete contains calcium nitrite, a widely used inorganic corrosion inhibitor that acts at the surface of the steel reinforcement to limit the electrochemical reaction involved in the corrosion process. The calcium nitrite is added in liquid form to the concrete at a rate of 3 to 5 gal/yd^3, depending on the quality of the concrete, the level of chlorides expected, and the life required for the concrete. (See *Manual for Corrosion Protection of Concrete Components in Bridges,* Task Force 32 Report, February 19, 1992, AASHTO, Washington, D.C.)

Lightweight Concrete. Although the concrete most often used in bridge construction is normal-weight (hardrock) concrete having a unit weight, reinforced, of 150 lb/ft^3, lightweight concrete can be produced from manufactured aggregate that is available from several sources around the United States. The coarse aggregate is produced by heating shale in a kiln, which expands it. Lightweight fine aggregate can also be produced but is not recommended for bridge concrete. Use of lightweight coarse aggregate can reduce the weight of reinforced concrete to 115 lb/ft^3. While the history of lightweight concrete for bridges includes premature failures, it also includes successful applications in both deck slabs and beams. It is important to recognize the different (greater) creep characteristics of lightweight concrete in structures where long-term deflections are a significant design factor.

4.5.2 Structural Steel

Steel for bridges is available in five different strength levels, each of which may be specified under ASTM A709, Standard Specification for Structural Steel for Bridges. The grade designations for the five strength levels are indicated in Table 4.1, as well as alternative specifications that may be more familiar. The grade designation

TABLE 4.1 Steels for Bridges

Type of steel	ASTM designations		AASHTO designations
	Bridge specification*	Structural steel specification	
Structural carbon	A709 grade 36	A36	M270 grade 36
High-strength, low-alloy	A709 grade 50	A572 grade 50	M270 grade 50
High-strength, low-alloy	A709 grade 50W	A588	M270 grade 50W
Quenched and tempered high-strength, low-alloy	A709 grade 70W	A852	M270 grade 70W
Quenched and tempered, high-strength alloy	A709 grade 100/ A709 grade 100W	A514	M270 grade 100/ M270 grade 100W

*When the supplementary requirements of A709 are specified, the steel exceeds the requirements of the listed structural steel specification. The supplementary requirements include toughness testing, grain size, and frequency of tension tests.

indicates the specified minimum yield stress in kips per square inch, and a "W" indicates that it is a weathering steel composition. ASTM A709 contains supplementary requirements for notch toughness and other items that are available but apply only when specified by the purchaser. When such supplementary requirements are specified, they exceed the requirements of the basic specifications such as A36 or A572.

Grades 36, 50, and 50W are available either as structural shapes or as plates. The other grades are available only as plates. Grades 36, 50, and 50W are the most frequently used materials. In general, compared with A36 steel, where other limitations such as deflection or stiffness do not override, the extra unit cost of the higher-strength grades (50 or 50W) is more than offset by the higher yield strength. Grades 70W and 100/100W have proven economical in some applications. Generally, these are longer-span structures, or the higher-stressed portions of medium-span structures.

Weathering grades (50W, 70W, and 100W) have chemical compositions that provide enhanced resistance to atmospheric corrosion. They can be used in the bare (unpainted) condition for bridges in many cases (see Art. 4.13). The savings on cost of painting and repainting frequently makes them an economical choice.

In reference to the unit price of grade 36 steel, the relative material price of the other steels in plate grades is approximately as follows:

Grade	Price relative to A36
36	1.00
50	1.12
50W	1.23
70W	1.52
100W	2.07

As indicated, these are only price factors and do not consider the reduced quantity of steel that may be required as the yield strength increases. For structural shapes, grade 50 steel can usually be obtained for about the same price as grade 36 steel, but there would usually be some additional cost for grade 50W.

The cost of fabrication and erection for members of grade 36 and grade 50 or 50W steel is approximately the same. Thus, in preliminary cost studies, only the cost of the mill material for the members selected need be compared. Fabrication costs for grade 70W, grade 100, and grade 100W tend to be higher than those of the as-rolled products, and thus, the cost comparisons must include those costs.

Steels with greater strength than grade 36 tend to be economical for beams and girders in many cases, and are particularly attractive under the following conditions:

- When dead load is a major part of total load
- When deflection limits do not control
- When deflections can be reduced (composite design, continuous structure, etc.)
- When weight reduction cuts cost of foundations, shipping, etc.
- When selection avoids use of built-up members (cover plates, fabricated girder versus rolled beam, etc.)

The higher-strength steels often show advantage for tension members of trusses because the higher strength is used more effectively (for the entire depth of the member, because there is no stress gradient). The same is true for compression members of trusses where the member slenderness ratio is small to moderate (ratio of length to radius of gyration of about 80 or less, depending on grade).

4.5.3 Other Materials

Aluminum.   A few bridges, including highway plate girder bridges and arch-type pedestrian bridges, have been constructed of aluminum. These bridges have generally performed well and have not required much maintenance. The plate girder bridges do not seem to have experienced problems that one might anticipate due to the difference in thermal coefficient between the aluminum girders and the concrete deck. The main reason aluminum bridges have not captured a larger share of the market is high cost. Lack of availability of appropriate design specifications for aluminum bridges has also been a factor, but this has been alleviated by the publication, in 1991, of the AASHTO *Guide Specifications for Aluminum Highway Bridges.*

Aluminum railings, while not having the strength or ductility of steel, do not require maintenance painting. Aluminum posts are cast, and aluminum rail elements are extruded in shapes that are convenient for bolted assembly of the railing. Bolts to anchor aluminum railings in concrete parapets are generally stainless steel.

Rubber.   Rubber, sometimes natural but more often synthetic, is used in bridge bearings and expansion joint sealing devices. Reinforced rubber sheets are used to fabricate troughs to conduct storm water that is permitted to flow through open expansion joints.

Stone.   Stone is used in some states to face barriers and to provide waterline protection of piers. It is sometimes also used for aesthetic reasons.

4.6   BRIDGE DECK MATERIALS AND SYSTEMS

Bridge decks can be constructed of timber, concrete, or steel.

Timber Decks.   For bridges on unpaved roads or low-volume roads in rural environments, timber decks of modern construction, such as Glulam (glued laminated) decks, can be serviceable and durable. For high-traffic-volume highways, timber is at a great disadvantage because of high cost, difficulty of fitting to a variable support profile, lack of skid resistance if a separate wearing surface is not provided, and difficulty in maintaining adhesion of an asphalt wearing surface. The choice then is usually between concrete and steel, or a combination of both.

Cast-in-Place Concrete Decks.   Where light weight or speed of construction is not prerequisite, cast-in-place concrete decks prevail because they easily conform to the top of the supporting superstructure and to the required surface profile. Cast-in-place concrete decks also easily accommodate concrete sidewalks, median barriers, and outside safety barriers.

The durability of concrete decks became a matter of great concern after the use of roadway deicing salts became prevalent and decks began to develop spalls at an early age. Extensive investigation and development have resulted in adoption of improved design and construction requirements that show promise of extending deck life by preventing premature corrosion of reinforcing steel. Life-extending measures include greater design cover over the top reinforcing bars, tighter control of actual construction tolerances, use of lower water/cement ratio concrete, use of admixtures and special concretes such as silica fume concrete to reduce permeability, imposition of stricter curing requirements, use of epoxy-coated reinforcing bars, application of various types of waterproofing surface sealants, installation of membranes or protective overlays, and installation of cathodic protection.

Precast Concrete Deck Units.   For rehabilitation of existing bridges requiring deck replacement, the use of precast prestressed-concrete deck units placed transversely across existing beams permits deck replacement at night or during other hours of reduced traffic volume. Only as much of the existing deck slab as can be replaced in a work shift is removed, and the gap between the remaining slab and the new deck units is minimized so that it can be bridged by a steel plate to maintain traffic. The deck units can be fastened by welding studs to the beams through formed holes in the deck unit, and by filling the holes with a fast-setting concrete. Adjusting devices can be built into the deck units to control the deck profile. Longitudinal posttensioning can be used to ensure a tight deck at the grouted joints. The deck units can cantilever beyond the outside beams and have provision for barrier placement.

Steel Decks.   Sometimes the weight of the deck needs to be minimized. This is true when replacing the deck on an existing superstructure of limited strength. It is also important on movable bridge spans where every pound reduced in the movable span is accompanied by a similar reduction of counterweight. In these cases, steel decks can effectively reduce weight. The lightest-weight decks have been open steel-grid decks, but these decks often have an unpleasant riding quality when new, and can become slippery and unsafe with wear. Skid resistance can be restored by grinding grooves on the riding surface, which weakens the grid, or by welding studs to the surface, but the unpleasantness of the sound and overall sensation perceived by the traveling public remains. Another type of damage to which open-grid decks are vulnerable is breakage of bars when chains, dragged from passing vehicles such as car haulers, become lodged in the grid openings. Open-grid decks are also prone to fatigue failure at the welds. The location and nature of the welds create a severe condition for fatigue susceptibility. For all the reasons above, open steel-grid decks are falling from favor.

Concrete-filled, partially filled, and overfilled steel-grid decks (now referred to as *grid reinforced-concrete bridge decks*) are also available and have often provided many years of service under heavy traffic with minimum maintenance. Where the concrete is filled only to the surface of the steel grid, wear of the concrete between the grid members, called *cupping,* can result in an unpleasant and unsafe riding condition. Therefore, overfilling is recommended.

In a few cases, concrete-filled steel-grid decks have been known to "grow," breaking welds to the supporting members. After extensive testing and analysis, the cause was determined to be corrosion on the vertical interfaces between the steel and the concrete fill. Although a very small expansion occurred at each interface, the accumulated expansion measured several inches at the ends of units. This phenomenon emphasizes the importance of preventing corrosion by any suitable means. In this case, the use of a corrosion-inhibiting concrete admixture seems appropriate

An attribute of the steel-grid deck, whether filled or open, is that it is or can be fabricated off-site, complete with concrete fill and wearing surface if necessary. This can be advantageous for speedy redecking where downtime must be held to a minimum, and may be a reason for selecting this type of deck even if weight reduction is not necessary.

A more recent (1980) variation of the concrete-filled steel-grid deck is the patented "exodermic" deck, where a thin reinforced-concrete slab is constructed on top of and made composite with the steel grid.

Another type of steel deck is the orthotropic deck, where the steel plate that supports traffic, and its stiffeners, are a part of the longitudinal load-carrying member of the bridge. Some of these decks have experienced problems with wearing surface adhesion, but the main reason they are not used more extensively is their high cost of fabrication.

Corrugated-Steel Bridge Flooring. Corrugated-steel bridge flooring, like stay-in-place steel forms but thicker (up to $\frac{3}{8}$ in thick), can be used on bridges such as existing truss bridges where the tops of the stringers are at the same level transversely. The planks are usually galvanized. They extend the full width of the roadway but are narrow, and so can be erected without cranes. The planks are fastened to the stringer flanges by bolted clips or by plug welding in holes over the stringers, thus permitting installation by the owner's forces. The deck is then paved with asphalt concrete. The valleys are filled first, and then the entire deck is overlaid, building in crown if necessary. To promote longevity of the plank and wearing surface, drainage holes are placed in the valleys of the plank. However, leakage of salt-laden water can corrode supporting stringers. Measures can be taken to prevent leakage, including seal-welding the seams, and eliminating the drain holes and waterproofing the entire plank surface before paving, but these measures can make this floor system costly.

4.7 CONCRETE BRIDGE DECK DESIGN

AASHTO specifications for design of concrete bridge deck slabs on longitudinal beams are based on distribution of loads in the slab according to Westergaard theory and assume flexural action of the slab. On the basis of these specifications, many states have developed design tables and charts for quick determination of slab thickness and both primary (transverse) and secondary (longitudinal) reinforcement.

The main variables in the design of the deck slab are:

- Beam spacing
- Concrete strength
- Weight allowance for future paving
- Live load (generally HS 20 or HS 25)
- Continuity factor for dead load

Applying the specifications, the simple dead and live load moments per unit width of slab are calculated. Dead load includes the client-specified future paving allowance, weight of any separate wearing surface, and weight of the deck slab including any monolithic wearing surface. Live load is the wheel load(s) of the client-specified HS truck loading. The simple span moments are calculated for the design slab span length, and are then modified for continuity over the beams. For this factor, most states use 0.8 for both dead and live load, but some states use 1.0 for dead load. The moments are factored, and the slab is designed by the strength method using the slab thickness minus any monolithic wearing surface considered subject to loss due to traffic wear. Effective depth dimension d from the compressive face is usually different for the top and bottom steel, because a minimum cover of 1 in is permissible and generally adequate for the bottom of the slab but much greater cover, up to 3 in, is specified for the top steel to provide protection against intrusion of chlorides. The rebar diameter is usually different as well, since most agencies maintain the same spacing of top and bottom steel and vary the bar size. A uniform spacing makes bar placement and inspection easier and facilitates concrete placement. Secondary steel is provided in accordance with the specifications, with a lesser amount in the outer quarters compared with that in the middle of the distance between beams.

Slab overhang beyond outside beams is limited so that the reinforcement furnished for interior panels is adequate for the overhang, or extra reinforcement is provided if required. Slab overhang is sometimes also limited for construction reasons; the weight of fresh concrete on an excessive overhang, acting through a diagonal brace, can cause local buckling of unstiffened steel girder webs or can damage the web of a prestressed-concrete girder.

The Ohio DOT Bridge Bureau's deck slab design procedures based on the AASHTO load factor design method are shown in Figs. 4.1 and 4.2. The detailed calculations shown for the example in Fig. 4.1 can be omitted by using the chart given with the example in Fig. 4.2. This chart reflects many of the variables discussed above. Note that one exception from usual practice is that the top distribution reinforcement is placed above rather than below the primary steel. This practice was adopted in recognition of the fact that most deck slab cracking is transverse, and the distribution steel is more effective in resisting that cracking if placed closest to the surface.

Some states continue to use deck slab design tables developed using the allowable-stress design method. In addition, some states assign an allowable concrete stress that is less than AASHTO specifications would allow on the basis of the required 28-day strength of the specified concrete. These conservative practices reflect the prevalent attitude gained from a common experience of premature and extensive bridge deck deterioration, mostly in the form of spalling due to reinforcing bar corrosion. If the preventive measures now being taken prove to be effective in eliminating or greatly reducing this premature deterioration, those states will be more inclined to adopt less conservative design methods.

The design procedures described above have resulted in safe designs. However, research has determined that significant membrane action is present in interior panels,

Sample problem : Using load factor design procedures determine the slab thickness and main reinforcement for a deck slab with an 9'-6" stringer spacing and an HS20-44 loading.

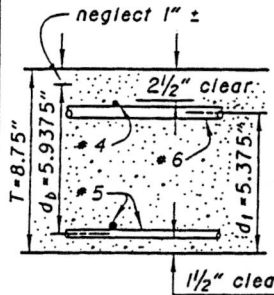

neglect 1" ±

$S = 9'\text{-}6''$ minus $6'' = 9'\text{-}0''$
$T_{min} = (17+S)/(36) = 0.72'$ or $8\frac{5}{8}''$ min. . use $8\frac{3}{4}''$
$f'_c = 4500$ psi
$f_y = 60,000$ psi
$\phi = 0.9$ (8.16.1.2.2)⊕
$Z = 130$ (top steel), 170 (bottom) (8.16.8.4)⊕
$n = 8$

Dead load W :
    Slab = (0.73)(0.15) = 0.110 ⎱ 0.170 kip/ft
    FWS = 60 p.s.f. = 0.060   ⎰

Design Moments :
DLM = $(0.125)(W)(S^2) \times (0.8) = (.125)(.170)(9.0^2)(.8) = 1.38$ ft-k   (3.24.3.1)⊕
LLM = $(S+2)(16)(1.3)(0.8)/32 = (11)(16)(1.3)(.8)/32 = 5.72$ ft-k   (3.24.3.1)⊕
Mu = $1.3[DL+1.67(LL)] = 1.3[1.38+1.67(5.72)] = 14.21$ ft-k   (3.22)⊕
Mw = Service load moment = DLM + LLM = 7.10 ft-k

Top Reinforcement	Bottom Reinforcement

$R = Mu/\phi bd^2$
  = $(14.21)(1000)/(.9)(1)(5.375^2)$
  = 547 psi
$\rho = 0.00987$, from chart
As = $(.00987)(12)(5.375) = 0.64$ In²/ft
 	
Try #6 bars at 8.25" In (As=0.64)
	$R = (14.21)(1000)/(.9)(1)(5.938^2)$
	  = 448 psi
	$\rho = 0.00796$, from chart
	As = $(0.00796)(12)(5.938) = 0.57²$In/ft
	
	Try #5 bars at 6½" In (As=0.57)

Check steel spacing (8.16.8.4)⊕	
f_s (all.) = $z/(d_c A)^{1/3}$	

f_s (all.) = $130/[(3.375^2)(2)(8.25)]^{1/3}$
  = 22.70 or 36.0 max.
	f_s (all.) = $170/[(1.81^2)(2)(6.5)]^{1/3}$
	  = 48.68 or 36 ksi

f_s (act.) = Mw/As Jd	

f_s (act.) = (7.10)(12)/(0.64)(0.89)(5.375)
  = 27.83 ksi (no good)
Reduce #6 bar spacing
As = 27.83(0.64)/22.70 = 0.79 In²/ft
Spaced @ 6.67"
	f_s (act.) = (7.10)(12)/(0.57)(0.91)(5.938)
	  = 27.66 ksi (o.k.)
	
	Spaced @ 6.5"

✿ Use #6 bars @ 6½" c/c (As=0.81)	✿ Use #5 bars @ 6½" c/c (As=0.57)

⊕ AASHTO Bridge Specifications   ✿ Top and bottom bars shall coincide.

FIGURE 4.1 Example of transverse slab design for HS 20 loading using calculations based on AASHTO load factor design. (*From* Bridge Design Manual, *Ohio DOT, Columbus, 1993, with permission.*)

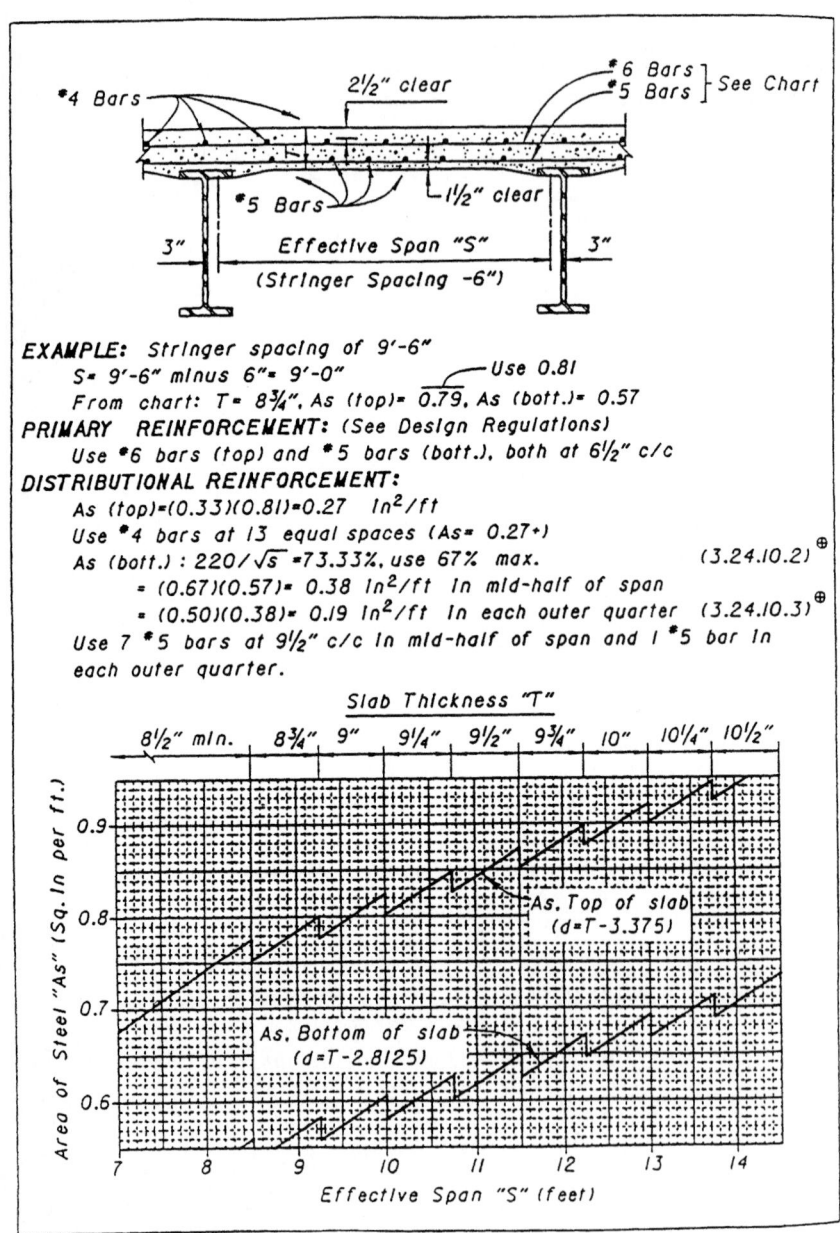

EXAMPLE: Stringer spacing of 9'-6"
   S = 9'-6" minus 6" = 9'-0"   — Use 0.81
   From chart: T = 8¾", As (top) = 0.79, As (bott.) = 0.57
PRIMARY REINFORCEMENT: (See Design Regulations)
   Use #6 bars (top) and #5 bars (bott.), both at 6½" c/c
DISTRIBUTIONAL REINFORCEMENT:
   As (top) = (0.33)(0.81) = 0.27 In²/ft
   Use #4 bars at 13 equal spaces (As = 0.27+)
   As (bott.) : 220/√s = 73.33%, use 67% max.   (3.24.10.2)⊕
      = (0.67)(0.57) = 0.38 In²/ft in mid-half of span
      = (0.50)(0.38) = 0.19 In²/ft in each outer quarter (3.24.10.3)⊕
   Use 7 #5 bars at 9½" c/c in mid-half of span and 1 #5 bar in
   each outer quarter.

⊕ *AASHTO Bridge Specification*

FIGURE 4.2 Example of transverse slab design for HS 20 loading using chart based on AASHTO load factor design. (*From* Bridge Design Manual, *Ohio DOT, Columbus, 1993, with permission.*)

and actual stresses are considerably lower than design stresses calculated on the basis of flexural action. Following laboratory testing, the province of Ontario and several states have constructed and tested full-scale bridges with so-called orthotropic deck slab reinforcement. In these designs, the reinforcement is the same size and spacing in both directions, and of a reduced total amount compared with designs by current AASHTO specifications. These experimental decks have performed well, in most cases.

A further experimental advancement of this concept is the complete elimination of top-slab reinforcement, except for longitudinal continuity reinforcement over piers. Load testing was conducted on an actual full-scale four-span continuous prestressed concrete girder bridge in Colorado, on which the deck slab in one span was built in this manner, i.e., without top reinforcement except continuity reinforcement in the composite slab. The bridge was test-loaded with a vehicle having five axles, with maximum axle loads of 28 kip. The research report abstract states, "...It was found that the peak transverse tensile strains developed in the top of the deck were less than 30% of the cracking strain....This study confirms the feasibility of eliminating most of the top reinforcement in bridge decks."* Additional testing with repeated heavier loads, combined with environmental effects, will no doubt be required before this major change in deck slab construction practice is adopted. The great potential saving in bridge deck construction cost that can be realized if the major source of distress—spalling due to rebar corrosion—can be eliminated by eliminating the steel itself warrants further experimentation and development.

Yet another experimental approach to concrete deck slab design has been taken at Canada's Technical University of Nova Scotia, where a 2-m span concrete slab was reinforced with polypropylene fibers to control shrinkage cracking, and with greatly reduced steel reinforcement. The design takes advantage of the slab's arching action. A flat steel bar was welded to the undersides of the girder top flanges to provide the reaction to the arch thrust. The slab failed in static punching shear at 94 tons. In rolling-load tests of an identical slab at Carleton University, Ottawa, after 4 million passes of a 12-ton load, the deck showed no distress.

4.8 CONCRETE BRIDGE DECK CONSTRUCTION

During construction, bridge deck concrete can be supported by reusable wood forms, permanent stay-in-place steel forms, or precast prestressed-concrete planks. Where permitted, contractors will generally use stay-in-place steel forms rather than removable wood forms. Allowance must be made in the design of the bridge for the extra weight of the steel forms, and for extra concrete where required. The forms are corrugated. Where the bottom transverse bar spacing can be made the same as the pitch of the corrugated form, the extra concrete in the valley below the nominal bottom of the slab line is compensated for by the concrete displaced by the peak of the corrugation above the bottom of the slab line, and the allowance can be for the weight of the forms

---

*"A Case Study of Concrete Deck Behavior in a Four-Span Prestressed Girder Bridge: Correlation of Field Test Results and Numerical Results," by Li Cao, John H. Allen, and P. Benson Shing, University of Colorado; and Dave Woodham, Colorado DOT; Report to the Sponsor, Colorado DOT, April 1994.

only. If the spacing is different from the pitch, a greater allowance will be required because extra concrete must provide the necessary cover. For long slab spans, the stay-in place forms are corrugated, but with a flat top plate. In this case, no extra concrete is required, and the extra weight allowance is for the forms only.

Prestressed concrete planks can also be used as support forms. In this case, the planks also serve as a component of the structural slab. Some agencies have used prestressed planks for years with success, but others have experienced problems—particularly, longitudinal cracking through the cast-in-place top slab over the ends of the planks at the supporting beams—and have discontinued their use.

When stay-in-place steel forms or prestressed-concrete planks are used, the slab overhang beyond the outside beam is generally formed separately using conventional removable wood forms.

4.9 CONCRETE BRIDGE DECK PROTECTION

Concrete bridge decks designed by the current AASHTO method described above have large amounts of reinforcing steel. (Some say it is enough to drive on, and that the concrete is provided just to make the ride smoother!) In the past, in areas where deicing salts are used, corrosion of the top steel caused extensive spalling that led to premature repair or replacement of many decks. In some coastal areas, saltwater spray on the bottom of deck slabs has caused similar corrosion of the bottom reinforcing steel.

For the foreseeable future, concrete bridge decks will continue to be reinforced with steel bars, even though revised design procedures may be adopted that permit lesser amounts. Therefore, it will continue to be necessary to protect those bars against corrosion. Reinforcing bar corrosion can be prevented or forestalled by a number of means, including:

- Making concrete more resistant to penetration of chlorides (less permeable)
- Preventing chlorides from penetrating the concrete by applying concrete sealers or waterproofing membranes
- Applying a physical coating to the bars to prevent contact between the chlorides and the bars
- Adding a corrosion-inhibiting admixture to the concrete mix
- Installing cathodic protection

Concrete Permeability. As discussed above, improvements have been made in the quality of concrete, and in the development of special concretes, in an effort to reduce the amount of chlorides reaching the reinforcing steel. These improvements themselves may be adequate to prevent premature corrosion, especially in areas where the application of deicing salt is moderate. In areas of greater rates of salt application, it may be necessary to provide supplementary protection of the types listed above.

Concrete Sealers. Sealers are available that can reduce the permeability of hardened concrete. Forms of silanes and siloxanes are among the best sealers. In some cases, however, the field performance of concrete sealers has not lived up to expectations based on laboratory testing. When selecting a sealer, one should avail oneself of the most current field evaluations of effectiveness over a reasonable period of time, and not rely solely on the claims of the manufacturer's representative.

Waterproofing Membranes. Where an asphalt concrete overlay is placed on a bridge deck in an area where deicing salt is used, the salt will penetrate through the overlay unless an impermeable membrane is installed on the concrete deck. Both hot and cold rubberized materials are available, as well as more labor-intensive built-up systems. Built-up systems, like roof systems, combine layers of fabric alternated with applications of a bituminous coating. Built-up systems may cause the asphalt overlay to slide on steep grades or superelevation. All kinds of membranes are subject to development of blisters due to entrapped water vapor if the membrane cures before the vapor escapes. This can generally be prevented by placing the membrane when the temperature in the deck is decreasing, that is, during the late afternoon or evening, rather than in the morning or midday.

Epoxy-Coated Reinforcing Steel. The coating that has received the most widespread acceptance for physical encapsulation of the reinforcing bars is fusion-bonded epoxy coating. Some agencies require epoxy-coated bars for the top mat only; others require them top and bottom.

The epoxy coating is applied electrostatically in powder form to cleaned and heated bars in a continuous operation, and rapidly quenched immediately after being applied. The coating is quite hard, but must be handled carefully to avoid damage. Nylon slings are used to lift the bundles, and padding is used within the bundles. Specifications that allowed a small but liberal percentage of openings in the coating have recently been reexamined and tightened. Following the widespread adoption of epoxy-coated reinforcing steel, some unfavorable experience in marine structures has put somewhat of a damper on its enthusiastic acceptance.

A disadvantage of epoxy coating is that the coating reduces the bond between the bars and the concrete, requiring longer lap splices.

Galvanized Reinforcing Steel. About the time when many states began to install epoxy-coated bars for experimental evaluation, some states experimented with galvanized bars. This was based partly on the contention of some corrosion experts that flaws in epoxy-coated bars would result in aggravated corrosion at those flaws. One state, Pennsylvania, adopted galvanized reinforcing steel for a time as its primary means of protecting the bars. Now at least one state has changed to that policy.

Galvanizing is not a permanent barrier, but is a sacrificial coating, and consequently would be expected to have limited life expectancy when exposed to sufficient quantities of chlorides over a period of time. Because of the electrochemical nature of the way galvanizing prevents corrosion, it should not be used on only one mat of reinforcement.

Corrosion-Inhibiting Admixtures. Another means of protecting against corrosion of reinforcing steel, without application of physical coating to the bars, is the incorporation of a corrosion-inhibiting admixture in concrete. The amount of chemical added to the concrete mix is proportioned to the amount of chlorides expected to penetrate to the reinforcing steel. Therefore, the degree of effectiveness of the inhibitor is related to the accuracy of that prediction. Higher dose rates will provide greater protection, but at greater cost. A lower dose rate may not provide the necessary protection.

Some inhibitors have undesirable effects on other properties of the concrete, but one admixture that is effective without side effects is calcium nitrite.

Available publications do not provide specifications or guidelines for the evaluation and comparison of corrosion-inhibiting admixtures, requiring users to rely on information provided by product manufacturers. However, a National Cooperative Highway Research Program project is planned to develop test procedures to evaluate and compare the effectiveness of corrosion inhibitors, and to recommend performance criteria for their acceptance.

Cathodic Protection. Since rebar corrosion is an electrochemical reaction, an effective means of preventing or arresting corrosion is cathodic protection. The two main types of cathodic protection are sacrificial anode and impressed current. In the sacrificial anode system, disks of metal are installed at intervals in the deck before placement of the deck or overlay concrete. Corrosion activity involves the consumption of this metal rather than rusting of the bars. The impressed-current method requires the input of electricity, and therefore requires an electric source and is dependent on the continued monitoring and maintenance of the system. Power consumption is low. One reason cathodic protection was late in being implemented is that it involves the expertise of electrical or corrosion engineers rather than the structural engineers who are normally responsible for bridge design and rehabilitation.

4.10 DECK SURFACES AND DECK OVERLAYS

New Construction. Some agencies use asphalt concrete overlays on new decks and protect the deck with a waterproofing membrane below the asphalt overlay. Currently, however, concrete wearing surfaces are more popular on new bridges than asphalt concrete. Concrete surfaces may be placed as an integral part of the structural bridge deck (monolithic), or placed as bonded overlays of various types including dense concrete, latex-modified concrete, and silica fume concrete (see Art. 4.5.1).

Repair of Existing Bridge Deck Surface. The concrete overlays discussed above can be used, in combination with patching of spalled areas, as a means of repairing deteriorated existing bridge decks. In addition to these overlays, which are at least $1\frac{1}{4}$ in thick and usually thicker, thin overlays $\frac{1}{2}$ in or less thick are available. Binder materials include epoxy, epoxy-urethane blends, and polyester resin. Because of their thinness and light weight, they are advantageous for bridges where weight reduction is desirable, or where thicker overlays would present problems with expansion joint or scupper modification, or where railing height would be reduced more than an acceptable amount by a thicker overlay.

The repair of a spalled bridge deck involves removal of the fractured or disintegrated concrete by some means. Mechanical methods include scabblers, scarifiers, and jackhammers. Because these methods all tend to create microfractures in the sound concrete, a better method is hydroblasting, or use of very-high-pressure water jets. This method is selective in that it automatically removes unsound concrete while leaving the sound concrete undamaged. The operation consumes large quantities of water and is noisy, and passing motorists must be protected from stray jets and flying debris. The muddy effluent must be disposed of properly, and not allowed to flow into catch basins.

After removal of unsound concrete, or concurrently with it, the surface of the remaining good concrete is removed to a depth of about $\frac{1}{4}$ in. The entire surface to be overlaid is dried or wetted to the required moisture condition, and the overlay placed. Where deep removal areas are present, it is generally preferable to patch these areas in a separate operation from the general overlay. After texture is applied, the fresh overlay concrete is then given an appropriate cure of the required duration. In cold weather the overlay must be prevented from freezing. For this reason specifications require placement at temperatures well above freezing.

4.11 SELECTION OF MATERIALS FOR MAIN SUPERSTRUCTURE MEMBERS

For the primary superstructure members of a bridge (not including the deck), concrete (reinforced and prestressed) and structural steel are the principal candidates. Concrete and steel both have desirable attributes and shortcomings as bridge materials. In general, bridges of both materials can be designed, constructed, and maintained to ensure long life. Claims of both steel and concrete industry associations, including references to national bridge inventory data used to support contentions of superiority of one material over the other, must be critically considered. One can find examples of both concrete and steel bridges that are old and in good condition, and conversely, relatively new and in poor condition. The trade associations do a service in countering each other's claims.

Some advantages of concrete bridges are:

- They do not require painting.
- They do not rust (but are susceptible to rebar corrosion).
- They can be formed to the desired shape (if of reinforced concrete).
- If of prestressed concrete, they may be fabricated more quickly than steel, although in some emergencies steel replacement structures have been fabricated and erected as quickly as prestressed members.
- They are not susceptible to fatigue failure (to date).

Some advantages of steel bridges are:

- Lighter weight permits smaller cranes for erection.
- Lighter weight permits reduction of substructure size, number of piles, etc.
- They are more readily dismantled and reused at the same or another site.
- Use of conventional erection and construction techniques may avoid construction cost overruns and litigation sometimes experienced with segmental concrete.
- Attachments to bridge are readily made by bolting or welding.
- Components are accessible and visible for inspection.
- Members damaged by vehicular collision may be more easily repaired than concrete members.

For short- to medium-span bridges, the selection of material will depend on which bridge type and material is the most economical for the particular site. This may be known by experience with bids received over a period of time, or can be determined by taking alternative bids on projects.

Until recently, the FHWA required that bridges with federal funding and an estimated cost over $10 million be designed in both materials, or in different framing systems of the same material, so that contractors bidding on both sets of plans can make the determination of which is less costly. Currently, alternative designs may be used at the option of the state. An increasingly common practice for bridges of all sizes is to allow the contractor to submit alternative designs, which must be designed by professional engineers and conform to the requirements of the owner.

4.12 *CORROSION PROTECTION OF NEW STEEL BRIDGES*

The application of protective coatings to steel bridges, and the maintenance reapplication of coatings, is costly and so alternatives to the use of coated steel should be sought. Where appropriate, unpainted weathering steel should be used instead (see Art. 4.13). If a coated bridge is still the best candidate for the particular location, a long-lasting coating system should be applied.

New Paint Systems.   The development of high-performance paint systems for new bridges has resulted mostly in two- or three-coat systems involving combinations of various materials including organic and inorganic zinc, epoxy, and urethane. The prime coats of these systems require cleaning the steel to a white or near white condition, which is an expensive operation even when done conveniently in a steel fabricating plant. The application of subsequent coats, especially field coats, is labor-intensive. Despite these factors, these new systems can provide acceptably economical protection.

Water-Based Paint.   The most recent emphasis in the development of paint systems has been on water-based paints. Because they do not contain volatile organic compounds, water-based paints can easily conform to the environmental restrictions placed on the levels of those compounds emitted during the painting process. Evaluation of this system and other systems continues. At this time, one can say that there is no one single paint system that is the best and most economical for all exposures. If a department of transportation or another agency representing an area with diverse geography, climate, or industrial development dictates a single paint system for all parts of that area, it is likely that some of the bridges will be overprotected and some underprotected.

Galvanizing.   Depending on local availability of galvanizing facilities of adequate size, steel members of limited length can be hot-dip galvanized. In addition to the deposition of zinc, the galvanizing process results in a change in chemistry of the surface of the steel, where an alloy is created, so that a degree of protection remains after the zinc coating is gone. The different stages of loss of coating and rusting that will eventually occur on galvanized steel can be seen on exposed highway hardware such as galvanized steel roadside barriers, luminaire supports, and traffic sign and signal supports. Since these structures are more exposed to salt spray than a bridge superstructure may be, unless the bridge is a grade-separation structure, the longevity of the protection may be expected to be greater on a bridge.

Fusion-Bonded Coating.   The coating of large structural members by fusion bonding with epoxy or other powders is now feasible in at least one coating plant, but this method of coating has not been used extensively.

Metallizing.   Another method of coating steel, which has been used on small components of new bridges, such as bearing plates, and on a few existing bridges, is application of a metallic coating by the flame spray method, or "metallizing." The existing steel is first prepared to a near white condition. Then a continuously fed wire is vaporized in a flame and sprayed onto the surface of the steel. Although results have been satisfactory, the cost on complete bridges has been extremely high compared with other methods of coating.

Selection of Protection System.   Environmental conditions and owners' experience may dictate the selection of a corrosion protection system. Where acceptable life of protection can be expected from galvanizing, painting, or use of unpainted weathering steel, the selection may be based on cost. Alternative bids should be encouraged.

4.13   WEATHERING STEEL

The cost of initial painting and periodic repainting of structural steel bridges can often be eliminated by the use of bare weathering steel. From an economic standpoint, the use of multicoat high-technology paint systems should be reserved to those bridges that are not suitable candidates for weathering steel.

To ensure successful long-term performance, the Federal Highway Administration (FHWA) has published "Guidelines for the Use of Unpainted Weathering Steel." Principal considerations are as follows:

- Consider with caution use in marine coastal areas; in areas of frequent high rainfall, high humidity, or persistent fog or condensing conditions; at grade separations in "tunnel-like" conditions; and at low-level water crossings.
- Eliminate expansion joints where possible.
- Use a trough under open expansion joints.
- Paint all steel within a distance of 1½ times the depth of girders from bridge joints.
- Seal box members where possible or provide weep holes to allow proper drainage and circulation of air.
- Seal overlapping surfaces exposed to water to prevent capillary penetration action.
- Implement maintenance and inspection procedures designed to detect and minimize corrosion.
- Divert roadway drainage away from the bridge.
- Clean troughs, reseal deck joints, and periodically clean and—when needed— repaint all steel in the vicinity of joints.
- Regularly remove all dirt, debris, and other deposits that trap moisture.
- Regularly remove all vegetation that can prevent natural drying of wet steel surfaces.

4.14   DEFLECTION AND EXPANSION JOINTS

Joints in bridges fall into two categories: deflection joints, and expansion joints.

4.14.1   Deflection Joints

Contrary to what the name implies, deflection joints, when placed in concrete barriers and parapets, are used primarily to minimize the vertical shrinkage cracking that would otherwise occur in long, unjointed panels. Some states permit a longitudinal spacing of joints as great as 30 ft in simple spans. Over piers of continuous bridges, the spacing is generally less, 7.5 ft or closer. Preformed joint filler is used to form the joints and is left in place. Sometimes the placement of parapet concrete is required to be done in two stages, with placement of alternate panels only in the first stage, to facilitate placement of the joint filler.

When barriers are permitted to be slipformed, the deflection joints are sawn an inch or so deep on the periphery of the barrier, and then caulked with a joint sealer. In this case the steel is not made discontinuous at the joints. Slipforming is a much faster way of constructing barriers, but the finished appearance, especially the straightness of the top, is sometimes rather crude compared with conventionally formed barriers.

Deflection joints can extend full depth of the barrier or parapet, or through only the top portion. Deflection joints in the New Jersey safety-shape barriers, when the concrete is placed in forms, are sometimes placed only above the curb portion of the barrier. In this case the longitudinal reinforcing steel is continuous in the curb, but discontinuous at the joints above the curb. This usually results in reflection cracks developing in the curb below the joints.

It is also common, in spite of the joints, to see one or more vertical cracks between the joints in long panels. These cracks may be aggravated by bridge deflection but are caused primarily by shrinkage. The development of these cracks illustrates a rule of thumb applied to slabs on grade, that there will be a tendency to crack if the slab is longer than twice its width. However, the likelihood of ultimate damage to the bridge resulting from these unwanted cracks is small, and so the cost to provide more closely spaced joints is not justified.

Deflection joints are also used in the deck slab at piers or over transverse floor beams where the slab is not continuous (and sometimes when it is continuous), and at abutments where the bridge slab abuts the approach slab. Since the amount of movement is small, due only to rotation, the joint can be sealed with a small compression seal or with liquid joint sealer.

4.14.2 Expansion Joints

Bridge roadway expansion joints are provided to accommodate the thermal changes in the superstructure, and, in the case of prestressed-concrete bridges, to accommodate creep shortening of the superstructure as well. They are required at abutments that are restrained against longitudinal movement and at the end of supported superstructures free to translate due to provision of expansion bearings. In some long-span steel bridges, expansion joints and expansion bearings must also accommodate change of length of span due to live load stresses. Expansion joints are not required in short bridges where movement is small—for example, in steel bridges with span less than 50 ft—or in longer bridges where the superstructure is fixed to the abutment (jointless bridges or integral construction). For these longer bridges, designs that eliminate or minimize bridge expansion joints, without introducing problems in the approach roadway or causing distress in the superstructure or substructure, are favored. (See Arts. 4.15 and 4.16.7.)

Expansion joints, or, more accurately, rotation joints, are also provided where the deck is made discontinuous, or a hinge is provided, in anticipation of settlement of the end of a span.

Where expansion joints are required, they should be sized to accommodate the anticipated movement with a liberal allowance. A joint-sealing device such as a strip seal can be destroyed by one occurrence of a record cold period. For deep simple-span girders, the joint movement due to live load rotation of the end of the span should be included. Specifications for installation of joints should take setting temperature into account. A table giving required joint opening dimensions for different ambient temperatures is preferred over an equation for adjustment of a fixed dimension that is applicable to a given temperature. In areas where roadway deicing salts are not applied, it may not be necessary to seal expansion joints. Even in this case, though, sealed expansion joints will prevent intrusion of foreign objects, which can damage the bridge by causing excessive local pressure, and will prevent accumulation of debris on bridge seats.

Large-capacity open expansion joints can be fabricated using steel plates with meshed fingers, the so-called finger joint. Finger joints have served well for many years on many bridges. The plates used in these joints must be thick to withstand the direct cantilever wheel loading to which the fingers are subjected. The two halves of finger joints are massive steel fabrications, but they can be gas-cut with accurate dimensional control. In snowplow areas, where the ends of the fingers may be snagged by the plow blade, the use of a finger joint should be avoided, or the ends of the fingers should be rounded downward. In areas where joints should be sealed, the finger joint surface may be left open, but an elastomeric trough should be installed beneath the joint.

Bicycles should not be permitted on bridges with finger joints, because the wheels can drop into the space between fingers, causing injury to the rider. Conversely, open finger joints should not be used on bridges on which bicycles are permitted.

In areas of salt application, expansion joints must be sealed. Northern states have incurred tremendous cost to repair damage to superstructures and substructures in the form of steel corrosion, prestressed-concrete beam deterioration, and concrete spalling due to salt drainage through open or inadequately sealed expansion joints.

4.14.3 Expansion Joint Sealers

Several types of expansion joint sealing devices are available. Properly sized and installed, they can greatly reduce, if not eliminate, drainage through the joint. Some of the available types are:

- Polymer-modified asphalt
- Compression seal
- Slab-type seal
- Strip seal
- Modular seal

Polymer-Modified Asphalt. For resurfacing projects where an asphalt concrete overlay or a portland cement concrete overlay is placed on an existing bridge deck, the approach slab is also overlaid, and the joint movement is moderate (1.5 in or less), an expansion joint seal using a poured liquid joint sealer and "armor" of polymer-modified asphalt concrete (elastomeric concrete) can be used. Construction is simple and requires only a minimum of removal of the existing structure, if any. The elastomeric concrete will bond to steel, concrete, or asphalt concrete, and also develops a tenacious bond with the liquid joint sealer, which is poured over backer rod installed between formed vertical faces of the elastomeric concrete. Figure 4.3 illustrates this seal.

Compression Seal. The compression seal (Fig. 4.4) is a rectangular elastomeric tube that has internal webbing and is manufactured by extrusion. It is installed between formed or sawn faces of concrete, or, more commonly in bridges, between steel armor.

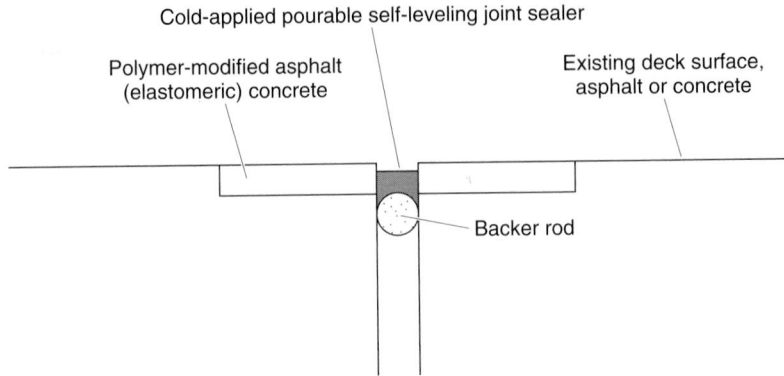

FIGURE 4.3   Cross section of polymer-modified asphalt concrete joint seal.

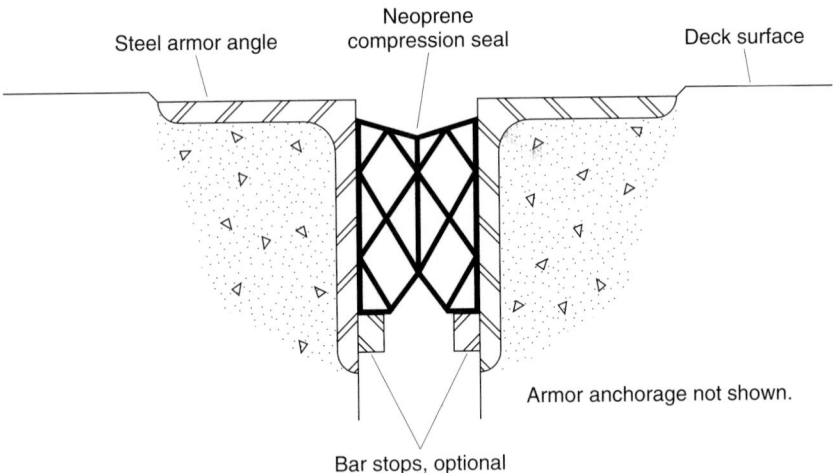

FIGURE 4.4   Cross section of compression seal.

A lubricant/adhesive is used to facilitate installation and prevent displacement in service.

Compression seals can be of the high-compression type, which rely more on their internal compression than on the adhesive to stay in place, or the low-compression type, which rely more on the adhesive. The high-compression type are more subject to loss of tight fit due to compression decay with age, and so are less desirable. Catalogs will not describe the seals in this manner, but manufacturers' representatives will know. A clue to the type of seal is the number of internal webs, which is greater for the high-compression type.

Compression seals are available for joint widths up to 5 in, but some agencies impose a 4-in limit. Some skew can be accommodated by using a larger seal than would be required for an unskewed joint. A maximum allowable skew of 15° (with respect to a line normal to the bridge centerline) is imposed by one state. Seals should be one piece for the entire length of the joint.

A variation of the compression seal is a proprietary seal that has no internal webbing. It is installed in the joint by air inflation, which presses the sides against the supporting surfaces, onto which an epoxy adhesive has been applied. The air pressure is released after an adequate curing period, and the adhesive is relied upon to maintain the seal in position.

Slab-Type Seal. The slab-type sealing device consists of an elastomer and internal steel plates that combine to provide a surface that bridges over the joint opening and supports traffic loads. There are notches in the slab that, along with the elasticity of the elastomer, permit it to change length. The sides of the slab are supported on horizontal steel or concrete surfaces, and a bedding adhesive is applied before the slab is fastened down. The slab is fastened to the bridge by closely spaced bolts.

A primary disadvantage of the slab-type seal is that large stresses are induced in the slab by temperature changes, which, along with pounding by traffic, tend to break it loose. Some users of slab-type seals have had satisfactory experience with them, but most users have changed to other types of seals.

Strip Seal. The strip seal (Fig. 4.5) is an elastomeric extrusion, called a *gland,* that spans between supporting steel armor. It is anchored by enlargements on the ends of the glands, which are inserted into grooves in the armor. The gland is generally only one layer thick, but some strip seals have two layers, the lower of which should act as a backup if the top layer is punctured. A lubricant/adhesive is used to facilitate installation. Like the compression seal, the gland should not be spliced. Special tools should be used to install the gland, and the gland should not be stretched during installation. Left by themselves, contractors may try to use inappropriate tools and brute force to install the gland.

Strip seals can accommodate skew somewhat better than compression seals and are favored by agencies for joint openings larger than can be accommodated by compression seals.

Modular Seal. At the ends of long bridges, or the ends of individual units of long bridges, the joint movement may be greater than can be accommodated by a single joint seal of the types described above. In this case a finger joint with an elastomeric

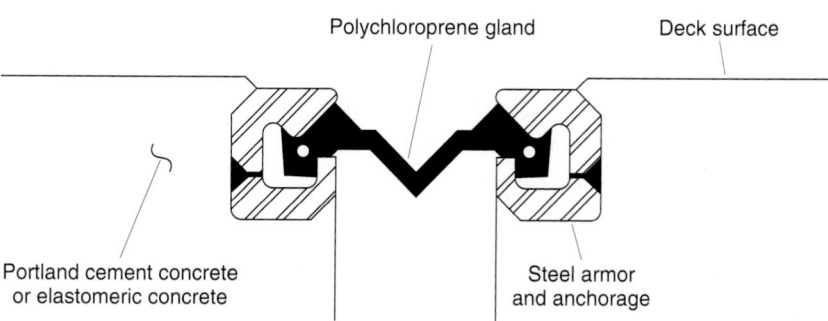

Polychloroprene gland   Deck surface

Portland cement concrete
or elastomeric concrete

Steel armor
and anchorage

FIGURE 4.5 Cross section of strip seal.

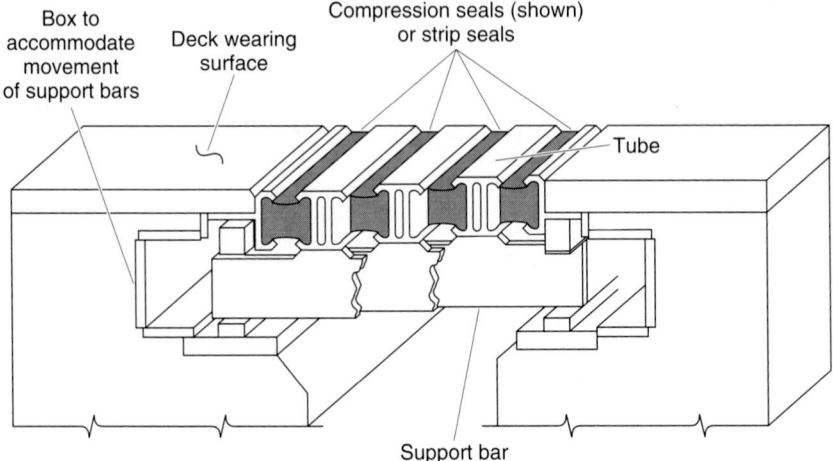

FIGURE 4.6   Cutaway view of modular expansion joint seal.

trough may be used, or a modular joint can be provided. The modular joint consists of multiple compression seals or strip seals separated by steel or aluminum structural members, which are in turn supported by bars that span transversely to the joint, parallel to the centerline of the roadway. Figure 4.6 illustrates this type of seal. The support mechanisms can become quite elaborate, with sliding bearings and components to ensure that uniform spacing between longitudinal seal elements is maintained. The designs must provide for joint rotation as well as translation. It is also important that the design of these joints allow for replacement of components.

The fact that many specifiers of sealed expansion joints do not expect them to be, or remain, watertight throughout their life is indicated by the practice recommended for weathering steel bridges (see Art. 4.13). That recommendation is that the steel be painted at the joints, and is applicable to sealed and unsealed joints alike.

Failure of an expansion joint can occur in the sealing mechanism itself, but in the past, failures have occurred as frequently in the anchorages. Some causes of anchorage failure have been:

- Inadequate consolidation of concrete below wide legs of armor angles
- Too small or too widely spaced welded stud anchors
- Vulnerability to snowplow damage because the sealing device was not recessed below the wearing surface
- Pressure exerted during thermal changes by overlapping steel angles because the joint design did not properly accommodate longitudinal grade
- Material used to bed or anchor the joint sealing device that was not shrinkage-resistant and broke under traffic

4.15 CONTINUITY AND JOINTLESS BRIDGES

Where possible, bridges should be made continuous. Continuous spans are less prone to catastrophic collapse from loss of substructure support due to stream erosion, earthquake, or vehicle or vessel collision. Bridges with multiple simple spans must have two lines of bearings and an expansion joint at each intermediate support. Two lines of bearings, each having the required capacity for the end of a simple span, will almost certainly be more expensive than the single line of bearings required for continuous spans. Expansion joints are expensive and in most geographic locations should be sealed against storm drainage and intrusion of debris, which further increases their cost. (Even the manufacturers of sealed expansion joints agree that the best joint is no joint.) Aesthetically, continuous bridges are generally superior, especially if constant depth is maintained, and do not require the cosmetic plates or other devices that have sometimes been used to conceal the gaps between simple spans.

Continuous bridges are generally more economical than simple-span bridges because of the reduction of midspan moments. Most bridges can be designed continuous for live load, and some bridges may be designed continuous for dead load as well. In the case of precast prestressed-concrete bridges, it is generally more convenient and economical to place the deck slab concrete while the beams are supported on their bearings, without temporary intermediate shoring, so that the beams are not continuous until the deck slab has acquired its strength and top longitudinal reinforcing bars are present in the composite section over the piers to resist negative moment. Therefore these bridges are designed continuous for live load and for superimposed dead loads (loads above the deck slab) only. Note that this type of construction unavoidably requires two lines of bearings at intermediate supports because practical prestressed-concrete design and construction requires that the spans be simple initially. This is called "made-continuous" construction.

There are situations where simple spans are preferable. Examples include situations where adjacent spans are unavoidably different in length and depth, or where adjacent spans have widely different geometrics with beam layouts that do not lend themselves to continuity, such as varying beam spacing or splayed framing. Simple spans may also be preferable where the bridge is part of a facility, such as an interchange, where stage construction will require future removal or addition of one or more spans. Simple spans are also desirable where substructure settlement is anticipated.

4.16 CHARACTERISTICS AND SELECTION OF BRIDGE TYPES

The type of bridge and the span layouts are interdependent. Bridge type cannot be selected without regard to the length of spans, the ratio of adjacent span lengths, and whether spans are to be made continuous.

Table 4.2 lists common types of bridges and the maximum span lengths below which they may be an economical choice. The maximum spans tabulated are approximate, and are presented as a guide only. They are subject to increase as technology advances. The economic competitiveness of a particular bridge type varies with regional availability and workload of fabricators and specialty contractors, yearly fluctuations of labor and material costs, and other factors. Thus, it is usually desirable to seek and permit bids on alternative bridge types.

TABLE 4.2   Approximate Maximum Span for Various Types of Bridges

Type	Approximate maximum span, ft
Reinforced-concrete flat slab, continuous	55
Composite steel beam (36-in series), simple	100
Precast prestressed-concrete voided box beam	120
Precast prestressed-concrete beams (bulb-tee), simple	120
Composite steel beam (36-in series), continuous	125
Precast prestressed-concrete beams (bulb-tee), made-continuous	140
Composite steel plate girder, simple	230
Cast-in-place (on falsework) posttensioned-concrete box girder, continuous	300
Precast posttensioned segmental concrete box girder, continuous, balanced cantilever	400
Composite steel plate girder, continuous, parallel flange	460
Composite steel plate girder, continuous, haunched	540
Cast-in-place posttensioned segmental concrete box girder, continuous, balanced cantilever	850
Steel arch (New River Gorge, Fayetteville, West Virginia, U.S.A.)*	1700
Steel cantilever truss (Pont de Québec, Canada)*	1800
Steel cable–stayed (Normandie Bridge, Le Havre, France)*	2808
Suspension (Humber Estuary, Hull, England, U.K.)*	4626

Note: No attempt has been made to include every bridge type in the above tabulation.
*Indicates record span.

Characteristics of some of the more common bridge types for short and intermediate spans and considerations in their selection are discussed in the following articles.

4.16.1   Reinforced-Concrete Flat-Slab Bridge

For short simple spans (30 ft or less) and for somewhat longer continuous spans (interior spans up to 55 ft), reinforced-concrete flat slabs provide a minimum-depth bridge. Figure 4.7 shows a schematic of this bridge type. At a slab depth of about 2 ft, the slab begins to become uneconomical, with too much of the section required to support itself.

Falsework is required to construct the slab. Where space is available beneath the structure, scaffolding may be used. If the bridge is over a stream, or over a highway or railroad where traffic must be maintained, the falsework must include support beams to span over the feature crossed. In that case, camber should be built into the falsework to compensate for its deflection. Also, the falsework must provide for the vertical geometry of the bridge and for deflection of the slab after removal of the falsework.

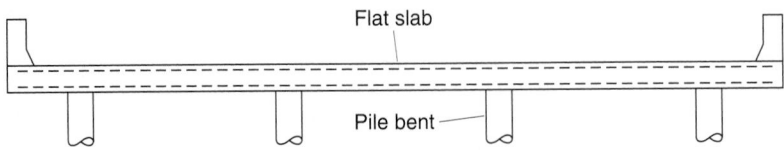

FIGURE 4.7   Cross section of reinforced-concrete flat-slab bridge on pile bent; pile caps may be required for shorter spans.

Longer continuous-slab spans can be constructed if the slab is haunched, that is, made deeper over the piers or bents. However, the cost and difficulty of constructing the forms, and bending and placing the longitudinal reinforcing bars, often negates the advantage of haunched construction.

Another type of construction that can be used to extend the span capability of slab bridges is voided construction. Voids, similar to those used to fabricate prestressed-concrete box beams, are used to replace the relatively ineffective concrete at mid-depth of the slab, thereby reducing the weight of the slab. However, where this type of construction has been used, it has generally been found to be more expensive than competitive types of bridges. A principal reason is the cost of providing adequate hold-down devices to prevent the voids from floating when the concrete is placed.

For balanced design of continuous-slab bridges, the usual rule that the end span should be shorter than the adjacent interior span may not apply. In the current design of a three-span continuous flat-slab bridge with three equal spans of 30 ft, designed for HS 25 live load (a load 25 percent greater than HS 20) and the AASHTO Alternate Military Loading, a good balance resulted between maximum positive and negative moments.

4.16.2 Prestressed-Concrete Box-Beam Bridges

The span range of a shallow bridge may be extended beyond the limits of a slab bridge by using precast prestressed-concrete box beams as illustrated in Fig. 4.8. The beams are prefabricated off-site. They are rectangular and, except for very shallow beams (12 in), which may be solid, have from one to three rectangular or circular voids. The void forms are either waterproofed cardboard or solid polystyrene foam and are left in the beams. Void drains must be provided to prevent entrapment of water. Prestressing strands are located on the bottom and in the sidewalls of the box, and may include debonded or deflected strands. The selection depends upon owner preference or, where the designer and owner allow the option, fabricator preference.

This type of bridge can be constructed using adjacent beams or spread beams. In adjacent box-beam construction, prefabricated box beams are placed side by side, abutting each other. The box beams are connected by transverse tie rods or posttensioned tendons, or by welded connection of tie plates to plates embedded in the tops of the beams. Shear keys between beams are grouted. The combination of transverse connection and grouted shear keys is intended to make the beams act together as a unit and prevent relative movement and cracking at the longitudinal joints. This type of bridge can be erected quickly, and temporary traffic can be maintained on partially completed portions of the bridge. These features have made this a popular type of bridge, despite at least one shortcoming discussed below, and sometimes cause it to be

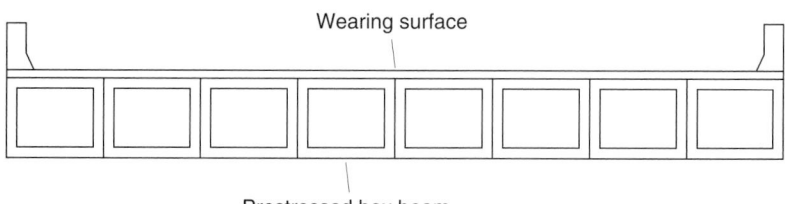

Wearing surface

Prestressed box beam

FIGURE 4.8   Cross section of prestressed-concrete box-beam bridge.

selected over a competitive type when both types are viable candidates for a bridge of a given span length.

For low-traffic-count roads, the tops of the beams may constitute the riding surface, but on most bridges a topping will be used. This may be a composite concrete slab, which adds to the strength of the bridge, or an asphalt concrete overlay, which is used to smooth out any irregularities between beams, to compensate for difference between roadway profile and final camber of the beams, and sometimes to maintain continuity of pavement type when the adjacent roadway is asphalt concrete. A waterproofing membrane should be used with this type of construction, with special attention to the joints.

The elimination of movement at the longitudinal joints and the maintenance of a waterproof condition has not always been achieved, even when a composite concrete slab has been used. Leakage of roadway drainage containing deicing salt through longitudinal joints has sometimes resulted in corrosion of the prestressing strands. In some cases, wires have broken.

In spread-box construction, the beams are spaced apart, and a reinforced-concrete slab is constructed on top. The slab between the beams is formed, and stay-in-place steel forms are frequently used. This has been an economical type of construction in Pennsylvania. In bridges with end spans shorter than interior spans, the beams can be the same depth for aesthetic reasons, with the spacing between beams varied to meet structural requirements.

4.16.3   Prestressed-Concrete I-Beam Bridge

Prestressed-concrete beams of the basic I-shape, but with variations, can be used over approximately the same range of spans as steel beams. The deepest prestressed beams (72 in) have a somewhat greater simple-span capacity than 36-in deep rolled steel beams. This type of bridge is illustrated in Fig. 4.9.

Prestressed-concrete beams are heavier to transport and erect than steel beams, and require more care in handling A prestressed-concrete beam can be destroyed if it is not maintained in an upright position.

I-beams may be standard AASHTO-PCI sections or conform to individual state standards. Depth varies from 28 in for the little-used AASHTO type I to 72 in for the AASHTO type VI and BT-72 bulb-tee. The basic difference between the AASHTO type V and type VI beams and the bulb-tee beams, all of which have 3.5-ft-wide top flanges, is that the bulb-tees have a thinner web (6 in instead of 8 in) and shallower top and bottom flanges. The bulb-tees have a flatter slope on the top of the bottom flange, as well. A variant of both is the modified AASHTO type VI, which uses the side forms for the AASHTO type VI beam but only a 6-in web. Individual analysis

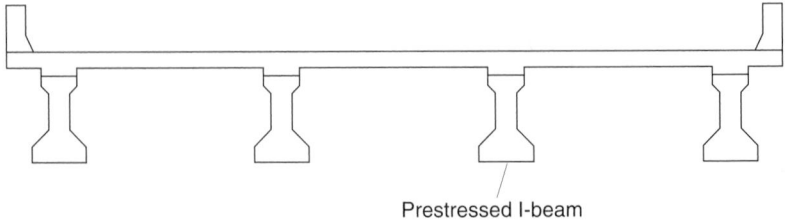

Prestressed I-beam

FIGURE 4.9   Cross section of prestressed-concrete I-beam bridge.

will determine which shape is best, but only shapes that are available from local precasters should be investigated unless the project is large enough to economically justify the purchase of special forms.

As with prestressed-concrete box-beam bridges, the prestressing strands may include deflected or debonded strands. When strands are deflected and a number of beams are cast in line on a casting bed, resulting in many hold-down or hold-up points, stressing procedures should be used and verified that limit the maximum prestress loss due to friction to the amount permitted by specifications.

For very long bridges with repetitive spans over water, and where there is a precasting plant at a site from which the bridge units can be delivered by barge, the option of precast deck units consisting of the beams, diaphragms, and deck slab cast monolithically should be considered.

4.16.4 Steel-Beam Bridge

The steel-beam bridge uses rolled steel beams as shown in Fig. 4.10. Until recent years, rolled beams were available only in depths up to 36 in. Now, rolled sections 40 in and 44 in deep are available. Because the 44-in beams are rolled outside the United States, and because federal law applicable to federally aided projects, as well as many state laws, prohibits the use of foreign steel, 44-in beams have not been used to any appreciable extent. However, there is currently one domestic producer that can roll certain of the 40-in sections. Rolled 40-in beams have been included in designs for bridges in Nebraska, Ohio, and Tennessee.

Steel beams may be made continuous by welding or bolting sections in the field. In the past, some states made welded connections at the piers, and currently at least one state makes welded connections at contraflexure points, supporting the field sections temporarily and providing enclosures to shield the joint from wind. More commonly, field sections are spliced by high-strength bolts, using web-and-flange splice plates. Bolts may be installed using calibrated wrenches, by the turn-of-nut method, or by use

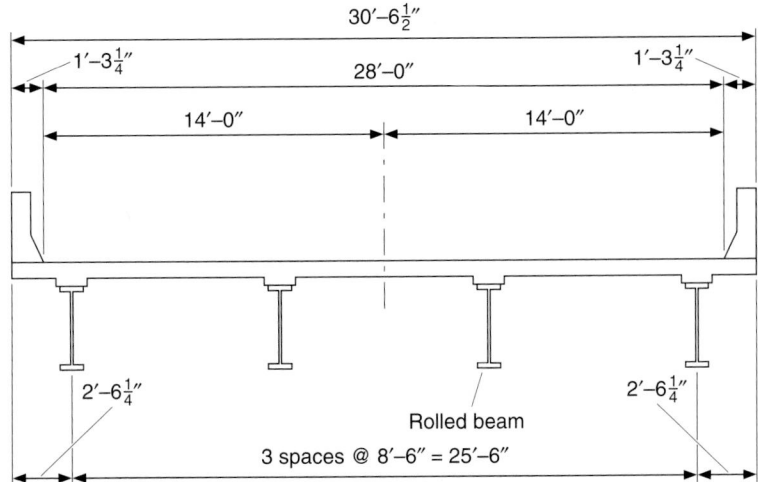

FIGURE 4.10   Cross section of bridge with rolled steel beams.

of tension-indicating washers, depending on what the designer allows and what the erector prefers to use. With all methods of bolting, it is important to use a procedure and sequence of bolting that will compact the joint and prevent a bolt initially adequately tightened from losing tension when subsequent bolts in the joint are tightened. Fasteners are generally ASTM A325 or A490 high-strength bolts.

To increase the span capacity of a rolled beam, or to permit a lighter beam to be used, cover plates may be added above the top flange and below the bottom flange in regions of high bending stress due to both positive and negative moments. The fatigue strength at the end of the cover plates, which is generally at a point of low maximum stress but high stress range, is much less than the fatigue strength of the unplated beam. Allowable fatigue stresses must not be exceeded, and this consideration may favor an unplated beam. However, an improved detail is available that uses bolts at the end of the plate, and the fatigue strength is somewhat higher.

4.16.5 Welded Steel Plate Girder Bridge

The welded steel plate girder bridge (Fig. 4.11) extends the span range of deck-type bridges (bridges having all the structural support below the deck slab) well beyond the range of rolled steel beams or precast prestressed-concrete beams.

Whereas haunched girders were economical in the past for long spans, the current practice, strongly advocated by the steel fabricating industry, is to use parallel-flange girders wherever possible. This is an economic consideration rather than an aesthetic one. Properly configured haunched girders are thought by many to be more pleasing. They permit a shallower structure depth at mid-span, which can result in a lower grade

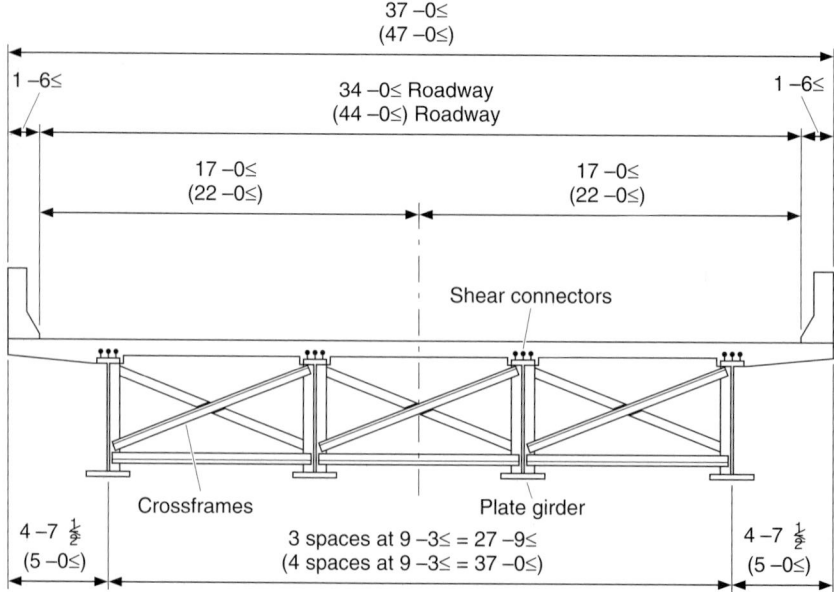

FIGURE 4.11 Cross section of bridge with steel plate girders.

line and consequent savings in roadway construction cost. However, parallel-flange girders can be fabricated more rapidly and economically than haunched girders. As an example of long-span parallel-flange steel girder construction, the Tennessee DOT has designed a continuous 1717-ft-long parallel-flange five-span steel plate girder bridge of girder–floor beam–stringer type construction, including two spans of 460 ft each. This design uses ASTM A36 and A572 steel in the webs, and A572 and A517 steel in the flanges. The A517 steel, which has a minimum yield strength of 100 kips/in², is used for the flange plates at points of maximum stress over the piers and at mid-span of one of the 460-ft spans.

In designing a steel plate girder bridge for economy, designing for minimum weight does not always result in the most economical girder. The cost saved by reducing web or flange plate width or thickness may be more than offset by the cost of making the welded splices. Cost data should be obtained from local fabricators to make this comparison. One rule of thumb is that the weight saved by a change of flange plate thickness should be at least 1500 lb. Also, it is generally desirable to use a constant-width flange plate to reduce fabrication and construction costs.

The use of excessively thin webs and narrow flanges, while saving weight, can result in flimsy sections that require special handling and erection equipment such as strongbacks. If such measures are not employed, the girder may be damaged in handling. Either consequence may more than offset the cost saved through weight reduction. For this reason many states have adopted minimum plate dimensions that are greater than minimum requirements of AASHTO or industry recommendations.

4.16.6 Composite Construction for Steel Beam and Plate Girder Bridge

The concrete deck for steel beam and girder bridges may be designed and constructed on the basis of either composite or noncomposite behavior. With composite construction, the effective area of the slab can be calculated and used in determining the moment resistance of the section in positive moment regions. In negative moment regions, tensile stresses can be resisted by the reinforcing steel. The required number of shear connectors must be calculated and furnished. These are generally headed studs that are welded to the top flange (Fig. 4.11). Overall economy depends upon the cost of the installed shear connectors and the reduction in steel weight that can be obtained. However, composite construction is frequently the economical choice.

4.16.7 Economical Design of Steel Beam and Plate Girder Bridge

Suggestions for maximum economy of steel beam and girder bridges may be summarized as follows:

- Develop and take bids on as many alternative designs as practical. Include rolled beams and plate girders, and alternative details such as cover plates for rolled beams.
- Consider higher-strength steels in general and weathering grades in particular. If a bare weathering steel such as grade 50W is used for the stringers, also use grade 50W for the diaphragms and cross frames; if coated grade 50 steel is used for the stringers, use grade 36 steel for the diaphragms and cross frames.
- Consider wider stringer spacings and develop realistic cost comparisons.

- Carefully evaluate changes in flange plate size, because flange splices are costly. Consider allowing the fabricator the option of continuing the heavier plate to eliminate the splice.

- Consider alternative systems for diaphragms or cross frames including rolled channels and bent plates for shallow sections, and angle frames for deeper sections. At expansion joints, a thickened bridge deck may be used to provide support for the discontinuous slab rather than developing special cross frame details.

- Give strong consideration to the design of a jointless bridge. Jointless bridges have been used successfully in many states to avoid the problems associated with deck joints. Both initial costs and maintenance costs can be reduced by keeping moisture away from the stringers and the substructure.

- Steel-reinforced elastomeric bearings usually provide the most economical system. Such bearings have proven suitable for both steel and concrete bridges under a variety of conditions.

4.16.8   Beam and Girder Spacing for Steel Beam and Plate Girder Bridge

In regard to efficiency in the number of lines of girders in bridges consisting of multiple girders connected by cross frames, cursory cost comparisons almost always conclude that the widest spacing of girders is the most economical. Savings result not only from the reduced number of main members but also from the reduced number of secondary elements (shear connectors, cross frames, stiffeners, and bearings). However, other costs must be considered. Wide girder spacing will generally be accompanied by a wide slab overhang over the outside girders, for a balance of load on interior and exterior girders. This may necessitate extra reinforcing steel in the top of the deck slab beyond the amount required for the slab span between girders. Use of three lines rather than four or more puts the bridge closer to a nonredundant condition. In some cases, a greater number of girders than the optimum for minimum material cost may be necessary or desirable to permit the bridge to be built in stages, or to have the deck replaced while maintaining traffic on a portion of the deck width. In general, beam and girder spacings of up to 10.0 or 12.5 ft should be investigated for typical bridges. For economy, the size of interior and exterior stringers should be the same.

4.16.9   Welded Steel Box Girder Bridge

The steel box girder bridge is depicted in Fig. 4.12. The steel elements are fabricated and erected as "tubs," and the composite concrete deck is placed in the field. This configuration has some advantages over plate girder construction. Visually, it is "cleaner," and it does not provide surfaces for birds to perch. For high-visibility bridges, such as

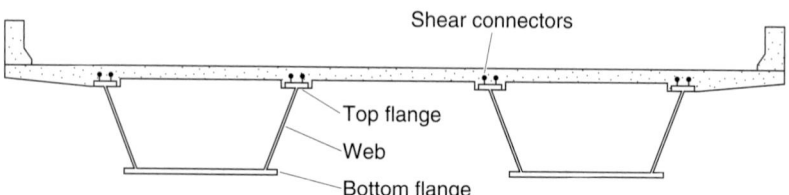

FIGURE 4.12   Cross section of bridge with steel box girders.

urban interchange bridges where motorists are constantly passing beneath the bridge, the enhanced appearance may be a deciding factor. Also, the cleaner surface areas tend to improve durability and reduce repainting costs. The bridge is torsionally stiff—especially beneficial for horizontally curved bridges. These advantages come at a price, however, because box girder bridges are generally more expensive than plate girder bridges. Sometimes the extra cost is knowingly borne for the aesthetic advantage.

4.17 DETERMINATION OF SPAN LENGTHS

Where the spans are not controlled by features crossed—such as roads, railroads, streams, or existing buildings—and there is freedom to locate piers, the lengths of spans will be controlled by aesthetic, economic, and structural requirements. Generally, from an aesthetic standpoint, spans should have a length at least 3 or 4 times the pier height.

The profile of the site crossed will influence the span proportions. On the uphill end of a crossed hillside, the end spans will be shorter than at the bottom of the valley. The type of bridge will also affect the selection of span ratios, from both aesthetic and structural standpoints. Where spans are continuous, the end span should not be made too short, because uplift may occur under live load, and loss of positive reaction at the abutment will occur sooner if the abutment settles.

The most economical bridge will generally not be either the one with the most economical superstructure or the one with the most economical substructure, but the one with the least combined cost. That determination is made by performing a cost study wherein a number of different span lengths are investigated, along with the cost of their substructures. To be meaningful, the superstructure and substructure designs should be fairly detailed. The superstructure and substructure costs are then plotted. The optimum span length will be at the low point of the combined cost curve. A typical cost study curve is shown in Fig. 4.13.

4.18 USAGE OF BRIDGE TYPES IN NEW CONSTRUCTION

The Federal Highway Administration annually publishes tables listing the number and cost of bridges of the basic material types authorized for construction with participation of federal funds in the preceding year. Data are given for reinforced-concrete, steel, prestressed-concrete, and "other" bridges. The information developed for 1993 is summarized in Table 4.3. The data show that for new bridges on all systems, in terms of percent of total cost, the most popular types of construction were prestressed concrete (56.5 percent), steel (33.4 percent), and reinforced concrete (4.1 percent). For major bridge rehabilitation, the order changes to steel (60.8 percent), prestressed concrete (17.3 percent), and reinforced concrete (10.6 percent).

4.19 BRIDGE WIDENING AND REHABILITATION

Shoulders were not always provided on bridges in the past. This in itself can be a reason for widening an existing bridge. More frequently, widening is necessitated by the addition of lanes to the highway, at which time a full shoulder can be provided.

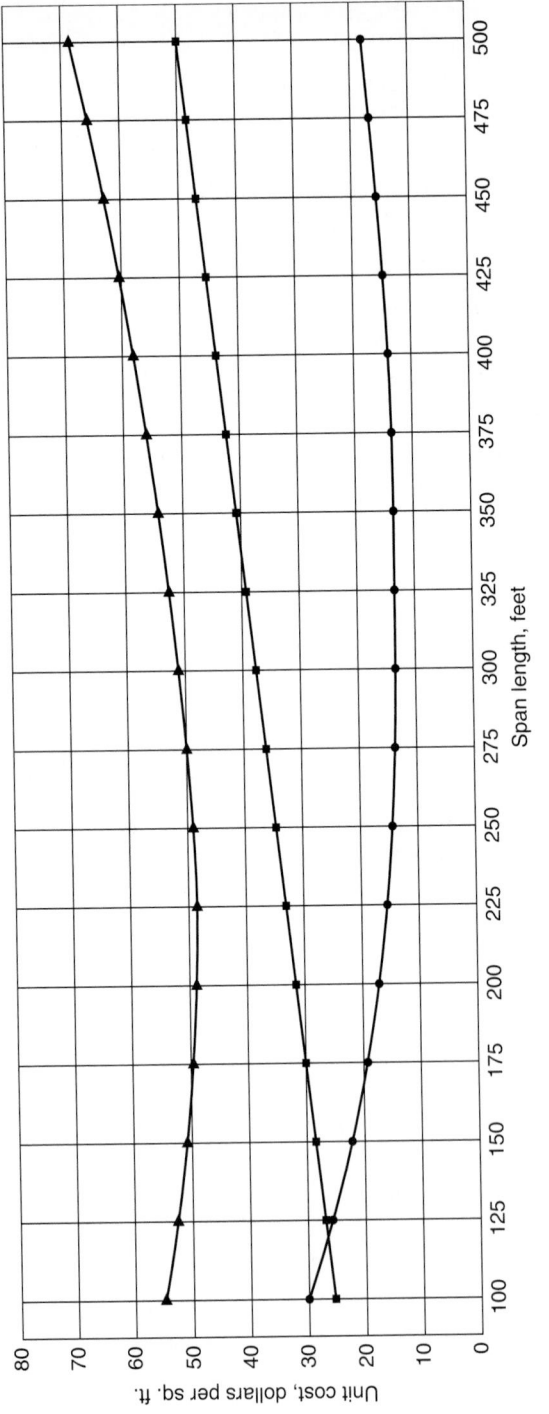

FIGURE 4.13 Example of cost study for optimum span length.

TABLE 4.3 Data Extracted from Compilation by FHWA of New Bridge, Bridge Replacement, and Bridge Rehabilitation Projects with Participation of Federal Funds in Calendar Year 1993

System	Number of bridges				Total cost of bridges, $				Percentage by cost			
	Reinforced concrete	Steel	Prestressed concrete	Other	Reinforced concrete	Steel	Prestressed concrete	Other	Reinforced concrete	Steel	Prestressed concrete	Other
Data classified by type of system for all improvement types (new, replacement, and major and minor rehabilitation)												
Interstate	61	271	157	4	36,933,376.00	370,438,185.00	279,678,910.00	495,043,134.00	3.1	31.3	23.7	41.9
Primary	244	293	384	5	83,410,908.00	408,618,032.00	310,407,853.00	1,480,393.00	10.4	50.8	38.6	0.2
Secondary	237	81	224	14	42,378,533.27	56,245,660.00	76,431,793.00	2,329,473.00	23.9	31.7	43.1	1.3
Urban	58	70	64	6	25,628,095.00	123,052,183.00	87,631,742.00	2,060,801.00	10.8	51.8	36.8	0.9
Off system	453	190	442	41	73,227,335.79	133,423,472.15	146,617,783.00	7,621,494.19	20.3	37	40.6	2.1
Grand total	1053	905	1271	70	261,578,248.06	1,091,777,532.15	900,768,081.00	508,535,295.19	9.5	39.5	32.8	18.4
Data classified by type of improvement for all systems												
New bridges, all systems	61	130	203	5	22,449,171.00	181,868,832.00	307,549,885.00	32,469,695.00	4.1	33.4	56.5	6
Bridge replacement, all systems	814	342	879	54	170,122,614.81	443,190,758.15	430,190,758.15	409,727,100.12	11.7	30.5	29.6	28.2
Major bridge rehabilitation, all systems	132	320	141	6	61,011,887.25	348,751,454.00	99,230,206.00	64,905,794.00	10.6	60.8	17.3	11.3
Minor bridge rehabilitation, all systems	40	91	35	1	4,659,624.00	73,159,747.00	6,776,191.00	1,016.07	5.5	86.5	8	—

4.37

The design and preparation of plans for bridge widening usually requires all the same elements as the preparation of the original plans for the structure, plus details and notes for partial removal of the existing bridge, rebar splice details, and notes on sequence of construction and maintenance of traffic. Therefore, it is a mistake to think of such a design project as "just a widening job" when estimating the hours required to design and prepare plans, or when reviewing such estimates for agency approval.

Bridges are generally widened in kind—that is, steel-beam bridges are widened with additional steel beams, prestressed-concrete beam bridges are widened using pre-stressed-concrete beams, etc. However, beam types different from the original have been used successfully in some widening projects.

It is sometimes possible to increase the design load capacity of a bridge when widening. If the bridge is steel and was originally designed and constructed noncompositely, it may be feasible to weld shear studs onto the top flange of the beam or girder if the deck slab must also be replaced. If the deck slab is good, another available technique is to carefully core holes in the slab over the beams, weld shear studs onto the beam flange, and fill the hole with high-strength concrete, thereby making the beam composite. One should always be cognizant of the effect of retrofits on fatigue life, just as one is conscious of the fatigue effect of structural details on new construction.

Another means of increasing load capacity is to space the existing beams closer together. Before this is done, a study should compare the cost of renovation with the cost of replacement with a new superstructure.

Posttensioning of members can also be used to increase the load capacity of existing bridges, or to correct deficiencies in the original design. External posttensioning of prestressed or posttensioned girders has been necessary on some bridges where the design did not adequately anticipate the magnitude of time-dependent deflections that occurred. In one case, a utility bridge developed a sag that trapped rainwater, further increasing the deflection. It was corrected by external posttensioning. Members of truss bridges can be posttensioned to increase the load capacity of the truss. A computer program is available from the BEST Center, University of Maryland, which allows analysis including the effect of posttensioning cables.

When determining the load capacity of an existing bridge, one should refer to the original plans, if available. These plans will generally state the design specifications used and the type and required strength of materials. For steel bridges for which plans are not available but the year of construction is known, the type of steel and the allowable stress may be obtained by reference to the AASHTO *Manual for Maintenance Inspection of Bridges.* If the bridge is a large or significant one, the type and strength of the steel should be determined by chemical and physical analysis performed on a coupon taken from the bridge. The chemistry, particularly the carbon equivalent, will be important if welding is proposed on the existing steel.

When evaluating the strength of an existing bridge for widening or rehabilitation, reference should be made to AASHTO publications dealing with evaluation and rating for strength and fatigue. Two such publications are the AASHTO *Guide Specifications for Fatigue Evaluation of Existing Steel Bridges* and the AASHTO *Guide Specifications for Strength Evaluation of Existing Steel and Concrete Bridges.*

Evaluation of an existing prestressed-concrete girder bridge for which plans are not available may be more difficult. The type of girder, whether a standard AASHTO shape or a state standard, may be determined by measurement. The number and size of strands may be apparent at an exposed end, but whether any strands are deflected or debonded, and whether the strands are stress-relieved or low-relaxation and what their strength is, are not easily determined. Full-scale load-deflection testing may provide some answers, but is very expensive. This illustrates the importance of maintaining and safeguarding the original plans and as-built drawings, and having complete design data on those plans.

4.20 REPAINTING OF EXISTING BRIDGES

Repainting of bridges over highways or railroads may necessitate use of protective covers, or require traffic lanes to be diverted or work interrupted during passage of trains, while existing paint is removed and new paint applied. These factors all favor use of a bridge material or protection system that does not require maintenance reapplication of a coating. (See Art. 4.13.)

Removal of Existing Paint. Complete removal of the existing paint on a bridge that is to be recoated with a paint system that requires it can be extremely expensive, particularly if the existing paint contains lead. Lead-based paints were used extensively in the past because they provided good protection. Because of the health issue involved, portions of bridges where lead-based paints are being removed are often required to be completely enclosed, and the paint particles contained and properly disposed of, often at great cost. Severe monetary penalties can be imposed if violations occur. (See Art. 1.4.)

Repainting. Maintenance paint should be applied with the same care as paint on new bridges, but it must be applied in a more difficult environment. Painting must be done under acceptable atmospheric and environmental conditions, particularly in regard to temperature, humidity, wind, and absence of dirt. Overspray onto vehicles and other objects must be prevented. Some painting contractors are consciously careless about this, preferring to take their chances and let their insurance company pay claims, rather than taking necessary precautions. This results in bad public relations between travelers and the owner.

Because of the cost associated with complete removal of existing paint, paint systems that do not require complete removal, but only removal of loose paint and minimal preparation of sound paint and exposed steel, are much desired. Some such systems are on the market. While longevity can be projected by accelerated testing, only real-time exposure will truly prove their worth.

Inspection. Thorough inspection during repainting contracts is essential to satisfactory performance of the work. Painting contractors often work during off-hours, and so the owner's inspectors should be prepared to work those same hours. Inspectors should follow closely behind the painters. On high bridges, the use of inspection devices such as high lifts, Reach-Alls, Snoopers, or cherry pickers, which permit inspectors to reach areas otherwise not readily accessible, will keep the painters on their toes just through awareness of their availability to the inspector.

4.21 DECK DRAINAGE

Adequate drainage of the deck is important for safe operation during rainstorms, to prevent accumulation of rainwater or snowmelt that could freeze and cause skidding, and to prolong the life of the deck by removing standing water, which would otherwise contribute the water element necessary for corrosion.

Design of the drainage system is frequently one of the last items listed in the scope of services for a bridge design project. This implies a secondary importance, but that is certainly not the case. On the contrary, the South Carolina DOT requires that scuppers of adequate size and spacing for high-intensity rainfall be shown in the preliminary bridge plan view. The DOT reviewers check the scupper design as part of the preliminary design review. This is the stage when deck drains should be designed. At this stage changes as major as span layout revision can be made with relative ease to ensure adequate drainage, including provision for longitudinal conductors with acceptable slopes, and downspouts, if necessary.

Scupper Size and Shape.   Bridge scuppers are unlike roadway drains, which can have large grates through which the storm water flows and drops into a large sump and then is conducted away by transverse or longitudinal drain pipes. Bridge scupper outlet pipes generally must be small circular, square, or rectangular pipes, and so any enlargement of the scupper surface opening must be limited to prevent debris from being trapped, clogging the scupper and making it ineffective. A minimum surface opening larger than the diameter of a beverage can is desirable and should be maintained to the outlet end. An open 4-in- or 6-in-diameter pipe meets this requirement and is economical compared with scuppers having large fabricated boxes with crossbars, so that more of them can be provided at the same cost. Where the discharge can be corrosive, such as where deicing salts are used, the pipe should extend below the bottom of an adjacent beam to prevent the discharge from being blown onto the beam. This is especially important on weathering steel bridges, and is also important on painted steel and concrete bridges. In areas that do not experience freezing, a formed opening in the deck may be acceptable, or the pipe need not extend below the beam. In this case one can expect a stain to develop on the beam onto which the discharge is blown.

Deck Drainage Criteria.   Deck drainage design is based on preventing storm water from spreading in the gutter more than an acceptable amount during a rainfall of a given intensity. For example, the maximum permissible spread may be the width of the shoulder, or the spread may be permitted to extend across half of the outside lane of a multilane directional roadway. One source of such drainage design criteria is the U.S. Department of Transportation document "Drainage of Highway Pavements," Hydraulic Engineering Circular 12, March 1984.

Scupper Design Procedure.   After the size of the scupper opening is established, the spacing may be determined. Rather than try to directly determine the spacing, it may be easier to select a trial spacing and then check the adequacy of that spacing by using a hydraulic analysis method acceptable to the client. The use of computer programs greatly expedites this task.

Collection of Runoff.   In areas such as reservoirs and sensitive wetlands, it may be necessary to provide a collector system and temporarily detain the first half inch or inch of rainfall from each storm. It is assumed that the initial rainwater that falls at the beginning of a rainstorm will contain most of the roadway pollutants that can be carried away with the runoff. The objective is to prevent these pollutants, especially petrochemicals (crankcase drippings, fuel spills, etc.), from polluting the area under the bridge. The collected storm water is treated and then discharged. In some cases this protection is extended to the roadway as well.

4.22   BRIDGE BEARINGS

For concrete-slab bridges where expansion is not provided, the slab is normally supported directly on the substructure, concrete on concrete. A "centerline of bearing (singular)" is denoted on plans at each support. (Some states do not identify a centerline of bearing at the end bent of a slab bridge. Instead, they measure the end span to the end of the slab.)

In other types of bridges, individual bearings are used to support the superstructure. The centerline of these devices is denoted as the "centerline of bearings (plural)."

AASHTO requires that steel bridges with spans of 50 ft or greater have a type of bearing employing a hinge, curved bearing plates, elastomeric pads, or pin arrangements for deflection (rotation) purposes. (This specification does not distinguish between simple and continuous spans. Presumably, it was written for simple spans, and so it would make sense that a span greater than 50 ft be allowed for pier bearings not providing for rotation when spans are continuous.)

Bearings consist of some or all of the following components:

- Masonry plate resting on the substructure bridge seat
- Rotation device
- Sliding device
- Movement-restraining devices, or "keepers"
- Sole plate attached to the superstructure

Bearings may be fixed bearings, providing for rotation only and preventing differential movement between superstructure and substructure, or expansion bearings. Sliding expansion bearings have a finite capacity depending on the length of the contact surface, or may employ keepers to limit the movement. The range of movement accommodated by the bearing should be greater than the calculated movement.

The type of bearing is typically denoted on the elevation view of the general plan and elevation sheet in the project plans, using "E" for expansion and "F" for fixed.

The main types of bearings are:

- Sliding plates
- Rockers and bolsters
- Pins
- Rollers
- Elastomeric bearings
- Disk bearings
- Pot bearings

Sliding Bearings. Sliding bearings will generally have a component made from a material that has a lower coefficient of friction than steel, and that is more corrosion-resistant. Bronze has been used in the past, but has not always maintained sliding capability over the life of the structure. When bearings "freeze," that is, lose their sliding capability, forces much greater than those anticipated in the design can be exerted on the substructure and ends of beams, doing great damage to both. To reduce friction and prolong the life span of bronze bearings, long-lasting lubrication can be forced under great pressure into trepanned rings on the surface of the bronze. These bearings are known by the brand name Lubrite.

Low friction can be achieved by use of polytetrafluoroethylene (TFE) sheets mated with stainless steel. This combination is included in many current bearing types to provide for expansion, while other components are used to accommodate rotation. The TFE can be in solid sheet form or woven fabric. During shipping and storage at the job site, the assembly should be banded to prevent dirt from contaminating the sliding surface. The configuration of the bearing should be such that the sliding surface will not easily become dirty in service.

Rockers and Bolsters.   Another means of allowing the superstructure to move, without sliding, is by use of rockers. Rockers, as the name implies, permit the superstructure to rock, like a person in a rocking chair, on the substructure. Rockers (Fig. 4.14*a*) consist of a masonry plate that is bolted to the concrete bridge seat; a rocker element, which is a heavy steel fabrication with a large-radius curved bottom surface and a small-radius semicircular convex pintle on top; and a sole plate, which has a mating concave surface. The height of the rocker is made proportional to the anticipated movement. Within the design range of movement, as the superstructure translates, the rocker tips, but the reaction to the base plate is maintained within the geometric limits of the rocker, so that the rocker does not tip over. To help prevent the rocker from tipping over, the space between the shoulder of the rocker and the bottom of the sole plate is limited, so that the assembly will bind as its capacity is reached.

Compared with sliding plates, rockers use more massive plates and are therefore less susceptible to severe corrosion and freezing. Because their inclination is readily visible, inspectors can easily determine whether they are functioning properly. Assuming that the rockers were properly installed to be vertical at a given average temperature, one can easily see whether they are inclined excessively or in the wrong direction. In cold weather, when the tops of the rockers should be tipped toward the center of the bridge, an opposite inclination indicates either that unexpected movement of the superstructure has occurred, or that the substructure has moved. Sliding plates can give similar indications, but they require closer observation.

Generally, when rockers are used for expansion bearings, bolsters are used for fixed bearings. The bolster (Fig. 4.14*b*) is a large steel fabrication consisting of a base plate, which is bolted to the concrete bridge seat; a pintle, which extends upward from the base plate and has at its top a machined convex semicircular shape that fits into a mating concave shape in the sole plate; and reinforcing plates on the sides of the pintle. These side plates are tapered, being wider at the base. This configuration of the mating surfaces allows the beam or girder to rotate, but fixes the superstructure against translation.

Rockers and bolsters, being taller than sliding plates, place the superstructure higher above the substructure. This is desirable for inspection and maintenance, including painting, and in fact is in line with AASHTO requirements that beams, girders, and trusses on masonry be so supported that the bottom flanges or chords will preferably be 6 in minimum above the bridge seat. However, if rockers or bolsters are tall and narrow, they may be aesthetically undesirable for overpass structures, giving the appearance of placing the superstructure on stilts.

Pin Bearings.   The pin bearing is used where a fixed condition is desired but rotation needs to be accommodated. As illustrated in Fig. 4.15, it consists simply of a masonry plate, a bottom plate welded to the masonry plate and machined to receive a pin, the pin, and a sole plate that is machined to bear on the pin. The pin is fabricated with shoulders to restrain it laterally within the plates. Clearance is provided between the ends of the plates and the inside faces of the shoulders to allow for lateral expansion of the bridge. For wide bridges, a greater clearance should be provided. A smooth finish is machined onto the pin and the mating surfaces. To prevent corrosion, all parts should be galvanized or metallized.

Rollers.   For long-span bridges with large reactions, rollers have been used, sometimes in combination with a geared rocker mechanism that is used to transmit the

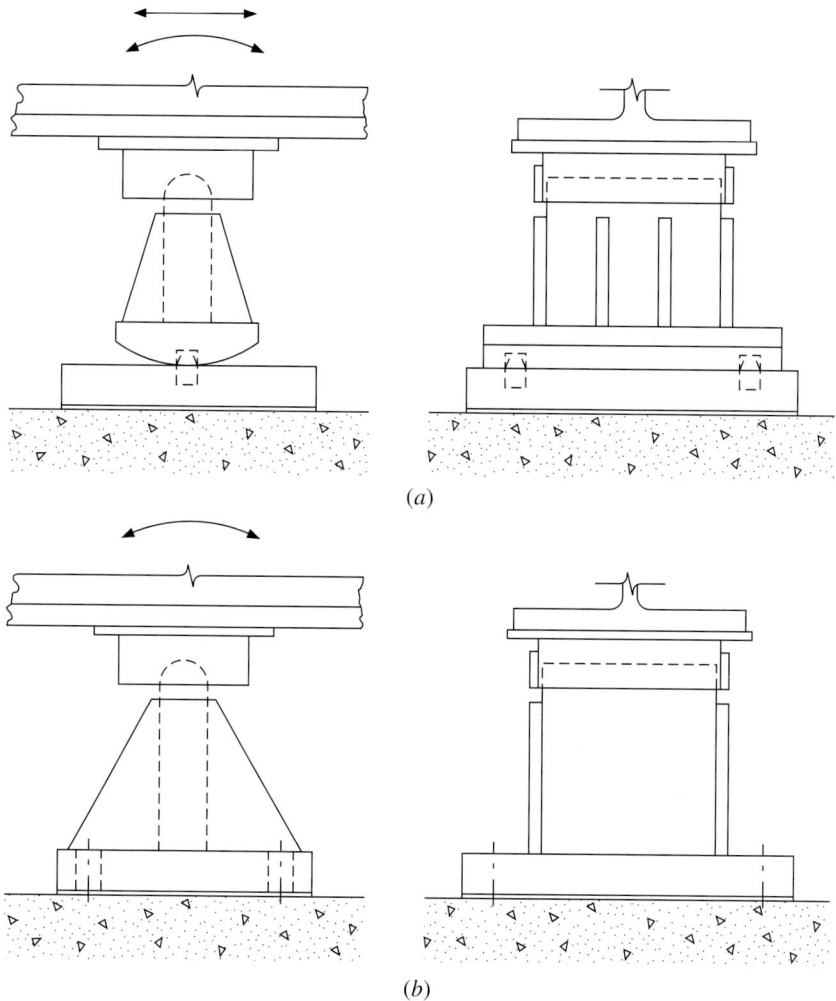

FIGURE 4.14   Bearings for structural steel showing (*a*) rocker (expansion) type and (*b*) bolster (fixed) type.

superstructure reaction to the rollers. Several rollers are usually installed in a roller nest, which is a box having (1) a bottom plate, on which the rollers bear, (2) end plates, and (3) substantial side bars by which the relative position of the rollers is maintained. Grease is placed in the box, and a skirt is added to shield the rollers from water and dirt. Unfortunately, the measures taken to prevent corrosion have often not been successful, and the rollers end up being piles of rusted steel, with all expansion capability lost.

Elastomeric Bearings.   The elastomeric bearing, in which superstructure translation can be accommodated by shear, is often the most economical type for both steel and

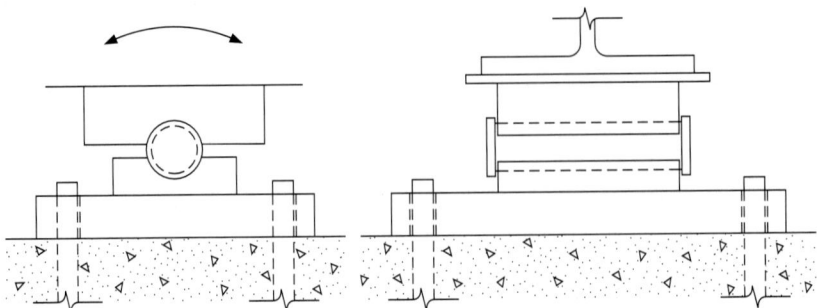

FIGURE 4.15   Fixed-pin bearing for structural steel.

concrete bridges. As shown in Fig. 4.16, this bearing consists of an elastomer such as natural or synthetic rubber (polychloroprene, or neoprene), with or without internal reinforcement, which may be steel plates or glass fiber fabric laminates. A steel-reinforced elastomeric bearing is cast as a unit in a mold and is bonded and vulcanized under heat and pressure; fabric-reinforced elastomeric bearings, popular in California, may be vulcanized in sheets and cut to size. Elastomeric bearings may have external steel load plates bonded to the upper or lower elastomer layer, or both. Natural rubber has better low-temperature properties than neoprene but is not as resistant to surface decay.

In addition to accommodating horizontal movement by deforming in shear, elastomeric bearings can accommodate superstructure rotation. AASHTO design specifications provide methods for properly designing elastomeric bearings, taking into account both translation and rotation requirements. Another desirable attribute of elastomeric bearings is that they can tolerate movements or rotations in directions other than longitudinal. This is not true of sliding plates, rockers and bolsters, and pin bearings. For structures with large skew or curvature, where it is known either qualitatively or quantitatively that such out-of-plane rotations exist, this is a desirable quality. Elastomeric bearings can be fixed bearings (shear prevented), in which case the allowable average compressive stress may be increased 10 percent over that permitted for bearings allowed to deform in shear. Shear is prevented by placing anchor bolts

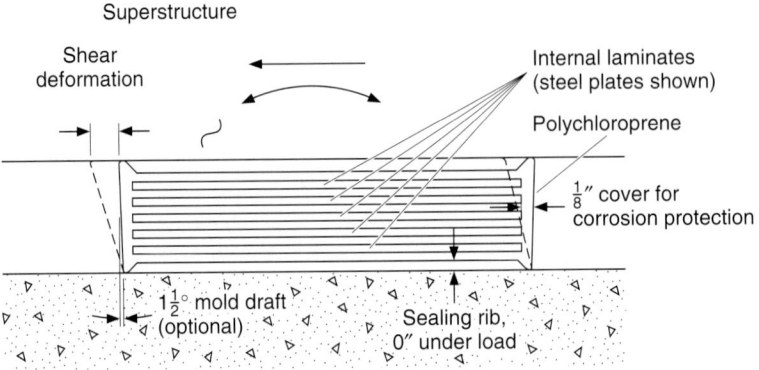

FIGURE 4.16   Elastomeric bearing for structural steel or concrete.

through holes in the bearing, the holes being only slightly larger than the anchor bolts. Steel reinforcement of elastomeric bearings is protected against corrosion by being contained in the elastomer. A minimum cover of $\frac{1}{8}$ in is maintained at the edges of the bearing, except at laminate restraining devices and around holes that are entirely closed in the finished structure.

Under load, elastomeric bearings will undergo a compressive deflection that is determined by the shape factor (loaded plan area divided by the perimeter area free to bulge) and the hardness of the elastomer. Where the dead load compressive strain is significant, allowance should be made for it when establishing bridge seat elevations. The compressibility of elastomeric bearings should also be considered in the design of expansion joints. Joints with overlapping steel elements should be avoided.

A significant shear force can be induced in an elastomeric bearing by movement of the superstructure. This force should be calculated and used in design of the substructure, taking into account also the flexibility of the substructure. For large-movement bearings, a tall and uneconomical elastomeric bearing would be required if the movement were taken entirely by shear. As an alternative, a sliding surface can be combined with an elastomeric bearing so that the bearing initially deforms in shear until the shear force exceeds the frictional resistance of the sliding surface, at which point the bearing slides.

Disk Bearings. Disk bearings are used where rotations occur in different planes. They consist of a polyether urethane disk confined by upper and lower steel bearing plates. Fixed-disk bearings provide for rotations in all directions but do not provide for longitudinal or transverse movements.

To permit expansion, a polytetrafluoroethylene (TFE) to stainless steel sliding surface is provided above the upper bearing plate. Expansion disk bearings may be guided or nonguided. In a guided bearing, a guide bar or keyway system is used to restrict transverse movement, with the sliding surfaces being TFE and stainless steel. Nonguided bearings allow rotation and longitudinal and transverse movement.

Pot Bearings. Pot bearings, like disk bearings, are used where rotations occur in different planes. In a pot bearing, the rotational motion is accommodated by compression of elastomeric material in a shallow steel base cylinder, or pot. The load is transmitted to the elastomer through a circular plate or piston, which is part of the upper load plate and which is just slightly smaller in diameter than the inner circumference of the pot. The surface of the elastomeric rotational element is lubricated or has TFE attached to it to facilitate rotation. Brass sealing rings are used between the steel piston and the elastomeric rotational element to prevent the elastomeric material from being squeezed out. This type of bearing is illustrated in Fig. 4.17. The elements of pot bearings that provide for guided or nonguided expansion are like those described for disk bearings. As can be concluded from the discussion of disk and pot bearings, these devices require expensive machining and demand high-quality materials. They are therefore expensive and will not likely be used where other, less costly bearings can serve adequately.

4.23 PROVISION FOR INSPECTION OF NEW BRIDGES

In the design of a new bridge, provision must be made for maintenance inspection. For example, plate girders can be provided with safety handrails, safety railings can be

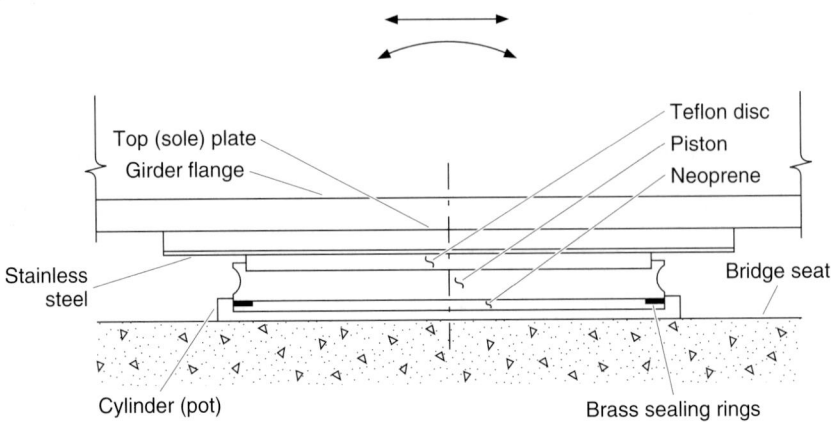

FIGURE 4.17   Pot bearing for structural steel or concrete.

specified on top of wide piers for inspectors to check bearings, and safety ladders can be installed to provide access to elements of the bridge otherwise difficult to reach. For deck-type bridges of moderate span and width, it will be possible to access the superstructure from special bridge inspection vehicles operating on the deck. For longer spans where the depth of girder exceeds the vertical capacity of a boom, and for wider bridges where the horizontal reach of the boom is not adequate, it may be necessary to provide catwalks or permanent movable inspection platforms. These devices are becoming increasingly popular as inspection and maintenance require-ments are given the attention they deserve in the design process.

4.24   SCOUR

Stream scour can undermine bridge piers or abutments, resulting in collapse of spans and loss of life. Several such incidents in recent years, including collapses in Alabama, New York (Schoharie Creek bridge on the New York Thruway, 1987), and Tennessee, have caused FHWA to mandate the evaluation of all highway bridges for scour vulnerability by 1997.

An insidious aspect of scour is that soil around a foundation can be removed and redeposited during a flood without leaving clear evidence that this has occurred, so that material may be present but may not provide the required support. Beyond sur-veying the stream bottom for local lowering of the flow line and inspecting around the pier by visual, manual, and remote means, current techniques for determining whether a loss of support has occurred are limited. They include physical probing and use of ground-penetrating radar.

Scour Study.   A scour study at an existing bridge will include some or all of the fol-lowing:

• A channel bottom physical inspection
• A channel bottom topographic inspection

- A localized scour evaluation conducted around each substructure element
- Photographic or video recording of observations
- Hydraulic analysis
- Soils investigation including laboratory testing

Hydraulic Analysis.   In the hydraulic analysis, depth of scour is calculated for 100-year and 500-year floods. Inclusion of the 500-year flood calculation reflects a change of thinking in regard to bridge hydraulics that has taken place in the last 20 or 30 years. Previously, it was thought acceptable to have a very small percentage of bridges wash out in a severe flood, and if this did not occur the hydraulic design requirements were considered excessive. The current thinking is that a complete washout should be avoided, even in very extreme floods. A difficulty in implementing this policy, as in earthquake engineering, is that the hydrologic database has been developed over a relatively short period of time in the United States.

A series of hydraulic analysis computer programs are available to assist in scour analysis. They include HEC-18, Evaluating Scour at Bridges; HEC-20, Stream Stability at Highway Structures; and FHWA's water surface modeling program, WSPRO. These programs are primarily for inland streams. In coastal areas, tidal velocities and hurricane surge velocities may also cause scour. To perform the hydraulic analyses for these conditions it may be necessary to obtain data from the Federal Emergency Management Agency (FEMA) and the National Oceanic and Atmospheric Administration (NOAA) and to use other analysis techniques.

Soils Investigation.   The objective of the soils investigation is to determine whether and to what degree the soils are subject to being eroded. Grain size is of particular interest.

Countermeasures.   Where a potential for undermining is found, countermeasures will be required to ensure the stability of the bridge. Countermeasures include riprap, poured-concrete protective aprons with keyed edges, cabled-concrete sections, pre-cast-concrete units, rock-filled basket mattresses, and protective piles. In the case of new bridges, where more opportunities for preventing scour exist, some of the available options are a larger waterway opening that reduces stream velocity, location of piers out of the scour-vulnerable zone, use of deeper piles, and selection of a different pier shape.

Design Information on Plans for New Bridges.   Information from the scour analysis for a new bridge should be placed on the construction drawings so that a permanent record of scour estimates, and their effect on design, is readily available for future inspections and for improvement of this design process.

4.25   SEISMIC DESIGN

In recognition of the serious potential destructive effects of earthquakes, AASHTO Specifications contain comprehensive provisions for seismic design. Although earlier specifications contained some provisions, the more comprehensive provisions were not adopted until the 1980s. They were based on a detailed study by consultants who were specialists in that field, with review and participation by bridge engineers and design firms. The standards developed apply to conventional steel and concrete girder and box girder construction with spans up to 500 ft, but do not cover suspension, cable-stayed, arch-type, and movable bridges.

Bridges and components designed to the AASHTO seismic provisions may suffer damage under severe seismic events, but should have a low probability of collapse due to ground shaking. The general philosophy adopted in the development was:

- Small to moderate events should be resisted elastically without significant damage.
- Realistic seismic ground motion intensities should be used in design.
- Large events should not cause bridge collapse, and damage that occurs should be readily detectable and repairable.

Seismic performance categories are assigned on the basis of a ground acceleration coefficient for the site determined from a contour map of the United States, and an importance classification of "essential" or "other." Different degrees of design complexity are specified, depending on the seismic performance category. Site coefficients are applied to approximate the effects of the site conditions (soil profile) on the response. Lateral forces and displacements may be determined from a single-mode spectral analysis, a multimode spectral analysis, or more rigorous procedures. Elastic response is assumed in the analysis, but forces are adjusted with response modification factors. The lateral forces are applied in orthogonal directions to account for the directional uncertainty of earthquake motions. An important requirement specifies the minimum length of the bearing seat supporting the expansion ends of girders, determined as a function of the span length and the height of the supporting columns. Foundation design is also treated.

Seismic retrofit is a major consideration for older structures, particularly in the western United States. During recent years, serious distress and collapse of some bridges in California during the Loma Prieta (1989) and Northridge (1994) earthquakes has received wide publicity. However, the problem structures were generally those designed and constructed to earlier standards. Where bridges were built according to modern methods, problems were minimal. Problems included failure of reinforced-concrete rigid-frame supports, failure of reinforced-concrete columns, columns punching through decks, and collapse of a structural steel span where the longitudinal displacement was excessive.

Active programs are in place to retrofit older structures to current criteria, but it is a massive undertaking that requires several years to accomplish. Some of the techniques being applied include (1) increasing the length of the seats for the bearings to provide a greater tolerance for longitudinal displacements, (2) adding cable restraints and hold-down devices at supports and hinges to restrict movement and keep members in place, (3) adding spiral reinforcing steel and steel jackets or composite overwraps to strengthen concrete column piers, and (4) adding foundation tie-down rods inserted into holes drilled into the soil. In the Northridge earthquake, several structures that had recently been retrofitted survived intact, giving confidence to the retrofit program. Retrofitting is not limited to the west but is under way in other parts of the United States as well.

CHAPTER 5
CULVERTS, DRAINAGE, AND BRIDGE REPLACEMENT

Paul W. Cotter, P.E.

Underground Structures Engineer
California Department of Transportation
Sacramento, California

A properly designed highway requires a well-designed drainage system. This requires a determination of the quantity of runoff reaching the drainage structures and an accurate analysis of water flow through the structures in order to properly size them. Also, a working knowledge of culvert structural characteristics and effects of environmental factors is necessary to provide for long-term performance. Timely inspection and maintenance of drainage facilities will ensure satisfactory service life. If all of these issues are properly addressed, an efficient drainage system can be developed. Because a large percentage of highway funds is spent on culverts and storm drains, it is incumbent upon the engineer to use funds wisely and create an efficient drainage system. Thus, the roadway and adjacent property will be protected without wasting taxpayers' money.

This chapter includes a review of fundamental hydrology considerations and runoff estimation, fundamentals of the hydraulics of open-channel flow, and design considerations and methods for the various components of highway drainage. The design, construction, and service life of both flexible and rigid pipe are addressed, as well as rehabilitation and maintenance. The range of products is broad, extending from small-diameter drainage pipe to long-span structures that may be used for the replacement of short-span bridges. Article 5.6 may be referred to for a general description of the major products available.

5.1 HYDROLOGY

The science of hydrology is concerned with the estimation of the intensity of rainfall, the distribution of the flow of the rainwater over the land, and the determination of the flow quantity (peak and total) that eventually reaches some specified point. Of primary concern to the highway engineer is the frequency of occurrence of the peak discharge. Although many methods for determining runoff have been proposed over the years, making an accurate prediction is difficult, because of the many and varying parameters that contribute to the complexity of the problem. These parameters include the affected drainage area, the rainfall intensity, the time of concentration of the rainfall, and the percent of the rainfall that will actually reach the point under consideration. In addition to the difficulty in forecasting flows due to the inaccuracies in measuring and predicting the above parameters, different techniques that are commonly used to predict flows may produce significantly different results for a specific site and situation.

The objective of hydrologic analysis is to estimate the quantity of runoff for which a specific hydraulic structure must be designed. The magnitude of the study must be proportional to the risks involved. Those risks include the potential for damage to the roadway and adjacent property and the importance of the roadway in the transportation system.

5.1.1   Watershed Characteristics

Characteristics of the watershed area directly affect the hydrologic analysis. Basic features of the watershed basin include size, shape, slope, land use, soil type, storage, and orientation.

The size of the watershed basin is the most important characteristic affecting the determination of the total runoff. It is generally measured in acres, square miles, or square kilometers and is defined by the limits of the topographic divide. A *topographic divide* is a line that separates water flow between basins, thus causing the rainfall that falls on one or the other side to flow into a particular watershed. The location of this divide, and thus the perimeter of the basin, may be determined from aerial photographs, topographic maps available from the U.S. Geological Survey (USGS), and field surveys.

The shape of the watershed primarily affects the rate of water flow to the main channel. Because the rainfall in narrow watersheds reaches the main stream relatively quickly, a narrow basin generally has a low peak discharge compared with a fan- or pear-shaped basin of otherwise similar characteristics.

The main effect the slope has on water flow is on the time of concentration, or the time it takes the rainfall to flow from the farthest point in the watershed to the point under consideration. Everything else being equal, steeper slopes cause a shorter time of concentration, and thus a higher peak discharge, than do flatter slopes.

The use of the land and the type of surface the precipitation falls upon have an obvious impact on the flow of water. Developed areas covered by asphalt or concrete will allow a much greater percentage of the rainfall to flow to the point under deliberation than will an undeveloped vegetated area. In addition, the study must take into account any reasonably anticipated changes in the use of the land. The vast majority of changes to the land cause an increase in the surface runoff due to an increased percentage of impervious surfacing.

Peak flows may be reduced by the effective storage of drainage water. Of the three main types of storage—interception, depression, and detention—detention storage has the major impact in determining runoff. *Interception* refers to storage on aboveground fixtures such as plants, and *depression* refers to storage in depressions in the ground surface. Interception storage will eventually evaporate, and depression storage will either evaporate or infiltrate into the ground. *Detention* storage is runoff that is either in transit to the main channel or in storage in a pond, swamp, basin, or constructed detention chamber prior to transmission.

The final characteristic of the watershed basin is orientation. Taking into account the slope of the basin, if it is north- or south-facing, the runoff may be affected. If the basin accumulates snow and faces north, the snow may not melt until the late spring. If the snow melt is caused by a spring rain, the total runoff will be increased. On the other hand, if the basin faces south, the snow melt may come much earlier in the year, and with evaporation and infiltration, it may not contribute as greatly to the runoff.

5.1.2 Flood Frequency

There are two accepted alternatives for determining the design flood frequency at a specific site: (1) by policy and (2) by economic assessment. An example of an establishment of a design flood frequency by policy is the Code of Federal Regulations, which specifies that the design flood for encroachment onto through lanes of interstate highways shall not be less than the 50-year discharge. Most state and local agencies have established guidelines for policy requirements of design flood frequencies. For example, whereas bridges are designed to convey a 50-year discharge with a specified freeboard and to convey the 100-year discharge with no freeboard, California has adopted the policy that culverts may be designed for a 10-year flood without headwater or to convey the base flood without damage to the facility or adjacent property. The base flood is defined as the flood or tide having a 1 percent chance of being exceeded in any given year, which is also defined as the 100-year flood. A design flood is a flood that will not inundate the highway—that is, will not cause the through lanes to be overtopped. An overtopping flood is a flood that will overtop the roadway, culvert, or bridge.

Blind adherence to the policy guideline to determine the design storm should be avoided. As a minimum, a range of peak flows should be considered and their potential effects on the traveling public, the potential damage to upstream and downstream properties, and the possibility of loss of life should be analyzed. This preliminary assessment will indicate whether the policy determination for the design flood frequency was applicable or whether further analysis is required. Additional studies could take the form recommended in the Federal Highway Administration (FHWA) Hydraulic Engineering Circular (HEC) 17 of providing the greatest flood hazard avoidance at the least total expected cost.

5.1.3 Estimation of Runoff by Statistical Methods

Estimating the peak discharge for which highway drainage structures are to be designed is one of the most common problems and biggest challenges faced by the highway engineer. The problem may be separated into two categories: (1) watersheds for which historical runoff data are available, those with gauged sites; and (2) areas for which no data are available. Gauged sites lend themselves to analysis of runoff by statistical methods, whereas ungauged sites rely upon hydrologic equations based on the hydrologic and physiographic characteristics of the watershed.

The runoff data necessary to utilize statistical methods are available through the USGS, which is the primary collector of such data. Appendix C of HEC 19, "Hydrology," consists of a list of local, state, and federal agencies that maintain records of peak discharges for specific sites. Provided that sufficient data are available for a specific site, a statistical analysis may be made that will result in a reasonable determination of the peak discharge. Water Resources Council Bulletin 17B, 1981, suggested that a minimum of 10 years of historic data are necessary to make an accurate estimation based on statistical methods. The USGS has no specific time requirements for historical hydrologic data collection. In the past, however, the recommended time period varied between 10 years for a 10-year design flood to 25 years for a 100-year design flood. HEC 19 should be referenced for different techniques available for determining the inferences of population characteristics from statistics.

Data collection can be categorized and arranged in groups that lend themselves to statistical analysis. The common groupings are by magnitude of peak annual discharge, by time of occurrence, and by geographic location. Of the three, magnitude of peak annual discharge is the most useful in determining peak discharge. Time of occurrence is the most useful in trend analysis or determining the effects of changing land use on runoff. Grouping by geographic location is the most useful when looking at sites that have insufficient flood data either because they are ungauged or because the historical time frame of the collected data is too short.

There are several standard frequency distributions that have been extensively studied in the statistical analysis of hydrologic data. Three of the most useful are (1) log-Pearson type III distribution, (2) lognormal distribution, and (3) Gumbel extreme value distribution. The log-Pearson type III distribution is popular largely because the distribution very often fits the available data closely and it is flexible enough to be used with many other types of distributions. Because of this flexibility, the US. Water Resources Council has recommended that it be used by all governmental agencies as the standard distribution for flood frequency studies. The characteristics of the lognormal distribution are the same as those of the classical normal or gaussian mathematical distribution except that the flood flow at a specified frequency is replaced with its logarithm and has a positive skew. *Positive skew* means that the distribution is skewed toward the high flows or extreme values. The characteristics of the Gumbel extreme value distribution (also known as the *double exponential distribution* of extreme values) are that the mean flood occurs at the return period T_r of 2.33 years and that it has a positive skew.

If runoff data are unavailable for a specific watershed area, one method that may be used to determine the peak stream discharge is a *regional flood-frequency* analysis. By using historical runoff records from similar drainage basins in the immediate area, estimates of peak discharges may be developed. (For the USGS procedure, see T. Dalrymple, "Flood Frequency Analysis," USGS Water Supply Paper 1543-A, 1960.)

5.1.4 Estimation of Runoff by Peak Flow Equations

Where there are no or insufficient stream gauging records available, peak flow methods such as the "rational method" and the Soil Conservation Service method may be used. The *rational method* is the most common procedure for determining the quantity of flow for the design of minor hydraulic structures. Its use in the United States dates back to the late 1800s. One of the basic design assumptions for its use is that the rainfall intensity is uniform throughout the watershed. This assumption limits its use to relatively small watershed areas (in the neighborhood of 200 to 300 acres, or 1 km²).

The rational method is based on the simple intensity-runoff equation

$$Q = KCiA \qquad (5.1)$$

where Q = design discharge, ft³/s (m³/s)
  C = runoff coefficient
  i = average rainfall intensity, in/h (mm/h) for selected frequency and duration equal to time of concentration
  A = drainage area, acres (km²)
  K = 1.0 for English units (0.278 for SI units)

If significant tributaries feed into the watershed area, the rational method should be applied separately to each one and the flows channeled to the main stream. The flow

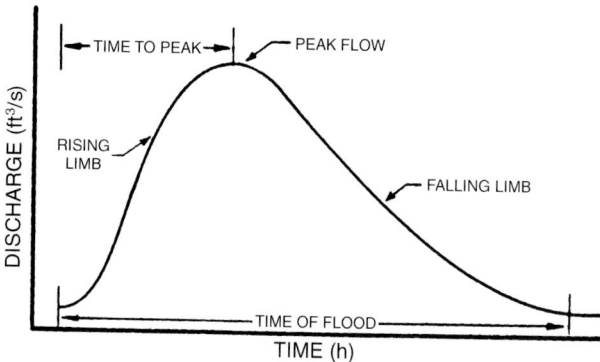

FIGURE 5.1   Typical flood hydrograph. (*From* Highway Design Manual, *California Department of Transportation, with permission*)

in the main stream may be considered through the construction of a hydrograph, which is merely a plot of stream discharge passing a selected point versus time. A schematic hydrograph is shown in Fig. 5.1. [Information and estimation methods for hydrographs may be found in HEC 19 and in *Modern Sewer Design*, American Iron and Steel Institute (AISI).]

After precipitation falls to the earth, it either is infiltrated into the earth, is evaporated back into the atmosphere, is subjected to depression or detention storage, or becomes runoff. The runoff coefficient C in Eq. (5.1) depicts the percent of precipitation that will run off the ground from the storm. Representative values of C for undeveloped and developed areas, respectively, are given in Tables 5.1 and 5.2. The application of the coefficient for any particular area should take into account possible future uses of that area. Also, if the watershed is made up of various surfaces, a weighted average should be used for C. This may be determined for surfaces with coefficients C_1, C_2, etc., and areas A_1, A_2, etc., as follows:

$$C = \frac{C_1 A_1 + C_2 A_2 + \cdots}{A_1 + A_2 + \cdots} \tag{5.2}$$

The intensity value i in Eq. (5.1) is dependent upon the time of concentration of the storm and the frequency of the design storm selected. Once these two parameters are selected, the rainfall intensity may be determined from an intensity-duration-frequency (IDF) curve. Such curves, which are derived from an accumulation of rainfall data recorded over the years, are available from both local and regional public agencies. (See, for example, U.S. Weather Bureau Technical Paper No. 25, "Rainfall Intensity-Duration-Frequency Curves for Selected Stations in the US.") A method for developing rainfall intensity curves and equations is shown in FHWA publication HEC 12. A typical IDF curve is shown in Fig. 5.2.

5.1.5   Estimation of Time of Concentration

The time of concentration or rainfall duration is equivalent to the length of time it takes for the runoff to travel from the most remote point of the watershed to the point under investigation. This assumes that there is a uniform rate of rainfall over the entire

TABLE 5.1 Runoff Coefficients for Undeveloped Areas

	Watershed type			
	Extreme	High	Normal	Low
Relief	0.28–0.35 Steep, rugged terrain with average slopes above 30%	0.20–0.28 Hilly, with average slopes of 10 to 30%	0.14–0.20 Rolling, with average slopes of 5 to 10%	0.08–0.14 Relatively flat land, with average slopes of 0 to 5%
Soil infiltration	0.12–0.16 No effective soil cover, either rock or thin soil mantle of negligible infiltration capacity	0.8–0.12 Slow to take up water, clay or shallow loam soils of low infiltration capacity, imperfectly or poorly drained	0.06–0.08 Normal; well-drained light or medium textured soils, sandy loams, silt and silt loams	0.04–0.06 High; deep sand or other soil that takes up water readily, very light well-drained soils
Vegetal cover	0.12–0.16 No effective plant cover, bare or very sparse cover	0.08–0.12 Poor to fair; clean cultivation crops, or poor natural cover, less than 20% of drainage area over good cover	0.06–0.08 Fair to good; about 50% of area in good grassland or woodland, not more than 50% of area in cultivated crops	0.04–0.06 Good to excellent; about 90% of drainage area in good grassland, woodland, or equivalent cover
Surface storage	0.10–0.12 Negligible surface depressions (few and shallow); drainageways steep and small, no marshes	0.08–0.10 Low; well-defined system of small drainageways; no ponds or marshes	0.06–0.08 Normal; considerable surface depression storage; lakes and pond marshes	0.04–0.06 High; surface storage, high; drainage system not sharply defined; large floodplain storage or large number of ponds or marshes

Given: An undeveloped watershed consisting of (1) rolling terrain with average slopes of 5%, (2) clay-type soils, (3) good grassland area, and (4) normal surface depressions.

Find: The runoff coefficient C for the above watershed.

Solution:
Relief	0.14
Soil infiltration	0.08
Vegetal cover	0.04
Surface storage	0.06
	$C = 0.32$

Source: From *Highway Design Manual*, California Department of Transportation, with permission.

TABLE 5.2 Runoff Coefficients for Developed Areas

Type of drainage area	Runoff coefficient
Business:	
Downtown areas	0.70–0.95
Neighborhood areas	0.50–0.70
Residential:	
Single-family areas	0.30–0.50
Multiunits, detached	0.40–0.60
Multiunits, attached	0.60–0.75
Suburban	0.25–0.40
Apartment dwelling areas	0.50–0.70
Industrial:	
Light areas	0.50–0.80
Heavy areas	0.60–0.90
Parks, cemeteries	0.10–0.25
Playgrounds	0.20–0.40
Railroad yard areas	0.20–0.40
Unimproved areas	0.10–0.30
Lawns:	
Sandy soil, flat, 2%	0.05–0.10
Sandy soil, average 2–7%	0.10–0.15
Sandy soil, steep, 7%	0.15–0.20
Heavy soil, flat, 2%	0.13–0.17
Heavy soil, average, 2–7%	0.18–0.25
Heavy soil, steep, 7%	0.25–0.35
Streets:	
Asphaltic	0.70–0.95
Concrete	0.80–0.95
Brick	0.70–0.85
Drives and walks	0.75–0.85
Roofs	0.75–0.95

Source: From *Highway Design Manual,* California Department of Transportation, with permission.

watershed resulting in the maximum flow at the point being investigated. The total time of concentration is made up of three components: overland flow time, channel flow time, and culvert flow time. The overland flow time is typically the longest of the three. The flow time in the culvert has such a small effect on the total that it may generally be included in the calculation for channel flow time.

The overland flow time may be approximated by the curves in Fig. 5.3. It is based on the following equation:

$$T_o = \frac{1.8(1.1 - C)(L)^{1/2}}{[S(100)]^{1/3}} \qquad (5.3)$$

where T_o = overland flow travel time, min
C = runoff coefficient
L = overland travel distance, ft
S = slope

CHAPTER FIVE

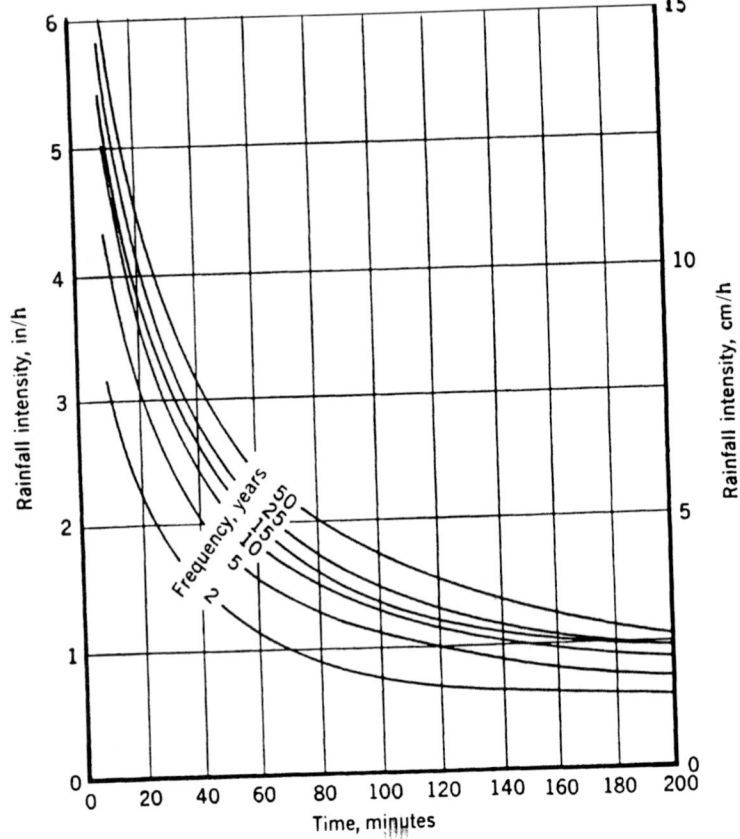

FIGURE 5.2 Typical rainfall intensity-duration-frequency curves. (*From* Design and Construction of Storm and Sanitary Sewers, *ASCE, 1986, with permission*)

The overland flow time can also be calculated by the *kinematic wave equation:*

$$T_o = \frac{K(L^{0.6})(n^{0.6})}{(i^{0.4})(S^{0.3})}$$

(5.4)

where T_o = overland flow travel time, min
  K = 0.93 for English units (6.92 for SI units)
  L = length of overland flow path, ft (m)
  S = slope of overland flow
  n = Manning's roughness coefficient
  i = rainfall intensity, in/h (mm/h)

The open-channel flow time may be determined from Manning's equation (see Art. 5.3.3). The total length of the channel is divided by the flow velocity from Manning's

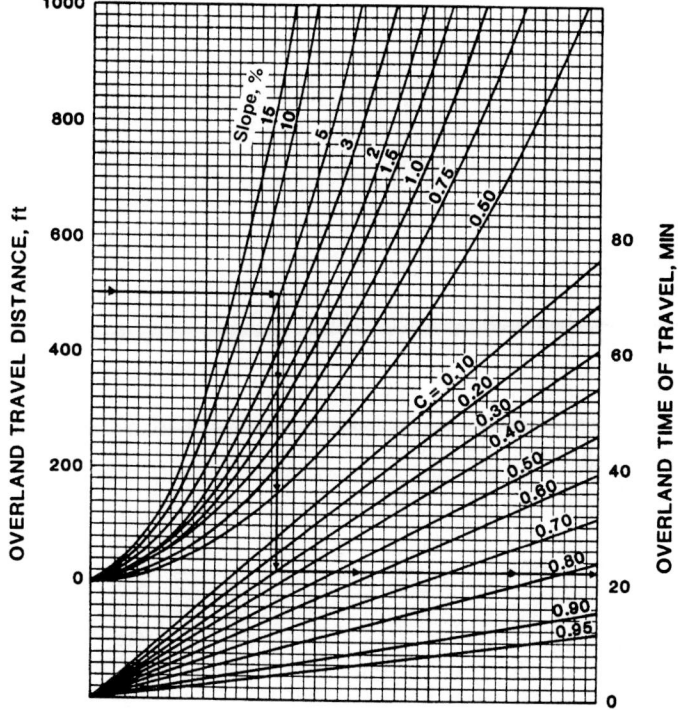

FIGURE 5.3 Overland time of concentration curves. (*From* Highway Design Manual, *California Department of Transportation, with permission*)

equation to arrive at the channel flow time. The overland flow time is added to the channel flow time to arrive at the total time of concentration.

As an alternative to the above procedure, where the channels are well defined and the overland flow is generally over bare ground, the total time of concentration may be estimated from the Kirpich equation:*

$$T_c = K\left(\frac{L}{S^{0.5}}\right)^{0.77} \tag{5.5}$$

where T_c = time of concentration, min
K = 0.0078 for English units (3.97 for SI units)
L = maximum flow length, ft (km)
S = total slope = total change in elevation divided by L

The value of T_c should be multiplied by 2 where the surfaces are grassy, by 0.4 where they are asphalt or concrete, or 0.2 for concrete channels. (See *Modern Sewer Design,* AISI.)

---

*Z. P. Kirpich, "Time of Concentration in Small Agricultural Watersheds," *Civil Engineering,* vol. 10, p. 362, 1940.

The total time of concentration may also be calculated from the following modified form of the Williams equation:*

$$T_c = KLA^{-0.1}S^{-0.2} \qquad (5.6)$$

where T_c = time of concentration, min
K = 21.3 for English units (14.6 for SI units)
L = maximum flow length, mi (km)
A = total watershed area, mi^2 (km^2)
S = slope, ft/ft (m/m)

A minimum time of concentration of 5 min is recommended by the FHWA. (See E. Maidment, *Handbook of Hydrology,* McGraw-Hill, 1993.)

Another common and simple method for determining the runoff is the Soil Conservation Service method. The determination of the peak discharge is dependent upon the time of concentration, the cumulative rainfall, and the soil and cover classifications. (See the following from the Soil Conservation Service: *National Engineering Handbook,* 1985; and "Urban Hydrology for Small Watersheds," TR-55, 1986.)

5.1.6 Computer Models

Many computer models have been developed in recent years for calculating rainfall runoff. Examples include the U.S. Army Corps of Engineers HEC-1 model, the Soil Conservation Service TR-20 model, and the FHWA-funded HYDRAIN system. As with all computer models, the accuracy and validity of the output can be only as accurate and valid as the input. The input and output data must be carefully inspected by a capable and practiced user to ensure valid results. (See E. Maidment, *Handbook of Hydrology,* McGraw-Hill, 1993; and *Highway Drainage Guidelines,* Vol. 2, AASHTO, 1992.)

Example: Time of Concentration, Rainfall Intensity, and Design Discharge. A grassy roadside channel runs 500 ft from the crest of a hill. The area contributing to the flow is 324 ft wide and is made up of 24 ft of concrete pavement and 300 ft of grassy backslope. The distance from the channel to the ridge of the drainage area is 200 ft. The channel has a grade of 0.4 percent, and the edge of the contributing area is 5 ft above the channel. Determine the time of concentration, rainfall intensity, and design discharge based on a 10-year-frequency rainfall.

Assume the grassy backslope is similar to the watershed described by the example in Table 5.1 with $C = 0.32$. From Table 5.2, assume for the pavement $C = 0.90$. Then, from Eq. (5.2), the weighted average value of the runoff coefficient is

$$C = \frac{0.90 \times 24 + 0.32 \times 300}{24 + 300} = 0.36$$

---

*The *modified* Williams equation is found in Maidment, cited below. The original Williams reference is G. B. Williams, "Flood Discharge and the Dimensions of Spillways in India," *The Engineer,* vol. 121, pp. 321–322, September 1922.

The length of travel is

$$L = 500 + 200 = 700 \text{ ft}$$

The difference in elevation between the most remote point on the basin and the outlet is

$$H = 5 + (0.004 \times 500) = 7 \text{ ft}$$

The slope is

$$S = \frac{H}{L} = \frac{7}{700} = 0.01$$

From Eq. (5.5), the total time of concentration is

$$T_c = 0.0078 \left[\frac{700}{(0.01)^{0.5}} \right]^{0.77} = 7.1 \text{ min}$$

Because this is an unmowed grassy area, multiply by 2.0 to get $T_c = 14$ min. Figure 5.2 shows that the rainfall intensity for this time of concentration and a 10-year-frequency rainfall is 4 in/h. Finally, the design discharge from Eq. (5.1) is

$$Q = 1 \times 0.36 \times 4 \times 3.7 = 5.3 \text{ ft}^3/\text{s}$$

5.2 DESIGN OF OPEN CHANNELS

As the name implies, open-channel flow is concerned with the conveyance of water with a free surface. This article primarily concerns lined and unlined channels such as encountered along roadways in highway design.

5.2.1 General Considerations

The parameters to consider in choice of channel cross section include hydraulics, safety, maintenance, economics, and the environment. These considerations are usually so interdependent that optimizing one can have detrimental effects on the others. The hydraulic engineer's objective is to achieve a reasonable balance among the competing criteria.

Safety is always of primary concern to the highway engineer. If the channel is located far enough away from the traveled way, an adequate recovery zone may be available for vehicles accidentally leaving the roadway. Additionally, with regard to safety, a channel with flattened sideslopes and a curved transition to the bottom is preferred to allow time for recovery for the errant vehicle. (See Chap. 6, Safety Systems.)

Periodic maintenance is required of hydraulic channels regardless of the cross-sectional design chosen. Access should be planned and provided for maintenance personnel and equipment. The proliferation of sediment and debris and the growth of vegetation can cause erosion or reduction of the capacity of the channel. The channel design should balance the cost of preventing these restrictions against the anticipated increased costs of removing them as they accumulate.

The proposed channel location and shape affect the economics of the project. A channel located away from the traveled way may be safer for the traveling public and more aesthetically pleasing; however, these considerations must be balanced against the potential increase in right-of-way costs as well as other associated costs. The shape also affects the cost of the channel. A channel with vertical sidewalls will typically be more expensive than one with sloping sides; the vertical walls must not only maintain flow within the channel but must also be designed to retain the earth outside the channel.

Proposed channel improvements must take into account the possible effects the project will have with regard to erosion, sedimentation, water quality, aesthetics, and fish and wildlife. Local, state, and federal resources and flood control agencies have an interest in drainage improvements and environmental impacts and should be contacted early in the planning process for input, cooperation, and assistance. A partial list of these agencies may be found in the AASHTO *Highway Drainage Guidelines.*

The necessary hydraulic parameters should be determined early in the design phase. As previously mentioned, the scope of the hydrologic study should be proportional to the importance of the hydraulic structure involved, the type of highway, the impacts on the local property, and potential risks involved. The hydraulic design of the channel involves selecting the cross section and lining to maintain the flow predicted from the hydrologic study. The capacity of the channel is affected by its size, shape, roughness, and slope.

The slope is generally controlled by the existing terrain, and the engineer has little control over this. As much as is practical, however, the engineer should avoid sudden changes in the slope as well as the alignment of the channel. Abrupt changes in channel alignment can lead to unintentional channel changes by aggradation and avulsion. Abrupt changes in slope can cause either erosion, if the grade is steepened; or an accumulation of buildup, if it is flattened.

Erosion and deposition may also be limited by controlling the velocity of the flow. The velocity of the water is dependent upon the size, shape, roughness, and slope of the channel as well as the quantity of flow. Recommended flow velocities for unlined channels are shown in Table 5.3. Velocities in lined channels can generally be much greater. To minimize deposition of sediment, the minimum gradient should be about 0.5 percent for earth-lined and grass-lined channels and 0.35 percent for paved channels. Also, decreasing gradients should be avoided.

5.2.2   Channel Realignment

At times it will be advantageous or necessary to realign or change the hydraulic characteristics of the channel. Reasons for altering the channel include improving culvert alignment, protecting roadways from erosion damage, reducing maintenance requirements, and eliminating hydraulic structures where the roadway recrosses the channel.

Plans for channel modifications must include a determination of what effect the change will have on the stream and the surrounding environment. Long- and short-term effects must be considered. The impact on the stream of the realignment or change in slope will vary from one site to another. At some sites, minor changes will have significant impacts, while at others the opposite may be true. Regardless of the magnitude of the effect on the stream and its environment that the change may have, plans should be developed to mitigate those effects.

Changes to a channel usually cause a decrease in the roughness and an increase in the slope. The resultant higher velocity may lead to increased scour and sedimentation buildup at the downstream end of the channel improvement, and may result in changes

TABLE 5.3 Recommended Permissible Velocities for Unlined Channels

	Permissible velocity, ft/s	
Type of material in excavation section	Intermittent flow	Sustained flow
Fine sand (noncolloidal)	2.5	2.5
Sandy loam (noncolloidal)	2.5	2.5
Silt loam (noncolloidal)	3.0	3.0
Fine loam	3.5	3.5
Volcanic ash	4.0	3.5
Fine gravel	4.0	3.5
Stiff clay (colloidal)	5.0	4.0
Graded material (noncolloidal)		
Loam to gravel	6.5	5.0
Silt to gravel	7.0	5.5
Gravel	7.5	6.0
Coarse gravel	8.0	6.5
Gravel to cobbles (under 6 in)	9.0	7.0
Gravel and cobbles (over 8 in)	10.0	8.0

Source: From *Highway Design Manual*, California Department of Transportation, with permission.

that affect the habitat in and around the stream. Any changes to existing streams that support fish or wildlife must be coordinated with the appropriate resource agencies early in the planning phase.

5.2.3 Channel and Shore Protection

Highways are often located adjacent to streams, lakes, and coastal areas. Channel and shore protection must be provided wherever the need is apparent or the risk is high. In other circumstances, where the possibility of damage to the roadway or adjacent land is not clear or risk is low, it may be acceptable to delay construction of embankment stabilization measures until a problem actually develops.

There are a number of methods of protecting the roadway from damage due to erosion. The simplest and surest of these is to locate the highway away from the erosive forces. This should always be considered, although it is rarely the most economical alternative. The most common method used to protect the roadway is to line the roadway embankment with a material that is resistant to erosion such as concrete or rock. Another method is to reduce the force of the water that would cause the erosion. Such bank protection structures retard the flow of the water while at the same time allowing a sedimentation buildup to reverse the trend of erosion and replace material that may have been lost. A final method of protection that should be considered is redirecting the eroding force away from the embankment. This may be done by the use of jetties or baffles, or even by creating a new channel.

Any combination of the above methods may be used to achieve the desired protection. The design of the protective features should be commensurate with the importance of the roadway being protected and with the risks involved. (See *Highway Drainage Guidelines*, Vol. III, *Erosion and Sediment Control*, AASHTO; "Design of Riprap Revetment," HEC 11, FHWA; "Hydraulic Design of Energy Dissipaters," HEC 14, FHWA; and "Design of Roadside Channels with Flexible Linings," HEC 15, FHWA.)

5.3 FUNDAMENTALS OF OPEN-CHANNEL FLOW

The fundamental relationships for hydraulic flow are the same for channels that are physically open at the top, such as roadway channels and curbs and gutters, and for pipes and culverts that have a free water surface. In both cases, hydraulic design is based on *open-channel flow*. An understanding of these relationships is important for comprehending various design aids subsequently presented.

5.3.1 Types of Flow

Open-channel flow may be categorized by three characteristics: the flow may be (1) steady or unsteady, (2) uniform or nonuniform, and (3) either subcritical, critical, or supercritical. This discussion will begin with the first two categories, and the third will be discussed later.

Steady flow means that at a particular point, there is no change in depth with respect to time. By extension, this means that there is no change in the quantity of flow. *Unsteady flow* means that the depth does change with time.

Uniform flow assumes that there is no change in depth or quantity of water at any section along the length of the channel (or culvert) under investigation. This requires that there be no change in velocity of the flow, and it is possible only if the slope, roughness, and cross section all remain constant along the length of the channel. This state is evidenced by the fact that the water surface is parallel to the channel bottom. Nonuniform flow assumes a change in depth or velocity along the length of the channel. This type of flow may be further classified as rapidly varying or gradually varying flow.

For most highway applications, the flow is steady and the changes in the section are so gradual that the flow may be considered uniform. The equations for open-channel flow are based on that assumption. Where the change in the cross section of the channel is dramatic, nonuniform flow should be assumed. (For analysis of nonuniform flow, see E. F. Brater and H. W. King, *Handbook of Hydraulics,* McGraw-Hill, 1976.)

5.3.2 Continuity Equation

The continuity equation is based on the basic and fundamental concept that the quantity of flow passing any cross section remains constant throughout the length of the stream flow:

$$Q = AV \qquad (5.7)$$

where Q = discharge, ft³/s (m³/s)
  A = area, ft² (m²)
  V = velocity, ft/s (m/s)

5.3.3 Manning's Equation

Manning's equation assumes uniform, turbulent flow conditions and computes the mean flow velocity for an open channel:

$$V = \left(\frac{1.486}{n} \right) R^{2/3} S^{1/2} \text{ in English units} \qquad (5.8a)$$

$$V = \frac{(R^{2/3}S^{1/2})}{n} \text{ in SI units} \tag{5.8b}$$

where V = mean velocity, ft/s (m/s)
n = Manning coefficient of roughness
R = hydraulic radius = A/WP, ft (m)
A = cross-sectional flow area, ft^2 (m^2)
WP = wetted perimeter = total perimeter of cross-sectional area of flow minus free surface width, ft (m)
S = channel slope

Manning's equation may be solved directly or obtained from the nomograph in Fig. 5.4. Typical Manning's n values are given in Table 5.4. For shallow flows, the effective n values should generally be increased, because the wetted perimeter will have a greater effect on the flow.

The continuity equation and Manning's equation may be used in conjunction to directly compute channel discharges. Substitute Eq. (5.8) into Eq. (5.7) and rearrange terms to obtain

$$AR^{2/3} = \frac{Qn}{1.486S^{1/2}} \text{ in English units} \tag{5.9a}$$

$$AR^{2/3} = \frac{Qn}{S^{1/2}} \text{ in SI units} \tag{5.9b}$$

R is a function of A. Thus, for a given slope, flow quantity, and n value, $AR^{2/3}$ may be determined and the normal depth of flow calculated by trial and error.

5.3.4 Energy Equation

The energy equation is based on the principle that energy must be conserved; that is, the energy at any one cross section on a stream is equivalent to the energy at any other section plus any intervening energy losses. This relationship, a form of the Bernoulli equation, may be used wherever there is a change in the size, shape, or slope, of the channel and is useful in determining the depth of flow.

$$z_1 + d_1 + \left(\frac{V_1^2}{2g}\right) = z_2 + d_2 + \left(\frac{V_2^2}{2g}\right) + h_L \tag{5.10}$$

where z_n = distance above some datum, ft (m)
d_n = depth of flow, ft (m)
V_n = flow velocity, ft/s (m/s)
g = acceleration of gravity, 32.2 ft/s^2 (9.8 m/s^2)
h_L = head loss between the two sections, ft (m)

Subscripts 1 and 2 refer to two sections along the flow line as depicted in Fig. 5.5. The velocity head is given by $V^2/2g$ and the specific energy is defined as $d + V^2/2g$. The plots in Fig. 5.5 illustrate the head at points along the length of the channel. The line drawn through points of static head is known as the *hydraulic grade line,* and the line drawn through points of total head is known as the *energy grade line.* The head loss between sections includes losses due to flow friction along the channel and losses due to turbulence at junctions and bends.

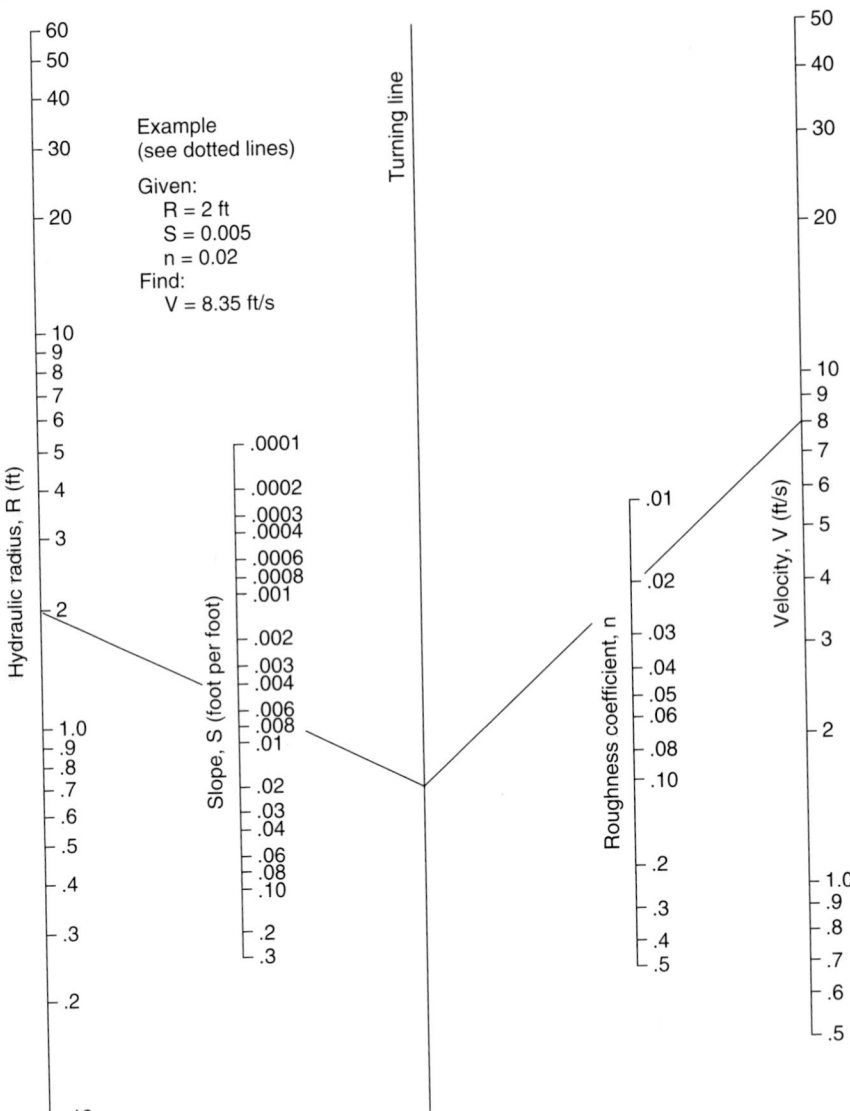

Example
(see dotted lines)

Given:
  R = 2 ft
  S = 0.005
  n = 0.02
Find:
  V = 8.35 ft/s

FIGURE 5.4 Nomograph for solution of Manning equation. (*From* Highway Design Manual, *California Department of Transportation, with permission*)

TABLE 5.4 Values of the Roughness Coefficient n for Use in the Manning Equation

	Min	Avg	Max
A. Open-channel flow in closed conduits			
1. Corrugated-metal storm drain	0.021	0.024	0.030
2. Cement-mortar surface	0.011	0.013	0.015
3. Concrete (unfinished)			
a. Steel form	0.012	0.013	0.014
b. Smooth wood form	0.012	0.014	0.016
c. Rough wood form	0.015	0.017	0.020
B. Lined channels			
1. Metal			
a. Smooth steel (unpainted)	0.011	0.012	0.014
b. Corrugated	0.021	0.025	0.030
2. Wood			
a. Planed, untreated	0.010	0.012	0.014
3. Concrete			
a. Float finish	0.013	0.015	0.016
b. Gunite, good section	0.016	0.019	0.023
c. Gunite, wavy section	0.018	0.022	0.025
4. Masonry			
a. Cemented rubble	0.017	0.025	0.030
b. Dry rubble	0.023	0.032	0.035
5. Asphalt			
a. Smooth	0.013	0.013	
b. Rough	0.016	0.016	
C. Unlined channels			
1. Excavated earth, straight and uniform			
a. Clean, after weathering	0.018	0.022	0.025
b. With short grass, few weeds	0.022	0.027	0.033
c. Dense weeds, high as flow depth	0.050	0.080	0.120
d. Dense brush, high stage	0.080	0.100	0.140
2. Dredged earth			
a. No vegetation	0.025	0.028	0.033
b. Light brush on banks	0.035	0.050	0.060
3. Rock cuts			
a. Smooth and uniform	0.025	0.035	0.040
b. Jagged and irregular	0.035	0.040	0.050

Source: From F. S. Merritt, ed., *Standard Handbook for Civil Engineers,* McGraw-Hill, 1983, with permission.

5.3.5 Critical Flow Depth

When the depth of flow is plotted against the specific energy, the specific energy diagram may be obtained and the critical depth found as illustrated in Fig. 5.6. The critical depth is defined as that depth where the specific energy is a minimum. The flow velocity at the critical depth is called the *critical velocity.* The channel slope that causes the critical depth and critical velocity is termed the *critical slope.* If the depth is greater than the critical depth, the flow is said to be subcritical and the velocity head reduces. Where the depth is less than the critical depth, the flow is said to be supercritical and the velocity head increases. For any particular energy level, except where

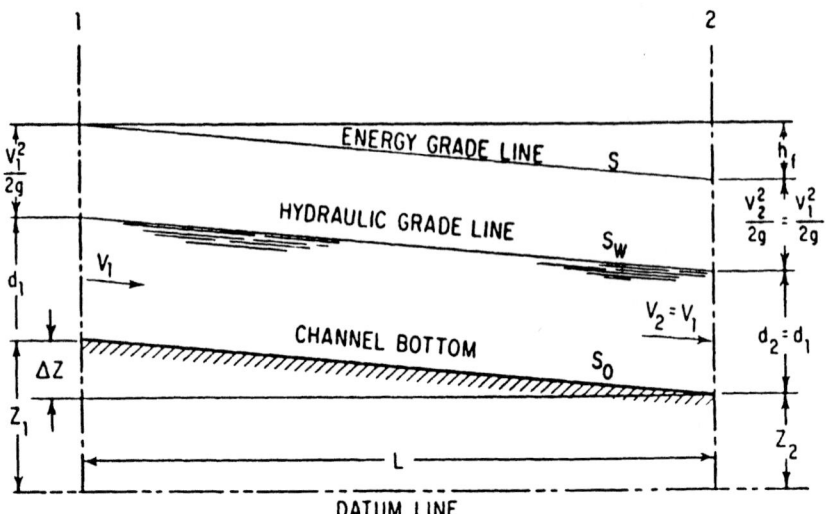

FIGURE 5.5 Flow characteristics for uniform open-channel flow. (*From F. S. Merritt, ed.,* Standard Handbook for Civil Engineers, *McGraw-Hill, 1983, with permission*)

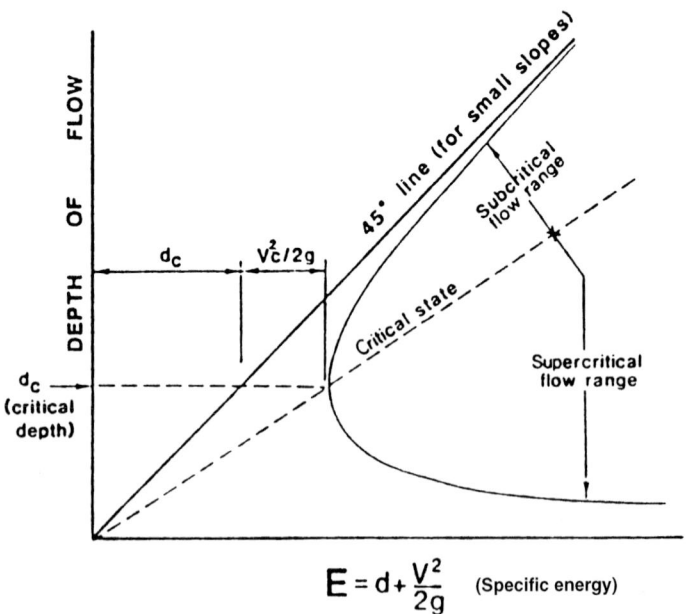

$$E = d + \frac{V^2}{2g} \quad \text{(Specific energy)}$$

FIGURE 5.6 Specific energy diagram. (*From* Highway Design Manual, *California Department of Transportation, with permission*)

the depth is critical, there are two corresponding depths that may occur. However, the depth may not alternate between these two values without a change in the channel configuration or slope.

Although the critical depth gives the greatest discharge, flow that causes the depth to be close to critical should be avoided, and thus the critical slope should be avoided. Flows near the critical depth may be turbulent. Inspection of the specific energy diagram reveals that where the depth is close to the critical depth, it takes little energy to change the flow from subcritical to supercritical or the reverse. If the flow does change from subcritical to supercritical, a hydraulic jump will occur. If placing the depth of flow near critical is unavoidable, it is advisable to assume the least favorable type of flow for design purposes. The critical depth may be determined from the following relationship:

$$\frac{A^3}{T} = \frac{Q^2}{g} \tag{5.11}$$

where A = cross-sectional flow area, ft^2 (m^2)
$\quad$ T = top width of channel flow, ft (m)
$\quad$ Q = discharge, ft^3/s (m^3/s)
$\quad$ g = acceleration of gravity, 32.2 ft/s^2 (9.8 m/s^2)

For a channel with vertical walls, the velocity corresponding to the critical depth is given by

$$V_c = \left(\frac{gA}{T} \right)^{1/2} \tag{5.12}$$

where V_c = critical velocity, ft/s (m/s). Also, for a channel with vertical walls, the flow area at a point of critical depth d_c is

$$A = T(d_c) \tag{5.13}$$

Substitution in Eq. (5.11) leads to

$$d_c = \left(\frac{Q^2}{gT^2} \right)^{1/3} \tag{5.14}$$

It can be seen from this relationship that for a given flow, as the width of the channel changes the critical depth also changes. Such locations should be investigated for a hydraulic jump.

Points of control are locations where the depth of flow may be easily determined. The critical depth is one point of control and may be found in several typical locations. As discussed above, one of these locations may be where there is a change in the channel section. Other typical locations are where the slope changes abruptly from flat (subcritical) to steep (supercritical), at the crest of an overflow dam or weir, and at the outlet of a culvert on a subcritical slope discharging into a basin or wide channel.

The Froude number (Fr) may also be used in determining whether the channel is under supercritical, critical, or subcritical flow:

$$\text{Fr} = \frac{V}{(gd_h)^{1/2}} \qquad (5.15)$$

where $d_h = A/T$. If Fr < 1.0, the channel flow is subcritical; if Fr = 1.0, the channel flow is critical; and if Fr > 1.0, the channel flow is supercritical.

Water surface profiles for the gradually varying flow condition may be determined by either the direct step method or the standard step method. The former method is applicable only to straight prismatic channel sections with gradually varying areas of flow. The standard step method may be used in nonprismatic channel sections and channel alignments that are not straight. Where the flow is subcritical, the analysis for determination of the water profile begins at the control point and proceeds upstream. Where the flow is supercritical, the opposite is true. (See V. T. Chow, *Open-Channel Hydraulics,* McGraw-Hill, 1959; and F. S. Merritt, ed., *Standard Handbook for Civil Engineers,* McGraw-Hill, 1983.)

Example: Critical Depth and Critical Velocity.   A channel has a width of 10 ft and vertical sides. Determine the critical flow depth and critical velocity for a flow of 1000 ft³/s.

From Eq. (5.14), $d_c = (Q^2/gT^2)^{1/3} = [(1000)^2/32.2(10)^2]^{1/3} = 6.77$ ft. From Eq. (5.13), $A = T(d_c) = 10(6.77) = 67.7$ ft². From Eq. (5.12), the critical velocity is $V_c = (gA/T)^{1/2} = (32.2 \times 67.7/10)^{1/2} = 14.8$ ft/s.

5.4   DESIGN OF ROADWAY DRAINAGE

Roadway drainage includes the entire system from pavement drainage through storm drains. Drainage features that make up the system include curbs, gutters, drop inlets, median drains, overside drains, roadside ditches, and storm drains. The basic design procedure for roadway drainage includes hydrology, surface water removal, and disposal. A properly designed system must adequately accommodate the design runoff by removing it from the roadway surface and conveying it to the outfall, avoiding damage to adjacent property and roadway hazards from overflowing and ponding.

5.4.1   General Considerations

Pavement may be drained in one of two ways. The runoff may be allowed to sheet-flow across the roadway surface and into roadside ditches. This may not always be possible or cost-effective, because of right-of-way constrictions. Alternatively, a curb and gutter section is used to channel the flow.

An appropriate design storm must be selected so that the drainage facilities may be properly designed. This design storm must relate to an acceptable level of flooding of the roadway with regard to both area and frequency. The acceptable level of flooding is termed the *design water spread* (Fig. 5.7) and is defined by the acceptable amount of encroachment on the roadway surface that is assumed to have a certain probability of occurrence. It may not be economically feasible to completely prevent encroachment on the roadway. Alternatively, it is unwise to allow spread that results in unsafe driving conditions. Greater water spread produces greater splash and spray effect and an accompanying decrease in visibility and vehicle control by the users of the facility. The amount and frequency of encroachment should vary with the type of roadway

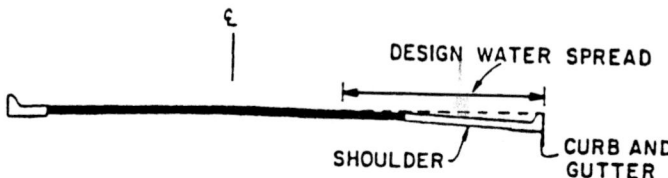

FIGURE 5.7   Illustration of design water spread. (*From* Highway Drainage Guidelines, *Vol. IX, American Association of State Highway and Transportation Officials, Washington, D.C., 1992, with permission*)

being designed, because roads with higher volumes and speeds can tolerate less loss of visibility than local and collector roads.

AASHTO has developed general guidelines on highway drainage that may be used to formulate roadway surface drainage criteria. Table 5.5 shows the suggested AASHTO procedure for relating the road classification, the frequency of the design storm, and the design spread. However, more specific local or regional guidelines are usually developed and should be referenced for highway drainage design. An example of a regional guideline developed by the California Department of Transportation (Caltrans) is shown in Table 5.6. It is apparent that a more severe storm (25-yr versus 10-yr mean recurrence interval) is used for roadways with higher volumes and speeds, as well as a more limited design water spread.

TABLE 5.5   Minimum Design Frequency

	Design frequency, years			Design spread	
Road classification	<10	10	50	Shoulder or parking	Partial driving lane (¼, ½, ¾)
1. High-volume divided highway					
a. <45 mi/h		X			X
b. >45 mi/h		X		X	
c. Sag point			X		X
2. High-volume bidirectional					
a. <45 mi/h		X			X
b. >45 mi/h		X		X	
c. Sag point			X		X
3. Collector					
a. <45 mi/h	X				X
b. >45 mi/h		X		X	
c. Sag point		X			X
4. Local streets					
a. Low ADT*	X				X
b. High ADT		X			X
c. Sag point		X			X

*Average daily traffic.

Source: From *Highway Drainage Guidelines,* Vol. IX, American Association of State Highway and Transportation Officials, Washington, D.C., 1992, with permission.

TABLE 5.6 Desirable Roadway Drainage Guidelines

	Design storm		Design water spread		
Highway type/category/feature	4% (25 yr)	10% (10 yr)	Parking lane	Shoulder or outer lane	½ local standard
Freeways					
Through traffic lanes, branch connections, and other major ramp connections	x	—	x	—	—
Minor ramps	—	x	x	—	—
Frontage roads	—	x	—	—	x
Conventional highways					
High volume, multilane, speeds over 45 mi/h	x	—	x	—	—
High volume, multilane, speeds 45 mi/h and under	—	x	—	x	—
Low volume, rural, speeds over 45 mi/h	x	—	x	—	—
Urban, speeds 45 mi/h and under	—	x	—	—	x
All state highways					

Depressed sections that require pumping: Use a 2% (50-yr) design storm for freeways and conventional state highways. Design water spread at depressed sections should not exceed that of adjacent roadway sections. A 4% (25-yr) design storm may be used on local streets or road undercrossings that require pumping.

Source: From *Highway Design Manual,* California Department of Transportation, with permission.

5.4.2 Curbs, Gutters, and Inlets

The roadway surface water can be removed by a series of drains that carry the water into a collection and disposal system. The curb, gutter, and inlet design must keep flooding within the parameters established in roadway drainage guidelines. The hydraulic efficiency of inlets is related to the roadway grade, the cross grade, the inlet geometry, and the design of the curb and gutters.

Curbs are divided into two classes: barrier and mountable. Barrier curbs are steep-faced and generally 6 to 8 in (150 to 200 mm) high. Mountable curbs are generally 6 in (150 mm) high or less with relatively flat sloping faces to allow vehicles to cross them when required. Neither barrier curbs nor mountable curbs should be used on high-speed roadways. (See Chap. 6, Safety Systems.)

Gutters begin at the bottom of the curb and extend toward the roadway a varying distance, usually 1 to 6 ft (300 to 1800 mm). They may or may not be constructed with the same material as the roadway.

The longitudinal grade of the gutter is controlled by the highway grade line. For drainage purposes, it is important to maintain some minimum longitudinal slope to ensure that runoff does not accumulate in ponds. Gutter cross slopes of 5 to 8 percent should be maintained for a distance of 2 to 3 ft (600 to 900 mm) for that portion of the gutter adjacent to the curb.

The following modification of Manning's equation may be used to determine the spread of the gutter flow as well as the maximum depth at the curb face. This applies to a section with a single cross slope. (For additional information, nomographs, and flow solutions for gutters with composite cross slopes, see "Drainage of Highway Pavements," HEC 12, FHWA.)

$$Q = \left(\frac{K}{n} \right) S_X^{1.67} S^{0.5} T^{2.67} \tag{5.16}$$

where Q = rate of discharge, ft³/s
  K = 0.56
  n = Manning's coefficient of roughness
  S_X = cross slope
  S = longitudinal slope
  T = spread or top width of flow in gutter = d/S_x, ft
  d = depth of flow at face of curb, ft

Example: Gutter Flow Spread and Depth.   A concrete gutter for a roadway with a grade of 0.05 and a cross slope of 0.04 must accommodate a flow of 1.4 ft³/s. Determine the spread of the flow and its depth at the curb face. Assume $n = 0.15$.
  Substitute in Eq. (5.16) and solve for the spread T as follows:

$$1.4 = \left(\frac{0.56}{0.015} \right)(0.04)^{1.67}(0.05)^{0.5} T^{2.67}$$

$$T^{2.67} = 36.23$$

$$T = 3.84 \text{ ft}$$

It follows that the depth at the curb is $d = TS_X = 3.84 \times 0.04 = 0.15$ ft.

5.4.3   Inlet Location and Type

One of the major objectives in the design of the roadway drainage system is to limit the encroachment of the flow to that developed in the roadway drainage guidelines. However, this spread cannot be determined until the inlet is located. After the inlet is located, the drainage area contributing to the flow into that inlet is determined. Discharge based on the rational method is then calculated, and finally the spread is determined based on that discharge and the gutter characteristics. If the spread is found to be too great (leading to possible unsafe conditions) or too small (possibly indicating an inefficient design), the inlet should be relocated and the process repeated. As can be seen, this design is an iterative process. The process is also controlled by surface features that restrict possible location of inlets, such as streets, driveways, and utilities.

There are also areas where inlets are nearly always required. These include sag points, points of superelevation reversal, street intersections, and at bridges. Where an inlet is required in the vicinity of a driveway, it should always be located upstream of the driveway. If it is located downstream, the driveway may affect the flow and cause a significant portion to bypass the inlet.

Finally, the type and size of the inlet has a direct affect on location and spacing. Similarly, designing for greater spread and allowing some bypass of the upstream inlets to occur with the residual being intercepted by those farther downstream (carry-over flow) will result in fewer inlets.

The basic types of inlets are the curb opening inlet and the grate inlet as shown in Fig. 5.8. Two other types frequently used are the slotted drain inlet and the combination inlet (grate plus curb opening) shown in Fig. 5.9.

Curb-opening inlets, which have the drainage opening in the face of the curb, are very durable and are comparatively free from blockage by debris. This type generally relies heavily on the bordering depression to be effective at intercepting the water flow and is relatively inefficient when located in an on-grade situation. It is probably the most efficient inlet type at sag points because of its tendency to remain free of clogging by debris and its large, hydraulically efficient opening. In addition, this type of inlet opening offers little interference to vehicular traffic, pedestrians, or bicyclists.

The length of the opening required for total interception of the gutter flow can be determined by the following equation:

$$L_t = 0.6Q^{0.42}S^{0.3}(nS_X)^{-0.6} \qquad (5.17)$$

where L_t = length of curb opening for total interception of flow, ft
 Q = discharge, ft³/s
 S = longitudinal slope of gutter
 n = Manning's roughness coefficient
 S_X = transverse slope of gutter

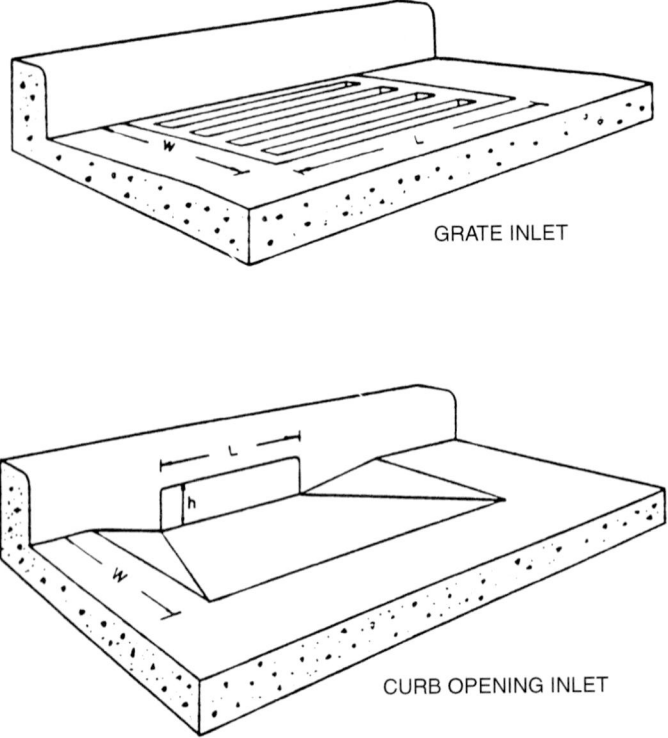

GRATE INLET

CURB OPENING INLET

FIGURE 5.8 Perspective views of grate inlet and curb-opening inlet. (*From "Hydraulic Charts for the Selection of Highway Culverts," Hydraulic Engineering Circular No. 12, FHWA, with permission*)

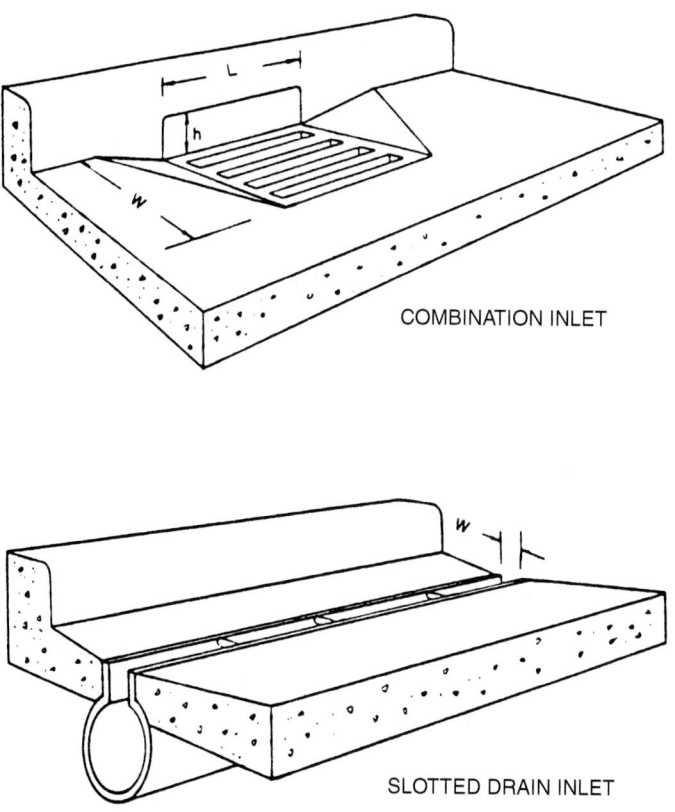

COMBINATION INLET

SLOTTED DRAIN INLET

FIGURE 5.9   Perspective views of combination and slotted drain inlets. (*From "Hydraulic Charts for the Selection of Highway Culverts," Hydraulic Engineering Circular No. 12, FHWA, with permission*)

Where there is a depression, the equivalent transverse slope S_e must be determined and used for S_x. (See "Drainage of Highway Pavements," HEC 12, FHWA, for a complete discussion of this and flow at sag points.)

Grate inlets come in a variety of shapes and sizes and are efficient where debris is not a problem. However, most inlets are subject to varying amounts of debris, and the selection of grate inlets, especially those located at sag points, must take this possibility into account.

For greatest hydraulic efficiency, grate inlets should be oriented parallel to the surface flow. However, grate bars oriented parallel with traffic can cause problems where bicycles are present, and specifically designed "bicycle-proof" grates with additional transverse bars should be used. Other factors influencing the hydraulic capacity of this type of inlet include the longitudinal and cross slope of the gutter, the width and length of the gutter, and the size and shape of the bars. The grate inlet will intercept all of the flow that passes over the top of the grate as long as the grate is long enough. In addition, a portion of the side flow, or the flow that is located above the grate toward the roadway centerline, will be intercepted. The amount of the intercepted side flow depends upon the velocity of the flow, the length of the inlet, and the cross slope of the gutter.

Combination grates—generally, curb opening and grate inlets—are desirable at sag points. The curb opening will generally keep the inlet from clogging. At grade locations, however, the efficiency of the combination inlet approaches that of the grate inlet.

Slotted drains can provide continuous interception of the flow when used on grades. However, because of the possibility of clogging, they should be used only in combination with other types of inlets at sag points. Slotted drains are also useful to supplement the existing drainage system where the roadway needs to be widened.

Inlets at grade sags deserve additional deliberation, since any blockage of the inlet will typically lead to flooding. Typical design considerations are to provide additional inlets or base the design on a relatively high assumption of debris blockage.

5.4.4 Culverts and Storm Drains

The conduits used to convey water from one side of the roadway through the embankment to the other side are typically referred to as *culverts*. A network or system of such conduits to carry storm water is referred to as a *storm drain system*. Conduits for culverts and storm drains are available in many different shapes, sizes, and materials, as discussed subsequently. Available shapes include circular, elliptical (horizontal or vertical), pipe-arch, arch, and box shapes. Factors that affect the shape at a particular site include the fill height, construction costs, and potential for clogging by debris. Where the cover over the conduit is limited, pipe-arch, arch, elliptical (horizontal), or box shapes may be more applicable. Where the fill height is great, circular shapes tend to be structurally and economically more favorable. Factors involved in the selection process include hydraulic, structural, construction, maintenance, and durability requirements. (See Art. 5.5 for hydraulic design of culverts.)

A system of closed conduits (storm drains and culverts) to convey the runoff from the inlets to the outfall must be designed starting at the upstream end and proceeding downstream. Each section of pipe that extends from inlet to inlet, or from an inlet to the final outfall, is called a *run*. Each run requires a separate analysis because of the change in flow at each, and possible changes in slope, pipe size, and type. After all runs are initially sized, the hydraulic grade line is developed (Art. 5.3.4). Unlike the sizing of the conduits, the calculations for this proceed in an upstream direction. In addition to head loss from friction along the length of the culvert, the hydraulic grade line must account for the effects of losses caused by turbulence at junctions and bends. Once the hydraulic grade line is established, it may be compared with the grade line of the system to ensure that it does not exceed an allowable high-water elevation. If it should extend above these allowable elevations, then the initial design must be adjusted.

In addition to system sizing based on hydraulic requirements, conduits should generally not be smaller than 12 to 18 in (300 to 450 mm) in diameter, to reduce the potential for debris clogging. Greater minimum diameters may be appropriate in some cases, particularly under high fills.

Flow in storm drains is assumed to be steady uniform flow. With this assumption, one of two hydraulic design approaches for sizing the run may be used, either open-channel flow or pressure flow. Open-channel flow assumes the flow in the conduit is open to atmospheric pressure; that is, the depth of the flow must be less than the height of the conduit. Pressure flow assumes the conduit is full with the wetted perimeter equal to the complete perimeter of the conduit. In this case, unlike open-channel flow, a pressure head will be above the conduit.

Storm drain systems based on open-channel flow will have larger conduits than those based on pressure flow. This allows for a slight factor of safety when there is an unanticipated increase in runoff, which is desirable because the determination of the flow entering the system is not an exact science. However, initial construction costs will be somewhat higher if the conduit is oversized.

If the design is based on pressure flow, the inlet and access hole elevations will be the allowable high-water elevations and should not be exceeded. Additionally, existing systems may need to be analyzed assuming pressure flow in order to accommodate new design flows.

The storm drain system can outfall into a body of water, a stream or river, an existing storm drain system, or a channel. Conformance to National Pollutant Discharge Elimination System (NPDES) and local water quality regulations may be necessary whenever discharging pavement runoff. (See Chap. 1.) Regardless of the type of outfall, the flow line of the outfall should be lower than the elevation of the outlet. The outlet should be positioned so that the flow of the outfall is directed downstream, thus limiting erosion. (See *Highway Drainage Guidelines,* Vol. IX, AASHTO, 1992; and *Design and Construction of Storm and Sanitary Sewers,* ASCE, 1986.)

5.3.5 Subsurface Drainage

Saturation of the structural section under the roadway (subgrade and base course) and the foundation materials is a primary cause of early roadbed failure because of decreased ability to support heavy truck loads. Saturated conditions can lead to piping of fines and frost damage or icing of the roadway surface. Designs to prevent water from infiltrating beneath the pavement will lead to longer-lasting and more economical roadbed sections. Designs typically include *subsurface drainage* (subdrains) to intercept and reroute encroaching groundwater and *subgrade drainage* to handle surface water inflow.

The design of subsurface drainage begins with flow determination. Although this may be determined by analytical methods, it is usually cumbersome and unsatisfactory to do so. Field explorations will generally yield better results. These investigations should include soil and geological studies, borings to find the elevation and extent of the aquifer, and measurements of the groundwater discharge. The investigation should be thorough and should be conducted during the rainy season or during snow melt if the region has snow cover. It may involve digging a trench or pit to aid in estimating flow. After the design flow is established, the pipe may be sized using Manning's equation, Eq. (5.8).

The standard underdrain consists of a perforated pipe near the bottom of a narrow trench. The trench is filled with a permeable material and may be lined with filter fabric if the trench is excavated in erodable soils. Figure 5.10 illustrates an underdrain used to intercept sidehill seepage.

The following considerations apply to the design of subsurface drainage:

1. Surface drainage should not be allowed to discharge into the subsurface drainage system.

2. Outlets for the underdrain system should be provided for at intervals not exceeding 1000 ft (300 m). Outlet may run into the storm drain system as long as there is no possibility of backflow due to a buildup of hydrostatic pressure.

3. Pipe underdrains should be placed on grades steeper than 0.5 percent if possible. Minimum grades of 0.2 percent are acceptable.

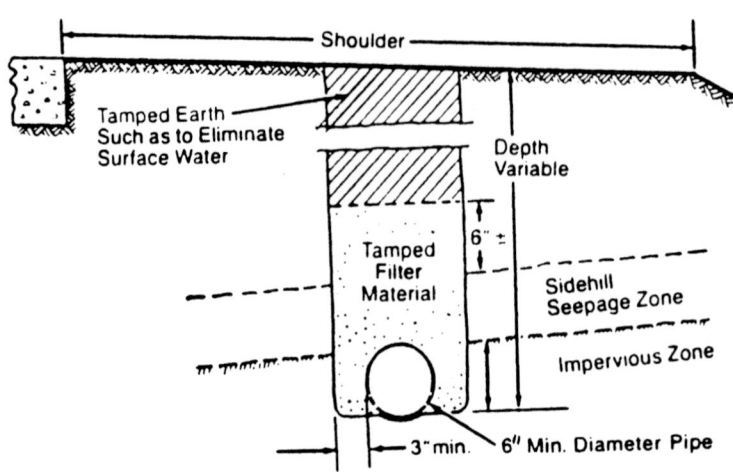

FIGURE 5.10   Intercepting drain in impervious zone for keeping free water out of roadway and subgrade. (*From* Handbook of Steel Drainage and Highway Construction
Products, *American Iron and Steel Institute, 1994, with permission*)

4. The depth of the underdrain will depend upon the permeability of the soil, the elevation of the aquifer, and the amount of necessary drawdown to achieve stability.
5. Pipes for underdrains may be made of metal, plastic, concrete, clay, asbestos
   cement, or bituminous fiber. Two types of openings are used to allow the groundwater into the pipe: perforated and open-jointed. Open-jointed pipes such as clay
   and concrete drain tiles are limited to areas where the admission of excessive
   solids through the joints may be avoided.

(See "Highway Subdrainage Design," FHWA-TS-80-224, August 1980, and
"Pavement Subsurface Drainage Systems," NCHRP Synthesis 96, TRB, November
1982.)

5.4.6   Subgrade Drainage

As indicated previously, subgrade drainage is designed to handle surface water inflow,
whereas subdrains are designed to accommodate encroaching groundwater. Surface
water can enter the pavement subsection through joints, cracks, and infiltration of the
pavement. Rapid drainage of the pavement structural section is necessary to minimize
piping and swelling of the subgrade material, and the subsequent increased deflections
and cracking of the pavement surface. This rapid drainage can best be achieved by
placing a highly permeable drainage layer under the full width of the pavement and
allowing it to drain the infiltration to an edge drain. Figure 5.11 illustrates edge drain
designs using either a pipe (perforated or slotted) in a trench filled with a permeable
material, or a drainage mat (geotextile).

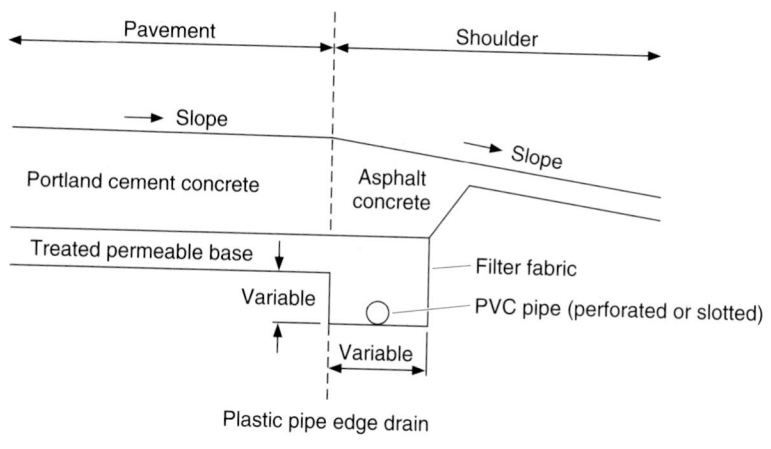

Plastic pipe edge drain

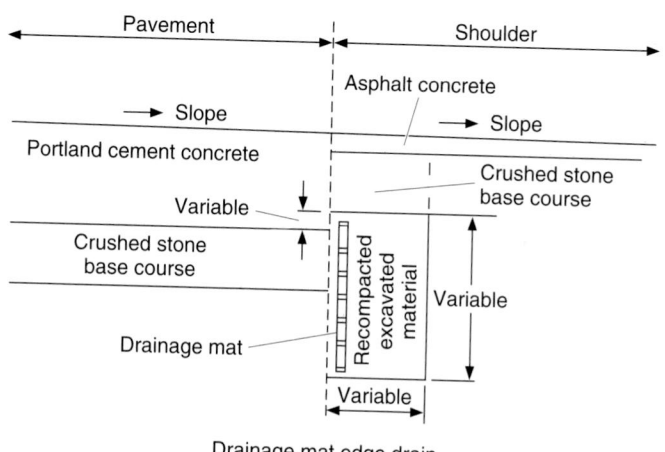

Drainage mat edge drain

FIGURE 5.11   Typical pavement edge drains. (*From* Highway Drainage Guidelines, *Vol. IX, American Association of State Highway and Transportation Officials, Washington, D.C., 1992, with permission*)

5.5   HYDRAULIC DESIGN OF CULVERTS

Culverts convey surface flow from one side of the roadway to the other. Although the hydraulics of the site is not generally the controlling criterion for determination of the shape of the culvert, culvert size is directly dependent upon the hydraulic requirements determined from the hydrologic study. Sizing the culvert on the basis of the hydraulic requirements necessitates that the alignment, slope, shape, and material of the culvert already be established.

5.5.1   General Considerations

The most common materials used are concrete, steel, aluminum, and plastic. The material used may affect the hydraulic capacity of the culvert, as different materials and wall configurations have different coefficients of roughness. The choice of the material is often controlled by structural and durability considerations.

The inlet configuration generally has a direct effect on the hydraulic capacity of the culvert and the backwater upstream from the site. The natural channel approaching the culvert is usually wider than the culvert, and thus the inlet operates as a flow contraction and can be the control for determining the hydraulic capacity. In many instances, the culvert is designed to operate hydraulically with the inlet submerged. This is one advantage that culverts have over bridges, which are designed for freeboard between the high-water elevation and the soffit. If the inlet provides for a gradual transition from the wider natural channel to the narrower culvert barrel, energy losses can be limited. Figure 5.12 depicts some common transitions used to improve culvert hydraulics. Some of the common end treatments used at inlets and outlets include projecting ends, mitered ends, flared ends, and headwalls and wingwalls.

Projecting ends exist when the barrel of the culvert extends out from the face of the embankment. This is probably the least expensive and most hydraulically inefficient of the listed end treatments. In addition it is unsightly and potentially hazardous to traffic. For these reasons its use should be limited to smaller culverts.

Mitered ends exist where the culvert is formed or manufactured to be in the same plane as the embankment. Mitered ends, when compared with projected ends, are more aesthetically pleasing. However, the projected end is structurally more stable and the mitered end may require the addition of a headwall to compensate for this instability. The hydraulic efficiency of both the mitered and the projected inlets is approximately the same.

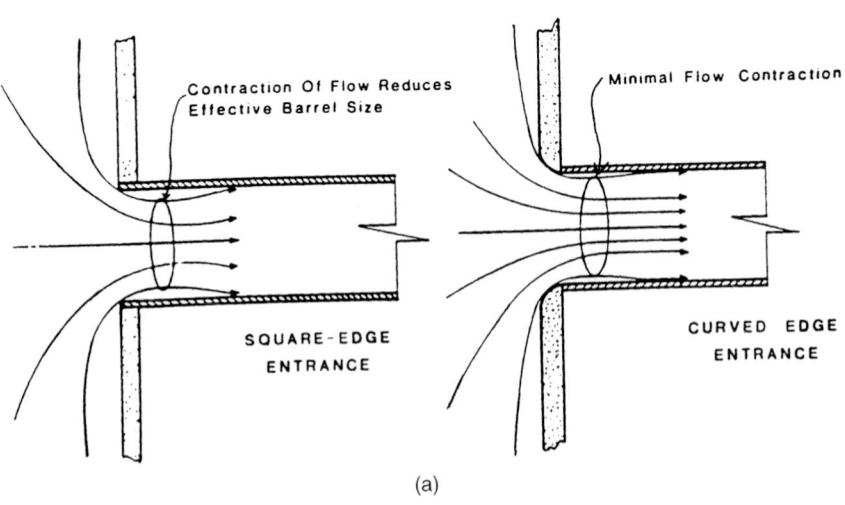

(a)

FIGURE 5.12 Illustration of common transitions that improve culvert hydraulics. (*a*) Entrance contraction. (*b*) Side-tapered inlet. (*c*) Slope-tapered inlet. (*From "Hydraulic Design of Highway Culverts," Hydraulic Design Series No. 5, FHWA, with permission*)

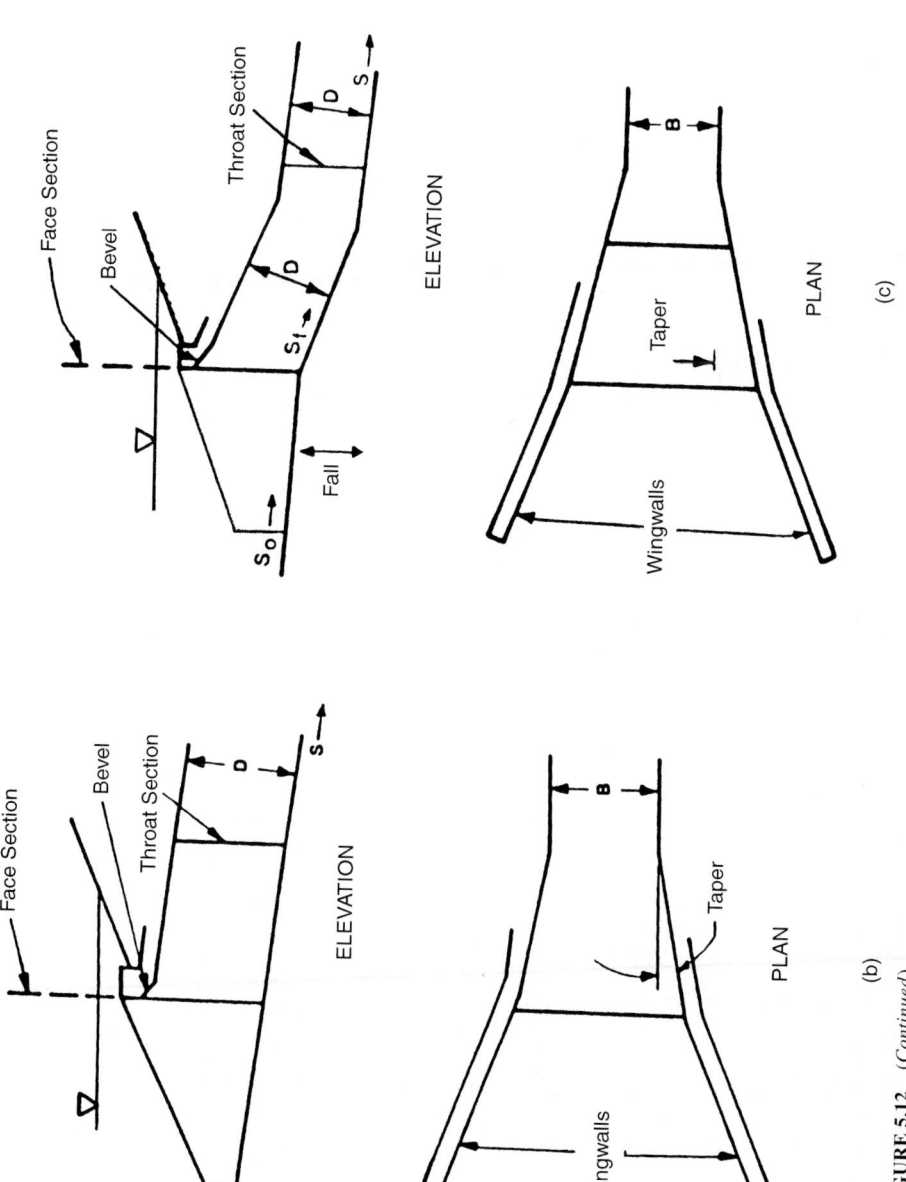

FIGURE 5.12 (*Continued*)

Flared ends are generally precast or prefabricated for use with concrete, corrugated steel or aluminum, and plastic pipes. They are used to retain the earth embankment and provide a hydraulic efficiency comparable to that of a headwall.

Headwalls and wingwalls are usually cast-in-place structures. They are designed to retain the embankment, improve hydraulics, prevent erosion, and, in larger-diameter flexible structures, provide support at the inlet and outlet ends. Retaining the earth has an economic benefit in that the culvert may be shortened, thereby providing cost savings. The hydraulics may be improved by skewing or warping the wingwalls to provide for a smooth transition between the wider channel and the narrower barrel.

The preferred location of the culvert is in the natural streambed. This alignment usually provides for efficient inlet and outlet configurations and keeps construction costs to a minimum by limiting excavation and backfill work. Aligning the culvert in this manner can result in an inordinately long structure if the natural channel is on a high skew with respect to the roadway. This may be avoided by realigning the channel so that the culvert is placed perpendicular to the highway, but this may lead to erosion and siltation problems. Erosion may occur where the channel is angled to provide for the perpendicular crossing. Siltation may occur as the slope is necessarily reduced because the flow travels a longer distance to traverse the roadway.

5.5.2   Inlet and Outlet Control

There are two types of flow in culverts: inlet control and outlet control. Accurate prediction of the condition of flow is difficult, and an assumption of the most conservative control may at times be warranted. Figures 5.13 and 5.14 depict several conditions of inlet and outlet control.

For inlet control, the discharge capacity is controlled at the upstream or inlet end. Factors that have an effect on the culvert performance under this condition are the headwater elevation, the inlet area of the barrel, and the inlet configuration. For outlet control, the discharge is controlled at the downstream end. Additional factors affecting performance under this condition include the tailwater elevation, characteristics of the culvert barrel (slope, length, roughness, shape, and cross-sectional area), and the outlet configuration.

With inlet control, the culvert usually flows only partially full; the roughness, slope, length, and outlet condition of the culvert do not affect the discharge capacity. The headwater depth is measured from the invert. The inlet area is generally the same as the cross-sectional area of the barrel. However, when tapered or beveled inlets are utilized, the face area is enlarged and the control area is at the throat. The efficiency of a culvert is greatly affected by the inlet configuration and may be heightened by the use of beveled edges and tapered inlets, which reduce the contraction of the flow, thereby effectively enlarging the face area. Bevels are large chamfers or rounded corners at the inlet. They improve the hydraulics at minimal additional cost and should always be considered as an improvement. Tapered inlets may be tapered either at the sides or at the bottom (slope tapers). Either type will increase the flow capacity or, conversely, decrease the headwater elevation for a given capacity.

The coefficient k_e, which represents the efficiency of the culvert inlet, is listed in Table 5.7 for many different designs. It may be used to calculate the head loss at the entrance from the equation

$$H_e = k_e \left(\frac{V^2}{2g} \right) \tag{5.18}$$

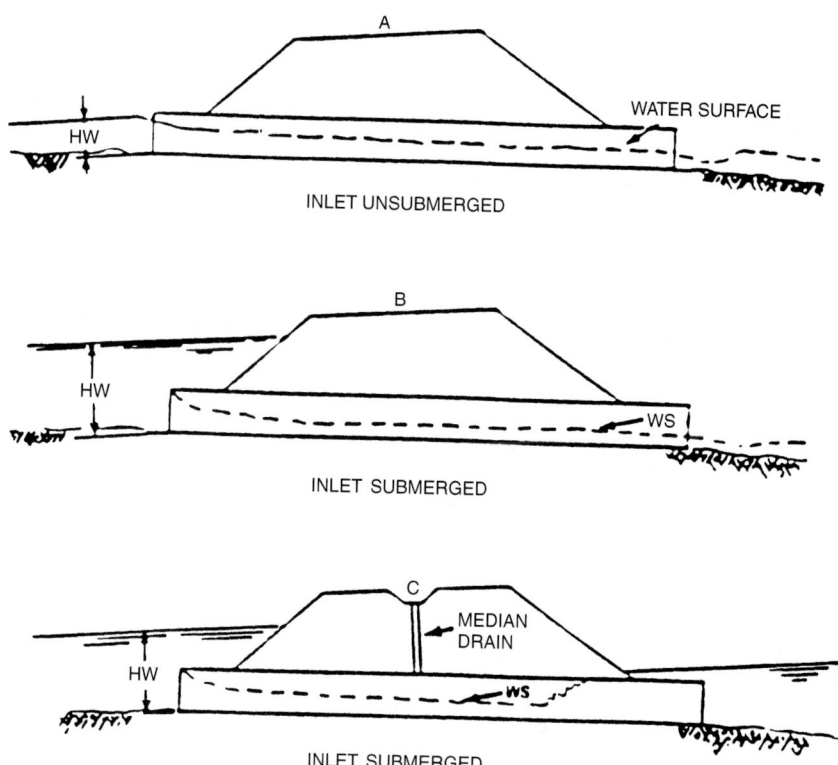

FIGURE 5.13   Illustration of culvert under inlet flow control. (*From* Highway Drainage Guidelines, *Vol. IV, American Association of State Highway and Transportation Officials, Washington, D.C., 1992, with permission*)

where H_e = entrance head loss, ft (m)
     k_e = energy coefficient
     V = velocity, ft/s (m/s)
     g = acceleration of gravity, 32.2 ft/s^2 (9.8 m/s^2)

5.5.3   Culvert Size Determination by Performance Curves

Once the design discharge and the allowable headwater are determined and the culvert alignment and slope decided upon, an efficient culvert size may be found through the use of either performance curves or nomographs as shown in FHWA publications. Performance curves are plots of headwater versus discharge for different sizes and types of culverts. They have been developed for nearly all common sizes and shapes used in culvert construction and are available from the FHWA. (For conditions of low headwater, see "Capacity Charts for the Hydraulic Design of Highway Culverts," HEC 10, FHWA. For conditions of high headwater, or where the outlet is submerged, see "Hydraulic Design of Highway Culverts," Hydraulic Design Series 5, FHWA.)

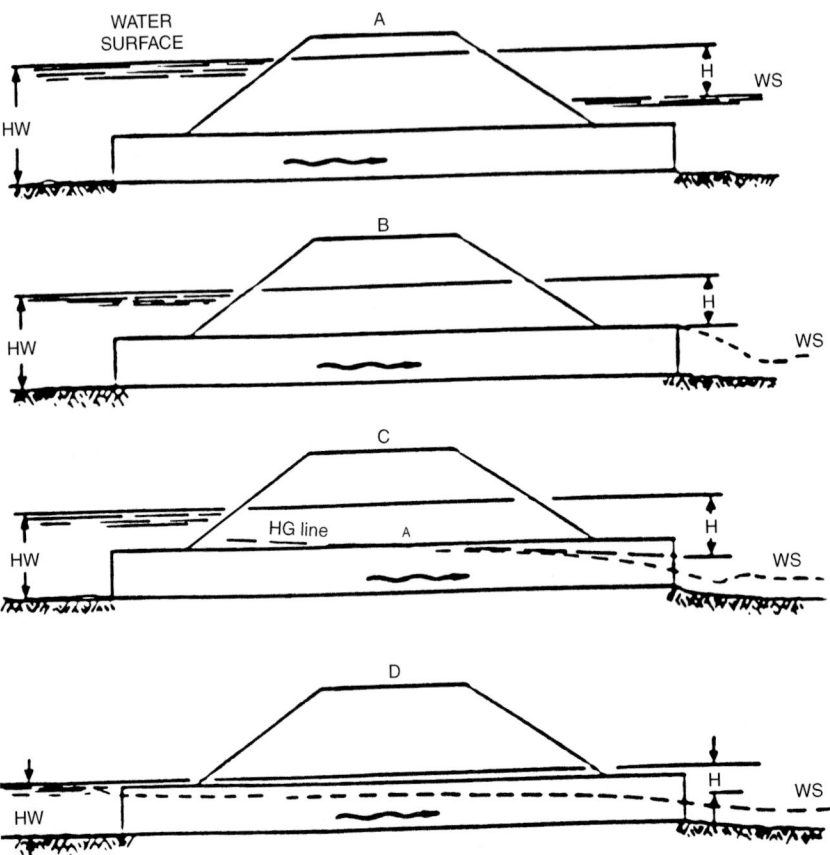

FIGURE 5.14 Illustration of culvert under outlet flow control. (*From* Highway Drainage Guidelines, *Vol. IV, American Association of State Highway and Transportation Officials, Washington, D.C., 1992, with permission*)

The charts in HEC 10 are arranged in groups according to shape and material. In addition, two sets of charts for shape are given, for two differing inlet conditions (projecting inlet and inlet with a headwall). The charts include performance curves for the types, sizes, and inlet configurations listed below. (See HEC 10 for additional inlet configurations which apply to those listed.)

1. Concrete boxes, with a maximum height of 6 ft (1.8 m) in a rectangular shape and 12 ft (3.6 m) in a square shape, all with headwalls or wingwalls at the inlet
2. Circular concrete pipe with a maximum diameter of 15 ft (4.6 m) with both square-edged and grooved-edged inlets
3. Oval concrete pipe with the maximum long horizontal axis of 91 in (2300 mm) with both square-edged and grooved-edged inlets
4. Oval concrete pipe with the maximum long vertical axis of 83 in (2100 mm) with both square-edged and grooved-edged inlets

TABLE 5.7   Entrance Loss Coefficients for Culverts under Outlet Control, Flowing Full or Partly Full

Type of structure and design of entrance	Coefficient k_e
Pipe, concrete	
Projecting from fill, socket end (groove end)	0.2
Projecting from fill, square-cut end	0.5
Headwall or headwall and wingwalls	
Socket end of pipe (groove end)	0.2
Square edge	0.5
Rounded, radius = $(\frac{1}{12})D$	0.2
Mitered to conform to fill slope	0.7
End section conforming to fill slopes*	0.5
Beveled edges, 33.7° or 45° bevels	0.2
Side- or slope-tapered inlet	0.2
Pipe, or pipe-arch, corrugated metal	
Projecting from fill (no headwall)	0.9
Headwall or headwall and wingwalls, square edge	0.5
Mitered to conform to fill slope, paved or unpaved slope	0.7
End section conforming to fill slope*	0.5
Beveled edges, 33.7° or 45° bevels	0.2
Side- or slope-tapered inlet	0.2
Box, reinforced concrete	
Headwall parallel to embankment (no wingwalls)	
Square-edged on 3 edges	0.5
Rounded on 3 edges to radius of ½ barrel	
dimension, or beveled edges on 3 sides	0.2
Wingwalls at 30° to 15° to barrel	
Square-edged at crown	0.4
Crown edge rounded to radius of ½ barrel	
dimension, or beveled top edge	0.2
Wingwall at 10° to 25° to barrel	
Square-edged at crown	0.5
Wingwalls parallel (extension of sides)	
Square-edged at crown	0.7
Side- or slope-tapered inlet	0.2

*"End section conforming to fill slope," made of either metal or concrete, is the section commonly available from manufacturers. From limited hydraulic tests it is equivalent in operation to a headwall in both *inlet* and *outlet* control. Some end sections, incorporating a *closed* taper in their design, have a superior hydraulic performance. These latter sections can be designed using the information given for the beveled inlet.

Source: From *Highway Drainage Guidelines,* Vol. IV, American Association of State Highway and Transportation Officials, Washington, D.C., 1992, with permission.

5. Corrugated metal pipe with 1/2-in (13-mm) corrugations to a maximum span of 10 ft (3 m) with both projecting and headwall inlets

6. Corrugated metal pipe-arch with 1/2-in (13-mm) corrugations to a maximum span of 6 ft (1.8 m) with both projecting and headwall inlets

7. Structural-plate corrugated metal pipe with 2-in (50-mm) corrugations to a maximum span of 15 ft (4.5 m) with both projecting and headwall inlets

8. Structural-plate corrugated metal pipe-arch with 2-in (50-mm) corrugations and 18-in (450-mm) corner radii to a maximum span of 16.6 ft (5 m) with both projecting and headwall inlets

An example of the performance charts is shown in Fig. 5.15. Charts such as these assume free-flowing water; if substantial debris is present, there will be a decrease in culvert capacity. To use the chart, the design discharge, allowable headwater depth, culvert length, and slope must be known. To find the required culvert size, find the intersection between the allowable headwater depth on the vertical axis and the design discharge on the horizontal axis; the smallest culvert providing that headwater depth or less, as indicated by the appropriate curve, can be selected. The solid line on the chart represents the culvert performance under inlet control, and the dashed line depicts the same culvert under outlet control. Where only one curve is shown, the performance of the culvert is almost identical under inlet and outlet control. To determine which curve applies for a particular size culvert, the index number represented by the value of the length of the culvert (ft) divided by the slope times 100 must be calculated ($L/100S_o$). If the index number is less than that shown on a solid line curve, the culvert is under inlet control and the solid line is used. If the index number falls between the numbers indicated on the solid and dashed lines, interpolate to determine the amount of headwater.

The dashed horizontal line represents a headwater depth equal to approximately twice the depth of the culvert. This is the limit for unrestricted use of the chart. If the headwater depth exceeds this value, the depth should be checked using the nomographs available in "Hydraulic Charts for the Selection of Highway Culverts," HEC 5, FHWA.

Additionally, if the culvert is not on a constant slope, if the index number exceeds those values shown on the dashed lines, or if the tailwater depth equals or exceeds the critical depth, use the nomographs available in "Hydraulic Design of Highway Culverts," HDS 5, FHWA.

The charts for rectangular sections in HEC 10 are limited where the span is wide with heights of 5 ft (1.5 m) or more. In this range, because resistance losses are a small part of the headwater depth, the capacity of the culvert is approximately proportional to the span for any given height. Therefore, the capacity of a rectangular box per linear foot of span may be found and extrapolated for longer spans.

The tailwater depth may be calculated from the continuity equation and Manning's equation, Eqs. (5.7) and (5.8).

5.5.4   Size Determination for Culverts with Submerged Outlets

The headwater depth for a culvert with a submerged outlet may be determined with the aid of the outlet control nomographs in HEC 5. An example of these nomographs for box culverts is shown in Fig. 5.16. The following procedure may be used to determine the head H from the nomographs. The length L (ft), entrance coefficient k_e, and design discharge must be known. Locate L on the appropriate k_e curve, and connect this point with the proposed culvert size. Locate the design discharge and extend a line from that point through the turning point intersection of the previous line to read the value of the head H (ft) on the right. For example, assume $L = 306$ ft, $k_e = 0.5$, a 2-ft×2-ft box, and $Q = 40$ ft³/s. The nomograph shows that $H = 7.3$ ft. The headwater depth, HW, may then be determined by geometry from the equation

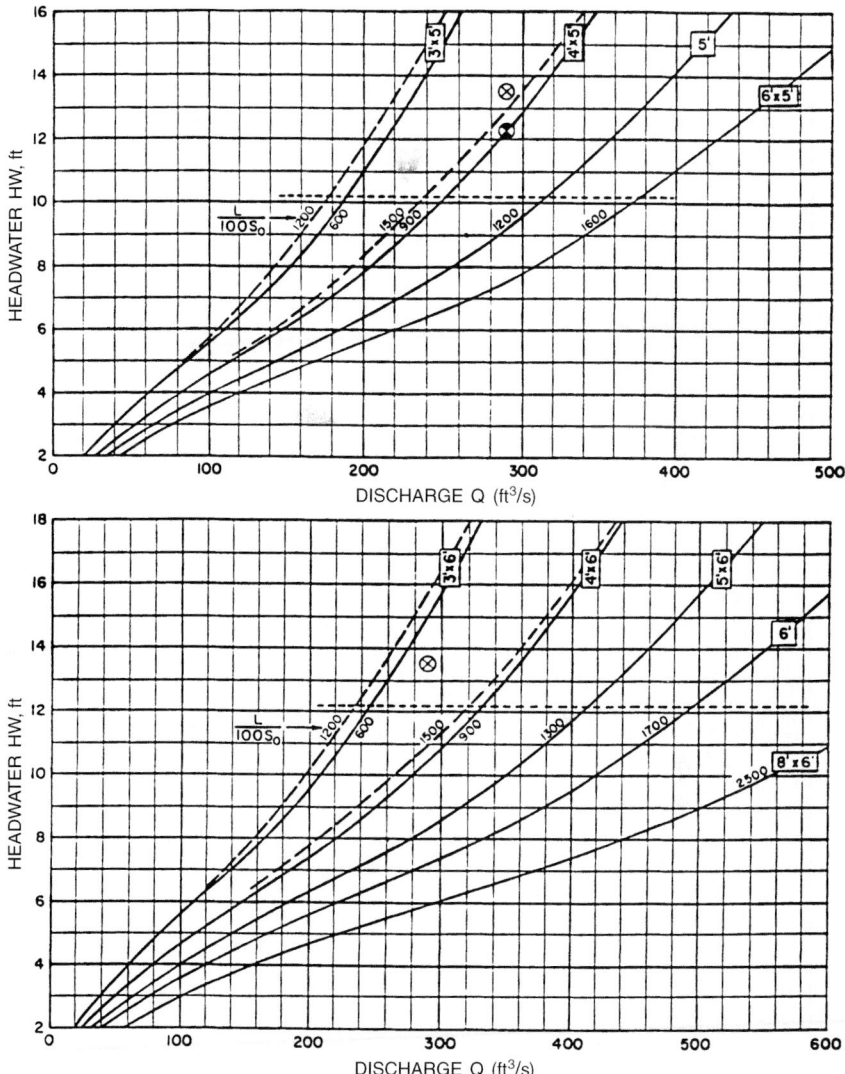

FIGURE 5.15 Capacity chart for hydraulic design of rectangular box culvert, 90° and 15° wingwall flare, 5- and 6-ft heights. (*From "Hydraulic Charts for the Selection of Highway Culverts," Hydraulic Engineering Circular No. 10, FHWA, with permission*)

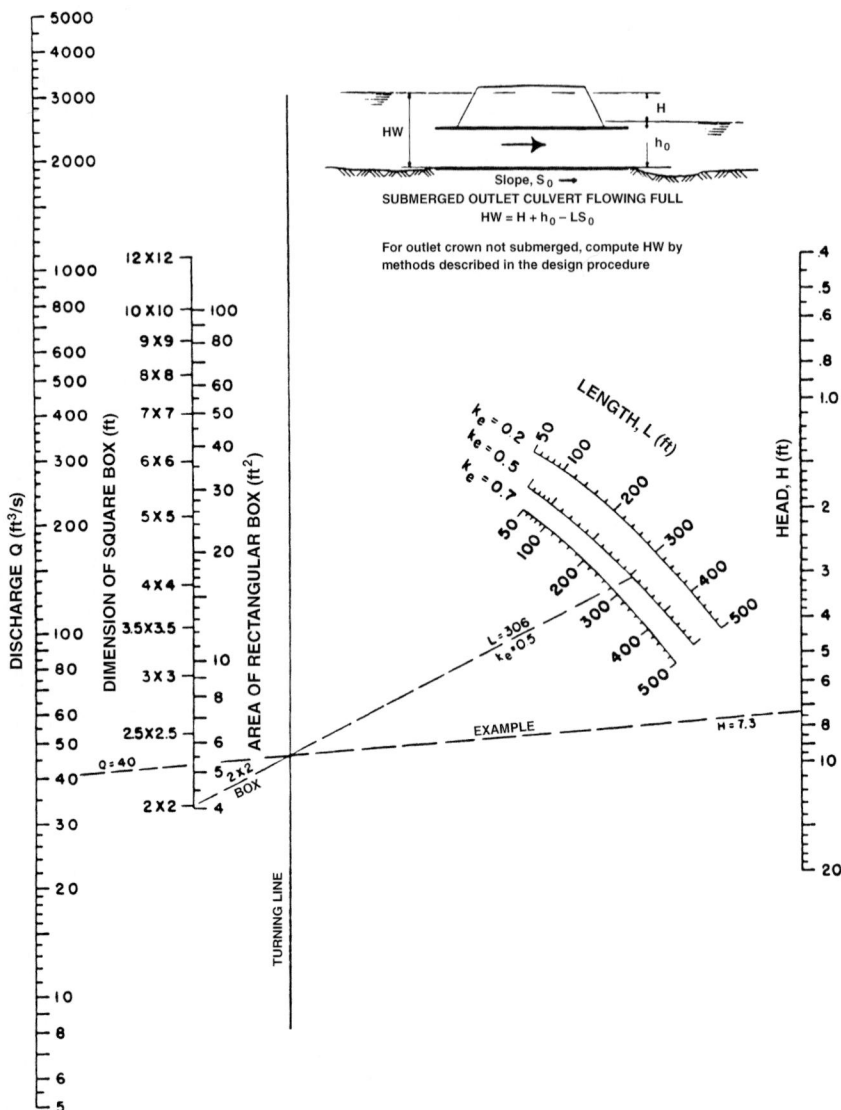

FIGURE 5.16   Flow nomograph for box culvert under outlet control with $n = 0.012$. (*From "Hydraulic Charts for the Selection of Highway Culverts," Hydraulic Engineering Circular No. 5, FHWA, with permission*)

$$HW = H + h_o - LS_o \qquad (5.19)$$

where the terms are defined by the inset figure in the nomograph. Where the outlet is submerged, h_o is equal to the tailwater depth just downstream of the outlet and may be calculated from Manning's equation as applied to the channel. Where the outlet is not submerged, h_o is equal to the greater of the tailwater depth or one-half of the sum of the culvert height plus the critical depth. By examining different alternatives, a culvert can be selected that provides the required flow within the allowable headwater depth.

5.5.5 Critical Depth Determination

The critical depth for various culvert cross sections may be found from charts in HEC 5. An example is given in Fig. 5.17 for a rectangular section. In this case the ratio of the flow Q (ft³/s) to the width B (ft) is used to find the critical depth d_c (ft). Of course, d_c cannot exceed the depth of the box section.

5.5.6 Size Determination for Long-Span Structures

Because culvert shapes are so numerous and new shapes are often developed, design charts showing performance curves are not available for all culvert sizes and shapes. One example is long-span corrugated-metal sectional plate structures. Although the product is available in several cross-sectional shapes, performance curves are available only for circular or elliptical cross sections (Fig. 5.18) and high- and low-profile arches (Fig. 5.19). (See Art. 5.6.2 for description.) These charts, which are for inlet control only, address four different inlet configurations ranging from mitered to beveled-edge ends. Because long-span structures are commonly used when headroom is low, they generally do not flow under head at design discharge but flow partly full.

The first step in using these charts is to obtain information on available sizes, including cross-sectional area A (ft²) and vertical height D (ft). For the design discharge Q (ft³/s), calculate $Q/AD^{0.5}$ and read the value of HW/D at the intersection of the appropriate edge condition curve. Multiply by the depth (height) of the structure (D) to obtain the headwater depth HW and compare with the allowable design value. To consider a long-span structure under outlet control, an analysis including pressure flow and backwater calculation can be made. (See "Hydraulic Design of Highway Culverts," HDS 5, FHWA.) The inlet and outlet control headwater elevations are then compared. The higher value is compared against the allowable elevation to determine if the size is satisfactory or if the process should be repeated.

5.5.7 Discharge Velocity and Energy Dissipation

Because of its hydraulic characteristics, the outlet velocity of a culvert is usually higher than the velocity in the discharge channel. The outlet velocity may be calculated either using Manning's equation, Eq. (5.8), if the culvert is under inlet control, or by dividing the discharge by the cross-sectional area of the flow if under outlet control. Under outlet control, if the tailwater is above the crown of the pipe, or if the discharge is high enough to result in a critical depth equal to the depth of the culvert barrel, then the flow area may be taken as the area of the barrel. If the tailwater depth is low, the

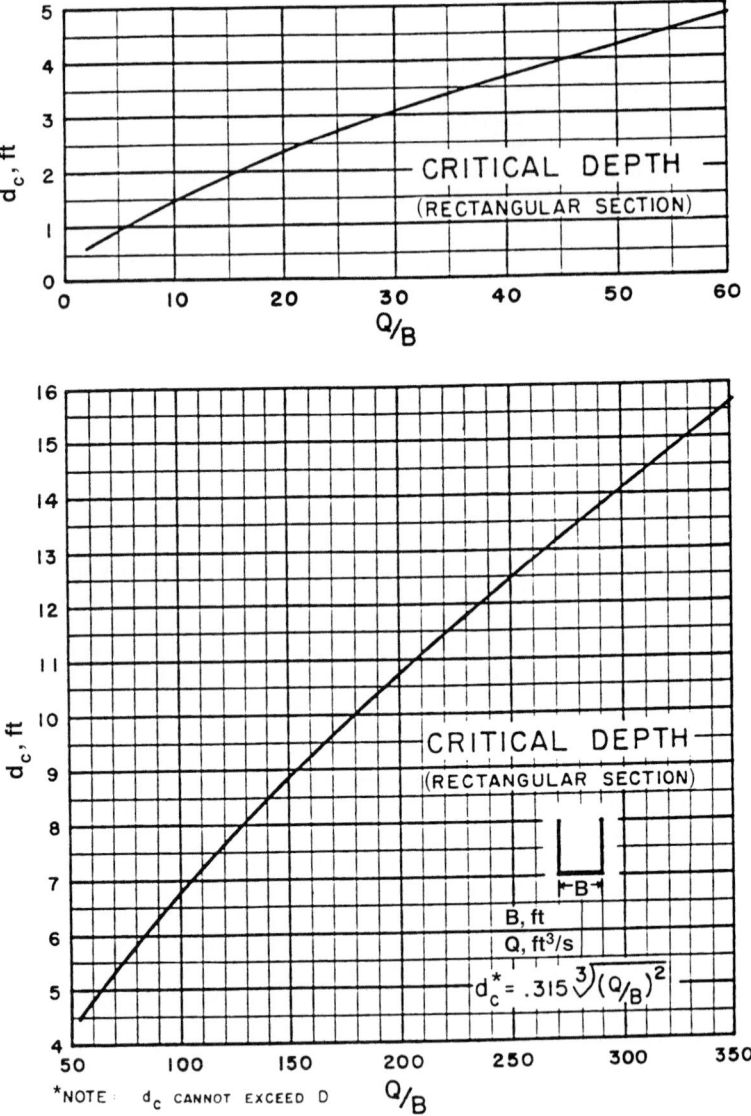

FIGURE 5.17　Critical depth for flow in rectangular channel. (*From "Hydraulic Charts for the Selection of Highway Culverts," Hydraulic Engineering Circular No. 5, FHWA, with permission*)

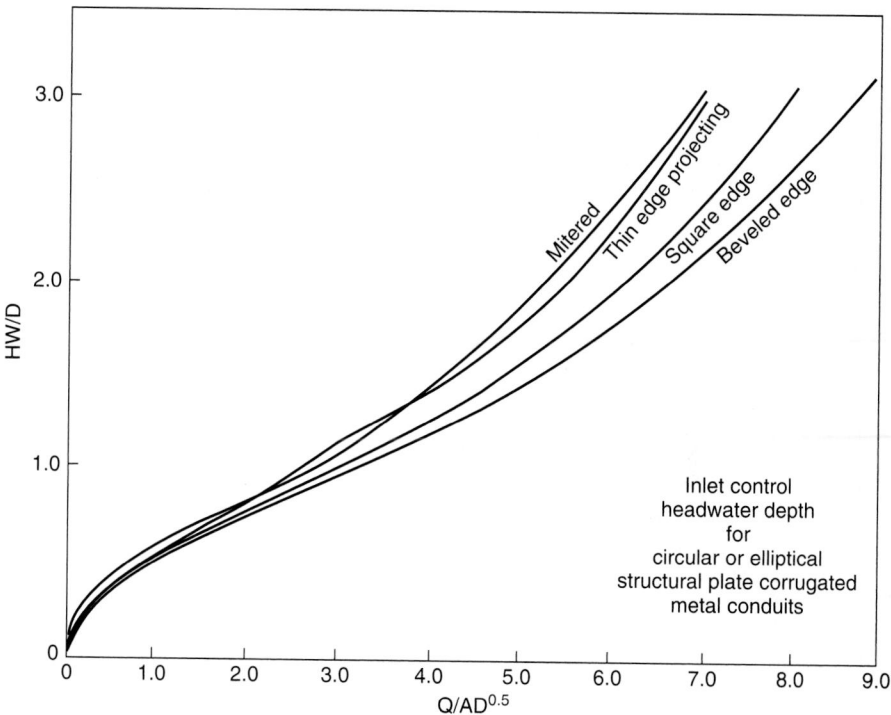

FIGURE 5.18 Performance charts for circular or elliptical structural-plate conduits under inlet flow control. (*From "Hydraulic Design of Highway Culverts," Hydraulic Design Series No. 5, FHWA, with permission*)

area of flow, and thus the velocity, may be determined using the chart in Fig. 5.20 or 5.21. To use these charts, first calculate the normal depth or tailwater TW (ft) in the channel; the ratio TW/D, where D is the structure height (ft); and the flow parameter $Q/BD^{3/2}$, where B (ft) is the width of the barrel and Q (ft^3/s) is the discharge. Enter the chart with TW/D and find Y_o/D at the intersection of the appropriate curve. Multiply by D to determine the depth of flow at the outlet end of the culvert, Y_o. The flow area is then calculated for Y_o and the velocity for the flow Q from the continuity equation, Eq. (5.7).

Recommended maximum channel velocities were presented in Table 5.3. The velocity at the outlet should be kept at or below these values, or, if this is not possible, the channel should be protected from erosion. The controlling parameters for the culvert velocity are its slope and roughness. If the recommended velocity is exceeded, consider decreasing the slope or using a culvert with a greater roughness coefficient. If the velocity at the outlet cannot be reduced by these means, channel protection or energy dissipaters should be used to protect against erosion. Channel protection may consist of treatments such as concrete aprons or cutoff walls. In some cases, concrete or rock riprap may be required. These types of protection do not necessarily dissipate the energy, but protect against erosion. Energy-dissipating devices may be necessary either separately or in conjunction with channel protection where flow velocities are high. Dissipation devices, if used, are generally located at the outlet end or in the interior near the end of the culvert. If such devices are used, consideration must be given

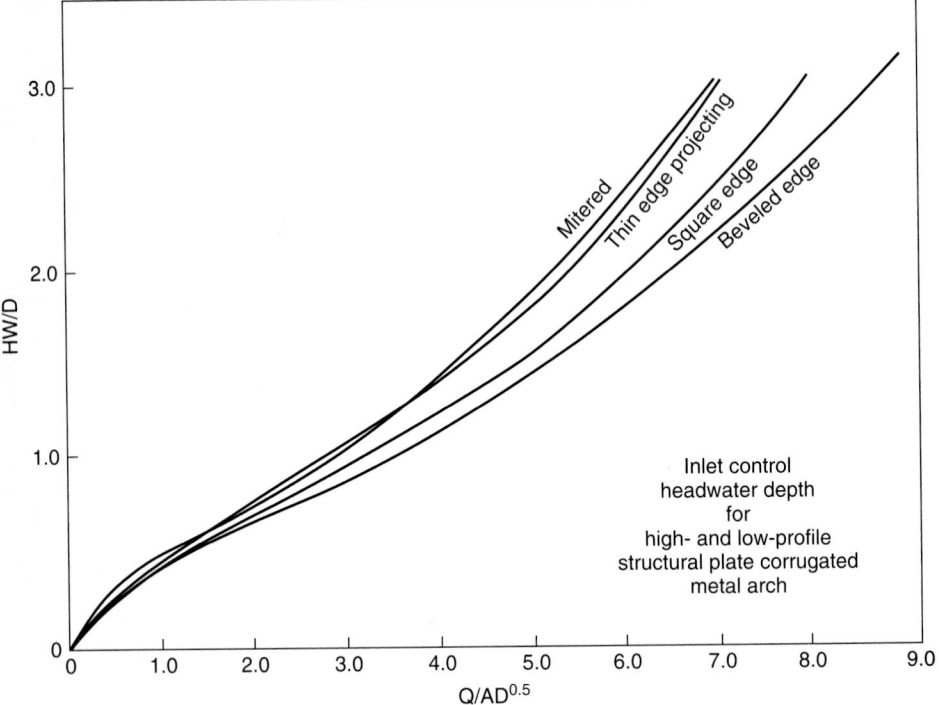

FIGURE 5.19   Performance charts for high- and low-profile structural-plate arches under inlet flow control. (*From "Hydraulic Design of Highway Culverts," Hydraulic Design Series No. 5, FHWA with permission*)

to the effects on possible debris collection. (See "Hydraulic Design of Energy Dissipators for Culverts and Channels," HEC 14, FHWA.)

5.6   CULVERT TYPES AND MATERIALS

The main types of pipe used in highway construction are concrete pipe, metal pipe (steel or aluminum), and plastic pipe (high-density polyethylene and polyvinyl chloride). They are available in a wide array of sizes, shapes, and properties. Some of the characteristics of these pipes are reviewed below.

5.6.1   Concrete Pipe

Concrete pipe is manufactured as nonreinforced, reinforced, or cast-in-place pipe; as box culverts and special shapes; and as field-constructed pipe.

Factory-Made Pipe.   Nonreinforced pipe is used for smaller diameters, whereas pipe with steel reinforcement is used for larger diameters and greater loads. Both are manufactured in a plant, cured, and shipped to the job site. They are furnished in relatively

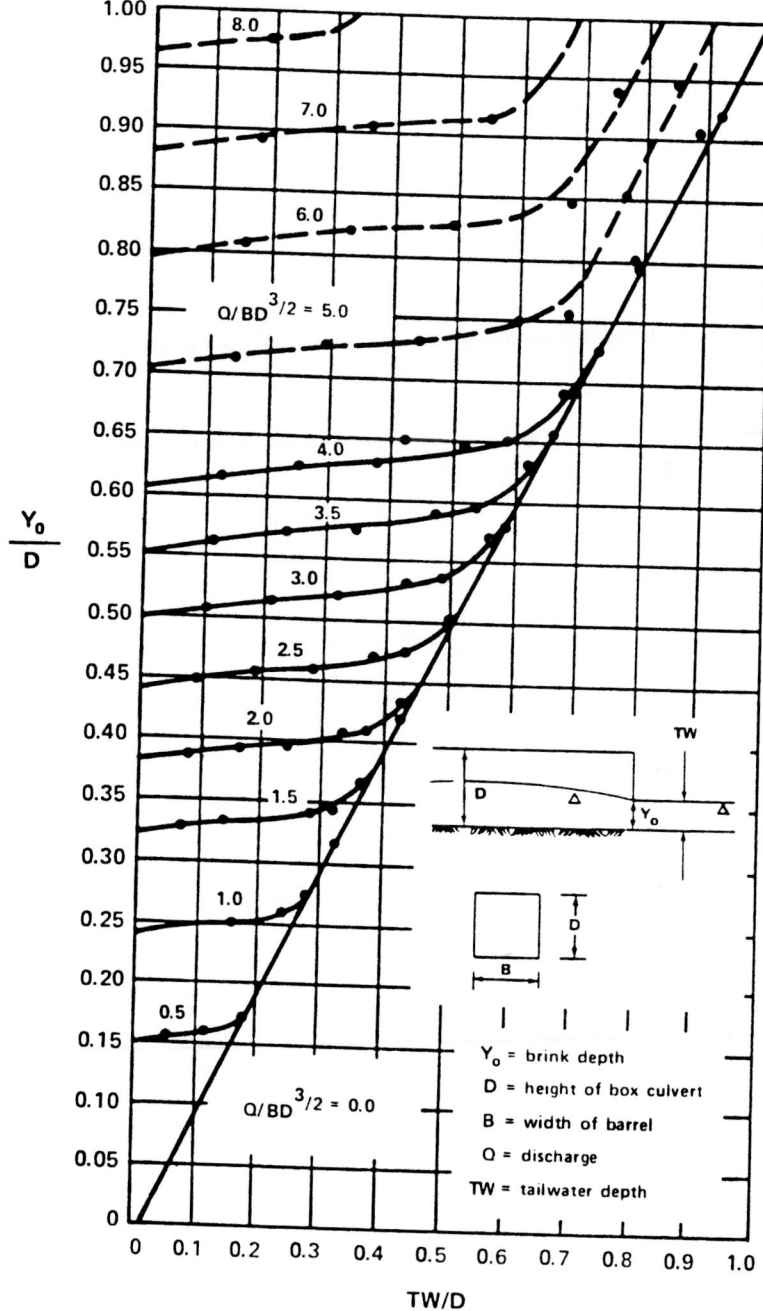

FIGURE 5.20 Dimensionless rating curves for outlets of rectangular culverts on horizontal and mild slopes. (*From "Hydraulic Design of Energy Dissipators for Culverts and Channels," Hydraulic Engineering Circular No. 14, FHWA, with permission*)

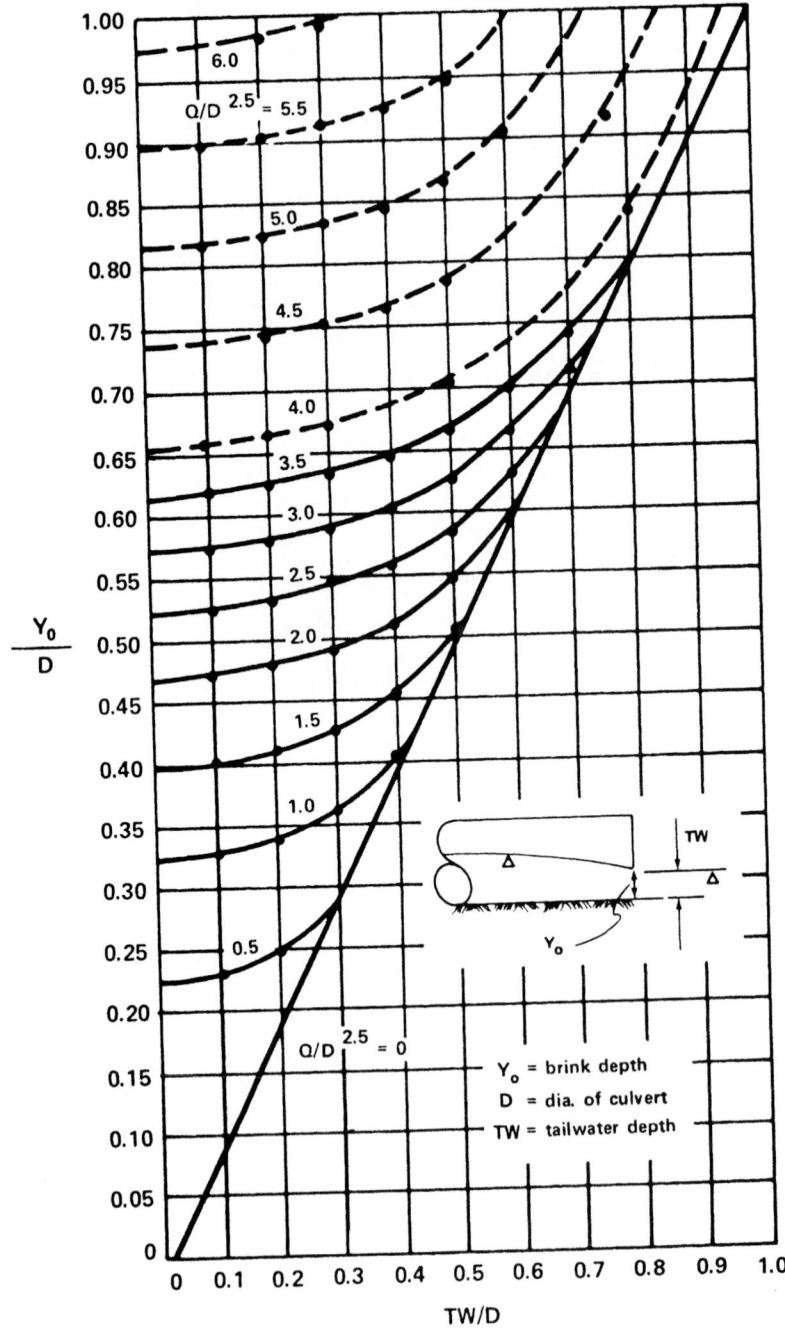

FIGURE 5.21 Dimensionless rating curves for outlets of circular culverts on horizontal and mild slopes. (*From "Hydraulic Design of Energy Dissipators for Culverts and Channels," Hydraulic Engineering Circular No. 14, FHWA, with permission*)

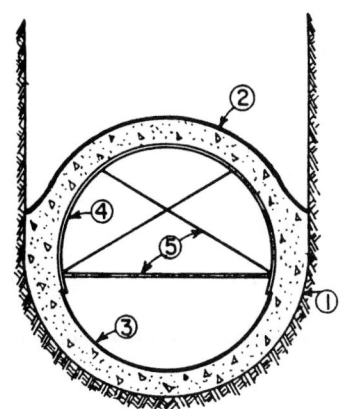

FIGURE 5.22   Cross section of cast-in-place concrete pipe showing form components. (*From Lynch Manual,* Cast-in-Place Concrete Process, *W. M. Lynch Co., Dixon, Calif., 1990, with permission*)

short lengths and coupled with a bell-and-spigot type joint. Shapes include round, horizontal ellipse, vertical ellipse, and arch configurations.

Cast-in-Place Pipe.   This type of nonreinforced pipe is formed in a trench using a continuous process. First a trench is excavated so that it has a semicircular bottom and vertical or near vertical sidewalls, which serve as the outer form for the bottom and sides. The upper portion of the pipe is cast against an inner arch form as illustrated in Fig. 5.22. The form is pulled along the trench while concrete is poured into a hopper located above. Powered spading mechanisms and variable-speed vibrators aid the flow of the concrete.

Box Culverts.   Box culverts are rectangular shapes with flat sides, top, and bottom. These shapes are constructed with steel reinforcement. Factory-made boxes are shipped in sections 6 to 8 ft long and joined in the field to make a structure of the required length.

Special Shapes.   Other shapes are also manufactured. One example is a section made up of an arch top and vertical sidewalls. It is furnished in segments that are joined in the field to make up the required length. With a span of 12 to 36 ft, it can be used as a replacement structure for short-span bridges. Also, segmental tunnel liners can be furnished as precast concrete segments.

Field-Constructed Pipe.   Large reinforced structures may be constructed at the job site using appropriate formwork. Large arches and box culverts are often constructed in this manner.

5.6.2   Steel and Aluminum Pipe

Numerous drainage products are available in steel with protective coatings and in aluminum. These include corrugated pipe, spiral-rib pipe, sectional-plate pipe, box culverts, and, where a tunnel is required, tunnel liner plates. Figure 5.23 shows the variety of profiles available for the wall cross section of steel drainage products. The arc-and-tangent profiles shown with depths of ¼ through 1 in are wall profiles for pipe factory-corrugated to the full pipe cross section. The 2-in-deep profile, which is used for sectional-plate pipe and box culverts, is corrugated and curved into arc segments that can be bolted together in the field. The 5½-in-deep profile is a similar product used for longer-span structures. The ¾- and 1-in-deep rectangular profiles are for factory-corrugated spiral-rib pipe. Figure 5.24 illustrates the shapes of the products, the range of sizes available in steel, and common uses. Some corrugation profiles and size ranges vary for aluminum products. The larger sizes of sectional-plate products and box culverts in steel or aluminum are often used as replacements for short-span bridges.

Corrugated Steel.   Most of the metal pipe used is corrugated from coils of coated sheet steel. Coatings, which are applied by the continuous hot-dip process in the production of the steel coil, include zinc (galvanizing), aluminum, 55 percent aluminum-zinc alloy, and zinc–5 percent aluminum misch metal alloy. In addition, coils are available precoated with a polymer (on one or both sides) to provide extra protection against corrosion and/or abrasion. Most corrugated pipe has a continuous helical lockseam, but some manufacturers use a continuous helical welded seam, or a longitudinal riveted or spot-welded seam. Wall profiles from $1½ \times ¼$ in to 5×1 in are factory-corrugated to the full pipe cross section. The pipe is furnished in lengths (typically 20 ft or more) and joined in the field by coupling bands. Diameters through 120 in are available, depending on the wall profile. Pipe-arch shapes for installations with low cover are formed to shape from lengths of round pipe.

Corrugated Aluminum.   Corrugated aluminum pipe is usually furnished with one of the following wall profiles: $1½ \times ¼$ in, $2⅔ \times ½$ in, or 3×1 in. The pipe may have a helical lockseam or a riveted seam. It is furnished in lengths similar to steel pipe and joined in the field by coupling bands. Diameters through 120 in are available, depending on the wall profile, and pipe-arch shapes are formed to shape from lengths of round pipe.

Spiral-Rib Pipe.   This is a newer type of steel pipe that is helically corrugated to the rectangular profiles shown in Fig. 5.23. The cross-section profile has been developed so that flow characteristics are similar to that of a smooth-walled pipe. It is available in either coated steel or aluminum, as either round pipe through 108-in diameter, or as pipe-arch.

Sectional-Plate Pipe.   This product type is available in either zinc-coated steel or aluminum.

Steel.   The 6- $\times$ 2-in profile used for sectional-plate pipe and box culverts is corrugated and curved into arc segments. The segments provide an arc length of up to about 86 in, in lengths of 10 or 12 ft. The segments are joined together with high-strength bolts in a sequential manner during construction. All of the shapes illustrated in Fig. 5.24 can be constructed with this product. The 15- $\times$ 5½-in profile can be used for the larger structures. With spans up to about 50 ft, sectional-plate structures can provide an economical alternative for replacing short-span bridges. Field coatings can be applied to enhance durability.

Aluminum.   The 9-in-wide by 2½-in-deep profile is used for the aluminum sectional-plate pipe and box culvert structures. Product characteristics are generally similar to those of the steel product.

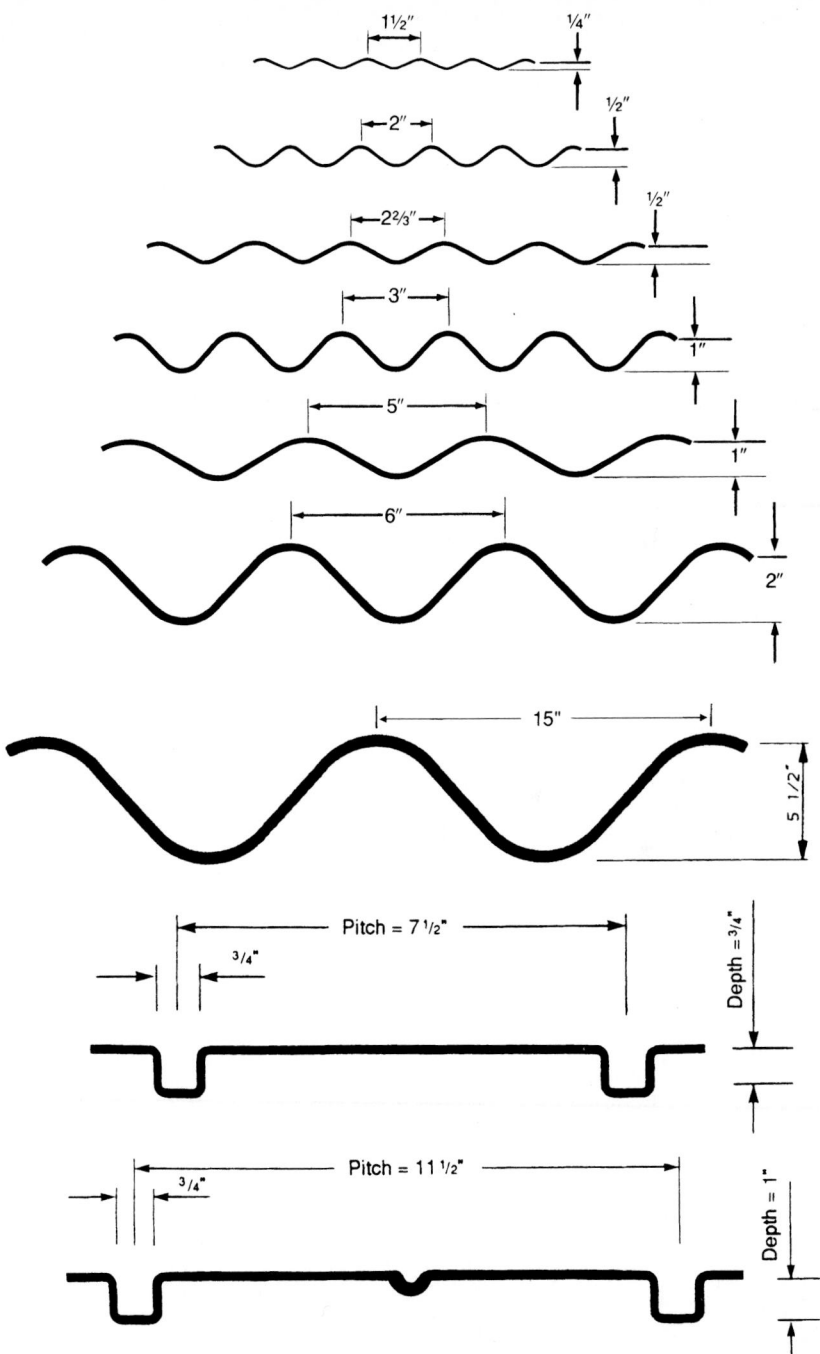

FIGURE 5.23 Profiles for corrugated steel pipe and spiral-rib pipe. (*From* Handbook of Steel Drainage and Highway Construction Products, *American Iron and Steel Institute, 1994, with permission*)

Shape		Range of Sizes	Common Uses
Round	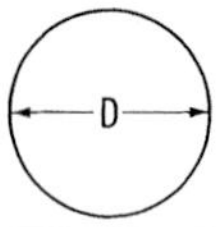 D	6 in.–26 ft	Culverts, subdrains, sewers, service tunnels, etc. All plates same radius. For medium and high fills (or trenches).
Vertical ellipse 5% nominal	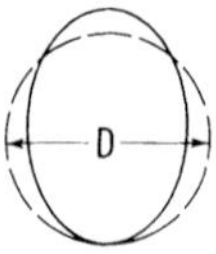 D	4–21 ft nominal; before elongating	Culverts, sewers, service tunnels, recovery tunnels. Plates of varying radii; shop fabrication. For appearance and where backfill compaction is only moderate.
Pipe-arch	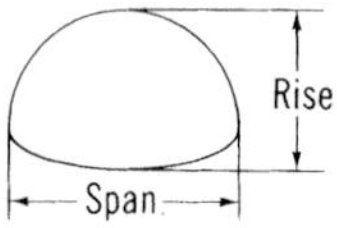 Rise / Span	Span x Rise 17 in. x 13 in. to 20 ft 7 in. x 13 ft 2 in.	Where headroom is limited. Has hydraulic advantages at low flows. Corner plate radius, 18 inches or 31 inches for structural plate.
Underpass*	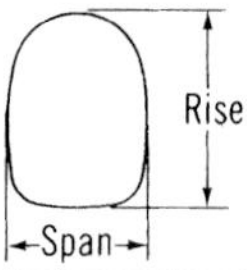 Rise / Span	Span x Rise 5 ft 8 in. x 5 ft 9 in. to 20 ft 4 in. x 17 ft 9 in.	For pedestrians, livestock or vehicles (structural plate).
Arch	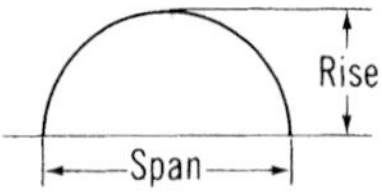 Rise / Span	Span x Rise 6 ft x 1 ft 9½ in. to 25 ft x 12 ft 6 in.	For low clearance large waterway opening, and aesthetics (structural plate).
Horizontal Ellipse	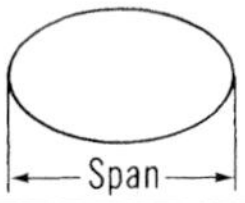 Span	Span 7-40 ft	Culverts, grade separations, storm sewers, tunnels.
Pear	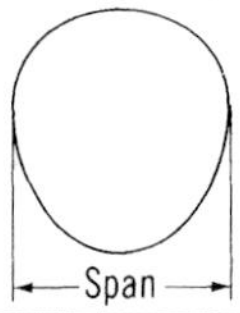 Span	Span 25–30 ft	Grade separations, culverts, storm sewers, tunnels.
High Profile Arch	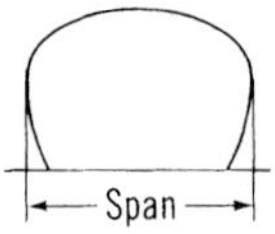 Span	Span 20-45 ft	Culverts, grade separations storm sewers, tunnels. Ammunition magazines, earth covered storage.
Low Profile Arch	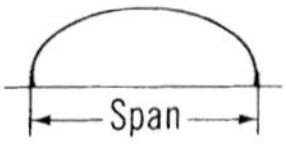 Span	Span 20-50 ft	Low-Wide waterway enclosures, culverts, storm sewers.
Box Culverts	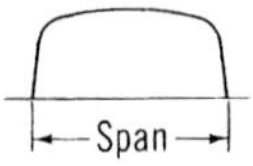 Span	Span 10-26 ft	Low-wide waterway enclosures, culverts, storm sewers.
Specials		Various	For lining old structures or other special purposes. Special fabrication.

*For equal area or clearance, the round shape is generally more economical and simpler to assemble.

FIGURE 5.24 Shapes, range of sizes, and common uses of corrugated steel drainage products. (*From* Handbook of Steel Drainage and Highway Construction Products, *American Iron and Steel Institute, 1994, with permission*)

Long-Span Structures. Long-span structural-plate structures are defined as having either special shapes that involve a relatively large radius in the crown or side plates, or a span that exceeds certain structural design criteria as specified in AASHTO *Standard Specifications for Highway Bridges.* These structures generally have spans in the range of 20 to 50 ft (6 to 15 m). They are advantageous where headroom is restricted and can often provide the required waterway area at a lower cost than building a short-span bridge. Long-span structures are made up of a structural-plate barrel of coated steel or aluminum and integral special features that enable the structure to reach long spans. Special features include either (1) continuous longitudinal stiffeners of metal and/or reinforced concrete attached to the plates at the sides of the top arc, or (2) circumferential reinforcing ribs curved from structural shapes and attached to the plates to provide additional stiffness. Typical sections of each are illustrated in Fig. 5.25. They may be constructed to most of the shapes shown in Fig. 5.24 except box culverts.

Box Culverts. This product type is available in either zinc-coated steel or aluminum.

  Steel. Box culverts are available in three types, including (1) 6- × 2-in corrugated plate shell with 6- × 3-in corrugated rib stiffeners (inside, outside, or both), (2) 6- × 2-in corrugated plate shell with 3- × 5-in hot rolled angle rib stiffeners, and (3) 15- × 5.5-in corrugated plate shell without stiffeners. Sizes range from 2 ft 6 in × 9 ft 2 in (rise × span) to 10 ft 6 in × 24 ft 9 in. The structures usually have an open bottom and are supported on a base channel or corrugated footing pads, on either a concrete footing or compacted soil, depending on size and other factors. They are also available with full invert plates.

  Aluminum. Box culverts have a 9- × 2½-in corrugated shell plate with extruded bulb angle rib stiffeners, 3 × 2½ in or 3⁵⁄₁₆ × 2½ in. Sizes range from 2 ft 6 in × 8 ft 9 in (rise × span) to 10 ft 2 in × 25 ft 5 in. Figure 5.26 shows a typical section and rib cross sections.

Tunnel Liners. Tunnel liners are press-formed from steel in an arc segment 16 or 18 in long. A corrugated profile is pressed in to make the wall cross section, and flanges are formed on the sides. Two styles are available: (1) two-flange plates that are bolted through the flanges on the two longitudinal sides and lap-bolted on the other two sides, and (2) four-flange plates that are bolted together through flanges on all four sides. Installation and assembly can be done entirely from the inside as the tunnel is constructed. The assembled liner plates may then act as a temporary structure that is lined by concrete, or may act alone as a permanent conduit. In addition to tunneling, the liner plates can be used in rehabilitation work, such as for lining a deteriorated concrete box culvert or a concrete or masonry arch bridge.

5.6.3 Plastic Pipe

Both high-density polyethylene (PE) and polyvinyl chloride (PVC) are used for drainage pipe. PE pipe may be single-wall corrugated, smooth-wall (double-wall), or ribbed. Common diameters are 4 to 24 in for single wall, 4 to 48 in for double wall, and 18 to 96 in for ribbed pipe. Single-wall pipe has a deep corrugation, whereas a smooth internal liner is added for double-wall pipe. Wall profile details vary with the manufacturer. PVC pipe may be either smooth-wall or ribbed, with diameters ranging up to 48 in. Plastic pipe is furnished in lengths (typically about 20 ft) and joined in the field by coupling bands. It is available only as round pipe.

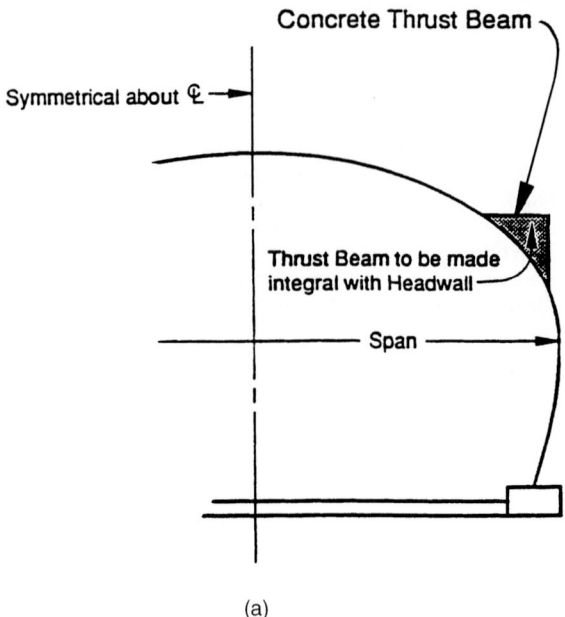

(a)

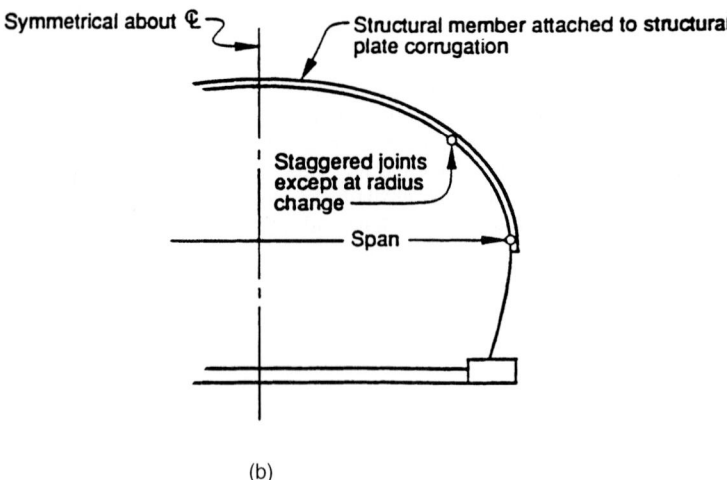

(b)

FIGURE 5.25 Typical sections of long-span structural-plate structures. (*a*) Longitudinally stiffened with concrete thrust beam. (*b*) Transversely stiffened with structural members. (*From* Highway Design Manual, *California Department of Transportation, with permission*)

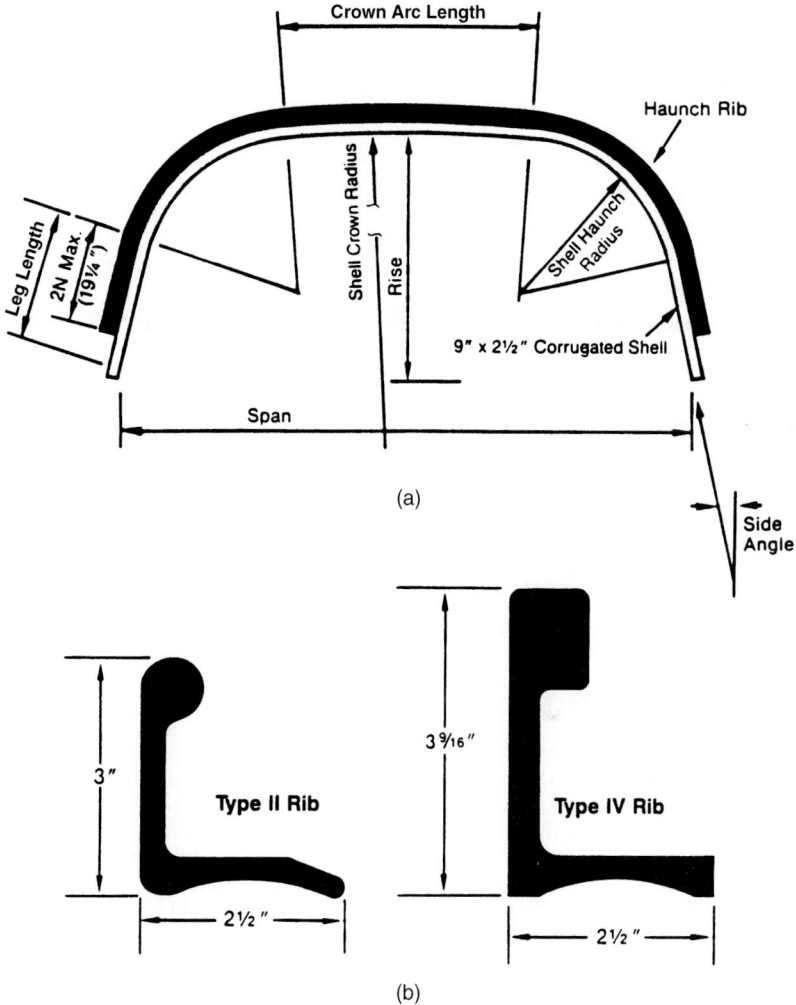

FIGURE 5.26 Corrugated aluminum box culvert. (*a*) Typical cross section. (*b*) Details of stiffening ribs. (*From* Aluminum Box Culverts, *Contech Construction Products, with permission*)

5.7 CULVERT SERVICE LIFE

The prediction of service life of drainage facilities is difficult because of the wide range of environments encountered and the various protective measures available. Service life and durability are directly related to resistance to corrosion, abrasion, and other modes of deterioration.

5.7.1 Design Service Life

Drainage facilities are usually designed for a specific service life. The design service life is sometimes defined as the expected period for which they are relatively free from maintenance. However, it can be defined to include a planned rehabilitation after a given number of years to reach the required service life as part of a value analysis approach. (See Art. 10.19.1.)

For a metal culvert, the design service life can be based on the number of years between the time it is installed and the time a perforation from either corrosion or abrasion occurs at any location in the culvert. However, this is a rather conservative approach because the consequences of small perforations are usually minimal and a single perforation can occur long before there is a general thinning of the metal. Thus, service life charts are often based on an average service life that extends life past first perforation by 25 percent or more. For a concrete culvert, the design service life is usually defined as the time between installation and when deterioration reaches the point of exposed reinforcement anywhere in the culvert.

The selection of design service life—whether 10, 25, 50, or more years—is dependent upon the use, importance, and ease of replacement of the culvert. A culvert located under a high fill or a roadway with high traffic volumes will be expensive to replace, and the replacement will disrupt traffic. Thus, such culverts are often assigned a design service life of 50 years or more. In contrast, a culvert parallel to the main road—for example, a pipe underneath an access road—will be relatively easy to replace and can be replaced with little disruption. Thus, such culverts, including those under low fill or on a minor roadway, are often assigned a shorter service life.

5.7.2 Environmental Factors

Important environmental factors that affect culvert durability include the acidity (pH) of the effluent and the soil, the electrical resistivity of the effluent and the soil, and the concentration of sulfates and chlorides. Data on these factors should be obtained at each pipe location, unless a random sampling plan is justified by establishing that the samples are uniform throughout a given length of the project. Water samples should be taken only during times of typical flows. If corrosive conditions are found to be present in the soil but not in the water samples, consideration should be given to using a better backfill material.

Concrete Pipe. Environmental factors that can affect the deterioration of concrete culverts include freeze-thaw, acids, sulfates, and chlorides. Freeze-thaw damage can occur if water penetrates the concrete interstices and then freezes and expands, causing cracking. Such damage would occur only at exposed ends of culverts, and low water-cement ratios can increase resistance. Continuous exposure to severe acidity is detrimental to concrete pipe; a pH below 5.0 is considered aggressive and below 4.0 highly aggressive. Improved resistance to acid attack can be attained by selecting aggregate that increases the total alkalinity of the concrete, increasing concrete cover over reinforcement, or adding barrier linings. Sulfates in the soil, groundwater, or effluent can be aggressive to concrete. Such problems, which are generally limited to arid regions with alkali soils, may be addressed with special cements and mix design. Chloride attack can potentially result from use of deicing salts and subsequent runoff, but problems of this type have not been generally reported.

Metal Pipe. Environmental factors that affect the corrosion of metal culverts include the acidity (pH) and the resistivity of the soil and water, and the moisture content, soluble salt content, oxygen content, and bacterial activity of the soil. These corrosion processes all involve the flow of current from one location to another. The current flows from an anodic area to a cathodic area through moist soil acting as an electrolyte, and this system is known as a corrosion cell. Thus, durability increases with increasing resistivity. Acid soils, those with low pH, tend to be more corrosive. Also, soils with high moisture content, such as loams and clays, tend to be more corrosive. High levels of chlorides and sulfates increase corrosion, as do increasing levels of dissolved oxygen and carbon dioxide. Numerous field studies have shown that the culvert invert is the portion most susceptible to corrosion, because it is generally exposed to water for a greater length of time. Thus, design charts are usually based on service life observed in the invert.

Plastic Pipe.   PE and PVC pipe are not affected by acid conditions, or by sulfates or other alkalis. These materials can become embrittled from ultraviolet radiation as a result of prolonged exposure to direct sunlight, such as at culvert ends, but inhibitors are added to the composition of the material to reduce this effect. If problems are encountered, ends can be shaded, covered with a coupling, or painted.

5.7.3   Abrasion

Abrasion causes a loss of section thickness due to impacts by the aggregate carried by stream flow. Protection from abrasion generally takes the form of providing a sacrificial thickness of the structural material, whether it be a thicker sheet of steel for metal pipe or more concrete cover over the reinforcement for reinforced concrete pipe. Alternatives to providing for a thicker section include using debris control structures to prevent the abrasive material from reaching the culvert, and providing metal planking longitudinally along the invert as a separation between the bedload and the bottom of the culvert.

Abrasion can be considered in four levels of severity as categorized by streambed velocity and general aggregate size. Protective measures, particularly in the invert, should increase with increasing levels of abrasion as discussed subsequently. (See *Project Development and Design Manual,* Federal Lands Highway, FHWA.)

Level 1, termed *nonabrasive,* has very low flow velocities and no bed load.

Level 2, *low abrasive,* has flow velocities of 5 ft/s (1.5 m/s) or less and light bed load consisting of sand.

Level 3, *moderately abrasive,* has flow velocities of between 5 and 15 ft/s (1.5 and 4.5 m/s) and moderate bed loads consisting of sand and gravel.

Level 4, *severely abrasive,* has flow velocities exceeding 15 ft/s (4.5 m/s) and heavy bed loads consisting of sand, gravel, and rock.

The projected velocities should be based upon a typical flow and not upon the design flood for which the culvert has been designed. The bed load size may be determined by visual inspection of the surrounding environment and the upstream channel. Sampling of the aggregate for a gradation analysis is not necessary.

5.7.4   Guidelines for Culvert Selection

The following general guidelines from the FLH manual should assist in determining appropriate culvert material types and necessary coatings. Other methods are available, and a materials engineer should be consulted for important applications. Of course, the final selection must provide for structural requirements as discussed in Art. 5.8.

Concrete Pipe.   Where the pH is less than 3.0 and the resistivity is less than 300 $\Omega \cdot$ cm, reinforced concrete pipe should not be specified. If the sulfate concentration exceeds 0.2 percent in the soil or water, type V cement should be specified. If the sulfate concentration exceeds 1.5 percent in the soil or water, an increased cement ratio using type V cement should be specified. The concrete cover over the reinforcement or the cement factor should be increased where there is severe abrasion.

Steel Pipe.   Figure 5.27 shows a chart for determining the service life of a galvanized steel culvert under nonabrasive and low abrasive conditions. The average service life of culvert with a wall thickness of 0.064 in (1.62 mm) is displayed in terms of pH and resistivity in Fig. 5.27*a*. For culverts with other wall thicknesses, obtain the service life from the chart and multiply by the factors in Fig. 5.27*b*. Use the chart for both the outside conditions and the inside (water side) conditions and base the design on the worst case. Generally, the inside condition controls.

For steel with a type 2 aluminum coating, the FLH manual assigns a greater service life under certain conditions. For nonabrasive and low abrasive flow, where the resistivity is equal to or greater than 1500 $\Omega \cdot$ cm and the pH is between 5 and 9, aluminized steel is considered to provide a service life twice that of galvanized steel as determined from Fig. 5.27.

Protective Coatings on Steel Pipe.   Under nonabrasive and low abrasive conditions, the service life of galvanized steel culvert can be extended by application of protective coatings. For example, when the water side environment controls the pipe thickness, application of a bituminous coating (a postfabrication coating by the pipe manufacturer) can add 10 years of service life to the culvert, and an application of a bituminous paved invert in addition to the coating will add a total of 25 years. If the soil side controls, application of the bituminous coating will add 25 years of life. Concrete lining will add 25 years of service life. Ethylene acrylic acid film coatings (a polymer precoat on the galvanized coil) with a 10-mil (0.25-mm) thickness can be expected to provide an additional 30 years of service life. Protective coatings of this type are not suitable for water side corrosion protection where the abrasion is moderate or severe, but special concrete pavings can be designed to add service life.

Aluminum Pipe.   Under nonabrasive and low abrasive conditions, where the resistivity is equal to or greater than 500 $\Omega \cdot$ cm and the pH is between 4 and 9, aluminum culverts can be assumed to have a service life of 50 years when the metal thickness is appropriately sized for structural adequacy.

Design for Abrasion.   In moderate abrasive environments, the sheet thickness for both steel and aluminum pipes should be increased by one nominal thickness, or the invert should be protected. In severe abrasive conditions, the sheet thickness should be increased by one nominal thickness and the invert should be protected. Invert protection under severe abrasive conditions may consist of metal rails or energy-dissipating devices at the inlet. Under moderate abrasive conditions, invert protection may consist of (1) paving with portland cement concrete or (2) bituminous coating and invert paving with bituminous concrete.

Plastic Pipe.   Under nonabrasive or low abrasive conditions, polyethylene and polyvinyl chloride plastic pipes may be specified without regard to the pH and resis-

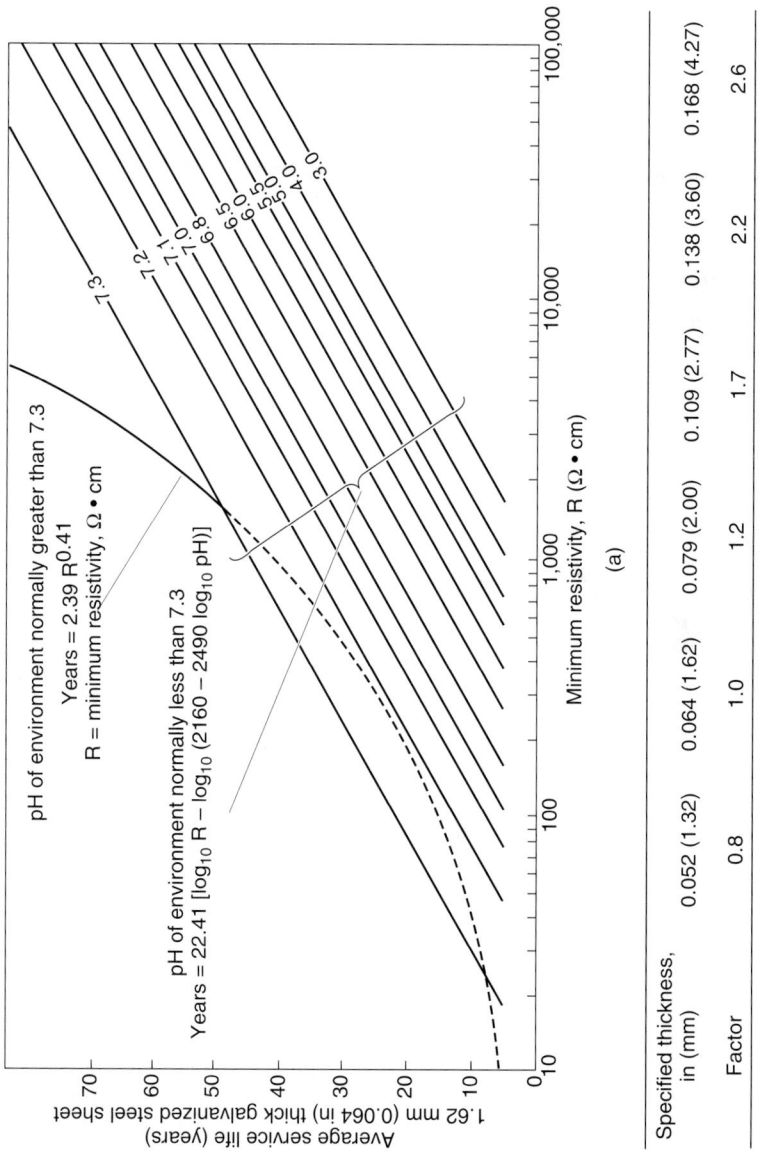

FIGURE 5.27 Method for estimating service life of plain galvanized steel culverts. (*a*) Service life chart for 0.064-in thickness based on invert performance. (*b*) Conversion factors for other thicknesses. (*From Project Development and Design Manual, FHWA, with permission*)

The following appears within image (a):

pH of environment normally greater than 7.3

$$\text{Years} = 2.39\, R^{0.41}$$

R = minimum resistivity, $\Omega \cdot$ cm

pH of environment normally less than 7.3

$$\text{Years} = 22.41\, [\log_{10} R - \log_{10} (2160 - 2490 \log_{10} pH)]$$

Vertical axis: Average service life (years) — 1.62 mm (0.064 in) thick galvanized steel sheet

Horizontal axis: Minimum resistivity, R ($\Omega \cdot$ cm)

(a)

The table (b):

Specified thickness, in (mm)	0.052 (1.32)	0.064 (1.62)	0.079 (2.00)	0.109 (2.77)	0.138 (3.60)	0.168 (4.27)
Factor	0.8	1.0	1.2	1.7	2.2	2.6

(b)

5.55

tivity of the site. Under moderate or severe abrasive conditions, they should not be used without invert protection.

Example: Minimum Thickness of Galvanized Steel Culvert. The design service life for the culvert has been set at 50 years. A site investigation of a potential location shows that the soil has a pH of 7.2 and a resistivity of 5000 $\Omega \cdot$ cm. The water flow shows a pH of 6.8 and a resistivity of 4000 $\Omega \cdot$ cm. Determine the minimum sheet thickness for durability.

> *Outside condition.* In Fig. 5.27a, find the intersection of the vertical line for 5000 $\Omega \cdot$ cm with the inclined line for 7.2 pH, and read the average service life of 52 years from the vertical scale at the left.

> *Inside condition.* In like manner, for a resistivity of 4000 $\Omega \cdot$ cm and a pH of 6.8, find the average service life of 42 years.

In this example, the inside conditions control the design, and the thickness must be increased. For the 0.064-in sheet thickness, the ratio of the design service life to the anticipated service life is 50/42 = 1.2. From Fig. 5.27b, the multiplying factor is 1.2 for a thickness of 0.079 in. Therefore, a thickness of 0.079 in should provide the desired service life of 50 years.

An alternative is the application of a bituminous coating, which can add 10 years of service life when the inside condition controls. For the 0.064-in sheet thickness, 42 + 10 = 52 years. Therefore, consider an 0.064-in sheet thickness with a bituminous coating.

5.8 STRUCTURAL DESIGN OF CULVERTS

5.8.1 General Considerations

Loads acting on buried structures include the dead load of the structure itself, the dead load of the earth over the structure, the weight of the fluid within the structure, and live loads from vehicles. The structural capacity of an underground structure and the method of determining that capacity are dependent upon the material properties of the structure, its physical configuration, and the characteristics of the surrounding soil. Buried structures, particularly flexible pipes, rely heavily upon the surrounding soil for their ability to withstand loads. For this reason, essential elements related to the installation of culverts are reviewed in some of the paragraphs that follow.

Culverts are usually classified as rigid or flexible, depending on their bending stiffness. For a round pipe under load, deflection due to bending is proportional to D^3/EI where D is the diameter, E is the modulus of elasticity, and I is the moment of inertia of the wall cross section. EI is the wall bending stiffness. Concrete pipe usually has a relatively thick wall and a high bending stiffness, and is referred to as *rigid* pipe. Corrugated metal pipe and plastic pipe tend to have a relatively thin wall and a low bending stiffness, and are referred to as *flexible* pipe. Rigid pipe, unless designed by an empirical method, is generally designed for moment, thrust, and shear. Flexible pipe, however, deflects laterally under vertical loads, mobilizes the passive earth pressure of the compacted sidefill, and can generally be designed for thrust alone.

Pressure Distributions. The pressure distribution usually assumed for the design of a concrete box culvert is shown in Fig. 5.28. Possible pressure distributions for round

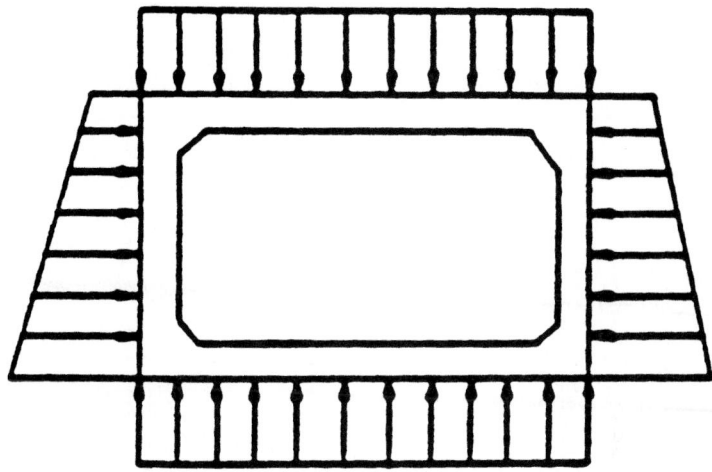

FIGURE 5.28   Pressure distribution used for design of concrete box culverts. (*From "Hydraulic Design of Highway Culverts," Hydraulic Design Series No. 5, FHWA, with permission*)

concrete pipe as given in the AASHTO *Standard Specifications for Highway Bridges* are shown in Fig. 5.29, based on work by Olander (Fig. 5.29*a*) and Paris (Fig. 5.29*b*). Pressure distributions under consideration for inclusion in the AASHTO specifications as part of a revised design approach based on studies by Heger are shown in Fig. 5.30. In contrast to these patterns, pressures on flexible pipe are assumed to act radially because of the ability of the structure to deflect under load. For round flexible pipe, the pressures are approximately uniform radial pressures. For other curved shapes, such as pipe arches, the pressures are inversely proportional to the radius. Thus, the pressure at the corners of a pipe arch (P_c) is approximately equal to the pressure at the top (P_t) multiplied by the ratio of the top radius (R_t) to the corner radius (R_c), which can be expressed as $P_c = P_t R_t / R_c$.

Rigid Box Culvert Design.   Moments and shears for rigid box culverts may be determined by standard methods of structural analysis. In contrast to rigid pipe, thrust is generally ignored for rigid boxes because bending dominates. For reasons of economy, if the span is 8 ft (2.4 m) or less, the box is generally designed as having pinned corners. For greater spans, it is designed as a rigid frame.

Rigid Pipe Design.   Rigid pipe may be designed by either the empirical "*D*-load" method, termed *indirect design,* or by an analytical method of *direct design.* If the pipe is designed by the *D*-load method, moments, thrusts, and shears need not be determined. All that is necessary for the analysis is the total load on the pipe, the bedding factor as determined from the proposed bedding condition, and the proposed inside and outside diameters of the pipe. If the pipe is designed by an analytical method, the moments, thrusts, and shears must first be determined. Those values may either be determined from figures and charts developed by H. C. Olander or J. M. Paris, or they may be found by more exact finite element methods using available computer programs. Developments in this area have been made by F. J. Heger. After the moments, thrusts, and shears are determined, the required pipe wall thickness, concrete strength, and area of reinforcing steel may be determined. (See Art. 5.8.6.)

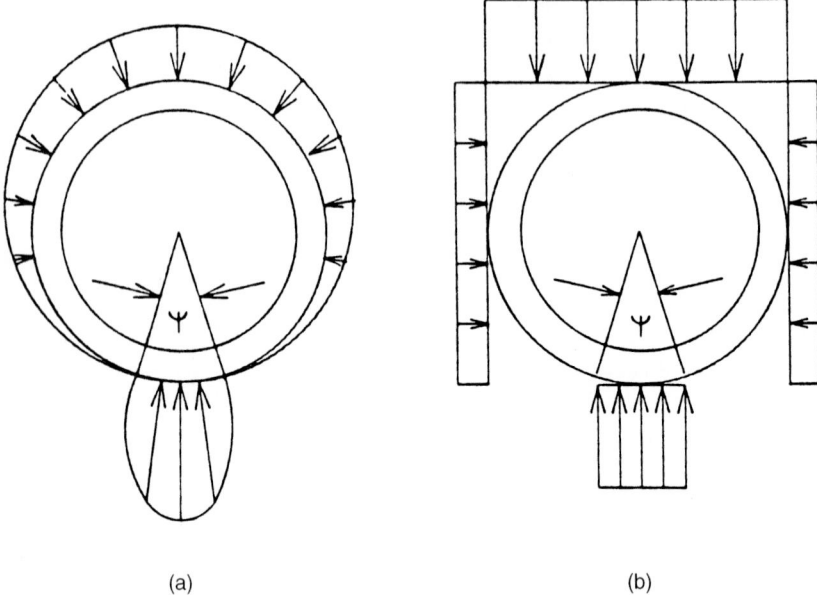

(a)   (b)

FIGURE 5.29   Pressure distribution for round concrete pipe for direct design. (*a*) Radial pressures. (*b*) Equivalent horizontal and vertical pressures. (*From* Standard Specifications for Highway Bridges, *American Association of State Highway and Transportation Officials, Washington, D.C., 1992, with permission*)

Flexible Pipe Design.   Flexible structures are generally designed for ring thrust by a semiempirical method that includes checks for wall area, buckling, and seam strength. A check is also made to ensure the structure has sufficient rigidity for installation and handling. Flexible pipes may also be designed using finite element computer programs that model both the structure and the soil. The design procedures for metal box culverts and long-span structures differ somewhat from those for other flexible structures.

Bedding and Backfill.   Figure 5.31 shows nomenclature generally used for culvert installations. The supporting soil beneath the culvert is the foundation, and the bedding is that portion of the foundation in contact with the bottom of the culvert. The springline of the culvert is located at the point where the span is a maximum. For a circular or elliptical pipe, this occurs at mid-height. The haunch is the zone between the springline and the invert. The soil placed and compacted around the culvert is known as the backfill, or sometimes as the sidefill.

Proper foundation, bedding, and placement and compaction of backfill are critical to the performance of underground structures. They are also essential factors for achieving an accurate structural analysis of the system. Bedding and backfill of underground structures should be done in accordance with standards established by local and state transportation agencies. These standards vary from one region to another, but the important aspects of typical practices are reviewed below. (Also see Art. 5.9.)

Installation of Flexible Pipe.   Flexible pipe may be installed in an embankment or in a trench. The foundation must provide relatively uniform resistance to loads. If rock is encountered, it should be excavated and replaced with soil. If soft material is encoun-

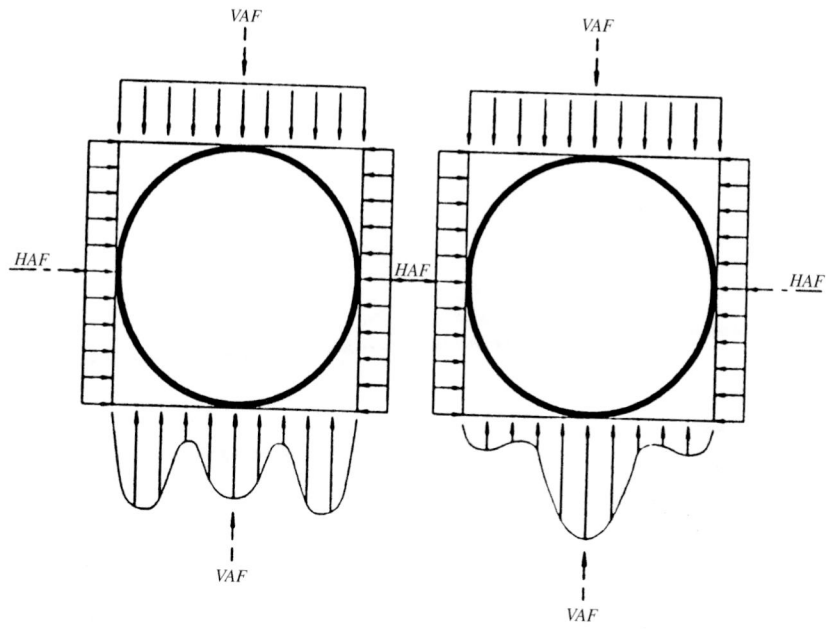

(a) Lower Haunch Highly Compacted Soil   (b) Lower Haunch Loose Soil

FIGURE 5.30   Pressure distribution for round concrete pipe for direct design based on finite element model SPIDA (soil-pipe interaction design and analysis). (*a*) Distribution with highly compacted soil under haunch. (*b*) Distribution with loose soil under haunch. (*From* Concrete Pipe Technology Handbook, *American Concrete Pipe Association, 1994, with permission*)

tered, it should be removed for a width of three pipe spans and replaced with suitable material. Care must be taken to ensure that the foundation under the pipe is not stiffer than the adjacent zones, because this will attract additional loads. Pipe-arch structures require excellent soil support at the corners, because pressures are higher there. Most pipe can be placed directly on the fine-graded foundation without shaping the bedding, provided that care is taken to compact the soil under the haunches. Exceptions are pipe arches, which should be placed on a bedding shaped to the contour of the invert or to a slight V shape. Also, large structures (those with a span over about 15 ft or 45 m) should have a shaped bedding. Some agencies require a shaped bedding for all pipe, over a width equal to half the pipe diameter, because of the difficulty in properly compacting the backfill in the area of the haunch. The backfill should be placed in 6- to 8-in (15- to 20-mm) compacted layers around the structure. Each layer must be compacted to a minimum of 90 percent of standard density (AASHTO T99 method) before the next is added, and the backfill must be kept in balance on each side of the pipe. A granular material free of organic content and with little or no plasticity makes good backfill. The assumption of a uniform radial pressure distribution for flexible structures is applicable only if the soil surrounding the structure is capable of supporting the lateral loads. The compacted backfill must be of sufficient width to dissipate the horizontal pressures from the structure.

Installation of Rigid Pipe.   Standard installations for circular concrete pipe in an embankment or in a trench are shown in Figs. 5.32 and 5.33. Similar figures for ellip-

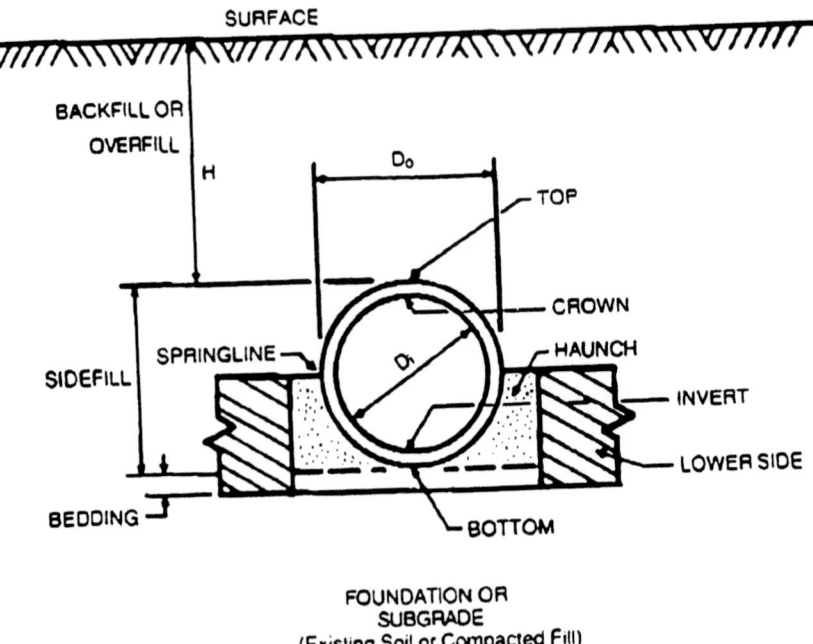

FIGURE 5.31   Pipe installation terminology. (*From Design Data 40, American Concrete Pipe Association, 1992, with permission*)

tical and arch pipes may be found in the *Concrete Pipe Handbook,* American Concrete Pipe Association (ACPA). The specified backfill installation corresponds to a particular bedding factor, B_f noted in the figures. For a given fill height, the bedding factor directly affects the required strength of the concrete pipe when designed by the D-load method. Higher bedding factors are required for the lower pipe strengths. Additionally, each of these installations is associated with an assumed bedding angle that may be used to determine the moments, thrusts, and shears acting on the pipe using the methods of Olander and Paris. The backfill installations with higher bedding factors have higher corresponding bedding angles. These higher bedding angles lead to more uniform distribution of loads and accompanying lower stress values in the pipe. Aside from the bedding conditions, most of the preceding discussion on installation of flexible pipe constitutes good practice for rigid pipe as well.

5.8.2   Dead Loads on Underground Structures

As previously mentioned, loads acting on buried structures include the dead load of the structure itself, the dead load of the earth over the structure, the weight of the fluid within the structure, and live loads from vehicles. The structure dead load is significant only for rigid structures. Flexible structures are manufactured from plastic—either polyethylene or polyvinyl chloride—or corrugated metal, either aluminum or steel. In each case, the weight of the material is insignificant when compared with the

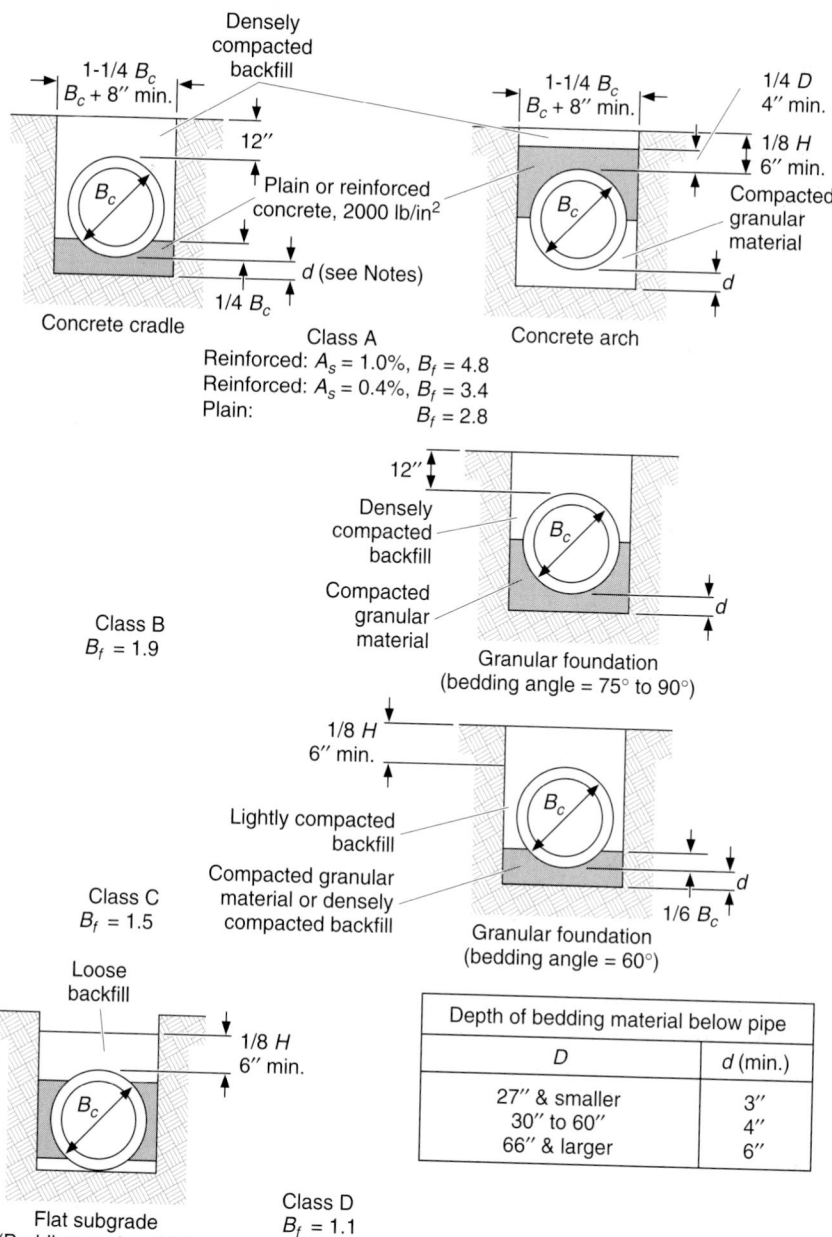

FIGURE 5.32 Trench installations of round concrete pipe showing bedding factors B_f and bedding angles for class A, B, C, and D installations. Symbols: B_c = outside diameter, H = backfill cover above top of pipe, D = inside diameter, d = depth of bedding material below pipe, A_s = area of transverse steel in the cradle or arch expressed as a percentage of area of concrete at invert or crown. Notes: (1) For class A beddings, use as depth of concrete below pipe unless otherwise indicated by soil or design conditions. (2) For class B and C beddings, subgrades should be excavated or overexcavated if necessary, so a uniform foundation free of protruding rocks may be provided. (3) Special care may be necessary with class A or other unyielding foundations to cushion pipe from shock when blasting can be anticipated in the area. (*From* Concrete Pipe Technology Handbook, *American Concrete Pipe Association, 1994, with permission*)

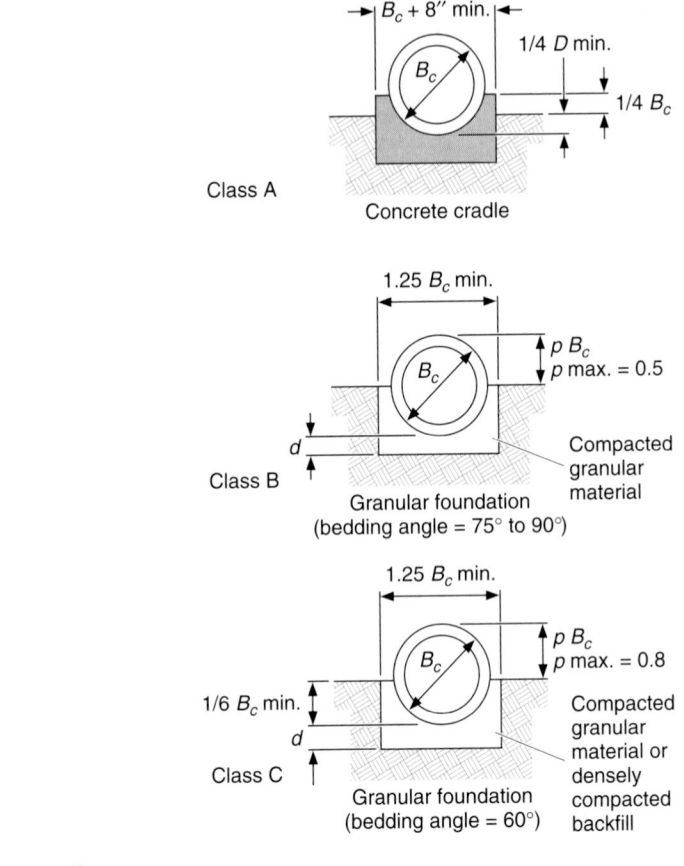

FIGURE 5.33 Embankment installations of round concrete pipe showing bedding angles for class A, B, C, and D installations. Symbols: B_c = outside diameter, D = inside diameter, d = depth of bedding material below pipe. Notes: (1) For class B and C beddings, subgrades should be excavated or overexcavated if necessary, so a uniform foundation free of protruding rocks may be provided. (2) Special care may be necessary with class A or other unyielding foundations to cushion pipe from shock when blasting can be anticipated in the area. (*From* Concrete Pipe Technology Handbook, *American Concrete Pipe Association, 1994, with permission*)

total load on the structure. For rigid structures, however, because the material is generally concrete and because the culvert wall thickness is considerable, the weight of the material should be included in the determination of the total load applied on the structure. The application of the earth load on the culvert varies with the type of structure and the installation conditions.

Earth Load on Rigid Structures.   For rigid structures, the earth load applied to the culvert varies with the type of installation. In the early 1900s, Anson Marston of Iowa State University developed a theory, still in wide use today, by which loads on underground structures may be determined for almost every installation condition. Basically, what Marston discovered was that the load on a pipe was equal to the weight of the soil prism located directly above the pipe, adjusted for the difference between the settlement of that prism area and the soil adjacent to it. Figure 5.34 illustrates the unadjusted prism load. The Marston adjustment is necessary because as one area settles more relative to the other, it will impart shearing friction forces on that area and cause a subsequent transfer of load. Depending upon the conditions, the earth load may be more or less than the unadjusted prism load.

Although there are numerous different installation conditions, the loading cases have been simplified in the AASHTO *Standard Specifications for Highway Bridges* so that design loadings adequate for most purposes may be easily determined. AASHTO allows for the load to be determined for trench and embankment conditions from the following equations:

$$W_E = F_e w B_c H \qquad \text{(embankment installations)} \qquad (5.20)$$

$$W_E = F_t w B_c H \qquad \text{(trench installations)} \qquad (5.21)$$

where W_E = total unfactored earth load, kips/ft
  B_c = outside diameter of pipe or out-to-out dimension of box culvert, ft
  H = depth of backfill, ft
  w = unit weight of the backfill, kips/ft^3
  F_e = soil-structure interaction factor for embankment installations
  F_t = soil-structure interaction factor for trench installations

The soil-structure interaction factors may be found from the following equations:

$$F_e = 1 + 0.20 \left(\frac{H}{B_c} \right) \qquad (5.22)$$

≤ 1.2 (pipe installations with compacted backfill)

≤ 1.5 (pipe installations with uncompacted backfill)

≤ 1.15 (box installations with compacted backfill)

≤ 1.4 (box installations with uncompacted backfill)

$$F_t = \frac{(C_d B_d{}^2)}{(HB_c)} \leq F_e \qquad (5.23)$$

where C_d = trench load coefficient from Fig. 5.35
  B_d = horizontal width of trench, ft

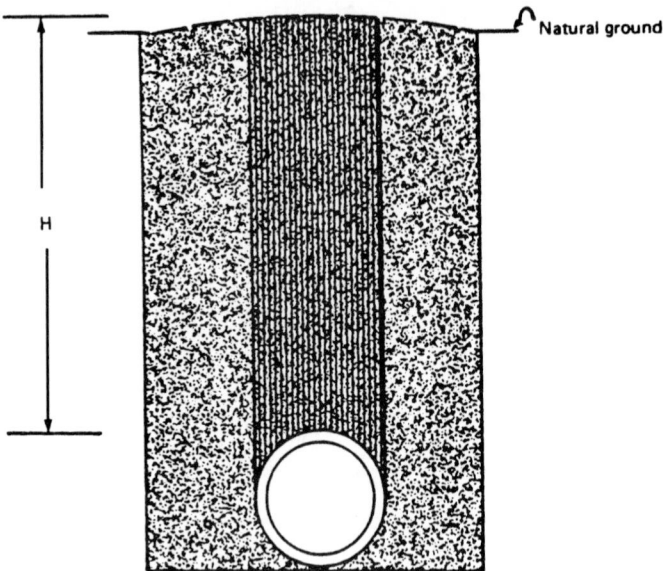

FIGURE 5.34   Illustration of prism load acting on pipe. (*From A. P. Moser,* Buried Pipe Design, *McGraw-Hill, 1990, with permission*)

The essential features of several different installation conditions for trench, embankment, and jacked or tunneled construction are illustrated in Fig. 5.36. Culverts in embankments may be installed as positive- or negative-projecting. A positive-projecting culvert is one installed in a shallow bedding with its crown projecting above the natural ground, and then covered with an embankment. A negative-projecting culvert is one installed in a relatively narrow trench with its crown below natural ground, and then covered with an embankment. An induced trench condition is achieved when a culvert is installed in a positive-projecting condition and then compressible material is placed over the crown to relieve a portion of the load. The earth load for a negative projecting embankment, an induced trench condition, or a jacked or tunneled installation is generally less than for a positive-projecting condition. However, increased bending moments may be developed when using an induced trench. The AASHTO load calculation method for embankments cited above assumes the positive-projecting condition. (For additional information and more refined analysis methods, see A. P. Moser, *Buried Pipe Design,* McGraw-Hill, 1990; *Design and Construction of Sanitary and Storm Sewers,* ASCE, 1969; and *Concrete Pipe Technology Handbook,* American Concrete Pipe Association.)

Earth Load on Flexible Structures.   For flexible structures, whether in an embankment or a trench, it is usually assumed that the complete prism of earth over the pipe acts on the structure. This is a conservative analysis, since flexible structures tend to deflect under the earth load and induce arching action in the overlying soil. Thus,

$$W_E = wB_cH \tag{5.24}$$

where the terms are as defined above.

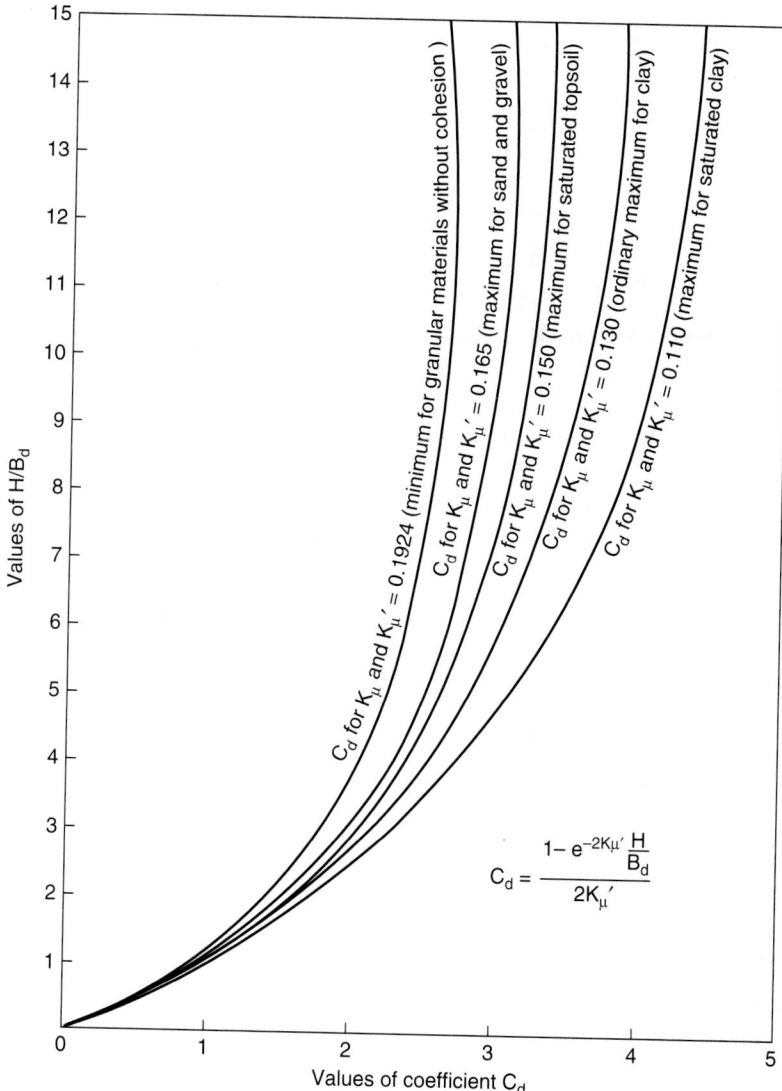

FIGURE 5.35   Coefficient C_d for load calculations for concrete pipe in trench installations. (*From* Standard Specifications for Highway Bridges, *American Association of State Highway and Transportation Officials, Washington, D.C., 1992, with permission*)

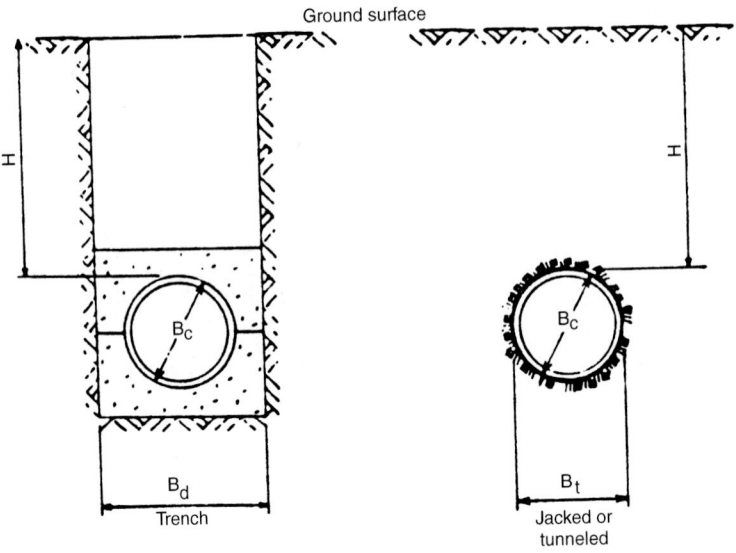

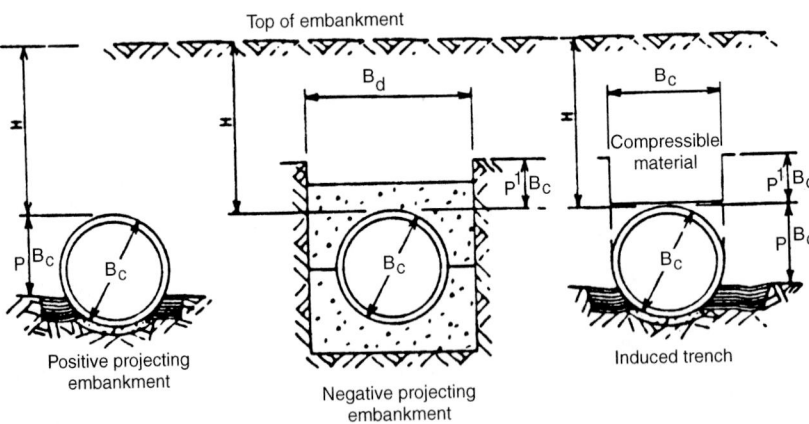

FIGURE 5.36 Essential features of types of installation of round concrete pipe. The multiplier P or P^1 is the projection ratio defined by AASHTO. (*From* Standard Specifications for Highway Bridges, *American Association of State Highway and Transportation Officials, Washington, D.C., 1992, with permission*)

Steel tunnel liner plate and steel ribs and lagging are flexible structures placed by a tunneling operation. Like other flexible structures, they are designed to deflect vertically under load so that the lateral side pressure will be established and uniform radial pressure will develop around the perimeter of the structure. Because these structures are used in tunneling operations, however, under most circumstances it is not necessary to design for the complete prism load. The AASHTO *Standard Specifications for Highway Bridges* state that the earth pressure on a tunnel liner may be determined from the following equation:

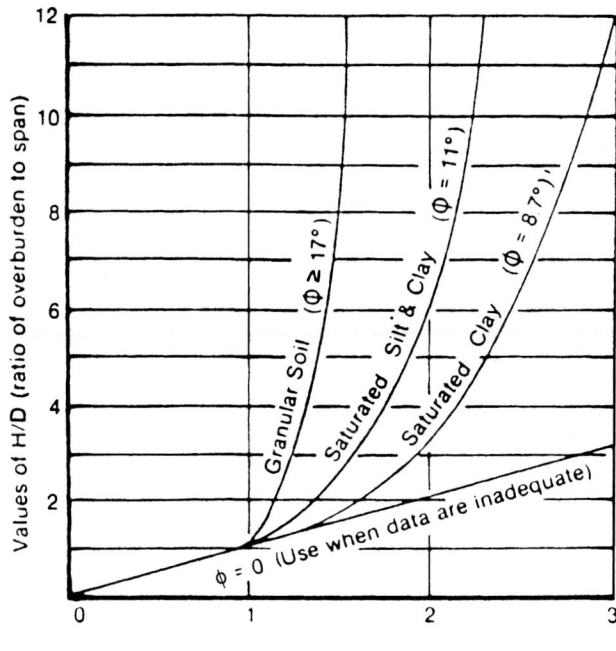

FIGURE 5.37   Diagram for coefficient C_d for load calculations for tunnels. ϕ is soil friction angle. (*From* Standard Specifications for Highway Bridges, *American Association of State Highway and Transportation Officials, Washington, D.C., 1992, with permission*)

$$W_E = C_{dt} v_s S \qquad (5.25)$$

where W_E = earth pressure at the crown, kips/ft^2
       C_{dt} = load coefficient for tunneling, from Fig. 5.37
       v_s = unit weight of the soil, kips/ft^3
       S = tunnel diameter or span, ft

Example: Earth Load on Concrete Pipe.   A 60-in-diameter concrete pipe with 6-in walls is to be installed in an 8-ft-wide trench under 12 ft of cover. The fill material is granular material without cohesion, and will be compacted to a density of 120 lb/ft^3. Determine the design earth load acting on the pipe.

Calculate the ratio H/B_d = 12/8 = 1.5. From Fig. 5.35, find trench load coefficient C_d = 1.3. Next, from Eq. (5.23), calculate the soil-structure interaction factor for trench installations as

$$F_t = \frac{(C_d B_d^{\,2})}{HB_c} \le F_e$$

$$= \frac{(1.3)(8)^2}{12 \times 6} \le 1.2$$

$$= 1.16 \leq 1.2$$

$$= 1.16$$

The design load is then calculated from Eq. (5.21) as

$$W_E = F_t w B_c H = 1.16 \times 0.120 \times 6 \times 12 = 10.0 \text{ kips/ft, or } 10,000 \text{ lb/ft}$$

Example: Earth Load on Flexible Pipe. A 60-in-diameter flexible pipe is to be installed in a trench under 12 ft of cover. The fill material will be compacted to a density of 120 lb/ft^3. Determine the design earth load acting on the pipe.

Because the wall thickness is small, the nominal diameter is usually used for the load calculation for a flexible pipe. Therefore, from Eq. (5.24),

$$W_E = w B_c H = 0.120 \times 5 \times 12 = 7.22 \text{ kips/ft, or } 7220 \text{ lb/ft}$$

The load is less than that on the concrete pipe in the preceding example because the flexible pipe effectively has a soil-structure interaction factor of 1.0 and the outside diameter is less.

5.8.3 Live Load and Impact on Underground Structures

Live Load. Culverts are usually designed for the live load generated by an AASHTO HS 20 truck. The controlling loading for culverts consists of two axles spaced 14 ft (4.3 m) apart, each weighing 32 kips (145 kN), with wheels on the axle spaced 6 ft (1.8 m) apart transversely. The 16-kip wheel load is the same as for an H 20 loading. The live load applied to underground structures under load factor criteria is either a standard HS truck, or a live load lane. Where the culvert has a span of 20 ft (6.1 m) or greater, it is classified as a bridge and must also be investigated for an Alternate Military Loading of two axles 4 ft (1200 mm) apart with each axle weighing 24 kips (107 kN). The live load lane consists of a uniform load applied in conjunction with a concentrated load. The concentrated load is distributed across the design lane of 10 ft (3000 mm), as is the uniform load. Because of this and because of the relatively short spans associated with culverts, the standard HS truck usually controls as the critical loading.

Impact Load. An impact factor is added to the highway live loading. The factor is equal to 30 percent for a soil cover of 1.0 ft (300 mm) or less and decreases to 20 percent for a cover up to 2.0 ft (600 mm) and to 10 percent for a cover up to 2 ft 11 in (900 mm). There is no impact applied when the cover is greater.

5.8.4 Distribution of Loads through Fill

Where the fill over the culvert is 2 ft (600 mm) or more, the wheel live load of 16 kips (73 kN) is applied as a concentrated load acting on the wheel print area and uniformly distributed over a square with sides equal to $1\frac{3}{4}$ times the depth of the fill. If areas from several concentrated loads overlap, the total load is uniformly distributed over an area as defined by the outside limits of those individual areas. For flexible structures and for

TABLE 5.8 Highway and Railway Live Loads

Highway loading*			Railway E 80 loading*	
Depth of cover, ft	Load, lb/ft²		Depth of cover, ft	Load, lb/ft²
	H 20	H 25		
1	1800	2280	2	3800
2	800	1150	5	2400
3	600	720	8	1600
4	400	470	10	1100
5	250	330	12	800
6	200	240	15	600
7	175	180	20	300
8	100	140	30	100
9	—	110	—	—

*See ASTM A796. Neglect live load when less than 100 lb/ft²; use dead load only.

Source: From *Handbook of Steel Drainage and Highway Construction Products,* American Iron and Steel Institute, 1994, with permission.

rigid structures with 2 ft (600 mm) or more of fill, the live load pressures given in Table 5.8 are often used instead. The table also includes pressures for an H 25 wheel load, which is 25 percent greater than an H 20 load, and an E 80 railway loading. All of the pressures in the table include an impact allowance for shallow covers.

The combination of dead and live loads causes variable pressures on the underground structure depending upon the amount of fill. As illustrated in Fig. 5.38, the dead load pressure increases uniformly with the increase in fill height, whereas the live load pressure decreases at a variable rate with the increase in fill height. For highway loadings, this results in a minimum load on the structure when there is approximately 5 ft (1.5 m) of fill over the structure. Standard designs for underground structures may be found in industry publications with minimum and maximum fill heights indicated. However, when a structure is designed for a site-specific fill height, the designer should be aware that future changes in roadway elevations may cause increased loading conditions. This may be of particular concern for a metal box culvert or a structural-plate long-span structure.

For rigid structures, when the fill is less than 2 ft (600 mm), the wheel load is applied as a concentrated load and there is no assumed distribution due to the fill. Since most concrete pipes have 2 ft or more of cover, this generally applies only to reinforced concrete boxes. The distribution length longitudinally along the top slab of reinforced concrete box culverts of wheel loads applied as concentrated loads is defined as the distance E given by

$$E = 4 + 0.06S \text{ (English units)} \tag{5.26a}$$

$$E = 1220 + 0.06S \text{ (SI units)} \tag{5.26b}$$

where S is the span in feet (mm). The live load may be distributed to the bottom slab of a reinforced concrete box culvert over a longitudinal distance equal to the width of the top slab strip increased by twice the box height.

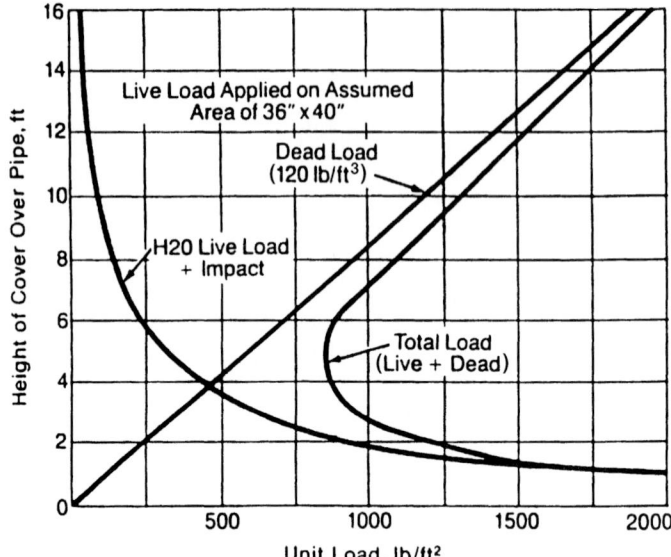

FIGURE 5.38   Design pressures for metal culverts under earth dead load and H 20 highway live load. (*From* Handbook of Steel Drainage and Highway Construction Products, *American Iron and Steel Institute, 1994, with permission*)

5.8.5   Design of Flexible Underground Structures

Flexible underground structures include corrugated metal and plastic pipes, corrugated metal pipe arches, structural plates of varying shapes, and metal box culverts. Present design specifications allow for design of these structures by either service load or load factor design. Service load design is implemented by the use of safety factors that may be applied to yield stress, buckling stress, or seam strength to determine an allowable stress. Load factor design is utilized by applying load factors (ß and γ) to the dead and live loads and allowing the design stress to approach the yield stress of the material adjusted by a capacity reduction factor (ф).

The calculations include checks for wall area, buckling, and installation strength. In addition, the seam strength must be checked for metal pipe with annular corrugations and all structural-plate structures except box culverts. Although structural-plate box culverts are erected with longitudinal bolted seams, the AASHTO *Standard Specifications for Highway Bridges* states that "moment requirements govern the design in all cases. Effects of thrust were found to be negligible when combined with moment."

Values for moment of inertia and wall area of steel products are given in Table 5.9, and minimum longitudinal seam strengths for steel structural plate are given in Table 5.10. Data for plastic pipe are given in Table 5.11. Data for aluminum pipe may be found in the AASHTO specifications. (See *Standard Specifications for Highway Bridges,* AASHTO; *Handbook of Steel Drainage and Highway Construction Products,*

TABLE 5.9 Moment of Inertia and Cross-Sectional Area of Corrugated Steel Pipe

Corrugation pitch × depth, in	Specified thickness, in*									
	0.052	0.064	0.079	0.109 / 0.111	0.138 / 0.140	0.168 / 0.170	0.188	0.218	0.249	0.280
	Moment of inertia I, in⁴ per ft of width									
1½ × ¼	0.0041	0.0053	0.0068	0.0103	0.0145	0.0196				
2 × ½	0.0184	0.0233	0.0295	0.0425	0.0586	0.0719				
2⅔ × ½	0.0180	0.0227	0.0287	0.0411	0.0544	0.0687				
3 × 1	0.0827	0.1039	0.1306	0.1855	0.2421	0.3010				
5 × 1		0.1062	0.1331	0.1878	0.2438	0.3011				
6 × 2				0.725	0.938	1.154	1.296	1.523	1.754	1.990
¾ × ¾ × 7½†		0.0431	0.0569	0.0858	0.1157					
¾ × 1 × 11½†		0.0550	0.0730	0.1111						
	Cross-sectional wall area A, in² per ft of width									
1½ × ¼	0.608	0.761	0.950	1.331	1.712	2.093				
2 × ½	0.652	0.815	1.019	1.428	1.838	2.249				
2⅔ × ½	0.619	0.775	0.968	1.356	1.744	2.133				
3 × 1	0.711	0.890	1.113	1.560	2.008	2.458				
5 × 1		0.794	0.992	1.390	1.788	2.196				
6 × 2				1.556	2.003	2.449	2.739	3.199	3.658	4.119
¾ × ¾ × 7½†		0.511	0.715	1.192	1.729					
¾ × 1 × 11½†		0.374	0.524	0.883						

*Where two thicknesses are shown, top is corrugated steel pipe and bottom is structural plate.
†Ribbed pipe. Properties are effective values.
Source: From *Handbook of Steel Drainage and Highway Construction Products*, American Iron and Steel Institute, 1994, with permission.

TABLE 5.10   Minimum Longitudinal Seam Strength for
6- × 2-in Steel Structural Plate

	Minimum seam strength (kips/ft) for indicated bolt pattern		
Specified thickness, in	4 bolts/ft	6 bolts/ft	8 bolts/ft
0.111	43		
0.140	62		
0.170	81		
0.188	93		
0.218	112		
0.249	132		
0.280	144	180	194

Source: Based on *Standard Specifications for Highway Bridges,* AASHTO.

TABLE 5.11   Design Properties for Plastic Pipe

A. PE corrugated pipes					
Nominal size, in	I.D., in (min.)	O.D., in (max.)	A, in^2/ft (min,)	c, in (min.)	I, in^4/in (min.)
12	11.8	14.7	1.50	0.35	0.024
15	14.8	18.0	1.91	0.45	0.053
18	17.7	21.5	2.34	0.50	0.062
24	23.6	28.7	3.14	0.65	0.116
30	29.5	36.4	3.92	0.75	0.163
36	35.5	42.5	4.50	0.90	0.222

B. PE ribbed pipes					
					I, in^4/in (min.)	
Nominal size, in	I.D., in (min.)	O.D., in (max.)	A, in^2/ft (min.)	c, in (min.)	Cell Class 335433C	Cell Class 335434C
18	17.80	21.0	2.96	0.344	0.052	0.038
21	20.80	24.2	4.05	0.409	0.070	0.051
24	23.80	27.2	4.66	0.429	0.081	0.059
27	26.75	30.3	5.91	0.520	0.125	0.091
30	29.75	33.5	5.91	0.520	0.125	0.091
33	32.75	37.2	6.99	0.594	0.161	0.132
36	35.75	40.3	8.08	0.640	0.202	0.165
42	41.75	47.1	7.81	0.714	0.277	0.277
48	47.75	53.1	8.82	0.786	0.338	0.277

TABLE 5.11 Design Properties for Plastic Pipe *(Continued)*

C. Profile wall PVC pipes						
					I, in^4/in (min.)	
Nominal size, in	I.D., in (min.)	O.D., in (max.)	A, in^2/ft (min.)	c, in (min.)	Cell Class 12454C	Cell Class 12364C
12	11.7	13.6	1.20	0.15	0.004	0.003
15	14.3	16.5	1.30	0.17	0.006	0.005
18	17.5	20.0	1.60	0.18	0.009	0.008
21	20.6	23.0	1.80	0.21	0.012	0.011
24	23.4	26.0	1.95	0.23	0.016	0.015
30	29.4	32.8	2.30	0.27	0.024	0.020
36	35.3	39.5	2.60	0.31	0.035	0.031
42	41.3	46.0	2.90	0.34	0.047	0.043
48	47.3	52.0	3.16	0.37	0.061	0.056

Source: From *Standard Specifications for Highway Bridges,* 1992, and *Interim Specifications,* 1995 (American Association of State Highway and Transportation Officials, Washington, D.C.), with permission.

American Iron and Steel Institute; J. B. Goddard, "Plastic Pipe Design," Tech. Report 4.103, Advanced Drainage Systems; and *Handbook of PVC Pipe,* Uni-Bell Corp.)

Load Calculation. AASHTO calculations for load (pressure) are as follows. The design load pressure P for service load design is the sum of the applicable live load plus impact and earth load. The earth load is the height of the fill above the pipe multiplied by the unit weight of the soil. For load factor design, the design load is calculated by multiplying the dead load by a ß factor of 1.5 and the live load by a ß factor of 1.67. The summation of the dead load and the live load is then multiplied by a γ factor of 1.3.

Service Load Design for Steel or Aluminum. Calculations for factory-corrugated or structural-plate structures proceed as follows. The thrust in the pipe wall is

$$T = P \times \frac{S}{2} \qquad (5.27)$$

where T = thrust, lb/ft
  P = design load, lb/ft^2
  S = diameter or span, ft

The required wall area to resist the thrust is

$$A = \frac{T}{f_a} \qquad (5.28)$$

where A = required wall area, in²/ft
  f_a = allowable stress, lb/ft²
     = f_y/SF = specified minimum yield stress (lb/in²) divided by safety factor (2.0)

For steel, f_y = 33,000 lb/in²; for aluminum, f_y = 24,000 lb/in².
  After selecting a corrugation profile and sheet thickness, check for possible buckling. If the buckling stress f_{cr} is less than the yield stress, recalculate the area using f_{cr}/SF in lieu of f_y. The buckling stress is given by the following equations:

$$\text{If} \quad S < \frac{r}{k}\sqrt{\frac{24E_m}{f_u}} \quad \text{then} \quad f_{cr} = f_u - \frac{f_u^2}{48E_m}\left(\frac{kS}{r}\right)^2 \tag{5.29}$$

$$\text{If} \quad S \geq \frac{r}{k}\sqrt{\frac{24E_m}{f_u}} \quad \text{then} \quad f_{cr} = \frac{12E_m}{(kS/r)^2} \tag{5.30}$$

where f_{cr} = critical buckling stress, lb/in²
  f_u = specified minimum tensile stress, lb/in²
     = 45,000 lb/in² (steel culvert pipe)
     = 31,000 lb/in² (aluminum culvert pipe)
     = 35,000 lb/in² (aluminum structural plate, 0.100–0.175 in thick)
     = 34,000 lb/in² (aluminum structural plate, 0.176–0.250 in thick)
  k = soil stiffness factor = 0.22
  S = diameter or span, in
  r = radius of gyration of corrugation = $\sqrt{I/A}$, in
  E_m = modulus of elasticity of metal, lb/in²
     = 29,000,000 lb/in² (steel) or 10,000,000 lb/in² (aluminum)
  I = moment of inertia of corrugation, in⁴/in
  A = cross-sectional area of corrugation, in²/in

Pipe with annular corrugations is fabricated with longitudinal seams, and a seam strength check is required; helically corrugated pipe has no longitudinal seams, and therefore such a check is not required. For pipe fabricated with longitudinal seams, the required seam strength is

$$SS = T(SF) \tag{5.31}$$

where SS = required seam strength, lb/ft
  T = thrust in pipe wall, lb/ft
  SF = safety factor = 3.0

Check handling and installation rigidity by calculating the flexibility factor FF, in/lb:

$$FF = \frac{S^2}{E_m I} \tag{5.32}$$

where terms are as defined above. Limit FF to the values listed in Table 5.12.

TABLE 5.12 Maximum Flexibility Factors for Metal Culverts

Material	Corrugation	Flexibility factor, in/lb
Steel	¼ and ½ in deep	4.3×10^{-2}
Steel	1 in deep	3.3×10^{-2}
Aluminum	¼ and ½ in deep	3.1×10^{-2} (0.060 in thick)
		6.1×10^{-2} (0.075 in thick)
		9.2×10^{-2} (>0.060 in thick)
Aluminum	1 in deep	3.3×10^{-2}
Steel and aluminum	Spiral rib pipe	Varies with I and with type of installation. See AASHTO *Standard Specifications for Highway Bridges.*
Steel	6 × 2 in	2.0×10^{-2} (pipe)
		3.0×10^{-2} (pipe-arch)
		3.0×10^{-2} (arch)
Aluminum	9 × 2½ in	2.5×10^{-2} (pipe)
		3.6×10^{-2} (pipe-arch)
		3.6×10^{-2} (arch)

Source: Based on *Standard Specifications for Highway Bridges,* AASHTO.

Load Factor Design for Steel or Aluminum.   Calculations proceed as follows. The thrust in the pipe wall is

$$T_L = P_L \times \frac{S}{2}$$ (5.33)

where T_L = load factor thrust, lb/ft
  P_L = factored load, lb/ft²
  S = diameter or span, ft

The required wall area to resist the thrust is

$$A = \frac{T_L}{\phi f_y}$$ (5.34)

where A = required wall area, in²/ft
  f_y = specified minimum yield stress, lb/in²
  ϕ = capacity modification factor
   = 1.00 for helical pipe with lockseam (including spiral rib pipe) or fully welded seam
   = 0.67 for annular pipe with spot-welded, riveted, or bolted seam (including structural-plate pipe)

After selecting a corrugation profile and sheet thickness, check for possible buckling. If the buckling stress f_{cr} is less than the yield stress, recalculate the area using f_{cr} in lieu of f_y. The buckling stress is given by Eqs. (5.29) and (5.30). For pipe with longitudinal seams, the required seam strength SS, lb/ft, is

$$SS = \frac{T_L}{\phi} \tag{5.35}$$

where ϕ is as given for Eq. (5.34). Check the flexibility factor by Eq. (5.32) and Table 5.12.

Design for Plastic Pipe.   Design calculations for service load design and for load factor design are similar to the respective calculations for steel or aluminum, with the following exceptions.

In calculating the required area for service load design by Eq. (5.28), the allowable stress $f_a = f_u/\text{SF}$, where f_u is the specified minimum tensile strength. Similarly, in calculating the required area for load factor design by Eq. (5.34), use f_u instead of f_y. There is no longitudinal seam strength required, and the capacity modification factor is $\phi = 1.00$.

The buckling stress is given by

$$f_{cr} = 9.24\left(\frac{R}{A}\right)\sqrt{\frac{BM_s EI}{0.149R^3}} \tag{5.36}$$

where f_{cr} = critical buckling stress, lb/ft^2
  R = effective radius = c + (inside diameter)/2, in
  c = distance from inside surface to neutral axis, in
  A = cross-sectional area of pipe wall, in^2/ft
  B = water buoyancy factor = $1 - 0.33h_w/h$
  h_w = height of water surface above top of pipe, ft
  h = height of ground surface above top of pipe, ft
  M_s = soil modulus = 1700 lb/in^2
  E = long-term (50-yr) modulus of elasticity of plastic, lb/in^2
  I = moment of inertia of pipe wall, in^4/in

The flexibility factor for plastic pipe may be calculated from Eq. (5.32), but substitute the effective diameter (inside diameter + $2c$) for the span. The limit for flexibility factor for PE and PVC pipe is 9.5×10^{-2} in/lb.

PE and PVC materials have stress-strain relationships that are nonlinear and time-dependent. Values for initial and long-term (50-yr) loadings are given in Table 5.13. AASHTO requires that the long-term properties be used for buckling but leaves to the judgment of the engineer what values are appropriate in other calculations.

AASHTO also cautions against excessive long-term tensile fiber strains in pipes that are significantly deflected. Allowable long-term strains are included in Table 5.13. However, AASHTO indicates that such values would not likely be reached in pipes designed and constructed in accordance with the specification.

Pipe-Arch Design.   The pipe-arch type of steel or aluminum culvert exerts high pressures at the corner radii as illustrated in Fig. 5.39. For this reason, in addition to the need for designing a pipe to withstand the imposed loads, the soil at the corner radii must be able to withstand the high bearing pressures applied to it. The corner pressure may be considered to be approximately equal to the thrust divided by the radius of the pipe-arch corner. However, the live load portion of the corner pressure may be reduced slightly to account for the longitudinal distribution of thrust in the pipe wall as the stresses from the wheel load flow from the crown toward the corner.

TABLE 5.13   Minimum Mechanical Properties for Design of Plastic Pipe

Type of pipe	ASTM cell class	Initial tensile strength, lb/in²	Initial modulus of elasticity, lb/in²	50-yr tensile strength, lb/in²	50-yr modulus of elasticity, lb/in²
PE, smooth wall	D3350, 335434C	3000	110,000	1440	22,000
PE, corrugated	D3350, 315412C	3000	110,000	1440	22,000
PE, ribbed	D3350, 334433C	3000	80,000	1125	20,000
PE, ribbed	D3350, 335433C	3000	100,000	1440	22,000
PVC, smooth wall	D1784, 12454C	7000	400,000	3700	140,000
PVC, smooth wall*	D1784, 12364C	6000	440,000	2600	158,400
PVC, ribbed	D1784, 12454C	7000	400,000	3700	140,000
PVC, ribbed*	D1784, 12364C	6000	440,000	2600	158,400

*Allowable long-term net tensile bending strain = 3.5%; for all other PE and PVC pipe, = 5%. Allowable long-term strains should not be reached in pipes properly designed and constructed.

Source: Based on *Standard Specifications for Highway Bridges,* AASHTO.

Pipe Deflection.   Deflection of flexible pipes is not a design criterion in AASHTO, because if pipes are properly installed with approved soil and compaction level, deflections will be within normal limits. However, the deflection for given loading and backfill conditions can be approximated for a round pipe. The traditional method of predicting deflection is the Iowa formula introduced by M. G. Spangler and modified by R. K. Watkins:

$$\Delta X = \frac{D_L K W_c r^3}{EI + 0.061E'r^3} \qquad (5.37)$$

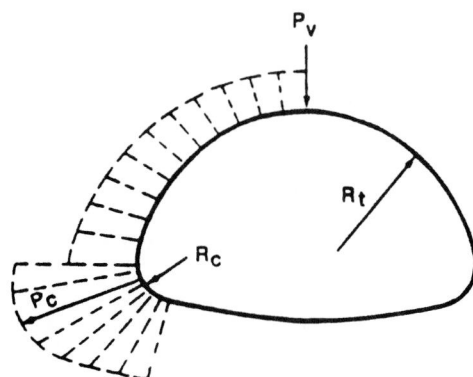

FIGURE 5.39   Pressure distribution assumed in design of metal pipe arches. (*From* Handbook of Steel Drainage and Highway Construction Products, *American Iron and Steel Institute, 1994, with permission*)

TABLE 5.14 Values of
Bedding Constant K for
Deflection Calculations for
Flexible Pipe

Bedding angle, °	K
0	0.110
30	0.108
45	0.105
60	0.102
90	0.096
120	0.090
180	0.083

Source: From A. P. Moser,
Buried Pipe Design, McGraw-
Hill, 1990, with permission.

where ΔX = total horizontal deflection, in
$\quad D_L$ = deflection lag factor
$\quad K$ = bedding constant
$\quad W_c$ = vertical load on pipe, lb/in
$\quad r$ = mean radius of pipe, in
$\quad E$ = modulus of elasticity of pipe material, lb/in^2
$\quad I$ = moment of inertia of pipe wall, in^3
$\quad E'$ = modulus of soil reaction, lb/in^2

Values for the bedding constant may be found in Table 5.14. Because the bedding
constant does not vary greatly and the bedding angle is generally not well known, it is
often taken as 0.10. Values of the modulus of soil reaction are given in Table 5.15. As
used in this table, the bedding material refers to the soil surrounding the pipe, not just
the bedding layer on which the pipe rests. The deflection lag factor accounts for the
tendency for deflections to increase over time, particularly if the soil is not well compacted or if the soil has a significant plastic content. The value of D_L used ranges from
1.0 to 1.5. Generally, reverse curvature of a round flexible pipe occurs when the
deflection reaches approximately 20 percent. Traditionally, a factor of safety of 4 is
used, so deflections are limited to 5 percent.

Design of Steel and Aluminum Long-Span Structures. Structural-plate structures
that cannot, because of their long span length, meet the design requirements for structural-plate pipe structures are defined as long-span structural-plate structures. These
structures, which often serve as short-span bridges, are not required to meet buckling
or flexibility requirements but must have certain special features. (See Art. 5.6.2.) The
required area of the metal plate is determined by the same method as for other pipe—
Eqs. (5.28) or (5.34)—but the span in the formulas is replaced by twice the radius of
the top arc. In addition, certain minimum thicknesses that have been found satisfactory
through experience are specified by AASHTO for the top-arc plate. Also, the structures must exhibit special features accepted by AASHTO. The design requirements for
long-span structures have been based not on an analytical analysis of the soil-structure
interaction system, but upon experience with successful installations. There is ongoing
research to provide for an analytical and reasonably simple method for the design of
these structures.

TABLE 5.15 Average Values of Modulus of Soil Reaction E' for Deflection Calculations for Flexible Pipe

Soil type—pipe bedding material (Unified Classification System)	E' for degree of compaction of bedding, lb/in^2			
	Dumped	Slight, <85% proctor, <40% relative density	Moderate, 85–95% proctor, 40–70% relative density	High, >95% proctor, >70% relative density
Fine-grained soils (LL>50)† Soils with medium to high plasticity CH, MH, CH-MH	No data available; consult a competent soils engineer; otherwise use $E' = 0$			
Fine-grained soils (LL<50) Soils with medium to no plasticity CL, ML, ML-CL, with less than 25% coarse-grained particles	50	200	400	1000
Fine-grained soils (LL<50) Soils with medium to no plasticity CL, ML, ML-CL, with more than 25% coarse-grained particles Coarse-grained soils with fines GM, GC, SM, SC contains more than 12% fines	100	400	1000	2000
Coarse-grained soils with little or no fines GW, GP, SW, SP‡ contains less than 12% fines	200	1000	2000	3000
Crushed rock	1000	3000	3000	3000
Accuracy in terms of percentage deflections§	±2	±2	±1	±0.5

Note: Values applicable only for fills less than 50 ft (15 m). Table does not include any safety factor. For use in predicting initial deflections only; appropriate deflection lag factor must be applied for long-term deflections. If bedding falls on the borderline between two compaction categories, select lower E' value or average the two values. Percentage proctor based on laboratory maximum dry density from test standards using about 12,500 ft · lb/ft^3 (598,000 J/m^3) (ASTM D698, AASHTO T-99, USBR Designation E-11). 1 lb/in^2 = 6.9 kN/m^2.

*ASTM Designation D2487, USBR designation E-3.

†LL = liquid limit.

‡Or any borderline soil beginning with one of these symbols (i.e., GM-GC, GC-SC).

§For ±1% accuracy and predicted deflection of 3%, actual deflection would be between 2% and 4%.

Source: From American Society of Civil Engineers, *J. Geotech. Eng. Div.,* January 1977, pp. 33–43, with permission. (Based on Amster K. Howard, "Soil Reaction for Buried Flexible Pipe," U.S. Bureau of Reclamation, Denver, Colo.)

The backfill gradation, placement, and compaction for structural-plate structures of any type or span are critical to the adequate performance of the structure. This is even more true for long-span structures. The backfill design requirements as given in AASHTO or those of the manufacturer of the product—whichever are more severe—should be followed. To make a determination of the extent of the backfill, which is dependent upon the quality of the adjacent embankment, a geotechnical investigation is advised. AASHTO recommends that a minimum of 6 ft (1.8 m) beyond the structure be backfilled with suitable material. In addition, since most long-span structures require concrete footings, the geotechnical report should provide allowable bearing pressures by which they may be designed.

Design of Steel and Aluminum Box Culverts.   The effects of moments on structural-plate box culverts control over those of thrust. The design moments may be calculated from simple tables provided in AASHTO and compared against allowable moments supplied by the manufacturer of the product. Unlike long-span structures, for which design methodology is based more on experience then analysis, design procedures for metal box culverts have been developed from finite element analyses.

Design of Tunnel Liners and Ribs and Lagging.   Tunnel liners act in compression caused by ring thrust. If a structural member with significant stiffness is used, the effects of ring flexure must be included because the flexure stress will reduce the capacity of the member to carry load. The loads on tunnel liners have been previously discussed (Art. 5.8.2), and the same loads may be used for steel ribs with lagging. The design of tunnel liners generally consists of designing the liner for joint strength, wall buckling, and minimum stiffness for installation. The analysis of steel ribs with lagging is a fairly straightforward procedure. The steel ribs are generally placed at 4-ft intervals on centers. The lagging must carry the load between these and is designed for moment and shear over the 4-ft span. The load per linear foot may be taken as that for liner plates. The ribs must be designed to withstand the load transferred from the lagging. The stress in the steel ribs should include the effects of both flexure and thrust. The use of precast-concrete tunnel liners as an initial support is rare. The analysis is complex, but may be aided by the use of moment-thrust interaction diagrams. (See *Standard Specifications for Highway Bridges,* AASHTO; R. V. Proctor and T. L. White, *Earth Tunneling with Steel Supports,* Commercial Shearing, Inc., 1977; and T. D. O'Rourke, *Guidelines for Tunnel Lining Design,* ASCE, 1984.)

Example: Service Load Design of Steel Culvert.   A 48-in-diameter culvert is required for a site with 6 ft of cover and an HS 20 live load. Determine a suitable corrugation and sheet thickness for a steel culvert using service load design procedures. Use factory-corrugated pipe with a helical seam.

First calculate the design load as follows. From Table 5.8, the live load pressure for the 6-ft cover is 200 lb/ft^2. The earth load pressure is the product of the fill density and the fill height: $120 \times 6 = 720$ lb/ft^2. The design load is the sum of these pressures: $P = 200 + 720 = 920$ lb/ft^2. Then, from Eq. (5.27), the thrust in the pipe wall is

$$T = P \times \frac{S}{2} = 920 \times \frac{4}{2} = 1840 \text{ lb/ft}$$

From Eq. (5.28), the required wall area to resist this thrust is

$$A = \frac{T}{f_a} = \frac{1840}{33,000/2} = 0.112 \text{ in}^2/\text{ft}$$

From Table 5.9, make a tentative selection of corrugation profile and sheet thickness as follows: $2\frac{2}{3}$- $\times$ $\frac{1}{2}$-in profile, 0.064-in thickness. Properties per foot of length are $A = 0.775$ in^2/ft (> 0.112 in^2/ft required) and $I = 0.0227$ in^4/ft. Divide by 12 to get properties per inch of length: $A = 0.0646$ in^2/in and $I = 0.00189$ in^4/in. Next, check the buckling stress. The radius of gyration of the corrugation is $r = \sqrt{I/A} = \sqrt{0.00189/0.0646} = 0.171$ in. To determine whether Eq. (5.29) or (5.30) applies, compare the span ($S = 48$ in) with

$$\frac{r}{k}\sqrt{\frac{24E_m}{f_u}} = \frac{0.171}{0.22}\sqrt{\frac{24 \times 29 \times 10^6}{45,000}} = 96.7 \text{ in}$$

Because $S < 96.7$ in, Eq. (5.29) applies. The buckling stress is

$$f_{cr} = f_u - \frac{f_u^2}{48E_m}\left(\frac{kS}{r}\right)^2 = 45,000 - \frac{(45,000)^2}{48 \times 29 \times 10^6}\left(\frac{0.22 \times 48}{0.171}\right)^2 = 39,500 \text{ lb/in}^2$$

Compare this with the yield stress, $f_y = 33,000$ lb/in^2. Because $f_{cr} > f_y$, buckling does not control. Also, because there are no longitudinal seams, the seam strength check does not apply. Finally, check handling and installation rigidity by calculating the flexibility factor. From Eq. (5.32),

$$FF = \frac{S^2}{E_m I} = \frac{(48)^2}{29 \times 10^6 \times 0.00189} = 4.2 \times 10^{-2} \text{ in/lb}$$

Table 5.12 gives the maximum value of FF for this profile as 4.3×10^{-2} in/lb. Therefore, the design is satisfactory. Select the $2\frac{2}{3}$- $\times$ $\frac{1}{2}$-in profile with a specified minimum thickness of 0.064 in.

Example: Load Factor Design of Steel Culvert. For the 48-in-diameter culvert in the preceding example, determine a suitable corrugation and sheet thickness for a steel culvert using load factor design procedures.

First calculate the factored load as follows:

$$P_L = \gamma(\beta_E P_{EL} + \beta_E P_{LL+I}) = 1.3(1.5 \times 720 + 1.67 \times 200) = 1840 \text{ lb/ft}^2$$

Then, from Eq. (5.33), the factored thrust in the pipe wall is

$$T_L = P_L \times \frac{S}{2} = 1840 \times \frac{4}{2} = 3680 \text{ lb/ft}$$

From Eq. (5.34), the required wall area to resist the thrust is

$$A = \frac{T_L}{\phi f_y} = \frac{3680}{1.0 \times 33,000} = 0.111 \text{ in}^2/\text{ft}$$

The remaining checks are similar to those for service load design. It may be seen that a satisfactory design is provided with the $2\frac{2}{3}$- $\times$ $\frac{1}{2}$-in profile with a specified minimum thickness of 0.064 in.

Example: Deflection of Plastic Pipe.   A 36-in-diameter plastic corrugated PE pipe is installed under a fill of 20 ft having a density of 120 lb/ft³. The soil surrounding the pipe is described under ASTM D2487 as type GM, and it will be compacted to 95 percent of AASHTO T-99 density. The pipe has a moment of inertia of $I = 0.222$ in⁴/in, and the c distance is 0.90 in. Estimate the expected deflection. Assume a bedding constant of 0.10 and a deflection lag factor of 1.25.

The load on the pipe is $W_c = 120 \times 20 \times 36/144 = 600$ lb/in. The mean radius can be taken as $18 + 0.90 = 18.90$ in. Because this will be a long-term loading, use the long-term (50-yr) modulus of elasticity from Table 5.13 of 22,000 lb/in². From Table 5.15, $E' = 2000$ lb/in². From Eq. (5.37), calculate the deflection as

$$\Delta X = \frac{D_L KW_c r^3}{EI + 0.061 E' r^3}$$

$$= 1.25 \times 0.10 \times 600 \times \frac{(18.90)^3}{22,000 \times 0.222 + 0.061 \times 2000 \times (18.90)^3}$$

$$= \frac{506,345}{4880 + 823,655} = 0.61 \text{ in}$$

As a percent of diameter, the deflection is $(0.61/36)100 = 1.7$ percent.

5.8.6   Design of Rigid Underground Structures

There are two general types of rigid underground structures—those with a curvilinear shape, and those made up of straight walls and flat slabs. A reinforced concrete pipe is an example of the former, while a reinforced concrete box is an example of the latter. Rigid structures built in a curvilinear shape tend to act in compression. Because of their limited deflection capability, they develop moment as well as compressive stresses. The effect of moment is reduced, however, because the curvilinear shape increases the compression in the member. Structures built with straight members act very differently. The effect of moment on the individual members is so great that it is not unusual for the engineer to completely ignore the small benefit obtained from the compression of the member.

Design Methods for Curvilinear Concrete Pipe.   The simplest curvilinear structure is the round pipe. Reinforced concrete pipes presently may be designed according to AASHTO by one of two methods: the direct design method, and the indirect (or *D*-load) design method. An additional method that has received positive support from many engineers and is being considered for inclusion in the next revision of AASHTO is design based on the finite element model SPIDA (soil-pipe interaction design and analysis).

Indirect Design.   The indirect design method is an empirical method in which the pipe is tested by the three-edge bearing test (Fig. 5.40), and is subjected to a previously calculated load. If the pipe supports the application of the load without exceeding a crack width criterion of 0.01 in (2.5 mm), it is considered acceptable for the application for which it was manufactured.

For both the direct and the indirect design methods, the load on the pipe is determined in substantially the same manner. As discussed in Art. 5.8.2, the dead load is determined by adjusting the prism load by the soil-structure interaction factor, F_e or

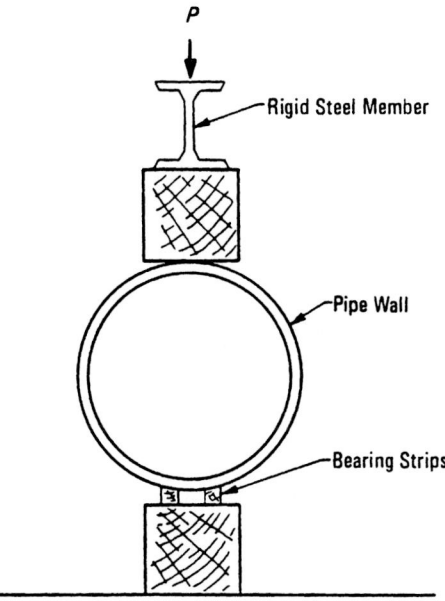

FIGURE 5.40   Three-edge bearing test for concrete pipe. (*From* Concrete Pipe Technology Handbook, *American Concrete Pipe Association, 1994, with permission*)

F_t. The dead load is then combined with the live load and impact load to obtain the total load. The appropriate load to be applied to the pipe for the three-edge bearing test may be found by dividing the total load by the appropriate bedding factor as determined from the installation conditions. This final value is termed the *required minimum three-edge bearing load* (*TEB*). If the pipe is nonreinforced, a safety factor of between 1.25 and 1.5 should be applied to the load. This is because as soon as the nonreinforced pipe cracks, it has reached its ultimate strength. However, a reinforced concrete pipe has significant reserve strength after cracking.

$$\text{TEB} = \frac{W_E + W_L}{B_f} \qquad (5.38)$$

where W_E = dead load due to the earth, lb/ft
W_L = live load, lb/ft
B_f = bedding factor for the specified installation (see Figs. 5.32 and 5.33)

The bedding factor is a ratio of the supporting strength of the buried pipe to the strength of the pipe as determined from the three-edge bearing test. For any given load, with a better installation, a lower-strength pipe is needed than would be required with an installation of poorer quality Where the pipe is jacked, it maintains good contact in the area of the invert. In addition, it is common practice to grout outside the pipe after jacking operations are completed. Because of this, a bedding factor as high as 3 may be used for jacked pipe.

Concrete pipe may be specified on the contract plans by its D-load value (lb/ft/ft), defined as follows:

$$D \text{ load} = \frac{\text{required TEB}}{S} \tag{5.39}$$

where S = inside diameter of the pipe in feet. The installation type by which the three-edge bearing value was determined must be specified along with the D-load value.

Direct Design. The direct design procedure for analyzing a concrete pipe is based not on empirical methods, but on engineering analysis. Consequently, the engineer may determine precisely what strength of concrete, wall thickness, and reinforcement are necessary. However, the method is more complicated.

The load on the pipe is determined in the same manner as for indirect design. Then, the bedding angle must be determined. Just as for indirect design, different installations have different bedding angles. The better the bedding condition, the larger the bedding angle that is assigned to it. Larger bedding angles are accompanied by smaller stresses in the pipe. Although recommended bedding factors are given in Figs. 5.32 and 5.33, the determination of a proper bedding angle for each installation is left to the judgment of the engineer.

After determining the bedding angle, the moments, shears, and thrusts may be calculated. There are two common calculation methods, which involve two separate applications of distributing pressures to pipes. As illustrated in Fig. 5.41, one method applies the pressures in a radial load system while the other employs a uniform or linear load system. The radial distribution method, developed by H. C. Olander, provides diagrams with coefficients for calculating moments, thrusts, and shears at any point around the circumference of the pipe with a given bedding angle. The uniform distribution method, developed by J. M. Paris, provides coefficients for calculating moments, thrusts, and shears at the locations where they are highest—the crown, invert, or springline. As indicated in Fig. 5.42, the C_m coefficients are multiplied by the load and the mean radius to obtain the moment, and the C_v and C_n coefficients are multiplied by the load to obtain shears and thrusts.

Actually, the location of critical shear generally occurs at between 15 and 45° from the invert. For a more precise calculation, the designer may refer to the diagrams by Olander, which will give the shear at any point around the circumference of the pipe. Alternatively, the Federal Highway Administration (FHWA) has developed a computer program titled PIPECAR from which the moments, thrusts, and shears may be obtained. The program will also design the pipe utilizing either indirect or direct design procedures. After determining the forces acting on the pipe, the wall thickness, concrete strength, and amount of reinforcement may be calculated following the detailed design provisions in the AASHTO *Standard Specifications for Highway Bridges.* (See H. C. Olander, "Stress Analysis of Concrete Pipe," *Engineering Monographs* No. 6, U.S. Department of the Interior, Bureau of Reclamation, October 1950; J. M. Paris, "Stress Coefficients for Large Horizontal Pipes," *Engineering News Record,* vol. 87, no. 19, November 10, 1921; and *Concrete Pipe Technology Handbook,* American Concrete Pipe Association, 1994. Also, for the design of concrete arches, refer to "Analysis of Arches, Rigid Frames and Sewer Sections," Concrete Information Bulletin No. ST 53, Portland Cement Association.)

Design of Box Culverts. The application of vertical loads on reinforced concrete box culverts (RCBs) has been previously discussed (Arts. 5.8.2 through 5.8.4). AASHTO specifies that RCBs shall be designed for two lateral earth pressures, 30 and

Load Case	Radial Load System	Uniform Load System
1. Pipe Weight		
2. Soil Weight		
3. Fluid Load		
4. Live Load		

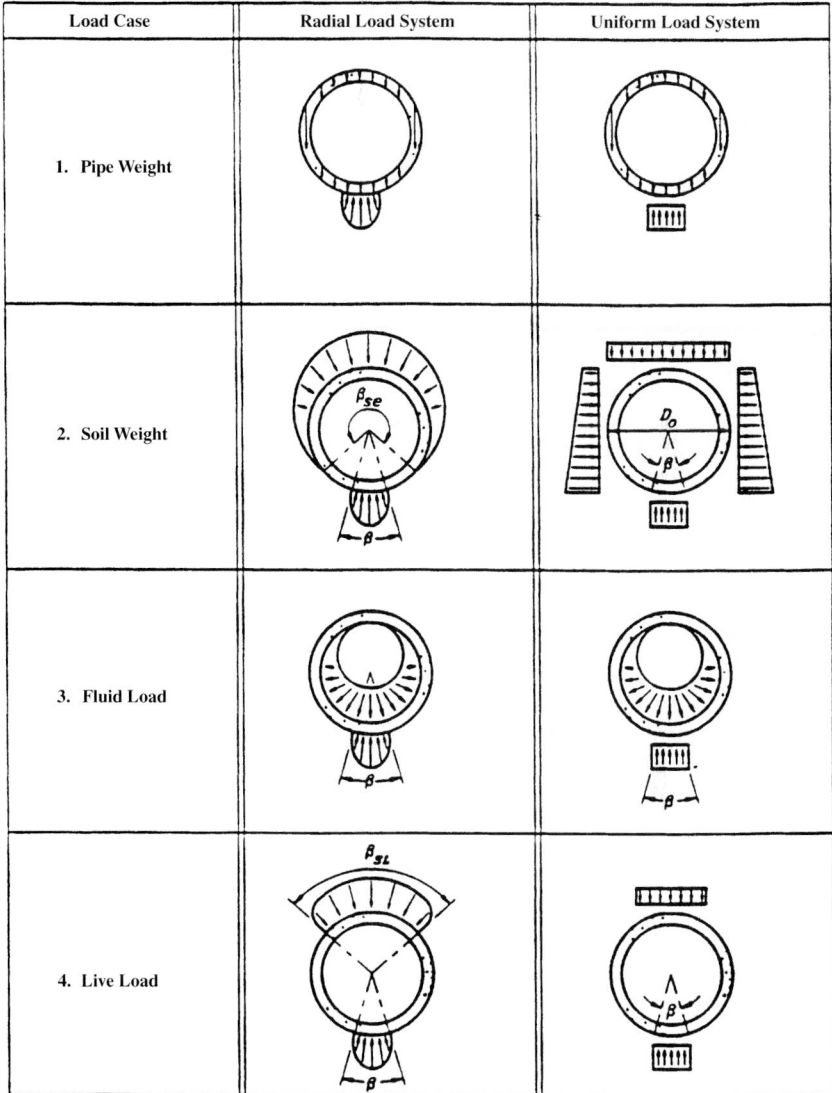

FIGURE 5.41 Pressure distributions for four load cases used in design of round concrete pipe. (*From* Concrete Pipe Technology Handbook, *American Concrete Pipe Association, 1994, with permission*)

60 lb/ft³. Box culverts may be designed with either fixed or pinned corners. All pre-cast RCBs and cast-in-place RCBs with spans greater than about 8 ft are designed with fixed corners. Those with pinned corners, cast-in-place RCBs with shorter spans, are designed for simple beam moments, and the lateral earth pressure of 60 lb/ft³ is the only condition for which the sidewall need be designed. Because fixed corners trans-fer moments around the corners, applying 60 lb/ft³ lateral pressure will reduce the

UNIFORM LOAD ON 180° TOP

	Conc. Support at Invert			θ = 60°			θ = 90°			θ = 120°			θ = 180°		
	C_m	C_n	C_v	C_m	C_n	C_v	C_m	C_n	C_v	C_m	C_n	C_v	C_m	C_n	C_v
TOP	+.1495	-.0530	0	+.1435	-.0400	0	+.1368	-.0268	0	+.1304	-.0132	0	+.1250	0	0
SIDE	-.1535	+.5000	+.0530	-.1465	+.5000	+.0400	-.1401	+.5000	+.0268	-.1327	+.5000	+.0132	-.1250	+.5000	0
INVERT	+.2935	+.0530	±.5000	+.1885	+.0400	0	+.1572	+.0268	0	+.1376	+.0132	0	+.1250	0	0

UNIFORM LOAD ON 90° TOP

	Conc. Support at Invert			θ = 60°			θ = 90°			θ = 120°			θ = 180°		
	C_m	C_n	C_v	C_m	C_n	C_v	C_m	C_n	C_v	C_m	C_n	C_v	C_m	C_n	C_v
TOP	+.1817	.0262	0	+.1757	-.0132	0	+.1690	0	0	+.1627	+.0136	0	+.1572	+.0269	0
SIDE	-.1683	+.5000	+.0262	-.1613	+.5000	+.0132	-.1549	+.5000	0	-.1475	+.5000	-.0136	-.1398	+.5000	-.0269
INVERT	+.3055	+.0262	±.5000	+.2005	+.0132	0	+.1690	0	0	+.1496	-.0136	0	+.1370	-.0269	0

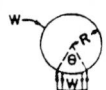

LOADING DUE TO WEIGHT OF RING

	Conc. Support at Invert			θ = 60°			θ = 90°			θ = 120°			θ = 180°		
	C_m	C_n	C_v	C_m	C_n	C_v	C_m	C_n	C_v	C_m	C_n	C_v	C_m	C_n	C_v
TOP	+.0796	-.0796	0	+.0736	-.0666	0	+.0669	-.0534	0	+.0606	-.0389	0	+.0551	-.0266	0
SIDE	-.0909	+.2500	+.0796	-.0839	+.2500	+.0667	-.0775	+.2500	+.0536	-.0701	+.2500	+.0399	-.0624	+.2500	-.0267
INVERT	+.2389	+.0796	±.5000	+.1339	+.0666	0	+.1025	+.0534	0	+.0829	+.0389	0	+.0704	+.0266	0

LOADING DUE TO WATER; PIPE FULL, ZERO PRESSURE HEAD ON SOFFIT

	Conc. Support at Invert			θ = 60°			θ = 90°			θ = 120°			θ = 180°		
	C_m	C_n	C_v	C_m	C_n	C_v	C_m	C_n	C_v	C_m	C_n	C_v	C_m	C_n	C_v
TOP	+.0796	-.2389	0	+.0736	-.2257	0	+.0669	-.2124	0	+.0606	-.1991	0	+.0551	-.1859	0
SIDE	-.0909	-.0680	+.0797	-.0836	-.0680	+.0667	-.0775	-.0680	+.0532	-.0701	-.0680	+.0399	-.0624	-.0680	+.0267
INVERT	+.2389	-.3981	±.5000	+.1337	-.4109	0	+.1025	-.4243	0	+.0829	-.4379	0	+.0704	-.4511	0

	C_m	C_n	C_v
TOP	-.1250	+.5000	0
SIDE	+.1250	0	0
INVERT	-.1250	±.5000	0

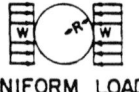
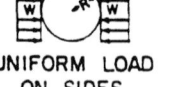
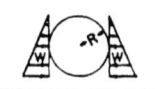

UNIFORM LOAD ON SIDES   TRIANGULAR LOAD ON SIDES

	C_m	C_n	C_v
TOP	-.1042	+.3125	0
SIDE	+.1250	0	-.0625
INVERT	-.1458	+.6875	0

FIGURE 5.42   Coefficients for calculating moment (M), thrust (N), and shear (V) in concrete pipe under linear pressure distributions. The coefficients used are as follows: $M = C_m WR$, $N = C_n W$, $V = C_v W$, where W is total load in each case and R is mean radius. *Sign convention:* $+M$ is tension on inside face, $+N$ is compression, $+V$ is positive shear for left side. (*From J. M. Paris, "Stress Coefficients for Large Horizontal Pipes," Engineering News Record, vol. 87, no. 19, November 10, 1921, with permission*)

TABLE 5.16 Loading Conditions That Generally Control Design of
Concrete Box Culverts for Moment and Shear

Member	Live load position	Lateral pressure, lb/ft³	Type of moment
Top slab	0.5 of span length	30	Positive
Top slab and sidewall	0.3 of span length	60	Negative
Sidewall	No live load	60	Positive

positive moments in the adjacent members when compared with the application of 30 lb/ft³. For this reason, RCBs with fixed corners must be designed for both lateral load conditions to ensure that all maximum moments in all members are found. Table 5.16 indicates the loading conditions that generally control the design of the top slab and sidewalls of concrete box culverts for moment and shear. After determining maximum moments and shears, the designer may utilize standard principles of concrete design to size the members and the reinforcement.

As is evident from the foregoing, and as is true for any structural design, many different loading conditions must be investigated to ensure that all maximum stresses are found. To reduce design time, the FHWA has developed the computer program BOX-CAR, which may be used for structural analysis and determination of stresses.

Design of Cast-in-Place Pipe. Because cast-in-place concrete pipe has no reinforcement, its flexural strength is limited. In this type of structure, the effects of the compression on the member keep the effects of the moment below the modulus of rupture of the concrete, and no reinforcing is necessary. Also, because it is cast against the ground at the invert and walls, it has excellent bedding conditions, and this contributes to a reduction in moments, thrusts, and shears. After determining the moments and thrusts, at different locations (crown, invert, and springline) in the pipe by the application of the Paris coefficients (Fig. 5.42, $\theta = 180°$), normal stresses may be calculated from the fundamental equation

$$f_c = \frac{P}{A} \pm \frac{Mc}{I}$$

where f_c = concrete stress, lb/in²
P = thrust, lb/in
A = wall area, in²/in
M = bending moment, in · lb/in
c = distance from neutral axis to extreme fiber, in
I = moment of inertia of wall, in⁴/in

Because the concrete is unreinforced, the stress may not exceed the modulus of rupture of the concrete adjusted by an appropriate safety factor. The FHWA recommends that the use of cast-in-place pipe on federal-aid highway projects be monitored through experimental projects in locations under roadways or with moderate to high fills.

Example: Three-Edge Bearing Load for Reinforced Concrete Pipe. For the 60-in-diameter pipe in the example given in Art. 5.8.2, determine the required minimum three-edge bearing load for a class C bedding.

For this example, the earth load was calculated as 10,000 lb/ft. For a cover of 12 ft, live load effects are negligible for highway loadings. The bedding factor for a class C bedding is obtained from Fig. 5.39 as $B_f = 1.5$. Therefore, from Eq. (5.38), the three-edge bearing load is

$$\text{TEB} = \frac{W_E + W_L}{B_f} = \frac{10,000}{1.5} = 6670 \text{ lb/ft}$$

or $6670/5 = 1334D$ lb/ft/ft, the D load where D is the diameter in feet.

5.9 CONSTRUCTION METHODS

Underground structures may be built by a variety of means including embankment construction, open-trench construction, jacking, tunneling, and microtunneling. Bedding and backfill requirements are extremely important for embankment and trench construction and vary with the type of structure being placed. Further details are given in Art. 5.8.1 and in the AASHTO *Standard Specifications for Highway Bridges,* Division II, "Construction."

5.9.1 Embankment Construction

Where a pipe is required as part of an embankment construction, it may be installed by compacting layers of fill uniformly on either side. It is important to bring the layers up uniformly on either side of the pipe. After a sufficient layer is compacted over the top of the pipe, ordinary embankment construction may proceed. Alternatively, some agencies require that the embankment be constructed first, then a trench dug for the installation of the pipe.

5.9.2 Trench Construction

The open-trench method is commonly used for culvert construction. It is more cost-effective than tunneling except when a pipe must be constructed in an existing high fill. Shoring may be necessary, particularly if the installation is under a traveled way. This will keep the limits of excavation to a minimum and, by the use of steel cover plates, allow the roadway to remain open during nonworking hours. Where it is necessary to use an open-trench method of construction in urban areas, it is wise for the designer to make available to the contractor options for the type of structure to be placed. For example, if a box culvert is deemed necessary by the engineer because of hydraulic considerations and physical constraints, a precast concrete or a prefabricated metal box, as alternatives to cast-in-place construction, should be permitted. In this manner, the traveling public experiences a minimum of disruption of service when open-trench construction is used. AASHTO recommends a trench width equal to 1.25 times the outside diameter of the pipe for concrete pipe and a width to provide for 2 ft minimum on each side of the pipe for flexible culverts. However, some states simply recommend a constant clearance between the outside of the pipe and the trench wall to ensure that there is room for compaction and compaction testing equipment.

5.9.3 Camber Requirements

Where high embankments are placed on original ground, the fill may compress and consolidate the foundation soil. Thus, culverts constructed on or near the original ground surface tend to undergo some settlement. The amount of settlement varies with fill height and the consolidation characteristics of the foundation soil. Because the amount of settlement varies with the fill height, the culvert will tend to settle more toward the center than at the ends. If the culvert is built upon a straight grade between the inlet and outlet elevations, a sag will develop. The sag may create a low point in the culvert, or may cause accumulation of debris and silt and opening and leaking of joints. These in turn may lead to a reduced waterway capacity and the possibility of loss of stability to the embankment through piping of fines at the joint. As illustrated in Fig. 5.43, these dangers may be avoided by cambering the culvert so that after settlement occurs, the culvert grade line will be at or close to that desired. Almost any type of culvert that is not cast in place may be cambered. These include precast concrete pipes and box culverts, corrugated metal pipes, structural-plate steel or aluminum pipes, and plastic pipes. The amount of camber required can be determined by a soils engineer.

5.9.4 Jacking and Tunneling

Should open-trench construction prove uneconomical or the disruption to the traveling public too great, either jacking or tunneling may prove to be more efficient. Either method removes from consideration the possible disruption of traffic. In addition, for deep fills, these methods can be economically competitive with the open-trench method. The designer should be cautioned that when jacking or tunneling is used, small differences in anticipated geologic conditions may lead to large changes in the method by which the contractor solves the problem. For example, the difference between "running" and "flowing" ground can be not only very costly, but disastrous as well. If unanticipated geologic conditions are encountered by the tunneling or jacking contractor, the cost of the contract could increase dramatically. For this reason, if geologic conditions are in doubt, the designer is advised to obtain adequate geotechnical information through borings.

Jacking. Jacking of underground structures requires that the structure being jacked be able to withstand the large compressive forces acting on it. This generally limits the possibilities to reinforced concrete pipe, reinforced concrete boxes, and steel pipes. The first step is to adequately provide for a jacking pit, or to design a thrust wall if the jacking is to take place above ground. The jacking force and the adequacy of the structure itself to withstand that force are often left to the contractor. The jacking force required is dependent upon the type and diameter or span of the structure, the type of soil, the amount of overfill, and the jacking distance. Table 5.17 provides values of frictional resistance on reinforced concrete pipe determined from past jacking projects. These values may be reduced if a lubricant such as bentonite slurry is injected into the void created by the overcut. If the frictional resistance is too high for the thrust blocks or the jacks, intermediate jacking stations may be necessary.

Figure 5.44 shows the sequence of work in a jacking operation. An articulated shield is placed on the first unit to be jacked, and all excavation occurs within this shield. The unit is pushed forward, material is excavated, and spoils are removed. The excavation may be made with a sophisticated boring machine, a small front end loader, or manual labor. Spoils are removed by augers, conveyor belts, or small carts. Pipe sec-

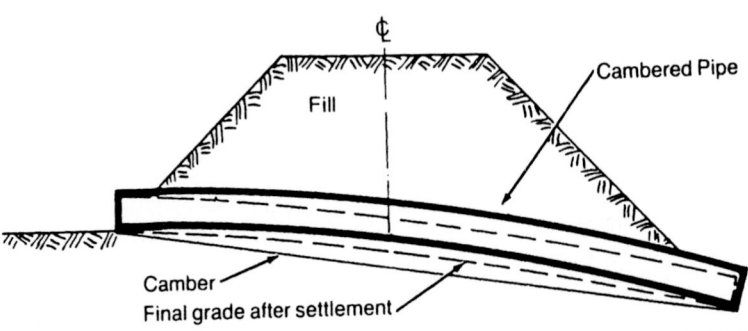

FIGURE 5.43 Illustration of camber to allow for settlement of culvert under high fill. (*From* Handbook of Steel Drainage and Highway Construction Products, *American Iron and Steel Institute, 1994, with permission*)

TABLE 5.17 Frictional Resistance of Reinforced Concrete Pipe for Jacking Projects

Soil condition	Frictional resistance, lb/ft²	Frictional resistance, kN/m²
Rock	40–60	2–3
Firm clay, silt	100–400	5–20
Wet sand	200–300	10–15
Dry loose sand	500–900	24–45

Source: From S. J. Klein, "Geotechnical Aspects of Pipe Jacking Projects," *Pipeline Crossing Proceedings,* Special Conference, Pipeline Division, American Society of Civil Engineers, Denver, March 25–27, 1991, with permission.

tions tend to remain in a straight line when jacked, either through the stiffness provided by the bell-and-spigot joints of concrete pipe or the welded joints of steel pipe. On the other hand, precast reinforced concrete boxes tend to rotate when jacked unless furnished to full length. This precludes their use when a jacking pit is necessary.

Tunneling. Tunneling through soft ground is accomplished by pushing a shield forward and erecting a liner inside of it. The shield is then pushed off the liner as the tunneling progresses, so that there is no limit to the length that may be tunneled. The initial liner may consist of precast concrete sections, steel tunnel liner plates, or steel ribs with either wood or steel lagging. After the liner is erected within the shield and the shield is jacked forward, the void created between the liner and the ground due to overcut may or may not need to be grouted. The grouting of this area depends upon the judgment of the engineer and the type of liner. Tunnel liner plates may not be expanded once they are erected. Because of this, the void caused by the overcut is generally grouted. Precast concrete sections and steel ribs may be expanded to contact the earth once the shield is jacked forward. In this case it is left to the judgment of the engineer whether or not grouting is necessary. After the tunnel is completed, the carrier pipe is placed inside the liner and the void between the two is generally filled with either sand or grout.

Microtunneling. Microtunneling is a term used to describe a method of horizontally boring pipes approximately 36 in (900 mm) in diameter and smaller, using highly

Pits are excavated on each side. The jacks will bear against the back of the left pit, so a steel or wood abutment is added for reinforcement. A simple track is added to guide the concrete pipe sections. The jacks are positioned in place on supports.

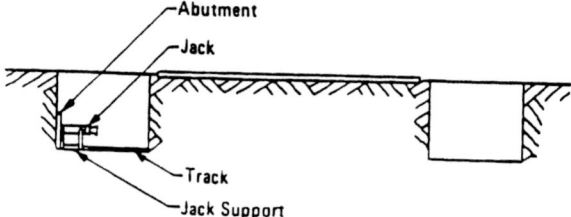

A section of concrete pipe is lowered into the pit.

The jacks are operated, pushing the pipe section forward.

The jack rams are retracted and a "spacer" is added between the jacks and pipe.

The jacks are operated and the pipe is pushed forward again.

It may be necessary to repeat the last two steps several times until the pipe is pushed forward enough to allow room for the next section of pipe. It is extremely important, therefore, that the stroke of the jacks be as long as possible to reduce the number of spacers required and thereby reduce the amount of time and cost. The ideal situation would be to have the jack stroke longer than the pipe to completely eliminate the need for spacers.

The next section of pipe is lowered into the pit and the above steps repeated. The entire process above is repeated until the operation is complete.

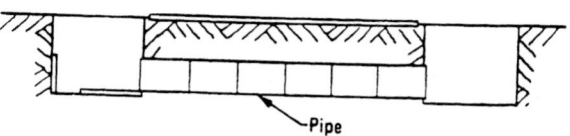

FIGURE 5.44 General procedures for jacking concrete pipe. Procedures for steel pipe are similar. (*From* Concrete Pipe Handbook, *American Concrete Pipe Association, 1980, with permission*)

sophisticated remotely controlled equipment. The use of lasers allows for extremely accurate placement of the pipe in both grade and alignment. The pipe is jacked from a jacking pit as the tunnel is being bored and the spoils are removed.

Directional Drilling. Directional drilling is similar to microtunneling except that where microtunneling is a one-stage process, the directional drilling method consists of first drilling a pilot hole, reaming it to the proper diameter, and then pulling the pipe through. Because of this methodology, no jacking pit is required. This method has a high degree of precision in location of grade and may be used where the pipe diameter is 42 in (1050 mm) or smaller and the length to be placed is less then 5000 ft (1.5 km).

Stabilization Methods for Tunneling. As previously stated, tunneling may be required where it is necessary to keep a roadway or rail line open. This may occur where there is little fill over the crown of the excavation, or where there is adequate fill but it is lacking in stiffness or cohesive strength. When this happens, the soil above the excavation cannot, by itself, develop an arching effect that will adequately support the roadway. This situation necessitates unusual solutions such as chemical grouting, compaction grouting, ground freezing, and the use of spiles. The applicable method of increasing the support depends upon the site and soil conditions. Chemical grouting, compaction grouting, and ground freezing are all methods of stabilizing the soil. Spiles are horizontally drilled small-diameter holes extending from one side of the proposed tunnel to the other and surrounding the tunnel, generally in an arch shape. The holes, after being drilled, have a steel pipe placed in them, which is subsequently filled with concrete. The spile diameter is commensurate with the size of opening to be excavated, and the spacing is reliant upon the amount of coverage and cohesiveness of the soil. After the spiles are in place, the tunnel excavation may begin with steel arch supports placed as necessary.

5.10 INSPECTION

Many storm drains and highway culvert systems have in the past been and are presently designed for a 50-year life span. The local roadway and state highway and interstate systems have in large part reached this age or soon will. Consequently, rehabilitation and repair of existing storm sewers and highway drainage culverts is presently requiring more and more attention and resources from the responsible agencies. It is generally less expensive to rehabilitate or repair an existing underground structure than to replace it. In addition, the cost of repair to the facility after a catastrophic failure greatly exceeds the cost of rehabilitating the structure and preventing that failure. The key, of course, is being able to identify those structures that are in jeopardy of failing.

5.10.1 General Considerations

Failure of a culvert can be defined as any condition that could reasonably lead to the collapse of the roadway above. Failure of the roadway above may be a direct result of the collapse of the structure, or may be caused by a loss of the fill due to piping and the infiltration of fines. Excessive seepage through open joints can cause loss of the backfill material as illustrated in Fig. 5.45.

Fortunately, the complete collapse of a culvert is a rare occurrence. Culverts that are overstressed, either because of loss of the surrounding soil support or because of overloads, tend to redistribute those stresses in many cases. For example, the loss of

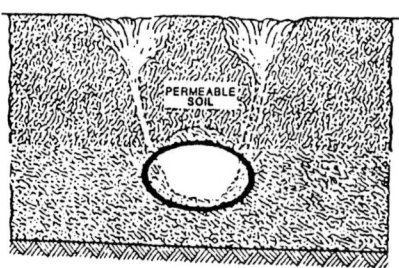
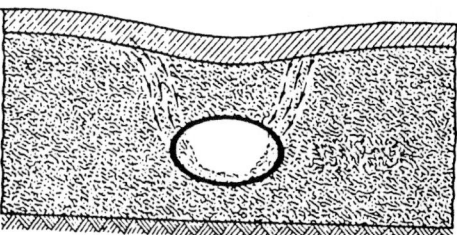

EFFECT ON UNPAVED AREAS

EFFECT ON PAVEMENT

FIGURE 5.45 Schematic of surface indications of infiltration. (*From "Culvert Inspection Manual," Report No. FHWA-IP-86-2, FHWA, 1986, with permission*)

support or the effect of excessive live loads may not occur over the complete length of the structure. Consequently, as one section becomes overstressed, it may deflect more than the adjacent sections and transfer loads to those stiffer sections. In addition, underground structures that show distress, such as a concrete pipe that cracks excessively or a flexible pipe that deflects excessively, may reduce the loads upon themselves by the very act of failing. For instance, flexible pipe that overdeflects may have a reduced overburden load on it because the complete prism of earth above the structure may not necessarily move downward with the deflection; competent soils will have a tendency to arch over the pipe and support some of the load. The concrete pipe that cracks may form hinges and redistribute loads within the structure; the concrete pipe may now have more of a tendency to act as a flexible structure with reduced moments and increased compression forces. However, this discussion should not give the false impression that structural distress can be ignored. Catastrophic failures have occurred and caused fatalities when vehicles plunged into the void left by the collapse. Large structures with low covers are probably the most susceptible to structural failures and should be evaluated carefully.

Even if complete collapse does not occur, structural distress can affect the adjacent soil and accelerate failure. Piping and infiltration that cause loss of adjacent soil support may proceed at an increasing rate and cause failure of the roadway above. In some cases, enough fill may be lost through piping to create a sinkhole with the structure below showing no signs of severe structural distress.

The National Bridge Inspection Program requires that all structures with span greater than 20 ft (6000 mm) be inspected every 2 years. That is, all structures with spans greater than 20 ft when measured along the centerline of the roadway are classified as bridges for purposes of inspection. Two important points should be mentioned here. First, the measured distance is along the centerline of the roadway. That means a structural-plate pipe or a reinforced concrete box culvert with a 15-ft (4500-mm) span on a 42° skew will be classified as a bridge for inspection purposes and included in the bridge inspection program, even though the span is 15 ft for hydraulic and structural design purposes. The second point is that multiple pipes are considered to be a bridge for inspection purposes when the out-to-out distance between the first and last pipes is 20 ft or greater and there is a maximum of one-half diameter of the smaller pipe between them. For example, two 102-in-diameter pipes separated by 51 in would qualify as a bridge (102 + 51 + 102 = 255 in, or 21.25 ft).

Culverts that do not qualify for inspection under the bridge program should nevertheless be given consideration for inclusion in a regular inspection program. To avoid repetition of inspections, some coordination between the engineers responsible for the two programs is necessary. Although the ideal would be to inspect all culverts, obvious constraints, with regard to both physically inspecting the culverts and the costs of doing so, place limits on any program of culvert inspection. It may be less expensive to replace small culverts that are located beneath lightly traveled roads and have little fill on them than it would be to maintain them in an inspection program with rehabilitation prior to failure as a goal. Conversely, where some culverts may not warrant inspections absent obvious signs of distress, others may require frequent inspections. Large structures that carry high flows during major storms or have a history of structural deficiencies, such as cracking (in concrete) or corrosion (in metal), should be inspected more frequently and especially after periods of storms.

5.10.2 Elements of Inspection

An inspection of the culvert should include the approach roadway, the embankment, the headwalls and wingwalls, the waterway, and the culvert barrel.

Roadway. The roadway over the culvert should be inspected for sags and cracks in the pavement that are the result of settlement. These may be evident in both the roadway itself and adjacent guiderail. The settlement may be the result of poorly compacted material adjacent to the culvert piping (infiltration or transportation of fines by water flowing through the backfill), or settlement of the culvert itself. The structural integrity of the culvert itself may or may not have been compromised. An inspection of the culvert must be made.

Embankment, Headwalls, and Wingwalls. The embankment, headwalls, and wingwalls at the inlet and outlet ends of the culvert should be inspected for signs of erosion, undermining, and settlement. If there is erosion at the ends, the structural integrity of the culvert will not necessarily be immediately compromised, but the hydraulic capacity will be affected. Any erosion or undermining will only worsen, and corrective action should be scheduled. If there is separation between the culvert and the endwalls, there could be a loss of supporting soil somewhere along the length of the culvert, which would affect structural capacity.

Waterway. The waterway should be inspected directly upstream and downstream for changes in the drainage. The culvert may have effected the changes in this drainage, and conversely, the changes in the drainage may have an effect on the culvert. An example of the former is where the velocity of the water is increased because of the channeling effect of the culvert. This velocity change could then cause either scour or accretion downstream. An example of the latter is accretion affecting the backwater up to the culvert, which can alter the subsequent performance of the culvert. In addition, the waterway should be inspected for accumulations of debris and sediment at both the inlet and the outlet and within the culvert itself.

Culvert Barrel. The barrel or structure of the culvert should be inspected for defects, distortions, and deflections. The nature of these will depend upon the type of culvert being inspected.

5.10.3 Inspection of Flexible Structures

A flexible structure should be checked to ensure that the cross-sectional shape it was designed for is intact. If the flexible culvert, whether it is a round pipe, a pipe arch, an arch, a horizontal ellipse, or any other structural shape, deflects from its design shape, it is not receiving the required support from the backfill. It is assumed in the design of flexible structures that moment in the structure is negligible and that due to the thrust forces, the structure is in compression throughout. If the deflection is large enough to cause a flattening of the structure, these assumptions will not hold true and the structure may collapse. Larger structures with large top radii, such as long-span structures, can withstand a smaller percentage of deflection before reverse curvature occurs than can round structures.

Visual observations of the culvert shape may reveal only large distortions and deflections; deformations may not be readily apparent until they reach approximately 10 percent. For this reason, if excessive deflections in the cross-sectional shape are suspected, physical measurements should be taken and documented with changes over time. Reference points should be permanently marked, and for a corrugated structure, measurements should be taken to inside corrugations for consistency. General deflections of round pipe greater than 5 percent should be investigated and monitored; reversal of curvature is expected at 20 percent for a metal culvert, but it may occur at a lesser value for a large structure. Localized flat spots or reversals of curvature are matters of special concern. A computer program is available to aid in the investigation and evaluation of multiple-radius metal structures. (See D. C. Cowherd et al., "Application of the Program MULTSPAN/SOILEVAL to Analyze Problem Structures," *Proceedings of the Second Conference on Structural Performance of Pipes,* Ohio University, Athens, Ohio, 1993, A. A. Balkema, Rotterdam, 1993.)

All metal culverts should be investigated for evidence of corrosion and erosion. With a general loss of section there will be an accompanying loss of structural capacity. Wear will first be noted by a loss of the galvanized or other coating. If this occurs, then the unprotected metal may be expected to deteriorate more rapidly because of the erosive effects of the bedload.

Bolted longitudinal seams of structural-plate culverts should be inspected for cocking, cracking, and bolt tipping. Cocking occurs where the structure deflects inward at the seam, causing a significant change in the structure's shape or appearance. This may be caused by improper erection or fabrication of the plates and can result in loss of backfill due to piping and a reduced allowable compression strength of the structure due to the distortion. Cracking may occur where there is excessive deflection at the seam. This could ultimately lead to a disjointing, which would result in loss of ring thrust. Bolt tipping is rare; it occurs where the plates slip because of high compressive forces. However, if the structure is under high fill and the plates slip, the bolt holes could become elongated, with the result that the bolt is eventually pulled through the plate.

Plastic pipe should be inspected for excessive deflection and for cracking.

5.10.4   Inspection of Rigid Structures

Inspection of reinforced concrete pipe should focus on problems with alignment, joints, and the wall.

The alignment of the culvert may be inspected visually. Misalignment may be caused either by poor installation practices or by subsequent settling of the pipe or the backfill. In any case, the pipe should be periodically monitored to ensure that the condition does not worsen. Close inspection of the joints may reveal conditions that will lead to an increase in the misalignment of the structure.

Joints should be inspected for cracks, separation, exfiltration, and infiltration. Cracks and separation of joints are detrimental to the culvert only insofar as they increase the possibility of infiltration and exfiltration. Infiltration is the inflow of water and the accompanying fines during times of high groundwater when the flow in the pipe itself is low. If the inspection is made during this time period and infiltration is occurring, it will be evident. If the inspection is made during a period when high groundwater is not present, but infiltration has occurred, there may be evidence of residual fines and silt at the joints. Infiltration can cause the loss of backfill and eventually lead to a failure of the roadway above as shown in Fig. 5.45.

Exfiltration is the outflow of water from the pipe into the surrounding backfill. This may cause piping, a loss of backfill material carried away by the outflowing water. This can create problems both with the roadway above and with the culvert itself, which can lose structural integrity because of the loss of side support. If exfiltration is occurring, it may be observed when the flow is relatively low by inspection of the joints. In addition, there may be some evidence of piping at the outlet end of the culvert, where undermining and the deposition of fines may be present.

Whereas loss of backfill support would be evidenced by excessive deflection in a flexible culvert, rigid culverts will not exhibit this condition. Despite the loss of backfill support, there may be little or no sign of distress in the wall of the culvert.

The walls of concrete pipe should be inspected for longitudinal and transverse cracks and spalls and wearing of the invert. Longitudinal cracks (cracks that run lengthwise down the culvert) are indicative of high flexural stresses in the pipe. As the pipe is loaded, it tends to deflect downward and outward. These deflections cause the inside of the pipe at the crown and invert to be in tension as well as the outside of the pipe at the springlines. If the pipe is subjected to a high load, longitudinal cracks may develop at these locations. Because the pipe is buried, inspection of the longitudinal cracks located at the springline on the outside of the pipe is not possible. However, the longitudinal cracks at the crown and the invert will be evident if they exist. Cracks 0.01 in (0.25 mm) or less in width are considered to be hairline cracks and are of minor importance. Larger cracks should be noted and monitored.

Transverse cracks (cracks extending around the circumference of the pipe) are caused by differential settlement along the length of the pipe. This can be caused by either unsuitable foundation material or poor installation practices. These cracks are usually not structural in nature but can lead to spalling or subsequent corrosion of the reinforcing steel. Figure 5.46 illustrates transverse cracking resulting from improperly prepared bedding.

Invert wear on a reinforced concrete pipe or box culvert will be indicated by rutting of the surface or rust stains on the surface. In the extreme case, there will be exposed reinforcement. All of these conditions lead to a reduction in the structural adequacy of the culvert. Where the reinforcing is exposed, the bond is broken between it and the concrete and the reinforcing is not able to carry the intended stresses.

Unreinforced concrete pipe, whether cast in place or precast, should be inspected for invert wear and cracking. Because the concrete itself must take the flexural stresses, any reduction in thickness due to abrasive wear is of concern. For that reason, if rutting of the invert is evident, an attempt should be made to determine the amount of loss of section. The culvert should be reanalyzed for its structural capacity using this changed section to determine whether or not rehabilitation or replacement is necessary. If longitudinal cracks are present in unreinforced concrete pipe, the modulus of rupture has been met or exceeded and the flexural capacity of the pipe has been reached. As previously mentioned, only those cracks at the crown and the invert may be easily detected.

(See "Culvert Inspection Manual," Report No. FHWA-IP-86-2, Federal Highway Administration.)

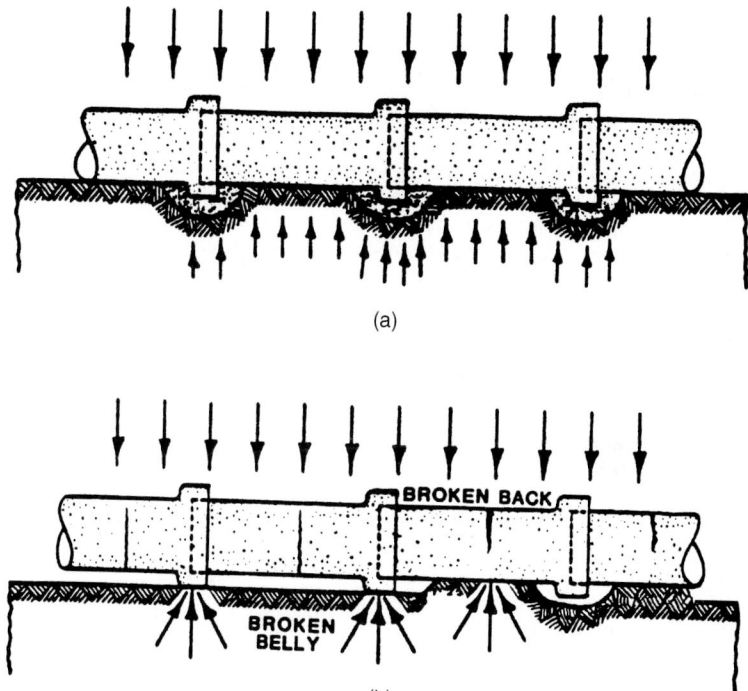

FIGURE 5.46   Illustration of transverse (circumferential) cracking in concrete pipe because of differential settlement. (*a*) Properly prepared bedding evenly distributes loads. (*b*) Improperly prepared bedding results in stress concentrations. (*From "Culvert Inspection Manual," Report No. FHWA-IP-86-2, FHWA, 1986, with permission*)

5.11 REHABILITATION

The appropriate method to be used for culvert rehabilitation depends upon the type and size of the culvert, its condition, and hydraulic and economic considerations.

Hydraulic and economic considerations bear on the issue of repair versus replacement. If the hydraulic capacity of the culvert is in question, or if a rehabilitation method that would reduce its capacity—either by reducing the waterway or by increasing its roughness—is under consideration, a hydraulic analysis would be prudent. In addition, if there will be additional highway construction in the area or if there are plans to widen the roadway in the future, these considerations should be included in the decision of rehabilitation versus replacement.

Pipe replacement is the only method applicable to all pipe types regardless of defects. It is also the most disruptive to the traveling public if done using an open-trench method of construction. As has been previously discussed, jacking or tunneling, at an increased cost, may eliminate this disruption. The advantage is that the hydraulic capacity may be increased and, at the present time, replacement is comparable in cost to relining. However, other methods of rehabilitation will often suffice, as discussed below.

5.11.1 Rehabilitation of Rigid Structures

Rigid culverts with invert wear may be rehabilitated by paving the lower quadrant of the culvert. Where there is no reduction in the structural capacity of the culvert, the invert may be protected from further erosion by placing portland cement concrete or by using shotcrete. Welded wire mesh may be used to strengthen the culvert where it is necessary to do so. For an unreinforced concrete pipe, this will be the case where there is either significant invert wear or longitudinal cracking. Strengthening of a reinforced concrete pipe may be deemed necessary where there is significant longitudinal cracking, invert wear, or spalling. Dowels should be drilled into the member to be repaired, to provide anchorage for the welded wire fabric.

Cracks and spalls may be repaired in rigid culverts by sealing and patching. Spalls may be patched with a mortar- or cement-based material, a procedure that is inexpensive and requires little resource allocation. Cracks may be sealed with either a flexible or a nonflexible sealant. If the crack is continuing to move, and if there will be no loss in the structural capacity of the culvert if it continues to do so (circumferential cracks may be an example), a flexible sealant may be used. If the crack has stabilized, or if additional movement is not acceptable, a nonflexible sealant such as a cement mortar may be appropriate. However, the sealant itself will not prevent additional movement. The underlying cause of the cracking must be discovered and appropriate measures such as pressure grouting applied.

If reinforced concrete pipes separate at the joints and infiltration or exfiltration occurs, not only must the joint be repaired, but the surrounding embankment must be stabilized. The concrete pipe joint may be sealed by the use of an expansion ring gasket and band to prevent further infiltration or exfiltration. Stabilizing the embankment may be accomplished by pressure grouting.

5.11.2 Rehabilitation of Flexible Structures

Flexible metal pipes may need rehabilitation wherever there is a loss of section or where large deflections (greater than 5 percent) are present. Where the culvert has undergone a loss of its structural section due to corrosion or erosion, the amount of loss should be noted and a determination made whether the culvert needs to be strengthened or only protected. The loss of section in metal culverts usually occurs at the invert due to the abrasive conditions of the water flow and/or the corrosive effects of the water. If the loss of section is not significant, it may be adequate to protect the invert with a coating to prevent future erosion or corrosion. The reason for the loss of section should be determined. If the loss is due to corrosion, the application of a bituminous material should provide protection against future corrosion. However, the bituminous coating does not withstand abrasion well. If the loss is due to erosion, paving the lower quadrant with portland cement concrete will be adequate. Either of these methods is applicable as long as there is no significant loss of structural section that would reduce the structural capacity of the culvert. Where the loss of section is considerable, the structural integrity may be maintained by the addition of welded wire fabric to the concrete paving of the invert. The wire mesh may be welded to the invert corrugations of the metal culvert and then the portland cement concrete placed to provide a smooth channel for the water. Figure 5.47 illustrates rehabilitation with concrete paving. Should the culvert have major structural defects, it may be necessary to replace or reline it or place reinforced concrete around the complete periphery.

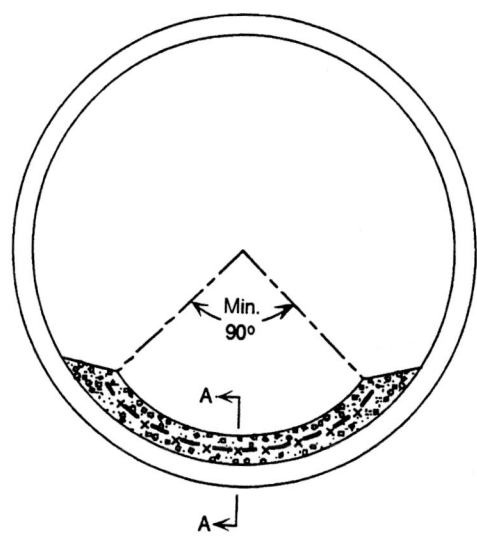

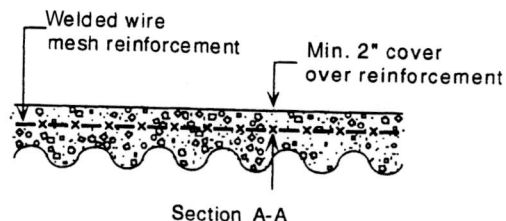

Section A-A

FIGURE 5.47   Example of invert paving of metal culverts
with reinforced concrete. (*From* Highway Design Manual,
California Department of Transportation, with permission)

5.11.3 Relining

Relining of either rigid or flexible culverts may be done either by slip lining or by
installing a flexible liner. Slip lining is merely the insertion of a prefabricated pipe
inside an existing pipe. The most common insertion pipes are either corrugated metal
or plastic. Obviously, the cross-sectional area of the pipe will be reduced. This will
likely affect the hydraulic capacity of the culvert. If a smooth plastic pipe is utilized,
the velocity of the flow may be increased, in part offsetting the reduction in the capacity due to the decreased area. Should this be the case, the downstream end of the culvert should be investigated to prevent additional erosion. Regardless of the type of
pipe inserted, the annular space between the inside of the existing pipe and the outside
of the inserted pipe should be grouted. The final product is comparable to a new culvert, placed with little disruption to the traveling public.

Another method of relining consists of inserting a flexible tube inside the damaged pipe. The flexible tube will generally consist of a resin liner, which, after being inserted inside the subject pipe, is expanded to fit the full cross section of the pipe. It spans irregularities such as joints that may have opened. However, it supplies little additional structural support.

5.11.4 Shotcrete Lining

Welded wire fabric may be used in conjunction with shotcrete to rehabilitate deteriorated portions of either metal or concrete culverts. The welded wire should be anchored to the in-place pipe either through the use of drilled dowels, if the pipe is concrete; or by welding to either the corrugations (of metal pipe) or to previously welded studs. The shotcrete can then be placed by the use of high-pressure hoses. The repair can be designed to restore structural integrity, with little loss of hydraulic capacity.

5.11.5 Grouting Soil Voids

Regardless of the method of repair or rehabilitation chosen, the possible need for grouting potential voids in the soil envelope surrounding the pipe should be addressed. Portland cement grout may be pressure-injected from the interior of the culvert through drilled holes located toward the bottom of the suspected voids. Drilled holes located toward the top of the pipe then allow for the trapped air and water to exit prior to the grout. Grouting of the voids is necessary to complete the structural rehabilitation of the culvert and to reduce the possibility of any future piping of the culvert backfill.

Where conventional grouting will fill the large voids adjacent to the culvert created by infiltration or piping, compaction grouting will density the soil. The equipment, method of injection, and makeup of the grout will all be different from what is required for conventional grouting. Compaction grouting in addition to or in lieu of conventional grouting may be necessary. However, the benefit of knowing that there is a well-compacted, stabilized soil in the vicinity of the culvert may not outweigh the expense involved in the process of compaction grouting.

CHAPTER 6
SAFETY SYSTEMS

Roger L. Brockenbrough, P.E.

R. L. Brockenbrough & Associates, Inc.
Pittsburgh, Pennsylvania

One of the most important and most challenging aspects of highway engineering is designing to enhance life safety. This chapter focuses on roadside safety, which encompasses the safety of vehicles that leave the roadway and shoulder.

This material is based largely on the publication of the American Association of State Highway and Transportation Officials (AASHTO), *Roadside Design Guide* (1996), which was developed by the AASHTO Subcommittee on Design, Task Force for Roadside Safety, under the chairmanship of Wayne F. Cobine. Made up of about 25 highway engineers with diverse experience, the task force prepared a synthesis of current information and operating practices to serve as a comprehensive guide to individuals and agencies in developing standards and policies. Their contribution to promoting highway safety is gratefully acknowledged.

6.1 CONCEPTS AND BENEFITS OF ROADSIDE SAFETY

The roadside is defined as that area beyond the traveled way and shoulder. Thus, roadside safety is concerned with treatments that minimize the likelihood of serious injuries when a vehicle runs off the roadway.

Roadside safety design has received particular emphasis over the last 25 years or so. The increased awareness of its importance, and the development of improved safety concepts and devices, have contributed significantly to improved safety. As shown in Figure 6.1, the traffic fatality rate expressed in terms of driven distance has declined by more than 50 percent since the mid-1960s. Many factors have contributed to the declining rate, including safer vehicles (occupant restraints, door beams, crash energy management, etc.) and improved roadways (intersection geometry, superelevation, grade separation, etc.). However, roadside improvements have played a key role in reducing fatalities.

The safety effort has definitely been worthwhile, but most of this effort has been directed at new construction. Now, many highway projects constructed before the era of safety emphasis are becoming candidates for major reconstruction. The opportunity therefore presents itself to incorporate cost-effective roadside safety concepts and features in this reconstruction to the benefit of all concerned.

Roadside safety must be addressed because a significant number of vehicles inevitably leave the roadway. There are a variety of reasons for this, such as:

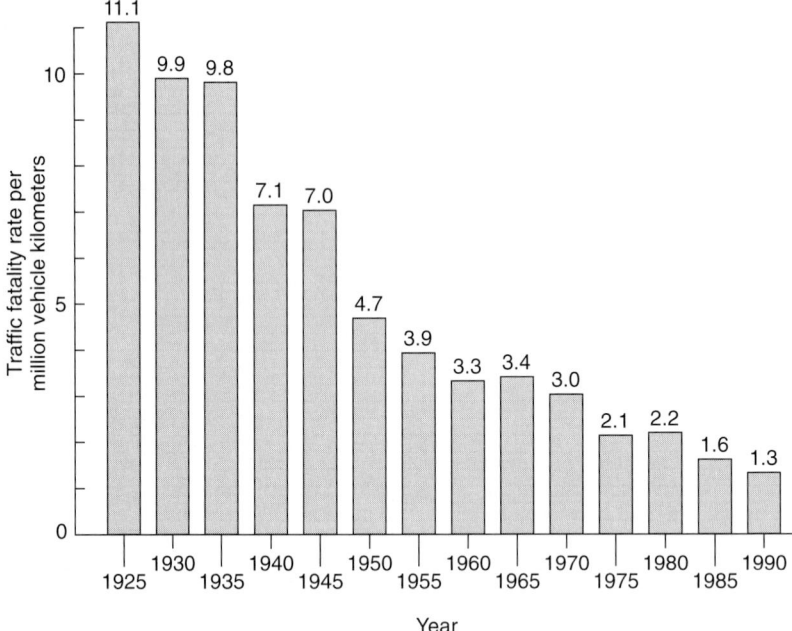

FIGURE 6.1   Traffic fatality rate continues to decline. (*From* Roadside Design Guide, *AASHTO, Washington, D.C., 1996, with permission*)

- Driver fatigue or inattention
- Excessive speed
- Driving under the influence of alcohol or drugs
- Collision avoidance
- Roadway condition (ice, snow, rain)
- Poor visibility
- Mechanical failure

To reduce the severity of accidents involving these errant vehicles, the roadside should have relatively flat slopes and be free of fixed objects. What is known as the *forgiving roadside concept* has generally become an integral part of highway design criteria.

   Obstacles most often responsible for roadside fatalities include:

- Trees and shrubs
- Utility poles
- Culverts and ditches
- Curbs and walls
- Sign and luminaire supports
- Bridge piers and abutments

Design options often employed for addressing a roadside obstacle include:

- Removing the obstacle
- Redesigning the obstacle so it can be safely traversed
- Relocating the obstacle
- Using breakaway devices
- Shielding the obstacle with a barrier or crash cushion
- Delineating the obstacle

As with virtually all highway construction, funds for safety improvements are limited, and thus, emphasis must be given to improvements that are cost-effective and offer the greatest opportunities for safety enhancement. Some features such as breakaway supports and bridge railings are routinely included on the basis of a subjective analysis of obvious benefits. In other cases, where alternatives exist, benefit-cost and value engineering studies should be used to aid in rational decisions. Benefits include expected reduction in accident costs, including the cost of personal injuries and property damage, based on the expected reduction in number and severity of accidents associated with the improvement. Costs include direct construction cost and maintenance. The study must be based on a specific project life so that benefits and costs can be annualized. This involves the application of discount rates and life cycle costs as discussed in Chap. 10. The computer program ROADSIDE is available to aid in the selection process.

6.2 APPLICATION OF CLEAR ZONE CONCEPT TO SLOPE AND DRAINAGE DESIGN

The clear roadside concept has a direct and obvious application to the selection of slopes and design of drainage features such as ditches, curbs, culverts, and drop inlets. A traversable, unobstructed roadside zone should extend beyond the edge of the driving lane for an appropriate distance so that the motorist can generally stop or slow the vehicle and return to the roadway safely.

The width of the zone depends on the traffic volume, the design speed, and the roadside slope. Vehicles on high-volume, high-speed routes obviously require more room to recover than those on less congested routes. A suggested guide for determining the width of the clear zone is presented in Fig. 6.2. The clear zone distance (width) is given in terms of the range of design average daily traffic (ADT) or vehicles per day (VPD), the design speed, and the roadside slope. Enter the chart from the left with the slope, intersect the appropriate design speed curve, and project down to the appropriate scale at the bottom to read the suggested width. The width should be used as a guide and may be adjusted for site-specific conditions and practicality. The AASHTO guide gives modification factors (1.1 to 1.5) that can be applied to increase the clear distance on horizontal curves where accident histories or site investigations show a need. Increased superelevation may be another option, depending on climatic conditions.

6.2.1 Roadside Geometry

Except for flat roadsides, a motorist leaving the roadway may encounter a negative grade (such as on an embankment), a positive grade (such as in a cut section), or a change from negative to positive grade (such as for a roadside channel).

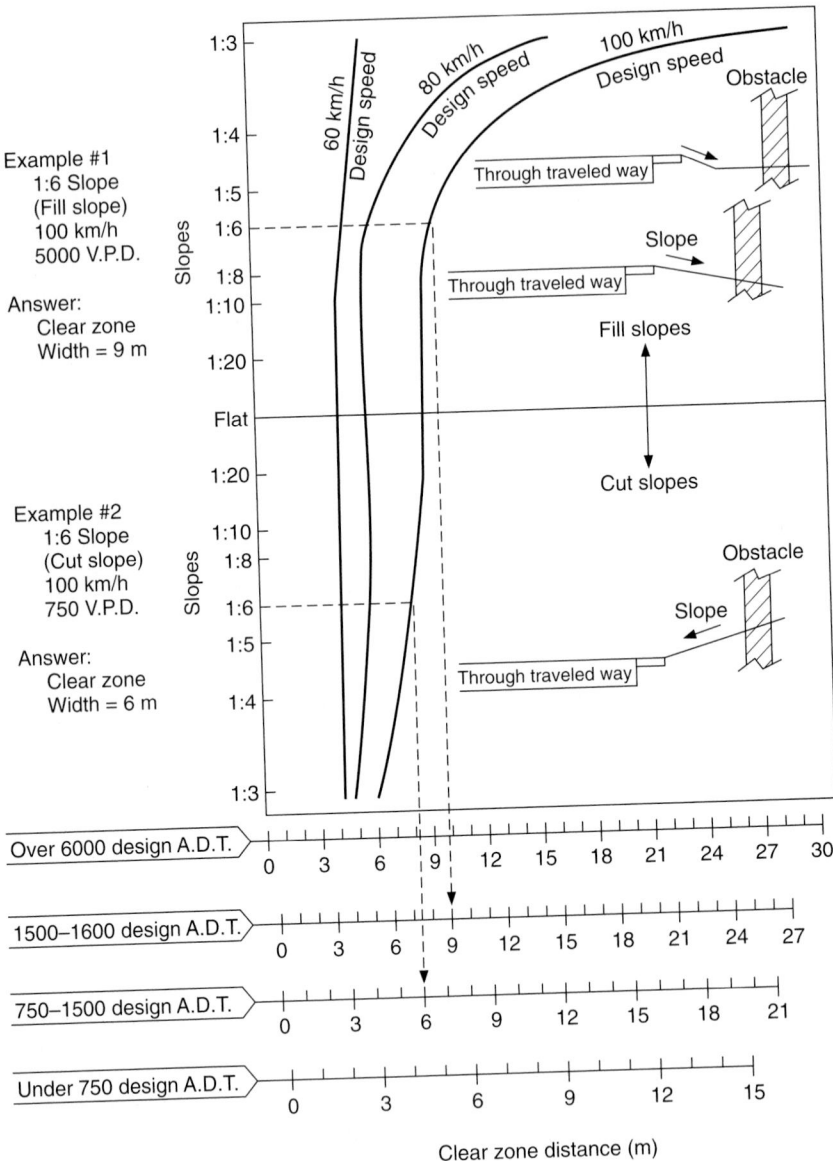

FIGURE 6.2 Clear zone distance curves. (*From* Roadside Design Guide, *AASHTO, Washington, D.C., 1996, with permission*)

Parallel embankment slopes, those oriented with the traffic flow, may be categorized as recoverable, nonrecoverable, or critical. Recoverable slopes are 1:4 or flatter, and the clear zone distance from Fig. 6.2 applies directly. Fixed obstacles such as culvert headwalls should not extend above the embankment in this zone. Nonrecoverable slopes, generally between 1:4 and 1:3, are traversable, but most motorists will reach the bottom of the slope and not be able to stop or return to the roadway easily. Fixed obstacles should not be constructed along such slopes, and a clear runout area at the bottom of the slope is desirable. Critical slopes, generally steeper than 1:3, are those on which a vehicle is likely to overturn. A barrier might be warranted in such cases. Figure 6.3 discusses alternatives that might be considered on critical parallel slopes.

Many highway agencies construct so-called barn roof sections in embankment conditions as illustrated in Fig. 6.4. A relatively flat slope is provided adjacent to the roadway, followed by a steeper slope and a clear runout area at the bottom. This is more economical than a continuous flat slope and apparently safer than a continuous steeper slope from the edge of the shoulder. In applying the clear zone concept, side slopes ranging from flat to 1:4 may be averaged to produce a composite clear zone distance. Slopes that change from negative to positive should be treated as channel sections. Changes in slope and toes of slopes should generally be rounded to keep vehicles in contact with the ground and enhance traversability.

In a cut section, the traversability of the backslope depends on its relative smoothness and the presence of fixed obstacles. If traversable (1:3 slope or flatter) and obstacle free, it may be acceptable. Conversely, a steep rough-sided rock cut (one that will cause excessive vehicle snagging) should be shielded unless it is outside the clear zone.

Embankment cross slopes may be created by median crossovers, intersecting side roads, or driveways. These generally create a more serious condition than parallel slopes because they can be struck head on by errant vehicles. To minimize the effect, slopes of 1:10 or flatter are desirable where practical, and a maximum of 1:6 for high-speed, high-volume routes. Steeper slopes may be suitable for low-speed facilities. Drainage pipes should be located as far from the roadway as practical. Also, where a vehicle could be led into the culvert inlet or outlet by a drainage channel, consideration should be given to special inlet or outlet treatment, as subsequently discussed.

Roadside channels are open flow areas generally paralleling the highway embankment within the right-of-way. They serve to collect surface runoff that drains from the highway and convey it to outlets. In addition to providing drainage functions, channels should be proportioned so that they are traversable. The shaded areas in Figs. 6.5 and 6.6 show preferred (traversable) slopes for the sides of channels. Where practical, channel sections outside the shaded areas may be reshaped, converted to a closed system (culvert), or shielded by a barrier. For all channels, roadside hardware (for example, sign supports) should not be located in or near channel bottoms or slopes because vehicles leaving the roadway may be funneled along the channel and impact the obstacle. Breakaway hardware may not function properly if impacted by airborne or sideways-sliding vehicles.

6.2.2 Drainage Features

The drainage system should be designed, constructed, and maintained with considerations for both the hydraulic function and roadside safety. (See Chap. 5.) In addition to channels, elements of the system include curbs, cross-drainage (transverse) structures (pipes and culverts), parallel drainage structures, and drop inlets. The following three options, listed in order of preference, are applicable to each:

Example 1

Design ADT: 12000
Design speed: 110 km/h
Recommended clear zone distance for 1:6 slope: 9–10.5 meters

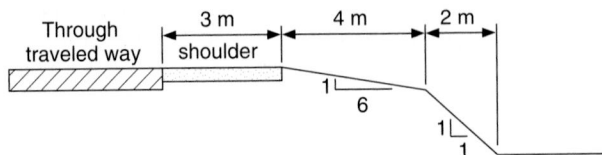

Discussion: Since the critical slope is only 7 meters from the traveled way, instead of the suggested 9–10.5 meters, it should be flattened if practical or considered for shielding. However, if this is an isolated obstacle and the roadway has no significant accident history, it may be appropriate to do little more than delineate the drop-off in lieu of slope flattening or shielding.

Example 2

Design ADT: 350
Design speed: 60 km/h
Recommended clear zone distance for 1:5 slope: 2–3 meters

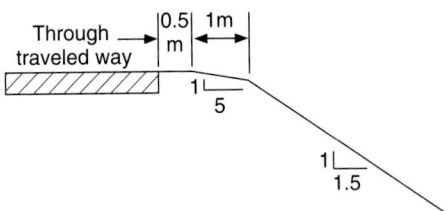

Discussion: The available 1.5 meters is 0.5 to 1.5 meters less than the recommended recovery area. If much of this roadway has a similar cross section and no significant run-off-the-road accident history, neither slope flattening nor a traffic barrier would be recommended. On the other hand, even if the 1:5 slope was 3 meters wide and the clear zone requirement was met, a traffic barrier might be appropriate if this location had noticeably less recovery area than the rest of the roadway and the embankment was unusually high.

FIGURE 6.3 Examples of application of clear zone concept to critical parallel embankment slopes on (Example 1) high-volume and (Example 2) low-volume Washington, D.C., highways. (*From* Roadside Design Guide, *AASHTO, 1996, with permission*)

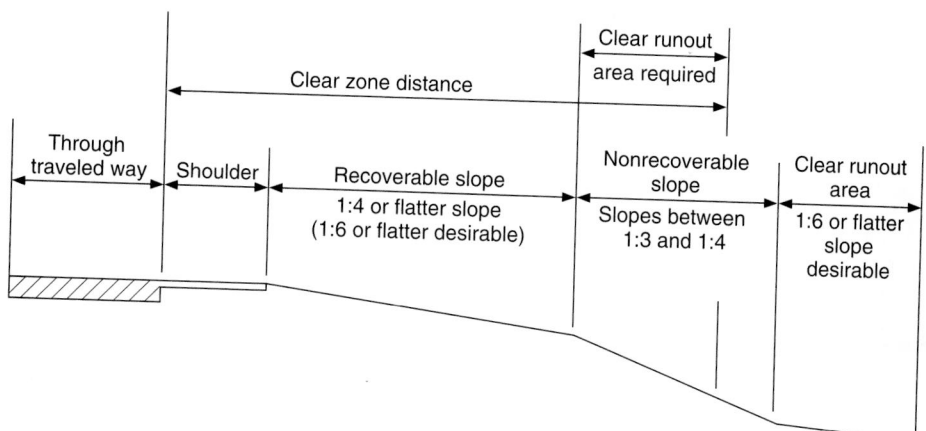

FIGURE 6.4 Example of "barn roof" section for parallel embankment slope design; recoverable slope is followed by a nonrecoverable slope. (*From* Roadside Design Guide, *AASHTO, Washington, D.C., 1996, with permission*)

- Eliminate nonessential drainage structures.
- Design or modify drainage structures so they are traversable or present minimal hazard to errant vehicles.
- If relocation or redesign is impractical, shield with a traffic barrier if in a vulnerable location.

Curbs may be classified as barrier or mountable types. Barrier curbs are nearly vertical, 150 mm (6 in) or more in height, and discourage motorists from leaving the highway. Mountable curbs have lower heights with sloping faces that can be easily traversed. Heights of 100 mm (4 in) or less are preferred for the latter to avoid dragging the underside of vehicles. Neither type of curb is desirable on high-speed highways, because either may cause overturning, particularly if the vehicle is spinning or slipping. In urban areas, a minimum horizontal clearance of 0.5 m beyond the face of the curb should be provided to obstacles. On high-speed roadways, curbs should not be used in front of traffic barriers, because unpredictable postimpact trajectories can result. If a curb must be used, locate it flush with the face of the railing or behind it. Curb-barrier combinations for bridge railings should be crash-tested unless data are available.

Cross-drainage structures carry streams or drainage water transversely underneath the embankment. They may range in size from 450 mm (18 in) to 6 m or more, may be constructed of concrete, metal, or plastic (in some sizes), and may be furnished as round pipe, elliptical shapes, or boxes. Typically, inlets and outlets of larger structures have concrete headwalls and wingwalls, and those of smaller structures are beveled to match the slope. Pipe may also be furnished with square-cut ends. These designs may result in a fixed object protruding above an embankment or an opening into which a vehicle can drop, causing an abrupt stop. Options available to minimize such obstacles include:

- A traversable design
- Extension of the structure so it is less likely to be hit
- Shielding the structure
- Delineating the structure (when other measures are not feasible)

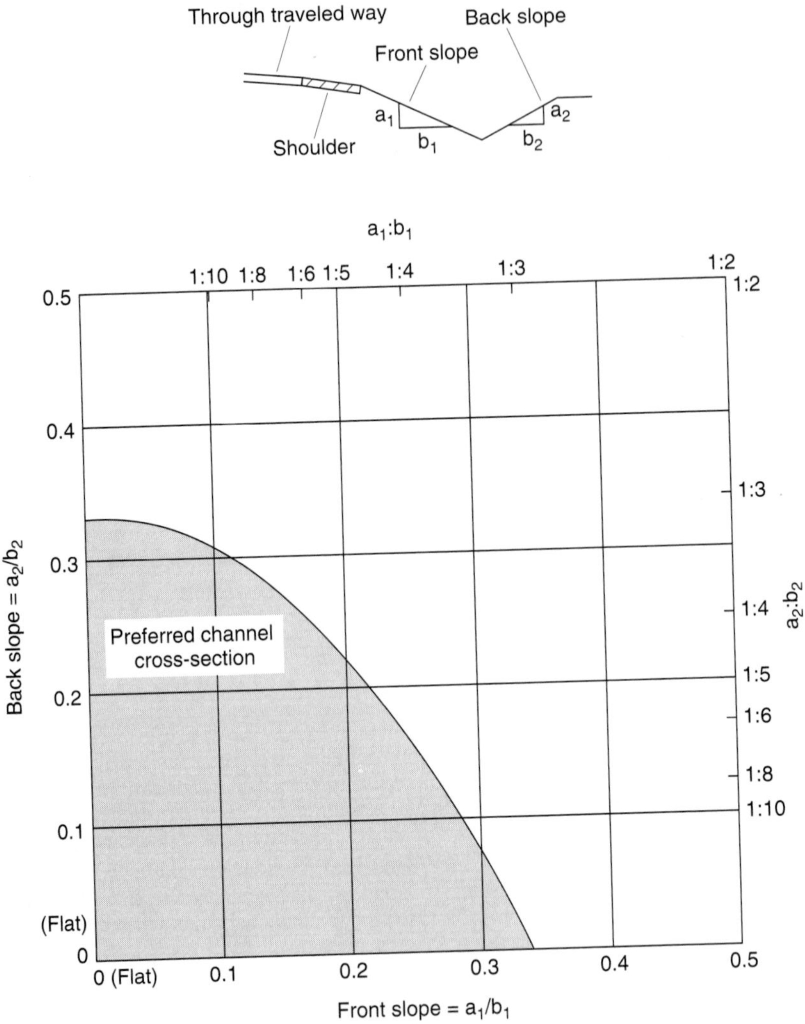

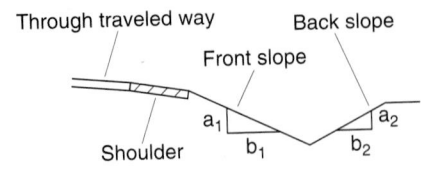

$a_1{:}b_1$

FIGURE 6.5   Preferred cross sections for channels with abrupt slope changes. Chart is applicable to all vee ditches, rounded channels with a bottom width less than 2.4 m, and trapezoidal channels with bottom widths less than 1.2 m. (*From* Roadside Design Guide, *AASHTO, Washington, D.C., 1996, with permission*)

Traversable design.   If a slope is generally traversable, the preferred treatment is to extend or shorten the structure and match the inlet or outlet shape to the embankment slope. Further treatment should not be required for small culverts, defined as a single pipe with a diameter of 1000 mm or less or multiple pipes with diameters of 750 mm or less each. Single structures and end treatments wider than 1000 mm can be made traversable for passenger-size vehicles by using bar grates or pipes to reduce the clear opening width. To maintain hydraulic efficiency, it may be necessary to apply bar grates to flared wingwalls, flared end sections, or culvert extensions larger than

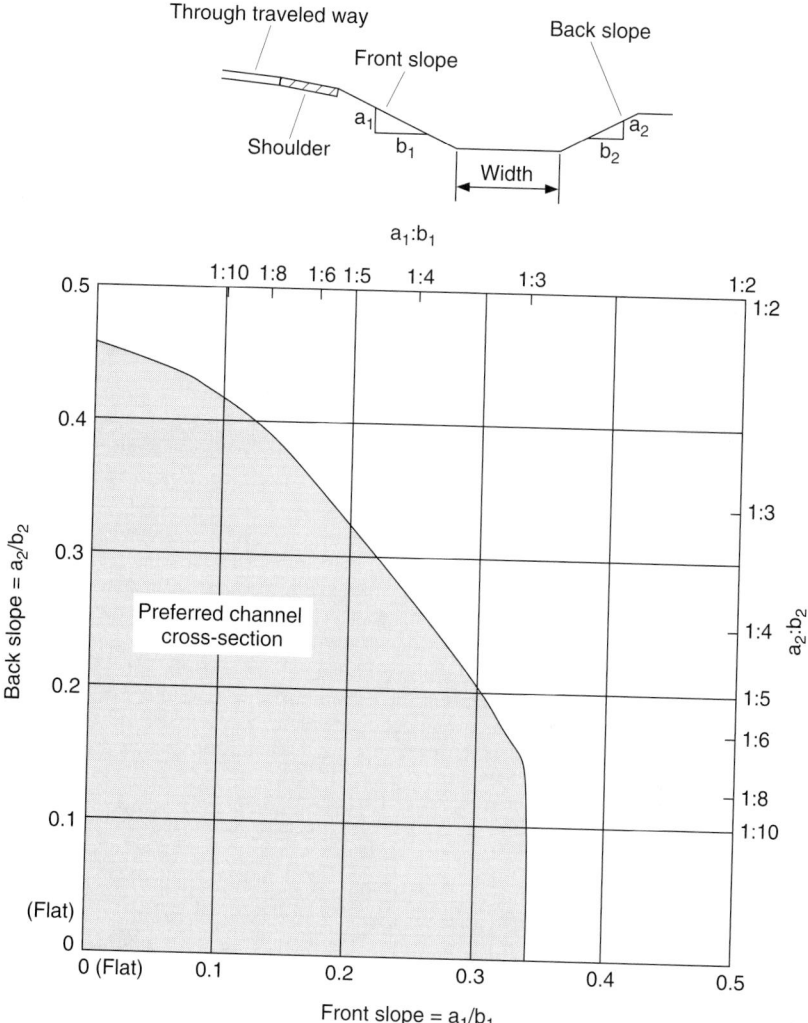

FIGURE 6.6 Preferred cross sections for channels with gradual slope changes. Chart is applicable to rounded channels with bottom widths of 2.4 m or more and trapezoidal channels with bottom widths equal to or greater than 1.2 m. (*From* Roadside Design Guide, *AASHTO, Washington, D.C., 1996, with permission*)

the main barrel. Crash tests have shown that automobiles can cross grated culvert end sections on slopes as steep as 1:3, at speeds from 30 to 100 km/h (20 to 60 mi/h), when steel pipe on 750-mm centers is used. This spacing does not significantly affect flow unless debris accumulates and causes clogging.

Design recommendations for safety treatment of culvert ends are summarized in Fig. 6.7. Where debris accumulation is not a concern and mowing operations are frequent, smaller openings may be tolerated and grates similar to those for drop inlets may be appropriate. In median areas, consider making culverts continuous and adding

Span length (m)	Inside diameter (mm)
Up to 3.65	75
3.65 to 4.90	87
4.90 to 6.10	100
6.10 or less with center support	75

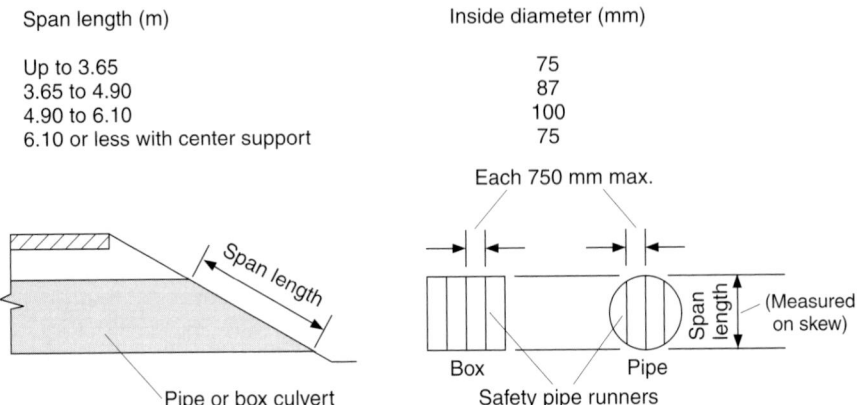

FIGURE 6.7   Safety treatment for ends of culverts showing diameters of grating (schedule 40 steel pipe) for various span lengths. (*From* Roadside Design Guide, *AASHTO, Washington, D.C., 1996, with permission*)

a median drainage inlet. Figure 6.8 shows a design that can be used where cover is insufficient to construct a drop inlet over the pipe.

Extension of structure.   For larger-sized drainage structures with inlets or outlets that cannot be readily made traversable, the structure can be extended so the obstacle is located at the edge or beyond the clear zone. This reduces but does not eliminate the possibility of hitting the pipe end. If the culvert headwall remains the only fixed object at the edge of the zone, simply extending the opening to the edge may not be the best solution. However, if there are numerous obstacles along the edge of the zone on the section under consideration, extension of the pipe might be an appropriate solution.

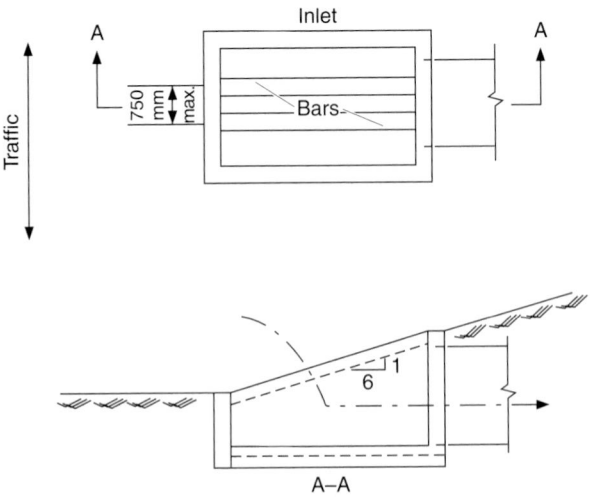

FIGURE 6.8   Safety treatment for drainage inlet where cover is insufficient for drop inlet. (*From* Roadside Design Guide, *AASHTO, Washington, D.C., 1996, with permission*)

Shielding of structure. An appropriate traffic barrier should be considered for shielding a drainage structure that cannot be reasonably made traversable or extended outside the clear zone. Because the barrier will be closer to the roadway and longer than the obstacle, it is more likely to be hit. However, if properly designed, constructed, and maintained, the barrier should provide an increased level of safety.

Parallel drainage structures are those that are oriented parallel to the traffic flow to carry water under driveways, entrances, ramps, side roads, and median crossovers. Such structures may represent a significant hazard if they can be struck head on by an errant vehicle. Options for safety treatment are similar to those for cross-drainage structures. If entrances are closely spaced, consider converting the open channel into a closed storm drain, backfilling areas between entrances, and eliminating multiple obstacles. Research has shown that, for parallel drainage structures, wheel snagging can be significantly reduced with pipe grates oriented perpendicular to the traffic direction and having a spacing of 600 mm or less. Single pipes of 600 mm diameter or less generally do not require a grate, but multiple small pipes may require one. In situations such as intersecting ramps, consider relocating the culvert farther back from the main road, out of the clear zone.

Drop inlets include on-roadway and off-roadway structures. On-roadway inlets, which are located along the shoulder to intercept surface runoff, include curb opening inlets, grated inlets, and slotted drains. If installed flush with the pavement, they do not cause a significant safety problem. Off-roadway drop inlets are used in medians of divided roadways or in roadside ditches. The hazard can be minimized by making the inlets flush with the drainage surface. Safety treatment should be such as to prevent a vehicle from dropping into the inlet, snagging, and losing control.

6.3 SIGN AND LUMINAIRE SUPPORTS AND SIMILAR FEATURES

Approximately 25 percent of all fixed-object fatalities involve sign and luminaire supports or utility poles. The options available to the highway engineer to improve on this record are similar to those presented earlier: remove or redesign, relocate, use a breakaway device, shield, or delineate. Although it is desirable to have an unobstructed roadside, it is not always possible to relocate appurtenances such as signing and lighting supports, because they must remain near the roadway to fulfill their intended purpose. Thus, emphasis is given to the use of breakaway hardware—selection of the most appropriate device and installing it to ensure acceptable performance. (See Chap. 7.)

Breakaway supports include all types of sign, luminaire, and traffic signal supports designed to yield when hit by a vehicle. Typical release mechanisms include slip planes, plastic hinges, and fracture elements. Criteria for breakaway supports are given by AASHTO in *Standard Specifications for Structural Supports for Highway Signs, Luminaires, and Traffic Signals.* The criteria require that a breakaway support fail in a predictable manner when struck head on by an 820-kg vehicle, or its equivalent, at speeds of 35 and 100 km/h. It is desirable to limit the occupant impact velocity to 3.0 m/s, but values as high as 5.0 m/s are acceptable. Also, the maximum stub height is set at 100 mm to avoid snagging the undercarriage after impact. The crash vehicle must remain upright with no significant deformation or intrusion of the passenger compartment.

Full-scale crash tests, tests with bogie vehicles (reusable, adjustable surrogate vehicle), and tests with pendulums (having special nose sections to model vehicles) are used for acceptance. Pendulum tests are the least expensive, but are used mostly for luminaire support hardware and are limited to 35 km/h. National Cooperative Highway Research Program (NCHRP) Report 350 discusses acceptance testing. Tests are run in a standard soil, but weak soil should be used in addition for any feature whose impact performance is sensitive to soil-structure interaction.

Many general practices are similar to those previously discussed. Supports should not be placed in drainage ditches, because vehicles may be channeled into the obstacle and freezing might interfere with proper functioning of the breakaway device. Also, breakaway supports must not be located near ditches or on steep slopes where a vehicle is likely to be partially airborne at impact, because breakaway devices may bind and not function properly when hit in this manner. They have been developed to be struck about 500 mm above the ground.

Locate supports where they are least likely to be hit, such as behind roadway barriers (beyond design deflections of the barriers) or on existing structures. In general, only when the use of breakaway supports is not feasible should a traffic barrier or crash cushion be used for shielding. Breakaway supports should be considered when design speeds are moderate to high. An occupant in a vehicle striking a fixed object at only 40 km/h can sustain serious injuries. However, in areas where pedestrians and bicyclists may be struck by falling supports after a crash, breakaway supports are not usually considered.

6.3.1 Sign Supports

Roadway signs include overhead signs, large roadside signs (area over 5 m^2), and small roadside signs.

Overhead signs include sign bridges and cantilevered signs. Their supports are generally too large to adapt to a breakaway design. When possible, install overhead signs on existing bridges or other structures. Otherwise, supports within the clear zone should be shielded with a traffic barrier.

Large roadside signs typically have two or more supports, each of which is of the breakaway type. Figures 6.9 and 6.10 show the loading conditions and the breakaway features. Note the hinge joint with fuse plate just below the sign and the breakaway base (shear plate). The supports must resist ice and wind loads and also meet the following criteria:

- The hinge must be at least 2100 mm above ground to prevent windshield penetration.
- A single post 2100 mm or more from another post should have a mass less than 65 kg/m; total mass below the hinge but above the shear plate should not exceed 270 kg. Two posts spaced less than 2100 mm apart should have a mass less than 25 kg/m.
- Supplementary signs should generally not be placed below the hinges.

The base joint activates when two parallel plates slide apart as bolts are pushed out under impact. As shown in Fig. 6.11, the designs may be of the unidirectional or multidirectional type. The upper hinge design includes a saw cut through the front flange and web of the plate, and a fuse plate on the front flange (impact side). The fuse plate has slotted bolt holes, and the bolts must be torqued to specified values for proper functioning. Alternatively, the fuse plate may have a line of open holes at the cut line, with the plate designed to rupture at the required load, negating the need for the specific values of bolt torque. Even with the breakaway design feature, it is good practice to locate large signs outside the clear zone where feasible.

Small roadside signs may be driven directly into the soil, set in drilled earth holes, or mounted on a base. U-shaped steel posts driven into the ground can generally bend

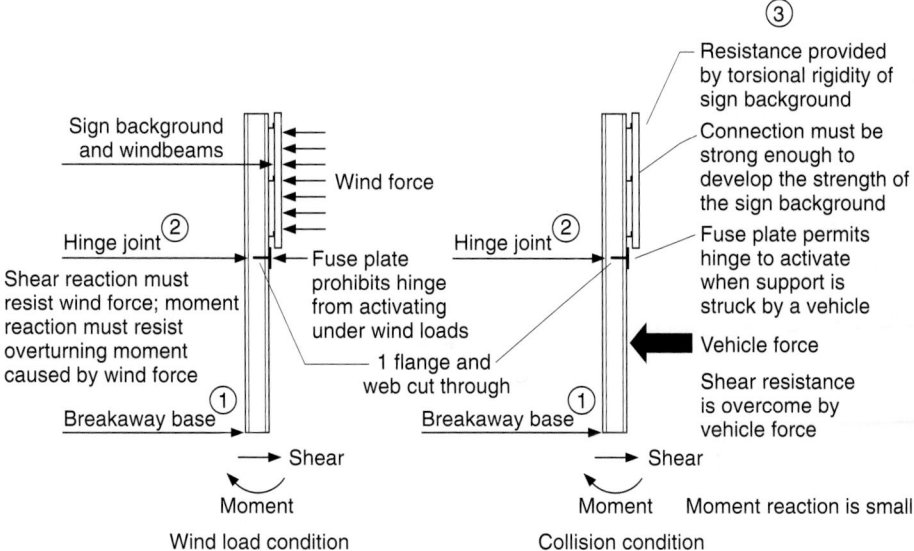

FIGURE 6.9 Wind and impact loads on large roadside sign. (*From* Roadside Design Guide, *AASHTO, Washington, D.C., 1996, with permission*)

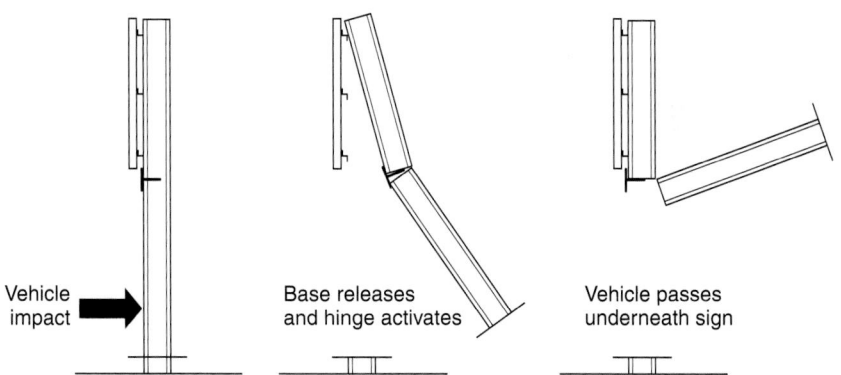

FIGURE 6.10 Impact performance of large roadside sign. (*From* Roadside Design Guide, *AASHTO, Washington, D.C., 1996, with permission*)

and yield at the base without special devices. Splicing the posts is not usually recommended, because performance is not predictable. Wood posts set in drilled holes can fracture at the base, as well as steel pipes connected to anchors driven into the ground. Also, small sign supports may be mounted on slip bases of the unidirectional or multidirectional type. A typical unidirectional design uses a four-bolt slotted slip base, inclined in the direction of traffic by 10 to 20°. This angle allows the sign to move up so the impacting vehicle can pass underneath. A hinge in the top of the post is not needed. Multidirectional bases are usually triangular and release when struck in any

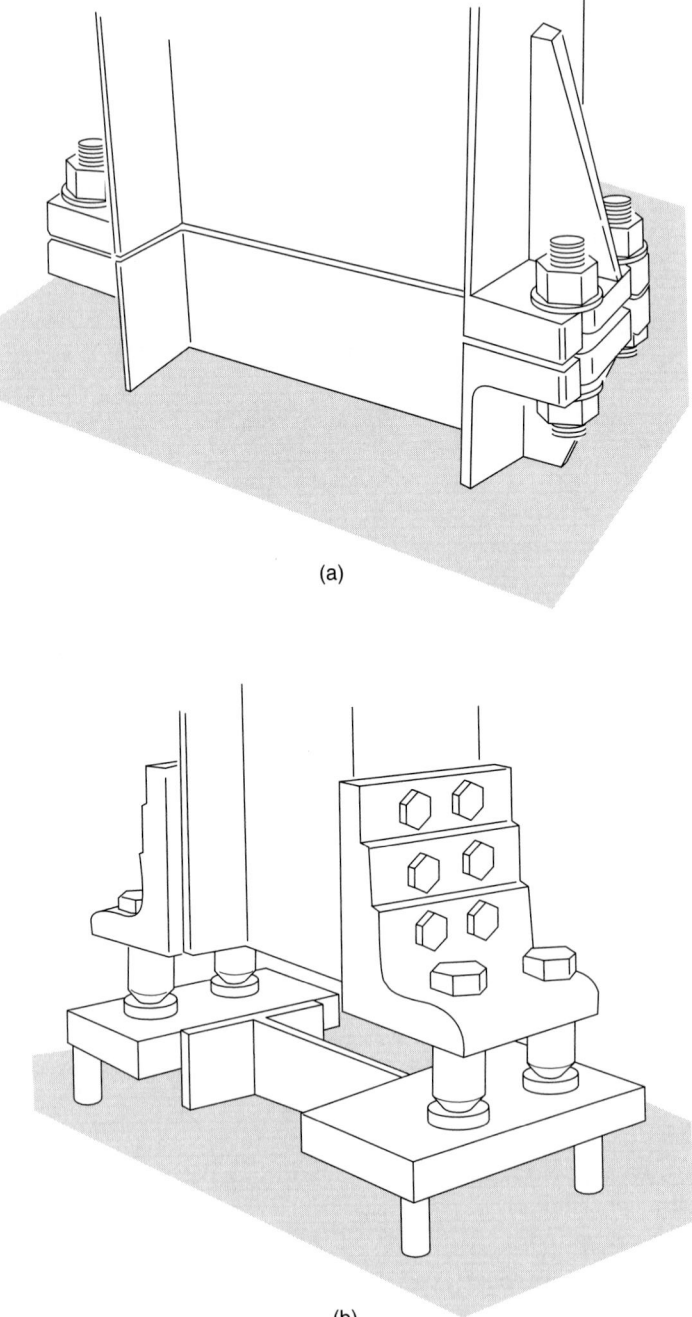

(a)

(b)

FIGURE 6.11 Breakaway bases for large overhead signs may provide for (*a*) unidirectional or (*b*) multidirectional impacts. (*From* Roadside Design Guide, *AASHTO, Washington, D.C., 1996, with permission*)

direction. They are often used in medians and at ends of ramps and similar locations. Because torque requirements for slip base bolts are low, wind vibrations have caused supports to "walk" from the slots under wind vibrations, but this can be prevented by using a sheet metal keeper plate. Overtorquing must be prevented, because this causes high friction between the slip base elements and prevents the support from releasing as intended.

6.3.2 Luminaire Supports

Breakaway supports for luminaires are usually a cast-aluminum transformer-type frangible base, a slip base, or frangible couplers. These devices have been developed to activate when loaded in shear by impacts at a bumper height of about 500 mm. If the supports are located so that they may be impacted at a greater height, the performance may not be as intended. Thus, negative side slopes between the roadway and the support should be limited to 1:6 or flatter. The mast arm of a falling support will usually rotate away from the roadway. However, the danger of falling poles striking pedestrians, bicyclists, and other motorists should be considered.

Breakaway supports are suitable for poles that do not exceed 17 m in height and 450 kg in mass. Foundations must be designed for the surrounding soils to prevent the foundation from pushing through the soil. A preferred method for lighting major intersections is to use high-mast lighting, because fewer supports are required and they can be located farther from the roadway. Supports located in the clear zone should be protected with a suitable traffic barrier.

6.3.3 Supports for Traffic Signals and Service Devices

Supports for traffic signals are not usually of the breakaway type, because of the potential consequences of the loss of the signal at an intersection. Supports in the clear zone should be shielded. Call boxes can often be located behind existing barriers, but a breakaway support is an option. The call box should be securely attached to its support to prevent windshield penetration if dislodged, and the top of the support should be at least 2700 mm above ground to avoid penetration after impact. Warning devices at railroad-highway crossings are usually the responsibility of the railroad, but highway officials should cooperate in proper selection and safety treatments. Fire hydrants have not been tested to current criteria, but at least one breakaway design is available that includes immediate water shutoff after impact. Mailbox supports should be embedded no more than 600 mm in the ground and not set in concrete, the mailboxes should be attached to the supports so that they will not separate after impact, and multiple mailboxes should be spaced apart by a distance of three-fourths of their height.

6.3.4 Supports for Utility Poles

Utility supports represent a serious hazard that accounts for about 10 percent of all fixed-object fatal crashes. Elimination, relocation, and burying the lines are preferred options. Increased spacings or multiple use may reduce the number of poles. A breakaway device has been tested and may be considered for vulnerable locations. A breakaway device for utility pole guy wires has also been developed. As with other obstacles, shielding is also an option.

6.3.5 Trees

Collisions of single vehicles with trees account for nearly 25 percent of fixed-object fatal crashes and result in about 3000 deaths each year. Most of these are along county and township roads, which tend to have narrow recovery zones. Certainly, trees should not be planted in the clear zone for new construction, and mowing should discourage growth of seedlings. For existing situations, the hazard should be evaluated. Generally, a single tree with a trunk diameter of 150 mm or more is considered a fixed object. For small trees close together, calculate an equivalent diameter based on the combined cross-section area. Large trees should be removed where possible. Warning signs and roadway delineators can be used to indicate where extra caution is advised. Roadside barriers should generally be used only where the severity of striking the tree is greater than that of striking the barrier.

6.4 WARRANTS FOR ROADSIDE BARRIERS

Longitudinal roadside barriers are used to shield motorists from natural or human-made obstacles located along either side of the traveled way, and sometimes to protect pedestrians and bicyclists. Median barriers and barrier end treatments are discussed separately in Arts. 6.9 and 6.12.

Barriers must contain and redirect vehicles. Because of the complicated dynamic behavior involved, the most effective way to ensure performance of new designs is through full-scale crash testing. Standard crash tests are presented in NCHRP Report 350, "Recommended Procedures for the Safety Performance Evaluation of Highway Features." To match barrier performance to service needs, a series of six test levels are recommended to evaluate occupant risk, structural integrity, and postimpact vehicle behavior. Various vehicle masses, velocities, and impact angles are included.

Barriers typically go through an experimental phase in which a barrier that has passed crash test evaluation is subjected to an in-service evaluation, and an operational phase in which a barrier that has proven acceptable in the in-service evaluation is used while its performance is further monitored. Barriers are also considered operational if they are used for extended periods and demonstrate satisfactory performance in construction, maintenance, and accident experience.

The criteria by which the need for a safety treatment or improvement can be determined are termed *warrants*. Barrier warrants are based on the premise that traffic barriers should be installed only where they reduce the probability or frequency of potential accidents. Warrants may be based on a subjective analysis of roadside conditions or a benefit-cost study (life cycle cost analysis). The latter can be used to rationally analyze factors such as design velocity and traffic volume in relation to barrier needs and associated costs and accident costs.. Three options may be evaluated:

- Remove or reduce the area of concern so that it does not require shielding.
- Install an appropriate barrier.
- Leave the area unshielded.

The last of these options would usually be cost-effective only where the accident probability is low.

The main uses of roadside barriers are to shield either embankments or obstacles, as discussed below. Barriers may also be used to protect pedestrians, school yards, or bicyclists. There are no firm criteria for these applications, and each must be evaluated on its own merits.

6.4.1   Embankments

As indicated in Fig. 6.12, the main factors considered in determining the need for barriers are the embankment height and the side slope. These criteria are based on studies of the severity of encroachments on embankments as compared with impacts with roadside barriers. The figure does not include the probability of an encroachment or

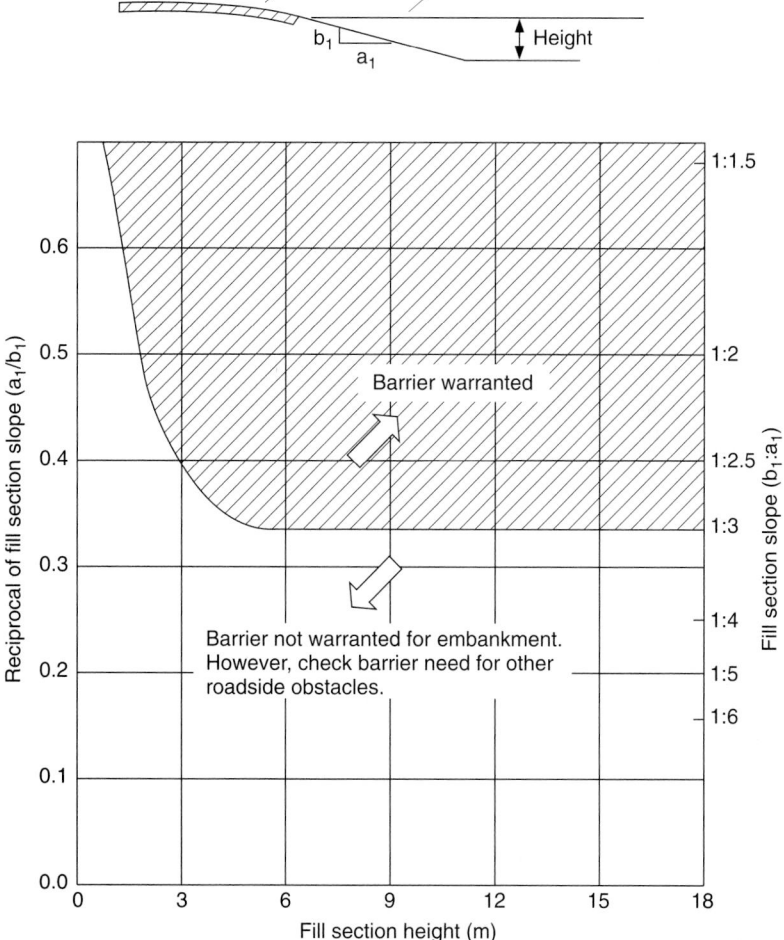

FIGURE 6.12   Embankment warrants based on comparative risk analysis. (*From* Roadside Design Guide, *AASHTO, Washington, D.C., 1996, with permission*)

relative costs. Some states have made their own studies and developed charts having a series of curves for different traffic densities.

6.4.2 Obstacles

Roadside obstacles include nontraversable terrain and fixed objects, either constructed (such as culvert headwalls or structural supports) or natural (such as trees). Such obstacles account for over 30 percent of highway fatalities. The need for a barrier depends on both the nature of the obstacle and the probability that it will be hit. Table 6.1 lists the major types of obstacles and considerations for barrier warrants. Refer to the clear zone chart (Fig. 6.2) as a guide in determining whether the location of an obstacle constitutes a significant threat.

TABLE 6.1 Barrier Warrants for Nontraversable Terrain and Roadside Obstacles*,†

Bridge piers, abutments, and railing ends	Shielding generally required
Boulders	A judgment decision based on nature of fixed object and likelihood of impact
Culverts, pipes, headwalls	A judgment decision based on size, shape, and location of obstacle
Cut slopes (smooth)	Shielding not generally required
Cut slopes (rough)	A judgment decision based on likelihood of impact
Ditches (parallel)	Refer to Figs. 6.5 and 6.6
Ditches (transverse)	Shielding generally required if likelihood of head-on impact is high
Embankment	A judgment decision based on fill height and slope (see Fig. 6.12)
Retaining walls	A judgment decision based on relative smoothness of wall and anticipated maximum angle of impact
Sign and luminaire supports‡	Shielding generally required for nonbreakaway supports
Traffic signal supports§	Isolated traffic signals within clear zone on high-speed rural facilities may warrant shielding
Trees	A judgment decision based on site-specific circumstances
Utility poles	Shielding may be warranted on a case-by-case basis
Permanent bodies of water	A judgment decision based on location and depth of water and likelihood of encroachment

*Shielding nontranversable terrain or a roadside obstacle is usually warranted only when it is within the clear zone and cannot practically or economically be removed, relocated, or made breakaway and it is determined that the barrier provides a safety improvement over the unshielded condition.

†Marginal situations, with respect to placement or omission of a barrier, will usually be decided by accident experience, either at the site or at a comparable site.

‡Where feasible, all sign and luminaire supports should be a breakaway design regardless of their distance from the roadway if there is reasonable likelihood of their being hit by an errant motorist.

§In practice, relatively few traffic signal supports, including flashing light signals and gates used at railroad crossings are shielded. If shielding is deemed necessary, however, crash cushions are sometimes used in lieu of a longitudinal barrier installation.

Source: From *Roadside Design Guide,* AASHTO, Washington, D.C., 1996, with permission.

6.5 CHARACTERISTICS OF ROADSIDE BARRIERS

Depending on their deflection characteristics upon impact, roadside barriers can be classified as flexible, semirigid, or rigid. Table 6.2 lists the most widely used barriers in each classification. Details of these operational barriers are presented along with other available information in Figs. 6.13 through 6.23 (all dimensions are in mm). The dynamic deflection listed is that observed during the standard test defined by NCHRP 230, which involves a 2000-kg car, a 25° impact angle, and a 100 km/h velocity. (Maximum available data are for an 1800-kg vehicle in some cases.) Other characteristics of the barriers are discussed below.

6.5.1 Flexible Systems

The *three-cable system* (Fig. 6.13) is made up of three steel cables mounted on weak posts. Each of the design variations will redirect vehicles in the 820- to 2000-kg range. The design with an S75 × 8.5 steel post has successfully redirected a low-front-profile vehicle and an 1800-kg van. The cable barrier redirects impacting vehicles after the cable deflects and develops tension, with the posts offering little direct resistance. Several states allow a backslope as steep as 1:2 behind the rail. If the barrier is placed on the inside of a curve, additional deflection will occur before tension develops in the cable, and thus it may be desirable to limit the radius. New York installs the barrier having S75 × 8.5 steel posts for radii of 220 m or more with the standard 4.9-m post spacing, and for radii of 135 m or more with a 3.7-m post spacing. Advantages of the three-cable barrier include low initial cost, effective vehicle containment and redirection over a wide range of vehicle sizes and installation conditions, low deceleration forces, and functionality in snow or sand areas because the open design prevents drifting. Disadvantages include the long lengths that are nonfunctional and must be repaired after an impact, the clear area behind the barrier needed to accommodate the design deflection distance, reduced effectiveness on the inside of curves, and sensitivity to correct height installation and maintenance.

The *W-beam (weak-post) system* (Fig. 6.14) behaves much like a cable system, but the deflection is much less. Thus, the required clear area behind the barrier is less. The barrier can redirect vehicles in the 800- to 1800-kg range. Lateral deflection was 2.2 m in an 1800-kg test at 28° and 95 km/h. The barrier has not been crash-tested with a 2000-kg vehicle, but field performance over many years has not indicated a problem. However, in a test with a 2100-kg van at 24° and 95 km/h, the barrier failed to keep the van upright. The system is sensitive to mounting height and irregularities in terrain.

The *thrie-beam (weak-post) system* (Fig. 6.15) is similar to the W-beam system but has a deeper rail element to accommodate a greater range of vehicle sizes and minor terrain irregularities. Field experience is limited, but the system is classified as operational because it is an improvement over an existing operational system. The barrier can redirect vehicles in the 800- to 2000-kg range. To prevent twisting, the thrie beam should be mounted by alternating upper and lower bolts.

6.5.2 Semirigid Systems

The *box-beam (weak-post) system* (Fig. 6.16) achieves its resistance through the combined flexural and tensile resistance of the box beam. Posts near the impact point are designed to break or tear away and distribute the impact force to adjacent posts. The barrier can redirect vehicles in the 800- to 1800-kg range. The barrier has not been

TABLE 6.2 Classification of Roadside Barriers

Flexible systems
  Three-cable
  W-beam (weak post)
  Thrie beam (weak post)

Semirigid systems
  Box beam (weak post)
  Blocked-out W-beam (strong post)
  Blocked-out thrie beam (strong post)
  Modified thrie beam
  Self-restoring barrier (SERB)
  Steel-backed wood rail

Rigid systems
  Concrete safety shape
  Stone masonry wall

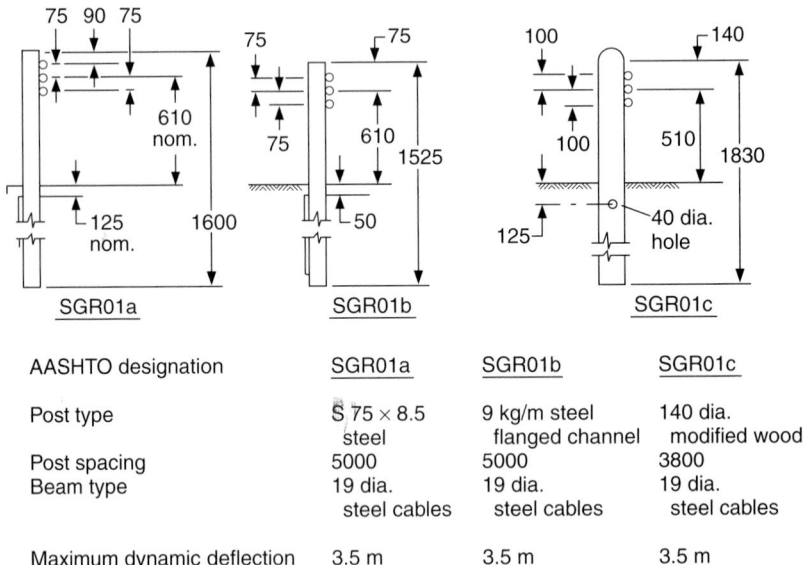

AASHTO designation	SGR01a	SGR01b	SGR01c
Post type	S 75 × 8.5 steel	9 kg/m steel flanged channel	140 dia. modified wood
Post spacing	5000	5000	3800
Beam type	19 dia. steel cables	19 dia. steel cables	19 dia. steel cables
Maximum dynamic deflection	3.5 m	3.5 m	3.5 m

Remarks: For shallow angle impacts, barrier damage is usually limited to several posts, which must be replaced. Cable damage is rare except in severe crashes. A crashworthy end terminal is critical in each of the cable systems, both to provide adequate anchorage to develop full tensile strength in the cable and to minimize vehicle decelerations for impacts on either end of an installation.

FIGURE 6.13 Three-cable roadside barrier. (*From* Roadside Design Guide, *AASHTO, Washington, D.C., 1996, with permission*)

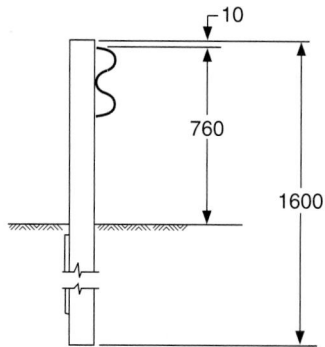

AASHTO designation: SGR02

Post type   S 75 × 8.5 steel
Post spacing   4100
Beam type   2.67 W-section

Maximum dynamic   Approximately 2 m
deflection

FIGURE 6.14   W-beam (weak-post) roadside
barrier. (*From* Roadside Design Guide, *AASHTO,
Washington, D.C., 1996, with permission*)

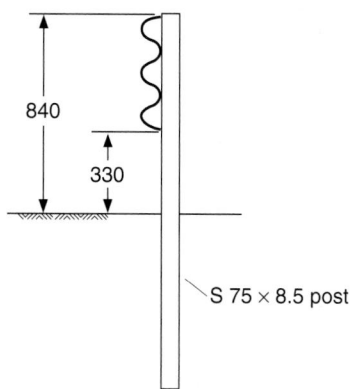

AASHTO designation: None

Post type   S 75 × 8.5 steel
Post spacing   4100
Beam type   3.43 thrie-beam

Maximum dynamic   Approximately 1.2 m
deflection

FIGURE 6.15   Thrie-beam (weak-post) roadside
barrier. (*From* Roadside Design Guide, *AASHTO,
Washington, D.C., 1996, with permission*)

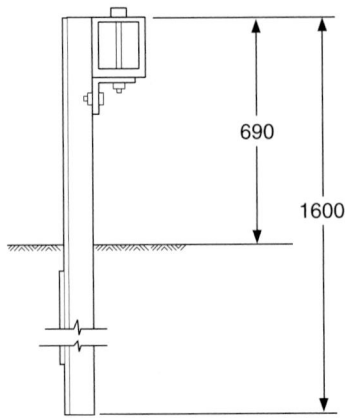

AASHTO designation: SGR03

Post type	S 75 × 8.5 steel
Post spacing	1830
Beam type	152 × 152 × 4.78 steel tube
Maximum dynamic deflection	Approximately 1.5 m

FIGURE 6.16 Box-beam (weak-post) roadside barrier.
(*From* Roadside Design Guide, *AASHTO, Washington, D.C.,*
1996, with permission)

crash-tested with a 2000-kg vehicle, but field performance over many years has not
indicated a problem. However, in a test with a 2100-kg van, the barrier failed to keep
the van upright. The system is sensitive to mounting height and irregularities in terrain.

The *blocked-out W-beam (strong-post) system* (Fig. 6.17) is the most common barrier. The blockout or offset of the rail from the post minimizes vehicle snagging and
reduces likelihood that a vehicle would vault over the barrier. As with all strong-post
systems, resistance is developed by a combination of the tensile and flexural resistance of the rail and the flexural and shear strength of the post. Dynamic deflections
are less than those of flexible systems. Bolt washers on the posts can be eliminated on
this and other strong-post systems; they are not needed for strength, and it is desirable
for the rail to break away as the post rotates downward. The barrier can redirect vehicles in the 800- to 2000-kg range. It has also redirected a 2100-kg van at 21° and 95
km/h. Strong post systems tend to remain functional after moderate collisions, so that
immediate repairs are not necessary.

The *blocked-out thrie-beam (strong-post) system* (Fig. 6.18) is similar to the preceding system, but it has a deeper, stiffer, three-corrugation rail. This makes it less
prone to damage during impact, allows higher rail mounting, and is better able to contain larger vehicles. A top railing height of 900 mm is considered optimum. In crash
tests the barrier has redirected vehicles from 820 to 1990 kg.

The *modified thrie-beam system* (Fig. 6.19) has a triangular notch cut from its web.
This allows the lower part of the beam and the face of the spacer block to bend in during impact, causes the rail face to remain nearly vertical as the post is bent back, and

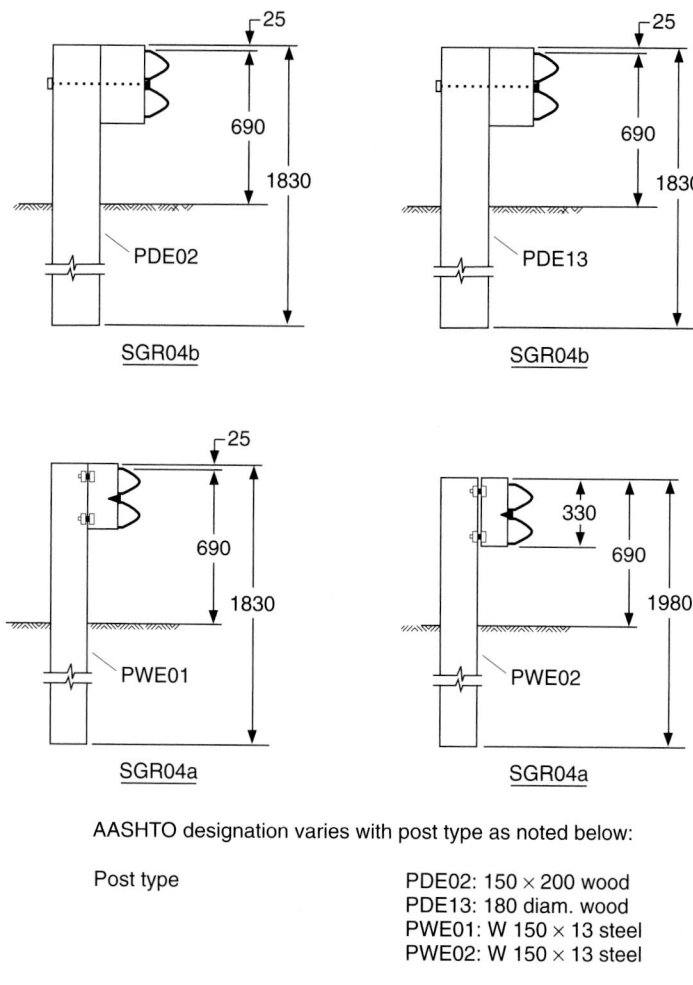

AASHTO designation varies with post type as noted below:

Post type	PDE02: 150 × 200 wood
	PDE13: 180 diam. wood
	PWE01: W 150 × 13 steel
	PWE02: W 150 × 13 steel
Post spacing	1905
Beam type	2.67 W-beam
Maximum dynamic deflection	Approximately 0.9 m

FIGURE 6.17   Blocked-out W-beam (strong-post) roadside barrier. (*From* Roadside Design Guide, *AASHTO, Washington, D.C., 1996, with permission*)

reduces likelihood that a vehicle would roll over the barrier. Also, bolt washers have been eliminated on the posts, as discussed previously. Successful crash tests have been conducted with an 820-kg vehicle, a 9100-kg school bus (15°, 90 km/h), and a 14,500-kg intercity bus (14°, 97 km/h). Repair costs of either thrie-beam system should be considerably less than for other metal barrier systems, because damage tends to be slight in shallow-angle impacts. Also, it is considered easier to install and maintain than a W-beam system with rub rail.

The *self-restoring barrier (SERB) system* (Fig. 6.20) is a high-performance barrier designed to be maintenance-free for most impacts and capable of containing and redi-

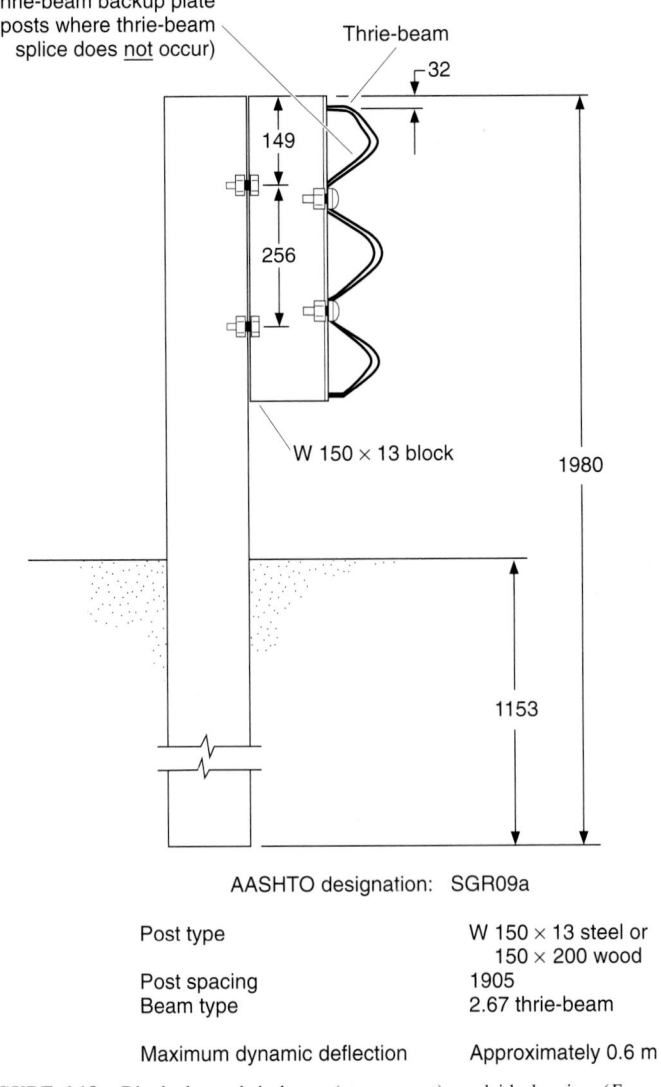

AASHTO designation: SGR09a

Post type	W 150 × 13 steel or 150 × 200 wood
Post spacing	1905
Beam type	2.67 thrie-beam
Maximum dynamic deflection	Approximately 0.6 m

FIGURE 6.18   Blocked-out thrie-beam (strong-post) roadside barrier. (*From Roadside Design Guide, AASHTO, Washington, D.C., 1996, with permission*)

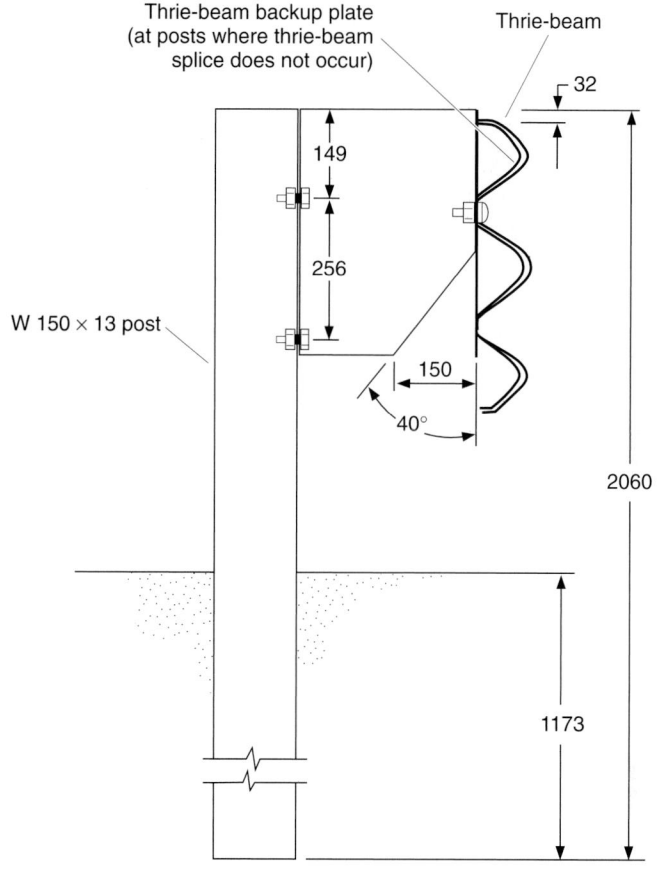

AASHTO designation: SGR09b

Post type	W 150 × 13 steel or 150 × 200 wood
Offset block	M 369 × 26 steel
Post spacing	1905
Beam type	2.67 thrie-beam
Maximum dynamic deflection	Approximately 0.9 m with a 9000-kg school bus (90 km/h, 15°)

Remarks: Modified thrie-beam was first installed in Rhode Island, Colorado, Nebraska, and Michigan as an experimental barrier. Since that time, it has been reclassified as an operational system, requiring virtually no repair for shallow-angle hits. This barrier can accommodate vehicles ranging in size from 800-kg subcompacts to a 15,000-kg intercity bus.

FIGURE 6.19 Modified thrie-beam roadside barrier. (*From* Roadside Design Guide, *AASHTO, Washington, D.C., 1996, with permission*)

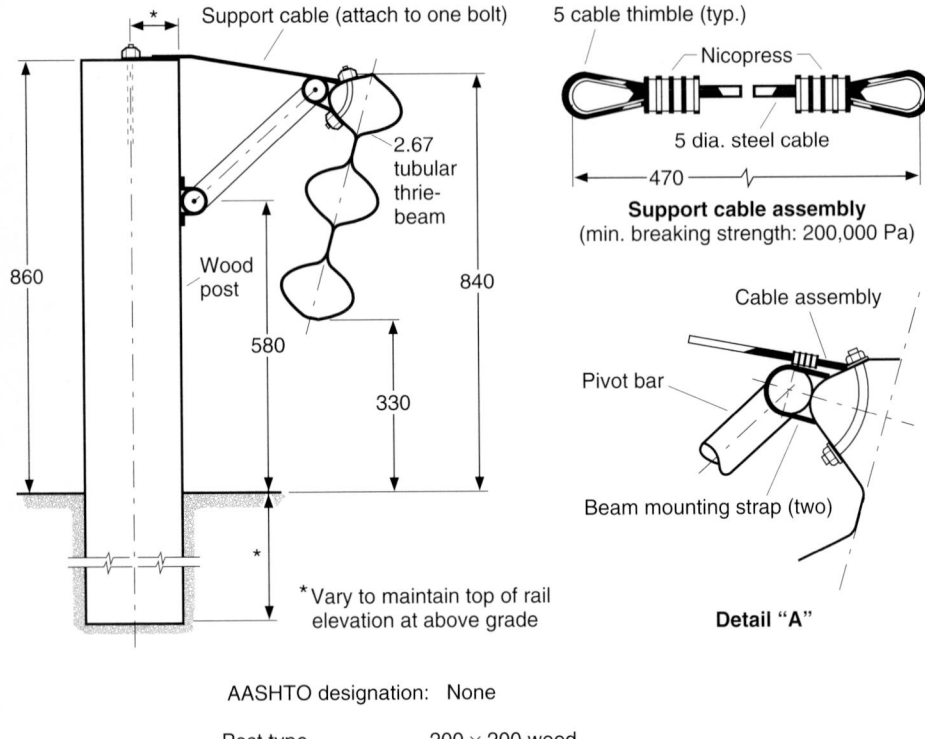

FIGURE 6.20 Self-restoring roadside barrier (SERB). (*From* Roadside Design Guide, *AASHTO, Washington, D.C., 1996, with permission*)

recting larger vehicles. The tubular thrie-beam element is supported at each wood post by a steel pivot bar and a cable. The rail can deflect backward and up during impact and return to its original position after redirecting the vehicle. Successful crash tests have been conducted on a range of vehicles from an 820-kg auto to a 18,000-kg intercity bus (16°, 92 km/h). Field performance indicates very low maintenance costs, but the initial cost is high (about 2 times that of the concrete safety shape). It is particularly suited for sites with high accident frequency.

The *steel-backed wood-rail system* (Fig. 6.21) is an aesthetic alternative to conventional systems, often selected for use along park roads. Successful crash tests have been conducted with an 820-kg vehicle (20°, 81 km/h) and a 2000-kg vehicle (25°, 81 km/h).

6.5.3 Rigid Systems

The *concrete safety shape system* (Fig. 6.22), which has a sloping front face, is similar to the concrete median barrier (Art. 6.9.1) but usually has a vertical back face. The

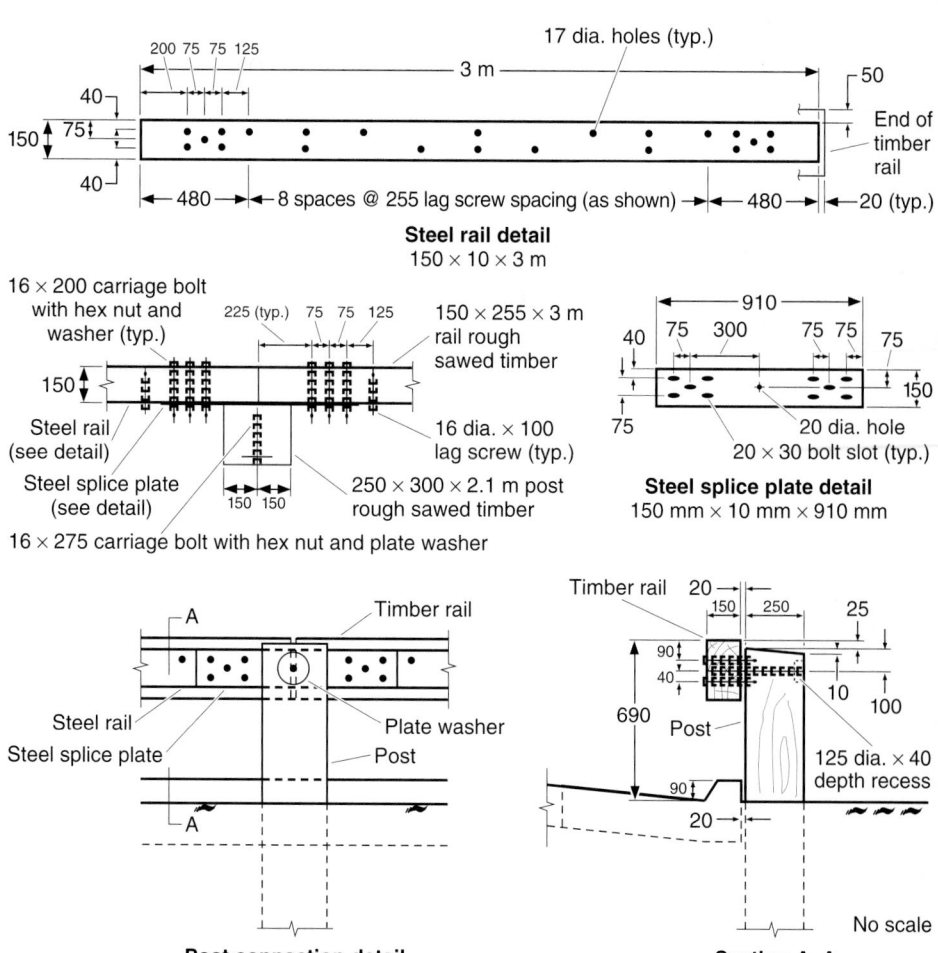

FIGURE 6.21 Steel-backed wood-rail roadside barrier. (*From* Roadside Design Guide, *AASHTO, Washington, D.C., 1996, with permission*)

reduced cross section of the roadside barrier version makes it more vulnerable to overturning, thus requiring more reinforcing steel and/or a more elaborate footing design. The barrier height of the basic design is 810 mm, but higher designs have been tested and constructed to redirect heavy vehicles. For example, the version shown in Fig. 6.22*a* is 2290 mm high; also, it is sloped on both faces. It has contained impacting tractor-trailers but has not completely eliminated rollovers. The version shown in

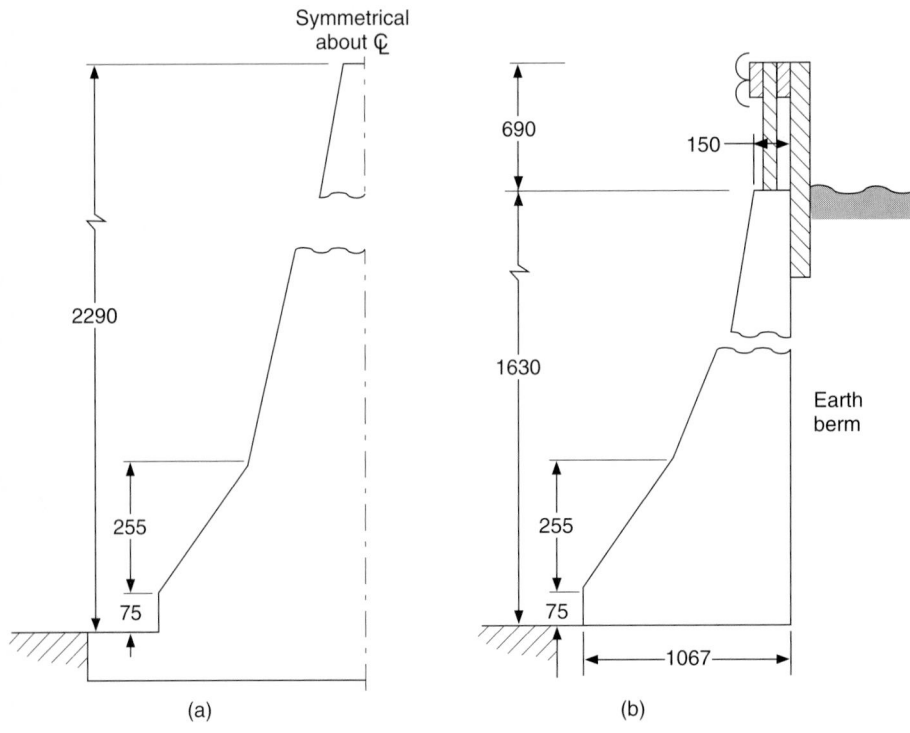

(a)   (b)

AASHTO designation: MB5

Note: The 810-mm-high concrete safety shape was
       initially installed primarily as a median barrier, but
       has become commonly used as both a bridge railing
       and a roadside barrier. Most of these barriers use
       the standard New Jersey shape; any extension in
       barrier height occurs above the slope break point.
       Several states extend the upper stem to serve as a
       maintenance-free glare screen. The two designs
       shown above are the extreme heights to which
       roadside barriers have been constructed, both along
       ramps with a history of truck accidents.

FIGURE 6.22 Variations of concrete safety shape for roadside barrier in severe applications showing (*a*) symmetrical form and (*b*) earth-backed installation. (*From* Roadside Design Guide, *AASHTO, Washington, D.C., 1996, with permission*)

Fig. 6.22*b* is 1630 mm high. It is buttressed by an earth berm on the back side and topped with a W-beam barrier. The New Jersey shape at a height of 810 mm has successfully redirected vehicles in the 820- to 2000-kg range and occasionally redirected buses up to 18,000 kg during moderate impacts. At a height of 1070 mm, it has redirected a 36,300-kg tractor-trailer (15° and 84 km/h).

The *stone masonry wall system* (Fig. 6.23) with a reinforced concrete core and a facing of stone and mortar, offers another aesthetic alternative for parks and similar applications. Successful crash tests have been conducted with an 820-kg vehicle (15°, 97 km/h) and a 1950-kg vehicle (25°, 97 km/h).

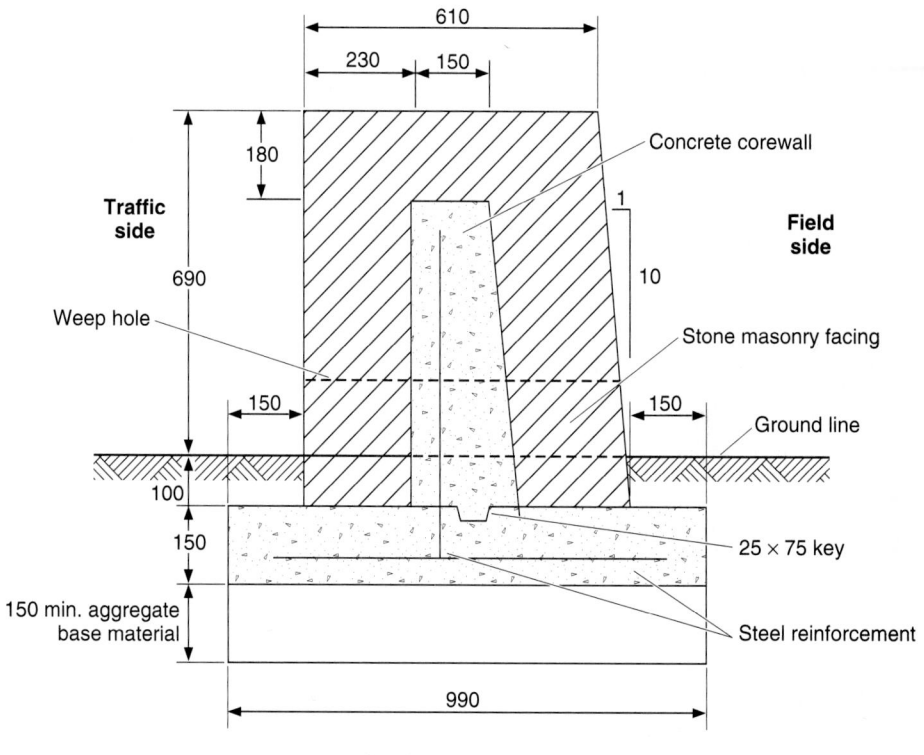

AASHTO designation: None

Note: This barrier consists of a reinforced concrete core faced with stone rubble masonry. Designed for use in scenic areas, its natural appearance and low height combine to make it an effective barrier for use on parkways and similar facilities.

FIGURE 6.23 Stone masonry roadside barrier. (*From* Roadside Design Guide, *AASHTO, Washington, D.C., 1996, with permission*)

6.6 SELECTION OF ROADSIDE BARRIERS

In most cases the selection of a roadside barrier should be made on the basis of the system that will provide the required degree of shielding at the lowest cost. The lowest cost should be based on a life cycle cost analysis, considering initial and maintenance costs and project life. Factors that should be considered in making the selection are summarized in Table 6.3. Most barriers have been developed and tested for passenger cars and offer marginal protection for heavier vehicles. However, some have been tested for more severe applications, as noted in the preceding discussions. In general, standard barriers are satisfactory for most locations, but higher-performance barriers should be considered for locations with poor geometries, high traffic volumes and speeds, and concentrations of heavy truck traffic. The deflection characteristics of the barrier must be considered in relation to the available space. Some systems can be modified to decrease deflections by decreasing post spacing or increasing the number of rails. A computer program, Analysis of Roadside Design (NARD), is available for predicting maximum deflections for blocked-out W-beam and thrie-beam systems with different post spacings and single or double rails. On all systems, data on impact performance and maintenance costs should be tabulated and made available to provide better information for the selection of roadside barriers.

6.7 PLACEMENT OF ROADSIDE BARRIERS

Factors to consider in specifying the exact layout of a barrier at a given location include lateral offset from the edge of the traveled way, terrain effects, flare rate, and length of need. (See also Art. 6.10.)

6.7.1 Lateral Offset

Roadside barriers should generally be placed as far from the traveled way as conditions permit, to allow motorists the best chance of regaining control and to provide better sight distance. It is desirable to maintain a uniform clearance between traffic and roadside features such as bridge railings, retaining walls, and roadside barriers. The distance beyond which a roadside object will not be perceived as an obstacle and cause a motorist to reduce speed or change position is known as the *shy line offset*. According to the AASHTO *Roadside Guide,* this distance varies with design speed as follows:

Design speed km/h	Shy line offset, m
130	3.7
120	3.2
110	2.8
100	2.4
90	2.2
80	2.0
70	1.7
60	1.4
50	1.1

TABLE 6.3 Selection Criteria for Roadside Barriers

Criterion	Comments
Performance capability	Barrier must be structurally able to contain and redirect design vehicle.
Deflection	Expected deflection of barrier should not exceed available room to deflect.
Site conditions	Slope approaching the barrier, and distance from traveled way, may preclude use of some barrier types.
Compatibility	Barrier must be compatible with planned end anchor and capable of transition to other barrier systems (such as bridge railing).
Cost	Standard barrier systems are relatively consistent in cost, but high-performance railing can cost significantly more.
Maintenance	
Routine	Few systems require a significant amount of routine maintenance.
Collision	Generally, flexible or semirigid systems require significantly more maintenance after a collision than rigid or high-performance railings.
Materials storage	The fewer different systems used, the fewer inventory items or the less storage space required.
Simplicity	Simpler designs, besides costing less, are more likely to be reconstructed properly by field personnel.
Aesthetics	Occasionally, barrier aesthetics is an important consideration in selection.
Field experience	The performance and maintenance requirements of existing systems should be monitored to identify problems that could be lessened or eliminated by using a different barrier type.

Source: From *Roadside Design Guide*, AASHTO, Washington, D.C., 1996, with permission.

Place the barrier beyond the shy line offset when possible, particularly for short, isolated installations. Uniform alignment reduces the possibility of snagging. Proper transition where a barrier connects to other features is essential. Short gaps between barriers should be avoided; make the barriers continuous instead. The barrier-to-obstacle distance must be greater than the expected dynamic deflection of the barrier. Where shielding an embankment, the distance from the barrier to the beginning of the down slope should generally be at least 0.6 m, but this may vary with local conditions for soil support of the post.

6.7.2 Terrain Effects

Ideally, at the moment of impact, a vehicle should have all wheels on the ground and the suspension system in a neutral state. Thus, terrain conditions between the traveled way and the barrier are very important. For example, curbs should be avoided and should be no higher than 100 mm if used. In many cases, they can be located behind the barrier. Barriers are usually tested on level terrain. If installed on slopes steeper than 1:10, vehicles may go over standard barriers or impact them too low, and thus not perform as anticipated.

6.7.3 Flare Rate

Roadside barriers must be flared (must have variable offset from the traveled way) to locate the barrier terminal back from the roadway and thus to minimize drivers' reaction to a perceived hazard near the road when approaching a bridge parapet or railing, for example. The greater the flare rate, the greater the potential impact angle and the severity of an accident if the barrier is hit. Also, the chance that a vehicle would be redirected across the roadway increases. Maximum flare rates depend on design speed, barrier type, and location relative to the shy line as shown in Table 6.4. Adjustment to a flatter rate is sometimes made to avoid extensive grading.

6.7.4 Length of Need

The total length of a longitudinal barrier needed to shield an area of concern is referred to as the *length of need*. Figure 6.24 illustrates the variables that must be considered, particularly the *runout length* L_R and the *lateral extent of the area of concern* L_A. The runout length is the theoretical distance needed for a vehicle that has left the road to come to a stop, measured as shown. Suggested values are given in Table 6.5 in terms of the traffic volume and the design speed. The lateral extent of the area of concern is the distance from the edge of the traveled way to the far side of the fixed object, or the outside edge of the clear zone L_C of an embankment or fixed object that extends past the clear zone. After major variables are established, the length of the barrier will then depend on the tangent length L_1, the distance from the traveled way L_2, and the flare rate $a:b$. If a semirigid railing is connected to a rigid barrier, the tangent length should be at least as long as the transition section to reduce pocketing and increase likelihood of redirection. After variables have been selected, the required length of need X in advance of the area of concern, for essentially straight sections of roadway, can be calculated from

$$X = \frac{L_A + (b/a)(L_1) - L_2}{(b/a) + (L_A/L_R)} \tag{6.1}$$

The lateral offset Y from the edge of the traveled way to the beginning of the length of need is

$$Y = L_A - \frac{L_A}{L_R}X \tag{6.2}$$

The amount of rail installed should be a multiple of 3.8 or 7.6 m, because metal-beam barriers are furnished in these lengths. A crashworthy end treatment must be added if the end treatment is located within the clear zone or in a location where it is likely to be struck. If the end treatment permits vehicle penetration, it must be extended upstream to preclude a vehicle from penetrating and striking the shielded feature.

Figure 6.25 shows the definition of variables of an approach barrier for opposing traffic. In this case, lateral dimensions are measured from the edge of the traveled way of the opposing traffic. This would be the centerline for a two lane roadway or the

TABLE 6.4 Suggested Maximum Flare Rates for Roadside Barriers

Design speed, km/h	Flare rate for barrier inside shy line	Flare rate for barrier beyond shy line	
		Rigid systems	Semirigid systems
110	30:1	20:1	15:1
100	26:1	18:1	14:1
90	24:1	16:1	12:1
80	21:1	14:1	11:1
70	18:1	12:1	10:1
60	16:1	10:1	8:1
50	13:1	8:1	7:1

Source: From *Roadside Design Guide, AASHTO,* Washington, D.C., 1996, with permission.

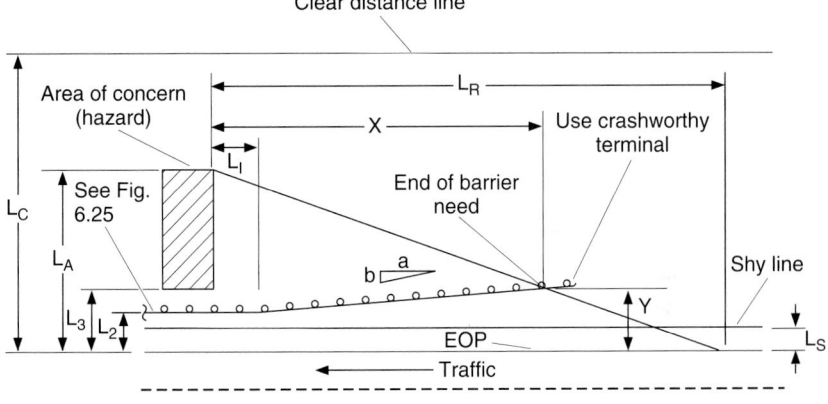

FIGURE 6.24 Layout of barrier approach. (*From* Roadside Design Guide, *AASHTO, Washington, D.C., 1996, with permission*)

TABLE 6.5 Suggested Runout Lengths for Barrier Design

Design speed, km/h	Runout length L_R for indicated traffic volume, m			
	>6000 ADT	6000–2000 ADT	2000–800 ADT	<800 ADT
110	145	135	120	110
100	130	120	105	100
90	110	105	95	85
80	100	90	80	75
70	80	75	65	60
60	70	60	55	50
50	50	50	45	40

Source: From *Roadside Design Guide,* AASHTO, Washington, D.C., 1996, with permission.

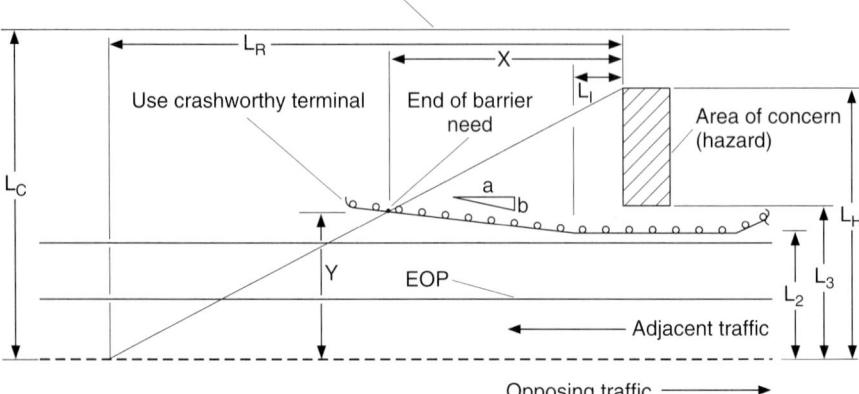

FIGURE 6.25 Layout of barrier approach. (*From* Roadside Design Guide, *AASHTO, Washington, D.C., 1996, with permission*)

edge of the driving lane next to the median for a two-way divided roadway. There are three ranges of clear zone width L_C to consider for an approach barrier for opposing traffic:

- If the barrier is beyond the clear zone, no additional barrier or crashworthy end treatment is required.
- If the barrier is within the clear zone but the area of concern is beyond it, no additional barrier is required but a crashworthy end treatment should be used.
- If the area of concern extends well beyond the clear zone, consider shielding only that portion that lies within the clear zone (set L_A equal to L_C).

The lateral placement of the approach rail should satisfy the criterion for embankment slopes. If steeper than 1:10, consider flattening the slope or decreasing the flare rate so the embankment criterion is not violated.

6.8   UPGRADING ROADSIDE BARRIER SYSTEMS

Table 6.6 provides a checklist that can be used to review existing barrier installations and determine adequacy for either structural or functional (design or placement) causes. Factors to be considered in determining the scope and extent of upgrading include the nature and extent of the deficiency, past accident history, and the cost-effectiveness of the recommended improvement. Remember to always consider the cost-effectiveness of eliminating or relocating the shielded feature.

6.9   MEDIAN BARRIERS

Longitudinal median barriers are used to separate opposing traffic on divided highways, to separate local and through traffic, or to separate traffic in designated lanes.

TABLE 6.6   Roadside Barrier Inspection Checklist

I. Structural adequacy*
    A. Longitudinal section
       1. Standard barrier design†
       2. Adequate post spacing
       3. Rail element blocked out on strong-post system
       4. Adequate splices in rail element
    B. Terminal
       1. Standard terminal design†
       2. Adequate anchorage strength
    C. Transition section
       1. Standard transition design†
       2. Adequate anchorage strength
       3. Adequate stiffening in advance of rigid system
       4. Adequate blockout and/or rubrail

II. Functional adequacy‡
    A. Longitudinal section
       1. Adequate length to shield area of concern
       2. Proper height of rail§
       3. Proper flare rate
       4. Barrier to object distance exceeds barrier deflection distance
       5. Placement behind curb consistent with vehicle trajectory data
       6. Placement on flat slopes (1:10) or on slopes up to 1:6 consistent with vehicle trajectory data
       7. Beam backup plates present on steel strong-post system

    B. Terminal
       1. Adequate clear recovery area behind yielding terminal
       2. Adequate offset of terminal end

---

*Structural adequacy is inherent in the barrier itself, rather than resulting from design, placement, or maintenance.

†Standard systems or elements are those which are currently an approved agency standard or have been successfully crash tested. Certain barriers that fall outside these categories may be left in place depending on the characteristics of the barrier and the results of an engineering analysis of the site.

‡Functional adequacy results from barrier design or placement and is essential for barrier effectiveness.

§Generally, a 75-mm variation from the nominal height is acceptable.

Source: From *Roadside Design Guide*, AASHTO, Washington, D.C., 1996, with permission.

Median barriers designed to redirect vehicles striking from either side require some different considerations from those for roadside barriers. However, performance requirements are the same as given in NCHRP 350 for roadside barriers.

Median barriers should be installed only if the consequences of striking the barrier are less severe than those of striking the feature in question. Figure 6.26 provides suggested warrants for median barriers on high-speed controlled-access roadways with relatively flat, traversable medians. The median width and the traffic volume dictate the need. Site-specific data should also be considered where available. Also, special consideration should be given to barrier needs for medians separating roadways at different elevations.

The information presented in Arts. 6.6, 6.7, and 6.8 on selection, placement, and upgrading of roadside barriers applies generally to median barriers as well. Some

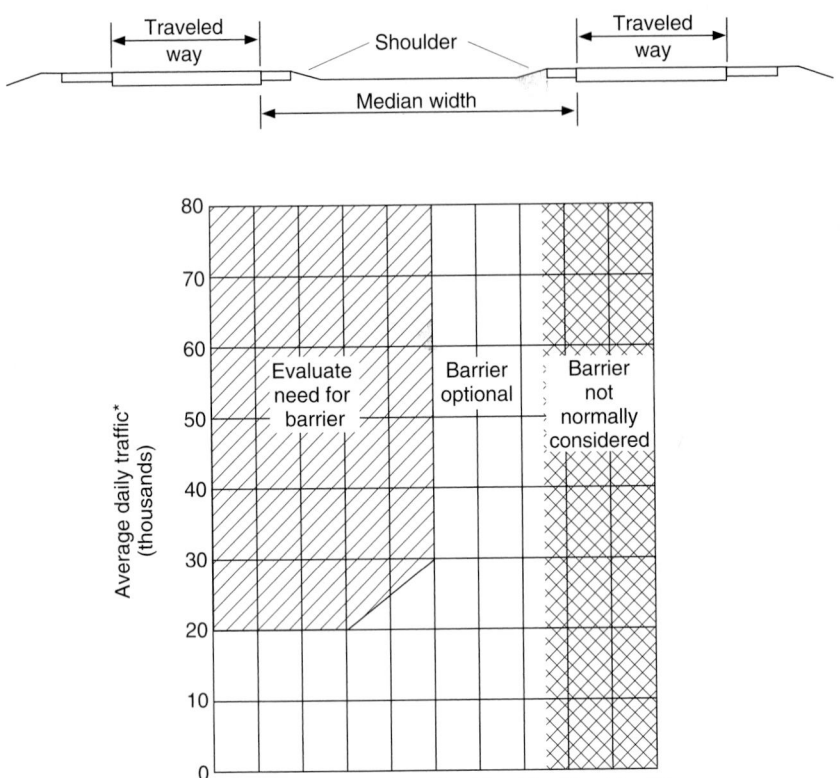

*Based on a 5-year projection

FIGURE 6.26 Median barrier warrants for freeways and expressways. (*From* Roadside Design Guide, *AASHTO, Washington, D.C., 1996, with permission*)

additional information on transitions and placement follows in Arts. 6.9.2 and 6.10. End treatments are discussed in Art. 6.12.

6.9.1 Characteristics of Median Barriers

Like roadside barriers, median barriers can be classified as flexible, semirigid, or rigid as indicated in Table 6.7. Figures 6.27 through 6.39 show details of these various types of median barriers and factors to be considered in selection and application. Additional comments on several of the systems follow. In many of their characteristics they are similar to their roadside barrier counterparts.

Cable systems (Figs. 6.27 and 6.28) should be used only if there is adequate deflection distance, about 3.5 m in each direction. Performance is sensitive to mounting height. Proper end anchorage is critical. They are not well suited for areas hit frequently, on sharp curves, and on facilities with high truck volumes.

TABLE 6.7 Classification of Median
Barriers

Flexible systems
  Three cable
  Six cable
  W beam (weak post)

Semirigid systems
  Box beam (weak post)
  Blocked-out W beam (strong post)
  Blocked-out thrie beam (strong post)
  Modified thrie beam
  Self-restoring barrier (SERB)

Rigid systems
  Concrete safety shape
  Single-slope concrete barrier
  Movable concrete barrier
  Earth berm

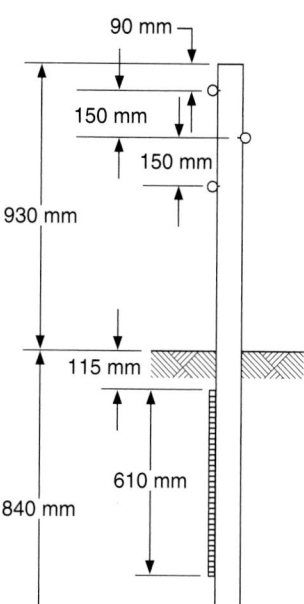

AASHTO designation: None
Post type   S 75 × 8.5
Post spacing   4880 mm
Beam type   19-mm-diam steel cable
Nominal barrier height 760 mm
Maximum dynamic deflection 3500 mm

Remarks: Because of high dynamic deflections,
this system is not recommended for use in
medians narrower than approximately 7 m.

FIGURE 6.27 Three-cable median barrier. (*From* Roadside Design Guide, *AASHTO, Washington,
D.C., 1996, with permission*)

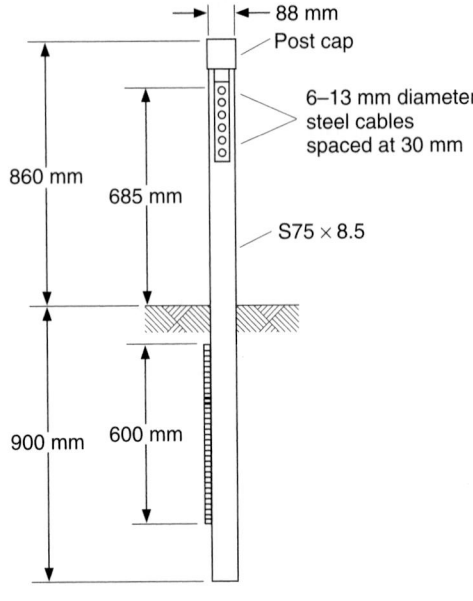

AASHTO designation:	None
Post type	S75 × 8.5
Post spacing	4000 mm
Beam type	13-mm-diam steel cable
Nominal barrier height	535–685 mm
Maximum dynamic deflection	3000 mm

Remarks: Because of high dynamic deflections, this system is not recommended for use in medians narrower than approximately 6 m. The six-cable guide rail has been successfully used by the Ontario Ministry of Transportation.

FIGURE 6.28 Six-cable median barrier. (*From* Roadside Design Guide, *AASHTO, Washington, D.C., 1996, with permission*)

The *W-beam (weak-post) system* (Fig. 6.29) is even more sensitive to mounting height than the cable type. Proper end anchorage is essential. It is not well suited where terrain irregularities exist or where frost heave or erosion is likely to alter the mounting height by more than 50 mm.

The *concrete safety shape* (Fig. 6.34) is the most common rigid median barrier because of low cost, effective performance, and low maintenance. Field performance has shown successful redirection of truck combinations at approaches less than 10°. Variations in the face of the barrier can have a significant effect on barrier performance. The critical dimension is the height of the break (change in slope of face) above the road surface. The New Jersey and F shapes (Fig. 6.34) are the preferred shapes. The latter performs better for small vehicle impact with regard to vehicle roll. Foundation requirements do not appear critical, and there are many variations. Some

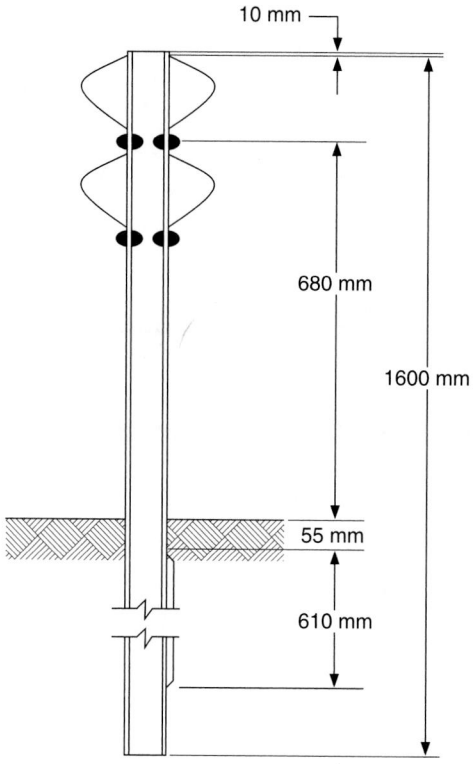

10 mm

680 mm

1600 mm

55 mm

610 mm

AASHTO designation	SGM02
Post type	S75 × 8.5
Post spacing	3810 mm
Beam type	Two steel W-sections
Offset brackets	None
Nominal barrier height	760–840 mm
Maximum dynamic deflection	Approximately 2100 mm

Remarks: This barrier system is suitable for wide, flat medians where sufficient space is available to accommodate deflections. In order to place rigid objects within the median, the SGM02 must be divided into parallel SGR02 barriers with the object centered in 6.7 m plus gap, or be transitioned to a semirigid system.

FIGURE 6.29   W-beam (weak-post) median barrier. (*From* Roadside Design Guide, *AASHTO, Washington, D.C., 1996, with permission*)

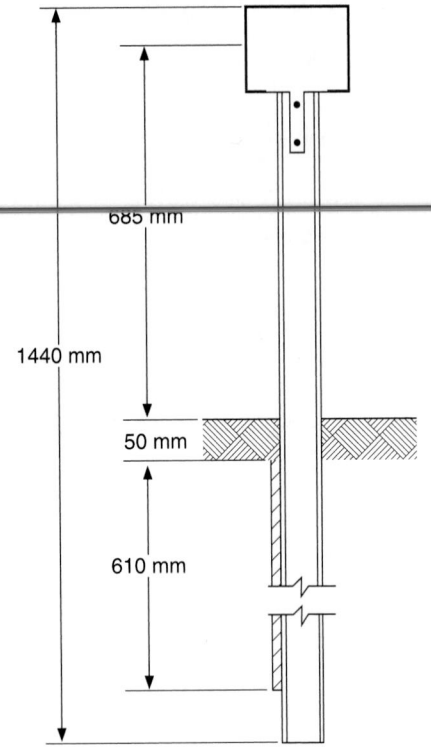

685 mm

1440 mm

50 mm

610 mm

AASHTO designation: SGM03

Post type	S75 × 8.5
Post spacing	1830 mm
Beam type	TS-203 × 152 × 6.4
Offset brackets	None
Mountings	Steel paddles
Nominal barrier height	760 mm
Maximum dynamic deflection	1700 mm

Remarks: This barrier system is suitable for both wide and narrow medians and locations where the terrain is moderately irregular. Even moderate vehicle impacts cause a large number of posts to be damaged. Temporary supports may be used to maintain beam height until posts are replaced.

FIGURE 6.30   Box-beam median barrier. (*From* Roadside Design Guide, *AASHTO, Washington, D.C., 1996, with permission*)

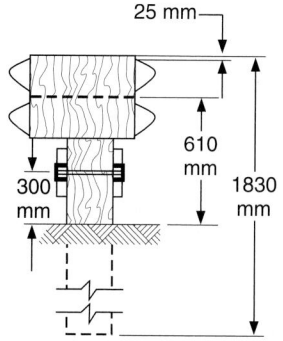

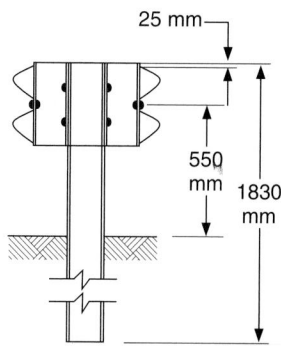

SGM04a		SGM06b
AASHTO designation:	SBM06b	SGM04a
Post type	150-mm × 200-mm Douglas fir*	W150 × 13
Post spacing	1905 mm	1905 mm
Beam type	Two steel W-sections Two C150 × 12 rub rails	Two steel W-sections
Offset brackets	Two 200-mm × 200-mm 360-mm Douglas fir	Two W 150 × 13 × 360 mm
Nominal barrier height	760 mm	700 mm
Maximum dynamic deflection	Approximately 600 mm	Approximately 600 mm

Remarks: SGM06b and SBM04a systems are semirigid and are satisfactory for use in narrow medians. After typical impacts, the system remains serviceable. Some states use a W-section as a rub rail, centered 250 mm above grade. This modification is appropriate for the SGM06b and a higher SGM04a. By dividing an SGM06b or SGM04a into parallel SGR04a-b rails (stiffened if deflection criteria cannot be met), objects in the median can be accommodated.

* 150-mm × 200-mm post and blockout acceptable

FIGURE 6.31 W-beam (strong-post) median barrier. (*From* Roadside Design Guide, *AASHTO, Washington, D.C., 1996, with permission*)

states use unreinforced concrete. A sand-filled metal version has been used in several states on an experimental basis.

Concrete median barriers can be slipformed, precast, or cast in place. Although the barrier does not deflect when hit, passenger vehicles may become airborne and even reach the top in high-angle, high-speed impacts. Fixed objects on top of the barrier such as luminaire supports can cause snagging. Also, even for shallow-angle impacts, the roll angle of a high-center-of-gravity vehicle may be great enough to permit contact of the cargo box with objects on or just behind the barrier. Taller barriers (Figs. 6.35 and 6.36) offer improved characteristics in this regard.

The *movable concrete barrier* is an F-shaped barrier furnished in lengths of 940 mm and arranged in a chain fashion with ends joined by pins. The T segment at the

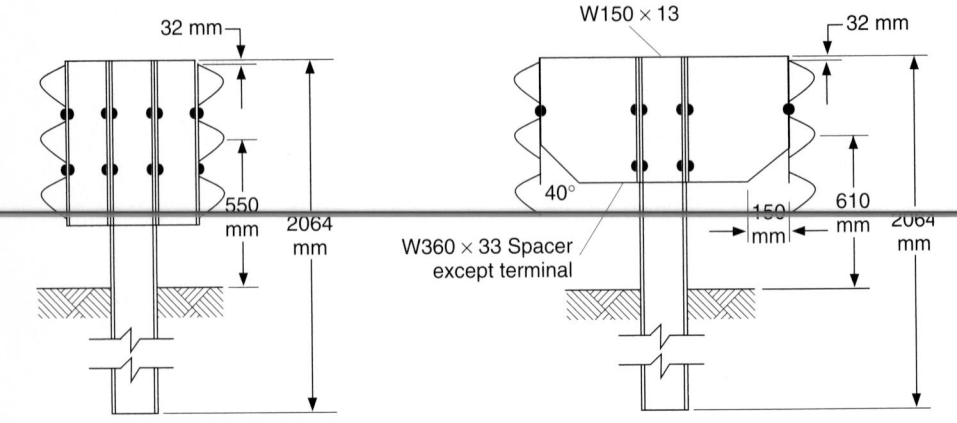

AASHTO designation	SGM09a	SGM09b
Post type	W150 × 13*	W150 × 13*
Post spacing	1905 mm	1905 mm
Beam type	Two thrie-beams	Two thrie-beams
Offset brackets	W150 × 13*	M360 × 26 or W360 × 33
Nominal barrier height	810 mm	870 mm
Maximum dynamic deflection	Approximately 500 mm	Approximately 500 mm

Remarks: SGM09 systems are satisfactory for use in narrow medians. Normal impacts do little damage to the rail. Under severe impact conditions, the rail of an SGM09b system remains upright and has the capability to redirect 18,000-kg vehicles impacting at 80 km/h and 15°.

* 150-mm × 200-mm wood post acceptable alternative (AASHTO designation SGM09c)

FIGURE 6.32   Thrie-beam (strong-post) median barrier. (*From* Roadside Design Guide, *AASHTO, Washington, D.C., 1996, with permission*)

top facilitates lifting. The system is often used in construction zones where traffic lanes are opened and closed frequently.

The *earth berm* can be used to mitigate obstructions in medians. It is critical that the berm be introduced gradually so it does not become an obstacle. Redirectional capabilities are minimal, and thus it should not be used where high-angle impacts are likely, such as on the outside of a curve, or for other critical conditions.

6.9.2   Median Barrier Transitions

Transition sections are used between adjoining median barriers having significantly different deflection characteristics, between a semirigid median barrier and a rigid barrier (such as a bridge rail), and in similar situations. The transition sections should provide impact performance similar to standard sections, and emphasis should be

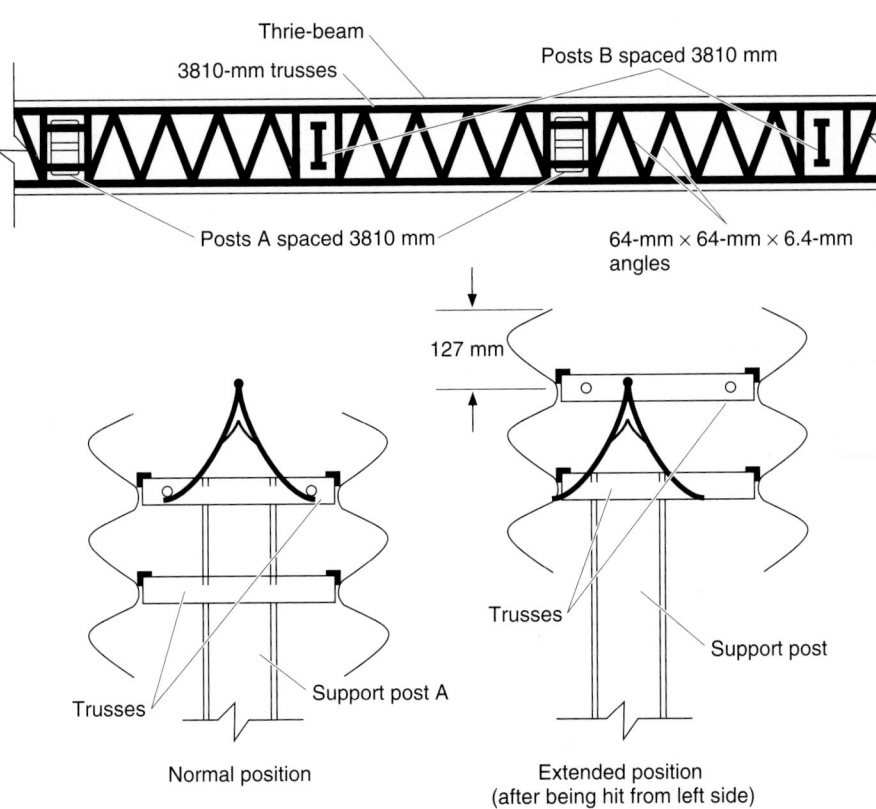

Section A-A

AASHTO designation	None
Post type	W150 × 29.8 × 2.1 m (A posts); W150 × 29.8 × 2.4 m (B posts)
Post spacing	1900 mm
Beam type	Two thrie-beams spaced at 500 mm by two steel trusses
Nominal barrier height	865 mm
Maximum dynamic deflection	730 mm (18,000-kg bus at 100 km/h and 15°)

Remarks: This is a medium-performance semiflexible median barrier designed for use in narrow medians. The thrie-beam can deflect 90 mm laterally and 150 mm vertically and return to its original configuration without major damage. Transitions to end treatments have also been crash-tested.

FIGURE 6.33 Self-restoring median barrier. (*From* Roadside Design Guide, *AASHTO, Washington, D.C., 1996, with permission*)

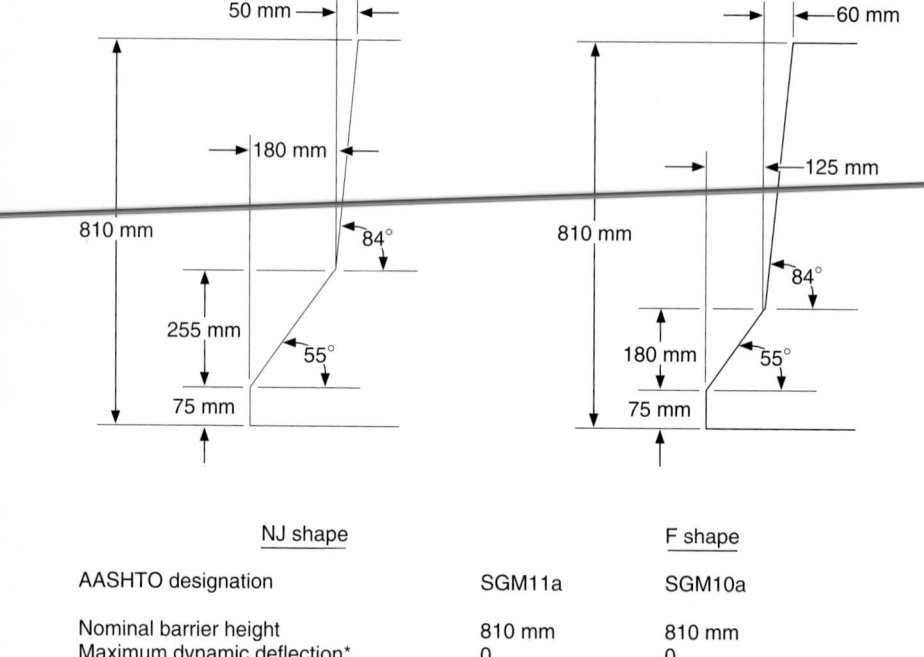

	NJ shape	F shape
AASHTO designation	SGM11a	SGM10a
Nominal barrier height	810 mm	810 mm
Maximum dynamic deflection*	0	0

Remarks: The lower-sloped face redirects vehicles without damage under low impact conditions. During moderate to severe impacts, some energy is dissipated when the vehicle is lifted off the pavement. The loss of tire contact with the pavement also aids redirection. In crash tests the F shape has proven to be more successful in preventing rollover of smaller vehicles.

The details of the shape are critical. The distance from the pavement to the break between the upper and lower slopes should be kept at 330 mm or below. Barrier performance under moderate to severe impact conditions is not significantly affected by overlays on the lower-sloped face. The overall height of the barrier, however, needs to be maintained at a minimum of 740 mm.

The safety shape barrier is suitable for narrow medians. Both faces can be flared away from the centerline to provide room for rigid objects to be installed in the medians. Since this barrier requires a paved approach, its application in wide medians is less cost-effective.

* Very severe hits may destroy the barrier. Reinforcing is recommended to prevent shattering of concrete where the top of the barrier has a width less than 300 mm.

FIGURE 6.34   Concrete safety-shape median barrier. (*From* Roadside Design Guide, *AASHTO, Washington, D.C., 1996, with permission*)

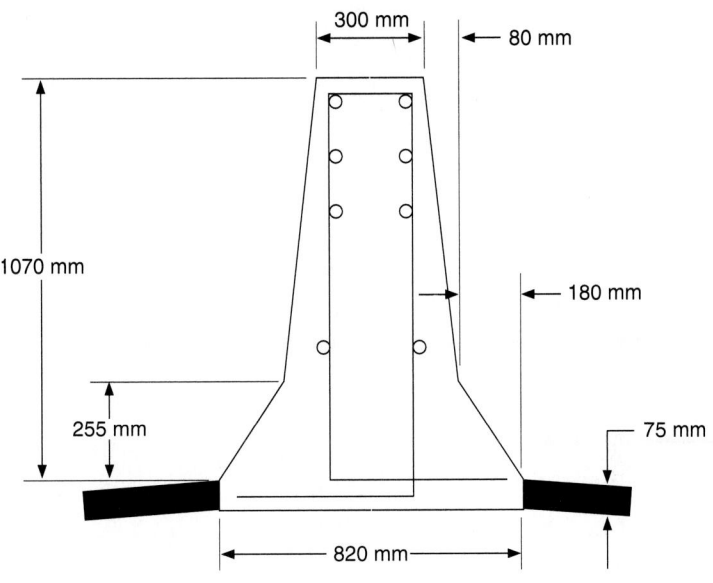

AASHTO designation   SGM12

Nominal barrier height   1070 mm
Maximum dynamic deflection   0

Remarks: This tall-wall concrete safety-shaped barrier is used by the
New Jersey Turnpike Authority.

FIGURE 6.35   Tall-wall concrete safety-shape (reinforced) median barrier. (*From
Roadside Design Guide, AASHTO, Washington, D.C., 1996, with permission*)

placed on designs to avoiding vehicle snagging. Structural details of special importance include the following:

- Rail splices should develop the tensile strength of the weaker rail.
- Use a flared or sloped connection if the connection could snag an opposite-direc-
  tion vehicle. Use a standard terminal connector to attach a W-beam or thrie-beam
  rail to a rigid bridge railing or parapet, or provide a recessed area in the parapet
  wall to receive the rail.
- In cases such as a strong-post W-beam transition to the concrete safety shape, use a
  blockout design and consider adding a rub rail or using a thrie beam instead.
- Use a transition length 10 to 12 times the difference in lateral deflection of the two
  systems under consideration.
- Increase stiffness gradually from the weaker to the stronger system by means such
  as decreasing the post spacing, increasing the post size, and using nested sections of
  W beam or thrie beam.

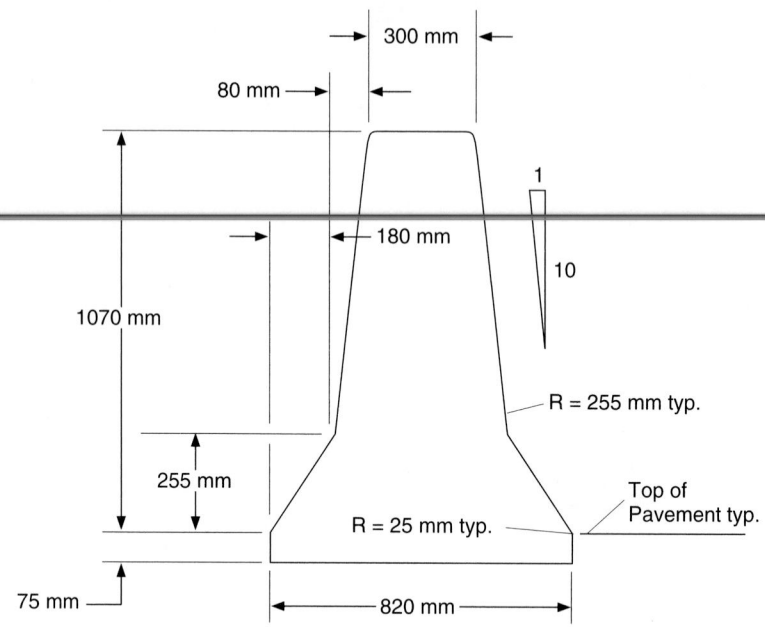

AASHTO designation	None
Nominal barrier height	1070 mm
Deflection	0

Remarks: This tall-wall concrete safety-shaped barrier is used by the Ontario Ministry of Transportation.

FIGURE 6.36   Tall-wall concrete safety-shape (unreinforced) median barrier. (*From Roadside Design Guide, AASHTO, Washington, D.C., 1996, with permission*)

6.10   PLACEMENT OF BARRIERS ON SLOPED MEDIANS

Either roadside barriers or median barriers may be appropriate for sloped medians, depending on conditions. If a relatively flat median (slope of 1:10 or flatter) free of rigid objects is available, a median barrier can be placed at the center. When such desirable conditions are not available, some additional guidelines should be considered. Figure 6.40 shows three basic types of median sections. Section I (illustrations 1–3) represents a depressed median or one with a ditch; section II (illustrations 4–6) represents a stepped median or a median that separates traveled ways with significant differences in elevation; and section III (illustration 7) applies to a raised median.

Section I.   Check to see if the slopes warrant a barrier. If both slopes require shielding (illustration 1), place a *roadside barrier* near the shoulder on each side of the median. If only one slope must be shielded, place a *median barrier* near the shoulder on that side; use a rigid or semirigid barrier, and install a rub rail on the ditch side

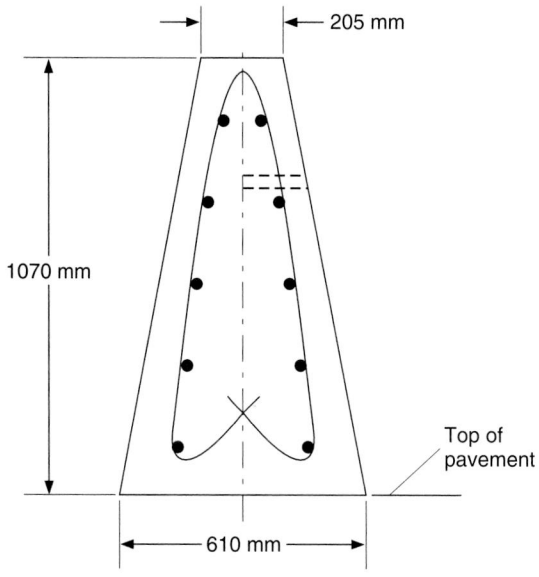

AASHTO designation	None
Nominal barrier height	1070 mm
Deflection	0

Remarks: This barrier is suitable for both permanent and temporary applications. Its primary advantage is that the adjacent pavement can be overlaid several times without affecting the performance of the barrier. Its disadvantage is that greater vehicle damage occurs at shallower impact angles as compared to other safety-shaped barriers.

FIGURE 6.37   Single-slope concrete median barrier. (*From* Roadside Design Guide, *AASHTO, Washington, D.C., 1996, with permission*)

of the barrier to prevent snagging of a vehicle that has crossed the ditch. If neither slope requires shielding but one is steeper than 1:10 (illustration 2), place a rigid or semirigid median barrier on the side with the steeper slope when warranted. If both slopes are relatively flat (illustration 3), place a median barrier (any type with dynamic deflection not greater than half median width) at or near the center of the median if vehicle override is not likely.

Section II.   If the embankment slope is steeper than 1:10 but traversable (illustration 4), place a median barrier near the shoulder on the high side of the slope. If the slope is not traversable (such as a rough rock cut, illustration 5), place a *roadside barrier* at the top and bottom of the slope. If a retaining wall is located at the bottom of the slope, contour the base of the wall to the exterior shape of a concrete safety shape. If the slope is flatter than 1:10 (illustration 6), place a median barrier near the center.

Section III.   If the median is sufficiently high and wide (illustration 7), vehicles may be redirected without a barrier. If the slopes are relatively flat and traversable,

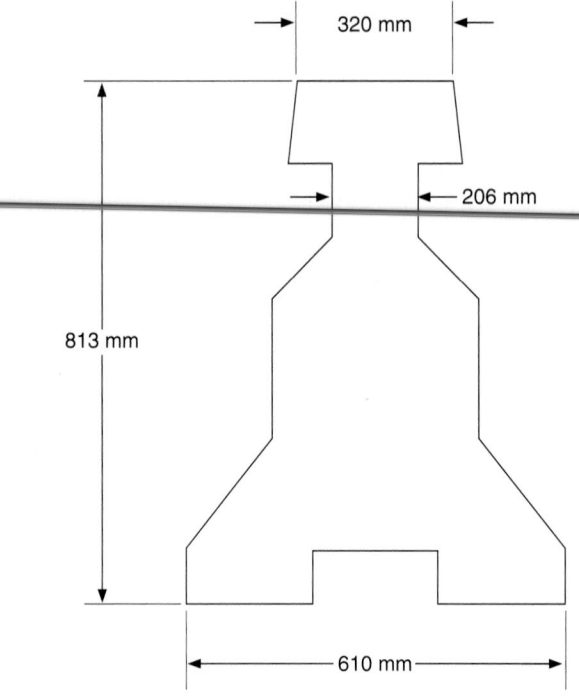

AASHTO designation	SWC01
Nominal barrier height	813 mm
Deflection	1.2 m

Remarks: This proprietary portable barrier system is suitable for both permanent (unbalanced traffic flow) and temporary applications. It is composed of a chain of safety-shaped concrete barrier segments 940 mm in length which can be shifted laterally. Even though cost is relatively high, the system becomes cost-effective when frequent lateral movement of the temporary barrier is required while maintaining traffic.

FIGURE 6.38   Movable concrete median barrier. (*From Roadside Design Guide, AASHTO, Washington, D.C., 1996, with permission*)

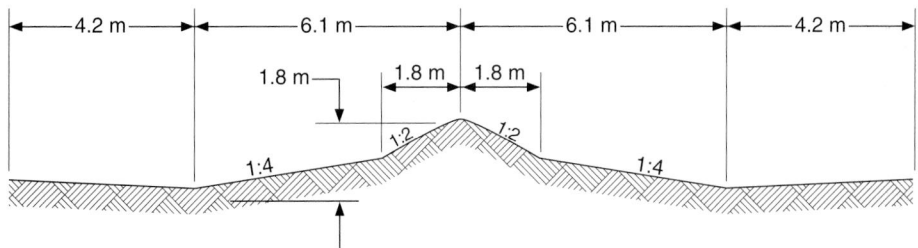

AASHTO designation   None

Median is shaped to provide redirection:

Height   1.2 m minimum
Slope varies to suit field conditions:
                            1:2, desirable and maximum
                            1:3, maximum for mowing
                            1:10, minimum approach grade
Length   Minimum of full berm in advance of hazard

Remarks: This system provides redirection at low impact angles (10° or less) or low speeds (60 km/h or less). It may also be used in wide medians outside of the clear zone distance for both directions of travel.

FIGURE 6.39   Earth berm median barrier. (*From* Roadside Design Guide, *AASHTO, Washington, D.C., 1996, with permission*)

place a semirigid median barrier at the apex. If the slopes are not traversable, place a *roadside barrier* on either side.

When a median barrier is warranted, it is best to use the same barrier throughout the length of need. In cases where a roadside barrier is required on both sides of the median for some length and a centrally located median barrier is situated upstream and downstream, use a gradual transition between the systems proceeding in the direction of traffic.

6.11 BRIDGE RAILINGS AND TRANSITIONS

Bridge railings are longitudinal barriers intended to prevent vehicles from running off the edge of a bridge. A metal post-and-rail system, a concrete safety shape, and various combinations have been used. Bridge railings are attached to the structure and designed to have minimal deflection under impact. Design loadings and procedures have been traditionally based on *Standard Specifications for Highway Bridges* (AASHTO). Railings that were successfully tested were acceptable even though they did not fully comply with the specifications. More recently, AASHTO has developed *Guide Specifications for Bridge Railings,* which requires full-scale crash testing on all railing systems used for new construction. Low-speed low-volume roadways are often not held to such high standards. Most railings have been developed to redirect passen-

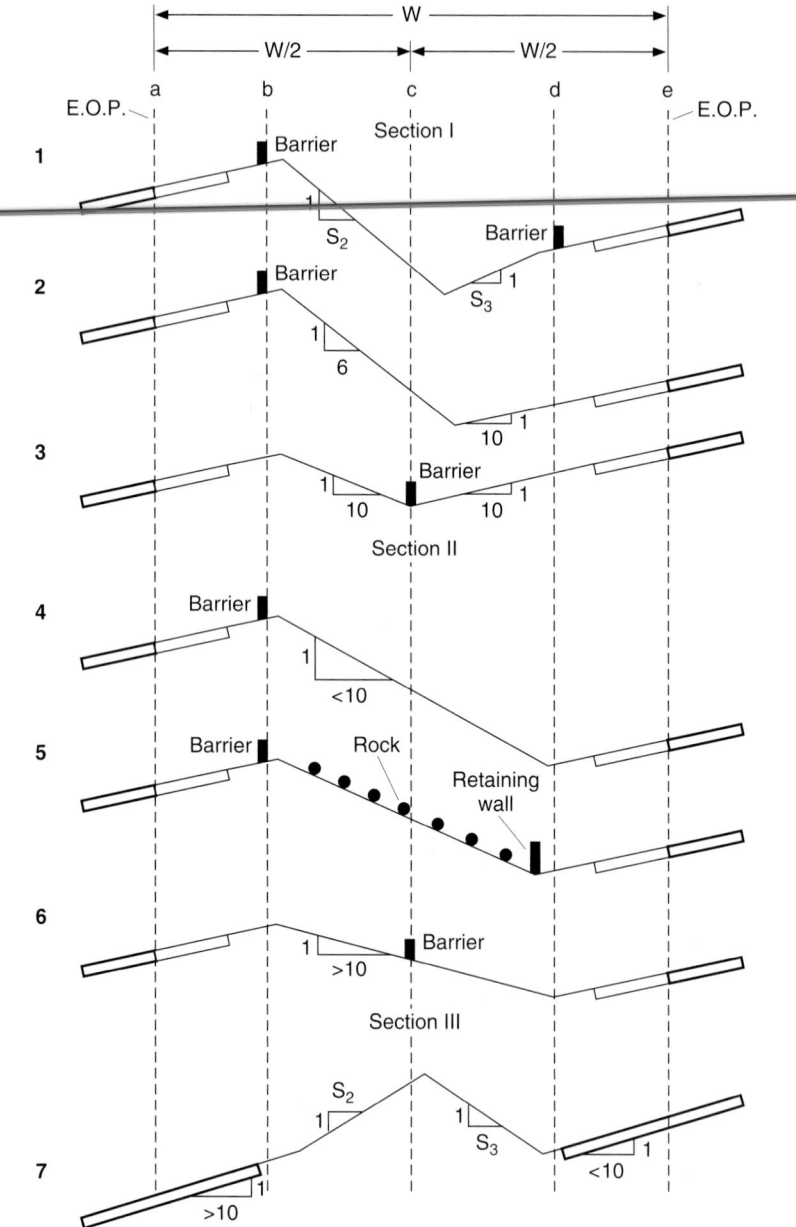

FIGURE 6.40 Barrier placement for sloped medians. (*From* Roadside Design Guide, *AASHTO, Washington, D.C., 1996, with permission*)

ger-size vehicles, but railings have also been developed for larger vehicles. In addition to strength, height and the shape of the face are key factors that affect performance.

6.11.1 Selection Considerations

The selection of a railing should include consideration of the following five factors:

Railing performance. There must be evidence that the system will provide the desired performance level.

Compatibility. A crashworthy transition section is required if the approach barrier differs in strength, height, or deflection characteristics.

Cost. Life cycle cost analysis is desirable to compare alternatives. Initial costs, maintenance costs, and the costs of accidents must be considered.

Field experience. Review in-service performance of existing systems to evaluate effectiveness and cost.

Aesthetics. Appearance is particularly important in scenic areas, but systems must be selected that meet required performance levels.

6.11.2 Placement Considerations

Bridges should provide a full, continuous shoulder that maintains uniform clearance with approaching roadside elements. However, if the bridge is narrower than the approaching roadway and shoulder, the appropriate flare rate should be provided where the railing is within the shy distance (Art. 6.7.3). In general, it is desirable to avoid curbs. If a sidewalk is present, use a bridge railing between the traffic and the sidewalk to protect pedestrians, and a pedestrian railing along the outside of the bridge. End treatment of the bridge railing is difficult under these circumstances. If a crash cushion or other barrier cannot be used, a vertically tapered end section may be the best solution. The location and extent of the taper must be carefully considered for the conditions present.

6.11.3 Upgrading Bridge Railing Systems

The first step in an upgrading project is to identify potentially deficient systems. Bridge railing designs prior to 1964 are particularly suspect. Strength and performance should be documented. Verify critical details such as base plate connections, anchor bolts, material (strength, toughness, and condition), welding details, reinforcement development, etc. Open-faced railings may cause snagging. Curbs or sidewalks adjacent to a railing may cause an impacting vehicle to vault or roll over. Approach transitions may be inadequate.

Retrofits can be developed to address inadequacies. When possible, use crash-tested designs in such updating. One common improvement is to rebuild the approach barrier and transition to current standards, continuing the metal-beam rail element, for example, across the structure to provide continuity. If a narrow curb is in place, a retrofit railing can often be blocked out to minimize rollover and ramping. Some specific retrofit concepts are discussed in the following.

Concrete retrofit (safety shape or vertical). The concrete safety shape can be used most effectively when it can be constructed in front of an existing railing that can remain in place. A vertical-faced concrete shape creates an effective barrier when added on top of and flush with an existing safety curb. The structure must be evaluated for the extra dead load imposed, and for the development of the required anchorage to resist impact forces.

W-beam and thrie-beam retrofits.   A partial solution sometimes used is to continue an approaching W-beam or thrie-beam roadside barrier across the bridge. It may not bring the bridge into full compliance with AASHTO criteria, but may be satisfactory as an interim solution, particularly on low-volume roadways. Adequate anchorage is provided by the continuous system. Gradual stiffening in the transition area is advised to avoid snagging.

Metal post-and-beam retrofit.   Where a sidewalk is present, a steel post (S shape or channel shape) can be anchored to the top and a pair of steel tubes attached to the roadway side to provide a smooth traffic barrier between the sidewalk and the roadway. The tube elements must be in line with the face of the curb. The post attachment can be designed to resist the impact loads, or a yielding design can be developed to minimize possible bridge deck damage. The existing bridge railing on the outside of the sidewalk can be adapted to a pedestrian railing.

Tubular thrie-beam retrofits.   Figure 6.41 shows two typical thrie-beam retrofits used to strengthen a substandard concrete bridge railing. Rectangular tubes are used to block out the tubular thrie beam from the existing railing so that its face is in line with the curb. The system is readily adaptable to different geometries. When set at a height of 960 mm, this retrofit successfully redirected an 18,000-kg intercity bus (15° and 80 km/h).

Self-restoring bridge rail (SERB) retrofit.   A retrofit with the SERB system is shown in Fig. 6.42. It is installed in a manner similar to the thrie-beam retrofit but has the advantage of requiring minimal maintenance for most impacts. This design has passed a full range of crash tests, including a subcompact (20° and 100 km/h) and an 18,000-kg intercity bus (15.5° and 85 km/h).

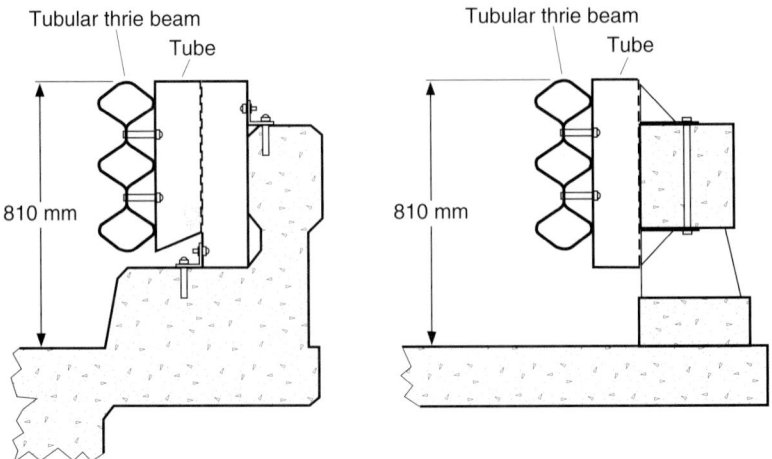

FIGURE 6.41   Tubular thrie-beam retrofits. (*From* Roadside Design Guide, *AASHTO, Washington, D.C., 1996, with permission*)

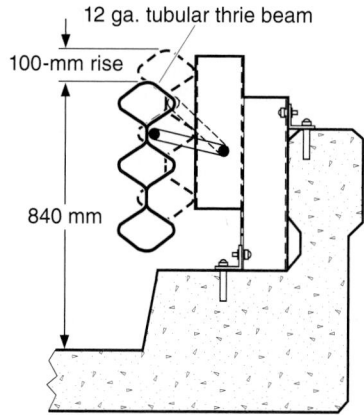

FIGURE 6.42 Self-restoring bridge rail (SERB) retrofit. (*From* Roadside Design Guide, *AASHTO, Washington, D.C., 1996, with permission*)

6.11.4 Crash-Tested Bridge Railings

Figures 6.43 through 6.52 show a variety of bridge railing systems that have been successfully crash-tested in accord with NCHRP 230. The railings include such types as W beams, thrie beams, rectangular tubes, and concrete safety shapes. The mass of the test vehicle, the impact speed, and the impact angle are given for each system. They have been tested with vehicles ranging from a car of about 900 kg to a large truck of over 36,000 kg. Thus, a railing may be selected from this group to provide the performance level needed for a particular application.

6.11.5 Transitions to Bridge Railings

Most of the principles previously discussed for median transitions (Art. 6.9.2) apply here as well. Transition designs should gradually stiffen the approach system to avoid vehicle pocketing, snagging, or penetration. Some considerations of importance follow. The concepts are appropriate for both new construction and retrofits.

- The splice between the rail of the approach barrier and the bridge rail should develop the tensile strength of the approach rail.
- Strong-post systems, or combination normal-post and strong-beam systems, can be used for transitions. These systems normally should be blocked out to avoid snagging. Also, a rub rail may be desirable with W-beam or tube-type transitions. Tapering the rigid bridge railing end behind the transition members may also be desirable. The rub rail and railing taper are specially appropriate when the approach transition is recessed into the end of a concrete railing or other rigid hazard.
- Use a gradual transition, typically 10 to 12 times the difference in lateral deflection of the two systems. Gradually stiffen by decreasing post spacing, increasing post size, and strengthening the rail (nested W beams or thrie beams, for example).

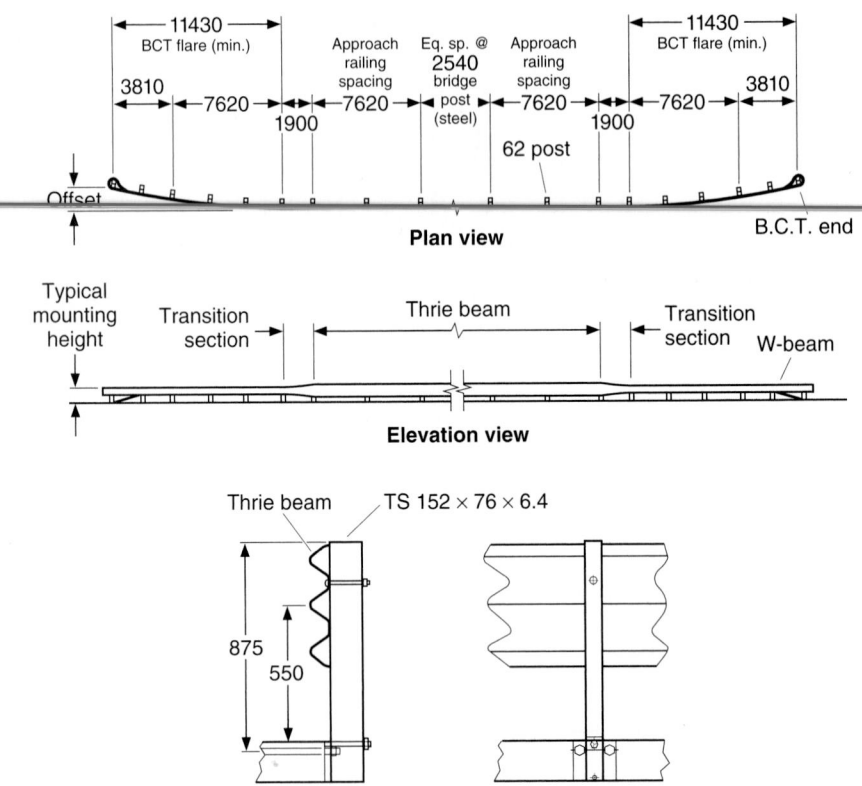

FIGURE 6.43 Side-mounted thrie-beam bridge railing. (*From* Roadside Design Guide, *AASHTO, Washington, D.C., 1996, with permission*)

- Eliminate curbs, inlets, and other drainage features in front of the barrier. Keep the slope between the edge of the driving lane and the barrier to 1:10 or less.
- When possible, relocate roads that intersect near the end of the bridge and interfere with a proper transition. Crash cushions may provide an option in some cases.
- A thrie-beam transition may be preferred over a W-beam transition because it is stiffer, better matches the bridge railing geometry in most cases, and eliminates the need for a rub rail.

(*Text continues on page 6.64.*)

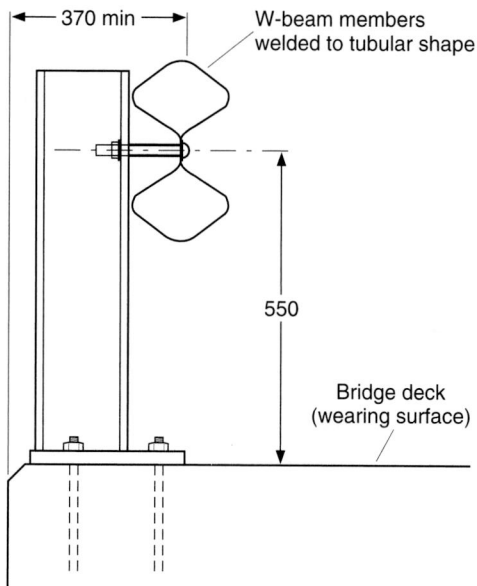

Elevation of post

Rail height: 690 mm

Test vehicle	1035-kg car	2043-kg car
Impact speed, km/h	93	99
Impact angle, degrees	14.0	27.5

FIGURE 6.44 Tubular W-beam bridge railing. (*From* Roadside Design Guide, *AASHTO, Washington, D.C., 1996, with permission*)

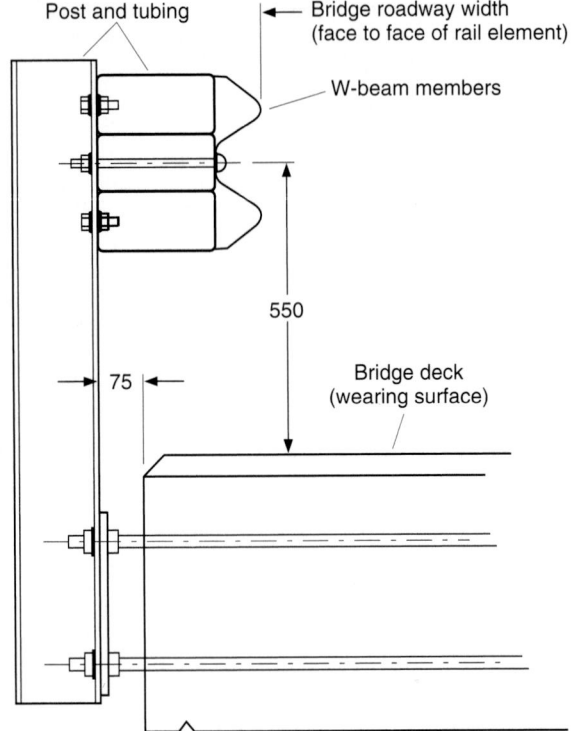

All dimensions shown are mm, unless otherwise noted.

Type-1 post

Rail height: 690 mm

Test vehicle	899-kg car	2175-kg car
Impact speed, km/h	98	97
Impact angle, degrees	19.6	25.0

FIGURE 6.45 Side-mounted rectangular-tube bridge railing. (*From Roadside Design Guide, AASHTO, Washington, D.C., 1996, with permission*)

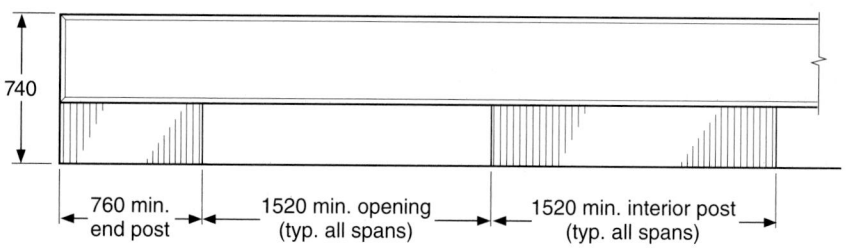

Elevation of traffic rail

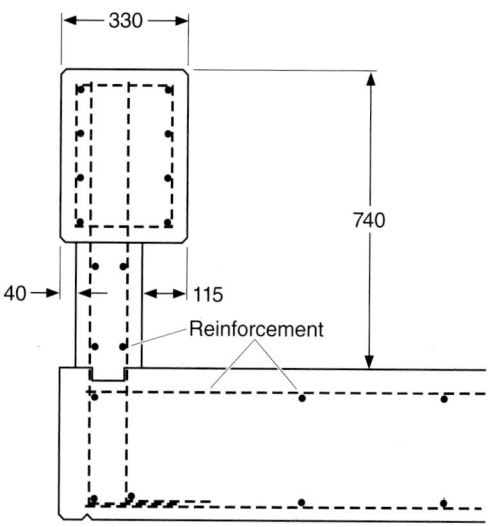

All dimensions shown are mm, unless otherwise noted.

Section

Rail height: 740 mm

Test vehicle	899-kg car	2116-kg car
Impact speed, km/h	95	95
Impact angle, degrees	18.9	25.4

FIGURE 6.46 Open concrete beam-and-post bridge railing (Oklahoma modified TR-1). (*From* Roadside Design Guide, *AASHTO, Washington, D.C., 1996, with permission*)

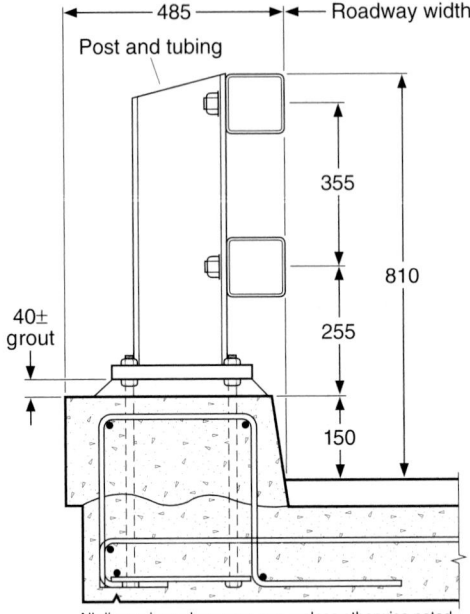

All dimensions shown are mm, unless otherwise noted.

Curb mount-post detail

Rail height: 810 mm

Test vehicle	905-kg car	2107-kg car
Impact speed, km/h	94	97
Impact angle, degrees	18.8	25.0

FIGURE 6.47 Curb-mounted tubular bridge railing. (*From* Roadside Design Guide, *AASHTO, Washington, D.C., 1996, with permission*)

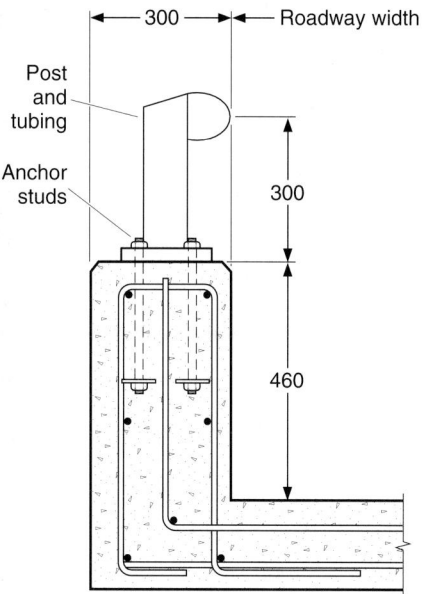

All dimensions shown are mm, unless otherwise noted.

Section through parapet and rail

Rail height: 810 mm

Test vehicle	903-kg car	2116-kg car	9044-kg bus
Impact speed, km/h	96	96	92
Impact angle, degrees	18.8	25.0	14.8

FIGURE 6.48   Aluminum post-and-tube bridge railing (BR1 type C). (*From* Roadside Design Guide, *AASHTO, Washington, D.C., 1996, with permission*)

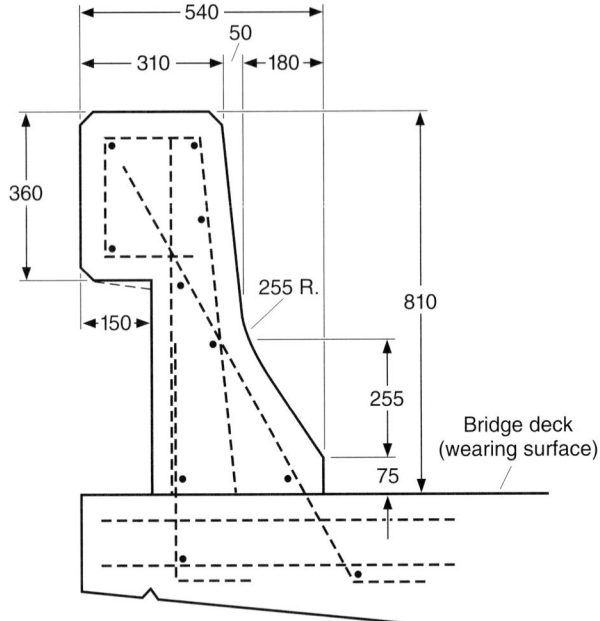

All dimensions shown are mm, unless otherwise noted.

Rail height: 810 mm

Test vehicle	2061-kg car	2061-kg car	2061-kg car
Impact speed, km/h	61	105	101
Impact angle, degrees	7.0	7.0	25.0

FIGURE 6.49 Safety-shape concrete bridge railing. (*From* Roadside Design Guide, *AASHTO, Washington, D.C., 1996, with permission*)

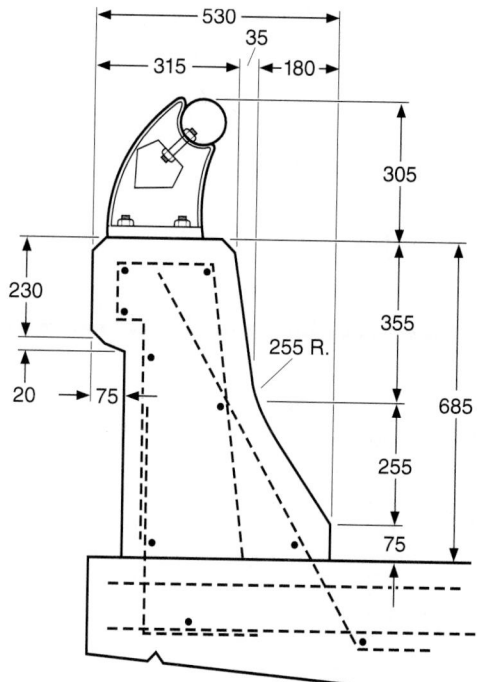

All dimensions shown are mm, unless otherwise noted.

Rail height: 990 mm

Test vehicle	868-kg car	2111-kg car	18,160-kg bus
Impact speed, km/h	98	99	95
Impact angle, degrees	19.3	24.9	16.4

FIGURE 6.50 Nevada concrete safety-shape bridge railing. (*From* Roadside Safety Design Guide, *AASHTO, Washington, D.C., 1996, with permission*)

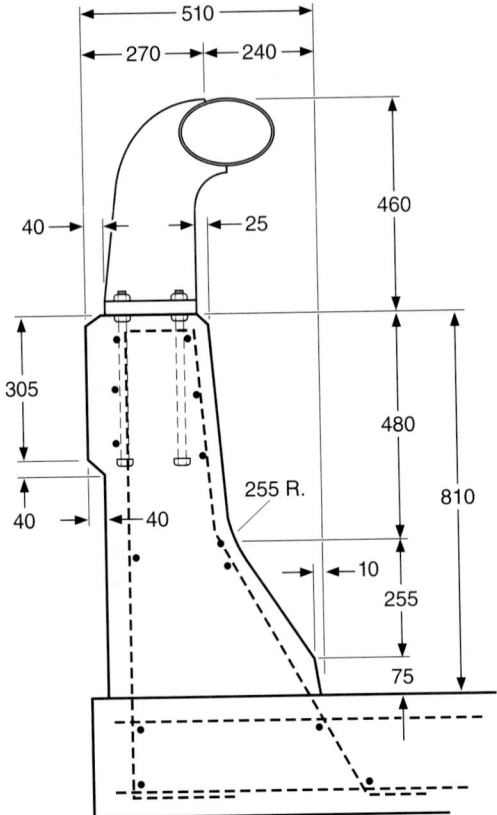

All dimensions shown are mm, unless otherwise noted.

Rail height: 1270 mm

Test vehicle	36,356-kg truck
Impact speed, km/h	78
Impact angle, degrees	14.5

FIGURE 6.51 Texas concrete safety-shape bridge railing (type HT). (*From* Roadside Design Guide, *AASHTO, Washington, D.C., 1996, with permission*)

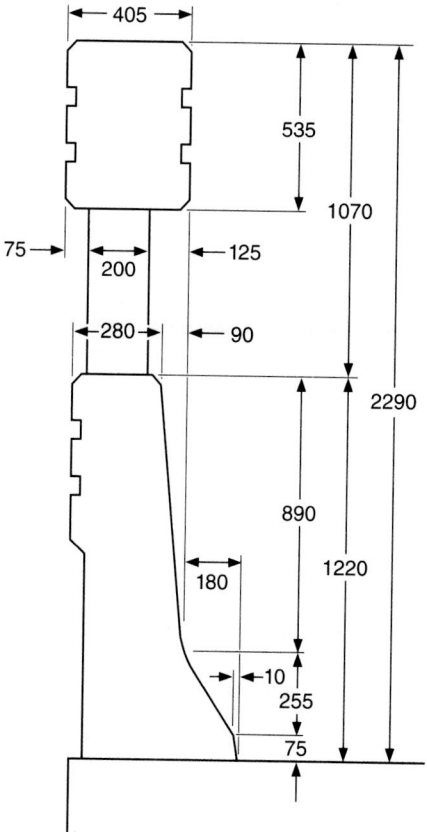

All dimensions shown are mm, unless otherwise noted.

Rail height: 2290 mm

Test vehicle	36,374-kg tank-type tractor trailer
Impact speed, km/h	83
Impact angle, degrees	15.0

FIGURE 6.52 Texas concrete safety-shape bridge railing (type TT). (*From* Roadside Design Guide, *AASHTO, Washington, D.C., 1996, with permission*)

6.12 BARRIER END TREATMENTS AND CRASH CUSHIONS

Barrier terminals and crash cushions are developed to gradually decelerate an impacting vehicle to a stop or to suitably redirect it. Otherwise, untreated ends of barriers and fixed objects can cause severe accidents. A crashworthy end treatment is essential if a barrier terminates within the clear zone or other area where it is likely to be hit by an errant vehicle. Recommendations for testing and performance are contained in NCHRP 350. Suitable devices must be able to perform under both head-on and side impacts, with no objects penetrating the passenger compartment or encroaching on other traffic. The vehicle should remain upright and not be redirected into adjacent traffic lanes. Occupant deceleration levels must be within target values. For longitudinal barriers that depend on the tensile strength of the elements, the end treatment must develop the full tensile strength of the rail, whether or not a crashworthy end treatment is employed.

6.12.1 Characteristics of End Treatments

Many types of end treatments are available. A description of a number of them and information on their performance follows. As indicated, a number of the systems are proprietary.

The *breakaway cable terminal (BCT)* was developed for a W-beam system (weak or strong post). However, as with other systems, it can be used with a thrie-beam system if placed after a thrie-beam/W-beam section transition. The BCT uses a looped metal end section that can buckle under impact and two posts (the first two) that can fracture or slip so the vehicle can pass behind the terminal. Tensile strength is developed by a cable that transfers forces from the W beam to the base of the end post. A parabolic flare (1220-mm end offset) in location over the 12-m length of the BCT is essential for proper performance. The approach and the back area should have a slope of 1:10 or flatter. At a 700-mm top-of-rail height, crash tests were successful for a 2000-kg and a 1020-kg vehicle, but not with 820-kg vehicles. Thus, this terminal should not be used on high-speed high-volume highways.

The *Extruder Terminal (ET-2000)* is a proprietary system for installation at the end of a W-beam rail element. No flare is required. The terminal consists of an extruder shoe and a modified cable anchor similar to that of the BCT. When impacted, the shoe travels along the rail, flattens it, and bends it out of the way. The basic parts are generally reusable, but components must be kept on hand and maintenance people must be trained. The terminal performed well under head-on and side impact tests (NCHRP 230) with 820-kg and 2000-kg vehicles at 100 km/h.

The *turned-down rail terminal* reduces the height of a W-beam rail from full value to ground level over a distance of 8 m or more . One of the first efforts to eliminate vehicle impalement, the turned-down rail is intended to collapse on impact so the vehicle can safely pass over the rail. However, there is a tendency for the vehicle to vault and roll. Flaring the turned-down terminal away from the roadway reduces the likelihood of its "capturing" a vehicle. Some design variations are more successful than others, but none of the conventional designs can accommodate an 820-kg vehicle that impacts near the buried end. On the basis of crash-test results, the turned-down terminal should not be used on high-speed high-volume highways.

A *sloped concrete* end treatment is used to terminate a concrete safety shape in locations where speeds are 60 km/h or less and space is limited, or where end impacts are unlikely. The height of the barrier is tapered over a minimum length of 6 m, but 10 m to 13 m is desirable. The end of the taper must not exceed 100 mm. Crash-test data are not available.

The *Crash Cushion Attenuating Terminal (CAT)* is a proprietary device designed to provide anchorage for a W-beam barrier. The system, which is 14 m long, includes energy-absorbing beams, breakaway wooden posts, and a cable anchorage. It is designed to crush in stages, depending on the severity of the impact. It is well suited for use with a semirigid median barrier or in a narrow gore area. If used as an attenuation device, an additional transition section is required before the rigid object (bridge pier or concrete median barrier). It has been successfully crash-tested with 820-kg and 2000-kg vehicles.

The *Safety End Treatment (SENTRE)* terminal is a proprietary device used with a W-beam or thrie-beam barrier. The 7-m-long system is made up of telescoping thrie-beam fender panels, slip-base support posts, and sand-filled plastic containers, mounted on a common concrete pad. The basic parts are generally reusable, but components must be kept on hand and maintenance people must be trained. The terminal performed well with 820-kg and 2000-kg vehicles, installed parallel to the rail or flared with a 1220-mm end offset.

The *Transition End Treatment (TREND)* is a proprietary system used to shield rigid barrier ends and other fixed objects where the available longitudinal distance is limited. This 6-m-long device is similar to the SENTRE except that a steel tension strap connects the upper rear of all posts to the rigid barrier. It is intended for use at locations where there is insufficient length for other transition systems, such as at bridge rails or similar fixed objects. The area behind the terminal remains vulnerable to penetration. The terminal performed well with 820-kg and 2000-kg vehicles.

The *three-cable* terminal has been developed and modified to accommodate three-cable barriers with S75 × 8.5 posts. The cable barrier is flared backward at full height to a slip-base end post that is offset 1070 mm from the barrier. The three cable strands are then turned down at 45° and anchored to a concrete block in the ground. A clear area must be provided in back of the terminal to allow an impacting vehicle to pass through. The modified terminal performed well with 820-kg and 2000-kg vehicles.

The *Brakemaster* terminal is a proprietary device used as a terminal for a W-beam barrier or as a crash cushion for protecting narrow objects. The system is made up of an anchor assembly, a cable-brake assembly, and W-beam panels. It is intended for use in low-frequency impact areas. Testing in accord with NCHRP 230 was reportedly satisfactory.

The *Advanced Dynamic Impact Extension Module (ADIEM)* is a proprietary device used at the end of a concrete barrier. It is made up of two perlite concrete, lightweight crushable modules, mounted on a 9-m-long base structure of standard concrete. It has been successfully crash-tested with 820-kg and 2000-kg vehicles in accordance with NCHRP 230.

A backslope anchorage is sometimes used to terminate a traffic barrier, such as in roadway cut sections or in a cut-to-fill transition. By flaring the barrier into the backslope area, the exposed portion of the barrier can redirect impacting vehicles and the hazards of an untreated end are mitigated. Previous criteria given for flare rate apply in the clear zone. Grade the approach to 1:10 or flatter and avoid ditches. The anchorage must develop the strength of the rail element. Crash testing with 820-kg and 2000-kg vehicles has shown mixed results, depending upon whether a ditch was present, height in the flared region, and the sharpness of the turnback into the backslope.

An *earth berm* can be ramped in front of a barrier end to provide protection. The berm must be designed with appropriate slopes and must be constructed to start beyond the length of need for the barrier. In a typical installation, an overall taper of 1:20 was used for the berm, with side slopes varying from 1:4 at the beginning to 1:2 at the barrier. Earth berms are typically used with rigid and semirigid median barriers, and end anchorage for the latter must still be provided. A vehicle hitting the berm head on may well straddle the barrier, but the consequences should be more favorable than an end-on crash into an unprotected barrier. There are no crash-test data available.

The *Wyoming box beam end terminal (WYBET)* consists of a nosepiece welded to a square tube, which is inserted into a larger tube that contains a crushable two-stage composite fiberglass tube, a wood support post, and a cable anchorage. The terminal may be installed parallel or flared out at a rate of up to 1:10. The device performed well when crash-tested with 820-kg and 2000-kg vehicles at 100 km/h, in both end-on and angled impacts.

6.12.2 Crash Cushions

Crash cushions are impact attenuators developed to prevent errant vehicles from impacting fixed obstacles. The crash cushion should either decelerate the vehicle to a safe stop, such as in a head-on hit; or redirect it safely away from the obstacle, in the case of a side hit. Crash cushions are typically used where fixed objects cannot be removed, relocated, converted to a breakaway design, or shielded by a longitudinal barrier. Examples of application sites include exit ramp gores where a bridge rail end or bridge pier presents a hazard, and the ends of longitudinal barriers. Most crash cushions are patented systems developed and tested by the manufacturer, who can also provide design charts for selection of appropriate designs.

Most crash cushions perform their function by the principle of kinetic energy absorption or transfer of momentum. In the first case, energy is absorbed by materials or devices that crush or plastically deform, or by hydraulic devices. A rigid backup support is required. In the second case, the momentum of the vehicle is transferred to an expendable mass, such as containers filled with sand. No rigid backup support is needed for such "inertial" barriers. Some crash cushions use a combination of these principles.

Table 6.8 provides a list of some of the most common crash cushions in use today, a brief description of how they function, and information on capabilities. All were developed primarily for passenger car attenuation, and some were discussed previously in Art. 6.12.1. One facility specially developed for runaway trucks on grades is the gravel-bed attenuator. For information and design guidelines, see the AASHTO publication *A Policy on Geometric Design of Highways and Streets.*

6.12.3 Selection of Crash Cushions

Selection of the most appropriate crash cushion depends on site characteristics, performance of the systems, maintenance characteristics, and life cycle cost. Both the geometrical conditions encountered and the space requirements for the different systems vary widely. Obstacles greater than 5 m wide can be shielded by systems such as arrays of sand-filled barrels, the CIAS, or the bullnose attenuator. Where space is limited, narrow systems such as the CAT or the GREAT may be appropriate. The structural and safety characteristics of alternative systems must be carefully reviewed and matched with needs. Items to consider include impact deceleration, redirection capability, impact debris, and anchorage and backup requirements. Table 6.9 has been prepared to compare the maintenance requirements of the different systems. Agency maintenance records should be used to establish associated costs. After potential systems have been identified for a given site, the final selection should be based on a life

TABLE 6.8 Characteristics of Operational Crash Cushions

Designation	Description	Mechanism	Performance
Hi-Dro Sandwich System*	Nest of water-filled plastic tubes with plastic fender panels, cable restraints, rigid backup.	Discharge of water from collapsible plastic tubes, movement of cushion mass, cushion drag.	Crash tests 1967–1973. Tests per NCHRP 230 or 350 not available.
Hi-Dro Cell Cluster*	Smaller version of Hi-Dro Sandwich for speeds ≤70 km/h.	Same.	Same.
Hex-Foam Sandwich System*	Similar to Hi-Dro Sandwich but uses cartridges instead of water-filled tubes.	Crushing of expendable cartridges of polyurethane foam matrix, etc.	Extensive test results from manufacturer.
Guardrail Energy Absorbing Terminal (GREAT)*	Hex-Foam cartridges held in place by telescoping corrugated steel-rail panels.	Same.	Same.
Sand-filled plastic barrels*	Barrels are clustered in front of the obstacle.	Momentum transfer, no backup required.	No side impact redirection. Extensive testing for 820-kg and 2000-kg vehicles.
Connecticut impact attenuation system (CIAS)	Cluster of thin-wall steel cylinders, steel tension straps, compression pipes, rigid backup.	Collapse of steel cylinders.	Not available.
Bullnose attenuator	W-beam barrier in oval-type pattern around obstacle; some use cable anchorages.	Rail tension, post bending.	Not available.
Dragnet*	Chain-link fence, steel-tape reels, anchor posts.	Steel tape is deformed as it is pulled through a set of rollers.	2000-kg vehicle stopped in 20 m at average deceleration of 2g.
CAT*	Slotted telescoping W beams, breakaway wooden posts, cable anchorage.	Beams crush in stages.	Successfully crash-tested with 820-kg and 2000-kg vehicles.
Brakemaster*	Anchor assembly, cable-brake assembly, and W-beam panels.	Frictional resistance of brake-cable assembly.	Testing in accord with NCHRP 230 was satisfactory.

TABLE 6.8 Characteristics of Operational Crash Cushions (*Continued*)

Designation	Description	Mechanism	Performance
Low-Maintenance Attenuator (LMA)*	Modular bays of elastomeric cylinders, steel diaphragms, thrie-beam rail.	Compression of cylinders, telescoping of thrie beam.	No side impact redirection. Extensive testing for 820-kg and 2000-kg vehicles.
Advanced Dynamic Impact Extension Module (ADIEM)*	Perlite concrete, lightweight modules, mounted on a base structure of standard concrete.	Crushing of lightweight concrete.	Testing with 820-kg and 2000-kg vehicles (NCHRP 230) was satisfactory.

*Proprietary system.

cycle cost analysis. (See Chap. 10.) Costs to consider are the initial cost of the device, site preparation and installation costs, and maintenance costs, as well as the cost of accidents.

6.12.4 Placement of Crash Cushions

For proper performance, crash cushions should be placed on level terrain with a clear path between the roadway and the attenuator so the vehicle can strike at normal height, with the suspension system in a neutral state. Avoid curbs or slopes in front of the device. Install the attenuator on a smooth surface (usually concrete) so it can compress uniformly. Conspicuous, well-delineated crash cushions are less likely to be hit than those that blend into the background. If the system is not reflective, install standard object markers to improve visibility at night and during inclement weather.

TABLE 6.9 Comparative Maintenance Requirements for Crash Cushions

Type of unit	Regular maintenance	Collision repair	Material storage
Hi-Dro Sandwich System,* Hi-Dro Cell Cluster*	Must be inspected on-site to determine if fluid level is adequate; unit is susceptible to vandalism; unit must be inspected more frequently during extreme high or low temperatures.	Unit is generally reusable after a collision; liquid on roadway may present a temporary problem in some cases.	Extra cells should be stored to replace any damaged by vandals or by collisions; antifreeze solution or liquid calcium chloride needed to prevent freezing in cold climates.
Hex-Foam Sandwich System,* GREAT System*	Can normally be inspected on a drive-by; missing or displaced Hex-Foam cartridges can be readily noted.	Unit is generally reusable after a collision; expendable Hex-Foam cartridges must be replaced after unit is repositioned.	Hex-Foam cartridges and other replacement parts per manufacturer's recommendations.
Sand barrels	Can be inspected on drive-by for external damage. If lids are not riveted on, sand content should be checked periodically.	Individual sand and barrels must be replaced after a collision; units damaged by nuisance hits must also be replaced. Debris must be removed from site.	Spare barrels, sand support inserts, and lids; supply of sand.
CIAS	Can be inspected on drive-by.	Crushed units must be removed from site; minor damage can be repaired on-site by jacking.	Spare cylinders to replace badly damaged units.
Bullnose attenuator	Can be inspected on drive-by.	Similar to roadside barrier repair.	Standard barrier hardware, including posts.
Dragnet*	Can be inspected on drive-by.	Energy-absorbing reels must be replaced; remainder of unit usually reusable. Spent reel casings can be refurbished and reused.	Energy-absorbing reels of steel tape.
LMA*	Can be inspected on drive-by.	Unit generally reusable after collision.	Side fenders and other replacement parts per manufacturer's recommendation.
CAT*	Can be inspected on drive-by, except for cable tension, which should be checked periodically.	Nose, rail elements, and wood posts must be replaced. Foundation tubes are normally reusable.	Rail elements and wood posts.
Brakemaster*	Can be inspected on drive-by. Should be inspected on-site periodically.	Most aboveground components can be damaged and need replacement.	Braking mechanism, fender panels, diaphragms, etc., per manufacturer's recommendation.

*Proprietary system.
Source: From *Roadside Design Guide*, AASHTO, Washington, D.C., 1996, with permission.

6.69

CHAPTER 7
SIGNING AND ROADWAY LIGHTING

PART 1

SIGNING

Brian L. Bowman, P.E.

Associate Professor
Auburn University
Auburn, Alabama

Part 1 of this chapter presents a comprehensive review of the design, construction, and maintenance of highway signs. Both single- and multiple-mounted sign supports are addressed, with an emphasis on highway safety. Breakaway supports with various types of slip bases, frangible bases, and post hinging systems are explained and illustrated. Commercially available devices and alternatives are identified and discussed. Guidelines on use and construction are summarized. An extensive list of references, which are noted in the text, concludes the section. Much of this material was derived from studies made by the author under a Federal Highway Administration project, "Design, Construction and Maintenance of Highway Safety Features and Appurtenances."

7.1 TRAFFIC SIGNING NEEDS

The capability of roadways to safely and efficiently serve vehicular traffic is dependent to a large extent on the adequacy of traffic control devices. The majority of motorists drive in an orderly and safe manner, provided they are given reliable regulatory, warning, and guide information. Motorists, through training and experience, develop expectations on when and in what manner they will be provided necessary information for safely controlling their vehicles. Motorists expect that similar traffic control devices will always have the same meaning and will require the same motorist action regardless of where they are encountered. This expectation has been enhanced by the use of uniform traffic control devices which enable motorists to consistently interpret the general intent of a device by its message, shape, and color.

The advantages of traffic control device uniformity were recognized long ago. The American Association of State Highway Officials published specifications of road markers and signs for rural roadways in 1925. A manual for urban roadways was published in 1929 by the National Conference on Street and Highway Safety. The unification of the standards applicable to the different classes of roadways was addressed by a joint committee of the American Association of State Highway Officials and the

National Conference on Street and Highway Safety. The joint committee developed, and printed in 1935, the first *Manual on Uniform Traffic Control Devices for Streets and Highways (MUTCD)* [1]. That joint committee, although subsequently reorganized and named the National Committee on Uniform Traffic Control Devices (NCUTCD), has been in continuous existence and contributes to periodic revisions of *MUTCD*.

The benefits of traffic control device uniformity include increasing safety by providing the road user with required information for vehicle guidance or control at the right time and place and in the proper manner. Signs should be installed only where warranted. This can include locations where special regulations apply at specific places or specific times or where hazards are not self-evident. They also provide information of highway routes, directions, destinations, and points of interest. The general standards for signs provided in Sec. 2A of the *MUTCD* and those sections pertaining to the particular type of sign being installed should be followed to ensure proper placement and message uniformity.

7.1.1 Uniformity Considerations and Necessary Deviations

While the advantages of uniformity far outweigh the disadvantages, there are some undesirable effects when complete uniformity is maintained. One of the principal disadvantages is that strict uniformity may result in the failure to adopt an improved device or procedure simply because it is not in common use. In addition, total uniformity would require the specification of a separate traffic control device for every conceivable roadway geometric and traffic operational condition. This would be a monumental task that undoubtedly would still not cover every situation, while simultaneously increasing the size of *MUTCD* with devices of limited application.

This difficulty is recognized in *MUTCD*, which indicates that warning signs other than those specified in the manual may be required under special conditions [2, Sec. 2C-1]. *MUTCD* requires exercising good engineering judgment in determining the need for other warning devices. It also mandates that the innovative devices be understood easily by the motorist. Ensuring that warning signs are easily understood necessitates that they be of standard shape and color and that the legends be unambiguous and brief. Establishing the need for distinct warning devices can be accomplished by identifying when standard devices do not properly address unusual conditions. While these conditions are unusual, they can typically be classified into the same use categories that are appropriate for standard warning signs. The *Traffic Control Devices Handbook* [3] identifies the following uses of warning devices:

- To indicate the presence of geometric features with potential hazards
- To define major changes in roadway character
- To mark obstructions or other physical hazards in or near the roadway
- To locate areas where hazards may exist under certain conditions
- To inform motorists of regulatory controls ahead
- To advise motorists of appropriate actions

The need to provide advance warning for unusual roadway, roadside, operational, and environmental conditions has resulted in the development of a wide diversity of devices. The majority of these devices can be categorized as warning signs containing different symbols and legends. Other warning devices include flashing beacons, rumble strips, pavement surface treatments, and pavement markings. Device complexity ranges from simple passive warning signs to devices that are activated by vehicle speed, headway, or presence on one or more approaches to a potentially hazardous roadway element. Further information on supplemental warning and rumble strips can be obtained from the National Cooperative Highway Research Program (NCHRP) publications *Synthesis of Highway Practice 186: Supplemental Advance Warning Devices* and *Synthesis of Highway Practice 191: Use of Rumble Strips to Enhance Safety* [4, 5].

7.1.2 Legal Responsibility

Estimates by the FHWA indicate that there are an average of 15 signs per mile on the nation's 3.8 million miles of streets and roadways [6]. The resultant 57 million traffic signs represent a huge investment in materials, labor, equipment, and maintenance costs. While this is a significant investment, improvements using standard traffic control signing are reported in the "1988 Annual Report on Highway Safety Improvement Programs" as having the highest benefit-cost ratio of any highway safety improvement [7]. Properly designed, located, and maintained standard traffic signs and other carefully conceived devices can be an effective method of increasing traffic and operational efficiency and subsequently decreasing the tort liability exposure of roadway agencies.

The concerns about tort liability judgments are valid, as the number of cases is steadily increasing. In almost every state, the shield of sovereign immunity either has been abolished by judicial decisions or has been eroded by legislative modifications to governmental immunity. In one state, for example, the legislature was instructed to enact comprehensive tort claim procedures in the near future or the doctrine of immunity would be abrogated by the State Supreme Court. In another state, the concept of sovereign immunity was declared unconstitutional [8].

A tort is a civil wrong or injury. The purpose of a tort action is to seek compensation for damages to property and individuals. The following elements must exist for a valid tort action:

- The defendant must owe a legal duty to the plaintiff.
- There must be a breach of duty; that is, the defendant must have failed to perform a duty or performed it in an improper manner.
- The breach of duty must be a proximate cause of the accident that resulted.
- The plaintiff must have suffered damages as a result.

In highway-related tort cases, the first element is relatively easy to establish. Roadway authorities have been vested with the responsibility of providing reasonably safe travel opportunity for roadways under their jurisdiction. The failure of the roadway agency to properly perform that duty, and that this breach of duty was the proximate cause of an accident, are more difficult to establish. In most instances, establishing that a breach of the legal duty occurred becomes a major issue in tort liability cases. Plaintiffs typically will attempt to establish that the agency having roadway jurisdiction was negligent in its duty and/or a physical condition was permitted to exist that was a hazard.

Negligence is the failure to exercise such care as a reasonably prudent and careful person would use under similar circumstances. Roadway agencies can be judged negligent in two ways: (1) wrongful performance (misfeasance), or (2) the omission of performance when some act should have been performed and was not (nonfeasance).

Roadway agencies can, therefore, be judged negligent either by addressing a safety problem incorrectly or by ignoring it. The critical issue in highway tort liability is the care with which highway agencies perform their responsibilities. If it is judged that a reasonable standard of care was not exercised, then the responsible persons and/or organizations may be held liable for injuries and damages that resulted.

In an attempt to familiarize roadway agencies and their employees with the potential liability, and to make them aware of their duties and responsibilities to the traveling public, the NCHRP published *Synthesis of Highway Practice 106: Practical Guidelines for Minimizing Tort Liability* [8]. In particular, this publication advises agencies to supply a consistent highway environment for motorists. The use of standard design features and uniform traffic control devices is also emphasized.

All states are to adopt the standards of *MUTCD* as the basis for designing and installing traffic control devices. Some states adopt *MUTCD* in its entirety, while other states incorporate additional devices and practices into their manuals which address their specific roadway design and driver expectancy needs. *MUTCD* provides minimal requirements, and states that do prepare their own manuals are required to conform to the national standard. Additional devices not included in either the federal *MUTCD* or those of the states are frequently developed to provide motorist warning of roadway hazards which, ideally, should be eliminated. The reasons for not eliminating the hazard can include geometric constraints, planned improvements, usefulness of the condition for other purposes, burden of removing the condition, and the lack of a method to correct the situation. When the need to warn motorists involves commonly encountered hazards, such as a stop sign ahead on a rural roadway, then an appropriate warning device can be found in *MUTCD*. When the hazard is posed by unusual or unique conditions, however, the highway engineer is placed in the difficult position of identifying, or often designing, a warning device that provides a clear message to the motorist of the potential hazard. It should be emphasized that the installation of a warning device does not remove the agency from liability, especially if it can be shown that it was reasonably possible to eliminate the hazard.

7.1.3 Design of Supplemental Warning Devices

Designing a warning device that provides a clear, unambiguous message to the motorist can be a difficult task. The difficulty is due in part to the concern of the engineer to act in a "reasonable and prudent" manner. Increasing motorist safety and minimizing liability requires that the device provide a readily understood and unambiguous message.

In the design of warning signs, it is important to remember that signs are designed to draw attention to themselves through contrast, color, shape, composition, reflectorization, and illumination, with a simple message providing a clear and understandable instruction to the motorist. Sign size, symbol size, lettering size, and placement should be such to allow adequate time for proper response. Uniform and reasonable instructions to the motorist will instill respect and develop willing compliance with the sign message. For these reasons, the majority of general warning signs should be designed as diamond shapes with black letters on a yellow background. Standard sign letters are prescribed in the *Standard Alphabets for Highway Signs,* which should be used to develop lettering size and style [9]. Sections 2C-1, 2C-2, and 2C-40 of *MUTCD* contain information that must be followed in the design of warning signs. In addition, Sec. 1A-7 of *MUTCD* lists additional publications and documents that provide requisite information for the proper design of warning signs.

7.1.4 Identifying Need for Supplemental Advance Warning Devices

Locations which would benefit from the installation of supplemental advance warning devices typically exhibit safety and/or operational problems. Establishing the need for supplemental devices, therefore, requires identifying the problem locations and performing a safety and/or operational analysis. Deficient locations can be identified by a traffic safety management system, citizen complaints, employee observations, and by safety analysis during a planned resurfacing, restoration, and rehabilitation (RRR) project.

Accident-based studies are used to identify locations that can be considered hazardous due to a large number of accidents. These studies involve the review and analysis of systemwide accident information. To compare the accident experience of several locations, the length of time over which accidents are counted, the traffic volumes, and the length of roadway section involved should be the same at each location. If not, accident rates may be compared between locations, provided that a common unit of exposure (e.g., accidents per million vehicle miles for longer roadway sections, or accidents per million entering vehicles for spot locations and intersections) is used.

Potential locations can also be identified by complaints received from citizens and by observations made by employees. Often a combination of accident analysis and an investigation of complaints and observations are required for low-volume roadways. Complaints about "near misses" and observations of hazardous roadway elements can be considered indicators of site deficiencies. This type of information is treated by some agencies with the same importance as a documented accident history. Such treatment has the advantage of reducing the number of accidents required to identify the hazardous roadway locations.

It should be recognized that maintaining a complaint and employee observation file requires that the agency be responsive to these inputs. Complaints and observations are notifications of hazards that become a matter of public record and are available as evidence should an accident result in litigation. This alone is not a valid reason to fail to maintain a complaint and observation file. If a defect is allowed to remain for an unreasonable period of time, even if no complaints or observations were received, the courts can consider it as constructive notice and assign liability. Complaint and observation files should, therefore, be maintained and a program established to respond to all complaints and to document facts and engineering decisions to minimize the possibility of lawsuit losses.

An opportune time to identify the need for a device is during the design phase of projects primarily intended to upgrade the physical and operational characteristics of the roadway. This opportunity can be used to detect safety and operational deficiencies and to select appropriate improvements that can be incorporated into the upgrading project.

The identification of potential locations for each of the previous methods should include a field inspection to help establish the cause of the deficiency and appropriate countermeasures. If the site inspection indicates that the deficiency cannot be readily corrected due to cost or physical constraints, then an advance warning device should be installed. If the site conditions are sufficiently unusual that an appropriate warning device is not contained in the federal or appropriate state *MUTCD*, then a supplemental device may need to be used or developed until it is feasible to take care of the underlying problem.

For example, consider a situation where a sag vertical curve was constructed to provide sufficient vertical bridge clearance on a roadway with a posted speed of 70 km/h. Analysis of the areawide accidents indicated that there is a higher than expected occur-

FIGURE 7.1   Example of supplemental advance warning sign.

rence of intersection-related and rear-end accidents at a signalized intersection immediately downstream of the bridge. A visit to the site indicated that the signal faces were not visible to approaching drivers until they were 120 m from the stop line. Since this distance is less than the minimum visibility distance of 140 m specified by Sec. 4B-12 of *MUTCD*, a Signal Ahead sign (W3-3) was installed [2]. The engineer determined that, although the minimum recommendations of *MUTCD* were being achieved, safety improvements could be achieved by providing real-time warning that a stop will be required at the intersection. Since removing the sight obstruction was not possible, the engineer considered lowering the speed limit and/or providing additional motorist warning. Experience with lowering speed limits indicated that this countermeasure was not an effective long-term solution. The engineer decided to install an active supplemental advance warning device with the legend "Prepare to Stop When Flashing" configured as shown in Fig. 7.1. The device was installed over the roadway, 150 m in advance of the stop bar, and interconnected with the traffic signal controller. The horizontally mounted beacons were timed to flash yellow 8 s prior to the red indication so that drivers passing the beacon at the legal speed limit would have advance warning of the required stop at the intersection. The yellow beacons continued to flash until 3 s before the end of the red indication to allow the start of queue dissipation. Motorists not encountering the flashing lights could expect not needing to come to a complete stop at the signal, while still having the signal presence reinforced by the overhead sign. The engineer plans to continue monitoring the location to determine if the active advance warning device is effective in reducing accidents.

7.1.5   Concerns on Use of Supplemental Advance Warning Devices

A large number of supplemental advance warning devices have been used by roadway agencies to inform motorists of unusual geometric, operational, or traffic control features. The use of a device by an agency does not imply that it is a viable or desirable device to use for identified deficiencies. The following concerns should be considered prior to the installation of any device not specified in *MUTCD*.

- Many warning devices are attempts at political, inexpensive, and/or quick solutions to totally inappropriate roadway conditions. The proper countermeasure for many of these conditions is to correct the fault rather than installing an additional motorist warning. Installing a supplemental warning device should be considered a temporary countermeasure until the inadequate roadway conditions can be corrected.

- *MUTCD* provides guidance on the proper placement of traffic control devices to provide adequate time for motorists to perceive, identify, decide upon, and perform any necessary maneuver. Section 2C-3 provides guidelines for the minimum placement distances of warning signs, while Sec. 4B-12 specifies the minimum continuous visibility distances that should be present for motorists approaching a traffic signal. The inability to provide the minimum visibility distance is one indication of the need to install an advance warning sign. Guidelines on the height and lateral location of signs are summarized in Fig. 2-1 in *MUTCD* [2]. The guidelines of Pts. 1, 2A, and 2C of *MUTCD* should be followed for the installation of all traffic signs.

- Section 2C-2 of *MUTCD* states that warning signs shall consist of a black legend and border on a yellow background [2].

- Section 2C-40 of *MUTCD* permits the design of warning signs for special conditions [2]. These signs should, however, be constructed with clear and concise verbal messages. Letter legibility and size, combined with placement, must provide a clear meaning and provide ample time for response. Sections 1A-2, 1A-6, 2A-13, and 2C-40 of *MUTCD* provide an approval process for new symbols and do not permit the use of symbols that are new or unique and, thereby, not readily understandable by the motorist [2]. The only exception to the provision of nonstandard symbols is where minor modifications to *MUTCD* symbols are necessary to adequately describe specific design elements of the roadway (Sec. 1A-2). An example of a permitted symbol modification is displaying a curve on a "Side Road" sign (W2-2) if the side road occurs in the vicinity of a horizontal curve. Devices that use symbols not contained in *MUTCD,* or in *Standard Highway Signs,* are nonstandard devices [2, 10].

- Warning devices should have the same silhouette shape as the device shape. For example a 915-mm×915-mm diamond warning sign mounted on a 1220-mm×1220-mm square piece of plywood would not satisfy the shape requirement. Dawn and dusk light conditions, fog, and other poor-visibility situations can result in interpreting the warning sign as a guide sign.

- Section 2A-20 of *MUTCD* permits the use of hazard identification beacons to supplement an appropriate warning sign or marker [2]. The hazard identification beacon consists of one or more sections of the circular yellow traffic signal head indication with a visible diameter of not less than 205 mm. *MUTCD* prohibits the placement of the beacons within the border of the sign except when used with a School Speed Limit sign [2, Sec. 4E-1]. Section 4E-5 states that when flashing beacons are used to supplement warning signs, they should be horizontally or vertically aligned. When the beacons are used in conjunction with a sign that is longer horizontally than vertically, they can be horizontally aligned similar to the provision of Sec. 4E-2. If two beacons are used, they should be alternately flashed at a rate of not less than 50 nor more than 60 times per minute.

- Unique situations in the roadway environment can result in the need for changes or additions to *MUTCD*. Section 1A-6 provides the procedure to be followed for consideration of a new device to replace a present standard device, for additional devices to be added to the list of standard devices, or for revisions to recommended application. Agencies that encounter the frequent need of a unique application are encouraged to request permission to experiment from the Federal Highway Administration, Office of Traffic Operations (HTO-30), 400 Seventh Street S.W., Washington, DC 20590.

7.2 CRASHWORTHY CONCERNS OF ROADSIDE FEATURES

The need for traffic signs, roadway illumination, utility service, and postal delivery results in roadside features frequently placed within the roadway right-of-way. (Also see Chap. 6, Safety Systems.) The presence and location of these obstacles varies by roadway type and location. Rural freeways, for example, can be designed where traffic signs are the only obstacles that are added to the roadside. Signs, light pole standards, utility poles, and mailboxes are all frequently encountered on rural collectors. These obstacles, when present, perform a necessary function, but are also potential fixed objects for an errant vehicle. To reduce accident severity it is important that signs, roadway illumination supports, utility poles, and mailboxes be properly designed, located, and placed when necessary within the right-of-way. As a general rule, there are a number of options that can be used by design engineers to provide a safe design. In order of preference these options are

- Do not install the obstacle.
- Install it on existing overhead structures, where it does not become an additional fixed object hazard.
- Locate the feature away from the traveled way or behind existing barriers where it will be less likely to be struck.
- Reduce impact severity by using appropriate breakaway or yielding design.
- Shield the feature with a properly designed longitudinal barrier or crash cushion if it cannot be eliminated, relocated, or redesigned.
- Delineate an existing feature if other measures are not practical. Putting up hazard markers is a cost-effective method for alerting motorists to an existing hazard. Obviously, delineators will not make any difference if a driver hits the object, but they might help a driver avoid running off the road at that spot.

Yielding or breakaway supports are recommended on all types of sign, luminaire, traffic signal, and mailbox supports that are located within the desirable clear zone. The clear zone is the total roadside area, starting at the edge of the traveled way, that is available for safe use by a vehicle. The desirable width of the clear zone is dependent upon traffic volume, speed, and the roadside geometry. The transversable area is the roadside border area that permits a motorist to maintain vehicle control including being able to slow and stop safely. The transversable area can exceed the desirable clear zone called for in the *Roadway Design Guide* [11]. Only yielding or breakaway supports should be permitted in the transversable roadside, even if it is located beyond the clear zone. In those instances where yielding or breakaway supports are not possible, such as large cantilever sign installations, shielding with crash cushions or guardrail should be used.

Yielding supports refer to those supports that are designed to remain in one piece and bend at the base upon vehicle impact. The anchor portion remains in the ground and the upper assembly passes under the vehicle. The term *breakaway support* refers to support systems that are designed to break into two parts upon vehicle impact. The release mechanism for a breakaway support can be a slip plane, plastic hinges, fracture elements, or a combination of these.

The technology of yielding and breakaway support systems has experienced dramatic improvements. These improvements were prompted by an increased emphasis on roadside safety and by the large reduction that has occurred in the weights of automobiles. Many foreign and domestic automobiles on our roadways weigh less than 1020 kg, which was at the bottom of the domestic weight range in 1975. By 1983 the trend to more fuel-efficient automobiles had resulted in approximately 40 percent of auto sales being vehicles weighing less than 1020 kg. Automobiles of 725 kg and less are now operating on U.S. highways. The typical family automobile weighs somewhere between 900 kg and 1800 kg, with only the luxury, and a few station wagons, weighing more. A survey of high-level automotive industry leaders, conducted by the University of Michigan, indicates that the total vehicle weight will remain fairly constant through the year 2000 [12].

The evolving safety feature environment and the change to the vehicle fleet weights have resulted in a number of revised standard specifications for the testing and acceptance of yielding and breakaway support systems. Requirements for yielding and breakaway support systems were introduced by AASHTO in 1975 and revised in 1985 to keep abreast of new research and development. Two of the most significant changes in the 1975 and 1985 specifications are the reduction in weight of the design vehicle from 1020 kg to 820 kg and the change from measures of momentum to measures of change in velocity. These changes, however, do not imply that safety features that satisfied the old specifications do not satisfy revised specifications. For example, the 1985 standard testing guidelines require that supports should impart a preferred vehicle change in velocity of 3.1 m/s or less, but not more than 4.6 m/s. A support that would cause a 1020-kg vehicle (i.e., 1975 design vehicle weight) to experience a 3.4 m/s change in vehicle velocity at a test speed of 32 km/h would likely result in 4.6 m/s change in velocity when tested under the same conditions with an 820-kg vehicle (i.e., 1985 design vehicle weight) [13]. These values compare favorably with the change in momentum requirements cited in the 1975 specifications. Supports that had acceptance test numbers near the preferred values for the old specification can, therefore, be expected to meet the new specification requirement.

Some of the changes in the 1985 AASHTO standard specifications were due to testing guidelines contained in NCHRP Report 230 [14]. NCHRP Report 350 establishes current testing guidelines for vehicular tests to evaluate the impact performance of permanent and temporary highway features, and supersedes those contained in NCHRP Report 230 [15, 14]. These guidelines include a range of test vehicles, impact speeds, impact angle, point of impact on the vehicle, and surrounding terrain features for use in evaluating impact performance. Acceptance testing of yielding and breakaway supports requires evaluation in terms of the degree of hazard to which occupants of the impacting vehicle are exposed, the structural adequacy of the support, the hazard to workers and pedestrians that may be in the path of debris from the impact, and the behavior of the vehicle after impact. The guidelines include requirements for:

- The structural adequacy of the device to determine if detached elements, fragments, or other debris from the assembly penetrate, or show potential for penetrating, the passenger compartment or present undue hazard to other traffic.

- A range of preferable and maximum vehicle changes in velocity resulting from impact with the support system. The preferable change in vehicle velocity is 3.0 m/s or less. The maximum acceptable change in vehicle velocity is 5.0 m/s. Note that due to conversion to the SI system the limiting velocity changes were rounded and consequently are not precisely the same as those in NCHRP Report 230 [14].

- The impacting vehicle to remain upright during and after the collision.
- The vehicle trajectory and final stopping position after impact to intrude a minimum distance, if at all, into adjacent or opposing lanes.

It is important to use only those support assemblies that have been tested, using the standard specifications, and subsequently approved for use by the FHWA. This is true for city and county jurisdictions where roadway speeds are generally less than what can be expected on state and rural roadways. Impacts with supports can be hazardous even at lower speeds, especially for occupants of a small vehicle. It should be noted that many supports can be more hazardous at low speeds (25 to 40 km/h) than at high speeds (90 to 100 km/h). For example, sign supports that fracture or break away can be more hazardous at low speeds, where the energy imparted to the support is not sufficiently large to make the device swing up and over the vehicle. The result can be intrusion of the lower portion of the support into the passenger compartment. Similarly, devices designed to yield are generally more hazardous at high speed, due to the reduced time available for deformation and subsequent passage beneath the vehicle.

The acceptance testing guidelines are intended to enhance experimental precision while maintaining cost within acceptable bounds. The wide range of vehicle speed, impact angle, vehicle type, vehicle condition, and dynamic behavior with which vehicles can impact the support cannot be economically replicated in a limited number of standardized tests. The use of an approved device does not, therefore, guarantee that it will function as planned under all impact conditions. However, the failure or adverse performance of a highway safety feature can often be attributed to improper design or construction details. The incorrect orientation of a unidirectional breakaway support, or something as simple as a substandard washer, are major contributors to improper function. It is important for proper device function that the safety feature has been properly selected, assembled, and erected and that the critical materials have the specified design properties.

When possible, and appropriate, the placement of traffic signs, luminaires, and utility and mailbox supports should take advantage of existing guiderail, overhead structures, and other features that will reduce their exposure to traffic. Care should be taken to ensure that supports placed behind existing, otherwise required barriers are outside the maximum design deflection standards of the barrier. This will prevent damage to the support structure and help ensure that the barrier functions properly if impacted. The design deflections are based on crash tests using a 2000-kg vehicle impacting the barrier at 100 km/h and an angle of 25°. The crash tests are conducted under optimum conditions. Other conditions such as wet, frozen, rocky, or sandy soil may result in deflections greater or less than the design values. Typical anticipated deflections are presented in Table 7.1.

7.2.1 Need Determination and Placement of Traffic Signs

Estimates on the number of signs present on our roadways vary drastically. An NCHRP synthesis indicated 58 million signs, while a study for the FHWA estimated that there are approximately 250 million sign assemblies on the U.S. roadway system [16, 17]. Signs contribute an important role in increasing the safety of the roadway by providing regulatory, warning, control, and guidance information to the driver. Every sign that is installed on its own support system, however, provides a fixed object for a potential collision. Even a relatively small sign on an apparent weak support can have

TABLE 7.1 Design Dynamic Deflections of
Various Barrier Types

Barrier type	Design deflection,* mm
Concrete safety shape	Minimal deflection
W-beam strong shape	900
W-beam weak post	1980
Box beam	1500
Cable guiderail	3350

*Deflections are measured from the back of the post at the top.

severe consequences when struck at high speed. The first step in the design of any sign is to determine if a sign is really needed and where it should be placed. The following points should be considered:

- *MUTCD* provides information on when traffic signs should be installed. In the case of regulatory signs, and in most cases for warning signs, there are specific warrants that should be met prior to installation [2]. Installing unnecessary signs increases operating and maintenance costs, increases the potential of fixed-object collisions, and reduces sign credibility to the motorist.

- *MUTCD* provides guidelines on the height and lateral placement of typical sign installations. These guidelines, which are presented in Fig. 7.2, provide the maximum practice lateral clearance from the edge of the traveled way for motorist safety. In rural areas, signs should be no closer than 1.8 m from the edge of the shoulder and no closer than 3.7 m from the edge of the traveled way if no shoulder exists. In urban areas, 600 mm is recommended as a minimum, with 300 mm from the curb face permissible where sidewalk width is limited.

7.2.2 Sign Assemblies

The sign panel, the support, and the embedment or anchorage system are the three components of a sign assembly. Each component contributes to the effectiveness, structural adequacy, and safety upon impact of the device. The sign assembly must be structurally adequate to withstand its own weight and the wind and ice loads subjected to the sign panel. In some northern climates, this requirement includes the forces created by snow ejected by snowblowers or the lateral forces resulting from snowplow activity. The majority of the design guides for each state contain recommendations on the size, number, and type of support required in different regions of the state. These guidelines are based on the size of the sign panel and the recurrent wind intensity. Average wind loads for 10-, 25-, and 50-year recurrence intervals are also contained in AASHTO's standard specification for structural supports [13]. Roadside sign structures that are considered to have a relatively short life expectancy may be designed using wind speeds based on the 10-year mean recurrence interval.

Opposing the need for structural adequacy is the requirement that the sign assembly provide a safe driving environment. This safety requirement necessitates that the complete sign assemblies are able to pass rigid testing criteria prior to approval for use

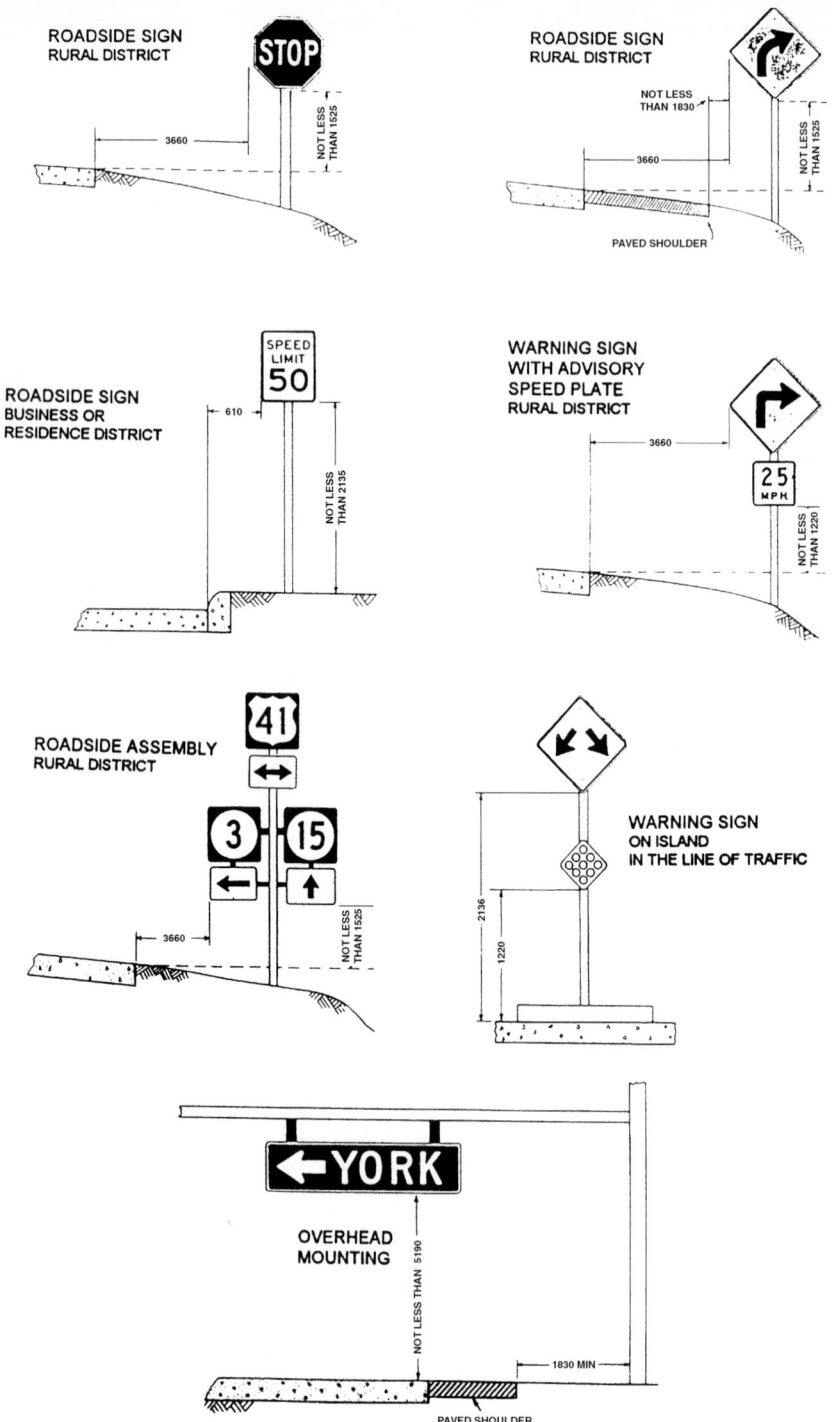

FIGURE 7.2 Height and lateral placement of signs. (*Adapted from* Manual on Uniform Traffic Control Devices, *FHWA, 1988*)

by the Federal Highway Administration. These standards are established with consideration to the specific purpose of the device and its performance as a unit. Installing a breakaway device that is designed to perform with impact from one direction (unidirectional) at a location which can experience impact from a number of directions (multidirectional) can result in improper performance. Only those devices that have been approved for use by the Federal Highway Administration should be installed. The installation of approved assemblies must be performed with proper design and construction to achieve the required performance.

7.2.3 Sign-Support Considerations

There are a variety of systems used to support ground-mounted traffic signs. These systems were often categorized by whether they were intended to support small or large signs. Small signs were arbitrarily defined as those having a total panel area of less than 4.7 m² [18]. This designation is, however, arbitrary and not effective in identifying the characteristics of the support used. An alternative method of categorizing sign types is by designating them as single- or multiple-mount systems. Multiple mounts include two or more supports that are separated by 2100 mm or more. Sign panels supported by a single support or by multiple supports less than 2100 mm apart are considered single mounts. The 2100-mm separation criterion allows for the possibility that a vehicle, leaving the roadway at an angle, can impact more than one support. Signs supported by more than one support, in addition to being separated by more than 2100 mm, must also be designed for each support to independently release from the sign panel. Multiple-support systems, therefore, must have sign panels with sufficient torsional strength to ensure proper release from the impacted support while remaining upright on the support(s) that were not impacted. This also requires that the remaining support(s) have sufficient strength properties to prevent the sign panel from breaking loose and entering the passenger compartment or becoming a projectile.

Metal supports that yield upon impact have been used for many years to provide effective economical supports for traffic signs. The U-channel post design is the most widely used support for both single-and multiple-support designs [18]. Yielding supports are designed to bend at the base and have no built-in breakaway or weakened design. Systems in this category include the full-length steel U-channel, aluminum shapes, aluminum X-posts, and standard steel pipes. For successful impact performance, the support must bend and lie down or fracture without causing a change in vehicle velocity of more than 5 m/s. Tests have shown that supports that fracture offer much less impact resistance, especially at high-speed impacts, than yielding supports of equal size.

The impact behavior of base-bending supports depends upon a number of complex variables including cross-sectional shape, mechanical properties, energy-absorption capabilities under dynamic loading, chemical composition, type of embedment, and characteristics of the embedment soil. The wide number of variables related to the properties of the support itself require that full-scale crash testing be performed to evaluate the impact behavior of base-bending supports. Tests are performed on categories of support types that need to be specified during their purchase. For example, U-channel posts, while of the same shape, will have different impact characteristics depending upon their unit weight and whether they are cold-rolled or hot-shaped.

The impact performance of base-bending supports depends upon the interaction between the structure and the soil in which it is embedded. Soil conditions vary drastically with location, even within small geographic locations. Due to this variability, NCHRP 350 has established *standard* soil conditions (previously referred to as "strong soil") and *weak* soil for testing. Weak soil consists of relatively fine aggregates that provide less resistance to lateral movement than that provided by a standard soil.

The rules on weak soil versus strong soil are, however, in question. The FHWA has insisted that yielding supports be qualified in both soils in order to be eligible for federal aid. However, recently completed crash testing yielded very few acceptable supports in weak soil. FHWA considers that it may be too restrictive to forbid all use of those supports that failed in weak soil. The standard soil in NCHRP Report 350 is the "strong soil." If a state has potential sites where the device will be installed in weak soils and believes that the device may not behave as well as in strong soil, then weak soil testing is called for. Otherwise, a device that has been found acceptable only in strong soil may be used only in strong soil.

The proper performance of some base-bending supports requires that they do not pull out of the soil upon low-speed impact. Placing these base-bending devices in weak soil, when they have been approved for use only in standard soil, or at an improper embedment depth will not provide acceptable low-speed performance. If the device was installed on a narrow median, for example, it can pull out of the ground upon impact and become a lethal trajectory to opposing traffic. Consideration must be given to the soil acceptance criteria of the post planned for use, the soil condition present, sign location, and the safety performance needs of the sign assembly.

Breakaway supports are designed to separate from the anchor base upon impact. Breakaway designs include supports with frangible couplings, supports with weakened sections, bolted sections, and slip base designs. Breakaway supports are classified by their ability to properly separate from the base upon impact from one direction (unidirectional) or from any direction (multidirectional). Large signs, requiring multiple supports separated by 2100 mm or more, often use a hinged breakaway mechanism with a horizontal slip base. The use of slotted hinge plates, on both sides of the upper beam, and a horizontal slip base results in proper device function from either the front or the back. The action of the hinged breakaway is illustrated in Fig. 7.3.

In addition to the yielding and breakaway sign support, overhead and fixed-base support systems may be used. Overhead sign support systems include the use of existing structures, such as bridges, that span the traffic lanes. Fixed-base support systems include those that do not yield or break away upon impact. Fixed-base systems are made of materials that will not fracture upon impact and are firmly embedded in or rigidly attached to a foundation. Fixed-base systems are often used to support overhead signs on roadway facilities with three or more lanes or for traffic signal supports. The large mass of these support systems and the potential safety consequences of the systems falling to the ground necessitate a fixed-base design. Fixed-base systems are rigid obstacles and should not be used in the clear zone area unless shielded by a barrier.

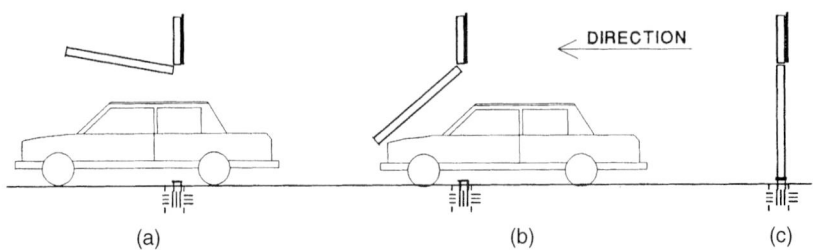

FIGURE 7.3   Illustration of hinged breakaway action. (*a*) Hinge activation. (*b*) Slip plate release. (*c*) Sign prior to impact.

The total combination of support systems and methods of embedment is large. Considering the following factors can assist in selecting the most appropriate sign support system.

- Large ground-mounted signs can be located 15 m or more from the edge of pavement on high-speed facilities. These substantial lateral clearances increase the roadside recovery zone while still meeting motorist viewing needs.
- The performance of any sign assembly is influenced by the surrounding terrain. Terrain that will cause the vehicle to impact the sign assembly at a higher or lower point than the design impact height can cause unpredictable and often hazardous results.
- The height of post-mounted signs is determined by drivers' need of a legible message and by the functional requirements of the support system. A breakaway support system designed with a hinge, for example, will not function properly if the sign is mounted so low on the support system as to interfere with the hinge action.
- Efforts should be exerted to keep the top of the sign panel at a height of 2700 mm. Placing the sign at this height reduces the possibility that the top of the sign will break the windshield and intrude into the passenger compartment during impact. If the top of the sign panel is at least 2700 mm high, then the sign will hit the vehicle's roof and reduce the probability of vehicle intrusion. The majority of signs that meet the *MUTCD* standards for clearance to the bottom of the sign will also meet the minimum height to the top of the sign panel. Exceptions to this include rural installations with mounting heights less than 2130 mm to the bottom of the sign with sign panels less than 1200 mm in size. The FHWA is working on changes to *MUTCD*, which include raising the minimum clearance for a sign panel to 2.1 m, which will accomplish the same end result.
- Traffic signs should not be considered permanent solutions to inappropriate or hazardous roadway conditions. Installing a warning sign, for example, to warn of a shoulder dropoff does not eliminate the dropoff problem and presents an additional fixed object.

7.2.4 Influence of Surrounding Terrain on Proper Breakaway Performance

Breakaway supports are designed and evaluated to operate safely on the basis of the characteristics of the vehicle fleet. One of the primary characteristics included in discussions of the impacting vehicle is its weight. While weight is very important, the bumper height is equally important, since it establishes where the vehicle weight is first concentrated on the breakaway support. The majority of the safety evaluation tests are conducted on level terrain. This implies that the impacting design vehicles are striking the breakaway supports at a known height—typically, about 500 mm above the ground. Roadside safety could, therefore, be enhanced if wide, level areas are provided along the roadside.

Providing this level roadside is not practical or possible in the majority of roadside situations. Side slopes, ditches, cross slopes, curbs, and other drainage and terrain features are necessary roadside design features. How these features can interact with and influence vehicle trajectory and device performance must be considered prior to device installation.

Breakaway support devices are designed to function properly when the slip base is subjected to shear forces. If the point of impact is at a significantly higher point than the design height of 500 mm, then sufficient shearing forces may not be transmitted to

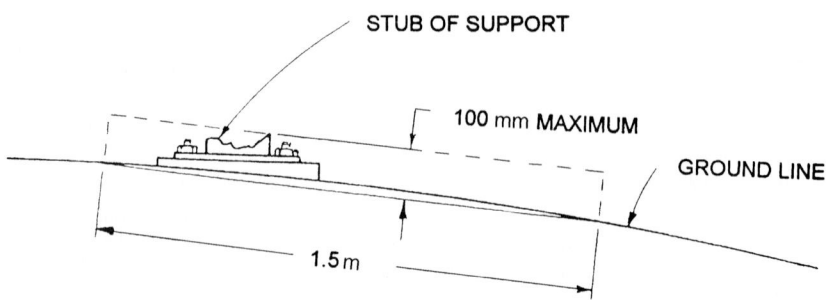

FIGURE 7.4   Breakaway support stub height measurement.

the base. The result can be binding of the mechanism and nonactivation of the break-away device. It is critical, therefore, that breakaway supports not be located near abrupt changes in elevation, superelevation transitions, changes in slope, or curbs that will cause vehicles to become partially airborne at the time of impact. As a general rule, if negative side slopes are limited to 6:1 or flatter between the roadway and the breakaway support, then vehicles will usually strike the support at an acceptable height.

Supports should not be placed in locations where the terrain features can possibly impede their proper operation. Placing supports in drainage ditches can result in erosion and freezing, which can affect the operation of the breakaway support. In addition, vehicles entering the ditch can be inadvertently guided into the support.

Supports should not be installed closer than 2100 mm to other fixed objects. If the supports are placed closer than 2100 mm to other objects that by themselves are considered acceptable, such as a 76-mm-diameter tree, then a vehicle will be able to strike both the support and the object simultaneously. The combined effect of both the tree and the support on the change of velocity can be much higher when impacting both objects simultaneously.

Terrain in the vicinity of the support base must be graded to allow vehicles to pass over portions of the support that remain in the ground or that are rigidly attached to a foundation. Remaining portions of the support that protrude more than 100 mm above the ground line over a horizontal span of 1.5 m, as presented in Fig. 7.4, can snag the vehicle undercarriage.

7.3   DESIGN OF SINGLE-MOUNT SIGN SUPPORTS

Traffic signs are a primary source of information to motorists. The majority of traffic signs consist of sign panels held in place by a single support. Single supports can usually be used for signs as large as 1.7 m^2 in area. The only purpose of the sign support is to hold the sign at the proper position for driver visibility. This requires that this support be strong enough to maintain the sign panel in its intended position while subjected to wind, ice, and snow loads. The magnitude of these forces increases as the sign panel becomes larger in size, until the panel is so large that multiple supports are required. Single sign supports are made of different materials, of various sizes and configurations, each capable of withstanding different environmental loads.

Considering only the environmental loads and selecting a support system to hold the sign panel at the proper position can result in severe vehicular damage and occupant injury upon impact. Proper sign installation requires that the sign assembly be able to hold its proper position and give way under impact to minimize severity to an errant vehicle and its occupants. This requires the proper design in sign system selection and placement.

Sign supports are classified as single-support and multiple-support systems. *Single sign support* refers to a support that has no other support, or fixed object, within a 2100-mm radius [19]. *Multiple supports* refers to installations that are spaced less than 2100 mm from each other, or from other fixed objects. With the closer spacing, it is possible for a vehicle, leaving the roadway at an angle, to impact more than one fixed object or support at a time. Support systems that provide acceptable performance when struck alone can result in severe occupant injury when struck simultaneously with another support. The discussion of this article pertains to single sign supports (i.e., supports installed no closer than a 2100-mm radius to other sign supports or fixed objects).

7.3.1 Sign Components

Sign assemblies consist of four components:

- The sign panel on which the message is displayed
- The signpost
- Mounting hardware and fasteners
- The base for the post

Sign Panels. The majority of sign panels in use today are made from sheet aluminum stock [20]. The thickness of the stock varies depending upon the sign size but is generally not less than 4.0 mm (8 gauge). Plywood is occasionally used by some agencies as the blank material for the reflective sheeting face in areas of frequent vandalism due to gunshots. Wooden sign blanks deform less from gunshots, are easier to repair, and are not as attractive a target as aluminum sign blanks. Plywood, however, does not weather as well as aluminum and, if the edges are not sealed correctly, has a relatively short life. More important, the plywood is heavier than aluminum, thus requiring a stronger post system and increasing the probability of intrusion into the passenger compartment upon impact. Composites such as fiberglass have also been used as sign blank materials with limited success. Early problems with composites included separation of the material and problems with the reflective sheeting adhering to the sign blank. A relatively new sign blank manufactured from recycled thermoplastic soft drink bottles is available from Composite Technologies [21]. These sign panels are molded with sealed edges, will not bend like aluminum, offer excellent bonding to adhesive sheeting, are weather and corrosion resistant, and are cost-effective compared with current aluminum pricing.

Sign Posts. The sign support must be strong enough to resist the wind and other loads yet safely give way when struck by a vehicle [22]. The loading conditions for which the support must be designed are illustrated in Fig. 7.5. The required size of a signpost is dependent upon the surface area of the sign it is supporting and the prevailing environmental conditions. Each state has a series of tables and/or graphs that spec-

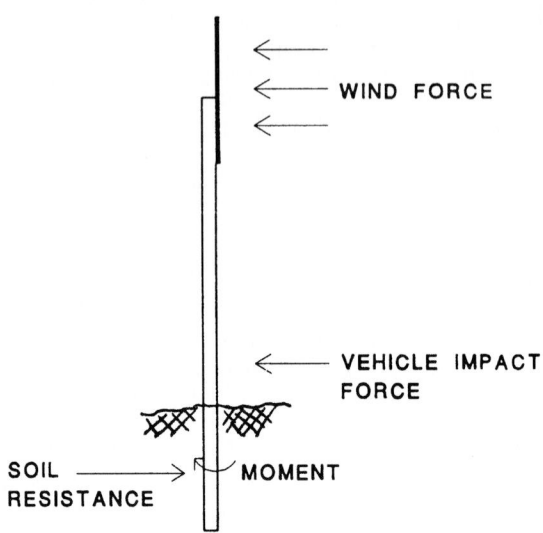

FIGURE 7.5   Wind and impact forces.

ify support post requirements based on prevailing wind and ice loads, sign size, and the height of the sign from the ground. These tables provide the information on the support size, embedment depth, and the support type that is required to withstand the environmental loads. The ability of the sign support to operate safely upon impact is dependent on the sign location, features of the surrounding terrain, and the intended method by which the support will give way. All give-way sign support systems operate by (1) complete or partial fracture of the support post, (2) failure of intentionally weakened (frangible) bolts or splices, and (3) mechanical release methods. These designs allow the support system to either bend at the base (base-bending) or break away into more than one piece. Sign support systems that do not give way upon impact are fixed-base supports which must be shielded with an appropriate barrier when placed within the transversable area.

Base-Bending Support Types.   A base-bending support (Fig. 7.6) is designed to bend over, lie down, and pass beneath the impacting vehicle. How effectively it performs is dependent upon the type of support and the velocity of impact. These supports tend to perform better at lower-speed impacts, which provide sufficient time for them to function as designed. Impacts at high speeds will frequently result in the support's partially fracturing or being pulled out of the ground. The performance of base-bending supports is more difficult to predict than that of other support types. Their behavior upon impact is influenced by variations in the depth of embedment, the soil resistance, stiffness of the sign support, mounting height of the sign, and the method of effecting the yielding action. One-piece assemblies are typically either driven directly into the ground or set in drilled holes and backfilled. Instead of a one-piece support, the yielding action is often effected by constructing an anchor system and connecting the sign support to the anchor assembly. The connection can be by direct splicing or the use of commercially available couplers that are designed to bend (fracturing) or break partially (frangible). The advantage of the two-piece assembly is that the anchor system is often not damaged during impact, thereby reducing replacement time. Base-bending supports provide

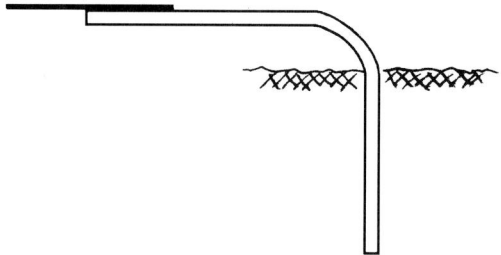

FIGURE 7.6   Example of base-bending support type.

a relatively inexpensive support system that reduces the probability that the sign assembly will become a deadly projectile to other traffic, pedestrians, and bicyclists.

Breakaway Support Types.   Breakaway sign-support systems (Fig. 7.7) are designed to have the system separate, at or near ground level, into more than one piece upon impact. This is accomplished by complete fracture of the support or by the separation of weakened splice parts. Wood is the most common material used for complete fracture designs. Weakened splice parts can be field-assembled splices, commercially available splices, or frangible couplings. Frangible couplings are usually made out of aluminum bolts that are necked down to provide a reduced cross section. Frangible couplings can be used for single sign supports but are generally used for large, multiple-support systems. Breakaway support systems typically work best for high-speed impacts where the vehicle has sufficient energy to both break the support and propel it away or over the vehicle.

FIGURE 7.7   Example of breakaway single sign support.

FIGURE 7.8   Example of mechanical release support type.

Mechanical Release Support Types.   Mechanical support types include slip base designs (Fig. 7.8), which have flat plates welded to both the sign support and the anchor piece. Upon impact, the plates slide against each other allowing the connecting bolts to release.

7.3.2 Sign-Support Selection

The only type of sign-support systems that should be used are those that have been approved for use by the FHWA. The following concerns should be addressed in the selection of an appropriate single–sign-support system.

- The specifications for support size provided by many states provide information on the maximum sign panel area to be mounted on the support. The shape of the sign as well as the area should be considered when determining the type and number of supports required. For example, a 1525-mm × 610-mm guide sign will have less area than a 1220-mm × 1220-mm warning sign. The wide dimension of the guide sign, however, will result in excessive vibration from wind loads if it is placed on a single sign support without bracing. As a general rule, signs over 1000 mm wide should be placed on multiple supports.

- Sign-support systems that are not placed in concrete foundations perform better in strong soils than in weak soils, such as sand. When the system is directly placed in weak soils, an anchor plate, a proper concrete footing, or embedment to a greater depth than used for strong soils may be required. This will hold the post firmly in the ground, preventing rotation due to wind loads, and help ensure proper operation during impact.

- The embedment depth is important for proper sign assembly operation. One-piece sign assemblies will pull out of the ground if not buried sufficiently deep. If buried too deep, it is difficult to remove the buried segment. Similarly, proper embedment depth for assemblies that use an anchor piece is important to prevent damage to the anchor piece on impact and to prevent rotation due to wind loads. The proper embedment depth varies by type of support system.

- Do not use sign-support sizes larger than required to support the sign or larger than approved for single-support types. For example, a slip base assembly should be used rather than a 9 kg/m U-channel post.

- Do not combine supports, such as square tube inside of pipe, or double the supports, such as back-to-back U-channels.

- Do not use diagonal bracing to strengthen a damaged or improperly designed support system.

- Sign-support assemblies are categorized as unidirectional, bidirectional, and multidirectional. Unidirectional supports will function properly only when impacted from one direction, and bidirectional, from two directions. Multidirectional supports will function properly when impacted from any direction.

- The same type of support post can be configured to operate in different ways upon impact. For example, the U-channel post is basically a unidirectional, base-bending support when buried directly in the ground. It can also be spliced to an anchor piece to provide breakaway characteristics or installed with a frangible coupling to provide multidirectional capability.

- Whenever an anchor system design is used, the anchor stub should not extend more than 100 mm above the ground. Extensions above the ground more than 100 mm can snag the vehicle undercarriage.

- A minimum mounting height of 2740 mm from the ground to the top of the sign panel is recommended for all single–sign-support installations. Mounting the signs at this minimum height will reduce the possibility of windshield penetration by a sign that bends or yields into the vehicle upon impact.

7.3.3   Steel U-Channels

The steel U-channel support is the most common type of single sign support used in the United States [20]. The steel U-channel is a unidirectional support available in different sizes and stiffnesses from a variety of manufacturers. The most popular steel U-channel sizes are 3, 3.7, 4.5, and 6.0 kg/m (weight is prior to making the fastening holes). The channel is constructed with 9.5-mm holes on 25.4-mm centers to eliminate the need for drilling to mount the sign panel. The posts are available with baked alkyd resin, with gloss enamel paint, or hot-dipped galvanized to inhibit corrosion. The stiffness of U-channel posts is a function of the material from which they are made, and the method by which they are shaped. The literature refers to billet steel or rail steel as the material from which U-channel is constructed. Rail steel is old railroad track— which has a high carbon content—that has been rerolled into the U-channel shape. The high carbon content results in a steel that is strong but relatively brittle. Billet steel is newly formulated steel. The most common grade of billet steel is A36, which is a relatively low-carbon "mild" steel, but billet steel can be manufactured with sufficient carbon to equal or exceed the strength characteristics of rail steel. For years the FHWA required "rerolled rail steel" instead of billet steel, since such a specification helped ensure a high carbon content. High-carbon billet steel U-channel posts are available from manufacturers, however. U-channel posts should be made from steel with sufficient carbon content to provide a minimum yield strength of 415 MPa and a minimum tensile strength of 550 MPa.

TABLE 7.2 Maximum Allowable Sign Area for One-Piece U-Channel Installation,* m^2

Post size, kg/m	Height from ground to center of sign for 113 km/h wind, m†						
	1.8	2.1	2.4	2.7	3.0	3.4	3.7
3.7	0.7	0.6	0.5	0.5	0.4	0.4	0.3
4.5	0.9	0.8	0.7	0.7	0.6	0.5	0.5
6.0	1.2	1.1	1.0	0.9	0.8	0.7	0.7

*Sign sizes are for typical U-channel with a minimum yield strength of 415 MPa.
†Height measured from ground to center of sign panel.

Crash tests indicate that, under some test conditions, high-carbon steel U-channel sign posts perform differently from those made of steel having a lower carbon content. The reason for this performance difference is that high-carbon steel posts, because of their low fracture toughness, break under high-speed impacts [23]. Lower-carbon steel posts bend and shape themselves to the front of the vehicle, thereby forming a tethering hook. Billet steel posts that have carbon content similar to that found in rail steel posts match the performance of the rail steel posts when crash-tested.

Table 7.2 presents the maximum allowable sign area for one-piece 415-MPa minimum yield strength U-channel installation. The table is for a maximum allowable pressure resulting from a 113 km/h wind velocity. State guidelines should be followed for the expected wind velocities for different regions of the state and to obtain support sizes for wind velocities different from 113 km/h.

Base-Bending Installation.   One-piece base-bending U-channel post installations are usually obtained by driving the post directly into the ground with a sledge hammer, manual post driver, or air-operated post driver. Drive caps should be used to protect the top end of the post while it is being driven into the ground. U-channel posts should not be encased in concrete unless a breakaway design is used. Typical embedment depth is 920 mm, and for ease in removing damaged posts, the driven depth should generally not exceed 1070 mm. A typical installation and cross section of a steel U-channel post is provided in Fig. 7.9. Also included in Fig. 7.9 is the patented RIB-BAK design, which has a ribbed back and flange. This design provides extra strength, a flush back-to-back sign-mounting surface, and a ridge for mounting channel locking clips. An alternative to the direct burial system is the V-loc anchor from Foresight Industries. The V-loc anchor is currently the only alternative method of anchoring an unspliced U-channel post that has been found acceptable by FHWA. This anchor system uses locking inserts to hold the U-channel securely into the V-shaped anchor piece. Upon impact, the post will bend at the ground line and may pull completely out of the V-loc anchor.

Breakaway Installations.   The repair and performance of large U-channel posts can be eased by using a breakaway design. Breakaway design in U-channel installations is obtained by developing splices at ground level. The splice consists of attaching the signpost to an anchor piece that is embedded in the soil or a concrete foundation. Splice designs include the Eze-Erect by Franklin Steel, the Minute-Man by Marion Steel, and various lap designs [24]. The intent of the splice designs is for the splice to fail upon impact. The commercially available splices are designed so that the signpost remains attached to the embedded anchor piece and passes beneath the impacting

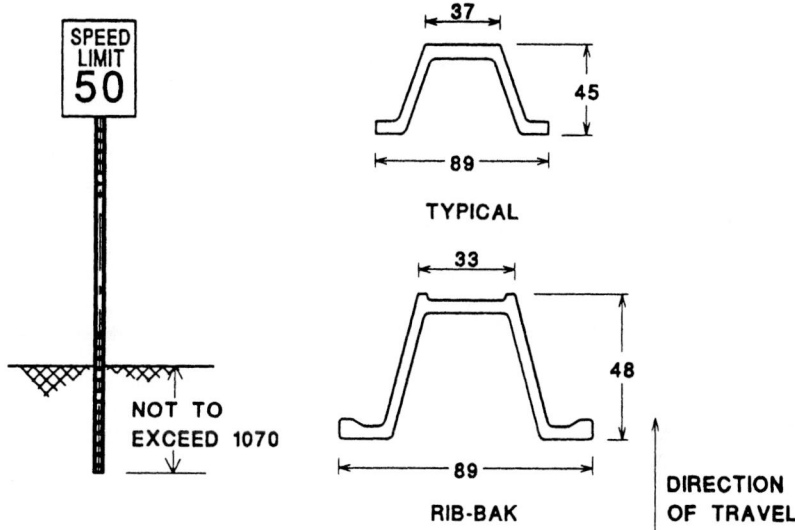

FIGURE 7.9 Installation and cross section of typical steel U-channel and RIB-BAK 3 lb/ft (4.5 kg/m) posts.

vehicle. This is accomplished by designing the splice device to partially fracture or to completely fracture a frangible coupling. To prevent vehicle snagging, the anchor piece should not extend more than 100 mm above the ground. Two commercially available breakaway designs are presented in Fig. 7.10. The Minute-Man consists of a frangible coupling with a backup plate, to hold the anchor and sign support pieces together. The Minute-Man coupler makes the U-channel a multidirectional support system.

The generic splice (Fig. 7.11) does not require special hardware [25]. It is acceptable for use on 6 kg/m U-channel, or less, installed in strong soil. The generic splice consists of an overlap of 150 mm and uses two 8-mm bolts spaced 100 mm center to center. Spacers, 8 mm thick, are used to separate the U-channel signpost and the anchor piece. The spacer must be strong enough to transfer the load between the webs of the signpost and the anchor piece. The signpost should be mounted behind the stub.

The anchor piece of all breakaway devices should be the same size as the signpost and must not extend more than 100 mm above the ground. A splice configuration, as in Fig. 7.12, does not provide protection for the anchor and increases the probability of snagging or of the sign's entering the passenger compartment. Breakaway devices improve the safety characteristics of the post and generally reduce maintenance costs. They should always be used when the sign support is placed in concrete areas. If the sign can be impacted from different directions, then a breakaway device similar to that shown in Fig. 7.10 should be used. Splicing the signpost to the anchor piece with bolts, with or without the splice breakaway device of Fig. 7.10*a,* does not make the U-channel support a multidirectional sign support.

Mounting Concerns. The U-channel post is approved for use in strong soils when impacted from a frontal direction. Installing the support in weak soils or in locations where it can be impacted from more than one direction requires more than direct bur-

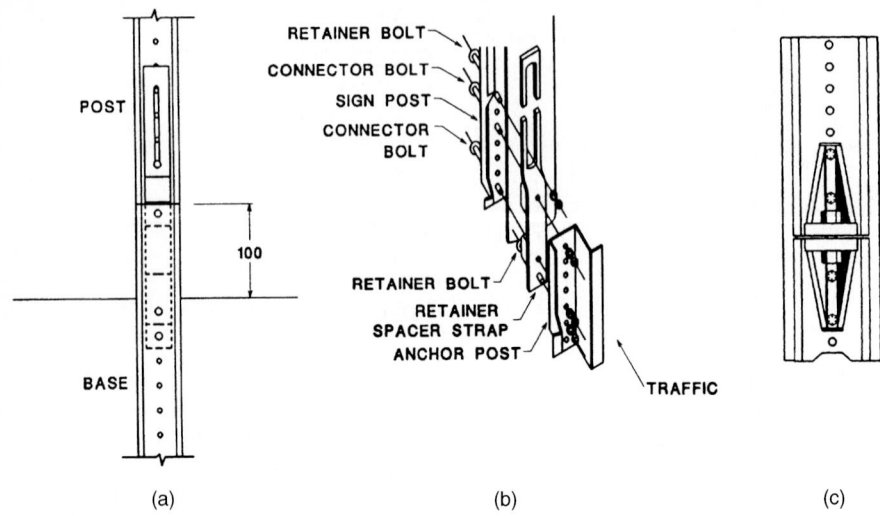

FIGURE 7.10   Breakaway devices for U-channel posts. (*a*) Eze-Erect splice joint. (*b*) Details of Eze-Erect splice. (*c*) Minute-Man coupler.

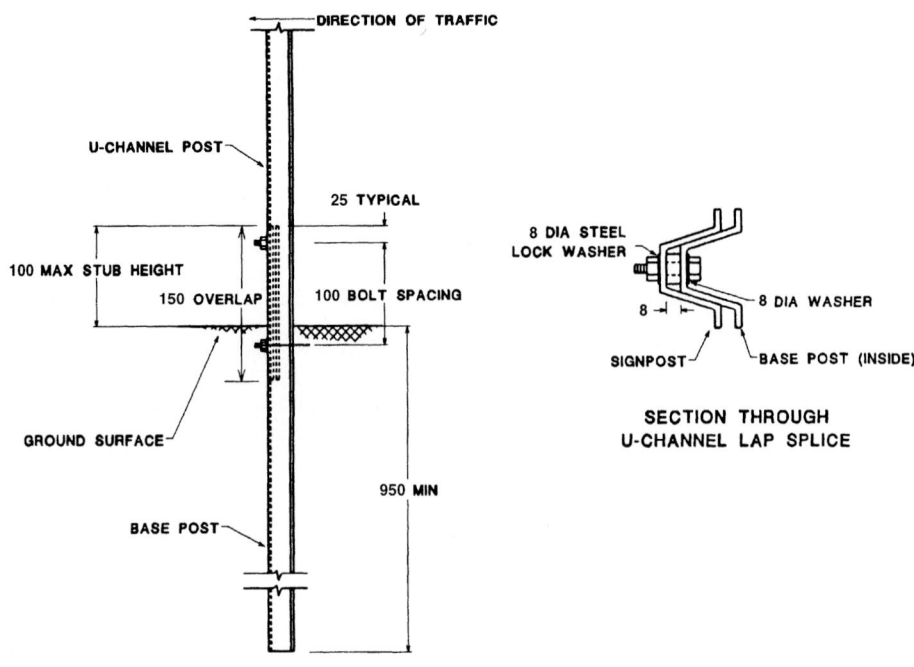

FIGURE 7.11   Details of generic splice configuration.

FIGURE 7.12   Improper splice of U-channel post system.

ial to make the U-channel perform as required. If the U-channel is installed in weak soil, an anchor plate, similar to that shown in Fig. 7.13, can be used to hold the sign in its proper position and to help ensure proper operation upon impact. In addition, the generic splice can allow the signpost to separate from the base. The possible consequences of this separation, and the trajectory of the sign assembly, should be considered prior to use of the generic splice.

7.3.4   Wooden Support Posts

Wooden support posts are available in shaped sizes, as engineered products, and as timber posts. The shaped sizes are described by their nominal dimensions, such a 100 mm $\times$ 100 mm. This is their size prior to the surfacing required to provide smooth and straight posts. Their actual size is typically 10 mm less than the nominal size. A 100-mm $\times$ 100-mm post will therefore have an actual size of 90 mm $\times$ 90 mm. The engineered products are made from laminated or pressure-glued wood and nonwood recycled products. Timber posts are round in shape.

All wooden posts are of breakaway design, with the intended fracture of the post near the base and less than 100 mm above the ground. The post features that influence fracture include the size of the post, effective cross-sectional area, embedment depth, type of soil, and the species of wood. The majority of wood post tests have been conducted using grade 2 southern yellow pine posts.

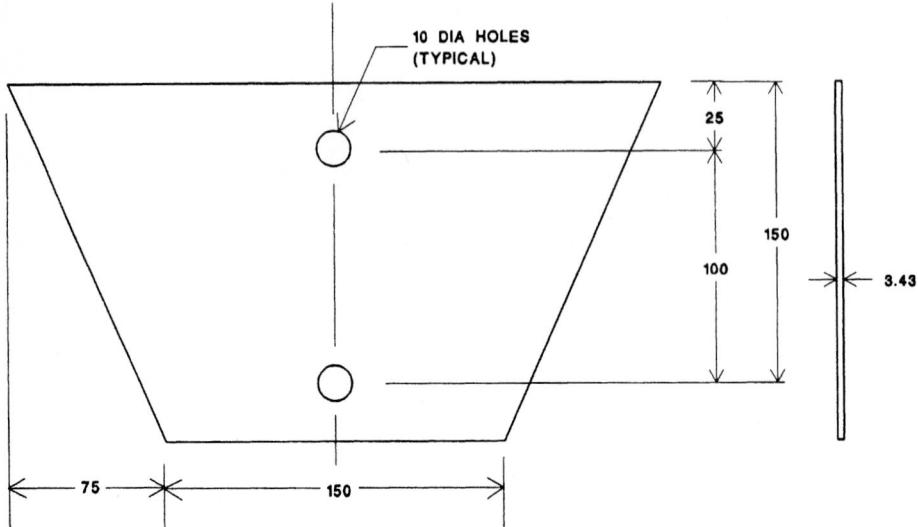

FIGURE 7.13   Large anchor plate for use with U-channels in weak soil locations.

Shaped Wood Posts.   The most common size wood post used for single sign installations is the 90-mm × 90-mm support. This support should be buried directly in strong soil to a depth of at least 920 mm. The cross-sectional area of the 90-mm×90-mm post is sufficiently small that drilled holes are not needed to provide a weakened section.

A 90-mm × 140-mm post installed in strong soil will provide acceptable performance upon impact without reducing the cross section. Tests have shown, however, that the 90-mm × 140-mm, when installed in loose or sandy soil, is unacceptable when impacted by a small vehicle. Single 90-mm × 140-mm wood posts installed in weak soil should be modified with 40-mm-diameter holes to be crashworthy. The holes should be centered at 100 mm and 460 mm above the ground line and perpendicular to the roadway centerline.

Table 7.3 provides the modifications required for acceptable performance of various shaped wood post sizes. The holes in each case are drilled at 100 mm and 460 mm above the ground line and perpendicular to the roadway centerline.

Typical details for installation of shaped wood posts by direct burial and concrete methods are presented in Fig. 7.14. The 140-mm × 140-mm post should be set in unreinforced concrete to help ensure that the post fractures upon impact. To make it easier to remove a broken stub, a post can be wrapped with 13-mm-thick Styrofoam prior to filling with concrete.

Some states have used larger shaped wooden posts, such as 140 mm × 215 mm, with appropriately sized holes to reduce the cross-section area. These holes provide a weak section that appears acceptable, but the increased mass of these posts, and lack of testing, result in unpredictable impact performance. Shaped wooden posts larger than those that have been crash-tested should not be used. If larger posts are required, then a multiple-post configuration, slip base design, or other alternatives should be used.

TABLE 7.3 Required Modifications to Shaped Wood Posts

Post size, mm	Hole size at 100 and 460 mm above ground level, mm
90 × 90	None
90 × 140	Strong soil, none; weak soil, 40
140 × 140	50
140 × 190	75

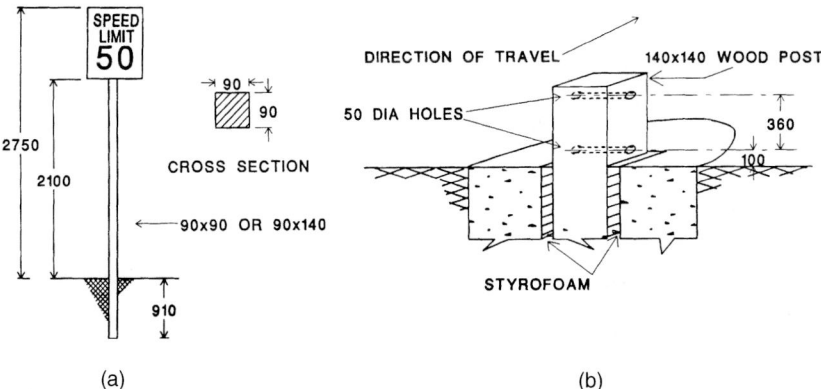

(a)   (b)

FIGURE 7.14 Typical installation of wooden sign supports. (*a*) Direct burial. (*b*) In concrete.

Engineered Wood Posts. A number of relatively new products have been developed for use as sign supports. These include engineered wood product posts made from recycled plastics and wood chips, and laminated veneer lumber posts. The Microllam laminated posts in 200 mm × 200 mm and in 380 mm × 200 mm have been accepted for use. These posts, manufactured by the Trus Joist MacMillan Corporation, have a wall thickness of 32 mm and mitered 45° corners. The post is placed in predrilled holes and backfilled. The posts require four 25-mm-diameter holes drilled on the two sides parallel to the direction of travel. Two of the holes are at 76 mm, and the other two holes are at 533 mm above ground height. A saw cut parallel to the ground that connects each set of holes is required as presented in Fig. 7.15.

Timber Poles. The majority of wooden sign-support systems consist of square or rectangular shapes. However, round timber poles, up to 190 mm diameter of southern pine, grade 2, have recently been accepted for use by the FHWA [26, 27]. The acceptable sizes and required holes to provide acceptable breakaway performance are presented in Table 7.4.

TYPICAL CROSS SECTION   TYPICAL CROSS SECTION

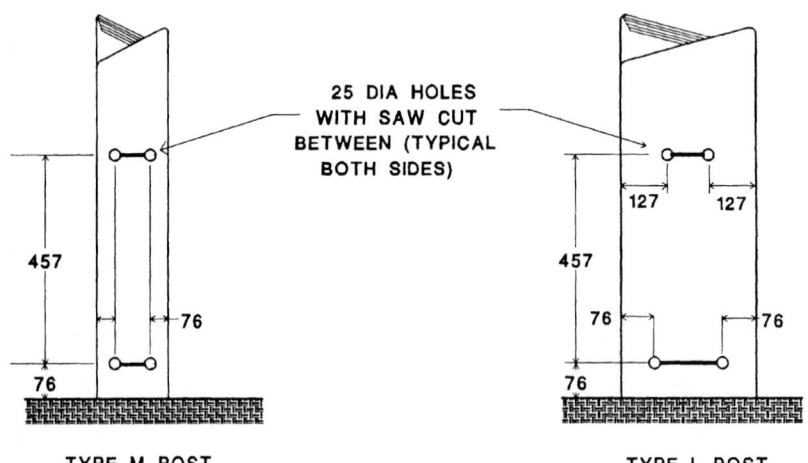

25 DIA HOLES
WITH SAW CUT
BETWEEN (TYPICAL
BOTH SIDES)

TYPE M POST   TYPE L POST

FIGURE 7.15 Laminated Microllam wood posts.

TABLE 7.4 Timber Wood Post Requirements

Timber post diameter, mm	Required hole size, mm*	Effective area, mm²
100	None	81
115	None	103
127	None	127
150	19	154
165	32	162
178	51	159
190	70	155

*Holes are centered at approximately 100 mm and 460 mm above ground, with their axes horizontal and in a plane parallel to the sign face.

7.3.5 Square Steel Tubes

Square steel-tube sign supports are used in many localities. They provide four flat surfaces for mounting sign panels, facing different directions, without special hardware as required by some support types. Square-tube supports can be purchased from a number of manufacturers and are available with 11-mm holes or knockouts at 25-mm centers on all sides [27, 28]. The square tubing is available in 6.4-mm incremental sizes from 38 mm × 38 mm to 64 mm × 64 mm and in strength factors of 275 and 415 MPa minimum yield strengths. Maximum sign areas for various square-tube sizes and strengths are illustrated in Table 7.5. Notice that the higher-strength posts allow for greater load-carrying capacity.

Square tubing can be driven directly into the ground using a drive cap with sledge or power equipment. The performance of the support assembly upon impact, and subsequent repair, are enhanced by using an anchor base. Three common methods of installing a single square-tube sign support are presented in Fig. 7.16. Figure 7.16a shows a direct burial installation. Square tube up to a maximum size of 57 mm × 57 mm has been approved for installation by direct burial. The performance of square-tube sign supports upon impact is more predictable, and easier to repair, by the use of an anchor base system [30]. Figure 7.16c shows an anchor base system where a 900-mm-long piece of square tube, one size larger than the anchor piece, is driven into the ground. This anchor piece is left protruding 25 to 50 mm above the ground to permit bolting of the signpost. The signpost is inserted 150 to 200 mm into the anchor piece and bolted in place. Figure 7.16b shows a device similar to the anchor base installation except that it uses an outer stiffener sleeve one size larger than the 900-mm-long anchor base piece. The stiffener sleeve provides a double-walled thickness that reduces damage to the anchor piece. Upon impact, the post yields at the top of the anchor assembly, normally leaving it undamaged as in Fig. 7.17.

Square steel tubes with perforations on all four sides have been found to provide acceptable crash performance for sizes as large as 64 mm × 64 mm when embedded directly into the soil. They are acceptable in both strong and weak soil when embedded to a depth of 1220 mm. Repairing direct-embedment supports, however, is more

TABLE 7.5 Maximum Sign Area for Square-Steel-Tube Single-Support Posts (113 km/h Wind), m²

Yield strength, MPa, and post size, mm	Height from ground to center of sign, mm						
	1830	2135	2440	2745	3050	3350	3360
275 MPa							
44 × 44 (12 ga)	0.7	0.6	0.5	0.4	0.4	0.3	0.2
51 × 51 (12 ga)	0.8	0.7	0.5	0.5	0.4	0.3	0.3
57 × 47 (12 ga)	1.1	1.0	1.0	0.9	0.8	0.7	0.5
64 × 64 (12 ga)	1.4	1.2	0.9	0.9	0.7	0.7	0.6
64 × 64 (10 ga)	1.7	1.4	1.2	1.1	0.9	0.8	0.7
415 MPa							
51 × 51 (14 ga)	1.0	0.9	0.8	0.7	0.6	0.5	0.5
51 × 51 (12 ga)	1.4	1.2	1.0	0.9	0.8	0.7	0.6
64 × 64 (12 ga)	2.5	2.1	1.9	1.6	1.4	1.3	1.1

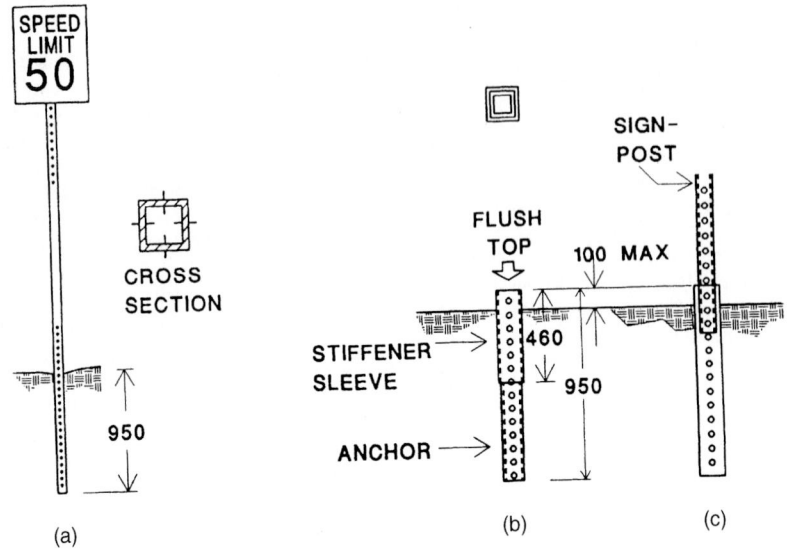

FIGURE 7.16 Square-tube sign-support system. (*a*) Direct burial. (*b*) Stiffener sleeve anchor. (*c*) Anchor assembly.

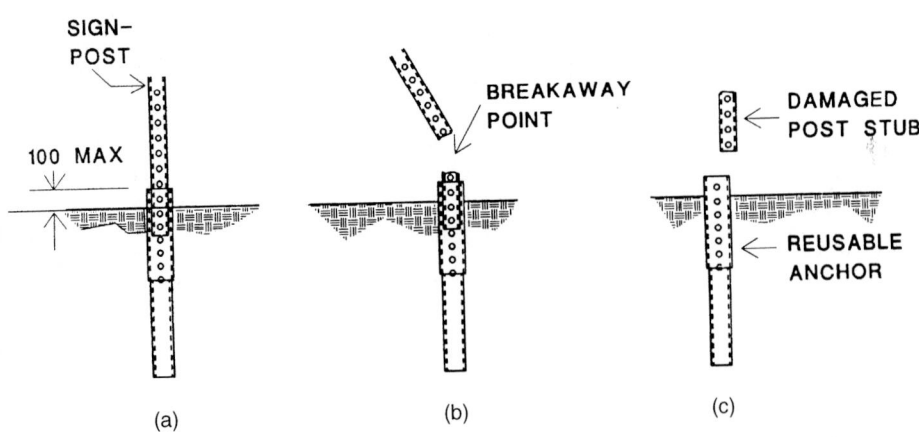

FIGURE 7.17 Typical square-tube damage with stiffener sleeve anchor assembly. (*a*) Prior to impact. (*b*) Breakaway action. (*c*) Removal of broken stub.

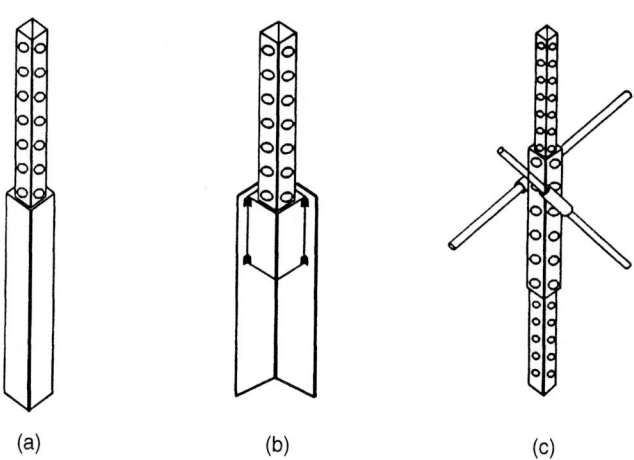

(a)   (b)   (c)

FIGURE 7.18 Anchor systems manufactured by Unistrut Corp. (*a*) Heavy-duty anchor. (*b*) Anchor post. (*c*) Stabilization anchor.

difficult than repairing the yielding breakaway system. The V-loc system from Foresight Industries can also be used as an anchor system for square-tube supports.

Figure 7.18 shows anchor systems for square tubing that are manufactured by Unistrut Corporation. Figure 7.18*a* shows a heavy-duty breakaway anchor for use with 51-mm and 64-mm square tubes. It consists of a 4.8-mm-thick wall that eliminates the need for a stiffness sleeve and allows the signpost to break away on impact without damaging the anchor wall. Figure 7.18*b* shows an anchor post that can be driven directly into extremely hard or rocky soil conditions. It is made from 6.4-mm × 102-mm steel angle section that can help stabilize the sign assembly in soil conditions that provide poor resistance to lateral and torsional forces. Figure 7.18*c* shows a stabilization anchor sleeve that helps adjust for inconsistent roadside gradients. The anchor rods help resist the environmental loads that can cause the signpost to lay over or twist in soft or shoulder dropoff conditions. The stabilization sleeve is installed over an anchor piece and the two rods inserted at a 45° angle to increase stability.

Figure 7.19 presents a soil stabilization anchor manufactured by Xcessories Squared [31]. The stabilizer is attached with a corner bolt, through the lower slots, to an anchor piece of square tube. The tops of the stabilizer piece and anchor are aligned and the assembly is driven into the ground until only 50 mm remains above the ground surface. After the bottom end of the signpost is inserted 205 mm into the anchor assembly, it is secured with a corner bolt from the back side, through the stabilizer, anchor, and signpost.

7.3.6 Steel-Pipe Posts

Steel-pipe posts are frequently used in urban areas and have the advantage of being readily available. They require special fastening hardware, and an earth plate when directly embedded, to prevent the post from rotating from its intended position. Standard steel pipe, schedule 40, galvanized, is readily available from plumbing sup-

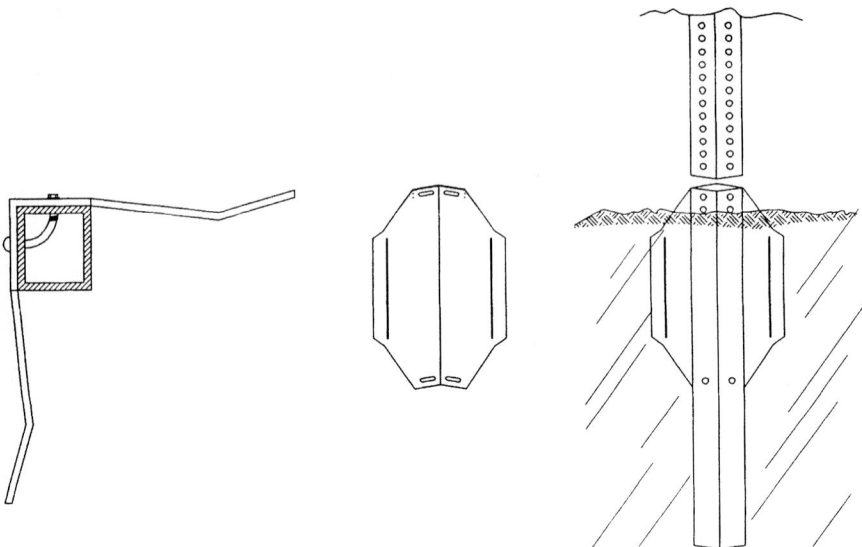

FIGURE 7.19 Anchor system manufactured by Xcessories Squared.

TABLE 7.6 Maximum Sign Area for Standard Steel-Pipe
Single-Support Posts (113 km/h Wind), m^2

Internal diameter post size, mm	Maximum sign area, m^2
51	0.6
64	1.1

ply wholesalers. The maximum sign panel areas that can be mounted on the 51-mm
and 64-mm internal diameter standard steel pipe are listed in Table 7.6.

Round steel supports, made from standard schedule 40 pipe, that have an internal
diameter (ID) of less than 51 mm can be embedded directly into the ground (Fig.
7.20a) to a depth of at least 1070 mm and provide acceptable performance upon
impact. A steel earth plate measuring 100 mm $\times$ 310 mm $\times$ 6 mm should be welded
or bolted to the pipe to prevent support rotation due to the wind.

Standard schedule 40 pipe, 51 mm and larger, is no longer approved for direct burial installation and must be installed with a weakening device [32]. A breakaway collar assembly (Fig. 7.20b) is required for standard schedule 40 pipe sizes, equal to or
greater than 51 mm ID, and also for smaller pipe sizes when the device is likely to be
hit. A regular pipe coupling or reducing coupling will provide acceptable breakaway
performance. The use of a pipe coupling will, however, frequently result in damage to
the anchor piece. Therefore the reducing coupling is the preferred breakaway device.
The anchor assembly consists of a concrete footing (usually 760 mm deep by 300 mm
in diameter) and a 610-mm-long piece of anchor pipe. The anchor pipe is usually one
size larger than the signpost to prevent damage to the anchor and to allow use of the
reducing coupling. The possibility of damage to the anchor post can be further
reduced by embedding the reducing coupling halfway into the concrete footing.

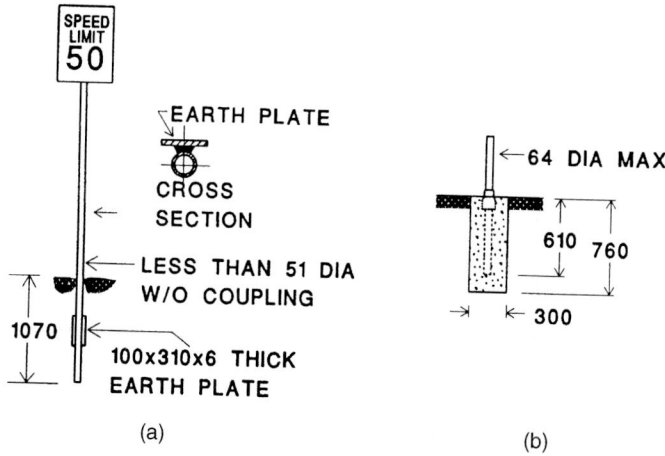

FIGURE 7.20   Standard steel-pipe sign supports. (*a*) Direct burial. (*b*) Breakaway collar assembly.

Round steel tube, in wall thicknesses of 12 gauge or less, can be used with anchor systems instead of standard schedule 40 pipe. These tubes are available from a number of manufacturers. Southwestern Pipe Inc. is one tubing manufacturer that also markets the Poz-Loc anchor system. This system consists of a tubular anchor socket 64 mm ID and 686 mm long constructed of 12 gauge steel. The socket, presented in Fig. 7.21, is pointed to facilitate driving into the ground and accepts a 51-mm ID steel round tube as the sign support. The sign support is held in place by driving a post wedge between the socket and the support. Should the post be damaged, the wedge can be removed, another post inserted, and the wedge replaced without disturbing the anchor socket. The wedge requires the use of a special puller for removal, which reduces vandalism to the sign system.

7.3.7 Aluminum Shapes

A limited number of single sign supports are constructed of aluminum. Shapes include U-channel, round and square tubing, and various X shapes. Examples of the U-channel and X shapes are presented in Fig. 7.22.

The aluminum shapes come in various sizes. The U-channel size shown is appropriate only for small or temporary sign installations. The X shape has been tested and found acceptable up to the 76-mm × 79-mm size and is commercially available form different suppliers in slightly different configurations. All of the X-shape supports are extruded and have a cross-sectional shape similar to that of back-to-back steel U-channels.

A thin-walled aluminum tube shape of 89 mm diameter, configured as presented in Fig. 7.23, has been determined as marginally acceptable. At the low-velocity test speed (32 km/h), the aluminum tube provided an acceptable deceleration rate but failed to yield to the vehicle. The test vehicle pushed the sign support over and came to rest on the support with some of its wheels off the ground. This suggests that the support is close to causing vehicle instability when impacted by small vehicles.

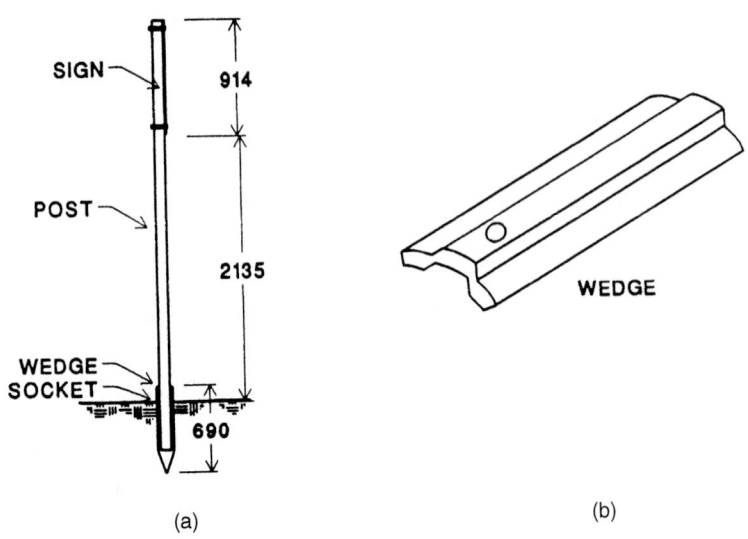

FIGURE 7.21   POZ-LOC standard pipe support system. (*a*) Socket assembly. (*b*) Locking wedge.

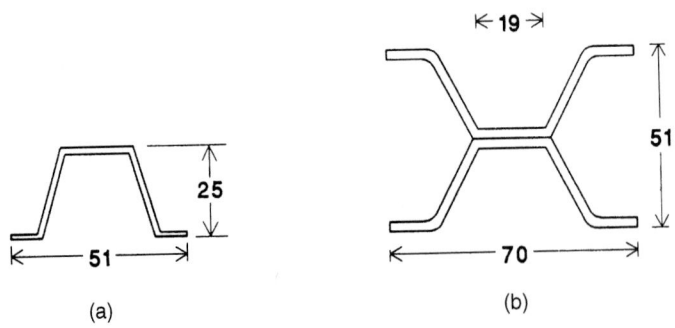

FIGURE 7.22   Examples of aluminum post shapes. (*a*) U-channel. (*b*) Typical X shape.

7.3.8 Fiberglass Support Posts

Fiberglass-reinforced pipe (FRP) support posts are available and have been accepted for use in single–sign-support systems [33]. The 76-mm outside diameter FRP post, manufactured by Hwycom and consisting of polyester resin and glass fiber reinforcement, is an example of an acceptable support. One method of installing the 3-mm-thick-walled FRP post is to insert it into a 686-mm-long steel tube that has a wall thickness of 1.6 to 2.4 mm. The steel anchor tube is coated with ceramic to reduce corrosion and flattened for 305 mm to prevent rotation. Self-tapping screws secure the post in place. A diagram of a fiberglass sign-support assembly, with the steel-tube anchor, is presented in Fig. 7.24.

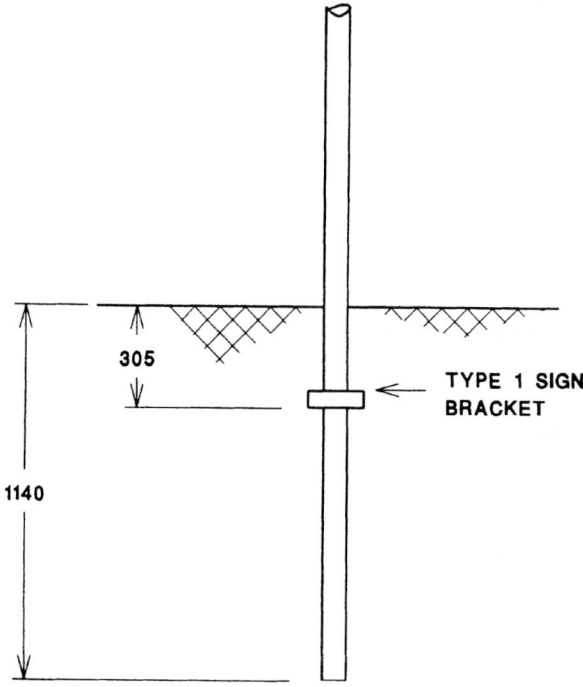

FIGURE 7.23 Aluminum tube sign installation.

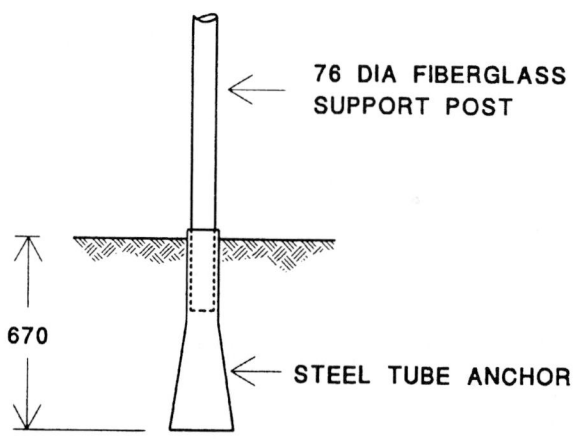

FIGURE 7.24 Fiberglass post installation with steel-tube anchor.

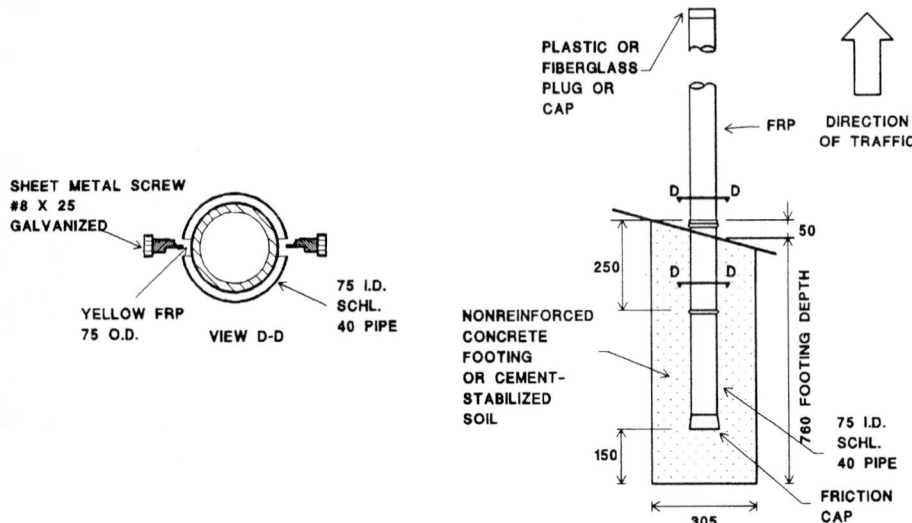

FIGURE 7.25   Details of concrete foundation design approved for dual installation of fiberglass sign supports.

The Hwycom fiberglass support can also be installed by inserting it into a steel anchor with a concrete foundation [34]. When the steel anchor and foundation system is used, the installation of two posts within a 2100-mm path has been approved. Details on the concrete foundation installation are provided in Fig. 7.25.

7.4   SLIP BASE DESIGNS

Slip base designs for small sign supports consist of two components: (1) the anchor assembly up to the bottom of the slip base, and (2) the sign support, containing the top of the slip base on the lower end and the sign panel on the upper end. Small sign slip bases are categorized as unidirectional or multidirectional.

Slip base designs allow the use of stronger sign supports than can safely be achieved by base-bending or fracture designs. The anchor piece of slip base designs is fixed into a foundation and should remain immovable during an impact. The sign support is connected to the anchor piece with bolts through a plate, which are attached to a similar plate on the anchor piece. The holes in the plates are slotted. When a vehicle impacts the sign support, the top plate, which is attached to the sign support, slides along the bottom plate until the bolts slide free of the slots. Inclined slip base designs, or designs with raised center cones, cause the sign support to move upward to allow the impacting vehicle to pass under the sign without being hit on the windshield by the sign during high-speed impact.

When slip base designs were first used, problems were encountered with assemblies that came apart without an impact. This was due to the wind and ice loads vibrating the assembly and causing the bolts to "walk" out of the slots, as in Fig. 7.26. This problem was solved by using a thin (20 to 28 gauge) keeper plate to ensure that the bolts remain properly located in the slots. During an impact, the bolts tear through the thin keeper plate as they slide free of the slots.

FIGURE 7.26   Loose slip base caused by vibration.

7.4.1   Unidirectional Slip Bases

Unidirectional slip bases for small sign supports consist of inclined slip bases, as shown in Fig. 7.27. The upper support piece is made from rolled-steel shapes, standard pipe, or structural tube. The base of the support assembly is inserted into a concrete footing to prevent movement of the anchor assembly.

The upward thrust obtained from the inclined slip base design is important to the proper action of a single-support sign system. The upward thrust causes the sign panel and support to rise and rotate when vehicle impact separates the mechanism. The sign panel and support stay together as a unit, which passes up and over the vehicle and lands behind it. This action is obtained only when the support is impacted from one direction. An impact from the opposite direction actually pulls the sign support downward, causing the support and sign panel to rotate toward the vehicle. Inclined slip bases should not be used where impact from more than one direction is expected. Horizontal slip bases will separate when impacted from the front or the rear but will not provide the uplift capability obtained from inclined-base designs. A typical design for an inclined slip base is provided as Fig. 7.28.

Horizontal slip bases, discussed in Art. 7.5.2, are not recommended for single sign supports. When impact can be expected from more than one direction, a multidirectional slip base design should be used.

7.4.2   Multidirectional Slip Base

The multidirectional slip base design operates on the same principle as the inclined slip base design. The multidirectional design consists of a triangular slip base employ-

FIGURE 7.27   Installation of unidirectional slip base.

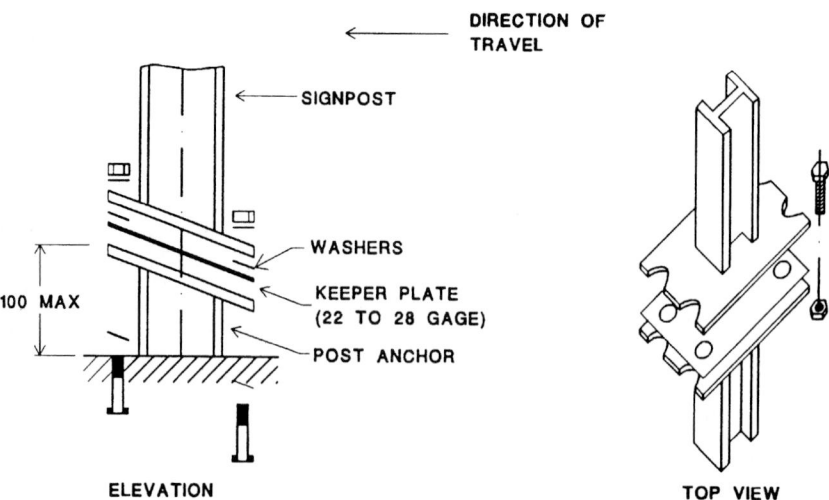

DIRECTION OF TRAVEL

SIGNPOST

WASHERS

KEEPER PLATE
(22 TO 28 GAGE)

POST ANCHOR

100 MAX

ELEVATION

TOP VIEW

FIGURE 7.28   Typical inclined, unidirectional slip base.

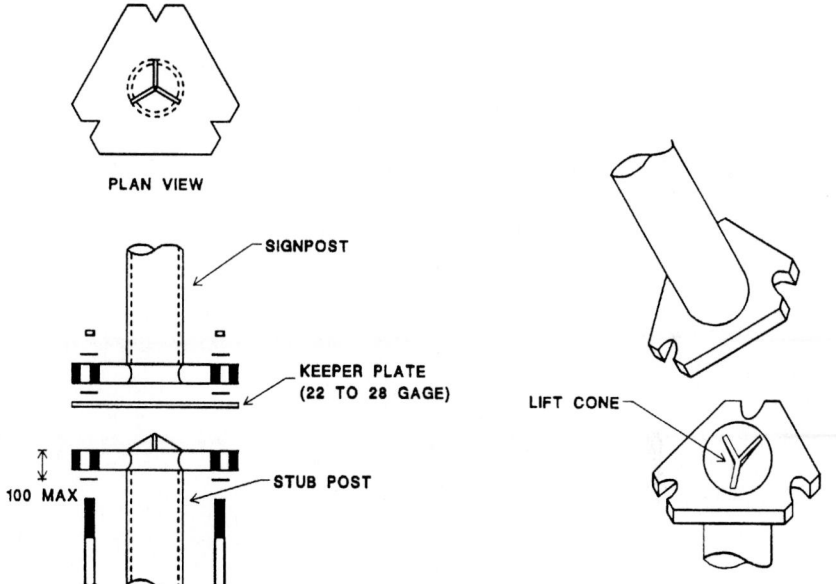

PLAN VIEW

FIGURE 7.29 Details of multidirectional triangular slip base.

ing only three slotted bolt holes, as presented in Fig. 7.29. The bolts are positioned at the flattened corners of the triangular plate. An impact from any direction slides the bolts out of the slots and allows the signpost to separate from the anchor piece. The desired lifting action is obtained by a lift cone located on the bottom plate. The sign support is tubular and beveled at the top triangular slip plate to help the lift cone push the support off the anchor plate during impact. The anchor piece is encased in concrete to prevent movement. The pipe generally used for multidirectional slip bases ranges from 76 to 127 mm in diameter. The design specification for each size must be checked, since the required bolt size, torque requirements, and lift cone design are dependent upon the size of the sign support.

7.4.3 Frangible Couplings

Acceptable single–sign-support performance can be achieved with the use of frangible couplings and load concentration couplers (Fig. 7.30). These couplings are either fabricated from die cast aluminum or extruded from an alloy. The couplers are used as inserts that bolt the support post plate to the anchor piece plate. They present a weak point on the sign-support assembly that fractures upon impact. The majority of applications for frangible couplings are for multiple sign supports. These couplings are, therefore, discussed more fully in Art. 7.5.2.

FIGURE 7.30   Frangible-coupling sign support.

7.4.4   Considerations in Design of Slip Bases

Failure of slip base designs to release properly can be due to the bolt torque, the gauge or thickness of the keeper plates, or the weight of the support. The following should be adhered to in the design of slip base supports:

- Horizontal and inclined slip bases can be constructed with wide-flange, standard-shape, and round signposts. Multidirectional designs are usually constructed with round signposts to enable the multidirectional rising action of the lift cone.
- The post should not weigh more than 67 kg/m, and the total weight of the support post, hardware, and sign panel should not be more than 270 kg.
- The bolts clamping the top and bottom portions of the slip base together should not be tightened more than the specifications. Overtorquing creates high friction between the slip base elements and may prevent the post from releasing properly. The clamping force must be controlled by installing the bolts with a torque wrench, using torque-limiting nuts, or using designs that are not dependent upon specific torque requirements.
- Washers used with the clamping bolts must be of sufficient strength to prevent the washers from deforming into the plate slots when the bolts are tightened to specification.
- The stub height must be no more than 100 mm above ground level at the highest point of the slip plate assembly.
- No bolts from the anchor piece should project into the upper support assembly.
- The choice of the proper sign support type for slip base designs is dependent upon the wind load, sign panel size, and the criterion that the weight of the sign support

TABLE 7.7 Round Sign Support Sizes
for Slip Base Designs Based on Sign Area

Round post internal diameter, mm	Total size area, m^2
51	0 to 0.37
64	0.37 to 0.74
89	0.74 to 1.9
100	1.9 to 3.3

and sign not exceed 270 kg. As a general rule, the maximum sign area presented in Table 7.7 can be used in selecting the appropriate size of wind sign support. Determining the maximum sign size for areas with high wind loads, or for the selection of post sizes other than round shapes, should be performed with reference to state requirements.

7.5 DESIGN OF MULTIPLE-MOUNT SIGN SUPPORTS

Multiple-mount sign support assemblies (Fig. 7.31) are required whenever the surface area or width of the sign is too large to withstand the wind and ice loads. Each state has guidelines, in the form of tables and graphs, that are used to select the size and numbers of supports required to withstand the prevalent environmental loads in differ-

FIGURE 7.31 Multiple-mount sign support.

ent parts of the state. These guidelines should be used to determine the required size and number of supports. The design of multiple–sign-support assemblies requires considerations that in some instances differ from single–sign-support assemblies. These considerations include the following:

- Tests have demonstrated that vehicles leaving the roadway at an angle can strike more than one support if supports are spaced closer than 2100 mm. If two supports are spaced less than 2100 mm apart, they must pass a crash test as a dual support assembly. Installing two acceptable single sign supports does not guarantee acceptable multiple-support performance.
- For multiple supports, the sign panel itself is an important part of the sign structure during impact. Depending upon the design, the sign panel must carry the weight of the impacted support and/or provide sufficient rigidity to enable the hinge mechanisms to activate. The sign panel must be made of material of sufficient thickness that it does not break into pieces when a support is impacted.
- Acceptable performance in multiple-support systems requires the sign panel to remain attached to the support(s) that are not impacted. This intended performance can be destroyed by:

  The use of bolts to fasten the sign panel that are too small

  The absence of washers, which allows the bolt head to pull through the sign panel

  Sign panel bracing that will twist or break and therefore not transfer the sign weight to the undamaged support(s)

- Slip base mechanisms must be designed with the proper sized bolts and washers. Bolts that are too small may not withstand the wind load forces. Oversized bolts can result in binding or friction forces between the base plates. Washers that are too small can deform into the slots and bind the plates together.
- Large multiple-support signs have a hinge mechanism that allows the support to swing upward upon impact. The hinge height should be at least 2100 mm above the base plate to allow vehicles to pass beneath the hinge point. The sign panel, or any auxiliary sign panels, should not be mounted below the hinge mechanism.
- Hinged multiple sign supports are generally designed to operate safely when impacted from one direction (they are unidirectional). They can be made bidirectional by selecting the proper hinge arrangement.
- Two posts within a 2100-mm path should each have a mass that does not exceed 27 kg/m.
- Supplemental signs or horizontal members between the supports and below the hinge should not be used.
- Multiple-support systems that are designed with anchor bases should have a maximum of 100 mm from the ground to the highest part of the anchor. This will prevent small vehicles from snagging the undercarriage on the anchor.
- Selection of unidirectional, bidirectional, or multidirectional support assemblies should be based on the possible directions from which the signs can be impacted. Unidirectional assemblies will not function correctly unless impacted from the front along the longitudinal axis of the slotted bolt holes. Bidirectional assemblies will function properly when struck from the front or the back. Impacts can be expected to occur from both travel directions in all cases except roadside sign supports located on the right side of divided roadways that have wide medians or positive median

barriers. Bidirectional or multidirectional support assemblies should be considered for:

Signs placed in the median that are within the clear recovery area of the opposite direction of travel

Signs placed on two-lane roadways or undivided multilane roadways

Signs placed near ramp terminals or intersections where impact could occur from any approach.

The majority of support types approved for use as single sign supports are approved for multiple installation. The approved usage as multiple supports, however, often requires the use of breakaway designs and a limit on the number of supports allowed within a 2100-mm distance of each other. Dual and triple installation refers to installing two and three supports, respectively, within a 2100-mm radius distance of each other. Acceptance of multiple–sign-support systems is based on the same vehicle deceleration characteristics used for single sign supports except that all of the supports within the 2100-mm path are impacted. Selection of approved multiple sign supports, therefore, requires knowledge of the number of supports required and the associated systems approved for use by FHWA.

Multiple–sign-support assemblies are required for signs with large surface areas but also for wide signs. For example, a guide sign 1525 mm × 610 mm contains less than 1.0 m² of sign area but will need more than one support to prevent the sign assembly from being damaged due to environmental loads. Sign panels that have relatively small surface areas but require multiple supports because of their shape can generally use two smaller-size supports than would be required if they were installed with a single support.

7.5.1 Approved Single Supports for Multiple-Support Assemblies

There are few single-support systems that can be buried directly and provide acceptable multiple-support performance upon impact. Two such systems are dual 4.5 kg/m U-channel and dual 90-mm × 90-mm shaped wooden posts. The majority of single-support adaptations to multiple-support assemblies require the use of anchor pieces and breakaway designs. Triple supports consisting of 45 mm × 45 mm square perforated tube and triple 3.7 kg/m U-channel are acceptable when installed with an anchor and breakaway design. Manufacturers are developing devices that enable the use of heavier supports for acceptable multiple-support systems. Figure 7.32 presents a slip base breakaway assembly for square-tube supports manufactured by Unistrut Corporation, which is acceptable for three 64-mm × 64-mm supports within a 2100-mm path [35]. The bottom subassembly is inserted into a 760-mm anchor piece and placed in a 200-mm-diameter, 760-mm-deep concrete foundation.

Multiple supports for large signs are often constructed as slip base designs with galvanized steel wide-flange (W) or American Standard (S) shapes for the sign support. These shapes, depicted in Fig. 7.33, are designated by their depth, in millimeters, and mass, in kilograms per meter. For example, a W150 × 18 is a wide-flange shape with a depth of 150 mm and a mass of 18 kg/m.

Multiple-support sign assemblies that are constructed of W and S shapes are frequently designed with frangible or load concentration couplers. The behavior of these designs is similar to slip bases except that, instead of the base slipping from between the bolts, the couplers, which are used in place of the bolts, break at impact.

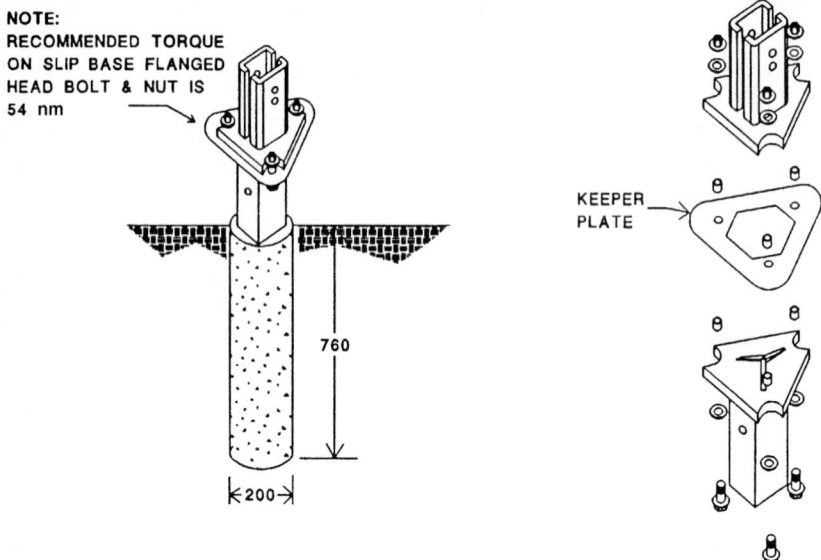

FIGURE 7.32 Acceptable slip base breakaway device for multiple–square-tube sign assemblies.

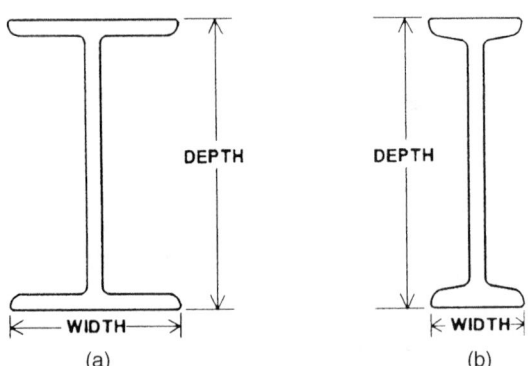

FIGURE 7.33 Examples of rolled-steel shapes for sign supports. (*a*) Wide-flange (W) shape. (*b*) American Standard (S) shape.

7.5.2 Multiple-Support Slip Base–Coupler Designs

Inclined slip base designs, commonly used for single sign supports, cause the sign to rise up upon impact and allow the vehicle to pass beneath the sign. In a multiple–sign-support system, each support is fastened to the other supports by the sign panel and any existing sign panel bracing. When an inclined slip base is used on multiple-support assemblies and only one support is struck, the sign panel stops the impacted support from moving upward. As a result, the slip base can become locked, or the sign

FIGURE 7.34   Installation of horizontal slip base.

panel torn from the other supports, causing intrusion of the panel or support into the vehicle. Inclined slip bases should be used only for multiple-support assemblies when all supports are within 1800 mm of each other. The horizontal slip base (Fig. 7.34) and the frangible coupler (Fig. 7.30) are the most frequently used designs for multiple-support systems. The horizontal slip base design, details of which are shown in Fig. 7.35, operates through separation of the top plate from the anchor plate.

Frangible coupling designs, presented in Fig. 7.36, are designed to effect separation from the anchor plate by fracturing the couplings. Figure 7.36*a* presents a load concentration design in which the small cross-sectional area, at the necked-down portion of the coupling, breaks at impact. Figure 7.36*b* presents a frangible aluminum coupling, by Transpo Industries, designed to break upon impact. The couplings are available in different sizes, designs, and resistance to fracture. Figure 7.37 presents a frangible coupler application for a large sign support. Notice the low profile of this design. The only portions of the sign assembly above ground level are the frangible couplings, so that the possibility of snagging the vehicle undercarriage is practically eliminated.

Horizontal slip base and coupler designs are intended to safely operate by allowing the vehicle to pass under the sign and support assembly upon impact, as presented in Fig. 7.38. This is accomplished by providing a hinge at least 2100 mm from the bottom anchor plate to allow the support to swing away.

7.5.3   Hinge Plate Designs

There are three basic types of hinge designs. One type, illustrated in Fig. 7.39*a*, develops a hinge by cutting through all but the back flange. The front flange is connected

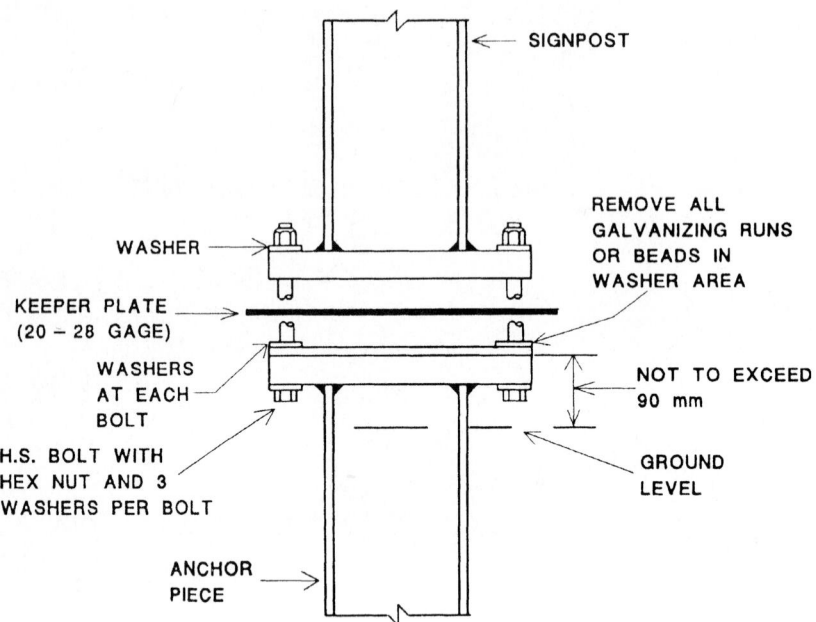

FIGURE 7.35   Design details of horizontal slip base.

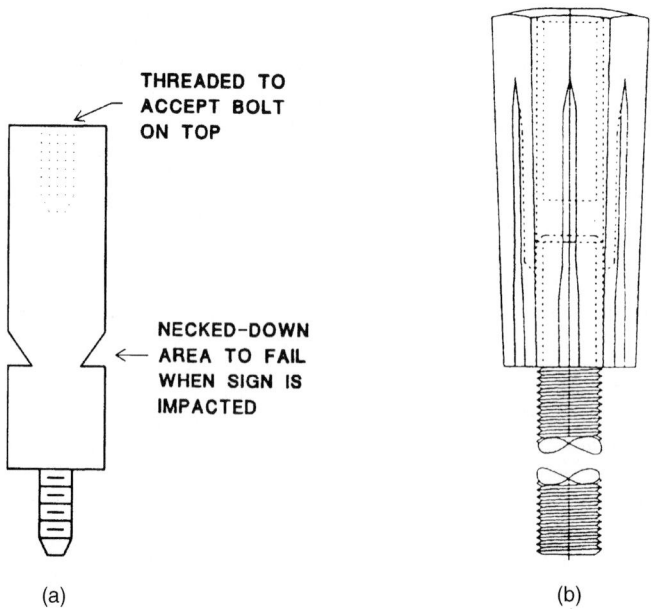

FIGURE 7.36   Examples of couplers. (*a*) Load concentration type. (*b*) Frangible type.

FIGURE 7.37 Base with load concentration couplers.

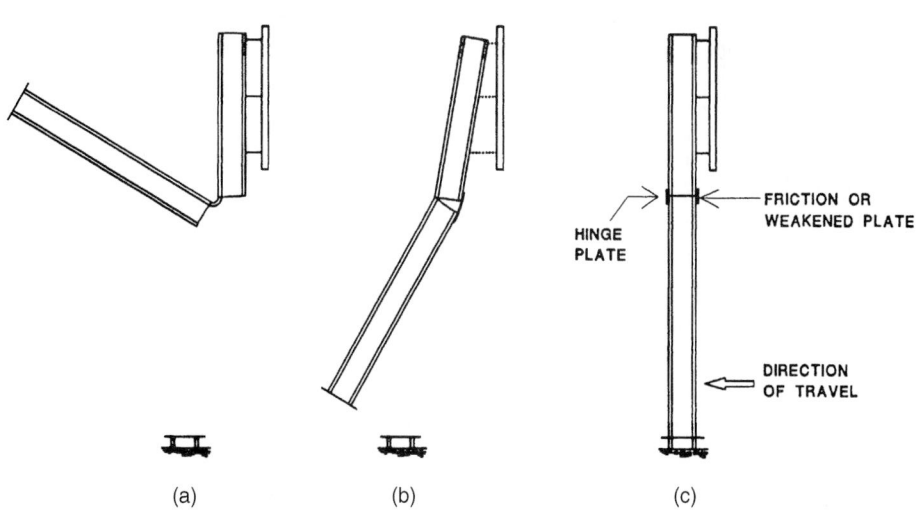

(a) (b) (c)

FIGURE 7.38 Illustration of hinge action for large multiple-support sign. (*a*) Vehicle passes under. (*b*) Hinge activates. (*c*) Vehicle impact.

DIRECTION OF TRAVEL ⟶

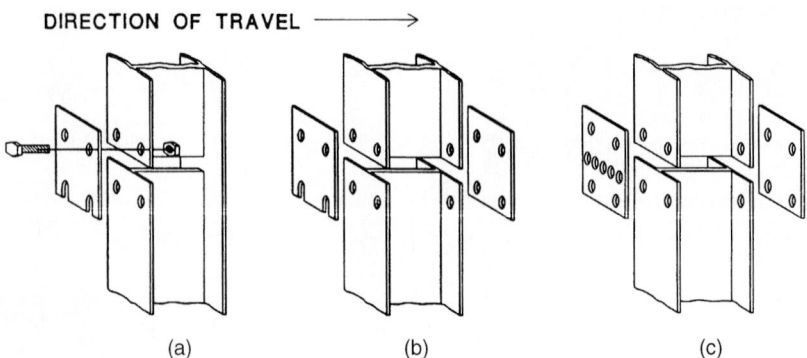

| (a) | (b) | (c) |

FIGURE 7.39 Common hinge designs used for large multiple supports. (*a*) Saw-cut support with front plate. (*b*) Saw-cut support with front plate and rear hinge plate. (*c*) Saw-cut support with weakened front plate and rear hinge plate.

with a slotted plate known as a friction plate. When the post is struck, the friction plate separates from the slotted bolt holes as the back flange bends. This type of hinge creates a maintenance problem, since the post is destroyed and must be replaced after each impact. It is also more difficult to predict the resistance of the hinge, which is dependent upon the post size and depth of cut.

Another type (Fig. 7.39*b*) utilizes a rear hinge plate. This plate is similar to the friction plate but does not have slotted bolt holes. With this type of hinge, the sign support is completely cut in two pieces, with the hinge plate bolted on the back and the friction plate on the front. When impacted, the friction plate releases through the slotted bolt holes and the hinge plate bends back. Maintenance after impact is simplified, since the hinge plate can be removed and the upper and lower support pieces reused with a new hinge plate. Proper operation of the friction plate design is dependent upon proper bolt size and torque. If the bolts are too small, or not torqued sufficiently, wind loads will cause the friction plate to become loose and the top of the sign to fall back. If the bolts are too large or torqued too much, the support will not separate properly upon impact [48].

The third hinge type (Fig. 7.39*c*) utilizes a rear hinge plate and a front hinge plate with a weakened section. When impacted, the section fractures through the plane of the holes, thus permitting the back hinge plate to bend. This design has an advantage over the friction-hinge plate design while remaining easy to repair. The advantage is that the torquing requirements on the friction plate are not critical for proper operation. The front hinge plate in Fig. 7.39*c* is weakened by drilling holes so that only 33 percent of the plate material remains. Figure 7.40 shows commercially available frangible hinge plates available from Transpo Industries [36]. The three hinge systems presented in Fig. 7.39 are unidirectional and should not be used in areas requiring bidirectional performance. Only the Transpo hinge system offers bidirectional performance.

7.5.4 Improved Base Devices

A number of manufacturers have developed products so that the anchor piece can be placed almost flush with the ground. The products can be used either to retrofit exist-

FIGURE 7.40   Commercially available hinge plates by Transpo Industries.

ing slip base designs or for new installations. One such manufacturer is Transpo, which markets the Breaksafe breakaway system for ground-mounted signs [36]. These devices use breakaway couplings and brackets designed for different support types and sizes. Included are back-to-back concrete and direct buried U-channel, 76- to 114-mm round pipe, 76- to 127-mm square tube, and various sizes of wide-flange and standard beam shapes. Figure 7.41 illustrates a typical before-after retrofit of a slip base design to a frangible coupler design. The advantage of the retrofit is that proper torquing, to prevent blowdown or walking due to environmental loads, yet permitting slip during vehicle impact, is not required with the frangible coupling retrofit. Figure 7.42 presents a new installation of the Breaksafe system for a 150- or 200-mm-wide flange support [36].

7.6   MAINTENANCE AND CONSTRUCTION OF SIGN SUPPORTS

An important element of a safe highway environment is the proper construction and maintenance of traffic signs. Good designs and the best of materials will not be effective in reducing accident potential or severity if the traffic signs are improperly placed or installed. This requires that field crews be knowledgeable of proper installation techniques and that they report and correct any possible problems instead of merely placing the signs at the roadside. (See also Art. 7.1.2.)

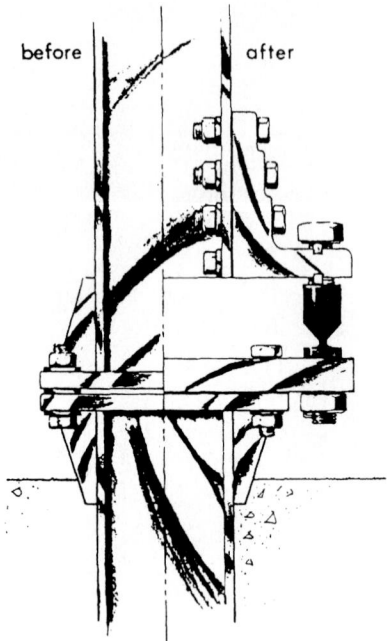

before   after

FIGURE 7.41 Typical before and after retrofit with Breaksafe.

7.6.1 Proper Placement

Important considerations for proper placement include the following:

- Warning signs should be placed sufficiently in advance that the driver has adequate time to perceive, identify, decide, and perform any necessary maneuver. A guide for the placement distance of warning signs is contained in *MUTCD* [2].
- The distance signs should be set back from the traveled way for urban and rural roadways is presented in Fig. 7.43 and for expressways and freeways, in Fig. 7.44.
- All sign assemblies located within the transversable area must be capable of giving way safely upon impact. This requires that the maximum vehicle deceleration does not exceed 5 m/s and that the sign assembly does not protrude into the passenger compartment [15].
- The height from the ground to the top of the sign should be 2750 mm. In the majority of cases, following the minimum 2100-mm requirement of *MUTCD* to the bottom of the sign will achieve the 2750-mm top of sign requirement. The 2750-mm requirement should, however, be the governing criterion, since tests have demonstrated that signs mounted with their tops at this height will hit the roof rather than the windshield upon impact.
- Sign supports installed with anchor systems must have a maximum height of 100 mm from ground level to the topmost part of the anchor.

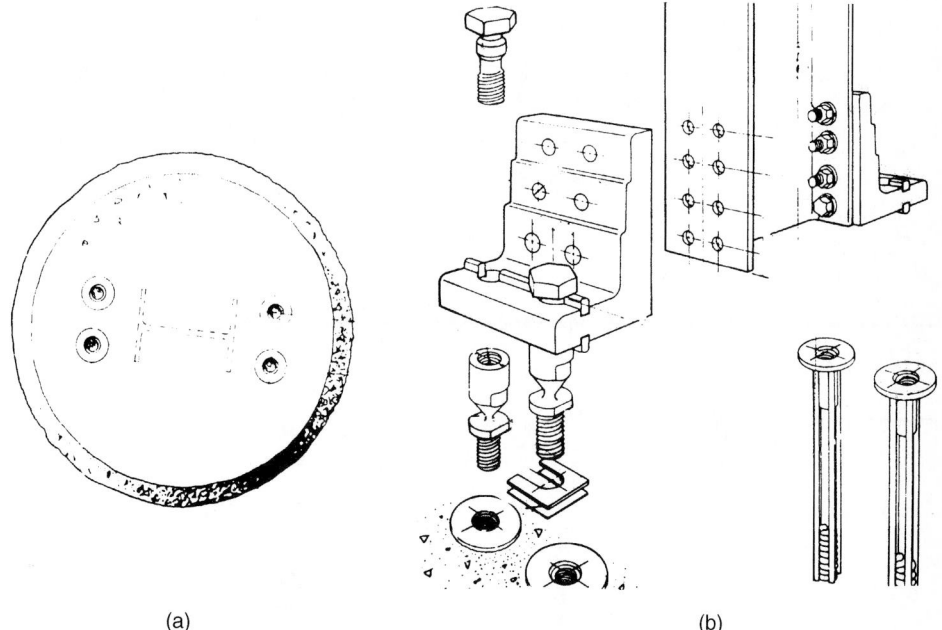

(a)   (b)

FIGURE 7.42 Breaksafe ground mount sign-support system. (*a*) Plan of footing. (*b*) Ground mount sign support.

- Most sign-support assemblies are designed to function properly when impacted at bumper height, typically about 500 mm above the ground. If impacted at a higher point, the assembly may bind at the planned shear point resulting in nonactivation of the breakaway mechanism. For this reason, it is critical that breakaway sign assemblies not be located near ditches or on steep slopes or other locations where the vehicle can become partially airborne at the time of impact.

- Sign supports should not be placed in ditches. The water in the ditch can erode the soil around the base of the support, cause premature deterioration of the post, and freeze, resulting in unpredictable performance during impact. The ditch can also act as a guideway that directs errant vehicles into the sign assembly.

- Sign-support assemblies are tested in both strong and weak soils. Supports that are designed to yield, or fracture, upon impact generally perform better in strong soil. Strong soil holds the buried portion in position, providing sufficient resistance for the sign support to break near ground level. Weak soils do not provide this resistance, but permit movement within the ground and unpredictable results. Yielding or fracturing supports that are embedded less than 1000 mm in weak soil will often pull out of the soil. While this may provide acceptable impact performance, the force of the wind and ice loads may cause the sign assembly to rotate or fall down. The actual soil type that is present may not be known until the start of installation. Weak soils are those that offer relatively little resistance to driving the signpost. If weak soils are encountered, there are measures that can be taken to maintain sign orientation in the face of environmental loads and still result in proper operation during impact. These include embedding the signpost to 1000 mm, and the use of anchor plates, concrete footings, and commercially available anchor systems [37].

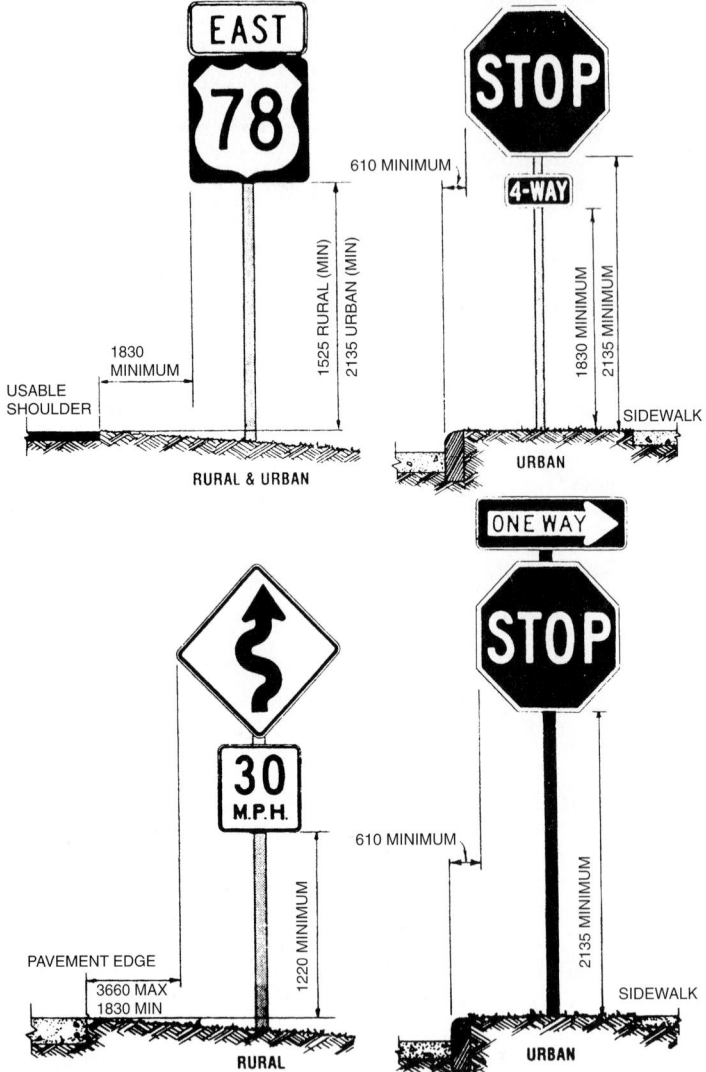

FIGURE 7.43   Height and lateral clearance of signs on rural and urban roadways.

- Single–sign-support systems are designed to operate safely when only one support is struck upon impact. Tests have shown that an errant vehicle, leaving the roadway at an angle, can impact more than one support if supports are not separated by more than 2100 mm. This separation applies to other fixed objects as well as signposts. For example, a 75-mm-diameter tree is sufficiently small to provide acceptable impact performance. Installing a sign support 2000 mm from this tree, however, can result in an errant vehicle's impacting both the tree and the signpost. The combined effect of the tree and sign can provide unacceptable impact performance.

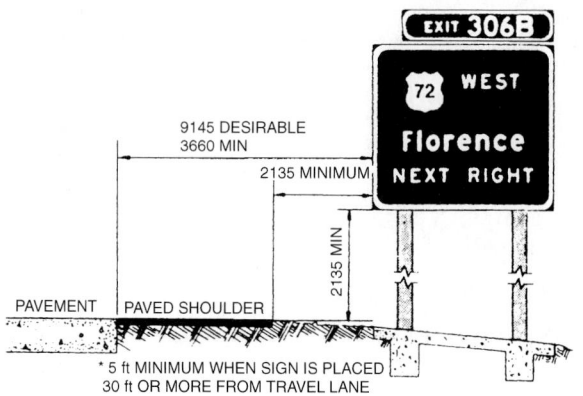

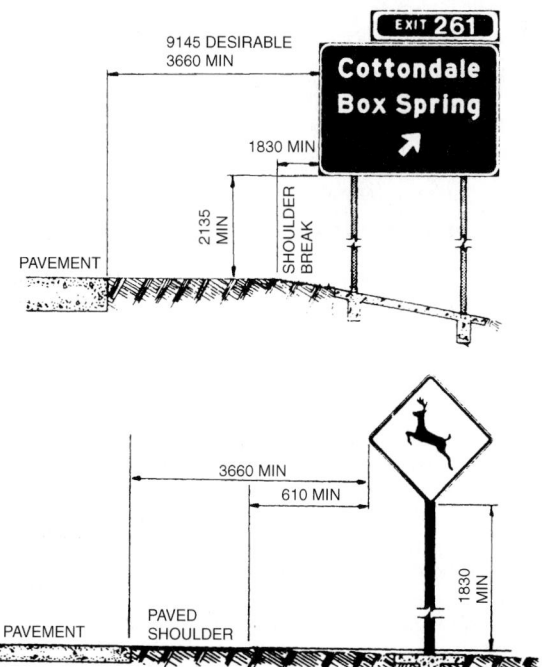

FIGURE 7.44 Height and lateral clearance of signs on expressways and freeways.

- Multiple-mount sign supports are required to support sign panels that are too large to withstand wind and ice loads with the use of only one support. Multiple-mount sign supports are designed to provide acceptable performance upon impact when the supports are placed 2100 mm or closer to each other. This close spacing results in the possibility that a vehicle leaving the roadway at an angle will impact two or more supports simultaneously. This possibility means that some supports approved for use in a single-support system are not approved for multimount designs. Support configurations that have not been approved for use as multiple-mount systems should not be used for multimount sign supports until they have been approved by the FHWA.

- Study the traffic patterns and surrounding geometrics prior to installing any sign. If the sign assembly can be expected to be struck from more than one direction, then a unidirectional slip base design is an improper choice. Two-lane rural roadways should use roadside supports that function safely when impacted from two directions. Installations on freeways, with wide medians or positive median barriers, can be expected to be impacted from only one direction.

- In summary, be aware of what is required for sign installations to function properly for both the environmental loads and vehicle impact. Do not install a device simply because it has been specified on the sign order. The actual site conditions may have been unknown, or different from what was expected by the designer who specified the type of sign assembly. If problems are identified, contact a supervisor to determine if changes should be made.

7.6.2 Single-Support Installation

The correct installation of sign support assemblies is dependent upon the type of support post, the type of soil present, and the impact performance design of the sign assembly. Installation instructions are contained in standard state drawings and, for proprietary devices, from signpost manufacturers. Proper installation practices for the most common types of single support assemblies are presented in subsequent sections.

U-Channel Posts.   The most common method of installing U-channel posts is by direct burial. The burial can be achieved by mechanical post drivers, by sledgehammer, or by digging a hole and backfilling. If the post is to be placed by driving into the ground, then a driving cap should be used to prevent damage to the end of the U-channel. Drive or place the posts at least 900 mm but no more than 1100 mm into the ground to make it easier to pull out damaged posts.

U-channel posts can also be installed as two-piece assemblies consisting of an anchor base and the post support. The advantage of the two-piece assembly is that the post will break off from the anchor piece upon impact. This often improves safety upon impact, makes repairs easier, and makes it possible to salvage portions of a damaged U-channel post. An anchor base assembly is especially advantageous when the post is placed in a paved area, such as a concrete median. The anchor piece should not extend more than 100 mm above the ground to prevent snagging the vehicle undercarriage.

The anchor piece can be directly driven, buried 900 mm in the ground, or embedded 610 mm in a concrete foundation that is 200 mm in diameter and 760 mm deep. The signpost can be attached to the anchor piece by a generic splice or the use of commercially available devices.

Figure 7.45 presents a generic method of attaching the signpost to the anchor piece. The signpost overlaps the anchor piece by 150 mm to provide stability against the

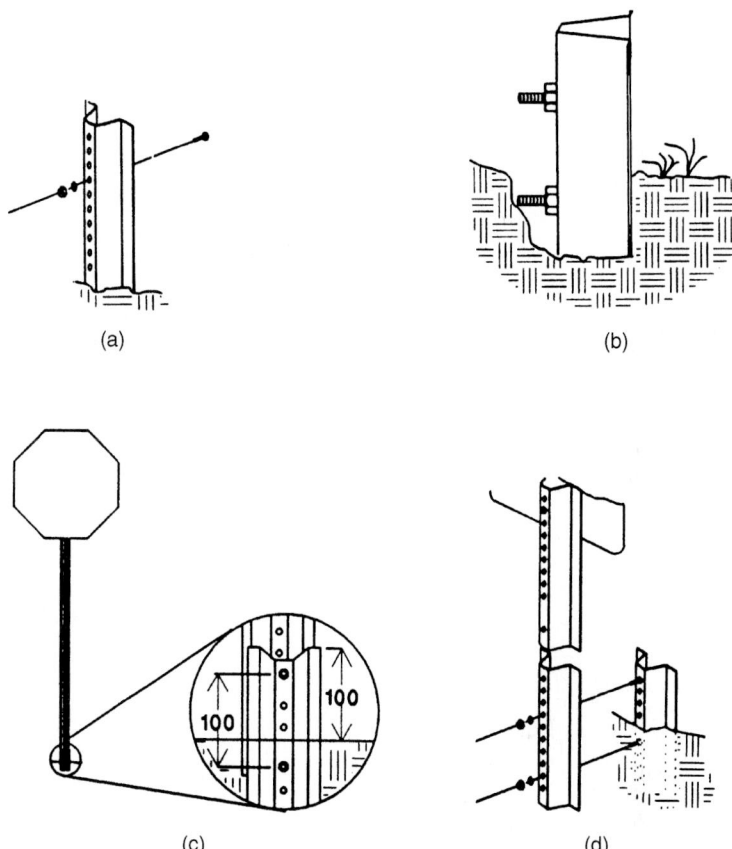

FIGURE 7.45 Generic method of splicing sign support to anchor piece. (*a*) Drive anchor post to within approximately 300 mm from top of ground and install bolt with lock washer in fifth hole from top. (*b*) Drive post to 100 mm or less from ground, and install bolt in first hole from back of post to allow room for sign post to be attached. (*c*) Install bolts 100 mm apart with ground stub no higher than 100 mm above ground. (*d*) Place signpost behind anchor stub, place bolts through first and fifth hole of sign post, use cut washers, and tighten securely.

environmental loads. Since the anchor piece cannot extend more than 100 mm above the ground, this means that the signpost is at least 50 mm below ground level. The signpost is placed behind the anchor stub, and the posts are attached together with two 8-mm bolts spaced 100 mm apart. Extra 8-mm nuts are used as spacers between the two post pieces to prevent binding during impact.

A number of commercial splicing devices for installing two-piece U-channel assemblies are also available. Figure 7.46 provides installation information on the Eze-Erect system available from Franklin Steel, and Fig. 7.47 is information on the Minute-Man coupling from Marion Steel [38, 39].

The following guidelines should be followed for the installation and use of U-channel posts:

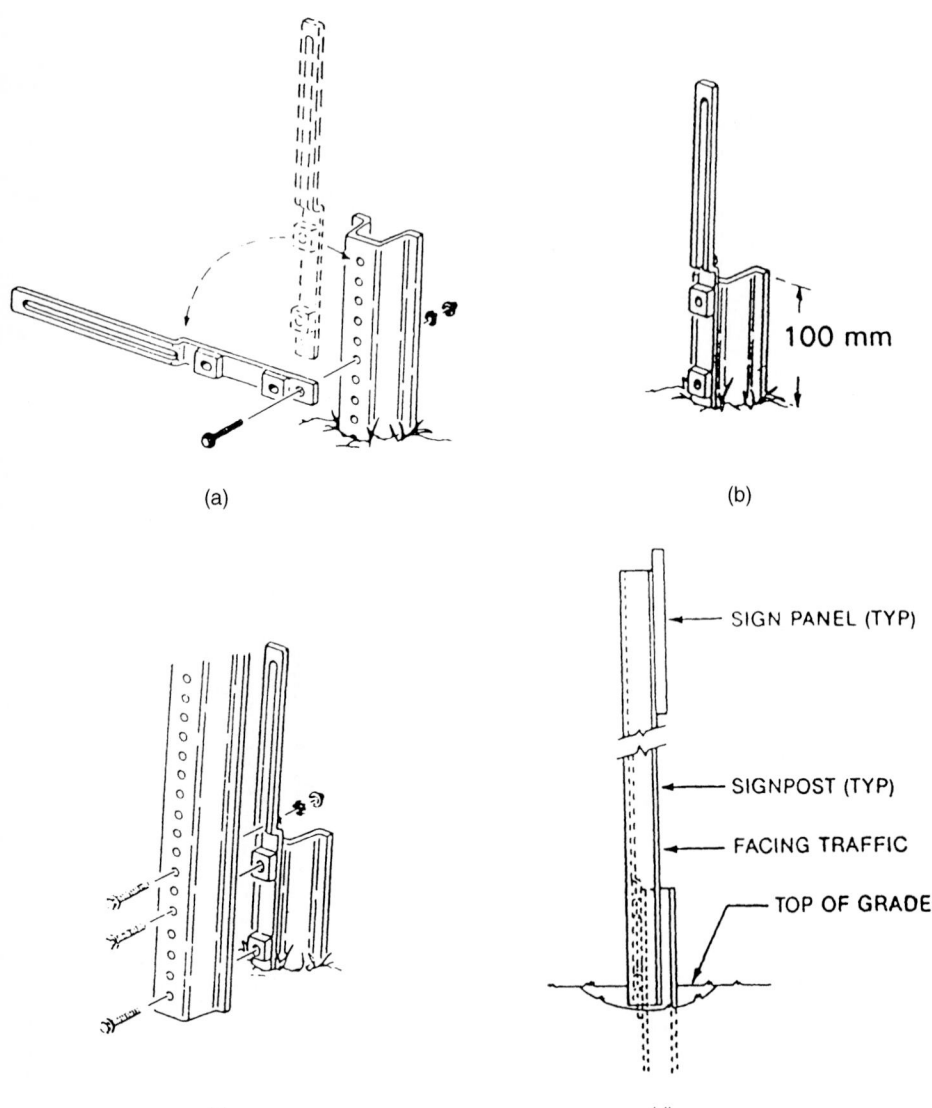

(a)

(b)

(c)

(d)

100 mm

SIGN PANEL (TYP)

SIGNPOST (TYP)

FACING TRAFFIC

TOP OF GRADE

FIGURE 7.46 Installation with Eze-Erect U-channel coupling. (*a*) Drive anchor post to within 300 mm of ground level, attach retainer spacer strap through bottom hole of strap and sixth hole of anchor post, and rotate strap to the side. (*b*) Drive anchor post to within 100 mm of ground level and rotate strap to vertical position. (*c*) Attach signpost with two bolts, nuts, and lock washers in bottom and fifth hole; insert one bolt through signpost and bottom of long slot in strap; and tighten all nuts snugly before completely tightening assembly. (*d*) Finished assembly.

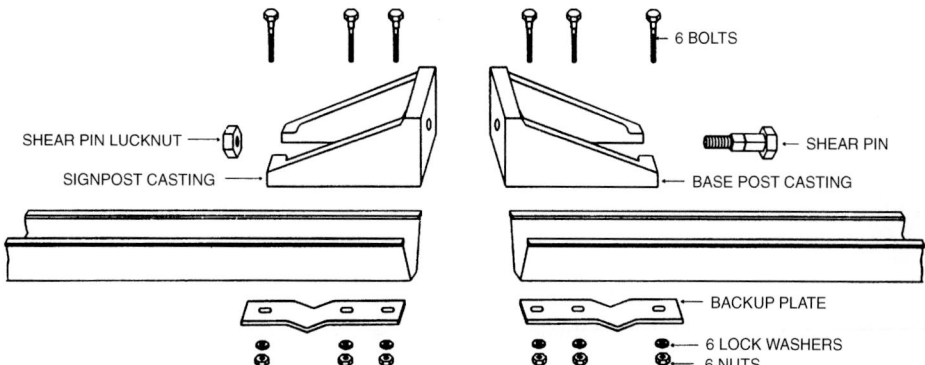

FIGURE 7.47   Minute-Man coupling for use with RIB-BAK U-channel signposts. Erection steps: (1) Bolt couplers to both Minute-Man groundpost and accompanying sign support using backup plates for reinforcement. (2) Drive groundpost into the ground until only 3 in remain above ground (1 in of bottom coupler is buried). (3) Raise sign and connect Minute-Man's top and bottom sections by inserting shear pin. To finish, simply tighten shear pin bolt.

- If U-channel posts are driven into the ground, they should not be embedded more than 1100 mm, to make it easier to pull out damaged posts.

- Use a drive cap to drive the U-channel into the ground to prevent damage to the post end.

- If an anchor base is used, do not leave the anchor stub protruding more than 100 mm above ground level.

- The generic splice should provide an overlap of 150 mm with the anchor base. This results in 50 mm of the signpost extending below ground level. The signpost is fastened to the anchor stub with 8-mm grade 9 bolts spaced 100 mm apart. The signpost and anchor piece should be separated with a 16-mm spacer to prevent possible binding of the posts upon impact.

- Anchor pieces one size larger than the sign post will help prevent damage to the anchor piece upon impact.

- The Florida splice requires an overlap of 200 mm. This results in embedding of 100 mm of the signpost below ground. The splice is secured with A307 9.5-mm bolts, 50 mm long, spaced at 150 mm center to center. A 16-mm spacer is placed between the anchor piece and the signpost. The use of 9.5-mm-diameter bolts requires that the post holes be reamed in order to insert the bolts. Reaming destroys the corrosion protection of the hole, necessitating the application of zinc-rich paint paste to prevent corrosion.

- If commercial splices are used, the manufacturer's installation instructions must be closely followed for proper impact performance.

- The frangible bolt provided with the Minute-Man must be used for proper impact performance. Do not replace this bolt with a regular steel bolt.

- It is not recommended to interchange signposts and anchor stubs of different manufacturers when there is variation in cross section between the two sections. No crash tests have been done on mixed anchor stubs and signposts. The difference in cross section may be sufficient to cause problems in nesting under some splice orientations.

- The signpost should be placed behind (on the nonimpact side of) the anchor stub for U-channel anchor base assemblies.
- Splices that are performed above the anchor piece to extend short pieces of U-channel or to piece together salvaged U-channel are not recommended. One-piece U-channel posts perform better under impact than posts that have been spliced above the anchor stub. A splice in the impact zone can strengthen the post and degrade its impact performance. Splices above the impact zone can open, allowing the sign panel to take an unpredictable and potentially hazardous trajectory. The splice can also open with the lower end of the upper post section penetrating the impacting vehicle. If splices above the anchor piece are used with U-channel, it is important that the following conditions are met [40]:

The splice does not extend below ground level.

The overlap is approximately 460 mm fastened by four 8-mm bolts, with two bolts, through the holes nearest the ends, at each end of the splice. Spacers 16 mm thick should be placed over the bolts between the spliced pieces of U-channel.

A splice that is mostly below a vehicle bumper height should have a maximum top elevation of 500 mm, and a splice that is mostly above the bumper should have a bottom elevation of 400 mm or above. A diagram of these recommendations is presented in Fig. 7.48.

Square Steel Tubes.   Square steel tubes are available from a number of manufacturers in perforated, and punched but not perforated, styles [41, 42, 43]. Two of the

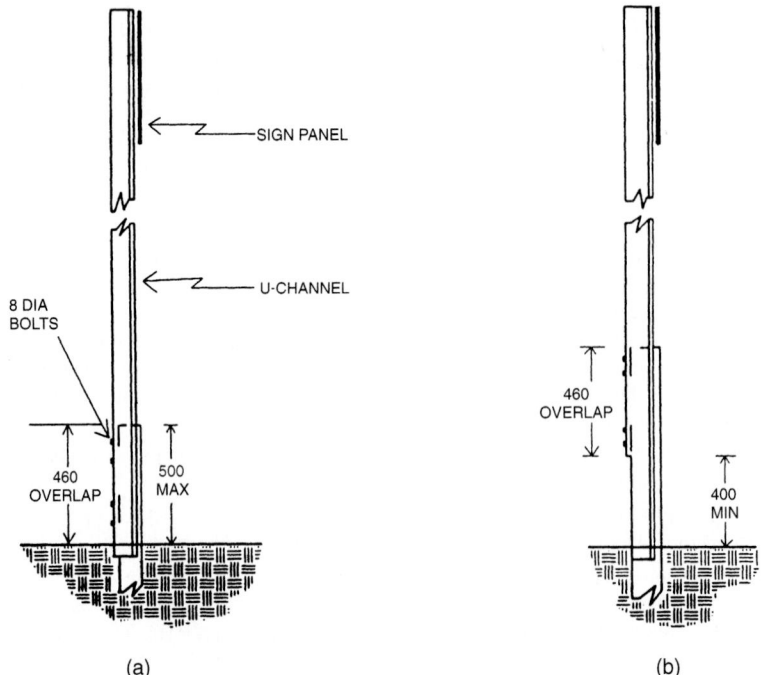

FIGURE 7.48   Allowable but not desirable splicing of U-channel sign supports. (*a*) Limits on lower splice. (*b*) Limits on upper splice.

major manufacturers of square-tube posts are Unistrut, with the brand name Telespar, and Allied Tube and Conduit, with the perforated Square Fit and the nonperforated Quick-Punch tubes [37, 38, 39]. Square-tube sign supports can be installed as one-piece direct burial assemblies and with anchor pieces. The anchor piece assemblies have the advantages of more predictable performance upon impact, a larger range of permissible sizes, and reduced maintenance required for repair after impact. Figure 7.16 shows types of installation.

Square steel-tube sign supports up to 57 mm × 57 mm in size can be installed by direct burial. Sizes larger than 57 mm × 57 mm require an anchor base assembly to provide acceptable impact performance characteristics. The most common method of direct burial is by driving directly into the ground, using a driving cap to protect the end, by mechanical drivers or a sledgehammer. Drive or place the square tube at least 900 mm deep but no more than 1100 mm into the ground to make it easier to pull out damaged posts.

Repair of damaged square tube is easier to perform when an anchor base assembly is used. The anchor base assembly for square tube usually consists of a 760-mm-long anchor piece, one size larger than the signpost, and a 450-mm-long stiffening sleeve, one size larger than the anchor piece. The sleeve provides a double-walled anchor base that helps prevent damage to the anchor assembly and makes the breakaway characteristics of the signpost more predictable. Acceptable impact performance can also be obtained by the use of only the anchor piece, but damage to the anchor piece and increased maintenance are more likely to occur than when using a stiffening sleeve. Sizes larger than 64 mm × 64 mm should not be used for breakaway performance with the anchor breakaway design. The anchor piece must not extend more than 100 mm above ground level. The installation procedures for the square-tube anchor base system are provided in Fig. 7.49.

In addition to the telescoping anchor bases, made from larger sizes of square tubing, there are heavy-duty anchor bases commercially available. These bases can be used in hard or rocky soil conditions that can present problems for driving the regular-sized tubing as anchor pieces.

The following guidelines should be followed for the installation and use of square steel-tube signposts:

- Do not directly bury square steel tubing that is larger than 57 mm × 57 mm. If a sign larger than 57 mm × 57 mm is required, use an anchor base system.

- Repair of the square steel-tube sign assembly is much easier if an anchor base system is used. The stiffening sleeve helps reduce damage to the anchor and provides a strengthened base for reliable impact performance.

- The anchor assembly should be driven or placed into the ground with only 25 to 50 mm protruding above ground level. This will expose one to two holes for fastening the sign assembly, reduce vehicle sagging, and ease repair.

- If driving the post or anchor base into the ground, use a drive cap to protect the exposed end. If a drive cap is not used, the exposed end will become distorted, inhibiting insertion of the telescoping tube.

- Do not install a two-piece anchor assembly if the top of the anchor piece and sleeve is not flush or if the holes are misaligned. The bolts will be difficult to insert and the higher piece may bend upon impact, damaging the anchor assembly.

- Do not overtighten the bolts that fasten the signpost to the anchor assembly. Tightening the bolt too much will distort the tubing and hinder the removal or insertion of the signpost into the anchor assembly.

- Sections of square steel tube can be spliced together to allow the reuse of damaged posts. The splice is made by using a 300-mm-long section of tubing one size small-

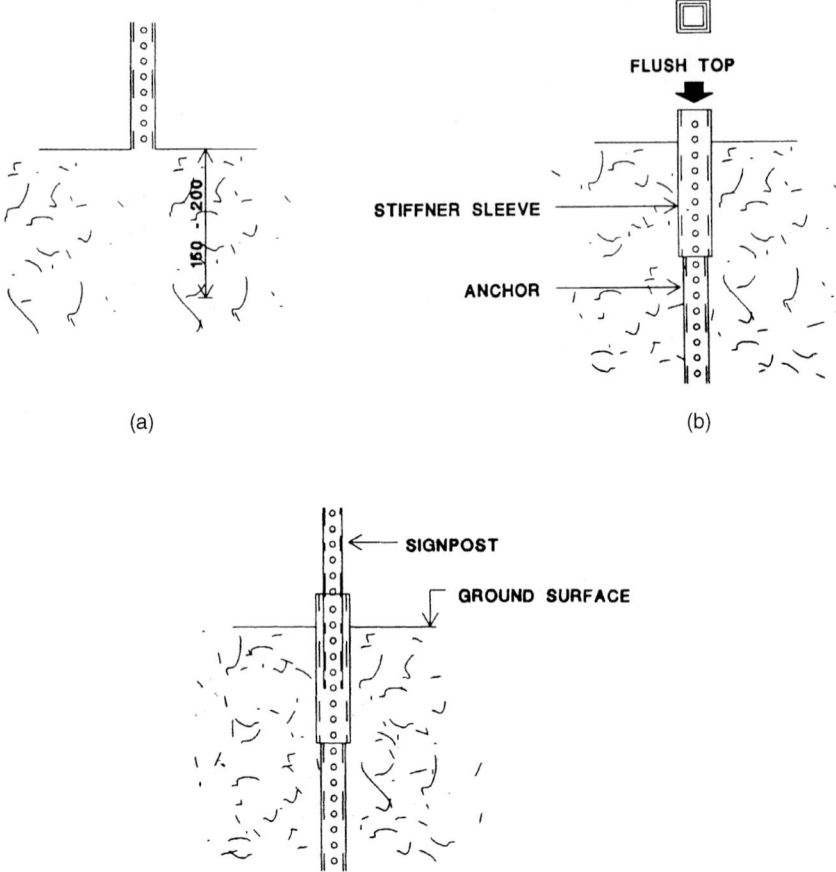

FIGURE 7.49 Installation procedure for square-tube anchor base assemblies. (*a*) Drive the anchor post 150 to 200 mm into the ground, remove post, and knock out soil from post end. (*b*) Reinsert post into hole and drive with stiffer sleeve to 25 to 50 mm above ground level. (*c*) Attach sign to signpost, insert 150 to 200 mm into anchor, and fasten to base.

er than the tubing to be repaired. The 300-mm section is inserted halfway into one of the tubes and secured with two drive rivets or one bolt. The second tube is then slipped over the free end of the 300-mm section and fastened in place.

- Square tube can be used to install signs in areas of concrete or asphalt by drilling or chipping through the surface and driving an anchor assembly in place. An anchor base is recommended in concrete or asphalt areas to make repair easier in case of impact.

Wooden Posts. The most common wooden supports for single signpost installation are the 90-mm × 90-mm shaped and the 100-mm-diameter round posts. These posts should be directly buried to a depth of at least 910 mm (Fig. 7.14*a*). Deeper burial is often performed to reduce vandalism. Posts larger than the 90-mm × 90-mm and the 100-mm-diameter posts require drilled holes to reduce the cross section and embedment in concrete so as to safely break away during impact. The requirements presented in Tables 7.8 and 7.9 should be followed for the installation of shaped and timber posts.

The use of 13-mm-thick Styrofoam for the concrete foundation (Fig. 7.14*b*) eases the removal of broken stub pieces [44]. An example of hole placement to achieve a weakened cross section is also presented in Fig. 7.14*b* [45]. The bottom hole should never be centered more than 100 mm above the ground, because the stub piece must remain at 100 mm or less after impact. Rectangular shaped posts are placed with the long post dimension parallel to the direction of travel. The holes of the proper size for the post are drilled perpendicular to the expected direction of impact.

Steel-Pipe Posts. Steel-pipe (schedule 40) posts smaller than 51 mm internal diameter (I.D.) can be directly buried and still provide acceptable impact performance. As indicated in Art. 7.3.6, a plate 102 mm × 305 mm × 6 mm, or two sign clamps, should be bolted or welded to the pipe, beneath ground level, to prevent rotation due to wind. Schedule 40 steel-pipe supports should be direct buried, with the attached earth plate, to a depth of at least 1070 mm to provide acceptable performance upon impact.

TABLE 7.8 Installation of Shaped Wooden Posts for Single Sign Installation

Post size, mm	Embedment type and depth	Comments and required post modifications
90 × 90	Direct burial to a minimum of 920 mm.	No holes required.
90 × 140	Direct burial to a minimum of 1 m.	Tests have shown that in loose or sandy soil, this post is not safe for a small car. In weak soils, therefore, holes perpendicular to probable impact path must be used: one 40-mm hole at 100 mm and one 40-mm hole at 140 mm above ground level.
90 × 140	Set in a concrete foundation of 460 mm diameter and 760 mm deep to a depth of 610 mm with a steel sleeve.	No holes required when placed in a concrete foundation.
140 × 140	Set in unreinforced concrete foundation. Wrap 13-mm-thick sheet Styrofoam around post before setting it in concrete to ease removal of broken stub.	Holes must be drilled perpendicular to probable impact path: one 50-mm hole at 100 mm and one 50-mm hole at 460 mm above ground level.
140 × 190	Placed in 610-mm-diameter, 760-mm-deep concrete foundation with steel sleeves to a depth of 610 mm.	Holes must be drilled perpendicular to probable impact path: one 75-mm hole at 100 mm and one 75-mm hole at 460 mm above ground level.

TABLE 7.9 Installation of Timber Wooden Posts

Post diameter, mm	Embedment type and depth	Comments and required post modifications
100 115 127	Direct burial to at least 920 mm	No holes required.
127	Placed in soilcrete foundation of 460 mm diameter and 1100 mm deep	Holes must be drilled perpendicular to probable impact path: one 50-mm hole at 100 mm and one 50-mm hole at 460 mm above ground level.
150	Direct burial to 1500 mm	Holes must be drilled perpendicular to probable impact path: one 50-mm hole at 100 mm and one 50-mm hole at 460 mm above ground level.
165	Direct burial to 1500 mm	Holes must be drilled perpendicular to probable impact path: one 32-mm hole at 100 mm and one 32-mm hole at 460 mm above ground level.
178	Diret burial to 1500 mm	Holes must be drilled perpendicular to probable impact path: one 51-mm hole at 100 mm and one 51-mm hole at 460 mm above ground level.
190	Direct burial to 1500 mm	Holes must be drilled perpendicular to probable impact path: one 70-mm hole at 100 mm and one 70-mm hole at 460 mm above ground level.

A breakaway collar assembly is required for schedule 40 standard pipe that is equal to or greater than 51 mm I.D. The breakaway collar can be made by the use of a regular pipe coupling or reducing coupling. The reducing coupling is recommended since it reduces the probability of damage to the anchor piece, thereby easing repair. The anchor piece is usually one size larger than the signpost. The anchor assembly consists of a 610-mm-long anchor piece placed in a concrete footing that is 760 mm deep and 300 mm in diameter.

In addition to standard steel pipe, there are round steel-tube sign supports available from a number of manufacturers, with a wall thickness of 12 gauge or less and designed for use in an anchor system. Commercial anchor systems, such as the Poz-Loc, can be used for the round steel tubes and for standard pipe 51 mm or less in size [47]. The use of commercial anchor systems requires closely following the manufacturer's instructions for proper performance.

A summary of steel-pipe sign-support installation recommendations is provided in Table 7.10. Also consider the following guidelines:

- Standard steel pipe (schedule 40) that is less than 51 mm I.D. can be direct buried for use as sign supports. Direct burial supports should have an anchor plate, or sign brace, bolted or welded to the buried portion to prevent rotation.

TABLE 7.10   Installation of Steel-Pipe Posts

Post type and diameter	Embedment type and depth	Comments
Standard (schedule 40) steel pipe Less than 51 mm I.D.	Direct burial to at least 1070 mm	An earth plate measuring 100 mm × 300 mm × 6 mm must be bolted or welded to the buried end to prevent rotation.
Equal to or less than 51 mm I.D.	Commercial anchor system such as Poz-Loc	Follow manufacturer's instructions.
Equal to or greater than 51 mm I.D.	Breakaway design with concrete anchor base	The concrete base is 760 mm deep by 300 mm diameter. The anchor piece is embedded 610 mm in the concrete base. Use an anchor piece one size larger than the anchor post and a reducing coupling for the breakaway action. Top of coupling should not be more than 100 mm above ground level.
Round steel tube 60 mm O.D. or less with 12 gauge walls	Commercial anchor system such as Poz-Loc	Follow manufacturer's instructions.

- Standard steel pipe (schedule 40) that is equal to or greater than 51 mm I.D. must be of breakaway design with anchor base.
- Anchor pieces should be placed in a concrete foundation and be one size larger than the signpost. The breakaway mechanism can be achieved by the use of a reducing coupling. The top of the coupling should not be more than 100 mm above ground level.

Aluminum Shapes.   A number of aluminum shapes, including U-channel, round and square tubing, and various X shapes, are available for sign supports. The available U-channel sizes are appropriate only for small or temporary sign installations. Round aluminum shapes require an anchor base to prevent rotation due to wind loads.

A summary of installation recommendations for aluminum shapes is presented in Table 7.11. Also consider the following:

- Do not install aluminum round signposts larger than 90 mm diameter. Recent tests show that the larger aluminum post sizes fail in weak soil conditions.
- Anchor plates or two sign clamps configured to encircle the post should be used below ground level to prevent rotation due to wind loads.

TABLE 7.11 Installation of Aluminum Post Shapes

Post type and size	Embedment type and depth	Comments
X shape: up to 76 mm × 88 mm	Drill and backfill to a depth of 1000 mm	
Round shapes: up to 90 mm I.D. by 4.8 mm wall thickness (often called thin wall tubing)	Direct burial to a depth of 1000 mm	Sizes larger than 90 mm are not acceptable for use in all soil types.

Fiberglass Support Posts. Fiberglass sign supports have been approved for use up to 76 mm diameter with a 3-mm-thick wall. An anchor system is necessary to provide acceptable breakaway performance and prevent rotation due to wind loads. The anchor systems are available from the same suppliers that provide the fiberglass post. Dual installation (two supports within a 2100-mm path) has been approved for fiberglass supports manufactured by Hwycom when installed with a steel pipe sleeve in a concrete foundation. A diagram of the steel sleeve and concrete foundation design was shown in Fig. 7.25. A summary of the fiberglass post installation requirements is provided as Table 7.12.

7.6.3 Slip Base Designs

Slip base designs for single sign supports provide the opportunity to use stronger sign supports than would be possible without the slip base design. The purpose of the slip base is to provide a separation plane between the sign support and the anchor system. The two pieces are fastened together with bolts that must be properly tightened, or torqued. If the bolts are not torqued enough, they will be loosened by vibration from environmental loads, causing the sign assembly to separate. If the bolts are torqued too much, the friction between the base of the signpost and the anchor piece will be too large to permit proper separation upon impact. A 20 to 28 gauge metal "keeper

TABLE 7.12 Installation of Fiberglass Single Sign Supports

Post size	Embedment type and depth	Comments
76-mm O.D., 3-mm-thick wall	Inserted into a 710-mm-long commercially available steel-tube anchor system	Self-tapping screws are installed above ground level to prevent post rotation.
	Inserted into 76-mm schedule 40 pipe embedded in 305-mm × 760-mm-deep concrete foundation	
76-mm O.D., 3-mm-thick wall	Inserted into 76-mm I.D., schedule 40 pipe, embedded in 305-mm-diameter × 760-mm-deep concrete foundation	Two supports within a 2100-mm radius can be used with concrete foundation design.

plate" should be inserted between the faces of the top and bottom slip bases to prevent the bolts from migrating out of the assembly (Art. 7.4).

There are three basic types of slip base designs for single sign supports. The horizontal slip base design (Fig. 7.34) will operate correctly when impacted from either the front or the back. Horizontal slip base designs do not provide the lift capability available from inclined or multidirectional designs. Horizontal slip bases when used as single sign supports therefore do not function as well upon impact as the other slip base designs.

The inclined slip base (Fig. 7.27) is the recommended type of slip base for single sign supports when impact can be expected from only one direction. Its performance upon impact is designed to cause the upper sign support and sign panel to raise up, thus allowing the vehicle to pass completely under the support assembly. The anchor piece of the inclined slip base must be installed so that approaching vehicles encounter the lower edge before the high edge.

The multidirectional slip base is fastened together with three bolts and has a lift cone fastened to the bottom plate. The sign support is tubular with a maximum size of 127 mm diameter.

All of the slip base designs require a firm foundation for proper operation. Concrete foundations should be used for all slip bases, since direct burial may result in base movement and improper release of the slip base. To prevent snagging of the vehicle undercarriage, no part of the anchor piece and its attached slip base may extend more than 100 mm above ground level. Horizontal and inclined slip base designs can be constructed with wide-flange, standard-shape, and round sign supports. The concrete footing sizes for wide-flange and standard-shape signposts should be constructed to state specifications. The concrete foundation and anchor stub sizes listed in Table 7.13 are appropriate for round signposts with slip base designs.

Torque Requirements. Specifications for bolt tightness must be followed so that the sign assembly (1) remains intact under normal environmental loadings and (2) separates correctly upon impact. The specifications can be given in a number of ways, such as residual tension, clamping force, or torque. *Torque* refers to the amount of force used in tightening the nut to the bolt. The result of the nut-to-bolt tightening places the bolt in tension and exerts the clamping force. Measuring the torque is the most convenient method of obtaining a specified tightness. Providing a specified torque, however, does not guarantee a certain clamping force. Irregularities in the threads, heavy coating deposits on galvanized parts, or irregularities on the mating surfaces of the nut and plate faces result in friction forces. These friction forces cause an increase in torque to move the nut without a resulting increase in clamping pressure. To help ensure that the torque specification provides the required clamping force, the nut should be tested for thread irregularities by being threaded on the bolt

TABLE 7.13   Slip Base Anchor Piece Installation
Requirements for Round Signposts

Internal diameter size, mm	Anchor piece stub length, mm	Concrete foundation dimensions, mm
50	920	1000 deep × 300 diameter
64	920	1000 deep × 300 diameter
90	1200	1400 deep × 300 diameter
100	1500	1700 deep × 460 diameter

TABLE 7.14 Connection Requirements for Horizontal and Inclined Slip Bases

Support shape and size	Number of bolts	Bolt diameter, mm	Required torque, N · m
Round, by internal diameter, mm			
51	4	13	11 to 16
64	4	13	11 to 16
76	4	13	11 to 16
100	4	13	11 to 16
127	4	16	26 to 39
Shapes, by kg/m			
0–5.4	4	14	10 to 15
5.5–13.4	4	16	23 to 35
13.5–20.2	4	20	40 to 58
720.2	4	27	60 to 75

TABLE 7.15 Connection Requirements for Multidirectional Slip Bases

Round support size, mm	Number of bolts	Bolt diameter, mm	Required torque, N · m
76	3	16	26 to 39
89	3	16	26 to 39
100	3	19	42 to 63
127	3	19	42 to 63

by hand. Also, the proper size flat washers should be used beneath the nut and the head of the bolt. The proper size bolts and recommended torque requirements are provided in Tables 7.14 and 7.15. The tables show requirements for round sign supports by diameter and for other support shapes by kilograms per meter. Remember that an S150 × 18.6 member is a standard shape with a depth of 150 mm and a mass of 18.6 kg/m.

Slip Base Orientation.   The proper operation of slip bases is also dependent upon proper assembly and correct orientation to the expected direction of impact. The parts, and orientation to the primary direction of travel, for the three types of slip bases are presented in Figs. 7.50 and 7.51. The horizontal slip base is generally not used for single sign supports. Where impact from more than one direction is expected, the multidirectional slip base provides better performance because of its design for lift upon impact.

Guidelines for Slip Base Installation.   The following guidelines should be followed for slip base installation:

- Slip base installations require a firm foundation to operate properly upon impact. Slip base installations should always include a concrete base and never be directly buried or drilled and backfilled.

- Use proper size bolts for the slip base fastening. Bolts that are too small may not be able to be sufficiently tightened and may fail under environmental loads. Bolts that

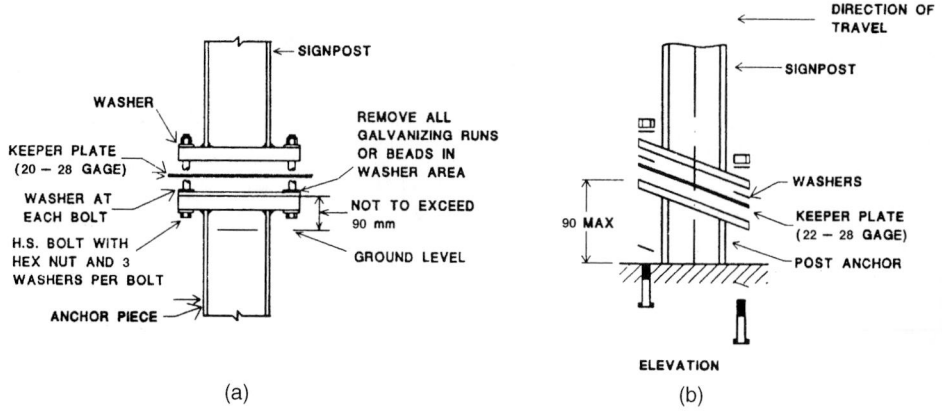

FIGURE 7.50 Components of horizontal and inclined slip bases. (*a*) Horizontal slip base. (*b*) Inclined slip base.

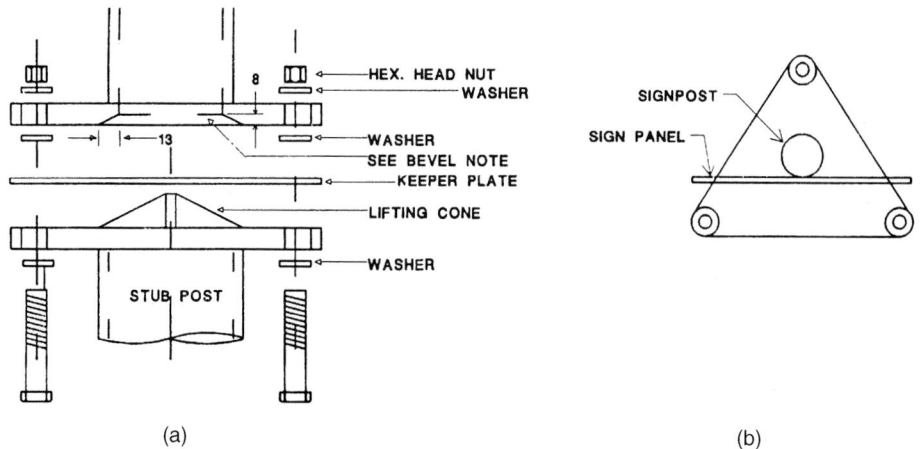

FIGURE 7.51 Components of multidirectional slip base. (*a*) Side view. (*b*) Top view.

are too large may become stuck in the release slots upon impact and prevent proper separation.

- Use proper size and strength washers. The washers beneath the nut and head surfaces should be sufficiently strong to withstand the torque requirements without deforming into the release slots of the base.

- Three washers should be used on each bolt—one each beneath the nut and bolt head, and one between the upper and lower faces of the slip base. The purpose of the washer between the two slip faces is to prevent binding upon impact. All galvanizing runs or beads should be removed from both the upper and lower faces in the washer areas.

- The nut should be run by hand down the bolt to find thread irregularities that will provide inaccurate torque readings.

FIGURE 7. 52   Improper installation of slip base anchor piece.

- Torque each base bolt to the required specifications.
- Remember that the top portion of the slip base must be attached to the anchor piece. Therefore it is recommended that the anchor piece be installed so that its highest portion extends no more than 90 mm above the ground. This will help ensure that the addition of the top plate will not result in a height that can snag the undercarriage of an impacting vehicle. The installation shown in Fig. 7.52 is improper and can snag the undercarriage of an impacting vehicle.
- The bolts must be sufficiently long that they can extend approximately 10 mm beyond the nut after complete assembly.
- Do not install an inclined slip base where impact from more than one direction is expected.

7.7 FASTENING SIGN BLANKS ON SINGLE–SIGN-SUPPORT SYSTEMS

Regardless of the type of support that is being used, there are three general rules that must be followed: (1) the top of the sign should be 2750 mm above ground level to reduce the possibility of intrusion into the passenger compartment upon impact, (2) the retaining bolts must be snug but not so tight as to distort the sign face, and (3) the bolts must be of the proper size and length to prevent the sign blank from separating from the support.

Fastening to a U-Channel.   Signs are normally mounted on the front face of the U-channel. This is the widest face of the U-channel. Signs can also be mounted on the narrow face, as in back-to-back sign installations, for example; but the decreased sur-

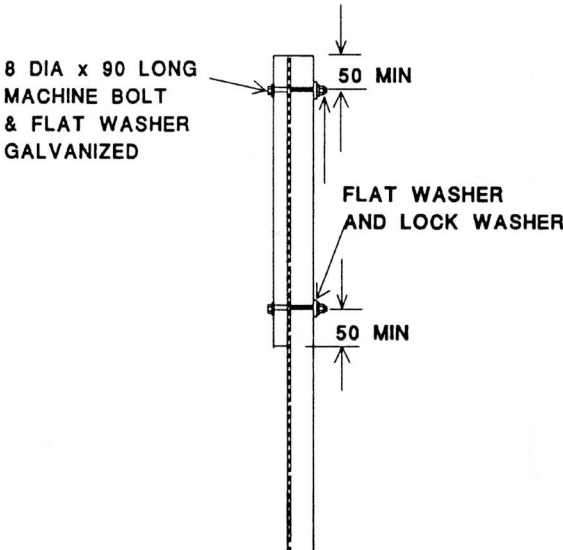

FIGURE 7.53 Typical sign blank fastening to U-channel.

face contact with the sign can result in damage to the sign face due to environmental loads. Side mounting on a U-channel requires channel brackets, and brackets for mounting street name signs to the top of the post are available. Typical sign blank installations for a U-channel are presented in Fig. 7.53.

Fastening to Wood Supports.   Sign installation on wood supports requires a bolt completely through the post and fastened with a nut. Lag bolts are not recommended for fastening sign blanks to the posts. Fastening with lag bolts is unpredictable upon impact because of the size of the bore hole, possible post splitting, the presence of knots, and variable characteristics of the wood.

Fastening to Square Tubes.   Sign installations on square tubes can be accomplished by 8-mm bolts and nuts or by the use of rivets. Figure 7.54 presents typical sign panel fastening details for shaped wooden and square-tube sign supports.

Fastening to Round Steel Shapes.   Signs should not be fastened directly to standard steel pipe or to light standards. The contact area between the back of the sign and the support due to the round support shape is too small to withstand the wind and other environmental loads. A number of different fastening methods can be used, including B-clamps, U-bolts, and stainless steel band clamps. Figure 7.55 presents the configuration of the B and stainless steel band clamps. The dimensions shown for the B-clamp are typical and vary slightly by manufacturer. The B-clamp is also available in other sizes than shown in Fig. 7.55.

U-bolts can also be used to fasten signs to round supports. The U-bolt is attached to the sign by the use of Z-bar aluminum channel or pieces of 0.9-kg U-channel. The U-bolt is purchased with an anchor chair to grip the post. Fastening details for U-bolts with aluminum Z-bar are presented in Fig. 7.56 and with aluminum channel or U-channel in Fig. 7.57.

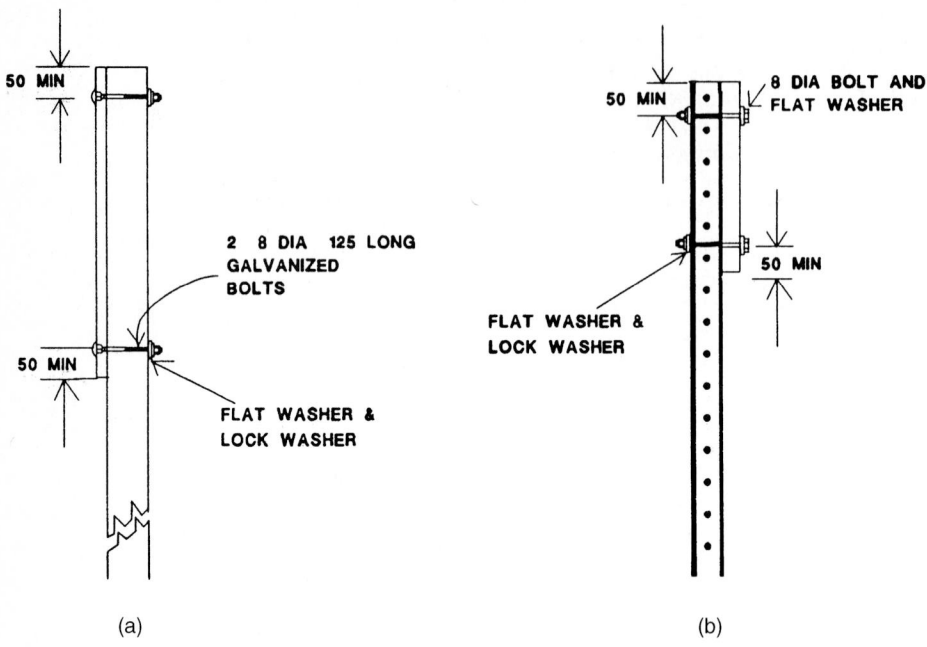

FIGURE 7.54   Typical sign blank fastening to shaped wood and square-tube supports. (*a*) Shaped wooden posts. (*b*) Square tube.

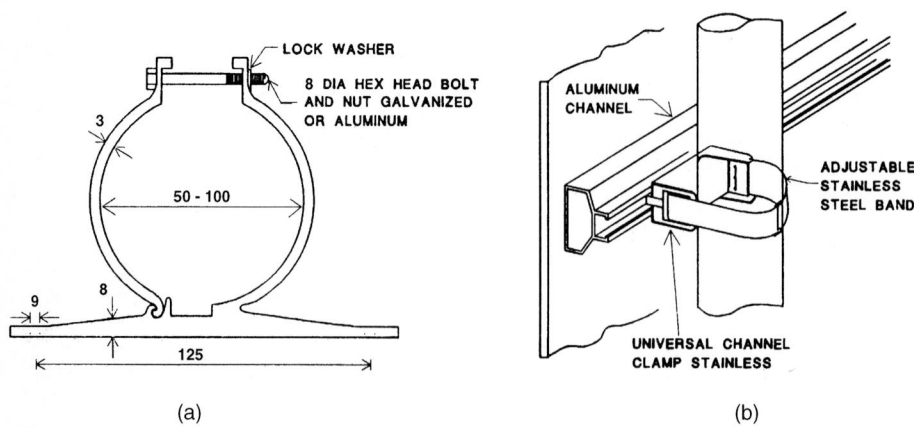

FIGURE 7.55   Typical design of B and stainless band clamps for round sign supports. (*a*) B-type clamp. (*b*) Stainless steel band clamps for round sign supports.

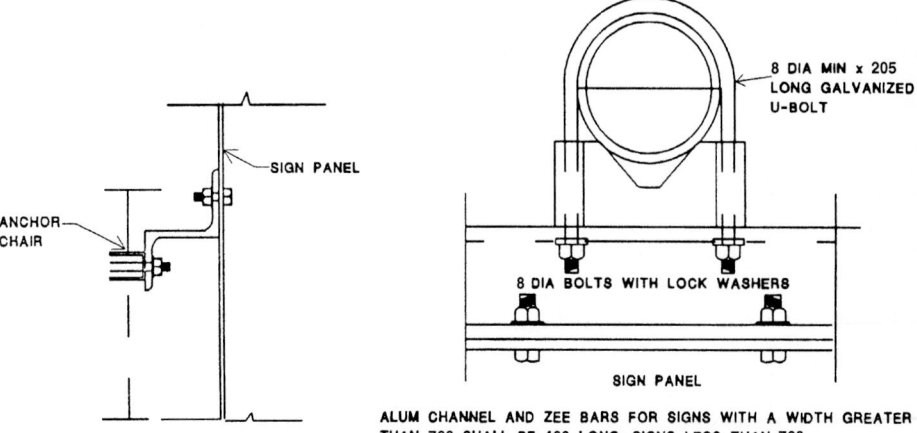

SIGN PANEL

ANCHOR CHAIR

8 DIA MIN x 205 LONG GALVANIZED U-BOLT

8 DIA BOLTS WITH LOCK WASHERS

SIGN PANEL

ALUM CHANNEL AND ZEE BARS FOR SIGNS WITH A WIDTH GREATER THAN 760 SHALL BE 460 LONG; SIGNS LESS THAN 760, ALUMINUM CHANNEL AND ZEE BARS SHALL BE 355 LONG

(a)   (b)

FIGURE 7.56   Fastening details for aluminum Z-bar and channel. (*a*) Z-bar. (*b*) Aluminum channel.

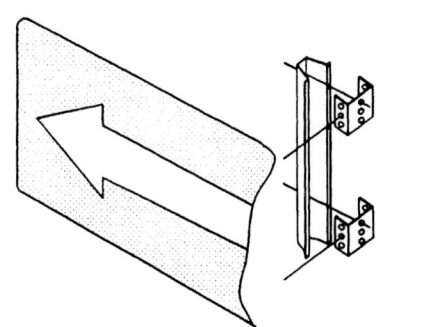

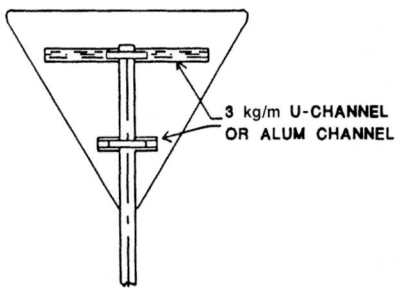

3 kg/m U-CHANNEL OR ALUM CHANNEL

FORMED CHANNEL (11 GAUGE) TO BE USED FOR MOUNTING SIGN PANELS ON MORE THAN 2 SIDES OF U-CHANNEL POST

MOUNTED WITH 2 B-2 CLAMPS, UNIVERSAL CHANNEL CLAMPS, OR 2 U-BOLTS

(a)   (b)

FIGURE 7.57   Typical variations for installing signs on round posts. (*a*) With formed channel. (*b*) With clamps or bolts.

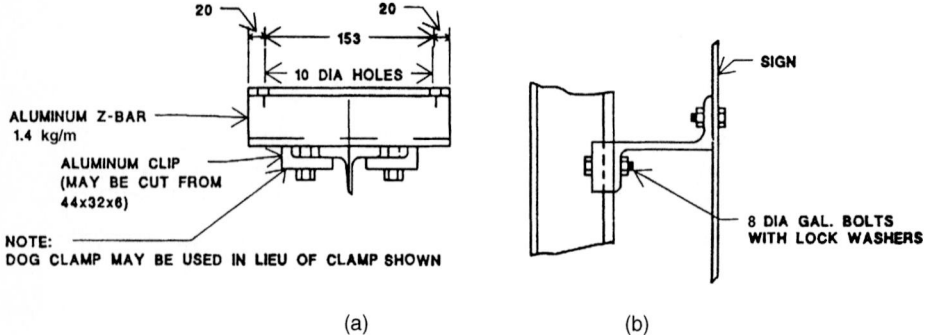

FIGURE 7.58   Aluminum Z-bar for attaching sign panels to beam post. (*a*) Top view. (*b*) Side view.

Fastening to Structural Steel Shapes.   Fastening signs to S- or W-shaped beam posts, often used with slip base designs, should be accomplished by using a stiffener. Clamps are used to fasten the stiffeners to the support, eliminating the need to drill into the support itself. Fastening details using aluminum Z bar as the stiffener are presented in Fig. 7.58. Figures 7.59 and 7.60 present fastening details for commercially available aluminum stiffeners specifically designed for mounting signs to beam supports.

7.7.1   Guidelines for Fastening Sign Panels to Single-Support Systems

The following guidelines should be followed for fastening sign panels to supports:

- Bolts smaller than 8 mm should not be used to fasten sign panels to the support. The bolts must be long enough to provide for bolt extension beyond the fastening nut.
- Carriage bolts, or hex bolts with washers between the hex head and sign face, should be used to reduce the possibility that the sign might separate from the sup-

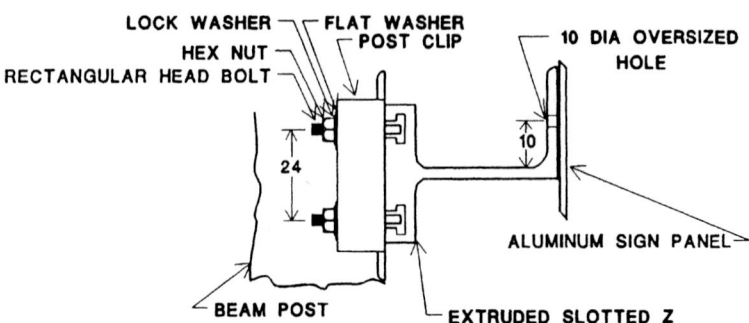

FIGURE 7.59   Aluminum slotted Z-bar for attaching sign panels to beam post.

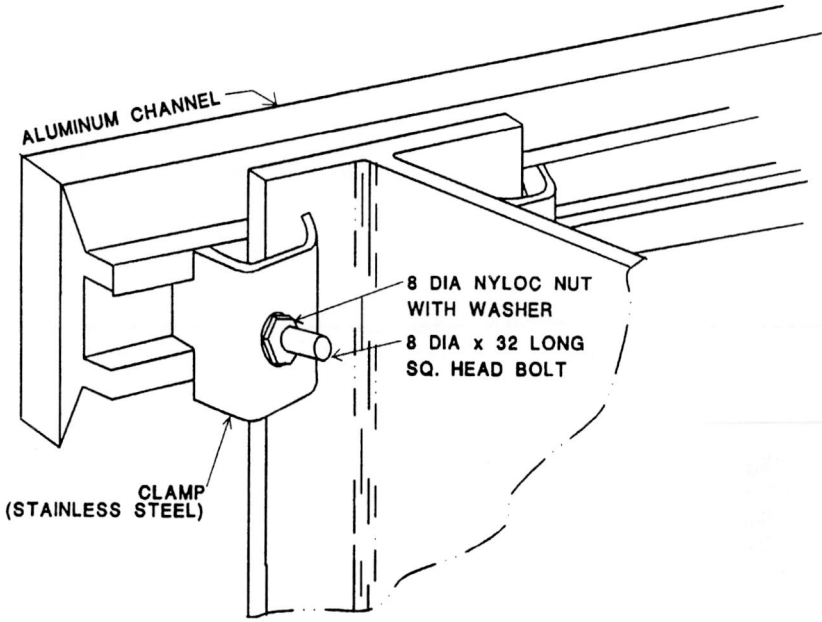

FIGURE 7.60 Extruded aluminum channel for attaching sign panels to beam post.

port upon impact. Flat washers and lock washers should be used at the nut end of the bolt.

- The bolts must be tightened sufficiently to prevent loosening, but not so tight as to distort the sign face.
- Do not allow the bolt to twist when tightening. This will often cause ripping of the reflective sign face material.
- Signs should be mounted on the supports so that the top of the sign is at least 2750 mm above ground level.
- For slip base designs, the sign panel stiffeners, mounting hardware, and the upper sign support itself must not weigh more than 270 kg.
- Follow the manufacturer's instructions when using commercially available fastening hardware.

7.8 MULTIPLE–SIGN-MOUNT INSTALLATION

Multimount sign supports have two or more support posts of breakaway design. The breakaway mechanism is either a fracture or a slip-base type. Fracture mechanisms consist of frangible couplers or frangible one-piece posts. Figures 7.31 and 7.61 show examples of multiple-mount slip base and frangible post designs.

FIGURE 7.61   Multiple-mount frangible wood post design.

7.8.1   One-Piece Multiple-Mount Sign Supports

Direct burial assemblies that are approved for use include dual 4.5 kg/m U-channel and dual 90-mm × 90-mm wooden posts that have been modified with two 38-mm holes placed at 100 mm and 460 mm above the ground line. Other than these exceptions, multiple-mount sign supports require the use of anchor pieces, sleeves, slip bases, or frangible couplings for acceptable impact performance. Only those devices approved for use by the FHWA should be used for multiple-mount supports. Since more than one support can be simultaneously struck by an errant vehicle, it is important that proper installation procedures be followed. Figure 7.62 illustrates the holes drilled in a direct burial wooden post to provide the weakened cross section required for multiple wooden post installation.

7.8.2   Frangible Coupler Designs

Transpo Industries manufactures a series of breakaway systems for ground-mounted sign supports, marketed under the trade name Break-Safe. The Break-Safe system uses frangible couplers and is available for U-channel, both concrete footing and direct burial; 76- to 114-mm round pipe; 76- to 127-mm square tube; and wide-flange and standard beam shapes. Schematics of Transpo Industries concrete base and the direct burial system for back-to-back U-channel are presented in Fig. 7.63. Figure 7.64 shows the Break-Safe system for square-tube and round-pipe supports [36].

The Break-Safe has a number of advantages over slip base designs. One advantage is that the critical torque requirements of the slip base bolts are eliminated by the use of frangible couplings. There are retrofit kits available, for wide-flange and standard

FIGURE 7.62 Weakened wooden post for multiple supports.

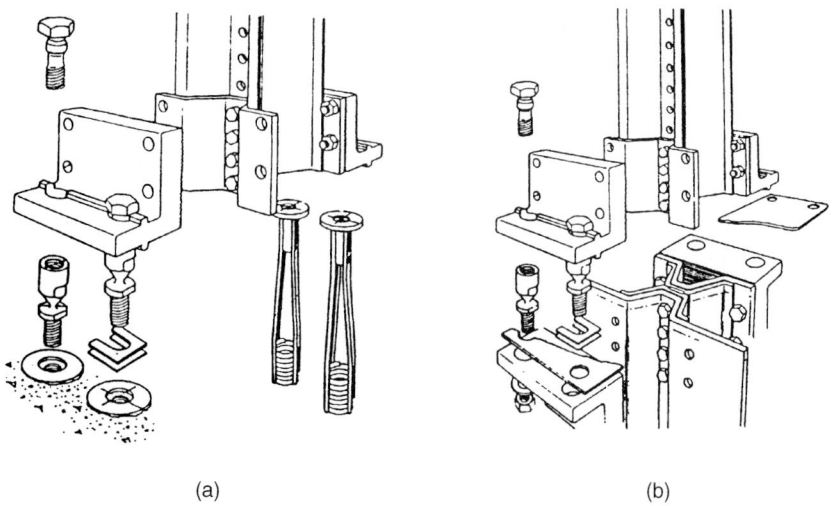

(a)   (b)

FIGURE 7.63 Transpo Break-Safe system for back-to-back U-channel supports. (*a*) Concrete footing. (*b*) Direct burial.

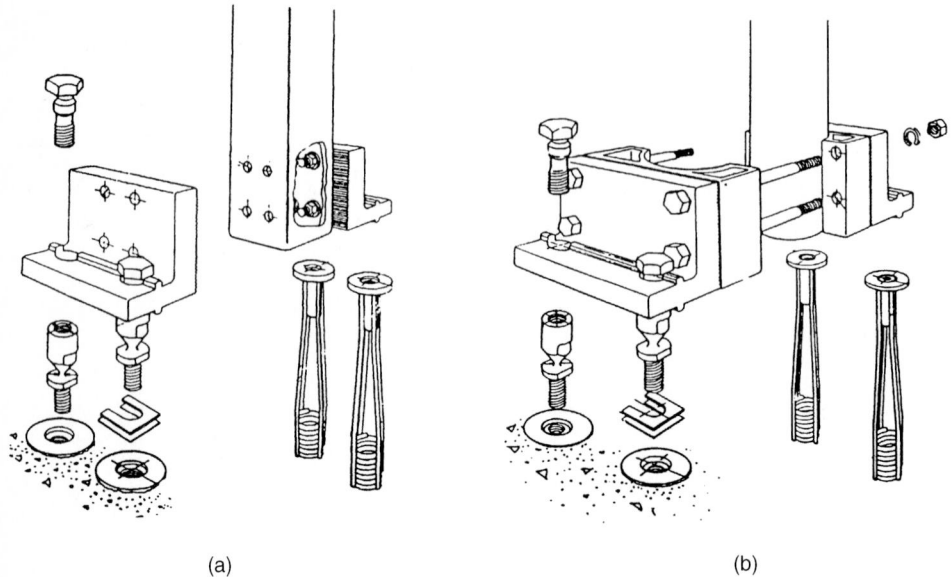

(a)   (b)

FIGURE 7.64 Transpo Break-Safe system for square-tube and round-pipe supports. (*a*) Square-tube support. (*b*) Round-pipe support.

beam supports, that use the existing slip base anchor to convert to a frangible coupler design. Another advantage of the Break-Safe system is that the concrete base installations do not require a protruding stub. This decreases the probability of snagging the undercarriage of an impacting vehicle, and damage to the anchor system itself. The protruding stub is eliminated by bolting the frangible coupling into anchors placed in the concrete footing. Proper assembly requires that the anchors be accurately placed for each type of support system. Accurate anchor placement requires the use of an installation jig similar to Fig. 7.65 [36]. The anchor spacings, dimensions A and B in Fig. 8.65*b*, vary with the type and size of sign support being installed. Frangible and load concentration couplers (Fig. 7.31) usually perform satisfactorily when struck from any direction (they are multidirectional).

7.8.3 Slip Base Designs

Slip base designs for multiple sign supports are usually of horizontal design as shown in Figs. 7.32 and 7.35. Horizontal slip bases, when used in multiple–sign-support systems, operate satisfactorily when impacted from only one direction. Horizontal slip bases should not, therefore, be used for multiple sign supports where there is a high probability of impacts from more than one direction. In Fig. 7.35, the keeper plate prevents the bolts from "walking" out of the assembly as a result of wind vibration (Art. 7.4). The washers should separate the upper and lower slip plates by at least 3 mm, but not more than 7 mm, to prevent mating of the surfaces and possible binding due to friction. Proper size washers must also be used under the nut and bolt head to prevent the washers from deforming into the slots of the slip plates and binding the mechanism.

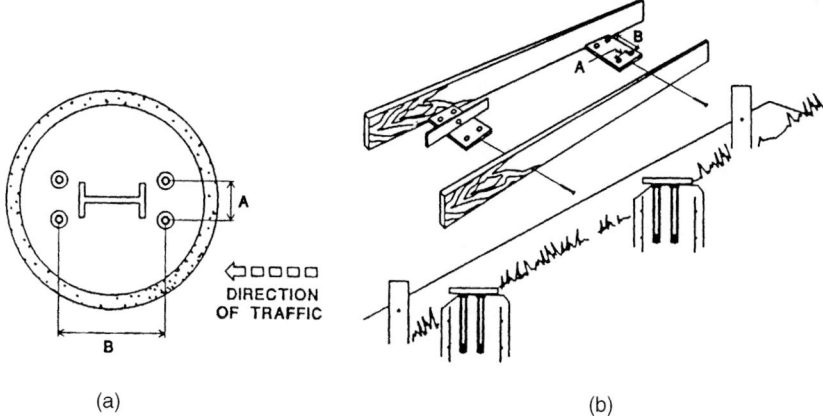

FIGURE 7.65 Concrete footing and installation jig for Break-Safe installations. (*a*) Plan view of footing. (*b*) Installation jig.

A typical concrete foundation detail is shown in Fig. 7.66, and specifications for the anchor piece of slip base designs are presented in Table 7.16. Notice that the foundation design includes eight reinforcing bars spaced around the anchor piece. This is a typical installation which is effective in maintaining the integrity of the foundation under vibrations resulting from environmental loads. State specifications should be consulted to determine if local requirements deviate from details shown in Fig. 7.66.

Proper functioning of the slip base requires correct selection of bolt size and torque. Table 7.17 gives typical design specifications for large roadside sign slip bases and concrete foundations.

7.8.4 Hinge Requirements

Multiple sign supports are designed to operate correctly when either one or all of the supports within a 2100-mm radius are impacted. When only one support is impacted, the remaining signpost should support the sign and prevent it from penetrating the windshield. The desired impact performance of slip base and frangible coupler designs for large sign supports is depicted in Fig. 7.38. The base releases upon impact and the impacted support rotates up, allowing the vehicle to pass underneath the sign. This requires that the post be cut, at least 2100 mm above the ground, to provide a hinge for rotation.

As shown in Fig. 7.39 and discussed in Art. 7.5.3, hinges for large sign supports consist of three basic designs: (1) partially cut post with front friction plate, (2) completely cut post with front friction and rear hinge plate, and (3) completely cut post with weakened front plate and rear hinge plate. Proper performance of the hinge requires the correct selection of plate size, bolt size, and torque. Figure 7.67 and Table 7.18 present the design values for friction plates. Figure 7.68 and Table 7.19 present the design values for hinge plates. The bolt torque values for both friction and hinge plates are the same as presented as Table 7.17 for slip bases. Proper sized flat washers should be used under each nut and the head of each bolt.

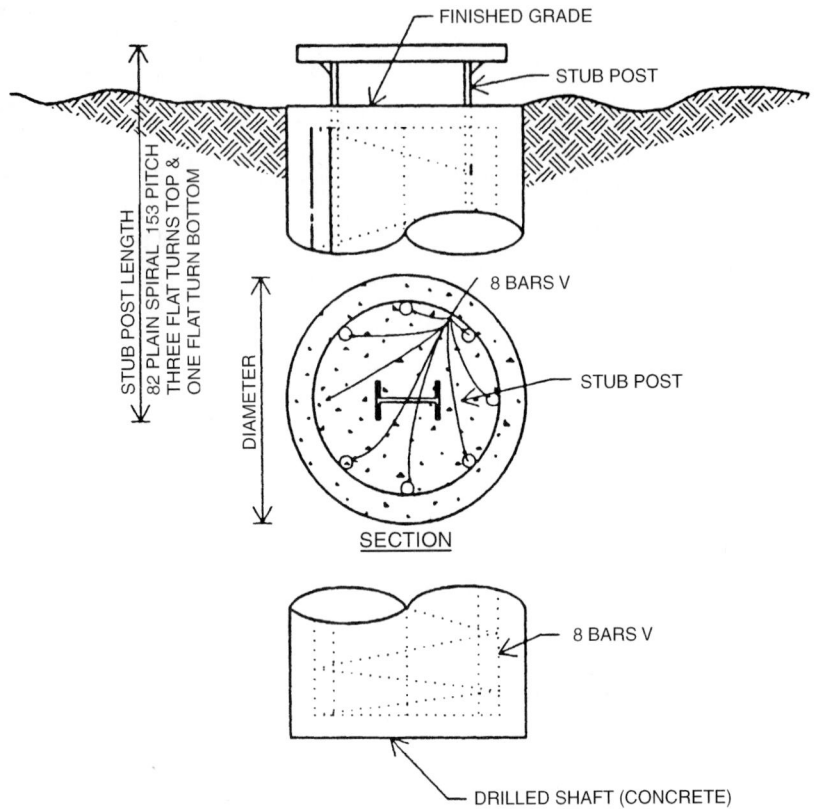

FIGURE 7.66 Horizontal base and concrete foundation detail.

TABLE 7.16 Details of Concrete Foundation Design for Large Slip Bases

Post size, mm × kg/m	Stub length, mm	Stub projection, mm	Drilled shaft diameter, mm	A615M bar size, no.
S76 × 8	460	90	460	15
S100 × 11	460	90	460	15
W150 × 13	600	75	610	15
W150 × 18	600	75	610	15
W150 × 23	760	75	610	20
W200 × 25	760	75	610	25
W200 × 30	910	65	610	25
W254 × 31	910	65	610	30
W254 × 37	910	65	610	35
W305 × 40	910	65	610	35

TABLE 7.17 Bolt-Tightening Specifications for Slip Base Design

Post size, mm × kg/m	Bolt size, mm	Clamping force, N	Torque, N · m
S76 × 8	14	4092–6139	10–15
S100 × 11	14	4092–6139	10–15
W150 × 13	16	7740–11,832	23–25
W150 × 18	16	7740–11,832	23–25
W150 × 23	16	7740–11,832	23–25
W200 × 25	16	7740–11,832	23–25
W200 × 30	16	10,676–16,014	23–25
W254 × 31	20 or 22	10,676–16,014	42–62
W254 × 37	20 or 22	10,676–16,014	42–62
W305 × 40	20 or 22	10,676–16,014	42–62
> 45 kg/m	27	10,676–16,014	60–75

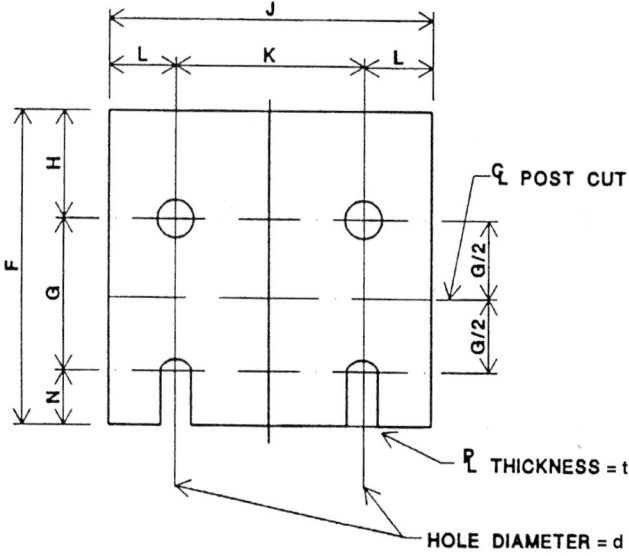

FIGURE 7.67 Design of friction plate. (See Table 7.18 for dimensions.)

TABLE 7.18 Specification of Friction Plate Design

Dimensions in mm; see Fig. 7.67

Post size, mm × kg/m	Bolt diameter	Bolt length	F	G	H	J	K	L	N	d	t
S76 × 8	13	38	80	38	29	67	35	16	13	15	6
S100 × 11	13	38	80	38	29	67	35	16	13	15	6
W150 × 13	13	38	92	51	29	102	57	23	13	15	6
W150 × 18	13	38	92	51	29	120	57	23	13	15	6
W150 × 23	16	38	112	64	32	152	89	32	16	18	10
W200 × 25	16	38	112	64	32	134	70	32	16	18	10
W200 × 30	19	45	121	54	38	134	70	32	19	21	13
W254 × 31	19	45	134	76	38	146	70	38	19	21	13
W254 × 37	19	48	134	76	38	146	70	38	19	21	13
W305 × 40	19	48	134	76	38	165	89	38	19	21	13

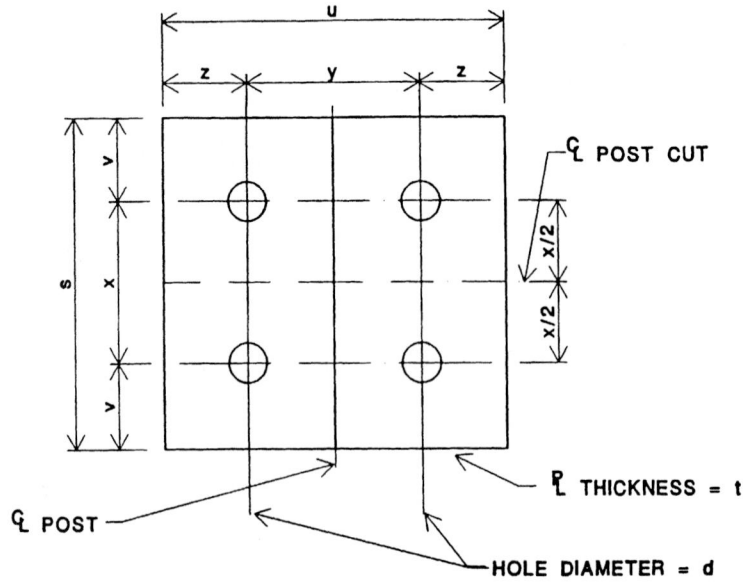

FIGURE 7.68 Design of hinge plate. (See Table 7.19 for dimensions.)

TABLE 7.19 Specification of Hinge Plate Design

Dimensions in mm; see Fig. 7.68

Post size, mm × kg/m	Bolt diameter	Bolt length	S	U	V	X	Y	Z	d	t
S76 × 8	13	38	95	67	29	38	35	16	15	8
S100 × 11	13	38	95	67	29	38	35	16	15	8
W150 × 13	13	38	108	102	29	51	57	23	15	6
W150 × 18	13	38	108	102	29	51	57	23	15	8
W150 × 23	16	38	127	152	32	64	89	32	18	8
W200 × 25	16	38	127	133	32	64	70	32	18	8
W200 × 30	19	45	140	133	38	64	70	32	21	10
W254 × 31	19	45	152	146	38	76	70	38	21	10
W254 × 37	19	48	152	146	38	76	70	38	21	11
W305 × 40	19	48	152	165	38	76	89	38	21	11

The hinge systems shown in Fig. 7.39 are all unidirectional designs and should not be used in areas requiring bidirectional breakaway performance. Only the Transpo hinge system shown in Fig. 7.40 offers bidirectional breakaway capability.

7.9 FASTENING SIGN BLANKS ON MULTIPLE–SIGN-SUPPORT SYSTEMS

The sign blank and its mounting hardware become a structural component of the sign assembly upon impact. Slip base and frangible coupler designs of multiple–sign-support systems require the sign panel hardware, and the upright signposts, to provide the rigidity necessary for proper operation. This includes providing sufficient resistance to activate the hinge and to prevent intrusion of the sign and impacted support into the passenger compartment. Proper hinge activation also requires that no portion of the primary sign, or any supplemental signs, be attached to the support posts below the hinge. In addition, no portion of the sign panel should extend lower than 2100 mm above ground level.

Fastening of sign panels to multisupport sign systems usually requires the use of stiffeners to provide the required rigidity. The exception to this is for relatively small surface area signs, which require multiple supports because of their shape, and for wooden signs. Consult state specifications for installation requirements. Clamps are used to fasten the stiffeners to S- or W-shaped beam posts, eliminating the need to drill into the post itself. Fastening details using aluminum Z-bar as the stiffener were presented in Fig. 7.58, with other common methods presented in Figs. 7.59 and 7.60. U-channel posts can also be used as stiffeners for large signs. When U-channel is used, it should be galvanized and should weigh no more than 3.7 kg/m. U-channel of 3 kg/m is sufficiently strong to withstand wind loads of 130 km/h.

Figure 7.69 presents the configuration of stiffeners for various sign sizes on a dual multiple-support system. Signs with a height of 600 mm or more should be reinforced with two stiffeners placed a distance of one fourth the sign height from the top and bottom of the sign. The stiffeners should not extend closer than 25 mm to the sign edge. Supplementary signs, added to the bottom of the primary sign, should be attached to the sign stiffeners and not to the posts. Signs should never be allowed to

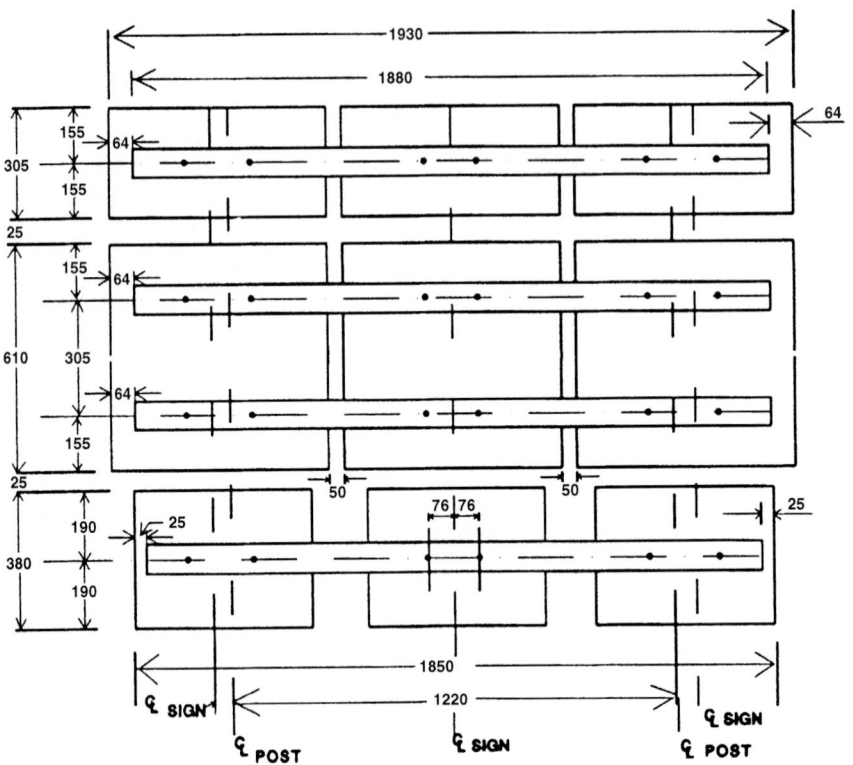

FIGURE 7.69 Example showing placement of sign stiffeners.

extend above and below the hinge at the post. Extending the sign at the hinge can cause the hinge to bind and improper operation upon impact.

7.10 GUIDELINES FOR MULTIPLE–SIGN-SUPPORT CONSTRUCTION

The following guidelines should be followed for multiple–sign-support construction

- Multimount sign supports are designed to function properly when more than one support is struck by an errant vehicle. There will be occasions, however, where only one support will be impacted. When this occurs, it is necessary that the sign panel be properly fastened and have sufficient rigidity that the post(s) that are not impacted will support the sign panel, preventing intrusion into the vehicle.
- The hinge should be located at least 2100 mm above the ground to prevent the upper section of the support from penetrating the windshield.
- No portion of the primary sign, additional signs, or bracing should be attached to the supports below the hinges. Fastening below the hinge will interfere with the breakaway performance of the support post. Signs that are mounted to the primary

sign panel and are less than 2100 mm above the ground can intrude into the passenger compartment even when the hinge operates correctly. Supplemental sign panels should not, therefore, be less than 2100 mm above the ground.

- Two posts within a 2100-mm path should each have a mass that does not exceed 27 kg/m.

- Slip base mechanisms must be constructed with the proper size bolts and washers. Oversized bolts can result in bending between the upper and lower base plates. Washers that are too thin can deform into the slots and bind the plates together.

- The torque specifications must be followed when assembling slip bases and hinges. With insufficient torque, wind and ice loads can cause the bolts to become loose, with subsequent "blowdown" from hinge release or "walking" at the slip base. Applying too much torque can result in binding between the mating surfaces, with subsequent improper operation upon impact.

- Crash tests, performed on level terrain, indicate that breakaway designs perform satisfactorily upon impact. When installed on slopes, however, there is the possibility that they may not function as planned. This is due to the slope's changing the trajectory of the impacting vehicle from the test conditions achieved with level ground. Multiple-mount signs should be installed on level ground when possible and outside the clear zone, in a location where they will be least likely to be hit. Some state agencies routinely require multimount signs to be installed 12 m or more from the edge of the traveled way.

- Follow the installation plans of multiple-mount supports for both construction and maintenance. Do not make temporary maintenance repairs using wrong size bolts or shear plates. Temporary repairs often become permanent, or at any rate can be subjected to an impact prior to correction.

- Do not install any sign supports in a ditch line. The water funneled in the ditch will cause premature corrosion and can freeze, preventing proper operation. The ditch can also channel errant vehicles and guide them into the support.

- Multiple-mount sign support systems are often classified as dual and triple installations. This classification refers to the number of posts permitted within a 2100-mm radius. Approval of support types for dual installation, for example, indicates that no more than two of these supports are permitted within a 2100-mm radius of each other. Acceptable impact performance can be achieved by reducing, but never increasing, the number of supports. A support type approved for dual use can be installed as a single-mount post but not as a triple installation.

- Multimount supports installed with slip base and/or frangible coupler designs must have a maximum height of 100 mm, over a span of 1.5 m, from the ground to the topmost part of the anchor. This is necessary to prevent the anchor piece from snagging the undercarriage of impacting vehicle.

- Each post of a hinge design should be fabricated from a continuous piece of material. The holes for the friction and hinge plates should be drilled and sections match-marked before cutting and weatherproofing. The match marks must be visible after weatherproofing.

- Supports, posts, and anchor pieces should be fabricated and assembled in a shop to ensure proper alignment and match of base plates. Any dismantling that may be required necessitates the placement of match marks to ensure reassembly in the original manner.

- Each post should be installed as a unit to ensure proper alignment of the post and anchor piece assemblies.

- Proper functioning of the slip base feature requires that the interior washers, between the post slip plate and the anchor piece slip plate, transfer the bearing pressures equally. After assembly, the upper and lower slip plates should have a clearance between them of at least 3 mm but not in excess of 7 mm.

- All bolts for attaching the signs to the stiffeners should be 8 mm placed in bolt holes of 10 mm. Flat washers should be used beneath the head of hex head bolts. Fiber washers should be used beneath the head of carriage bolts to prevent possible damage to the reflective sheeting when tightening. All bolts should be sufficiently long to allow the bolt to extend beyond the nut when tightened correctly.

7.11 SIGN VANDALISM PROBLEMS AND COUNTERMEASURES

Sign vandalism costs millions of dollars each year in increased maintenance costs and is a contributing cause to many accidents as well. In addition to the accident itself, vandalized signs can expose the roadway agency and municipality to tort liability cases. Surveys of state and local agencies indicate that an average of 30 percent of all sign replacement and repair is due to vandalism and that an average of 30 percent of the sign maintenance budget is required for vandalized signs. Acts of sign vandalism are categorized as destruction, mutilation, and theft [17].

7.11.1 Destruction

Destruction occurs when the sign support or sign face is physically damaged to the extent that it no longer serves its intended purpose. Destruction vandalism includes damage from:

- Gunshot
- Thrown projectiles such as rocks and bricks
- Sign bending
- Sign or support burning
- Deliberate sign or support knockdown
- Sign cutting with snips or saw
- Support twisting that results in improper orientation
- Support cutting

7.11.2 Mutilation

Sign mutilation occurs when the installation is altered or defaced in such a manner that the sign is illegible or loses its nighttime retroreflectivity characteristics. Examples of sign mutilation include:

- Application of paint by spray or brush
- Application of unauthorized stickers or decals
- Contamination by caustic substances
- Alteration of sign legend by crayon, lipstick, or ink markers

- Reorientation of the sign panel
- Scratching the sign surface
- Peeling or removing reflective sheeting

7.11.3 Theft

Theft is the unauthorized removal of a sign assembly or any of its parts. Some common reasons for theft include:

- Home decoration
- Relationship of the sign legend to an individual's name or interests
- Construction or scrap value of the wood, aluminum, or metal parts
- Firewood
- Uniqueness of the sign legend

7.11.4 Techniques to Reduce Vandalism

Techniques to reduce incidents of sign vandalism include steps that address the reasons for vandalism, enable the prosecution of offenders, ease maintenance, and make it more difficult to perform the vandalism. Consider the following to reduce vandalism:

- The theft and damage to many street name signs is due to the similarity to someone's name. Vandalism to signs can often be reduced by adding St., Ave., or Blvd. to the sign.
- Use only standard signs. Signs that have an unusual message experience a higher vandalism rate.
- Use sign blank materials that are less susceptible to specific types of vandalism. Thicker gage aluminum sign blanks can be used in areas that are subject to damage by bending. Plywood sign blanks are less susceptible to gunshots. Aluminum signs, when struck by gunshot, are indented over a 12.5-mm diameter circle per bullet hole, resulting in severe chipping and loss of reflectivity and legibility. Plywood signs remain legible even with numerous bullet holes. Plywood signs are also a less attractive target than aluminum signs, since they provide less noise and movement when used for target practice.
- Place an agency identification sticker on the back of each sign. This sticker should have a unique number for each sign, the agency name, whom to contact if the sign is found, and a warning about the legal consequences of stealing or damaging the sign. The identification sticker enables law enforcement officials to prosecute individuals stealing or vandalizing the sign. The date of installation can also be placed on the sticker for maintenance information.
- Apply protective coatings to the sign face to ease the removal of foreign substances. Clear coatings, such as product number 711 or 731 from the 3M Company, can be applied by spraying, roll coating, or hand brushing. Transparent overlay films such as Scotchlite brand graphic overlay (GOF™) from the 3M Company are also available. The clear coatings and overlays allow the removal of crayon, paint, lipstick, and other contaminants with the use of strong solvents that would normally harm uncoated sign face material.

- Support twisting or removal can be reduced by installing approved supports of a heavier gauge and using anchor plates. Driven sign supports, as opposed to those installed by drilling and backfilling, are less susceptible to twisting.
- Use commercially available antitheft fasteners that make it difficult for vandals to remove signs. These fasteners include Tufnet, Teenut, aluminum fluted nuts, blind aluminum rivets, and Vandalgard nuts as illustrated in Figs. 7.70 through 7.72.

7.12   MAINTENANCE OF TRAFFIC SIGNS

Continuing maintenance is required to ensure that traffic signs function for their intended purpose. Proper maintenance of all signs is important since the condition of the signs is a visual statement on the competency of the roadway agency. Regulatory and warning signs that are missing or in poor condition pose safety hazards to motorists and can result in tort liability. Regulatory and warning signs must be repaired as soon as a defect is noticed. All of the signs on an agency's roadway system should be inspected periodically to determine that their orientation and retroreflectivity properties are adequate for nighttime visibility.

Damage to traffic signs can occur as the result of environmental and wind load, accidents, improper installation, end of effective service life, and vandalism. Repairs can be required for the sign panel, the sign support, or both.

7.12.1   Repair and Replacement of Sign Panels

The decision on the appropriate action for damaged sign panels is a field judgment. Minor bending of a sign will prevent headlights from illuminating the sign at night. Signs with minor bends can be repaired by removing the sign from the post and straightening the sign. Signs that are badly bent cannot be properly repaired in the field. Attempts to straighten badly bent signs result in cracking and peeling of the sign face material. Many agencies consider it more economical to replace rather than repair signs that maintenance workers judge to be badly worn or damaged [49].

There are field repair kits available with pressure-sensitive reflective background sheeting and die-cut pressure-sensitive prespaced letters, borders, and symbols. It is often difficult, however, to properly apply these materials under field conditions. In addition to the difficulty of field repairs, a regulatory or warning sign should be placed on the post while repairs are being made. This sign may as well be a replacement sign and a more economical and durable repair made in the controlled environment of a shop operation. Do not take down a sign without immediately positioning a replacement. Extra signs should be placed in the service truck prior to leaving the garage. If field replacement of reflective sheeting is performed, however, the proper procedure provided by the manufacturer must be followed.

7.12.2   Sign Cleaning

Sign legibility can be restored or improved by general cleaning and removal of foreign substances from the sign face. Sign cleaning products are commercially available for use in removing common soil and severe contaminants such as paint and adhesives. In

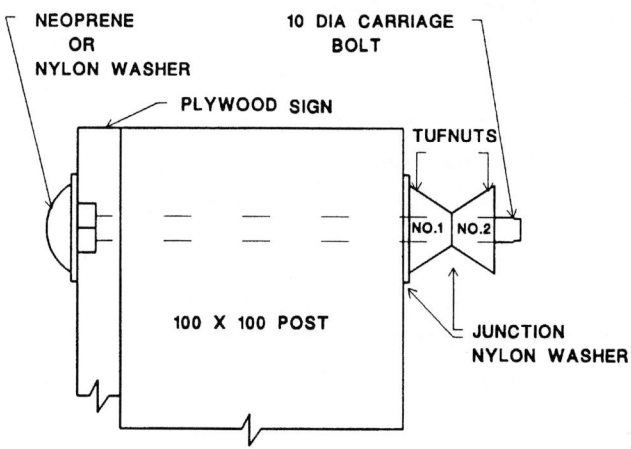

TYPICAL INSTALLATION PROCEDURE

STEP 1: INSTALL FIRST TUFNUT (NO. 1) FINGER TIGHT AS SHOWN.
STEP 2: INSTALL SECOND TUFNUT (NO. 2) FINGER TIGHT AS SHOWN.
STEP 3: INSTALL WRENCH AT JUNCTION TO TIGHTEN (OR LOOSEN) AS NECESSARY.
STEP 4: REMOVE TUFNUT NO. 2; THEN INSTALLATION IS COMPLETE.

SINGLE TUFNUT IS DIFFICULT TO REMOVE BECAUSE OF ITS SHAPE.
ALWAYS USE FOUR TUFNUTS FOR EACH SIGN INSTALLATION.

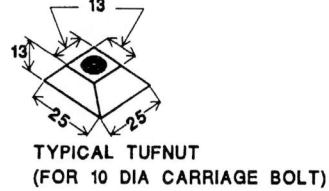

TYPICAL TUFNUT
(FOR 10 DIA CARRIAGE BOLT)

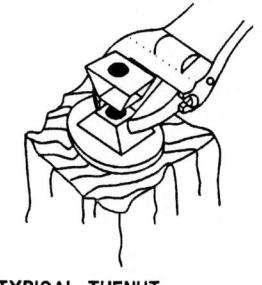

TYPICAL TUFNUT
(FOR 10 DIA CARRIAGE BOLT)

FIGURE 7.70 Tufnut sign fasteners.

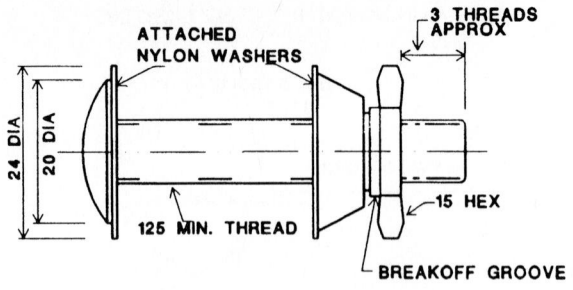

ATTACHED NYLON WASHERS

3 THREADS APPROX

24 DIA
20 DIA

125 MIN. THREAD

15 HEX

BREAKOFF GROOVE

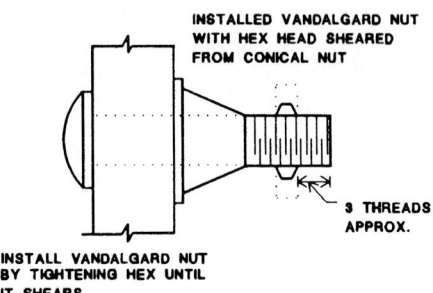

INSTALLED VANDALGARD NUT WITH HEX HEAD SHEARED FROM CONICAL NUT

3 THREADS APPROX.

INSTALL VANDALGARD NUT BY TIGHTENING HEX UNTIL IT SHEARS.

INSTALLATION

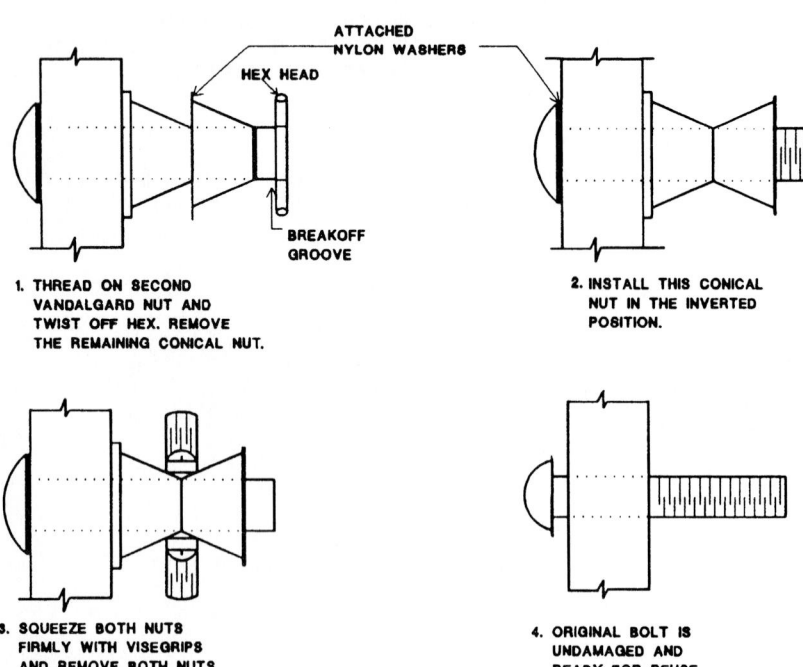

ATTACHED NYLON WASHERS

HEX HEAD

BREAKOFF GROOVE

1. THREAD ON SECOND VANDALGARD NUT AND TWIST OFF HEX. REMOVE THE REMAINING CONICAL NUT.

2. INSTALL THIS CONICAL NUT IN THE INVERTED POSITION.

3. SQUEEZE BOTH NUTS FIRMLY WITH VISEGRIPS AND REMOVE BOTH NUTS TOGETHER.

4. ORIGINAL BOLT IS UNDAMAGED AND READY FOR REUSE.

REMOVAL

FIGURE 7.71   Vandalgard sign fastener.

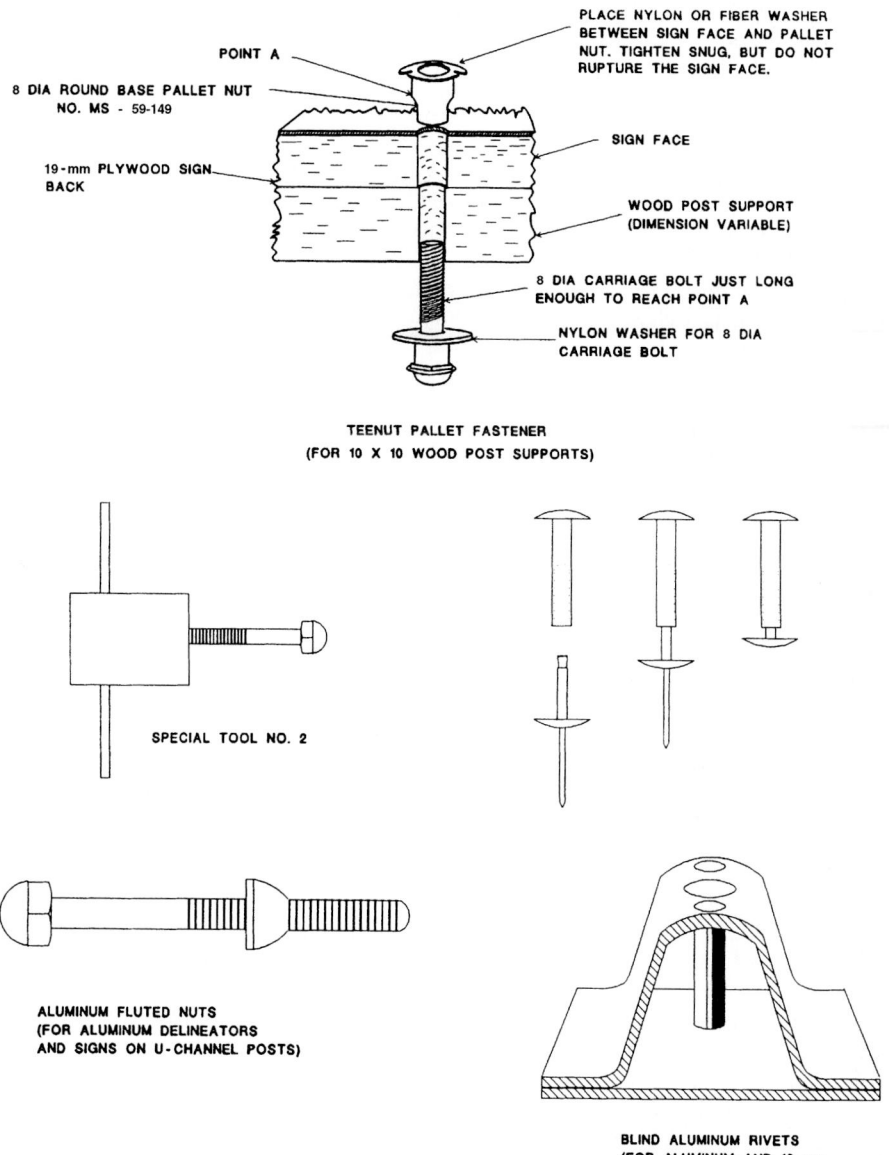

PLACE NYLON OR FIBER WASHER BETWEEN SIGN FACE AND PALLET NUT. TIGHTEN SNUG, BUT DO NOT RUPTURE THE SIGN FACE.

POINT A

8 DIA ROUND BASE PALLET NUT NO. MS - 59-149

19-mm PLYWOOD SIGN BACK

SIGN FACE

WOOD POST SUPPORT (DIMENSION VARIABLE)

8 DIA CARRIAGE BOLT JUST LONG ENOUGH TO REACH POINT A

NYLON WASHER FOR 8 DIA CARRIAGE BOLT

TEENUT PALLET FASTENER
(FOR 10 X 10 WOOD POST SUPPORTS)

SPECIAL TOOL NO. 2

ALUMINUM FLUTED NUTS
(FOR ALUMINUM DELINEATORS
AND SIGNS ON U-CHANNEL POSTS)

BLIND ALUMINUM RIVETS
(FOR ALUMINUM AND 13-mm
PLYWOOD SIGNS ON
U-CHANNEL POSTS)

FIGURE 7.72 Teenut, aluminum fluted nut, and blind aluminum rivet sign fasteners.

the majority of cases the required cleaning products can be obtained from local hardware stores. Strong solvents and incomplete removal of the contaminant can damage the sign reflectivity. Strong solvents should be trial-tested prior to application on the sign. The following steps can be used to clean signs of common soil and some contaminants [50]:

- General sign cleaning: Mild, nonabrasive cleaners and detergents suitable for painted or enameled surfaces are recommended for removal of common soil. Cleaners should be free of strong aromatic solvents or alcohols and be chemically neutral (pH of 6 to 8 is recommended).
- Pollen and fungus can be removed by washing the surface with 3 to 5 percent solutions of laundry bleach (sodium hypochlorite). This should be followed with detergent wash and a clear water rinse.
- Lipstick, crayon, tar, oil, bituminous materials, and some oil-based paints can often be removed with mild solvents such as mineral spirits (toluene), kerosene, heptane, or naphtha. Wipe the contaminated area lightly with a soft cloth saturated with the solvent. Continue wiping lightly until the contaminant is removed. If this does not work, then try the next step.
- Wipe the contaminated area with a soft cloth moistened with lacquer thinner. Continue wiping lightly until the contaminant is removed.
- A nighttime reflectivity check or a nighttime visual inspection of all signs from which contaminants have been removed should be conducted.

7.12.3   Patching Holes and Punctures

It is not necessary to repair each hole in a sign. When a hole does not damage the message or symbol and does not create the impression of a sloppy sign, then repair may not be needed. The following procedures can be used to make field repairs on signs.

Retroreflective Aluminum Sign Panels

- Remove all damaged background sheeting and legend. Usually this means about one inch from the edge of the hole. A retractable-blade knife is a useful tool for this.
- Straighten the sign (flatten out the hole puncture nipple area) using a ball peen hammer and a flat surface (truck bed, trailer bed, or a fender dolly).
- Remove any additional sheeting damaged during straightening.
- Clean the entire area with xylol; then apply varnish maker's and painter's (VM&P) naphtha.
- Patch the hole or puncture on both sides of the sign backing material using 3M Company No. 425 UAL aluminum foil tape or equal. Use a squeegee to apply firm pressure on both sides of the sign. On large holes, start placing the foil at the bottom of the hole, overlapping each strip about ¼ inch in shingle fashion as you move up, and cover the hole area.
- Apply retroreflective background sheeting, extending it at least 13 mm beyond the foil tape strips.
- Replace damaged legend with die-cut, pressure-sensitive, prespaced letters, bor-

ders, or symbols and firmly squeegee them into place.

- Seal edge of new background sheeting and legend with 3M Company No. 700 edge sealer or equal. If the sign is subject to snow burial and replacement sheeting extends to the edge of sign, place 3M Company transparent film (No. 639 or equal) along that top edge.

Instead of making small patches to signs with holes, a portable double-roller unit for applying a full-sized sign face to a sign blank in the field can be used. After patching the holes, remove the paper material protecting the adhesive backing. Carefully align the new sign face sheet with one edge of the sign blank and spread the new sign face over the sign blank as smoothly as possible by hand. Then crank the sign blank with new sign face through the portable roller unit to properly pressure-seat the new sign face. Seal the edge of the new sheeting if necessary.

Retroreflective Plywood Panel Signs

- Remove all loose wood on both sides of the sign and all damaged sheeting.
- Fill holes with wood filler, let the surface set, and sand smooth if you think the holes need to be filled for a field repair. Allow filler to harden. Small holes can be covered by foil tape without filling.
- Wipe areas with clean cloth.
- Cover holes on both sides of the plywood sign blank with 3M Company No. 425 UAL aluminum foil tape or equal. Apply firm pressure to the tape on both sides of the plywood sign back using a squeegee. On large holes, start placing the foil at the bottom of the hole, overlapping each strip about ¼ inch in shingle fashion as you move up and cover the hole area.
- Apply retroreflective background sheeting, extending it at least 13 mm beyond the foil tape strips on the face of the sign.
- In the area covered by the patching, replace any damaged legend with die-cut, pressure-sensitive, prespaced letters, borders, or symbols and firmly squeegee them in place.
- Seal edge of new background sheeting and legend with 3M Company No. 700 edge sealer or equal. If the sign is subject to snow burial and replacement sheeting extends to the top edge of the sign, place 3M Company transparent film (No. 639 or equal) along the top edge.
- Lightly spray a sealing film of flat black enamel paint (use an aerosol can) over the aluminum foil tape covering the holes on the back of the sign panel. Be careful to keep paint off the front sign face, because paint will destroy the night retroreflection. If your agency paints plywood sign backs some color other than black, use an appropriate color if possible.

7.12.4 Sign-Support Straightening

A tool such as that shown in Fig. 7.73 can be constructed out of pipe to straighten twisted U-channel posts [51]. Similar devices with a metal U-shape at the end of a pipe handle can be constructed to realign shaped wood posts and square tubing. A large pipe wrench can also be used to realign U-channel and square-tube supports. Small signs should be mounted at 90° to the road.

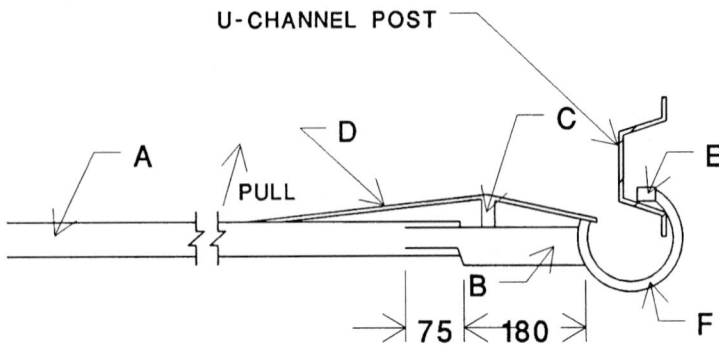

A. 1 Pc - 38 I.D. BLACK PIPE - 1270 LONG

B. 1 Pc - 16 x 75 x 255 LONG

C. 1 Pc - 20 x 20 x 15 LONG

D. 1 Pc - 6 x 20 x 510 LONG

E. 1 Pc - 10 x 10 x 75 LONG

F. 1 - CLEVIS SLIP HOOK (REMOVE EYE)

FIGURE 7.73   Shop-fabricated tool to straighten twisted U-channel.

7.13   REFERENCES ON SIGNING

1. American Association of State Highway Officials and National Conference on Street and Highway Safety, *Manual on Uniform Traffic Control Devices,* Washington, D.C., 1935.

2. *Manual on Uniform Traffic Control Devices,* Federal Highway Administration, U.S. Department of Transportation, Washington, D.C., 1988.

3. *Traffic Control Devices Handbook,* Federal Highway Administration, U.S. Department of Transportation, Washington, D.C., 1983.

4. Bowman, Brian L., *NCHRP Synthesis of Highway Practice 186: Supplemental Advance Warning Devices,* Transportation Research Board, National Research Council, Washington, D.C., 1993.

5. Harwood, Douglas W., *NCHRP Synthesis of Highway Practice 191: Use of Rumble Strips to Enhance Safety,* Transportation Research Board, National Research Council, Washington, D.C., 1993.

6. Cunard, Richard A., *NCHRP Synthesis of Highway Practice 157: Maintenance Management of Street and Highway Signs,* Transportation Research Board, National Research Council, Washington, D.C., 1990.

7. "1988 Annual Report on Highway Safety Improvement Programs," Report of the Secretary of Transportation to the United States Congress, Report No. FHWA-SA-88-0003, Federal Highway Administration, U.S. Department of Transportation, Washington, D.C., 1988.

8. Lewis, Russel M., *NCHRP Synthesis of Highway Practice 106: Practical Guidelines for Minimizing Tort Liability*, Transportation Research Board, National Research Council, Washington, D.C., 1983.

9. *Standard Alphabets for Highway Signs and Pavement Markings*, Federal Highway Administration, U.S. Department of Transportation, Washington, D.C., 1977.

10. *Standard Highway Signs*, Federal Highway Administration, U.S. Department of Transportation, Washington, D.C., 1979.

11. *Roadside Design Guide*, American Association of State Highway and Transportation Officials, Washington, D.C., 1996.

12. *Delphi V—Forecast and Analysis of the U.S. Automotive Industry through the Year 2000*, University of Michigan Transportation Research Institute, Ann Arbor, July 1989.

13. *Standard Specifications for Structural Supports for Highway Signs, Luminaires and Traffic Signals*, American Association of State Highway and Transportation Officials, Washington, D.C., 1985.

14. Mickie, J. D., "Recommended Procedures for the Safety Performance Evaluation of Highway Appurtenances," NCHRP Report 230, Transportation Research Board, Washington, D.C., March 1987.

15. Ross, Hayes, Jr., Dean L. Sicking, Richard A. Zimmer, and Jarvis D. Mickie, "Recommended Procedures for the Safety Performance Evaluation of Highway Features," NCHRP Report 350, Transportation Research Board, National Research Council, Washington, D.C., 1993.

16. Cunard, Richard A., *NCHRP Synthesis of Highway Practice 157;* Maintenance Management of Street and Highway Signs, Transportation Research Board, National Research Council, Washington, D.C., 1990.

17. Perkins, David, D., "Manual on Countermeasures for Sign Vandalism," Report No. FHWA-IP-86-7, Federal Highway Administration, Washington, D.C., September 1986.

18. Texas Transportation Institute, "State of the Practice in Supports for Small Highway Signs," Technology Sharing Report 80-222, Federal Highway Administration, Washington, D.C., 1980.

19. Heinz, Ronald E., "Request for Technical Assistance: Sign Support Design," Memorandum, Federal Highway Administration, Highway Design Division, Washington, D.C., July 1985.

20. Hayes, E. Ross, Jr., Jesse L. Buffington, et al., "State of the Practice in Supports for Small Highway Signs," Federal Highway Administration, Washington, D.C., 1980.

21. Composite Technologies Company, *Signs Manufactured from 100% Landfill-Destined Plastic*, undated brochure, Composite Technologies Company, Dayton, Ohio.

22. *Standard Specifications for Highway Signs, Luminaires and Traffic Signals*, American Association of State Highway and Transportation Officials, Technical Committee, Washington, D.C., 1985.

23. Phillips, David K., "Steel Flanged Channel Posts for Small Highway Sign Supports," technical advisory, Federal Highway Administration, Office of Engineering, Washington, D.C., September 27, 1983.

24. Staron, L. A., "Minute Man Breakaway Device," letter, Federal Highway Administration, Federal Aid and Design Division, Washington, D.C., January 1987.

25. Heinz, Ronald E., "Splicing of U-Channel Steel Posts," memorandum, Federal Highway Administration, Highway Design Division, Washington, D.C., October 1984.

26. Noel, Leon M., "Timber Sign Supports," letter, Federal Highway Administration, Highway Design Division, Washington, D.C., August 1982.

27. Hove, R.W., "Ground Mounted Signs: Timber Sign Supports," memorandum, Federal Highway Administration, U.S. Department of Transportation, Washington, D.C., May 1984.

28. Allied Tube and Conduit, *Qwik-Punch and Qwik-Coat Systems,* Harvey, Ill., January 1991.

29. Noble, Glen, "Support System Acceptance," memorandum, Unistrut Corporation, Wayne, Mich., November 1992.

30. Staron, L. A., "Quick-Punch Sign Supports," letter, Federal Highway Administration, Federal Aid and Design Division, Washington, D.C., October 3, 1986.

31. Xcessories Squared, *Soil Stabilizer* (brochure), Auburn, Ill., February 23, 1995.

32. Staron, Lawrence A., Geometric and Roadway Acceptance Letter No. SS-25, Federal Highway Administration, Federal Aid and Design Division, Washington, D. C., June 4, 1991.

33. Staron, Lawrence A., "Fiberglass Sign Post System," letter, Federal Highway Administration, Federal Aid and Design Division, Washington, D.C., May 11, 1989.

34. Hwycom, *Hwycom's New FRP Signpost System* (brochure), Hwycom Inc., Big Springs, Tex., September 8, 1993.

35. Staron, L. A., Geometric and Roadside Design Letter No. SS-38, Federal Highway Administration, Federal Aid and Design Division, Washington, D.C., November 1993.

36. *Break-Safe™ Breakaway System for Sign Supports,* Transpo Industries, New Rochelle, N.Y., August 1993.

37. Staron, L. A., ACTION: Breakaway Sign Supports, memorandum, Federal Highway Administration, Federal Aid and Design Division, Washington, D.C., September 1993.

38. *Franklin Steel Eze-Erect™ Sign Posts,* Franklin Steel, Franklin, Pa., May 1989.

39. *The Minute Man™ U-Channel Breakaway Signpost System,* Marion Steel Co., Marion, Ohio, 1992.

40. Staron, L. A., "Splicing of Steel U-Channel Posts on Small Sign Supports," memorandum, Federal Highway Administration, Federal Aid and Design Division, Washington, D.C., September 1991.

41. *Unistrut-Telespar™ Sign Support System,* Unistrut Corporation, Wayne, Mich., 1986.

42. *Allied Square Tube Signposts,* Allied Tube and Conduit, Harvey, Ill., February 1992.

43. *14 Ga. Qwik-Punch System,* Allied Tube and Conduit, Harvey, Ill., September 1991.

44. *Roadside Improvements for Local Roads and Streets,* Office of Highway Safety, Federal Highway Administration, Washington, D.C., October 1986.

45. Staron, L. A., Geometric and Roadside Acceptance Letter No. SS-27, Federal Highway Administration, Federal Aid and Design Division, Washington, D.C., May 1992.

46. Staron, L. A. Geometric and Roadside Acceptance Letter No. LS-23, Federal Highway Administration, Federal Aid and Design Division, Washington, D.C., January 1991.

47. Van Ness, Norman, Acceptance Letter to Southwest Pipe Inc., Highway Design Division, Federal Highway Administration, Washington, D.C., July 1986.

48. Hanna, Howard, "Evaluation of the Upper Hinge Mechanism of Multiple Leg Breakaway Sign Supports," memorandum, Federal Highway Administration, Program Development Division, Washington, D.C., October 1987.

49. "Maintenance of Small Traffic Signs—A Guide for Street and Highway Maintenance Personnel," Federal Highway Administration, Report No. FHWA-RT-90-002, Washington, D.C., 1991.

50. Nettleson, Tom, *Signs Maintenance Guide,* Forest Service, U.S. Department of Agriculture, October 1979.

51. McGee, H. W., et al., "Sign Fabrication, Installation and Maintenance," Federal Highway Administration, Report No. FHWA-SA-91-033, Washington, D.C., May 1992.

CHAPTER 7
SIGNING AND ROADWAY LIGHTING

PART 2

ROADWAY LIGHTING

C. Paul Watson, P.E.

State Electrical Engineer
Alabama Department of Transportation
Montgomery, Alabama

Brian L. Bowman, P.E.

Associate Professor
Auburn University
Auburn, Alabama

Part 2 of this chapter presents considerations in the selection of lighting for freeways and other types of roadways. Both standard and high mast lighting are addressed. Roadside safety and the application of various types of bases are discussed and illustrated. Information on construction, acceptance testing, and maintenance is presented. An extensive list of references, which are noted in the text, concludes the section. Portions of this material were derived from studies made under a Federal Highway Administration Project, "Design, Construction and Maintenance of Highway Safety Features and Appurtenances."

7.14 BENEFITS AND FUNDAMENTALS OF LIGHTING

Properly designed and installed roadway lighting can result in significant reductions in nighttime traffic accidents, act as a deterrent to crime, increase commercial activity, and improve aesthetic value. Roadway lighting increases traffic safety by enhancing the visibility of potential roadway hazards, other vehicles, pedestrians, and roadway geometrics. Pedestrians are among the largest beneficiaries of lighting installed on urban streets. Studies indicate reductions of up to 80 percent in pedestrian accidents and reductions ranging from 20 to 40 percent for all types of night accidents [1]. Another study identified a 40 percent reduction in the ratio of night accidents to day accidents resulting from the installation of roadway lighting on freeways [2]. While these figures are significant, it is anticipated that the safety benefits derived from the installation of roadway lighting will become even more pronounced in the future. This is due to the increasing age of the driving population and the significantly reduced visual abilities of persons over 65 years of age. The savings realized by accident prevention alone can often justify the costs of a modern lighting system [3].

Although much progress has been made in improving lighting system efficiency and effectiveness, there are still many streets, particularly in small communities, that are not lighted in accordance with present guidelines. This is primarily due to the scarcity of local funds, which can be mitigated by the use of federal funds on qualifying projects. Roadway lighting has been recognized as a viable countermeasure for increasing traffic safety since 1966, when federal legislation enabled federal aid expenditures for construction and maintenance of roadway lighting [4].

The benefits of providing roadway lighting include enhancing traffic safety, improving pedestrian visibility, deterring crime, improving commercial interests, and promoting community pride. The actual benefits obtained are dependent upon the type of facility and area in which the lighting will be installed. Only the traffic operational and safety benefits obtained from the proper design and installation of roadway lighting are discussed in this chapter. It should be noted that properly designed and installed roadway lighting can result in roadway facilities operating almost as efficiently and safely at night as during the daytime. Lighting cannot, however, be expected to achieve the same safety levels as daytime operation, because of the influence of other factors, such as fatigue, higher speeds, and intoxication, which make a greater contribution to nighttime accident frequency.

7.14.1 Visibility: Luminance versus Illuminance

The requirement of adequate visibility is essential for safe traffic operations during both day and night operation. Visibility can be separated into at least three classifications when applied to highway driving: perception, recognition, and decision making [5]. Perception involves the condition of our eyes, the quantity and the direction of the available light, size of the object being viewed, contrast of the object against its background, and the time available for viewing the object. Effective roadway lighting can aid in these tasks by providing the quality of light required by the human eye to increase its visual acuity.

Illuminance and luminance are two methods used to quantify, or measure, the quality of lighting on a roadway. Prior to 1983, horizontal illuminance was the undisputed method for evaluating roadway lighting in North America [6]. Illuminance is generally described as the light falling on a surface and is expressed in footcandles (fc) or in lux (lx) in the International System. While illuminance is the most common measure, it fails to adequately describe what is actually perceived by the human eye. It is still frequently used as a design tool, since it is easy to measure by the use of light meters to determine if the installation meets design criteria. Another reason for its common use is that the required calculations are easy to perform without the use of a computer. Pavement luminance, and veiling luminance (glare) criteria, however, provide a more realistic method for calculating and predicting effective roadway lighting.

Luminance measures surface reflected light and is expressed in candelas per square meter (cd/m^2). Although the 1983 ANSI/IES American National Standard Practice for Roadway Lighting recognizes both illuminance and luminance, the currently preferred method is luminance, since it more accurately describes what is perceived by the human eye. The introduction of computerized calculation programs should increase the use of the more accurate luminance calculations. In fact, it has been shown that a lighting project can be designed that meets both the illuminance and the luminance criteria by being conscious of the location of the light source and adjusting the mounting height and/or the setback from the roadway.

7.14.2 Warranting Conditions

The potential traffic safety benefits of lighting are due to an increase in driver comfort and confidence resulting from enhanced vision. This reduces driving stress and tension, increases roadway capacity, and reduces the potential for traffic accidents. The economic return of roadway lighting is greatest in urban and suburban areas with high traffic volumes. Rural locations can also benefit from full, or partial, lighting of decision points such as at isolated intersections, on- or off-ramps, and ramp terminals [3]. For some lighting applications there are warranting criteria that can be used to help determine when lighting should be installed.

Warranting conditions are based on minimum conditions which signify that providing lighting would be beneficial. Satisfying the warrants does not obligate an agency to provide lighting, since warrants are not the only criteria that should be considered. Local conditions such as frequent fog, ice, snow, roadway geometry, ambient lighting, sight distance, and signing could justify modifying the warrants either positively or negatively [3]. Judgments on lighting need should include an assessment of the anticipated benefits, traffic volume, speed, road use during the night, night accident rate, road geometrics, and general night visibility. Some agencies justify lighting based on an economic analysis. This requires placing monetary value on the expected reductions in personal injuries, fatalities, and property damage accidents, in addition to other societal benefits estimated to be realized from illumination.

Warranting conditions have been established for freeways. However, due to the wide diversity of conditions that can exist, there are no established warrants to assist the designer in determining when lighting should be provided for urban streets, highways, walkways, and bikeways. The justification for urban lighting is left to engineering judgment, coupled with perceived user needs and user benefits.

7.15 FACILITY AND AREA CLASSIFICATIONS

The following descriptions of facility types and area classifications are used to describe the warranting conditions and design needs of roadway lighting.

7.15.1 Roadway and Walkway Classifications [3]

- *Freeway.* A divided major highway with full control of access and no crossings at grade.
- *Expressway.* A divided major arterial highway for through traffic with full or partial control of access and generally with interchanges at major crossroads. Parkways are expressways for noncommercial traffic within parks and parklike areas.
- *Major.* The part of the roadway system that serves as the principal network for through traffic flow. Major roadways connect areas of principal traffic generation with important rural highways entering the city.
- *Collector.* The distributor and collector roadways serving traffic between major and local roadways. These are roadways used mainly for traffic movements from residential to commercial and industrial areas.
- *Local.* Roadways used primarily for direct access to residential, commercial, industrial, or other abutting property. They do not include roadways carrying

through traffic. Long local roadways will generally be divided into short sections by collector roadway systems.

- *Alleys.*   A narrow public way within a block, generally used for vehicular access to the rear of abutting properties.
- *Sidewalks.*   Paved or otherwise improved areas for pedestrian use, located within public street rights-of-way that also contain roadways for vehicular traffic.
- *Pedestrian ways.*   Public sidewalks for pedestrian traffic generally not within rights-of-way for vehicular traffic roadways. Included are skywalks (pedestrian overpasses), subwalks (pedestrian tunnels), walkways giving access to park or block interiors, and crossings near centers of long blocks.
- *Bicycle lanes.*   Any facility that explicitly provides for bicycle travel.

7.15.2   Area Classifications

- *Commercial.*   That portion of a municipality in a business development where there are large numbers of pedestrians and a heavy demand for parking space during periods of peak traffic; or a sustained high pedestrian volume and a continuously heavy demand for off-street parking space during business hours. This definition applies to densely developed business areas outside of, as well as those that are within, the central part of a municipality.
- *Intermediate.*   That portion of a municipality which is outside of a downtown area but generally within the zone of influence of a business or industrial development, often characterized by moderately heavy nighttime pedestrian traffic and a somewhat lower parking turnover than found in commercial areas. This definition includes densely developed apartment areas, hospitals, public libraries, and neighborhood recreational centers.
- *Residential.*   A residential development, or a mixture of residential and commercial establishments, characterized by few pedestrians and low parking demand or turnover at night. This definition includes areas with single-family homes, townhouses, and/or small apartments. Regional parks, cemeteries, and vacant lands are also included.

7.16   *FREEWAY LIGHTING CONSIDERATIONS*

Freeway lighting can substantially reduce accident frequency and increase capacity. Lighting is of additional benefit in freeway operations by providing motorists additional warning time of stalled or disabled vehicles on the roadway. A listing of the specific conditions considered by most authorities to warrant lighting is included below.

7.16.1   Warranting Conditions for Continuous Freeway Lighting

Continuous lighting along the freeway is warranted in the following cases:

Case CFL-1.   On those sections in or near cities where the current average daily traffic (ADT) is 30,000 or more.

Case CFL-2.   Where three or more successive interchanges are located with an average spacing of 1½ mi or less and adjacent areas outside the right-of-way are substantially urban.

Case CFL-3.   Where for a length of two or more miles the freeway passes through a developed suburban or urban area in which at least one of these conditions exists: (1) local traffic operates on a complete street grid having some form of street lighting, parts of which are visible from the freeway; (2) the freeway passes through a series of developments that are lighted; (3) separate cross streets with or without connecting ramps occur with an average spacing of ½ mi or less, some of which are lighted as part of the local street lighting system; (4) the freeway cross-section elements, such as median and borders, are substantially reduced in width.

Case CFL-4.   Where the ratio of night accidents to day accidents is at least 2.0 or higher than the statewide average for all unlighted similar sections and lighting may be expected to result in a significant reduction in the night accidents. [3]

7.16.2   Warranting Conditions for Complete Interchange Lighting

Complete lighting of freeway interchanges is warranted under the following circumstances:

Case CIL-1.   Where total current ADT ramp traffic entering and leaving the freeway within the interchange area exceeds 10,000 for urban conditions, 8000 for suburban conditions, or 5000 for rural conditions.

Case CIL-2.   Where current ADT on the crossroad exceeds 10,000 for urban conditions, 8000 for suburban conditions, or 5000 for rural conditions.

Case CIL-3.   Where existing substantial commercial or industrial development, which is lighted, is located in the immediate vicinity of the interchange; or where the crossroad approaches are lighted for at least ½ mi on each side of the interchange.

Case CIL-4.   Where the ratio of night accidents to day accidents within the interchange area is at least 1.5 times higher than the statewide average for all unlighted similar interchanges and lighting may be expected to result in a significant reduction in the night accident rate.

7.16.3   Warranting Conditions for Partial Interchange Lighting

Partial lighting of freeway interchanges (Fig. 7.74) is warranted when the following circumstances exist:

Case PIL-1.   Where the total current ADT ramp traffic entering and leaving the freeway within the interchange area exceeds 5000 for urban conditions, 3000 for suburban conditions, or 1000 for rural conditions.

Case PIL-2.   Where the current ADT on the freeway through traffic lanes exceeds 25,000 for urban conditions, 20,000 for suburban conditions, or 10,000 for rural conditions.

Case PIL-3.   Where the ratio of night accidents to day accidents within the interchange area is at least 1.25 times higher than the statewide average for all unlighted similar sections and lighting may be expected to result in a significant reduction in the night accident rate.

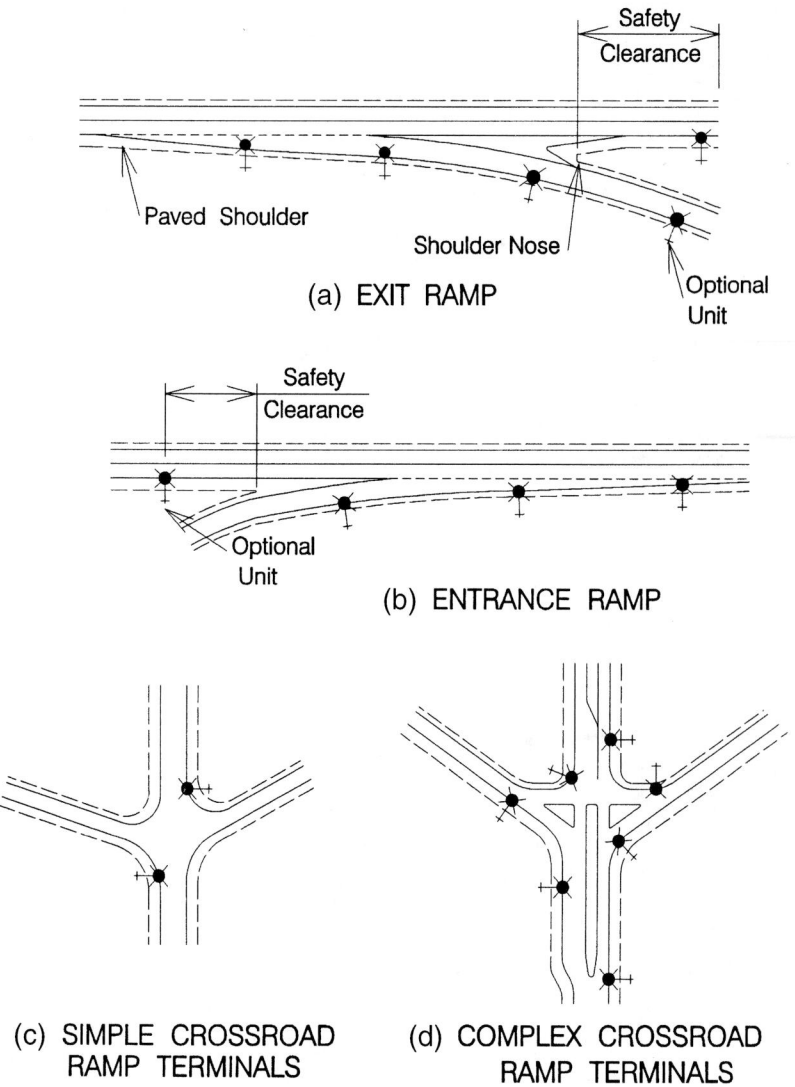

FIGURE 7.74   Typical luminaire locations for partial interchange lighting.

7.16.4   Special Conditions That Warrant Lighting [3]

Continuous, complete, or partial interchange lighting is considered to be justified in the following situations:

- In general, lighting is warranted where the local government agency finds sufficient benefit in the forms of convenience, safety, policing, community promotion, public relations, etc., to pay an appreciable percentage of the cost of the installation, maintenance, and operation of the lighting system.
- Where there is continuous freeway lighting, there should be complete interchange lighting.
- Where complete interchange lighting is warranted but not initially fully installed, a partial lighting system that exceeds the normal partial installation in number of lighting units is considered to be justified.
- Lighting of crossroad ramp terminals is warranted regardless of volumes where the design requires the use of channelizing or divisional islands, and/or where there is poor sight distance.

7.16.5   Freeway Lighting Design Values [6, 7]

The following should be satisfied for proper design when using the illuminance method:

- Continuous freeway and complete interchange lighting should be designed to provide an average maintained horizontal illuminance in the range of 0.6 to 0.8 fc (6 to 9 lx) on the traveled roadway.
- The ramps should be lighted to the same level as the main roadway.
- The point of least illuminance on the roadway should not be less than 0.2 horizontal fc (2 lx).
- An average to minimum uniformity ratio of 3:1 or 4:1 is reasonable. The more uniform design is preferred.
- The higher levels of illuminance should be at the gores and intersections.
- Situations such as high ambient brightness near the roadway or closed circuit television surveillance equipment may justify higher levels of illuminance.

The following criteria should be satisfied for proper design when using the luminance method:

- The average maintained luminance should be in the range of 0.4 to 0.6 cd/m^2 (0.12 to 0.17 foot-lambert, or ft · L).
- The ratio of average luminance to minimum luminance should not exceed 3.5 to 1.
- The ratio of maximum luminance to minimum luminance should not exceed 6.0 to 1.
- The ratio of veiling luminance to average luminance should not exceed 0.3 to 1.

7.16.6   Transition Lighting

Rapid changes in lighting levels which occur, especially when leaving a continuously lighted section of roadway, may be compensated for by using transition lighting or adaptation techniques. Lighting levels as recommended above should be reduced to approximately one-half the recommended level for about 15 s to allow the eye to adapt.

7.16.7 Bridges and Overpasses

Lighting on bridges and overpasses should be at the same level as the roadway. It may be desirable to provide lighting on long bridges in urban and suburban areas even though the approaches are not lighted, since lighting enhances both the safety and utility of bridges. Where bridges are provided with sidewalks, lighting is warranted to increase pedestrian safety and security.

7.17 STREETS AND HIGHWAYS OTHER THAN FREEWAYS

Fixed roadway lighting systems increase night visibility, as well as improve safety, traffic movement, and general roadway use. Traffic volume, numbers of pedestrians, at-grade intersections, turning movements, signalization, and unusual geometrics are some elements that make lighting of streets and highways desirable. Lighting, in addition to its safety benefits, is a crime deterrent and a valuable aid to law enforcement agencies and often contributes to the pride of a community.

7.17.1 Warranting Conditions for Roadways Other Than Freeways [3]

It is not practical to establish specific warrants for the installation of roadway lighting to satisfy all prevailing or anticipated conditions. In general, lighting may be considered for those locations where the respective governmental agencies concur that lighting contributes substantially to the efficiency, safety, and comfort of vehicular or pedestrian traffic. Lighting may be provided for all major arterials in urbanized areas. It may also be provided for locations or sections of streets and highways where the ratio of night accidents to day accidents is higher than the statewide average for similar locations and a study indicates that lighting may be expected to significantly reduce the night accident rate. Determinations to install lighting that have been made on the basis of accident experience at a particular site can be applied to other similar highway locations. The latter should include similar geometric layouts on which experience or accident data are not available and also on highway sections where anticipated increase in vehicular and pedestrian traffic will present problems within a few years. Lighting may be considered at locations where severe or unusual weather or atmospheric conditions exist. In other situations, lighting may be considered where the local governmental agency finds sufficient benefit in the form of convenience, safety, policing, community promotion, or public relations to pay an appreciable percentage of the cost of, or wholly finance, the installation, maintenance, and operation of the lighting facilities [3].

Lighting has been successfully used on rural conventional highways at hazardous locations to reduce the number of accidents. Lighting of spot locations in rural areas should be considered whenever the driver is required to pass through a section of road with complex geometry and/or raised channelization as well as at intersections with higher than normal accident rates [3]. Isolated lighting of railroad grade crossings has been used to help the driver identify when a train is present in the crossing.

7.17.2 Lighting Design Values

Average maintained luminance or illuminance levels are contained in Tables 7.20 and 7.21. The values given are minimum levels when the output of the lamp and luminaire are diminished by the maintenance factors.

7.17.3 Other Considerations

In using Tables 7.20 and 7.21, there may be conditions for which different luminance and illuminance levels are desirable or necessary. The lighting designer should use all available pertinent information in reaching a decision regarding the level to be used for any specific street or highway.

There are many locations where very high levels of luminance or illuminance are provided for streets in the central city business district. This is usually a commercial consideration directed toward making the downtown business area more appealing to shoppers. Levels considerably higher than the levels in the table must be justified on some basis other than solely for the safe and efficient flow of traffic. If higher than recommended levels are desired, the lighting designer should consider using a white light source, such as metal halide, rather than a monochromatic source, such as high-pressure sodium (HPS). Visibility tests have shown there is a lack of contrast with

TABLE 7.20 Average Maintained Luminance Design Values for Roadways Other Than Freeways

Roadway classification		Luminance				Veiling luminance ratio, $L_{v, max}/L_{avg}$
		L_{avg}		Uniformity		
		cd/m^2	Foot-lamberts	L_{avg}/L_{min}	L_{max}/L_{min}	
Expressway*	Commercial	1.0	0.29	3:1	5:1	0.3:1
	Intermediate	0.8	0.23	3:1	5:1	
	Residential	0.6	0.17	3.5:1	6:1	
Major*	Commercial	1.2	0.35	3:1	5:1	0.3:1
	Intermediate	0.9	0.26	3:1	5:1	
	Residential	0.6	0.17	3.5:1	6:1	
Collector*	Commercial	0.8	0.23	3:1	5:1	0.4:1
	Intermediate	0.6	0.17	3.5:1	6:1	
	Residential	0.4	0.12	4:1	8:1	
Local*	Commercial	0.6	0.17	6:1	10:1	0.4:1
	Intermediate	0.5	0.15	6:1	10:1	
	Residential	0.3	0.09	6:1	10:1	
Alleys	Commercial	0.4	0.12	6:1	10:1	0.4:1
	Intermediate	0.3	0.09	6:1	10:1	
	Residential	0.2	0.06	6:1	10:1	

For design of pedestrian walkways and bicycle lanes, use values as shown in Table 7.21.

*Adapted from "American National Standard Practice for Roadway Lighting," ANSI/IES RP-8, 1983, Illuminating Engineering Society of North America [6]. Used by permission.

TABLE 7.21 Average Maintained Horizontal Illuminance for Roadways Other Than Freeways, Walkways, and Bicycle Lanes

Roadway and walkway classification		R1		R2 and R3		R4		Uniformity, avg/min
		Footcandles	Lux	Footcandles	Lux	Footcandles	Lux	
Expressway*†	Commercial	0.9	10	1.3	14	1.2	13	3:1
	Intermediate	0.7	8	1.1	12	0.9	10	
	Residential	0.6	6	0.8	9	0.7	8	
Major†	Commercial	1.1	12	1.6	17	1.4	15	3:1
	Intermediate	0.8	9	1.2	13	1.0	11	
	Residential	0.6	6	0.8	9	0.7	8	
Collector†	Commercial	0.7	8	1.1	12	0.9	10	4:1
	Intermediate	0.6	6	0.8	9	0.7	8	
	Residential	0.4	4	0.6	6	0.5	5	
Local†	Commercial	0.6	6	0.8	9	0.7	8	6:1
	Intermediate	0.5	5	0.7	7	0.6	6	
	Residential	0.3	3	0.4	4	0.4	4	
Alleys	Commercial	0.4	4	0.6	6	0.5	5	6:1
	Intermediate	0.3	3	0.4	4	0.4	4	
	Residential	0.2	2	0.3	3	0.3	3	
Sidewalks	Commercial	0.9	10	1.3	14	1.2	13	3:1
	Intermediate	0.6	6	0.8	9	0.7	8	4:1
	Residential	0.3	3	0.4	4	0.4	4	6:1
Pedestrian ways and bicycle lanes‡		1.4	15	2.0	22	1.8	19	3:1

Average illuminance on the traveled way or on the pavement area between curb lines of curbed roadways.

Pavement classifications are as follows. R1: Portland cement concrete road surface; asphalt road surface with minimum of 15 percent of aggregates composed of artificial brightener (e.g., Synopal) or with naturally bright aggregates (e.g., labradorite, quartzite). R2: Asphalt road surface with an aggregate composed of a minimum of 60 percent gravel, of size greater than 19 mm; asphalt road surface with 10 to 15 percent artificial brightener in an aggregate mix (not normally used in North America). R3: Asphalt road surface, regular and carped teal, with dark aggregates (e.g., trap rock, blast furnace slag), having a rough texture after some months of use. R4: Asphalt road surface with a very smooth texture.

*Both mainline and ramps. Expressways with full control of access are covered in Art. 7.16.

†Adapted from "American National Standard Practice for Roadway Lighting," ANSI/IES RP-8, 1983; Illuminating Engineering Society of North America [6]. Used by permission.

‡This assumes a separate facility. Facilities adjacent to a vehicular roadway should use the illuminance or luminance levels for that roadway.

high levels of HPS. The lack of contrast reduces the ability to distinguish an object from its background, the details of an object, or the color of the object.

7.18 TUNNEL LIGHTING

A tunnel is defined as a structure over a roadway that restricts the normal daytime illumination of a roadway section so that the driver's visibility is substantially diminished. Design of tunnel lighting requires adaptation for driver needs in the approach, and the threshold, transition, and interior zones, as presented in Fig. 7.75. Tunnels are classified by structure length and geometric alignment (visibility through the structure). A straight tunnel having an overall length from portal to portal equal to or less than the safe stopping sight distance (SSSD; Table 7.22) is considered to be a short tunnel. A tunnel with an overall length greater than one SSSD, or having an alignment or curvature that prevents motorists from seeing through the structure to the exit end, is considered to be a long tunnel. Overpasses and underpasses are those structures in which the length does not exceed one width of the roadway over (or under) which they are constructed.

Underpasses with a length-to-height ratio of approximately 10:1 or less will not normally require daytime lighting. When the length-to-height ratio exceeds 10:1, it is necessary to analyze the specific conditions, including vehicular and pedestrian activity, to determine the need for daytime lighting. Roadways that are not continuously lighted warrant underpass lighting in areas having frequent nighttime pedestrian traffic or where unusual or critical roadway geometry occurs under or adjacent to the underpass area. On roadways with continuous lighting, favorable positioning of luminaires adjacent to the underpass can often provide adequate lighting without supplemental luminaires. Lighting levels and uniformities should match the values on the adjacent roadway when practical. Because of limited mounting height, when lights are placed within a tunnel, special consideration should be given to glare and uniformity. Raised lighting levels may be achieved by using closely spaced low-wattage luminaires. Such increased levels should not exceed twice that of the adjacent roadway.

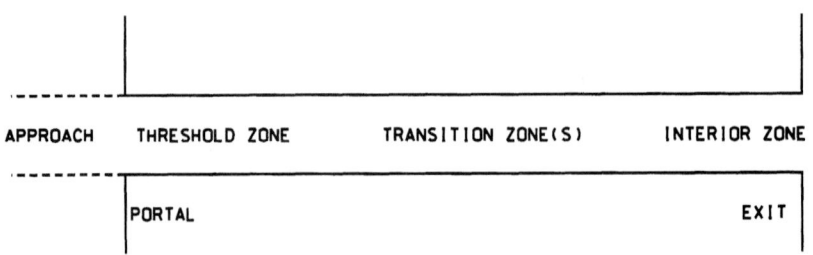

FIGURE 7.75 Lighting adaptation zones for tunnel lighting. Approach: The external roadway area leading to the tunnel. Portal: The plane of entrance into the tunnel. Threshold zone: The area inside the tunnel where a transition is made from the high natural lighting level to the beginning of the transition zone. Transition zone(s): Areas that allow the motorist to achieve appropriate eye adaptation by incrementally reducing the level of luminance required in the threshold zone to the luminance of the interior zone. Interior zone: Area within the tunnel after eye adaptation has been completed.

TABLE 7.22 Safe Stopping Sight Distances in Tunnels

Traffic speed		Minimum safe stopping sight distance (SSSD)*	
km/h	mi/h	m	ft
50	30	60	200
60	40	90	300
80	50	140	450
90	55	160	530
100	60	190	620
110	65	220	720

Source: A Policy on Geometric Design of Highways and Streets, American Association of State Highway and Transportation Officials, Washington, D.C., 1990, with permission.

*Assumes average prevailing speeds in a straight and level tunnel approach roadway are at, or near, the posted speed limit of the facility. For other geometric conditions, refer to the AASHTO documents.

7.18.1 Warrants for Tunnel Lighting

The use of artificial daytime lighting is warranted when user visibility requirements are not satisfied by the natural sunlight. Overall tunnel visibility varies considerably with such factors as geometry of the tunnel and its approaches, traffic characteristics, roadway and environmental reflective surfaces, the climate and orientation of the tunnel, and visibility objectives. Comprehensive literature is available on the technical aspects of visibility and lighting of tunnels [8]. Information on lighting levels for tunnels requires a detailed analysis of the tunnel approach characteristics. Tunnel lighting requires considerable experience to achieve proper design.

7.19 ROADWAY REST AREAS

The design of lighting for rest areas requires consideration of both vehicle and pedestrian needs. Properly designed rest area lighting will enhance the architectural and landscape features of the facility, promote safety by easing the task of policing, and contribute to the rest and relaxation of motorists by adequately lighting the driving, parking, and walking areas. In areas with landscaping or in natural settings, the lighting designer often attempts to make the light poles less noticeable by causing them to blend with the environment. One cost-effective method uses colored fiberglass reinforced poles that blend with the surrounding environment. These poles are usually of the direct burial type that can be installed with or without breakaway devices.

The lighting system designer should be mindful of motorists on the travelway by not allowing glare or spill light from the rest area luminaires to adversely affect their vision. The motorist on the main roadway should be able to see any vehicles leaving the rest area as well as traffic along the main route. The lighting concerns for rest areas can be divided into several distinct areas:

- *Entrance and exit.*   The deceleration and acceleration lanes adjacent to the main roadway can be lighted so that a motorist can safely transition into and out of the rest area. When the main roadway is not lighted, an average illumination of 6 lx should be maintained on the deceleration with three to five luminaires along the speed change lanes. On the exit gore and acceleration lane, 6 lx is recommended to a point where the motorist can merge onto the main roadway. If the main route is lighted, the entrance and exit lanes should be lighted to a level equal to that of the main route.

- *Interior roadways.*   These are the roads for the entrance gore to the parking areas and from the parking areas to the exit gore. The recommended illumination is 6 lx with a uniformity of 3:1 to 4:1.

- Parking areas.   The recommended average maintained lighting level is 11 lx for both automobile and truck parking areas with a uniformity of 3:1 to 4:1 over the entire area. Special areas that should have the higher levels are handicap ramps, sanitary disposal stations, and other features that require detail viewing.

- *Activity areas.*   The major pedestrian activity areas are restrooms, information centers, and walkways to and from the buildings and the parking lot. Minor activity areas include picnic tables, dog walks, and other walk areas. The recommended lighting level for major areas is 11 lx with a 3:1 to 4:1 uniformity ratio. Minor activity areas should be lighted to 5 lx with a uniformity ratio of 6:1 [3].

Rest areas are often located in remote areas that are not readily accessible by bucket trucks or other special maintenance equipment. This requires that lighting system components be selected that provide maximum protection against vandalism and require minimal maintenance. One device that has been used to allow pole-mounted luminaires to be maintained, and lamps changed, without using a bucket truck is an individual lowering device (ILD) which allows the pole-mounted luminaires to be lowered to ground level, one at a time, for servicing. This is done with a hand-operated winch that is lightweight and easily portable by one person. One such ILD (Fig. 7.76) that has been designed to DOT requirements and used successfully since 1991 is manufactured by Lighting and Lowering Systems, Chicago.

7.20   ANALYTICAL APPROACH TO LIGHTING WARRANTS [3]

An analytical approach to determining if roadway lighting is warranted was developed through the National Cooperative Highway Research Program (NCHRP Report 152) using four comprehensive evaluation forms. The four forms relate to non-controlled-access roadways, intersections, freeways, and interchanges, and are presented in Figs. 7.77, 7.78, 7.79, and 7.80, respectively [9]. The forms are used by multiplying the rating of a characteristic by the difference in its unlighted and lighted weight to obtain a quantitative measure of the effect of that characteristic on driver visual information needs. After all the characteristics are rated, the scores are summed to obtain an overall measure of driver information needs.

There is an established "minimum warranting condition" of a given number of points for each of the four forms. The exact number of points is determined by assuming a rating of 3 for each of the characteristics. It should be emphasized that the mini-

FIGURE 7.76 Lowering device manufactured by Lighting and Lowering Systems.

mum warranting condition is not firm, but merely a starting point. This method is flexible and permits modifications to fit local needs. This procedure also provides a method for administrators or planners to prioritize lighting projects by using objective standards to determine where lighting would be most beneficial.

7.21 TYPES OF LUMINAIRES

Conventional roadway lighting has been the cobra head luminaire mounted on a support arm and positioned at the edge of the roadway or, on some cases, out over the roadway. The base of the pole when a breakaway device is present should be a minimum of 4570 mm from the travelway, but 6100 mm is preferred on roadway sections without a curb. The travelway is defined as being a continuous traffic lane and does not include an acceleration or deceleration lane merging with a through lane. When a curb is present, the pole with its breakaway device is preferred to be 3050 mm from the face of the curb. If this is not possible, the pole should be closer than 610 mm. This will ensure that an impacting vehicle strikes the pole at the designed impact height for proper breakaway operation. Breakaway devices should not be used on any pole located where the pedestrians are likely to be present, because of the danger to them if the pole falls.

CLASSIFICATION FACTOR	RATING					UNLIT WEIGHT (A)	LIGHTED WEIGHT (B)	DIFF. (A-B)	SCORE [RATING X (A-B)]
	1	2	3	4	5				
GEOMETRIC FACTORS									
No. of Lanes	4 or less	-	-	-	8 or more	1.0	0.8	0.2	
Lane Width (m)	> 3.6	3.6	3.3	3.0	<3.0	3.0	2.5	0.5	
Median Openings per Mile	<4.0 or one-way operation	4.0-8.0	8.1-12.0	12.0-15.0	>15.0 or no access control	5.0	3.0	2.0	
Curb Cuts	< 10%	10-20%	20-30%	30-40%	>40%	5.0	3.0	2.0	
Curves	<3.0°	3.1-6.0°	6.1-8.0°	8.1-10.0°	>10°	13.0	5.0	8.0	
Grades	< 3%	3.0-3.9%	4.0-4.9%	5.0-6.9%	7% or more	3.2	2.8	0.4	
Sight Distances (m)	>215	150-215	90-150	60-90	<60	2.0	1.8	0.2	
Parking	Prohibited both sides	Loading zones only	Off-peak only	Permitted one side	Permitted both sides	0.2	0.1	0.1	
OPERATIONAL FACTORS								GEOMETRIC TOTAL	
Signals	All major intersections signalized	Substantial majority of intersections signalized	Most major intersections signalized	About half the intersections signalized	Frequent nonsignalized intersections	3.0	2.8	0.2	
Left Turn Lane	All major intersections or one-way operation	Substantial majority of intersections	Most major intersections	About half the major intersections	Infrequent turn bays or undivided streets	5.0	4.0	1.0	
Median Width (m)	9	6-9	3-6	0-1.2	0-1.2	1.0	0.5	0.5	
Operating Speed (km/h)	40 or less	50	55	65	70 or greater	1.0	0.2	0.8	
Pedestrian Traffic at Night (peds/mi)	Very few or none	0-50	50-100	100-200	> 200	1.5	0.5	1.0	
ENVIRONMENTAL FACTORS								OPERATIONAL TOTAL	
% Development	0	0-30%	30-60%	60-90%	100%	0.5	0.3	0.2	
Predominant Type development	Undeveloped or backup design	Residential	Half residential &/or commercial	Industrial or commercial	Strip industrial or commercial	0.5	0.3	0.2	
Setback Distance (m)	> 60	45-60	30-45	15-30	< 15	0.5	0.3	0.2	
Advertising or Area Lighting	None	0-40%	40-60%	60-80%	Essentially continuous	3.0	1.0	2.0	
Raised-Curb Median	None	Continuous	At all intersections	At all signalized intersections	A few locations	1.0	0.5	0.5	
Crime Rate	Extremely low	Lower than city average	City average	Higher than city average	Extremely high	1.0	0.5	0.5	
ACCIDENTS								ENVIRONMENTAL TOTAL	
Ratio of Night-to-Day Accident Rates	< 1.0	1.0-1.2	1.2-1.5	1.5-2.0	2.0*	10.0	2.0	8.0	

	GEOMETRIC TOTAL =		ACCIDENT TOTAL
	OPERATIONAL TOTAL =		
*Continuous Lighting Warranted	ENVIRONMENTAL TOTAL =		
	ACCIDENT TOTAL =		
	SUM =	POINTS	
	WARRANTING CONDITION = 85 POINTS		

FIGURE 7.77 Evaluation form for non-controlled-access facility lighting.

CLASSIFICATION FACTOR	RATING					UNLIT WEIGHT (A)	LIGHTED WEIGHT (B)	DIFF. (A-B)	SCORE [RATING X (A-B)]
	1	2	3	4	5				
GEOMETRIC FACTORS									
Number of Legs		3	4	5	6 or more (including traffic circles)	3.0	2.5	0.5	
Approach Lane Width (m)	>3.6	3.6	3.3	3.0	<3.0	3.0	2.5	0.5	
Channelization	No turn lanes	Left turn lanes on major legs	Left turn lanes on all legs, right turn lanes on major legs	Left and right turn lanes on major legs	Left and right turn lanes on all legs	2.0	1.0	1.0	
Approach Sight Distance (m)	>215	150-215	90-150	60-90	<60	2.0	1.8	0.2	
Grades on Approach Streets	< 3%	3.0-3.9%	4.0-4.9%	5.0-6.9%	7% or more	3.2	2.8	0.4	
Curvature on Approach Legs	< 3.0°	3.0°-6.0°	6.1°-8.0°	8.1°-10.0°	> 10°	13.0	5.0	8.0	
Parking in Facility	Prohibited both sides	Loading zones only	Off-peak only	Permitted one side only	Permitted both sides	0.2	0.1	0.1	
OPERATIONAL FACTORS							GEOMETRIC TOTAL		
Operating Speed on Approach Legs (km/h)	40 or less	50	55	65	70	1.0	0.2	0.8	
Type of Control	All phases signalized (incl. turn lane)	Left turn lane signal control	Through-traffic signal control only	4-way stop control	Stop control on minor legs or no control	3.0	2.7	0.3	
Channelization	Left and right turn lane signal control	Left turn lane signal control on major legs	Left turn lane signal control on all legs	Left turn lane signal control on major legs	No turn lane control	3.0	2.0	1.0	
Level of Service (Load Factor)	A 0.0	B 0-0.1	C 0.1-0.3	D 0.3-0.7	E 0.7-1.0	1.0	0.2	0.8	
Pedestrian Volume (peds/h)	Very few or none	0-50	50-100	100-200	> 200	1.5	0.5	1.0	
ENVIRONMENTAL FACTORS							OPERATIONAL TOTAL		
Percent Adjacent Development	0	0-30%	30-60%	60-90%	100%	0.5	0.3	0.2	
Predominant Development near Intersection	Undeveloped	Residential	50% residential 50% industrial or commercial	Industrial or commercial	Strip industrial or commercial	0.5	0.3	0.2	
Lighting in Immediate Vicinity	None	0-40%	40-60%	60-90%	100%	3.0	1.5	1.5	
Crime Rate	Extremely low	Lower than City average	city average	Higher than city average	Extremely high	1.0	0.5	0.5	
ACCIDENTS							ENVIROMENTAL TOTAL		
Ratio of Night-to-Day Accident Rates	1.0	1.0-1.2	1.2-1.5	1.5-2.0	2.0*	10.0	2.0	8.0	
*Intersection Lighting Warranted	GEOMETRIC TOTAL =						ACCIDENT TOTAL		
	OPERATIONAL TOTAL =								
	ENVIRONMENTAL TOTAL =								
	ACCIDENT TOTAL =								
			SUM =			POINTS			
			WARRANTING CONDITION = 75 POINTS						

FIGURE 7.78 Evaluation form for intersection lighting.

CLASSIFICATION FACTOR	RATING					UNLIT WEIGHT (A)	LIGHTED WEIGHT (B)	DIFF. (A-B)	SCORE [RATING X (A-B)]
	1	2	3	4	5				
GEOMETRIC FACTORS									
Number of Lanes	4		6		≥8	1.0	0.8	0.2	
Lane Width (m)	>3.6	3.6	3.3	3.0	≤2.75	3.0	2.5	0.5	
Median Width (m)	>12	7-12	3.7-7.0	1.2-3.3	0	1.0	0.5	0.5	
Shoulders (m)	3.0	2.4	1.8	0.9	0	1.0	0.5	0.5	
Slopes	≥8:1	6:1	4:1	3:1	2:1	1.0	0.5	0.5	
Curves	0-1/2°	1/2-1°	1-2°	2-3°	3-4°	13.0	5.0	8.0	
Grades	< 3%	3-3.9%	4-4.9%	5-6.9%	> 7%	3.2	2.8	0.4	
Interchange Frequency	6.4 km	4.8 km	3.2 km	1.6 km	>1.6 km	4.0	1.0	3.0	
OPERATIONAL FACTORS						GEOMETRIC TOTAL			
Level of Service (any dark hour)	A	B	C	D	E	6.0	1.0	5.0	
ENVIRONMENTAL FACTORS						OPERATIONAL TOTAL			
% development	0%	25%	50%	75%	100%	3.5	0.5	3.0	
Offset to Development (m)	60	45	30	15	<15	3.5	0.5	3.0	
ACCIDENTS						ENVIRONMENTAL TOTAL			
Ratio of Night-to-Day Accident Rates	1.0	1.0-1.2	1.2-1.5	1.5-2.0	2.0+	10.0	2.0	8.0	

	GEOMETRIC TOTAL =	
	OPERATIONAL TOTAL =	ACCIDENT TOTAL
*Continuous Lighting Warranted	ENVIRONMENTAL TOTAL =	
	ACCIDENT TOTAL =	
	SUM = POINTS	
	WARRANTING CONDITION = 95 POINTS	

FIGURE 7.79   Evaluation form for controlled-access facility freeway lighting.

CLASSIFICATION FACTOR	RATING					UNLIT WEIGHT (A)	LIGHTED WEIGHT (B)	DIFF. (A-B)	SCORE [RATING X (A-B)]
	1	2	3	4	5				
GEOMETRIC FACTORS									
Ramp Types	Direct	Diamond	Buttonhooks cloverleafs	Trumpet	Scissor and left-side	2.0	1.0	1.0	
Crossroad Classification	None		Continuous		At interchange intersections	2.0	1.0	1.0	
Frontage Roads	None		One-way		Two-way	1.5	1.0	0.5	
Freeway Lane Widths (m)	>3.6	3.6	3.3	3.0	<3.0	3.0	2.5	0.5	
Freeway Median Widths (m)	>12	10-12	3.6-7.3	1.2-3.6	<1.2	1.0	0.5	0.5	
Number of Freeway Lanes	4 or less		6		8 or more	1.0	0.8	0.2	
Main Lane Curves	< 1/2°	1-2°	2-3°	3-4°	> 4°	13.0	5.0	8.0	
Grades	3%	3-3.9%	4-4.9%	5-6.9%	7% or more	3.2	2.8	0.4	
Sight Distance Crossroad Intersection	>304	210-300	150-210	120-150	<120	2.0	1.8	0.2	
OPERATIONAL FACTORS						GEOMETRIC TOTAL			
Level of Service (any dark hour)	A	B	C	D	E	6.0	1.0	5.0	
ENVIRONMENTAL FACTORS						OPERATIONAL TOTAL			
% development	None	1 quad	2 quad	3 quad	4 quad	2.0	0.5	1.5	
Setback Distance (m)	>60	45-60	30-45	15-30	<15	0.5	0.3	0.2	
Crossroad Approach Lighting	None		Partial		Complete	3.0	2.0	1.0	
Freeway Lighting	None		Interchanges only		Continuous	5.0	3.0	2.0	
ACCIDENTS						ENVIRONMENTAL TOTAL			
Ratio of Night-to-Day Accident Rates	< 1.0	1.0-1.2	1.2-1.5	1.5-2.0	> 2.0•	10.0	2.0	8.0	

GEOMETRIC TOTAL =	
OPERATIONAL TOTAL =	
ENVIRONMENTAL TOTAL =	
ACCIDENT TOTAL =	
SUM = POINTS	ACCIDENT TOTAL
COMPLETE LIGHTING WARRANTING CONDITION = 90 POINTS	
PARTIAL LIGHTING WARRANTING CONDITION = 60 POINTS	

•Continuous Lighting Warranted

FIGURE 7.80 Evaluation form for interchange lighting.

Cobra head luminaires are available in a wide range of cutoff, semicutoff, and non-cutoff beam patterns. All cobra head luminaires have a horizontal lamp position that causes them to produce a large amount of light directly under the luminaire. This requires the designer to closely inspect the calculated average to minimum light level ratios to ensure compliance with values given in the illuminance tables.

Other luminaires that can be used in the same locations as cobra heads utilize a vertical lamp position. These produce a more uniform light pattern, since a smaller portion of their lumens are directed straight down, thus providing a more uniform light level. These luminaires are not available in cutoff types. The two major manufacturers of this vertical lamp luminaire are Holophane and McGraw-Edison.

High mast luminaires are designed to be mounted on the lowering ring of a high mast pole. High mast luminaires are produced primarily in 400- and 1000-watt sizes in a wide variety of beam patterns. New lamps are being developed that have lumen to watt ratios equal to or better than the 1000-W that do not demonstrate the same fragile tendencies [10]. Manufacturers use different designations for their own luminaires, but generally type 2 and type 5 beam patterns are most popular. The beam patterns are also referred to as *long* and *narrow,* and *symmetrical* and *nonsymmetrical.* Cutoff and noncutoff types are used, although not all beam patterns are made in each category. The designer must be concerned with light trespass when using high mast luminaires and should locate them so as not to interfere with adjacent property usage. A technique used by some designers, when high mast poles cannot be located in the middle of the area to be lighted, is to specify an offset-type luminaire mounted on the high mast lowering ring in lieu of the "traditional" high mast luminaire. This produces a more directional pattern that can reduce the amount of off-premise light.

Offset luminaires are manufactured by several companies under names such as Vector, Turnpike, Multimount, and Interstate. These luminaires are specifically designed for roadway use and resemble a floodlight in appearance but not in beam pattern. The offset luminaires are intended to be mounted well away from the roadway edge and aimed at an approximate 45° angle. This design was originally conceived in the 1960s, and a test installation along I-55 south of Memphis was very satisfactory. The original design was large and difficult to handle, but perhaps the greatest handicap that prevented widespread use was the resistance among maintenance personnel due to the difficulty in getting to the pole location when servicing was required. From a safety aspect, the offset was, and still is, a very good choice, since it can be located well away from the travelway and the beam pattern allows a wide spacing between the poles. The development of an individual lowering device has increased the number of locations where the offset can be installed. The individual lowering device (ILD) provides each luminaire with its own lowering cable and latch assembly, as distinguished from the high mast lowering device, which has all luminaires mounted on the same ring and lowered together. The cost of the ILD is much less than that of the high mast device when one to four luminaires are located on the pole. Four ILDs is the maximum number used on a single pole, but one or two per pole is most commonly used.

Segmented reflectors are special-purpose luminaires that have been successfully used on top of concrete median barriers. The top of concrete safety shape barriers can be as wide as 300 mm. This limits the anchor bolt spacing, for attaching a luminaire pole, at 150 mm in order to provide a minimum of 75 mm of concrete around each bolt. The resultant anchor bolt spacing places a height restriction on the pole due to the structural needs required to counteract the overturn moments. Two options are to increase the width of the concrete barrier at the luminaire post, as in Fig. 7.81, or to use segmented reflector luminaires, which require less height to provide proper lighting for multiple lanes.

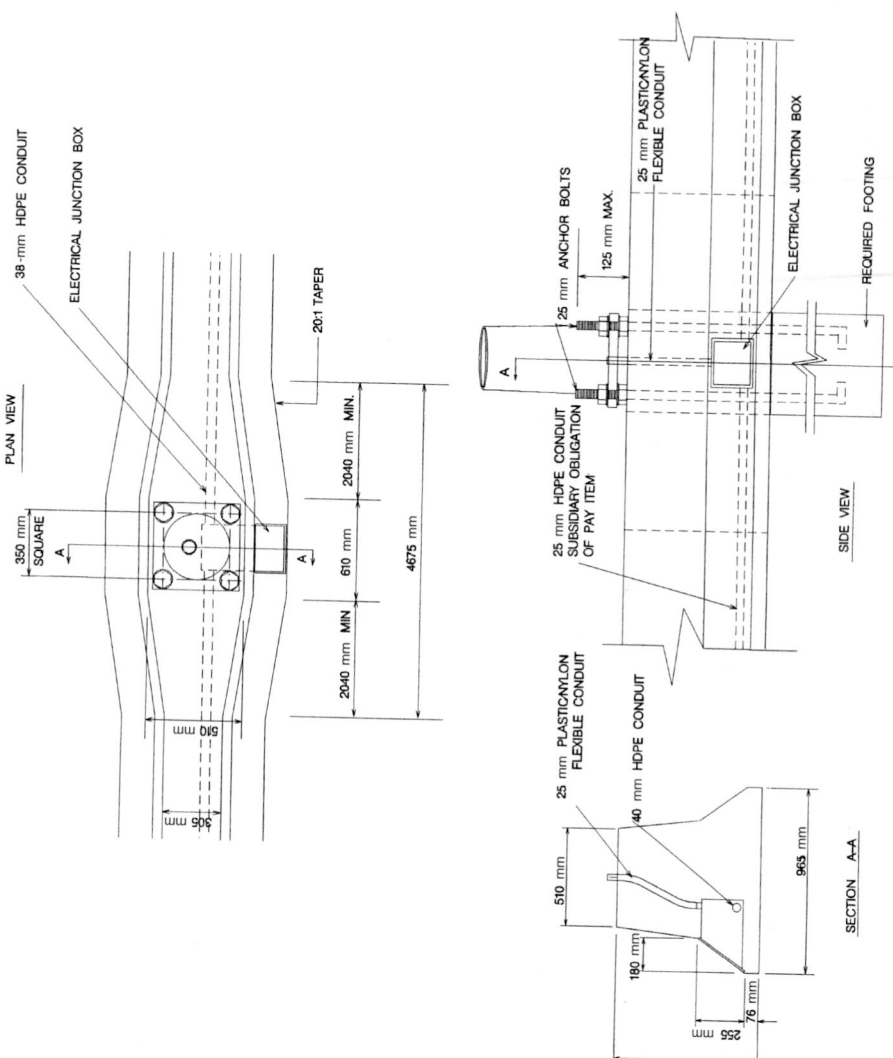

FIGURE 7.81 Example of widening concrete median safety shape to accommodate luminaire pole.

Two of the segmented reflector luminaires when mounted 12 m high can light up to six or eight lanes on each side of the barrier depending on the width of the inside shoulder. This luminaire was originally developed for use in parking decks and uses a vertically mounted lamp with only a small portion of its lumens directed straight down. Exceptional uniformity ratios and the cutoff pattern make these luminaires a good choice when veiling luminance (glare) and light trespass are concerns. Other types of luminaires are also used on top of median barriers. The cobra head is often used, and a traditional high mast luminaire has been found to be very effective in this application, although veiling luminance is a potential problem.

Poles mounted on top of median barriers have a number of advantages and disadvantages. One advantage is cost, since one pole in the middle of the lighted area can replace two roadside luminaires, requiring only one set of circuit conductors and conduits. The disadvantages include problems of traffic and maintenance safety. Placing the poles on top of the concrete median increases the probability of a pole being struck and landing in the opposing traffic lanes, when compared with offset luminaire pole locations. Maintenance crews are required to work with bucket trucks on the inside shoulder, requiring the closure of the inside traffic lane. Experience of over 10 years with median barrier–mounted poles indicates that few poles are actually struck and the majority of strikes that do occur take place late at night when traffic levels are low. The poles that were struck did not become detached from the anchor bolts, since breakaway devices are not used on the barrier rail poles. The use of bucket trucks to service these luminaires is a potential problem, because the knuckle of the boom can extend over an adjacent lane. The solution in at least one metropolitan area is to use the ILD (Fig. 7.76) with the luminaire, eliminating the need for a bucket truck. This facilitates traffic control and requires a smaller-size crew. Some maintenance departments prefer the ILDs and request their installation on barrier lighting projects.

There have been several types of floodlights and sports lights used in roadway applications over the years as lighting designers have attempted to cope with the increasing numbers of lanes and the confining rights-of-way. In most cases, the high level of accuracy required for proper aiming and the need for special glare shields have limited their usefulness.

7.22   HIGH MAST LIGHTING

The value of high mast lighting has been highly controversial since its introduction in the early 1960s. Proponents suggested that high mast lighting offered considerable enhancements to visibility. Opponents, on the other hand, argued that high mast lighting was expensive to build, offered little improvement to visibility, and often resulted in light trespass and light pollution. By the early 1980s, new data became available which suggested the superiority of high mast light over conventional systems. The reasons cited were:

- An improved visual field negating the "tunnel effect" caused by a limited lateral dimension when using conventional mounting heights. The tunnel effect prevents the driver's eyes from reaching a reasonable level of retinal stability—a failure believed by some to be the cause of a substantial number of accidents at night [11].

- Improved luminance uniformity within the principal visual field. The uniformity eliminates the need for the eye to adapt to a wide range of luminances, which adversely affects visual acuity. Many experienced engineers are of the opinion that luminance can be reduced when using high mast lighting because of the compensating factors of improved uniformity and reduced veiling glare. The IESNA RP-8 "American National Standard Practice for Roadway Lighting" has a special table for illuminance levels recommended for high mast lighting design. This table permits the reduction of the lighting level for class A freeways by approximately 35 percent compared with the recommended levels for conventional lighting [11].

- Disability veiling brightness negatively affects a driver's visual performance. In practice, luminaires spaced at long distances require large light sources with high beam intensity in the upper angles of the vertical plane. Light emitted from a luminaire above 75° can be considered a contributor to glare. High mast luminaires confine their distribution within the limits of 60 to 65° and practically eliminate the disability veiling brightness. Brightness from high mast systems is also reduced through geometric arrangement. Increased mounting height and greater offset remove the luminaires from the driver's active viewing area [11].

- The location of the high mast pole contributes to a clear roadside and results in a reduction in the number of vehicular collisions with fixed objects [11].

- Studies done using the older mercury vapor light sources indicated that on both diamond and cloverleaf type interchanges, high mast lighting systems utilized fewer luminaires and less energy than conventional lighting [11].

The growth in vehicular traffic combined with the continuous search by transportation authorities for safer and more cost-effective roadway design has resulted in a shift toward multilane roadways. The new freeway designs cannot be effectively illuminated by conventional methods using low mounting heights and light sources of limited lumen output. Because of these requirements, high mast systems offer a distinct advantage over alternative systems. High mast systems also offer advantages in cases where future road widening is expected. The poles can be located 15 m or more from the traffic lanes, enabling future road widening without the need for changes in the lighting system [11].

7.22.1 High Mast Design Methods

The recommended maintained illuminance design levels for high mast lighting are presented in Table 7.23. Two methods are available for formulating a design for high mast lighting [7]:

- *Utilized lumens method.*   Pole locations are established, and design values are calculated from standard utilized lumens formulas by first overlaying the isofootcandle (or isolux) curve transparencies on drawings of the area to be lighted, as in Fig. 7.82. This method can also be accomplished by using direct calculations made from the coefficient of utilization charts.

- *Average point method.*   Readings are determined at points designated in an established grid pattern on the roadway and then averaged.

7.23 ROADSIDE SAFETY

The primary purpose of roadway illumination is to increase safety by enhancing nighttime visibility. The net safety benefit from increased visibility is influenced by the hazard posed by the roadway lighting or luminaire support acting as a fixed object. If

TABLE 7.23   Recommended Maintained Illuminance Design
Levels for High Mast Lighting [6]

| Road classification | Horizontal illuminance, lx | | |
	Commercial area	Intermediate area	Residential area
Freeways	6	6	6
Expressways	10	8	6
Major	12	9	6
Collector	8	6	6

Recommended uniformity of illumination is 3:1 or better (average-to-minimum ratio for all road classifications at the recommended illuminance levels).

These design values apply only to the traveled portions of the roadway. Interchange roadways are treated individually for purposes of uniformity and illuminance level analysis.

Source: *American National Standard Practice for Roadway Lighting,*
ANSI/IES RP-8, Illuminating Engineering Society of North America, 1983,
with permission.

roadway illumination is not warranted, or if it is installed wrong, there is a strong possibility that traffic hazards will be increased rather than reduced by providing illumination. The AASHTO publication *Roadside Design Guide* requires the lighting designer not only to produce an effective, efficient lighting system but also to consider removing the hazards inherent in such a system [12]. The *Roadway Design Guide* stresses that safety should be enhanced by considering the following, in decreased order of desirability:

• Remove the hazard from the right-of-way
• Locate the hazard in a place less likely to be struck
• Provide a breakaway support
• Provide a barricade

The most common approach to meeting the safety requirement has been to provide a breakaway structure for the light poles. There are a number of devices that have been tested and approved by the Federal Highway Administration for this purpose, including transformer bases, frangible couplings, slip bases, and various schemes applicable to a particular type of pole such as fiberglass and aluminum. All these devices will perform as prescribed, but it is up to the designer to use the proper device in the particular situation encountered for the project. The FHWA approval process evaluates only the structural breakaway performance of a tested device, not the structural strength or the possible electrical hazard introduced when a pole is struck. The lighting designer must become familiar with the structural load limitations of each tested device in order to match the weight, height, and wind loading demands of the luminaires with the strength of the device being considered. The designer should also consider methods to mitigate or eliminate the possibility that damaged electrical wires will be exposed after a pole is knocked down. In urban areas or other locations where pedestrians or cyclists may be in the area where a breakaway pole would fall if struck, breakaway supports are not recommended.

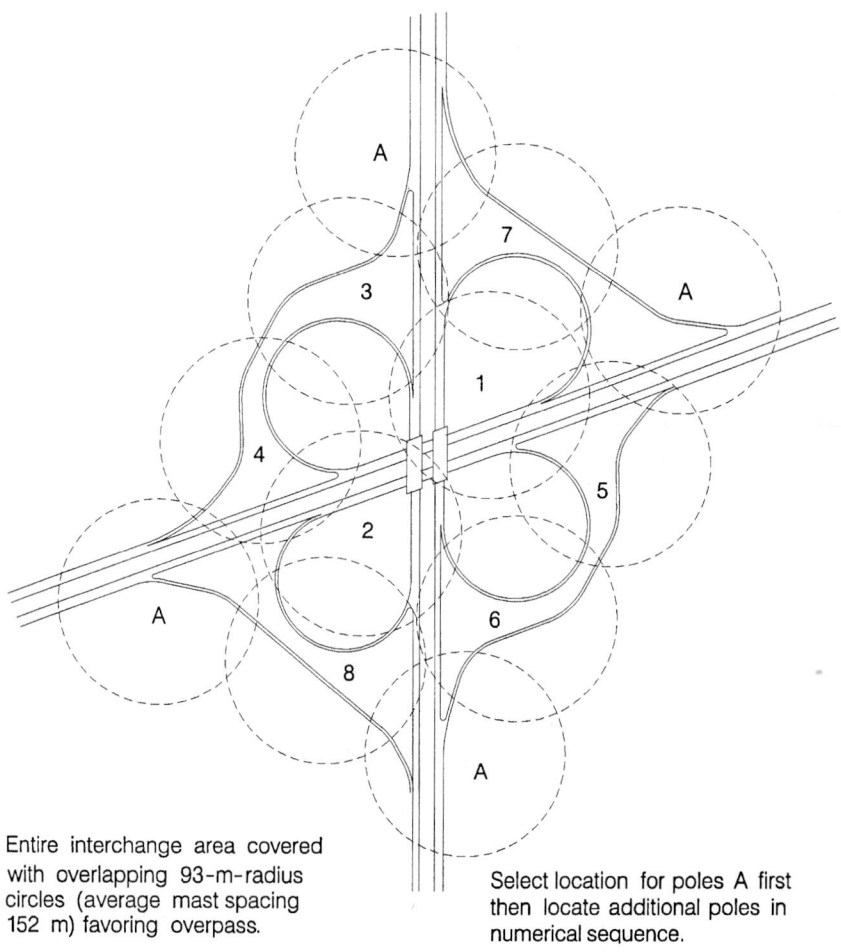

Entire interchange area covered with overlapping 93-m-radius circles (average mast spacing 152 m) favoring overpass.

Select location for poles A first then locate additional poles in numerical sequence.

FIGURE 7.82 Highway lighting design at typical cloverleaf interchange.

7.23.1 Location of Poles

The location of a lighting pole is partially dictated by the lighting scheme selected by the designer for a section of roadway. Using the conventional (cobra head) type luminaires requires the pole to be close to the travelway and therefore, unless it is behind a barrier, most likely to be struck by an errant vehicle. A median barrier–mounted pole is less likely to be struck, but occasionally an out-of-control vehicle will get high enough on the barrier to impact the pole. When this occurs, the danger to oncoming traffic will be increased if the pole is of a breakaway design. Because of this possibility, median-mounted poles are normally not designed as a breakaway type. The lighting scheme that incorporates offset and/or high mast luminaires is the least likely to create a hazard on the roadside, since the poles can be located 12 to 15 m, or farther, from the travelway. In addition to reduced accident rates, this type of lighting reduces maintenance costs due to pole knockdowns.

Pole locations are influenced by the location of sign structures, overpasses, guardrail, roadway curvature, gore clearances, overhead power lines, drainage pipes, drainage structures, underground utilities, and the shoulder slopes, in addition to the luminaire capabilities. The lighting designer must evaluate the eventual consequences of safety, aesthetics, maintenance, and economics when selecting the pole locations. Safety considerations for lighting pole locations include:

- Poles should be placed outside the clear zone whenever practical.
- Pole locations should consider the hazards in servicing the lighting equipment.
- Poles should be located to provide adequate safety clearance in the gore areas of exit and entrance ramps.
- Poles should be placed to minimize interference with motorists' view of the sign, and the luminaire brightness should not seriously detract from sign legibility at night.
- Poles should not be placed where overhead signs will cast distracting shadows on the roadway surface at night.
- Poles on the inside radius of superelevated roadways should have sufficient clearance to avoid being struck by trucks.
- Poles should never be placed on the traffic side of guardrail or any natural or manufactured deflecting barrier.
- Where poles are located in exposed areas, they should have an approved breakaway feature or device.
- Poles along the freeway should be located at least 4.6 m and preferably 6.1 m or more from the edge of the travelway and include a breakaway device unless located behind a barrier or guiderail or otherwise protected.
- Poles behind flexible or yielding type rails or barriers should provide the necessary clear distance for rail or barrier deflection. The design deflection distance of the particular barrier being used should be checked to ensure that vehicles impacting the barrier will not continue into the lighting support.
- Installing poles on the median, instead of the roadside, should be considered where median width is sufficient (on landscaped medians) and on top of properly designed concrete safety shapes present on narrow medians. Among the advantages with median-mounted poles are that one-half the number of poles are required, the quantities of conduit and cable are reduced, house sidelight is minimized, and visibility on the high-speed lanes is improved.

Clear zone is not a constant distance but varies on the basis of the design ADT, the design speed, and the slope, either positive or negative, of the shoulder. Clear zone dimensions are given in the AASHTO *Roadside Design Guide* [12]. (See Chap. 6.)

7.24 POLE TYPES

Poles are available in a number of materials. The advantages and disadvantages of each follow.

Steel.   Steel poles are available in galvanized, painted, powder-coated, and weathering types, plus a combination of powder coating over galvanizing. Galvanized is the most popular of the steel types because of the comparatively low cost and extended life. Painted poles are used primarily when a color is desired, but they require continual maintenance. The powder coating over galvanizing serves the same purpose and requires little maintenance. Weathering steel poles offer enhanced aesthetics but provisions must be made for the rusty runoff.

Aluminum.   Aluminum poles are popular because of their resistance to corrosion and the resultant low maintenance cost. They have an added advantage of being lighter in weight than most other types. Aluminum poles operate well as breakaway designs when impacted at the design height. Since they are less rigid than steel posts, however, aluminum poles can result in an increased probability of improper breakaway operation when impacted higher than the design height. Aluminum poles are also considerably more expensive than most other types.

Stainless Steel.   Stainless steel poles are corrosion-resistant and relatively lightweight. Their high rigidity results in dependable breakaway operation upon impact. They are, however, considerably more expensive than the other pole types.

Fiberglass.   Fiberglass-reinforced plastic (FRP) poles are approved for breakaway use both in the anchor base and in the direct burial series. Shaft lengths are currently limited to 14.3 m, which means 12 m height for the direct burial series and the full 14.3 m height for the anchor base series. Advantages of FRP poles include no rust, no corrosion, no rot, light weight, no additional breakaway device required, no maintenance, no electrical shock, and, for the direct burial series, no need for concrete foundation. FRP poles come in many decorative styles and several standard colors.

Wood.   Wood is perhaps the least expensive of pole types, particularly in areas where trees are plentiful. They can be treated to resist deterioration from the environment and damage due to insects. The use of existing utility poles for luminaire placement has the advantage of reducing the number of poles on the roadside. The huge mass of wood poles, however, makes it difficult to design them as breakaway, and thus, wooden poles should not be installed on high-speed facilities.

Concrete.   Concrete poles are popular in regions where cement and concrete aggregates are plentiful. One advantage to concrete poles is that they can be economical. Concrete poles cannot, however, be designed effectively to safely break away upon impact. They are extremely heavy even when made by prestressing concrete. Impacts with concrete poles result in extensive damage to vehicles and severe injury to occupants. Prestressed concrete poles, therefore, should not be used within the traversable area, unless shielded, on facilities with design speeds over 50 km/h. Concrete posts can be a functional and economical type of support on local urban streets if proper consideration is given to placement.

7.25   ELECTRICAL HAZARD

One problem that has recently been identified is the potential deadly threat posed by the electric circuits after pole impact by an errant vehicle. There are many documented deaths of motorists who survived the impact with a luminaire pole only to be subsequently killed from the resulting explosion and fire. The explosion and fire are usually caused when the fuel tank ruptures, the vehicle having been caught on an improperly constructed foundation, and the electrical system sparks repeatedly until the fuel explodes. In other incidences, medical personnel have been delayed from attending victims because of the risk of electrical shock from exposed conductors near or under a vehicle.

Past research efforts have concentrated on evaluating the structural breakaway characteristics of luminaire poles. In addition to the need for the pole itself to have breakaway ability, it is recognized that the underground wiring system should also be capable of properly separating. There are a number of reasons for requiring proper separation of the wiring system. One of these reasons is that the size, and associated tensile strength, of the wire cable is sufficient to significantly increase the deceleration rate of impacting vehicles and to also change the trajectory of the falling pole. Another reason is that improper separation of the electrical cabling can result in bare conductors that are still energized, posing an electrical and a possible fire hazard at the accident scene.

Early efforts to reduce electrical hazard concentrated on providing line fuses placed in a breakaway device. However, these widely used "breakaway fuse holders," which for years have been the standard, have not been certified by testing. Prior experience indicates that they frequently perform improperly during an accident situation. Rather than properly separating, the device frequently pulls off the wire, leaving an exposed end that is potentially deadly. Part of the problem with the breakaway fuse holder is the location of the device in the pole or T-base and the 610 to 910 mm of distribution cable inside the base. This extra length of wire is placed in the pole to allow service crews the ability to pull the wire out of the pole and make the connections to the luminaires. Upon impact this extra length of wire obstructs proper separation of the breakaway fuse, and allows the wiring insulation to be damaged by the fractured pole. The resulting bare electrical conductor poses a safety hazard because of the relatively large voltages used in underground roadway illumination systems.

Most luminaire underground wiring systems operate on 480 V. The reason for using 480 V is that the voltage drop in the copper conductors that supply a given load is only one-fourth the value of the voltage drop when using 120 V and one-half that of 240 V. In addition, luminaires are designed to perform within a certain percent of the rated voltage. Thus for a given percent, such as 10 percent, the allowable drop would be 4 times greater for a 480-V circuit than for a 120-V circuit (48 versus 12 V) or twice that of a 240-V circuit. These factors are additive, so a 480-V circuit requires a much smaller copper wire to deliver the necessary amount of energy over a long distance. Using 480 V is desirable, but proper precautions and installation techniques must be used to reduce the inherent hazard on the public right-of-way.

A modular cable system initially developed by MG^2 Inc. and Duraline Inc. eliminates a number of problems presented by the current wiring method [13]. This cable system is a submersible, modular plug and cable system that allows the circuit components (i.e., the low-amperage, fast-acting, current-limiting fuses; the surge arrester where desired; and the conductor splices) to be placed in an underground junction box adjacent to the pole foundation. The circuit breakaway connector can be positively positioned at the top edge of the conduit inside the pole base. Since the stiff, typically no. 4 or no. 6 copper, conducting cables never enter the pole, the system unplugs at ground level. The impact that knocks down the pole will not put stress on the electrical cables and will not weaken splices in adjacent poles. Most important, with the modular cable assembly, there is no exposed electrical hazard upon knockdown as can exist with the conventional wiring method. When this system (Fig. 7.83) is combined with a properly installed foundation, the possibility of fire and explosion or electrical shock is significantly reduced if not eliminated. Recent developments have shown that the splices, the surge arresters, the fuse holders, and the ground rod must be placed underground in a junction box adjacent to the pole base to provide the greatest possible degree of safety. This requires that all components be submersible. This design will positively place a breakaway connector in the wiring system at the top edge of the foundation; the fuses are underground, where no damage can occur on the supply side.

FIGURE 7.83   MG^2 Duraline modular pole cable system.

The Modular Cable System developed by MG^2/Duraline was the first of these submersible wiring systems on the market and has proven to be very reliable [13]. By using fast-acting, current-limiting fuses installed below ground, the potential for electrical shock and fuel explosions is greatly reduced if not eliminated.

7.26 FOUNDATIONS

The foundation for a luminaire pole must provide sufficient resistance to overturning moments caused by the static load of the mast arm plus a wind and/or an ice load. It must be capable of maintaining the correct alignment of the luminaire and able to withstand the impact should the pole be struck. For breakaway poles, the foundation must be rigid enough to allow the breakaway device to operate while not becoming a hazard itself.

Luminaire foundations are perhaps one of the most dangerous constructed hazards on the right-of-way. This is due to their placement or location, structural design, and unsafe wiring systems. Historically, pole foundations have been poured-in-place concrete with steel reinforcing rods and anchor bolts. The requirement that upon breakaway nothing shall project more than 100 mm above a chord line drawn between two points 1.5 m apart has caused redesign of concrete foundations. It has been recognized for several years that a problem exists when a foundation is placed on a slope. As early as 1985, a memorandum was issued stating that designers should not allow the slope between the travelway and the foundation to be greater than 6:1. This is even more effective when the diameter of the foundation is as small as possible, thus limiting the concrete protruding above the grade line. Eliminating the transformer base allows the foundation to be sized to accommodate the pole bolt circle, which in most

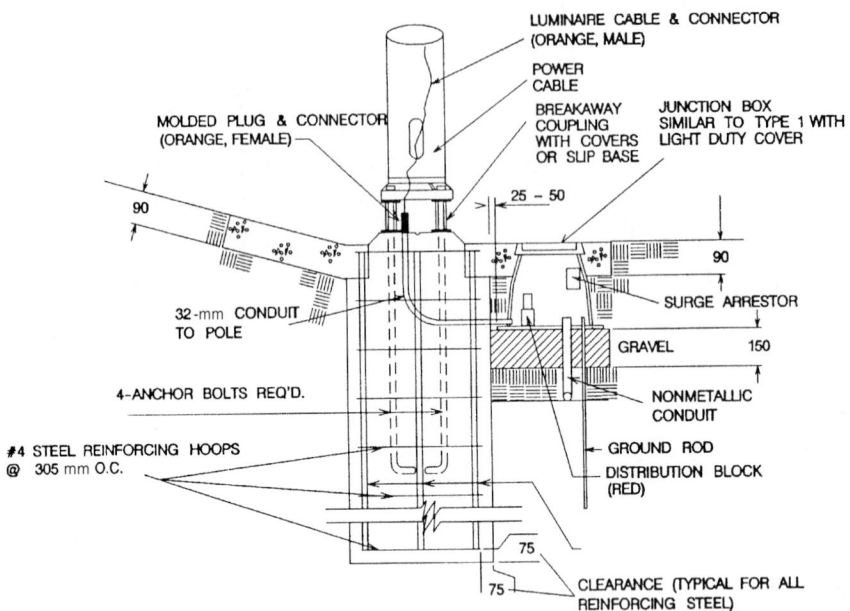

FIGURE 7.84 Small-base concrete luminaire foundation system with underground modular cable distribution system.

roadway size poles is considerably less than the bolt circle for a transformer base. In order to do this, the electrical circuit elements that were formerly housed in the T-base—i.e., splices in the conductors, fuse holders, and surge arresters—must be relocated underground. This requires that the electrical components be capable of being fully submerged and remain watertight. Figure 7.84 provides an example of a small-diameter concrete foundation and adjacent underground electrical junction box design.

Auger bases are an effective method of reducing diameter of the foundation. Many states use a galvanized steel auger base foundation instead of concrete. Most concrete foundations require 75 mm of concrete outside the anchor bolts to provide the necessary strength. Even if the T-base is eliminated, a concrete foundation is 150 mm larger in diameter than the pole base it serves. The flat steel top of the auger base foundation can be the same size as the pole base, which minimizes foundation size. Another advantage of the auger base foundation, with the circuit elements underground, is the resistance to damage when an accident breaks the pole. When a concrete foundation, with its anchor bolts poured in place, has one bolt damaged, the entire foundation should be replaced. The auger base foundation uses relatively short bolts, which are replaceable if damaged. Auger base foundations are easily installed by the electrical crew using the same auger trucks used to drill the hole for the concrete shaft. Electrical crews are not called upon to tie reinforcing steel, set and properly align anchor bolts, or finish the concrete—all tasks that require skill to perform properly. It has been reported that a two-worker crew can install 8 to 10 auger base foundations per day, resulting in significant labor savings. A diagram of the auger base foundation with underground electrical junction box is provided in Fig. 7.85.

Pole foundations cannot always be installed on a 6 to 1 slope, as described in many publications by the FHWA as being necessary to ensure the proper operation of a pole

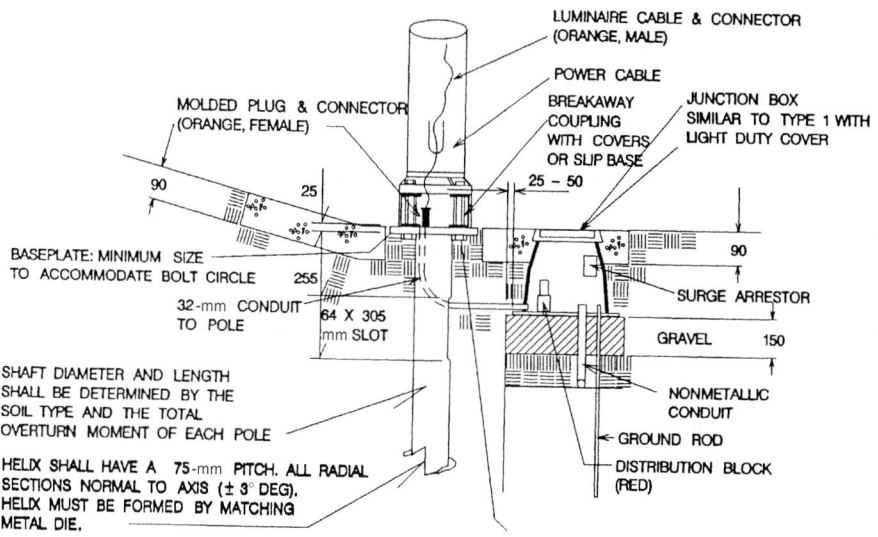

FIGURE 7.85 Auger base luminaire foundation system with underground modular cable distribution system.

breakaway device. To meet the requirement that no part of the foundation or remaining stub of a breakaway device extend no more than 100 mm above a theoretical line connecting possible tire tracks 1525 mm apart, it is desirable to reduce the diameter of the foundation. If a concrete foundation is 760 mm in diameter, then on a 6 to 1 slope, which is the very best condition expected, there will be 125 mm of concrete protruding above the ground plus the anchor bolts and the remaining portion of the breakaway device after breaking. Figure 7.86 shows a conventional transformer base installed on a concrete foundation. A 430-mm bolt circle is normal for the current transformer base design and 75 mm of concrete is needed outside the bolt circle for strength, resulting in a foundation diameter of 580 mm.

The foundation design shown in Fig. 7.87 incorporates the flush-mounted junction box made possible by the submersible wiring system. This allows the bolt circle to be reduced to that required by the pole base plate, usually in the range of 280 to 330 mm. The smallest foundation that can be provided is a steel plate that is the size of the pole base. This is attached to a steel shaft that extends into the soil approximately 1.5 to 2.5 m deep. This foundation design including the submersible wiring is shown on a 3 to 1 slope in Fig. 7.88. These foundations are available from several sources including Dixie Division of Aluma-Form [14] and the A. B. Chance Company [15]. Such a foundation is combined with the flush grade mounted junction box and the submersible Modular Cable System and either a frangible coupling or a slip base pole to achieve an effective breakaway lighting pole installation. Lighting designers cannot always influence the design of a roadway's shoulders and front slopes, but by using this foundation, designers do all within their power to ensure requirements of the AASHTO *Roadside Design Guide* are satisfied.

The Modular Cable System mentioned is recommended for use in both breakaway and nonbreakaway pole bases. Because there is only one splice to be made at each pole, with the other connections made by plug-in connectors, the time required to install and the skill level needed of the installer are minimal. Also, when troubleshooting a defec-

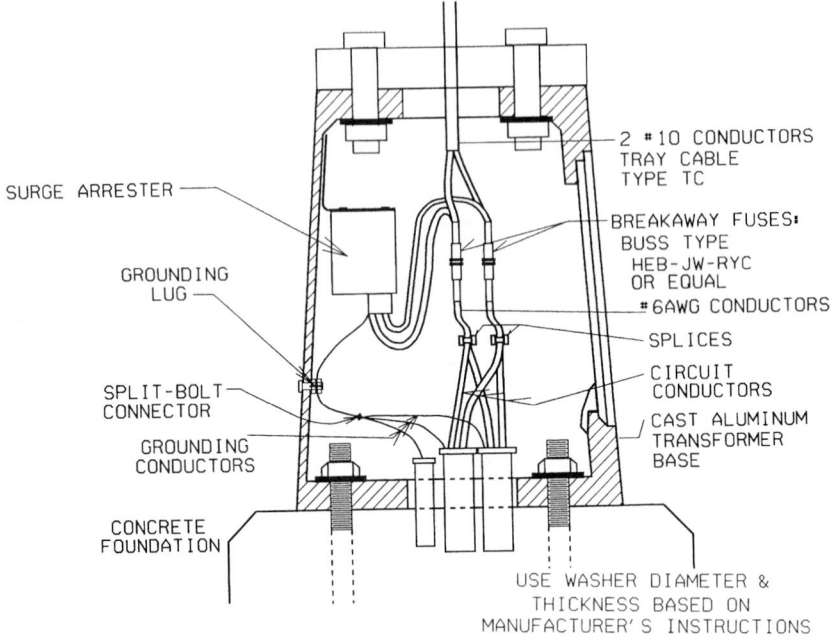

FIGURE 7.86   Details of conventional transformer base installed on concrete foundation.

tive circuit, it is a great advantage to be able to unplug the various circuit elements rather than deal with permanent splices.

7.27 BASES

Breakaway luminaire poles are designed to yield at their base attachment to the foundation. There are numerous types of bases currently in service. Some of these are designed for breakaway operation and others are not designed to yield. The nonyielding types have application where vehicle speeds are low and the danger from a falling pole is greater that the hazard of hitting the rigidly mounted pole. A description of the most common base types follows. Not all of these bases are crashworthy.

Direct Burial Base.   The direct burial base allows the pole to be directly embedded in the soil. It is the most economical, since it eliminates the need for a foundation. It is the common type of base for wood and is used frequently with concrete and fiberglass reinforced plastic (FRP) poles. FRP poles are the only direct burial poles currently approved for breakaway use. The other types are normally limited to low-speed facilities or should be located out of the recoverable area.

Flange Base.   Most steel and aluminum poles are fitted with a plate or flange at the base of the pole. With steel poles, this usually involves welding a steel plate to the bottom of the pole. With aluminum poles, a cast-aluminum shoe base is usually fitted

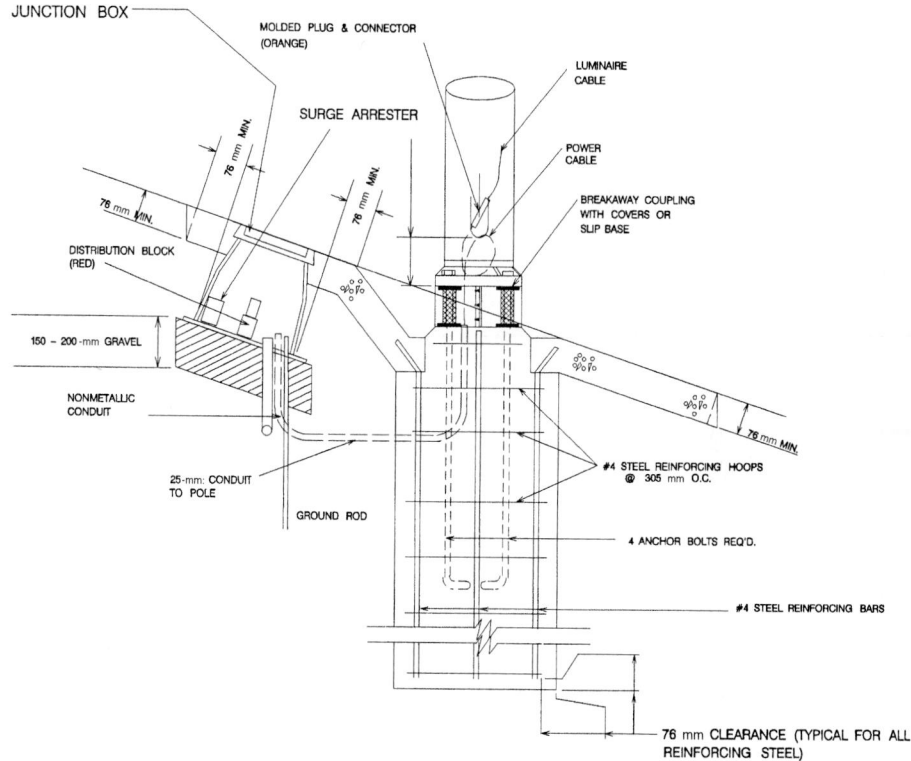

FIGURE 7.87 Flush-mounted electrical junction box.

to the bottom of the pole. The use of a flange base implies that the flange is to be fastened directly to the anchor bolts embedded in the foundation or to some type of breakaway device. When a flange is in direct contact with the concrete, some method needs to be employed that will allow water to flow out and not be trapped in the base of the pole. Trapped water can cause premature failure of the pole due to corrosion on the inside. Flange base designs without breakaway features are not crashworthy and should be restricted to where the hazard from a falling pole is greater than the hazard of impacting the rigidly mounted pole. A flange base is illustrated in Fig. 7.89.

Cast-Aluminum Transformer Base (T-base).   T-bases may be steel or cast aluminum and were originally devised to house the transformer. The T-base (Fig. 7.86) proved unacceptable for storage of the ballast because of moisture and insect damage to the electrical components. However, the cast base proved to have safety advantages, since it yielded and broke apart upon impact. The ballast is rarely stored in the base anymore, but the T-base is still frequently installed because it serves as an electrical junction box and because of its breakaway characteristics.

Frangible Couplings.   A number of manufacturers have developed cast and extruded aluminum frangible couplings. The typical coupling (Fig. 7.90) is a short connector attached to the foundation on the bottom and the flange of the pole on the top. Upon impact, the coupling fractures, separating the pole from the foundation. The proper

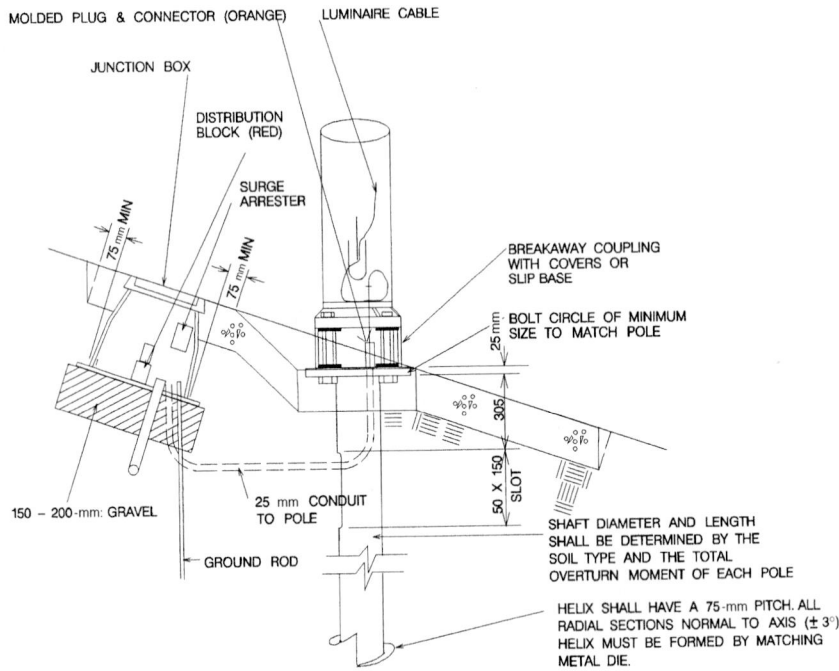

MOLDED PLUG & CONNECTOR (ORANGE)   LUMINAIRE CABLE

JUNCTION BOX

DISTRIBUTION
BLOCK (RED)

SURGE
ARRESTER

75 mm MIN

75 mm MIN

BREAKAWAY COUPLING
WITH COVERS OR
SLIP BASE

25 mm

BOLT CIRCLE OF MINIMUM
SIZE TO MATCH POLE

305

50 X 150
SLOT

150 – 200-mm: GRAVEL

25 mm CONDUIT
TO POLE

GROUND ROD

SHAFT DIAMETER AND LENGTH
SHALL BE DETERMINED BY THE
SOIL TYPE AND THE TOTAL
OVERTURN MOMENT OF EACH POLE

HELIX SHALL HAVE A 75-mm PITCH. ALL
RADIAL SECTIONS NORMAL TO AXIS (± 3°).
HELIX MUST BE FORMED BY MATCHING
METAL DIE.

FIGURE 7.88   Foundation design on 3 to 1 slope with submersible wiring.

FIGURE 7.89   Flange-type steel base.

FIGURE 7.90   Frangible-coupling luminaire support.

performance of frangible couplings requires proper matching of coupling and pole. Stiffer poles work best with frangible couplings, since the stiffness of the pole results in impact forces remaining in the direction of impact (shear). Flexible poles, such as aluminum poles, bend upon impact, resulting in translation of some of the impact force to vertical forces. This places the couplings in compression and tension, forces the couplings are specifically designed to resist. Frangible couplings often need to be enclosed in skirts to keep dirt and water from entering the conduit and to keep rodents from eating the wire insulation.

Slip Base.   Luminaire support slip bases are designed to resist wind and vibration loads while safely releasing upon impact from any direction. A typical base consists of two triangular plates, one welded to the support pole and the other welded to the foundation attachment. The plates accommodate three anchor bolts and are slotted to allow release upon impact. If installed correctly, the foundation part of the slip base will be reusable with minor repairs after impact of the support pole. The following criteria are necessary to ensure that the slip base operates correctly:

- Any bolts used to anchor the foundation piece to the foundation must be lower than the plane of the slip base.
- The upper surface of the foundation piece must be no more than 100 mm above the surface of the surrounding terrain.
- An 18 to 22 gauge keeper plate must be placed between the surfaces of the slip base to prevent the device from slipping apart in response to wind loads.
- Washers of sufficient strength to prevent deformations into the vee slots must be used between the plates and on the top and bottom.
- The bolts must be torqued to the specified level.

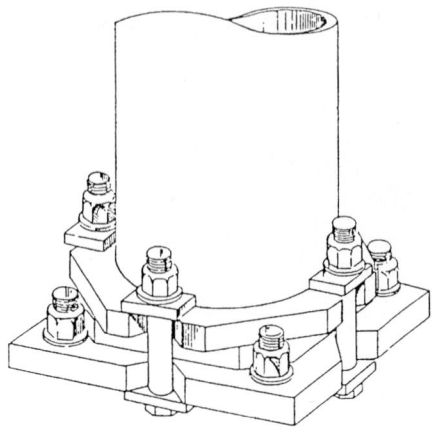

FIGURE 7.91   Four-bolt roadway lighting support slip base by Valmont Industries.

A four-bolt slip base (Fig. 7.91) is also available. Developed by Valmont Industries, it provides added structural resistance to environmental loads and is used extensively by some western states.

Shear Base.   The design concept for the shear base is to load the rivets or welds that secure the base to a foundation plate. When struck by a vehicle, the rivets or welds are sequentially sheared and the support breaks away. Typical designs for shear bases are thin-walled stainless steel bases and a family of cast-aluminum bases (not T-bases).

Other.   There are other breakaway methods that relate to specific materials used for the pole. These include a fiberglass-reinforced plastic pole with an anchor base that will break above a cast-aluminum base, and several schemes approved for use with aluminum poles.

7.28   *CONSTRUCTION CONSIDERATIONS*

7.28.1   Conduit on Bridges and Median Barriers

The current design standards for bridge rail and median barrier rail are very similar. Consequently, the method for installing conduit for one is applicable to the other. In the past, the most common method for installing bridge conduit was to attach it to the underside of the bridge deck. This required both the installer and the maintenance crews to work outside the bridge rail and underneath the slab by constructing special scaffolding or using expensive trucks with articulating booms. A method has recently been developed that places the conduit and junction boxes within the bridge rail. The conduit may be galvanized rigid, PVC, or high-density polyethylene (HDPE). HDPE is available on large reels in lengths up to 1500 m. The junction box is a curb box (Fig. 7.92) and may be a standard galvanized cast iron, or a box with reinforced fiber-glass sides and a polymer concrete ring and cover as manufactured by CDR [16]. The CDR box has the advantage of not requiring the cover to be grounded for safety rea-

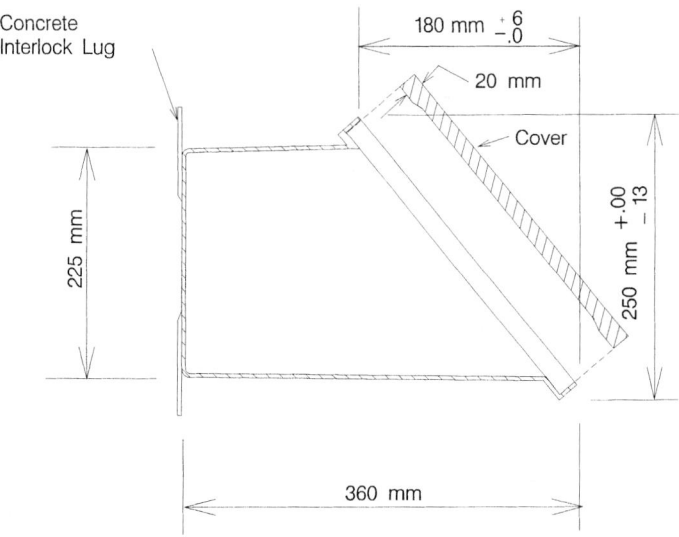

FIGURE 7.92   Curb box and cover section used for electrical junctions within bridge rail.

sons. Details of this wiring method with junction box and luminaire foundation are shown in Fig. 7.93.

This same method is also used for installing conduits for median-mounted lighting systems. In some areas, whenever a median of a divided highway is closed and a median barrier installed, an empty HDPE conduit is installed inside the barrier for future use. The advantage of using the HDPE in this case is that there are no joints required other than at bridge ends. No special skill is required to install the HDPE, so the concrete barrier construction is not delayed. When the conduit is for a future lighting system, no foundations are installed at the time the rail is constructed. A section of the barrier can be removed at a later time when the lighting system is installed to allow the conduit ends to be connected to the junction box placed at each luminaire location. Details of this foundation are shown in Fig. 7.94. Luminaire poles can be easily installed on existing barriers without the encased conduit by using a modification that places the conduit in the roadway shoulder below grade and connected to the barrier rail–mounted junction box.

7.28.2   High Mast Power Connection

The conventional method of supplying electric energy to the high mast ring, with the luminaires at the top of the pole, uses a normal twist-lock plug and cap set. Some of these have a neoprene cover to provide some degree of waterproofing, but oftentimes this cover is either poorly installed or missing entirely. The result has been premature failure of the plug and cap set due to moisture. In some cases, the connectors will short-circuit between phases, and in others a large hole is burned through the connector body where the electricity arcs to a grounded structure. This latter case is particularly dangerous to the maintenance technician who attempts to reset a tripped circuit breaker while standing directly in front of the power cord connector. A method to

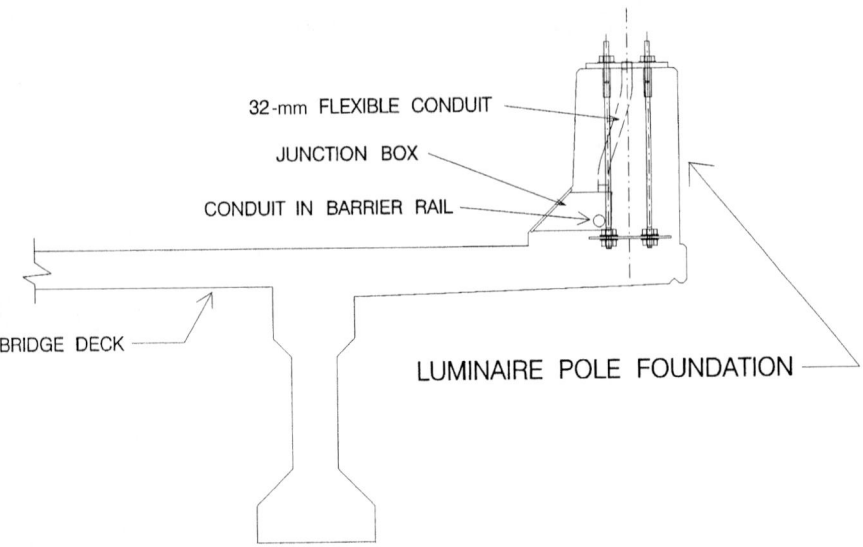

32-mm FLEXIBLE CONDUIT

JUNCTION BOX

CONDUIT IN BARRIER RAIL

BRIDGE DECK

LUMINAIRE POLE FOUNDATION

FIGURE 7.93   Details of wiring system with conduit in bridge barrier rail.

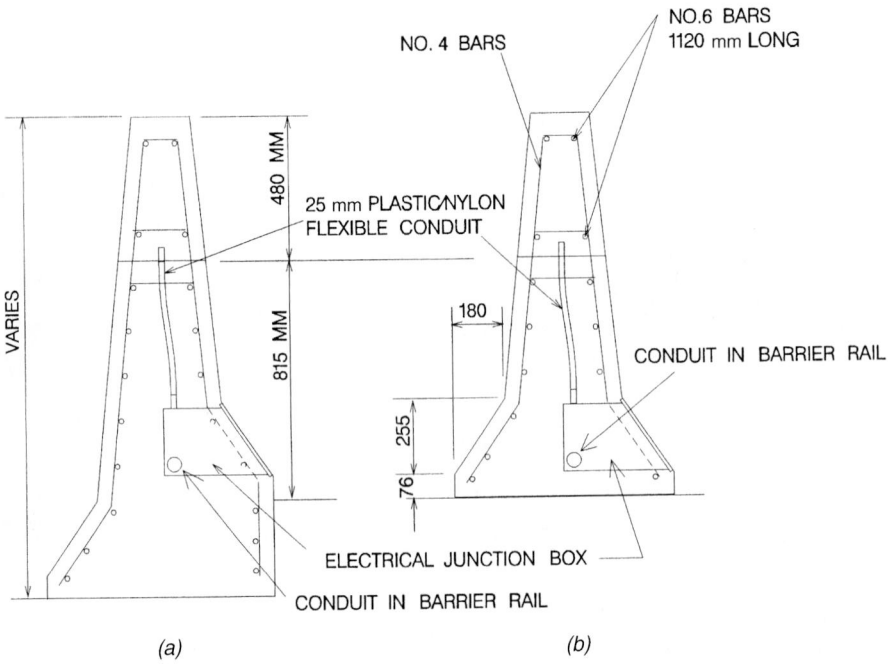

NO.6 BARS
1120 mm LONG

NO. 4 BARS

480 MM

25 mm PLASTIC/NYLON
FLEXIBLE CONDUIT

180

CONDUIT IN BARRIER RAIL

VARIES

815 MM

255

76

ELECTRICAL JUNCTION BOX

CONDUIT IN BARRIER RAIL

(a)   *(b)*

FIGURE 7.94   Details for luminaire foundation installed in concrete median barrier. (*a*) With super-elevation. (*b*) Without superelevation.

avoid this has been developed by MG²/Duraline by utilizing a submersible cable connector that is impervious to moisture [13].

7.28.3 Foundation Installation

A sometimes overlooked hazard on the roadside is a poorly installed foundation. Any foundation, regardless of the design, can be a hazard unless care is taken to ensure that neither any portion of the foundation nor the predicted breakline of the breakaway device is above the 100-mm line. Strict adherence to details, such as discussed in Art. 7.26, will produce a safe foundation even on a steep slope.

7.29 ACCEPTANCE TESTS

Before any lighting system is accepted as complete, or preferably before the electricity is turned on, several tests should be conducted to ensure the quality of the components:

Insulation tests.   The contractor should measure the conductor insulation resistance to ground of each lighting circuit using a 500-V megohm-range type instrument. A record should be made of each phase conductor's resistance to ground. The circuits should measure a minimum of 250,000 Ω resistance to ground before the power is turned on. The test should be arranged to test splices and all components of the circuit.

Ground resistance test.   Using an instrument designed for the purpose, the contractor should measure the resistance of each ground rod. A written record of the value should be signed and given to the inspector. Any ground rod with a resistance of 25 Ω or less is acceptable. Additional ground rods, up to a maximum of three at each location, should be installed to reach the 25 Ω.

High mast lowering test.   Each high mast lighting assembly should be tested by completely raising and lowering the luminaire ring once. Further testing of the latching operation for top latch devices is necessary. Each luminaire ring should be unlatched, lowered a minimum of 2 m, then raised and relatched a total of five times to demonstrate its acceptability.

Photocontroller test.   The control circuit of the lighting system should be demonstrated to show it operates properly in both manual and automatic modes. The lighting level at which the photocontroller actuates the system should be observed. Adjustment of the direction of the photocontroller should be made as necessary.

Voltage tests.   The supply voltage at the lighting control center should be measured and recorded. With the luminaires energized and at full brightness, the voltage at the last luminaire of the circuit should be measured to ensure no more than a 10 percent voltage drop is present.

7.30 MAINTENANCE CONSIDERATIONS

Maintenance must be considered from the earliest design stages of a lighting project. Top-quality materials should be specified and then arranged or located to protect the components from the potential hazards of the environment, whether these be rain, moisture, ultraviolet degradation, or threat of vehicular impact. After a system is installed and tested for operation and for component integrity, proper maintenance procedures can produce continued high performance of the roadway lighting system. If the lighting system is not properly maintained, the responsible authorities may expose themselves to potential liability—plus increased costs if expendable items are not replaced as they reach the end of their service life, because they can cause other components to fail.

7.30.1 Maintenance Operations

There are many reasons to routinely maintain a lighting system. The first reason is that only through good maintenance can the system continue to perform as designed. No matter how much knowledge and skill goes into the design, and how much care is put into the installation inspections and final system testing, the system will not provide the performance expected of it if regular maintenance is not performed. In addition to the legal liabilities of a substandard lighting system, the condition of the system reflects the civic concern of the responsible agency. A lighting system containing faults such as burned out lamps, dirty luminaires, or knocked-down poles reflects a poor attitude that is very noticeable. Another factor is that the electrical energy costs are more or less constant even though the light on the roadways may be significantly reduced, so the economic efficiency is decreased.

7.30.2 Routine versus Demand Maintenance

Maintenance activities usually fall into one of two categories: demand or routine. Demand-responsive maintenance is response to random occurrences such as luminaire failures—i.e., lamps, fuses, ignitors, ballasts—or pole knockdowns. Routine maintenance is scheduled activities such as group lamp replacement or luminaire cleaning that are intended to produce a certain level of performance of the lighting system and eliminate some of the demand maintenance [17].

7.30.3 Maintenance Guidelines

A comprehensive discussion of roadway lighting maintenance is presented in "Design Guide for Roadway Lighting Maintenance," IESNA DG-4-1993. In addition to the factors affecting maintenance, this guide includes information for establishing a maintenance management system that will be helpful to agencies attempting to upgrade maintenance activities.

7.31 IMPACT PERFORMANCE CRITERIA

The following criteria are necessary to ensure satisfactory impact performance of luminaire supports.

- Use only designs that have been approved as crashworthy by the FHWA.
- The FHWA has established upper limits on the support mass and height of luminaire supports. These limits are applicable even when the breakaway characteristics have proven acceptable by crash testing. The maximum acceptable support mass is 454

kg, and the maximum luminaire support height is 18.3 m. These values are increased from the limits of 272 kg and 15.2 m cited a few years ago. Any further increases in these limits should be based on full-scale crash testing and an investigation of vehicle characteristics beyond those recommended in NCHRP Report 350 [12, 18].

- Breakaway devices are designed to operate by being subjected to horizontal forces (device placed in shear). The devices are designed for this to occur when impacted at a typical bumper height of about 510 mm. Locating luminaire supports where they will be impacted at a different height will result in forces directed parallel to the support and thereby loading the devices in tension and compression. This results in improper operation of the breakaway device and possibly severe impacts and injuries to vehicle occupants. Superelevation, slope rounding, offset side slopes, curves, curbs, vehicle departure angle, and speed can all influence the striking height of a typical bumper. Negative side slopes should be limited to 1:6 between the roadway and the luminaire to help ensure that errant vehicles strike the support at an acceptable height [12].

- Use a wiring system that allows all circuit components to be shielded from impact, preferably underground, and that ensures that all electrical energy potentially available at the pole foundation surface is limited by the current-limiting fuses. Conductors protected only by a circuit breaker should be not be accessible in the pole base.

- The major cost of a luminaire assembly is the pole, foundation, and breakaway devices. Select luminaires for performance and for a design flexibility that allows more selection of pole locations to produce a lighting system with fewer potential hazards.

- As a general rule, a pole will fall in line with the path of an impacting vehicle. The mast arm usually rotates so that it is pointing away from the roadway when resting on the ground. Consideration, however, must be given to the fact that falling poles may endanger pedestrians and may pose a danger to other motorists.

- A maximum 100-mm stub height must be maintained to prevent vehicle snagging. Quick-disconnect electrical circuitry should also be used to facilitate the breakaway mechanism, to reduce the hazard of electrical shock from exposed wiring after impact, and to ease repairs.

- Foundations should be properly sized for surrounding soil conditions. Foundations that move through the soil upon impact place the breakaway mechanisms in bending rather than shear, resulting in improper actuation.

- Curbs, regardless of their shape or height, will elevate an impacting vehicle. The rise in height begins approximately 460 mm from the curb and can extend as far as 3050 mm. When possible, therefore, luminaire supports should be placed 3050 mm from the curb. If this is not possible, then they should be located closer than 610 mm from the curb. Luminaire poles placed between 610 and 3050 mm behind curbs increase the chances of improper breakaway operation.

- If a luminaire support is placed behind a barrier, it may not be necessary to provide a breakaway feature. In general, if the support is within the design deflection distance of the barrier, then either the barrier should be stiffened or a breakaway pole support should be used.

- Some agencies place luminaire assemblies on top of concrete median barriers. High-angle impacts, or impacts by large trucks or buses, can cause a luminaire mounted on top of a barrier to be struck. Breakaway design is not recommended for this type of installation because of the risk that a downed pole might pose to opposing traffic.

FIGURE 7.95   Failure of luminaire support without impact.

- If a luminaire support is to be placed on top of a concrete barrier, then the barrier must be adapted to fit the pole base. Concrete safety-shape types are typically designed with an approximately 150-mm-wide top surface. Since luminaire bases are typically 200 to 305 mm in width, it is necessary to either widen the barrier top to 305 mm, or flare the barrier in the area of the luminaire.

- Design alternatives should be investigated with the goal of reducing the number of luminaires used along a section of roadway. Higher mounting heights can significantly reduce the total number of supports needed. Tower or high mast lighting can be used to effectively illuminate major interchanges. This method reduces the number of poles and locates the supports much farther from the roadway.

- It should be noted that some agencies are experiencing problems with the failure of aluminum T-bases due to environmental loads. It is believed that this kind of failure, such as shown in Fig. 7.95, is initially caused by minor impacts from mowing units and other maintenance equipment. This causes small cracks at the bottom flange of the T-base, which grow under environmental loads. The result is eventual separation of the bottom flange from the T-base as in Fig. 7.96.

7.32   STRUCTURAL DESIGN

Detailed provisions for the structural design of supports for lighting, signs, and traffic signals are available from AASHTO [19].

FIGURE 7.96   Separation of bottom flange from T-base.

7.33 REFERENCES ON LIGHTING

1. Joint Committee of the Institute of Traffic Engineers and the Illuminating Engineering Society of North America, "Public Lighting Needs," *Illuminating Engineering,* vol. 61, no. 9 (September 1966), p. 585.

2. Box, Paul C., "Relationship between Illumination and Freeway Accidents," *Illuminating Engineering,* vol. 66, no. 5 (May–June 1971), p. 365.

3. Federal Highway Administration, *Roadway Lighting Handbook—Implementation Package 78-15,* U.S. Department of Transportation, Washington, D.C., 1978.

4. *Value of Public Roadway Lighting,* Illuminating Engineering Society of North America, New York, 1989.

5. Odle, H. A., *Light and Sight,* Holophane, Inc., 218 Oakwood Ave, Newark, Ohio.

6. "American National Standard Practice for Roadway Lighting," ANSI/IES RP-8-1983, Illuminating Engineering Society of North America, New York, 1983.

7. AASHTO, *An Informational Guide to Roadway Lighting,* American Association of State Highway and Transportation Officials, Washington, D.C., 1984.

8. IESNA Roadway Lighting Committee, "American National Standard Practice for Tunnel Lighting RP-22," Illuminating Engineering Society of North America, New York, 1995.

9. Walton, N. E., and N. J. Rowan, "Warrants for Highway Lighting," NCHRP Report 152, National Cooperative Highway Research Program, Transportation Research Board, Washington, D.C., 1974.

10. Ketvirtis, A. P., *High Masts Meet Modern Highway Illumination Needs,* Fenco Engineers, Inc., Toronto, Ont., Canada.

11. Ketvirtis, A. P., *High Mast Lighting Design: Concepts and Practice,* Fenco Engineers, Inc., Toronto, Ont. Canada.

12. *Roadside Design Guide,* American Association of State Highway and Transportation Officials, Washington, D.C., 1996.

13. Mg^2, 1718 28th Avenue South, Birmingham, AL 35209.

14. Dixie Division of Aluma-Form, Inc., P.O. Box 170040, Birmingham, AL 35317-0040.

15. A. B. Chance Company, 210 North Allen Street, Centralia, MO 65240.

16. CDR Division of Homac Manufacturing Company, P.O. Box 1118, Ormond Beach, FL, 32175.

17. "Roadway Lighting Maintenance," IESNA DG-4-1993, Illuminating Engineering Society of North America, New York, 1993.

18. Ross, Hayes, Jr., Dean L. Sicking, Richard A. Finimer, and Jarvis D. Mickie, "Recommended Procedures for the Safety Performance Evaluation of Highway Features," NCHRP Report 350, Transportation Research Board, Washington, D.C., 1993.

19. *Standard Specifications for Structural Supports for Highway Signs, Luminaires and Traffic Signals,* American Association of State Highway and Transportation Officials, Washington, D.C., 1985.

CHAPTER 8
RETAINING WALLS

A. J. Siccardi, P.E.

Formerly, Staff Bridge Engineer
Colorado Department of Transportation
Denver, Colorado

Retaining walls are an important element in highway construction. They are most frequently constructed in the highway environment to retain a mass of earth. They are also used to enable the highway designer to establish grade lines for roadways at differing elevations when such roadways are in close proximity to one another and are to be constructed within limited rights-of-way, as is generally the case in densely populated urban locations.

Until 1972, when the first Reinforced Earth wall in the United States was built in California, retaining walls utilized in highway construction were usually plain gravity or reinforced concrete walls. There are now more than 5000 Reinforced Earth–type wall systems in place. Because early walls included metal strap reinforcement as the primary mechanism for stabilizing the soil, corrosion of the reinforcement and lack of long-term durability were a major impediment to immediate acceptance. Currently, the utilization of metal reinforcing requires the addition of sacrificial galvanizing materials selected to ensure the design life of the structure.

More recent earth reinforcement systems utilize geosynthetic materials, which are deemed inert to attack by deicing salts used on the highways. Salts are a primary inducer of corrosion in metal reinforcement. Long-term creep characteristics of geosynthetic reinforcements, however, result in their cautious use by designers. There are also increasingly more specialty-type walls, such as the soil nail type for both temporary and permanent wall locations, especially for slope stabilization where slope materials are appropriate for nailing. Each of these wall types is discussed briefly in this chapter.

The material in this chapter is drawn from many sources, including personal experience, but primarily from the following sources: (1) Section 5, "Retaining Walls," American Association of State Highway and Transportation Officials (AASHTO), *Standard Specifications for Highway Bridges,* 15th ed., 1992; and (2) Subsection 5 of the *Colorado Bridge Design Manual,* "Earth Retaining Wall Design Requirements," Colorado Department of Transportation. The latter document was prepared under the author's general supervision by Dr. T. C. Wang, P.E. A list of references is given at the end of the chapter.

8.1 EARTH RETAINING WALL CLASSIFICATION

A classification system is an essential part of the description and selection of different earth retaining wall types. Figure 8.1 indicates the many types of walls that are possible.

Earth pressure walls can be classified logically into three categories according to their basic mechanisms of retention, or into three categories based on their source of support. The retention mechanisms include internally stabilized, externally stabilized, and hybrid systems. The sources of support are described as gravity, semigravity, and nongravity.

An externally stabilized system uses a physical structure to hold the retained soil. The stabilizing forces of this system are mobilized either through the weight of a morphostable structure or through the restraints provided by the embedment of the wall into the soil, if needed, plus the tieback forces of anchorages.

An internally stabilized system involves reinforced soils to retain fills and sustain loads, adding reinforcement either to the selected fills as earth walls or to the retained earth directly to form a more coherent stable slope. These reinforcements can either be layered reinforcements installed during the bottom-to-top construction of selected backfill material, or be driven piles or drilled caissons built into the retained soil. All this reinforcement must be oriented properly and must extend beyond the potential failure plane of the earth mass.

A hybrid or mixed system is one that combines elements of both externally and internally stabilized systems.

Regarding sources of support, gravity walls derive their capacity through the dead weight of the wall itself or through an integrated mass that can be either externally or internally stabilized. They can further be classified into four types. The first is an internally stabilized system: earth walls with either facing covered cuts in situ doweled with uniformly spaced top-to-bottom constructed nails or selected fills reinforced with tensile reinforcements, which can be either metal (inextensible) reinforcements or geotextile (extensible) reinforcements. The second type is an externally stabilized system, either modular precast concrete walls or prefabricated metal bin walls. Third is an externally stabilized system—generic walls such as masonry, stone, dumped-rock, and gabion walls. The fourth type is an externally stabilized cast-in-place mass concrete wall or low-cost cement-treated soil wall system with anchored precast concrete facings.

Semigravity walls derive their capacity through the combination of dead weight and structural resistance. Semigravity walls designed with different shapes can be further classified into two groups: first is the conventional cast-in-place cantilever concrete wall, and second is a prefabricated system wall with cast-in-place base and many kinds of innovative precast or posttensioned stems. Semigravity walls are, in general, externally stabilized systems. They can be constructed either on spread footings or on deep foundations, such as caissons or piles, as foundation conditions may demand.

Nongravity walls derive their capacity through lateral resistance, either by embedment of vertical wall elements into firm ground, by anchorages provided by tiebacks, by dowel actions provided by piles, or by caissons drilled into a stabilized zone. They can be classified into, first, an externally stabilized system with embedded cantilever wall elements, sheet piles, drilled shafts, or slurries; second, similar embedded walls utilizing multiple anchorage tieback systems; and third, internally stabilized systems such as creeping slopes externally covered with multianchored facings and internally doweled with pile or caisson inclusions.

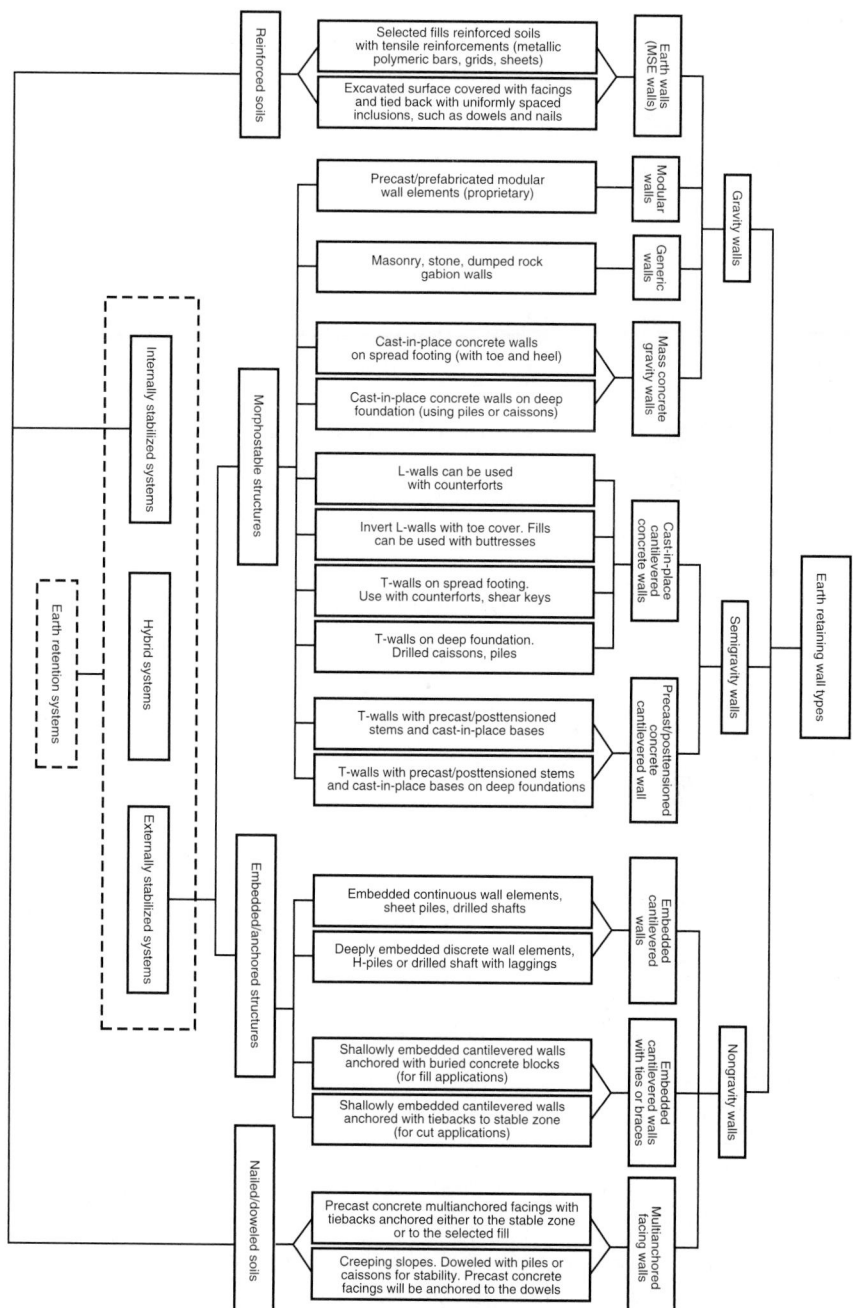

FIGURE 8.1 Types of earth retaining walls. (*From* Bridge Design Manual, *Section 5, Colorado Department of Transportation, Denver, Colo., with permission*)

Wall selection is an iterative process that involves cycles of preliminary design and cost estimation. The first and most important step is to define the design problem with design objectives and constraints. The objective of almost all design problems is least cost, although there will be many cases, particularly in urban areas, where objectives will include aesthetic and environmental considerations as well. Costs such as those for materials and construction are much easier to quantify than are aesthetic and environmental costs. In the latter instances, it is sometimes difficult to verify which one of the feasible solutions is the best. In order to find solutions that are at least feasible, constraints such as serviceability requirements (wall horizontal movement, vertical differential settlement, etc.) and spatial limitations (rights-of-way, underground easements, etc.) should be defined as comprehensively as possible. Designs (wall types) that meet the prescribed constraints are all *feasible* solutions. A ranking of these feasible solutions (wall types) is required. The ranking should include spatial behavior, and economic factors as discussed later in this article. Ideally, the wall with the highest rank should be adopted for detailed design; the rest can be used as design alternatives or discarded if the selected wall is confidently lowest cost, or is the only wall that satisfies all the established design requirements.

At the beginning of the selection process, rough sketches labeled with wall types should be adequate to screen out unfeasible types. As the selection process proceeds, a conceptual design with preliminary dimensions should be generated. Factors affecting the selection of an earth retaining structure may be grouped into three categories: spatial constraints; behavioral constraints; and environmental, aesthetic, and economic considerations. Factors to be considered for each of these categories are listed below.

1. Spatial constraints
   A. Functions of wall
      (1) Provide room for roadway at front of wall
      (2) Retain roadway at back or top of wall
      (3) Provide for grade separation, landscaping, or noise control
      (4) Provide for ramp or underpass wall support
      (5) Provide for temporary shoring of an excavation
      (6) Ensure stability of steep side slope
      (7) Flood control
      (8) Serve as bridge abutment
      (9) Other
   B. Space limitations and site accessibility
      (1) Right-of-way boundaries
      (2) Geological boundaries
      (3) Access of material and equipment
      (4) Temporary storage of material and equipment
      (5) Maintaining existing traffic lanes or widening
      (6) Temporary and permanent easement
      (7) Other
   C. Proposed finished profile (using combinations of different wall types along the wall alignment may be the optimal solution)
      (1) Limit of radius of wall horizontal alignment
      (2) Cut or fill with respect to original slope
      (3) Minimal site disturbance:
         (a) Anchored wall with minimal cut
         (b) Stepped-back wall on terrace profile
         (c) Superimposed or stacked low walls
         (d) Mechanically stabilized earth (MSE) wall with truncated base or trapezoidal reinforced zone

    D. Check available space versus required dimensions
       (1) Working space in front of wall (shoring, formwork, etc.)
       (2) Wall base dimension
       (3) Wall embedment depth
       (4) Excavation behind wall
       (5) Underground easement
       (6) Wall front face battering
       (7) Superimposed walls or trapezoidal profile of wall back
2. Behavioral constraints
    A. Earth pressure estimation (magnitude and location)
       (1) The magnitude of the earth pressure exerted on a wall is dependent on the amount of movement that the wall undergoes.
       (2) The vertical component of earth pressure is a function of the coefficient of friction and/or relative displacement (settling) between wall (stem, facing, and reinforced earth mass) and retained fill.
       (3) Compaction of confined soil may result in developing of earth pressure greater than active or at-rest condition.
       (4) For complex or compound walls such as bridge abutments, battered-faced walls, superimposed walls, and walls with trapezoidal backs, a global limit equilibrium analysis is required.
       (5) For embedded cantilever walls, profiles of lateral pressures acting on both sides of a wall are affected by the location of the center of wall rotation (pivot point), which is construction-dependent.
       (6) For multianchored embedded cantilever walls using a minimum penetration depth where there is no static pivot point, the soil pressure profile is anchorage design–dependent and should be developed with the recognition of beam-on-elastic foundation principles.
       (7) At the ultimate limit state, the location of the horizontal earth pressure resultant moves up from 0.33 to 0.40 of the wall height.
    B. Groundwater table
       (1) Reduce hydrostatic pressure if possible by an appropriate drainage system.
       (2) Introduce special precautions to reduce corrosion.
       (3) Prevent soil saturation; an appropriate groundwater drainage system is required except when the water table level must be maintained to prevent settlement of adjacent structures.
    C. Foundation pressure estimation
       (1) Uniform average pressure by Meyerhof effective width method for mechanically stabilized earth wall systems
       (2) Maximum toe pressure by flexural formula method for unreinforced or reinforced concrete–type walls
    D. Allowable bearing capacity estimation
       (1) Allowable bearing capacity is limited by and related to preset settlement or differential settlement criteria.
       (2) Earth walls integrated with wider flexible bases are allowed higher bearing capacity and tolerate more settlement than rigid walls on spread footings.
    E. Allowable differential settlement
       (1) Settlement is a time-dependent behavior.
       (2) Top-of-wall settlement is a sum of settlement from wall and from subsoil strata.
       (3) Allowable settlement should be evaluated by considering tolerable movement of superstructure and wall precast facings.
       (4) Simple-span bridges tolerate more angular distortion between adjacent footings than continuous-span bridges.

    (5) Tolerable (vertical and horizontal) movement of a wall facing is a function of panel joint width and pattern of connection.
- **F.** Earth pressure on wall facing
  - **(1)** The rigidity and slope of a wall facing affect the development of lateral pressure and displacement at facing.
  - **(2)** The earth pressure is reduced with a decrease in facing stiffness, while the facing deformation is only slightly increased for a decrease in stiffness.
- **G.** Settlement and bearing capacity improvement techniques
  - **(1)** Surcharge (two-phase construction) to hasten anticipated settlement
  - **(2)** Drainage (wick drain) to hasten anticipated settlement in fine-grain silt and clay substructure materials
  - **(3)** Excavation and compaction of a portion of weak foundation material
  - **(4)** Addition of reinforcement to subsoil
  - **(5)** Use of lightweight fill material to minimize loads beyond existing precompression of foundation materials
- **H.** Methods of reducing settlement on reinforced mass
  - **(1)** Increasing compaction of fill material
  - **(2)** Using more reinforcements (length, area, and spacings of reinforcements)
  - **(3)** Cement treatment of fills
  - **(4)** Reducing clay content of fill
  - **(5)** Using high-density in situ micronails
- **I.** Earth pressure applied at facing
  - **(1)** *High:* facing with posttensioned anchors
  - **(2)** *Medium-high:* mechanically stabilized earth wall with full-height panels
  - **(3)** *Medium:* rigid concrete facing with inextensible reinforcements
  - **(4)** *Medium-low:* concrete panel facing with extensible reinforcements
  - **(5)** *Low:* concrete panel facing with nailed soil
- **J.** Wall base width
  - **(1)** Wall types, foundation types
  - **(2)** Allowable bearing capacity of spread footing
  - **(3)** No tension allowed at heel of spread footing
  - **(4)** Internal and external stability of wall
  - **(5)** Reinforcement length to control lateral movement of reinforced earth wall
  - **(6)** Hybrid walls to reduce wall base width
- **K.** Toe penetration depth of embedded cantilever wall
  - **(1)** Water cutoff consideration
  - **(2)** Heave in front of wall
  - **(3)** Bearing capacity
  - **(4)** Stability of passive toe kickout
  - **(5)** Slope of ground in front of wall
- **L.** Wall sensitivity to differential settlement
  - **(1)** *High:* cast-in-place concrete retaining walls
  - **(2)** *Medium:* earth walls with inextensible reinforcements, geogrid walls with facings, precast modular walls
  - **(3)** *Medium-low:* geofabric walls without facing
  - **(4)** *Low:* gabion walls, crib walls, embedded cantilever walls, multianchored cantilever walls
- **M.** Potential settlement of retained mass
  - **(1)** *High:* embedded cantilever walls
  - **(2)** *High-medium:* some concrete modular walls, geofabric walls
  - **(3)** *Medium:* cast-in-place concrete retaining wall, concrete modular walls, geogrid walls

        (4) *Medium-low:* earth walls with inextensible reinforcements

        (5) *Low:* multianchored embedded cantilever walls

  N. Relative construction time

        (1) *Long:* cast-in-place concrete walls

        (2) *Medium:* earth walls with reinforcements

        (3) *Short:* embedded cantilever walls, multianchored embedded cantilever walls, precast modular walls

  O. Wall design life

        (1) Structural integrity

        (2) Color and appearance

  P. Load-carrying capacity and settlement of deep foundation

        (1) Maximum frictional resistance along the pile shaft will be fully mobilized when the relative displacement between the soil and the pile is about $\frac{1}{4}$ in irrespective of pile size and length.

        (2) Maximum point resistance will not be mobilized until the pile tip has gone through a movement of 10 to 25 percent of the pile width (or diameter). The lower limit applies to driven piles, and the upper limit is for bored piles.

        (3) The ultimate load-carrying capacity is the sum of pile point and total frictional resistance.

        (4) Pile-to-cap compatibility should be considered, especially with battered piles and semirigid pile-cap connection.

        (5) For the estimation of group efficiency in vertical and horizontal displacement, calculation of pile group, pile diameter, spacing, soil type, and total number of piles should be considered.

  Q. Fill material properties

        (1) The lower the soil friction angle, the higher the internal earth pressure restrained by the wall.

        (2) The lower the soil friction angle, the lower the apparent friction coefficient for frictional reinforcing systems.

        (3) The higher the plasticity of the backfill, the greater the possibility of creep deformation, especially when the backfill is wet.

        (4) The greater the percentage of fines in the backfill, the poorer the drainage and more severe the potential problem from high water pressure.

        (5) The more fine-grained and plastic the fill, the more potential there is for corrosion of metallic reinforcement.

  R. Fill retention versus cut retention

        (1) Fill retention (bottom-to-top construction)

            (a) Earth walls (extensible and inextensible tensile reinforcements)

            (b) All semigravity walls

            (c) Modular walls, generic walls, and mass concrete walls

        (2) Cut retention (top-to-bottom construction)

            (a) Earth walls, soil nails

            (b) All nongravity walls

3. Environmental, aesthetic, and economic considerations

  A. Environmental constraints

        (1) Ecological impacts on wetlands

        (2) Effect of corrosive environment on structural durability

        (3) Water pollution, sediment, or contaminated material

        (4) Noise or vibration control policy

        (5) Stream encroachment

        (6) Fish and wildlife habitat or migration routes

    (7) Unstable slope
    (8) Other
  B. Aesthetic constraints
    (1) Urban versus rural
    (2) Design policy of scenic routes
    (3) Acoustic or aesthetic properties of wall facing
    (4) Antigraffiti wall facing
    (5) Avoiding valley effect of long or high wall
    (6) Other
  C. Economic considerations
    (1) Construction schedule
    (2) Availability of fill material
    (3) Supply of laborers
    (4) Heavy equipment requirements
    (5) Formwork, temporary shoring
    (6) Dewatering requirements
    (7) Available standard designs
    (8) Temporary versus permanent wall and future widening
    (9) Cost of drainage system
    (10) Design and installation of wall attachments
    (11) Negotiated bidding and design/build on nonstandard projects
    (12) Maintenance cost, readjustment, and remodeling
    (13) Uncertainty of site and wall loads
    (14) Complexity of project
    (15) Height differences in finished or base grades
    (16) Number of wall turning points
    (17) Scale of project
    (18) Length or height of wall—quality control of fill material
    (19) Posttensioning, grouting, trenching, slurry
    (20) Pile driving, caisson drilling
    (21) Precasting, transportation, and inspection
    (22) Quantity of excavation
    (23) Quantity of backfill material
    (24) Experience and equipment of local contractor
    (25) Proprietary product and quality assurance
    (26) Other

    The logical consequence of considering these factors is to reduce the number of feasible wall types. The first stage of the decision process eliminates obviously inappropriate walls through spatial and behavior constraints before considering economic factors. The behavior constraints involve the properties of the earth the wall must retain and the ground it rests on. A detailed geological investigation and soil property report is needed in the second stage of the decision process. At this stage, conceptual designs with dimensioned wall sections and subsoil strata are required. In the third stage, behavior constraints and economic constraints should be repeatedly or simultaneously considered.

    After identification of the feasible set of wall types (a subset of the available walls), work proceeds on the more refined or detailed preliminary designs. Then a rating of these feasible designs should be made.

    To consider the various factors during the selection process, use the worksheets shown in Figs. 8.2, 8.3, and 8.4, along with the properly defined design problems (objectives and constraints) and cost requirements (Fig. 8.5). These sheets form a part of the documentation in support of the final selection(s).

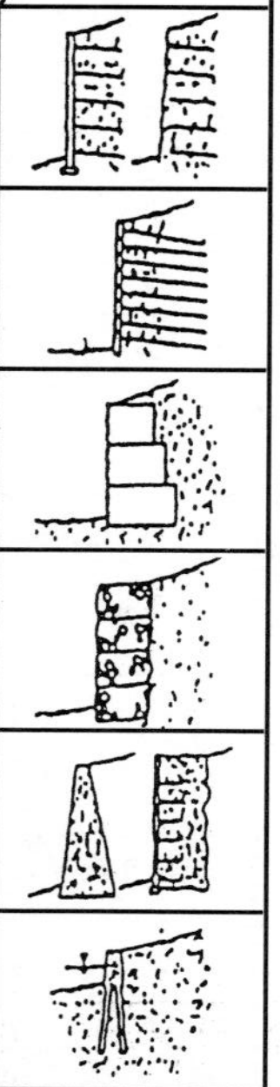

FIGURE 8.2 Gravity wall worksheet. (*From Bridge Design Manual, Section 5, Colorado Department of Transportation, Denver, Colo., with permission*)

Measurement indicators:

L Large	H High	Y Yes
M Medium	M Medium	? ?
S Small	L Low	N No

System names and descriptions

Wall type	System name	Description
Earth walls	Reinf. soil walls	Selected fills reinforced with tensile reinforcements [either metal (inextensible) or geotextile (extensible) bars, mats, grids, etc.]
Earth walls	Nailed soil walls	Facing covered cuts with uniformly spaced, top-to-bottom constructed nails.
	Modular walls	Precast/prefabricated modular walls. Most are proprietary products. Some patents have run out.
	Generic walls	Prefabricated wall elements such as masonry or concrete blocks. Rough elements such as dumped rocks, gabions.
Mass concrete walls		Cast-in-place solid concrete walls or precast concrete facings anchored in cement stabilized soil zones.
Mass concrete walls		C.I.P. reinforced concrete wall on deep foundation, either drilled caissons or piles.

Spatial factors:
- Fig. width (emb. depth) to height ratio
- Maximum economical wall height (ft)
- Excavation behind wall
- Working space behind wall
- Fronting face battering
- Trapezoidal wall back
- Sensitivity of marginal bearing capacity
- Sensitivity of differential settlement
- Lateral movement
- Marginal backfill material
- Unstable slope

Behavior factors:
- Scour and flood
- High water table/seepage
- Load-carrying structural facing
- Active earth pressure condition
- Construction-dependent loads
- Noise/water pollution loads
- Quantity of backfill material
- Fill compaction and control
- Construction time
- Project scale affects cost

Economic factors:
- Cost of maintenance
- Availability of standard design
- Labor usage
- Facing as an extra cost
- System durability problem

Typical illustrations

Measurement indicators:

L Large	H High	Y Yes
M Medium	M Medium	? ?
S Small	L Low	N No

System names and descriptions

Factor groups: Economic factors · Behavior factors · Spatial factors

Column headers (left to right):
System durability problem · Facing as an extra cost · Labor usage · Availability of standard design · Cost of maintenance · Project scale affects cost · Construction time · Construction and control · Fill compaction material · Quantity of backfill material · Noise/water pollution · Construction-dependent loads · Active earth pressure condition · Construction structural facing · Load-carrying structural facing · High water table/seepage · Scour and flood · Unstable slope · Unsuitable backfill material · Marginal differential settlement · Lateral movement · Sensitivity of marginal bearing capacity · Sensitivity of wall back · Trapezoidal wall back · Front face battering · Working space behind wall · Excavation behind wall · Maximum economical wall height (ft) · Fig. width (emb. depth) to height ratio · Typical illustrations

Cast-in-place

- L-walls can be used with counterforts.
- Invert L-walls can be used with buttresses. Use with toe cover fills.
- T-walls on spread footing. Use with counterforts and shear key if applicable.
- T-walls or L-walls on deep foundations. Use either drilled caissons or piles.

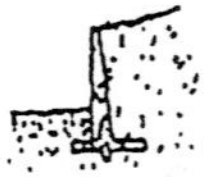

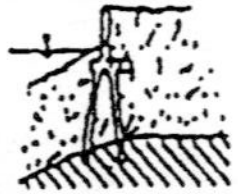

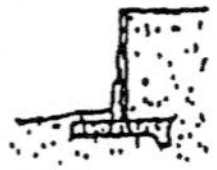

Precast posttension

- T-walls with precast posttensioned stem and C.I.P. base on spread footing.
- T-walls with precast posttensioned stem and C.I.P. base on deep foundation, either drilled caissons or piles.

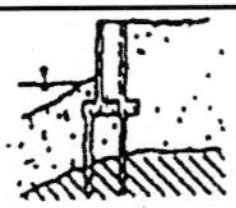

FIGURE 8.3 Semigravity wall worksheet. *(From Bridge Design Manual, Section 5, Colorado Department of Transportation, Denver, Colo., with permission)*

System names and descriptions	Spatial factors						Behavior factors										Economic factors										
Measurement indicators: L Large, H High, Y Yes M Medium, M Medium, ? S Small, L Low, N No	Ftg. width (emb. depth) to height ratio	Maximum economical wall height (ft)	Excavation behind wall	Working space of wall	Front face battering	Trapezoidal wall back	Sensitivity of marginal bearing capacity	Sensitivity of differential settlement	Lateral movement	Marginal backfill material	Unstable slope	Scour and flood	High water table/seepage	Load-carrying structural facing	Active earth pressure condition	Construction-dependent loads	Noise/water pollution	Quantity of backfill material	Fill compaction and control	Construction time	Project scale affects cost	Cost of maintenance	Availability of standard design	Labor usage	Facing as an extra cost	System durability problem	Typical illustrations
Generic walls anchored with geofabric grid reinforcements. Gabion walls anchored with geogrids.			L M S	L M S	Y ? N	Y ? N	H M L	H M L	L M S	Y ? N	Y ? N	Y ? N	Y ? N	Y ? N	Y ? N	Y ? N	Y ? N	L M S	H M L	L M S	Y ? N	H M L	Y ? N	H M L	Y ? N	Y ? N	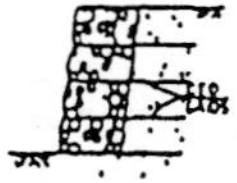
Modular precast L-walls anchored with geofabric grid reinforcements.																											
Geofabric wall(s) stacked on top of T-wall.																											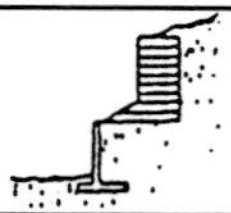
Either inverted L-wall stacked on MSE wall for bridge abutment applications, or L-wall with rail stacked on top of earth wall for roadway applications.																											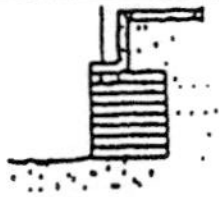
T-wall with anchors added to stabilized zone. Used for wall remodeling or rehabilitation, and for roadway-widening applications.																											
T-wall with precast/posttensioned modular stem elements. Anchored with geogrid or with reinforcements.																											

FIGURE 8.4 Hybrid wall worksheet. (*From* Bridge Design Manual, *Section 5, Colorado Department of Transportation, Denver, Colo., with permission*)

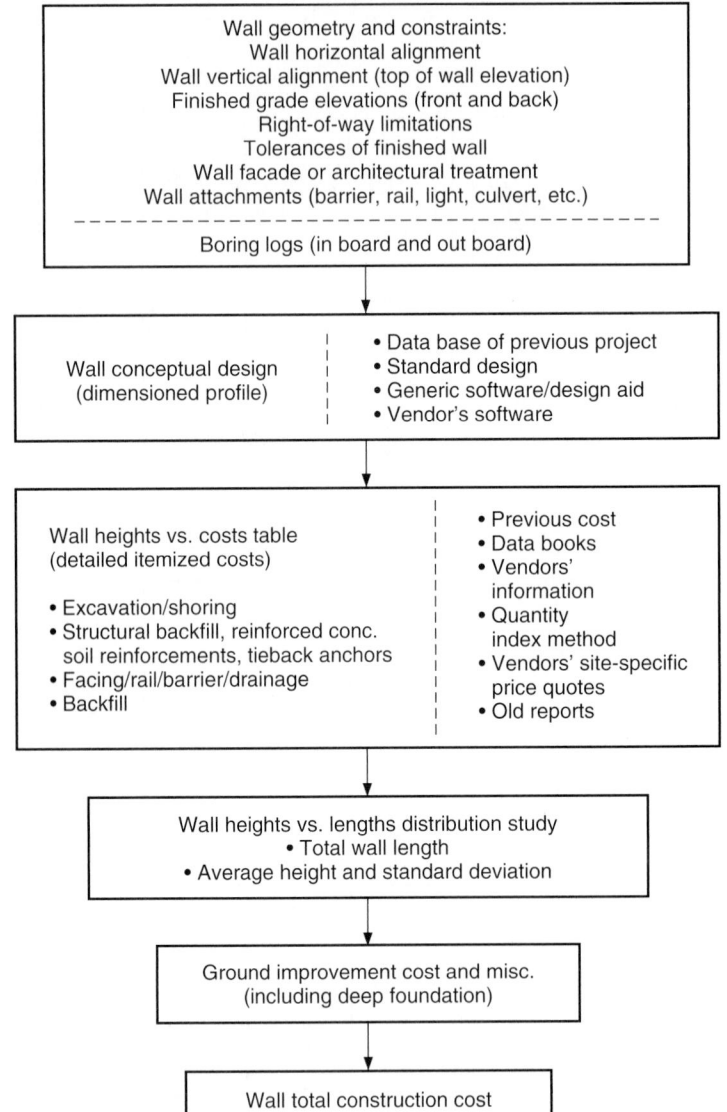

FIGURE 8.5 Requirements for wall cost study. (*From* Bridge Design Manual, *Section 5, Colorado Department of Transportation, Denver, Colo., with permission*)

After the worksheets are completed, a list of selected wall types with conceptual designs should be generated. A rating matrix can then be developed for a qualitative evaluation of the selected alternatives. On the basis of each evaluation factor, a qualitative rating between 1 and 5 can be given each alternative. The qualitative ratings are usually multiplied by weight factors reflecting the importance of the factors; usually, cost- and durability-related factors are given higher weights than the rest. The alternative(s) with the highest score is (are) then selected for final design and detailed cost estimation.

The intent of this procedure is to identify equally satisfactory alternative wall types. The plans or specifications will provide the opportunity for the contractor to select from the acceptable alternatives, should the designer make the decision to permit alternative walls. The specifications will outline the acceptable alternatives with dimensioned conceptual designs and indicate the requirements for the contractor to submit final site-specific details. These submitted (design/build) shop drawings should clearly establish that the design criteria are satisfied. They may include aesthetic features, bearing capacity and stability requirements, design computations for the alternative site-specific selection signed and sealed by a licensed professional engineer, and other data as may be necessary to document compliance with project needs.

8.1.1 Evaluation Factors

Evaluation factors that can be used on selected conceptual wall designs include the following:

- Constructibility
- Maintenance
- Schedule
- Aesthetics (appearance)
- Environment
- Durability or proven experience
- Available standard designs
- Cost

The sum of all weight factors should be 100 points. To simplify the selection process, minor factor(s) may be removed from the rating matrix. This is readily achieved by assigning the same score for minor factors on all the selected feasible wall types.

8.1.2 Notes on Using the Worksheets (Figs. 8.2, 8.3, and 8.4)

1. Factors that can be evaluated in percentage of wall height
   A. Base dimension of spread footing
   B. Embedded depth of wall element into firm ground
2. Factors that can be described as "large (high)," "medium (average)," or "small (low)"
   A. Quantitative measurement
      (1) Amount of excavation behind wall
      (2) Required working space during construction
      (3) Quantity of backfill material
      (4) Effort of compaction and control
      (5) Length of construction time
      (6) Cost of maintenance
      (7) Cost of increasing durability
      (8) Labor usage
      (9) Lateral movement of retained soil

    B. Sensitive measurement
       (1) Bearing capacity
       (2) Differential settlement
3. Factors that can be appraised with "yes," "no," or "question" (insufficient information)
    A. Front face battering
    B. Trapezoidal wall back
    C. Using marginal backfill material
    D. Unstable slope
    E. High water table or seepage
    F. Facing as load-carrying element
    G. Active (minimal) lateral earth pressure condition
    H. Construction-dependent loads
    I. Project scale
    J. Noise or water pollution
    K. Available standard designs
    L. Facing cost
    M. Durability
4. Factors that can be approximated from recorded height
    A. Maximum wall height
    B. Economical wall height

8.2 EARTH PRESSURE CONSIDERATIONS AND DETERMINATION

Once a proper selection has been made of feasible wall types that satisfy the necessary constraints, design consists of determining the earth pressure against the back of the wall and then proportioning the wall so that it will be structurally sufficient to satisfy a number of traditional checks. These checks include stability against sliding and overturning, and foundation bearing pressure limits. Clearly, satisfying the traditional checks would be of no value if the entire structure were to move because of some condition not related to any of these three checks. Therefore, it is also important that the designer be assured that the wall is globally stable—i.e., that no deep-seated slide or slip surface exists.

    An important and essential part of the design of retaining walls consists of determining the earth pressure on the back of the wall. The earliest theory of earth pressure traces back to Charles-Augustin de Coulomb, who published his work in 1773. Coulomb's theory presented a method by which a designer could determine the pressure that dry, granular, cohesionless material would exert upon the back of a wall constructed to restrain the material. His work was based on the theory that failure is characterized by a wedge-shaped mass of the supported sand material that slides down along a sloping plane such as is shown in Fig. 8.6.

    The Coulomb theory assumes a hydrostatic distribution of pressure such that the resultant forces R (reaction needed to hold wedge in equilibrium) and P (summation of normal pressure times area) act at the lower third point of the planes upon which they act, planes *ab* and *ac*, respectively. The force R acts at an angle of friction of soil on concrete, ordinarily 25°, while P acts at an angle of friction of soil on soil, generally assumed to be 34°. This latter angle will vary significantly from 34° to 40° or more. Because of the different angles of friction, the theory produces an error in the result; however, the error is generally accepted as negligible. In essence, if it is assumed that no friction exists between the earth and the wall, the pressure determined from the

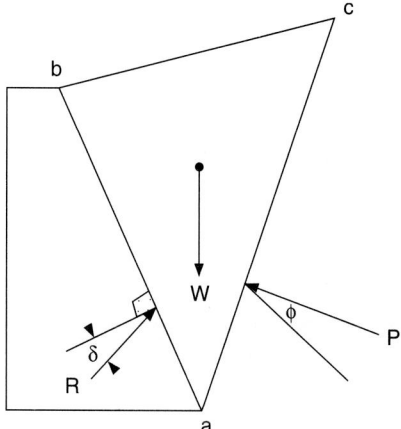

FIGURE 8.6 Forces acting on walls according to Coulomb's theory. W = weight of wedge *abc*; P = intensity of normal pressure; ϕ = angle of internal friction of backfill; δ = angle of friction of resultant force that accounts for difference in friction angle between backfill itself and backfill against concrete wall; R = pressure needed to hold wedge in equilibrium.

Coulomb theory is the same as that determined from the Rankine theory. Thus, because of its simplicity, the tendency is to use the Rankine theory. See Art. 8.2.3 for an example of active pressure calculation.

It is evident that the theory as expressed in Fig. 8.6 does not suggest a particular plane of failure. Thus, the pressure determination of the Coulomb theory is traditionally left to graphical methods, in particular those first developed by J.-V. Poncelet, and later by a German engineer, Culmann. These constructions, which allow for the complete determination of lateral pressure acting on the wall (i.e., magnitude, direction, and point of application), are not further discussed herein. However, several failure planes are usually assumed, pressure from each assumption is graphically determined, and an envelope line of pressure is developed from these pressure points from which the maximum pressure can be determined. The methods are laborious but straightforward and may again gain in popularity with the increasing use of computers.

At approximately the same time as Culmann's construction was developed, a Scottish physicist, W. J. M. Rankine, presented his theory in a work called *On the Stability of Loose Earth,* a theory that remains in active use. Rankine assumed a mass of loose earth of infinite extent, and a planar top surface subjected to its nonweight. The theory assumed granular backfill material without cohesion, but was adapted in 1915 by a British engineer to allow for cohesion.

8.2.1　States of Earth Pressure

Lateral earth pressure loadings are applied in various states—specifically, active, at-rest, and passive states. The state of pressure to be considered varies with the wall type.

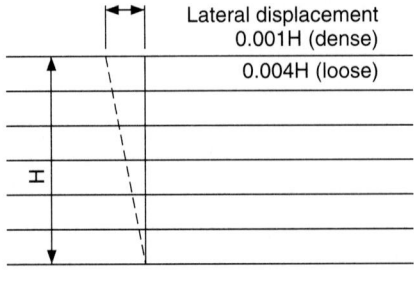

FIGURE 8.7   Lateral displacement to develop active state in cohesionless soils for yielding walls.

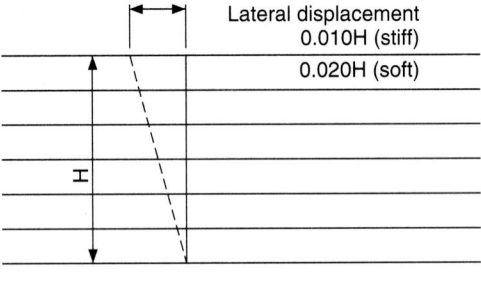

FIGURE 8.8   Lateral displacement to develop active state in cohesive soils for yielding walls.

Yielding Walls.   Yielding walls are free to translate or rotate about their top or base. For such walls, the lateral earth pressure may be computed assuming active conditions and wedge theory. In general, the lateral displacement at the top of a rigid wall of height H necessary to develop the active state varies from $0.001H$ in dense cohesionless soils to as much as $0.004H$ in loose cohesionless soils. For clay soils, a greater displacement on the order of $0.01H$ to $0.02H$, for stiff and soft soils, respectively, is necessary to develop an active state. See Figs. 8.7 and 8.8.

Thus, it is noted that the amount of displacement necessary to develop active pressure can vary, say, for a 20-ft-high wall, from less than ¼ in, in a dense cohesionless material, to as much as 5 in, in a soft clay. Clearly, the backfill material selected at any location plays a major role in the earth pressure for which a wall must be designed.

Restrained Walls.   Restrained walls are walls that are fixed or partially restrained against translation or rotation. Lateral earth pressures are computed assuming at-rest conditions using the following relationship:

$$P_0 = \frac{\gamma H^2 K_0}{2} \tag{8.1}$$

where P_0 = resultant of at-rest earth pressure, kips/ft
γ = unit weight of soil or rock, kips/ft^3
H = wall height, ft
K_0 = at-rest pressure coefficient

This latter condition may occur naturally at walls that are not totally freestanding—for example, at the junction of the wingwall at bridge abutments—or the condition may occur by design. Examples include locations where the lateral deflection cannot be tolerated because it retains a structure; or a heavily reinforced concrete counterfort wall, which is sensitive to settlement, located on material susceptible to settlement, especially differential settlement. In the latter case the designer must evaluate options that may include, depending upon the depth of the material that will settle, (1) removal and replacement, (2) deep foundations to adequate bearing material, or (3) selection of a different wall type, if conditions permit, that will be more tolerant to the potential for the differential settlement.

Should some force be present that tends to push the wall into the earth mass it is intended to retain—which therefore develops a resistance to slip on the failure plane or a resistance to the lateral displacement needed to mobilize the active pressure state—a condition known as *passive pressure* develops. The lateral earth pressure for which the wall must be designed increases significantly, as much as 10 times, and requires special attention. See Fig. 8.9 for a qualitative depiction of the relative lateral displacement.

Rigid Walls. For the case of rigid walls, which involves wall translation or rotation "into the backfill," the movement necessary to develop passive earth pressure behind the wall varies from $0.020H$ to $0.060H$ for cohesionless soils, dense to loose, respectively. Also, for stiff to soft cohesive soils, the lateral displacement will vary from $0.020H$ to $0.040H$. It is obvious that passive earth pressures can be developed in these defined conditions. Certainly, the best way for the designer to account for these pressures is to avoid them wherever possible and practical, alleviating the conditions under which such pressures develop. This brief discussion is intended only to generate an awareness in the designer that such conditions can be created. An example would

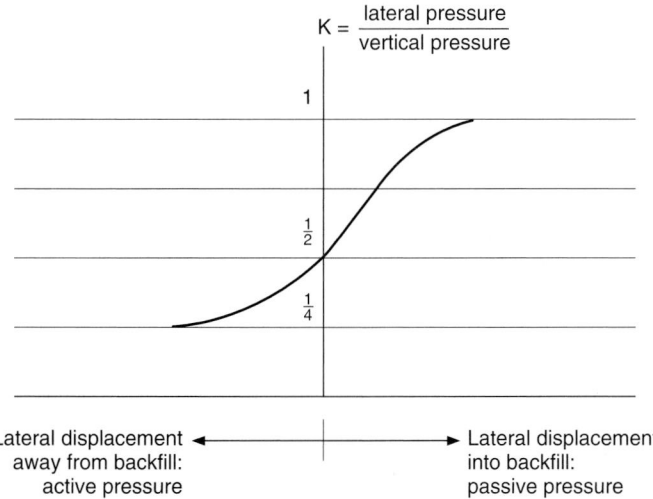

FIGURE 8.9   Active and passive pressures on restrained walls.

be dead-man type anchorages tying the wall top into solid materials or outside the failure plane of the wedge, thus preventing the movement necessary for development of the active pressure state.

8.2.2   Earth Pressure Calculations

For yielding walls, lateral earth pressures can be computed assuming active conditions and wedge theory, using a planar surface of sliding defined by the Rankine theory. Table 8.1 provides soil properties for computing active earth pressures for five types of soil. Table 8.2 provides friction factors and adhesion for dissimilar materials. See Figs. 8.10 and 8.11 for the magnitude and location of resultant forces on retaining walls considering various types of soil backfill and backslide geometries. The pressures presented in these figures assume mobilization of the soil shear strength along the entire Rankine active failure plane, extending uninterrupted from the ground surface at the base of the wall or to the location on the wall at which the total earth load is being computed. Figure 8.11 shows the failure surface geometry and associated earth pressure distributions for various design conditions. If the soil behind the wall consists of more than one soil type, the design earth pressure should be determined using the weighted average of the properties of the soil types within and along the theoretical failure plane.

AASHTO provides that, for yielding walls, lateral earth pressures should be computed assuming active stress conditions and wedge theory using a planar surface of sliding defined by Coulomb theory. The computational procedures for active pressures are given below. An alternative to the computation for active pressures using the Coulomb theory for yielding walls and for cohesionless soils is to define the lateral pressures utilizing the Rankine theory.

The above procedures for developing the design pressures on yielding walls are based on the following assumptions:

1. The backfill soils are compacted with lightweight hand-compaction equipment.
2. The soil within the theoretical failure zone is made up entirely of the backfill soil.
3. No point or line loads act on the backfill surface.
4. The retaining wall deflections are consistent with the deflections required to develop the design active earth pressure.

8.2.3   Example of Active Pressure Calculations

The active pressure coefficient K_a is given by Coulomb theory as

$$K_a = \frac{\sin^2 (\theta + \phi')}{\sin^2 \theta \sin (\theta - \delta) \left[1 + \sqrt{\dfrac{\sin (\phi' + \delta) \sin (\phi' - \beta)}{\sin (\theta - \delta) \sin (\theta + \beta)}} \right]^2} \qquad (8.2)$$

where θ = angle of slope of back wall to horizontal, °
   ϕ' = effective angle of internal friction, °
   δ = angle of wall friction, °
   β = angle of back slope, °

TABLE 8.1 Soil Properties for Rankine Active Earth Pressure Computation

Soil type number	Soil description	USCS symbol[a]	Unit horizontal soil pressure[b] k_h, (lb/ft²)/ft	Rankine active earth pressure coefficient[b] K_a	Total soil unit weight γ, lb/ft³	Effective angle of friction ϕ', °
1	Sands and gravels with little or no fines	GW, GP, SW, SP (AASHTO A7)	30[c]	0.25	120	37
2	Sands and gravels with some silt	GM-GP, GM-GW, SM-SP, SM-SW	35	0.29	120	33
3	Silty and clayey sands and gravels	GM, GC, SM, SC	45	0.45	100	22
4	NOC[d] to LOC[e] silts and clays	ML, MH, CL, CH	100	0.80	125	0[g]
5	HOC[f] clays which can become saturated	CL, CH	120	1.00	120	0[g]

[a]Unified Soil Classification System (see Fig. 8.14).

[b]At β = 0°, representing a horizontal backslope behind the wall. For a sloping backfill (β > 0), refer to Figs. 8.10 and 8.11. $k_h = K_a \gamma$.

[c]The minimum value of k_h for design should be 35 lb/ft³.

[d]Normally overconsolidated (OCR = 1).

[e]Lightly overconsolidated (OCR = 1 to 2).

[f]Heavily overconsolidated (OCR > 2).

[g]Undrained shear strength.

Source: From *Design Manual,* Part 4, Pennsylvania Department of Transportation, Harrisburg, Pa., with permission.

TABLE 8.2 Ultimate Friction Factors and Adhesion for Dissimilar Materials

Interface materials	Friction factor, tan δ	Friction angle δ, °
Mass concrete on the following foundation materials:		
Clean sound rock	0.70	35
Clean gravel, gravel-sand mixtures, coarse sand	0.55–0.60	29–31
Clean fine to medium sand, silty medium to coarse sand, silty or clayey gravel	0.45–0.55	24–29
Clean fine sand, silty or clayey fine to medium sand	0.35–0.45	19–24
Fine sandy silt, nonplastic silt	0.30–0.35	17–19
Very stiff and hard residual or preconsolidated clay	0.40–0.50	22–26
Medium stiff and stiff clay and silty clay	0.30–0.35	17–19
(Masonry on foundation materials has same friction factors.)		
Steel sheet piles against the following soils:		
Clean gravel, gravel-sand mixtures, well-graded rock fill with spalls	0.40	22
Clean sand, silty sand-gravel mixture, single-size hard rock fill	0.30	17
Silty sand, gravel or sand mixed with silt or clay	0.25	14
Fine sandy silt, nonplastic silt	0.20	11
Formed concrete or concrete sheet piling against the following soils:		
Clean gravel, gravel-sand mixture, well-graded rock fill with spalls	0.40–0.50	22–26
Clean sand, silty sand-gravel mixture, single-size hard rock fill	0.30–0.40	17–22
Silty sand, gravel or sand mixed with silt or clay	0.30	17
Fine sandy silt, nonplastic silt	0.25	14
Various structural materials:		
Masonry on masonry, igneous and metamorphic rocks:		
Dressed soft rock on dressed soft rock	0.70	35
Dressed hard rock on dressed soft rock	0.65	33
Dressed hard rock on dressed hard rock	0.55	29
Masonry on wood (cross grain)	0.50	26
Steel on steel at sheet pile interlocks	0.30	17

Interface materials*	Adhesion C_a, lb/ft^2
Very soft cohesive soil (0–250 lb/ft^2)	0–250
Soft cohesive soil (250–500 lb/ft^2)	250–500
Medium stiff cohesive soil (500–1000 lb/ft^2)	500–750
Stiff cohesive soil (1000–2000 lb/ft^2)	750–950
Very stiff cohesive soil (2000–4000 lb/ft^2)	950–1300

*Cohesion values are shown in parentheses.

Source: From *Design Manual,* Part 4, Pennsylvania Department of Transportation, Harrisburg, Pa., with permission.

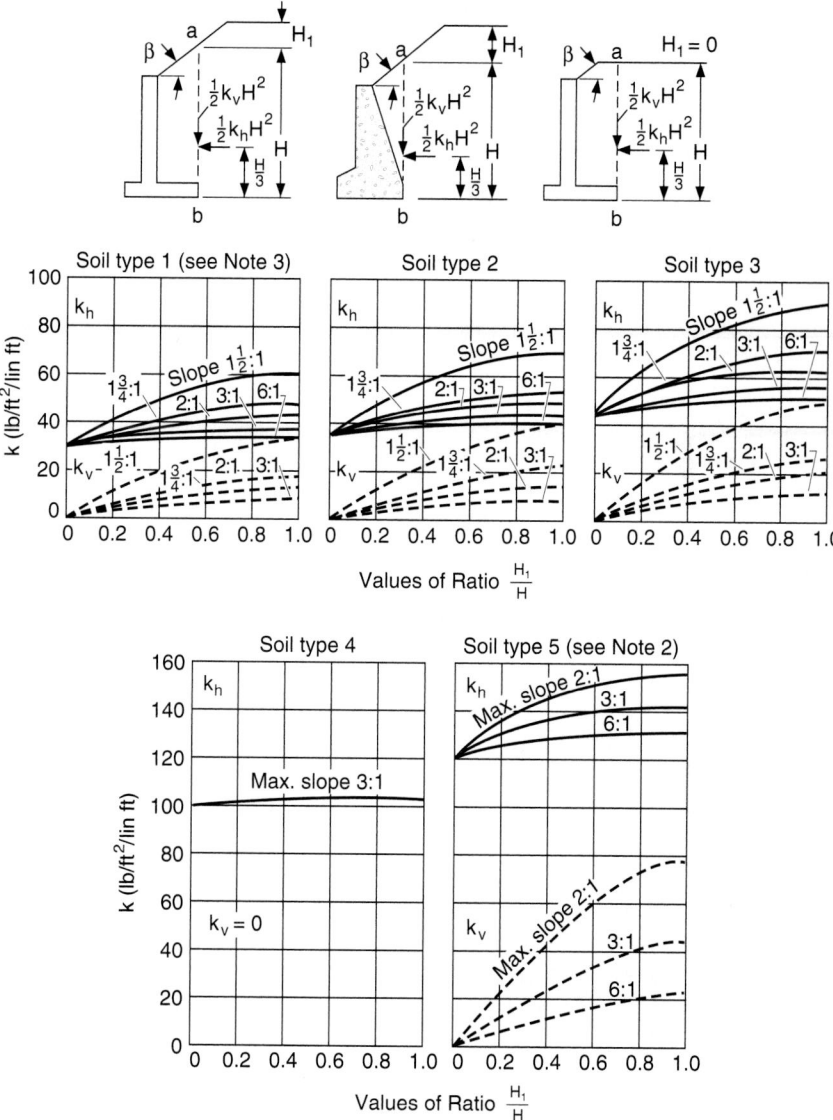

FIGURE 8.10 Charts for estimating Rankine active earth pressures against retaining walls supporting sloped ground of limited height. *Notes:* (1) Soil types shown on curves correspond to soil types described in Table 8.1. (2) For soil type 5, computations of soil pressure may be based on a value of H 4 ft less than the actual value. (3) The minimum value of k_h for design should be 35 lb/ft^2/lin ft. (4) Add pressures due to water and surcharge (including 2 ft minimum soil surcharge) to the active earth pressures from these charts. (*From* Design Manual, *Part 4, Pennsylvania Department of Transportation, Harrisburg, Pa., with permission*)

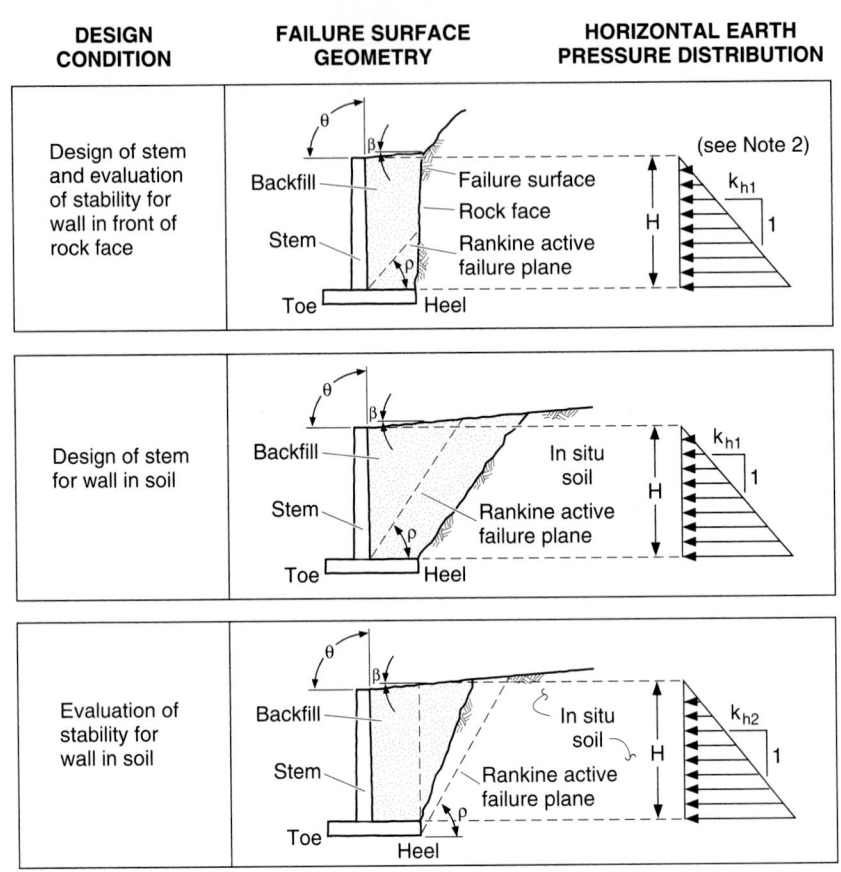

DESIGN CONDITION	FAILURE SURFACE GEOMETRY	HORIZONTAL EARTH PRESSURE DISTRIBUTION

LEGEND

k_{h1} = unit horizontal soil pressure due to backfill

k_{h2} = unit horizontal soil pressure due to in situ soil

ρ = angle between Rankine active failure plane and horizontal

ϕ' = weighted average effective stress angle of internal friction along failure plane

$$\rho = \tan^{-1}\left(\tan\phi' + \sqrt{1 + \tan^2\phi' - \frac{\tan\beta}{\sin\phi'\cos\phi'}}\right)$$

NOTES

(1) Obtain values of k_{h1}, k_{h2}, and vertical component of soil pressure.

(2) The earth pressure resultant for this condition can be more accurately determined by Culmann's graphical construction.

(3) Add pressures due to water and surcharge (including 2 ft minimum soil surcharge).

FIGURE 8.11 Assumed failure surfaces and horizontal earth pressure distributions. (*From* Design Manual, *Part 4, Pennsylvania Department of Transportation, Harrisburg, Pa., with permission*)

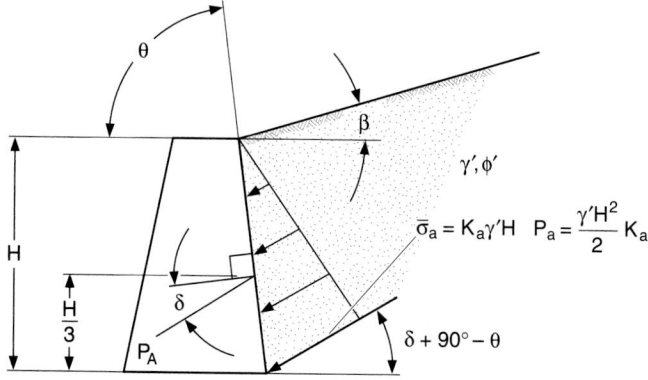

FIGURE 8.12   Forces for active pressure calculations. (*From* Design Manual, *Part 4, Pennsylvania Department of Transportation, Harrisburg, Pa., with permission*)

Refer to Figure 8.12 for the force diagram. The resultant horizontal earth force is to be determined for a design case wherein the following assumptions apply:

Design assumptions

$$\phi' = 34°$$

$$\delta = 25°$$

$$\beta = 0°$$

$$\theta = 90°$$

$$\gamma' = 125 \text{ lb/ft}^3$$

$$H = \text{height of wall} = 20 \text{ ft}$$

$$\text{Soil type} = 1 \text{ (see Table 8.1)}$$

Computations

$$\sin (\theta + \phi') = \sin (90° + 34°) = \sin 124° = 0.8290$$

$$\sin^2 (\theta + \phi') = \sin^2 (90° + 34°) = \sin^2 124° = 0.6873$$

$$\sin (\phi' + \delta) = \sin (34° + 25°) = \sin 59° = 0.8572$$

$$\sin (\phi' - \beta) = \sin (34° - 0°) = \sin 34° = 0.5592$$

$$\sin (\theta - \delta) = \sin (90° - 25°) = \sin 65° = 0.9063$$

$$\sin (\theta + \beta) = \sin (90° + 0°) = \sin 90° = 1.0000$$

$$\sin^2 (\theta) = \sin^2 (90) = 1.0000$$

$$K_a = \cfrac{0.6873}{1 \times 0.9063 \left[1 + \sqrt{\cfrac{0.8572 \times 0.5592}{0.9063 \times 1.00}} \right]^2}$$

$$= \frac{0.6873}{2.7041} = 0.2542$$

k_a = horizontal active pressure = $K_a \delta' H$ = 0.2542(125)20 = 635.5 lb/ft²

P_a = force resultant due to horizontal active pressure

$$= \frac{\gamma' H^2 K_a}{2} = \frac{125(20)^2(0.2542)}{2} = 6355 \text{ lb/ft}$$

Alternate calculation. Figure 8.10 gives the horizontal and vertical components of active earth pressure, k_h and k_v, for the five soil types listed in Table 8.1. The pressures are given in terms of the ratio H_1/H, when H_1 is the surcharge height and H is the height of the fill from the base, both as defined by the sketches in Fig. 8.10.

From Fig. 8.10, soil is type 1, $H_1/H = 0$, $k_h = 30$ (lb/ft²)/ft. Use 35, per note 3, Fig. 8.10.

$$P_a = \tfrac{1}{2} k_h H^2 = \tfrac{1}{2} \times 35(20)^2 = 7000 \text{ lb/ft}$$

8.3 FOUNDATION INVESTIGATIONS AND SOILS ANALYSIS

8.3.1 General Considerations

Since the stability and safety of a structure—more specifically, retaining wall structures—depend upon the proper performance of the foundation, it is important that an adequate foundation investigation be made. The purpose of the investigation is to provide the designer with information concerning the engineering properties of the subsurface conditions. Generally, a retaining wall extends for a considerable length. Accordingly, the amount and type of foundation investigation that should be made and/or which the owner can afford must be considered. The owner must understand that once an exploration crew is dispatched to the site of a proposed wall, the investigation should be sufficiently complete to allow for the selection of an appropriate wall type.

When a rigid concrete retaining wall is to be used, the designer must consider that such a wall can tolerate only minimal differential settlement. If differential settlement is predicted, the designer may have to accommodate this situation by vertical joints in the wall and other systems of articulation in the wall. In many instances, this type of wall, under situations where differential or excessive settlement is anticipated, will require deep foundations such as caissons or piling driven to firm supporting material. Alternatively, subexcavation and replacement of poor material at the base of the wall may be appropriate. When a mechanically stabilized wall is selected under conditions of poor foundation soils, the wall is more tolerant to such a foundation condition. It is important for the owner to realize that while the wall is more tolerant to this condition, the end result as viewed from the finished top surface of the wall may be decidedly different. Therefore, it is important for the owner to set out the requirements of, and acceptance criteria for, the wall prior to the selection process. All alternative wall types evaluated should meet those criteria. Otherwise, the owner is not evaluating equal alternatives.

Subsurface Exploration Plan.   Retaining walls are often viewed as subsidiary structures not worthy of any substantial expenditure for subsurface exploration. To the contrary, retaining walls can cost anywhere from $10 to $40 per square foot of wall face. Thus, they are costly structures. Further, the ultimate cost of most walls is quite sensitive to the foundation material.

The subsurface exploration plan can include obtaining subsurface data through the use of geophysical methods, such as seismic and electrical resistivity methods. More often the subsurface exploration effort is a simple and traditional boring program.

The boring program can be a simple auger drilling program with an experienced geologist classifying the soil on the basis of the auger cuttings. Clearly, if a physical examination of the type, nature, and characteristics of the subsurface materials is desired, samples will be necessary. The samples can be disturbed or undisturbed. The disturbed sample is generally taken in cohesionless soils and is used for classification and for moisture determination and compaction tests. More commonly, such samples may be taken by driving a heavy walled sampler into a clean hole. The size of the sampler or spoon varies from 2 in O.D. to $4\frac{1}{2}$ in O.D. When a standard penetration test (SPT) is required, the sample is obtained by driving a 2-in O.D. by $1\frac{3}{8}$-in I.D. sampler.

Where it is necessary to evaluate the structural properties of the subsurface material in its natural condition, an undisturbed sample is taken. This type of sample will produce a core sample that can be used for such laboratory tests as the triaxial shear, unconfined compression, and consolidation tests. This type of sample is more frequently taken in cohesive soils that contain little or no granular materials. They are often taken with thin-walled tube samplers (Shelby type).

The SPT results may be used to describe soil density and clayey soil consistency as shown in the following table:

Granular soil		Clay	
Blows	Density	Blows	Consistency
0–4	Very loose	0–1	Very soft
5–10	Loose	2–4	Soft
11–24	Medium dense	5–8	Medium stiff
25–50	Dense	9–15	Stiff
Over 50	Very dense	16–30	Very stiff
		31–60	Hard
		Over 60	Very hard

The blows are for the test procedures given in AASHTO Test Designation T-206.

Whenever rock is encountered, core drilling is done to advance the boring and to sample the rock in order to determine the profile and nature of the underlying rock strata. A general method by which the quality of the rock at a site is related to the amount of fracturing and alteration is known as the *rock quality designation (RQD)*. The procedure consists of summing the total length of core recovered by counting only those pieces of hard and sound core that are 4 in or greater in depth. The ratio of this modified core recovery length to the total core run is the RQD. Rock quality is related to the RQD as follows:

Rock quality designation (RQD)	Rock quality
0–25	Very poor
25–50	Poor
50–75	Fair
75–90	Good
90–100	Excellent

Soil Properties. Soils include matter in three states: solid, liquid, and gas. Figure 8.13 shows a diagram of a soil block and presents the fundamental weight-volume relationships among the terms. The following sample problem illustrates application to a soil sample. Refer to Fig. 8.13 for nomenclature.

- Data:

  Clay sample with water content of 31.2 percent by weight.

  Specific gravity of soil particles is 2.80.

  Sample is 98 percent saturated.

- Determine void ratio e and soil unit weight γ; assume 1 cm^3 of solids for calculations; $V_s = 1.00$ cm^3:

$$V_w/V = 0.98$$

$$W_s = 2.80 \times 1 \text{ g/cm}^3 = 2.80 \text{ g}$$

$$W_w = 0.312 \times 2.80 = 0.874 \text{ g}$$

$$W = W_s + W_w = 2.80 + 0.874 = 3.674 \text{ g}$$

$$V_w = W_w/\gamma_w = 0.874 \text{ g}/(1 \text{ g/cm}^3) = 0.874 \text{ cm}^3$$

$$V_v = V_w/0.98 = 0.874/0.98 = 0.892 \text{ cm}^3$$

Soil block	Volumes	Weights
Gas (air)	V_g	$W_g = 0$
Water	V_w	$W_w = V_w\gamma_w$
Solids	V_s	$W_s = V_s\gamma_sG_s$

$V = V_V + V_S$ = total volume $\qquad V_V = V_g + V_W$ = volume of voids
W = total weight = $V\gamma$
$V/V_V = V_V/V_V + V_S/V_V \quad$ or $\quad 1/n = 1 + 1/e \quad$ and $\quad n = e/(1 - n)$
$e = V_V/V_S$ = void ratio $\qquad n = V_V/V$ = porosity

where G_S = specific gravity of soil
$\quad \gamma_w$ = unit weight of water
$\quad \gamma_S$ = unit weight of solids
$\quad \gamma$ = unit weight of soil

FIGURE 8.13 Weight-volume relationships for soils.

$$V_g = V_v - V_w = 0.892 - 0.874 = 0.018 \text{ cm}^3$$

$$V = V_s + V_v = 1.00 + 0.892 = 1.892 \text{ cm}^3$$

$$e = V_v/V_s = 0.892/1.00 = 0.892$$

$$\gamma = W/V = 3.674/1.892 = 1.94 \text{ g/cm}^3$$

8.3.2 Soils Analysis

Retaining wall design engineers not fully trained in soil mechanics need to be acquainted with certain basic principles, in order to understand the data developed by the geotechnical engineer or geologist responsible for the subsurface exploration. Soil is a nonhomogeneous earthen material that varies laterally and vertically in mineral context, grain size, density, grain shape, moisture content, strength, consistency, and compressibility. For the design of retaining walls and other structure-type foundations, the engineering properties of the soil must be evaluated. Such an evaluation will always require consideration of foundation soil classification, bearing capacity, and compressibility.

Soil Classification. Since the types of soils are so numerous and variable, a classification system is important. The Unified Soil Classification System (USCS) has been generally accepted by engineers. It is based upon the sizes of the particles, the distribution of the particle sizes, and the properties of the fine-grained portion. Only particle sizes of 3 in or less are included in the USCS. Materials greater in size than 3 in are generally indicated in the log of borings as cobbles or boulders. Figure 8.14 shows the Unified Soil Classification Chart. The basic classifications include coarse-grained and fine-grained soils.

Coarse-Grained Soils. Coarse-grained soils are classified as either gravels or sands, dependent upon the fraction of the material retained on a no. 200 sieve. The classification threshold is 50 percent; i.e., if more than 50 percent of the fraction retained on a no. 200 sieve is retained on a no. 4 sieve, the soil is classified a gravel. If more than 50 percent passes the no. 4 sieve, the soil is classified a sand. There are many groupings of these coarse-grained soils, as indicated in the chart.

Fine-Grained Soils. Fine-grained soils are subdivided by plasticity and compressibility rather than by grain size. Fine-grained soils are classified as silt or clay, and as lowly or highly compressible. Criteria for classification are based upon the relationship between the liquid limit and the plasticity index. The relationship is given in the form of a plasticity chart shown by the inset in Fig. 8.14 and reproduced in Fig. 8.15. The "A" line on the chart divides clays from silts. Soils whose Atterberg limits plot above the line are clays, designated C; limits that plot below the line are silts, designated M.

8.3.3 Bedrock

Bedrock is divided by geologists into three large groups, namely (1) igneous, (2) metamorphic, and (3) sedimentary. Igneous rocks are those that have resulted from the cooling and crystallization of molten masses of mineral matter and gases either at or below the earth's surface. Sedimentary rocks consist of the transported and subse-

UNIFIED SOIL CLASSIFICATION CHART

MAJOR DIVISIONS	GROUP SYMBOLS	TYPICAL NAMES	FIELD IDENTIFICATION PROCEDURES (excluding particles larger than 3 in. and basing fractions on est. weights)	INFORMATION REQUIRED FOR DESCRIBING SOILS	LABORATORY CLASSIFICATION CRITERIA
COARSE-GRAINED SOILS — More than half of material is *larger* than No. 200 sieve size. (The No. 200 sieve size is about the smallest particle visible to the naked eye.) → **GRAVELS** — More than half of coarse fraction is *larger* than No. 4 sieve size. (For visual classifications, the 1/4" size may be used as equivalent to the No.4 sieve size.) → **CLEAN GRAVELS** (Little or no fines)	GW	Well-graded gravels, gravel-sand mixtures, little or no fines.	Wide range in grain size and substantial amounts of all intermediate particle sizes.	Give typical name; indicate approximate percentages of sand and gravel, max. size; angularity, surface condition, and hardness of the coarse grains; local or geologic name and other pertinent descriptive information, and symbol in parentheses. For undisturbed soils add information on stratification, degree of compactness, cementation, moisture conditions and drainage characteristics. EXAMPLE: *Silty sand, gravelly; about 20% hard, angular gravel particles 1/2 in. max. size rounded and subangular sand grains coarse to fine; about 15% non-plastic fines with low dry strength; well compacted and moist in place; alluvial sand; (SM).*	$C_u = D_{60}/D_{10}$ greater than 4; $C_g = (D_{30})^2/D_{10} \times D_{60}$ between 1 and 3
	GP	Poorly graded gravels, gravel-sand mixtures, little or no fines.	Predominantly one size or a range of sizes with some intermediate sizes missing.		Not meeting all gradation requirements for GW.
(CLEAN GRAVELS) **GRAVELS WITH FINES** (Appreciable amount of fines)	GM	Silty gravels, poorly graded gravel-sand-silt mixtures.	Non-plastic fines (for identification procedures see ML below).		Atterberg limits below "A" line, or PI less than 7. / Above "A" line with PI between 4 and 7 are *borderline* cases requiring use of dual symbols.
	GC	Clayey gravels, poorly graded gravel-sand-clay mixtures.	Plastic fines (for identification procedures see CL below).		Atterberg limits above "A" line with PI greater than 7.
SANDS — More than half of coarse fraction is *smaller* than No. 4 sieve size. → **CLEAN SANDS** (Little or no fines)	SW	Well-graded sands, gravelly sands; little or no fines.	Wide range in grain sizes and substantial amounts of all intermediate particle sizes.		$C_u = D_{60}/D_{10}$ greater than 6; $C_g = (D_{30})^2/D_{10} \times D_{60}$ between 1 and 3
	SP	Poorly graded sands, gravelly sands; little or no fines.	Predominantly one size or a range of sizes with some intermediate sizes missing.		Not meeting all gradation requirements for SW.
SANDS WITH FINES (Appreciable amount of fines)	SM	Silty sands, poorly graded sand-silt mixtures.	Non-plastic fines (for identification procedures see ML below).		Atterberg limits below "A" line or PI greater than 7. / Above "A" line with PI between 4 and 7 are *borderline* cases requiring use of dual symbols.
	SC	Clayey sands, poorly graded sand-clay mixture.	Plastic fines (for identification procedures see CL below).		Atterberg limits above "A" line with PI greater than 7.

Use grain size curve in identifying the fractions as given under field identification.

Determine percentages of gravel and sand from grain size curve. Depending on percentage of fines (fraction smaller than No. 200 sieve size), coarse-grained soils are classified as follows:

Less than 5% .... GW, GP, SW, SP
More than 12% .... GM, GC, SM, SC
5% to 12% .... *Borderline* cases requiring use of dual symbols

FINE-GRAINED SOILS — More than half of material is *smaller* than No. 200 sieve size.

Identification procedures on fraction smaller than No. 40 sieve size

MAJOR DIVISIONS	GROUP SYMBOLS	TYPICAL NAMES	DRY STRENGTH (crushing characteristics)	DILATANCY (reaction to shaking)	TOUGHNESS (consistency near plastic limit)	INFORMATION REQUIRED FOR DESCRIBING SOILS
SILTS and CLAYS — Liquid limit *less than* 50	ML	Inorganic silts and very fine sands, rock flour, silty or clayey fine sands with slight plasticity.	None to slight	Quick to slow	None	Give typical name; indicate degree and character of plasticity, amount and max. size of coarse grains; color in wet condition, odor if any, local or geologic name, and other pertinent descriptive information, and symbol in parentheses. For undisturbed soils add information on structure, stratification, consistency in undisturbed and remolded states, moisture and drainage condition. EXAMPLE: *Clayey silt, brown; slightly plastic; small percentage of fine sand; numerous vertical root holes; firm and dry in place; loss; (ML).*
	CL	Inorganic clays of low to medium plasticity, gravelly clays, sandy clays, silty clays, lean clays.	Medium to high	None to very slow	Medium	
	OL	Organic silts and organic silt-clays of low plasticity.	Slight to medium	Slow	Slight	
SILTS and CLAYS — Liquid limit *greater than* 50	MH	Inorganic silts, micaceous or diatomaceous fine sandy or silty soils, elastic silts.	Slight to medium	Slow to none	Slight to medium	
	CH	Inorganic clays of high plasticity, fat clays.	High to very high	None	High	
	OH	Organic clays of medium to high plasticity.	Medium to high	None to very slow	Slight to medium	
HIGHLY ORGANIC SOILS	PT	Peat and other high organic soils.	Readily identified by color, odor, spongy feel, and frequently by fibrous texture.			

PLASTICITY CHART — For laboratory classification of fine-grained soils.
(A line; CH; OH & MH; CL; ML; CL-ML. Comparing soils at equal liquid limit: Toughness and dry strength increase with increasing plasticity index. Plasticity index vs. Liquid limit.)

1. **Boundary classifications:** Soils possessing characteristics of two groups are designated by combinations of group symbols; *for example* GW-GC, well-graded gravel-sand mixture with clay binder.
2. All sieve sizes on this chart are U.S. standard.

FIGURE 8.14 Unified soil classification chart. (*Adopted by U.S. Army Corps of Engineers and Bureau of Reclamation, January 1952*)

8.28

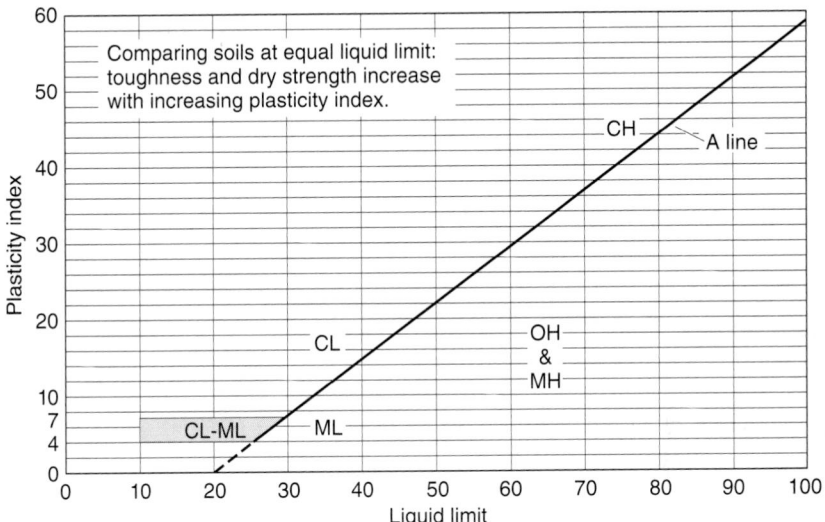

FIGURE 8.15 Plasticity chart for laboratory classification of fine-grained soils. (*Adopted by U.S. Army Corps of Engineers and U.S. Bureau of Reclamation, January 1952*)

quently indurated products of weathering of previously existing rock types, while metamorphic rocks are frequently defined as those having characteristic textures and mineral compositions that have resulted from high temperatures and pressures and/or hot mineralizing solutions acting on a parent rock. Figures 8.16, 8.17, and 8.18 indicate easily recognizable descriptions for field classification of igneous, metamorphic, and sedimentary rock, respectively.

8.3.4 Soils Laboratory Tests

Grain size, shape, and gradation are generally established by sieve analysis. For the finer clays, a hydrometer analysis is necessary. Figure 8.19 depicts a classification of sediment based on grain size.

Atterberg limit tests are performed on fine-grained soils and represent the amount of water present in the voids. The liquid limit (LL), plastic limit (PL), and plasticity index (PI) constitute the Atterberg limits.

The *triaxial shear test* is used to find the shear strength of a soil for the determination of pile lengths and of bearing capacity for spread footings or drilled shafts. Triaxial shear test results are also needed to give soil parameters for the design of retaining walls. High-quality, undisturbed samples are required for triaxial shear tests. Poor samples should be discarded rather than tested, as they will give misleading results.

The *direct shear* test is sometimes performed in lieu of other shear tests, and the use of its results is the same as that noted above for the triaxial shear test. It is important to remember that direct shear test results are usually less reliable than those obtained from the triaxial shear test, since the failure line in the direct shear test is imposed by the method of testing, whereas the triaxial method allows the sample to fail in its weakest plane. On occasion, it is desirable to shear soil or rock along a par-

FIELD CLASSIFICATION OF IGNEOUS ROCKS				
ROCK TEXTURE	ROCK COLOR AND ESSENTIAL MINERALS			
	Light gray, white, or pink contains orthoclase and quartz	Dark gray or black contains plagioclase, hornblende, and/or biotite	Dark gray or black contains plagioclase and pyroxene	Black or green contains augite and/or olivene and/or hornblende
Granular or (course-grained)	Granite	Diorite	Gabbro	Peridotite
Porphyritic and aphanitic (coarse and fine)	Rhyolite	Andesite	Basalt	
Aphanitic (fine-grained)	Felsite (light-colored)		Basalt (dark-colored)	
Glassy (amorphous)	Obsidian (black)	Pitchstone (red or brown)		Pumice (a glass froth)

FIGURE 8.16 Field classification of igneous rock. *Note:* Consolidated volcanic ash is called *tuff* if no large fragments are present. If large fragments are present, it is called *breccia.* (*From C. H. Harned,* Some Practical Aspects of Foundation Studies for Highway Bridges, *U.S. Bureau of Public Roads, January 1959*)

ticular plane. In these cases, a direct shear test may be used. High-quality, undisturbed samples are needed for this test.

The *unconfined compression test* of a soil is a uniaxial compression test in which the test specimen is provided with no lateral support while undergoing vertical compression. The test measures the unconfined, compressive strength of a cylinder of cohesive or semicohesive soil, which, indirectly, may be indicative of the shearing strength. The test is usually performed on an undisturbed sample of soil at its natural moisture content. It may also be performed on a remolded sample to evaluate the effects of disturbance and remolding upon the shearing strength.

Unconfined compression tests are relatively quick to perform and relatively inexpensive. When used in conjunction with the triaxial test, the unconfined compression test is of value. Also, it is sometimes used as an index test because it is easy to conduct.

8.3.5 Engineering Properties of Soils

The equation for shearing strength S (lb/ft^2) of a soil may be taken as follows:

$$S = c + \sigma \tan \phi \tag{8.3}$$

where c = cohesion, lb/ft^2
  σ = confining pressure or normal stress, lb/ft^2
  ϕ = angle of internal friction of the soil, degrees

FIELD CLASSIFICATION OF METAMORPHIC ROCKS	
NONFOLIATED (no parallel alignment of minerals)	FOLIATED (parallel alignment of minerals)
Quartzite conglomerate (from conglomerate) Quartzite (from sandstone) Marble (from limestone) Serpentine (from basic igneous rocks) Anthracite (from bituminous coal)	Gneiss (individual foliation planes are easily distinguishable with the naked eye) Schist (individual foliation planes are distinguishable with a hand lens) Slate (microfoliated)

ESSENTIAL MINERAL COMPOSITION OF THE COMMON METAMORPHIC ROCKS			
NONFOLIATED		FOLIATED	
ROCK	MINERALS	ROCK	MINERALS
Quartzite	Quartz	Gneiss	Quartz, feldspar, muscovite, biotite, pyroxenes, amphiboles
Marble	Calcite or dolomite		
Serpentine	Serpentine	Schist	Garnet, staurolite, talc, muscovite, biotite, chlorite, epidote
Coal	No minerals		
		Slate	Microscopic quartz, muscovite, biotite, chlorite
Note: The names of foliated rocks are frequently modified by designating the conspicuous minerals present, e.g., chlorite schist, talc schist, mica schist, staurolite schist.			

FIGURE 8.17   Field classification of metamorphic rock. (*From C. H. Harned,* Some Practical Aspects of Foundation Studies for Highway Bridges, *U.S. Bureau of Public Roads, January 1959*)

The shearing strength of the soil should account for the effect of pore water pressure when present. Equation (8.3) can be modified as:

$$S = c + (\sigma - u) \tan \phi \qquad (8.4)$$

where u = pore water pressure, lb/ft^2

Soil consolidation is produced by load and is associated with changes in soil moisture. It is also a function of time. The time required for drainage to occur, which results from the change in soil moisture, is a function of the permeability of the soil and the distance the water must travel in the material to be released. It is clear that consolidation of coarse-grained materials will occur fairly rapidly. This explains the often used assumption that consolidation of such materials under applied load, for example, the load of a retaining wall, generally occurs during the construction of the

CLASTIC SEDIMENTARY ROCKS		
RESIDUAL AND /OR MECHANICAL SEDIMENT		
GRAIN SIZE	UNCONSOLIDATED SEDIMENT	CONSOLIDATED ROCK
Coarse	Boulders, cobbles, gravel, and coarse sand	Conglomerate (rounded particles) Breccia (angular particles) Sandstone (coarse)
Medium	Sand	Sandstone Arkose = +25% feldspar Graywacke = dark colored
Fine	Silt and clay	Siltstone and shale
NONCLASTIC SEDIMENTARY ROCKS		
CHEMICAL SEDIMENTS		ORGANIC SEDIMENTS
Gypsum Salt Dolomite Clauconite Some chert (flint) Some iron ores Some phosphate rock		Some limestone Some chert (flint) Some phosphate rock Peat Coal

Note: 1. The cementing agents for sedimentary rocks are calcite, quartz, limonite, hematite, and chalcedony. Clay minerals may also function as binder or semicementing material.

2. Compositional descriptive adjectives such as siliceous, argillaceous, arenaceous, calcareous, carbonaceous, ferrugenous, feldspathic, opaline, and cherty are frequently used.

3. Other descriptive adjectives such as massive, laminated, stratified, varved, cross-bedded, concretionary, and fissile are also used.

FIGURE 8.18 Field classification of sedimentary rock. (*From C. H. Harned,* Some Practical Aspects of Foundation Studies for Highway Bridges, *U.S. Bureau of Public Roads, January 1959*)

wall. Thus, long-term settlement is not normally considered to occur. On the contrary, clays and/or silts are relatively impermeable, so that long-term settlement should be anticipated in the design. The designer must consider various options to accommodate this projected long-term settlement. For example, the designer may (1) require pre-loading to effect the settlement before the wall is constructed, (2) accelerate the consolidation by drilling for and placing sand drains, and (3) decide to build the structure with pile or caisson support systems that are independent of the consolidation.

SIZE RANGE (mm)	CLASS NAME
Over 256	Boulder
256–64	Cobble
64–4	Pebble
4–2	Granule
2–1	Very coarse sand
1–0.5	Coarse sand
0.5–0.25	Medium sand
0.25–0.125	Fine sand
0.125–0.0625	Very fine sand
0.0625–0.002	Silt
Less than 0.002	Clay

(a)

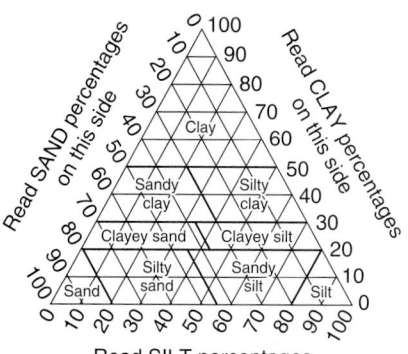

Read SILT percentages
on this side

GROUPING USED IN TRIANGULAR CLASSIFICATION			
CLASS NAMES	PERCENT OF SIZES PRESENT		
	SAND	SILT	CLAY
Sand	80–100	0–20	0–20
Silty sand	45–80	0–55	0–20
Sandy silt	0–45	35–80	0–20
Silt	0–20	80–100	0–20
Clayey sand	38–80	0–42	20–30
Clayey silt	0–38	32–80	20–30
Sandy clay	30–70	0–40	30–50
Silty clay	0–30	20–70	30–50
Clay	0–50	0–50	50–100

If gravel is present in appreciable amounts, the term "gravelly" may be added to the class name, vis. "gravelly sand". The terms "coarse", "medium", and "fine", when used to describe gravel, sand, and silt, refer to standard grade size limits.

(b)

FIGURE 8.19   Classification of soil based on (*a*) grain size of sediment and (*b*) standard grain size limits. (*From C. H. Harned,* Some Practical Aspects of Foundation Studies for Highway Bridges, *U.S. Bureau of Public Roads, January 1959*)

8.4   *RIGID RETAINING WALLS*

8.4.1   General Criteria

Rigid retaining walls are those that develop lateral resistance primarily from their own weight. Figure 8.20 shows the terms used in the design of this type of wall. On the basis of their overall cross sections, those walls may be referred to as L walls or T walls. (See insets, Fig. 8.3.)

Examples of rigid structures typically include concrete gravity walls, thick concrete slurry walls, and gabion walls. Additionally, some reinforced earth walls, if

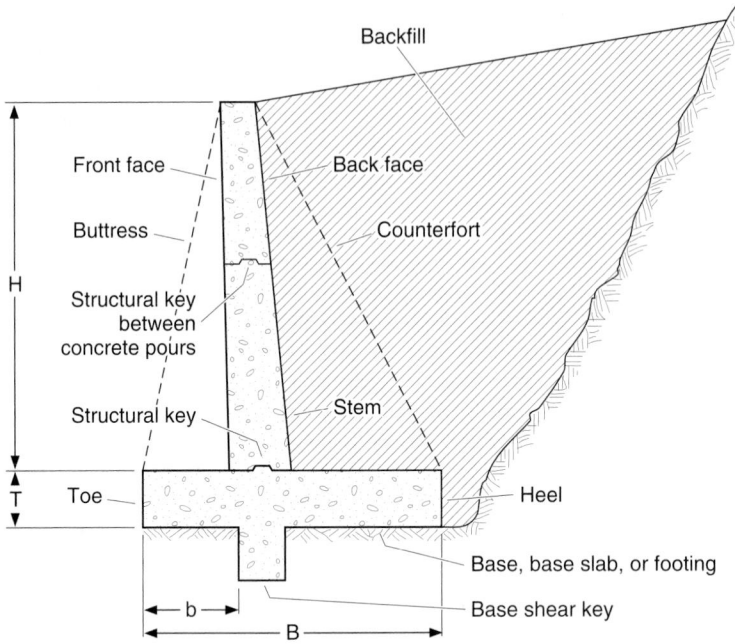

FIGURE 8.20   Terminology for rigid retaining walls. (*From* Design Manual, *Part 4,
Pennsylvania Department of Transportation, Harrisburg, Pa., with permission*)

designed to be reinforced in such a way that limited lateral movement will occur, can
also be categorized as rigid walls. In fact, a wall may have considerable flexibility in
its vertical dimension and nevertheless be classified and designed as a "rigid" wall
system. Requirements for resistance of these wall types include sliding stability, overturning, bearing pressure evaluation, and settlement considerations. Design criteria for
rigid retaining walls are summarized in Fig. 8.21. Overall, or global, stability is an
important consideration in that, while the wall itself may adequately retain a soil
mass, the soil mass may be unstable because, for example, of a deep-seated failure
plane. This type of consideration is evaluated by *slip circle analysis.*

8.4.2   Sliding Stability

To provide adequate resistance against sliding, the base of the wall should be at least 3
ft below ground surface in front and below the depth of frost action, depth of seasonal
volume change, and depth of scour. Sliding stability should be adequate without
including passive pressure at the toe. If insufficient sliding resistance is available, the
designer may increase base width, provide a pile foundation, or lower the base of the
wall and consider passive resistance below frost depth. If the wall is supported on rock
or very stiff clay, a key may be installed below the foundation to provide additional
resistance to sliding. Considerations of the need for the 3-ft depth when dealing with
reinforced earth walls should be evaluated in that such walls are not as susceptible to
frost action as more rigid concrete walls. In any event, it is recommended that some
nominal depth below ground line be provided to accommodate changes in natural terrain over the anticipated life of the structure, often 75 to 100 years. Such changes

Type of wall	Load diagram	Design factors
Gravity	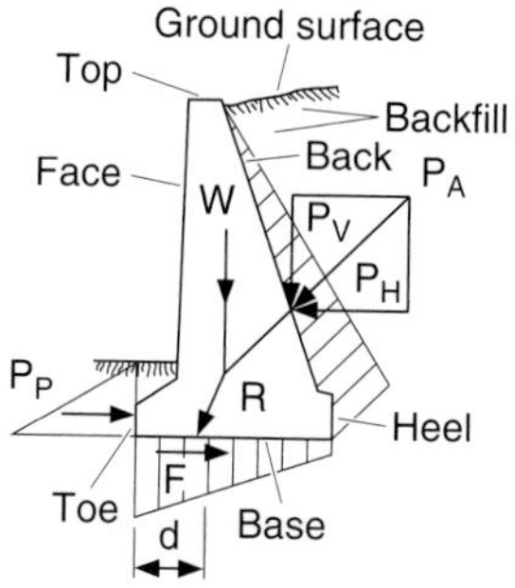	**LOCATION OF RESULTANT** Moments about toe: $$d = \frac{W_a + P_Ve - P_Hb}{W + P_V}$$ Assuming $P_P = 0$ **OVERTURNING (GRAVITY AND SEMIGRAVITY)** Moments about toe: $$F_S = \frac{W_a}{P_Hb - P_Ve} \geqq 1.5$$ Ignore overturning if R is within middle third (soil), middle half (rock). Check R at different horizontal planes for gravity walls.
Semigravity		**RESISTANCE AGAINST SLIDING** $$F_S = \frac{(W + P_V)\,\text{TAN}\,\delta + C_aB}{P_H} \geqq 1.5$$ $$F_S = \frac{(W + P_V)\,\text{TAN}\,\delta + C_aB + P_P}{P_H} \geqq 2.0$$ $$F_S = (W + P_V)\,\text{TAN}\,\delta + C_aB$$
Cantilever		For coefficients of friction between base and soil see Table 8.2. C_a = Adhesion between soil and base TAN δ = Friction factor between soil and base W = Includes weight of wall and soil in front for gravity and semigravity walls. Includes weight of wall and soil above footing, for cantilever and counterfort walls.
Counterfort		

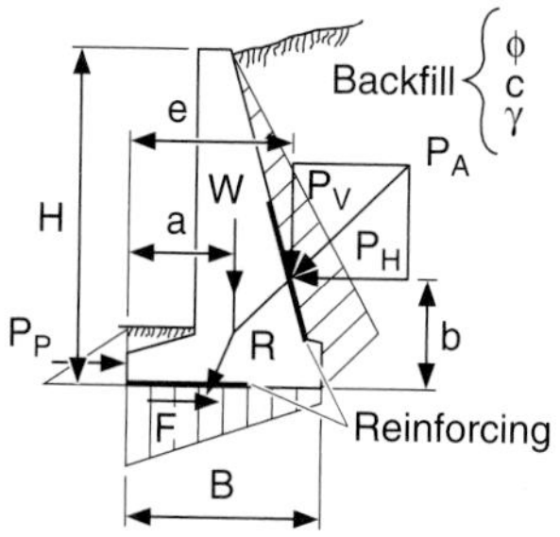

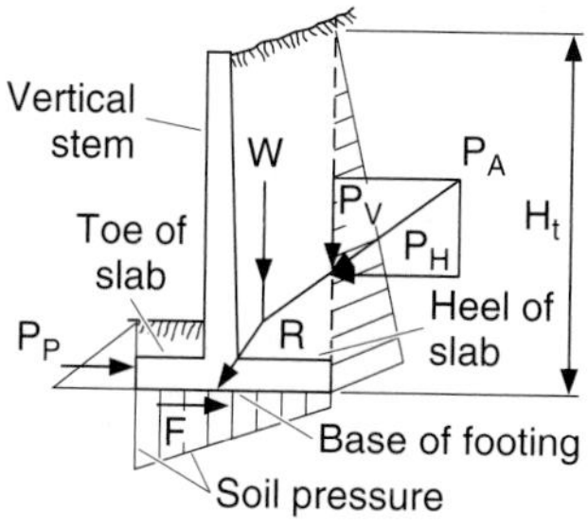

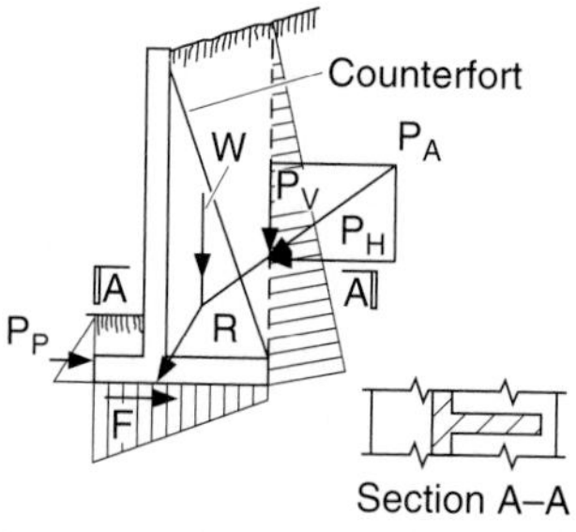

FIGURE 8.21   Design criteria for rigid retaining walls. (*From* Design Manual, *Part 4, Pennsylvania Department of Transportation, Harrisburg, Pa., with permission*)

TABLE 8.3 Relationship between Soil Backfill Type and Wall Rotation to Mobilize Active and Passive Earth Pressures behind Rigid Retaining Walls

	Wall rotation, Δ/H	
Soil type and condition	Active	Passive
Dense cohesionless	0.001	0.020
Loose cohesionless	0.004	0.060
Stiff cohesive	0.010	0.020
Soft cohesive	0.020	0.040

Source: From *Design Manual*, Part 4, Pennsylvania Department of Transportation, Harrisburg, Pa., with permission.

occur as a result of normal soil erosion caused by wind, rainfall, and other natural processes. Of course, in situations where scour may occur, hydrologic and hydraulic evaluations of scour depth must be made.

8.4.3 Settlement and Overturning

For walls on relatively incompressible foundations, apply the overturning criteria of Fig. 8.21. If the foundation is compressible, compute settlement by available methods previously referred to and estimate tilt of a rigid wall from the settlement. If the consequent tilt is anticipated to exceed acceptable limits, proportion the wall to keep the resultant force at the middle third of the base. If a wall settles so that the resulting movement forces it into the soil it supports, then the lateral pressure on the active side increases substantially. Table 8.3 shows the magnitudes of wall rotation required to mobilize active and passive earth pressures for different types of soil.

8.4.4 Overall Stability on Weak Soils

Where retaining walls are underlain by weak soils, the overall stability of the soil mass containing the retaining wall should be checked with respect to the most critical surface of sliding. A minimum safety factor of 2.0 is desirable but may not always be achievable. A technique known as *slip circle analysis* can be used to check for global stability. Refer to standard texts on soils engineering.

8.4.5 Design Procedures for a Cantilever Retaining Wall

A typical cantilever retaining wall is illustrated by the insert sketch in Fig. 8.21. This rigid-type wall can be constructed with or without a base shear key (see Fig. 8.20) depending on an analysis for resistance to sliding, as discussed later.

The specifications of the owner will govern the selection and use of backfill materials behind retaining walls. In most cases, clean backfill materials having an internal friction angle of at least 34° are assumed in the design of retaining walls, subject to the following considerations:

1. With a proper drainage system and with backfilling controlled so that no compaction-induced lateral loads are applied to the wall, the above-noted or better material may be used in construction. A minimum lateral earth pressure of 30 (lb/ft^2)/ft (equivalent fluid weight) for level backfills, or 40 (lb/ft^2)/ft for 2:1 sloped fills, should be assumed.

2. Backfill is assumed as on-site inorganic material; however, if it is of a lower class designation, the wall must be designed for an equivalent fluid weight lateral pressure suitable for that class. Therefore, should the designer select a backfill material of lower classification, it will be necessary to clearly specify the backfill material by a supplemental project special provision and to use an appropriate equivalent fluid weight lateral pressure for design.

The design aids provided in Figs. 8.22 and 8.23 may be used for preliminary dimensions in the design of a cantilever cast-in-place retaining wall. On the basis of the Rankine theory of earth pressure, final design may proceed with the following steps:

1. Obtain soil parameters for both backfill and foundation. Usually the cohesionless backfill is slightly larger than Rankine zone. This enables the designer to use the properties of backfill material to estimate earth loads; otherwise the properties of retained material must be used.

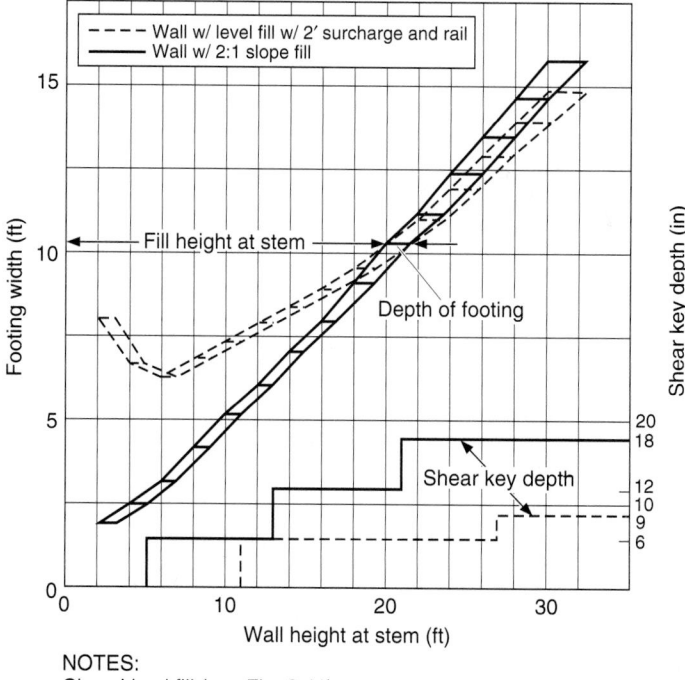

NOTES:
Class I backfill (see Fig. 8.41)
Class D concrete
Coef. of friction (soil to soil = 0.67, soil to concrete = 0.42)

FIGURE 8.22 Aid for preliminary design of cast-in-place concrete retaining walls showing wall and footing dimensions. (*From* Bridge Design Manual, *Section 5, Colorado Department of Transportation, Denver, Colo., with permission*)

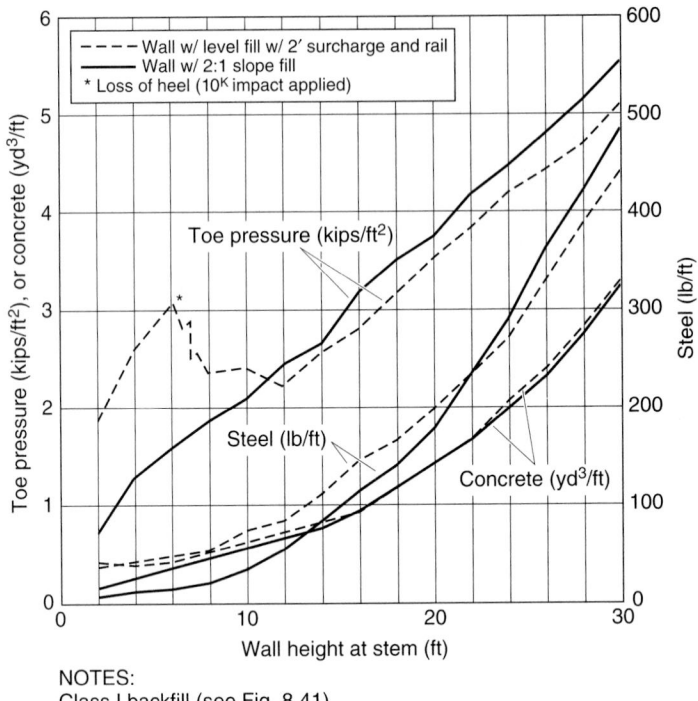

FIGURE 8.23 Aid for preliminary design of cast-in-place concrete retaining walls showing toe pressure and steel and concrete quantities. (*From* Bridge Design Manual, *Section 5, Colorado Department of Transportation, Denver, Colo., with permission*)

2. Determine the appropriate design cases and load combinations. Load types are designated as follows: D, dead load; E, earth load; SC, surcharge; RI, rail impact; W, wind load. Typical load combinations are as follows: sloped or leveled fill without rail, $D + E$; leveled fill without rail, $D + E + $ SC; leveled fill with rail, $D + E + $ RI; leveled fill with rail and fence, $D + E + $ SC $+ W$.

3. Determine the overall design height including footing thickness T and stem height H, and select a trial footing width dimension B. (See Fig. 8.20.) Usually the toe width b is approximately one-third to one-half of B. The ratio of footing width to overall height should be in the range from 0.4 to 0.8 for T-shaped walls as shown by the design aids in Figs. 8.22 and 8.23. In these preliminaries, wide-base L-shaped walls (footing width to height ratios larger than 0.8) are used for low wall heights (less than 10 ft), and the factor of safety with respect to overturning is relaxed from a minimum of 2.0 to 1.5 when considering the case of $D + E + $ RI.

4. Draw a vertical line from the back face of the footing to the top of the fill. This line serves as the boundary of the free body to which the earth pressure is applied. The applied active earth pressure can be estimated by Rankine theory, and the direction assumed parallel to the backfill surface. Compute the resultant P of the applied earth pressure and associated loads. Resolve P into horizontal and vertical components P_h

and P_v, and apply at one-third the total height H_t of the imaginary boundary from the bottom of the footing. (See Fig. 8.21.)

5. Take a free body of the stem and compute the loads applied at the top of the stem as well as loads along the stem (height H), and find the moment and shear envelope to meet all the design cases at several points along the height. The working stress design method and the concept of shear friction can be used to calculate the shear strength at the joint between footing and stem.

6. Calculate the weight W, which is the sum of the weight of concrete and the weight of soil bounded by the back of the concrete wall and the vertical line defined by step 4 above. Find the distance from the extremity of the toe to the line of action of W, which is the stabilizing moment arm a.

7. Calculate the overturning moment M_o applied to the wall free body with respect to the tip of the toe as:

$$M_o = P_h\left(\frac{H_t}{3}\right) \tag{8.5}$$

Calculate the resisting moment M_r with respect to the tip of the toe as:

$$M_r = Wa + P_v B \tag{8.6}$$

The safety factor SF against overturning is

$$\text{SF (overturning)} = \frac{M_r}{M_o}$$

$$= \frac{Wa + P_v B}{P_h H_t /3} \tag{8.7}$$

The required safety factor (overturning) should be equal to or greater than 2.0 unless otherwise accepted and documented by the engineer (see step 3).

8. Compute the eccentricity e of the applied load with respect to the center of the footing based on the net moment:

$$e = \frac{B}{2} \times \frac{M_r - M_o}{W} \tag{8.8}$$

The resultant should be within the middle third of the footing width; i.e., the absolute value of e should be less than or equal to $B/6$ to avoid tensile action at the heel.

9. The toe pressure q can be evaluated and checked by the following equation:

$$q = \frac{W}{B}\left[1 + 6\left(\frac{e}{B}\right)\right] \tag{8.9}$$

The toe pressure must be equal to or less than the allowable bearing capacity based on the soils report. Toe pressure is most effectively reduced by increasing the toe dimension.

10. The footing, both toe and heel, can be designed by working strength design. Soil reactions act upward and superimposed loads act downward. The heel design loads should include the portion of the vertical component P_v of earth pressure that is applied to the heel. For the toe design loads and stability, the weight of the overburden should not be used if this soil could potentially be displaced at some time during the life of the wall.

11. Check the factor of safety against sliding without using a shear key. The coefficient of friction between soil and concrete is approximately $\tan(\frac{2}{3}\phi)$, where ϕ is the internal friction angle of the soil in radians. Neglect the passive soil resistance in front of the toe. The sliding resistance SR can be evaluated as:

$$SR = (W + P_v) \tan\left(\frac{2}{3}\phi\right) \qquad (8.10)$$

The SF (sliding), which is SR/P_h, should be equal to or greater than 1.5. If SF (sliding) is less than 1.5, then either the width of the footing should be increased or a shear key should be installed at the bottom of the footing.

If a shear key is the choice, the depth of the inert block d_c is computed by the sum of the key depth KD and the assumed effective wedge depth, which is approximately half the distance between the toe and the front face of the shear key ($b/2$). Using the inert block concept, knowing the equivalent fluid weight (γ_p) of passive soil pressure, and neglecting the top 1 ft of the toe overburden T_o, the toe passive resistance P_p is

$$P_p = 0.5\gamma_p[(T_o + T + d_c - 1)^2 - (T_o + T - 1)^2] \qquad (8.11)$$

Total sliding resistance F from friction is the sum of the horizontal component of the resistance from toe to shear key and the resistance from shear key to heel. Therefore:

$$F = [\cos^2(\frac{2}{3}\phi)\, R_1 \tan\phi] + R_2 \tan(\frac{2}{3}\phi) \qquad (8.12)$$

where ϕ = internal friction angle of base soil
R_1 = soil upward reaction between toe and key, lb/ft
R_2 = soil upward reaction between key and heel, lb/ft

Sliding resistance is:

$$SR = F + P_p \qquad (8.13)$$

The SF (sliding), which is SR/P_h, should be equal to or greater than 1.5.

12. Repeat steps 3 through 11 as appropriate until all design requirements are satisfied.

Figure 8.24 represents typical values for equivalent fluid pressures of soils. These values are suggested for use in the absence of a more detailed determination.

8.5 MECHANICALLY STABILIZED EARTH (MSE) WALLS

8.5.1 Types of MSE Walls

MSE walls are made up of several elements—specifically, the reinforcement of a soil mass through the use of steel strips, steel or polymeric grids, or geotextile sheets, capable of withstanding tensile forces, and a facing material. Figure 8.25 depicts different types of geosynthetic reinforced walls. The walls depicted range from a sloping geotextile wrapped face, usually used for the more temporary conditions; to stabilized soil masses faced with more long-term cast-in-place concrete or masonry block facings.

Structural backfill class designation	Type of soil (compaction conforms with AASHTO 90–95% T180)	Typical values for equivalent fluid unit weight of soils, lb/ft[3a,b,c]	
		Level backfill	2:1 (H:V) backfill
Class I[d]: borrowed, selected, coarse-grained soils	Loose sand or gravel	40 (active) 55 (at rest)	50 (active) 65 (at rest)
	Medium dense sand or gravel	35 (active) 50 (at rest)	45 (active) 60 (at rest)
	Dense[e] sand or gravel, 95% T180	30 (active) 45 (at rest)	40 (active) 55 (at rest)
Class IIA[f]: on-site, inorganic, coarse-grained soils, low percentage of fines	Compacted, clayed, sand gravel	40 (active) 60 (at rest)	50 (active) 70 (at rest)
	Compacted, clayed, silty gravel	45 (active) 70 (at rest)	55 (active) 80 (at rest)
Class IIB: on-site, inorganic LL < 50%	Compacted, silty/sandy gravelly, low/medium plasticity lean clay	Site-specific material, use with special attention; see geotechnical engineer. Soils report on workmanship of compaction, drainage design, and waterstop membrane is required.	
Class IIC: on-site, inorganic LL > 50%	Fat clay, elastic silt that can become saturated	Not recommended	

[a]At rest, pressure should be used for earth that does not deflect or move.
[b]Active pressure state is defined by movement at the top of wall of 1/240 of the wall height.
[c]The effect of additional earth pressure that may be induced by compaction or water should be added to that of earth pressure.
[d]Class I: 30 percent or more retained on no. 4 sieve and 80 percent or more retained on no. 200 sieve.
[e]Dense: No less than 95 percent density per AASHTO T180.
[f]Class IIA: 50 percent or more retained on no. 200 sieve.

FIGURE 8.24 Typical values for equivalent fluid pressure for soils. (*From* Bridge Design Manual, *Section 5, Colorado Department of Transportation, Denver, Colo., with permission*)

The advantages of MSE walls over the more conventional reinforced concrete walls include:

1. Inherent flexibility to accommodate reasonable differential settlements
2. Lower total cost
3. Less construction time
4. Inherent capability to provide drainage to avoid buildup of hydrostatic forces

The reinforcement elements are characterized as extensible or inextensible. Extensible reinforcements can deform without rupture to develop deformations greater than can the soil in which they are placed. Such reinforcements include polymeric geotextiles and geogrids. Inextensible reinforcements cannot deform to deformations

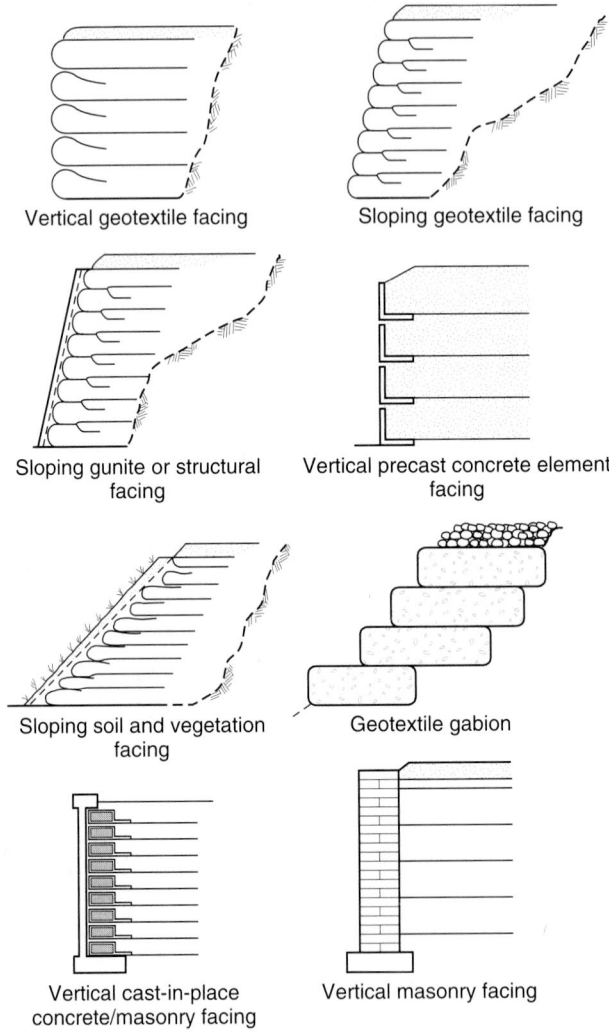

Vertical geotextile facing

Sloping geotextile facing

Sloping gunite or structural facing

Vertical precast concrete element facing

Sloping soil and vegetation facing

Geotextile gabion

Vertical cast-in-place concrete/masonry facing

Vertical masonry facing

FIGURE 8.25   Facings for geotextile-reinforced walls.

greater than the soil they reinforce. Metallic-strip or grid reinforcements are included in this category.

A summary of the available MSE systems in terms of the reinforcement and facing panel details is included in Table 8.4. The summary includes the major proprietary systems available. Figure 8.26 includes data regarding the geometries and some mechanical properties of the different reinforcement types available for use in MSE walls with geotextile reinforcements.

Reinforced Earth was invented by Henri Vidal, who first published results of his studies in 1963. After a brief period of skepticism, the first significant projects were constructed in 1967. The use of Reinforced Earth then spread rapidly, and by the early

TABLE 8.4 Reinforcement and Face Panel Details for Several Reinforced Soil Systems Used in North America

System name	Reinforcement detail	Typical face panel detail*
Reinforced Earth (The Reinforced Earth Company, 1700 N. Moore St., Arlington, VA 22209-1960	Galvanized ribbed steel strips, 0.16 in (4 mm) thick; 2 in (50 mm) wide. Epoxy-coated strips also available.	Facing panels are cruciform shaped precast concrete 4.9 ft × 4.9 ft × 5.5 in (1.5 m × 1.5 m × 14 cm). Half-size panels used at top and bottom.
VSL Retained Earth (VSL Corporation, 101 Albright Way, Los Gatos, CA 95030)	Rectangular grid of W11 or W20 plain steel bars, 24 in × 6 in (61 cm × 15 cm) grid. Each mesh may have 4, 5, or 6 longitudinal bars. Epoxy-coated meshes also available.	Precast concrete panel. Hexagon shaped, 59½ in high, 68⅜ in wide between apex points, 6.5 in thick (1.5 m × 1.75 m × 16.5 cm).
Mechanically stabilized embankment (Calif. Dept. of Transportation, Div. of Engineering Services, 5900 Folsom Blvd., P.O. Box 19128, Sacramento, CA 95819)	Rectangular grid, nine ⅜-in (9.5-mm) diameter plain steel bars on 24 in × 6 in (61 cm × 15 cm) grid. Two bar mats per panel (connected to the panel at four points).	Precast concrete; rectangular 12.5 ft (3.81 m) long, 2 ft (61 cm) high, and 8 in (20 cm) thick.
Georgia stabilized embankment (Dept. of Transportation, State of Georgia, No.2 Capitol Square, Atlanta, GA 30334-1002)	Rectangular grid of five ⅜-in (9.5-mm) diameter plain steel bars on 24 in × 6 in (61 cm × 15 cm) grid, 4 bar mats per panel.	Precast concrete panel; rectangular 6 ft (1.83 m) wide, 4 ft (1.22 m) high with offsets for interlocking.
Hilfiker retaining wall (Hilfiker Retaining Walls, PO Drawer L, Eureka, CA 95501); and Lane retaining wall (Lane Enterprises, Inc., P.O. Box 345, Pulaski, PA 16143)	Welded wire mesh, 2 in × 6 in grid (5 cm × 15 cm) of W4.5 × W3.5 (0.24 in × 0.21 in diameter), W7 × W3.5 (0.3 in × 0.21 in), W9.5 × W4 (0.34 in × 0.23 in), and W12 ×W5 (0.39 in × 0.25 in) in 8-ft-wide mats.	Welded wire mesh, wraparound with additional backing mat and 1.4 in (6.35 mm) wire screen at the soil face (with geotextile or shotcrete, if desired).
Reinforced Soil Embankment (The Hilfiker Company, 3900 Broadway, Eureka, CA 95501)	6 in × 24 in (15 cm × 61 cm) welded wire mesh: W9.5 × W20—0.34 in to 0.505 in (8.8 mm to 12.8 mm) diameter.	Precast concrete unit 12 ft 6 in (3.8 m) long, 2 ft (61 cm) high. Cast-in-place concrete facing also used.
Tensar Geogrid system (The Tensar Corporation, 1210 Citizens Parkway, Morrow, GA 30260)	Nonmetallic polymeric grid mat made from high-density polyethylene or polypropylene.	Nonmetallic polymeric grid mat (wraparound of the soil reinforcement grid with shotcrete finish, if desired), precast concrete units.
Miragrid system (Mirafi, Inc., P.O. Box 240967, Charlotte, NC 28224)	Nonmetallic polymeric grid made of polyester multifilament yarns coated with latex acrylic.	Precast concrete units or grid wrap around soil.

*Many other facing types are possible with any specific system.

TABLE 8.4 Reinforcement and Face Panel Details for Several Reinforced Soil Systems Used in North America (*Continued*)

System name	Reinforcement detail	Typical face panel detail*
Maccaferri Terramesh system (Maccaferri Gabions, Inc., 43A Governor Lane Blvd., Williamsport, MD 21795)	Continuous sheets of galvanized double-twisted woven wire mesh with PVC coating.	Rock fill gabion baskets laced to reinforcement.
Geotextile reinforced system	Continuous sheets of geotextiles at various vertical spacings.	Continuous sheets of geotextiles wrapped around (with shotcrete or gunite facing). Others possible.

*Many other facing types are possible with any specific system.

Source: From J. K. Mitchell and B. R. Christopher, "North American Practice in Reinforced Soil Systems," *Proceedings, Specialty Conference on Design and Performance of Earth Retaining Structures*, Geotechnical Division, American Society of Civil Engineers, 1990, with permission.

1970s many significant projects were in place in several countries. These included the 23-m-high Peyronnet wall on the Nice-Menton Highway and the coal and ore loading facility at the port of Dunkirk, in France; the major retaining walls built along California Route 39 and along Interstate 70 through Vail Pass in the Colorado Rocky Mountains, in the United States; the Henri Bourassa Interchange in Quebec City, Canada; the several retaining walls on the Bilbao-Behobia Expressway in Spain; and the 11-km-long wall built on the St. Denis coastal road on Reunion Island in the Indian Ocean. Subsequently, Reinforced Earth has been accepted by civil engineers in all of the world's industrialized nations, and its uses have been greatly diversified. By the end of 1989, more than 13,500 projects had been completed throughout the world, including 4500 in the United States. Predominant applications are highway and railway retaining walls and bridge abutments.

As indicated in Table 8.4, several other systems have been used since the introduction of Reinforced Earth. The Hilfiker retaining wall, which uses welded wire reinforcement and facing, was developed in the mid-1970s, and the first experimental wall was built in 1975 to confirm its feasibility. The first commercial use was on a wall built for the Southern California Edison Power Company in 1977 for repair of roads along a power line in the San Gabriel Mountains. In 1980, the use of welded wire wall expanded to larger projects, and to date about 1600 walls have been completed in the United States.

Hilfiker also developed the Reinforced Soil Embankment (RSE) system, which uses continuous welded wire reinforcement and a precast-concrete facing system. The first experimental Reinforced Soil Embankment system was constructed in 1982. The first use of RSE on a commercial project was in 1983, on State Highway 475 near the Hyde Park ski area northeast of Santa Fe, New Mexico. At that site, four reinforced soil structures were constructed totaling 17,400 ft^2 (1600 m^2) of wall face. More than 50 additional RSE systems have been constructed since.

A system using strips of steel grid (or "bar mat") reinforcement, VSL Retained Earth, was first constructed in the United States in 1981 in Hayward, California. Since then, 150 VSL Retained Earth projects containing over 600 walls totaling some 5 million ft^2 (465,000 m^2) of facing have been built in the United States.

The mechanically stabilized embankment, a bar mat system, was developed by the California Department of Transportation on the basis of its research studies starting in

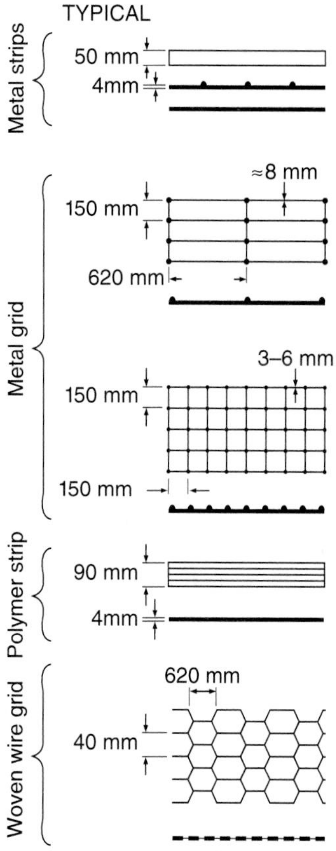

	TYPE	TYPICAL MECHANICAL PROPERTIES		
		J MODULUS* (kN/m)	TENSILE[†] CAPACITY (kN/m)	STIFFNESS RATIO ($S_v = 0.6$m) (kN/m/m)
Metal strips — TYPICAL 50 mm / 4mm	Ribbed smooth	830,000	540	90,000
Metal grid — 150 mm, 620 mm, ≈8 mm	Bar mat (strip)	80,000– 90,000	60	40,000– 60,000
Metal grid — 150 mm, 150 mm, 3–6 mm	Welded wire mesh (continuous sheet)	10,000– 40,000	10–30	17,000– 67,000
Polymer strip — 90 mm / 4mm		20,000	300	3,000
Woven wire grid — 620 mm, 40 mm	Woven wire mesh (continuous sheet)	2,000– 10,000	20–30	3,000– 16,000[‡]

J represents the modulus in terms of force per unit width of the reinforcement.

* $J = E(A_c/b)$ where: A_c = total cross section of reinforcement material
b = width of reinforcement
E = modulus of material

[†] Allowable values with no reduction for durability considerations

[‡] Confined

FIGURE 8.26 Types of reinforcement and mechanical properties. (*From J. K. Mitchell and B. R. Christopher, "North American Practice in Reinforced Soil Systems,"* Proceedings, Specialty Conference on Design and Performance of Earth Retaining Structures, *Geotechnical Division, American Society of Civil Engineers, 1990, with permission*)

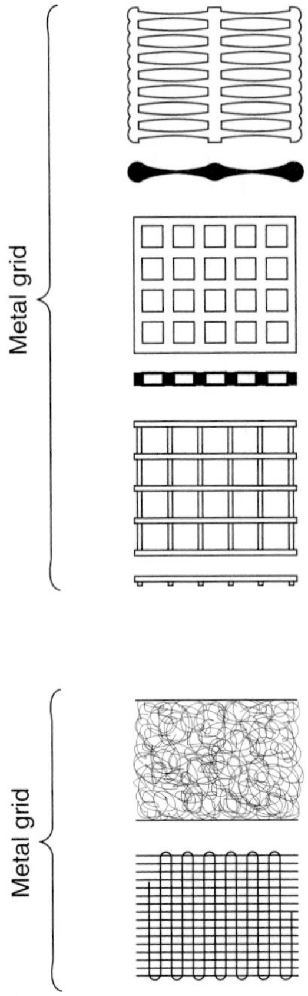

TYPE	MECHANICAL PROPERTIES		
	J MODULUS* (kN/m)	TENSILE† CAPACITY (kN/m)	STIFFNESS RATIO ($S_v=0.6$m) (kN/m/m)
Extruded (15–100-mm openings)			
Punched (5–50-mm openings)	75–2,000	5–50	125–3,000
Connected strips or strands (5–50-mm openings)			
Nonwoven	2–800‡	2–25	3–1,300
Woven	75–10,000	5–40	125–17,000

Metal grid

Metal grid

FIGURE 8.26   (*Continued*)

1973 on Reinforced Earth walls. The first wall using this bar mat type of reinforcement system was built near Dunsmuir, California, about two years later. Here, two walls were built for the realignment and widening of highway I-5. Since then, California has built numerous reinforced soil walls of various types.

Another bar mat reinforcing system, the Georgia stabilized embankment system, was developed more recently by the Georgia Department of Transportation, and the first wall using its technology was built for abutments at the I-85 and I-285 interchange in southwest Atlanta. Many additional walls have been constructed using this system.

Polymeric geogrids for soil reinforcement were developed after 1980. The first use of geogrid in earth reinforcement started in 1981. Extensive marketing of geogrid products in the United States was started about 1983 by the Tensar Corporation. Since then, over 300 wall and slope projects have been constructed using this type of reinforcement.

The use of geotextiles in reinforced soil walls started after the beneficial effect of reinforcement with geotextiles was noticed in highway embankments over weak subgrades. The first geotextile reinforced wall was constructed in France in 1981, and the first structure of this type in the United States was constructed in 1974. Since about 1980, the use of geotextiles in reinforced soil has increased significantly, with over 80 projects completed in North America.

The highest vertical reinforced soil wall constructed in the United States to date is about 100 ft. Polymeric reinforced soil walls have been constructed to a height of 40 ft.

8.5.2 Facing Systems

The types of facing elements used in the different reinforced soil systems control their aesthetics, since they are the only visible parts of the completed structure. A wide range of finishes and colors can be provided in the facing. In addition, the facing provides protection against backfill sloughing and erosion, and provides drainage paths. The type of facing influences settlement tolerances. In multianchored structures, the facing is a major structural element. Major facing types include the following:

1. *Segmental precast-concrete panels.* Examples of these are found in Reinforced Earth, the Georgia stabilized embankment system, the California mechanically stabilized embankment system, the VSL Retained Earth system, the Hilfiker Reinforced Soil Embankment, Tensar GeoWall, the American Geo-Tech system, the Stress Wall systems, the TRES system, the WEBSOL system, the Tensar system, and the York system of the Department of Environment, United Kingdom. (See Fig. 8.27.)

2. *Cast-in-place concrete, shotcrete, or full-height precast panels.* This type of facing is available in the Hilfiker and Tensar systems. Shotcrete is the most frequently used system for permanent soil nailed retaining structures. (See Fig. 8.28.)

3. *Metallic facings.* The original Reinforced Earth system had facing elements of galvanized steel sheet formed into half cylinders. Although precast concrete panels are now usually used in Reinforced Earth walls, metallic facings are still used in structures where difficult access or difficult handling requires lighter facing elements. Preformed metallic facings are also used in some soil nailing systems.

4. *Welded wire grids.* Wire grid can be bent up at the front of the wall to form the wall face. This type of facing is used in the Hilfiker and Tensar retaining wall systems. Welded wire grid facing is also commonly used with soil nailing in fragmented rocks or intermediate soils (chalk, marl, shales).

5. *Gabion facing.* Gabions (rock-filled wire baskets) can be used as facing with reinforcing strips consisting of welded wire mesh, welded bar mats, polymer geogrids, or the double-twisted woven mesh used for gabions placed between the gabion baskets.

6. *Fabric facing.* Various types of geotextile reinforcement are looped around at the facing to form the exposed face of the retaining wall. These faces are susceptible to ultraviolet light degradation, vandalism (e.g., target practice), and damage due to fire.

7. *Plastic grids.* A plastic grid used for the reinforcement of the soil can be looped around to form the face of the completed retaining structure in a similar manner to welded wire mesh and fabric facing. Vegetation can grow through the grid

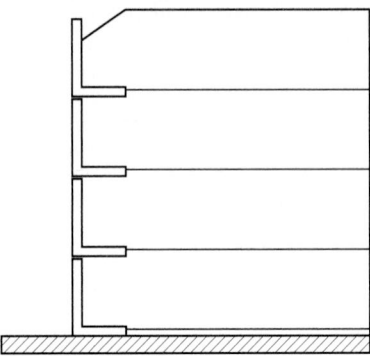

FIGURE 8.27   Sloping or vertical wall with reinforcement attached to precast-concrete facing elements.

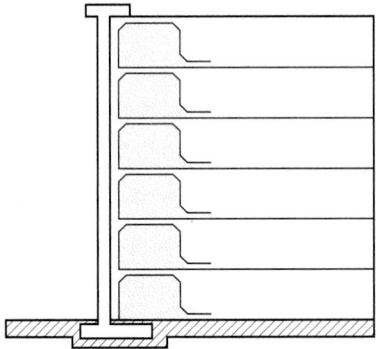

FIGURE 8.28   Vertical wall with cast-in-place concrete facing. Reinforcement is wrapped around fill used for drainage.

structure and can provide both ultraviolet light protection for the polymer and a pleasing appearance.

8. *Postconstruction facing.*   For wrapped faced walls, whether geotextiles, geogrids, or wire mesh, a facing can be attached after construction of the wall by shotcreting, guniting, or attaching prefabricated facing panels made of concrete, wood, or other materials. Shotcrete is the most frequently used system for permanent soil nailed retaining structures.

Precast elements can be cast in several shapes and provided with facing textures to match environmental requirements and to blend aesthetically into the environment. Retaining structures using precast-concrete elements as the facings can have surface finishes similar to any reinforced concrete structure. In addition, the use of separate panels provides the flexibility to absorb differential movements, both vertically and horizontally, without undesirable cracking, which could occur in a rigid structure.

Retaining structures with metal facings have the disadvantage of shorter life because of corrosion unless provision is made to compensate for it.

Facings using welded wire or gabions have the disadvantages of an uneven surface, exposed backfill materials, more tendency for erosion of the retained soil, possible shorter life from corrosion of the wires, and more susceptibility to vandalism. These can, of course, be countered by providing shotcrete or hanging facing panels on the exposed face and compensating for possible corrosion. The greatest advantages of such facings are low cost; ease of installation; design flexibility; good drainage (depending on the type of backfill), which provides increased stability; and possible treatment of the face for vegetative and other architectural effects. The facing can easily be adapted and well blended with the natural environment in the countryside. These facings, as well as geosynthetic wrapped facings, are especially advantageous for construction of temporary or other short-term design life structures.

8.5.3   Structure Dimensions

MSE walls should be dimensioned as required by AASHTO to have a minimum base width approximately 70 percent of the wall height. The soil reinforcement length is also required to meet this same minimum length criterion, but not less than 8 ft for both strip and grid type reinforcement. AASHTO requires the reinforcement length to be uniform throughout the entire height of the wall. The specification does allow deviation from this uniform length requirement, subject to the availability of substantiating evidence.

MSE walls must be designed for both external stability and internal stability. The recommended minimum factors of safety in various areas of external stability are noted in AASHTO as follows:

External stability	Factor of safety
Overturning	≥ 2.0
Ultimate bearing capacity	≥ 2.0
Sliding	≥ 1.5
Deep-seated failure	≥ 1.5
Seismic stability	$>75\%$ static safety factor

Settlement should be investigated based on a geologic study. In regard to internal stability, AASHTO notes the following:

Internal stability	Factor of safety
Pullout resistance	≥ 1.5
Rupture strength reinforcement	Allowable tension
Durability	According to design life
Seismic stability	1.1

Figures 8.29 and 8.30 indicate the two basic failure modes for internal stability analysis—specifically, rupture or creep failure of the reinforcement and a pullout failure mode. These failure modes suggest the use of the tied-back wedge analysis approach

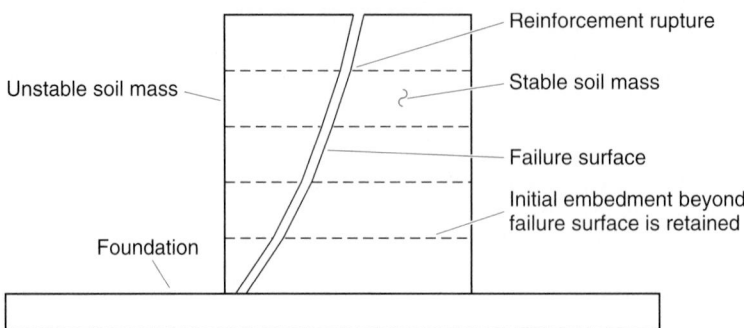

FIGURE 8.29 Reinforcement rupture or creep failure mode for internal stability evaluation.

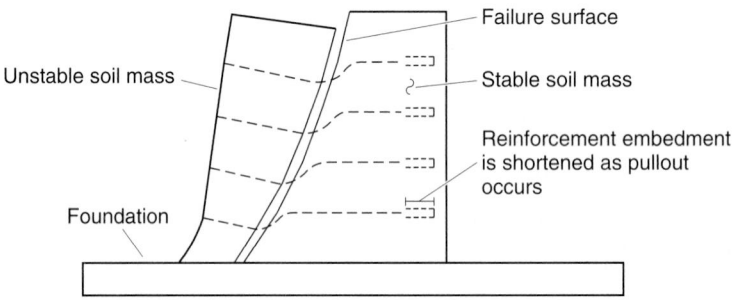

FIGURE 8.30 Reinforcement pullout failure mode for internal stability evaluation.

depicted in Fig. 8.31, which represents the basic method of analysis included in the AASHTO specifications.

8.5.4 Reinforced Fill Materials

Well-graded, free-draining granular material is usually specified for permanent-placed soil reinforced walls. Lower-quality materials are sometimes used in reinforced embankment slopes. Experience with cohesive backfills is limited. However, low strength, creep properties, and poor drainage characteristics make their use undesirable. Some current research is focused on the use of cohesive soil backfills.

The following gradation and plasticity limits have been established by the AASHTO-AGC-ARTBA* Joint Committee Task Force 27 for mechanically stabilized embankments:

*AGC, Associated General Contractors; ARTBA, American Road and Transportation Builders Association.

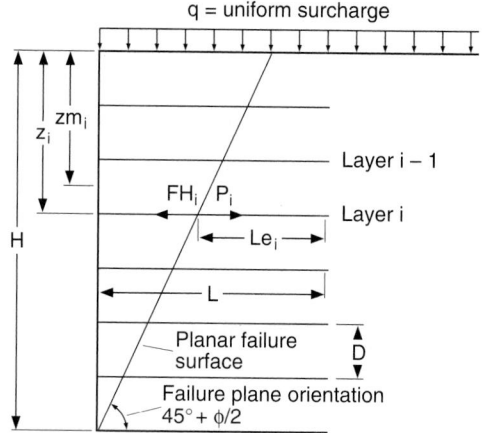

FIGURE 8.31 Parameters for tied-back wedge analysis.

U.S. sieve size	Percent passing
4 in	100
No. 40	0–60
No. 200	0–15

Plasticity index (PI) less than 6 percent

It it recommended that the maximum particle size be limited to $\frac{3}{4}$ in for geosynthetics and epoxy-coated reinforcements unless tests show that there is minimal construction damage if larger particle sizes are used.

Metallurgical slag or cinders should not be used except as specifically allowed by the designer. Material should be furnished that exhibits an angle of internal friction of 34° or more, as determined by AASHTO T-236, on the portion finer than the No. 10 sieve. The backfill material should be compacted to 95 percent of AASHTO T-99, method C or D, at optimum moisture content. See Art. 8.5.7 for backfill requirements that are important in relation to the durability of the steel reinforcement.

On-site or local material of marginal quality can be used only with the discretion and approval of the designer.

8.5.5 Design Methodology for MSE Walls

Figure 8.32 shows the general design equations given by AASHTO for MSE walls with a horizontal backslope and a traffic surcharge. Included is the calculation of safety factors for overturning and sliding, and the maximum base pressure. Inclusion of a traffic surcharge is required only in those instances where traffic loadings will actually surcharge the wall. Separate surcharge diagrams are applied for the two conditions shown. For stability of the mass, the traffic surcharge should act at the end of the reinforced zone so as to eliminate the "stabilizing" effect of this loading. However, for purposes of determining horizontal stresses, which are increased as a result of this surcharge, the loading is deemed to apply over the entire surface of the wall backfill. Figure 8.33 shows the AASHTO equations for the sloping backfill case.

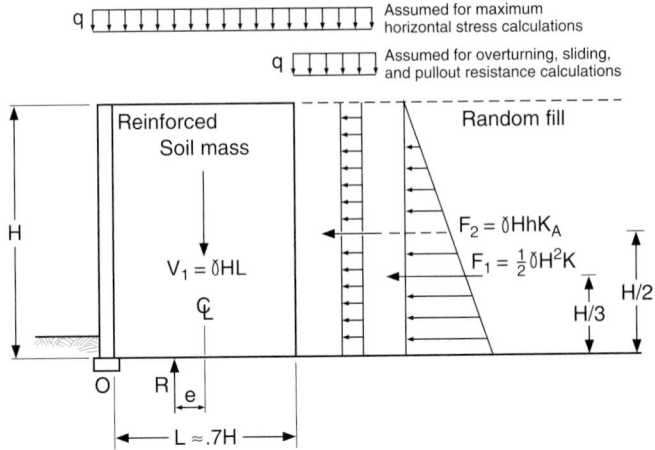

FIGURE 8.32 General design requirements for MSE walls with horizontal backfill and traffic surcharge. (*From* Standard Specifications for Highway Bridges, *American Association of State Highway and Transportation Officials, Washington, D.C., 1992, with permission*)

SAFETY FACTOR AGAINST OVERTURNING (MOMENTS ABOUT POINT O):

$$\text{S.F.(O)} = \frac{\Sigma \text{ moments resisting } (M_r)}{\Sigma \text{ moments overturning } (M_o)} = \frac{V_1(L/2)}{F_1(H/3) + F_2(H/2)} \geq 2.0$$

SAFETY FACTOR AGAINST SLIDING:

$$\text{S.F.(S)} = \frac{\Sigma \text{ horizontal resisting force(s)}}{\Sigma \text{ horizontal driving force(s)}} = \frac{(V_1 + V_2) \tan \phi}{F_1 + F_2} \geq 1.5$$

ϕ = friction angle of backfill or foundation, whichever is lower

$$e = \frac{L}{2} - \frac{M_r - M_o}{R} \leq \frac{L}{6} \qquad \sigma_v = \frac{R}{L - 2e}$$

where: e = eccentricity
q = traffic surcharge
σ_v = maximum base pressure
R = resultant of vertical forces $(V_1 + V_2)$
K = varies from K_o to K_a

While the conventional analysis of a mechanically stabilized earth wall assumes a rigid body, field evaluation has shown that the variation and magnitude of the foundation loading exerted by the wall on the underlying soil differ from the traditional trapezoidal pressure distribution assumed under reinforced-concrete cantilever walls. Tests were performed by placing pressure cells under the base of an MSE wall. The wall was the Fremersdorf wall constructed in Germany, which is depicted in Fig. 8.34 along with the bearing pressure recorded from the pressure cells. Tests on that structure demonstrated that loading is greater toward the front of the structure because of earth pressure imposed by the retained fill behind the wall. In addition, the total load was slightly greater than the total weight of the wall, indicating that the thrust behind the structure was inclined. The difference between total loading and weight, and the location of the resultant, made it possible to compute the thrust angle β.

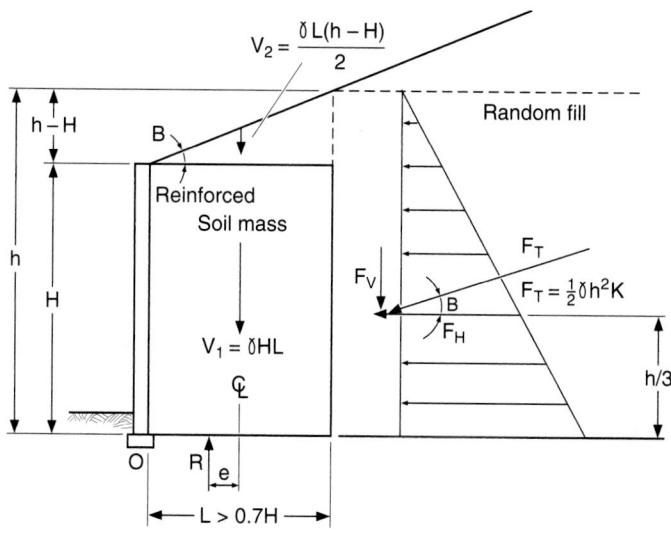

SAFETY FACTOR AGAINST OVERTURNING (MOMENTS ABOUT POINT O):

$$S.F.(O) = \frac{\Sigma \text{ moments resisting } (M_r)}{\Sigma \text{ moments overturning } (M_o)} = \frac{V_1(L/2) + V_2(2L/3) + F_V(L)}{F_H(h/3)} \geq 2.0$$

SAFETY FACTOR AGAINST SLIDING:

$$S.F.(S) = \frac{\Sigma \text{ horizontal resisting force(s)}}{\Sigma \text{ horizontal driving force(s)}} = \frac{R \tan \phi}{F_H} \geq 1.5$$

ϕ = friction angle of backfill or foundation, whichever is lower

$$e = \frac{L}{2} - \frac{M_r - M_o}{R} \leq \frac{L}{6} \qquad \sigma_v = \frac{R}{L - 2e}$$

where: e = eccentricity
σ_v = maximum base pressure
R = resultant of vertical forces $(V_1 + V_2 + F_V)$

FIGURE 8.33   General design requirements for MSE walls with sloping backfill. (*From* Standard Specifications for Highway Bridges, *American Association of State Highway and Transportation Officials, Washington, D.C., 1992, with permission*)

The bearing pressure distribution from the Fremersdorf wall is idealized in the AASHTO equation for soil pressure (σ_v) shown in Fig. 8.35. A uniform pressure (Meyerhof distribution) is calculated over a width equal to the length of the soil reinforcement elements minus 2 times the eccentricity of the vertical force.

8.5.6   Superimposed versus Terraced Structures

There are instances when one MSE wall is built on top of another. In certain instances, these walls can be considered to be two independent structures, each requiring its own internal design and external stability. The global stability of the slope must be sufficiently stable so as not to undermine the stability of the entire embankment.

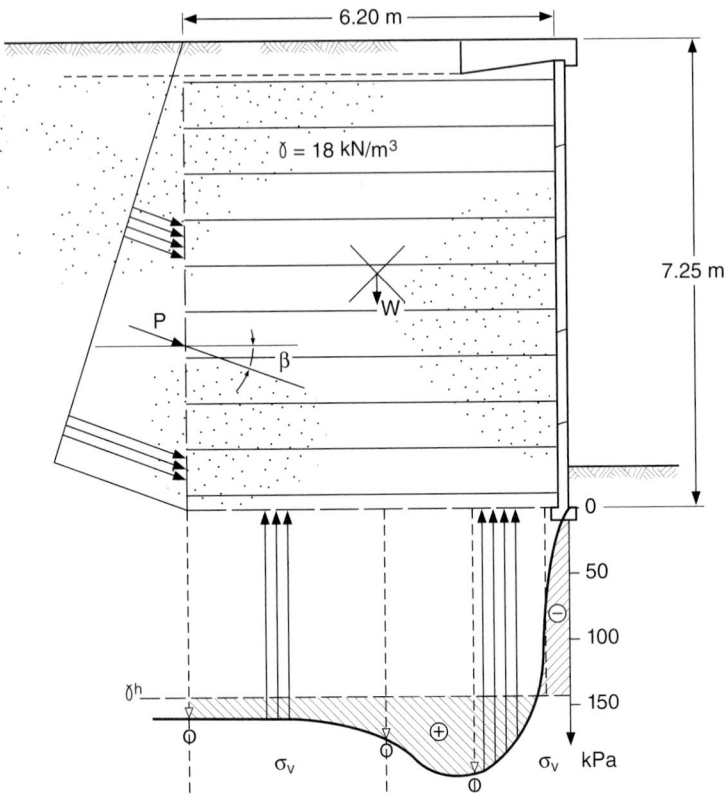

FIGURE 8.34   Fremersdorf MSE wall with foundation pressures from pressure cell readings. (*From the Reinforced Earth Co., with permission*)

Figure 8.36 shows a superimposed structure. The walls are such that the load of the upper wall level serves as a surcharge load on the lower wall. Each wall is independently designed.

This design approach does not hold when the MSE structures are directly superimposed, one on another, as shown in Fig. 8.37. Such terraced arrangements are sometimes used for high walls. These offset structures are obviously similar to a single embankment with a sloping face. They exhibit essentially the same overall behavior, and are designed as sloping faced walls.

8.5.7   Durability Considerations for MSE Walls with Metal Reinforcement

Where metallic reinforcement is used, the life of the structure will depend on the corrosion resistance of the reinforcement. Practically all the metallic reinforcements used in construction of embankments and walls, whether they are strips, bar mats, or wire mesh, are made of galvanized steel. Epoxy coating can be used for additional corrosion protection, but it is susceptible to construction damage, which can significantly reduce its effectiveness. PVC coatings on wire mesh also provide corrosion protec-

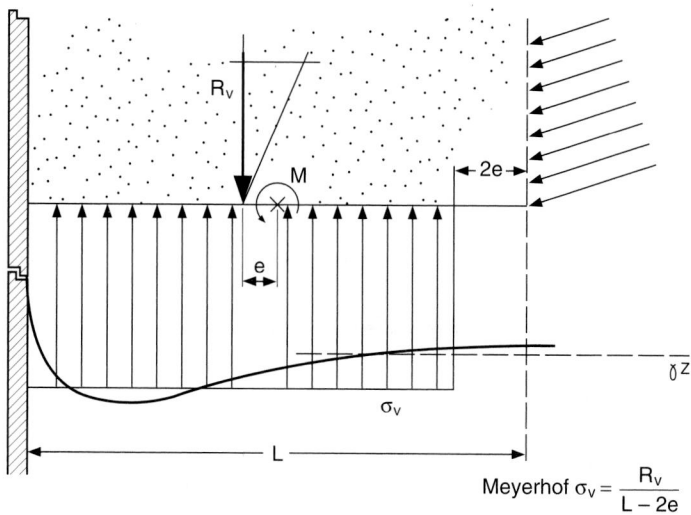

Meyerhof $\sigma_v = \dfrac{R_v}{L - 2e}$

FIGURE 8.35   Foundation pressure for MSE wall calculated by the AASHTO method based on Meyerhof. (*From the Reinforced Earth Co., with permission*)

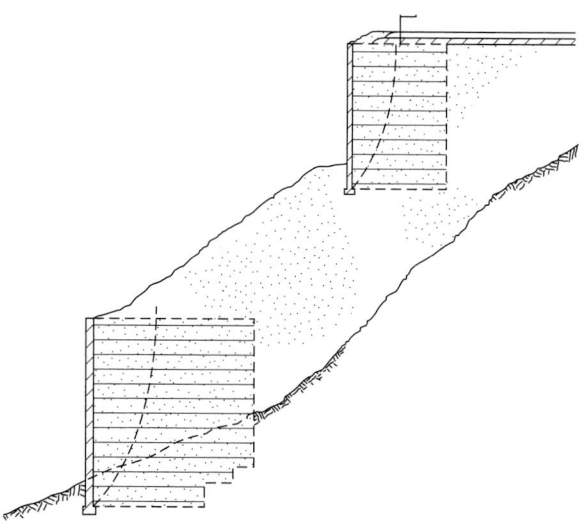

FIGURE 8.36   Superimposed MSE walls. (*From the Reinforced Earth Co., with permission*)

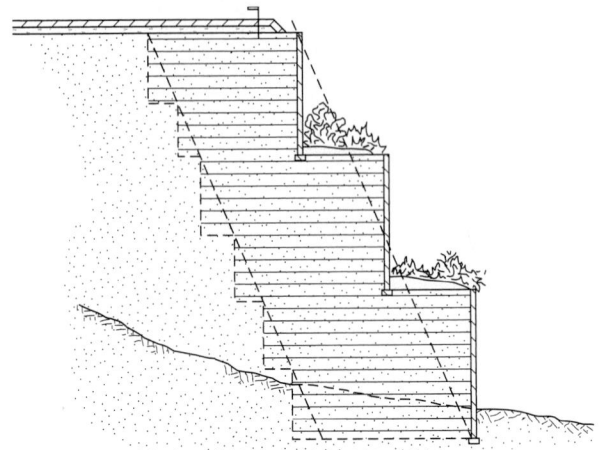

FIGURE 8.37   Terraced MSE wall. (*From the Reinforced Earth Co., with permission*)

tion, provided again that the coating is not significantly damaged during construction. When PVC or epoxy coatings are used, the maximum particle size of the backfill should be restricted to ¾ in or less to reduce the potential for construction damage.

For the purpose of determining the sacrificial metal required (corrosion allowance), the following design life is provided, pursuant to recommendations of Task Force 27 of AASHTO-AGC-ARTBA:

Structure classification	Design life, yr
Permanent structure	75
Abutments	100
Rail supporting structures	100
Marine structures	75

The required cross-sectional area of steel reinforcement is calculated using the relationships given in Fig. 8.38 for the selected type of reinforcement (strips or grids). The corrosion loss assumed is based on the following.

In 1985, an FHWA study was initiated to develop practical design and construction guidelines from a technical review of extensive laboratory and field tests on buried metals. The results of this research were published in December 1990 in the Federal Highway Administration report FHWA-RD-89-186, "Durability/Corrosion of Soil Reinforced Structures":

> For structures constructed with carefully selected and tested backfills to ensure full compliance with the electrochemical requirements, the maximum mass presumed to be lost per side due to corrosion at the end of the required service life may be computed by assuming a uniform loss model which considers the following loss rates:
>
> **1.** Zinc corrosion rate for first 2 years: 15 μm/yr
> **2.** Zinc corrosion to depletion: 4 μm/yr
> **3.** Carbon steel rate: 12 μm/yr

STEEL STRIPS

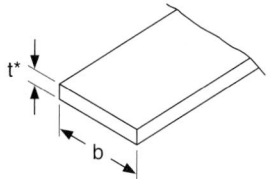

$A_c = b \cdot t^*$

Where t^* = thickness corrected for corrosion loss
A_c = cross section area

STEEL GRIDS

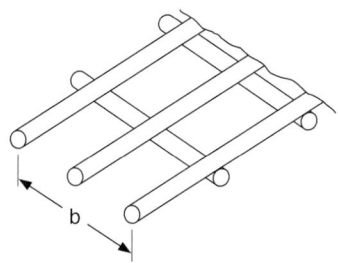

A_c = number of bars $\cdot \pi \dfrac{(d^*)^2}{4}$

Where d^* = diameter of bar or wire corrected for corrosion loss

FIGURE 8.38   Metallic reinforcement for MSE walls showing correction for corrosion loss. (*From the Reinforced Earth Co., with permission*)

The resulting sacrificial thickness for a 75-year life based on initial galvanization of 2 oz/ft^2 (86 μm) is approximately 1.5 mm of total sacrificial thicknesses. Since this is a *maximum* loss rate, it is presently assumed that the reduced minimum thickness remains proportional to tensile strength and therefore no further reduction is necessary. (See Fig. 8.39.)

The select backfill materials shall meet the following requirements:

Internal friction angle.   The material shall exhibit an internal friction angle of not less than 34 degrees as determined by the standard direct shear test, AASHTO T-236, utilizing a sample of the material compacted to 95 percent of AASHTO T-99, Methods C or D (with oversize correction), at optimum moisture content. Internal friction angle testing is not required for backfill materials that have at least 80 percent of the material greater than or equal to the ¾-in size.

Soundness.   The material shall be substantially free of shale or other soft, poor durability particles. The material shall have a magnesium sulfate soundness loss of less than 30 percent after four (4) cycles, as determined by AASHTO T-104.

Electrochemical requirements.   The material shall conform to the electrochemical requirements as described in Table 8.5.

The Contractor shall furnish to the Engineer a Certificate of Compliance certifying that the select granular backfill material complies with this section of the specification. A

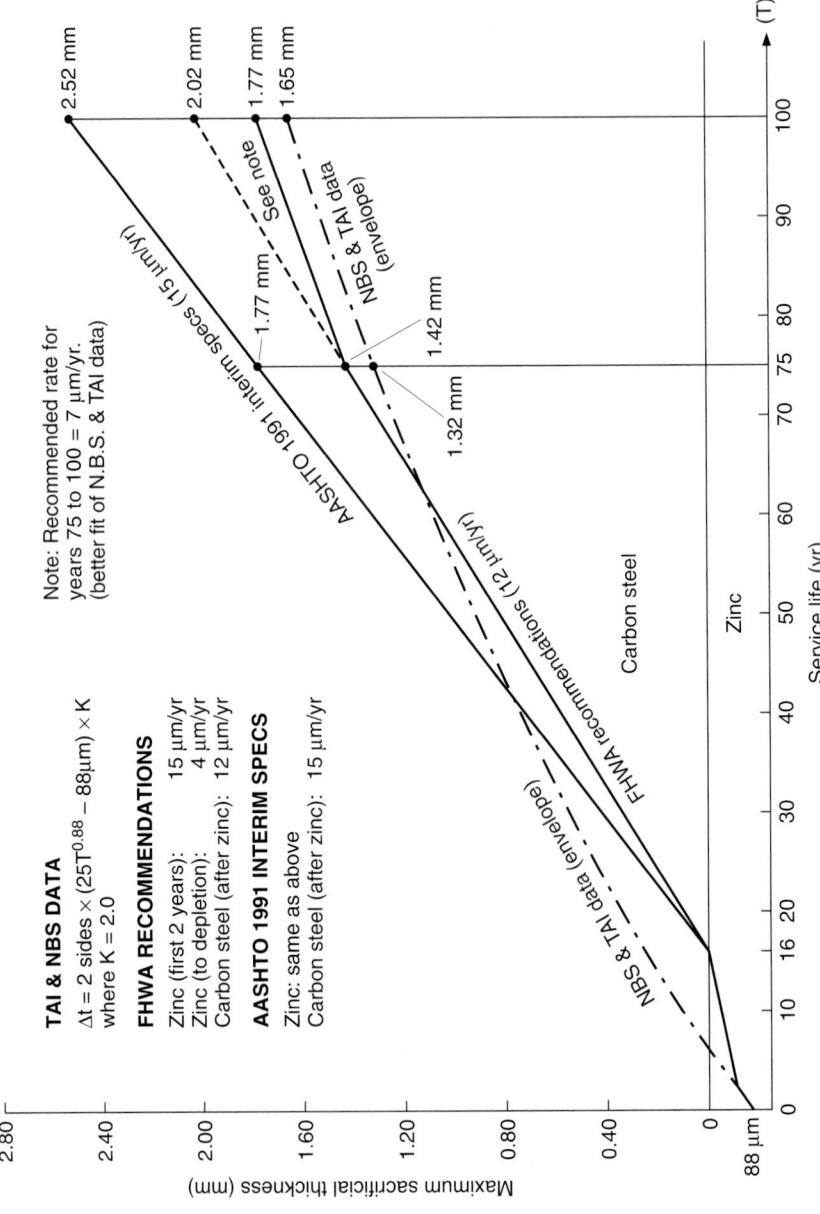

FIGURE 8.39 Maximum sacrificial thickness for calculating corrosion loss. *(From the Reinforced Earth Co., with permission)*

8.58

TABLE 8.5 Backfill Requirements Related to Durability of Steel
Reinforcement

Property	Requirement	Test method
Resistivity	Minimum 3000 $\Omega \cdot$ cm, at 100% saturation	California DOT 643
pH	Acceptable range 5–10	California DOT 643
Chlorides	Maximum 100 ppm	California DOT 422
Sulfates	Maximum 200 ppm	California DOT 417

Source: From the Reinforced Earth Co., with permission.

copy of all test results performed by the Contractor, which are necessary to assure compliance with the specifications shall also be furnished to the Engineer.

Backfill not conforming to this specification shall not be used without the written consent of the Engineer.

The frequency of sampling of select granular backfill material, necessary to assure gradation control throughout construction, shall be as directed by the Engineer.

8.5.8 Durability Considerations for MSE Walls with Polymeric Reinforcement

The durability of polymeric reinforcements is influenced by time, temperature, mechanical damage, stress levels, microbiological attack, and changes in the molecular structure due to radiation or chemical exposure. The effects of aging and of chemical and biological exposure are highly dependent on material composition, including resin type, grade, and additives; manufacturing process; and final product physical structure.

Polymeric reinforcement, although not susceptible to corrosion, may degrade as a result of physicochemical activity in the soil, such as hydrolysis, oxidation, and environmental stress cracking. In addition, it is susceptible to construction damage, and some forms may be adversely affected by prolonged exposure to ultraviolet light. The durability of geosynthetics is a complex subject, and research is ongoing to develop reliable procedures for quantification of degradation effects. Moderate-strength geosynthetics have tensile strengths of about 100 lb/in; some are now available that have strengths well over an order of magnitude higher. Current procedure to account for strength loss due to construction damage, and as a result of aging and chemical and biological attack, is to decrease the initial strength of the intact, unaged material for design.

8.5.9 Design Example of MSE Retaining Wall with Steel Reinforcement

The following design example is provided with the permission of the Reinforced Earth Company. Typical calculations are shown, including the determination of allowable reinforcement tension for galvanized steel reinforcing strips. Figure 8.40 shows a cutaway view of a typical Reinforced Earth retaining wall. Refer to Fig. 8.41 for illustration of calculation steps.

Geometry

$$\text{Height of wall } H = 20 \text{ ft}$$

$$\text{Strip length } B = 20 \text{ ft (AASHTO minimum} = 0.7H = 14 \text{ ft)}$$

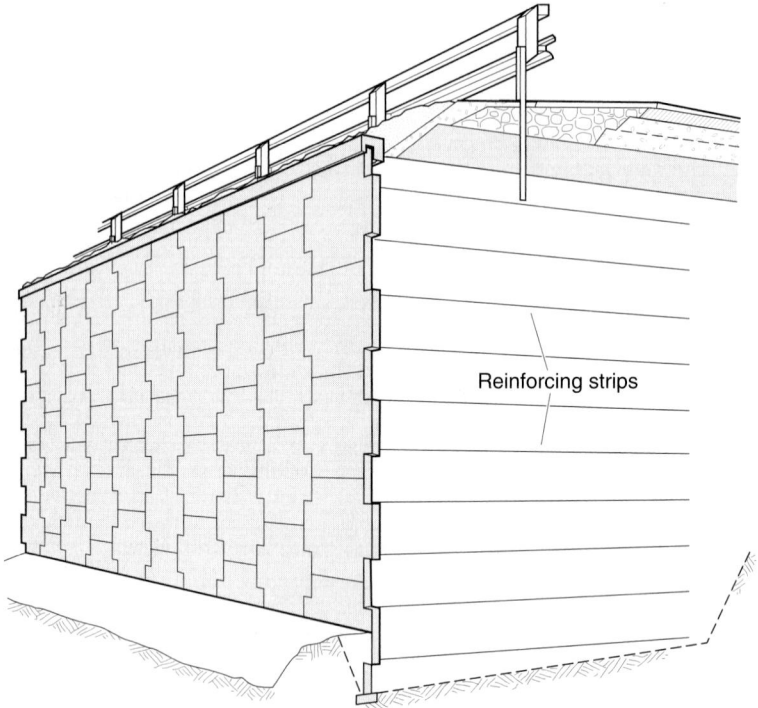

FIGURE 8.40   Cutaway view of typical Reinforced Earth retaining wall. (*From the Reinforced Earth Co., with permission*)

Soil Properties

	ϕ	Cohesion, c	Unit weight, γ
R.E. material	34.00°	—	0.125 kip/ft³
Random fill	25.00°	0.100 kip/ft²	0.120 kip/ft³
Foundation	25.00°	0.300 kip/ft²	—

Other Properties

- Equivalent fill height for traffic surcharge of 0.25 kip/ft² =

$$\frac{0.25 \text{ kip/ft}^2}{0.120 \text{ kip/ft}^3} = 2.08 \text{ ft}$$

- Maximum value of apparent coefficient of friction (bond) = 1.50.
- Coefficient of friction at foundation level (sliding) = 0.47.
- Surface area of one "A" panel = 24.2 ft².
- Maximum reinforcement tension = 7.20 kips per strip.
- Stress at connection = 100 percent of maximum tie tension.

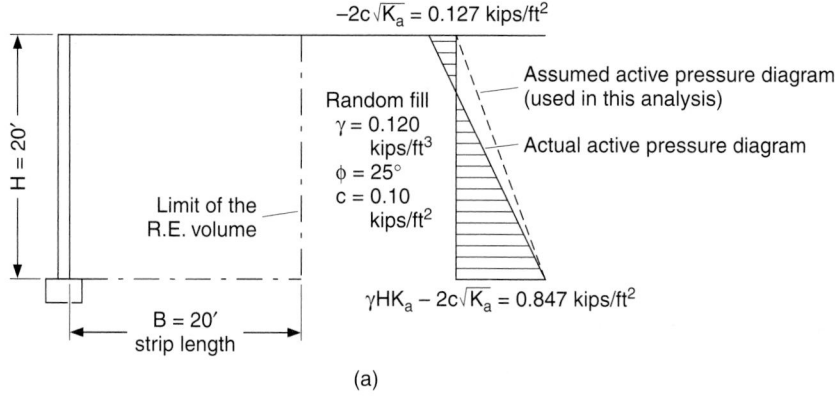

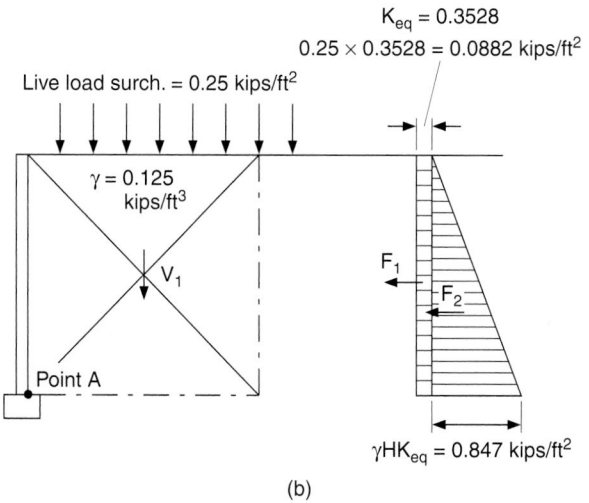

FIGURE 8.41 Design of Reinforced Earth retaining wall showing (*a*) active earth pressure, (*b*) addition of pressure from surcharge, (*c*) analysis at intermediate level, and (*d*) effective strip length. (*From the Reinforced Earth Co., with permission*)

General Calculations. Random fill is used outside the zone filled with R.E. (Reinforced Earth) material.

Pressure Coefficient for Random Fill. For the case of level ground at the top of the wall, a vertical backface, and neglecting the effect of wall friction, the pressure coefficient for the fill is given by

$$K_a = \tan^2\left(45° - \frac{\phi}{2}\right)$$

Substituting $\phi = 25°$ gives

$$K_a = \tan^2\left(45° - \frac{25°}{2}\right) = 0.4059$$

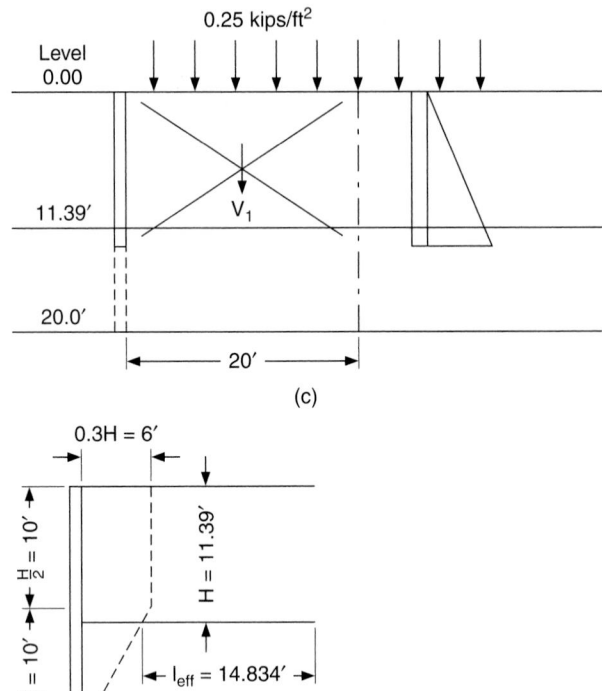

(c)

(d)

FIGURE 8.41 (*Continued*)

To allow for the effects of cohesion in the fill (see Fig. 8.41*b*), define an equivalent pressure K_{eq} such that

$$\gamma H K_{eq} = \gamma H K_a - 2c\sqrt{K_a}$$

Thus

$$K_{eq} = K_a - \frac{2c\sqrt{K_a}}{\gamma H} = \tan^2\left(45° - \frac{\phi_{eq}}{2}\right)$$

Solving for ϕ_{eq}, the equivalent soil friction angle can then be calculated as

$$\phi_{eq} = 2\left(45° - \arctan\sqrt{K_a - \frac{2c\sqrt{K_a}}{\gamma H}}\right)$$

$$= 2\left(45° - \arctan\sqrt{0.4059 - \frac{2 \times 0.10 \text{ kip/ft}^2 \times \sqrt{0.4059}}{0.12 \text{ kip/ft}^3 \times 20 \text{ ft}}}\right)$$

$$= 28.58°$$

The calculation of the equivalent pressure coefficient follows as

$$K_{eq} = \tan^2\left(45° - \frac{25.58°}{2}\right) = 0.3528$$

This coefficient is subsequently used to calculate F_1, the horizontal force on the wall caused by the surcharge, and F_2, the horizontal force on the wall caused by the fill.

Vertical Loads and Resisting Moment. The vertical loads to be considered are the weight of the reinforced fill, V_1, and of the surcharge, V_{surch}. These loads are calculated and multiplied by their horizontal moment arm from the base (point A in Fig. 8.41b), and the results are summed to determine the resisting moment M_r. The sum of the vertical loads is designated R_V.

Load, kips/ft	Moment arm, ft	Resisting moment M_r, kips · ft/ft
$V_1 = 0.125$ kip/ft$^3 \times$ 20 ft $\times$ 20 ft $= 50.0$ kips/ft	10 ft	500 kips · ft/ft
$V_{surch} = 0.25$ kip/ft$^2 \times$ 20 ft $= 5.0$ kips/ft	10 ft	50 kips · ft/ft
$R_v = V_1 + V_{surch} = 50 + 5 = 55$ kips/ft		Total $M_r = 550$ kips · ft/ft

Horizontal Forces and Overturning Moment. The horizontal force due to the surcharge, F_1, and that due to the random fill, F_2, are illustrated in Fig. 8.41b. They are calculated using the value of K_{eq} determined previously and multiplied by their vertical moment arm from the base, and the results are summed to determine the overturning moment M_o.

Load, kips/ft	Moment arm, ft	Overturning moment M_o, kips · ft/ft
$F_1 = 0.3528 \times 0.250$ kip/ft$^2 \times$ 20 ft $= 1.764$ kips/ft	10 ft	17.64 kips · ft/ft
$F_2 = 0.3528 \times 0.120$ kip/ft$^3 \times$ (20 ft)$^2 \times$ (½) $= 8.47$ kips/ft	20 ft/3 $= 6.67$ ft	56.44 kips · ft/ft
$F_1 + F_2 = 10.23$ kips/ft		$M_o = 74.08$ kips · ft/ft

Eccentricity e *(without Surcharge).* The eccentricity without surcharge must be calculated to make sure it is less than one-sixth of the base dimension B, which is the length of the reinforcing strip.

$$e = \frac{B}{2} - \frac{M_r - M_o}{V_1} = \frac{20}{2} - \frac{500 - 74.08}{50} = 1.4816 \text{ ft}$$

$$e < \frac{B}{6} = \frac{20}{6} = 3.33 \text{ ft} \quad \text{OK}$$

Safety Factors. The safety factor against overturning is the ratio of the resisting moment to the overturning moment. The safety factor against sliding is the ratio of the horizontal resisting forces (weight of reinforced fill times friction factor plus foundation cohesion force) to the horizontal active forces. These safety factors must be calculated to make sure they are within limits.

$$\text{SF (overturning)} = \frac{M_r}{M_o} = \frac{500}{74.08} = 6.75 > 2.0 \quad \text{OK}$$

$$\text{SF (sliding)} = \frac{V_1 \tan 25° + c \times B}{F_1 + F_2}$$

$$= \frac{50 \times \tan 25° + 0.3 \times 20}{10.23} = 2.87 > 1.50 \quad \text{OK}$$

Eccentricity e *(with Surcharge; Use Total* M_r*).* The eccentricity calculated with the surcharge should also be less than $B/6$. This value of e will be used to calculate the bearing pressure.

$$e = \frac{B}{2} - \frac{M_r - M_o}{R_v} = \frac{20}{2} - \frac{550 - 74.08}{55} = 1.347 \text{ ft}$$

Bearing Pressure. The bearing pressure σ_V under the reinforced fill can be calculated from Myerhof's equation. The pressure must be within the allowable value for the site.

$$\sigma_v = \frac{\Sigma V}{B - 2e}$$

$$= \frac{55 \text{ kips}}{20 \text{ ft} - 2 \times 1.347 \text{ ft}} = 3.18 \text{ kips/ft}^2$$

Design at Intermediate Level. Design is illustrated for a level 11.39 ft below the top of the wall. (See Fig. 8.41c and d.) The same procedure is used for other levels.

$$\Sigma V_1 = 0.25 \text{ kip/ft}^2 \times 20 \text{ ft} + 0.125 \text{ kip/ft}^3 \times 11.39 \text{ ft} \times 20 \text{ ft}$$
$$= 33.475 \text{ kips/ft}$$

$$\text{Resisting moment} = 33.475 \text{ kips} \times 10 \text{ ft} = 334.75 \text{ kips} \cdot \text{ft/ft}$$

$$\text{Overturning moment } \Sigma M_o = 0.3528 \times [0.25 \text{ kip/ft}^2 \times (11.39)^2 \times \tfrac{1}{2} +$$
$$0.120 \text{ kip/ft}^3 \times (11.39)^3 \times \tfrac{1}{6}] = 16.15 \text{ kips} \cdot \text{ft/ft}$$

$$\text{Safety factor for overturning} = \frac{334.75}{16.15} = 20 > 2.0 \quad \text{OK}$$

$$e = \frac{20}{2} - \frac{334.75 - 16.15}{33.475} = 0.482 \text{ ft}$$

$$e \leq \frac{20}{6} = 3.33 \text{ ft} \quad \text{OK}$$

$$\sigma_v = \frac{\Sigma V_1}{B - 2e} = \frac{33.475 \text{ kip}}{20 \text{ ft} - 2 \times 0.482 \text{ ft}} = 1.759 \text{ kips/ft}^2$$

The pressure coefficient K is assumed to vary linearly between K_0 (the coefficient of earth pressure at rest) at the top of the wall and K_a (the coefficient of active earth pressure) at a depth of 20 ft. Below a 20-ft depth, $K = K_a$. The distance below the top of the wall is d.

Maximum horizontal pressure $\sigma = K\sigma_v$

$$K_0 = 1 - \sin 34° = 0.4408$$

$$K_a = \tan^2\left(45° - \frac{34°}{2}\right) = 0.2827$$

$$K = K_0 - \frac{K_0 - K_a}{20 \text{ ft}} \times d$$

$$= 0.4408 - \frac{0.4408 - 0.2827}{20 \text{ ft}} \times 11.39 \text{ ft}$$

$$K = 0.3508$$

$$\sigma_h = 0.3508 \times 1.759 \text{ kips/ft}^2 = 0.617 \text{ kip/ft}^2$$

The area of a standard "A" panel is 24 ft^2. Use four strips per panel.

$$\text{Reinforcing strip tension} = \frac{0.617 \times 24}{4} = 3.73 \text{ kips per strip}$$

3.73 kips/strip < 7.2 maximum tension for 75-yr design life   OK

(See subsequent calculations for maximum tension allowable for strip and connections.)
Check length of strip:

$$V = 0.125 \text{ kip/ft}^2 \times 11.39 \text{ ft} \times 20 \text{ ft} = 28.475 \text{ kips (with-}$$

out surcharge)

Resisting moment = 28.475 kips $\times$ 10 ft = 284.75 kips $\cdot$ ft/ft

Overturning moment $\Sigma M_o = 16.15$ kips $\cdot$ ft

$$\text{Safety factor for overturning} = \frac{284.75}{16.15} = 18 > 2.0 \qquad \text{OK}$$

$$e = \frac{20}{2} - \frac{284.75 - 16.15}{28.475} = 0.567 \text{ ft}$$

$$e \leq \tfrac{20}{6} = 3.33 \text{ ft} \qquad \text{OK}$$

$$\sigma_v = \frac{28.475 \text{ kips}}{20 \text{ ft} - 2 \times 0.567 \text{ ft}} = 1.509 \text{ kips/ft}^2$$

$$\sigma_h = 0.3508 \times 1.509 \text{ kips/ft}^2 = 0.5295 \text{ kip/ft}^2$$

$$T = \text{tension on an "A" panel} = \sigma_h \times A = 0.5295 \text{ kip/ft}^2$$

$$\times 24.2 \text{ ft}^2$$

$$= 12.81 \text{ kips}$$

$$R = \text{frictional resistance of reinforcing strips}$$

$$= 2b \times l_{\text{eff}} \times H \times \delta \times f^* \times N$$

where $2b = \dfrac{2 \times 1.97}{2} = 0.328$ ft = width of top and bottom surface of one strip

$H = 11.39$ ft = overburden
$l_{\text{eff}} = 14.834$ ft = effective strip length
$\delta = 0.125$ kip/ft^3
$f^* = 1.5 - [(1.5 - \tan 34° \times 11.39 \text{ ft})/20 \text{ ft}] = 1.03$
     = coefficient of apparent friction
$N = 4$ = number of strips per panel

$$R = 0.328 \text{ ft} \times 14.834 \text{ ft} \times 11.39 \times 0.125 \text{ kip/ft}^3 \times 1.03 \times 4 = 28.54 \text{ kips}$$

$$\text{Effective length safety factor} = \frac{R}{T} = \frac{28.54}{12.81} = 2.23 > 1.5 \qquad \text{OK}$$

Design Summary at Intermediate Levels

Level, ft	Maximum horizontal stress,	Stress at facing, kips/ft^2	Straps per panel	Reinforcing strip tension, kips	Horizontal stress (bond), kips/ft^2	Effective length safety factor	Strip length, ft
2.00	0.21	0.21	4	1.29	0.11	2.52	20.00
4.01	0.31	0.31	3	2.50	0.21	1.84	20.00
6.47	0.42	0.42	3	3.39	0.32	1.76	20.00
8.93	0.52	0.52	3	4.21	0.43	1.67	20.00
11.39	0.62	0.62	4	3.73	0.53	2.23	20.00
13.85	0.70	0.70	4	4.26	0.62	2.28	20.00
16.31	0.79	0.79	4	4.76	0.71	2.29	20.00
18.77	0.86	0.86	4	5.22	0.79	2.25	20.00

Calculation of Allowable Reinforcement Tension. The following calculations show the determination of the allowable reinforcement tension for galvanized reinforcing strips in permanent mechanically stabilized earth structures. Allowable stresses in strips and components are based on the AASHTO *Bridge Specifications*. The allowable reinforcement tension is based on maintaining allowable hardware stresses to the end of a 75-year service life. After 75 years, the structure will continue to perform with reinforcement stresses that may or may not exceed allowable levels, depending on the soil environment and the applied reinforcement loads. The calculations are based on the following mechanical properties of the reinforcement components.

- *Reinforcing strips*

    50- $\times$ 4-mm ribbed (1.97 $\times$ 0.16 in)

    ASTM A572 grade 65

    F_u = 80 kips/in^2 (minimum tensile strength)

    F_y = 65 kips/in^2 (minimum yield point)

- *Tie strips*

    50 $\times$ 3.0 mm (1.97 $\times$ 0.12 in)

    ASTM A570 grade 50

    F_u = 65 kips/in^2

    F_y = 50 kips/in^2

- *Bolts*

    ½-in-diameter $\times$ 1¼ inch long

    ASTM A325

To begin, consider the tie strips at a section where there are no bolt holes. (Section A-A, Fig. 8.42). There are two 50- $\times$ 3-mm tie strip plates with 2 oz/ft^2 (86 μm) of zinc. Calculate the life of the zinc coating (see Art. 8.5.7):

$$T = 2 \text{ yr} + \frac{86 \ \mu\text{m} - 2 \ \text{yr}(15 \ \mu\text{m/yr})}{4 \ \mu\text{m/yr}} = 16 \text{ yr}$$

No carbon steel is lost until after depletion of the zinc.

Next, calculate the carbon steel loss in the subsequent 59 years. (See Art. 8.5.7.) The thickness of the carbon steel loss on one side is determined as follows:

$$\Delta e = 59 \text{ yr} \times 12 \ \mu\text{m/yr} = 708 \ \mu\text{m on each exposed side}$$

The outside surfaces of the tie strip plates are in contact with soil; the inside surfaces are not in contact with soil. Therefore, use one-half the carbon steel loss rate for the inside surfaces. The sacrificial thickness of reinforcement during service life is determined from:

$$E_S = 708 \ \mu\text{m} + 354 \ \mu\text{m} = 1062 \ \mu\text{m per plate}$$

The thickness of the reinforcement at end of service life is the nominal thickness minus the sacrificial thickness:

$$E_C = E_n - E_S = 3000 \ \mu\text{m} - 1062 \ \mu\text{m} = 1938 \ \mu\text{m per plate}$$

The cross-sectional area at end of service life is found from:

$$A_S = \frac{2 \text{ plates} \times 50 \text{ mm} \times 1938 \ \mu\text{m/plate}}{25.4 \text{ mm/in} \times 25,400 \ \mu\text{m/in}} = 0.300 \text{ in}^2$$

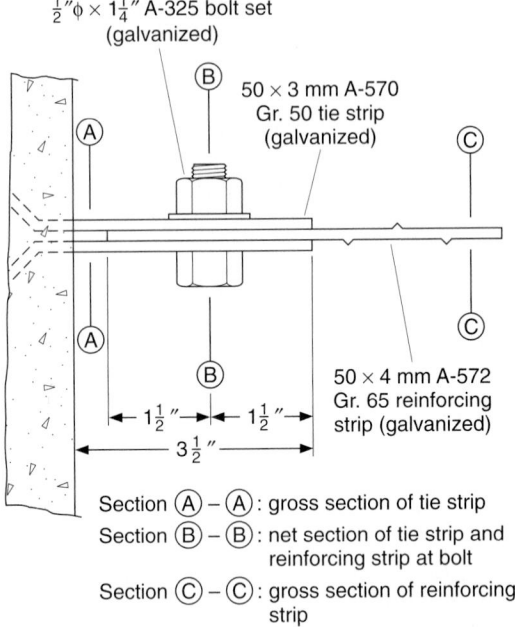

Section (A) – (A) : gross section of tie strip
Section (B) – (B) : net section of tie strip and
                    reinforcing strip at bolt
Section (C) – (C) : gross section of reinforcing
                    strip

FIGURE 8.42   Structural connection of reinforcing strip to facing panel. (*From the Reinforced Earth Co., with permission*)

The allowable tensile stress is found from:

$$F_T = 0.55F_y = 0.55(50 \text{ kips/in}^2) = 27 \text{ kips/in}^2$$

The allowable tension on reinforcement is:

$$T_{AL} = F_T A_S = 27 \text{ kips/in}^2 \times 0.300 \text{ in}^2 = 8.10 \text{ kips per connection}$$

Now, consider the tie strips at a section through the bolt holes (Section B-B, Fig. 8.42). There are two 50- × 3-mm tie strip plates with 2 oz/ft² (86 μm) of zinc. The diameter of each bolt hole is ⁹⁄₁₆ in (14.3 mm). The life of the zinc is 16 years, as found in the calculation for Section A-A.

Calculate the thickness of carbon steel loss over the subsequent 59 years:

$$\Delta e = 708 \text{ μm per exposed side}$$

(See the preceding calculation for Section A-A.) Corrosion does not occur on the inside surfaces of the plates, because of protection provided by sandwiching the reinforcing strip. Thus,

$$E_S = 708 \text{ μm per plate}$$

Proceed with calculations for thickness at end of service life, cross-sectional area, allowable tensile stress, and allowable tension force:

$$E_C = E_n - E_S = 3000 \ \mu m - 708 \ \mu m = 2292 \ \mu m \text{ per plate}$$

$$A_S = \frac{2 \text{ plates} \times (50 \text{ mm} - 14.3 \text{ mm}) \times 2292 \ \mu m/\text{plate}}{25.4 \text{ mm/in} \times 25,400 \ \mu m/\text{in}} = 0.254 \text{ in}^2$$

$$F_T = 0.50F_u = 0.50(65 \text{ kips/in}^2) = 32 \text{ kips/in}^2$$

$$T_{al} = F_T A_S = 32 \text{ kips/in}^2 \times 0.254 \text{ in}^2 = 8.13 \text{ kips per connection}$$

Now, consider the reinforcing strip at a section through the bolt holes (Section B-B, Fig. 8.42). The reinforcing strip is 50 × 4 mm with 2 oz/ft² (86 μm) of zinc. The diameter of each bolt hole is $\%_{16}$ in (14.3 mm). No carbon steel is lost from reinforcing strip surfaces at the net section, because of the sandwiching protection by the tie strip. Thus,

$$E_S = 0$$

$$E_C = E_n = 4000 \ \mu m \text{ or } 4 \text{ mm}$$

$$A_S = \frac{50 \text{ mm} - 14.3 \text{ mm}}{(25.4 \text{ mm/in})^2} \times 4 \text{ mm} = 0.221 \text{ in}^2$$

$$F_T = 0.50F_u = 0.50(80 \text{ kips/in}^2) = 40 \text{ kips/in}^2$$

$$T_{al} = F_T A_S = 40 \text{ kips/in}^2 \times 0.221 \text{ in}^2 = 8.84 \text{ kips per connection}$$

The shear strength of each bolt is found as follows. Each bolt is ½ in × 1¼ in, ASTM A325, galvanized. It is assumed that no carbon steel is lost from the bolt shank, because of sandwiching protection by the strips. The bolt head, nut, and washer have more than adequate metal for loss to corrosion.

The allowable shear stress on the bolt (with threads excluded from the shear plane) is:

$$F_V = 1.4 \times 19 \text{ kips/in}^2 = 27 \text{ kips/in}^2 \text{ allowable}$$

The nominal cross-sectional area of the ½-in-diameter bolt is 0.196 in². The allowable force on each bolt, considering two shear planes, is

$$T_{al} = F_V A_S = 27 \text{ kips/in}^2 \times 0.196 \text{ in}^2 \times 2 = 10.60 \text{ kips per connection}$$

A check shows that bearing strength does not control for this case.

Next, consider the reinforcing strip at a section where there are no bolt holes (Section C-C, Fig. 8.42). The reinforcing strip is 50 × 4 mm, with 2 oz/ft² (86 μm) of zinc. The life of the zinc is 16 years, from previous calculations.

Calculate the thickness of carbon steel loss over the subsequent 59 years.

$$\Delta e = 708 \ \mu m \text{ per exposed side (see previous calculations)}$$

$$E_S = 2 \text{ sides} \times 708 \ \mu m/\text{side} = 1416 \ \mu m$$

Calculations follow for the thickness of each reinforcing strip at the end of service life, cross-sectional area, allowable tensile stress, and allowable tensile force:

$$E_C = E_n - E_S = 4000 \ \mu m - 1416 \ \mu m = 2584 \ \mu m$$

$$A_S = \frac{50 \text{ mm} \times 2584 \ \mu m}{25.4 \text{ mm/in}} \times 25,400 \ \mu m/\text{in} = 0.200 \text{ in}^2$$

$$F_T = 0.55F_Y = 0.55(65 \text{ kips/in}^2) = 36 \text{ kips/in}^2$$

$$T_{al} = F_T A_S = 36 \text{ kips/in}^2 \times 0.200 \text{ in}^2 = 7.20 \text{ kips per connection}$$

Design Summary for Allowable Reinforcement Tension

Component	Section	Allowable force, kips
Tie strip	Main	8.10
Tie strip	Through bolt holes	8.13
Reinforcing strip	Main	7.20
Reinforcing strip	Through bolt holes	8.84
Bolt	Shear planes	10.60

The least value controls the design. In this case, the allowable reinforcement tension (7.20 kips) is governed by the strength of the reinforcing strip at a section where there are no bolt holes.

8.5.10 Material Properties of Polymeric Reinforcement

The tensile properties of polymeric reinforcement are subject to creep under load because properties of the materials are both time- and temperature-dependent. Also, the materials are subject to damage during the construction process and are affected by durability considerations such as aging. Furthermore, characteristics of geosynthetic products made from the same base polymer exhibit the normal variation of most manufactured products.

As a result, selection by the designer of a long-term tension capacity is a complex matter. AASHTO requires that long-term stress-strain-time behavior be determined from the results of controlled laboratory creep tests conducted on samples of the finished product for a minimum duration of 10,000 hours (approximately one year) for a range of load levels. Samples are required to be tested in either a confined or an unconfined mode in the direction in which the load will be applied in use. The test procedure is outlined in ASTM D5262, "Standard Test Method for Evaluating the Unconfined Tension Creep Behavior of Geosynthetics." The results of this test procedure are extrapolated to the required design life, usually 75 to 100 years, using procedures identified in ASTM D2837, "Standard Test Method for Obtaining Hydrostatic Design Basis for Thermoplastic Pipe Materials." Clearly, the designer cannot depend upon tests of this relatively short duration to establish ultimate tensile strength values for design purposes, nor can manufacturers use such tests for purposes of quality control in the manufacturing process.

In the absence of sufficient test data, the designer can calculate long-term tension capacity T_{al} from the following equation:

$$T_{al} = \frac{T_{ult}(\text{CRF})}{\text{FD} \times \text{FC} \times \text{SF}} \leq T_s \tag{8.14}$$

where T_{al} = long-term tension capacity of the geosynthetic material at a selected design strain (usually 5% or less)

T_{ult} = ultimate strength (tensile strength) from wide-strip tensile strength tests (ASTM D4595)

FD = durability factor of safety (It is dependent on the susceptibility of the geosynthetic material to attack by microorganisms and chemicals, thermal oxidation, and environmental stress cracking. It typically ranges from 1.1 to 2.0. In the absence of product-specific durability information, use 2.0.)

FC = construction damage factor of safety (Typical range is 1.1 to 3.0. In the absence of product-specific construction damage tests, use 3.0.)

SF = overall factor of safety to account for uncertainties in the geometry of the structure, fill properties, reinforcement properties, and externally applied loads; for permanent, vertically faced structures, it should be a minimum of 1.5

CRF = creep reduction factor, T_1/T_{ult}, where T_1 is creep limit strength obtained from creep test results (If the CRF value for the specific reinforcement is not available, consider AASHTO-AGC-ARTBA Task Force 27 recommendations: for polyester, use 0.4; polypropylene, 0.2; polyamide, 0.35; polyethylene, 0.2.)

ASTM designation D4595, "Standard Test Method for Tensile Properties of Geotextiles by the Wide-Width Strip Method," serves as a quality control test from which a "minimum average roll value (MARV)" is determined and certified by the manufacturer to the user of the product. The MARV value is a measure of the ultimate tensile strength of the polymeric material under the stated test conditions. For purposes of design, the value is reduced by a creep reduction factor that varies according to the basic polymer used in the manufacturing process as presented above.

As noted, the manufacturing process is subject to variation. The minimum value the manufacturer certifies must therefore meet or exceed the design minimum value. The manufacturer must also be able to meet this minimum value at a specific confidence level. The ASTM and the industry have adopted a 95 percent confidence level. A normal distribution of the test results is assumed.

ASTM Designation D5262. The test method prescribed in ASTM D5262 is intended for use in determining the unconfined tension creep behavior of geosynthetics at constant temperature and humidity when subjected to a sustained tensile loading. The load is applied in one step, and the total elongation of the test specimen is measured as a function of time. The test results can be used in conjunction with interpretive methods to evaluate creep strain potential at design load. AASHTO requires that the interpretive methods be extrapolated using procedures outlined in ASTM D2837.

The test is a controlled test and is performed in an unconfined environment. The results, therefore, are deemed to be conservative with regard to the behavior of the material, in that the creep in service is likely to be reduced in soil because of load transfer to the soil.

The full load is applied rapidly and smoothly at a specified strain rate. The extension of the specimen is measured in accordance with a 1000-h time schedule. In design, it is generally accepted that creep data should not be extrapolated beyond one order of magnitude. Thus, a 1000-h test is not deemed to be sufficient to accurately reflect the long-term creep behavior. For such cases, creep tests should be conducted for a minimum duration of 10,000 h, which is specified by AASHTO.

ASTM Designation D2837. The test described in ASTM D2837 is identified by AASHTO as the procedure for extrapolation. D2837 describes a procedure for obtaining a hydrostatic design basis for thermoplastic pipe materials by evaluating stress rupture data derived from testing pipe made from thermoplastic material.

The procedure for estimating long-term hydrostatic strength is essentially an extrapolation with respect to time of a stress-time regression line based on data obtained in accordance with Test Method D1598, "Test Method for Time-to-Failure of Plastic Pipe under Constant Internal Pressure." Stress-failure time plots are obtained for the selected temperature and environment, and the extrapolation is made in such a manner that the long-term hydrostatic strength is estimated for these conditions.

ASTM Designation D4595.   Test method ASTM 4595, which is prescribed by AASHTO, covers the measurement of tensile properties of geotextiles using a wide-width strip specimen. The test is also applied to geogrids. A relatively wide specimen is gripped across its entire width in the clamps of a constant-rate-of-extension (CRE) type tensile testing machine operated at a prescribed rate of extension, applying a longitudinal force to the specimen until the specimen ruptures. The distinctive feature of this test is that the width of the specimen is greater than the length, and this tends to minimize the contraction (neck-down) effect that is present with other test methods for measuring strip tensile properties of geotextiles. It is believed that the test will provide a closer relationship to expected geotextile behavior in the field. Tensile strength, elongation, initial and secant modulus, and breaking toughness of the test specimen can be calculated from the results.

The determination of the wide-width strip force-elongation properties of geotextiles provides design parameters for reinforcement applications such as reinforced MSE walls. D4595 may be used for acceptance testing of commercial shipments of geotextiles, although an individual owner may specify other acceptance criteria.

This test method is generally used by manufacturers, but when it is not, it should be required by owners in order to provide supporting data for the manufacturer's stated MARV. To the end user, MARV is a minimum value that exceeds design requirements. To account for testing variation, the manufacturer is required to take a sufficient number of specimens per fabric swatch that the user may expect, at the 95 percent probability level, that the test result will not be more than 5.0 percent of the average above or below the true average of the swatch for both the machine and the cross-machine direction.

The number of tests required to establish a MARV depends upon whether a reliable estimate of the coefficient of variation v of individual observation exists, in the laboratories of either the manufacturer or the end user. Specifically, when there is a reliable estimate of v based upon extensive past records for similar materials tested as directed in the method, the required number of specimens is calculated using the equation:

$$n = \left(\frac{tv}{A}\right)^2 \tag{8.15}$$

where n = number of specimens (rounded upward to a whole number)
   v = reliable estimate of coefficient of variation of individual observations on similar materials in user's laboratory under conditions of single-operator precision, %
   t = value of Student's t for one-sided limits (see Table 8.6), a 95% probability level, and degrees of freedom associated with the estimate of v
   A = 5.0% of average, the value of allowable variation

When there is no reliable estimate of v for the manufacturer's or user's laboratory, the equation should not be used directly. Instead, specify the fixed number of six spec-

TABLE 8.6 Values of Student's t for One-Sided Limits and 95% Probability

df	One-sided	df	One-sided	df	One-sided
1	6.314	11	1.796	22	1.717
2	2.920	12	1.782	24	1.711
3	2.353	13	1.771	26	1.706
4	2.132	14	1.761	28	1.701
5	2.015	15	1.753	30	1.697
6	1.943	16	1.746	40	1.684
7	1.895	17	1.740	50	1.676
8	1.860	18	1.734	60	1.671
9	1.833	19	1.729	120	1.658
10	1.812	20	1.725	∞	1.645

df = degrees of freedom = number of samples − 1.

Source: From *Geotextiles* Magazine, with permission.

imens each for the machine direction and the cross-machine direction tests. The number of specimens is calculated using $v = 7.4$ percent of the average. This value for v is somewhat larger than usually found in practice. When a reliable estimate of v for the user's laboratory becomes available, the above equation will usually require fewer than the fixed number of specimens.

D4595 specifically includes formulas for determining the initial tensile modulus and the offset tensile modulus. Additionally, the formula for breaking toughness is included. The appendix to the designation contains graphical representations for the determination of the modulus values.

8.5.11 Design Example of MSE Retaining Wall with Geogrid Reinforcement

The following design example (provided courtesy of Tensar Earth Technologies) illustrates an application of AASHTO specifications and the tieback wedge method of analysis.

Step 1: Qualify Design Assumptions. Review plans, specifications, and available information to confirm feasibility, to determine if the information is adequate to continue with design, and to ascertain that the wall layout is clearly understood.

Step 2: Define Parameters for Soil, Reinforcement, Geometry, and Loading. On the basis of the information provided, clearly state the design parameters and factors of safety that will be used for design. Provide a diagram for the geometry of the wall that will be designed indicating slopes above and below the wall, any surcharge loadings and their locations, magnitude and direction of application, and hydrostatic and seismic loading conditions.

For this example, refer to Fig. 8.43 for geometry. Design parameters are as follows:

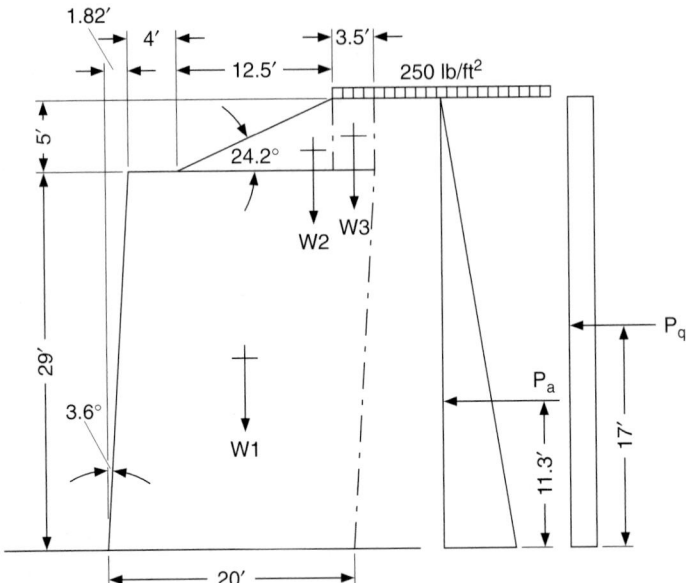

FIGURE 8.43 Design example of MSE retaining wall with geogrid reinforcement. (*From Tensar Earth Technologies, with permission*)

1. Soil

Zone	ϕ', °	c	γ, lb/ft^3
Reinforced fill	34	0	120
Retained fill	30	0	120
Foundation	30	0	120

Allowable foundation bearing stress is 6000 lb/ft^2.

2. Groundwater: none
3. Surcharge: 250 lb/ft^2 uniform
4. Seismic loading: none

Step 3: Calculate External Stability. First calculate the coefficient of active earth pressure, K_a. The slope angle β is zero above the wall because the slope levels before reaching the end of the reinforcement. Had the slope extended beyond the tail of the reinforcement, a trial wedge solution or infinite slope calculation would be required, depending on the distance of the slope extension.

For the following calculation, refer to Art. 8.2.3 for equation and nomenclature:

For $\phi' = 30°$, $\beta = 0$, $\theta = 93.6°$ (face has 3.6° batter), $\delta = 0$: $K_a = 0.31$.

Minimum embedment length $L \approx 0.7H = 0.7(29) = 20.3$ ft. Use 20 ft.

Sum moments and forces about the toe of the wall and solve for external safety factors (SF) as follows.

Item, Fig. 8.43	Force, lb	Moment arm, ft	Moment, ft · lb
W_1	69,600	10.91	759,336
W_2	3,750	14.15	53,075
W_3	2,099	20.07	42,127
W_4	875	20.07	17,661
P_a	21,502	11.33	243,613
P_q	2,635	17.00	44,795

$R_v = W_1 + W_2 + W_3 = 76,325$ lb
$R_h = P_a + P_q = 24,137$ lb
Resisting moment $= M_1 + M_2 + M_3 = 854,538$ ft · lb
Overturning moment $= 243,613 + 44,795 = 288,408$ ft · lb
SF overturning $= 854,538/288,408 = 2.96 \geq 2.0$ OK
SF sliding $= R_v C_i \tan 30°/R_h = 76,325 \times 0.5774/24,137$
$\quad = 1.82 \geq 1.5$ OK

The safety factor for sliding should be calculated in at least two locations: at the interface of the foundation and the reinforced fill, and at the lowest geogrid. In this case, C_i, the coefficient of interaction between the geogrid and the reinforced fill, is 1.0 according to test data supplied by the geogrid manufacturer. Because the reinforced fill is stronger than the foundation soils, the lowest safety factor for sliding is at the foundation interface.

Next check bearing. The eccentricity of the vertical reaction is:

$$ e = \frac{L}{2} - \frac{M_r - M_{ot}}{R_v} = \frac{20}{2} - \frac{872,188 - 288,408}{76,325} = 2.35 \leq \frac{L}{6} $$

The maximum bearing stress is then:

$$ \sigma_v = \frac{R_v}{L - 2e} = \frac{76,325}{20 - 2 \times 2.35} = 4990 \text{ lb/ft}^2 \leq 6000 \text{ lb/ft}^2 \quad \text{OK} $$

All external safety factors are satisfied. Next, calculate internal safety factors for geogrid tension, pullout at face, and pullout past the Rankine failure plane.

Step 4: Calculate Internal Stability. The calculation of K_a for this check is similar to the external calculation, except that the slope angle above the wall (if any) is always assumed to be zero. Thus, $K_a = 0.31$ in this example. The additional forces contributed by the sloping surface are accounted for in the summation of forces and moments in determining bearing stress. Calculation of internal stability and tension in reinforcements is similar to the preceding calculations. At each level of reinforcement, the vertical stress, σ_i, is calculated on the basis of the resultant of the forces and moments of both the reinforced fill and the external forces. This stress is then multiplied by K_a and the vertical tributary area, v_i, to calculate the tension in the reinforcement. If the calculated tension T exceeds the allowable tension T_{al}, either a stronger reinforcement or a reduced vertical spacing must be adopted.

The allowable design stress for the geogrids is determined from AASHTO criteria, considering both ultimate strength and serviceability. Both the geogrid and the connection of the grid to the face must be considered. In this case the following allowable tension values have been determined for two geogrids:

$$\text{Geogrid UX1500: } T_{al} = 1267 \text{ lb/ft}$$

$$\text{Geogrid UX1600: } T_{al} = 1731 \text{ lb/ft}$$

The calculations for tension in Table 8.7 can now be made; the last column indicates the reinforcement selected.

Check pullout in the top geogrid layer. Geogrids must extend beyond the failure plane ($45° - \phi/2$) by at least 3 ft.

$$L_e = 20 - [26.67 \tan (45° - \text{¾}) + 26.67 \tan (3.6°)]$$

$$= 7.50 \text{ ft} \geq 3.0 \quad \text{OK}$$

Calculate pullout resistance by friction (two grid sides) based on weight acting beyond the failure plane:

$$\text{Minimum pullout capacity} = 2[7.5 \text{ ft} \times 2.33 \text{ ft} \times 120 \text{ lb/ft}^3 + W_3]C_i \tan \phi$$

$$= 2(2097 + 2099)1.00 \tan 34°$$

$$= 5660 \text{ lb/ft}$$

$$\text{FS} = 5660/508 = 11.1 \geq 2.0 \quad \text{OK}$$

8.6 NONGRAVITY CANTILEVERED WALL DESIGN

Nongravity cantilevered walls are those that provide lateral resistance through vertical elements embedded in soil, with the retained soil between the vertical elements usually supported by facing elements. Such walls may be constructed of concrete, steel, or timber. Their height is usually limited to about 15 ft, unless provided with additional support anchors.

8.6.1 Earth Pressure and Surcharge Loads

Lateral earth pressure can be estimated assuming wedge theory using a planar surface of sliding as defined by Coulomb theory. For permanent walls, effective stress methods of analysis and drained shear strength parameters for soils can be used for determining lateral earth pressures. Alternatively, the simplified earth pressure distributions shown in Figs. 8.44 and 8.45 can be used. Nomenclature and notes for Fig. 8.44 are given in Table 8.8.

For temporary applications in cohesive soils, total stress methods of analysis and undrained shear strength parameters apply. The simplified earth pressure distributions shown in Figs. 8.46 and 8.47 can alternatively be used with the following limitations:

1. The ratio of overburden pressure to undrained shear strength must be less than 3. This ratio is referred to as the stability number $N = \gamma H/c$.

2. The active earth pressure must not be less than 0.25 times the effective overburden pressure at any depth.

TABLE 8.7 Calculations for Tension in Geogrid Reinforcement of MSE Retaining Wall

No.	Height, ft	Depth, ft	W_1, lb	W_2, lb	W_3, lb	W_4, lb	P_a, lb/ft²	P_q, lb/ft²	σ_v, lb/ft²	v_z, ft²	T, lb/ft	Grid, UX-
13	26.67	2.33	6,692	3750	2099	875	999	568	492	3.99	509	1500
12	23.35	5.65	13,560	3750	2099	875	2,108	826	895	3.34	774	1500
11	20.00	9.00	21,600	3750	2099	875	3,643	1084	1326	3.34	1148	1500
10	16.67	12.33	29,592	3750	2099	875	5,582	1342	1782	3.00	1387	1600
9	14.00	15.00	36,000	3750	2099	875	7,434	1549	2175	2.67	1504	1600
8	11.34	17.66	42,384	3750	2099	875	9,543	1755	2598	2.33	1570	1600
7	9.34	19.66	47,184	3750	2099	875	11,302	1910	2940	2.00	1525	1600
6	7.34	21.66	51,984	3750	2099	875	13,210	2065	3309	1.67	1433	1600
5	6.00	23.00	55,200	3750	2099	875	14,571	2168	3573	1.34	1237	1600
4	4.67	24.33	58,392	3750	2099	875	15,988	2271	3851	1.33	1328	1600
3	3.34	25.66	61,584	3750	2099	875	17,471	2374	4146	1.34	1436	1600
2	2.00	27.00	64,800	3750	2099	875	19,032	2478	4463	1.34	1545	1600
1	0.67	28.33	67,992	3750	2099	875	20,647	2581	4801	1.34	1662	1600
0	0.67	29.00	69,600	3750	2099	875	21,502	2635	4989			

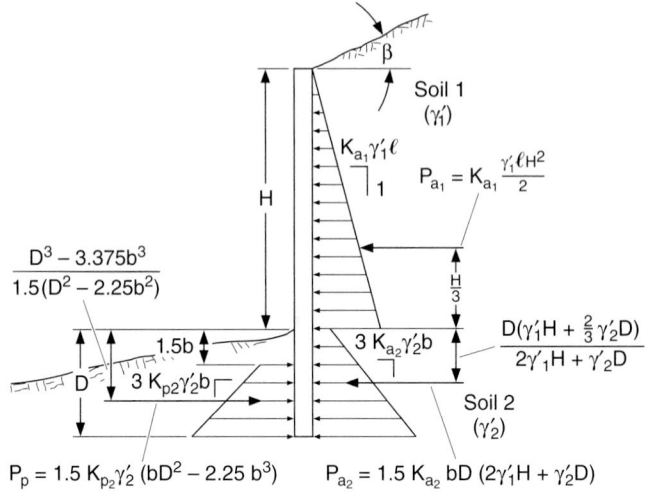

$$\frac{D^3 - 3.375b^3}{1.5(D^2 - 2.25b^2)}$$

$P_p = 1.5 \, K_{p2} \gamma_2' \, (bD^2 - 2.25 \, b^3)$ $P_{a_2} = 1.5 \, K_{a_2} \, bD \, (2\gamma_1'H + \gamma_2'D)$

(a) Embedment in soil

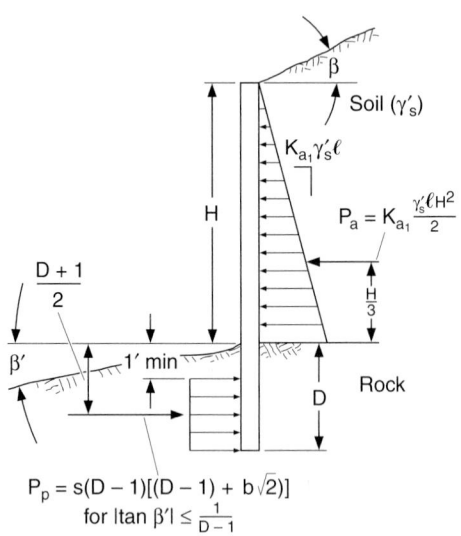

$P_p = s(D-1)[(D-1) + b\sqrt{2})]$
for $|\tan \beta'| \le \dfrac{1}{D-1}$

(b) Embedment in rock

Note: Refer to Table 8.8 for general notes and legend.

FIGURE 8.44 Simplified earth pressure distributions for permanent flexible cantilevered walls with discrete vertical wall elements. (*From* Standard Specifications for Highway Bridges, *American Association of State Highway and Transportation Officials, Washington, D.C., 1992, with permission*)

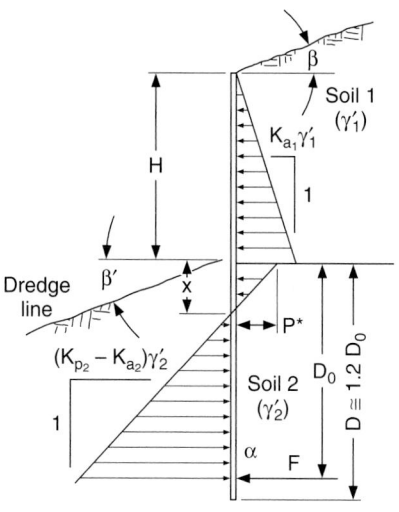

1. Determine the active earth pressure on the wall due to surchage loads, the retained soil, and differential water pressure above the dredge line.

2. Determine the magnitude of active pressure at the dredge line (P^*) due to surcharge loads, retained soil, and differential water pressure, using the earth pressure coefficient K_{a2}.

3. Determine the value of $x = P^*/[(K_{p2} - K_{a2})\gamma_2']$ for the distribution of net passive pressure in front of the wall below the dredge line.

4. Sum moments about the point of action of F to determine the embedment (D_0) for which the net passive pressure is sufficient to provide equilibrium.

5. Determine the depth (point α) at which the shear in the wall is zero (i.e., the point at which the areas of the driving and resisting pressure diagrams are equivalent).

6. Calculate the maximum bending moment at the point of zero shear.

7. Calculate the design depth, $D = 1.2\,D_0$ to $1.4\,D_0$, for a safety factor of 1.5 to 2.0.

(a) Pressure distribution   (b) Simplified design procedure

Notes: (1) Surcharge and water pressures must be added to the above earth pressures.

     (2) Forces shown are per horizontal foot of vertical wall element.

FIGURE 8.45 Simplified earth pressure distributions and design procedures for permanent flexible cantilevered walls with continuous vertical wall elements. (*From* Standard Specifications for Highway Bridges, *American Association of State Highway and Transportation Officials, Washington, D.C., 1992, with permission*)

Nomenclature and notes for Fig. 8.46 are given in Table 8.8.

Where discrete vertical wall elements are used for support, the width of each vertical element should be assumed to equal the width of the flange or diameter of the element for driven sections, and to equal the diameter of the concrete-filled hole for sections encased in concrete.

For permanent walls, Figs. 8.44 and 8.45 show the magnitude and location of resultant loads and resisting forces for discrete vertical elements embedded in soil and rock. The procedure for determining the resultant passive resistance of a vertical element assumes that net passive resistance is mobilized across a maximum of 3 times the element width or diameter (reduced, if necessary, to account for soft clay or discontinuities in the embedded depth of soil or rock). Also, a depth of 1.5 times the width of an element in soil, and 1 ft for an element in rock, is ineffective in providing passive lateral support.

The design lateral pressure must include lateral pressure due to traffic, permanent point and line surcharge loads, backfill compaction, or other types of surcharge loads, as well as the lateral earth pressure.

TABLE 8.8 General Notes and Legend for Simplified Earth Pressure Distributions for Permanent and Temporary Flexible Cantilevered Walls with Discrete Vertical Wall Elements, Figs. 8.44 and 8.46

Legend:
γ' = effective unit weight of soil
b = vertical element width
l = spacing between vertical wall elements, center to center
S_u = undrained shear strength of cohesive soil
s = shear strength of rock mass
P_p = passive resistance per vertical wall element
P_a = active earth pressure per vertical wall element
β = ground surface slope behind wall   } + for slope up from wall
β' = ground surface slope in front of wall } − for slope down from wall
K_a = active earth pressure coefficient; refer to Art. 8.2.3
K_p = passive earth pressure coefficient; refer to Figs. 5.5.2D and 5.5.2E in *Standard Specifications for Highway Bridges,* AASHTO, 1992.
ϕ' = effective angle of soil friction

Notes:

1. For temporary walls embedded in granular soil or rock, refer to Fig. 8.44 to determine passive resistance and use diagrams on Fig. 8.46 to determine active earth pressure of retained soil.

2. Surcharge and water pressures must be added to the indicated earth pressures.

3. Forces shown are per vertical wall element.

4. Pressure distributions below the exposed portion of the wall are based on an effective element width of $3b$, which is valid for $l \geq 5b$. For $l < 5b$, refer to Figs. 8.45 and 8.47 for continuous wall elements to determine pressure distributions on embedded portions of the wall.

Source: From *Standard Specifications for Highway Bridges,* American Association of State Highway and Transportation Officials, Washington, D.C., 1992, with permission.

8.6.2 Water Pressure and Drainage

Flexible cantilevered walls must be designed to resist the maximum anticipated water pressure. For a horizontal static groundwater table, the total hydrostatic water pressure can be determined from the hydrostatic head by the traditional method. For differing groundwater levels on opposite sides of the wall, the water pressure and seepage forces can be determined by net flow procedures or other methods. Seepage can be controlled by installation of a drainage medium. Preformed drainage panels, sand or gravel drains, or wick drains can be placed behind the facing with outlets at the base of the wall. It is important that drainage panels maintain their function under design earth pressures and surcharge loadings. AASHTO requires that they extend from the base of the wall to a level 1 ft below the top of the wall.

Where thin drainage panels are used behind walls, saturated or moist soil behind the panels may be subject to freezing and expansion. In such cases, insulation can be provided on the walls to prevent soil freezing or the wall can be designed for the pressures that may be exerted on it by frozen soil.

8.6.3 Structure Dimensions and External Stability

Flexible cantilevered walls should be dimensioned to ensure stability against passive failure of embedded vertical elements using a factor of safety of 1.5 based on unfactored loads. Vertical elements must be designed to support the full design earth, surcharge, and water pressures between the elements. In determining the depth of embed-

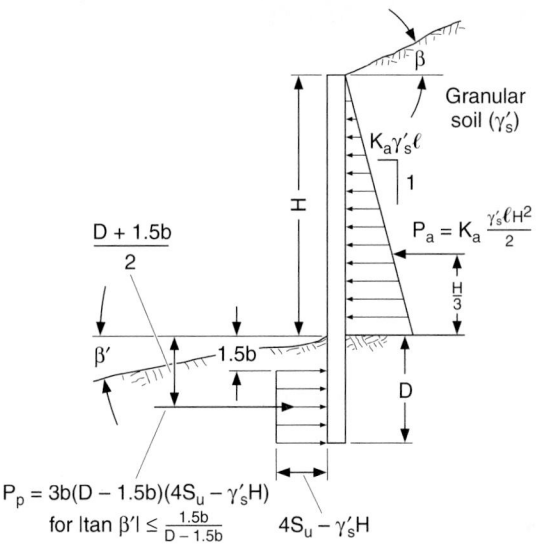

$$P_a = K_a \frac{\gamma'_s \ell H^2}{2}$$

$K_a \gamma'_s \ell$

$$P_p = 3b(D - 1.5b)(4S_u - \gamma'_s H)$$
$$\text{for } |\tan \beta'| \le \frac{1.5b}{D - 1.5b}$$

$$4S_u - \gamma'_s H$$

(a) Embedment in cohesive soil retaining granular soil

Note: For sloping backfill
use effective shear strength
parameters (c = 0).

Consider possible condition
of water in tension crack

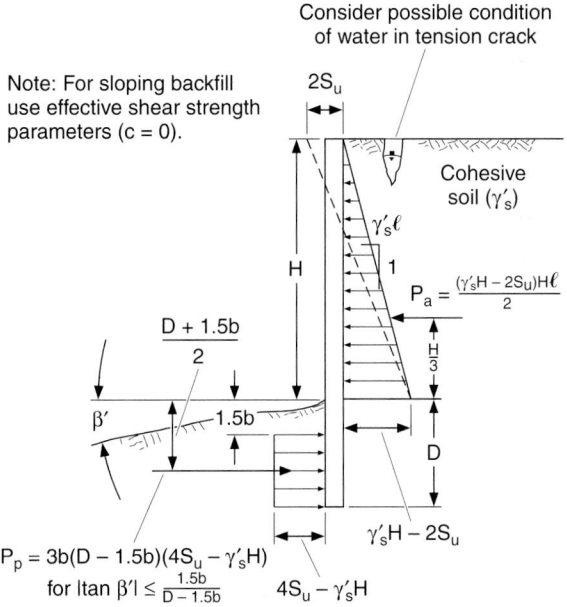

$2S_u$

Cohesive
soil (γ'_s)

$\gamma'_s \ell$

$$P_a = \frac{(\gamma'_s H - 2S_u)H\ell}{2}$$

$$P_p = 3b(D - 1.5b)(4S_u - \gamma'_s H)$$
$$\text{for } |\tan \beta'| \le \frac{1.5b}{D - 1.5b}$$

$$4S_u - \gamma'_s H$$

$$\gamma'_s H - 2S_u$$

(b) Embedment in cohesive soil retaining cohesive soil

Note: Refer to Table 8.8 for general notes and legend.

FIGURE 8.46 Simplified earth pressure distributions for temporary flexible cantilevered walls with discrete vertical wall elements. (*From Standard Specifications for Highway Bridges, American Association of State Highway and Transportation Officials, Washington, D.C., 1992, with permission*)

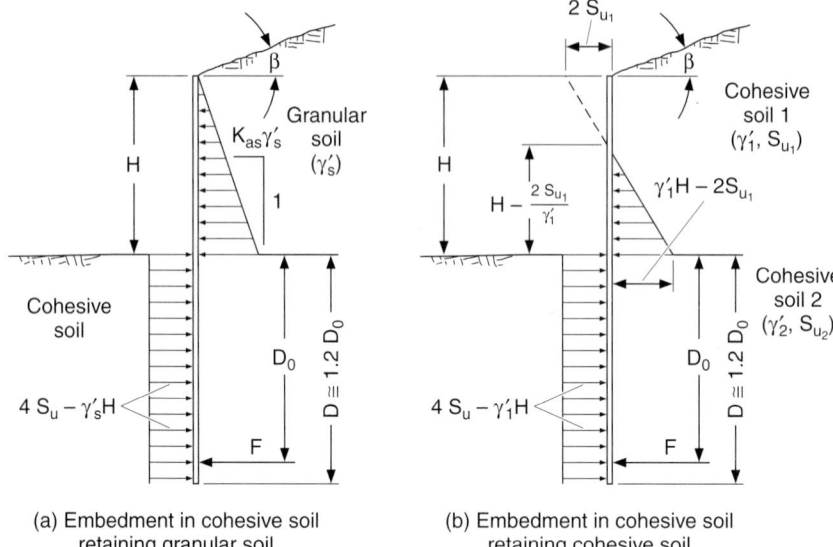

(a) Embedment in cohesive soil
retaining granular soil

(b) Embedment in cohesive soil
retaining cohesive soil

Notes: (1) For walls embedded in granular soil, refer to Fig. 8.45 and use above diagram
for retained cohesive soil when appropriate.

(2) Surface and water pressures must be added to the above earth pressures.

(3) Forces shown are per horizontal foot of vertical wall element.

FIGURE 8.47 Simplified earth pressure distributions for temporary flexible cantilevered walls with continuous vertical wall elements. (*From* Standard Specifications for Highway Bridges, *American Association of State Highway and Transportation Officials, Washington, D.C., 1992, with permission*)

ment to mobilize passive resistance, consideration should be given to planes of weakness (such as "slickensides," bedding planes, and joint sets) that could reduce the strength of the soil or rock from that determined by field or laboratory tests. AASHTO recommends that for embedment in intact rock, including massive to appreciably jointed rock, which should not be allowed to fail through a joint surface, design should be based on an allowable shear strength of 0.10 to 0.15 times the uniaxial compressive strength of the intact rock.

8.6.4 Structure Design

Structural design of individual wall elements may be performed by service load or load factor design methods.

The maximum spacing L between vertical supporting elements depends on the relative stiffness of the vertical elements and facing, the design pressure P_a, and the type and condition of soil to be supported. Design the facing for the bending moment M_{max} at any level, as determined by the following equations:

Simple span (no soil arching):

$$M_{max} = \frac{P_a L^2}{8} \tag{8.16}$$

Simple span (soil arching):

$$M_{max} = \frac{P_a L^2}{12} \tag{8.17}$$

Continuous:

$$M_{max} = \frac{P_a L^2}{10} \tag{8.18}$$

Equation (8.16) is applicable for simply supported facings where the soil will not arch between vertical supports (e.g., in soft cohesive soils or for rigid concrete facing placed tightly against the in-place soil). Equation (8.17) is applicable for simply supported facings where the soil will arch between vertical supports (e.g., in granular or stiff cohesive soils with flexible facing, or rigid facing behind which there is sufficient space to permit the in-place soil to arch). Equation (8.18) is applicable for facings that are continuous over several vertical supports (e.g., reinforced shotcrete).

AASHTO gives the following recommendations for timber facings:

> Timber facings should be constructed of stress-grade lumber. If timber is used where conditions are favorable for the growth of decay-producing organisms, wood should be pressure treated with a wood preservative unless the heartwood of a naturally decay-resistant species is available and is considered adequate with respect to the decay hazard and expected service life of the structure.

8.6.5 Overall Stability

The overall stability of slopes in the vicinity of walls is considered part of the design of retaining walls. The overall stability of the retaining wall, retained slope, and foundation soil or rock can be evaluated for all walls using limiting equilibrium methods of analysis. AASHTO gives the following requirements:

> Where soil and rock parameters and ground water levels are based on in-situ and/or laboratory tests, the minimum factor of safety shall be 1.3 (or 1.5 where abutments are supported above a retaining wall). Otherwise, the minimum factor of safety shall be 1.5 (or 1.8 where abutments are supported above a retaining wall). It must be noted that, even if overall stability is satisfactory, special exploration, testing and analyses may be required for bridge abutments or retaining walls constructed over soft sub-soils where consolidation and/or lateral flow of the soft soil could result in unacceptable long-term settlements or horizontal movements.

8.6.6 Corrosion Protection

Prestressed anchors and anchor heads must be protected against corrosion that would result from ground and groundwater conditions at the site. The level of corrosion protection depends on both the ground environment and the potential consequences of an anchor failure. Also, anchors for permanent walls require a higher level of corrosion protection than those for temporary walls.

8.7 ANCHORED WALL DESIGN

Anchored walls are made up of the same elements as cantilevered walls but are furnished with one or more tiers of anchors for additional lateral support. Anchors may be either prestressed or dead-man type. Tendons or bars extend from the wall face to a region beyond the active zone where they are grouted in place or mechanically anchored. Such walls are typically constructed from the top down in cut situations rather than fill conditions. Figure 8.48 illustrates an anchored wall and defines terminology.

8.7.1 Earth Pressure and Surcharge Loadings

The choice of lateral earth pressures used for design should take into account the method and sequence of construction, rigidity of the wall-anchor system, physical characteristics and stability of the ground mass to be supported, allowable wall deflections, space between anchors, anchor prestress, and potential for anchor yield. For stable ground masses, the final lateral earth pressures on a completed wall with two or

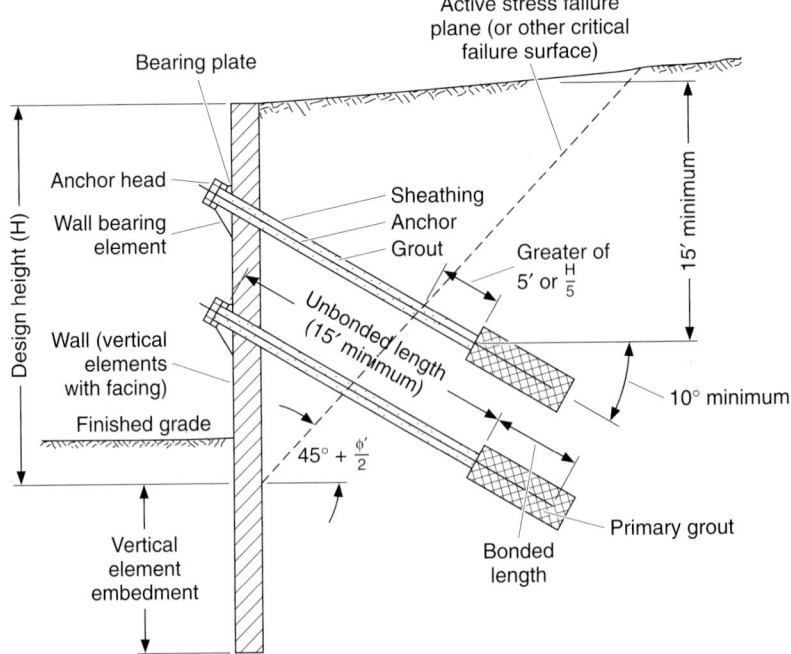

FIGURE 8.48   Terms used in flexible anchored wall design. (*From* Standard Specifications for Highway Bridges, *American Association of State Highway and Transportation Officials, Washington, D.C., 1992, with permission*)

more levels of anchors constructed from the top down can be calculated using the apparent earth pressure distributions shown in Fig. 8.49. For unstable or marginally stable ground masses, design earth pressures will be greater than those shown in Fig. 8.49. Therefore, loads should be estimated using methods of slope stability analysis that include the effects of anchors, or that consider "interslice" forces.

In developing design earth pressures, consideration should be given to wall displacements that may affect adjacent structures or underground utilities. Rough estimates of settlement adjacent to braced or anchored flexible walls can be made using Fig. 8.50. If wall deflections estimated using Fig. 8.50 are excessive, a more detailed analysis can be made using beam-on-elastic-foundation, finite element, or other meth-

SOIL TYPE	APPARENT EARTH PRESSURE DISTRIBUTION	NOTATION

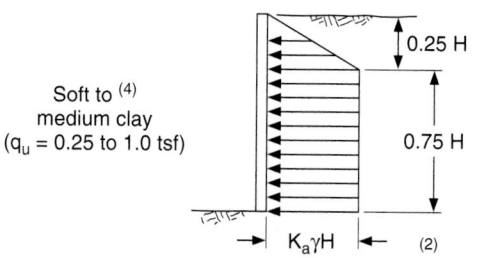

NOTATION

H = final wall height
K_a = active earth pressure coefficient
γ' = effective soil unit weight
γ = total soil unit weight
m = reduction factor
q_u = unconfined compressive strength

NOTES

(1) $K_a = \tan^2\left(45 - \frac{\phi'}{2}\right)$

(2) $K_a = 1 - m\,(2q_u/\gamma H)$ but not less than 0.25

   $m = 1$ for overconsolidated clays
   $m = 0.4$ for normally consolidated clay

(3) Value of 0.4 should be used for long-term excavations; values between 0.4 and 0.2 may be used for short-term conditions.

(4) Surcharge and water pressures must be added to these earth pressure diagrams. The two lower diagrams are not valid for permanent walls or walls where water level lies above bottom of excavation.

FIGURE 8.49 Guidelines for estimating earth pressures on walls with two or more levels of anchors constructed from the top down. (*From* Standard Specifications for Highway Bridges, *American Association of State Highway and Transportation Officials, Washington, D.C., 1992, with permission*)

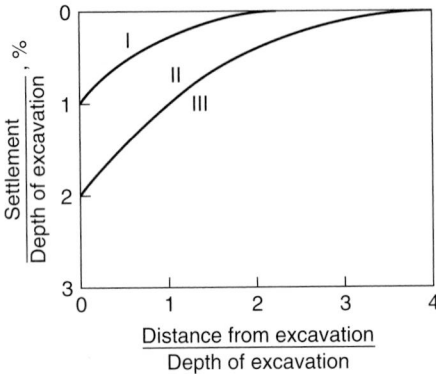

FIGURE 8.50 Settlements adjacent to braced or anchored flexible walls. Zone I: sand and soft-to-hard clay. Zone II: very soft to soft clay when the depth of clay below excavation is limited or when $S_u > 0.25$ $\gamma'H$. Zone III: very soft to soft clay when $S_u < 0.25$ $\gamma'H$ to a significant depth below the wall base. *Note:* Curves are based on data from excavations using standard soldier piles or sheet piles braced with cross-bracing or tiebacks. (*From* Standard Specifications for Highway Bridges, *American Association of State Highway and Transportation Officials, Washington, D.C., 1992, with permission*)

ods of analysis that consider soil-structure interaction effects. Where a structure or utility particularly sensitive to settlement is located close to a wall, wall deflections should be calculated on the basis of the loading, soil properties, anchor spacing, and wall element stiffness.

The distribution of earth pressure loading for anchored walls with one level of anchors can be assumed to be triangular and to be based on a lateral earth pressure coefficient (i.e., K_a, K_0, or K_p) consistent with the expected wall deflection. To consider the case where excavation has advanced down to the first anchor level but the anchors have not yet been installed, the wall can be treated as a nongravity cantilevered wall and the earth pressure distribution assumed triangular. Overstressing of anchors should be avoided, because excessive anchor loads, relative to the capacity of the retained ground mass, can cause undesirable deflections, or passive failure of the wall into the retained soil. As with other walls, design lateral pressures for walls constructed from the top down must include the lateral pressure due to traffic or other surcharge loading.

Where there is no anchor level or only one, the magnitude and distribution of lateral resisting forces for embedded vertical elements in soil or rock can be determined as described in Art. 8.6.1. When two or more levels of anchors have been installed, the lateral resistance provided by embedded vertical elements will depend on the element stiffness and deflection under load.

Earth pressures on anchored walls constructed from the bottom up (fill construction) are affected by the construction method and sequence. These must be well specified, and the basis for lateral earth pressures fully documented. For walls with a single anchor level, consider a triangular distribution, defined by $K_a\gamma$ per unit length of wall height, plus surcharge loads. For walls with multiple anchor levels, consider a rectangular pressure distribution, derived by increasing the total force from the triangular pressure distribution just described by one-third and applying the force as a uniform pressure distribution.

Drainage considerations for anchored walls are similar to those discussed in Art. 8.6.2.

8.7.2 Structure Dimensions and External Stability

The design of anchored walls involves a determination of several factors. Included are the size, spacing, and depth of embedment of vertical wall elements and facing; the type, capacity, spacing, depth, inclination, and corrosion protection of anchors; and the structural capacity and stability of the wall, wall foundation, and surrounding soil mass for all intermediate and final stages of construction. The bearing capacity and settlement of vertical wall elements under the action of the vertical component of the anchor forces and other vertical loads must also be evaluated.

AASHTO provides the following guidance:

> For walls supported in or through soft clays with $S_u < 0.3\gamma'H$, continuous vertical elements extending well below the exposed base of the wall may be required to prevent heave in front of the wall. Otherwise, the vertical elements are embedded several feet as required for stability or end bearing. (Where significant embedment of the wall is required to prevent bottom heave, the lowest section of wall below the lowest row of anchors must be designed to resist the moment induced by the pressure acting between the lowest row of anchors and the base of the exposed wall, and the force $P_b = 0.7(\gamma HB_e - 1.4cH - \pi cB_e)$ acting at the mid-height of the embedded depth of the wall.)

In the above, the following definitions apply:

$$B_e = \text{width of excavation perpendicular to wall}$$

$$c = \text{cohesion of soil}$$

$$H = \text{design wall height}$$

$$S_u = \text{undrained shear strength of cohesive soil}$$

$$\gamma = \text{soil unit weight}$$

$$\gamma' = \text{effective unit weight of soil}$$

8.7.3 General Design Procedures for Anchored Walls

For a typical wall with two or more rows of anchors constructed from the top down, the general procedure is to (1) design for the final condition with multiple rows of anchors and (2) check the design for the various stages of construction. The required horizontal component of each anchor force can be calculated using apparent earth pressure distributions such as given in Fig. 8.49. Any other applicable forces such as horizontal water pressure, surcharge, or seismic forces must be included where applicable. The anchor inclination must be considered in calculating the anchor force. The horizontal anchor spacing and anchor capacity must provide the required total anchor force.

Vertical wall elements must be designed to resist all applicable forces such as horizontal earth pressure, surcharge, water pressure, and anchor and seismic loadings, as well as the vertical component of earth pressure due to wall friction and the vertical component of anchor loads and any other vertical loads. In the analysis, supports may be assumed at each anchor location and at the bottom if the vertical element extends below the bottom of the wall.

All components should be checked for the various earth pressure distributions and other loading conditions that may exist during construction.

8.7.4 Anchor Design

Anchor design includes the selection of a feasible anchor system, estimation of anchor capacity, determination of unbonded length, and consideration of corrosion protection. In determining the feasibility of employing anchors at a particular location, considerations include the availability of underground easements, proximity of buried facilities to anchor locations, and the suitability of subsurface soil and rock conditions within the anchor stressing zone.

Ultimate anchor capacity per unit length may be estimated from Tables 8.9 and 8.10 for soil and rock, respectively. The values are based on straight-shaft anchors installed in small-diameter holes using a low grout pressure. Other anchor types and installation procedures may result in different anchor capacities. Allowable anchor capacity for small-diameter anchors may be estimated by multiplying the ultimate anchor capacity per unit length by the bonded (or stressing) length and dividing by a factor of safety. AASHTO suggests 2.5 for anchors in soil and 3.0 for anchors in rock.

Bearing elements for anchors must be designed so that shear stresses in the vertical wall elements and facing are within allowable limits. The capacity of each anchor should be verified as part of a stressing and testing program.

TABLE 8.9 Ultimate Values of Load Transfer for Preliminary Design of Anchors in Soil

Soil type	Relative density or consistency*	Estimated ultimate transfer load, kips per lineal foot
Sand and gravel	Loose	10
	Medium dense	15
	Dense	20
Sand	Loose	7
	Medium dense	10
	Dense	13
Sand and silt	Loose	5
	Medium dense	7
	Dense	9
Silt-clay mixture with minimum LL, PI, and LI restrictions, or fine micaceous† sand or silt mixtures	Stiff	2
	Hard	4

*Values corrected for overburden pressure.

†The presence of mica tends to increase the volume and compressibility of sand and soft deposits due to bridging action and subsequent flexibility under increased pressures.

Source: From *Standard Specifications for Highway Bridges,* American Association of State Highway and Transportation Officials, Washington, D.C., 1992, with permission.

TABLE 8.10 Ultimate Values of Load Transfer for Preliminary Design of Anchors in Rock

Rock type	Estimated ultimate transfer load, kips per line foot
Granite or basalt	50
Dolomitic limestone	40
Soft limestone	30
Sandstone	30
Slates and hard shales	25
Soft shales	10

Source: From *Standard Specifications for Highway Bridges,* American Association of State Highway and Transportation Officials, Washington, D.C., 1992, with permission.

Determination of the unbonded anchor length should consider the location of the critical failure surface farthest from the wall, the minimum length required to ensure minimal loss of anchor prestress due to long-term ground movements, and the depth to adequate anchoring strata. As shown in Fig. 8.48, the unbonded (or free) anchor length should not be less than 15 ft and should extend 5 ft or one-fifth of the design wall height, whichever is greater, beyond the critical failure surface in the soil mass being retained by the wall. For granular soils or drained cohesive soils, the critical failure surface is typically assumed to be the active failure wedge. This wedge is defined by a plane extending upward from the base of the wall at an angle of $45° + \phi'/2$ from the horizontal, where ϕ' is the effective angle of soil friction. Longer free lengths may be required for anchors in plastic soils or where critical failure surfaces are defined by planes or discontinuities with other orientations.

Selection of anchor inclination should consider the location of suitable soil or rock strata, the presence of buried utilities or other geometric constraints, and constructibility of the anchor drill holes. AASHTO suggests that anchors be located on a minimum inclination of 10° below horizontal and the bonded zone be located a minimum depth of 15 ft below the ground surface. The component of vertical load resulting from anchor inclination must be included in evaluating the end bearing and settlement of vertical wall elements.

AASHTO suggests that the minimum horizontal spacing of anchors be either 3 times the diameter of the bonded zone or 4 ft, whichever is larger. If small spacings are required, consideration can be given to different anchor inclinations between alternating anchors.

8.8 SOIL NAILED STRUCTURES

8.8.1 Development and General Considerations

Figure 8.51 shows a cross section of the first soil nailed wall, which was a temporary wall built in France (1972–1973) for a railroad project. Such walls are constructed from the top down during excavation. Reinforcing bars are either inserted in drilled holes and grouted into place, or driven into place. Then a facing of cast-in-place concrete or shotcrete is installed as the work progresses.

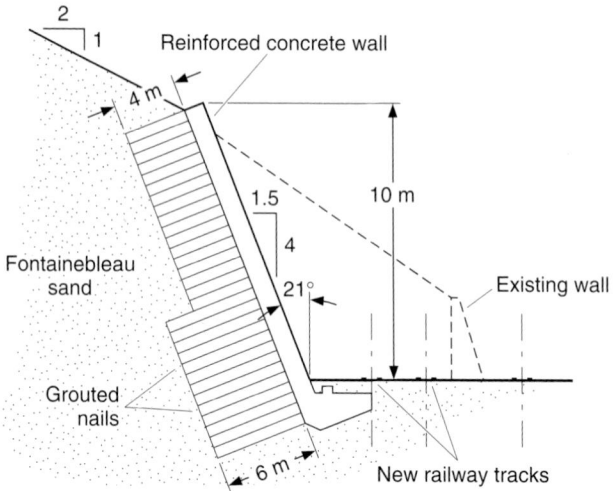

FIGURE 8.51   First soil nailed wall, constructed at Versailles, France, 1972–1973. (*F. Schlosser, Behavior and Design of Soil Nailing,* Proceedings of Symposium on Soil and Rock Improvement Techniques, *Bangkok, 1982, with permission*)

The wall in France was built in Fontainebleau sand using a high density of short nails of two different lengths: 13 ft (4 m) for nails in the upper portion of the wall and 20 ft (6 m) for those in the lower portion. The first full-size experimental wall was constructed in Germany in 1979 using grouted nails and loaded to failure. In 1981, a prefabricated-concrete-facing soil nailed wall was used in a commercial application in France. An extensive national research project conducted in France during the years 1986–1990 resulted in a noted publication titled *Recommendations Clouterre.*

A soil nailed wall is constructed as an integral part of the construction of an excavation as illustrated in Fig. 8.52. (See Art. 8.8.7.) The soil is reinforced as the slope excavation progresses. Reinforcement generally consists of bars inserted parallel to one another and placed at a downward-sloping angle. The bars are inserted in a passive state; however, as the skin friction between the soil and the nails is mobilized, the nails are placed into tension. Figure 8.53 compares the action of soil nails and ground anchors.

The work is carried out from the top downward in increments, gradually building up a reinforced soil mass. Some type of facing is generally necessary to keep the soil from caving between the soil nails. In the case of the Fontainebleau sand (effective friction angle of 38° and some cohesion), the distance for stability of the excavation between soil nails was about 6.6 ft (2 m) with failure occurring at about 9.8 ft (3 m). The sand between the nails will slough until an "arching" action occurs within the soil. The point where this action can no longer occur because the internal friction capacity of the soil has been exceeded defines the temporary facing limit.

The concept of soil nailing has not been patented, nor is it patentable; however, numerous technologies have been patented. For soil nail walls to be cost-effective, the ground must be capable of standing unsupported while the nails and shotcrete are installed. The success of soil nail walls depends upon:

1.   Selection of good applications in ground suitable for nailing

2.   Ability to quickly respond to changed ground conditions

3.   Use of a rational design procedure for the wall and each of its components

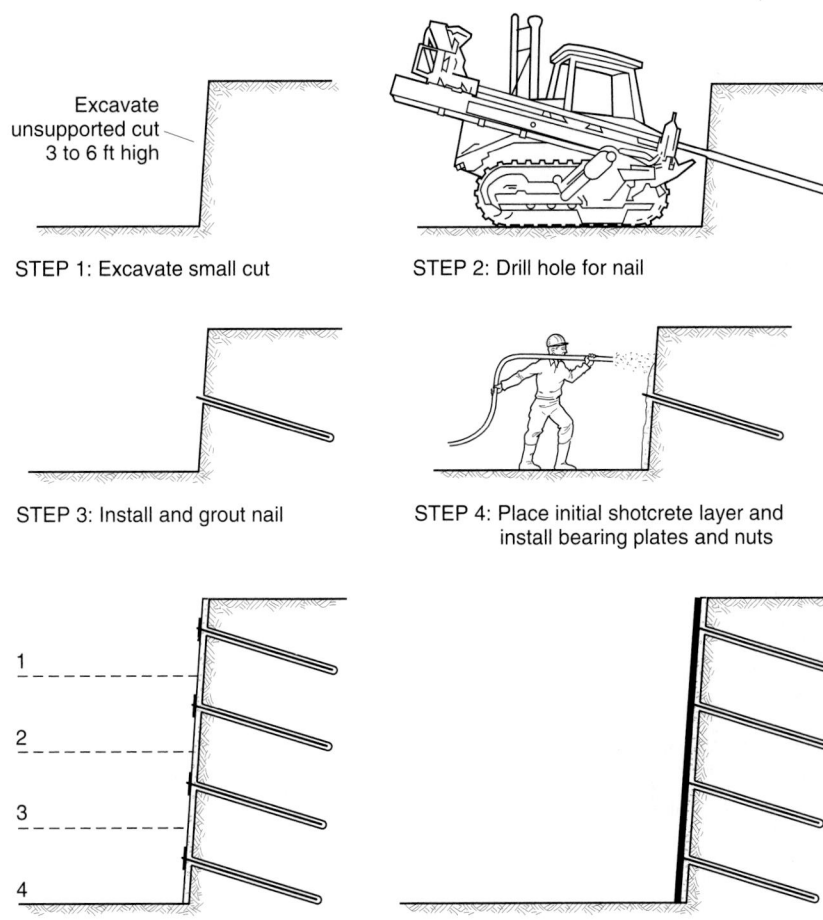

STEP 1: Excavate small cut

STEP 2: Drill hole for nail

STEP 3: Install and grout nail

STEP 4: Place initial shotcrete layer and install bearing plates and nuts

STEP 5: Repeat process to final grade

STEP 6: Place final facing

FIGURE 8.52 Typical construction sequence for soil nailed wall.

4. Use of good construction specifications
5. Ability of the owner and contractor to work together in a partnering concept
6. Handling of work performance on-site by knowledgeable personnel representing each of the parties, including the owner

8.8.2 Suitable Soils

Most research to date has been done in homogeneous soils. However, there is no reason why the concept cannot be applied to heterogeneous soil masses if proper consideration of soil properties is made and rationally applied to the selection of nail length and spacing.

To be economical, soil nailed walls should be constructed in ground that can stand unsupported on a vertical or steeply sloped cut of 3 to 6 ft (1 to 2 m) for one to two

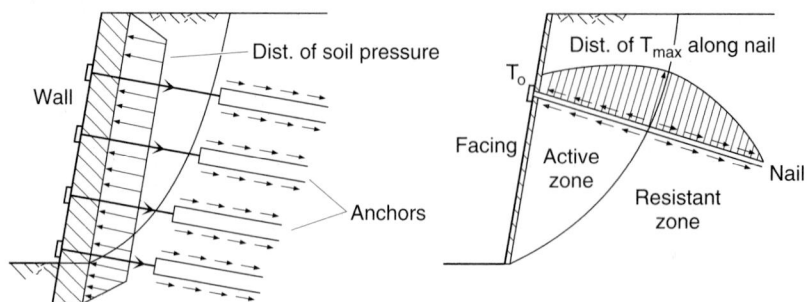

FIGURE 8.53 Resisting forces for (*left*) ground anchors and (*right*) soil nails.

days, and can maintain an open drill hole for a few hours. Soils considered favorable to soil nailing are as follows:

1. Naturally cohesive materials (silts and low-plasticity clays that are not prone to creep)
2. Naturally cemented sands and gravels
3. Weathered rock
4. Fine to medium, homogeneous sand with capillary cohesion of 60 to 100 lb/ft² associated with a water content of at least 5 to 6 percent

According to FHWA Report RD-89-108, soil nailing is generally *not* considered cost-effective or applicable in the following soils:

1. Loose granular soils with field standard penetration N values lower than about 10 or relative densities of less than 30 percent
2. Granular cohesionless soils of uniform size (poorly graded) with uniformity coefficient (D_{60}/D_{10}) less than 2, unless found to be very dense; nailing of these soils may be impractical because of the necessity of stabilizing the cut face (by grouting or another permanent technique) prior to excavation
3. Soft cohesive soils with undrained shear strengths of less than 500 lb/ft², because of the inability to develop adequate pullout resistance
4. Highly plastic clays (LL > 50 and PI > 20 percent), because of their potential for excessive creep deformation
5. Expansive (swelling) and highly frost-susceptible soils

Soil nailing is also *not* recommended for the following conditions:

1. In ground with water pressure present at the face
2. Below the groundwater table, unless the slope can be effectively dewatered prior to excavation
3. Loose fill, granular soil with no apparent cohesion

8.8.3   Comparison with MSE Walls

Soil nailed walls have some similarities with MSE walls but also some fundamental differences. The main *similarities* are:

1. The major mechanism in both MSE and soil nailed retaining structures is the development of tensile forces in the reinforcements due to frictional interaction and, consequently, restrainment of lateral deformations of the structures.

2. The reinforced soil mass is separated into two zones based on the points of maximum tension in the reinforcement (Fig. 8.53): an "active" zone close to the facing, where the shear stresses exerted on the surface of the reinforcement are directed outward and have a tendency to pull out the reinforcements; and a "resistant" zone, where the shear stresses are directed inward and prevent the sliding of the reinforcements.

3. The reinforcement forces are sustained by a frictional bond between the soil and the reinforcing element; the reinforced zone is stable and resists the thrust from the unreinforced soil it supports, much like a gravity retaining structure.

4. The facing of the retained structure is relatively thin, with prefabricated elements used for MSE walls and, usually, shotcrete for soil nailed walls.

The main *differences* are:

1. *The construction procedure.* Although at the end of construction the two structures may look similar, the construction sequence is radically different. Soil nailed walls are constructed "top down" by staged excavations, while MSE walls are constructed "bottom up." Thus, the wall deformation pattern is different during construction. This also results in differences in the distribution of the forces that develop in the reinforcement, particularly during the construction period. In an MSE structure (built bottom up), the working forces that develop in the reinforcement layers generally increase from top to bottom. In a nailed structure (built top down), the working loads that develop in the reinforcement layers are generally of uniform magnitude, similar to those in a braced excavation.

2. *Nature of the soil.* Soil nailing is an in situ reinforcement technique exploiting *natural ground,* the properties of which cannot be preselected and controlled as they are for MSE fills. MSE walls usually utilize clean, low-water-content granular backfills, which have a known friction angle and little to no cohesion. On the contrary, nails are installed into soil and rock (natural ground) whose strength properties (friction angle and cohesion) and water content can vary through a wide range.

3. *Soil-reinforcement bond.* Grouting techniques are usually employed to bond the nail reinforcement to the surrounding ground, with the load transferred along the grout to the soil interface. In MSE structures, friction is generated directly along the reinforcement-soil interface.

8.8.4 Wall Drainage Systems

Almost all shotcrete failures in slope stabilization applications have resulted from inadequate drainage. Therefore, drainage is a *critical* design and construction element. Drainage from behind the shotcrete face can be provided by the following methods:

1. *Surface interceptor ditch.* Excavate a shallow ditch along the crest of the excavation to lead away surface water. Drainage gutters or lined ditches are recommended immediately behind the top of the wall.

2. *Prefab geotextile drains.* Place 12-in (300-mm) wide prefabricated geotextile drain strips (Miradrain 6000, Amerdrain 200, etc.) vertically prior to applying the shotcrete. Typical spacings are the same as the horizontal nail spacing. Extend the

drain mats down the full height of the excavation and discharge into a collector pipe at the base.

3. *Weep holes and horizontal drains.* Install 2-in (50-mm) diameter PVC pipe weep holes on approximately 10-ft (3-m) centers through the shotcrete face where heavier seepage is encountered. Plug the pipe temporarily when shotcrete is applied. Longer PVC horizontal drain pipes can also be installed in heavy seepage areas.

8.8.5 Wall Facing Systems

Temporary walls are typically left with a rough shotcrete face—"gun" finish—with weep holes and protruding nail heads. For permanent walls, where the rough finish is aesthetically unacceptable, the following face options are available:

1. *Separate fascia wall.* As an alternative to the exposed shotcrete finish, the shotcrete can be covered with a separate concrete fascia wall, either cast in place (CIP) or constructed of precast panels. The CIP section is typically a minimum of 6 to 8 in (150 to 200 mm) thick. Precast face panels can be smaller modular panels or full-height fascia panels such as those used to cover permanent slurry walls. A disadvantage of the smaller modular face panels is difficulty of attaching the face panels to the nail heads and some proprietary patent restrictions. A disadvantage of full-height pre-cast panels is that, because of practical constructibility weight and handling limitations, their use is limited to wall heights less than 25 ft (8 m).

2. *Permanent exposed shotcrete facing.* Present technology for shotcrete placement is such that the final shotcrete layer can be controlled to close tolerances, and with nominal hand finishing, an appearance similar to a CIP wall can be obtained (if desired). The shotcrete, whether left in the natural gun finish or hand-textured, can also be colored either by adding coloring agent to the mix or by applying a pigmented sealer or stain over the shotcrete surface. *Only* experienced, well-qualified structural shotcrete specialty contractors should place and finish the permanent structural shot-crete. "Wet-mix" shotcrete should be specified instead of dry-mix because good quality control is easier with wet-mix. Also, wet-mix shotcrete can be air-entrained for improved freeze-thaw durability, whereas dry-mix cannot.

8.8.6 Design of Soil Nailed Retaining Structures

The stability of a soil nailed structure relies on (1) transfer of resisting tensile forces generated in the inclusions in the active zone into the ground in the resistant zone, through friction or adhesion mobilized at the soil-nail interface; and (2) passive resistance developed against the face of the nail. Ground nailing using closely spaced inclusions produces a composite coherent material. As shown in Fig. 8.53, the tensile forces generated in the nails are considerably greater than those transmitted to the facing.

The design procedure for a nailed retaining structure includes (1) estimation of nail forces and location of the potential sliding surface, (2) selection of the reinforcement type, cross-sectional area, length, inclination, and spacing, and (3) verification that stability is maintained during and after excavation with an adequate factor of safety. Methods for determining tensile, bending, and shear stresses in the nails are given by FHWA based on a limit equilibrium analysis.

The majority of soil nailed retaining structures constructed in France are based on two distinct technologies: (1) the method of Hurpin, with nails driven into the ground

TABLE 8.11   Typical Characteristics of Soil Nailed Walls with Vertical
Facing and Horizontal Earth Pressure

	Nails at close centers (method of Hurpin)	Widely spaced nails
Length of nails	0.5 to 0.7H	0.8 to 1.2H
Number of nails per m² of facing	1 to 2	0.15 to 0.4
Perimeter of nail	150 to 200 mm (6 to 8 in)	200 to 600 mm (8 to 24 in)
Tensile strength of reinforcing bar (nail)	120 to 200 kN (34 to 45 kips)	100 to 600 kN (22 to 135 kips)
Nailing density	0.4 to 1.5	0.13 to 0.6

Source: From *Recommendations Clouterre, French National Research,* 1991,
(English translation by Federal Highway Administration, 1993), with permission.

on close spacing, i.e., vertical and horizontal spacing equal to or less than 3 ft (1 m);
and (2) widely spaced grouted nails. With the method of Hurpin, the nails (generally
reinforcing bars) are relatively short and are driven into the ground by percussion or
vibratory methods. The relatively high nail density permits thinner wall facings. In
walls with widely spaced nails, the nails are generally longer. Typical data for soil
nailed walls with a vertical facing and a horizontal earth pressure are shown in Table
8.11. The "nailing density" listed in Table 8.11 is a dimensionless parameter representing soil nails placed in a uniform pattern. It is defined as

$$\chi = \frac{T_r}{\gamma S_h S_v L} \tag{8.19}$$

where T_r = ultimate tensile force that can be mobilized at head of nail
S_h = horizontal spacing between nails
S_v = vertical spacing between nails
L = length of nails
γ = total unit weight of soil

This parameter represents the maximum tensile force in a nail as it relates to the
weight of the soil reinforced with a chosen grid spacing.

A full set of preliminary design charts are included in the FHWA translation of
Recommendations Clouterre, 1991. Diagrams for an angle of installation of the nails
of $i = 20°$ are shown in Fig. 8.54 for illustration.

Figure 8.54 provides a preliminary chart for a soil nailed wall. It seeks to define in
approximate terms the lengths, spacings, and resistance values of the nails to ensure
internal and external stability. It may be used in an early evaluation stage based on
macro assumptions such as homogeneous soil, identical and evenly spaced loads in the
nails, and pure tension in the nails; the bending stiffness of the nails is neglected,
regardless of the angle of incidence of the potential failure surface. This approach is
based on the classic method of vertical slices with circular potential failure surfaces.
The charts are based on a system of coordinates that characterize the shear resistance
of the soil.

Consider the following example: height $H = 10$ m, $\gamma = 20$ kN/m³, $\phi = 35°$, $c =$
20 kPa. Assume $L/H = 0.8$. Calculate coordinates and plot as point M on the chart.

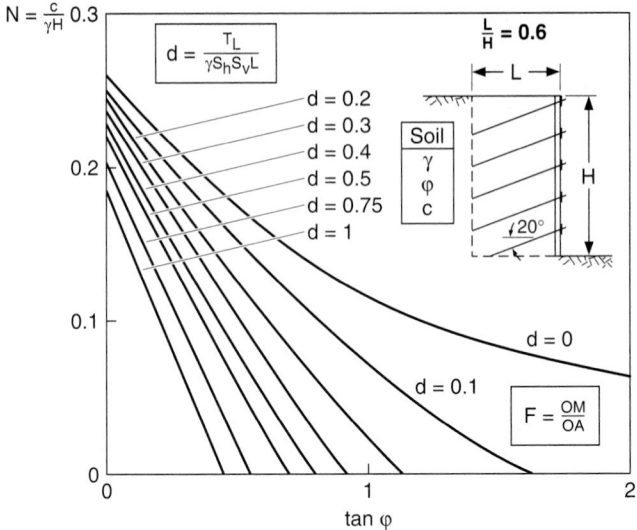

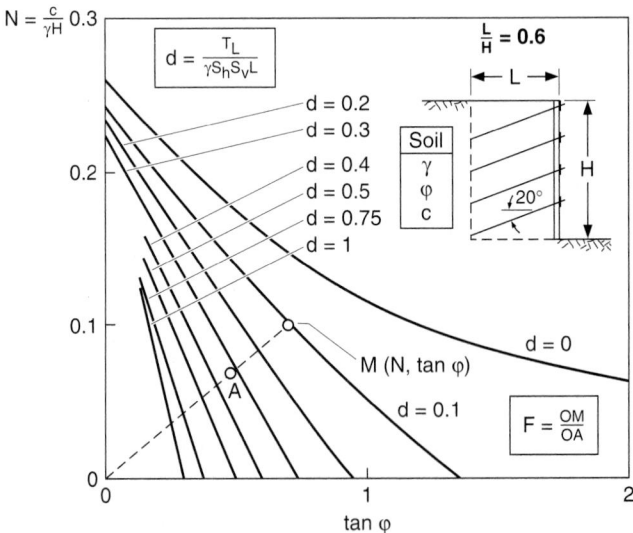

FIGURE 8.54 Preliminary design charts for soil nailed walls. (*From* Recommendations Clouterre, *French National Research, 1991, translation by Federal Highway Administration, 1993, with permission*)

$$N = \frac{c}{\gamma H} = \frac{20}{20 \times 10} = 0.10 \qquad \tan \phi = \tan 35° = 0.70$$

Draw a line from the origin O to M. The safety factor F is the ratio OM/OA. Therefore, for a safety factor of $\frac{3}{2}$, locate point A two-thirds of the distance along the line OM. Interpolation gives the required nailing density d as 0.33. Thus:

$$\frac{T_L}{\gamma S_h S_v L} = 0.33$$

$$\frac{T_L}{S_h S_v L} = 0.33 \times 20 \times 0.8 \times 10 = 52.8 \text{ kPa}$$

Thus, for a nail tensile force T_L, the spacings S_h and S_v can be determined. The result from the chart should be generally conservative and used only for preliminary evaluation.

The final design for stability of a soil nailed wall is analyzed either by calculating the deformations or by using limit equilibrium design. The first method uses finite element calculation and has not been refined to the point where there is an "acceptable" procedure. In Europe to date, there has been considerable diversity in some details among the various design approaches, both within and across national boundaries. A most significant factor is the postulated mechanism by which nails are considered to reinforce a soil mass. For nails installed nearly parallel to the direction of maximum soil tensile strain (e.g., near-horizontal nails and a near-vertical excavation face), the prevailing opinion is that the reinforcing action is predominantly related to tensile loading within the nails. Under service load conditions, the contribution of shear or bending is considered negligible. As failure conditions are approached, the contribution of shear or bending action is more significant but still small. From a practical point of view, however, it is recognized that the soil nails should exhibit ductile behavior in response to bending in order to minimize the potential for sudden failures related to brittleness. Where reinforcing elements are used as dowels and are oriented nearly perpendicular to the direction of maximum shear strain, the shearing, bending, and tensile action of the reinforcement should be considered.

All design methods are based on concepts of limiting equilibrium or ultimate limit states. Various types of potential slip surface are considered, including circular, log spiral, and bilinear wedge. In general, each of the methods appears to provide a satisfactory representation for design purposes. Consistent with the above, most design computer codes consider only the tensile action of the nails, but some also permit consideration of the shear or bending action of the nails. Almost all of the design methods do not explicitly consider the potential for pullout of the reinforcing nails within the active block between the facing and the slip surface. It is implicitly assumed that the nail-soil adhesion within this zone, together with the structural capacity of the facing, will be sufficient to prevent this type of failure. Some design approaches offer strict guidelines for the required structural wall capacity to prevent such active-zone failures, but others appear to rely on experience and do not directly address this issue.

A significant range of load and resistance factors or safety factors have been applied by individual European designers, in considering design for both ultimate limits and serviceability limits of soil nailed walls. It appears that Eurocode 7 requirements will impose a uniform approach in Europe for load and resistance factors and that these requirements will be similar to the recent recommendations from the Clouterre program.

On the basis of the overall reinforcing requirements determined from the limiting equilibrium design calculations, the reinforcing steel is empirically proportioned. In general, designers use nails of uniform length and cross-sectional area, on a uniform spacing. For drilled and grouted nails, the nail spacing is typically in the 3- to 6-ft (1- to 2-m) range. For driven nails, much higher densities (typically 1.5 to 2 nails per square meter) are used. The nail lengths are typically in the range of 60 to 80 percent of the height of the wall, but may be shorter in very competent rocklike materials and longer for heavy surface surcharge or high seismic or other operational loading.

As noted above, facing design requirements are empirically determined using a variety of techniques. German practice requires the use of a uniform facing pressure equivalent to 75 to 85 percent of the active Coulomb loading. The Clouterre recommendations require designing the facing and connectors to support between 60 and 100 percent of the maximum nail loading (for both ultimate and serviceability limit states), depending on the nail spacing.

8.8.7 Construction Considerations for Soil Nailed Walls

The construction sequence is typically to excavate, nail, and shotcrete the face in increments from the top down. Figure 8.55 shows a schematic of a sequence for underpass widening. Where face stability is a concern, a flashcoat of shotcrete may be applied before nail installation. The most common method of nail installation in Europe, as in the United States, is the drill-and-grout method. Most commonly, the steel tendon is installed prior to grouting, although this sequence is sometimes reversed. In France, however, very significant use is also made of driven nails without grouting. Other specialty techniques for installing nails include jet grout nailing (France), driven nails with an oversize head and subsequent grouting of the annulus (Germany), and compressed-air explosive injection of nails (United Kingdom).

Small hydraulic, track-mounted drill rigs of the rotary-percussive type are most commonly used to install nails. These rigs can work in relatively confined surroundings and are therefore compatible with many of the constraints associated with crowded urban environments. Open-hole drilling methods are predominant, with cased-hole methods used in particularly difficult ground conditions. The most common grouting method used with open-hole drilling is the low-pressure tremie method. Where extensive use of casing is required, alternative methods of construction are often more cost-competitive.

Steel tendons typically used for drill-and-grout soil nails usually consist of $\frac{3}{4}$- to 2-in (20- to 50-mm) diameter bars with a yield strength in the range of 60 to 70 kips/in^2 (420 to 500 N/mm^2). These steels exhibit ductile behavior under bending action. The driven nails used commonly in France are typically steel angle sections, which show better ability to deal with subsurface obstructions such as cobbles and small boulders than do circular steel sections.

Drainage is a critical aspect of soil nail wall construction. Face drainage is virtually always used with permanent walls, and very commonly used with temporary walls. Face drainage usually consists of synthetic drainage elements placed between the shotcrete and the retained soil, and may be typically 8- to 12-in (200- to 300-mm) wide synthetic strips or perforated pipes. Depending on the site groundwater conditions, face drainage may be supplemented with weep holes through the facing and longer horizontal perforated drain pipes. Control of surface water is also an important element of the drainage system.

Temporary soil nail wall facings generally consist of shotcrete 3 to 4 in (80 to 100 mm) thick and a single layer of wire mesh. Permanent shotcrete walls 6 to 10 in (150 to 250 mm) thick are very common in Germany, and these walls typically include a

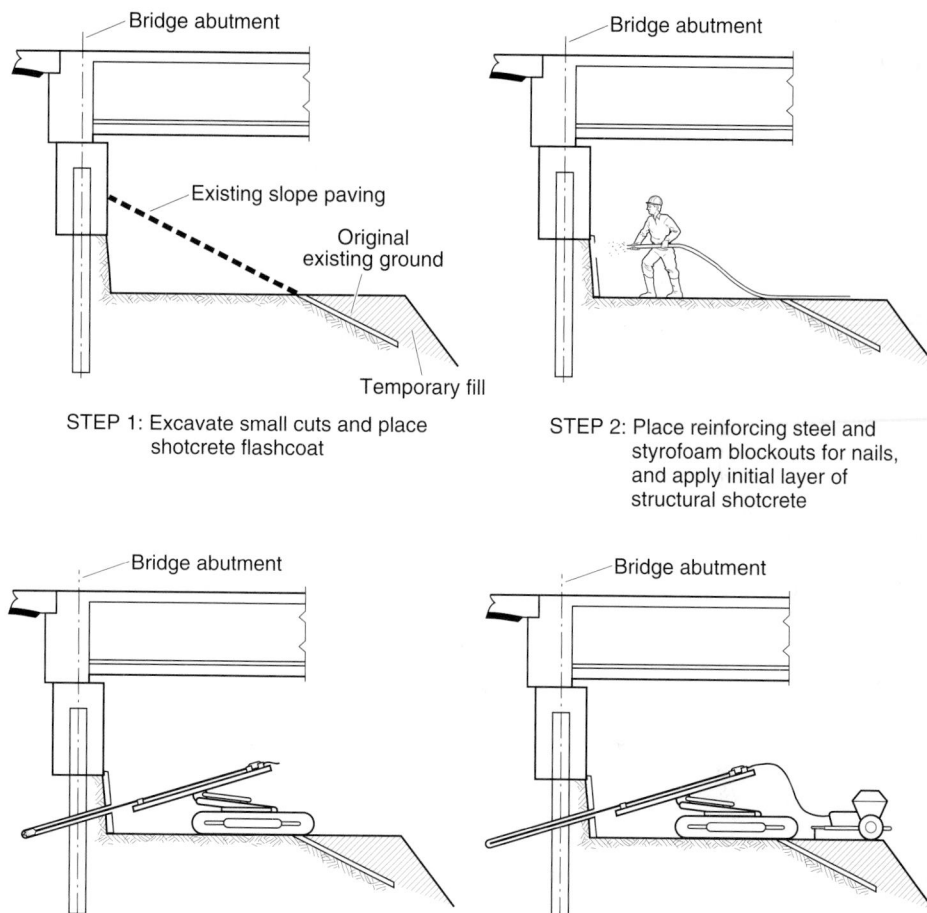

STEP 1: Excavate small cuts and place shotcrete flashcoat

STEP 2: Place reinforcing steel and styrofoam blockouts for nails, and apply initial layer of structural shotcrete

STEP 3: Drill hole for nail

STEP 4: Install nail and pressure grout while extracting casing

FIGURE 8.55   Construction sequence for soil nailed wall used in widening of underpass. (*From Oregon Department of Transportation, with permission*)

second layer of wire mesh. For architectural reasons, permanent walls of precast panels and cast-in-place concrete are also commonly used in France and Germany.

Testing and monitoring during construction are an important aspect of soil nail wall construction in Europe. Nail bond testing is almost universally performed, to confirm the assumptions made during design or to enable redesign in the event that the design assumptions cannot be realized. For relatively homogeneous sites, typically 3 to 5 percent of the nails will be tested, depending on the size of the job. Testing is also undertaken whenever changed geologic or construction conditions occur. Wall performance monitoring usually consists of measuring horizontal wall movements during construction. Some contractors make more use of inclinometers for displacement monitoring. Maximum horizontal displacements are typically in the range of 0.1 to 0.3

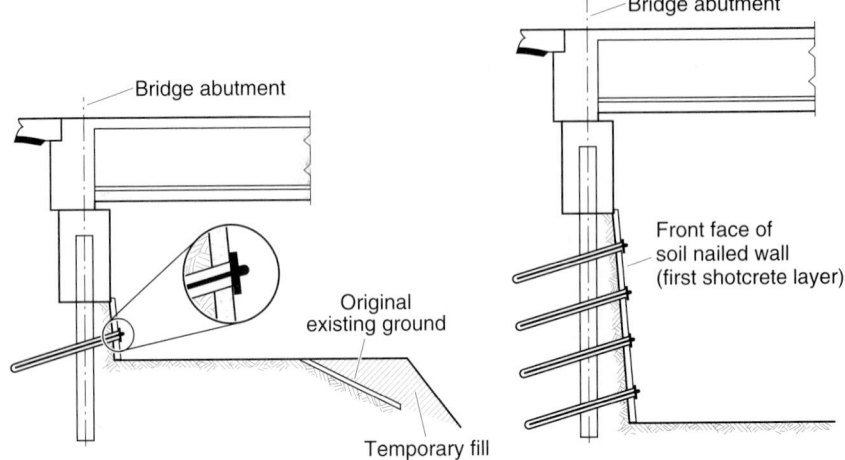

STEP 5: Install bearing plate and nut

STEP 6: Repeat process for
all nail layers

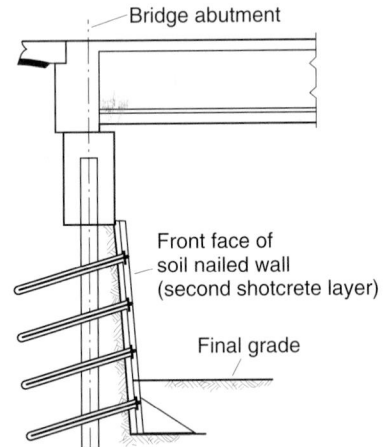

STEP 7: Place second structural application
to full height and architectural finish

FIGURE 8.55   *(Continued)*

percent of the height of the wall, depending on ground conditions. Strain gauging of nails, together with the use of load cells at the nail head, are usually reserved for experimental walls.

The level of quality assurance and control monitoring varies significantly. In Germany, for example, the QA-QC inspector might be on the job from 10 percent to full time.

Overall, soil nail wall performance in Europe has been very good. Problems during construction have typically been associated with encountering loose fill, granular soil with no apparent cohesion, water, and constructed obstructions such as utility trenches. Other problems have been associated with a contractor's failing to construct the wall in accordance with the plans and specifications (e.g., eliminating nails, overexcavation of lifts). Frost action on fully bonded nails has also resulted in development of very large loads near the head of the nail, where no insulating protection has been provided in the wall design.

8.8.8 Soil Nailed Wall Facing Design Procedure

The following typical details and design procedure are based primarily on Caltrans' method for use on highway construction, but the method is very similar to other methods presently in practice. Design facing pressures are based on the French Clouterre empirical method. The cast-in-place portion of the facing is designed for this pressure for permanent walls only. The strength of the shotcrete construction facing is ignored. Only the ultimate limit state is addressed; no serviceability calculations are made for cracking or deflections. Sample design calculations are illustrated following the presentation of the procedure.

Typical Details. (See Figs. 8.56 to 8.58.)

1. Use a shotcrete layer with a 4-in minimum thickness.
2. Include a single layer of welded wire fabric at mid-thickness:

   6×6-W4.0 $\times$ W4.0 (4 gauge wire; diameter, 0.225 in; cross-section area = 0.080 in^2/ft)

   4×4-W2.9 $\times$ W2.9 (6 gauge wire; diameter, 0.192 in; cross-section area = 0.087 in^2/ft)
3. Use two continuous horizontal no. 4 grade 60 reinforcing steel bars at each nail.
4. Use 1-ft-wide vertical geocomposite drain between nails; connect the geocomposite drain to a 2 in round plastic weep hole outlet drain just above finished grade near the bottom of the wall.
5. Place the nail bearing plate (1 in $\times$ 9 in $\times$ 9 in ASTM A36 steel) with wedge washer and nut on the outside face of the shotcrete; set into place before the shotcrete hardens. Add studs to this plate to engage permanent facings that are placed over this initial shotcrete layer.

Step 1: 4-in Shotcrete Construction Facing. This is the only facing required for temporary walls (service life less than 18 months) and the first portion required for permanent walls. It is placed immediately after each stage of excavation and nail placement. Current AASHTO and American Concrete Institute (ACI) codes do not address the loadings or the structural capacities for this facing. Therefore, many current designs rely on details that have shown good performance on previous projects rather than design calculations.

Step 2: 8-in CIP Permanent Facing—Compute Design Nail Load and Pressure at Facing. The design nail load at the facing is computed for the given nail size, steel grade, and nail spacing according to the French Clouterre empirical method. The French determined through field tests that the nail load at the facing (T_0) did not exceed about one-half the maximum nail load (T_{max}) near the soil failure surface. They established a design nail load for the facing that varies from 0.6 times T_{max} for closely

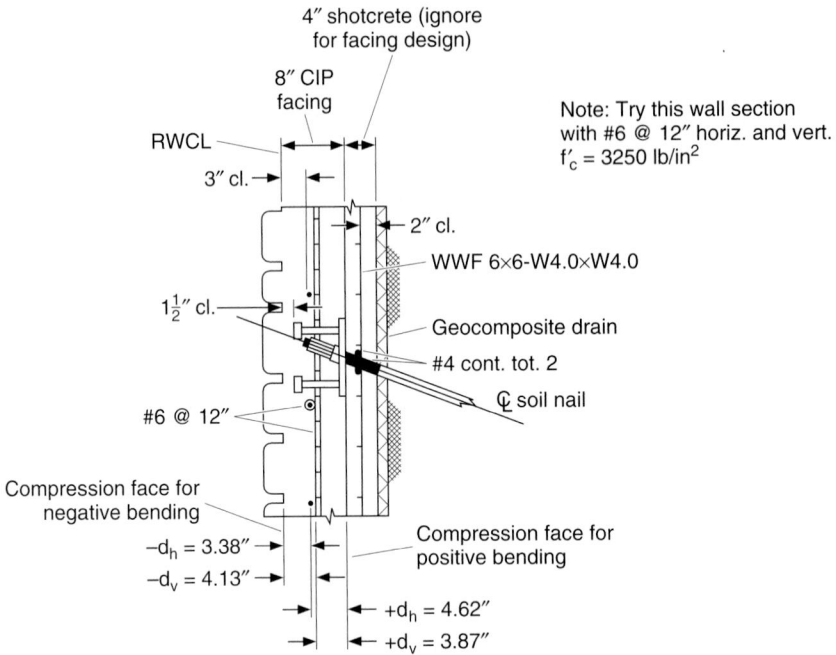

Negative steel

$$-d_h = 3 + \frac{0.75}{2} = 3.38'' \longleftarrow \text{Controls}$$

$$-d_h = 3 + 1.5 \times 0.75 = 4.13''$$

Positive steel

$$+d_h = 8 - 3.38 = 4.62''$$

$$+d_h = 8 - 4.13 = 3.87''$$

Note: d_h and d_v are effective horizontal distances to the horizontal and vertical steel, respectively.

FIGURE 8.56 Section through facing of soil nailed wall showing concrete reinforcement and soil nail connection. (*From J. W. Keeley,* Soil Nail Wall Facing: Sample Design Calculations, *Federal Highway Administration, 1993, with permission*)

spaced nails to 1.0 times T_{max} for larger nail spacings. T_{max} is the ultimate limit state established for the nail tension (Caltrans procedure) at $0.75 \times f_y$ ($T_{max} = A_{nail} \times 0.75 \times f_y$), where f_y is the yield strength of the nail. The design pressure for the facing is then simply the design nail load at the facing (T_0) divided by the nail tributary facing area (i.e., horizontal nail spacing times vertical nail spacing).

Step 3: 8-in CIP Permanent Facing—Design for Flexure. The cast-in-place facing is designed so that its ultimate strength is greater than the moments in the facing computed by simple continuous beam equations for the facing pressure from T_0. Only one layer of grade 60 reinforcing steel placed near the middle of the section is used. (See Fig. 8.56.) The controlling d is used for the ultimate strength computation.

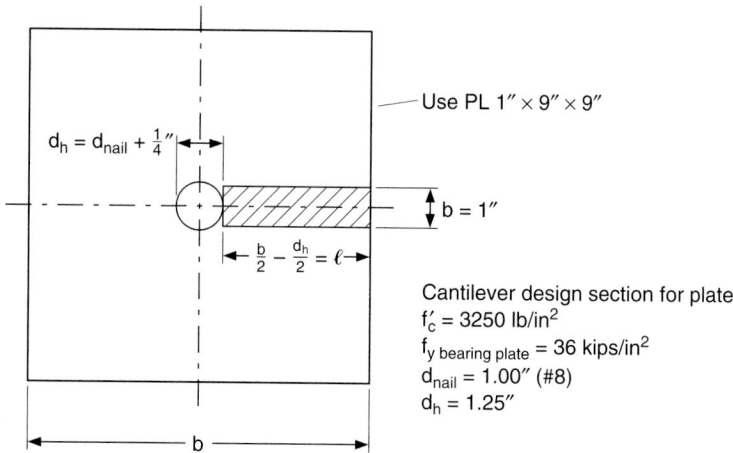

FIGURE 8.57 Bearing plate for soil nailed wall showing cantilever strip for calculations. (*From J. W. Keeley,* Soil Nail Wall Facing: Sample Design Calculations, *Federal Highway Administration, 1993, with permission*)

Step 4: 8-in CIP Permanent Facing—Nail Connection Design. The connection of the nail to the cast-in-place facing is designed to carry the nail's ultimate limit state in tension, T_{max}. The nail bearing plate is sized to limit the bearing pressure from T_{max} to the ultimate value allowed by AASHTO ($0.6f'_c$). The plate thickness is determined to provide sufficient bending strength for the moment from the bearing pressure about the nail nut. Studs are welded to the bearing plate to carry T_{max} entirely by the 8-in CIP permanent facing. The ultimate punching shear capacity of the steel embedment is computed according to American Concrete Institute specifications 349-90, Appendix B.

Sample Design Calculations for Soil Nail Wall Facing (Based on Caltrans Methods)

Step 1: 4-in Shotcrete Construction Facing. Details as previously described may be used.

Step 2: 8-in CIP Permanent Facing—Concrete Design Nail Load and Pressure at Facing. Given No. 8 nails; $f_y = 60$ kips/in²; $A_{nail} = 0.79$ in²; horizontal nail spacing $S_h = 6$ ft; and vertical nail spacing $S_v = 6$ ft.

Begin by calculating the design nail loads.

$$T_{max} = \text{design nail load at soil failure surface at ultimate limit state}$$

$$= 0.75 f_y A_{nail} = 0.75 \times 60 \times 0.79 = 35.6 \text{ kips}$$

$$T_0 = \text{design nail load at facing at ultimate limit state}$$

$$\frac{T_0}{T_{max}} = 0.5 + \frac{0.3 S_{max} - 0.5}{5} \quad \text{for} \quad S_{max} \text{ in ft}$$

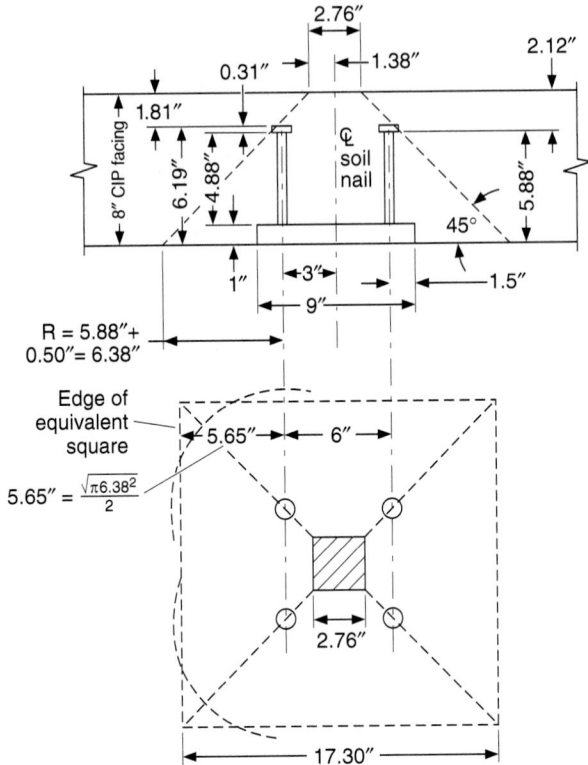

Effective stress area = A_{cp} = $17.30^2 - 2.76^2$ = 291.7 in^2

FIGURE 8.58 Bearing plate for soil nailed wall showing effective stress area for embedment design capacity. (*From J. W. Keeley*, Soil Nail Wall Facing: Sample Design Calculations, *Federal Highway Administration, 1993, with permission*)

(*Note:* This is the French Clouterre equation.)

$$T_0 = T_{max}\left(0.5 + \frac{0.3S_{max} - 0.5}{5}\right)$$

$$= 35.6\left(0.5 + \frac{0.36 - 0.5}{5}\right) = 27.0 \text{ kips}$$

The design nail load T_0 is then used to calculate the facing design pressure.

W_u = facing design pressure at ultimate limit state

$$= \frac{T_0}{S_h S_v} = \frac{27.0}{6 \times 6} = 0.75 \text{ kip/ft}^2$$

Step 3: 8-in CIP Permanent Facing—Design for Flexure.   *Note:* Try this wall section with No. 6 reinforcing bars spaced at 12 in horizontally and vertically; f'_c = 3250 lb/in².

Required ultimate moment per foot (horizontal and vertical; positive and negative)

$$M_u = W_u \frac{l^2}{10} = 0.75 \left(\frac{6^2}{10} \right) = 2.7 \text{ ft} \cdot \text{kips/ft}$$

Ultimate design capacity for trial section

8-in CIP permanent facing

b = 12 in per ft of design width

f'_c = 3250 lb/in²

f_y = 60 kips/in²

d = 3.38 in (see Fig. 8.56)

A_s = 0.44 in² (2 No. 6 @ 12 in) = area of steel

$$a = \frac{A_s}{0.85} \frac{f_y}{f'_c b} = \frac{0.44 \times 60}{0.85 \times 3.25 \times 12} = 0.80 \text{ in}$$

= depth of concrete compression

A strength reduction factor of ϕ = 0.90 is applied to the nominal moment strength M_n as follows:

$$\phi M_n = \phi A_s f_y \left(d - \frac{a}{2} \right)$$

$$= 0.9 \times 0.44 \times 60 \times \left(3.38 - \frac{0.80}{2} \right) \times \frac{1}{12} = 5.9 \text{ ft} \cdot \text{kips/ft} \geq 2.7 \qquad \text{OK}$$

Check of minimum steel requirements.   According to the AASHTO *Bridge Specifications*, the tension reinforcement must be equal to or greater than the lesser of that required to develop a moment (1) 1.2 times the cracking moment (based on the gross section modulus S_g) and (2) 1.33 times that required by analysis for the specified loading conditions. This leads to the following equations for the cross-section area of the reinforcement:

$$A_s = \frac{1.2 \times S_g \times 7.5 \sqrt{f'_c}}{\phi f_y \times 0.9d}$$

$$A_s = \frac{1.33 M_u}{\phi f_y \times 0.9d}$$

The strength reduction factor ϕ is 0.90. Substitution gives the following results:

$$A_s = \frac{1.2(12 \times 8^2/6)7.5\sqrt{3250}}{0.9 \times 60,000 \times 0.9 \times 3.38} = 0.40 \text{ in}^2/\text{ft}$$

and

$$A_s = \frac{1.33(2.7 \times 12,000)}{0.9 \times 60,000 \times 0.9 \times 3.38} = 0.26 \text{ in}^2/\text{ft}$$

Thus, A_s must be at least 0.26 in²/ft to meet these requirements. For temperature and shrinkage, minimum A_s is 0.13 in²/ft (No. 4 @18 in). The final selection is No. 6 reinforcing bars spaced at 12 in, which provides A_s of 0.44 in²/ft, a nominal level of reinforcing. For complete designs, also check cantilever sections at the top and bottom of the wall and any other special facing sections, such as at expansion or contraction joints.

Step 4: 8-in CIP Permanent Facing—Nail Connection Design

Design bearing plate at ultimate limit state for T_{max}

$$T_{max} = 0.75 f_y A_{nail} = 0.75 \times 60 \times 0.79 = 35.6 \text{ kips}$$

Calculate the ultimate concrete bearing strength under the plate using a strength reduction factor of $\phi = 0.70$. Therefore,

$$\text{Ultimate concrete bearing strength} = 0.85 \phi f'_c$$

$$= 0.60 f'_{c\,min}$$

$$\text{"A" bearing} = b^2 - \frac{\pi d_h 2}{4}$$

$$\text{Required "A" bearing} = \frac{35.6}{0.6 \times 3.25} = 18.26 \text{ in}^2$$

$$\text{Available "A" bearing} = 9^2 - \frac{\pi 1.25^2}{4} = 79.8 \text{ in}^2 \qquad \text{OK}$$

$$\text{Bearing stress under plate} = \frac{35.6}{79.8} = 0.45 \text{ kip/in}^2$$

$$\text{Allowable stress} = 0.6 \times 3.25 = 1.95 \qquad \text{OK}$$

$$l = \frac{1}{2}(b - d_h) = \frac{1}{2}(9 - 1.25) = 3.875 \text{ in}$$

$$t \text{ required} = l\sqrt{\frac{3.33w}{f_y}} = 3.875\sqrt{\frac{3.33 \times 0.45}{36}} = 0.79 \text{ in}$$

$$t \text{ available} = 1 \text{ in} \qquad \text{OK}$$

Note:

$$\frac{wl^2/2}{t^2/6} = 0.9f_y$$

Therefore,

$$t = l\sqrt{\frac{3.33w}{f_y}}$$

where w = ultimate bearing pressure.

Required steel area for studs to resist T_{max}. The design strength of the stud, f_{ds}, is determined as the lesser of (1) the stud yield strength (f_y = 50 kips/in²) multiplied by a strength reduction factor of ϕ = 0.90 and (2) 80 percent of the stud tensile strength (f_u = 60 kips/in²). Therefore,

$$f_{ds} = \min\,[\phi f_y;\,0.8f_u]$$

$$\phi f_y = 0.9 \times 50 = 45 \text{ kips/in}^2$$

$$0.8f_u = 0.8 \times 60 = 48 \text{ kips/in}^2$$

Therefore

$$f_{ds} = 45 \text{ kips/in}^2$$

$$\text{Required stud area } A_s = \frac{T_{max}}{f_{ds}}$$

$$= \frac{35.6}{45} = 0.79 \text{ in}^2$$

Try $4\frac{1}{2}$ in ϕ studs.

$$A_s = 4 \times 0.196 = 0.79 \text{ in}^2 \qquad \text{OK}$$

Check of anchor head bearing for ½ in ϕ stud.   Determine head area A_h.

$$A_h = \pi\left(\frac{d_h}{2}\right)^2 = 0.79 \text{ in}^2 \text{ (manufacturer's data, } d_h = 1)$$

$$\frac{A_h}{A_s} = \frac{0.79}{0.196} = 4 > 2.5 \qquad \text{OK}$$

$$T_h = 0.312 \text{ in} > 0.25 \text{ in}$$

Head thickness is OK.

Ultimate connection embedment design capacity P_d

$$P_d = 4\phi\sqrt{f'_c}\,A_{cp}$$

$$f'_c = 3250 \text{ lb/in}^2$$

$$\phi = 0.65 = \text{strength reduction factor in this case}$$

$$A_{cp} = 291.7 \text{ in}^2 = \text{effective stress area (see Figure 8.58)}$$

$$P_d = 4 \times 0.65 \times \sqrt{3250} \times 291.7 \times \frac{1}{1000} = 43.2 \text{ kips}$$

$$P_d > T_{max} = 35.6 \text{ kips} \qquad \text{OK}$$

8.8.9　Global Stability

Evaluation of a global safety factor that includes the nailed soil and the surrounding ground requires determination of the critical sliding surface. This surface may be located totally inside, totally outside, or part inside and part outside the nailed zone. Limit equilibrium methods are usually used, and the Davis method is recommended because of its simplicity and availability in the public domain (C. K. Shen et al., "Field Measurements of an Earth Support System," *Journal of the Geotechnical Division, American Society of Civil Engineers,* vol. 107, no. 12, 1981). The Davis method has been modified (V. Elias, and I. Juran, "Soil Nailing," Report for FHWA, DTFH 61-85-C, 1988) to permit input of interface limit lateral shear forces obtained from pullout tests, separate geometric and strength data for each nail, facing inclination, and a ground slope at the top of the wall. The concrete facing elements (shotcrete, cast-in-place concrete, or prefabricated panels) are considered for design to be analogous to a beam or raft of a unit width equal to the nail spacing supported by the nails.

8.8.10　Contracting Practices

Although procurement and contracting practices vary among the European countries, there are some common elements that tend to distinguish European practices from those in the United States. These include (1) strong industry, academic, and government cooperation in research and development and the introduction of new technologies; (2) a partnering approach among all parties involved in a particular project; (3) less litigation; and (4) a high level of contractor involvement in the conceptualization and design phases, as well as during construction.

In France, the contractor design-build approach appears to be dominant. For public agency work, a prequalified group of contractors are typically asked to prepare a final design and bid, based on a preliminary design prepared by the owner or the owner's consultant. Alternative designs may also be prepared by the contractor at this time, and may be selected if they are technically and financially viable and meet the overall performance and scheduling requirements of the project. French contractors tend to be much larger and stronger than their U.S. counterparts, and the major groups tend to support significant research and development efforts. Contractor-consultant-academic-government cooperation in areas requiring major research and development is particularly well developed in France.

In Germany, public agency work is again usually bid on a conceptual or preliminary design prepared by or for the agency, with the contractor required to submit a bid on the original design and also encouraged to submit any alternative design that will provide an equivalent wall at a reduced price. Ultimately, award is made for the lowest-cost responsive bid. Soil nailing in Germany requires the involvement of one of a small group of prequalified or "licensed" contractor organizations. As in France, these contractors tend to be technically and financially very strong. Private work, like public work, tends to be awarded on the basis of low bid.

Based on the European, and particularly the French, experience, two main recommendations are offered for encouraging the development of innovative construction methods and improving the construction performance for such methods. First, stronger and more formal government-academic-industry cooperation should be established to develop new technologies and disseminate the information in a nonproprietary manner. This should also include participating in corresponding European programs when the opportunity arises and when the information will be of mutual benefit. Second, alternative bidding, including contractor design-build alternatives, performance-oriented specifications, and the use of carefully prequalified specialty contractors, should be encouraged.

8.9 PREFABRICATED MODULAR WALLS

There are also a number of prefabricated modular wall systems in use. Such systems are generally composed of modules or bins filled with soil, and function much like gravity retaining walls. The bins may be of concrete or steel, and can be used in most cases where conventional gravity, cantilever, or other wall systems are considered. AASHTO indicates that such walls should not be used on curves less than 800 ft in radius, unless a series of chords can be substituted; or where the calculated longitudinal differential settlement along the face of the wall is excessive. Also, durability considerations must be addressed, particularly where acidic water or deicing spray is anticipated.

REFERENCE MATERIAL

The following reference sources were helpful in developing this chapter:

Bridge Design Manual, Colorado Department of Transportation, Section 5, "Earth Retaining Wall Design Requirements."

Claybourn, A. F., *Comparison of Design Methods for Geosynthetic-Reinforced Soil Walls,* Woodward-Clyde Federal Services.

Design Manual, Part 4, "Structures," Section 5, "Retaining Walls," Commonwealth of Pennsylvania, Department of Transportation.

"Durability/Corrosion of Soil Reinforced Structures," Federal Highway Administration Report RD-89-186, December 1990.

Elias, V., and I. Juran, "Soil Nailing," report for Federal Highway Administration, DTFH 61-85-c, 1988.

Geosystems for Highways and Transportation Structures: Guide to Selection, Design and Construction, Colorado Department of Transportation, June 1992.

Harned, C. H., *Some Practical Aspects of Foundation Studies for Highway Bridges,* Bureau of Public Roads, January 1959.

Keeley, J. W., *Soil Nail Wall Facing: Sample Design Calculations,* Federal Highway Administration, 1993.

Mitchell, J. K., and B. R. Christopher, "North American Practice in Reinforced Soil Systems," *Proceedings, Specialty Conference on Design and Performance of Earth Retaining Structures,* Geotechnical Division, American Society of Civil Engineers, 1990.

Recommendations Clouterre, 1991, French National Research, Project Clouterre (English translation, 1993, U.S. Department of Transportation, Federal Highway Administration).

Shen, C. K., et al., "Field Measurements of an Earth Support System," *Journal of the Geotechnical Division, American Society of Civil Engineers,* vol. 107, no. 12, 1981.

Standard Specifications for Highway Bridges, 15th ed., 1992, American Association of State Highway and Transportation Officials, Section 5, "Retaining Walls."

CHAPTER 9
NOISE WALLS

James J. Hill, P.E.

Structural Engineer
Minnesota Department of Transportation
Roseville, Minnesota

Roger L. Brockenbrough, P.E.

R. L. Brockenbrough & Associates, Inc.
Pittsburgh, Pennsylvania

During recent years, there has been increasing concern over noise generated by highway traffic in urban areas. Noise abatement programs have been implemented by many agencies. Source control methods have included the development of quieter pavements, quieter tire tread patterns, and speed restrictions. In some regions, noise levels have been reduced by depressing roadways or building tunnels, or by special designs of adjacent buildings. In many cases, however, noise reduction has been achieved through controlling the noise path by the design and construction of noise walls. Sometimes referred to as sound barriers or noise barriers, these longitudinal walls are built specifically to reduce traffic noise. However, in addition to their primary purpose, noise walls are sometimes adopted to shield unsightly areas from the public and restore a feeling of visual privacy. A noise wall project involves many areas including acoustical evaluations, consideration of aesthetics, cost evaluations, roadway safety design, structural design, foundation design, and construction.

This chapter includes information from the following sources: S. H. Godfrey and B. Storey, *Highway Noise Barriers: 1994 Survey of Practice,* Transportation Research Board, Washington, D.C., 1995; D. Byers, "Noise Wall Aesthetics: New Jersey Case Study," presentation, Transportation Research Board, Washington, D.C., 1995; *Guide on Evaluation and Abatement of Traffic Noise,* American Association of State Highway and Transportation Officials (AASHTO), Washington, D.C., 1993; *Guide Specifications for Structural Design of Sound Barriers,* AASHTO, Washington, D.C., 1989; *Road Design Manual,* Minnesota Department of Transportation, Roseville, 1982.

9.1 ACOUSTICAL CONCEPTS

Figure 9.1 illustrates the fundamental function of a noise wall. The noise source is traffic, particularly large truck traffic, which generates noise by the action of tires on pavement, the drive train, the engine, and the exhaust. The receiver or receptor can be defined as the location where land use results in exposure to highway traffic noise for an hour or more per day. It may typically be set at 5 ft above ground or at window level. Acoustical design includes controlling noise that passes over the wall and is diffracted to the receiver, noise that is transmitted through the wall, and noise that is reflected from the wall.

Noise levels are expressed in dBA, decibels measured with a frequency weighting network corresponding to the A scale on a standard sound-level meter. The ease of attaining increasing levels of attenuation has been estimated as follows:

5 dBA: simple

10 dBA: attainable

15 dBA: very difficult

20 dBA: nearly impossible

Designs for reductions greater than 15 dBA are usually not considered feasible because of unpredictable and uncontrollable atmospheric and terrain surface effects, scattering from trees and buildings, and other unknowns.

Diffracted Noise.   The noise that passes over the wall, which is the most important of the three types of noise, depends on the location and height of the wall. Attenuation is directly related to the difference between the length of the path from the source to the receiver in the absence of a noise wall, and the length of the path from the source over the top of the wall to the receiver by diffraction. At a given distance from the roadway, increasing the wall height increases the attenuation achieved. However, this relationship is obviously nonlinear, and as the height of the wall increases above some reasonable value, the attenuation that can be achieved decreases rapidly. Assuming a wall height that just breaks the line of sight from the source to the receiver, and assuming that such a wall provides a 5-dBA attenuation, a rule of thumb is to assume that an attenuation of ½ dBA can be achieved with each additional foot of wall height. But because the relationship is actually nonlinear, this approximation holds for only a limited range. Sometimes it is possible to take advantage of local terrain and locate a

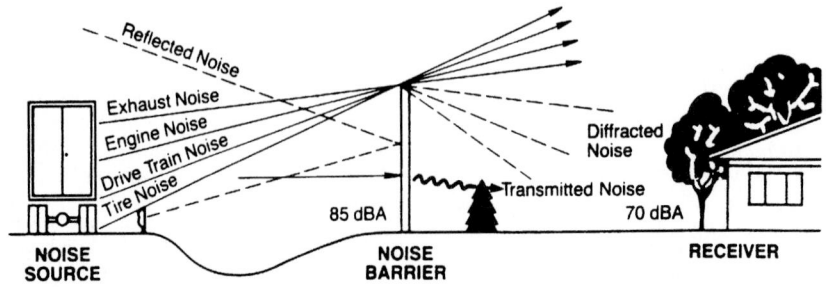

FIGURE 9.1   Acoustical concept of noise wall.   (*From* Handbook of Steel Drainage and Highway Construction Products, *American Iron and Steel Institute, Washington, D.C., 1994, with permission*)

noise wall on a stretch of land at a higher elevation. This reduces the required height and cost of the wall. Wall heights are generally in the range of 6 to 25 ft.

Traffic generates sound waves longitudinally as well as laterally. Thus, care must be taken to extend the length of the wall sufficiently to achieve the desired end result. A rule of thumb states that the noise wall should extend, in each direction from the boundaries of the receiver, 4 times the distance from the receiver to the noise wall. This length can be reduced by combining the ends of the wall with other features, such as natural knolls, or by flaring the wall toward the land use area to form a barrier to the longitudinal sound waves.

Transmitted Noise.   The noise that passes through the wall depends on the surface characteristics and composition (density) of the wall. Acoustical performance can be determined by testing in accordance with standards of the American Society for Testing and Materials (Test Designation E90). It is important that the wall not contain gaps or holes. Overlapping sections can be used to accommodate access through the wall for maintenance or other personnel when applicable. In such cases, the overlap should be at least 2.5 to 3 times the width of the opening.

Reflected Noise.   There is a possibility that noise wall effectiveness can be reduced by reflected noise, such as where longitudinal walls are located on either side of the roadway. To avoid this situation, it has been recommended that the width between two parallel barriers be at least 10 times the average height of the wall above the roadway.

9.2   ACOUSTICAL STANDARDS AND DESIGN

Acoustical design goals should be set in the initial project stages. Generally, this will be based on reducing the noise level to an acceptable level. A reference guide that should be consulted is the Federal Highway Administration publication *FHWA Noise Standards.* If a wall is being planned for an existing highway, it is desirable to measure noise levels before and after wall construction to document effectiveness and obtain data for use in other situations.

Noise levels can be predicted by mathematical models such as the computerized FHWA model known as STAMINA. Using the output from STAMINA, an interactive program known as OPTIMA can be used to design the most economical barrier configuration for a given site.

Local criteria may be more restrictive than federal criteria. In Minnesota, for example, daytime criteria in residential areas are an hourly L_{10} of 65 dBA and an hourly L_{50} of 60 dBA. L_{10} refers to the sound level that is exceeded 10 percent of the time over the period under consideration (1 h, in this case); L_{50} refers to the level exceeded 50 percent of the time. Noise abatement projects strive for a minimum reduction of 10 dBA in L_{10} and 6 dBA in L_{50} from existing traffic noise levels.

9.3   TYPES OF WALLS

Except for berms and brick or masonry construction, most noise walls are of post-and-panel construction, that is, vertical posts spaced several feet apart with horizontal or vertical panels running in between. Rails or girts may also run between the posts to support the panels. Posts are embedded in the foundation soil to design depth, which depends on the wind loading and soil properties. Brick and masonry walls generally require spread footings.

The main materials that have been used for noise wall construction, in order of usage, are the following:

- Concrete
- Earth berms
- Wood
- Brick or masonry
- Cold-formed steel sheet

Other materials sometimes used include aluminum, plastic, glass, composites, and gabions (rock-filled wire baskets). Glass and clear plastic are alternatives where it is desirable to not block scenic views.

Concrete.   Concrete is the most widely used material for noise walls. Users indicate that their selection has been based on cost, durability, low maintenance, surface treatments available, and acoustical properties. Concrete walls can be precast, cast in place, or of post-and-panel construction. Precast concrete panels may be of either prestressed or reinforced construction. Various surface finishes such as texturing are available and are relatively inexpensive. A 4-in-thick wall provides a relatively high transmission loss of 32 dBA.

Earth Berms.   Earth berms or mounds are preferred by many to serve the function of a noise wall. Natural appearance, favorable cost, ready availability of the material, low maintenance cost, and acoustical efficiency favor their selection. A disadvantage is the space needed for construction, particularly in view of safety requirements. Sometimes soil is used in combination with a noise wall of other material where space is limited. For example, if there is not enough space to achieve the full desired height with a berm, a noise wall can be located on top of a berm of lower height. Berm side slopes of 4:1 or flatter are desirable on the basis of considerations of safety (see Art. 6.2) and roadside maintenance. Some states permit up to 3:1, depending on lateral location. Both acoustics and aesthetics can be improved when the berm is combined with a dense planting of vegetation. Vegetation with a minimum depth of 100 ft (perpendicular to roadway), height of 15 ft, and density such that there is no clear path between the highway and the adjacent land use areas can result in a noise level reduction of up to 5 or 6 dBA.

Wood.   Wood is the third most widely used material for noise walls. Attributes that favor its selection include favorable cost, ease of construction, aesthetic appeal, and availability. Disadvantages include shrinkage, warpage, deterioration, difficulty of quality control, discoloration around fasteners, and low resistance to vandalism. Wooden walls have been constructed from timbers, planks, plywood, and laminated products. Often, these materials are used for the panels or facing and concrete or steel is used for the posts. Tongue-and-groove construction can be used for panels running between the posts. The durability of wooden walls can be enhanced by using materials that have received a pressure preservative treatment. Wood provides a transmission loss of 18 to 23 dBA per inch of thickness.

Brick or Masonry.   Brick and masonry construction is also popular, mainly because of its pleasing appearance and acoustical properties. However, initial cost is likely to be higher, as well as repair cost if damaged. Slump block, cinder block, stone, and brick have all been used. Units can be arranged to produce various patterns. The typical transmission loss is 33 dBA, and this can be improved by the addition of mineral wool or fiberglass to the wall interior.

Cold-Formed Steel Sheet.   Steel walls can be used as stand-alone walls or in combination with berms. Low cost, maintainability, and ease of construction favor use of steel. Disadvantages include vibration problems, denting, and ineffectiveness in the low-frequency range. For steel construction, the panels are fluted (have rectangular corrugations) vertically or horizontally, with a channel-shaped cap at the top. Prepainted galvanized sheet and weathering steel have been used, and other durability treatments are available. The transmission loss is generally between 10 and 22 dBA.

Proprietary Systems.   There are a number of proprietary systems on the market for sound walls. Some have been used with success, while others are still being evaluated by users. Some products have included recycled materials such as tire rubber, wood processing waste, and plastics. Of course, steel and aluminum products contain a high level of recycled metal. Some of the proprietary systems include Durisol, Evergreen wall, Fan wall, Sierra wall, Sound Lok, Sound Zero, and Soundtrap. Metal systems include the ConTech Bin Wall and Cameo Metal Panels.

9.4   WALL SELECTION

Presuming acoustical requirements are met, selection is usually based on cost and aesthetics. Costs that must be considered include site preparation, the wall material itself, foundations, fabrication, erection, and maintenance. Aesthetics should be judged with the involvement of personnel with diverse backgrounds, and public participation should be encouraged. However, there are numerous factors that go into the final selection. Some factors that should be considered in wall selection are summarized in Table 9.1.

The reasonableness of constructing a noise wall can be judged from a cost-benefit analysis. For example, Minnesota uses the following procedure. The benefit is based on the summed insertion loss (noise reduction) for each residence in the first two rows of homes nearest the noise wall where the insertion loss is greater than 5 dBA. The ratio of this sum in dBA to the cost of the barrier in thousands of dollars must be greater than 0.4 for the benefit to be considered reasonable.

9.5   AESTHETICS

Often, a detailed study is required to address the question of aesthetics. Alternative systems can be compared, with sketches, renderings, plan drawings, and other visual aids prepared to assist in the process. A multidiscipline team approach is desirable, including design engineers, planners, landscape architects, and environmental personnel. Public input to the selection system helps achieve acceptance of the final system. Designers should be concerned with the visual impact from both the driver's side and the land user's side of the wall.

Some of the important aspects of aesthetics include scale relationship, relationship to environmental setting, line form, color, and texture. A high wall alongside a row of single-story houses is not desirable, nor is one placed so close to the residences that unwanted shadows are created. A rule of thumb is to locate the wall a distance of at

TABLE 9.1 Factors to be Considered in Noise Wall Selection

Site
  Site geometry
  Right-of-way width
  Relation to source height
  Configuration, single or parallel

Noise source
  Traffic type and volume
  Noise frequencies
  Extraneous noise sources

Material
  Structural integrity
  Durability and maintenance
  Susceptibility to vandalism
  Acoustical properties

Cost
  Site preparation
  Wall material
  Foundations
  Fabrication
  Erection
  Maintenance

Aesthetics
  Scale relationship
  Environmental relationship
  Line form
  Color
  Texture
  Community preferences

least 4 times its height from the residences. Walls higher than 16 ft should be critically evaluated for potential unsightly impact.

Evergreens and other plantings are often used with noise walls to enhance appearance. Vines, encouraged to grow up the posts and across the top, have been appreciated by the public. Most agree that walls with extensive landscaping are the most visually appealing.

When the elevation changes along the length of the wall, it is generally considered more pleasing to step the wall rather than to taper it. Ordinarily, the wall will be constructed vertically. There has been some use of walls that have the top tilted away from the roadway in an effort to reduce echo, but such walls tend to give the appearance of instability when viewed from the back side.

If a wall is located in an area with dominant architectural features, this should be considered in the selection of wall material, texture, and color. On the other hand, if a wall is located near dominant roadside features such as bridges, there should be an effort to create a strong visual relationship to such features.

In most cases, there should be some consistency in color and surface treatment. For example, some agencies use color scheme and architectural treatment to distinguish between particular corridors.

In general, walls with darker colors are preferred to lighter ones because they tend to blend better with the background. Although it is usually desirable to avoid visual dominance, murals painted on noise walls have been well received in some urban regions. The murals tend to discourage graffiti, and in some cases, youth groups have been active in restoring murals that have been defaced by graffiti.

With concrete walls, a textured appearance can give the effect of shadows and is often considered desirable. Deep textures are more effective than shallow ones. Such treatments can be achieved by a raking technique on the surface of the newly placed concrete. Colors can be obtained with additions to the mix, or by applying a pigmented sealer after the wall is constructed. The latter technique helps take care of small color variations between panels and minor field problems. Also, coatings can aid in removing graffiti and restoring the intended surface.

For a pleasing visual effect, as well as for safety and acoustic considerations, walls should not begin or end abruptly. To achieve this, walls may be stepped down, flared, or tied into an earth berm, a hillside, a bridge abutment, or another feature. Tapering or stepping is particularly desirable where the height of the wall exceeds 6 ft.

Views of several noise walls are shown in Figs. 9.2 through 9.5 to illustrate some of the effects that can be achieved. Figure 9.2 shows concrete-block construction and deep texturing with vertical grooves. The wall is stepped rather than tapered in height. Figure 9.3 shows timber tongue-in-groove construction, with a natural finish and a stepped height. In Fig. 9.4, the alignment of the timber wall has been changed to a buttress configuration, and extensive plantings have been added. A much different effect has been obtained with concrete post construction, in Fig. 9.5, where the light posts make a distinct contrast with the darker timber.

FIGURE 9.2   Concrete-block noise wall with vertical groove treatment, stepped in height.

FIGURE 9.3 Timber noise wall with tongue-in-groove construction, stepped in height.

FIGURE 9.4 Timber noise wall with buttress-type alignment.

FIGURE 9.5 Timber noise wall with concrete posts showing effect of contrasting hues.

9.6 SAFETY CONSIDERATIONS

Care must be taken not to install a noise wall in such a way that it will be a safety hazard. The general considerations presented in Chap. 6, Safety Systems, apply here. Noise wall design should incorporate all of the safety design techniques used in the basic roadway design. Examples of features that should be considered include transverse location to provide required clear zone, slopes of berms, sight distances, wall ends, plantings, and transitions.

Ideally, noise walls should be located beyond the clear zone. If not, a traffic barrier may be warranted. It is usually best to design the barrier as part of the noise wall. If a wall is located at or near the edge of the shoulder, the portion of the wall above the traffic barrier should be capable of withstanding the force of an occasional vehicle that may ride up above the top of the barrier. Concrete or masonry construction would usually be used in this case.

At locations such as ramps, intersections, and merge areas, care must be taken to avoid blocking the line of sight between vehicles. The AASHTO *Guide on Evaluation and Abatement of Traffic Noise* gives the following suggestions for placement of noise walls (referred to therein as *noise barriers* or just *barriers*):

> For on and off ramps, the minimum set back of a noise barrier is based upon the stopping sight distance, which is a function of the design speed and radius of curvature of the ramp. For ramp intersections, proper barrier location is set by the sight distance corresponding to the time required for a stopped vehicle to execute a left-turn maneuver (approximately 7.5 s). For intersecting roadways, barrier placement is determined from stopping sight distance, which depends on driver reaction time and deceleration rate.

The AASHTO *Guide Specifications for Structural Design of Sound Barriers* indicates that, when locating a sound wall near a gore area, the wall should begin or end at least 200 ft from the theoretical curb nose location.

Protrusions that could constitute a hazard must be avoided near traffic lanes, as well as facings that could become missiles in the event of a crash. Also, surfaces must not create excessive glare.

Sometimes it is necessary to store plowed snow between the roadway and the walls over a width of 6 to 10 ft. In such cases, it should be removed as soon as practical to avoid blowing on the roadway and freezing. Also, there has been some occasional damage to wall panels from the pressure created by snowplows, and this should be avoided as well. Aside from snow storage, highway engineers should consider the potential for roadway icing problems resulting from deep shadows cast by walls.

The end of a noise wall or earth berm can be a hazard to approaching traffic. When exposed to approaching traffic within the clear zone area, it should be treated with protection similar to that for other fixed objects. Barrier rails or crash cushions may be appropriate. End slopes for earth berms should be 6:1 or flatter, with 10:1 or 15:1 desirable.

9.7 MAINTENANCE CONSIDERATIONS

It is wise to keep a stock of compatible replacement materials on hand to repair damage from impact or vandalism. Consideration should be given to keeping replacement materials where they can weather to match installed walls, such as for pressure-treated timber components. Also, if color is added to concrete panels during manufacture, it is desirable to make future replacement panels in the same operation.

The control of graffiti remains a problem in some urban areas. There are some antigraffiti surface treatments available, but they are generally costly. Power washing and repainting are current options.

Plantings should be tolerant of roadside environments and require little or no maintenance. Access must be provided to both sides of the wall for mowing, general maintenance, etc. Sometimes this may require backside access from city streets, or overlap openings along the length of the wall. In some cases, arrangements can be made with abutting property owners to maintain the area behind the wall. If the noise wall is over 5 ft high, the right-of-way fence can usually be eliminated.

9.8 PROJECT DEVELOPMENT STEPS

Table 9.2 outlines the major steps required in the development of final construction plans for a noise abatement project on an existing highway. Considerations in several of these steps are as follows.

Preliminary Engineering.   During the preliminary engineering step, the following actions should take place:

- Develop a basic noise abatement plan, and determine wall height and location.
- Develop alternative methods of abatement such as walls, earth berms, berm-wall combinations, etc.

TABLE 9.2 Project Development Steps for
Noise Walls for Existing Highways

1. Preliminary engineering
    a. Identify project limits
    b. Collect data
    c. Identify alternatives
2. Public and municipal involvement
    a. Discuss alternatives
    b. Decide on system
3. Preparation of preliminary plans
4. Preliminary approvals
    a. Municipal
    b. State DOT
    c. FHWA
5. Final design
6. Final approval and processing
7. Contract letting

- Develop alternative locations for abatement facilities.
- Develop alternative wall material types such as concrete, timber, masonry, or steel.
- Develop a conceptual landscaping plan for each alternative.
- Develop cost estimates for alternatives.
- Develop a general environmental plan.
- Make preliminary arrangements for public informational meetings.

Items to be considered in selecting proposed alternatives include aesthetics, traffic safety, sight distance, drainage, maintenance, existing utilities, lighting, signing, potential soil problems, compatibility with surrounding terrain and land use, and restrictions imposed by available right-of-way. Consider any requirements for snow storage, future construction of sidewalks, trails, etc.

Layouts, cross sections, and wall profiles should be prepared for each alternative. Aerial photography contour maps should provide sufficient accuracy for determining ground elevations. Supplementary field information may be required in problem areas. Drainage away from both sides of the noise wall should be provided, with a minimum slope of 0.04. Ditches or culverts may be required where walls or berms alter natural drainage patterns.

Public and Municipal Involvement.   Local officials and the affected public should be informed of the scope of the proposed work and the alternative methods being considered to achieve noise abatement. Work through these groups to achieve a consensus. Provide sketches, renderings, plan drawings, and other visual aids to assist in the process. With this input, a decision should be made on a system so that preliminary plans can be prepared.

Preparation of Preliminary Plans.   Preliminary plans must be prepared for design and safety review. The plans should include a layout with the wall placement and profiles of the ground line and the top of the wall. Supplemental layouts for sight distance requirements may be required.

Preliminary Approvals. Local approval of the preliminary plan developed is sought at this time. Where applicable, municipal acceptance of maintenance responsibility of back slopes or other areas outside the noise wall should be obtained. Subsequent approval by the state DOT and FHWA is then sought.

Final Design. Information on soil conditions at the final wall location should be obtained from the soils engineer. The required depth of the investigation should correspond to the depth of post embedment or depth of spread footings. For construction in new embankment areas, care must be taken to avoid excessive differential settlement, because of concern for wall tilting, rotation, or cracking (of rigid systems). If a combination wall and berm is to be constructed, consider specifying an embankment material that will result in an economical wall design. It may be desirable to use a cohesive material for the upper portion of the berm.

Wall alignment can be modified slightly when necessary to make adjustments for standard panel sizes or material sizes; to fit with existing features such as trees, signs, lights, or utilities; or to better meet safety or drainage requirements.

Often, wall designs are based on standard agency plans. Special designs may be required where a wall ties into a bridge abutment or retaining wall, where the wall height exceeds the standards, where lights or signs are constructed integrally with the wall, where the wall must also serve as a retaining wall, or where soil properties are outside the range of those anticipated in the design standards.

9.9 STRUCTURAL DESIGN

9.9.1 Sound Wall Design Loads

Wind Loads. In most cases, the wind load represents the main load. The design pressure depends upon the wind velocity, which should be based upon a 50-year mean recurrence interval (Fig. 9.6). The wind pressure is applied perpendicular to the wall surface to develop the design wind load. On the basis of AASHTO *Guide Specifications for Structural Design of Sound Barriers,* the pressure may be calculated from

$$P = 0.00256(1.3V)^2 C_d C_c \tag{9.1}$$

where P = wind pressure, lb/ft^2
V = wind velocity, mi/h
C_d = drag coefficient = 1.2 for noise walls
C_c = combined height, exposure, and location coefficient

The factor of 1.3 in Eq. (9.1) provides for wind gusts. Values of C_c and calculated wind pressures are given in Table 9.3. The following three conditions with increasing levels of wind pressure are included:

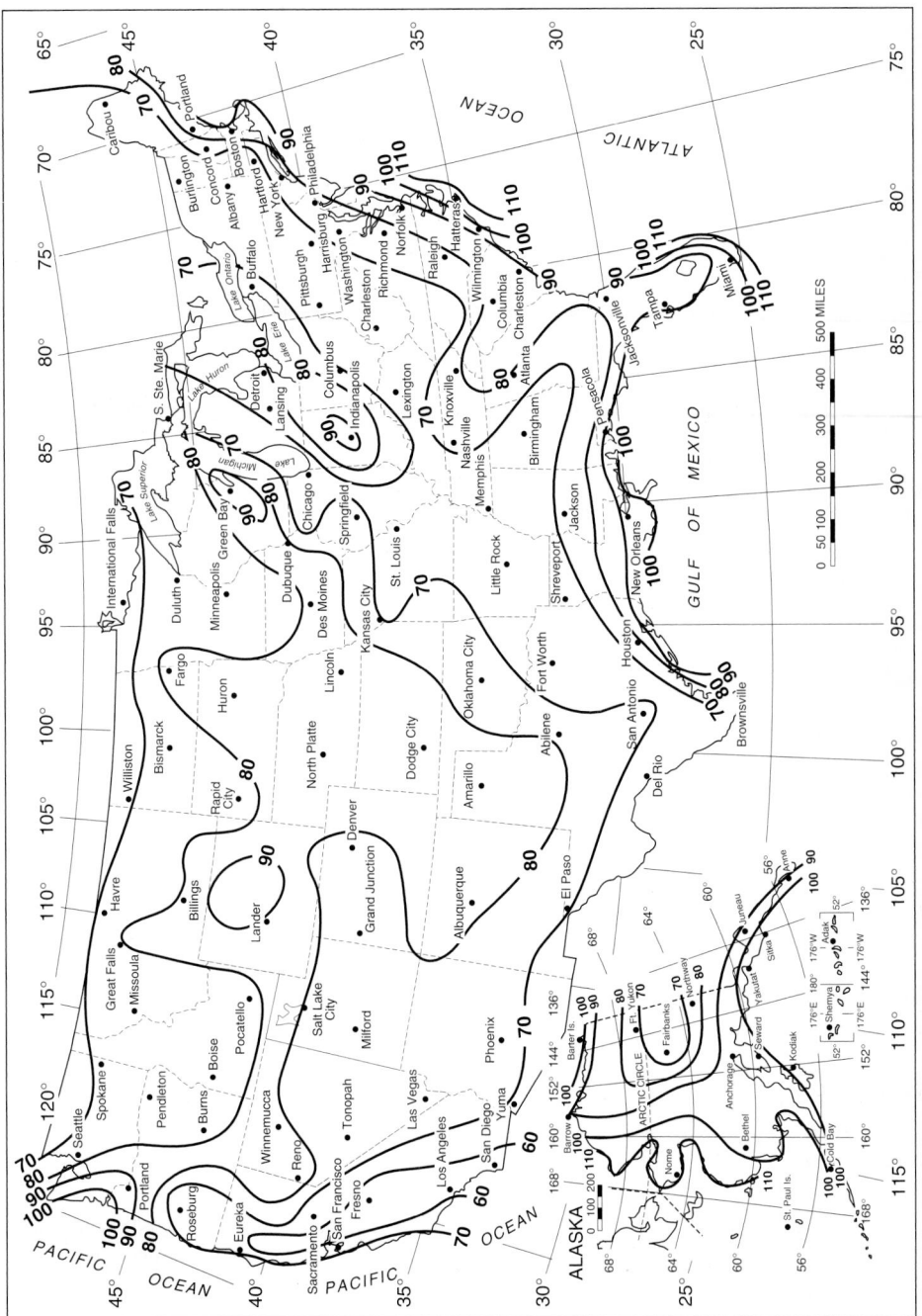

FIGURE 9.6 Wind velocities (mi/h) based on annual extreme-mile, 30 ft above ground, 50-year mean recurrence interval. (*From Guide Specifications for Structural Design of Sound Barriers, AASHTO, Washington, D.C., 1989, with permission*)

9.13

TABLE 9.3 Design Wind Pressures on Noise Walls

Location/ exposure	Height,* ft	Coefficient C_c	Pressure for indicated wind velocity, lb/ft²				
			70 mi/h	80 mi/h	90 mi/h	100 mi/h	110 mi/h
Ground/B1	≤14	0.37	9	12	16	19	23
	14–29	0.50	13	17	21	26	31
	>29	0.59	15	20	25	31	37
Ground/B2	≤14	0.59	15	20	25	31	37
	14–29	0.57	19	25	32	39	37
	>29	0.85	22	28	36	44	53
Structure†/C	≤14	0.80	20	27	34	42	50
	14–29	1.00	25	33	42	52	63
	>29	1.10	28	37	46	57	69

*Height refers to distance from average level of adjoining ground surface to centroid of loaded area in each height zone.

†Structure refers to noise walls on bridge structures, retaining walls, or traffic barriers.

Source: Adapted from AASHTO Guide Specifications for Structural Design of Sound Barriers, Washington, D.C., 1989.

1. *Noise walls not located on structures and having exposure B1.* This includes urban and suburban areas with numerous closely spaced obstructions having the size of single-family dwellings or larger that prevail in the upwind direction from the noise wall for a distance of at least 1500 ft.

2. *Noise walls not located on structures and having exposure B2.* This includes urban areas with more open terrain that does not meet exposure B1.

3. *Noise walls located on bridge structures, retaining walls, or traffic barriers (exposure C).* This is based on open terrain with scattered obstructions.

The interpretation of the surrounding terrain and identification of local conditions that may have increased effect on wind loads is left to the design engineer.

Seismic Loads. AASHTO requires that, where structures are designed for seismic load, noise walls also be designed for such. They define the seismic load (EQD) as

$$\text{EQD} = A \times f \times D \tag{9.2}$$

where A = acceleration coefficient (varies from 0.05 to 0.40 depending on geographical location; see AASHTO Guide Specifications, Fig. 1-2.1.3)
  D = dead load
  f = dead load coefficient (2.50, on bridges; 0.75, not on bridges; 8.0, connections of prefabricated walls to bridges; 5.0, connections of prefabricated walls to retaining walls)

The product of A and f must not be taken as less than 0.10.

Other Loads. In addition to dead load, other loads that might be encountered include earth load, live load surcharge, and ice and snow load. When encountered, these loads can be developed from information in the AASHTO *Standard Specifications for Highway Bridges.*

9.9.2 Load Combinations

Noise walls can be designed by working-stress design methods or load factor design. For the working-stress design method, the following load combinations should be considered:

$$\text{Group I: } D + E + SC$$

$$\text{Group II: } D + W + E + SC$$

$$\text{Group III: } D + EQD + E$$

$$\text{Group IV: } D + W + E + I$$

where
D = dead load
E = lateral earth pressure
SC = live load surcharge
W = wind load
EQD = seismic load
I = ice and snow load

For load combination I, the stresses are limited to 100 percent of the basic allowable stresses. For load combinations II, III, and IV, the stresses are limited to 133 percent.

9.9.3 Design Criteria

The AASHTO *Guide Specifications* state that, for the design of noise walls in concrete, timber, or steel, the design should conform to either the AASHTO *Bridge Specifications* or an industry-recognized design specification. Such sources may be referred to for allowable stress values and other details. For masonry walls, detailed design criteria are presented in the AASHTO *Guide Specifications*. Other materials can be designed using established engineering principles and appropriate industry specifications.

9.10 FOUNDATION DESIGN

The capacity of the foundation soil should be determined using accepted engineering principles and measurement of material parameters such as cohesion and angle of friction, or on the basis of field data such as the standard penetration test or the shear vane test. (See Chap. 8 for pertinent information.) One agency uses the following for default values:

1. Use angle of friction $\phi = 30°$ for granular soils and a cohesion value of $c = 1000$ lb/ft^2 for plastic soils to determine post embedment. Water encountered in soils above embedment depths will require special designs.
2. Use 2000 lb/ft^2 for allowable bearing capacity unless higher values are approved by the soils engineer.

3. A maximum of 2 ft of unbalanced fill on one side of the noise wall will be allowed.

The AASHTO *Guide Specifications* recommend the following safety factors for the design of spread footings that support noise walls:

Group	Overturning	Sliding
I	2.0	1.5
II	1.5	1.2
III	1.5	1.2
IV	1.5	1.2

For walls supported on two or more rows of piles, the design should follow procedures in *Standard Specifications for Highway Bridges* (AASHTO, Washington, D.C., 1992). For walls supported on a single row of piles, the pile must be designed as a column, considering both axial loads and bending. Also, the pile must be designed for the shear from the lateral loads.

For panel-and-post type walls, the embedment depth of the post can be determined using Rankine or Coulomb earth pressure theories. The following equation follows from static equilibrium analysis and applies for a pile or post on level ground:

$$0 = \frac{Rd^3}{12} - \frac{2Pd}{3} - \frac{P^2}{3Rd} - Ph \tag{9.3}$$

where $P =$ applied ultimate lateral load, lb
$h =$ vertical distance from lateral load to top of embedment, ft (disregard upper 6 in of soil at ground surface)
$R =$ allowable net horizontal ultimate lateral soil pressure, lb/ft² per ft of depth
$d =$ required depth of embedment, ft

Note that both P and R are ultimate values. The design load must be increased by an appropriate factor, and the resisting soil pressure decreased by an appropriate factor.

Example. $P = 200$ lb, $h = 6$ ft, $R = 600$ (lb/ft²)/ft. Determine d.
By trial and error, it is found that $d = 3.2$ ft satisfies Eq. (9.3). The final trial gives

$$0 = \frac{600(3.2)^3}{12} - \frac{2(200)(3.2)}{3} - \frac{200^2}{3(600)(3.2)} - 200(6)$$

$$0 = 1638 - 427 - 7 - 1200$$

$$0 \approx 4 \quad \text{(close enough; OK)}$$

The post should be embedded a distance of $3.2 + 0.5 = 3.7$ ft below the ground surface.

The depth of embedment can also be calculated by means of the following equation from the Uniform Building Code:

$$d = \frac{A}{2}\left(1 + \sqrt{1 + \frac{4.36h}{A}}\right) \tag{9.4}$$

where $A = \dfrac{2.34P}{S_1 b}$

P = applied ultimate lateral load, lb
b = width or diameter of post or pile, ft
S_1 = $Rd/3$, allowable ultimate lateral soil pressure, lb/ft^2, at one-third the depth of embedment
R = allowable net horizontal ultimate lateral soil pressure, lb/ft^2 per ft of depth
d = required depth of embedment, ft
h = vertical distance from lateral load to top of embedment, ft (disregard upper 6 in of soil at ground surface)

Note that both P and R are ultimate values. The design load must be increased by an appropriate factor, and the resisting soil pressure decreased by an appropriate factor.

Example. $P = 200$ lb, $h = 6$ ft, $R = 600$ (lb/ft^2)/ft, $b = 0.5$ ft. Determine d.
Try 3.2 ft:

$$S_1 = \frac{(600)(3.2)}{3} = 640 \text{ lb/ft}^2$$

$$A = \frac{(2.34)(200)}{(640)(0.50)} = 1.46$$

$$d = \frac{1.46}{2}\left(1 + \sqrt{1 + \frac{4.36 \times 6}{1.46}}\right)$$

$$d = 3.17 \text{ ft} \qquad \text{(close enough; OK)}$$

The post should be embedded a distance of $3.2 + 0.5 = 3.7$ ft below the ground surface.

The maximum moment in the pile or post can be expected to occur at a depth of $0.25d$. In this case, the maximum moment is

$$M = P(h + 0.25d)$$

$$= 200(6 + 0.25 \times 3.2)$$

$$= 1360 \text{ ft} \cdot \text{lb}$$

9.11 CONSTRUCTION

The following material is presented in the format of a typical specification used by one agency for the construction of noise walls. In addition to the type of wall included—timber wall with concrete posts—it can be adapted to walls of other types.

A. Miscellaneous Structure Removal

Abandoned structures and other obstructions shall be removed from the right-of-way and disposed of in accordance with DOT provisions except as modified below:

All debris resulting from the removal items and all other materials that become the property of the contractor and are not recycled into the project shall be disposed of outside the right-of-way in accordance with DOT provisions. This work shall be incidental to removal and salvage operations, and no direct compensation will be made therefor.

The contractor's attention is directed to possible privately owned appurtenances adjacent to the construction site that may need to be removed. If the private appurtenances are damaged, the contractor will be required to reinstate the appurtenances to satisfaction of owner. This work shall be considered incidental to the removal operations, and no direct compensation will be made therefor.

B. Clearing and Grubbing at Construction Site

The engineer shall have authority to limit the surface area of erodible earth material exposed by clearing and grubbing, excavation, and borrow and fill operations and to direct the contractor to provide immediate permanent or temporary control measures to prevent contamination of adjacent streams and other watercourses, lakes, ponds, and areas of water impoundment. Cut slopes shall be seeded and mulched as the excavation proceeds to the extent considered desirable and practicable.

The contractor will be required to incorporate all permanent erosion control features into the project at the earliest practicable time as outlined in his/her accepted schedules. Temporary pollution control measures will be used when needed to correct conditions that develop during construction but were not foreseen during the design stage, when needed prior to installation of permanent erosion control features, or when needed temporarily to control erosion that develops during normal construction practices; by definition, such temporary measures are not associated with the permanent control features on the project.

Where erosion is likely to be a problem, clearing and grubbing operations should be so scheduled and performed that grading operations and permanent erosion control features can follow immediately thereafter if the project conditions permit; otherwise, temporary erosion control measures may be required between successive construction stages. Under no conditions shall the surface area of erodible earth material exposed at one time by clearing and grubbing exceed 750,000 square feet without approval of the engineer.

C. Furnishing Concrete Post and Wood Noise Wall

This work shall consist of furnishing all materials for and constructing wood noise attenuator walls complete with concrete posts, and wood retaining wall, all in accordance with the plan details, the applicable DOT Standard Specifications, the required specifications for pigmented sealer and exterior wood surface stain, and the following:

1. General.   All thickness and width dimensions of solid sawn wood for timber facing material indicated in the plans for wood wall construction shall be construed to be nominal dimensions unless otherwise indicated in the plans or these special provisions.

2. Materials
    a. Concrete Posts.   Concrete posts shall be constructed as detailed in the plan and the required specification on pigmented sealer.

b. Wood Noise Walls. The facing lumber and battens shall be any species of southern pine conforming to the applicable provisions of DOT, modified to the extent that the lumber shall contain no holes and have tight knots. No intermixing of lumber species will be permitted within any continuous section of wall. If the wall abuts any earth fill greater than 2 ft, the facing planks installed below the top of the fill shall be 3- × 3-in, 1500 lb/ft² lumber with the 3-in dimension being rough-sawn. All facing lumber and battens shall be pressure preservative–treated with a chromated copper arsenate (CCA) waterborne preservative as provided hereinafter. Lumber treated with Millbrite will not be acceptable.

Facing boards shall be surfaced on two sides, and shall be tongue-and-grooved. All plank facing lumber shall be no. 1 structural grade or better. Facing lumber and battens shall be stamped with the appropriate grade mark.

c. Hardware. All hardware for noise wall shall be galvanized and meet the requirements of the American National Standards Institute (ANSI) and ASTM as to strength and testing.

D. Preservative Treatment

All lumber shall be pressure-treated with a preservative in accordance with the provisions of AASHTO M133 and AWPA P5.

1. All wall facings and battens shall be treated with chromated copper arsenate (CCA).

2. Wood materials shall be treated as required for aboveground installation, or for installation in contact with ground or water, in accordance with the applicable provisions of AASHTO M133 with a retention level of 0.60 lb/in · ft of wood.

3. All southern pine materials shall be free of sap stain (blue stain).

4. All wood members shall be kiln-dried to a moisture content of 15 percent or less after preservative treatment.

5. After completion of the preservative treatment, all lumber materials shall be protected from any increase in moisture content by covering or any other approved method until incorporated into the wall.

6. The same preservative treatment shall be used to treat bolt holes, saw cuts, etc., if any, and for any additional dressing deemed necessary by the engineer.

7. All treated wood members shall be cared for in accordance with the applicable provisions of AWPA *Standard for the Care of Preservative Treated Wood Products.*

E. Construction Requirements

1. Construction of wood noise attenuator walls, together with appurtenant posts, etc., shall be accomplished in accordance with the plan details, the applicable DOT Standard Specifications, these special provisions, or as otherwise approved by the engineer.

2. Nailing and fastening shall be accomplished in a manner that will avoid splitting boards. A 4-mil polyethylene sheeting may be placed between the planks and the earth for added protection when fill is being retained.

3. Joints shall be constructed in a manner that will completely arrest the passage of light. No daylight shall be visible through the joints 120 days after completion of the wall. The contractor is advised to take whatever measures are necessary to avoid excessive shrinkage or shifting that would cause the passage of light. Where passage of light does occur, the contractor shall take corrective action, in the form of caulking, to the satisfaction of the engineer, at his/her own expense.

4. Storage of materials within the right-of-way will be permitted only as approved by the engineer.

5. Debris shall be disposed of outside the right-of-way as specified by the engineer. Posts shall be plumb after installation.

6. The trench and trench backfill shall be compacted by the ordinary compaction method. The trench bottom shall be compacted to 90 percent of maximum density, and the bedding to 95 percent and 90 percent on each side of the footing. The density control shall not apply to the topsoil. The layers of material to be compacted shall be placed and compacted simultaneously so that the backfill material will be raised uniformly throughout the entire embedment depth.

F. Noise Wall Measurement and Payment

1. Concrete posts of each size will be measured separately by the length of the posts furnished and installed complete in place as specified. Payment will be made at the contract bid price per linear foot, which shall be compensation in full for all costs relative thereto.

2. Noise wall construction will be measured by the total front face area of the wall constructed (i.e., the area between the centers of end posts, and between the top of the wall and 6 in below the tabulated ground line).

3. Payment will be made for noise attenuator wall at the contract bid price per square foot, which price shall be compensation in full for all costs of constructing the wall complete in place as specified, except the appurtenant concrete posts, which shall be compensated for separately under the appropriate contract item provided.

4. Instead of the hand-driven "full-head" nail as shown in the plan, a reduced-head power-driven nail may be used to meet one of the following modifications:
   a. Use a nail one gauge heavier.
   b. Increase the number of nails used in each pattern by a minimum of 50 percent. For example, use 3 nails instead of 2, 5 instead of 3, 2 instead of 1.

5. In case of failure on the part of the contractor to control erosion, pollution, and siltation as ordered, the DOT reserves the right to employ outside assistance or to use its own forces to provide the necessary corrective measures. All expenses so incurred by the Department, including its engineering costs, that are chargeable to the contractor as his/her obligation and expense, will be deducted from any monies due or coming due the contractor.

9.12 TYPICAL DESIGNS OF NOISE WALLS

Typical designs for some of the many types of noise walls are shown in Figs. 9.7 through 9.15. These designs are generally based on a design wind pressure of 23 lb/ft^2

and a soil bearing value of 2 kips/ft^2. Included in the figures are noise walls constructed with concrete posts (reinforced or prestressed) and wood panels (plywood or timber), concrete blocks, self-supporting panels of prestressed concrete or wood glue-laminate, and wood posts with wood panels (glue-laminate, timber, or plywood). The designs are shown for illustration only; they should be checked against current standards and local requirements, and modified as necessary for particular applications. The height of the retained earth is denoted w. For w values above 2 ft, careful installation is required to avoid wall tilting. For some soil conditions, it will be necessary to limit w to 4 ft rather than 5 ft to avoid excessive wall tilting.

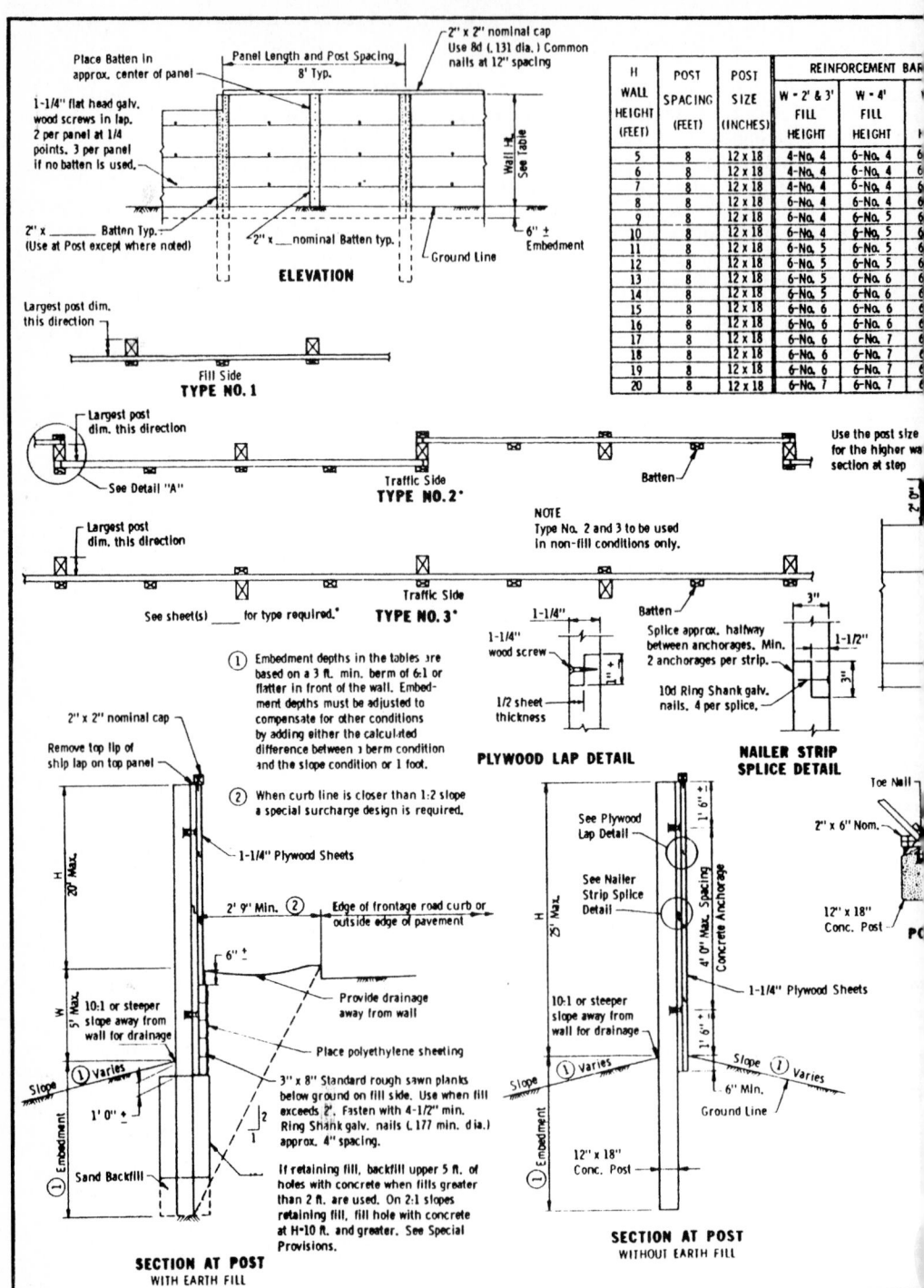

FIGURE 9.7 Typical design for noise wall using posts of reinforced or prestressed concrete and plywood panels. (*Adapted from Minnesota DOT design standards*)

NOISE WALL TABLE — Fed. Proj. No.

PRESTRESSED STRANDS			LEVEL GROUND φ=30°			2:1 SLOPE φ=30°			3:1 SLOPE φ=30°			4:1 SLOPE φ=30°		
W=2'&3' FILL HEIGHT	W=4' FILL HEIGHT	W=5' FILL HEIGHT	POST EMBEDMENT			POST EMBEDMENT			POST EMBEDMENT			POST EMBEDMENT		
			W=2'-3'	W=4'	W=5'	W=2'-3'	W=4'	W=5'	W=2'-3'	W=4'	W=5'	W=2'-3'	W=4'	W=5'
4	4	4	7'0"	8'0"	9'0"	13'0"	15'0"	15'0"	10'0"	11'0"	12'0"	9'0"	10'0"	11'0"
4	4	4	8'0"	9'0"	9'0"	14'0"	15'0"	16'0"	11'0"	12'0"	12'0"	10'0"	11'0"	11'0"
4	4	4	8'0"	9'0"	9'0"	15'0"	15'0"	16'0"	11'0"	12'0"	13'0"	10'0"	11'0"	11'0"
4	4	4	8'0"	9'0"	9'0"	15'0"	15'0"	17'0"	12'0"	12'0"	13'0"	11'0"	11'0"	12'0"
4	4	4	8'0"	9'0"	10'0"	15'0"	16'0"	17'0"	12'0"	12'0"	13'0"	11'0"	11'0"	12'0"
4	4	4	8'0"	9'0"	10'0"	14'0"	14'0"	15'0"	12'0"	13'0"	13'0"	11'0"	11'0"	12'0"
4	4	4	9'0"	9'0"	10'0"	14'0"	15'0"	16'0"	12'0"	13'0"	14'0"	11'0"	12'0"	13'0"
4	4	5	9'0"	10'0"	10'0"	15'0"	15'0"	16'0"	13'0"	13'0"	14'0"	11'0"	12'0"	13'0"
4	4	6	9'0"	10'0"	10'0"	15'0"	16'0"	16'0"	13'0"	13'0"	14'0"	12'0"	12'0"	13'0"
4	5	6	10'0"	10'0"	11'0"	15'0"	16'0"	17'0"	13'0"	14'0"	14'0"	12'0"	13'0"	13'0"
4	6	7	10'0"	10'0"	11'0"	15'0"	16'0"	17'0"	13'0"	14'0"	15'0"	12'0"	13'0"	14'0"
5	7	8	10'0"	11'0"	11'0"	16'0"	17'0"	17'0"	14'0"	14'0"	15'0"	13'0"	13'0"	14'0"
6	8	9	10'0"	11'0"	11'0"	16'0"	17'0"	18'0"	14'0"	15'0"	15'0"	13'0"	14'0"	14'0"
7	9	10	11'0"	11'0"	12'0"	17'0"	17'0"	18'0"	14'0"	15'0"	16'0"	13'0"	14'0"	14'0"
8	10	11	11'0"	11'0"	12'0"	17'0"	18'0"	18'0"	15'0"	15'0"	16'0"	14'0"	14'0"	15'0"
			11'0"	12'0"	12'0"	17'0"	18'0"	18'0"	15'0"	16'0"	16'0"	14'0"	14'0"	15'0"

- 8d Ring Shank galv. nails (.120" dia.) Approx. 3" sps. (Typ.)
- 3-1/2" min. Ring Shank galv. nails (.145 min. dia.) approx. 4" spacing (Typ.)

Plywood sheets may be 8' or 16' lengths and 2' or 4' widths — 2' or 4'

BATTEN DETAIL — **LEVEL DETAIL** — Typ. splice detail

FRONT ELEVATION AT WALL

DESIGN CRITERIA
φ = 30° (Granular)
Wind = 23 P.S.F.
f_b = 1500 P.S.I.
Wood Nailer Strip
Plywood Grade Stress
Level = S-2

1" standard sawn plank — Panel — Batten — Batten

ION FOR ANGLE TURNS

NOISE WALL TABLE

H WALL HEIGHT (FEET)	POST SPACING (FEET)	POST SIZE (INCHES)	REINF. BARS W=0 to 1' FILL HEIGHT	PRESTRESS STRANDS W=0 to 1' FILL HEIGHT	POST EMBEDMENT			
					Level Ground	2:1 Slope	3:1 Slope	4:1 Slope
5	8	12 x 18	4-No. 4	4	5'0"	8'0"	7'0"	6'0"
6	8	12 x 18	4-No. 4	4	6'0"	9'0"	8'0"	7'0"
7	8	12 x 18	4-No. 4	4	6'0"	9'0"	8'0"	8'0"
8	8	12 x 18	4-No. 4	4	7'0"	10'0"	9'0"	8'0"
9	8	12 x 18	4-No. 4	4	7'0"	11'0"	10'0"	9'0"
10	8	12 x 18	4-No. 4	4	8'0"	11'0"	10'0"	9'0"
11	8	12 x 18	4-No. 5	4	8'0"	12'0"	11'0"	10'0"
12	8	12 x 18	4-No. 5	4	8'0"	12'0"	11'0"	10'0"
13	8	12 x 18	4-No. 5	4	9'0"	13'0"	11'0"	10'0"
14	8	12 x 18	4-No. 5	4	9'0"	13'0"	11'0"	10'0"
15	8	12 x 18	6-No. 5	4	9'0"	13'0"	11'0"	11'0"
16	8	12 x 18	6-No. 5	4	9'0"	14'0"	12'0"	11'0"
17	8	12 x 18	6-No. 5	4	10'0"	14'0"	12'0"	11'0"
18	8	12 x 18	6-No. 6	4	10'0"	15'0"	13'0"	12'0"
19	8	12 x 18	6-No. 6	5	10'0"	15'0"	13'0"	12'0"
20	8	12 x 18	6-No. 6	5	10'0"	15'0"	13'0"	12'0"
21	8	12 x 18	6-No. 6	6	11'0"	16'0"	14'0"	13'0"
22	8	12 x 18	6-No. 7	6	11'0"	16'0"	14'0"	13'0"
23	8	12 x 18	6-No. 7	7	11'0"	17'0"	14'0"	13'0"
24	8	12 x 18	6-No. 7	8	11'0"	17'0"	15'0"	14'0"
25	8	12 x 18	6-No. 7	9	12'0"	17'0"	15'0"	14'0"

4   5   6   7   8   9   10   11
STRAND PLACEMENT

NOTES: The concrete shall be as per Spec. 2461.4A4b.

Slopes may be on both sides of wall for these embedments.

For slopes between 4:1 and 10:1 use interpolation to determine the embedments. Level ground assumed to be 10:1.

When the 10:1 berm dimension on each side of wall is equal to or greater than the post embedment length for level ground, use level ground embedments.

Embedment length is based on the water table being below embedment depth.

POST DESIGN CRITERIA

No. of Strands	f'c ①	f'c ②
6 or less	4000 psi	5500 psi
7 or more	4000 psi	6000 psi

① Min. concrete strength at time of prestress transfer

② Min. concrete strength when curing can be discontinued and post transported and installed

Reinforced Post: Min. concrete strength = 5000 psi when post may be moved and installed.

Prestressed steel strands are 1/2" dia. (area = 0.153 sq. in.) min. of 2" sps. 270 kip ultimate strength. Initial prestress equals 28,900 lbs./strand. Paint exposed ends with approved epoxy.

All reinforcement bars shall be grade 60.

POST DETAIL — PRESTRESSED — REINFORCED

- ₵ Post
- 3-1/2" Min. Ring Shank galv. nails
- Plywood Panel
- 3" x 10" standard rough sawn plank
- Additional bars when 6 used. May be omitted in upper 1/3 of post.
- 1/2" Chamfer or radius
- Batten as required
- 1" dia. flat head cap screw or 1/2" dia. Hex. Head bolt with 2" O.D. x 1/8" washer galv. per Spec. 3392 (Typ.) Countersunk to flush fit.
- No. 3 tie at 1' 6" sps. (typ.)
- Coil-spring conc. anchors for fasteners at 4' 0" Sps. alternate sides of ₵ post. Adjust to miss strands
- See table for strands required
- See Above for strand placement

CONCRETE POST TIMBER NOISE BARRIER (PLYWOOD)

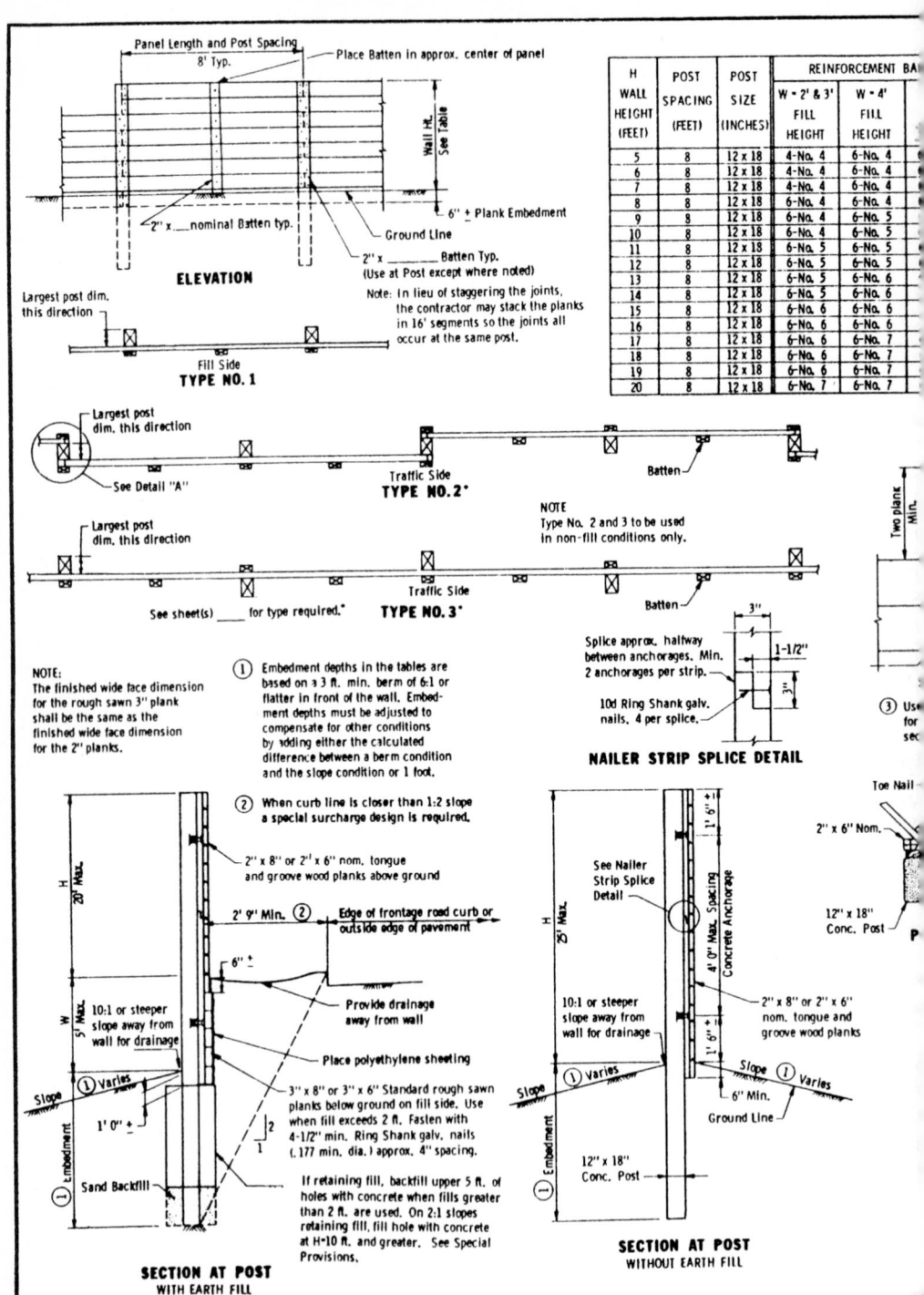

FIGURE 9.8 Typical design for noise wall using posts of reinforced or prestressed concrete and tongue-and-groove timber panels. (*Adapted from Minnesota DOT design standards*)

9.24

PRESTRESSED STRANDS			LEVEL GROUND Ø=30°			2:1 SLOPE Ø=30°			3:1 SLOPE Ø=30°			4:1 SLOPE Ø=30°		
2' & 3' FILL HEIGHT	W=4' FILL HEIGHT	W=5' FILL HEIGHT	POST EMBEDMENT			POST EMBEDMENT			POST EMBEDMENT			POST EMBEDMENT		
			W=2'-3'	W=4'	W=5'	W=2'-3'	W=4'	W=5'	W=2'-3'	W=4'	W=5'	W=2'-3'	W=4'	W=5'
4	4	4	7'0"	8'0"	9'0"	13'0"	15'0"	15'0"	10'0"	11'0"	12'0"	9'0"	10'0"	11'0"
4	4	4	8'0"	9'0"	9'0"	14'0"	15'0"	16'0"	11'0"	12'0"	12'0"	10'0"	11'0"	11'0"
4	4	4	8'0"	9'0"	9'0"	15'0"	15'0"	16'0"	11'0"	12'0"	13'0"	10'0"	11'0"	11'0"
4	4	4	8'0"	9'0"	9'0"	15'0"	15'0"	17'0"	12'0"	12'0"	13'0"	11'0"	11'0"	12'0"
4	4	4	8'0"	9'0"	10'0"	15'0"	16'0"	17'0"	12'0"	12'0"	13'0"	11'0"	11'0"	12'0"
4	4	4	8'0"	9'0"	10'0"	14'0"	14'0"	15'0"	12'0"	13'0"	13'0"	11'0"	11'0"	12'0"
4	4	4	9'0"	9'0"	10'0"	14'0"	15'0"	16'0"	12'0"	13'0"	14'0"	11'0"	12'0"	13'0"
4	4	4	9'0"	10'0"	10'0"	15'0"	15'0"	16'0"	13'0"	13'0"	14'0"	12'0"	12'0"	13'0"
4	4	5	9'0"	10'0"	10'0"	15'0"	15'0"	16'0"	13'0"	13'0"	14'0"	12'0"	12'0"	13'0"
4	4	6	10'0"	10'0"	11'0"	15'0"	16'0"	17'0"	13'0"	14'0"	14'0"	12'0"	13'0"	13'0"
4	5	6	10'0"	10'0"	11'0"	15'0"	16'0"	17'0"	13'0"	14'0"	15'0"	12'0"	13'0"	14'0"
4	6	7	10'0"	11'0"	11'0"	16'0"	17'0"	17'0"	14'0"	14'0"	15'0"	13'0"	13'0"	14'0"
5	7	8	10'0"	11'0"	11'0"	16'0"	17'0"	18'0"	14'0"	15'0"	15'0"	13'0"	14'0"	14'0"
6	8	9	11'0"	11'0"	12'0"	17'0"	17'0"	18'0"	14'0"	15'0"	16'0"	13'0"	14'0"	14'0"
7	9	10	11'0"	11'0"	12'0"	17'0"	18'0"	18'0"	15'0"	15'0"	16'0"	14'0"	14'0"	15'0"
8	10	11	11'0"	12'0"	12'0"	17'0"	18'0"	18'0"	15'0"	16'0"	16'0"	14'0"	14'0"	15'0"

NOISE WALL TABLE

H WALL HEIGHT (FEET)	POST SPACING (FEET)	POST SIZE (INCHES)	REINF. BARS W=0 to 1' FILL HEIGHT	PRESTRESS STRANDS W=0 to 1' FILL HEIGHT	POST EMBEDMENT			
					Level Ground	2:1 Slope	3:1 Slope	4:1 Slope
5	8	12 x 18	4-No. 4	4	5'0"	8'0"	7'0"	6'0"
6	8	12 x 18	4-No. 4	4	6'0"	9'0"	8'0"	7'0"
7	8	12 x 18	4-No. 4	4	6'0"	9'0"	8'0"	8'0"
8	8	12 x 18	4-No. 4	4	7'0"	10'0"	9'0"	8'0"
9	8	12 x 18	4-No. 4	4	7'0"	11'0"	10'0"	9'0"
10	8	12 x 18	4-No. 4	4	8'0"	11'0"	10'0"	9'0"
11	8	12 x 18	4-No. 5	4	8'0"	12'0"	11'0"	10'0"
12	8	12 x 18	4-No. 5	4	8'0"	12'0"	11'0"	10'0"
13	8	12 x 18	4-No. 5	4	9'0"	13'0"	11'0"	10'0"
14	8	12 x 18	4-No. 5	4	9'0"	13'0"	11'0"	10'0"
15	8	12 x 18	6-No. 5	4	9'0"	13'0"	11'0"	11'0"
16	8	12 x 18	6-No. 5	4	9'0"	14'0"	12'0"	11'0"
17	8	12 x 18	6-No. 5	4	10'0"	14'0"	12'0"	11'0"
18	8	12 x 18	6-No. 6	4	10'0"	15'0"	13'0"	12'0"
19	8	12 x 18	6-No. 6	5	10'0"	15'0"	13'0"	12'0"
20	8	12 x 18	6-No. 6	5	10'0"	15'0"	13'0"	12'0"
21	8	12 x 18	6-No. 6	6	11'0"	16'0"	14'0"	13'0"
22	8	12 x 18	6-No. 7	6	11'0"	16'0"	14'0"	13'0"
23	8	12 x 18	6-No. 7	7	11'0"	17'0"	14'0"	13'0"
24	8	12 x 18	6-No. 7	7	11'0"	17'0"	15'0"	14'0"
25	8	12 x 18	6-No. 7	9	12'0"	17'0"	15'0"	14'0"

NOTES: The concrete shall be as per Spec. 2461.4A4b.

Slopes may be on both sides of wall for these embedments.

For slopes between 4:1 and 10:1 use interpolation to determine the embedments. Level ground assumed to be 10:1.

When the 10:1 berm dimension on each side of the wall is equal to or greater than the post embedment length for the level ground, use the level ground embedments.

The embedment length is based on the water table being below the embedment depth.

POST DESIGN CRITERIA

No. of Strands	f'c (1)	f'ci (2)
6 or less	4000 psi	5500 psi
7 or more	4000 psi	6000 psi

(1) Min. concrete strength at the time of prestress transfer.

(2) Min. concrete strength when curing can be discontinued and the post can be transported and installed.

Reinforced Post: Min. concrete strength = 5000 psi when the post may be moved and installed.

Prestressed steel strands are 1/2" dia. (area = 0.153 sq. in.), min. of 2" spaces, 270 kip ultimate strength. initial prestress equals 28,900 lbs./strand. Paint exposed ends with approved epoxy.

All reinforcement bars shall be grade 60.

Left-side details

- 8d Ring Shank galv. nails (.120" dia.) 1 per plank (Typ.)
- 3-1/2" min. Ring Shank galv. nails (.145 min. dia.) approx. 4" spacing (Typ.)
- 2"x8" or 2"x6" nom. Tongue & Groove Planks
- BATTEN DETAIL
- LEVEL DETAIL
- Typ. splice detail alt. plank splices.

DESIGN CRITERIA
Ø = 30° (Granular)
Wind = 23 P.S.F.
f_b = 1500 P.S.I.
Wood Nailer Strip

ELEVATION AT WALL

x 10" standard gh sawn plank — Batten — lanking — Batten

ION FOR ANGLE TURNS

STRAND PLACEMENT   4 5 6 7 8 9 10 11

POST DETAIL

- 3-1/2" Min. Ring Shank galv. nails
- Wood Planking
- 3"x10" standard rough sawn plank
- Additional bars when 6 used. May be omitted in upper 1/3 of post.
- 1/2" Chamfer or radius
- Batten as required
- Post ℄
- 1" dia. flat head cap screw or 1/2" dia. Hex. Head bolt with 2" O.D. x 1/8" washer galv. per Spec. 3392 (Typ.) Countersunk to flush fit.
- No. 3 tie at 1'6" sps. (typ.)
- Coil-spring conc. anchors for fasteners at 4'0" Sps. alternate sides of ℄ post. Adjust to miss strands
- See table for strands required
- See Above for strand placement

REINFORCED   POST DETAIL   PRESTRESSED

Dimensions: 2-1/2" — 7" — 2-1/2" — 12" — 18" — 13" — 9" — 2-1/2" — 7" — 2-1/2" — 12"

CONCRETE POST TIMBER NOISE BARRIER (PLANKING)

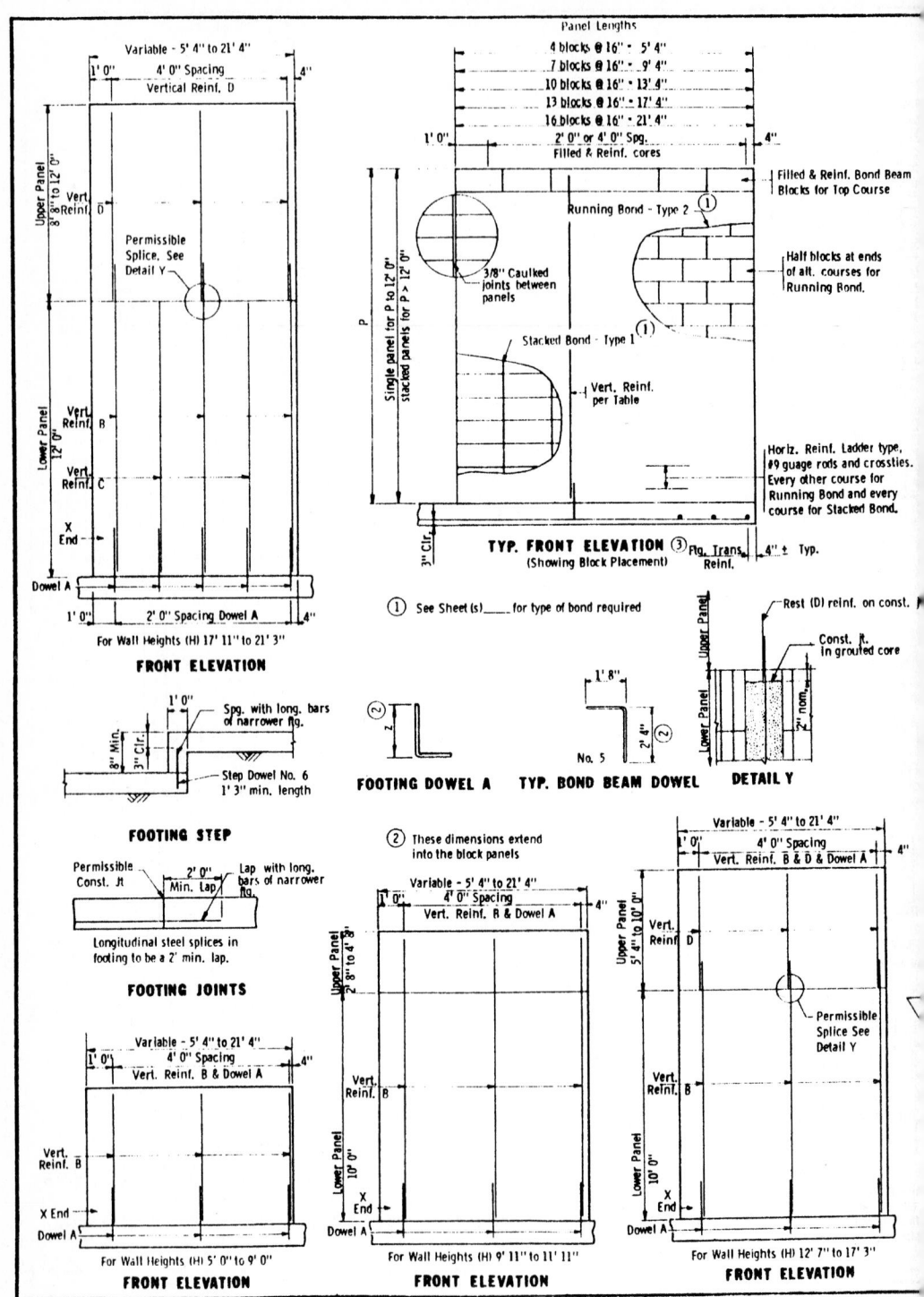

FIGURE 9.9 Typical design for noise wall using concrete block. (*Adapted from Minnesota DOT design standards*)

NOISE WALL TABLE
VERTICAL REINFORCEMENT

(H) WALL HEIGHT (FEET)	(P) Ftg. to Top of Wall	CONCRETE BLOCK SIZE (NOMINAL)	Dowel A SIZE	SPS.	DIM z	LENGTH	B SIZE	SPS.	LENGTH	C SIZE	SPS.	LENGTH	D SIZE	SPS.	LENGTH	FOOTING WIDTH (W)
5'0"	8'0"	8"x8"x16"	6	4'0"	3'6"	4'6"	6	4'0"	7'9"							2'8"
5'8"	8'8"	8"x8"x16"	6	4'0"	3'6"	4'6"	6	4'0"	8'5"							3'0"
6'4"	9'4"	8"x8"x16"	6	4'0"	3'6"	4'6"	6	4'0"	9'1"							3'4"
7'0"	10'0"	8"x8"x16"	6	4'0"	3'6"	4'6"	6	4'0"	9'9"							3'8"
7'8"	10'8"	8"x8"x16"	6	4'0"	3'6"	4'6"	6	4'0"	10'5"							3'8"
8'4"	11'4"	8"x8"x16"	6	4'0"	3'6"	4'6"	6	4'0"	11'1"							4'0"
9'0"	12'0"	8"x8"x16"	7	4'0"	4'4"	5'6"	6	4'0"	11'9"							4'4"
9'11"	12'8"	12"x8"x16"	7	4'0"	4'7"	5'9"	7	4'0"	12'5"				•			4'4"
10'7"	13'4"	12"x8"x16"	7	4'0"	4'7"	5'9"	7	4'0"	13'1"				•			4'4"
11'3"	14'0"	12"x8"x16"	7	4'0"	4'7"	5'9"	7	4'0"	13'9"				•			4'8"
11'11"	14'8"	12"x8"x16"	7	4'0"	4'7"	5'9"	7	4'0"	14'5"				•			4'8"
12'7"	15'4"	12"x8"x16"	7	4'0"	4'7"	5'9"	7	4'0"	13'8"				7	4'0"	5'2"	5'0"
13'3"	16'0"	12"x8"x16"	7	4'0"	4'7"	5'9"	7	4'0"	13'8"				7	4'0"	5'10"	5'4"
13'11"	16'8"	12"x8"x16"	7	4'0"	4'7"	5'9"	7	4'0"	13'8"				7	4'0"	6'6"	5'4"
14'7"	17'4"	12"x8"x16"	7	4'0"	4'7"	5'9"	7	4'0"	13'8"				7	4'0"	7'2"	5'8"
15'3"	18'0"	12"x8"x16"	8	4'0"	5'8"	7'0"	7	4'0"	13'8"				7	4'0"	7'10"	5'8"
15'11"	18'8"	12"x8"x16"	8	4'0"	5'8"	7'0"	7	4'0"	13'8"				7	4'0"	8'6"	6'0"
16'7"	19'4"	12"x8"x16"	8	4'0"	5'8"	7'0"	7	4'0"	13'8"				7	4'0"	9'2"	6'4"
17'3"	20'0"	12"x8"x16"	7	4'0"	5'8"	7'0"	7	4'0"	13'8"				7	4'0"	9'10"	6'4"
17'11"	20'8"	12"x8"x16"	6	2'0"	3'9"	4'9"	7	4'0"	15'8"	5	4'0"	11'9"	7	4'0"	8'6"	6'8"
18'7"	21'4"	12"x8"x16"	6	2'0"	3'9"	4'9"	7	4'0"	15'8"	5	4'0"	11'9"	7	4'0"	9'2"	6'8"
19'3"	22'0"	12"x8"x16"	7	2'0"	4'7"	5'9"	7	4'0"	15'8"	5	4'0"	11'9"	7	4'0"	9'10"	7'0"
19'11"	22'8"	12"x8"x16"	7	2'0"	4'7"	5'9"	7	4'0"	15'8"	5	4'0"	11'9"	7	4'0"	10'6"	7'0"
20'7"	23'4"	12"x8"x16"	7	2'0"	4'7"	5'9"	7	4'0"	15'8"	5	4'0"	11'9"	7	4'0"	11'2"	7'4"
21'3"	24'0"	12"x8"x16"	7	2'0"	4'7"	5'9"	7	4'0"	15'8"	5	4'0"	11'9"	7	4'0"	11'10"	7'8"

*Lower panel vert (B) extends into upper panel.

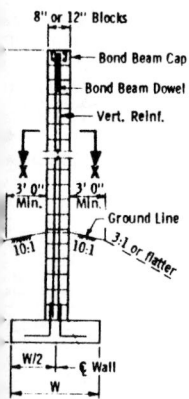

8" or 12" Blocks
- Bond Beam Cap
- Bond Beam Dowel
- Vert. Reinf.

3'0" Min. · 3'0" Min.
Ground Line
10:1 · 10:1 · 3:1 or flatter

W/2 · ℄ Wall · W

SECTION THRU WALL
...es For Additional Information

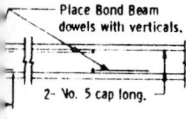

Place Bond Beam dowels with verticals.

2 - No. 5 cap long.

Plan View
BEAM REINFORCING

...this design on soils of
...ksf. allowable bearing
...fference in ht. of ground
...site faces of wall exceeds 2' 0".

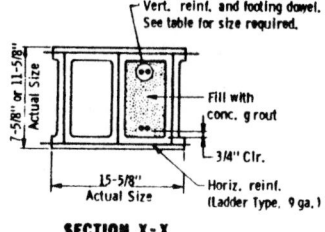

Vert. reinf. and footing dowel. See table for size required.

Fill with conc. grout

3/4" Clr.

Horiz. reinf. (Ladder Type, 9 ga.)

7-5/8" or 11-5/8" Actual Size

15-5/8" Actual Size

SECTION X-X
(Modular Block)

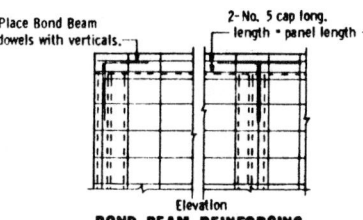

Place Bond Beam dowels with verticals.

2-No. 5 cap long. length = panel length - 4"

Elevation
BOND BEAM REINFORCING

DESIGN DATA
Wind Loading 23 p.s.f.
Min. required soil bearing 2 k.s.f.

LOAD FACTOR DESIGN METHOD
f_y = 60 ksi reinforcement
f'_m masonry = 2000 psi (Blocks Mortar, and Grout)
f'_c = 4000 psi (Footing)
Serviceability f_s = 36 Ksi at service load (23 psf wind)
Load factor for wind = 1.3

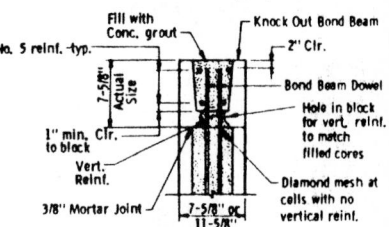

*Lower panel vert (B) extends into upper panel.

Fill with Conc. grout

2 No. 5 reinf. -typ.

Knock Out Bond Beam
2" Clr.
Bond Beam Dowel
Hole in block for vert. reinf. to match filled cores
Diamond mesh at cells with no vertical reinf.

1" min. Clr. to block
Vert. Reinf.
3/8" Mortar Joint

7-5/8" or 11-5/8" Actual Size

BOND BEAM CAP DETAIL

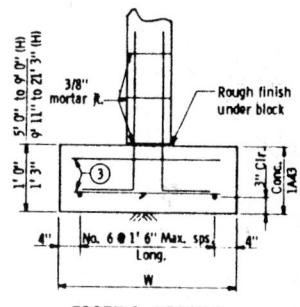

5'0" to 9'0" (H)
9'11" to 21'3" (H)

3/8" mortar R.

Rough finish under block

1'0" · 1'3"

③

No. 6 @ 1' 6" Max. sps. Long.

3" Clr.

4" · W · 4"

FOOTING DETAILS
(See Table)

③ Transverse footing bars are No. 6 @ 1' 6" max. sps. Length equals W - 6". Only bottom bars required for H = 15' 11" to 18' 7". Top and bottom bars are required for H = 19' 3" and above. No transverse bars required for H less than 15' 11".

CONCRETE BLOCK NOISE BARRIER

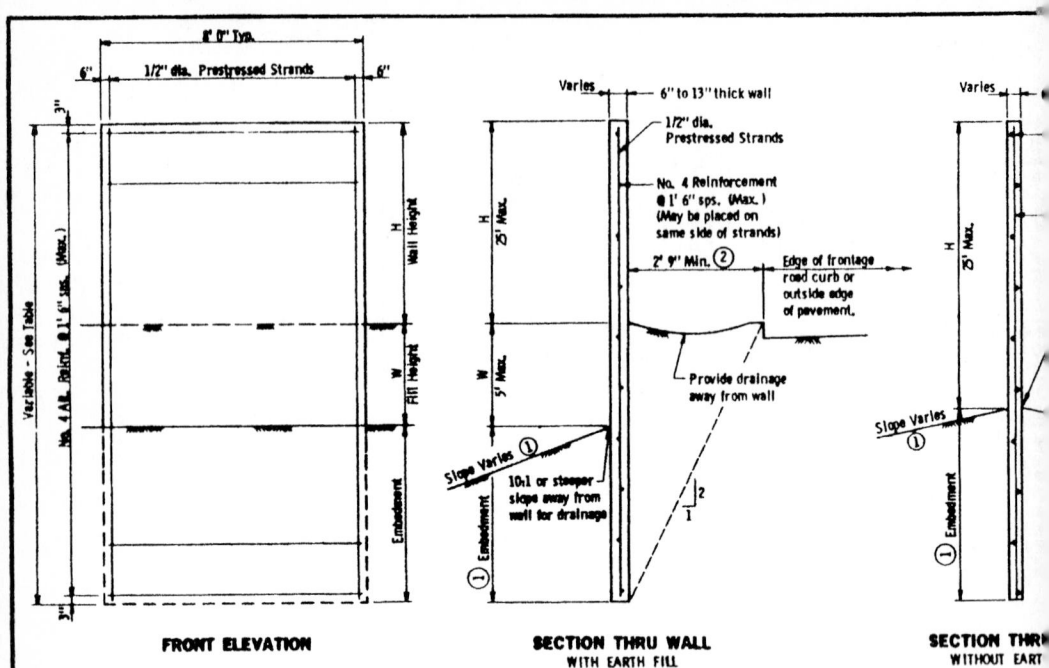

FRONT ELEVATION

SECTION THRU WALL
WITH EARTH FILL

SECTION THRU
WITHOUT EARTH

32' 2-1/4"

8' 0-3/4" | 8' 0-3/4" | 8' 0-3/4" | 8' 0"

3/4" Gap | 3/4" Gap | 3/4" Gap

TYPICAL PANEL LAYOUT

① Embedment depths in tables are based on a 3 ft. min. berm of 6:1 or flatter in front of the wall. Embedment depths must be adjusted to compensate for other conditions by adding either the calculated difference between a berm condition and the slope condition or 1 foot.

② When curb line is closer than 1:2 slope, a special surcharge design

NOISE WALL TABLE

"H" WALL HEIGHT (FEET)	PANEL WIDTH (FEET)	W=0 to 1' FILL HEIGHT WALL THICKNESS (IN.)	VERTICAL ROWS 1/2" dia. STRANDS	W=2' FILL HEIGHT WALL THICKNESS (IN.)	VERTICAL ROWS 1/2" dia. STRANDS	W=3' FILL HEIGHT WALL THICKNESS (IN.)	VERTICAL ROWS 1/2" dia. STRANDS	W=4' FILL HEIGHT WALL THICKNESS (IN.)	VERTICAL ROWS 1/2" dia. STRANDS	W=5' FILL HEIGHT WALL THICKNESS (IN.)	VERTICAL ROWS 1/2" dia. STRANDS	LEVEL GROUND EMBEDMENT W=0 to 1'	W=2'
5	8	6	5	6	6	6	6	6	7	6	12	4' 0"	4' 0"
6	8	6	5	6	6	6	6	6	7	6	12	4' 0"	4' 0"
7	8	6	5	6	6	6	6	6	7	6	12	4' 0"	4' 0"
8	8	6	5	6	6	6	6	6	7	6	12	4' 0"	4' 0"
9	8	6	5	6	6	6	6	6	8	6	14	4' 0"	5' 0"
10	8	6	5	6	6	6	6	6	10	7	11	4' 0"	5' 0"
11	8	6	5	6	6	6	8	6	12	7	13	4' 0"	5' 0"
12	8	6	5	6	6	6	10	7	10	7	15	5' 0"	5' 0"
13	8	6	5	6	8	6	12	7	12	8	12	5' 0"	5' 0"
14	8	6	5	6	10	6	14	7	14	8	15	5' 0"	6' 0"
15	8	6	5	6	13	7	12	8	11	9	12	5' 0"	6' 0"
16	8	6	6	6	15	7	14	8	13	9	14	5' 0"	6' 0"
17	8	6	8	7	13	7	16	9	11	10	11	6' 0"	6' 0"
18	8	6	10	7	15	8	13	9	13	10	13	6' 0"	6' 0"
19	8	6	12	8	12	8	16	10	10	10	15	6' 0"	7' 0"
20	8	6	14	8	15	9	13	10	12	11	13	6' 0"	7' 0"
21	8	7	11	9	12	9	15	10	15	11	15	7' 0"	7' 0"
22	8	7	13	9	14	10	13	11	12	12	12	7' 0"	7' 0"
23	8	8	10	10	12	10	15	11	14	12	14	7' 0"	7' 0"
24	8	8	12	10	14	11	12	12	12	12	16	7' 0"	7' 0"
25	8	8	14	10	16	11	15	12	14	13	14	7' 0"	8' 0"

FIGURE 9.10 Typical design for noise wall using self-supporting, vertical prestressed concrete panels. (*Adapted f*

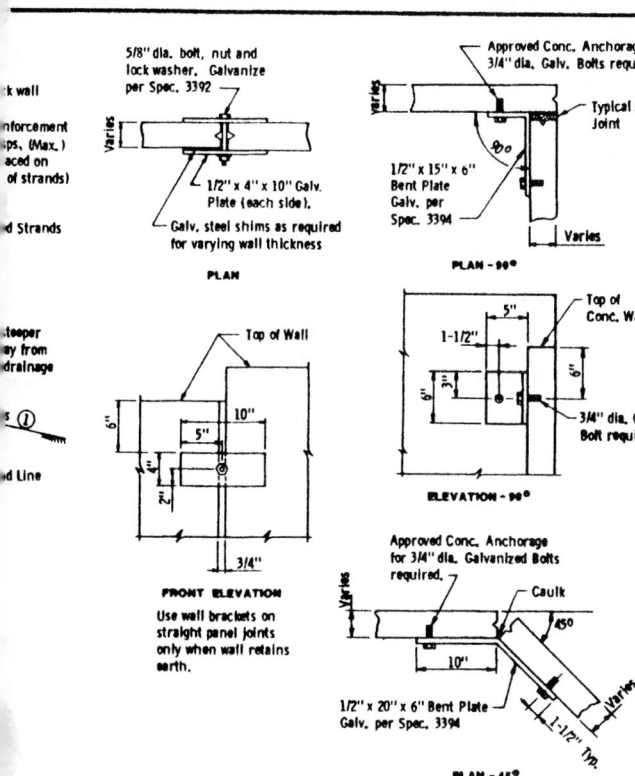

PLAN

5/8" dia. bolt, nut and lock washer. Galvanize per Spec. 3392

1/2" x 4" x 10" Galv. Plate (each side).

Galv. steel shims as required for varying wall thickness

Varies

FRONT ELEVATION

Top of Wall

10"

5"

6"

3/4"

Use wall brackets on straight panel joints only when wall retains earth.

PLAN - 90°

Approved Conc. Anchorage for 3/4" dia. Galv. Bolts required

Typical Panel Joint

90°

1/2" x 15" x 6" Bent Plate Galv. per Spec. 3394

Varies

ELEVATION - 90°

Top of Conc. Wall

5"

1-1/2"

6"

3"

6"

3/4" dia. Galv. Bolt required

PLAN - 45°

Approved Conc. Anchorage for 3/4" dia. Galvanized Bolts required.

Caulk

45°

10"

1/2" x 20" x 6" Bent Plate Galv. per Spec. 3394

1-1/2" Typ.

Varies

WALL BRACKET DETAILS

When wall brackets are used, grouting of top 2 feet shall be replaced with caulking.

3/4"

1/2 Wall thickness

3/4" V (Typ.)

Caulk top and sides to 6" below ground line. Grout top 2 ft.

TYPICAL JOINT

NOTES

The raked-surface finish of the precast concrete panels shall be randomly alternated between the neighborhood side and traffic side.

Wall shall be temporarily supported in a rigid manner until the concrete or compacted backfill material has sufficient strength for final support.

Finish edges of panels with small radius edger.

Prestressed steel strands shall be spaced symetrically and a min. of 2". Strands are 1/2" dia. (area - 0.153 Sq. in.) 270 kip ultimate strength. Initial prestress equals 28,000 lbs./strand.

Paint exposed ends of 1/2" strands with approved epoxy.

All reinforcement bars shall be grade 60 and have a 2" minimum cover.

Rustication grooves shall be 3" min. between centers and from edge of panel.

EMBEDMENT NOTES

Slopes may be on both sides of wall for these embedments.

For slopes between 4:1 and 10:1 use interpolation to determine the embedments. Level ground assumed to be 10:1

When the 10:1 berm dimension on each side of wall is equal to or greater than the post embedment length for level ground, use level ground embedments.

Embedment length is based on the water table being below embedment depth.

	3:1 SLOPE ∅ - 30°					4:1 SLOPE ∅ - 30°				
	EMBEDMENT ⓐ					EMBEDMENT ⓐ				
W - 5'	W-0 to 1'	W - 2'	W - 3'	W - 4'	W - 5'	W-0 to 1'	W - 2'	W - 3'	W - 4'	W - 5'
5'0"	4'0"	5'0"	5'0"	6'0"	7'0"	4'0"	4'0"	5'0"	6'0"	7'0"
6'0"	4'0"	5'0"	6'0"	7'0"	8'0"	4'0"	5'0"	5'0"	6'0"	7'0"
6'0"	4'0"	5'0"	6'0"	7'0"	8'0"	4'0"	5'0"	6'0"	6'0"	7'0"
6'0"	5'0"	6'0"	6'0"	7'0"	8'0"	4'0"	5'0"	6'0"	7'0"	7'0"
6'0"	5'0"	6'0"	7'0"	7'0"	8'0"	5'0"	5'0"	6'0"	7'0"	8'0"
6'0"	5'0"	6'0"	7'0"	8'0"	9'0"	5'0"	6'0"	6'0"	7'0"	8'0"
6'0"	6'0"	6'0"	7'0"	8'0"	9'0"	5'0"	6'0"	7'0"	7'0"	8'0"
7'0"	6'0"	7'0"	7'0"	8'0"	9'0"	6'0"	6'0"	7'0"	8'0"	8'0"
7'0"	6'0"	7'0"	8'0"	8'0"	9'0"	6'0"	7'0"	7'0"	8'0"	8'0"
7'0"	7'0"	7'0"	8'0"	9'0"	9'0"	6'0"	7'0"	7'0"	8'0"	9'0"
7'0"	7'0"	8'0"	8'0"	9'0"	10'0"	7'0"	7'0"	8'0"	8'0"	9'0"
7'0"	7'0"	8'0"	8'0"	9'0"	10'0"	7'0"	7'0"	8'0"	8'0"	9'0"
7'0"	8'0"	8'0"	9'0"	9'0"	10'0"	7'0"	8'0"	8'0"	9'0"	9'0"
8'0"	8'0"	8'0"	9'0"	10'0"	10'0"	7'0"	8'0"	8'0"	9'0"	9'0"
8'0"	8'0"	9'0"	9'0"	10'0"	11'0"	8'0"	8'0"	9'0"	9'0"	10'0"
8'0"	8'0"	9'0"	9'0"	10'0"	11'0"	8'0"	8'0"	9'0"	9'0"	10'0"
8'0"	9'0"	9'0"	10'0"	10'0"	11'0"	8'0"	9'0"	9'0"	10'0"	10'0"
8'0"	9'0"	9'0"	10'0"	11'0"	11'0"	8'0"	9'0"	9'0"	10'0"	10'0"
8'0"	9'0"	10'0"	10'0"	11'0"	11'0"	8'0"	9'0"	9'0"	10'0"	10'0"
9'0"	9'0"	10'0"	10'0"	11'0"	12'0"	9'0"	9'0"	10'0"	10'0"	11'0"
9'0"	10'0"	11'0"	11'0"	11'0"	12'0"	9'0"	9'0"	10'0"	10'0"	11'0"

ⓐ Use 2:1 slopes only for special cases, embedment for 2:1 slope shall be 85% greater than level ground embedment.

DESIGN CRITERIA

f'_{c1} - 4000 psi, minimum concrete strength at time of prestress transfer.

f'_c - 5000 psi, minimum concrete strength when curing can be discontinued and panel transported and installed.

Wind - 23 P. S. F.

PRESTRESSED CONCRETE NOISE BARRIER

| State Proj. No. | | Sheet No. of Sheets |

NOISE WALL TABLE

"H" WALL HEIGHT (FEET)	PANEL WIDTH (INCHES)	W = 2' & 3' FILL HEIGHT "T" WALL THICKNESS (INCHES)	W = 4' FILL HEIGHT "T" WALL THICKNESS (INCHES)	W = 5' FILL HEIGHT "T" WALL THICKNESS (INCHES)	LEVEL GROUND ∮ = 30° EMBEDMENT (a)			3:1 SLOPE ∮ = 30° EMBEDMENT (a)			4:1 SLOPE ∮ = 30° EMBEDMENT (a)		
					W = 2'-3'	W = 4'	W = 5'	W = 2'-3'	W = 4'	W = 5'	W = 2'-3'	W = 4'	W = 5'
5	22-1/4	2-11/16"	2-11/16"	3-7/16"	4' 0"	5' 0"	5' 0"	5' 0"	6' 0"	7' 0"	5' 0"	6' 0"	7' 0"
6	22-1/4	2-11/16"	2-11/16"	3-7/16"	4' 0"	5' 0"	6' 0"	6' 0"	7' 0"	8' 0"	5' 0"	6' 0"	7' 0"
7	22-1/4	2-11/16"	3-7/16"	3-7/16"	5' 0"	5' 0"	6' 0"	6' 0"	7' 0"	8' 0"	6' 0"	6' 0"	7' 0"
8	22-1/4	2-11/16"	3-7/16"	3-15/16"	5' 0"	5' 0"	6' 0"	6' 0"	7' 0"	8' 0"	6' 0"	7' 0"	7' 0"
9	22-1/4	2-11/16"	3-7/16"	3-15/16"	5' 0"	6' 0"	6' 0"	7' 0"	7' 0"	8' 0"	6' 0"	7' 0"	8' 0"
10	22-1/4	3-7/16"	3-7/16"	3-15/16"	5' 0"	6' 0"	6' 0"	7' 0"	8' 0"	9' 0"	6' 0"	7' 0"	8' 0"
11	22-1/4	3-7/16"	3-7/16"	3-15/16"	5' 0"	6' 0"	6' 0"	7' 0"	8' 0"	9' 0"	7' 0"	7' 0"	8' 0"
12	22-1/4	3-7/16"	3-15/16"	4-5/8"	6' 0"	6' 0"	7' 0"	7' 0"	8' 0"	9' 0"	7' 0"	8' 0"	8' 0"
13	22-1/4	3-7/16"	3-15/16"	4-5/8"	6' 0"	6' 0"	7' 0"	8' 0"	8' 0"	9' 0"	7' 0"	8' 0"	8' 0"
14	22-1/4	3-15/16"	3-15/16"	4-5/8"	6' 0"	6' 0"	7' 0"	8' 0"	9' 0"	10' 0"	7' 0"	8' 0"	9' 0"
15	22-1/4	3-15/16"	4-5/8"	5-1/4"	6' 0"	7' 0"	7' 0"	8' 0"	9' 0"	10' 0"	8' 0"	8' 0"	9' 0"
16	22-1/4	3-15/16"	4-5/8"	5-1/4"	6' 0"	7' 0"	7' 0"	8' 0"	9' 0"	10' 0"	8' 0"	8' 0"	9' 0"
17	22-1/4	4-5/8"	4-5/8"	5-1/4"	7' 0"	7' 0"	7' 0"	9' 0"	9' 0"	10' 0"	8' 0"	9' 0"	9' 0"
18	22-1/4	4-5/8"	4-5/8"	5-1/4"	7' 0"	7' 0"	8' 0"	9' 0"	10' 0"	10' 0"	8' 0"	9' 0"	9' 0"
19	22-1/4	4-5/8"	5-1/4"	6-3/16"	7' 0"	7' 0"	8' 0"	9' 0"	10' 0"	11' 0"	9' 0"	9' 0"	10' 0"
20	22-1/4	4-5/8"	5-1/4"	6-3/16"	7' 0"	8' 0"	8' 0"	9' 0"	10' 0"	11' 0"	9' 0"	9' 0"	10' 0"

NOISE WALL TABLE

"H" WALL HEIGHT (FEET)	PANEL WIDTH (INCHES)	W = 0 to 1' FILL HEIGHT "T" WALL THICKNESS	EMBEDMENT (a)		
			LEVEL GROUND	3:1 SLOPE	4:1 SLOPE
5	22-1/4	2-11/16"	4' 0"	4' 0"	4' 0"
6	22-1/4	2-11/16"	4' 0"	4' 0"	4' 0"
7	22-1/4	2-11/16"	4' 0"	4' 0"	4' 0"
8	22-1/4	2-11/16"	4' 0"	5' 0"	4' 0"
9	22-1/4	2-11/16"	4' 0"	5' 0"	5' 0"
10	22-1/4	2-11/16"	4' 0"	5' 0"	5' 0"
11	22-1/4	2-11/16"	4' 0"	6' 0"	5' 0"
12	22-1/4	2-11/16"	5' 0"	6' 0"	6' 0"
13	22-1/4	2-11/16"	5' 0"	6' 0"	6' 0"
14	22-1/4	3-7/16"	5' 0"	7' 0"	6' 0"
15	22-1/4	3-7/16"	5' 0"	7' 0"	7' 0"
16	22-1/4	3-7/16"	5' 0"	7' 0"	7' 0"
17	22-1/4	3-7/16"	6' 0"	8' 0"	7' 0"
18	22-1/4	3-15/16"	6' 0"	8' 0"	8' 0"
19	22-1/4	3-15/16"	6' 0"	8' 0"	8' 0"
20	22-1/4	3-15/16"	6' 0"	8' 0"	8' 0"
21	22-1/4	4-5/8"	7' 0"	9' 0"	8' 0"
22	22-1/4	4-5/8"	7' 0"	9' 0"	8' 0"
23	22-1/4	4-5/8"	7' 0"	9' 0"	8' 0"
24	22-1/4	5-1/4"	7' 0"	9' 0"	9' 0"
25	22-1/4	5-1/4"	7' 0"	10' 0"	9' 0"

WASHER PLATE
Structural Steel Spec. 3309

9/16"
3/4"
7/8"
2-3/4"
4-1/2"
1-3/4"
1/4"

DESIGN CRITERIA
∮ = 30° (Granular)
Wind = 23 P.S.F.

Glue Laminate Combination
Symbol = 20F

(a) Use 2:1 slopes only for special cases, embedment for 2:1 slope shall be 85% greater than level ground embedment.

NOTES

Slopes may be on both sides of wall for these embedments.

For slopes between 4:1 and 10:1 use interpolation to determine the embedments. Level ground assumed to be 10:1.

When the 10:1 berm dimension on each side of wall is equal to or greater than the post embedment length for level ground, use level ground embedments.

Embedment length is based on the water table being below embedment depth.

TOP VIEW
22-1/4"
Varies
1/2" to 3/4" max. Exterior
1-1/4" max. Interior
(3)

ELEVATION OF PANEL

21-1/4" 22-1/4"
1" Panel Overlap
(3)
1" nom. cap board to cover top of panel. Fasten with four 10d galv. nails per panel
12"
6" Max. Spacings
Varies — See Table
12"
T Varies

ELEVATION

FIGURE 9.11 Typical design for noise wall using self-supporting, vertical wood glue-laminate panels: type I, staggered panels. (*Adapted from Minnesota DOT design standards*)

9.30

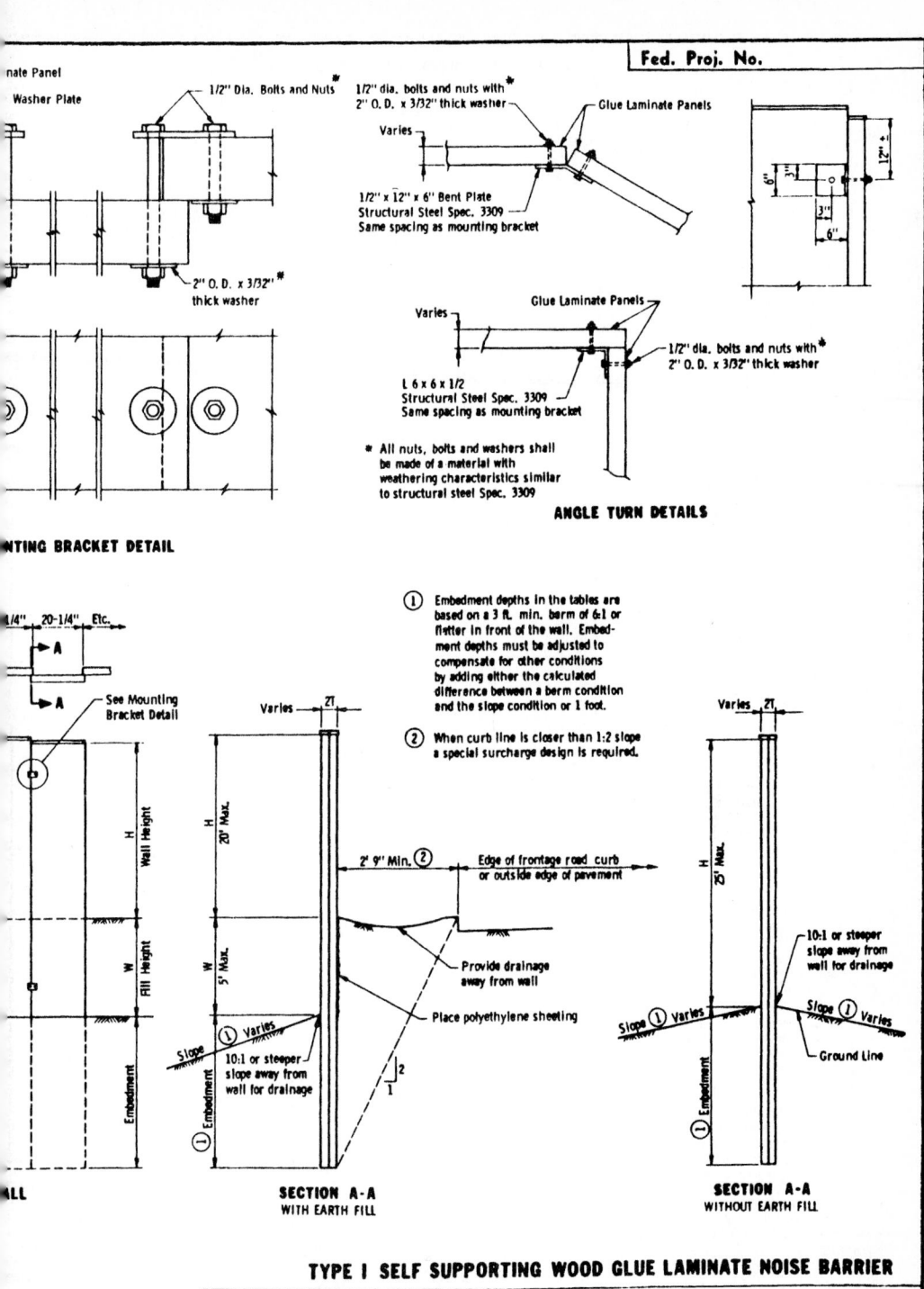

MOUNTING BRACKET DETAIL

ANGLE TURN DETAILS

1/2" Dia. Bolts and Nuts *

1/2" dia. bolts and nuts with *
2" O.D. x 3/32" thick washer

Glue Laminate Panels

Washer Plate

inate Panel

2" O.D. x 3/32" * thick washer

1/2" x 12" x 6" Bent Plate
Structural Steel Spec. 3309
Same spacing as mounting bracket

Glue Laminate Panels

Varies

1/2" dia. bolts and nuts with *
2" O.D. x 3/32" thick washer

L 6 x 6 x 1/2
Structural Steel Spec. 3309
Same spacing as mounting bracket

* All nuts, bolts and washers shall
be made of a material with
weathering characteristics similar
to structural steel Spec. 3309

① Embedment depths in the tables are
based on a 3 ft. min. berm of 6:1 or
flatter in front of the wall. Embedment depths must be adjusted to
compensate for other conditions
by adding either the calculated
difference between a berm condition
and the slope condition or 1 foot.

② When curb line is closer than 1:2 slope
a special surcharge design is required.

See Mounting
Bracket Detail

Wall Height
H
All Height
W
Embedment

WALL

Varies — 2T
20' Max.
H
5' Max.
W
Embedment ①

2' 9" Min. ② Edge of frontage road curb
or outside edge of pavement

Provide drainage
away from wall

Place polyethylene sheeting

Slope ① Varies

10:1 or steeper
slope away from
wall for drainage

SECTION A-A
WITH EARTH FILL

Varies 2T
25' Max.
H
Embedment ①

10:1 or steeper
slope away from
wall for drainage

Slope ① Varies   Slope ① Varies

Ground Line

SECTION A-A
WITHOUT EARTH FILL

TYPE I SELF SUPPORTING WOOD GLUE LAMINATE NOISE BARRIER

NOISE WALL TABLE

"H" WALL HEIGHT (FEET)	PANEL WIDTH (INCHES)	W = 2' & 3' FILL HEIGHT "T" WALL THICKNESS (INCHES) FB=16F	FB=20F	W = 4' FILL HEIGHT "T" WALL THICKNESS (INCHES) FB=16F	FB=20F	W = 5' FILL WEIGHT "T" WALL THICKNESS (INCHES) FB=16F	FB=20F	LEVEL GROUND Ø=30° EMBEDMENT (a) W=2'-3'	W=4'	W=5'	3:1 SLOPE Ø=30° EMBEDMENT (a) W=2'-3'	W=4'	W=5'	W=
5	22-1/4	2-11/16"	2-11/16"	3-7/16"	2-11/16"	3-7/16"	3-7/16"	4'0"	5'0"	5'0"	5'0"	6'0"	7'0"	5'
6	22-1/4	2-11/16"	2-11/16"	3-7/16"	2-11/16"	3-15/16"	3-7/16"	4'0"	5'0"	6'0"	6'0"	7'0"	8'0"	5'
7	22-1/4	2-11/16"	2-11/16"	3-7/16"	2-11/16"	3-15/16"	3-7/16"	5'0"	5'0"	6'0"	6'0"	7'0"	8'0"	6'
8	22-1/4	3-7/16"	2-11/16"	3-7/16"	2-11/16"	3-15/16"	3-7/16"	5'0"	5'0"	6'0"	6'0"	7'0"	8'0"	6'
9	22-1/4	3-7/16"	2-11/16"	3-15/16"	3-7/16"	4-5/8"	3-15/16"	5'0"	6'0"	6'0"	7'0"	7'0"	8'0"	6'
10	22-1/4	3-7/16"	3-7/16"	3-15/16"	3-7/16"	4-5/8"	3-15/16"	5'0"	6'0"	6'0"	7'0"	8'0"	9'0"	
11	22-1/4	3-15/16"	3-7/16"	3-15/16"	3-15/16"	4-5/8"	3-15/16"	5'0"	6'0"	6'0"	7'0"	8'0"	9'0"	
12	22-1/4	3-15/16"	3-7/16"	4-5/8"	3-15/16"	4-5/8"	4-5/8"	6'0"	6'0"	7'0"	7'0"	8'0"	9'0"	
13	22-1/4	3-15/16"	3-15/16"	4-5/8"	3-15/16"	5-1/4"	4-5/8"	6'0"	6'0"	7'0"	8'0"	8'0"	9'0"	7'
14	22-1/4	4-5/8"	3-15/16"	4-5/8"	4-5/8"	5-1/4"	4-5/8"	6'0"	6'0"	7'0"	8'0"	9'0"	9'0"	7'
15	22-1/4	4-5/8"	3-15/16"	5-1/4"	4-5/8"	5-1/4"	4-5/8"	6'0"	7'0"	7'0"	8'0"	9'0"	10'0"	
16	22-1/4	4-5/8"	4-5/8"	5-1/4"	4-5/8"	6-3/16"	5-1/4"	6'0"	7'0"	7'0"	8'0"	9'0"	10'0"	8'
17	22-1/4	5-1/4"	4-5/8"	5-1/4"	4-5/8"	6-3/16"	5-1/4"	7'0"	7'0"	7'0"	9'0"	9'0"	10'0"	8'
18	22-1/4	5-1/4"	4-5/8"	6-3/16"	5-1/4"	6-3/16"	5-1/4"	7'0"	7'0"	8'0"	9'0"	10'0"	10'0"	8'
19	22-1/4	5-1/4"	4-5/8"	6-3/16"	5-1/4"	6-3/16"	6-3/16"	7'0"	7'0"	8'0"	9'0"	10'0"	11'0"	9'
20	22-1/4	6-3/16"	5-1/4"	6-3/16"	5-1/4"	6-3/16"	6-3/16"	7'0"	8'0"	8'0"	9'0"	10'0"	11'0"	9'

NOISE WALL TABLE

"H" WALL HEIGHT (FEET)	PANEL WIDTH (INCHES)	W = 0 to 1' FILL HEIGHT "T" WALL THICKNESS FB=16F	FB=20F	EMBEDMENT (a) LEVEL GROUND	3:1 SLOPE	4:1 SLOPE
5	22-1/4	2-11/16"	2-11/16"	4'0"	4'0"	4'0"
6	22-1/4	2-11/16"	2-11/16"	4'0"	4'0"	4'0"
7	22-1/4	2-11/16"	2-11/16"	4'0"	4'0"	4'0"
8	22-1/4	2-11/16"	2-11/16"	4'0"	5'0"	4'0"
9	22-1/4	2-11/16"	2-11/16"	4'0"	5'0"	5'0"
10	22-1/4	2-11/16"	2-11/16"	4'0"	5'0"	5'0"
11	22-1/4	2-11/16"	2-11/16"	4'0"	6'0"	5'0"
12	22-1/4	3-7/16"	2-11/16"	5'0"	6'0"	6'0"
13	22-1/4	3-7/16"	3-7/16"	5'0"	6'0"	6'0"
14	22-1/4	3-7/16"	3-7/16"	5'0"	7'0"	6'0"
15	22-1/4	3-15/16"	3-7/16"	5'0"	7'0"	7'0"
16	22-1/4	3-15/16"	3-7/16"	5'0"	7'0"	7'0"
17	22-1/4	3-15/16"	3-15/16"	6'0"	8'0"	7'0"
18	22-1/4	4-5/8"	3-15/16"	6'0"	8'0"	7'0"
19	22-1/4	4-5/8"	3-15/16"	6'0"	8'0"	8'0"
20	22-1/4	4-5/8"	4-5/8"	6'0"	8'0"	8'0"
21	22-1/4	5-1/4"	4-5/8"	7'0"	9'0"	8'0"
22	22-1/4	5-1/4"	4-5/8"	7'0"	9'0"	8'0"
23	22-1/4	5-1/4"	5-1/4"	7'0"	9'0"	8'0"
24	22-1/4	6-3/16"	5-1/4"	7'0"	9'0"	9'0"
25	22-1/4	6-3/16"	5-1/4"	7'0"	10'0"	9'0"

DESIGN CRITERIA
Ø = 30° (Granular)
Wind = 20 P.S.F.
Glue Laminate Noise Barrier
Stress Level:
20F = 2000 PSI Allowable Bending Stress.
16F = 1600 PSI Allowable Bending Stress.

(a) Use 2:1 slopes only for special cases, embedment for 2:1 slope shall be 85% greater than level ground embedment.

NOTES

Slopes may be on both sides of wall for these embedments.

For slopes between 4:1 and 10:1 use interpolation to determine the embedments. Level ground assumed to be 10:1

When the 10:1 berm dimension on each side of wall is equal to or greater than the post embedment length for level ground, use level ground embedments.

Embedment length is based on the water table being below embedment depth.

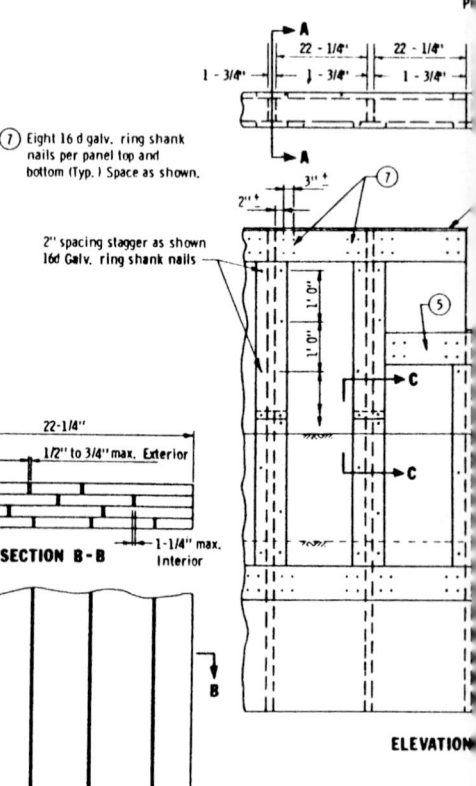

(7) Eight 16 d galv. ring shank nails per panel top and bottom (Typ.) Space as shown.

2" spacing stagger as shown 16d Galv. ring shank nails

22-1/4"
Varies
1/2" to 3/4" max. Exterior
1-1/4" max. Interior

SECTION B-B

ELEVATION OF PANEL

ELEVATION

(3) 1" nom. cap board to cover top of panel. Fasten to 2 x 8'... four 8d galv. nails per panel. Incidental to wall construction

FIGURE 9.12 Typical design for noise wall using self-supporting, vertical wood glue-laminate panels: type II, aligned panels. (*Adapted from Minnesota DOT design standards*)

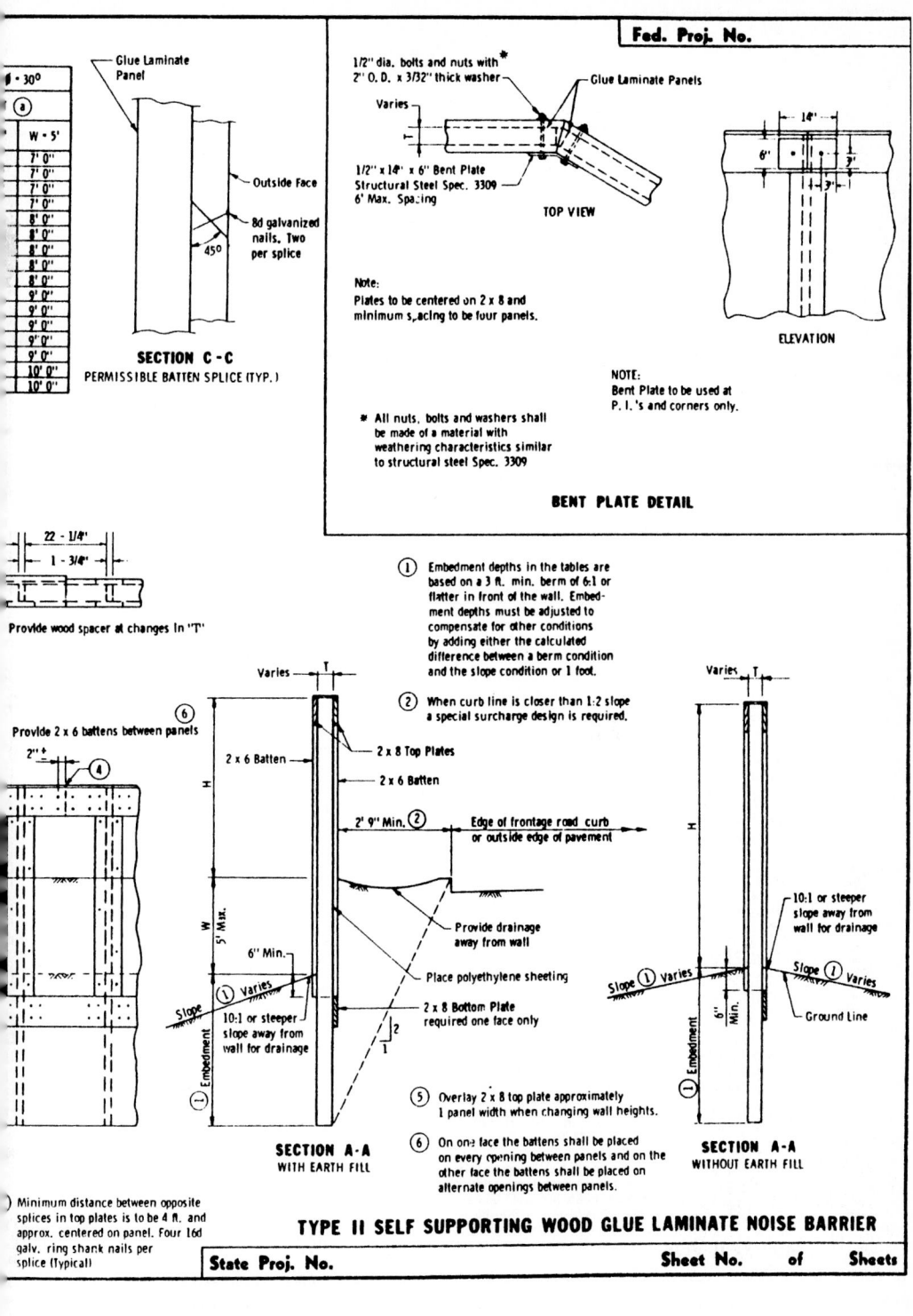

TYPE II SELF SUPPORTING WOOD GLUE LAMINATE NOISE BARRIER

9.33

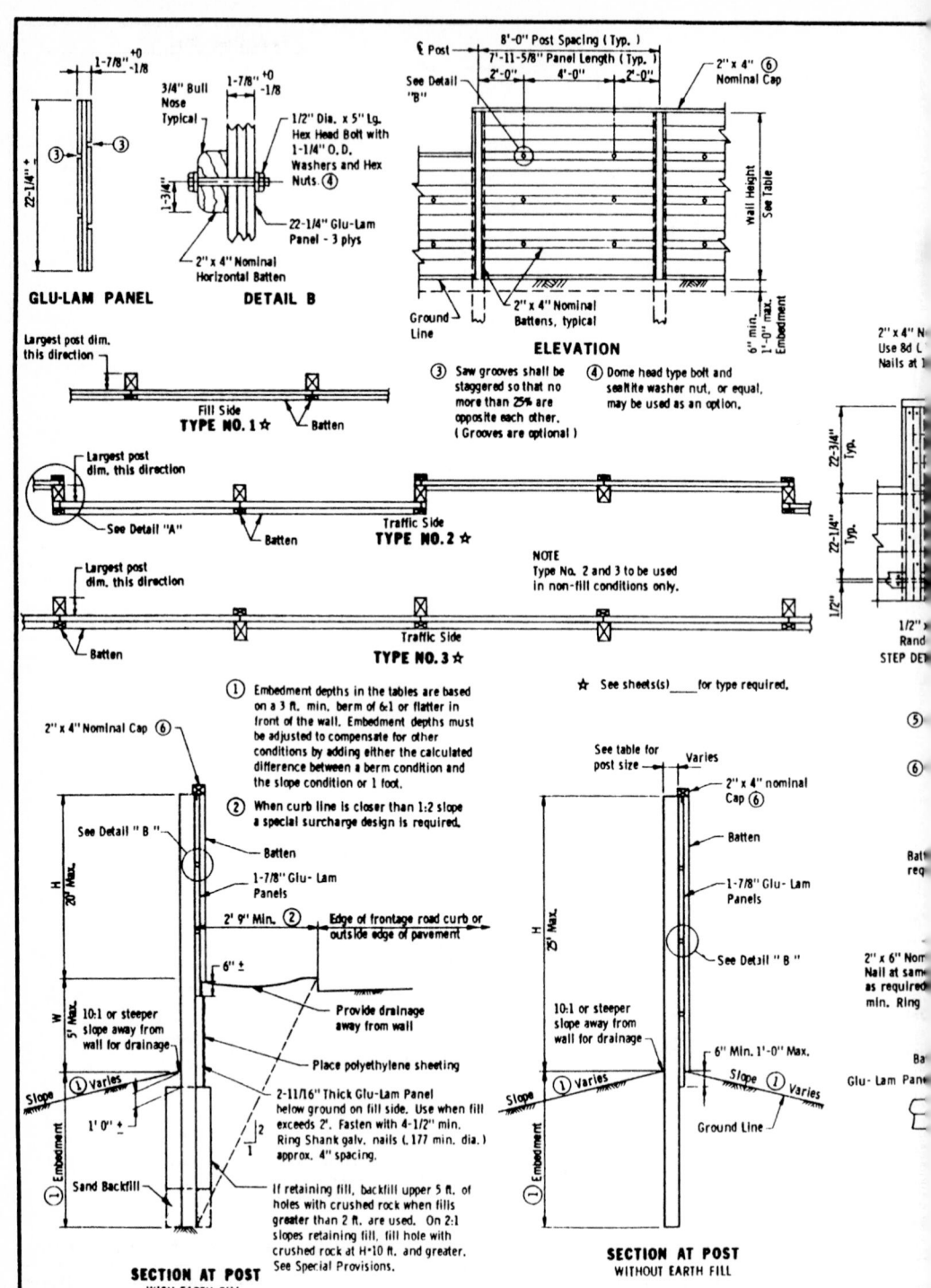

FIGURE 9.13 Typical design for noise wall using wood posts and horizontal wood glue-laminate panels. (*Adapted from Minnesota DOT design standards*)

9.34

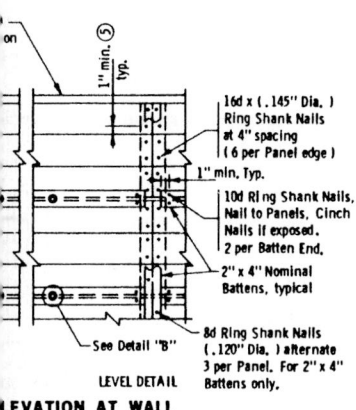

16d x (.145" Dia.)
Ring Shank Nails
at 4" spacing
(6 per Panel edge)

1" min. Typ.

10d Ring Shank Nails,
Nail to Panels, Cinch
Nails if exposed.
2 per Batten End.

2" x 4" Nominal
Battens, typical

See Detail "B"

8d Ring Shank Nails
(.120" Dia.) alternate
3 per Panel. For 2" x 4"
Battens only.

LEVEL DETAIL

ELEVATION AT WALL

stance from
cing groove
1/2"

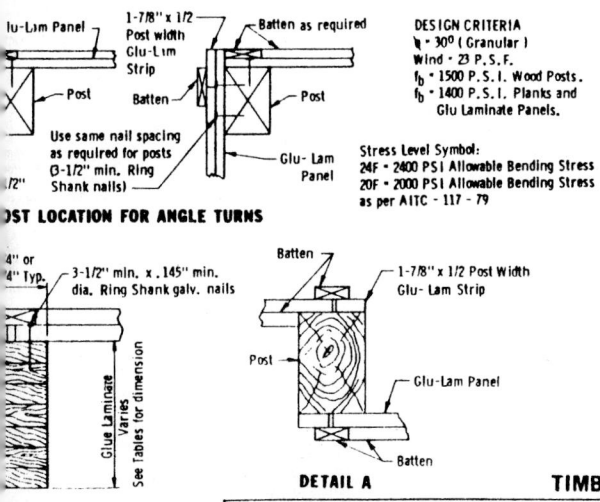

Iu-Lim Panel

1-7/8" x 1/2
Post width
Glu-Lim
Strip

Batten as required

Post

Batten

Post

Use same nail spacing
as required for posts
(3-1/2" min. Ring
Shank nails)

Glu- Lam
Panel

POST LOCATION FOR ANGLE TURNS

4" or
4" Typ.

3-1/2" min. x .145" min.
dia. Ring Shank galv. nails

Glue Laminate
Varies
See Tables for dimension

Batten

1-7/8" x 1/2 Post Width
Glu- Lam Strip

Post

Glu-Lam Panel

Batten

DETAIL A

MINATE POST DETAIL

DESIGN CRITERIA
ϕ = 30° (Granular)
Wind = 23 P.S.F.
f_b = 1500 P.S.I. Wood Posts.
f_b = 1400 P.S.I. Planks and
Glu Laminate Panels.

Stress Level Symbol:
24F = 2400 PSI Allowable Bending Stress
20F = 2000 PSI Allowable Bending Stress
as per AITC - 117 - 79

NOTES:

Slopes may be on both sides of wall for these embedments.

For slopes between 4:1 and 10:1 use interpolation to
determine the embedments. Level ground assumed to be 10:1.

When the 10:1 berm dimension on each side of wall is equal
to or greater than the post embedment length for level
ground, use level ground embedments.

Embedment length is based on the water table being below
embedment depth.

Posts shall be standard rough sawn dimensions as shown.

Galvanize nails per Mn/ D.O.T. 3392.

Bolts, nuts and washers shall be coated in accordance
with Military Specification Mil-P-16232, Type Z, Class 4.
The coating color shall be black.

**HORIZONTAL GLUE LAMINATE
TIMBER NOISE BARRIER (WOOD POSTS)**

State Proj. No.

Sheet No.   of   Sheets

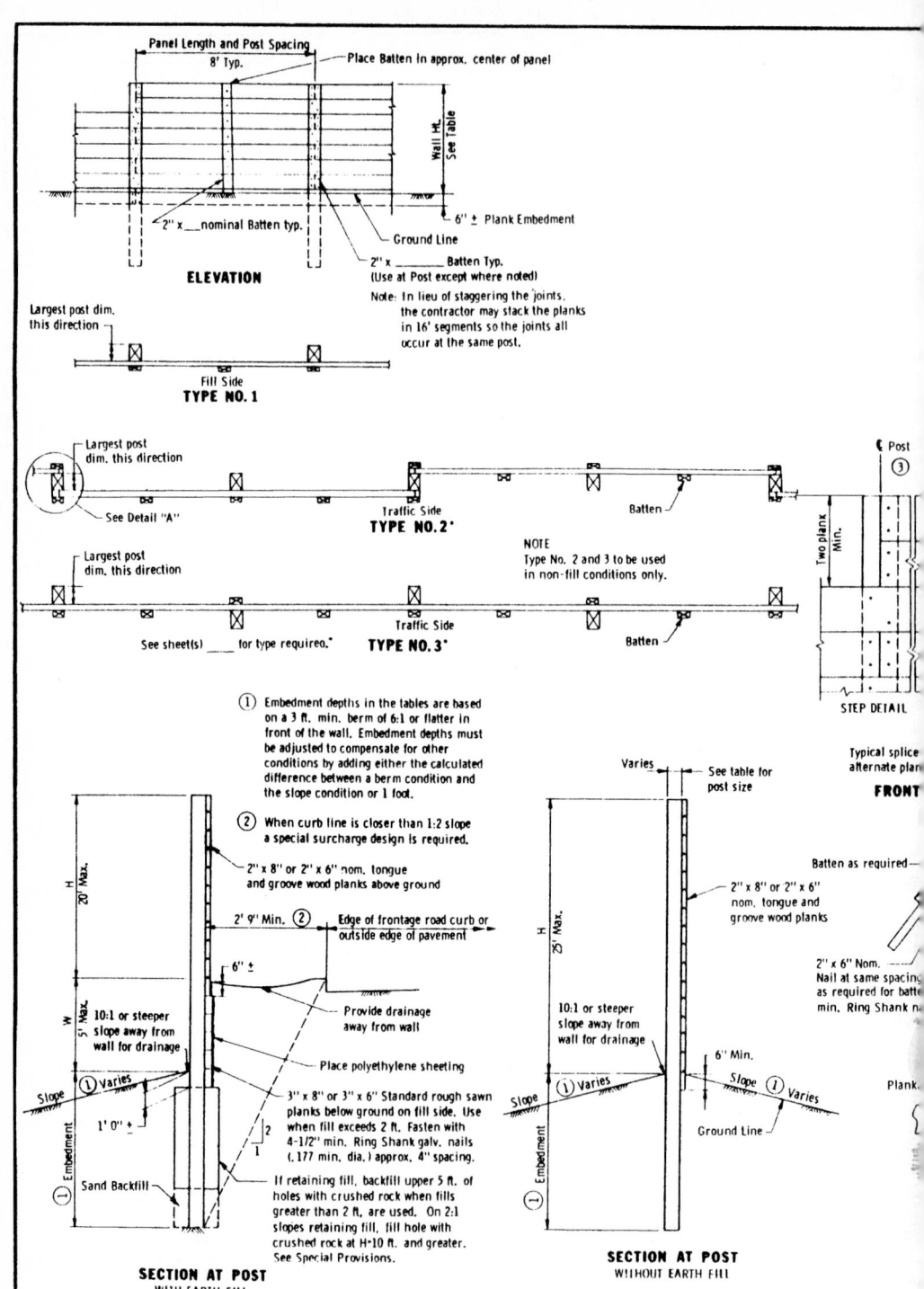

FIGURE 9.14 Typical design for noise wall using wood posts and horizontal tongue-and-groove timber panels. (*Adapted from Minnesota DOT design standards*)

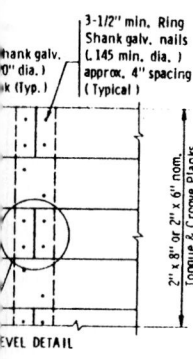

Fed. Proj. No.

hank galv.
0" dia.)
k (Typ.)
3-1/2" min. Ring Shank galv. nails (.145 min. dia.) approx. 4" spacing (Typical)
2" x 8" or 2" x 6" nom. Tongue & Groove Planks

EVEL DETAIL

③ Use the post size for the higher wall section at step

AT WALL

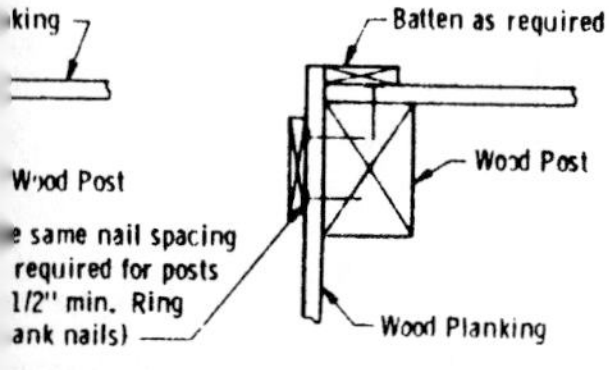

king
Batten as required
Wood Post
Wood Post
e same nail spacing required for posts 1/2" min. Ring ank nails)
Wood Planking

ATION FOR ANGLE TURNS

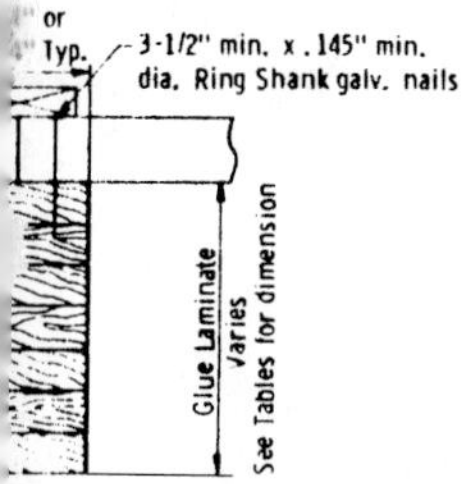

" or
" Typ.
3-1/2" min. x .145" min. dia. Ring Shank galv. nails
Glue Laminate
Varies
See Tables for dimension

MINATE POST DETAIL

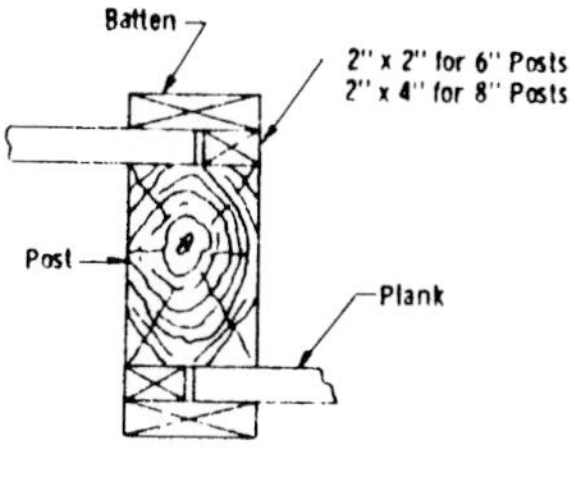

Batten
2" x 2" for 6" Posts
2" x 4" for 8" Posts
Post
Plank

DETAIL A

DESIGN CRITERIA
ϕ = 30° (Granular)
Wind = 23 P.S.F.
f_b = 1500 P.S.I. Wood Posts
Stress Level Symbol:
24F = 2400 PSI Allowable Bending Stress
20F = 2000 PSI Allowable Bending Stress
as per AITC-117-79

NOTES:

Slopes may be on both sides of wall for these embedments.

For slopes between 4:1 and 10:1 use interpolation to determine the embedments. Level ground assumed to be 10:1

When the 10:1 berm dimension on each side of wall is equal to or greater than the post embedment length for level ground, use level ground embedments

Embedment length is based on the water table being below embedment depth.

Posts shall be standard rough sawn dimensions as shown.

The finished wide face dimension for the rough sawn 3" plank shall be the same as the finished wide face dimension for the 2" planks.

WOOD POST TIMBER NOISE BARRIER (PLANKING)

State Proj. No. Sheet No. of Sheets

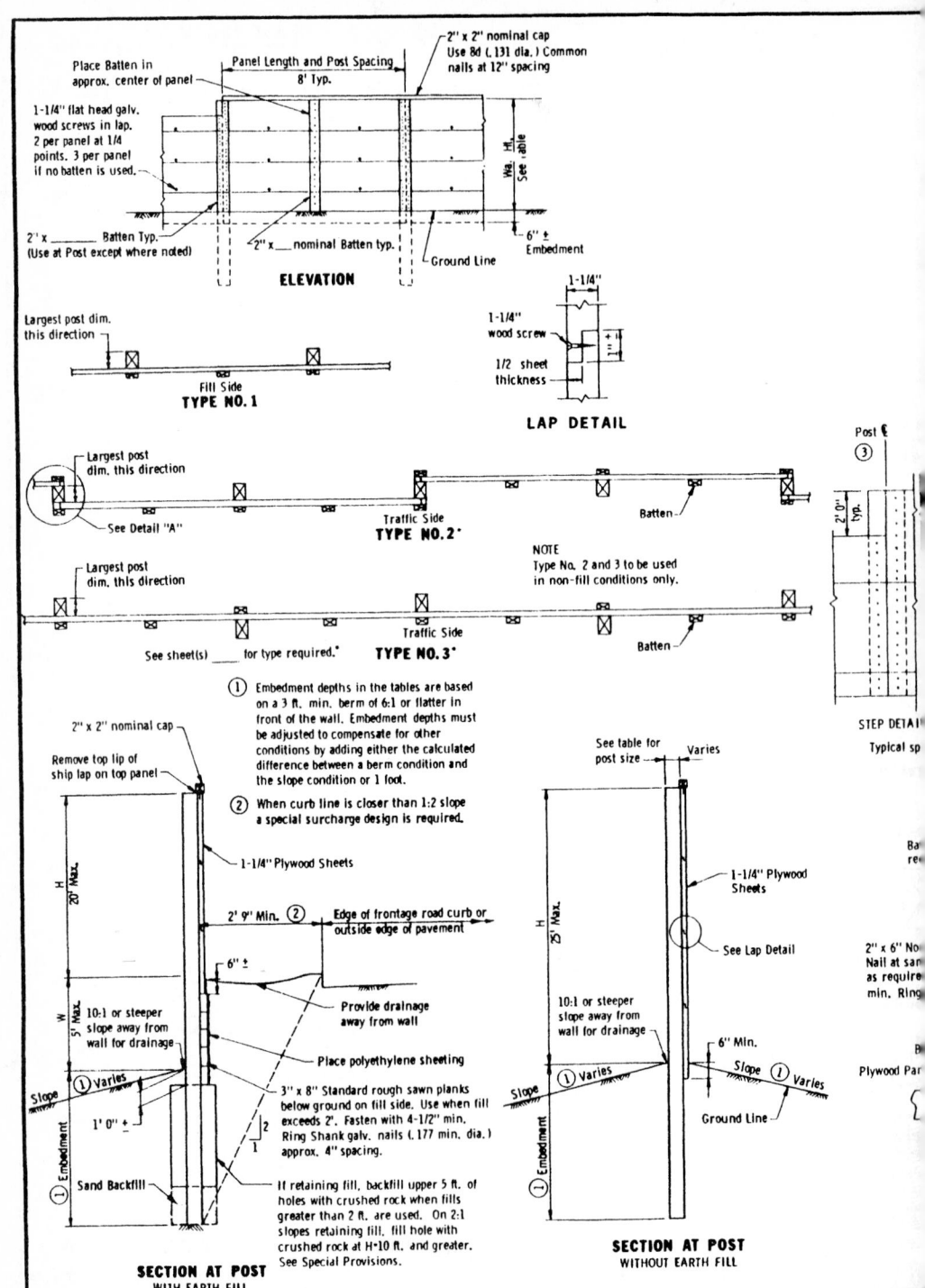

FIGURE 9.15 Typical design for noise walls using wood posts and horizontal tongue-and-groove plywood panels. (*Adapted from Minnesota DOT design standards*)

9.38

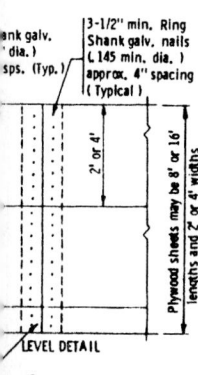

3-1/2" min. Ring
Shank galv. nails
(.145 min. dia.)
approx. 4" spacing
(Typical)

nk galv.
dia.)
sps. (Typ.)

2' or 4'

Plywood sheets may be 8' or 16'
lengths and 2' or 4' widths

LEVEL DETAIL

③ Use the post size for the
higher wall section at step

EVATION AT WALL

lywood Panel

Batten as required

Wood Post

Wood Post

Use same nail spacing
as required for posts
(3-1/2" min. Ring
Shank nails)

Plywood Panel

2"

ST LOCATION FOR ANGLE TURNS

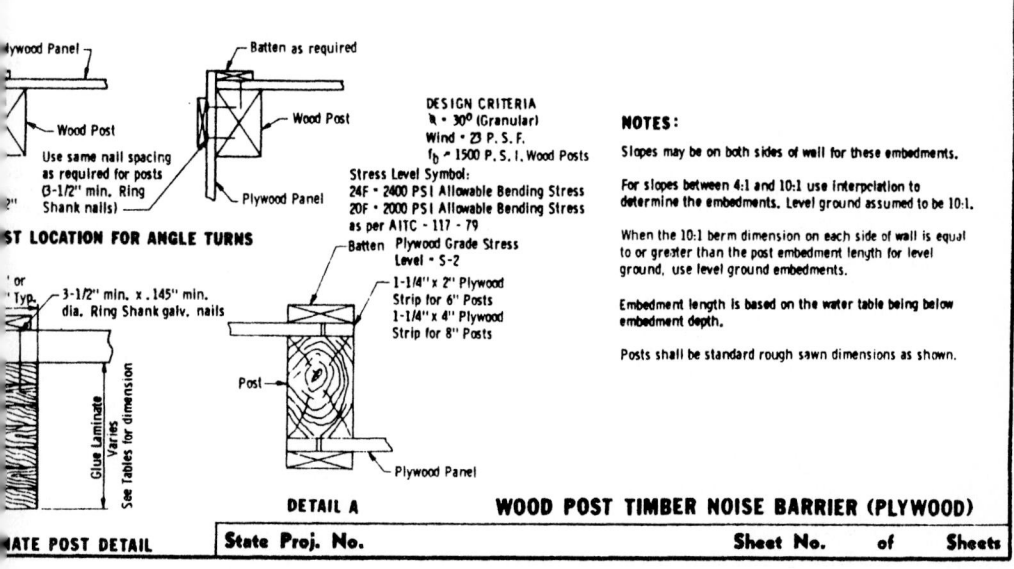

or
Typ.

3-1/2" min. x .145" min.
dia. Ring Shank galv. nails

Glue Laminate
Varies
See Tables for dimension

Post

Batten

1-1/4"x 2" Plywood
Strip for 6" Posts
1-1/4"x 4" Plywood
Strip for 8" Posts

Plywood Panel

DESIGN CRITERIA
Φ = 30° (Granular)
Wind = 23 P. S. F.
f_b = 1500 P. S. I. Wood Posts
Stress Level Symbol:
24F = 2400 PSI Allowable Bending Stress
20F = 2000 PSI Allowable Bending Stress
as per AITC - 117 - 79
Plywood Grade Stress
Level = S-2

NOTES:

Slopes may be on both sides of wall for these embedments.

For slopes between 4:1 and 10:1 use interpolation to
determine the embedments. Level ground assumed to be 10:1.

When the 10:1 berm dimension on each side of wall is equal
to or greater than the post embedment length for level
ground, use level ground embedments.

Embedment length is based on the water table being below
embedment depth.

Posts shall be standard rough sawn dimensions as shown.

DETAIL A

WOOD POST TIMBER NOISE BARRIER (PLYWOOD)

ATE POST DETAIL

State Proj. No.

Sheet No.   of   Sheets

9.39

CHAPTER 10
VALUE ENGINEERING AND LIFE CYCLE COST

Harold G. Tufty, CVS, FSAVE

Editor and Publisher
Value Engineering and Management Digest
Washington, D.C.

Value engineering has developed from its initial concepts over 50 years ago into a recognized method for analyzing the function of a product or service and developing an acceptable alternative with the lowest possible cost. It has been applied very successfully in industry, in the Department of Defense, and, in more recent years, to highway projects.

Life cycle costing, or least-cost analysis, is an integral part of value engineering. It provides a rational means of comparing the costs of alternatives in terms of today's dollars, including the effect of initial costs, maintenance costs, and rehabilitation.

This chapter reviews the application of value engineering on highway projects, explains the fundamentals of the process, and outlines methods for implementing value engineering programs.

VALUE ENGINEERING ON HIGHWAY PROJECTS

10.1 DEFINITION AND CONGRESSIONAL REQUIREMENTS

Value engineering (VE), as used in highways, has been defined by the Federal Highway Administration (FHWA) as "the systematic application of recognized techniques by a multi-disciplined team which identifies the function of a product or service; establishes a worth for that function; generates alternatives through the use of creative thinking; and provides the needed functions, reliably, at the lowest overall cost." Thus, it can be viewed as an organized application of common sense and technical knowledge to reduce the cost of highway projects. Value engineering should not be viewed as a duplication of the traditional engineering effort on a project.

Instead, it should be viewed as the use of a multidisciplined team, creative thinking, and function analysis to improve quality and productivity, foster innovation, eliminate unnecessary and costly design elements, and ensure that projects are cost-effective.

Although value engineering programs had been in effect in the FHWA and elsewhere for a number of years, the increased emphasis on the use of VE arose from the Intermodal Surface Transportation Efficiency Act (ISTEA) of 1991. This act required the secretary of transportation to study the effectiveness and benefits of value engineering review programs applied to federal-aid highway projects. Its intent was to utilize VE in all phases of highway design and construction to reduce costs and improve overall quality.

10.2 DOT POLICY

The FHWA's value engineering program is based on the policies of the U.S. Department of Transportation (DOT). Two categories of VE are recognized by the DOT. *Value engineering change proposals (VECPs)* are initiated by contractors working under DOT contracts. *Value engineering proposals (VEPs)* are developed by employees of the federal government or by VE personnel employed by the DOT to provide such services. In both cases, the proposals must result in measurable cost savings while maintaining equal or improved efficiency and quality.

The overall policy of the DOT was set forth in DOT Order 1395.1A as part of a program to obtain the maximum benefits for federally supported construction projects:

Each project authorized for direct construction by DOT shall use VE in its planning, design, and/or construction phases if the VE criteria established by the Operating Administrations can be applied. The VE criteria should take into account the overall complexity of the project, its estimated cost and other relevant design and/or construction factors. VE should generally be undertaken when there is potential for a significant ratio of savings to the cost of the study. This use of VE can also apply to design standards, construction procedures, and VE incentive clauses in construction contracts as deemed reasonable and appropriate by the Operating Administrations.

Each DOT Operating Administration should strongly encourage the use of VE in its grant awards or Federally assisted programs for major transportation projects throughout the planning, design, and/or construction phases. This may include the use of VEPs as a result of VE studies/analyses as well as VE incentive clauses in construction contracts. Major transportation projects are to be defined by each Operating Administration. Guidelines developed by the American Association of State Highway and Transportation Officials (AASHTO) may also be used in helping to identify projects that should be considered for VE. The Operating Administrations may exempt specified types of projects if the Administrator or the Commandant of the Coast Guard determines that the absence of complexity or potential cost savings makes the use of VE inappropriate. Design standards and construction procedures that are used in the planning, design, and construction of major transportation projects may also be considered for VE, as deemed reasonable and appropriate by the Operating Administrations. Internal goals or targets for including VE reviews in grant awards or some other measures should be established by each appropriate Operating Administration in order to evaluate the progress being achieved in promoting VE in Department grant projects.

Appropriate action should be taken to provide information, training, and technical assistance to grant award recipients in order to effectively promote the use of VE at the State and local levels.

10.3 VALUE ENGINEERING IN FEDERAL-AID AND FLH PROGRAMS

The FHWA value engineering programs are applied to both the federal-aid highway program, which provides reimbursements to the states, and the Federal Lands Highway (FLH) program, which involves highways built on land owned by the federal government. The federal-aid VE program is designed to (1) encourage the use of VE by the state highway authorities, (2) encompass a variety of activities focused on education and training, technical assistance, liaison with industry and states, promotional activities, and active participation in VE studies, and (3) focus primarily on training through the presentation of the National Highway Information VE workshop.

The FLH value engineering program is designed to address the direct expenditure of federal funds in design and construction activities, much like the VE programs that are implemented by the states. In the preconstruction phase, projects are identified early in the design phase for VE analysis by the FLH field divisions. In the construction phase, provisions for VECPs are included in each construction contract to encourage VE changes by construction contractors.

10.4 STATE VALUE ENGINEERING PROGRAMS

Through educational and technical assistance programs, the FHWA has strongly encouraged the development of state VE programs for highway development, construction, operation, and maintenance. Value engineering workshops are presented when requested by the states. The cost of conducting VE studies is an eligible federal-aid expenditure. Such studies may be done with state personnel or by a consulting firm contracted by the state.

The FHWA reported that, within a four-year period, over 1500 value engineering studies were performed by the states. During this time, $3.6 billion in savings were recommended and $615 million were being implemented.

The most active VE programs have been in Pennsylvania, Florida, Virginia, Illinois, California, Maryland, and Washington. In these states, VE programs have been in place for 10 to 15 years. These states typically conduct in-house training programs on a continuing basis, have a VE coordinator or VE office to manage the effort, and have action plans for carrying out the studies. These states have given the following main reasons for implementing a VE program:

- VE saves money and ensures that projects are cost-effective.
- VE can improve the quality of the project.
- VE can eliminate unnecessary design elements.
- VE can foster innovation and improve productivity.

In most states, VE programs are established by policy, but in Virginia, a comprehensive program has been mandated and implemented by state law. It applies to all transportation projects with a cost of $2 million or more. The legislation was adopted in 1990 after about 15 years' successful experience with VE. Virginia claims that over a recent three-year period, $39 million was saved, representing a return on investment of 30 to 1. Virginia divided the state into five regions with a coordinator for each. The coordinators were given extensive training in VE theory and application, as well as in the areas of interpersonal skills, group facilitation, problem solving, and computer applications. An automated tracking system was developed for monitoring recommendations and savings.

Virginia has found that it is beneficial to conduct VE studies at the initial stage of plan development and later when plans are 60 percent complete. Virginia has also used VE to examine its public review process and develop improved methods for reaching a wider audience. The public can now use less formal methods for presenting its views, such as recorded messages, cable television, direct mail, and toll-free telephone numbers.

Typical VE teams are made up of four to six members and are selected from different disciplines. By rotating assignments, the training and experience team members receive is carried into other parts of the state highway authority, thus encouraging the use of creative design techniques in everyday practice.

In general, state VE programs that have been successful have included the following characteristics:

- Top management and executive level recognition, support, and involvement
- A commitment of time, money, and personnel to perform the studies
- A commitment to train sufficient state staff in the VE technique
- An established statewide VE policy and implementation of a continuing statewide VE program
- A designated state VE coordinator

Although state VE programs have been optional so far as the federal government is concerned, it is likely that, in view of the proven worth of these programs, they may be mandated in the near future. For example, suggestions have been made to require the process on projects over a certain dollar amount, say $10 million. Current law sets the threshold at $25 million for mandatory VE on projects that are initiated as part of the designated National Highway System. The FHWA has proposed rules that would require its use on all federal-aid highway programs. Studies have shown that the cost of the VE effort is more than offset by the cost savings generated. For example, in 1993, California, Massachusetts, and Florida reported savings of over $100 million as a result of VE recommendations. These recommendations are often referred to as *VE opportunities*.

10.5 FEDERAL LANDS HIGHWAY (FLH) VALUE ENGINEERING PROGRAMS

The FLH program office is responsible for highway construction on federally owned land and administers a $125 to $175 million design and construction program yearly. Field offices are required to review highway programs to identify areas for VE studies and develop annual programs. The programs may include studies of projects, processes, procedures, specifications, or standards that hold the potential for achieving the greatest savings while maintaining high quality. A trained VE coordinator in each division monitors the progress of studies and follows up on study recommendations. VECP clauses are included in all contracts over $100,000. Evaluation and acceptance or rejection is required within 45 days.

During a four-year period, the FLH studied 28 design projects, which represented over 50 percent of the construction program. Value engineering studies saved $44 million, giving a return on investment of about 70 to 1. Also, 13 of 16 VECP submissions by contractors were approved, for an additional savings of $400,000.

Continuous improvements are being made to the VE program to:

- Ensure that VE studies are performed at the best stage to optimize savings
- Allow periodic VE team reviews of the planned design program to identify additional VE study opportunities
- Obtain services of certified VE consultants to overcome workload limitations
- Train and retrain engineering personnel on the benefits and uses of VE and VECP
- Require VE studies in contracts for engineering design services
- Emphasize VECP submissions during preconstruction and partnering conferences
- Publicize successful VE studies and recognize and reward VE teams for their work

10.6 VALUE ENGINEERING CHANGE PROPOSAL (VECP) PROGRAMS

The VECP program was developed to allow a construction contractor to submit a cost savings proposal after the construction contract has been awarded. The contractor has the opportunity to employ the VE process (formally or informally), develop a cost-effective alternative to the contract plans, and initiate a change proposal. As an incentive to submit proposals, the contractor receives a share (typically 50 percent) of the cost savings generated by any approved changes. VECP clauses should be included in most construction specifications. Although there have been significant savings through VECPs, the full potential of this process has not been realized in most state programs. Steps that can be taken to encourage contractor submittals include:

- Full support of the process by top management
- Guarding against pride of ownership and resistance to change
- Ensuring that the review and approval or rejection process is prompt

10.7 GENERAL CONSIDERATIONS

10.7.1 Selection of Projects for Value Engineering

State departments of transportation and other groups should have a program that defines specific criteria and guidelines for selecting projects for VE review. The FHWA has stated that "consideration should be given to projects that have shown recent substantial cost increases; projects with complex designs or construction phases; projects involving major structures; projects with unique specifications, standards or processes; multi-modal projects; projects with repetitive work elements; projects with high right-of-way costs; projects with unique or experimental features; projects with high maintenance, user impact, or traffic control costs; and projects specifically requested for review by State agency program offices or management." Increased emphasis is usually given to larger projects. Even so, it may well be worth the effort to perform focused, relatively short reviews on smaller projects. The FHWA has set a goal that VE studies address a minimum of 50 percent of the dollar volume of projects. Studies can be performed by trained personnel on staff or by consultants, depending on work loads and availability.

10.7.2 Level of Value Engineering Effort

The appropriate level of VE effort for a given project is a function of several factors, the major ones being project size, project complexity, constraints upon the scope of the VE study, and the degree of completion of the design. The Environmental Protection Agency (EPA) discourages any VE effort after 80 percent completion of design. The optimum opportunity for return on VE costs appears to be when VE is conducted at 10 to 30 percent of completion of the design. The priorities of the design team should recognize the VE effort as a project milestone at the 10 to 30 percent point in design and establish priorities that will ensure the availability of material described in the VE workbook. The subsequent discussions of level of effort are based upon conducting the VE study at the 10 to 30 percent point. The level of VE efforts for studies later than this should be reduced to reflect the reduced potential for return. The major elements to be determined for a given VE level of effort are total manpower required and the number of workshops.

Depending on the size and complexity of the project, the VE effort may vary from one team and one study to multiple teams and/or multiple studies in order to adequately review the project. In some larger and/or more complex projects, it may be desirable to schedule two VE workshops during the course of the design. The first may occur when 10 to 30 percent of the design is complete and would concentrate on basic factors such as project layout and systems to be used, general approach to electrical systems, etc. The second workshop would occur when the design is complete enough (approximately 50 to 60 percent complete) that a detailed review of the designs could be made.

The determination of the appropriate numbers of teams and studies must be made on a case-by-case basis. For example, a large project may readily justify separate teams, each with a study area such as bridges, pavements, drainage, or right-of-way. If the system in question is simple, the level of effort may be relatively small and readily handled in one VE review.

10.7.3 Team Composition

Typically, one VE team consists of about five members. In some cases, larger teams may be justified, and larger teams have been used on particularly large and complex projects. However, for most projects, the five-member team will be an appropriate size. Team members should be drawn from different disciplines such as design, construction, environmental science, traffic, maintenance, planning, and right-of-way. Experts in other project areas can be added, depending on the project being reviewed, as well as a representative from the public, when appropriate.

The team is typically led by a trained VE team coordinator who acts as a facilitator of the process for the team. The coordinator is responsible for the following functions:

1. Pre-workshop preparation: collecting project reports, drawings, specifications, quantity takeoffs, and cost data from the designer; also making arrangements with team members and distributing information to them.
2. Conducting the project review workshop
3. Preparing the preliminary VE report
4. Participating in the implementation phase, as required

The project manager, lead designer, or other person directly responsible for the planning or design should not be a team member. This avoids "pride of ownership" problems. However, that person should participate as an information resource. Typical duties would involve the following:

1. Pre-workshop preparation: working with the coordinator in assembling needed information
2. During workshop: providing answers to questions raised (any defensive reactions must not be allowed to interfere with the generation of alternative ideas)

10.7.4 Team-Building Suggestions

Experienced leaders in successful VE highway efforts have found the following suggestions to be beneficial in building VE teams:

1. Emphasize that VE is a positive process that will increase the value of the project to the owner and help the preliminary design team arrive at the "best" design.
2. Include a mixture of engineering and nonengineering personnel with different backgrounds on the team.
3. Develop trust and communication between the design team and the VE team.
4. Avoid political influence over design decisions; evaluate recommendations based on merit.
5. Keep an open mind to change.

10.7.5 Implementation

The VE program should include procedures to ensure that recommendations are reviewed promptly by applicable staff. A slow review process invites delays and can negate savings. Procedures should be established for monitoring implementation and documenting accepted and rejected recommendations. Cost savings generated, as well as costs of the VE process, must be maintained and reviewed regularly to judge the effectiveness of the program. Also, it is imperative that VE alternatives employed on one project be included early in the design process on similar projects.

("Value Engineering on Federal-Aid Projects," Report to Congress by the Secretary of Transportation, U.S. Department of Transportation, Federal Highway Administration, 1993; Proposed Rules, *Federal Register,* vol. 59, no. 220, 1994; R. W. Miller, "Value Engineering in Preliminary Design," in *Proceedings of the AASHTO Value Engineering Conference,* AASHTO, Washington, D.C., 1993.)

FUNDAMENTALS OF VALUE ENGINEERING

10.8 ORIGINS OF VALUE ENGINEERING

As previously stated, value engineering is the systematic application of various recognized engineering techniques by multidisciplinary team(s) that identifies the function of a product or service, establishes the worth of that function, generates alternatives through the use of creative thinking, produces recommendations in concept form for presentation to the owners or original designers, and provides the needed function at the lowest overall cost. Value engineering (VE) in the United States emerged from the industrial community during World War II, when many critical materials were difficult, if not impossible, to obtain. This problem forced the use of substitute materials and designs by many manufacturing concerns. The General Electric Company found that many of the substitute materials used were providing equal or better performance at less cost. It was felt that an effort should be made to improve product efficiency by intentionally selecting the best material the first time.

In 1947, the task of investigating this possibility was undertaken by Lawrence D. Miles, a staff engineer for General Electric. Miles developed a number of ideas and techniques to enable this type of change to be performed intentionally rather than accidentally. In effect, Miles took an old attitude about the search for value and developed from this attitude a successful methodology designed to ensure the development of value in a product. His prime contribution was to force designers to focus on the function required, not a particular material.

The concept quickly spread throughout private industry as the possibilities for large returns from relatively modest investments were recognized. Value engineering, whether called value analysis, value improvement, value management, or any of numerous other names, was, in 1961, formally implemented in the Department of Defense (DOD) in the United States. The DOD establishment applied the concept to procurement as well as other areas. Prior to the implementation of the contractual aspects of the DOD value engineering program, contractors had no specific financial incentive to propose specification or design changes in order to reduce costs. Now the DOD contractor is not only encouraged to make changes, but is offered an attractive opportunity to share in the savings.

In today's market, VE has proved itself to be one of the soundest economic ventures. Its overall record of performance where it has been intelligently applied, discreetly managed, and faithfully reported is impressive. It has often lowered manufacturing and procurement costs by 15 to 25 percent, including the costs of VE itself. In highway applications, its return on investment has been impressive.

The VE technique, or methodology, can be applied anywhere there is a function that can be identified and there is a way to measure that function. Value engineering is versatile because practically everything human beings do, in both public and private life, is done to accomplish something, some particular function. The measurement of the function can be dollars, time, weight, performance, safety, quality, or any other necessary feature.

10.9   VALUE ENGINEERING CONCEPT

Value engineering is an organized effort directed at analyzing functions such as the construction, maintenance, or rehabilitation of highways, bridges, or pavements; the installation, operation, maintenance, repair, or replacement of equipment; or the acquisition of facilities or supplies for the purpose of achieving the required function at the lowest total cost of effective ownership, consistent with requirements for performance, reliability, quality, and maintainability. In fundamental terms, it can be called an organized way of thinking about or looking at something through a functional approach. It involves an objective appraisal of functions performed by alternative designs for bridges, pavements, or roadway alignments, as well as parts, components, products, equipment, procedures, services, etc. Anything that costs money, or can be measured by other means, is a candidate for analysis. The VE approach is aimed at providing a necessary function for the lowest cost. It is designed to check the upward spiral of costs, while ensuring the essentials of reliability, performance, maintainability, and quality.

Value engineering is not primarily centered on a specific category of the physical sciences, but incorporates all available technologies, as well as the principles of economics and business management, into its procedures. When viewed as a management discipline, it utilizes the total resources available to an organization to achieve broad, top management objectives. Thus, VE may be seen as a systematic and creative approach for increasing the "return on investment" on highway facilities, operating procedures, or material acquisitions. In other words, VE is concerned with acquiring good value by investigating what the facility, product, or process does in relation to the money spent on it.

As indicated before, there are two types of VE applicable to highways:

1. Value engineering proposal (VEP): a specific proposal developed internally within an organization for total value improvement through the use of VE techniques

2. Value engineering change proposal (VECP): a specific cost-reduction proposal, developed and submitted by a contractor or supplier under VE contract provisions, that requires a change to the contract specifications, purchase description, or statement of work

10.10   VALUE ENGINEERING PROGRAM MANAGEMENT

A VE program is an organized set of definite tasks that support or apply the VE discipline to projects in major elements of an organization. An effective and sustained VE program will usually have:

1. Periodic top management attention to ensure implementation and continuing attention by middle management.

2. A key individual to manage the value engineering program. This individual should be well versed in VE principles, techniques, and contracting procedures.

3. A "master plan" to ensure that actions that may effectively contribute to a successful VE program are considered and acted upon.

4. Value engineering objectives, policies, responsibilities, and reporting requirements firmly established and implemented.

For most VE programs, the following tasks should also be included:

1. Close coordination with the contracting staff and marketing group

2. A strong VE training and executive briefing program

3. Management attention to ensure that the VE discipline is used to derive measurable benefits

Although there are many other specific tasks required to ensure that VE achieves its full potential, these items form the foundation upon which the structure of a successful VE program may be built.

A study conducted by a major federal agency determined the predominant sources of opportunity for VE. From a sample of 415 successful VE changes, the study identified seven factors that were responsible for about 95 percent of the savings realized. The seven factors in order of percent of total savings were:

1. Advances in technology
2. Elimination of excessive cost
3. Questioning of specifications
4. Additional design effort
5. Changes in user's needs
6. Feedback from test or use
7. Design deficiencies

However, the study revealed that a single factor was rarely the basis for a VE action.

VE is a fundamental approach that challenges everything and takes nothing for granted, including the necessity for the existence of a facility or product. It may be successfully introduced at any point in the design process. Some find it most effective to apply it in the initial stages and again in later stages.

VE requires an intensive review of requirements and the development of alternatives by the use of appropriate value techniques utilizing aspects of engineering, requirements analysis, the behavioral sciences, creativity, economic analysis, and the scientific method. Employed in an organized effort and properly managed, VE utilizes a systematic procedure for analyzing requirements and translating these into the most economical means of providing essential functions without impairment of needed performance, reliability, quality, or maintainability.

10.11 DETERMINATION OF VALUE AND FUNCTION

Value is a function of both performance and cost. The value of a project, product, or service can be described as "good" if it delivers the required basic function for the lowest cost, and "poor" if it costs too much or fails to perform the required function. Good value cannot be obtained by the simple procedure of cutting costs, particularly if there is any lowering of quality or reliability. Safety, appearance, and maintainability are important and cannot be sacrificed. Therefore, good value can truly be obtained only when the cost of all necessary features is brought into balance with the cost of attaining the essential function.

Top management, middle management, and even first-line supervisors have not been entirely successful in controlling the causes of higher costs. Costs frequently depend on the decisions of the design engineer. Further complications arise when no one individual is responsible for all phases of development. A project may be conceived by one group, designed by another, refined by a third, and implemented by a fourth. Every phase in the development involves new people, some of whom may not have an appropriate concern about costs.

In the search for value at a reasonable price, a user purchases an item or service because it will provide certain functions at a cost the user is willing to pay. If something does not do what it is intended to do, it is of no use to the user and no amount of cost reduction will improve its value. Actions that sacrifice needed utility of an item actually reduce its value to the user. On the other hand, expenditures to increase the functional capability of an item beyond that which is needed are also of little value to the user. Thus, anything less than the necessary functional capability is unacceptable; anything more is unnecessary and wasteful.

To achieve optimum value, functions must be carefully defined so that their associated costs may be determined and properly assigned. The VE approach requires the development of valid and complete answers to the following questions:

What is (the item)?

What *must* it do (its function)?

What does it cost?

What is it worth (the least possible cost to perform the function)?

What else would work?

What does that cost?

10.12 VALUE ENGINEERING JOB PLAN CONCEPT

10.12.1 Step 1: Evaluation of Function

When it has been decided that a VE study is to be conducted, the initial effort should be to determine the user's actual needs. The user's needs are those explicit performance qualities, traits, or characteristics that justify the existence of a system or an item, that is, the characteristics that must be possessed if the system or item is to be useful and efficient. They define what the item must be able to do in relation to overall goals, or what it must be able to do in relation to the whole of which it is a part. The manner in which the user's needs are expressed may imply a method of satisfying them, but it is the engineer's job to make the method tangible and explicit.

Therefore, the user's needs are the objectives; the design specifies the means by which the objectives are satisfied. The definition of the user's needs in explicit quantitative terms is a difficult task. Many times there is a temptation to look at an item and say that the function it performs is the required function, but this is not always true. By defining the function, one learns precisely which characteristics of the design are really required.

Determining Function. Attempts to identify and define the function(s) of an item can often result in several concepts being described in many sentences. While this method may conceivably describe the function(s) satisfactorily, it is neither concise nor workable enough for the VE functional approach. In VE, function is best expressed using two words—a verb and a noun.

1. The verb answers the question "What does it do?" The verb defines the item's required action (it may generate, control, pump, emit, protect, transmit, etc.).
2. The noun answers the question "What does it do it to?" The noun tells what is acted upon (electricity, temperature, liquids, light, surfaces, sound, etc.). This noun must be measurable or at least understood in measurable terms, since a specific value must be assigned to it during the latter evaluation process, that of relating cost to function.

3. The system of defining a function in two words, a verb and a noun, is known as *two-word abridgment*. This abridgment represents a skeletal presentation of relative completeness. Advantages of this system are that it:
   a. Forces consciousness
   b. Avoids combining functions and attempting to define more than one simple function at a time
   c. Aids in achieving the broadest level of disassociation from specifics
4. A function should be identified so as not to limit the ways in which it could be performed.
5. Identification of function should concern itself with how something can be used, not just what it is.
6. Identifying function in broadest possible terms provides the greatest potential for value improvement because it gives greater freedom for creatively developing alternatives. Further, it tends to overcome any preconceived ideas of the manner in which the function is to be accomplished.

Classifying Functions. Functions of items or systems may be divided into two kinds: (1) basic and (2) secondary. The definition of higher order applies to functions beyond the scope of the immediate study.

1. A *basic function* defines a performance feature that must be attained. It reflects the primary reason for an item or a system.
2. A *secondary function* defines performance features of a system or an item other than those that must be accomplished.
   a. A secondary function answers the question "What else does it do?"
   b. Secondary functions support the basic function but generally exist only because of the particular design approach that has been taken to perform the basic function.

10.12.2 Step 2: Evaluation of Cost

The consideration of cost is the second step in function analysis. In this application, it is the current, actual, as-designed cost of the method chosen to perform the function that is considered. Evaluation of cost serves several purposes:

1. High-cost elements will be identified. This is useful in determining the priority of individual VE studies to be undertaken.
2. Cost visibility is given to a function's performance where normally such costs are buried in unit or system estimates.
3. The validity of the claimed savings at the conclusion of a VE project depends upon the accuracy of the cost figures for the present design and the realism of cost estimates of the proposed design. Thus the cost figures obtained must be factual and realistic.

Determining Costs. The cost of the present design of a system, an item, or an operation should be determined in as great detail as practicable. However, in conducting functional analysis the following rules should always govern:

1. Where an item serves but one function, the cost of the item is equal to the cost of the function.

2. Where an item serves more than one function, cost of the item should be prorated to each function.

10.12.3   Step 3: Evaluation of Worth

The third step in a function analysis is to establish a monetary value of the worth for each function. Worth is the lowest possible cost to perform the function. It is done after all functions have been identified and classified. It is perhaps the most difficult step in VE, but it is an indispensable step. It is a highly creative endeavor because worth is a subjective rather than an absolute or objective measure. Skill, knowledge, and judgment play a major role in determining the quantitative aspect of worth in money terms. The worth of a basic function is usually determined by comparing the present design for performing the function with the least-cost method of performing essentially the same function.

Questions that might be asked during the evaluation include the following:

1. What is the least cost to achieve the basic function as the item is presently designed?

2. Do you think the present, as-designed cost of the basic function should be so high?

3. If not, what would you consider a reasonable amount to pay for the performance of the basic function (assuming for the moment that the function is actually required) if you were to pay for it out of your own pocket?

4. What is the cost of achieving this function if some other known item is used?

5. Is this a common, easily accomplished function, or one that is rare and difficult to achieve?

6. What is the price of some item that will almost, but not quite, perform the function?

Determining Worth.   The above guidelines for evaluating worth must be applied against the following rules for determining worth for a specific VE analysis:

1. The worth of all secondary functions is zero for VE purposes.

2. A monetary amount for the value of worth must be established for each basic function.

3. Worth is associated with the necessary function or functions and not with the present design of the item.

4. There must be no discrimination between a function that is definitely required and the consequences of failure to achieve that function. For example, if a bolt supporting a steel beam in a building fails, the building may collapse. Nevertheless, the worth of the bolt is the lowest cost necessary to provide a reliable fastening. Worth is not, therefore, affected by the consequence of failure.

10.12.4   Step 4: Evaluation of Value

The fourth step in function analysis is to make a determination of value. Value is the relationship of worth to cost as seen by the user (or customer) in the light of needs and resources in a given situation. The ratio of worth to cost is the principal measure of value. Thus, a value equation may be written to derive a *value index,* as follows:

$$\text{Value index} = \frac{\text{worth}}{\text{cost}} = \frac{\text{utility}}{\text{cost}}$$

Four key considerations apply.

1. Value may be increased by:
   a. Improving the utility of something with no change in cost
   b. Retaining the same utility for less cost
   c. Combining improved utility with a decrease in cost

   Optimum value is achieved when all utility criteria are met at the lowest overall cost. Although worth and cost can be expressed in monetary units, value is a dimensionless expression of the relationship of these two.

2. If something does not do what it is intended to do, it is not of any use to anyone and no amount of cost reduction will improve its value. Cost reduction actions that sacrifice the needed utility of something actually reduce its value to the user. On the other hand, expenditures to increase the functional capability of an item beyond that which is needed are also of little value to the user. These funds would be better spent providing additional usable capability. Thus, anything less than the necessary functional capability is unacceptable; anything more is unnecessary and wasteful.

3. Value engineering, overall, is also concerned with two other types of functions: the use function and the aesthetic function. Most simply stated, the use function of a product satisfies the user's desire for having an action performed, while an aesthetic function pleases either the user or someone the user wants to have pleased. These two functions are not mutually exclusive and are generally both present in some ratio in all products or services. Good value occurs when the user pays only the necessary costs to provide each type of function.

4. In function analysis, the value of an end item approaches its maximum when each function has been penetratingly studied with the user, and needs and wants have been accurately determined. Maximum value thus provides the best combinations of knowledge, skills, and materials the state of the art permits. A VE goal is to provide the maximum value through the control of use value and aesthetic value, and to eliminate costs associated with any other value not related to the performance of basic function.

   When correctly classified, all secondary-function costs can be challenged. Each can be eliminated if quality will not be downgraded.

   One of the first, most clearly defined, and most widely accepted adaptations of the function approach to problem solving is the *function analysis system technique* (*FAST*). FAST is particularly useful in complicated systems to identify the critical path of basic functions. A description of FAST with examples is included later in this chapter.

   The definition of worth in the function-cost-worth portion of the VE job plan is frequently a confusing concept to understand. One particularly effective tool to help in this area is a *cost model.* It will show the ratio between "should cost" and "does cost." The "should cost" figure helps define worth. This can be used in almost any VE application area and is particularly useful in the highway construction field. A sample of a cost model in construction is included later in this chapter.

10.13 VALUE ENGINEERING JOB PLAN CONCEPT IN DETAIL

A task that is accomplished in a planned and systematic manner is more likely to be fruitful than one that is unplanned and relies upon undisciplined ingenuity. Value engineering efforts generally follow a variation of the "scientific method" to ensure a planned, purposeful approach. This procedure is termed the *VE job plan*. It was conceived as a group undertaking because it is unlikely that an implemented VE proposal will be the product of the effort of a single individual.

The VE job plan outlines those tasks or functions necessary to properly perform a VE study. Adherence to a definite plan is essential to achieving optimum results. Good results come from a good system, and a good system is one that covers all aspects of a problem or situation to the necessary degree. Use of the job plan provides:

1. A vehicle to carry the study from inception to conclusion
2. A convenient way to maintain a written record of the effort as it progresses
3. Assurance that consideration has been given to facts that may have been neglected in the creation of the original design
4. A logical separation of the study into units that can be planned, scheduled, budgeted, and assessed
5. Assurance that proper emphasis is given to the essential creative work of a study and its analysis so that superior choices can be made for further development

The job plan attempts to generate, identify, and select the best-value alternative(s) by making specific recommendations supported with the proper data and identifying the actions necessary for implementation. Further, it provides a proposed implementation schedule and a summary of benefits to the user. The VE job plan is a planned program that has been tested, is being used, and has been proved to work.

Several versions of the VE job plan are described in current VE literature. Some sources list five phases, and others range up to nine. The number of phases into which the study procedure is divided is less important than the systematic approach employed. The seven-phase approach depicted in Fig. 10.1 is described below. A nine-phase approach often used in highway studies is shown in Table 10.1. Actually, there are no sharp lines of distinction between the phases; they tend to overlap in varying degrees. In actual practice it is frequently necessary to return to a previously completed phase for additional work prior to reaching a decision. A team may be working on two or more phases at the same time. Early steps, such as information gathering, may continue throughout most of the VE effort.

The VE effort must include all phases of the job plan. However, the proper share of attention given to each phase may differ from one application to another. Judgment is required in determining the depth to which each phase is performed, with consideration given to the resources available and the results expected.

An orientation phase is usually conducted by a VE manager prior to the assembling of a VE task team. This activity relates to the selection of ideas for VE projects and their planning and authorization. The VE team follows the VE job plan starting with the information phase after the item to be studied has been selected. The number of members of the VE team varies considerably, but usually the job plan is completed by a team of at least five persons.

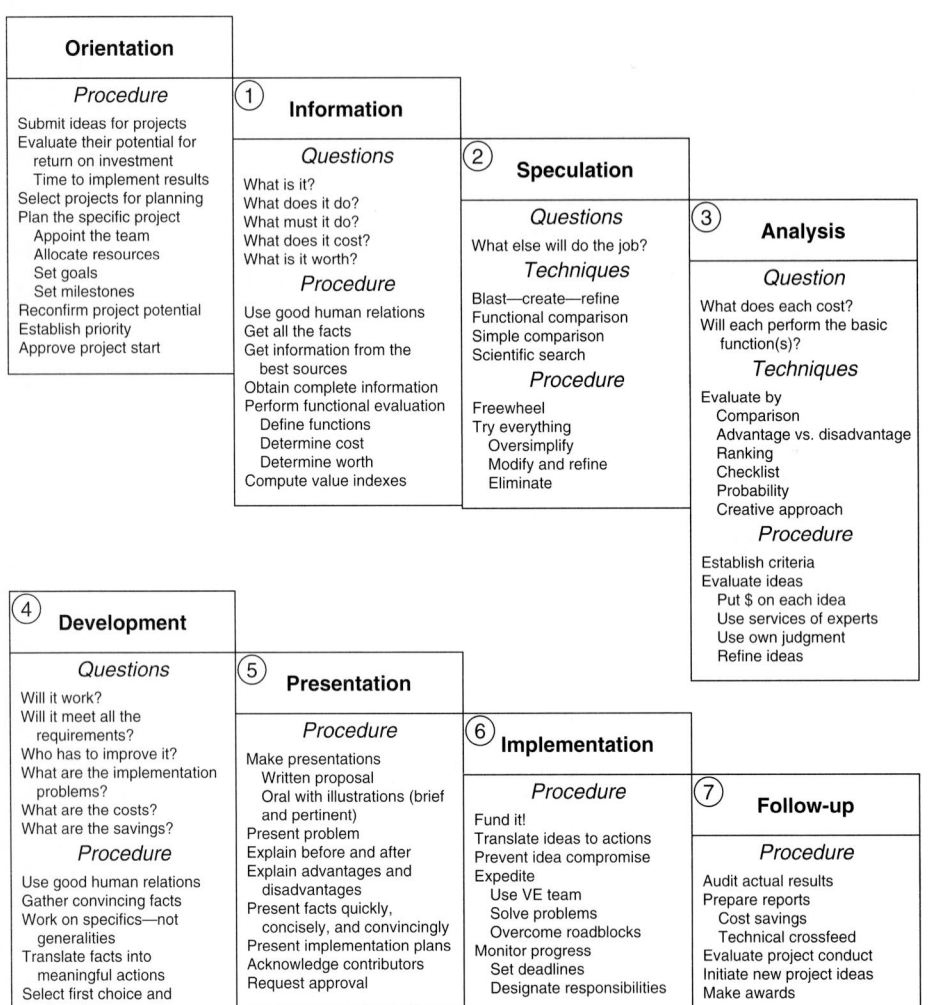

FIGURE 10.1   Seven-phase VE job plan.

10.13.1 Phase 1: Information

Objectives.   The first phase of the job plan has two basic objectives:

To obtain thorough understanding of the system, operation, or item under study by a rigorous review of all of the pertinent factual data

To define the value problem by means of functional description accompanied by an estimate of the cost and worth of accomplishing each basic function

TABLE 10.1 A Nine-Phase VE Job Plan

Phase	Objective	Key questions	Techniques	Tasks
Selection	Select projects	What is to be studied? Who is best able to study the problems? What must be known to start the study?	Solicit project ideas. Identify high-cost or low-value areas. Plan the project. Obtain authorization to proceed. Allocate resources.	Speculate on sources of projects. Develop plan to identify project. Analyze projects for applying VE. Evaluate projects for potential. Present projects to management. Select projects for VE study. Implement study plan.
Investigation	Investigate project	What is the project? What is the problem? What is the cost? What is now accomplished? What must be accomplished?	Get information from the best sources. Get all the facts. Work with specifics. Get all available costs. Challenge everything. Identify the function.	Speculate on sources of project data. Develop plan to gather project data. Implement data search plan. Investigate the project. Speculate on functions performed. Audit data.
Analysis	Analyze function and cost	What is the basic function worth? What are the secondary functions worth? What are the high-cost areas? Can any function be eliminated?	Evaluate by comparison. Put price on specs and requirements. Put price on key tolerances and finishes. Put price on key standards.	Analyze costs. Analyze functions. Evaluate function cost and worth. Evaluate project potential. Select specific study areas.
Speculation	Speculate on alternatives	What else will perform the function? Where else may the function be performed? How else may the function be performed?	List everything. Be imaginative. Use creative techniques. Defer judgment. Do not criticize. Be courageous.	Select techniques to be used. Speculate on alternatives. Speculate on parameters.
Evaluation	Evaluate alternatives	How might each idea work? What might be the cost? Will each idea perform the basic function?	Weigh alternatives. Choose evaluation criteria. Refine ideas. Put price on each main idea. Evaluate by comparison.	Speculate on evaluation criteria. Evaluate alternatives. Select best alternatives.

TABLE 10.1 A Nine-Phase VE Job Plan (*Continued*)

Phase	Objective	Key questions	Techniques	Tasks
Development	Develop alternative	How will the new idea work? How can disadvantages be overcome? What will be the total cost? Why is the new way better? Will it meet all the requirements? What are the life cycle costs?	Use search techniques. Get information from the best sources, specialists, and suppliers. Consider specialty materials, products, and processes. Consider standards. Use new information. Compile all costs. Work with specifics. Gather convincing facts.	Speculate on information needed. Speculate on information sources. Develop a plan of investigation. Develop selected alternatives. Select preferred alternative. Develop implementation plan. Audit data.
Presentation	Present alternative	Who must be sold? How should this idea be presented? What was the problem? What is the new way? What are the benefits? What are the savings? What is needed to implement the proposal?	Make recommendations. Use selling techniques. Be factual. Be brief. Give credit. Provide an implementation plan.	Develop a written proposal. Speculate on possible roadblocks to acceptance. Present recommended alternative.
Implementation	Implement alternative	Who is to implement the change? How will the contract be amended? Have needed resources been allocated?	Translate plan into action. Overcome problems. Expedite action. Monitor project.	Develop change documents. Implement approved alternative. Evaluate progress.
Audit	Audit results	Did the new way work? How much did it cost? How much money was saved? Did the change meet expectations? Who is to receive recognition?	Verify accomplishments. Make awards. Report to management.	Audit results of implementation. Evaluate project results. Present project results. Present awards.

Key questions. During this phase, the following key questions must be answered:

What is it?

What does it do?

What must it do?

What does it cost?

What is it worth? (What is the least the function could cost?)

Procedure

1. *Use good human relations.* The matter of human relations is of utmost impor-tance to the success of any VE study. "People" problems are sometimes more difficult to resolve than technical problems. The effectiveness of a VE team leader's efforts depends upon the amount of cooperation the leader obtains from the engineers, designers, estimators, managers, etc. If one is skillful in approach, diplomatic when resolving opposing viewpoints, and tactful in questioning a design requirement or specification, one will minimize the problems of obtaining the cooperation necessary to do the job effectively.

2. *Collect information.* All pertinent facts concerning the system, operation, or item must be drawn together. Getting all the relevant facts and getting them from the best sources are of paramount importance. The VE team should gather complete infor-mation consistent with the study schedule. All relevant information is important, regardless of how disorganized or unrelated it may seem when gathered. The data gathered should be supported by tangible evidence in the form of copies of all appro-priate documents. Where supported facts are not obtainable, the opinions of knowl-edgeable persons should be documented.

In addition to specific knowledge of the item, it is essential to have all available information concerning the technologies involved, and to be aware of the latest techni-cal developments pertinent to the subject being reviewed. Knowledge of the various manufacturing, fabrication, or construction processes that may be employed in pro-duction of the item is essential. The more information brought to bear on the problem, the more likely the possibility of a substantial cost reduction. Having all the above information would be the ideal situation, but if all of this information is not available, it should not preclude the performance of the VE effort. The checklists provided in Table 10.2 include types and sources of information.

3. *Determine functions.* The determination of functions is a requisite for all value studies. The decision to pursue the project through the remaining phases of the job plan can be made only by determining function, placing a worth on each required function, and then comparing worth against actual or estimated cost. The determina-tion of function should take place as soon as sufficient information is available to per-mit determination of true requirements. All members of the VE study team should par-ticipate in this exercise because the determination of required function is vital to subsequent phases of the job plan.

4. *Evaluate functions.* After the functional description has been developed, the next step is to estimate the as-designed cost and worth of performing each basic func-tion. The worth (the lowest possible cost to perform the function) determined will later be compared against the estimate of the item's cost. This comparison indicates whether the study should be terminated because cost and worth are approximately equal or pursued because cost greatly exceeds worth.

TABLE 10.2 Types and Sources of Information

A. Types of information

Physical data: Shape, dimensions, material, color, weight

Methods data: Information on how it is operated, constructed, made, shipped, fabricated, written, developed, installed, packaged, repaired, maintained, organized, replaced, etc.

Performance data: Information on what present performance is, and what performance requirements should be in areas of design, operation, maintenance, utilization, utility, etc.

Restrictional data: Information on required restrictions concerning tolerances, methods, performance, procedures, operations, time, schedule, cost, etc.

Cost data: Major breakdown of costs and labor, material, and markups for the elements of physical data, methods data, performance data, and restrictional data as they relate to design, acquisition, operation, maintenance, and disposal

Quantity data: The anticipated volume or repetition of use for the present and future

B. Sources of information

Government personnel: Project managers, design engineers, owners, users, managers, operators, maintenance personnel, specialists, etc.

Commercial personnel: Contractors, fabricators, suppliers, vendors, distributors, expert consultants, etc.

Highway users

Project data: Planning documents, drawings, specifications, sketches, computations, cost estimates, material lists, and financial records

Critical data: Handbooks, orders, standard specifications, commercial and government standards, published user regulations and requirements, engineering manuals, maintenance manuals, etc.

Experience data: Test reports, maintenance reports, user feedback, technical crossfeed reports, etc.

Commercial data: Catalogs, product literature, technical publications, magazines, etc.

Historical data: Libraries, previous studies, data files, management information systems, previous designs and contracts, conference and symposium proceedings, etc.

10.13.2 Phase 2: Speculation

Objective. The objective of this speculation is to generate, by creative techniques, numerous alternative means for accomplishing the basic function(s) identified. This phase, which is frequently called the *creative phase,* includes brainstorming.

Key Question. Accomplishing this phase should result in answering the question "What else will do the job, that is, perform the basic function(s)?" The completeness and comprehensiveness of the answer to this question determines to a very high degree the effectiveness and caliber of value work. The greater the number and quality of alternatives identified, the greater the likelihood of developing an outstanding solution. Additional alternatives that have not been considered will usually exist regardless of the skill and proficiency of the study team.

Procedure. Consideration of alternative solutions should not formally begin until the problem is thoroughly understood. All members of the VE study team should participate, for the greater the number of ideas conceived, the more likely that really effective, less costly alternatives will be among them. A proper frame of mind is important at this stage of the study; creative thinking should replace the conventional.

It should be a unique flight of the imagination, undertaken to generate numerous alternative methods of providing the necessary function(s).

Judicial thinking does not belong in this phase. As an aid to speculative thought, the techniques of creative thinking, such as brainstorming, should be employed. Every attempt should be made during this phase to depart from ordinary patterns, typical solutions, and habitual methods. Experience indicates that it is often the new, fresh, and radically different approach that uncovers the best-value solution. The individual or group members may supplement their ideas with those of others—everyone is expected to make a contribution.

The best solution may be complete elimination of the present function or item. This possibility should not be overlooked during the initial phases of this step. Perhaps some aspect can be modified which will permit elimination of the function under study. Only after determining that the function must remain should the study group look for alternative ways to perform the same function at the lowest conceivable cost. Free use of imagination is encouraged so that all possible solutions are considered.

A partial list of questions that can be used to stimulate and trigger ideas is given in Table 10.3. The questions shown can be rephrased by substituting terms like *project, system, item,* or *procedure* for the words *it* or *part* when appropriate.

Techniques

1. *Blast, create, and refine.* This theme has often been used by value engineers. *Blast*—get off the beaten path. *Create*—rally for an unusual idea; reach way out for another approach. *Refine*—strengthen or add to develop an idea to perform basic functions in a new or unique manner.

2. *Functional comparison.* Conduct a creative problem-solving session (brainstorming) in which new and unusual contributions of known things or processes are combined and/or rearranged to provide different ways to perform basic functions.

3. *Simple comparison.* Conduct a thorough search for other items that are similar in at least one significant characteristic to the study item. Determine whether they can be modified to satisfy basic functions.

4. *Scientific search.* Conduct a search for other scientific disciplines capable of performing the same basic function. This often involves interviewing specialists in disciplines that did not previously contribute to solving the problem. An industry (or its representatives) that specializes in some highly skilled technique can often make a substantial contribution when called upon for technical assistance.

10.13.3 Phase 3: Analysis

Objectives. The purpose of the analysis phase is to select the most promising alternatives from among those generated during the previous phase. During the speculation phase there is a conscious effort to prohibit any judicial thinking so as not to inhibit the creative process. In phase 3, all the alternatives must be critically evaluated because many of them may not be feasible. The alternatives are studied individually and/or grouped for the best solution. Identifying function may seem like a simple process—so simple, in fact, that it seems only a "simple" mind would be required to get the job done. In some ways this is true; a mind that can work in a simple, direct way is required—a mind with the ability to reduce concepts, ideas, and analyses to their best common denominators. The emphasis on function in this analysis phase is what makes the VE approach radically different from any other cost reduction effort.

TABLE 10.3 Questions to Stimulate Ideas

A. Idea stimulators

Eliminate—combine:
   Can it be eliminated entirely?
   Can part of it be eliminated?
   Can two parts be combined into one?
   Is there duplication?
   Can the number of different lengths, colors, types be reduced?

Standardize—simplify:
   Could a standard part be used?
   Would a modified standard part work?
   Does the standard contribute to cost?
   Does anything prevent it from being standardized?
   Is it too complex?
   Can connections be simplified?

Challenge—identify:
   Does it do more than is required?
   Does it cost more than it is worth?
   Is someone else buying it at lower cost?
   What is special about it?
   Is it justified?
   Can tolerances be relaxed?
   Have drawings and specifications been coordinated?

Maintain—operate:
   Is it accessible?
   Are service calls excessive?
   Would you like to own it and pay for its maintenance?
   Is labor inordinate to the cost of materials?
   How often is it actually used?
   Does it cause problems?
   Have users established procedures to get around it?

Requirements—cost:
   Are any requirements excessive?
   Can less expensive materials be used?
   Is it proprietary?
   Are factors of safety too high?
   Are calculations always rounded off on the high side?
   Would a thinner material work?
   Could a different finish be used?

B. Analysis techniques

Review all phases of the program being evaluated (speculation phase).
Designate the subordinate problems requiring solution (analysis phase).
Determine the data that might help with the evaluation (speculation phase).
Determine the most likely sources of data (analysis phase).
Conceive as many ideas as possible that relate to the problem (speculation phase).
Select for further study ideas most likely to lead to a solution (analysis phase).
Consider all possible ways to test the ideas chosen (speculation phase).
Select the soundest ways of testing the ideas (analysis phase).
Decide on the final idea to be used in the program (analysis phase).

TABLE 10.3 Questions to Stimulate Ideas (*Continued*)

C. Analysis criteria
Will the idea work?
Can it be modified or combined with another?
What is the savings potential?
What are the chances for implementation?
What might be affected?
Who might be affected?
Will it be relatively difficult or easy to make the change?
Will it satisfy all the user's needs?

Key Questions. The following questions must be answered about all alternatives being developed during this phase:

What does each alternative cost?

Will each perform the basic functions?

Techniques. Several techniques are available by which alternative ideas can be evaluated and judged. Comparisons can be made between the various advantageous and disadvantageous features of the alternatives under consideration. Advantages and disadvantages of each alternative can be listed and then the ideas sorted according to the relative numbers of advantages and disadvantages. A system of alternately using creative and judicial thinking processes for each basic idea to be evaluated can be applied according to the steps shown in the Analysis Techniques portion of Table 10.3.

Procedure. Evaluation may be accomplished either by the generating group or by an independent group. Authorities disagree upon which approach is better. The disagreement grows out of the question whether people who generate ideas can be objective enough in evaluating them.

1. *Establish criteria.* The first step is to develop a set of evaluation criteria or standards by which to judge the ideas. In developing these criteria, the team should try to anticipate all effects, repercussions, and consequences that might occur in trying to accomplish a solution. The resultant criteria should, in a sense, be a measure of sensitivity to problems (which might be inherent in changes caused by the new idea). In Table 10.4, three sets of criteria that could be used in the analysis phase are presented under Possible Ratings. Factors such as these are really the yardsticks by which the effectiveness of each idea can be tested.

2. *Screen ideas.* The next step in the procedure is the actual ranking, or rating, of ideas according to the criteria developed. No idea should be summarily discarded; all should be given this preliminary evaluation as objectively as possible. In Table 10.4, a three-part system that can be used to rate ideas is presented under Alternative Idea. Ratings and their weights are based on the judgment of persons performing the evaluation. This initial analysis will produce a shorter list of alternatives, each of which has passed the evaluation standards set by the team.

3. *Define alternatives.* The remaining alternatives can be ranked according to an estimate of their relative cost reduction potential. The ranking may be based on nothing more than relative estimates comparing the elements, materials, and processes of

TABLE 10.4 Typical Analysis Rating System

Alternative idea	Possible ratings
Ability to perform basic function	Excellent Good Fair Poor
Usability of the idea	Use now Modify Hold Reject
Ease of idea implementation considering complexity and schedule	Simple idea Moderately complex Complex idea

the alternatives and the original or present method of providing the function. The surviving alternatives are then developed further to obtain more detailed cost estimates. The cost estimating for each alternative proceeds only if the preceding step indicates it still to be a good candidate. Although the analysis phase is the responsibility of the VE team, authorities and specialists should be consulted in estimating the potential of these alternatives. Cost estimates must be as complete, accurate, and consistent as practicable to minimize the possibility of error in assessing the relative economic potential of the alternatives. Specifically, the method used to determine the cost of the original should also be used to cost the alternatives.

4. *Make final selection.* After the detailed cost estimates are developed for the remaining alternatives, one or more are selected for further study, refinement, testing, and information gathering. Normally, the alternative with the greatest savings potential will be selected. However, if several alternatives are not decisively different at this point, all should be developed further.

10.13.4 Phase 4: Development

Objective. In the development phase, the alternatives that have survived the selection process are developed into firm, specific recommendations for change. The process involves not only detailed technical and economic testing but also an assessment of the probability of successful implementation.

Key Questions. Several questions must be answered about each alternative during the development of specific solutions:

Will it work?

Will it meet all necessary requirements?

Who has to approve it?

What are the implementation problems?

What are the costs?

What are the savings?

Procedures

1. *General.*   To satisfy the questions above, each alternative must be subjected to:
   a. Careful analysis to ensure that the user's needs are satisfied
   b. A determination of technical adequacy
   c. The development of estimates of costs and implementation expenses, including schedules and costs of all necessary tests
   d. Consideration of changeover requirements and their impact
2. *Develop convincing facts.*   As in the information phase, the use of good human relations is of considerable importance to the success of the development phase. In developing answers to the questions above, the VE team should consult with personnel knowledgeable about what the item must do, within what constraints it must perform, how dependable the item must be, and under what environmental conditions it must operate. Technical problems related to design, implementation, procurement, or operation must be determined and resolved. Consideration must also be given to impact in areas such as safety, fire protection, maintenance, and supply support.
3. *Develop specific alternatives.*   Those alternatives that stand up under close technical scrutiny should be followed through to the development of specific designs and recommendations. Work on specifics rather than generalities. Prepare drawings or sketches of alternative solutions to facilitate the identification of problem areas remaining in the design, and to facilitate detailed cost analysis. Perform a detailed cost analysis for proposed alternatives to be included in the final proposal.
4. *Development implementation plans.*   Anticipate problems relating to implementation, and propose specific solutions to each. Particularly helpful in solving such problems are conferences with specialists. Develop a specific recommended course of action for each proposal that details the steps required to implement the idea, who is to do it, and the time required. Ask for ideas from the office that will approve or disapprove the recommendation.
5. *Testing.*   When testing is involved, the VE team may arrange the necessary testing and evaluation, although normally this will be done by other appropriate personnel in the organization. This testing and evaluation should be planned for and scheduled in the recommended implementation process.
6. *Select first choice.*   Finally, one alternative should be selected for implementation as the best-value (best overall cost reduction, usually) alternative, and one or more other recommendations selected for presentation in the event the first choice is rejected by the approval authority. The implementation schedule that will yield the greatest cost reduction should also be indicated.

10.13.5   Phase 5: Presentation

Objective.   The presentation phase involves the actual preparation and presentation of the best alternatives to persons having the authority to approve the VE proposals. This phase of the VE job plan includes the following steps:

1. Prepare and present the VE proposals.
2. Present a plan of action that will ensure implementation of the selected alternatives.
3. Obtain a decision of positive approval.

Discussion. A value engineering proposal (VEP) is almost without fail a challenge to the status quo of any organization. It is a recommendation for change. The recommendation was developed through a team effort, and its adoption is dependent upon another team effort. The success of a VE project is measured by the savings achieved from implemented proposals. Regardless of the effort invested and the merits of the proposal, the net benefit is zero, or is negative, if the proposals are not implemented. Presenting a proposal and subsequently guiding it to implementation often requires more effort than its actual generation. We review here some principles and practices that have been successfully used to facilitate the approval of VEPs:

1. *Form.* Presentation of a VEP should always be written. Oral presentation of study results is most helpful to the person who is responsible for making the decision; however, it should never replace the written report. A written report normally demands and receives a written reply, whereas oral reports can be forgotten and overlooked as soon as they are presented. In the rush to wrap up a project, promote a great idea, or save the laborious effort of writing a report, many proposals have fallen by the wayside because the oral presentation came first and was inadequate. The systematic approach of the VE job plan must be followed all the way through to include the systematic, meticulous, careful preparation of a written report. From this will evolve a more concise and successful oral presentation.

2. *Content.* Management responsible for review and approval must base its judgment on the documentation submitted with a proposal. The proposal and supporting documentation should provide all of the data the reviewer will need to reach a decision. Top management is primarily concerned with net benefit and disposition. A manager either may be competent in the areas affected by the proposal or may rely on the advice of a specialist. In either case, completely documented proposals are far more likely to be implemented. Generally, proposals should contain sufficient discussion to assure the reviewer that performance is not adversely affected, supporting technical information is complete and accurate, potential savings are based on valid cost analysis, and the change is feasible.

3. *VEP acceptance.* There are many hints that may be offered to improve the probability of and reduce the time required for acceptance and implementation of proposals. Those that appear to be most successful are as follows:

    a. *Consider the reviewer's needs.* Use terminology appropriate to the training and experience of the reviewer. Each proposal is usually directed toward two audiences. First is the technical authority, who requires sufficient technical detail to demonstrate the engineering feasibility of the proposed change. Second are the administrative reviewers, for whom the technical details can be summarized while the financial implications (implementation costs and likely benefits) are emphasized. Long-range effects on policies, procurement, and applications are usually more significant to the manager than to the engineer.

    b. *Prepare periodic progress reports—"no surprises."* The manager who makes an investment in a VE study expects to receive periodic progress reports with estimates of potential results. Reporting is a normal and reasonable requirement of management. It helps ensure top management awareness, support, and participation in any improvement program. There are very few instances where managers have been motivated to act by a one-time exposure at the "final presentation," no matter how "just" the cause. Therefore, it is advisable to discuss the change with the decision makers or their advisors prior to its submittal as a formal VEP. This practice familiarizes key personnel with impending proposals, and enables them to evaluate them more quickly after submittal. No manager likes to be surprised. Early disclosure may also serve to warn the originators of any objections to the proposal. This "early warning" will give the originators opportunity to incorporate modifications to overcome the objections. Often, the preliminary discussions produce additional suggestions that improve the proposal and enable the decision maker to contribute directly. If management has been kept informed of progress, the VEP presentation may be only a concise summary of final estimates and pro and con discussions, and perhaps trigger formal management approval.

c. Relate benefits to organizational objectives. The VEP that represents an advancement toward some approved objective is most likely to receive favorable consideration from management. Therefore, the presentation should exploit all of the advantages a VEP may offer toward fulfilling organizational objectives and goals. When reviewing a VEP, the manager normally seeks either lower total cost of ownership, or increased capability for the same or lesser dollar investment. The objective may be not only savings but also the attainment of some other mission-related goal of the manager.

d. Support the decision maker. The monetary yield of a VEP is likely to be improved if it is promptly implemented. Prompt implementation, in turn, is dependent upon the expeditious approval by the decision makers in each organizational component affected by the proposal. These individuals should be identified and the entire VE effort conducted under their sponsorship. The VE group becomes the decision maker's staff, preparing information in such a manner that the risk against the potential reward can be weighed. Like any other well-prepared staff report, each VEP should:

- Satisfy questions the decision maker is likely to ask
- Respect the decision maker's authority
- Permit the decision maker to preserve professional integrity
- Imply assurance that approval would enhance image
- Include sufficient documentation to warrant a favorable decision with reasonable risk factors (both technical and economic)

e. Minimize risk. If VE proposals presented to management are to be given serious consideration, they should include adequate evidence of satisfactory return on the investment. Often, current or immediate savings alone will ensure an adequate return. In other cases, life cycle or total program savings must be considered. Either way, evidence of substantial benefits will improve the acceptability of a proposal.

The cost and time spent in testing to determine the acceptability of a VE proposal may offset a significant portion of its savings potential. Committing such an investment with no guarantee of success constitutes a risk that could deter acceptance of a VEP. In some cases this risk may be reduced by prudent design and scheduling of test programs to provide intermediate assurances indicating the desirability of continuing with the next step. Thus, the test program may be terminated or the proposal modified when the concept first fails to perform at an acceptable level. Major expenditures for implementing proposed VE actions should not be presented as a lump sum aggregate, but rather as a sequence of minimum risk increments. A manager may be reluctant to risk a total investment against total return, but may be willing to chance the first phase of an investment sequence. Each successive investment increment would be based upon the successful completion of the previous step.

f. Combine testing. Occasionally, a significant reduction in implementation investment is made possible by concurrent testing of two or more proposals. Also, significant reductions in test cost can often be made by scheduling tests into other test programs scheduled within a desirable time. This is particularly true when items to be tested are part of a larger system also being tested. However, care must be exercised in instances of combined testing to prevent masking the feasibility of one concept by the failure of another.

g. *Show collateral benefits of the investment.* Often VE proposals offer greater benefits than the cost improvements specifically identified. Some of the benefits are collateral in nature and difficult to express in monetary terms. Nevertheless, collateral benefits should be included in the calculations. The likelihood of acceptance of the VEP is improved when all its collateral benefits are clearly identified and completely described.

h. *Acknowledge contributors.* An implemented VE proposal always results from a group effort. There is a moral obligation to identify all individuals and data sources contributing to a proposal. Identification of contributors also provides the reviewers with a directory of sources from which additional information may be obtained. Individuals, departments, and organizations should be commended whenever possible. This recognition promotes cooperation and participation essential to the success of subsequent VE efforts.

i. *Prepare the oral presentation.* The oral presentation can be the keystone to selling a proposal. It gives the VE team a chance to ensure that the written proposal is correctly understood and that proper communication exists between the parties concerned. Effectiveness of the presentation will be enhanced if:

- The entire team is present and is introduced
- The presentation is relatively short with time for questions at the end
- The presentation is illustrated through the use of visual aids such as mockups, models, slides, or flip charts
- The team is prepared with sufficient backup material to answer all questions during the presentation

10.13.6 Phase 6: Implementation

Objective. During the implementation phase, the VE manager must ensure that approved recommendations are converted into actions. Until this is done, savings to offset the cost of the study will not be achieved. Three major objectives of this phase are:

1. To provide assistance, clear up misconceptions, and resolve problems that may develop in the implementation process
2. To minimize delays encountered by the proposal in the implementation process
3. To ensure that approved ideas are not modified during the implementation process in such a manner as would cause them to lose their cost-effectiveness or basis for original selection

Implementation Investment. The need to invest in order to save must be emphasized when submitting VEPs. Some degree of investment is usually required if a VE opportunity is to become a reality. Funds and/or personnel for implementation have to be provided. The key to successful implementation lies in placing orders for the necessary actions into the normal routine of business. Progress should be reviewed periodically to ensure that any roadblocks that arise are overcome promptly.

Expediting Implementation. One of the fastest ways to achieve implementation of an idea is to effectively utilize the knowledge gained by those who originated it. Whenever possible, the VE team should be required to prepare first drafts of documents necessary to revise handbooks, specifications, change orders, drawings, and contract requirements. Such drafts will help to ensure proper translation of the idea into action and will serve as a baseline from which to monitor progress of final implementation. To further ensure proper communication and translation of the idea onto paper, the VE team should review all implementation actions prior to final release.

Monitoring Progress. Implementation progress must be monitored just as systematically as the VEP development. It is the responsibility of the management or the VE manager to ensure that implementation is actually achieved. A person should be designated by name with responsibility to monitor all deadline dates in the implementation plan.

Objective. The last phase of the job plan has several objectives; these might seem quite diverse, but when achieved in total, they will serve to foster and promote the success of subsequent VE efforts:

1. Obtain final copies of all completed implementation actions.
2. Compare actual results with original expectations.
3. Submit cost savings achievement reports to management. This will allow calculation of the total return on investment (ROI) of the VE effort.
4. Submit technical reports to management for possible use elsewhere.
5. Evaluate conduct of the project to identify problems that arose and recommend corrective action for the next project.
6. Initiate recommendations for potential VE study on ideas evolving from the study just completed.
7. Screen all contributors to the VEP for possible receipt of an award and initiate recommendations for appropriate recognition.

Discussion. A VE project is not completed with implementation of an idea. Full benefit is not derived from a VEP until the follow-up phase is completed. Until then, the records on a project cannot be closed. It is the responsibility of the VE manager to designate some individual to complete this phase of the job plan. Certain key questions must be answered to assess accomplishments:

1. Did the idea work?
2. Did it save money?
3. Would you do it again?
4. Could it benefit others?
5. Has it been forwarded properly?
6. Has it had proper publicity?
7. Should any awards be made?

10.13.7 Job Plan Summary

The VE job plan can be applied to any subject suitable for a VE study, usually anything with an identifiable function and a means to measure it. In serving as a vehicle to carry the study from inception to conclusion and in observing certain formalities, the VE job plan ensures that consideration is given to all necessary facets of the study. Although the job plan divides the study into a distinct set of work elements, judgment is necessary to determine the depth to which each phase is performed. In fact, each effort must be made in light of the resources available and the results expected. The VE job plan requires those making the study to clearly define the functions performed by the item under study.

Adherence to the job plan ensures that time is made available for the essential creative work and its necessary analysis so that best choices can be made for further development. The job plan leads to the establishment of an effective program aimed at the selection of the best-value alternatives. And finally, it concludes with specific recommendations, the necessary data supporting them, the identification of necessary implementing actions, a proposed implementation schedule, and a required follow-up procedure.

The job plan is normally followed in sequence, phase by phase. However, in actual practice it is often necessary to do additional work on a previously completed phase before reaching a decision. Thus, in practice, the phases may overlap broadly, and such early steps as information gathering may continue throughout most of the VE effort. Typically, two major documents are produced: a report summarizing the results of the effort, and a project book that contains all the detailed backup information.

The VE job plan is a planned program that has been tested through use and is being successfully used in highway applications to obtain real cost savings.

10.14 FAST DIAGRAMMING AND THE JOB PLAN

Function analysis system technique (*FAST*) is a diagramming technique to graphically show the logical relationships of the functions of an item, system, or procedure. FAST was developed in 1964 by Charles V. Bytheway at the UNIVAC Division of the Sperry Rand Corporation. Prior to the development of FAST, one had to perform a function analysis of an item by random identification of functions. The basic function had to be identified by trial and error, and one was never quite sure that all functions had been uncovered. FAST provides a system to do a better job in function analysis.

10.14.1 Purpose of the FAST Diagram

The FAST diagram should be created during the information phase of the VE job plan by the whole VE team. When used in conjunction with a value study, the FAST diagram serves the following purposes:

1. It helps organize random listing of functions. When answering the questions "What is it?", "What does it do?", "What must it do?", the study team develops many verb-noun function solutions at all levels of activity, which the FAST diagram can help sort out and interrelate.
2. It helps check for missing functions that might be overlooked in the above random function identification process.
3. It aids in the identification of the basic function or scope of the study.
4. It deepens and guides the involvement, visualization, and understanding of the problem to be solved and the proposed changes.
5. It demonstrates that the task team has completely analyzed the subject or problem.

6. It tests the functions through the system of determinate logic.

7. It results in team consensus in defining the problem in function terms and aids in developing more creative valid alternatives.

8. It is particularly helpful in "selling" the resulting changes to the decision makers.

10.14.2   Guidelines for FAST Diagrams

Figure 10.2 depicts the diagramming conventions to be used in preparing a FAST diagram. The relative positions of functions as displayed on the diagram are also levels of activity. The FAST diagram is a horizontal graphical display based on system functions rather than system flowcharting or components. Level 1 functions, the higher-level functions, appear on the left side of the FAST diagram, with lower-level activity successively graphed to the right as shown. In most cases, when conducting a VE study, various levels of activity of verb-noun functions will be automatically suggested as the basic function of an item or a system.

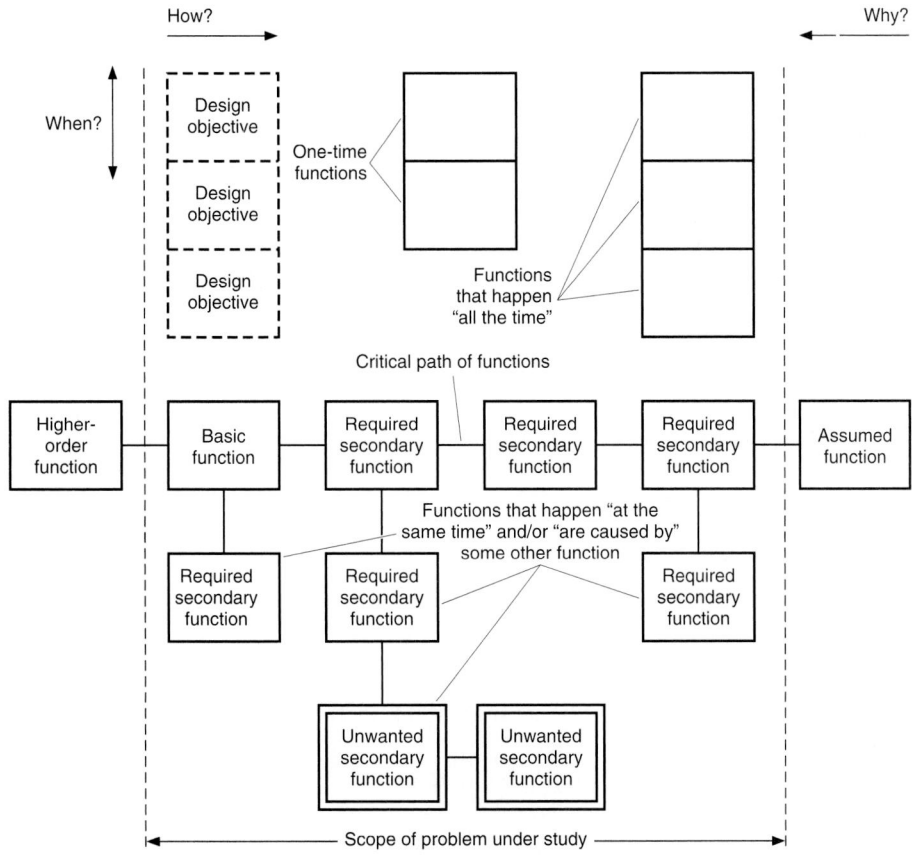

FIGURE 10.2   FAST diagram: functional analysis system technique.

The FAST diagram is just a tool. It is the process used in creating the diagram that is important, not the final diagram itself or its appearance. There is no such thing as a "right" or perfect schoolbook solution that each diagrammer should be able to create, if he or she had perfect knowledge of the technique and theory. Yet if the diagram logic is logical to the diagrammer, it will normally be logical to a reviewer. And if it is not, then the FAST diagram will have served another purpose—communication of a misunderstanding in statement of the problem. That is also valuable to know. With these things in mind, consider the following guidelines in preparing a diagram:

1. Show the scope of the problem under study by two vertical dashed lines, one to the extreme left and one to the extreme right of the diagram. Everything that lies between the two scope lines is defined as the problem under study.

2. Every FAST diagram will have a "critical path of functions" going from left to right across the scope lines.

3. On that critical path should be found only required secondary functions, the basic function(s), and the higher-order function.

4. The higher-order function will lie to the immediate left of the left scope line.

5. The basic function(s) will always lie to the immediate right of the left scope line.

6. All other functions on the critical path will lie to the right of the basic function and will be the required secondary functions (not normally aesthetic or unwanted secondary functions).

7. Any "assumed" functions lie to the right of the right-hand scope line.

8. All other secondary functions the item performs will lie either above or below the critical path of functions. These functions can be required secondary functions, aesthetic functions, or unwanted functions.

9. If the function "happens at the same time as" and/or "is caused by" some function on the critical path, place the function below that critical path function.

10. If the function happens "all the time" the system is doing its work, place it above the critical path function to the extreme right of the diagram.

11. If there are specific design objectives or general specifications to keep in mind as the diagram is constructed, place them above the basic function and show them as dotted boxes.

12. All "one-time" actions are placed above the critical path and in the center area of the diagram.

13. All functions that lie on the critical path must take place to accomplish the basic function. All other functions on the FAST diagram are subordinate to the critical path function and may or may not have to take place to accomplish the basic functions.

10.14.3 Steps in Construction of FAST Diagrams

The following steps are recommended in the construction of the FAST diagram:

1. *Function listing.* Prepare a list of all functions, by assembly or by system, using the verb-and-noun technique of identification of function. Do this by brainstorming the questions (*a*) "What does it do?" and (*b*) "What must it do?"

2. *The function worksheet.* Using lined paper, prepare a three-column function worksheet in the format shown in Fig. 10.3. Insert the listed functions from above,

Why?	Function	How?

FIGURE 10.3   Function worksheet for FAST.

one at a time, into the central column. Then, ask of each function the following questions:

a. How do I (verb) (noun)? Record the answer(s) in the right column.

b. Why do I (verb) (noun)? Record the answer(s) in the left column.

3. *The diagram layout.*   Next, write each function separately on a small card in verb-and-noun terminology. Select a card with the function that you consider to be the basic function. Determine the position of the next higher and lower function cards by answering the following logic questions:

a. Perform the "how" test by asking of any function the question, "How do I (verb) (noun)?" The function answer should lie to the immediate right. Every function that has a function to its immediate right should logically answer the "how" test. If it does not, either the function is improperly described or a function is in the wrong place.

b. The second test, "why," works in the same way, but in the opposite direction. Ask the question "Why do I (verb) (noun)?" The answer should be in the function to the immediate left and should read, "So that I can (verb) (noun)." The answer must make sense and be logical.

4. *The critical path.*   To determine whether a function belongs on the critical path, test the functions with these questions:

a. How is (verb) (noun) actually accomplished, or how is it proposed to be accomplished?

b. Why must (verb) (noun) be performed?

5. *The support logic block.*   A support logic block is a block immediately underneath a given block at the same general level of activity. This contains functions that "happen at the same time as" and/or "are caused by" some other function. They can be determined by answering these questions:

a. When is (verb) (noun) performed?

b. If (verb) (noun) is performed, what else must also happen?

6. *The locating scope lines.*   In determining where to place the scope lines, the choice is arbitrary. Actually, moving the left scope line from left to right lowers the level of activity of the problem to be studied. The basic function to be studied shifts, since it is always the function that lies to the immediate right of the left scope line. Locating the right scope line determines the assumptions and "givens" one is willing to accept before starting the study. Location of both scope lines is also subject to the point of view of the owner or user of the problem.

To better visualize the process, Fig. 10.4 shows a FAST diagram of a familiar system, a fire alarm system.

10.14.4 Diagramming Techniques

The following three considerations are general techniques that should be followed:

1. Usually only two FAST diagrams are of interest: the diagram that represents an existing plan, program, or design; and the diagram that represents the proposed concept. When diagramming something that exists, be sure not to slip off on a tangent and include alternatives and choices that are not present in the existing system.

2. When using a FAST diagram to design or propose a new concept, restrict it to a specific concept; otherwise the answers created in diagramming become meaningless. The "method selected" to perform a function brings many other functions into existence. Therefore, creation of several FAST diagrams during system design is a possibility.

3. The choice of the level of detail of functions to be used in the FAST diagram is entirely dependent on the point of view of the diagrammer, the purpose for which it is to be used, and to whom it will be presented. For presentation of VE study results to management, a very detailed FAST diagram should be simplified.

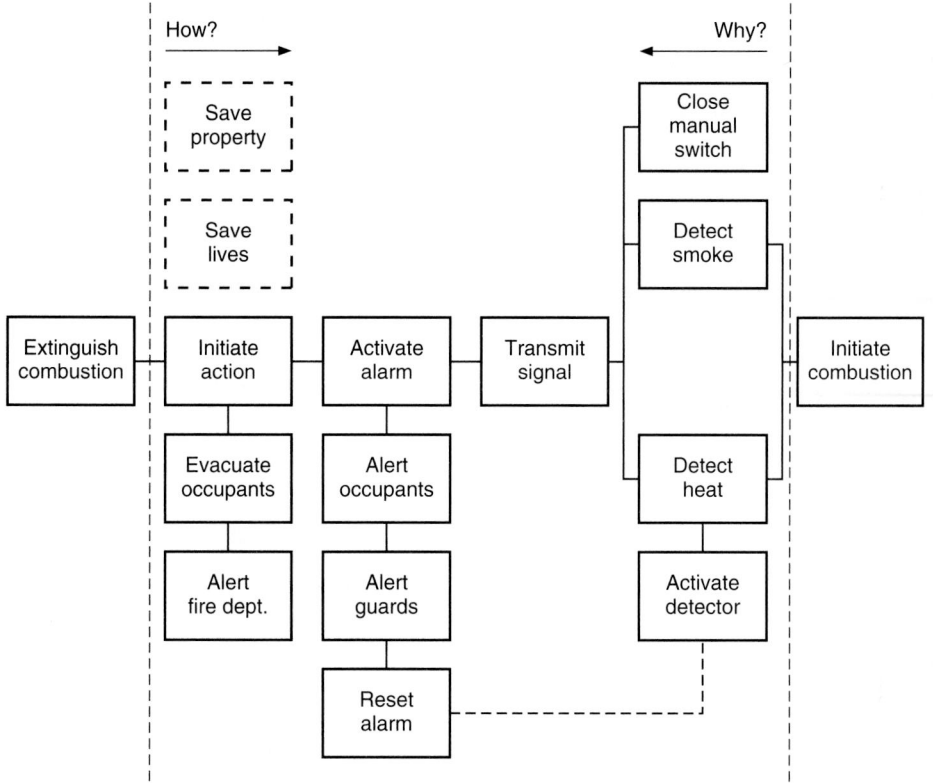

FIGURE 10.4 FAST diagram of fire alarm system.

10.14.5 Summary of FAST Diagramming

1. FAST is a structured method of function analysis that results in analyzing the basic function, establishing critical path functions and supporting functions, and identifying unnecessary functions.
2. FAST diagrams should be constructed at a level low enough to be useful, but high enough to be advantageous to the purpose of creatively seeking alternative methods.
3. FAST diagrams are used to communicate with subject matter experts; to understand the problems of specialists in their own profession; to define, simplify, and clarify problems; to bound the scope of a problem; and to show the interrelated string of functions needed to provide a product or service.
4. The FAST procedure will be useful only if thinking outlined in the steps to prepare a diagram is performed. The value of this technique is found not in recording the obvious, but in the extension of thinking beyond usual habits as the study proceeds.

5. A FAST diagram, as first constructed, may not completely comply with "how" and "why" logic. This is because it takes additional thinking to get everything to agree. However, when you are persistent and insist that the logic be adhered to, you will discover that your understanding has expanded and your creativity has led you into avenues that would not otherwise have been pursued. When the "how" and "why" logic is not satisfied, it suggests that either a function is missing or the function under investigation is a supporting function and not on the critical path.

6. A main benefit from using FAST diagramming and performing an extensive function analysis is to correct our ignorance factor, so that we can see the study in its true light. Once this function analysis is performed on a given topic, we can quickly see that the only reason a lower-level function has to be performed is because a higher-level function caused it to come into being. Essentially, whenever we establish one of these functional relationships that is visually presented by a FAST diagram, we correct our ignorance factor and open the door to greater creativity.

10.15 COST MODEL

A *cost model* is a diagrammatic form of a cost estimate. It is used as a tool in the VE process to provide increased visibility of the cost of the various elements of a system or an item, to aid in identifying the item's subelements most suitable for cost reduction attention, and to establish cost targets for comparison of alternative approaches. It also helps define the worth of an element.

A cost model is an expression of the cost distribution associated with a specific item, product, or system. In industry, it is often referred to as a *work breakdown structure*. A cost model is developed by first identifying assembly, subassembly, and major component elements or centers of work. From this, the model can be expanded to include a parts breakdown at more minute levels, as necessary. Next, the costs are developed (actual, estimated, or budgeted) for each of the above categories. These become the cost elements of the model and can be viewed as the cost building blocks of cost buildup from successive levels.

Figure 10.5 shows a general-purpose cost model for use in purchasing any manufactured commodity. The total cost consists of the cost of the purchase and the cost of logistics. These costs are further broken down into logical components as indicated. Shown in Fig. 10.6 are five common categories of cost for a government construction program. Some additional items that should be considered, particularly for a commercial project, include cost of land, financing charges, building permits, and taxes.

10.16 WORTH MODEL

The same form of model used to distribute cost of a system can be used to allocate worth. The cost model and the *worth model* should be identical in format. The procedures to follow in creating a worth model are as follows:

1. First, the VE team determines the necessary functions to be performed by each element of work at the lowest level of activity of the cost model.

2. The worth of each of these functions is determined as explained in the job plan.

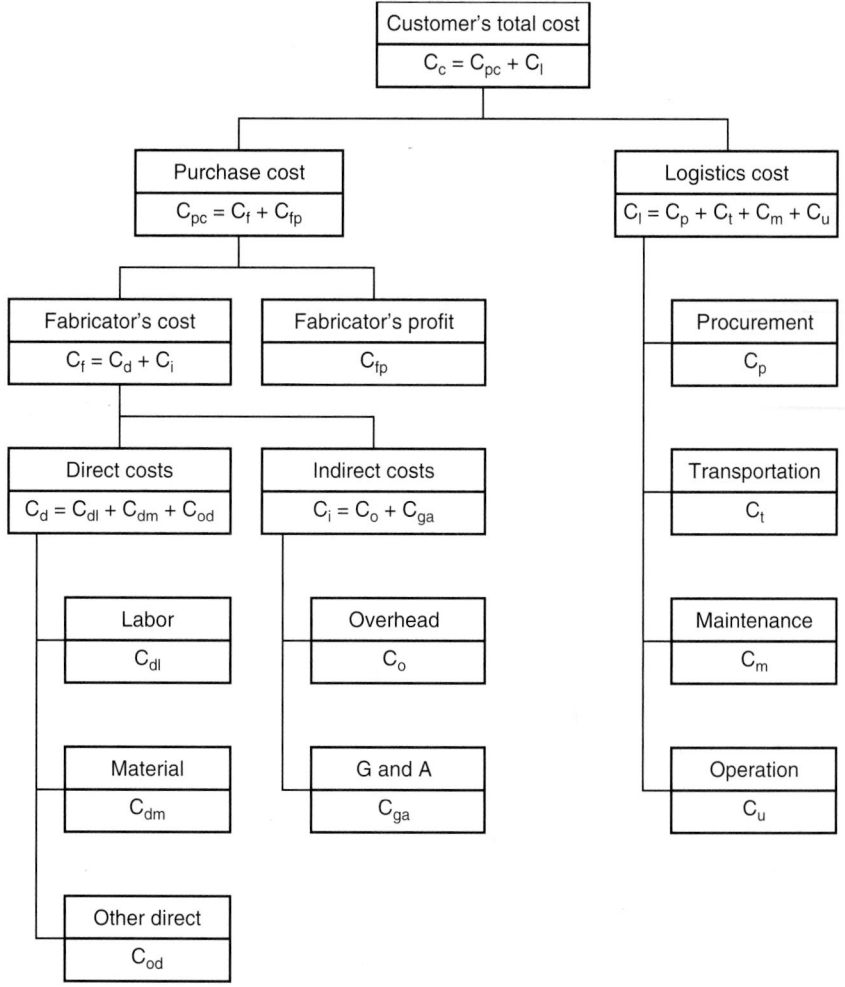

Formula: $C_c = ([(C_{dl} + C_{dm} + C_{ad}) + (C_o + C_{ga})] + C_{fp}) + (C_p + C_t + C_m + C_u)$

FIGURE 10.5   General-purpose cost model for a commodity.

3. The worth of all functions for each cost element is totaled and becomes the worth for that element.

4. The sum of the worth of all cost elements becomes the worth of the corresponding cost element at the next higher level.

Thus, the VE team develops the minimum costs it believes are possible for each block of the cost model. The result is a cost model representing minimum costs. These costs become targets to be compared with costs as reflected by the best estimates available.

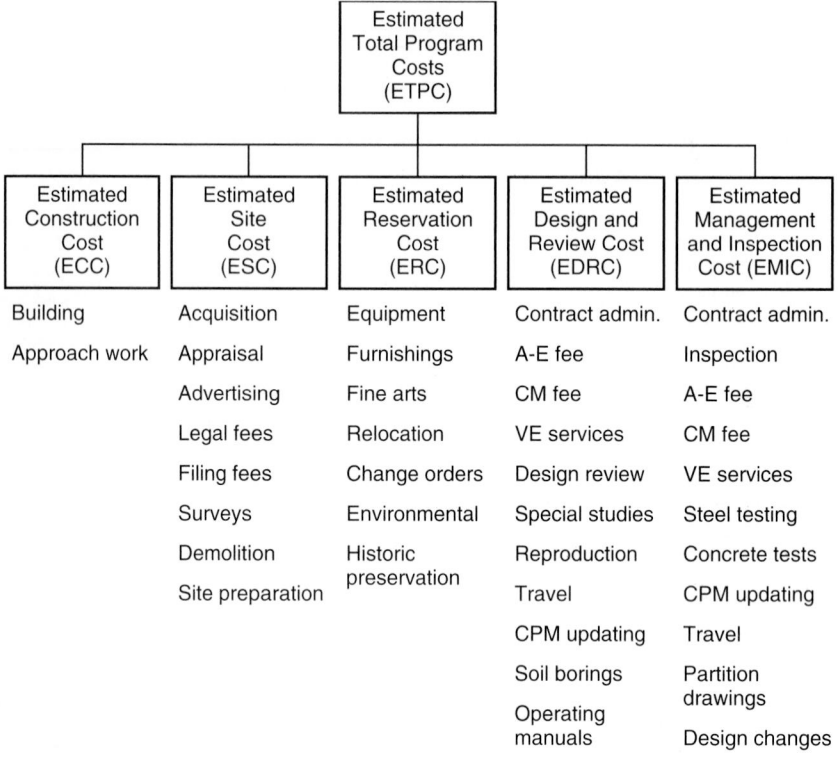

FIGURE 10.6 Cost model for construction program.

Cost blocks having the greatest differences between target and estimated costs are then selected for VE study.

(H. G. Tufty, *Compendium on Value Engineering,* Indo-American Society, Bombay, 1989.)

LIFE CYCLE COST

10.17 CONSIDERATIONS IN LIFE CYCLE COST ANALYSIS

Life cycle cost (*LCC*) is the total cost of ownership of an item, computed over its useful life. To rationally compare the worth of alternative designs, or different ways to do a job (accomplish a function), an LCC analysis is made of each. For those who follow the VE job plan, a life cycle cost analysis is very easy to perform because the total impact of each recommended VE alternative is an integral part of the total calculations. In reality, an LCC study uses VE techniques to identify all costs related to the subject (functional) area, and VE's special contribution can be the selection of the best alternatives to be "life cycle–costed."

LCC is the development of all significant costs of acquiring, owning, and using an item, a system, or a service over a specified length of time. LCC is a method used to compare and evaluate the total costs of competing solutions to satisfy identical functions based on the anticipated life of the facility or product to be acquired. In performing a value study, an LCC analysis is performed in the development phase of the value engineering (VE) job plan to determine the least costly alternative.

The value of an item includes not only consideration of what it costs to acquire it, but also the cost to use it or the cost of performance to the buyer for as long as the item is needed. The buyer, not the seller, pays the life cycle costs and therefore must determine value. One measure of value to the buyer is the calculated total cost of ownership.

Costs of repair, operations, preventive maintenance, logistic support, utilities, depreciation, and replacement, in addition to capital cost, all reflect on the total value of a product to a consumer. Calculation of the LCC for each alternative during performance of a value study is a way to judge whether product quality is being maintained in sufficient degree to prevent degradation of reliability, performance, and maintainability.

Life cycle cost analysis requires the knowledge of several economic concepts. One of these is the concept of equivalent costs in relation to time. Equivalent costs are typically developed by equating all costs to a common time baseline using interest rate to adjust for variable expenditure years. One must also hold the economic conditions constant while the cost consequences of each alternative are being developed. That is, the same economic factors are applied to each alternative using a uniform methodology.

10.17.1 Design Life

The first task one must accomplish in performing an LCC analysis is to determine the period of time for which the analysis of accumulated costs is to occur. This will usually be designated the project design life. The life span of the facility to be analyzed (a bridge, pavement, or culvert pipe) must be determined, together with the associated maintenance and rehabilitation costs. Another consideration that must be addressed is the realization that individual life spans of components of a system may be quite different. For example, in considering a highway system, the life of a bridge will likely be much longer than the life of a pavement. In considering a building, the life of the structural framework may well be 100 years or more, whereas the life of the roof may be only 20 years.

In performing a value study, the project design life or life span that should be selected is the period of time over which the owner or user of a product or facility needs the item. The user's need determines the life span when judging LCC and worth, and when comparing alternatives. The life span should be a realistic, reasonable time, and the same life span must be used for evaluating all choices. Assessment of obsolescence is part of a rational determination of design life. One must estimate how far in the future the functional capacity will be adequate. An unrealistically long design life may result in excessive expenditures on initial costs. On the other hand, an unrealistically short design life may lead to expensive replacement at a premature date.

The salvage or residual value at the end of the project design life must be determined and accounted for in the analysis. This may represent a net scrap value or the value associated with the reuse of a component, if that is feasible.

10.17.2 Discount Rates

The discount rate is used to convert costs occurring at different times to equivalent costs in present dollars. The selection of the discount rate to be used in the calculations is very important. If a low discount rate is selected, greater significance is given to future expenditures. If a high discount rate is selected, less significance is given to future expenditures. The discount rate should represent the rate of interest that makes the owner indifferent regarding whether to pay a sum now or at a future time. In government projects, the discount rate may be mandated by policy or law. The Office of Management and Budget prescribes rules for federal projects in Circular A-94. It states that the discount rate represents an estimate of the average rate of return on private investment, before taxes and after inflation. Thus, it may differ from the cost of borrowing. Guidelines on discount rates may be further amplified by federal agencies.

10.18   CATEGORIES OF COSTS

Costs that must be considered depend to some extent upon the system or project analyzed, but can generally be categorized as follows:

1. Initial costs
    a. *Item costs.*   These are costs to produce or construct the item.
    b. *Development costs.*   These are costs associated with conducting the value study, testing, building a prototype, designing, and constructing models.
    c. *Implementation costs.*   These are costs expected to occur after approval of the ideas, such as redesign, tooling, inspection, testing, contract administration, training, and documentation.
    d. *Miscellaneous costs.*   These costs depend on the item and include costs for owner-furnished equipment, financing, licenses and fees, and other one-time expenditures.

2. Annual recurring costs
    a. *Operation costs.*   These costs include estimated annual expenditures associated with the item such as for utilities, fuel, custodial care, insurance, taxes and other fees, and labor.
    b. *Maintenance costs.*   These costs include annual expenditures for scheduled upkeep and preventive maintenance to keep an item in operable condition.
    c. *Other recurring costs.*   These include costs for annual use of equipment associated with an item as well as annual support costs for management overhead.

3. Nonrecurring cost
    a. *Repair and replacement costs.*   These are costs estimated on the basis of predicted failure and replacement of major system components, predicted alteration costs for categories of space related to the frequency of moves, and capital improvements predicted necessary to bring systems up to current standards at given points in time. Each estimated cost is for a specific year in the future.
    b. *Salvage.*   Salvage value is often referred to as *residual value.* Salvage value is not really a cost, in that this factor is entered as a negative amount in the LCC calculation to reduce the LCC amount. Salvage value represents the remaining market value or use value of an item at the end of the selected LCC life span.

10.19 *METHODS OF CALCULATION*

The concepts of annualized cost and present worth are employed in LCC. Using the annualized cost method, all costs incurred are converted to equivalent annual costs using a baseline and a specified life span. For example, initial costs would be amortized over the life cycle and include principal and interest (similar to home mortgage payments). Replacement costs or rehabilitation costs at various points during the life cycle would also be converted to equivalent annual costs (sinking fund). The following steps can be employed:

1. *Annualized initial cost.* Tabulate all initial (acquisition) costs. These include the base cost of each of the alternative systems and any other initial cost. Total these initial expenditures to arrive at the total initial cost (IC). Next, amortize the initial costs (IC) by determining the annual payment necessary to pay off a loan equaling the total initial cost. Using a capital recovery table or the following equation, find the periodic payment (PP) necessary to pay off $1.00 at a discount rate of r over a period of n years. Each total initial cost is multiplied by this factor to determine the annualized cost for this element.

$$PP = \frac{r}{1 - (1 + r)^{-n}}$$

2. *Annual recurring cost.* The next step is to tabulate, for each alternative, the average annually recurring costs for operations, maintenance, and other known factors.

3. *Annualized nonrecurring cost.* Next, determine the replacement or rehabilitation costs for all major items, for each alternative, at appropriate times during the life span. Also determine the salvage value at the end of the life span. Each of the replacement costs is then discounted from the point in time where the funds are to be expended. Multiply each cost by the present worth factor (PW) from a table or calculated by the equation

$$PW = (1 + r)^{-n}$$

Then, the present worth of these replacement and salvage costs is reduced to a uniform series of payments by applying the same capital recovery periodic payment factor (PP) used in step 2. Salvage or residual values are treated similarly except that the resulting costs are negative.

4. *Total annual cost.* Finally, sum the annualized initial cost, annual recurring cost, and annualized nonrecurring cost for each alternative to determine total annual costs. These costs represent a uniform baseline of comparison for the alternatives over a projected life span at a selected interest rate. The annual differences are then determined and used for recommendations.

5. *Present worth of annual difference.* To determine the real value of an annual cost difference, calculate its present worth. Multiply each cost by the present worth annuity factor (PWA), which shows how much $1.00 paid out periodically is worth today in real dollars. The factor may be obtained from a table or calculated by the equation

$$PWA = \frac{1 - (1 + r)^{-n}}{r}$$

Thus, one may then compare the present worth of each alternative to assess the benefit derived.

6. *Effect of inflation.* The effect of inflation should be considered in the calculations when determining annual recurring cost, replacement cost, and salvage value, If inflation is constant at a rate i, costs at a future date of y years can be found by multiplying the cost by an inflation factor (IF) given by the equation

$$IF = (1 + i)^y$$

Thus, the calculations can be made using costs that allow for inflation. Using this procedure, different costs can be adjusted for different levels of inflation, if there is information to support such choices. More complex methods for handling inflation are also available.

If the items being compared do not involve different annual costs, it is more direct to make the present worth calculation directly. Future nonrecurring costs over the project design life can be reduced to their present worth value by multiplying by the PW factor given above, $PW = (1 + r)^{-n}$. These are added to the initial costs to determine total present worth of each system. The present worth of alternative systems can then be compared.

10.19.1 Example of Calculations

A simple example to illustrate the above calculation method is presented in Table 10.5. In this example, inflation is handled by using a net discount rate equal to the nominal discount rate (assumed as 10 percent) minus the rate of inflation (assumed as 5 percent). Two pipe materials are being considered for a drainage application where the project design life is 50 years. Initial costs associated with pipe A are $150,000, and those associated with pipe B are $180,000. Pipe A will require a $37,400 rehabilitation at the end of 40 years, and pipe B a $25,000 rehabilitation at the end of 45 years. Pipe A will have no salvage value, and pipe B will have a salvage value, of $30,000. For illustrative purposes, each is assumed to have an annual maintenance cost of $1000. Calculations in part A show that for the assumed conditions, pipe A will have the lower annualized cost and the present worth of the difference in annual cost is $24,900. Calculations in part B show the same difference in present worth, since the annual recurring costs are the same in this example. (For an example of LCC in pavements, see Art. 3.10.)

Life cycle costing is a technique to assess the total cost consequences between alternatives. The potential to optimize value through LCC is only as good as the alternatives being considered. It should be used in proper sequence as part of the VE effort.

(H. G. Tufty, *Compendium on Value Engineering,* Indo-American Society, Bombay, 1989; "Value Engineering and Least Cost Analysis," *Handbook of Steel Drainage and Highway Construction Products,* AISI, Washington, D.C., 1994.)

TABLE 10.5 Life Cycle Cost Calculations for Two Pipes

		A. Annualized cost method			
		Pipe A		Pipe B	
Type of cost	Equation for factor	Factor	Annualized cost, $	Factor	Annualized cost, $
Initial	PP $= r/[1 - (1 + r)^{-n}]$	0.0548 $n = 50$	$150,000 $\times$ 0.0548 = $8216	0.0548 $n = 50$	$180,000 $\times$ 0.0548 = $9864
Recurring	—	—	$1000	—	1000
Non-recurring (rehab.)	PW $= (1 + r)^{-n}$	0.1420 $n = 40$	$37,400 $\times$ 0.142 $\times$ 0.0548 = 291	0.1113 $n = 45$	$25,000 $\times$ 0.1113 $\times$ 0.0548 = $152
Non-recurring (salvage)*	PW $= (1 + r)^{-n}$	—	—	0.0872 $n = 50$	$30,000 $\times$ 0.0872 $\times$ 0.0548 = $143
Total	—	—	$8507	—	$9873

Pipe A has the lower annualized cost.

Annual difference = $9873 − 8507 = $1366.

Present worth of annual difference = (1/0.0548) $\times$ $1366 = $24,900.

*Note that salvage values are treated as negative numbers in the summations for annualized cost and present worth.

		B. Direct present worth method			
		Pipe A		Pipe B	
Type of cost	Equation for factor	Factor	Present worth, $	Factor	Present worth, $
Initial	—	—	$150,000	—	$180,000
Non-recurring (rehab.)	PW $= (1 + r)^{-n}$	0.1420 $n = 40$	$37,400 $\times$ 0.142 = $5311	0.1113 $n = 45$	$25,000 $\times$ 0.1113 = $2782
Non-recurring (salvage)*	PW $= (1 + r)^{-n}$	—	—	0.0872 $n = 50$	$30,000 $\times$ 0.0872 = $2616
Total	—	—	$155,311	—	$180,166

Pipe A has the lower cost based on present worth.

Difference in present worth = $180,166 − 155,311 = $24,900 (same result as in part A.)

*Note that salvage values are treated as negative numbers in the summations for annualized cost and present worth.

EXAMPLES OF SUCCESSFUL VE HIGHWAY STUDIES

10.20 VIADUCT REHABILITATION

This study concerns the Hoboken Viaduct, a two-level, 3380-ft roadway west of the Holland Tunnel in Jersey City, New Jersey. The multiple-span upper-level roadway is supported by a concrete-encased steel floor beam–stringer system. The viaduct is about 65 years old and is obsolete. The planned total rehabilitation included the following major items:

1. Mill and resurface lower-level roadway.
2. Improve drainage on lower level.
3. Make safety improvements to lower level with new barrier curb and median barrier.
4. Construct temporary roadway on structure upper level.
5. Replace upper deck and stringers.
6. Remove concrete encasement from floor beams.
7. Clean, paint, and repair floor beams.
8. Repair and rehabilitate substructure.

The project cost estimate is $51.15 million, including $49.55 million for construction and $1.60 million for utilities.

Evaluation. A seven-person VE team was organized consisting of personnel from bridges, construction, preliminary engineering, roadway design, and planning. Two major items were identified by the team in a VE workshop as constituting 40 percent of the estimated project cost:

> *Maintenance of traffic* in conjunction with construction of the temporary roadway to carry upper-level eastbound traffic is estimated to cost $6.5 million. Although necessary under the proposed plan, these funds do not contribute to physical rehabilitation. Continuous construction will be under way for six years. Revisions that could reduce the construction time would likely prove beneficial.
>
> *Floor beam rehabilitation* requires the removal of concrete encasement, then cleaning, repairing, and painting. New floor beams would be less expensive but are not feasible under the proposed plan. Also, the remaining service life of the rehabilitated floor beams will be less than that of new ones, and future replacement of floor beams could require replacement of stringers.

VE Recommendation. The VE team recommended that the eastbound traffic on the upper level of the viaduct be shifted south and placed on a partial embankment section with retaining walls on mostly vacant land. The upper roadway can then be replaced with new structures that meet current design requirements. Table 10.6 compares the original proposed work and the VE recommendation. The VE proposal offers an improved structure with a longer life, and still results in impressive initial savings. The project cost was reduced by $12.15 million, and the total present worth of the savings was about the same as the initial project cost estimate.

TABLE 10.6　VE Example: Viaduct Rehabilitation*

Original proposal	VE recommendation
Rehabilitate structure: 　Upper level: new deck and stringers 　Lower level: resurface and make safety 　improvements	Upper level: remove and place roadway on partial embankment, erect new structures Lower level: resurface and make safety improvements
Rehabilitate structure area of upper level (179,200 ft²)	New cross-street structures (58,700 ft²)
Construction stages: 6	Construction stages: 3
Construction time: 6 yr	Construction time: 5 yr
No right-of-way impacts	2.2 acres required
Major traffic impacts, all stages	Major traffic impacts, stage 2 only
Remaining service life: 50 yr	Remaining service life: 100 yr
Total project cost = $51.15 million	Total project cost = $39.00 million

Initial cost savings = $12.15 million

Present worth savings per LCC analysis = $39.70 million

Total present worth savings = $51.85 million

*Adapted from L. Ninesling, "Achieving Quality in Design through Value Engineering," in *Proceedings of the AASHTO Value Engineering Conference*, AASHTO, Washington, D.C., 1993.

10.21　BRIDGE WIDENING

A proposal was made to widen 1.9 miles of Route 9 near Woodbridge, New Jersey. The four lanes of the existing roadway would be increased to six with outside shoulders. The route includes the Edison Bridge, a 29-span, 4390-ft-long, 50-year-old structure. The bridge carries a divided roadway with two 12-ft-wide lanes, a 2.3-ft-wide sidewalk, and a 1-ft-wide shoulder in each direction. The proposed work would rehabilitate and symmetrically widen the structure to provide three 12-ft-wide lanes, an 8-ft-wide outer shoulder, and a 2-ft inner shoulder in each direction. The existing bridge is made up of two deck girders on a concrete pier, with transverse floor beams and longitudinal stringers between the girders. Cantilevered sections outside the floor beams provide the present 58-ft width of the structure. For the proposed widening project, a new concrete pier cap would be constructed by placing concrete around the existing pier cap and posttensioning new end sections on either side. Two additional deck girders would then be erected outside the existing floor beams to support floor beam extensions and stringers. Traffic would be handled in three stages: 1, widen structure on both sides while maintaining traffic; 2, shift outside lanes to new deck and work between split traffic; and 3, shift inside lane next to outside lane in completed area and work in center of bridge.

Evaluation.　A seven-person VE team was organized consisting of personnel from bridges, construction, project management, design, and right-of-way. The project was evaluated in a VE workshop, with the effort focusing on the Edison Bridge.

　Bridge superstructure and substructure must be considered together. It was noted that the widening could add considerable dead and live loads to the pier columns

TABLE 10.7 VE Example: Bridge Widening*

Original proposal	VE recommendation
Rehabilitate and widen existing bridge	Build new bridge and rehabilitate existing bridge
Shoulders: inside, 2 ft; outside, 8 ft	Shoulders: inside, 4 and 5 ft; outside, 10 and 12 ft
Construction stages: 3	Construction stages: 2
Wetlands impact: substantial	Wetlands impact: reduced
Traffic disruption: substantial	Traffic disruption: reduced
	Rehabilitation time reduced
	Accommodations for future improvements
Total project cost = $70.80 million	Total project cost = $72.30 million

Initial cost savings = − $1.50 million

Total present worth savings based on LCC = $24.10 million

*Adapted from B. Strizki, "Applying Life Cycle Cost Analysis in Value Engineering Bridge Projects," in *Proceedings of the AASHTO Value Engineering Conference,* AASHTO, Washington, D.C., 1993.

and foundations. Thus, to limit such loads, the proposed inside and outside shoulder widths are substandard. Even so, foundations will have to be modified for the extra loads, and some additional settlement and tilting is likely. The current proposal would result in major bridge components with dissimilar lives. Future replacement might require total replacement of the structure on a new alignment.

Traffic considerations add extra costs. With the current proposal, there is only one lane of traffic in each direction between 9 p.m. and 5 a.m. The lane reductions, extensive night work, complex staged construction, and traffic control are expensive and should be avoided. Costs spent in these areas contribute nothing to physical improvements.

VE Recommendation.   The VE team recommended that a new northbound structure be constructed and the existing structure be rehabilitated and used as a southbound structure. The new bridge would have three 12-ft-wide lanes, a 10-ft-wide outside shoulder, and a 4-ft inside shoulder. The rehabilitated bridge would have three 12-ft-wide lanes, a 12-ft-wide outside shoulder, and a 5-ft-wide inside shoulder. The rehabilitated bridge would not have to be widened, and thus, no additional loads would be imposed on its foundation system. Table 10.7 compares the original proposed work and the VE recommendation. The VE proposal offers an improved overall structural system with a longer life. The initial project cost is increased by $1.50 million. However, an LCC analysis based on a 110-year life indicates a present worth savings of $24.1 million. LCC calculations were based on a discount rate of 5.5 percent and an inflation rate of 4 percent. The following assumptions were made: original proposal, new structure and alignment at 60 years, major rehabilitation at 110 years; VE recommendation, replace the older bridge and rehabilitate the newer bridge at 60 years, and similarly at 110 years.

(L. Ninesling, "Achieving Quality in Design through Value Engineering," and B. Strizki, "Applying Life Cycle Cost Analysis in Value Engineering Bridge Projects," in *Proceedings of the AASHTO Value Engineering Conference,* AASHTO, Washington, D.C., 1993.)

INDEX

1

ABOUT THE EDITORS

ROGER L. BROCKENBROUGH, P.E., is president of R. L. Brockenbrough & Associates, Inc., a company that provides structural engineering services for a wide variety of special projects. He was formerly senior research consultant for U.S. Steel Group, USX. He has written numerous technical articles over the years and has also edited or co-edited many books, including the *Structural Steel Designer's Handbook,* Second Edition, published by McGraw-Hill.

KENNETH J. BOEDECKER, JR., P.E., is currently specifications engineer for W. R. Grace & Company. His long and distinguished career has included positions in highway marketing management and civil engineering, as well as chairmanship of various professional committees. He was honored with the prestigious ASTM Award of Merit in 1994.